Lecture Notes in Computer Science 10874

Commenced Publication in 1973
Founding and Former Series Editors:
Gerhard Goos, Juris Hartmanis, and Jan van Leeuwen

More information about this series at http://www.springer.com/series/7407

Sriram Chellappan · Wei Cheng
Wei Li (Eds.)

Wireless Algorithms, Systems, and Applications

13th International Conference, WASA 2018
Tianjin, China, June 20–22, 2018
Proceedings

Editors
Sriram Chellappan
University of South Florida
Tampa, FL
USA

Wei Li
Georgia State University
Atlanta, GA
USA

Wei Cheng
Virginia Commonwealth University
Richmond, VA
USA

ISSN 0302-9743 ISSN 1611-3349 (electronic)
Lecture Notes in Computer Science
ISBN 978-3-319-94267-4 ISBN 978-3-319-94268-1 (eBook)
https://doi.org/10.1007/978-3-319-94268-1

Library of Congress Control Number: 2018947327

LNCS Sublibrary: SL1 – Theoretical Computer Science and General Issues

Printed on acid-free paper

This Springer imprint is published by the registered company Springer International Publishing AG part of Springer Nature
The registered company address is: Gewerbestrasse 11, 6330 Cham, Switzerland

Preface

The 13th International Conference on Wireless Algorithms, Systems, and Applications (WASA 2018) was held during June 20–22, 2018, in Tianjin, Hebei, China. The conference is motivated by the recent advances in cutting-edge electronic and computer technologies that have paved the way for the proliferation of ubiquitous infrastructure and infrastructureless wireless networks. WASA is designed to be a forum for theoreticians, system and application designers, protocol developers, and practitioners to discuss and express their views on the current trends, challenges, and state-of-the-art solutions related to various issues in wireless networks.

The technical program of the conference included 59 regular papers together with 18 short papers, selected by the Program Committee from a number of 197 full submissions received in response to the call for papers. All the papers were peer reviewed by the Program Committee members or external reviewers. The papers cover various topics, including cognitive radio networks, wireless sensor networks, cyber-physical systems, distributed and localized algorithm design and analysis, information and coding theory for wireless networks, localization, mobile cloud computing, topology control and coverage, security and privacy, underwater and underground networks, vehicular networks, Internet of Things, information processing and data management, programmable service interfaces, energy-efficient algorithms, system and protocol design, operating system and middle-ware support, and experimental test-beds, models, and case studies.

We would like to thank the Program Committee members and external reviewers for volunteering their time to review and discuss conference papers. We would like to extend our special thanks to the steering and general chairs of the conference for their leadership, and to the publication, publicity, and local chairs for their hard work in making WASA 2018 a successful event. Last but not least, we would like to thank all the authors for presenting their works at the conference.

May 2018

Sriram Chellappan
Wei Cheng
Wei Li

Organization

Steering Committee

Xiuzhen Susan Cheng (Co-chair)	The George Washington University, USA
Zhipeng Cai (Co-chair)	Georgia State University, USA
Jiannong Cao	Hong Kong Polytechnic University, Hong Kong, SAR China
Ness Shroff	Ohio State University, USA
Wei Zhao	University of Macau, SAR China
PengJun Wan	Illinois Institute of Technology, USA
Ty Znati	University of Pittsburgh, USA
Xinbing Wang	Shanghai Jiao Tong University, China

General Chair

Keqiu Li	Tianjin University, China

Program Co-chairs

Sriram Chellappan	University of South Florida, USA
Wei Cheng	Virginia Commonwealth University, USA

Publication Co-chairs

Wei Li	Georgia State University, USA
Tom H. Luan	Xidian University, China

Publicity Co-chairs

Cheng Zhang	The George Washington University, USA
Zenghua Zhao	Tianjin University, China
Zaobo He	Miami University, USA

Local Chair

Xiaobo Zhou	Tianjin University, China

Program Committee

Ashwin Ashok	Georgia State University, USA
Yu Bai	California State University Fullerton, USA

Pratool Bharti	Communication Concepts Integration, USA
Ionut Cardei	Florida Atlantic University, USA
Mihaela Cardei	Florida Atlantic University, USA
Yacine Challal	Université de Technologie de Compiègne, France
Brijesh Chejerla	Missouri University of Science and Technology, USA
Changlong Chen	Microsoft, USA
Fei Chen	Shenzhen University, China
Songqing Chen	George Mason University, USA
Yu Cheng	Illinois Institute of Technology, USA
Lin Cui	Jinan University, China
Dezun Dong	National University of Defense Technology, China
Qinghe Du	Xi'an Jiaotong University, China
Neelanjana Dutta	Teradata Labs R&D, USA
Xiumei Fan	Xi'an University of Technology, China
Xinwen Fu	University of Central Florida, USA
Chunming Gao	University of Washington, USA
Xiaofeng Gao	Shanghai Jiao Tong University, China
Yong Guan	Iowa State University, USA
Meng Han	Kennesaw State University, USA
Gaofeng He	Nanjing University of Posts and Telecommunications, China
Liang He	University of Colorado Denver, USA
Zaobo He	Miami University, USA
Chunqiang Hu	Chongqing University, China
Baohua Huang	Guangxi University, China
Yan Huo	Beijing Jiaotong University, China
Anandhi Jayam	Virginia Commonwealth University, USA
Donghyun Kim	Kennesaw State University, USA
Hwangnam Kim	Korea University, South Korea
Yanggon Kim	Towson University, USA
Sanghwan Lee	Kookmin University, Korea
Feng Li	Indiana University-Purdue University Indianapolis, USA
Ming Li	University of Nevada, Reno, USA
Pan Li	Case Western Reserve University, USA
Qun Li	College of William and Mary, USA
Ruinian Li	The George Washington University, USA
Wei Li	Georgia State University, USA
Wenjia Li	New York Institute of Technology, USA
Yingshu Li	Georgia State University, USA
Zhenhua Li	Tsinghua University, China
Peixiang Liu	Nova Southeastern University, USA
Xiang Lu	Chinese Academy of Science, China
Zhihan Lu	University College London, UK
Yu Luo	South Dakota School of Mines and Technology, USA
Liran Ma	Texas Christian University, USA

Name	Affiliation
Jian Mao	National University of Singapore, Singapore
Bo Mei	The George Washington University, USA
Manki Min	Louisiana Tech University, USA
Hung Nguyen	Virginia Commonwealth University, USA
Linwei Niu	West Virginia State University, USA
Anupam Panwar	Yahoo Inc., USA
Anurag Panwar	University of South Florida, USA
Vamsi Paruchuri	University of Central Arkansas, USA
Winston Peng	National University of Defense Technology, China
Zheng Peng	City University of New York City College, USA
Jian Ren	Michigan State University, USA
Na Ruan	Shanghai Jiao Tong University, China
Sergio Salinas	Wichita State University, USA
BharathKumar Samanthula	Montclair State University, USA
Kewei Sha	University of Houston-Clear Lake, USA
Zhiguo Shi	Zhejiang University, China
Mark Snyder	Microsoft, USA
Junggab Son	Kennesaw State University, USA
Houbing Song	Embry-Riddle Aeronautical University, USA
Mukundan Sridharan	The Samraksh Company, USA
Guoming Tang	National University of Defense Technology, China
Srinivas Thandu	Amazon, USA
Xiaohua Tian	Shanghai Jiao Tong University, China
Chao Wang	North Chine University of Technology, China
Chaokun Wang	Tsinghua University, China
Guodong Wang	South Dakota School of Mines and Technology, USA
Honggang Wang	University of Massachusetts, USA
Li Wang	Beijing University of Posts and Telecommunications, China
Shengling Wang	Beijing Normal University, China
Tian Wang	Huaqiao University, China
Yu Wang	UNC Charlotte, USA
Yuexuan Wang	The University of Hong Kong, Hong Kong, SAR China
Zhaohui Wang	Michigan Technological University, USA
Sheng Wen	Swinburne University of Technology, Australia
Alexander Wijesinha	Towson University, USA
Yubao Wu	Georgia State University, USA
Yang Xiao	The University of Alabama, USA
Kai Xing	University of Science and Technology of China, China
Kaiqi Xiong	National Science Foundation, USA
Guobin Xu	Frostburg State University, USA
Kuai Xu	Arizona State University, USA
Wen Xu	Texas Woman's University, USA
Qingshui Xue	Shanghai Jiao Tong University, USA
Qiben Yan	University of Nebraska Lincoln, USA

Qing Yang	Montana State University, USA
Zhe Yang	Shanghai Jiao Tong University, USA
Jianguo Yao	Shanghai Jiao Tong University, USA
Qiang Ye	University of Prince Edward Island, USA
Jiguo Yu	Qufu Normal University, USA
Dongxiao Yu	Shandong University, USA
Wei Yu	Towson University, USA
Mira Yun	Wentworth Institute of Technology, USA
Sherali Zeadally	University of Kentucky, USA
Bowu Zhang	Marist College, USA
Lichen Zhang	Shanxi Normal University, China
Xu Zheng	University of Electronic Science and Technology of China, China
Jindan Zhu	Amazon Web Service, USA
Yuqing Zhu	California State University Los Angeles, USA
Haojin Zhu	Shanghai Jiao Tong University, USA

Contents

Hypergraph Based Radio Resource Management in 5G Fog Cell

Xingshuo An[1] and Fuhong Lin[1,2(✉)]

[1] School of Computer and Communication Engineering, University of Science and Technology Beijing, Beijing 100083, People's Republic of China
FHLin@ustb.edu.cn
[2] Beijing Engineering and Technology Research Center for Convergence Networks and Ubiquitous Services, University of Science and Technology Beijing, Beijing 100083, People's Republic of China

Abstract. 5G is a hot topic of current research in the field of wireless communication, micro base stations will be widely deployed in large quantities. The traditional cloud computing paradigm is unable to effectively solve the problem of 5G resource management, such as limited system capacity and low utilization rate of resource management. As a new paradigm, fog computing has the characteristics of low delay and geo-distribution. It can enable the resource management of 5G. Fog nodes are cooperative and geo-distribution. In order to improve the capacity of the system and the utilization of resources, we need to allocate fog nodes for each task. To address this issue, we propose a concept of 5G fog Cell network architecture that can be implemented by Macro-eNB and fog node. In this model, we use Hypergraph theory to establish a task model, and we design a Hypergraph clustering algorithm to manage and allocate radio resource. Simulations demonstrate that the 5G fog Cell network performs better than traditional macro cell network in radio resource utilization.

Keywords: Resource management · 5G · Fog computing · Hypergraph theory

1 Introduction

In 5G system, the radio network presents the diversification form with the coexistence of dense deployment, diversified commercial and heterogeneous networks [1]. 5G needs to support high concurrency of user equipment access and to guarantee its quality of service (QoS/QoE). This requires that the 5G network needs to meet the features of low latency, high mobility, high scalability, and real time execution, and so on [2]. The future 5G network virtualization technology can be separated from the constraints of the existing network architecture and protocol standards [3]. Through the abstraction and unification of network resources (including physical device resource and spectrum resource), the function of complex network of Management and Controlling can be decoupled from hardware. To construct a virtual network with higher flexibility, low cost and high efficiency, we can extract radio resource to the virtual resource pool uniformly [4]. The emerging wave of 5G virtualization technology requires seamless

S. Chellappan et al. (Eds.): WASA 2018, LNCS 10874, pp. 1–13, 2018.
https://doi.org/10.1007/978-3-319-94268-1_1

mobility support and geo-distribution in addition to location awareness and low latency. Conventional paradigm such as cloud computing will not be able to suit new requirement of 5G network virtualization [5]. In the process of 5G virtualization, the macro eNB as cloud server offer their services to the user equipments (UEs) from geo-spatially remote locations. This invokes high latency in service provisioning [6].

Fog Computing [7] provides a new idea for 5G virtualization. Fog extends cloud services to the edge of the network to provide low latency network services to user equipment (UE). Fog nodes [8] are service nodes with computing and storage capabilities closer to the user's device. In 5G network, high-frequency electromagnetic waves are used for communication in order to guarantee high-speed data transmission. The communication distance of high-frequency electromagnetic waves is limited. This requires the deployment of micro base stations at the edge of the network closer to users [9]. The deployment of micro base stations can be used to separate the user layer from the control layer (U/C) [10] and also provide the application of fog computing in 5G networks Necessary physical conditions. A typical application of the case shown in Fig. 1. The UE first establishes a connection with the macro base station, and the macro base station accesses the core network and may send a service request to the cloud server. Micro base station cluster constitutes a fog service layer, providing data services directly to users.

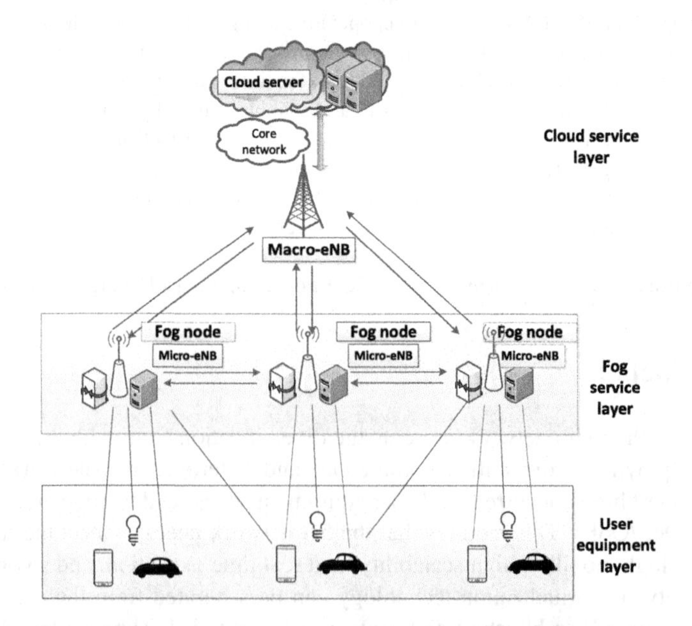

Fig. 1. 5G network architecture based on fog computing paradigm

In the 5G network, all baseband resources need to be managed and shared in virtual resource pool in order to provide the service to the user equipments. Through the wireless resource management, the network capacity can be improved under the limited physical equipment resources, and the resource utilization rate of the whole network can be improved. Key issues are how to accomplish this assignment efficiently and how to share or manage radio resources among resource nodes [11, 12]. The allocation mechanism of virtual resource pools requires multiple fog nodes to provide resources for UEs. When the user has access to the 5G network, the cloud server needs to calculate which fog nodes can provide services for the user's service requests. In this paper, a fog computing framework named 5G fog Cell for 5G is introduced. In this framework, we apply hypergraph theory to radio resource management modeling, and propose a radio resource partitioning algorithm based on Apriori clustering algorithm for 5G fog Cell.

The rest of this paper is organized as follows. Section 2 introduces some related work about fog computing in 5G and Hypergraph. Network model is built in Sect. 3. Section 4 proposes the network model using Hypergraph theory. The simulation is taken out in Sect. 5. In Sect. 6, a conclusion is drawn.

2 Related Works

In this Section, we review the related works. Some studies focused on the combination application of 5G and fog computing. In Ref. [9], the authors discuss the application prospect of fog calculation in 5G and they discussed on the potential research issues from the perspective of 5G networking. Reference [13] proposes and discusses a novel mobile edge computing design and architecture. It exploited many benefits of D2D by incorporating D2D functionalities into the proposed relay-gateways. Authors [14] evaluates Fog Computing service orchestration as a support mechanism for 5G Network in terms of round trip time. In Ref. [15], author evaluated fog computing as a support mechanism for 5G Network in terms of latency, throughput, and energy efficiency. Authors [16] presented the optimal tunable-complexity bandwidth manager (TCBM) in wireless fog computing for 5G networks.

Above mentioned works give us an insight of research about 5G radio networking mode. Some authors research on the definitions and architecture, and some authors work on 5G applications. Based on the above research, this paper put forward a kind of new 5G radio network architecture and carries on the research of using Hypergraph theory and clustering algorithm. In addition, we will review the related work of 5G resource management.

Authors in Ref. [17] integrated fog computing and MCPS to build fog computing supported MCPS (FC-MCPS) and proposed an LP-based two-phase heuristic algorithm to manage resource in fog. A tradeoff between bandwidth and energy consumption in the IoT is presented in Ref. [18]. A service providing model is built to find the relationship between bandwidth and energy consumption using a cooperative differential game model. Considering the emerging problem of joint resource allocation and minimizing carbon footprint problem for video streaming service in Fog computing, Authors [19] developed a distributed algorithm based on the proximal algorithm and alternating direction method of multipliers (ADMM).

3 Network Architecture of Fog Computing in 5G

High frequency electromagnetic wave signals must be used in the 5G wireless network in the future. Due to the characteristics of high frequency electromagnetic wave, the traditional Cell network architecture is no longer applicable. Considering that future heterogeneous hierarchical radio networks need to meet a series of design standards, such as capacity, support for mobility, effective flow share, spectrum allocation, design flexibility and operation as well as the throughput fairness. This Section will introduce a layered heterogeneous radio network enhancement scheme which is suitable for 5G combining with fog computing paradigm. This architecture blurs the boundaries of traditional Cell. In this framework, 5G fog Cell is no longer centered on the micro base station, instead of the user centered.

In the first Section, we introduce the 5G architecture based on fog computing. In this architecture, fog nodes provide users with access services and data services as service nodes. In 5G, the process of user communication in the network can be divided into three steps: (1) Users send access request and submit data traffic demand to Macro-eNB firstly. (2) Decision making by macro-eNB or cloud server to select fog nodes participating in the service. (3) Fog nodes participating in the service provide the service to the users according to the decision. In order to facilitate the research of resource allocation on fog computing in 5G architecture, we break the original physical layering and divide the resource allocation architecture of fog computing into three layers, as shown in Fig. 2.

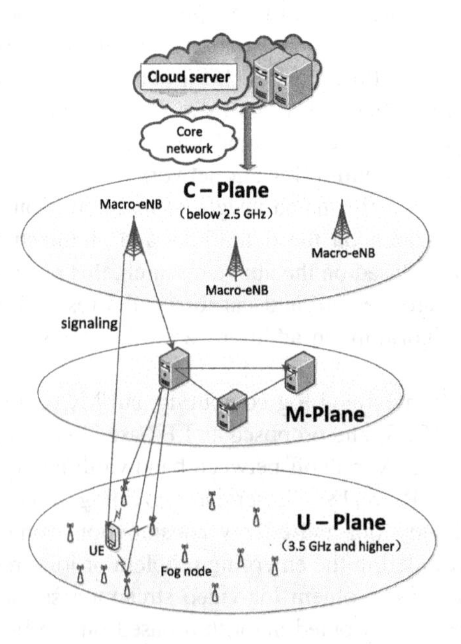

Fig. 2. 5G fog Cell framework

As shown in Fig. 2, this network architecture is composed of three layers, namely control plane (C-Plane), management plane (M-Plane) and user plane (U-Plane). U-Plane is a layer that provides services for users in fog nodes. It is equivalent to the fog service layer and the user layer in fog computing. M-Plane and C-Plane belongs to the cloud service layer. The C-Plane is mainly responsible for the virtual mapping for radio resource and radio resource management (RRM).

In C-Plane, the widely deployed Macro-eNB is composed as the control nodes. UE establishes a connection with the Macro-eNB at low-frequency stage. The macro Cell firstly guides UE to discovery fog node ID, and then UE utilizes the initial timing to carry out frequency search. UE will decode the discovered fog node ID and feedback the fog node ID measurement results to the macro Cell. The process is reflected in Fig. 3. The discovered signal must be synchronous with the Macro-eNB and orthogonal with each other.

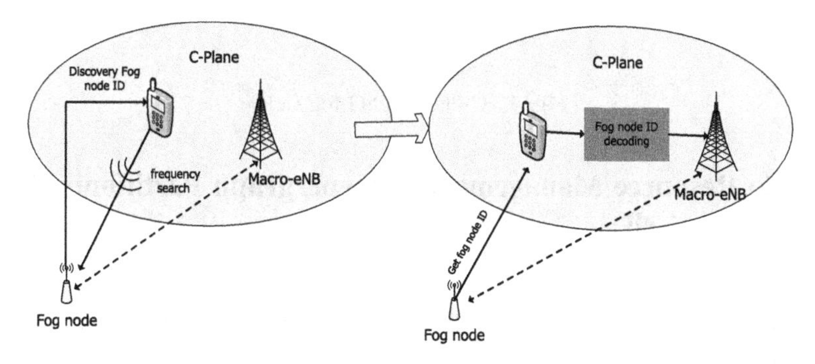

Fig. 3. Macro cell discovery in C-Plane

In M-Plane, the management of virtual resources by cloud server can be divided into three parts. The first part is information acquisition and abstraction, during which the cloud server can obtain various measurement information from the underlying nodes or users. The second part is the establishment of virtual resource pool. Virtual mapping is carried out for the related resource information of the communication link between fog node and UE to establish the corresponding radio resource management model and the Quality of Service (QoS) evaluation system focusing on the user. The third part is resource management decision. Cloud server utilizes all information in the resource pool to determine the allocation of resources among businesses and carry out dynamic adjustment and reconfiguration based on feedback results.

The U-Plane is composed of Fog nodes which provide high frequency electromagnetic wave signal and the UE which is connected to the network. Fog node is responsible for the digital analog conversion of the radio frequency transceiver function, and digital baseband processing function carried out in the M-Plane. In the U-Plane, it is no longer the center of a Cell with a fog node. The virtual baseband pool formed in M-Plane and it confuses the concept of Cell. Multiple fog nodes will carry out multi point cooperative communication (CoMP) for UE, and CoMP is initially used in the LTE system to eliminate and improve the interference between Cells. Figure 4 shows the U-Plane in 5G fog Cell. For the 5G fog Cell with fuzzy Cell boundary concept, the focus of the study is how the virtual resource pool will select and allocate fog nodes for UE to provide service of cooperative communication.

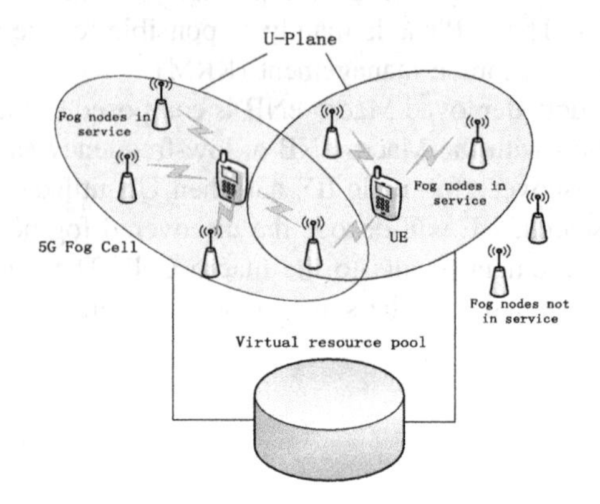

Fig. 4. C-Plane in 5G fog Cell

4 Radio Resource Management of Hypergraph Partitioning in 5G Fog Cell

In this model, fog node provides service for UE as a node, which will be provided with resource by virtual resource pool. Fog node resource network which has been deployed is regarded as a network composed of numerous nodes that can provide resources. In the communication services, UE firstly selects the specific resource pool, and then chooses the specific RHH in the resource pool to carry out the transmission of high frequency data services. The study proves that the Hypergraph theory is suitable for radio network modeling and radio resource management and allocation in such network.

4.1 Task Model

Assuming that UE of each access network is a task that needs to be processed. Each task is independent and can be handled by the fog nodes. Each task consists of a task ID, a task state, and an occupation amount of resource block (RB).

$\mathrm{T} = \{t_1, t_2, \cdots\cdots, t_n\}$ is a set of the resource management task. There is a total of n tasks within the system, among which t_i represents the ith task. $t_i = \{tID, tRB, tS\}$ is the attribute Set of the task, among which $i \in [1, n]$. tID represents the ID label of the task; tRB represents the resources amount required for the task communication; the tS represents the task state.

$tS = \{tmatch, tfree, tdoing, tdone\}$ represents the four task states of matching, idle, processing and processed respectively.

The task is idle before UE is accessed to the 5G fog Cell resource pool through the signaling dispatch of macro base station. After the macro base station monitoring

Channel State Information (CSI) of UE, it is shared with fog node of 5G fog Cell. According to CSI of UE, the matching between Hypergraph tree and communication node will be carried out. The communication business of UE on resource node is completed after the successful matching, and the task state switch from *tdoing* to *tdone*.

4.2 Hypergraph Model of 5G Fog Cell Resource Pool

When the connection is established by UE, it is divided into cluster and node selection. The Hypergraph is utilized to model the resource pool of 5G fog Cell. The Hypergraph is composed by the node and hyperedge provided by RB communication resources. Fog node which can provide RB communication resources is utilized to represent the nodes in the Hypergraph. The node set is composed of resource ID, bandwidth B, power P, current transmission rate R, fog node channel gain H corresponding to user UE, and resource state S. Resource description is as follows:

$G = (V, E)$ 5G fog Cell resource pool Hypergraph model

(1) Hypergraph node description

Definition 1: $V = \{v_1, v_2, \cdots\cdots, v_n\}$ is a set of fog nodes in the network, where v_i represents the *ith* fog node. $v_i = \{ID_i, B_i, B_{\mathrm{max}i}, P_i, P_{\mathrm{max}i}, H_i, Load_i, S_i, R_i, R_{\mathrm{max}i}\}$ ID is used to identify fog node; B_i is the bandwidth of the communication between UE and the node; P_i is the power when node is communicated with UE; H_i is the channel gain;

$$Load_i = \sum_{i \in Rmatch} \omega_{match} B_i + \sum_{i \in Rdoing} \omega_{doing} B_i \tag{1}$$

$Load_i$ represents the load of resource nodes.

S is the state of current node.

Definition 2: S is the current node state.

$S = \{Savailable, Sunavailable\}$ is the load of node, and the node state is determined by load rate φ_{Load}. When φ_{Load} is larger than resource threshold θ_{Load}, the node is in state of *Sunavailable*, and no longer participate in the allocation of nodes. On the contrary, the node is in the state of *Savailable*.

$$\varphi_{Load} = \frac{Load_i}{B_{\mathrm{max}}} \tag{2}$$

The maximum communication rate of a node before the connection between UE and a certain resource node is established:

$$R_{\max i} = B_{\max i} \log_2\left(1 + \frac{P_{\max i} H_i}{N}\right) \tag{3}$$

Where $B_{\max i}$ represents the maximum bandwidth reserved for the current resource node, $P_{\max i}$ is the maximum power provided by the node; H_i is the channel gain of the communication between UE and node, N is the noise.

(2) Hyperedge description

$E = \{e_1, e_2, \cdots\cdots, e_m\}$ is the set of hyperedge, and e_j is defined as $e_j = \{m_j, W_j\}$.

Where m_j is the quantity of nodes contained in the hyperedge. The weight value of the hyperedge W_j is defined as the average value of the system throughput of all nodes that are contained in the hyperedge.

$$W_j = \frac{\sum_{v_i \in e_j} R_{\max i}}{m_j} \tag{4}$$

4.3 Hypergraph Cluster and Resource Allocation

When calculating the collection of cooperative fog nodes, the resource remainder of each node is first judged. If the node load situation can allow nodes to participate inz resource allocation, then we give priority to the previously cooperative fog node combination to serve a UE. A double-layer cutting bandwidth allocation Hypergraph clustering algorithm is proposed based on the Apriori algorithm [20].

(1) Multiple groups of data set D composed of node ID is recorded with the apriori knowledge of fog node radio cooperation.
(2) Traverse the state value of S within the data set, and conduct ID_i of $S_i = Sunavailable_i$ with pruning to obtain new data set D.
(3) The hyperedge (candidate set) E_k of single ID node is generated and count the support degree of each ID in the data set D.
(4) Remove the ID set of which the support degree is less than the threshold value ψ in hyperedge E_k, and form the frequent item set L_{k-1}.
(5) L_{k-1} connect by itself to form the new hyperedge candidate set E_{k+1}, where the ID set has been removed in last round of searing for frequent item set is not included. Count the support degree.
(6) Carry out reiterative iterative operations to find all the ID sets larger than the threshold ψ. The selected nodes will form $V = \{v_1, v_2, \cdots\cdots, v_m\}$. The nodes in each Hypergraph can provide cooperative communication services for UE.
(7) Compare $W_j = \frac{\sum_{v_i \in e_j} R_{\max i}}{m_j}$, and select Max_{W_j} as aggregate of the fog node collaborative communication.

Algorithm 1 Hypergraph clustering algorithm based on the Apriori

L1 =find_frequent_1-itemsets(*D*);
For (k=2; L_{k-1} != null;k++){
// Produce a candidate and prune
C_k =apriori_gen(L_{k-1});
// D for candidates counting
For each *t* in *D* {
C_t =subset(C_k,*t*); // subset of *t*
For each $c \in C_t$
c.count++;
}
// Return an item set that is not less than minimum support
Lk ={$c \in C_k$ | c.count>=min_sup}
}
Return *L*= All frequent sets;
// First step: join.
Procedure apriori_gen (L_{k-1}: frequent(k-1)-itemsets)
For each $l_1 \in L_{k-1}$
For each $l_2 \in L_{k-1}$
If ((l_1 [1] =l_2 [1]) && (l_1 [2] =l_2 [2]) && ……&& (l_1 [k-2] =l_2 [k-2]) && (l_1 [k-1] $< l_2$ [k-1]))
then {
$c = l_1$ join to l_2

if has_infrequent_subset (*c*, L_{k-1}) then
delete *c*;
else add c to C_k;
}
Return C_k;
// Second step: prune.
Procedure has_infrequent_sub (*c*:candidate k-itemset; L_{k-1} :frequent(k-1)-itemsets)
For each (k-1)-subset *s* of *c*
If $s \notin L_{k-1}$, then
Return true;
Return false;

5 Numerical Simulation

In this Section, we combine the 5G radio network architecture with the corresponding radio resource allocation algorithm, and we simulate the radio communication system from the point of maximum UE access and radio resource utilization.

In the simulation environment, we assume that a radio communication system is composed of a Macro-eNB and a plurality of fog node (FN). We give a Transcendental cooperation data table in Table 1, and the main parameters for the simulations are shown in Table 2, which consults Ref. [7].

Table 1. Cooperation set data of the FN that once cooperated

ID	Transcendental cooperation data	Times
1	FN1, FN3, FN4, FN5	2
2	FN1, FN2, FN3	3
3	FN2, FN3, FN5	1
4	FN1, FN4, FN5	1
5	FN3, FN5	2
6	FN1, FN2, FN4	2

As shown in Fig. 5, we can see in 5G Fog Cell network, Due to the separation of U-Plane and C-Plane, with the number of FN in the macro cell increasing, the maximum number of UE access in the system can increase.

Figure 6 shows the radio resource utilization in 5G fog Cell by Hypergraph modeling. Because of the increase of the group size, the radio resource utilization grows rapidly in the range of 0 to 200, and the growth rate slows down gradually in the range of 200 to 400. By viewing the data table, we found that after the amounts of users reached 200, FN3 reached the *Sunavailable* state, no longer participate in allocating resource. Lastly, comparing with traditional macro cell network, 5G fog Cell network has better performance in radio resource utilization.

6 Conclusion

In the future, 5G radio mobile communication system presents new requirements as follows: the "user centric" should replace the "eNB centric" to construct the network, and how to allocate and manage radio resource. 5G fog Cell is one of the schemes to solve the mentioned problems above, and it can effectively utilize the frequency resource for data communication. In this paper, we propose a concept of 5G fog Cell network architecture which can be implemented by Macro-eNB and fog nodes. An important research point is how we efficiently allocate or manage resource in 5G fog Cell. We introduce a Hypergraph clustering algorithm to manage and allocate radio resource. Furthermore, we verify that the 5G fog Cell network has better performance than traditional macro cell network in radio resource utilization.

Table 2. Simulation parameters

Parameters	Values
Transmission power	43 dBm (Macrocell), 30 dBm (FN)
Path loss (macro eNB)	L = 35.3 + 37.6 log (d), d = distance in meters
Shadow fading	Log-normal, 8 dB standard deviation(Macrocell) and 10 dB standard deviation (FN)
Operating freq. of macro eNB	2 GHz
Operating freq. of FN	3.5 GHz
System bandwidth for eNBs	5 MHz
Average transmission rate of user demand	10 kbps–1000 kbps
Cell layout	Hexagonal grid, 3-sector sites
Maximum allowed time delay	50 ms–5 s

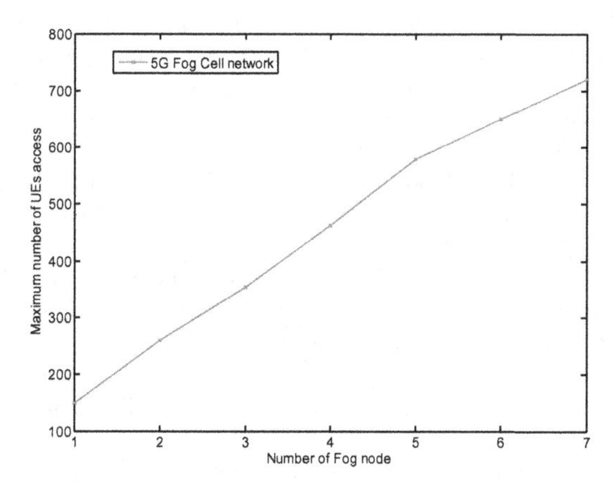

Fig. 5. Maximum number of UE access

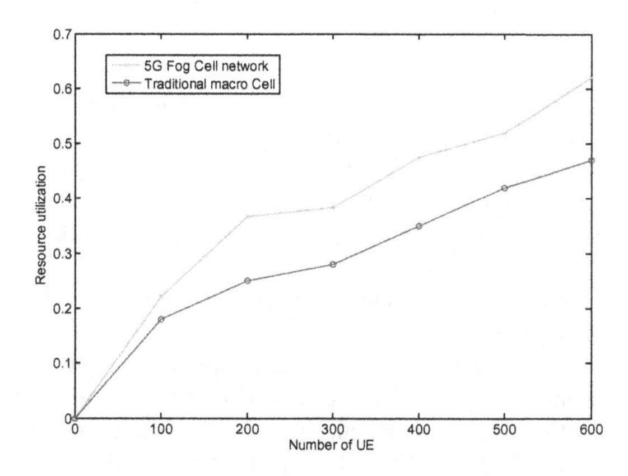

Fig. 6. Ratio of resource utilization

Acknowledgement. This work is supported by National Science Foundation Project of P. R. China (No. U1603116), the Foundation of Science and Technology on Information Assurance Laboratory (No. KJ-17-101), and the Foundation of Beijing Engineering and Technology Center for Convergence Networks and Ubiquitous Services.

References

1. Osseiran, A., et al.: Scenarios for 5G mobile and wireless communications: the vision of the METIS project. Commun. Mag. IEEE **52**(5), 26–35 (2014)
2. Zhang, X., Cheng, W., Zhang, H.: Heterogeneous statistical QoS provisioning over 5G mobile wireless networks. IEEE Netw. **28**(6), 46–53 (2014)
3. Palattella, M.R., et al.: Internet of things in the 5G era: enablers, architecture, and business models. IEEE J. Sel. Areas Commun. **34**(3), 510–527 (2016)
4. Tafazolli, R., et al.: Enabling 5G: energy and spectrally efficient communication systems. Trans. Emerg. Telecommun. Technol. **26**(1), 1–2 (2015)
5. Tran, T.X., et al.: Collaborative mobile edge computing in 5G networks: new paradigms, scenarios, and challenges. IEEE Commun. Mag. **55**(4), 54–61 (2017)
6. Fajardo, J.O., et al.: Introducing mobile edge computing capabilities through distributed 5G cloud enabled small cells. Mob. Netw. Appl. **21**(4), 1–11 (2016)
7. Bonomi, F., et al.: Fog computing and its role in the internet of things. In: Edition of the MCC Workshop on Mobile Cloud Computing, pp. 13–16. ACM (2012)
8. Jingtao, S.U., et al.: Steiner tree based optimal resource caching scheme in fog computing. China Commun. **12**(8), 161–168 (2015)
9. Gao, L., Luan, T.H., Liu, B., Zhou, W., Yu, S.: Fog computing and its applications in 5G. In: Xiang, W., Zheng, K., Shen, X. (eds.) 5G Mobile Communications, pp. 571–593. Springer, Cham (2017). https://doi.org/10.1007/978-3-319-34208-5_21
10. Yan, L., Fang, X., Fang, Y.: A novel network architecture for C/U-plane staggered handover in 5G decoupled heterogeneous railway wireless systems. IEEE Trans. Intell. Transp. Syst. **18**, 1–13 (2017)
11. Kamel, M.I., Le, L.B., Girard, A.: LTE multi-cell dynamic resource allocation for wireless network virtualization. In: Wireless Communications and Networking Conference, pp. 966–971. IEEE (2015)
12. Chen, L., et al.: Distributed virtual resource allocation in small-cell networks with full-duplex self-backhauls and virtualization. IEEE Trans. Veh. Technol. **65**(7), 5410–5423 (2016)
13. Singh, S., et al.: Mobile edge fog computing in 5G era: architecture and implementation. In: Computer Symposium, pp. 731–735. IEEE (2017)
14. Kitanov, S., Janevski, T.: State of the art: fog computing for 5G networks. In: Telecommunications Forum, pp. 1–4. IEEE (2017)
15. Kitanov, S., Janevski, T.: Fog computing as a support for 5G network. J. Emerg. Res. Solut. ICT **1**, 47–59 (2016)
16. Amendola, D., Cordeschi, N., Baccarelli, E.: Bandwidth management VMs live migration in wireless fog computing for 5G networks (2017)
17. Gu, L., et al.: Cost efficient resource management in fog computing supported medical cyber-physical system. IEEE Trans. Emerg. Top. Comput. **5**(1), 108–119 (2015)
18. Fuhong, L., et al.: Cooperative differential game for model energy-bandwidth efficiency tradeoff in the Internet of Things. China Commun. **11**(1), 92–102 (2014)

19. Do, C.T., et al.: A proximal algorithm for joint resource allocation and minimizing carbon footprint in geo-distributed fog computing. In: International Conference on Information Networking, pp. 324–329. IEEE (2015)
20. Liang, R., Guo, W., Yang, D.: Mining product problems from online feedback of Chinese users. Kybernetes **46**(3), 572–586 (2017)

A Novel Energy Harvesting Aware IEEE 802.11 Power Saving Mechanism

Yigitcan Celik(✉) and Cong Pu

Weisberg Division of Computer Science, Marshall University, Huntington, WV 25755, USA
{celik1,puc}@marshall.edu

Abstract. The spread of wirelessly connected computing sensors and devices and hybrid networks are leading to the emergence of an Internet of Things (IoT), where a myriad of multi-scale sensors and devices are seamlessly blended for ubiquitous computing and communication. However, the communication operations of wireless devices are often limited by the size and lifetime of the batteries because of the portability and mobility. To reduce energy consumption during wireless communication, the IEEE 802.11 standard specifies a power management scheme, called Power Saving Mechanism (PSM), for IEEE 802.11 devices. However, the PSM of IEEE 802.11 was originally designed for battery-supported devices in single-hop Wireless Local Area Networks (WLANs), and it does not consider devices that equip with rechargeable batteries and energy harvesting capability. In this paper, we extend the original PSM by incorporating with intermittent energy harvesting in the IEEE 802.11 Medium Access Control (MAC) layer specification, and propose a novel energy harvesting aware power saving mechanism, called *EH-PSM*. The basic idea of EH-PSM is that a longer contention window is assigned to a device in energy harvesting mode than that of a device in normal mode to make the latter access the wireless medium earlier and quicker. In addition, the device in energy harvesting mode stays active as far as it harvests energy and updates the access point of its harvesting mode to enable itself to be ready for receiving and sending packets, or overhearing any on-going communication. We evaluate the proposed scheme through extensive simulation experiments using OMNeT++ and compare its performance with the original PSM. The simulation results indicate that the proposed scheme can not only improve the packet delivery ratio and throughput but also reduce the packet delivery latency.

Keywords: Energy harvesting · Power saving mechanism
Medium access control · IEEE 802.11

1 Introduction

With recent technological advances in portability, mobility, low-power microprocessors, and high speed wireless Internet, embedded computing devices capable of

S. Chellappan et al. (Eds.): WASA 2018, LNCS 10874, pp. 14–26, 2018.
https://doi.org/10.1007/978-3-319-94268-1_2

wireless communication are rapidly proliferating. The growing presence of WiFi and 4G Long-Term Evolution (LTE) enable users to pursue seemingly insatiable access to Internet services and information wirelessly. It is predicted that 30 billion wirelessly connected devices will be available by 2020, nearly triple the number that exists today [1]. The spread of these devices and hybrid networks is leading to the emergence of an Internet of Things (IoT) [2], where a myriad of multi-scale sensors and devices are seamlessly blended for ubiquitous computing and communication. The prevalence of cloud, social media, and wearable computing and the reduced cost of processing power, storage, and bandwidth are fueling explosive development of IoT applications in major domains (i.e., personal and home, enterprise, utilities, and mobile) [3], which have the potential to create an economic impact of $2.7 trillion to $6.2 trillion annually by 2025 [4]. We envision that wirelessly connected smart devices under the IoT will not only enhance flexible information accessibility and availability but also improve our lives further.

To realize this vision, however, a limited lifetime of the battery to power wireless devices must be overcome. For example, the TMote^{TM} Sky node consumes 64.68 mW in receive mode [5]. Using two standard 3,000 mAh AA batteries, the network lifetime is only 5.8 days if nodes are heavily utilized [6]. In addition, rapidly proliferating wearable devices implanted to anywhere of user (e.g., glasses [7], clothes, shoes, accessories, or even under skin [8]) are directly affected by the lifetime of batteries. In order to extend the lifetime of the batteries, energy harvesting from surrounding environmental resources (e.g., vibrations, thermal gradients, lights, wind, etc.) has been given considerable attention as a way to either eliminate replacing the batteries or at least reduce the frequency of recharging the batteries. For example, ambient vibration-based energy harvesting has been widely deployed because of the available energy that can be scavenged from an immediate environment, such as the pulse of a blood vessel, or the kinetic motion of walking or running. Piezoelectric-based energy harvesting is favored when vibration is the dominant source of environmental energy and solar light is not always available. Though energy harvesting from photo-voltaic cells is popular and well-studied, it is inefficient because of the unpredictable availability of solar irradiation, such as installed location (e.g., a shaded area), initial position (e.g., seasonal variations between the sun's angle and solar panel), weather conditions (e.g., a rainy season), and harvesting period (e.g., daytime only).

Thus, we anticipate that energy harvesting will play a pivotal role in making possible self-sustainable wireless devices ranging from nano-scale sensors to hand-held mobile devices, and serve as a major building block for emerging IoT applications. Although environmental energy harvesting has been well studied in civil and mechanical engineering, research on energy harvesting sensitive communication algorithms and protocols embedded into link layer (i.e., IEEE 802.11 Medium Access Control (MAC)) is still in its infancy. Thus, key research motivations of this paper are summarized: (i) Prudent energy-efficient mechanisms have been proposed to extend the network lifetime. Despite the prior best effort-based approaches, manually replacing the batteries or recharging the batteries is ultimately unavoidable. Disposable batteries can address this issue but they may pose a potential environmental hazard; (ii) Energy harvesting from immediate environmental sources may radically shift the paradigm on energy management from reducing energy consumption to maximizing the utilization of opportunistic energy; (iii) With the increasing prevalence of wearable computing, the adoption of energy harvesting techniques to multi-scale wireless sensors and mobile devices is essential to the design of IoT applications.

In this paper, we design and propose a novel Power Saving Mechanism (PSM) of IEEE 802.11 for self-sustainable devices supported by intermittent energy harvesting. The proposed research will shift the paradigm of energy management from conserving limited battery energy to maximizing the utilization of harvested energy, and provide design considerations to the broader IoT community seeking new applications. Our major contribution is briefly summarized in two-fold:

- We design and propose a novel energy harvesting aware power saving mechanism of IEEE 802.11, called *EH-PSM*, incorporating with intermittent energy harvesting from environmental resources. The basic idea of EH-PSM is that a longer contention window is assigned to a device in energy harvesting mode than that of a device in normal mode to make the latter access the wireless medium and earlier and quicker. In addition, the device in energy harvesting mode stays awake as far as it harvests energy and updates the access point of its harvesting mode to enable itself to be ready for receiving and sending packets, or overhearing any on-going communication.
- We develop a customized discrete event-driven simulation framework by using OMNeT++ [9] and evaluate the proposed scheme through extensive simulation experiments in terms of packet delivery ratio, throughput, packet delivery latency, and total active time. We also revisit and implement the original PSM in energy harvesting environment for performance comparison.

The simulation results indicate that the proposed scheme can not only improve the packet delivery ratio and throughput but also reduce the packet delivery latency.

The rest of paper is organized as follows. The prior approaches are analyzed and presented in Sect. 2. A system model and the proposed energy harvesting aware power saving mechanism are presented in Sect. 3. Section 4 presents extensive simulation results and their analyses. Finally, we conclude the paper in Sect. 5.

2 Related Work

In [10], an energy saving algorithm is proposed to reduce power consumption of tethering smartphone, which plays a role of mobile access point temporarily. The basic idea is that smartphone turns off its WiFi interface when there is no traffic in order to conserve battery power without increasing the packet delay substantially. The [11] proposes a system, called *Percy*, to maximize the energy saving of static and dynamic PSM respectively while minimizing the delay of flow completion time. The Percy deploys a web proxy at the access point and suitably configures the PSM parameters, and is designed to work with unchanged clients running dynamic PSM, unchanged access points, and Internet servers. In [12], a detailed anatomy of the power consumption by various components of WiFi-based phones has been provided. Through a measurement-based study of WiFi-based phones, the power consumption for various workloads at various components have been analyzed.

WiFi continues to be a prime source of energy consumption in mobile devices, and energy optimization has conventionally been designed with a single access point. However, network contention among different access points can dramatically increase client's energy consumption. Thus, the [13] designs and proposes an approach to achieve energy efficiency by evading network contention. The basic idea is that the access points regular the sleeping window of their clients in a way that different access points are active/inactive during non-overlapping time windows. In [14], a micro power management is proposed to enable an IEEE 802.11 interface to enter unreachable power saving mode even between medium access control (MAC) frames, without noticeable impact on the traffic flow. To control data lost, the proposed scheme leverages the retransmission mechanism in IEEE 802.11 and controls frame delay to adapt to demanded network throughput with minimal cooperation from the access point. The [15] presents an experimental study of the IEEE 802.11 power saving mechanism on PDA in a Wireless Local Area Network (WLAN), where the power consumption of the PDA in both continuous active mode and power saving mode are measured under various traffic scenarios, beacon period, and background multicast traffic. In [16], the performance of different MAC schemes based on CSMA and polling techniques for wireless sensor networks which are solely powered by solar energy are studied. In [17], the feasibility of powering wireless metering devices (e.g., heat cost allocators) by thermal energy harvested from radiators is investigated.

In summary, there is a significant amount of research effort on the IEEE 802.11 power saving mechanism and its variants. However, to the best of authors' knowledge, the proposed research focusing on designing energy harvesting aware power saving mechanism and integrating it with the original PSM of IEEE 802.11 is new.

3 The Proposed Approach

In this section, we first review the IEEE 802.11 Power Saving Mechanism (PSM). Then we present the system model, and propose a novel energy harvesting aware PSM of IEEE 802.11 for self-sustainable devices supported by intermittent energy harvesting.

3.1 Overview of the IEEE 802.11 Power Saving Mechanism

As proposed in the IEEE 802.11 standard [18], an IEEE 802.11 based wireless network interface can choose to stay in either awake state or sleep state at any time. In the awake state, the device turns on its wireless interface and performs normal data communications, for example receiving or sending packets, or just stay in idle. In order to save the residual energy, the device can switch to the sleep state, where the radio of a device is turned off and the wireless interface cannot detect or sense any wireless communication. Wireless interface in awake state usually consumes an order of magnitude more power than that in sleep state [15].

To reduce the energy consumption of IEEE 802.11 devices during wireless communications, the [18] specifies a power management scheme, called Power Saving Mechanism (PSM). The basic idea of PSM is that the devices sleep most of the time and stay at a low power state (i.e., turn off wireless interface) but periodically wake up and switch to a high power state (i.e., turn on wireless interface) to receive the packets buffered at the access point (AP). In PSM, the AP buffers incoming packets destined for devices in low power state and periodically announces its buffering status through the traffic indication map (TIM) contained in the beacon frames. The device wakes up at beginning of beacon interval periodically to listen to the beacon frames. If the corresponding bit of the association ID (AID) of device is set in the TIM, the device will stay awake, and wait for the AP to initialize a PS-Poll packet to retrieve data packet from it and/or send the buffered packet to it. The AP can issue multiple PS-Poll packets until all outstanding packets from the devices have been retrieved. As opposed to the continuously awake mode, a device applying power saving mechanism can often have opportunities to turn off its wireless interface to save energy when it has no packets buffered at the AP, or no packets need to be sent to the AP. Here, a snapshot of the power saving mechanism of IEEE 802.11 is shown in Fig. 1. For example, the device S_A and S_B wake up at the beginning of beacon interval and listen to the TIM broadcasted by the AP. Suppose that the AP initializes a PS-Poll packet to S_A first and sends the buffered packet $DATA_a$ to it. After receiving the packet $DATA_a$, S_A replies an ACK packet to the AP after a short time period $SIFS$, and then switches to power saving mode and turns off its wireless interfaces to save energy.

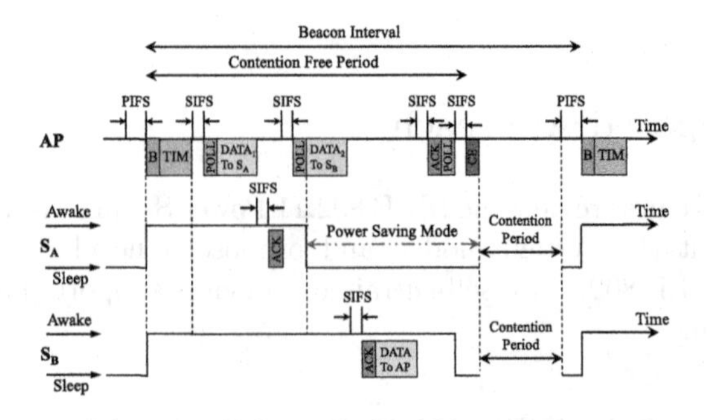

Fig. 1. A snapshot of the power saving mechanism of IEEE 802.11.

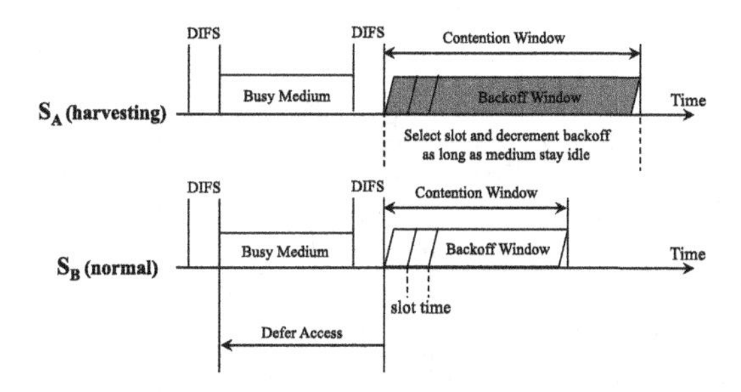

Fig. 2. An example of longer contention window for device in energy harvesting mode.

3.2 System Model

In this paper, we assume that each IEEE 802.11 device equips with an energy harvesting component to replenish its rechargeable battery. For example, a piezo-electric fiber composite bi-morph (PFCB) W14 based energy harvesting from an immediate environment (e.g., disturbance or typical body movements) can generate sufficient power (i.e., 1.3 mW–47.7 mW) for wireless sensors [19–21]. It is envisaged that multi-scale piezo devices and integrated self-charging power cells (SCPCs) [22] will enhance the efficiency of energy harvesting. An energy harvesting is modeled by a two-state Markov process with energy harvesting (S_h) and normal (S_n) modes [6]. A device stays in S_n mode for a random amount of time, which is exponentially distributed with a mean λ_n, and changes its mode into S_h mode. After harvesting energy for some amount of time in S_h mode, which is also assumed to be exponentially distributed with a mean λ_h, the device changes its mode back to S_n mode. Both S_n an S_h modes are repeated.

3.3 Energy Harvesting Aware Power Saving Mechanism

The PSM of IEEE 802.11 was originally designed for battery-supported devices in single-hop Wireless Local Area Networks (WLANs), and does not consider devices that equip with rechargeable batteries and energy harvesting component. In this paper, we will extend the original PSM by incorporating with intermittent energy harvesting in the IEEE 802.11 Medium Access Control (MAC) layer specification, and propose a novel energy harvesting aware power saving mechanism, called *EH-PSM*. The basic idea of EH-PSM is that a longer contention window is assigned to a device in energy harvesting mode than that of a device in normal mode to make the latter access the wireless medium earlier and quicker. In addition, the device in energy harvesting mode stays awake as far as it harvests energy and updates the AP of its energy harvesting mode to enable itself to be ready for receiving and sending packets, or overhearing any on-going communication. The rationale behind of the EH-PSM is to shift the paradigm of energy management from conserving limited battery energy to maximizing the utilization of harvested energy. Thus, we focus on (i) how to assign contention

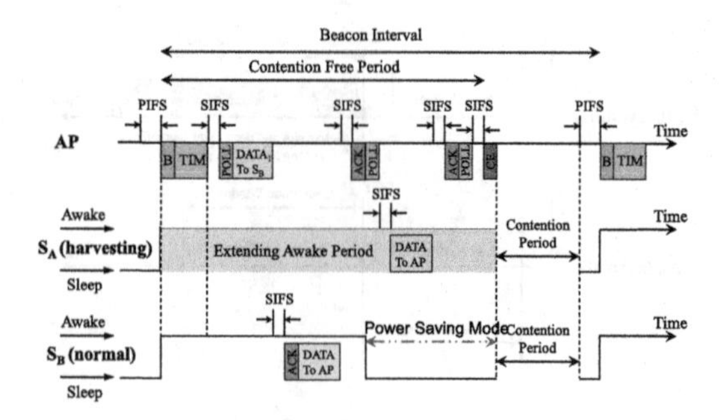

Fig. 3. An example of node in energy harvesting mode extends awake period.

window to a device in energy harvesting mode, and (ii) how to extend the awake time period of device in energy harvesting mode.

First, in the original PSM, each device uniformly chooses the contention window for backoff period before accessing the medium to avoid any potential collision. In the presence of energy harvesting, however, each device may need to adjust its contention window differently. The basic idea is to intentionally assign a longer contention window to a device in energy harvesting mode than that of a device in normal mode. Then a device containing the less amount of residual energy or staying in normal mode has more chances to choose a shorter backoff period to access the medium earlier and quicker, and then turn off its wireless interface, which finally results in lower energy consumption. Since a device in energy harvesting mode may experience a longer delay before initiating the communication, it will have a shorter contention window later to access the medium for fairness when it is in the normal mode. For example, as shown in Fig. 2, suppose that device S_A and S_B is in energy harvesting mode and normal mode, respectively. Since S_A is in energy harvesting mode, it intentionally sets a longer contention window and randomly chooses a backoff period. However, S_B is in normal mode, and it follows the original PSM and select the backoff period from a normal contention window that is shorter than that of S_A. Thus, S_B has more chances to select a shorter backoff period from normal contention window, accesses the medium earlier, and finally consumes less amount of residual energy.

Second, in the original PSM as shown in Fig. 1, a device immediately sleeps back after receiving the buffered packets from the AP or sending generated packets to the AP in order to reduce energy consumption. In the EH-PSM, however, the basic idea is that a device in energy harvesting mode stays awake as far as it harvests energy and updates the AP of its energy harvesting mode. This approach enables the device to be ready for receiving and sending packets, or overhearing any on-going communication. In addition, in order to reduce the energy consumption of device in normal mode, the AP can specify non-harvesting device as early polled device in the Traffic Indication Map (TIM) and send the PS-Poll packet and buffered packets immediately after broadcasting the TIM. After receiving buffered packets from AP or sending the outstanding packet to

Notations:

- B_{int}, T_{CFP}, T_{CP}, CW_i, t_i^{boff}, and δ: Beacon interval, contention-free period, contention period, contention window of device n_i, backoff period of device n_i, and extension of contention window.
- $pkt[src, des, type]$: A packet with source id, src, destination id, des, and packet type, $type$. Here, $type$ is *Data*, *PS-Poll* or *ACK*.
- *TIM*, S_h, S_n: Defined before.

```
◇ At the beginning of B_int, device n_i wakes up to listen to TIM:
  /* Contention free period T_CFP begins */
  if i ∈ TIM
    if n_i is in S_n
      /* n_i is in normal mode */
      Wait for pkt[AP, i, PS-Poll] from AP;
      Reply pkt[i, AP, ACK] to AP with potential pkt[i, AP, Data];
      Turn off wireless interface;
    else
      Wait for pkt[AP, i, PS-Poll] from AP;
      Reply pkt[i, AP, ACK] to AP with potential pkt[i, AP, Data];
      Keep wireless interface on; Overhear on-going communication;
  else
    if n_i is in S_n
      Turn off wireless interface;
    else
      Keep wireless interface on; Overhear on-going communication;
◇ When T_CFP ends, and T_CP begins:
  /* Contention period T_CP begins */
  if n_i is in S_n
    t_i^boff = rand(CW_i);
  else
    t_i^boff = rand(CW_i + δ);
```

Fig. 4. The pseudocode of EH-PSM.

AP, the device in normal mode can switch to power saving mode and turn off its wireless interface to save energy. For example, as shown in Fig. 3, S_A is in energy harvesting mode while S_B is in normal mode. In order to reduce the energy consumption of S_B, the AP first polls S_B's packet and sends the buffered packet to it after broadcasting the TIM. After receiving the buffered packet from AP and replying the *ACK* with the outstanding packet to AP respectively, S_B switches to power saving mode and turns off its wireless interface to reduce energy consumption. However, S_A stays awake as far as it harvests energy in energy harvesting mode, and extends awake time period to overhear any on-going communication, e.g., PS-Poll packet. Since S_A extends awake time period, whenever it has a newly generated packet for AP, it can directly send the packet after overhearing the PS-Poll packet. Here, major operations of the EH-PSM are summarized in Fig. 4.

4 Performance Evaluation

We develop a customized discrete-event driven simulator using OMNeT++ [9] to conduct our experiments. A 600×400 m^2 network area is considered, where 5 to 9 devices which belong to one access point (AP) are distributed in the network. The communication range of each device is 500 (m). The radio model simulates CC2420 with a normal data rate of 2 Mbps. The access point generates

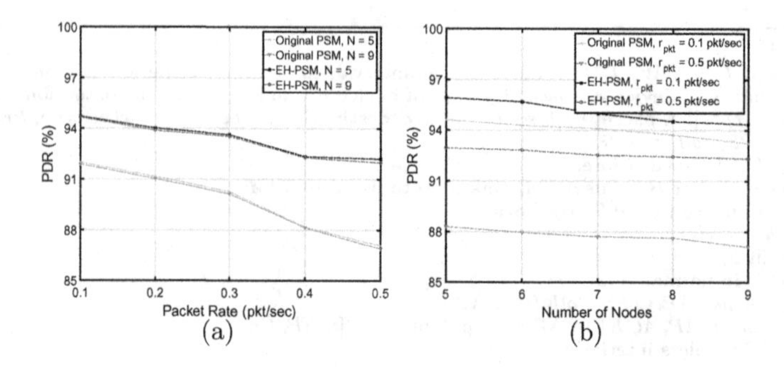

Fig. 5. The performance of packet deliver ratio (PDR) against packet rate and number of nodes.

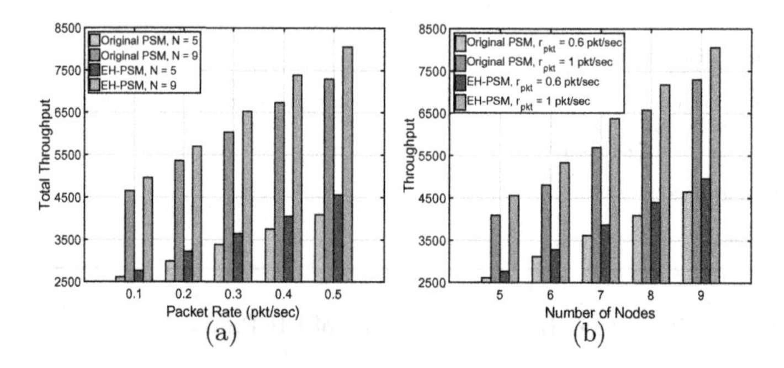

Fig. 6. The performance of throughput against packet rate and number of nodes.

data traffic with packet injection rate 0.1 to 1.0 pkt/s and the packet size is 60 Bytes. The periods of energy harvesting and normal states are assumed to be exponentially distributed with mean λ_h (50 s) and λ_n (25 s), respectively. The total simulation time is 5000 s. In this paper, we measure the performance in terms of packet delivery ratio (PDR), throughput, packet delivery latency, and total awake time by changing key simulation parameters, including number of nodes (N) and packet rate (r_{pkt}). For performance comparison, we compare the proposed scheme EH-PSM with the original PSM of IEEE 802.11.

First, we measure the packet deliver ratio (PDR) by varying packet rate and number of nodes in Fig. 5. In Subfig. 5(a), as the packet rate increases from 0.1 to 0.5 pkt/s, the PDR of EH-PSM and original PSM decreases from 95% and 92% to 93% and 87%, respectively. Since each node generates more packets with larger packet rate, packets could have more chances to collide with each other, and the PDR decreases. The EH-PSM shows the higher PDR than that of original PSM because each node stays awake as far as it is in energy harvesting mode and overhears on-going communication, it can forward the packets to AP directly with a shorter contention window based on the overhearing network traffic. However, the PDR is not sensitive to the number of nodes in the network, and thus, a slightly lower PDR is observed with larger number of nodes. In

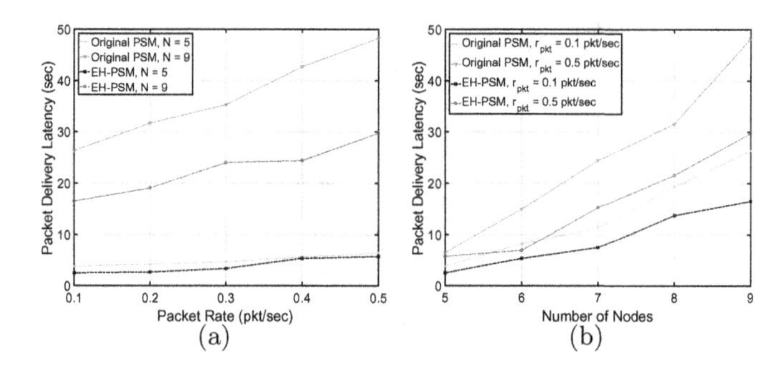

Fig. 7. The performance of packet transmission latency against packet rate and number of nodes.

Subfig. 5(b), the overall PDR of EH-PSM and original PSM decreases as the number of nodes increases. This is because more number of nodes will content for the medium for sending packets to the AP during the contention period, packets have more chances to collide with each other, the number of packets received by the AP is reduced. As the packet rate increases, a lower PDR is observed. This is because more number of packets are generated and forwarded to the AP, more packets collide with each other at the AP, which results in a lower PDR.

Second, Fig. 6 shows the throughput of EH-PSM and original PSM with varying packet rate and number of nodes. As shown in Subfig. 6(a), the overall throughput increases as the packet rate increases. This is because each node can generate more number of packets and send them to the AP. However, the EH-PSM shows a higher throughput than that of original PSM with different number of nodes, this is because each node in energy harvesting mode can extend their awake time period to be ready for receiving and sending more packets. When the number of nodes increases, a higher throughput is observed by EH-PSM and original PSM, respectively. This is because more number of nodes could send more packets to the AP, the throughput is increased. However, the EH-PSM still performs better than original PSM. This is because more nodes could stay awake for a longer time period in energy harvesting mode and overhear on-going communication, and more packets can be delivered to the AP when the medium is free. In Subfig. 6(b), as the number of nodes increases, the throughput of EH-PSM and original PSM increases, respectively. Since more number of nodes are associated with the AP and could generate and send more packets, finally a higher throughput is achieved. However, our scheme still achieves a better performance than original PSM. This is because more number of nodes can switch to energy harvesting mode and then stay awake as far as they harvest energy, more packets can be generated and sent to the AP, which result in a higher throughput.

Third, we measure the packet delivery latency by varying packet rate and number of nodes in Fig. 7. Overall, the packet deliver latency increases as the packet rate increases in Subfig. 7(a). With a larger packet rate, the AP can

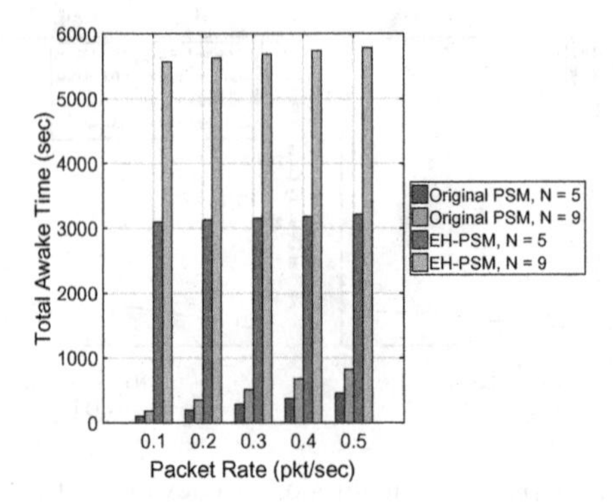

Fig. 8. The performance of total active time period against packet rate and number of nodes.

generate more packets for each node in the network. However, packet receiver could be in normal mode and switch to power saving mode (i.e., turn off wireless interface) after receiving all buffered packets from the AP. The newly generated packets at the AP have to be buffered until the next beacon period, thus, a higher packet delivery latency is observed. But, our scheme shows a lower packet delivery latency than original PSM, because each node could be in energy harvesting mode, extend awake time period, and then receive more newly generated packets from the AP quickly. As shown in Subfig. 7(b), the overall packet delivery latency of EH-PSM and original PSM increases as the number of nodes increases. This is because more number of nodes could be in normal node and switch to power saving mode after receiving all buffered packets from the AP, the newly generated packets have to be buffered at the AP and experience a longer packet delivery latency. However, the EH-PSM still provides the better performance than that of original PSM because the node in energy harvesting mode extends its awake time period and enables itself to be ready for receiving packets. Thus, the AP can directly send the newly generated packets quickly, which results in a lower packet delivery latency.

Fourth, we measure the total active time against packet rate and number of nodes in Fig. 8. Overall, the proposed EH-PSM achieves a much higher total active time than the original PSM. This is because the EH-PSM enables each node in energy harvesting mode to extend its awake time period, a larger total active time period can be observed compared to that of original PSM. As the number of nodes increases, the total active time is increased because more number of nodes could stay in energy harvesting mode and extend their active time period.

5 Conclusion

In this paper, we investigated the power saving mechanism incorporating with intermittent energy harvesting in the IEEE 802.11 Medium Access Control

(MAC) layer specification, and proposed a novel energy harvesting aware power saving mechanism, called *EH-PSM*. In the EH-PSM, a longer contention window is assigned to a device in energy harvesting mode than that of a device in normal mode to make the latter access the wireless medium earlier and quicker. In addition, the device in energy harvesting mode stays awake as far as it harvests energy and updates the access point of its energy harvesting mode to enable itself to be ready for receiving and sending packets, or overhearing any on-going communication. We evaluated the performance of the proposed scheme through extensive simulation experiments, compared it with the original PSM of IEEE 802.11. Extensive simulation results indicate that the proposed scheme achieves better performance in terms of packet delivery ratio, throughput, and packet delivery latency.

References

1. More than 30 billion devices will wirelessly connect to the internet of everything in 2020
2. Palattella, M., Dohler, M., Grieco, A., Rizzo, G., Torsner, J., Engel, T., Ladid, L.: Internet of things in the 5G era: enablers, architecture, and business models. IEEE J. Sel. Areas Commun. **34**(3), 510–527 (2016)
3. Gubbi, J., Buyya, R., Marusic, S., Palaniswami, M.: Internet of things (IoT): a vision, architectural elements, and future directions. Future Gener. Comput. Syst. **29**, 1645–1660 (2013)
4. Al-Fuqaha, A., Guizani, M., Mohammadi, M., Aledhari, M., Ayyash, M.: Internet of things: a survey on enabling technologies, protocols, and applications. IEEE Commun. Surv. Tutor. **17**(4), 2347–2376 (2015)
5. Raymond, D., Marchany, R., Brownfield, M., Midkiff, S.: Effects of denial of sleep attacks on wireless sensor network MAC protocols. In: Proceedings Workshop on Information Assurance, pp. 297–304 (2006)
6. Pu, C., Gade, T., Lim, S., Min, M., Wang, W.: Light-weight forwarding protocols in energy harvesting wireless sensor networks. In: Proceedings IEEE MILCOM, pp. 1053–1059 (2014)
7. Google Glass. http://www.google.com/glass/start/
8. Wearable computing is here already: how hi-tech got under our skin. http://www.independent.co.uk
9. Varga, A.: OMNeT++ (2014). http://www.omnetpp.org/
10. Jung, K., Qi, Y., Yu, C., Suh, Y.: Energy efficient Wifi tethering on a smartphone. In: Proceedings IEEE INFOCOM, pp. 1357–1365 (2014)
11. Ding, N., Pathak, A., Koutsonikolas, D., Shepard, C., Hu, Y.C., Zhong, L.: Realizing the full potential of PSM using proxying. In: Proceedings IEEE INFOCOM, pp. 2821–2825 (2012)
12. Gupta, A., Mohapatra, P.: Energy consumption and conservation in WiFi based phones: a measurement-based study. In: Proceedings IEEE SECON, pp. 122–131 (2007)
13. Manweiler, J., Choudhury, R.R.: Avoiding the rush hours: WiFi energy management via traffic isolation. In: Proceedings ACM MobiSys, pp. 253–266 (2011)
14. Liu, J., Zhong, L.: Micro power management of active 802.11 interfaces. In: Proceedings ACM MobiSys, pp. 146–159 (2008)

15. He, Y., Yuan, R., Ma, X., Li, J.: The IEEE 802.11 power saving mechanism: an experimental study. In: Proceedings IEEE WCNC, pp. 1362–1367 (2008)
16. Eu, Z., Tan, H., Seah, W.: Design and performance analysis of MAC schemes for wireless sensor networks powered by ambient energy harvesting. Ad Hoc Netw. **9**(3), 300–323 (2011)
17. Vithanage, M.D., Fafoutis, X., Andersen, C.B., Dragoni, N.: Medium access control for thermal energy harvesting in advanced metering infrastructures. In: Proceedings EuroCon, pp. 291–298 (2013)
18. Eu, Z., Tan, H., Seah, W.: IEEE Std 802.11: wireless LAN medium access control (MAC) and physical layer (PHY) specifications. In: IEEE Computer Society LAN MAN Standards Committee, no. 3, August 1999
19. Starner, T.: Human-powered wearable computing. IBM Syst. J. **35**(3 & 4), 618–629 (1996)
20. Starner, T., Paradiso, J.A.: Human generated power for mobile electronics, pp. 1–35. CRC Press (2004)
21. Wang, Z.L.: Nanogenerators for self-powered devices and systems. Georgia Institute of Technology, Atlanta, USA (2011)
22. Xue, X., Wang, S., Guo, W., Zhang, Y., Wang, Z.L.: Hybridizing energy conversion and storage in a mechanical-to-electrochemical process for self-charging power cell. Nano Lett. **12**(9), 5048–5054 (2012)

Interest-Aware Next POI Recommendation for Mobile Social Networks

Ming Chen[1], Wenzhong Li[1,2(✉)], Lin Qian[1,3], Sanglu Lu[1,2], and Daoxu Chen[1]

[1] State Key Laboratory for Novel Software Technology, Nanjing University, Nanjing, China
chenming@dislab.nju.edu.cn, lwz@nju.edu.cn
[2] Sino-German Institutes of Social Computing, Nanjing University, Nanjing, China
[3] NARI Group Corporation/State Grid Electric Power Research Institute, Nanjing, China

Abstract. Recommending the next point-of-interest (POI) to mobile users is an interesting topic for mobile social networks to provide personalized location-based services. In this paper, we propose an interest-aware next POI recommendation approach, which consider the location interest among similar users and the contextual information (such as time, current location, and friends preference) for POI recommendation. We develop a spatial-temporal topic model to describe users location interest, based on which we form comprehensive feature representations regarding user interest and contextual information. We propose a supervised learning prediction model for next POI recommendation. Experiments based on the Gowalla dataset verify the accuracy and efficiency of the proposed approach.

Keywords: Location-based service · POI recommendation
Mobile social network

1 Introduction

With the development of mobile Internet in the recent years, various location-based services (LBS) in mobile social networks are becoming more and more popular. In the LBS applications, users upload their locations to the LBS server, such as checking-in a restaurant using Foursquare [1]. The check-in locations are

This work was partially supported by the National Key R&D Program of China (Grant No. 2017YFB1001801), the National Natural Science Foundation of China (Grant Nos. 61672278, 61373128, 61321491), the science and technology project from State Grid Corporation of China (Contract No. SGSNXT00YJJS1800031), the Collaborative Innovation Center of Novel Software Technology and Industrialization, and the Sino-German Institutes of Social Computing.

S. Chellappan et al. (Eds.): WASA 2018, LNCS 10874, pp. 27–39, 2018.
https://doi.org/10.1007/978-3-319-94268-1_3

known as point-of-interests (POIs) representing the stores, bars, restaurants that maybe interested to the users. Next POI recommendation is an important topic in mobile social network, which has a broad range of applications such as Ad pushing, city plan, traffic prediction, emergency alert, etc.

Numerous works had been done on next POI recommendation. Some works simply recommended the most popular nearby locations to the users [2]. Such recommendation is not personalized and inaccurate. The Matrix Factorization approaches [9] adopted conventional recommendation algorithms such as Collaborative Filtering (CF) [12] and FPMC [14] to recommend POI based on the similarity of check-in behaviors among users. However, they neglected the context of the users. Some works proposed the usage of Markov Chain for POI prediction [22]. However, they failed to capture the temporal dependency and personalized location preferences. A few works predicted next POI based on neural network (NN), which is a black box and its factors are not easy to be explained. In this paper, we propose an interest-aware next POI recommendation approach, which comprehensively consider the location interest among similar users, the contextual information such as time and current location, and the location preferences of friends in mobile social network, to achieve high accurate POI prediction.

Intuitively, the historical check-in records reflect the location interest of a user, and the users with similar interests may have similar check-in preference. To extract location interest information, we propose a spatial-temporal topic model based on latent Dirichlet Allocation (LDA), which takes the historical POI check-in records as a document to represent users' interest as the distribution among a number of topics. We further apply a spectral clustering algorithm to find the group of users with similar interests. We combine the extracted interest information with contextual features (e.g., current location, time, friends) to build a supervised learning model for next POI recommendation. We test the performance of the proposed model using the dataset from Gowalla [5]. Extensive experiments show that the proposed interest-aware POI prediction approach outperforms the conventional approaches.

Our main contributions are summarized as follows.

- **A spatial-temporal topic model to extract users' interest.** We propose a spatial-temporal topic model based on LDA which is well used in NLP (natural language process), to extract users' interest on locations.
- **Comprehensive feature representations.** We comprehensively extract several features including the location interest among similar users, the contextual information of time and current location, and the location preferences of friends in mobile social network, which are helpful for POI prediction.
- **An interest-aware next POI recommendation approach.** We proposed an interest-aware POI prediction model based on supervised learning, which results are well visualized and well explained.

2 Related Work

In this section, we review related work about location recommendation models and location recommendation system.

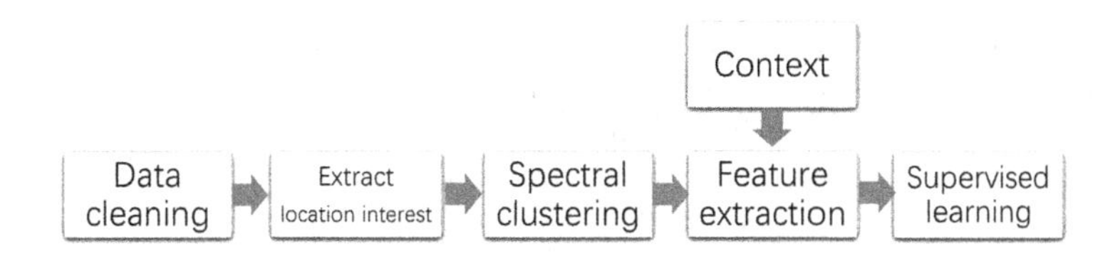

Fig. 1. The solution framework.

Location Recommendation Models: The trivial model is to recommend the most popular location as the next POI. Obviously, this recommendation is simple and inefficient. In the field of recommendation, Matrix Factorization (MF) method is a traditional way that factorize a user-item matrix into two low rank matrices, which product is the user-item matrix. [10] recommended POI using the conventional collaborative filtering. Zheng et al. [20] constructed location-activity, location-feature, and activity-activity matrices, and used a collective matrix factorization method to mine POI and activities. Tensor Factorization (TF) could add a time dimension to original Matrix Factorization. Zheng et al. [21] also used tensor factorization on the user-location-activity relationship to mine POI and activities. Markov chain is a traditional model that could recommend next POI based on user's past behavior. Zimdars et al. [22] used temporal data for making recommendations based on Markov chains. Rendle et al. [14] proposed Factorizing Personalized Markov Chains Model which subsumes Matrix Factorization and Markov chain together, learned an own transition matrix for each user, and outperformed both Matrix factorization and unpersonalized Markov chain model. There are also some work based on neural network. Zhang et al. [19] introduced a novel framework based on Recurrent Neural Networks (RNN) which directly models the dependency on users sequential behaviors into the click prediction process. Liu et al. [11] extended RNN and proposed ST-RNN to model continuous time interval and geographical distance for predicting the next location. Theses neural network [11,17–19], unlike the traditional algorithm, although it could achieve good performance, it is not easy to be explained.

Location Recommendation Systems: COMPASS [16] recommended locations to tourist based on weather, traffic conditions, tourist's profile, shopping list, schedule. MobyRec [15] could recommend restaurants based on users' preferences. These recommendation systems required user declare their interest such as which kind restaurant to eat. Our recommendation is different from above, while only using check-in log data, users do not need to specify their preferences.

3 Problem Formulation

We consider the scenario that people use mobile social network APPs such as Foursquare, Gowalla, etc., to check-in their locations. Assume the trajectories of users' check-in history are known. For a check-in event of a user, given his current location, the time stamp, the trajectory, and the friends of the user on the social network, the next POI recommendation problem is to predict the most probable next check-in location of the user. In another word, we want to find a mapping $f(current\ location, current\ time, trajectory, friends) \rightarrow nextPOI$, which predicts the next check-in POI to match the actual check-in location of the user as well as possible.

4 Interest-Aware Next POI Recommendation

4.1 Solution Framework

According to the description of next POI recommendation problem, directly taking the historical trajectory as input to predict the next POI is infeasible. Intuitively, we can extract location interests from the historical trajectory, and use them as features to build a prediction model. Location interest refer to the location preference of a user under some context. For example, some users would like to go to the gym after work. Some users tend to visit the bar after dinner. Exploring the location interest can help to build a prediction model for POI recommendation. The framework of the proposed interest-aware next POI recommendation contains the following processes, which are shown in Fig. 1.

- **Data cleaning.** Some of the collected could be unqualified due to some users may disable their GPS or barely check-in with the APPs. In the first step, we filter out the unqualified data which is unsuitable to be used for the prediction model.
- **Extract location interest.** We adopt a temporal-spacial Latent Dirichlet Allocation (LDA) model to extract users' location interests from their historical trajectory, and form a user-interest matrix.
- **Spectral clustering.** We apply a spectral clustering algorithm to group users into different clusters, where the users in the same cluster have similar location interests.
- **Feature extraction.** We further combine the extracted location interests with a Markov train model to form the interest features and context features representing by a number of transition matrices.
- **Supervised learning for next POI recommendation.** Using the extracted features as input, we propose a supervised learning approach to predict the most probable next POI of the user.

The details of the framework are explained below.

4.2 Data Cleaning

We use the mobile social network dataset from Gowalla [5]. Gowalla is a location-based social networking service where users share their locations by checking-in. The Gowalla dataset contains a total of 6,442,890 check-ins of 196,591 users over the period of Feb. 2009–Oct. 2010.

In our work, we extract the users from the city San Francisco from the dataset. Since some of the users are barely check-in their locations, we filter out the users whose check-in locations is less than 10. Since some of the locations are barely checked, we filter out the locations which have less than 10 check-ins. Before data cleaning, the extracted dataset has 6189 users, 11371 locations, and 152860 check-ins. After filtering out the unqualified users and locations, we obtained a dataset with 1995 users, 3251 locations, and 106,098 check-ins. The average check-in number for each user is 53 times, the minimum is 10, the maximum is 1604, and the median is 22. On average, each location is checked-in 32 times, the minimum is 10, the maximum is 948, and the median is 19.

4.3 Extract Location Interests

It's evidently that check-in a place at different time means different kind or different degree of location interest. For example, if a user checks-in a library at 10 am on weekend, he may want to read books. If a user regularly checks-in the library at 10 pm on weekday, he probably works at the library. In order to extract users' interests, both time and location factors should be considered. We propose a temporal-spacial latent Dirichlet Allocation (LDA) model to extract interests. In natural language processing, latent Dirichlet allocation (LDA) is a generative statistical model that assumes each document is associated with a topic distribution, and each topic is associated with a word distribution. Specifically, we treat the combination time and location of a check-in event as a word, and the historical trajectory os a user can be view as a document. An example is illustrated in Table 1, where time is represented by two values: a Boolean value indicating whether it is weekday or not and an integer representing the hour of the day. We apply the LDA model on all documents of all users, and convert their historical check-in trajectories into interest distribution represented by K topics. As illustrated in Fig. 2, we obtain a matrix of user-interest distribution to describe the interest of the users. The user-location matrix shows the user distribution on each location, which comes from the counting of the trajectory. The user-interest matrix shows the users' interest distribution, where the number of the i-th row j-th column shows how the i-th user be interested in the j-th interest topic. The interest-location matrix shows the interest distribution on each location. These latter two matrices come from the LDA process which use Gibbs sampling [6] as inference technique.

To determine the best number K of the topics of LDA, we use the cross validation method to show the likelihood. We varies the LDA topic number K from 5 to 95, and draw a figure of likelihood as shown in Fig. 3. It is observed that the maximum likelihood achieved for $K = 40$. Therefore 40 topics are chosen in our experiments.

To show the convergence of the LDA model, we show the change of likelihood with the number of iterations in Fig. 4. As shown in the figure, the LDA model converges after 30,000 iterations.

Table 1. The example of treating time and location as word.

Location	Time	Word
252	2010-06-10$T$19:52:54Z	(252, 1, 19)
245	2010-08-29$T$20:20:56Z	(245, 0, 20)
...	...	...

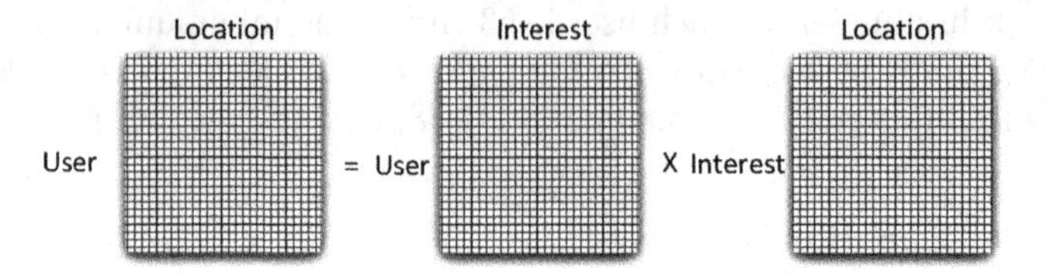

Fig. 2. The LDA approach.

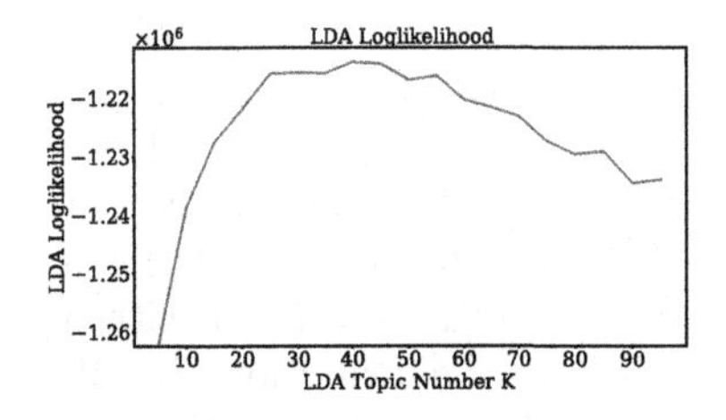

Fig. 3. Cross validation for K.

Fig. 4. Loglikelihood for K = 40.

4.4 Spectral Clustering

After obtaining the location interest of individual users, we group the users into clusters according to the similarity of interests. The intuition is that the check-in data of individual could be sparse and grouping similar users together can obtain statical preferences for next POI recommendation.

We adopt the spectral clustering [13] approach to form the user groups. The input is the extracted interests, and the output is the class ID for each user regarding to their interest.

We choose spectral clustering to partition user groups due to the following reasons. On one hand, spectral clustering requires only a matrix of similarity between the users and therefore it is very efficient for clustering sparse data. The traditional clustering algorithm such as K-Means is hard to achieve this. On the other hand, spectral clustering can reduce the dimensions of the input space, therefore the complexity in processing high-dimensional data is better than the traditional clustering algorithms. Further, the result of spectral clustering can be easily visualized, as will be demonstrated later.

First, we specify the input of spectral clustering. After applying the LDA process, we get two matrices: the user-interest matrix M^{UI} and the interest-location matrix M^{IL} which show the user-interest distribution and interest-location distribution. The user-interest matrix M^{UI} is our input of spectral clustering. The i-th row of user-interest matrix M^{UI} is a vector like $[M_{i,0}^{UI}, M_{i,1}^{UI}, \cdots, M_{i,K-2}^{UI}, M_{i,K-1}^{UI}]$ which is used to represent i-th user's interest. Each $M_{i,k}^{UI}$ is in the range [0, 1] representing how the i-th user interested in the k-th interest topic. We draw the user-interest matrix as shown in Fig. 5(a).

Second, we explain how spectral clustering works. Free software to implement spectral clustering is available in the open source projects like Scikit-learn [3]. Spectral clustering contains the three steps:

- **(1) Pre-processing.** Construct a similarity graph for all users' M_i^{UI}. Then build Laplacian matrix L of the graph.
  The adjacency matrix of similarity graph is as following:

$$A_{i,j} = \begin{cases} w_{i,j}, & : \text{weight of edge (i, j)} \\ 1, & : \text{if i = j} \end{cases} \tag{1}$$

  The adjacency matrix is constructed using the Gaussian kernel function of the Euclidean distanced $\|M_i^{UI} - M_j^{UI}\|_2$.

$$w_{i,j} = exp(-\gamma * \|M_i^{UI} - M_j^{UI}\|_2^2) \tag{2}$$

  where γ is the kernel coefficient, default γ is 1.0
  D is the diagonal matrix of degrees.

$$d_i = \sum_{j|(i,j)\in E} w_{i,j} \tag{3}$$

  The Laplacian matrix can be calculated by

$$L = D - A \tag{4}$$

- **(2) Decomposition.** Build a reduced space from multiple eigenvectors. Here we use the Ng-Jordan-Weiss algorithm for spectral clustering [13]. First, compute eigenvectors of the matrix L_{norm}.

$$L_{norm} = D^{-1/2} L D^{-1/2} \tag{5}$$

  Then find the first k eigenvectors $\boldsymbol{\nu}_1, \ldots, \boldsymbol{\nu}_k$ of the Matrix L_{norm}. Finally, let $U \in R^{n \times k}$ be the matrix containing the vector $\boldsymbol{\nu}_1, \ldots, \boldsymbol{\nu}_k$ as columns. In this step, we map each point to a low denominational representation based on eigenvectors.
- **(3) Grouping.** A classical clustering algorithm (here is k-means) is applied to matrix U to partition the users. The output of spectral clustering assigns each user i with a class ID $InterestClassID_i$. Figure 5(b) shows the user-interest matrix after spectral clustering with the number of clusters setting to 40. It is shown that the users are grouped together with similar interest distributions. Figure 6 shows the similarity between user before and after spectral clustering. It clearly shows that similar users are grouped together forming clusters.

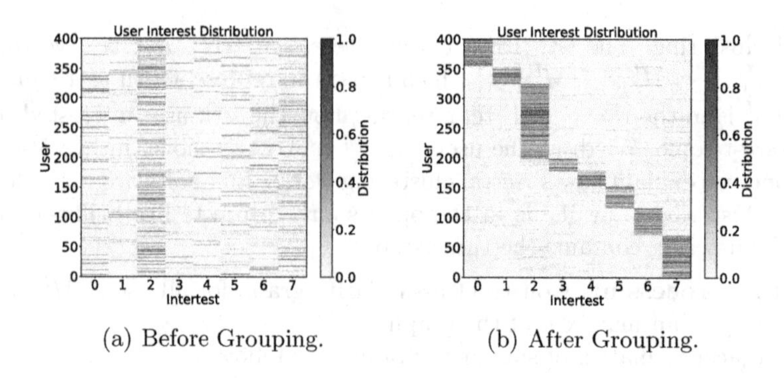

(a) Before Grouping. (b) After Grouping.

Fig. 5. Visualize user interest graph before and after grouping.

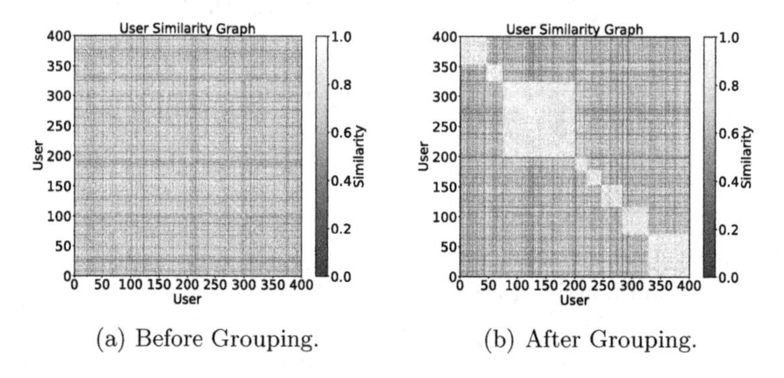

(a) Before Grouping. (b) After Grouping.

Fig. 6. Similarity graph before and after grouping.

4.5 Feature Extraction

Based on the obtained user interests and cluster IDs, we can form different types of features including interest features and context features, which will be used to build a prediction model. The feature extraction process including the following steps.

Step 1: We build a Markov transition matrix MC^{LOC} based on user trajectory, where each element $MC_{i,j}^{LOC,}$ in the matrix means the transition probability of a user transiting from POI i to POI j.

The user's trajectory can be represented by a sequence of POI IDs. We build a weighted graph G(V, E), where V is the set of nodes indicated by POI IDs, E is the set of edges. If a user moves from a node to another, there is an edge between them. Initially the weights on the edge $\omega'(E) = 0$ at the beginning. If we find a transit from POI ID_i to POI ID_j, let $\omega'_{ID_i,ID_j} = \omega'_{ID_i,ID_j} + 1$. After scanning the whole trajectory, we obtain G(V, E) with weight ω' on every edges. Then we normalize the weights ω' to ω: for each POI ID_i, set $\omega_{ID_i,ID_j} = \frac{\omega'_{ID_i,ID_j}}{\sum_{j=1}^{j=POI\ number} \omega'_{ID_i,ID_j}}$. We get the Matrix $MC_{i,j}^{LOC,}$, where $MC^{LOC}_{i,j} = \omega_{ID_i,ID_j}$ are the weights between two edges that satisfies $\Sigma_{j=1}^{j=POI\ number} \omega_{ID_i,ID_j} = 1$. Generally speaking, MC^{LOC} is a 2-dimension matrix where each row sums to 1.

Step 2: Now based on the spectral clustering result, we build a Markov transition matrix $MC^{Interest}$, where $MC^{Interest}_{c,i,j}$ means the transition probability of a user transiting from POI i to POI j on condition that the user's cluster ID satisfies $ClassID = c$.

The construction of matrix $MC^{Interest}$ is similar to that of MC^{LOC}, except that the trajectories are chosen from the set of users belonging to the same cluster c.

We may find the situation that some users in $ClassID_c$ never go from POI_i to anywhere. In this situation, in order to get locations to recommend, we set $MC^{Interest}_{c,i} = MC^{LOC}_i$. That means copying a whole row in MC^{LOC}_i to $MC^{Interest}_{c,i}$. It helps to fill the blank data. That also means that when it comes a new data with property c and current location i that we do not know which next POI to recommend (because the property c and current location i never occurred before), we can ignore the property c and simply use the current POI i to recommend next POI.

Step 3: Further, we extract the context features (such as time, friends, etc.) represented by Markov transition matrices. Besides interest, we could apply the similar method in Step 2 to obtain different kinds of context features such as: weekday, hour, log (friends number), and so on. Each feature could be represented by a transition matrix under the property condition.

The transition matrices we got will be used to predict the next POI location to be recommended.

4.6 Supervised Learning for Next POI Recommendation

After feature extraction, we obtain the interest features and context features represented by a number of transition matrices. Using the features as input, we can build a supervised learning model to predict the next POI. The process works as follows.

Step 1: Construction of training set. Here is how the training set is constructed. Based on the transition matrices regarding interest and context such as location, time, and friends, we choose the POIs with maximum transition probability as candidates for POI recommendation. Specifically, given the user ID, cluster ID c, and current POI i, we set $Loc_{loc} = Max(MC^{LOC}_i)$, which is the location with maximum transition probability according to location context. We set $Loc_{interest} = Max(MC^{Interest}_{c,i})$, which is the location with maximum transition probability under interest group c. Similar, we obtain Loc_{time}, Loc_{friend} from the transition matrices regarding the context of time and friends.

Using the candidate locations as features, we take the actual next POIs of the user as labels, a training set can be formed with the format like $[CurLoc, Loc_{loc}, Loc_{interest}, Loc_{time}, Loc_{friend}] => [nextPOI]$.

Step 2: Learning. We adopt the random forest [4,8] algorithm to train a prediction model. Random Forest is an ensemble classifier that consists of many decision trees. It outputs the class that is the mode of the class's output by individual trees. Here we choose no limit to the leaf node and tree depth. Learning random forest is fast, and could produces a highly accurate classifier.

5 Performance Evaluation

In this section, we conduct experiments based on the Gowalla dataset [5] to evaluate the performance of the proposed approach.

5.1 Baseline Algorithms

We compare the proposed approach with four traditional POI recommendation algorithms.

- Top. Choose the most popular locations as next POI for each user.
- Random. Choose random locations as next POI for each user.
- Markov Chain (MC) [22]. Based on the current location, choose the most likely locations from Markov transition matrix as next POI for each user.
- Factorizing Personalized Markov Chain (FPMC) [14]: It is a sequential prediction method which based on personalized Markov chains.

5.2 Default System Parameters

The default parameters are as follows. The number of decision trees in random forest is 10. The number of check-ins for each user and each location is at least 10 times. The LDA topic number is 40. The spectral clustering parameters N is 160. The training set is set to 80% of the dataset.

Table 2. Performance comparison.

	Recall	Precision	F1-score
Interest-aware	**0.101697**	0.192546	**0.125091**
FPMC	0.087795	**0.287373**	0.122852
MC	0.088313	0.211404	0.116741
Top	0.000484	0.003119	0.000720
Random	0.014722	0.000217	0.000427

Table 3. The impact of tree number.

	Recall	Precision	F1-score
Tree = 10	0.101697	0.192546	0.125091
Tree = 20	**0.101791**	**0.192579**	**0.125138**
Tree = 30	0.101744	0.192555	0.125107
Tree = 40	0.101744	0.192555	0.125107

Table 4. The impact of LDA parameter K

	Recall	Precision	F1-score
K = 20	0.102545	0.190029	0.125066
K = 40	0.101744	0.192555	0.125107
K = 60	0.102403	**0.195227**	0.126255
K = 80	**0.102828**	0.193236	**0.126832**

5.3 Comparison with Baselines

We adopt the commonly used performance metrics including precision, recall, and F1-score for performance evaluation. The definitions of the performance metrics can be found in [7]. We compare the performance of the proposed interest-aware approach with the baselines, which results are shown in Table 2. According to the table, the proposed approach performance much better than the baselines. The Random and Top approaches perform poor in precision and recall since they ignore the personalized location preference and contextual information. The MC approach works good in precision, but it performs poor in recall. FPMC which extent MC, also performs poor in recall too. As F1-score evaluate both recall and precision, our approach outperform the FPMC. By combining location interest and contextual information for prediction, the proposed approach achieves better performance than the baselines.

5.4 Sensitive Analysis

Next we conduct sensitive analysis by showing the performance under different system parameters.

The Impact of the Number of Decision Trees. The random forest algorithm relies on a set of decision trees for prediction, and the number of decision trees is an important system parameter for the model. In the experiments, we varies the number of decision trees from 10 to 40, and the results are shown in Table 3. The number of decision trees has effect on the performance. The best performance achieved when $Tree = 20$. While bigger tree number means more memory cost ($O(Tree)$), the improvement is not so much. It is not worthy use bigger tree number for a little performance improvement.

Table 5. The impact of parameter N.

	Recall	Precision	F1-score
N = 40	0.098021	0.186868	0.121204
N = 80	0.099717	0.187972	0.122556
N = 160	**0.101697**	0.192546	**0.125091**
N = 320	0.101037	**0.193670**	0.124512

Table 6. The impact of train set size.

	Recall	Precision	F1-score
Train = 60%	0.094910	0.179373	0.117071
Train = 70%	0.099749	0.186864	0.122339
Train = 80%	0.101697	0.192546	0.125091
Train = 90%	**0.102733**	**0.198377**	**0.126699**

The Impact of the Number of Topics in LDA. The number of topics K is an important parameter for the LDA model. In the experiments, we varies K from 20 to 80 and the results are shown in Table 4. The performance raise with the number of topics K raise. More topic number K, means more information about users, also means more memory and computation cost (both $O(K)$), while the performance improve not too much.

The Impact of the Number of Clusters. The number of clusters N is an important parameter for the spectral clustering algorithm. In the experiments, we vary N from 40 to 320, and the results are shown in Table 5. The performance raise with N varies from 40 to 160, and performance fall with N varies from 160 to 320. Smaller N means to recommend user some POIs from grouped similar users, which is not too personalized. While bigger N means to recommend user more personalized POI, meanwhile some POIs from grouped similar users could not be recommend. At the point N = 160, the model performs best.

The Impact of Train Set Size. We change the size of training set from 60% to 90%, and the results are shown in Table 6. The performance raise with the percentage of training data raise. More training set achieved better perform because of more information about users.

6 Conclusion

In this paper, we address the next POI recommendation for mobile social networks, and propose a novel interest-aware POI prediction model to solve the problem. We develop a spatial-temporal topic model based on LDA that takes the historical POI check-in records as input to extract users' location interests. We further combine location interest with contextual features to construct a prediction model based on supervised learning for next POI recommendation. We conduct experiments on the Gowalla dataset, which show that the proposed POI prediction approach outperforms the conventional approaches.

References

1. https://foursquare.com/
2. https://www.lifewire.com/location-apps-for-user-reviews-tips-3485920
3. http://scikit-learn.org/stable/modules/clustering.html
4. Breiman, L.: Random forests. Mach. Learn. **45**(1), 5–32 (2001)
5. Cho, E., Myers, S.A., Leskovec, J.: Friendship and mobility: user movement in location-based social networks. In: Proceedings of the 17th ACM SIGKDD International Conference on Knowledge Discovery and Data Mining, pp. 1082–1090. ACM (2011)
6. Darling, W.M.: A theoretical and practical implementation tutorial on topic modeling and Gibbs sampling. In: Proceedings of the 49th Annual Meeting of the Association for Computational Linguistics: HuMan Language Technologies, pp. 642–647 (2011)

7. Goutte, C., Gaussier, E.: A probabilistic interpretation of precision, recall and F-score, with implication for evaluation. In: Losada, D.E., Fernández-Luna, J.M. (eds.) ECIR 2005. LNCS, vol. 3408, pp. 345–359. Springer, Heidelberg (2005). https://doi.org/10.1007/978-3-540-31865-1_25
8. Ho, T.K.: Random decision forests. In: Proceedings of the Third International Conference on Document Analysis and Recognition 1995, vol. 1, pp. 278–282. IEEE (1995)
9. Koren, Y., Bell, R., Volinsky, C.: Matrix factorization techniques for recommender systems. Computer **42**(8), 30–37 (2009)
10. Levandoski, J.J., Sarwat, M., Eldawy, A., Mokbel, M.F.: LARS: a location-aware recommender system. In: 2012 IEEE 28th International Conference on Data Engineering (ICDE), pp. 450–461. IEEE (2012)
11. Liu, Q., Wu, S., Wang, L., Tan, T.: Predicting the next location: a recurrent model with spatial and temporal contexts. In: AAAI, pp. 194–200 (2016)
12. Mnih, A., Salakhutdinov, R.R.: Probabilistic matrix factorization. In: Advances in Neural Information Processing Systems, pp. 1257–1264 (2008)
13. Ng, A.Y., Jordan, M.I., Weiss, Y.: On spectral clustering: analysis and an algorithm. In: Advances in Neural Information Processing Systems, pp. 849–856 (2002)
14. Rendle, S., Freudenthaler, C., Schmidt-Thieme, L.: Factorizing personalized Markov chains for next-basket recommendation. In: Proceedings of the 19th International Conference on World Wide Web, pp. 811–820. ACM (2010)
15. Ricci, F., Nguyen, Q.N.: Acquiring and revising preferences in a critique-based mobile recommender system. IEEE Intell. Syst. **22**(3), 22–29 (2007)
16. van Setten, M., Pokraev, S., Koolwaaij, J.: Context-aware recommendations in the mobile tourist application COMPASS. In: De Bra, P.M.E., Nejdl, W. (eds.) AH 2004. LNCS, vol. 3137, pp. 235–244. Springer, Heidelberg (2004). https://doi.org/10.1007/978-3-540-27780-4_27
17. Xing, S., Liu, F., Zhao, X., Li, T.: Points-of-interest recommendation based on convolution matrix factorization. Appl. Intell. **47**, 1–12 (2017)
18. Yang, C., Bai, L., Zhang, C., Yuan, Q., Han, J.: Bridging collaborative filtering and semi-supervised learning: a neural approach for poi recommendation. In: Proceedings of the 23rd ACM SIGKDD International Conference on Knowledge Discovery and Data Mining, pp. 1245–1254. ACM (2017)
19. Zhang, Y., Dai, H., Xu, C., Feng, J., Wang, T., Bian, J., Wang, B., Liu, T.Y.: Sequential click prediction for sponsored search with recurrent neural networks. In: AAAI, vol. 14, pp. 1369–1375 (2014)
20. Zheng, V.W., Zheng, Y., Xie, X., Yang, Q.: Collaborative location and activity recommendations with GPS history data. In: Proceedings of the 19th International Conference on World Wide Web, pp. 1029–1038. ACM (2010)
21. Zheng, V.W., Cao, B., Zheng, Y., Xie, X., Yang, Q.: Collaborative filtering meets mobile recommendation: a user-centered approach. In: AAAI, vol. 10, pp. 236–241 (2010)
22. Zimdars, A., Chickering, D.M., Meek, C.: Using temporal data for making recommendations. In: Proceedings of the Seventeenth Conference on Uncertainty in Artificial Intelligence, pp. 580–588. Morgan Kaufmann Publishers Inc. (2001)

Approximate Minimum-Transmission Broadcasting in Duty-Cycled WSNs

Quan Chen[1(✉)], Tianbai Le[2], Lianglun Cheng[1], Zhipeng Cai[3,4], and Hong Gao[5]

[1] School of Computers, Guangdong University of Technology, Guangzhou, China
{quan.c,llcheng}@gdut.edu.cn

[2] Department of Computer Technology, Harbin Institute of Petroleum, Harbin, China

[3] Department of Computer Science, Georgia State University, Atlanta, USA
zcai@gsu.edu

[4] Department of Computer Science, Harbin Engineering University, Harbin, China

[5] School of Computer Science and Technology, Harbin Institute of Technology, Harbin, China
honggao@hit.edu.cn

Abstract. Broadcast is an essential operation for disseminating messages in multihop wireless networks. Unfortunately, the problem of Minimum Transmission Broadcast (MTB) in duty-cycled scenarios is not well studied. Existing works always have rigid assumption that each node is only active once per working cycle. Aiming at making the work more practical and general, MTB problem in duty-cycled network where each node is allowed to active multiple times in each working cycle (MTBDCA) is investigated in this paper. Firstly, the MTBDCA problem is proved to be NP-hard and unlikely to have an $(1 - o(1))ln\Delta$-approximation algorithm, where Δ denotes the maximum node degree. A novel covering problem is proposed based on an auxiliary graph, which is constructed to integrating the transmitting time slots into the network. Then, a $ln(\Delta + 1)$-approximation algorithm is proposed for MTBDCA. Finally, the theoretical analysis and experimental results verify the efficiency of the proposed algorithm.

1 Introduction

In multihop wireless networks, broadcast is an essential operation for many networking protocols, such as routing discovering, information dissemination, and network configuration [1–7]. Assume that there is one predefined source node s broadcasting a message to all other nodes in the network. By exploiting the broadcast nature of the wireless medium, the neighbors can receive the packet with a single transmission. Thus, the total number of transmissions in the broadcast process is defined as the number of senders in the network. Obviously, the smaller the number of transmissions, the more energy will be saved. Therefore,

S. Chellappan et al. (Eds.): WASA 2018, LNCS 10874, pp. 40–52, 2018.
https://doi.org/10.1007/978-3-319-94268-1_4

the problem of minimum-transmission broadcast (**MTB**) attracts many attentions of researchers. Many scheduling algorithms [8–16] has been proposed for the networks that nodes always keep awake. According to [8], the MTB problem is NP-hard, so that the aforementioned algorithms [8–16] are approximation ones, and they achieve high performance in terms of efficiency.

However, for many wireless networks, such as wireless sensor networks, the energy is quite limit. In order to conserve energy, the duty-cycled scheme is often adopted [17,18], in which each node switches between two states, i.e., active state and dormant state periodically, and all the functional modules are turned off in dormant state to save energy. Since the working mode of duty-cycled networks is completely different from the traditional wireless networks, the MTB problem should be reconsidered and the following two problems should be solved: (1) For a node which wants to broadcast a packet to all its neighbors, multiple transmissions may be required to guarantee all the neighbors can receive the packet; (2) Since a node may have multiple active times slots per working cycle, how to choose the optimal forwarding nodes to make use of all the active time slots to reduce the number of transmissions is also a big challenge.

Currently, the main works that studied the MTB problem in duty-cycled networks are [19–22]. By dividing the nodes into different sets according to their active time slots, several efficient algorithms for MTB are proposed. However, these methods assume that all the nodes have only one active time slot per working cycle. So they only provided a solution for the aforementioned problem 1. When each node has arbitrary active time slots [17,18], the problem 2 becomes more important for reducing the redundancy.

In order to overcome the above shortcomings and provide a more general solution, this paper studies the MTB problem in Duty-Cycled networks with Arbitrary active time slots allowed in a working cycle (**MTBDCA**). Firstly, MTBDCA is proved to be NP-hard and unlikely to have an $(1-o(1))ln\Delta$-approximation algorithm, where Δ denotes the maximum node degree. Secondly, a novel covering problem is proposed based on an auxiliary graph and an approximate algorithm is proposed to solve it. Then, a $ln(\Delta+1)$-approximation algorithm is proposed for MTBDCA. Finally, extensive experimental results verify the efficiency of the proposed algorithms.

The rest of this work is organized as follows. Section 2 surveys the related works. Section 3 presents the problem definition. Section 4 explains the proposed approximate algorithms for the MTBDCA problem. The simulation results are shown in Sect. 5. Section 6 concludes the paper.

2 Related Works

The MTB problem has attracted extensive attentions from researchers. It has been studied in both nodes always-awake networks and duty-cycled networks. In the nodes always-awake networks, the main works that studied the MTB problem are [8–16]. In [8], the MTB problem was proved to be NP-hard and a tree-based scheme is exploited to forward the message. By using the Minimum Connected

Dominate Set (MCDS) technique, the authors in [9–11] introduced the MCDS-based scheme to reduce the redundancy, where a virtual backbone is constructed for broadcasting. To further reduce the redundancy, Lou et al. improved the Dominant Pruning scheme with two-hop neighbors' information [12]. After that, Lou et al. proposed a quasi-local forward-node-set-based scheme which provides a better performance [13]. The authors in [14–16] introduced another scheme, i.e., the multipoint relay scheme, for the MTB problem and they provided a localized and optimized way for determining a small forwarding set. Additionally, to find a good approximation of MCDS, which is the core of solutions for the MTB problem, Wan et al. then proposed an distributed algorithm with an approximation ratio of 8 in [23] and then reduce the approximation ratio to 6.8 with a two-phase approach [24]. Recently, Shi et al. studies the CDS construction problem in the energy-harvesting networks and the weakly CDS problem [25–27]. However, all these approaches are not suitable for duty-cycled networks.

For the MTBDC problem in duty-cycled WSNs, several algorithms were proposed [19–22]. The MTBDC problem was first studied and proved to be NP-hard in [19]. By exploiting the set covering technique, a $3(ln\Delta+1)$-approximation algorithm is proposed. A distributed algorithm which has a constant approximation ratio of at most 20 was also proposed in [19]. To further reduce the number of transmissions, the authors in [20,21] proposed a level-based scheduling algorithm by constructing the backbone according to the level of the forwarding nodes. Recently, Le-Duc et al. revealed that the transmissions can be further reduced by using a depth-first traversal on the set of all forwarding nodes and proposed an algorithm which outperforms the previous algorithms [22]. However, these methods all assume that each node wakes up only once in a working cycle, which limits their application.

3 Problem Definition

Assume $G = (V, E)$ denote a multihop duty-cycled network, where $V = \{1, 2, \ldots, N\}$ is the set of sensor nodes and $E = \{(u, v) | 1 \leq v \neq u \leq N\}$ denotes the neighborhood relationship among sensor nodes. That is, node u and v can communicate with each other by one-hop messages if $(u, v) \in E$. According to the above discussion, each node has two states, *i.e.* the active state and the dormant state, and switches between them periodically. Let $W = \{0, 1, 2, \ldots, |W| - 1\}$ be a working cycle that contains $|W|$ time slots. Then the *working plan* of node u (denoted by $\mathcal{W}(u)$) can be defined as the set which contains all active time slots of u in a working cycle, i.e., $\mathcal{W}(u) = \{t_1, t_2, \ldots, t_k\} \subseteq \{1, 2, \ldots, |W|\}$. For any node u, it can only receive data at the time slots $t(t \in \mathcal{W}(u))$ in every working cycle. And if it wants to send a message, it can switch to the active state at any time slot when the receiving node is awake. The duty cycle is defined as the ratio of the working plan to a whole working cycle, which can be calculated as $|\mathcal{W}(u)|/|W|$.

Given a broadcast request, which includes a source node s, the MTB problem in the nodes always-awake sensor networks is to construct a broadcast tree that satisfies that: (1) it is rooted at the source node s and spanning all of the nodes in V; (2) the number of the non-leaf nodes is minimized since each node only needs to transmit once to inform its neighbors. However, in the duty-cycled network, the MTB problem becomes much complicated since one node may need to broadcast multiple times to inform the neighbors. In this case, one not only need to construct the broadcast tree carefully, but also take the *Transmitting Schedule* of the non-leaf nodes into consideration.

Before we give the formal definition of a transmitting schedule, we need to clarify some notations used in this paper. Let $NB(u)$ denote the set of one-hop neighboring nodes of u in G. As for a broadcast tree T, let $nl(T)$ denote the set of non-leaf nodes in T and the set of u's children is denoted by $ch(u)$.

Definition 1 (Transmitting Schedule). Given node $u \in V$, let $sch(u) = [u, t, ch_t(u)]$ denote the transmitting schedule of node u, where for any child node $v \in ch_t(u)$, we have $(u, v) \in E$ and $t \in \mathcal{W}(v)$.

The transmitting schedule $sch(u)$ means node u is scheduled to transmit its data to the child nodes in $ch_t(u)$ at time slot t, where t belongs to the working plan of each node $v \in ch_t(u)$. Since one node may need to transmit multiple times, node u may have multiple transmitting schedules and $ch_t(u)$ is only the part of node u's children in the broadcast tree. Let $\mathcal{S}(u)$ denote all the transmitting schedules of node u, then we can have $\bigcup_{sch(u)\in\mathcal{S}(u)} ch_t(u) = ch(u)$.

By exploiting the definition of the transmitting schedule, the MTB problem in duty-cycled sensor networks is then to construct a broadcast tree T and determine the transmitting schedules for each non-leaf node, i.e., $\mathcal{S}(T)$, while the total number of transmissions is minimized. The broadcast tree and the transmitting schedules is called a broadcast schedule.

Theorem 1. *The MTBDCA problem is NP-hard and there exists no polynomial-time approximation algorithm with performance ratio of* $(1-o(1))ln\Delta$ *for the MTBDCA problem unless* $NP \subseteq DTIME(n^{O(loglogn)})$, *where* Δ *is the maximum node degree of the duty-cycled network.*

This theorem can be proved since the problem of using the minimum number of active time slots to cover all the neighbors is equivalent to the Minimum Set Cover (MSC) problem [28]. Moreover, It has been proved that there exists no polynomial-time algorithm for the MSC problem with performance ratio of $(1 - o(1))lnn$ unless NP has quasi-polynomial time algorithms, where n is the size of the MSC problem.

4 Approximation Algorithm for MTBDCA

Since the above problem is NP-hard, we propose an algorithm with approximation ratio of $ln(\Delta + 1)$. The proposed algorithm includes four steps. Firstly, an auxiliary graph is constructed to integrate all the possible transmitting time slots into the network. Secondly, a novel kind of covering problem is designed based on the auxiliary graph and an approximation algorithm is proposed to solve it. Thirdly, a pseudo broadcast tree is constructed on the auxiliary graph. Finally, the pseudo broadcast tree is mapped into a broadcast schedule.

4.1 Constructing an Auxiliary Graph

The first step of our approach is to transform the duty-cycled network G into an auxiliary graph $\mathcal{G}$ to assist in building the broadcast tree. In the auxiliary graph, a new kind of node, i.e., *Schedule Node*, is introduced, which is used to determine the transmitting schedule for each non-leaf nodes.

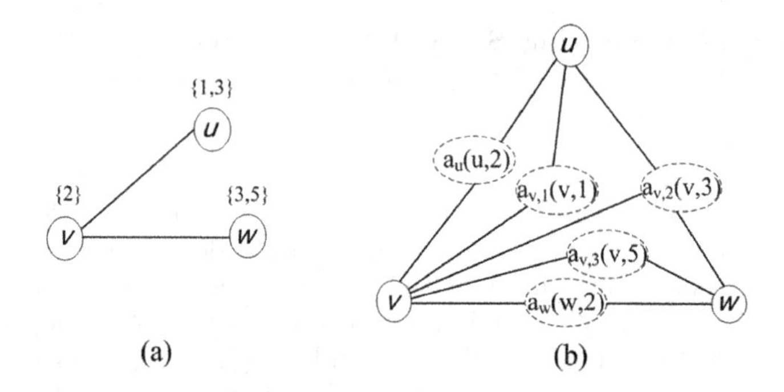

Fig. 1. The example of auxiliary graph.

For any node $u \in V$, we use a_u to denote a schedule node of u in the auxiliary graph $\mathcal{G}$. Each schedule node a_u owns two properties, *i.e.*, $(a_u.p,\ a_u.t)$, which denote its primary node u and transmitting time slot t respectively. In this paper, node u is called the primary node of the scheduling node a_u, i.e., $a_u.p = u$. Then, the auxiliary graph is constructed as follows.

Definition 2 (Auxiliary Graph). Given a duty-cycled sensor network $G = (V, E)$, its auxiliary graph $\mathcal{G} = (V', E')$ denotes the graph integrating the transmitting time slots, where V' and E' denote the set of nodes and edges. V' and E' are constructed as follows:

(i) Initially, $V' = V$, $E' = \emptyset$;

(ii) For each node $u \in V$ and each time slot $t \in \bigcup_{v \in NB(u)} \mathcal{W}(v)$, create a schedule node $a_{u,i} (1 \le i \le |\bigcup_{v \in NB(u)} \mathcal{W}(v))|$ with its two properties set as (u, t) (i.e., $a_{u,i}.p = u$ and $a_{u,i}.t = t$). Let the set of all of the schedule nodes of u be denoted by $\Upsilon(u)$. Then, we can have $V' = V' \bigcup\limits_{u \in V} \Upsilon(u)$;

(iii) For each node $u \in V$, we create an edge between u and each schedule node $a_{u,i} \in \Upsilon(u)$ in the auxiliary graph, which means node u can transmit at time slot $a_{u,i}.t$. Let $E'_u = \{(u, a_{u,i}) | a_{u,i} \in \Upsilon(u)\}$, then we can have $E' = \bigcup\limits_{u \in V} E'_u$;

(iv) Let v be a neighboring node of u in the original graph G. For any schedule node $a_{u,i} \in \Upsilon(u)$, we add an edge $(a_{u,i}, v)$ in $\mathcal{G}$ if $a_{u,i}.t \in \mathcal{W}(v)$, and we use $R(a_{u,i})$ to denote the set of such nodes v of $a_{u,i}$, which means the reaching nodes by $a_{u,i}.p$ at time slot $a_{u,i}.t$. After then, E' can be updated as $E' = E' \bigcup_{u \in V} \{ \bigcup_{a_{u,i} \in \Upsilon(u)} E'_{a_{u,i}} \}$, where $E'_{a_{u,i}} = \{(a_{u,i}, v) | v \in R(a_{u,i})\}$.

As the example in Fig. 1, there is an original graph in Fig. 1(a), where the number in the braces denotes the working plan of each node. As for the above original graph, we do as follows according to Definition 2, where the result is shown in Fig. 1(b). For example, there are three schedule nodes are created for node v, i.e., $a_{v,1}$, $a_{v,2}$ and $a_{v,3}$. And their properties are set $(v, 1)$, $(v, 3)$, and $(v, 5)$ respectively. Since time slot 1 is in node u's working plan, then the schedule node $a_{v,1}$ is connected to node u.

4.2 Minimum Schedule Node Covering

To reduce the number of transmissions, a new kind of problem, *Minimum Scheduling Node Covering*, is introduced. The minimum schedule node covering problem is aimed at covering all the primary nodes except the source node in the auxiliary graph with minimum schedule nodes.

Definition 3 (Minimum Schedule Node Covering Problem). Given an auxiliary graph $\mathcal{G}$ and a source node s, the minimum schedule node covering problem is to find a set of schedule nodes A_S which satisfy the following conditions:

1. The nodes in A_S are all schedule nodes;
2. For any primary node $u \in V \& u \neq s$, there exists a schedule node $a_v \in A_S$ and $u \in R(a_v)$, where $R(a_v)$ denote the set of nodes which can be reached by the primary node $a_v.p$ at time slot $a_v.t$;
3. The size of A_S is minimized.

To solve the minimum schedule node covering problem, we use a greedy set covering technique [29] to find a small node set that can cover all the primary nodes in $\mathcal{G}$. In each loop, we first find a schedule node a_v that has the maximum number of adjacent primary nodes which was uncovered. Then, we add a_v into the set A_S. This process repeats until there exist no uncovered primary nodes. It is called Algorithm 1 for simplicity. Since each schedule node can cover at most Δ nodes (which is shown in Lemma 1 and can be easily verified), then Algorithm 1 has an approximation ratio of $ln(\Delta + 1)$.

Lemma 1. *Any schedule node in the auxiliary graph $\mathcal{G}$ can be adjacent to at most Δ primary nodes.*

Theorem 2. *The approximation ratio of Algorithm 1 is $ln(\Delta + 1)$.*

4.3 Calculating the Pseudo Broadcast Tree

In this subsection, we will introduce the method to generate a *Pseudo Broadcast Tree* on the auxiliary graph to contain all the schedule nodes in A_S and their primary nodes. Let $L(v)$ denote the level of node v, which can be obtained by a Breadth-First Search initiated from the source node s on the auxiliary graph $\mathcal{G}$. It mainly works as follows.

First, a breadth-first search is conducted on from the source node s on the auxiliary graph $\mathcal{G}$.

Second, let $A_P = \{a_v.p \mid a_v \in A_S\} \cup \{s\}$ denote the primary nodes of all the schedule nodes in A_S. Note that, the source node is also contained.

Third, assume the node with the smallest level in A_P be u. Let T_u be the subtree rooted at node u and set $T_u = T_u \cup \{u\}$. The subtree T_u is constructed as follows:

1. Let m be a primary node in $T_u \cap A_P$ and $S_m = \{a_v | a_v \in A_S \& a_v.p = m\}$ denote all the schedule nodes which is in A_S and its primary node is m.
2. For each schedule node $a_v \in S_m$, add a_v in the subtree T_u, i.e., $T_u = T_u \cup \{a_v\}$ and connect a_v to node m. And then for any primary nodes in $R(a_v)$, i.e., n, if it is not included in any subtree then add it to subtree T_u, i.e., $T_u = T_u \cup \{n\}$ and connect n to schedule node a_v. Otherwise, ignore it.
3. Delete m from A_P, i.e., $A_P = A_P - \{m\}$.

The above three steps repeated until A_P is empty, which means all the schedule nodes in A_S and their primary nodes are included. Notice that, it may generate multiple subtrees in this step.

Next, we will introduce how to merge these subtrees into a pseudo broadcast tree. It mainly works as follows. Let TR denote the set of the root node of the generated subtrees. Obviously, the source node s is included in TR. The merging process is beginning from the subtree rooted at source node s, i.e., T_s.

- Case 1: If there exist a primary node in T_s, i.e., m, and a schedule node of m, i.e., a_m, that covers most primary nodes in TR, then connect a_m to m, and the covered root nodes in TR to node a_m.
- Case 2: Let the root node in TR with the smallest level be r. Then there exists a neighbor of r, i.e., $u \in NB(r)$, which is not included in T_s and can be reached by some node in T_s, i.e., x. Assume the schedule node of x that can reach u be a_x, i.e., $u \in R(a_x)$. Then connect x to node a_x and connect node a_x to node u. Additionally, let the schedule node of u that can reach r be a_u, then connect u to node a_u and connect a_u to node r. Let the subtree contains u be T_w, delete u from T_w.

Finally, all the subtrees except T_s can be added into the primary subtree T_s either by case 1 or case 2. This can be verified easily and we omit the proof here. Now, the pseudo broadcast is constructed.

4.4 Transform the Pseudo Broadcast Tree to a Broadcast Schedule

After obtaining the pseudo broadcast tree, we will show how to transform the pseudo broadcast tree to a broadcast schedule, which works as follows.

Firstly, the broadcast tree T in the duty-cycled network G is constructed by the following three steps:

Step 1. For any schedule node in the pseudo broadcast tree T_s, we create an edge between its father and all of its child nodes;
Step 2. Remove all of the schedule nodes from T_s;
Step 3. Replace all of the primary nodes with their corresponding nodes in the duty-cycled network G. Then, the broadcast tree T is obtained.

Secondly, we will show how to determine the transmitting schedules for each non-leaf in T, i.e., $\mathcal{S}(T)$. For each non-leaf node in the broadcast tree T, i.e., u, its transmission schedule can be just obtained according to its child node (i.e., schedule nodes) in the pseudo broadcast tree. That is, for each schedule node $a_u \in ch(u)$ in T_s, we add a transmitting schedule $[u, a_u.t, ch(a_u)]$ in $\mathcal{S}(T)$, where $ch(a_u)$ denote the child nodes of schedule node a_u in the pseudo broadcast tree.

Now, the complete Approximate Minimum Transmission Broadcasting algorithm, i.e., AMTB, is introduced. Next, we will present the theoretical analysis for the AMTB algorithm.

4.5 Performance Analysis of the Proposed Algorithm

Theorem 3 gives the correctness analysis of the proposed AMTB algorithm.

Theorem 3. *The broadcast schedule generated by AMTB is complete and correct.*

Proof. Firstly, we will prove all the nodes are included in the broadcast tree. According to Algorithm 1, all the nodes except the source node are covered by some schedule node, and the set of these schedule nodes is A_S. Then, during the construction of the pseudo broadcast tree, all the schedule nodes in A_S are included in a group of sub-trees and as a result, all the primary nodes are included in the pseudo broadcast tree. Finally, since all the sub-trees can be merged into the primary subtree, which is rooted at the source node, then all the nodes are included in the broadcast tree.

Secondly, we will prove the generated broadcast schedule for each non-leaf node is correct, i.e., all the nodes can receive the messages correctly. We prove it by contradiction. Assume there exist a contradiction, then there existed a node, i.e., u, that cannot receive the message from its parent. Assume the schedule node in the pseudo broadcast tree connect node u be a_v. According to the construction of auxiliary graph, node v can reach u at time slot $a_v.t$. Since the transmission schedule $[v, a_v.t, ch_{a_v.t}(v)]$ is added into the broadcast schedule, then u can receive the message from its parent, which is a contradiction.

Therefore, the result generated by AMTB is complete and correct.

Next, we will first give the lower bound analysis for MTBDCA in Theorem 4 and the approximation ratio analysis for AMTB in Theorem 5.

Theorem 4. *Let $|A_S|$ denote the size of obtained set of schedule nodes by Algorithm 1, then the lower bound on the number of transmissions of any optimal broadcast schedule for MTBDCA is at least $\frac{|A_S|}{ln(\Delta+1)}$, where Δ denotes the maximum degree in the duty-cycled network G.*

Proof. Let OPT denote the required number of transmissions of the optimal schedule for MTBDCA. Since each node except the source node need to be covered once, then $OPT \geq |A_S^{opt}|$, where A_S^{opt} denote the optimal schedule for the minimum schedule node covering problem. According to Theorem 2, $|A_S| \leq |A_S^{opt}| \cdot ln(\Delta+1)$. Thence, $OPT \geq |A_S^{opt}| \geq \frac{|A_S|}{ln(\Delta+1)}$. The theorem is proved.

Theorem 5. *The approximation ratio of AMTB is at most $ln(\Delta+1)$.*

Proof. Let B be the total number of transmissions of the AMTB algorithm. In AMTB, a pseudo broadcast tree is first constructed, which can then be transformed to a broadcast schedule. Apparently, B is equal to the number of the schedule nodes in the pseudo broadcast tree. The schedule nodes in the pseudo broadcast tree can be classified into two parts: (1) the computed scheduled nodes by Algorithm 1, which is equal to $|A_S|$; (2) the schedule nodes when merging the subtrees into the primary subtree. Let the number of subtrees generated by AMTB be k. Since at most two schedule nodes are added in the pseudo broadcast tree when merging the subtree, then we can get $B \leq |A_S| + 2k$.

According to Theorem 4, $|A_S| \leq OPT \cdot ln(\Delta+1)$, where OPT denote the required number of transmissions of the optimal schedule for MTBDCA problem, then $B \leq |A_S| + 2k \leq OPT \cdot ln(\Delta+1) + 2k$. Additionally, the generated subtrees in the network is a constant when the length and width of the network is fixed [22], then we can get $\frac{B}{OPT} \leq \frac{OPT \cdot ln(\Delta+1)+c}{OPT} \leq ln(\Delta+1)$, where c is a constant.

5 Simulation Results

In this section, the performance of AMTP is evaluated through extensive simulations, which is compared with the following algorithms.

First, the broadcasting algorithms designed for duty-cycled networks, i.e., CSCA [19] and BRMS [22], which assumes single active time slot per working cycle. In the experiments, these two algorithms are both evaluated to demonstrate our proposed algorithm can benefit from multiple active time slot scheduling.

Second, two baseline algorithms, i.e., CDS-based and SPT-based algorithms are also compared. In these two algorithms, a CDS-based and SPT-based broadcast tree is constructed respectively. Additionally, to reduce the transmissions, the transmitting schedule for each forwarding node is computed optimally by enumerating all the possible results.

In the simulations, we mainly focus on the number of transmissions of the proposed algorithms under various network topologies. As in [30], to test more complex networks, Networkx [31] was used to generate different network topologies with number of nodes set from 100 to 400. In all simulations, the duty cycle is set from 10% to 35% and the working plan of each node is generated randomly to test a wide range of configurations. In the simulation results, each plotted point represents the average of 100 executions.

5.1 Performance Under Different $|W|$

First, the number of transmissions of the proposed algorithms under different length of working cycle, i.e., $|W|$ and network size, i.e., N are evaluated. In this simulation, the duty cycle of each node is set 20% and the average number of neighbors is set 5.

Figure 2 shows the performance of the proposed algorithm when the number of nodes is set 100, 200, 300 and 400, and the results are shown in Fig. 2(a), (b), (c) and (d) respectively. The first observation from Fig. 2 is that the number of transmissions of AMTB is much less than other algorithms in all scenarios. Compared to BRMS, which is recently proposed, the number of transmissions is decreased by 50% on average. The effectiveness of AMTB is mainly attributed to it makes full use of the multiple active time slots among all neighbors to construct the broadcast tree. One can also observe that, although considering multiple active time slots in scheduling, the CDS-based and SPT-based method performs even worse than the BRMS method, which takes only one active time slot of each node into consideration. This demonstrates the importance of choosing the forwarding nodes in constructing the broadcast tree. If it is not constructed well, the number of transmissions can be still large even using the optimal scheduling.

Another finding is that, the performance of BRMS and CSCA is even worse when the length of the working cycle is increased, while the one of AMTB, SPT-based and CDS-based is improved. This is because BRMS and CSCA only use the first active slot of each node. When the length of working cycle is increased, the number of nodes sharing the same active time slot may be reduced and other methods can benefit from the increased active time slots in its working plan. Additionally, when N is increased, the number of transmissions of each method is also increased, which is shown in Fig. 2(a), (b), (c) and (d) respectively.

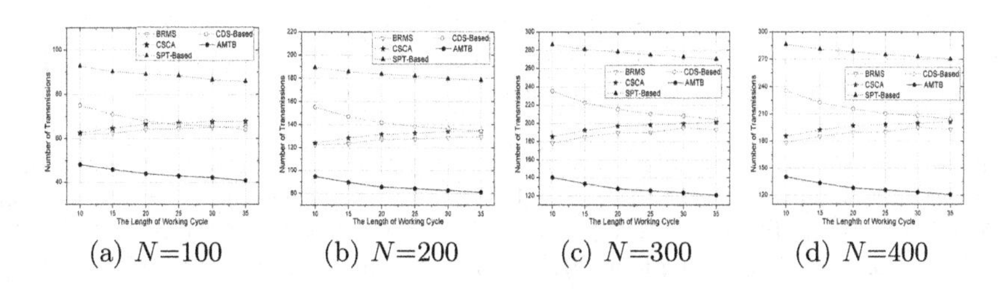

(a) N=100 (b) N=200 (c) N=300 (d) N=400

Fig. 2. The number of transmissions when duty cycle is set 20%.

5.2 Performance Under Different *Duty Cycle*

Second, the number of transmissions of the proposed algorithms under different duty cycle is evaluated. In this simulation, the length of a working cycle is set 20 and the number of nodes is set from 100 to 400.

Figure 3 shows the performance of the proposed algorithm when we vary the duty cycle, and the results are shown in Fig. 3(a), (b), (c) and (d) respectively. Similar as in Fig. 2, AMTB perform much better than the existing methods, *i.e.*, BRMS and CSCA, in terms of the number of transmissions. The performance of the SPT-based method is still much worse than other methods, which demonstrates that the SPT-based method is not suitable for the MTB problem in duty-cycled networks.

Different from Fig. 2, both the performance of BRMS and CSCA and the one of the proposed AMTB is improved when the duty cycle is increased. This is because the probability of the first active time slot of the neighboring nodes is same is increased when the duty cycle and the size of node's working plan is increased. In this case, the number of nodes share the same active time slots is increased. And the number of transmission of AMTB is much less than other methods, which demonstrates the efficiency of the proposed methods.

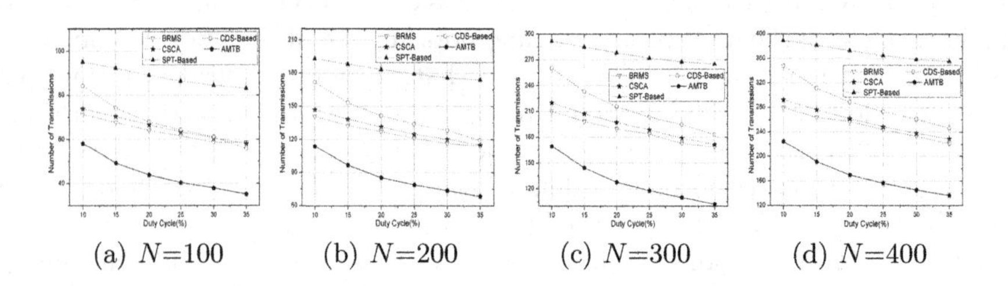

(a) N=100 (b) N=200 (c) N=300 (d) N=400

Fig. 3. The number of transmissions when $|W|$ is set 20.

6 Conclusion

In this work, the MTBDCA problem is investigated and is proved to be NP-hard and unlikely to have an $(1 - o(1))ln\Delta$-approximation algorithm unless $NP \subseteq DTIME(n^{O(loglogn)})$. A novel covering problem is proposed based on an auxiliary graph and a $ln(\Delta + 1)$-approximation algorithm is proposed for MTBDCA. Extensive simulation demonstrates the great efficiency of the proposed algorithm in terms of transmission redundancy.

Acknowledgement. This work is partly supported by the National Science Foundation (NSF) under grant NOs. 1252292, 1741277 and 1704287, the National Natural Science Foundation of China (NSFC) under Grant NOs. 61632010, 61502116, 61502110 and the major PSTP of Guangdong 2015B010104005 and 2017A010101017.

References

1. Cheng, S., Cai, Z., Li, J., Gao, H.: Extracting kernel dataset from big sensory data in wireless sensor networks. TKDE **29**(4), 813–827 (2017)
2. Li, J., Cheng, S., Cai, Z., et al.: Approximate holistic aggregation in wireless sensor networks. ACM Trans. Sens. Netw. **13**(2), 11 (2017)
3. Yu, J., Qi, Y., Wang, G., Gu, X.: A cluster-based routing protocol for wireless sensor networks with nonuniform node distribution. Int. J. Electron. Commun. **66**(1), 54–61 (2012)
4. Cheng, S., Cai, Z., Li, J.: Curve query processing in wireless sensor networks. IEEE Trans. Veh. Technol. **64**(11), 5198–5209 (2015)
5. He, Z., Cai, Z., Cheng, S., et al.: Approximate aggregation for tracking quantiles and range countings in wireless sensor networks. Theoret. Comput. Sci. **607**(3), 381–390 (2015)
6. Han, M., Li, L., Xie, Y., et al.: Cognitive approach for location privacy protection. IEEE Access **6**, 13466–13477 (2018)
7. Han, M., Wang, J., Duan, Z., et al.: Near-complete privacy protection: cognitive optimal strategy in location-based services. In: Proceedings of IIKI (2017)
8. Chlamtac, I., Kutten, S.: Tree-based broadcasting in multihop radio networks. IEEE Trans. Comput. **36**(10), 1209–1223 (1987)
9. Lou, W., Wu, J.: A cluster-based backbone infrastructure for broadcasting in MANETs. In: Proceedings of IPDPS (2003)
10. Das, B., Bharghavan, V.: Routing in ad-hoc networks using minimum connected dominating sets. In: Proceedings of ICC, pp. 376–380 (1997)
11. Wu, J., Li, H.: On calculating connected dominating set for efficient routing in ad hoc wireless networks. In: Proceedings of DIALM, pp. 7–14 (1999)
12. Lou, W., Wu, J.: On reducing broadcast redundancy in ad hoc wireless networks. IEEE Trans. Mob. Comput. **1**(2), 111–123 (2002)
13. Wu, J., Lou, W.: Forward-node-set-based broadcast in clustered mobile ad hoc networks. Wirel. Commun. Mob. Comput. **3**(2), 155–173 (2003)
14. Wu, J.: An enhanced approach to determine a small forward node set based on multipoint relays. In: Proceedings of VTC, pp. 2774–2777 (2003)
15. Adjih, C., Jacquet, P., Viennot, L.: Computing connected dominated sets with multipoint relays. Ad Hoc Sens. Wirel. Netw. **1**(1), 27–39 (2005)
16. Wu, J., Lou, W., Dai, F.: Extended multipoint relays to determine connected dominating sets in MANETs. IEEE Trans. Comput. **55**(3), 334–347 (2006)
17. Chen, Q., Gao, H., Cheng, S., et al.: Centralized and distributed delay-bounded scheduling algorithms for multicast in duty-cycled wireless sensor networks. IEEE/ACM Trans. Netw. **25**(6), 3573–3586 (2017)
18. Chen, Q., Gao, H., Cheng, S., et al.: Distributed non-structure based data aggregation for duty-cycle wireless sensor networks. In: Proceedings of IEEE INFOCOM, pp. 145–153 (2017)
19. Hong, J., Cao, J., Li, W., et al.: Minimum-transmission broadcast in uncoordinated duty-cycled wireless ad hoc networks. IEEE TVT **59**(1), 307–318 (2010)
20. Le-Duc, T., Le, D., Choo, H., et al.: On minimizing the broadcast redundancy in duty-cycled wireless sensor networks. In: ACM International Conference on Ubiquitous Information Management and Communication (2013)
21. Le-Duc, T., Le, D., Choo, H., et al.: Level-based approach for minimum-transmission broadcast in duty-cycled wireless sensor networks. Pervasive Mob. Comput. **27**, 116–132 (2016)

22. Le-Duc, T., Le, D., Zalyubovskiy, V.: Towards broadcast redundancy minimization in duty-cycled wireless sensor networks. Int. J. Commun. Syst. **30**(6), 1–21 (2017)
23. Wan, P.-J., Alzoubi, K.M., Frieder, O.: Distributed construction of connected dominating set in wireless ad hoc networks. Mob. Netw. Appl. **9**(2), 141–149 (2004)
24. Wan, P.-J., Wang, L., Yao, F.: Two-phased approximation algorithms for minimum CDS in wireless ad hoc networks. In: Proceedings of ICDCS, pp. 337–344 (2008)
25. Shi, T., Cheng, S., Li, J., et al.: Constructing connected dominating sets in battery-free networks. In: Proceedings of INFOCOM, pp. 1–9 (2017)
26. Yu, J., Wang, N., Wang, G.: Constructing minimum extended weakly connected dominating sets for clustering in ad hoc networks. J. Parallel Distrib. Comput. **72**(1), 35–47 (2012)
27. Yu, J., Ning, X., Sun, Y., et al.: Constructing a self-stabilizing CDS with bounded diameter in wireless networks under SINR. In: IEEE INFOCOM (2017)
28. Han, K., Liu, Y., Luo, J.: Duty-cycle-aware minimum-energy multicasting in wireless sensor networks. IEEE Trans. Netw. **21**(3), 910–923 (2013)
29. Feige, U.: A threshold of for ln n approximating set cover. J. ACM **45**, 634–652 (1998)
30. Chen, Q., Gao, H., Li, Y., et al.: Edge-based beaconing schedule in duty-cycled multihop wireless networks. In: Proceedings of INFOCOM (2017)
31. NetworkX. http://networkx.lanl.gov

Robust Network-Based Binary-to-Vector Encoding for Scalable IoT Binary File Retrieval

Yu Chen[1,2], Hong Li[1,2(✉)], Yuan Ma[1,2], Zhiqiang Shi[1,2], and Limin Sun[1,2]

[1] School of Cyber Security, University of Chinese Academy of Sciences, Beijing, China
{chenyu9043,lihong,mayuan2,shizhiqiang,sunlimin}@iie.ac.cn
[2] Beijing Key Laboratory of IoT Information Security Technology, Institute of Information Engineering, CAS, Beijing, China

Abstract. The goal of IoT binary file retrieval is to retrieve homologous binary files from a large IoT binary file database. Binary file retrieval has many applications, such as security analysis, OEM detection and plagiarism detection. However, traditional string-based approaches are hard to retrieve binary file which contains few or obfuscated strings. To solve this problem, we propose a novel neural network-based approach for encoding binary file into numerical vector based on non-string binary features. Moreover, by using this encoding method, the retrieval task can be accelerated by locality-sensitive hashing technique. For network training and testing, we compile 893 open source components into 71,129 labeled binary file pairs by using 16 different compilation configurations. We implement a prototype called B2V and compare it with IHB, a string-based approach, on both original and string obfuscated testing sets. The results show that the AUC of B2V is better than IHB (0.94 vs. 0.81) on the string obfuscated testing set, while still keeps comparable performance with IHB on the original testing set. Moreover, B2V can be easily retrained to adapt to string obfuscated scenarios with 15%–20% performance improvement. In the interest of open science, we also make our dataset publicly available to seed future improvements.

Keywords: Homologous binary retrieval · Binary feature encoding Function call graph · Graph embedding network

1 Introduction

Homologous binary file retrieval is very important for discovering IoT firmware with homologous vulnerabilities. Moreover, it also has some other significant applications such as OEM detection [1] and plagiarism detection [2]. However, performing homologous binary retrieval on IoT binaries faces the following four challenges. Firstly, the IoT firmware has more compilation configurations than the traditional software, such as more instruction sets and more factory-tailored

S. Chellappan et al. (Eds.): WASA 2018, LNCS 10874, pp. 53–65, 2018.
https://doi.org/10.1007/978-3-319-94268-1_5

compilers. As a result, even the same source code may generate quite different binary files. Second, dynamic analysis of IoT firmware is difficult because it is difficult to simulate the feedback of peripheral hardware Thirdly, there are millions of IoT firmware on market today, and the number of binary files included is even on the order of hundred millions. Therefore, there is a very high requirement for retrieval speed. Lastly, some manufactures may obfuscate strings in infringing binaries to avoid the detection by the string-based copyright detection tools, such as Binary Analysis Tool (BAT) [3]. In our previous work [4], IHB, a large-scale file retrieval scheme based on strings and Locality-Sensitive Hashing (LSH) [5], is proposed. IHB solves the above challenges except for the last one as the string-based scheme dose not apply to binaries with few or obfuscated strings. In recent years, deep learning [6] has been applied to many application domains, including binary analysis, and has shown better performance than other approaches. In this paper, we employ deep learning method to encode non-string binary features for giving the retrieval system more robustness against string absence or obfuscation. Then we use LSH and reverse indexing techniques to accelerate binary file retrieval as those used in IHB. In particular, a binary file is represented as a Attributed Function Call Graph (AFCG) which has a strong anti-obfuscation ability and then be converted into a numerical vector by a graph embedding network [7]. As there are no publicly labeled binary pairs as ground truth for training neural networks, we spend a lot of effort to generate the first binary pairs dataset which contains 71,129 labeled binary file pairs compiled from 893 open source components by using 16 different compilation configurations. We implement a prototype called B2V and compare it with IHB on both the original and obfuscated testing sets. The results show that the AUC of B2V is better than IHB (0.94 vs. 0.81) on the obfuscated testing set, whereas it still keeps comparable performance with IHB on the original testing set. Moreover, B2V can be easily retrained to adapt to string obfuscated scenarios with 15%–20% performance improvement. The contributions of this paper are summarized as follows:

- We propose a method for representing binary file using AFCG
- We propose the first neural network-based approach to encoding binary file into a numerical vector with cosine metric
- We implement a prototype called B2V. Our evaluation demonstrates that B2V can achieve a higher AUC than IHB (0.94 vs 0.81) on the obfuscated data set while can be comparable with B2V on data set without obfuscation
- We generate the first data set which contains 71,129 labeled binary file pairs compiled from 893 open source components by using 16 different compilation configurations and make it publicly available for other researchers.

2 Preliminary

A fundamental data-mining problem is to examine data for similar items. The problem arises when we search for similar items of any kind is that there may be far too many pairs of items to test each pair for their degree of similarity, even if computing the similarity of any one pair can be made very easy [8]. This motivates the LSH technique which focuses on retrieving pairs that are most likely to be similar. However, the precondition for using LSH technology to speed up the retrieval process is that the sample must be able to be encoded as a feature vector with a certain distance metric. The distance metric needs to have a LSH function family F that defined as follows:

Let Θ denote the sample space and F denote the function set. d_1 and d_2 are two distance values that satisfy $d_1 < d_2$ under the distance metric $\mathcal{D}$ in Θ. We call F a (d_1, d_2, p_1, p_2)-sensitive LSH function family if for $\forall f \in F$ satisfies the following conditions:

- f must be statistically independent of each other, that is, the joint probability of two or more functions is equal to the probability product of the independent events on each function
- if $\mathcal{D}(x, y) \leq d_1$, the minimal probability that $f(x) = f(y)$, namely the pair of items (x, y) to be chosen is p_1
- if $\mathcal{D}(x, y) \geq d_2$, the maximal probability that $f(x) = f(y)$, namely the pair of items (x, y) to be chosen is p_2.

Typical distance metrics with the LSH function family are jaccard distance, cosine distance, euclidean distance and hamming distance, etc. In our previous work [4], binary files were encoded as feature vectors with Jaccard distance metrics. However, in this paper, in order to make the retrieval process being able to resist the string confusion, we need to encode the binary file as a numerical vector with cosine distance metric which is more challenging than encode with Jaccard distance metrics.

3 Problem Definition

A homologous binary file family is a collection of binary files compiled from the same or similar source codes based on different compilation configurations. The compilation configurations include different instruction sets, different compilers and different compilation optimization levels, etc. Let b_1 and b_2 represent two binary files and π denotes the homology calculation operators. If b_1, b_2 belong to the same homologous binary file family, then $\pi(b_1, b_2) = +1$, otherwise $\pi(b_1, b_2) = -1$. by using a two-tuple to represent a data sample, the data set T can be defined as follows:

$$T := \{(<b_1, b_2>, \pi(b_1, b_2))\} \tag{1}$$

Our goal is to use the data set T to train, verify and test a graph embedding network that satisfies the following conditions:

$$Sim(g(b_1), g(b_2)) = cos(\phi(g(b_1)), \phi(g(b_2))) = \frac{\langle \phi(g(b_1)), \phi(g(b_2)) \rangle}{\| \phi(g(b_1)) \| \cdot \| \phi(g(b_2)) \|} \tag{2}$$

Where $g(b)$ is the graph representation of the binary file b, $Sim \in [-1, +1]$. The closer Sim is to -1, the lower the probability that b_1 and b_2 belong to the same homologous binary file family. Conversely, the closer Sim is to $+1$, the higher the probability that b_1 and b_2 belong to the same homologous binary file family.

4 Detailed Design

4.1 Graph Representation of Binary Files

In this work, we propose a method to represent binary files by using AFCG, which has strong anti-obfuscation capability. An example of AFCG is shown in the Fig. 1. The AFCG of a binary file is essentially a directed graph with functions as nodes. The attributes of the node (function) in the function call graph are shown in Table 1. The node has six statistical features and two structural features. The cross-platform and cross-compilation performance of statistical and structural features have been verified in our previous work [9] and in Feng's work [10], respectively.

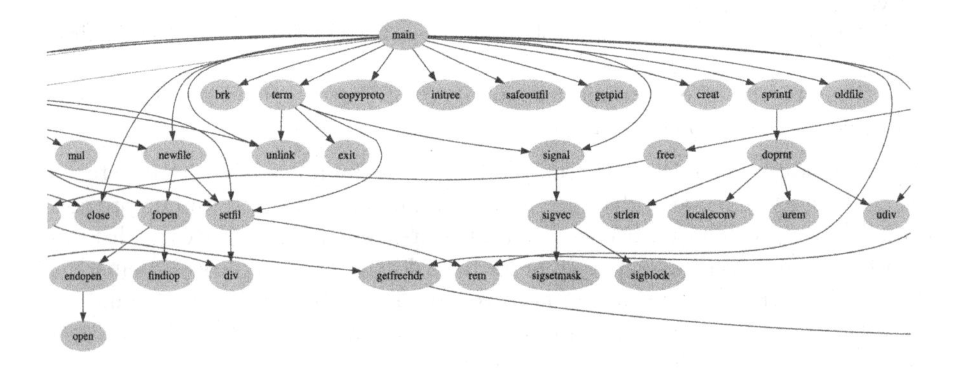

Fig. 1. An AFCG example

Table 1. Vector representation of binary function

Type	Feature name	Describe
Statistical features	size	Number of instructions
	nbbs	Number of basic blocks
	edges	Number of CFG edges
	cc	Cyclomatic complexity of CFG
	strnum	Number of string references
	strlen	Total length of refererad strings
Structural features	indegree	FCG indegree of the function
	outdegree	FCG outdegree of the function

4.2 Structure of B2V

The input of B2V is an AFCG, denoted as g, and the output is a numerical vector of length p, denoted as μ. The process of generating μ from g consists of two stages: the first stage is the iterations of the values of all the nodes in g, and the second stage is the synthesis of the result of the last iteration into the vector μ. Detailed steps for the T-round iteration are shown in Algorithm 1. **Lines 3–7** describe the process of node attribute iteration, and **Lines 8** describes how to synthesize the result of the last iteration into the result vector. In **Line 5**, $N(v_i) := \{v|\exists e \in \varepsilon, v \rightarrow v_i\}$. L_x represents a linear network, with no bias except L_5. Both tanh and ReLU in Algorithm 1 are the activation functions of neural networks. In order to facilitate the understanding of the above algorithm, a network structure diagram which is equivalent to the above algorithm process is shown in the Fig. 2. The arrows in the Fig. 2(a) describe the dependencies between nodes when updating their attributes. Taking node v_3 as an example, both v_1 and v_2 have edges that point to v_3. Therefore, in the k-th iteration, the update of the attribute value of the x_3 node is dependent on the attribute values of x_1, x_2, and x_3 itself in the k-1th iteration. Figure 2(b) describes the process of updating the node attributes in the k-th iteration, which is consistent with the formula described in **Line 5** in Algorithm 1.

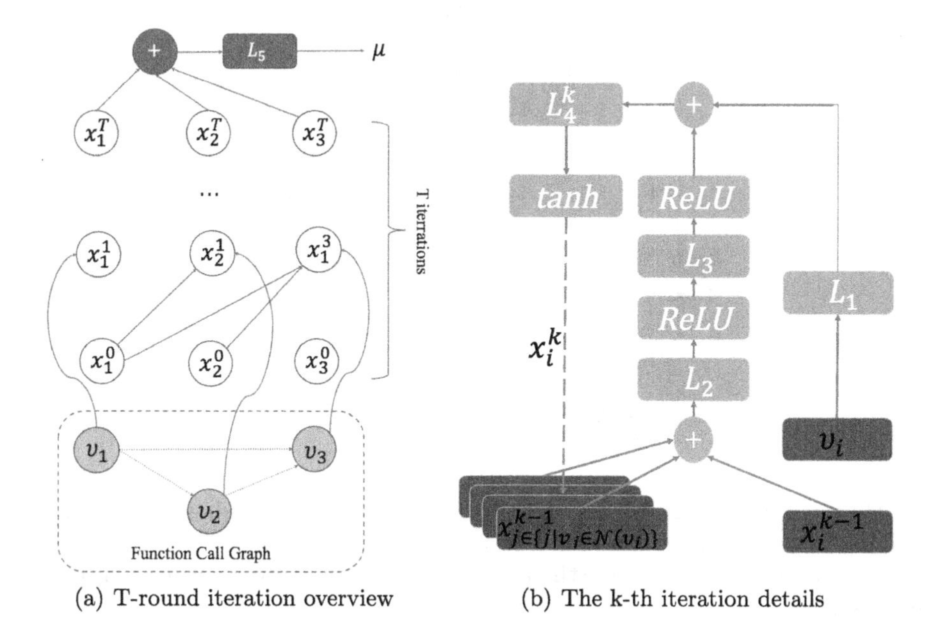

(a) T-round iteration overview (b) The k-th iteration details

Fig. 2. Network structure of B2V

4.3 Training

In reference to Xu's work [11], we use siamese architecture as shown in Fig. 3 to train the parameters of B2V. The two networks in the siamese architecture share the same parameters and use the following Loss Function.

$$\sum \left(cosine(\phi(g(b_1)), \phi(g(b_2))) - \pi(b_1, b_2)\right)^2 \tag{3}$$

Algorithm 1. T-round iteration process

1: Input: AFCG $g = \langle V, \varepsilon, X \rangle$
2: Initialize $x^0_{i \in [1, len(V)]} = \overline{0}$
3: **for** each $k \in [1, T]$ **do**
4: **for** each $i \in [1, len(\mathrm{V})]$ **do**
5: $x_i^k = tanh(L_4^k(ReLU(L_3(ReLU(L_2(\sum_{j \in \{j | v_j \in N(v_i)\}} x_j^{k-1} + x_i^{k-1})))) + L_1(v_i)))$
6: **end for**
7: **end for**
8: return $\phi(g) := L_5(\sum_i x_i^T)$

The **Minibatch** is set to 128 and the training environment is as follows: 48 core Intel Xeon E5-2650 v4 with 256G RAM, NVIDIA Tesla P100 GPU, NVIDIA Driver 384.90, CUDA-8.0, Pytorch-0.3.0.

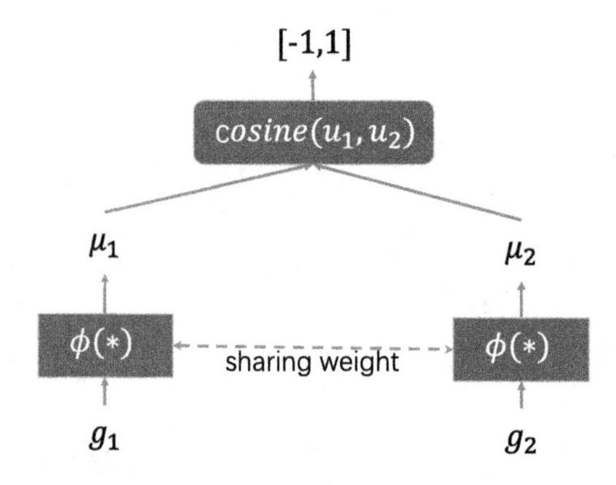

Fig. 3. Siamese architecture

During the training process, the Loss and AUC curves are shown in Fig. 4(a) and (b), respectively. As shown in Fig. 4, the performance of B2V stabilized after twenty epochs which takes less than ten minutes in the abovementioned training environment.

5 Evaluation

5.1 Data Set

Original Data Set. As described above, there are no publicly labeled binary pairs for training neural networks. In this work, we spend a lot of effort to generate the first binary pairs dataset as ground truth and make it publicly available for other researchers. Our dataset consists of 71,129 labeled binary file pairs

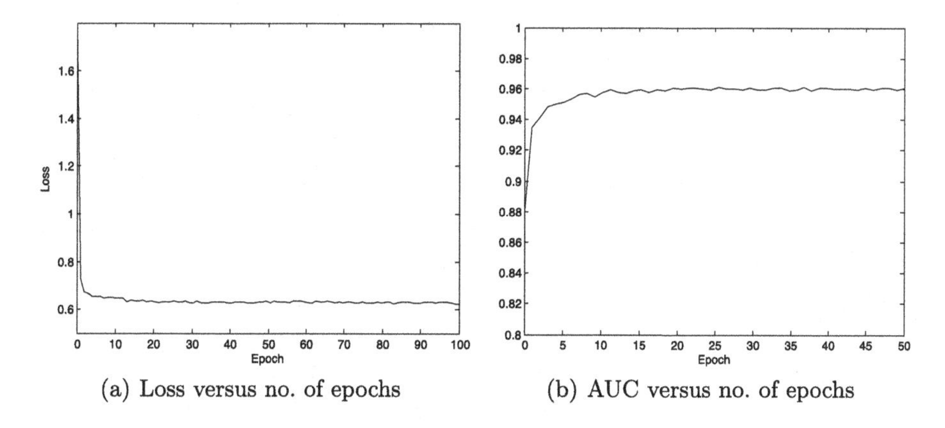

(a) Loss versus no. of epochs (b) AUC versus no. of epochs

Fig. 4. The training process

which are generated by compiling 893 different source codes with 16 different compilation configurations. The open source codes include common components in embedded devices such as openssl, binutils, libmad, boa web server, sudo, openssh, ntp, nginx, tcpdump, libgd and libxml2. The compilation configurations are combinations of different target platforms (ARM, MIPS, PowerPC, X86, X64), different compilers (gcc, clang, icc), different instruction set bits (32-bit, 64-bit) and different compiler optimization levels (-O0, -O1, -O2, -O3, -Os). We split 41,421 labeled binary files into three disjoint subsets of functions for training, validation, and testing, respectively. During the split, we guarantee that no two binary files compiled from the same source function are separated into two different sets among training, validation and testing sets. In doing so, we can examine whether the neural network can generalize to unseen binary files. Following the principle that the ratio of *pi* to $+1$ and -1 should be approximately equal, the binary files in each set are paired randomly to generate the corresponding two-tuples. The overview of the dataset is shown in the Table 2.

Obfuscated Data Set. In order to simulate the obfuscation that may exist in the actual application scenario, obfuscation processing is performed on the binary files in the testing set as follows:

- Randomly change the length of each string to 80%–100% of its original length
- Randomly replace characters in a string with other characters
- Randomly change the call relationship in the FCG with 20% probability
- Randomly change 0%–20% of the function basic attribute values with 20% probability.

5.2 Metric

The similarity of each two-tuple sample $(\langle b_1, b_2 \rangle, \pi(b_1, b_2))$ in the test set is calculated by the following equation:

Table 2. Binary code dataset

Instruction set	Instruction bits	Compiler	Optimization	TrainSet	ValidSet	TestSet
ARM	32	gcc	Default	2455	180	520
MIPS				3114	256	667
PPC				3280	253	634
X86				2685	192	586
X64	64		-O0	1639	293	190
			-O1	1486	258	175
			-O2	1479	41	157
			-O3	1433	39	151
			-Os	1303	253	138
		clang	-O0	1551	291	193
			-O1	1487	308	173
			-O2	1231	267	185
			-O3	1270	268	174
			-Os	1203	277	167
		icc	-O0	1422	293	167
			-O1	1078	258	128
			-O2	403	41	50
			-O3	398	39	50
			-Os	1084	253	123
Total				31695	4781	4945
Number of tuples ($<b_1, b_2>, \pi(b_1, b_2)$)				55253	7543	8333

$$Sim(g(b_1), g(b_2)) = cos(\phi(g(b_1)), \phi(g(b_2))) = \frac{\langle \phi(g(b_1)), \phi(g(b_2)) \rangle}{\parallel \phi(g(b_1)) \parallel \cdot \parallel \phi(g(b_2)) \parallel} \quad (4)$$

If threshold is donated by δ, the homology prediction of b_1 and b_2 is defined as follows:

$$Pred(b_1, b_2) = \big(Sim(g(b_1), g(b_2)) \geq \delta ? +1 : -1\big) \quad (5)$$

We use the ROC curve to evaluate the performance of the neural network. The ROC curve is an implicit function curve whose parameter is δ. The horizontal axis of the ROC curve is the false positive rate (FPR), and the vertical axis of the ROC curve is the true positive rate (TPR). Among them, FPR and TPR are defined as follows:

$$FPR = \frac{card(\{(<b_1, b_2>, \pi(b_1, b_2)) | (<b_1, b_2>, \pi(b_1, b_2)) \in TestSet, Pred(b_1, b_2) == 1 \cap \pi(b_1, b_2) == -1\})}{card(\{(<b_1, b_2>, \pi(b_1, b_2)) | (<b_1, b_2>, \pi(b_1, b_2)) \in TestSet, \pi(b_1, b_2) == -1\})} \quad (6)$$

$$TPR = \frac{card(\{(<b_1, b_2>, \pi(b_1, b_2)) | (<b_1, b_2>, \pi(b_1, b_2)) \in TestSet, Pred(b_1, b_2) == 1 \cap \pi(b_1, b_2) == 1\})}{card(\{(<b_1, b_2>, \pi(b_1, b_2)) | (<b_1, b_2>, \pi(b_1, b_2)) \in TestSet, \pi(b_1, b_2) == 1\})} \quad (7)$$

5.3 Performance Evaluation

In this section, we compare B2V with IHB [4], a string-based approach, on both the original and the obfuscated data sets. The experimental results are shown in Fig. 5. The specific meanings of the legend are as follows:

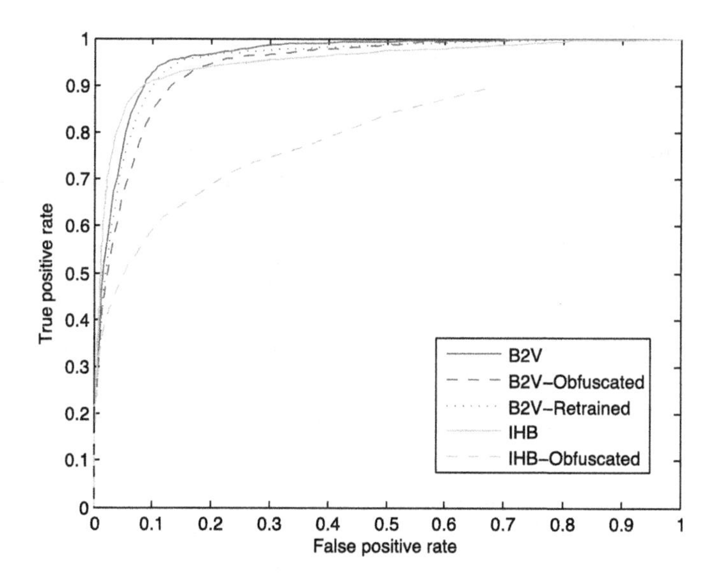

Fig. 5. ROC for different approaches evaluated on the original and obfuscated dataset

- **IHB** indicates the ROC curve of IHB on the original data set
- **B2V** indicates the ROC curve of B2V on the original data set
- **IHB-Obfuscated** indicates the ROC curve of IHB on the obfuscated data set
- **B2V-Obfuscated** indicates the ROC curve of B2V on the obfuscated data set
- **B2V-Retrained** indicates the ROC curve of B2V on the obfuscated data set after retraining with a small amount of obfuscated samples (10% of the number of testing set) on the basis of the original parameters.

We can have the following observations from the results in Fig. 5.

- On the original data set, the performance of IHB is slightly better than that of B2V only when FPR is less than 0.085 but the gap is not very significant.
- On the obfuscated data set, the performance of B2V is much better than IHB. Specifically, when the TPR reaches 0.9, the FPR of B2V and IHB is 0.12 and 0.67, respectively. In fact, B2V can achieve a higher AUC than IHB (0.94 vs. 0.81).

- On the obfuscated data set, the performance of B2V after retraining is about 15%–20% better than that without retraining.

In summary, on the obfuscated data set, the classification performance of B2V is much better than IHB while can be comparable with B2V on the original data set. Moreover, B2V can be retrained to achieve better performance with a small amount of obfuscated samples.

5.4 Hyperparameters

This section evaluates the impact of the model's hyperparameters, (the encoding bits p and the number of iterations T) on the performance of the neural network.

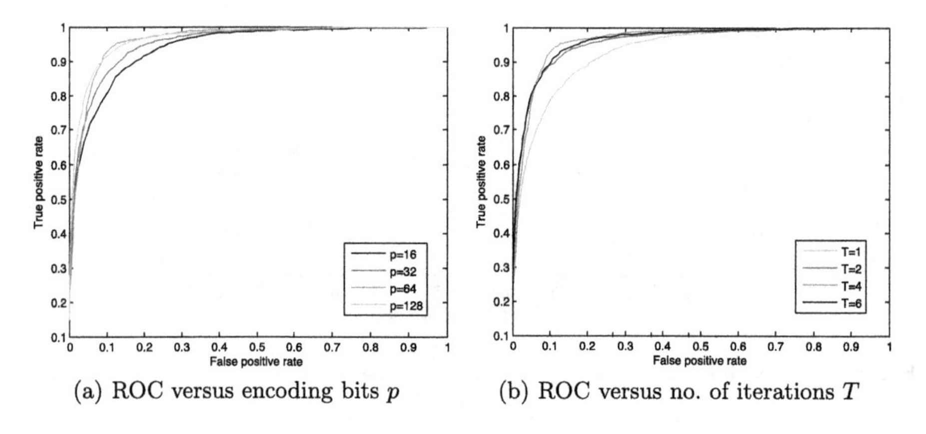

(a) ROC versus encoding bits p (b) ROC versus no. of iterations T

Fig. 6. The impact of the model's hyperparameter

As shown in Fig. 6(a), the longer the number of encoding bits, the better the classification performance. However, the gap between 64 bit and 128 bit is not very significant. Therefore for the sake of computational overhead, we choose 64 as the value of the hyperparameter p.

As shown in Fig. 6(b), when the value of T is 4, the classification performance is the best. Thus we choose 4 as the value of the hyperparameter T.

6 Related Work

Kornblum [12] proposed a fuzzy hash algorithm ssdeep, which implements automatic alignment and fuzzy matching based on the context-triggered rolling hash. Roussev [13] proposed another fuzzy hash algorithm sdhash which chooses the feature with the highest degree of differentiation based on entropy estimation. Li [14] validates the effectiveness of above fuzzy hash algorithms in security applications such as malware family classification. Bass [15] used ssdeep to perform clustering experiments on 10.7M files in the CERT Artifact Catalog database.

Costin [16] was the first one to perform a correlation analysis on more than 30,000 firmwares, but since the binary file association method they used is based on fuzzy hash [12,13] which requires the one-to-one calculation, it is not suitable for large-scale binary file retrieval applications. In our previous work [4], we designed, implemented, and evaluated a scalable and efficient homologous binary search scheme (termed as IHB) for IoT firmwares with time complexity $O(1)$. The main idea of our methodology is to leverage the readable strings in binaries to calculate the similarities between different IoT firmwares. Although the IHB solves the efficiency problems of large-scale retrieval, it performs poorly in scenes where strings are obfuscated.

In recent years, more and more machine learning methods have been used to solve problems in the field of binary analysis and IoT Privacy Policy [17–20]. Here are some of the representative work. In our another previous work [9], we proposed a function vulnerability association method which was based on the BP neural network [21] to extract the cross-platform features of the function. Shin [22] applies the Recurrent Neural Network (RNN) to function boundary recognition. Chua [23] uses three-layer RNN networks to identify the number of function parameters and data types. Feng [10] proposed to use an unsupervised learning model to encode a function expressed with ACFG into a numerical vector and finally using LSH technology to quickly retrieve vulnerability functions. Based on Feng's work, Xu [11] proposed the first graphical embedded network to encode the ACFG to improve retrieval performance. The idea of our work is inspired by the above research results, especially the work of Feng and Xu, but there are two significant differences: (1) Different encoding objects: the encoding object of our work is a binary file, but the encoding object of Feng and Xu is a binary function; (2) Original dataset: our work generates the first data set which contains 71,129 labeled binary file pairs compiled from 893 open source components by using 16 different compilation configurations for training, validating and testing the neural network.

7 Conclusion

In this paper, we design, implement and evaluate a robust binary encoding scheme for scalable IoT binary file retrieval. We evaluate our method and compared it with the state-of-the-art schemes on the original and obfuscated dataset. The results show that our scheme can achieve a higher AUC (0.94) than the state-of-the-art schemes (0.81).

Acknowledgment. This work was supported by National Key Research and Development Program of China (2016YFB0800202); National Natural Science Foundation of China under Grants No. U1636120; Fundamental Theory and Cutting Edge Technology Research Program of Institute of Information Engineering, CAS; SKLOIS (No. Y7Z0361104 and No. Y7Z0311104); Key Program of National Natural Science Foundation of China (U1766215); Key Research Program of Chinese MIIT under Grant No. JCKY2016602B001; Beijing Municipal Science & Technology Commission Grants No. Z161100002616032; The Science and Technology Project of State Grid Corporation of China (No. 52110418001K).

References

1. Hemel, A., Kalleberg, K.T., Vermaas, R., Dolstra, E.: Finding software license violations through binary code clone detection, pp. 63–72 (2011)
2. Jhi, Y.C., Jia, X., Wang, X., Zhu, S., Liu, P., Wu, D.: Program characterization using runtime values and its application to software plagiarism detection. IEEE Trans. Softw. Eng. **41**(9), 925–943 (2015)
3. Hemel, A., Coughlan, S.: BAT: binary analysis toolkit. http://www.binaryanalysis.org/en/home
4. Chen, Y., Li, H., Zhao, W., Zhang, L., Liu, Z., Shi, Z.: IHB: a scalable and efficient scheme to identify homologous binaries in IoT firmwares. In: 2017 IEEE 36th International Performance Computing and Communications Conference, IPCCC (2017)
5. Gionis, A., Indyk, P., Motwani, R.: Similarity search in high dimensions via hashing. In: International Conference on Very Large Data Bases, pp. 518–529 (1999)
6. Hinton, G.E., Osindero, S., Teh, Y.W.: A fast learning algorithm for deep belief nets. Neural Comput. **18**(7), 1527–1554 (2006)
7. Goyal, P., Ferrara, E.: Graph embedding techniques, applications, and performance: a survey (2017)
8. Leskovec, J., Rajaraman, A., Ullman, J.D.: Mining of Massive Datasets. Cambridge University Press, Cambridge (2014)
9. Chang, Q., Liu, Z., Wang, M., Chen, Y., Shi, Z., Sun, L.: VDNS: an algorithm for cross-platform vulnerability searching in binary firmware. J. Comput. Res. Dev. (2016)
10. Feng, Q., Zhou, R., Xu, C., Cheng, Y., Testa, B., Yin, H.: Scalable graph-based bug search for firmware images. In: Proceedings of the 2016 ACM SIGSAC Conference on Computer and Communications Security, pp. 480–491. ACM (2016)
11. Xu, X., Liu, C., Feng, Q., Yin, H., Song, L., Song, D.: Neural network-based graph embedding for cross-platform binary code similarity detection (2017)
12. Kornblum, J.: Identifying almost identical files using context triggered piecewise hashing. Digit. Investig. **3**(3), 91–97 (2006)
13. Roussev, V.: Data fingerprinting with similarity digests. In: Chow, K.-P., Shenoi, S. (eds.) DigitalForensics 2010. IAICT, vol. 337, pp. 207–226. Springer, Heidelberg (2010). https://doi.org/10.1007/978-3-642-15506-2_15
14. Li, Y., Sundaramurthy, S.C., Bardas, A.G., Ou, X., Caragea, D., Hu, X., Jang, J.: Experimental study of fuzzy hashing in malware clustering analysis. In: USENIX Conference on Cyber Security Experimentation and Test, p. 8 (2015)
15. Bass, L., Brown, N., Cahill, G.M., Casey, W., Chaki, S., Cohen, C., Niz, D.D., French, D., Gurfinkel, A., Kazman, R.: Results of SEI line-funded exploratory new starts projects (2012)
16. Costin, A., Zaddach, J., Balzarotti, D.: A large-scale analysis of the security of embedded firmwares. In: USENIX Conference on Security Symposium, pp. 95–110 (2014)
17. Cai, Z., Zheng, X.: A private and efficient mechanism for data uploading in smart cyber-physical systems. IEEE Trans. Netw. Sci. Eng. (2018)
18. Liang, Y., Cai, Z., Yu, J., Han, Q., Li, Y.: Deep learning based inference of private information using embedded sensors in smart devices. IEEE Netw. Mag. (2018)
19. Zheng, X., Cai, Z., Li, Y.: Data linkage in smart IoT systems: a consideration from privacy perspective. IEEE Commun. Mag. (2018)

20. Hu, C., Li, R., Mei, B., Li, W., Alrawais, A., Bie, R.: Privacy-preserving combinatorial auction without an auctioneer. EURASIP J. Wirel. Commun. Netw. **2018**(1), 38 (2018)
21. Li, J., Cheng, J., Shi, J., Huang, F.: Brief introduction of back propagation (BP) neural network algorithm and its improvement. In: Jin, D., Lin, S. (eds.) Advances in Computer Science and Information Engineering. AINSC, vol. 169, pp. 553–558. Springer, Heidelberg (2012). https://doi.org/10.1007/978-3-642-30223-7_87
22. Shin, E.C.R., Song, D., Moazzezi, R.: Recognizing functions in binaries with neural networks. In: USENIX Conference on Security Symposium, pp. 611–626 (2015)
23. Chua, Z.L., Shen, S., Saxena, P., Liang, Z.: Neural nets can learn function type signatures from binaries. In: USENIX Conference on Security Symposium (2017)

OSCO: An Open Security-Enhanced Compatible OpenFlow Platform

Haosu Cheng[1,2], Jianwei Liu[1,2], Jian Mao[1,2(✉)], Mengmeng Wang[2], and Jie Chen[2]

[1] School of Cyber Science and Technology, Beihang University, Beijing, China
maojian@buaa.edu.cn

[2] School of Electronic and Information Engineering, Beihang University, Beijing, China

Abstract. Software-defined Networking (SDN) is a representative next generation network architecture, which allows network administrators to programmatically initialize, control, change and manage network behavior dynamically via open interfaces. However, SDN brings new security problems, e.g., controller hijacking, black-hole, unauthorized data modification, etc. It is desirable to develop a unified platfom to enhance the security property and facilitate the security configuration and evaluation. In this paper, we propose OSCO (Open Security-enhanced Compatible OpenFlow) platform, a platform based on Raspberry Pi Single Board Computer (SBC) hardware and SDN network architecture, which supports highly configurable cryptographic algorithm modules, security protocols, flexible hardware extensions and virtualized SDN networks. Furthermore, we present an enhanced OpenFlow protocol to improve the security in SDN data plane. We implement and evaluate the prototype system and the experiment results show that our system conducted security functions with relatively low computational and networking performance overheads.

1 Introduction

The Software-Defined Networking (SDN) is a representative next generation network architecture with data plane decoupled from control plane that supports programmatical initialization and dynamic network behavior management. In the SDN architecture, the control plane is logically centralized and the data plane is employed simply to conduct decisions made by the control plane [14,16]. The application plane implements control functions invoking software-based logic in the control plane via the REST-APIs (Representational State Transfer Application Programming Interface), which is also called the *Northbound interface*. The deployed logic decisions are executed by the data plane through the *Southbound interfaces*, e.g. OpenFlow [16]. OpenFlow is a widely adopted implementation architecture of SDN. OpenFlow-based SDN applications, using the underlying network infrastructure, are developed to deploy various functions at run-time. The network control traffic is transferred from the infrastructure (Data plane) to

S. Chellappan et al. (Eds.): WASA 2018, LNCS 10874, pp. 66–77, 2018.
https://doi.org/10.1007/978-3-319-94268-1_6

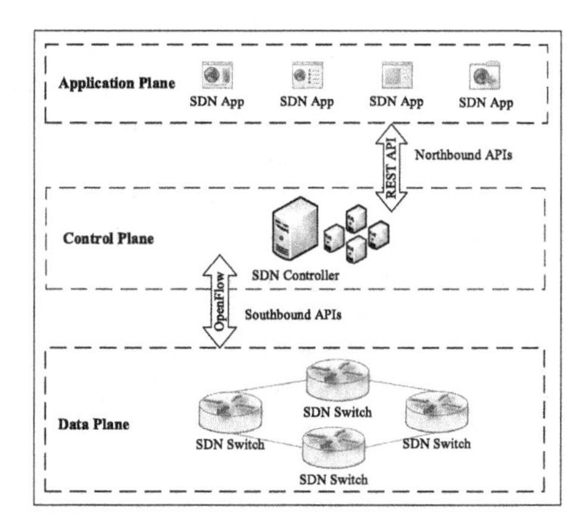

Fig. 1. The three planes in SDN architecture.

the controller. With the help of SDN apps, network operators may achieve distinguished properties of network control, automation and resource optimization. Figure 1 illustrates the overall architecture of an SDN framework.

Although SDN provides a new solution for the network architecture, it exposes many security problems [11]. The security vulnerabilities in SDNs are mainly distributed in application plane, control plane, and data plane. For instance, communication channels between isolated planes can be attacked by other compromised planes. Because of the control plane's visible nature, it is more attractive to attacker and vulnerable to DoS and DDoS attacks. The lack of authentication, authorization and accessing control mechanisms is another primary security limitation in the SDN application plane. The data plane is vulnerable to fraudulent flow rules, flooding, controller compromise and Man-in-the middle attacks.

Many security solutions are proposed targeting to the exposed security vulnerabilities of SDN. FRESCO [19] is proposed to enable development of OpenFlow security applications. The control plane security solutions consist of defending malicious applications, optimizing load balancing policies, DoS/DDoS attacks mitigation and reliable controller placement. Security-enhanced (SE) Floodlight [2] controller is an attempt towards an ideal secure SDN control layer by adding a secure programmable north-bound API to operate as an intercessor between applications and data plane. The fine-grained security enforcement mechanisms such as authentication and authorization are used for applications that can change the flow rules. FortNox [17] is a platform that enables the NOX controller to check flow rule contradictions real-time and authorize applications before they can change the flow rules.

However, these existing solutions only aim at some specified type of attacks or focus on specific planes. It is undesirable to develop and deploy security countermeasures in such an ad hoc manner to guarantee the data and network resource security. In addition, how to systematically test and verify the deployed security mechanisms is another challenge in SDN-based cyber-physical system (CPS) security enhancement, considering the high dependencies on the hardware/software implementation and the network topology. The current solutions to establish an SDN working environment are either to deploy hardware-based SDN switches with controllers or to simulate the SDN network topology and operations by using software simulators. In addition, the current hardware deployment or software simulation mostly focus on network function implementation. None of those solutions target at the security perspective. SDN switch hardwares are expensive to deploy massively and it is also difficult to modify the embedded protocols. Furthermore, commercial SDN switches cannot support customized protocols and interfaces. The software simulation (e.g. Mininet [15], NS3 [18]) is able to support large SDN network topology, but it lacks various hardware interfaces that enable some hardware-based security schemes, e.g., TPM, special encryption chip, random number generator, etc.

In this paper, we propose OSCO (Open Security-enhanced Compatible OpenFlow) platform, a platform based on Raspberry Pi Single Board Computer (SBC) hardware and SDN network architecture, which supports highly configurable cryptographic algorithm modules, security protocols, flexible hardware extensions and virtualized SDN networks. Furthermore, we present an enhanced OpenFlow protocol to improve the security in SDN data plane. We implement and evaluate the prototype system and the experiment results show that our system conducted security functions with relatively low computational and networking performance overheads.

Contributions. In summary, we make the following contributions in this paper:

- We propose an extensible security-oriented SDN network experiment platform, OSCO, allowing various security design schemes to experiment in an SDN environment. Our scheme provides an open framework with flexible hardware interfaces and extensible software modules that fully support standard OpenFlow protocol and its security enhancement solutions.
- To illustrate the extensibility, we improve the standard OpenFlow protocol and present a security enhanced OpenFlow scheme to protect the communication among switch nodes, which extend the crypto-based security mechanisums from *controller-to-switch* to *switch-to-switch*.
- We implement the OSCO platform in the Raspberry Pi hardware installed with the Ubuntu Linux operation system, which is acting as the OpenFlow switch. It supports physical SDN network deployment and simulation with physical network hybrid deployment. We evaluate the performance of our platform by conducting cryptographic operation test and network performance test. The results illustrate its efficiency.

Paper Organization. The rest of this paper is organized as follows: Sect. 2 presents the overall architecture and critical components of our platform, OSCO; Sect. 3 illustrates the proposed security enhanced open-flow protocol; Sect. 4 describes the implementation and evaluation results of OSCO platform and Sect. 5 concludes the paper.

2 Overall Architecture of OSCO Platform

In this section, we present the overall architecture of the OSCO platform and describe its core functions with auxiliary components in detail. The proposed architecture aims to bring the benefits of flexible functionality by providing well defined interfaces for control plane functions and enabling richer flexibility hardware interface in data plane secure traffic handling.

As shown in Fig. 2, the proposed platform, OSCO, includes two parts, the central part (enclosed by the dashed rectangle) and the peripheral systems. The central part consists of OSCO core function modules. The peripheral system consists of the REST-API formatted SDN Security applications, diversified peripheral security hardware, the SDN controller and the security-supported network simulator. The OSCO center part acts as an OpenFlow switch under the SDN architecture. The SDN Controller is able to communicate with applications and OSCO core. The OSCO is able to extend the security function through the peripheral security hardware. Besides, OSCO integrates the security-supported network simulator to create SDN network topology equipped with interal or exteral controllers.

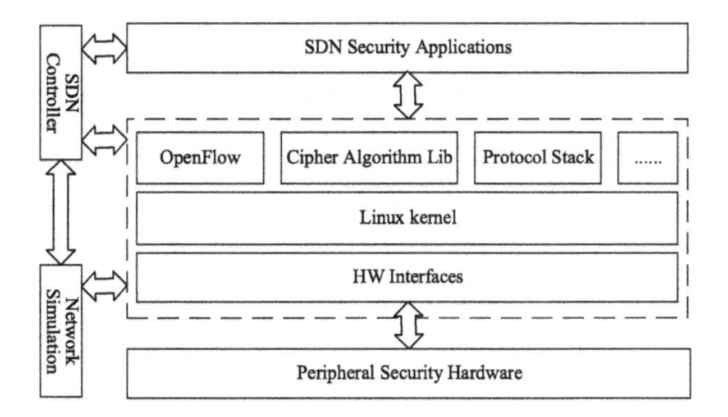

Fig. 2. The OSCO platform architecture.

2.1 OSCO Core Function Modules

The center part of the OSCO platform includes several core function modules, such as the *Hardware interface*, the *Linux kernel*, the *OpenFlow module*, the *Cipher algorithm library*, *Protocol stack*, etc. The core function modules act as a multi-function OpenFlow Switch. The security hardware is supported by the HW interfaces, such as the secure chip, random number generator, network interface and so on. The linux kernel is responsible for managing and invoking of software and hardware resources. The OpenFlow module is implemented above the Linux kernel, which is one of the most important protocols in the SDN architecture. OpenFlow is used for the interaction between a data plane constituted network switch and a control plane constituted controller. The cipher algorithm library contains the mainstream cryptographic algorithms, e.g., the MD5, SHA, DES, ASE, RSA algorithms, etc. The key management, key distribution and authentication protocols are stored in the protocol stack module. In addition, the OSCO platform is also extensible to employ other network functions, such as network package capture and analysis, etc.

2.2 Peripheral Systems

Another important part in OSCO is the peripheral systems that support the core function modules to fulfill their designed functionalities. The peripheral systems include four moduless: the first module is SDN Security Application that implements and invokes the security solutions; the second module is Peripheral Security Hardware that provides hardware-based security solutions; the third module is Network Simulation that enables the simulation of the complex network topology and SDN network functions; the last part is the SDN Controller that makes packages forwarding decisions, maintains the SDN network topology and coordinates network resources etc.

OpenFlow protocol links controller and OpenFlow switch, and it plays an important role in the SDN-based architecture. The OpenFlow protocol provides an optional SSL/TLS based secure channel between the SDN controller and switches. However, it is not a default configuration and just provides privacy and data integrity between SDN controller and switches. As a result, there might be information leakage among SDN switches in the regular OpenFlow protocol. To solve this problem and illustrate the extensibility of OSCO platform, we present a security enhanced OpenFlow protocol in Sect. 3.

3 OpenFlow Protocol with Security Enhanced Actions

OpenFlow is a standardized protocol [10] that is used for the interaction between a network switch and a controller. The switch performs packet forwarding according to the flow tables. Each table contains a set of rules with corresponding actions and measurement counters. The controller is able to choose a secure channel (SSL/TLS) to communicate with switches as an optional configuration. There is no such secure channel among switches, which makes the datalink between two switches vulnerable to the Man-in-the-middle, the Eavesdropping and the denial-of-service attacks [12].

3.1 Security Enhanced Actions

For the purpose of security, two actions (Encrypt and Decrypt) need to be added in the OpenFlow protocol implementation. The Encrypt action performs a symmetric encryption operation, which encrypts the packet data fields. The packet header will not be encrypted, as shown in Fig. 3. The Decrypt action will perform inverse operation to decrypt the encrypted packet. Encrypt and Decrypt actions are customized actions, and they will be carried out in the experimenting action structure under the OpenFlow version 1.3 standard. The packet header is not encrypted, therefore, the normal packet processing is not affected in the switch, as shown in Fig. 3.

The OpenFlow switch can perform data-link (L2) and IP layer (L3) packets forwarding, which means it is also able to perform L2 and L3 Encryption when two actions are implemented. In the security enhanced OpenFlow, it should not only enable two newly added customized actions, but also need to ensure the efficient key management and develop corresponding modification of controller side and security Apps as well. In this paper, we deploy a Kerberos-based [13] key management scheme to ensure the efficient and secure session key distribution among the switches.

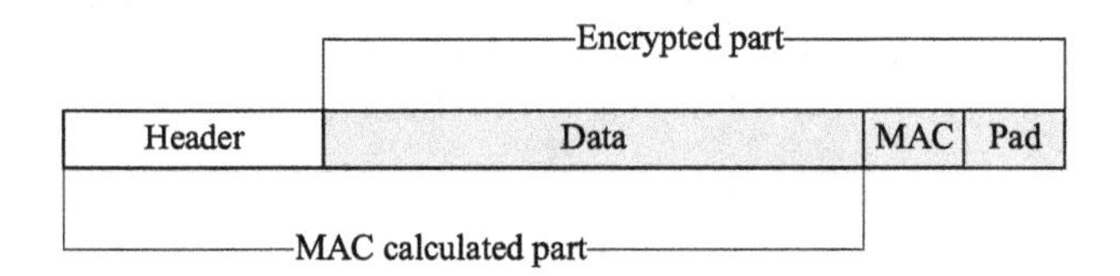

Fig. 3. The encrypted packet.

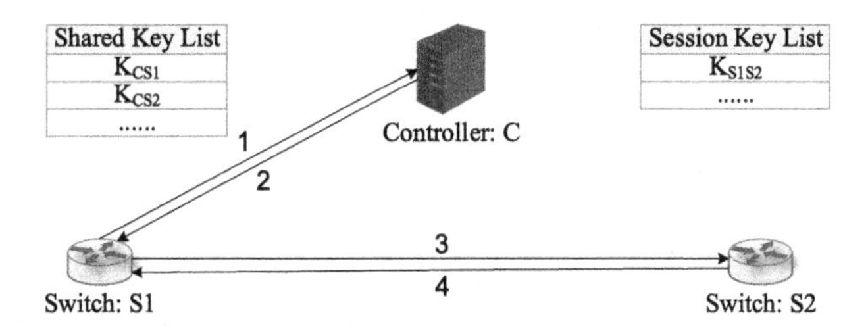

Fig. 4. The encryption key distribution.

Figure 4 shows the key management protocol between switch and controller. The protocol includes four phases:

- Phase one: Request a session key between switch 1 and 2 is indicated by step 1 in the figure.
- Phase two: Controller generates session key for switches and stores them in the key list inside controller, which is indicated by step 2.
- Phase three: Controller distributes the session key to switches, which is indicated by step 3 and 4 in the figure.
- Phase four: Switch 2 uses session key to establish a security channel and to verify the key distribution is complete, which is indicated by step 4 in the figure.

Table 1. Symbols in key management protocol

ID_C	Controller ID	T	Time stamp
ID_{S1}, ID_{S2}	ID for Switch 1 and 2	L	Life time
K_{S1S2}	Session key between Switch 1 and Switch 2	R	Random number
K_{CS1}	Shared key between Controller and Switch 1	E	Encryption operations
K_{CS2}	Shared key between Controller and Switch 2	M	Message

Switches pre-install shared key of the controller before joining in the network. Controller is responsible for session key generation, storage and distribution. The network time servers ensure the time synchronization between Controller and switches. Those are pre-conditions of our security enhanced Openflow protocol. The following text explains key distribution between controller C, switch S1 and S2. The symbols in the protocol are shown in Table 1.

① S1 → C:$M||H(M)$,
M = $ID_{S1}||ID_{S2}$
② C→S1:$M||H(M)$,
M = $M_1||M_2$,
$M_1 = E(ID_{S2}||K_{S1S2}||T||L, K_{CS1})$,
$M_2 = E(ID_{S1}||K_{S1S2}||T||L, K_{CS2})$,
$K_{S1S2} = H(ID_{S1}||ID_{S2}||R)$
③ S1 → S2:$M||H(M)$,
M = $M_1||M_2$,
$M_1 = E(ID_{S1}||T, K_{S1S2})$,
$M_2 = E(ID_{S1}||K_{S1S2}||T||L, K_{CS2})$
④ S2 → S1:$M||H(M)$, $M = E(T+1, K_{S1S2})$

First of all, S1 sends the request message M and a hash code H(M) to Controller. M contains switch 1 and 2 identifications ID_{S1} and ID_{S2}.

After receiving the request message from S1, C generates a Session key K_{S1S2} by performing H operation with ID_{S1}, ID_{S2} and a random number R. The outgoing message M contains two parts: the first part is the message M1, which includes ID_{S2}, T, L and the session key K_{S1S2}, M1 encrypts with K_{CS1}; the second part is the message M2, which includes ID_{S1}, T, L and the session key K_{S1S2}, M2 encrypts with K_{CS2}.

S1 uses K_{CS1} to decrypt the message M1 and gets the session key K_{S1S2} and M2. S1 creates a new M1 message by using the session key K_{S1S2} to encrypt S1's ID and time stamp. The new message M that includes M2 and the new M1 is sent to S2.

Finally, S2 decrypts the message M2 by shared key K_{CS2} and gets the session key K_{S1S2}. S2 decrypts the message M1 by session key K_{S1S2} and gets T and ID_{S1}. S2 sends new T to S1 by using the session key K_{S1S2} to encrypt the message M. All session keys and switch controller sheared keys are stored in Controller.

Here we share a basic idea on how to build an experiment platform capable of supporting such implementation and deployment. Furthermore, the other SDN-based security enhancement proposal [20] is evaluated on this platform.

4 Implementation and Evaluation

4.1 Open Pi Switch Implementation

We select the Single Board Computer (SBC) hardware Raspberry Pi3 model B [9] as computation and HW-interface platform in core module implementation. The Raspberry Pi is an open source hardware with a 1.2 GHz 64-bit quad-core ARM Cortex-A53 processor and 1 GB of RAM. It supports 10/100 Mbit/s Ethernet, 802.11n wireless, Bluetooth 4.1 On-board network, 4 USB2.0 ports and 17 General purpose input-output (GPIO) connectors including I2C, SPI, UART, PCM, and PWM interfaces. The Ubuntu MATE [7], which is a Linux-based implementation, is chosen as platform Operation System (OS). The rest of the core functions modules are deployed above the Linux level.

The three external USB ethernet adapters (10/100Mb/s) are mounted as OSCO physical ports. The Open Virtual Switch (OVS) [5] is installed as the implementation of OpenFllow protocol. We choose to use OpenSSL [4], Crypto++ [1] and PBC [6] libraries as the Cryptology Algorithms module in the Platform. Additionally, the Wireshark [8] is installed as OpenFlow packet inspector and monitor. SDN Network Setup and Topology.

4.2 SDN Network Setup and Topology

The POX/NOX [3] and Floodlight [2] SDN controllers are tested in our platform, and the Floodlight controller is used in this experiment. The SDN virtual network is created by the Mininet [15], which runs kernel, switch and application code on a single machine. In our experiment, we use Mininet simulate two switches and two hosts, with host1 connecting to switch 1, host 2 connecting to switch 2, and switch 1 connecting to switch 2. The Mininet configuration allows switch 2 to connect the controller via a physical OSCO switch. The OSCO switch 3 physically connects to OSCO switch 4, the virtual switch 2 and a controller. All switches connect (physically or logically) the controller via the OSCO switch 3 and switches work in OpenFlow-only mode with OpenFlow1.3 protocol. Figure 5 presents more detailed topology information.

4.3 Performance Evaluation

The Open Pi Switch platform executed two types of testing, including the cryptographic algorithms and the network performance testing. The testing parameters and results are shown below.

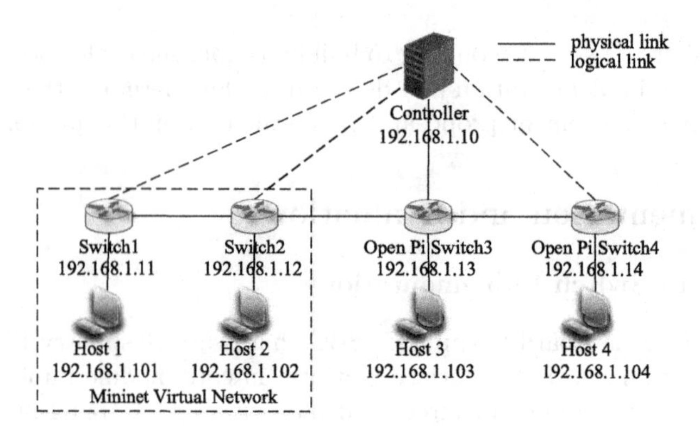

Fig. 5. SDN network topology.

(1) Hash algorithm performance

Hash function is used for assuring integrity of transmitted data and also can be used to form a hash-based message authentication code.

Figure 6 explains that the performance is influenced by the size of input. A bigger input size achieves the better performance, particularly for MD5, HMAC and SHA1 hash functions in OSCO Platform.

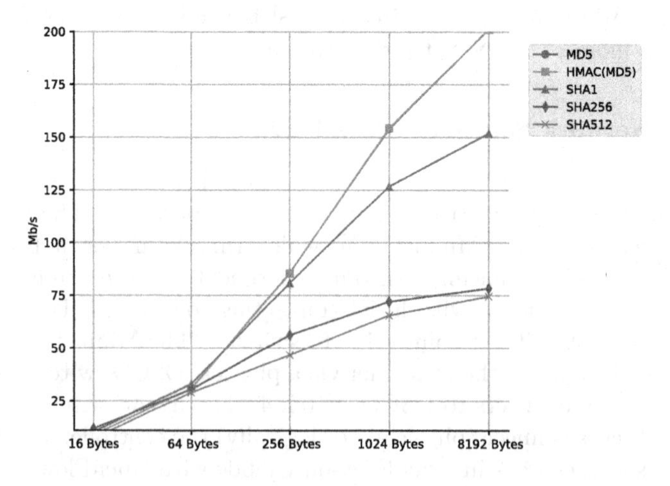

Fig. 6. Hash functions performance in OSCO platform.

(2) Stream/Block ciphers performance

Block ciphers operate as important elementary components in the design of many cryptographic protocols, and are widely used to implement encryption of bulk data.

The AES-128, AES-192 and AES-256 are commonly used block cipher algorithms. Besides, we also conducted the evaluation on another two classic stream/block ciphers, RC4 and DES.

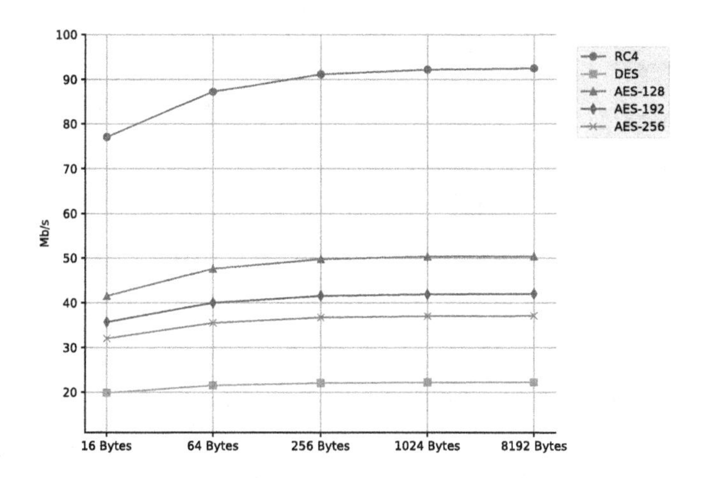

Fig. 7. Stream/Block ciphers performance in OSCO platform.

Figure 7 shows the performance in different input size configurations. The performance increases until 256 bytes input size in OSCO platform. And the implementation performances of the testing algorithms are consistent with their designed properties.

(3) OSCO platform network performance

In this experiment, the OSCO acts as an OpenFlow switch in the SDN network topology with some enhanced security abilities. Its data forwarding performance, which influences the whole SDN network performance, is one of the key factor besides computational power of the cryptosystem.

The experiment uses IPerf a Linux network tool to test the max bandwidth and minimum response time of two hosts which use the TCP/UDP network connection respectively. The UDP connection testing applies -ub 100 m as the input parameters and the rest part maintains using the default settings, e.g., UDP buffer size is 160 Kbyte and port is 5001 etc. The TCP connection uses default parameters, for example, setting the TCP window as 43.8 Kbyte. And we evaluate the performance of the network data transmission, p_5, according to Formula 1, where d_t is number of data being transmitted, t is transmission time.

$$p_5 = \frac{d_t}{t} \tag{1}$$

Figure 8 shows two different network connections in TCP and UDP mode. The 1st connection (the virtual Host 1 to the physical Host 3) is indicated by green bar, and the 2nd connection is the link between two physical hosts (Host 4 to Host 3), indicated by orange bar (see Fig. 5 for the network topology). The blue bar indicates the max theoretical network connection bandwidth (100 Mb/s). Both connections get a pretty good result comparing to max theoretical bandwidth. In addition, the minimum response time of the 1st connection is 7.902 ms, and the 2nd connection gets 0.029 ms (both use UDP connection).

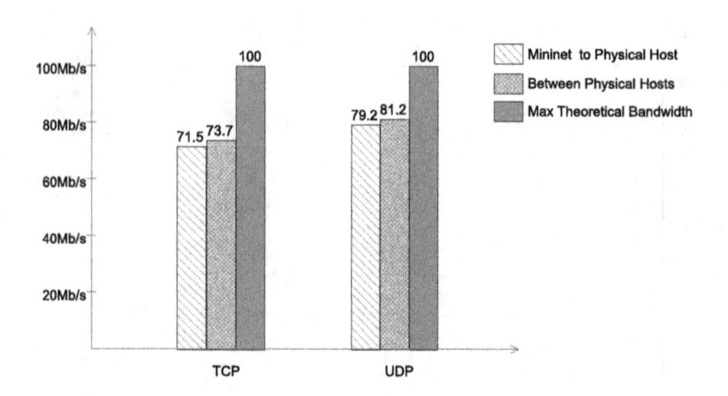

Fig. 8. Network performance in OSCO platform.

Overall, our experiments demonstrate that OSCO platform is not only capable of carrying out the various cryptographic algorithms but also provides good network bandwidth and speed. Furthermore, the OSCO platform has the lowest power consumption of 4 W (system power consumption) comparing to HTPC and Server platform which approximately consume 15 W and 80 W just only for the processors (Thermal Design Power from Intel) respectively.

5 Conclusion

In this paper, to effectively integrate security countermeasures and systematically test the developed security mechanisms in the SDN-based CPS environment, we propose OSCO (Open Security-enhanced Compatible OpenFlow) platform. The OSCO platform combines the Raspberry Pi Single Board Computer (SBC) hardware with security-supported network simulators, which supports highly configurable cryptographic algorithm modules, security protocols, flexible hardware extensions and virtualized SDN networks. Furthermore, we present an enhanced OpenFlow protocol to improve the security in SDN data plane. We implement and evaluate the prototype system and the experiment results (in a heterogeneous CPS environment) show that our system conducted security functions with relatively low computational and networking performance overheads.

Acknowledgment. This work was supported in part by the National Key R&D Program of China (No. 2017YFB1400700), the National Natural Science Foundation of China (No. 61402029, U17733115), the National Natural Science Foundation of China (No. 61379002, No. 61370190), and the Funding Project of Education Ministry for the Development of Liberal Arts and Social Sciences (No. 12YJAZH136).

References

1. Crypto++ Library. http://www.cryptopp.com/
2. Floodlight Controller. http://www.projectfloodlight.org/
3. NOX Controller. http://www.noxrepo.org/

4. OpenSSL. http://www.openssl.org/
5. Open vSwitch. http://www.openvswitch.org/
6. Pairing-Based Cryptography Library. http://crypto.stanford.edu/pbc/
7. Ubuntu MATE. http://ubuntu-mate.org/
8. Wireshark. http://www.woreshark.org/
9. Raspberry Pi Hardware Specification (2011). http://www.raspberrypi.org/documentation/hardware/
10. Open Networking Foundation (2015). https://www.opennetworking.org/
11. Ahmad, I., Namal, S., Ylianttila, M., Gurtov, A.: Security in software defined networks: a survey. IEEE Commun. Surv. Tutor. **17**(4), 2317–2346 (2015)
12. Kloti, R., Kotronis, V., Smith, P.: Openflow: a security analysis. In: 2013 21st IEEE International Conference on Network Protocols (ICNP), pp. 1–6. IEEE (2013)
13. Kohl, J.: The Kerberos network authentication service (V5). RFC **7**(3), 167 (1993)
14. Kreutz, D., Ramos, F.M., Verissimo, P.E., Rothenberg, C.E., Azodolmolky, S., Uhlig, S.: Software-defined networking: a comprehensive survey. Proc. IEEE **103**(1), 14–76 (2015)
15. Lantz, B., Heller, B., McKeown, N.: A network in a laptop: rapid prototyping for software-defined networks. In: Proceedings of the 9th ACM SIGCOMM Workshop on Hot Topics in Networks, p. 19. ACM (2010)
16. McKeown, N., Anderson, T., Balakrishnan, H., Parulkar, G., Peterson, L., Rexford, J., Shenker, S., Turner, J.: Openflow: enabling innovation in campus networks. ACM SIGCOMM Comput. Commun. Rev. **38**(2), 69–74 (2008)
17. Porras, P., Shin, S., Yegneswaran, V., Fong, M., Tyson, M., Gu, G.: A security enforcement kernel for openflow networks, pp. 121–126 (2012)
18. Riley, G.F., Henderson, T.R.: The *ns*-3 network simulator. In: Wehrle, K., Güneş, M., Gross, J. (eds.) Modeling and Tools for Network Simulation, pp. 15–34. Springer, Heidelberg (2010). https://doi.org/10.1007/978-3-642-12331-3_2
19. Shin, S., Porras, P., Yegneswaran, V., Fong, M., Gu, G., Tyson, M.: Fresco: modular composable security services for softwaredefined networks. In: Proceedings of Network and Distributed Security Symposium (2013)
20. Wang, M., Liu, J., Mao, J., Cheng, H., Chen, J., Qi, C.: Routeguardian: Constructing secure routing paths in software-defined networking. Tsinghua Sci. Technol. **22**(4), 400–412 (2017)

Hop-Constrained Relay Node Placement in Wireless Sensor Networks

Xingjian Ding[1], Guodong Sun[2], Deying Li[1], Yongcai Wang[1], and Wenping Chen[1](✉)

[1] School of Information, Renmin University of China, Beijing 100872, China
{dxj,deyingli,ycw,chenwenping}@ruc.edu.cn

[2] School of Information Science and Technology, Beijing Forestry University, Beijing 100083, China
sungd@bjfu.edu.cn

Abstract. Placing relay nodes in wireless sensor networks is a widely-used approach to construct connected network topology. Previous works mainly focus on the relay node minimization while achieving network connectivity but pay less attention on the path performance guarantee. In this paper we first investigate the hop-constrained relay node placement optimization which aims at using as few relays as possible to construct sensor-to-sink paths meeting the hop constraint given by the end user. We present a heuristic-based algorithm to solve the above optimization problem and evaluate its performance by extensive simulation. The experimental results demonstrate that the efficiency of our designs in comparison with two baselines.

Keywords: Wireless sensor networks · Connectivity
Relay node placement · Hop constraint

1 Introduction

Recent years have witnessed a growing interest of using wireless sensor networks (WSNs) to monitor the physical world [1,2]. A wireless sensor network contains lots of wireless sensor nodes, which are deployed in unattended filed to monitor the physical events, such as the temperature, the light strength, vibration, etc [3]. With the built-in radio chip, these wireless sensor nodes can be organized or self-organized into a wireless multihop network that delivers the sensory data to a sink node at the user end [3].

In real-world wireless sensor network systems, the on-board radio chip of sensors is the most energy hungry component. To reduce the energy consumption and achieve a long-term monitoring, researchers and engineers entirely have

W. Chen—This work was supported, in part, by the National Natural Science Foundations of China with Grants No. 11671400 and No. 61672524, the Fundamental Research Funds for the Central University, and the Research Funds of Renmin University of China with Grant No. 18XNH109.

S. Chellappan et al. (Eds.): WASA 2018, LNCS 10874, pp. 78–89, 2018.
https://doi.org/10.1007/978-3-319-94268-1_7

moved to low-power wireless communication [4]. One typical approach is to reduce the transmit power level of sensor nodes. In practice, therefore, the sensor node provides a limited coverage of wireless communication. In the other hand, some wireless sensor network applications are deployed in harsh environments, such as unattended forests, deserts, and so on. The propagation of low-power wireless signals often suffer nonnegligible reflection, scattering, and diffraction in these fields [5]. Consequently, the network connectivity of the initialized deployment cannot be sufficiently guaranteed in some cases, which leads to uneasy data collection. Of course, placing a great amount of sensor nodes in the monitoring area would offer us a connected network, however, such a redundant deployment is sure to increase the system cost, especially when the targets to be detected locate sparsely with respect to the communication coverage of sensor nodes. To obtain or recover connected network, a promising approach is to place extra nodes in the monitoring area. These nodes are also called *relay nodes* (or *relays*), which usually participate into the data propagation and do not acquisition any sensory data.

It is very challenging to achieve efficient relay placement which has been proven to be NP-Hard [6–8]. Existing works have focused on achieving connected network topology by including least extra relay nodes [6,7,9–11]. Some works study how to achieve connectivity with least relays [6,7,9]. Some articles [10,11] extend the previous researches: besides achieving the connectivity, they also guarantee the fault tolerance. However, they do not pay direct attention on the performance of forwarding paths—usually, these paths traverse count-uncontrollable hops before reaching the sink. As we know, the more the hops along a path are, the longer the delivery time is. Additionally, excessive hops increase the risk of losing data packets in transmission because the low-power links of wireless sensor network are time-varying.

In this paper, we first investigate the Hop-constrAined Relay Placement problem (HARP, in short), to the best of our knowledge, and we present a heuristic-based algorithm for the HARP problem. The major contributions of our work are as follows. First, we present some insightful observations that are beneficial to dissect the HARP problem and help us design better heuristics. Second, to solve the HARP problem, we propose two algorithms which can collectively find efficient solutions. Third, we conduct extensive simulations to evaluate our designs, and results show that our approach always outperforms the baseline algorithm with approximation ratio of 3 [6].

The rest of the paper is organized as follows. We introduces works related to ours in Sect. 2. We describes the problem to be addressed in Sect. 3. We presents the details of algorithm designs in Sect. 4. We evaluates the performance of our work in Sect. 5. Finally, We concludes this paper in Sect. 6.

2 Related Work

In the past few years, placing relay nodes for wireless sensor networks has been studied in various contexts [7–12]. Cheng et al. [6] present two algorithms of approximation ratios 3 and 2.5 to solve the general optimal relay node placement problem. Ma et al. [13] present a set-covering-based algorithm for delay constrained relay node placement problem in two-tiered wireless sensor networks. They analyzed the time complexity and the approximation ratio of the proposed algorithms, and demonstrate their performance through extensive simulations. Lee and Younis [7] propose an optimized relay placement algorithm under the minimum Steiner tree model with convex hull, and validate its performance through simulation experiments. In [14], the authors address the problem of placing least relays to establish multi-hop paths. They first derive an optimal solution for the case with three terminals and then present three heuristic algorithms to deal with the cases with more than three terminals; the simulation results demonstrate the performance of their algorithms. Ma et al. [9] present a local search approximation algorithm (LSAA) to solve the relay node single cover problem, aimed at improving fault tolerance; also, they extend LSAA to solve the relay node double cover problem.

To address the delay constrained relay node placement problem, [15] presents a set cover-based approximation algorithm (SCA), which deploys relays iteratively from the sink to the given sensor nodes; during the iteration, sensors are gradually connected to the sink with satisfying delay constraint. Authors of [16] propose a subtree and mergence based approach to addressing the delay constrained relay node placement problem. Different from our research, [15,16] assume that relay nodes can only be deployed at the positions specified in advance.

3 Problem

3.1 Network Model

In this paper we assume that the wireless sensor network consists of n sensor nodes and a sink node that collects the data acquired by the sensor nodes. All the sensor nodes and the sink are deployed in a two-dimensional area; the locations of these sensors can be obtained once being deployed. We assume that the sensor nodes and the relay nodes have the same transmit radius ρ. We consider such a scenario: the initial deployment of sensor nodes are sufficiently sparse so that any two sensor nodes, u and v, cannot directly communicate with each other, i.e., $L(u, v) > \rho$ where $L(u, v)$ denotes the physical distance between u and v.

3.2 Problem Description

Relay nodes can help establish a network that connects all the sensor nodes to the sink: offering a forwarding path from each sensor node to the sink. Intuitively, if we deploy relay nodes densely, it is easy to obtain a connected topology. However, the possible redundancy of relay nodes will lead to not only serious co-channel conflicts but also higher deployment cost. This paper studies a novel optimal relay placement problem (HARP) which is to place minimum number of relay nodes in the target area such that the network topology is connected and satisfies the hop constraint from every sensor to sink.

Figure 1 describes the motivation of the HARP problem, and Fig. 1(a) shows the original network topology that is unconnected. Figure 1(b) shows a case that uses only five relays—least relays—to connected all the segments. In Fig. 1(b), however, the path from X to the sink goes through eight hops. Figure 1(c) simply places the relays on the straight line connecting the segmented sensor and the sink. Figure 1(c) uses fourteen relays, more than those used in Fig. 1(b), and (c) constructs hop-minimum paths for sensors D, E, and X. Obviously, the trivial relay placement scheme shown in Fig. 1(c) results in high deployment cost, especially when the end user does not have a strict demand on the hop count. Figure 1(d) trades off the two relay placements shown in Fig. 1(b) and (c): it pursues not-too-long paths with as few relays as possible.

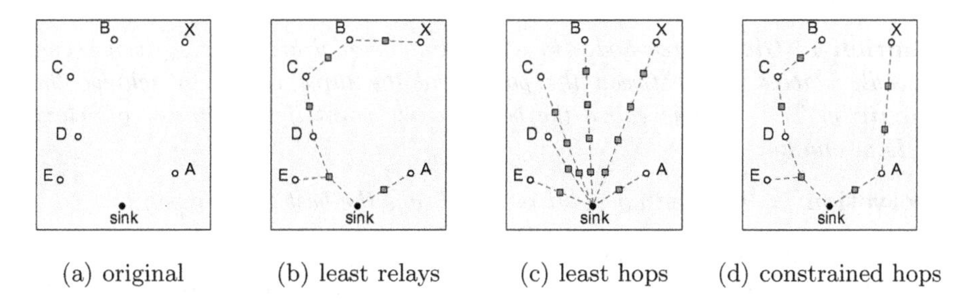

(a) original (b) least relays (c) least hops (d) constrained hops

Fig. 1. Illustration of three relay placement schemes. The blue square represents the relay node. (Color figure online)

To profile the path performance requirement, we set a hop constraint k for the sensor-to-sink path and $k > 1$. Specifically, the hop count of the path from sensor s_i to the sink s_0, denoted by $h(s_i, s_0)$, should not be greater than k times the hop count $\lceil L(s_i, s_0)/\rho \rceil$ which is the minimum hop count. Denote N_r as the number of relays used. Then, the HARP problem is formally modeled as Eq. (1). The requirement for guaranteeing the network connectivity is implicitly indicated by Eq. (2) and then is not given here for clearance.

$$\mathbf{min}: \; N_r \tag{1}$$

$$\mathbf{s.t.}: \; \frac{h(s_i, s_0)}{\lceil L(s_i, s_0)/\rho \rceil} \leq k, \qquad i \in \{1, 2 \ldots n\} \tag{2}$$

Theorem 1. *The HARP problem is NP-Hard.*

Proof. We consider a special case of HARP problem, in which the hop constraint k is large enough such that it will never pose a valid constraint to the minimization of relay nodes. That is the optimal relay node placement problem in [6]. It means that the HARP problem contains the optimal relay node placement problem as a particular case. Since the optimal relay node placement problem has been proven to be NP-Hard in [6], the HARP problem is also NP-Hard. □

4 Algorithm Designs

4.1 Benefical Observations on the HARP Problem

For the Euclidean Steiner tree problem [17], the goal is to connect given n nodes with minimum total length of all edges (wireless links). For the case of $n = 3$, if the maximum angle of the triangle formed by the given three nodes is less than 120°, the Steiner point locates at the Fermat point which is a point such that the total distance from the three vertices of the triangle to the point is the minimum possible; otherwise, the Steiner point locates at the vertex of the maximum angle. However, the HARP problem targets the reduction of relays with the hop constraint, other than the total length of all edges, and then, it is different from typical Euclidean Steiner tree problems. The following provable observations further profile the HARP problem and provide heuristics for our algorithm designs.

Definition 1. *Given three nodes in a two-dimensional area, for a given point, we evenly place relays between the point and the three vertex to achieve the connectivity. The point is called the best Steiner point if the number of relays used is minimum.*

Observation 1. *The Fermat point is not always the best Steiner point.*

Proof. For a Euclidean triangle Steiner tree, if the largest interior angle is smaller than 120°, then the Steiner point is just the Fermat point; otherwise, it is at the vertex of the angle that is 120° or larger. If each angle of the triangle is less than 120°, the sum of the distances between the Fermat point and the three vertices can then achieve the minimum. However, the HARP model does not target the Euclidean distance optimization. Therefore, Fermat point is not always the best Steiner point in the HARP model. □

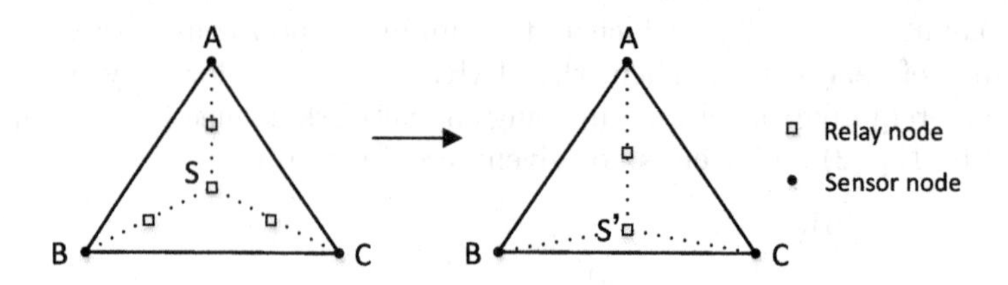

Fig. 2. Illustration of a triangle Steiner tree

As shown in Fig. 2, for example, $\triangle ABC$ is regular and its Fermat point is labeled with S, then we have $|SA| = |SB| = |SC|$. Suppose $|SA| = 1.1\rho$. To connect vertices A, B, and C, we will need four relays to build a triangle Steiner tree if the Fermat point is thought of as the best Steiner point. However, we will need only two relays if point S' is chosen. Obviously, the Fermat point is not optimal in this example.

Observation 2. *The best Steiner point definitely is in a triangle or on its edges.*

Proof. A rough proof of this theorem was given in [7]. Here we propose a strict proof. We need only to prove that every best Steiner point is not outside of the given triangle. Suppose P is a Steiner point of $\triangle ABC$, as shown in Fig. 3, and the three side lengths of $\triangle ABC$ are all greater than ρ, then, we can place relays on the three dashed lines starting from P and ending at A, B, and C, respectively. Now we choose a point Q on side BC such that $|BQ| = |BP|$. Since $|BP| + |CP| > |BQ| + |CQ|$, we have $|CP| > |CQ|$. Obviously, $|AP| > |AQ|$. Thus, we have

$$\left\lceil \frac{|AP|}{\rho} \right\rceil + \left\lceil \frac{|BP|}{\rho} \right\rceil + \left\lceil \frac{|CP|}{\rho} \right\rceil \geq \left\lceil \frac{|AQ|}{\rho} \right\rceil + \left\lceil \frac{|BQ|}{\rho} \right\rceil + \left\lceil \frac{|CQ|}{\rho} \right\rceil \tag{3}$$

By Eq. (3), we know that Q is better than P as a Steiner point; in other words, given a Steiner point outside the triangle, we can always find a better one that locates on some side of the triangle. □

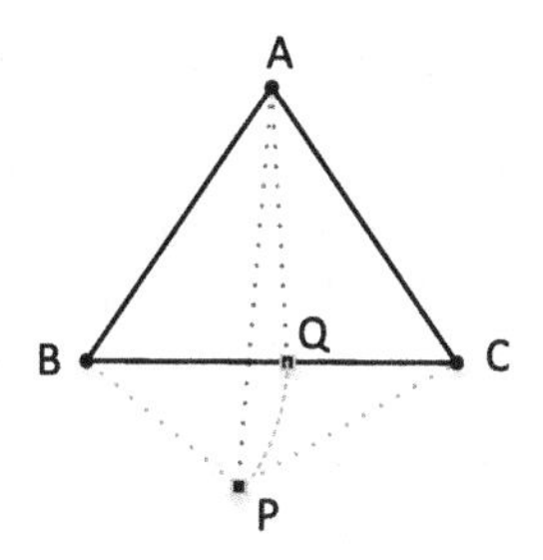

Fig. 3. Illustration for Observation 2

Observation 3. *Given a triangle whose side lengths are all greater than ρ, its circumcenter or midpoint of long edge is possibly the best Steiner point.*

Proof. Consider a special case that a triangle exactly needs one relay node to build a connected topology, then the choice of the best Steiner point can be quite tricky. As only one relay node is needed, we must confirm that the distances from the relay to the three vertices are not greater than ρ, then the triangle's circumcenter or midpoint of long edge that minimizes the maximum distance must be a suitable point. Thus this observation is true. □

4.2 Algorithm Description

Based on the three observations, we present a greedy algorithm for the HARP problem. The main idea of our algorithm (RPC) is heuristically selecting each sensor node and iteratively connecting them to the sink, at each iteration, RPC invokes a local optimization algorithm (FBS) to find the best Steiner point by searching grid cells, and deploy relays to achieve the connectivity.

4.2.1 Locally Optimal Determination of Best Steiner Point

In this subsection, we solve the local optimal relay deployment problem. Given a connected graph $G(V_G, E_G)$, and a separate sensor node u, and a hop constraint k, our goal is to construct a connected graph $G(V_G \cup \{u\}, E'_G)$ such that the total number of relay nodes used is minimized. We first mesh the network area, then greedily find the shortest edge(u, v_0) that meets the hop constraint where $v_0 \in V_G$, and then we form triangles with node u, v_0 and any v_j where v_j is a node incident to v_0 in G. After that, we search the grid cells to find the best Steiner point that satisfies the hop constraint.

The local optimization algorithm is shown in Algorithm 1. Figure 4 shows an walk-through example of Algorithm 1.

Algorithm 1. Finding Best Steiner point (FBS)

Input: $G(V_G, E_G)$, sensor node u, and hop constraint k
Output: a locally optimized graph $G(V_G \cup \{u\}, E'_G)$

1: For each node $v_i \in V_G$ that has a path to sink s_0, compute $h(v_i, s_0)$ and $L(u, v_i)$
2: Find node $v_0 \in V_G$ that is nearest to u, such that

$$h(v_0, s_0) + \lceil L(u, v_0)/\rho \rceil \leq k \times \lceil L(u, s_0)/\rho \rceil$$

3: Form a triangle $\triangle_j$ with nodes u, v_0, and any v_j, where v_j is a node incident to v_0 in G. For each $\triangle_j$, search the grid cells to find the best Steiner point that meets the hop constraint; and calculate n_j^r for each $\triangle_j$
4: **if** $\exists\, n_j^r > 0$ **then**
5: Determine triangle $\triangle^* = \arg\max_{\triangle_j}\{n_j^r\}$
6: Place a relay at the best Steiner point of $\triangle^*$ to build a triangle Steiner tree
7: **else**
8: Connect u with v_0 directly
9: **end if**
10: Place least relays on the edges whose length is greater than ρ
11: Update $h(v_i, s_0)$ for each node $v_i \in V_G$
12: Return $G(V_G \cup \{u\}, E'_G)$

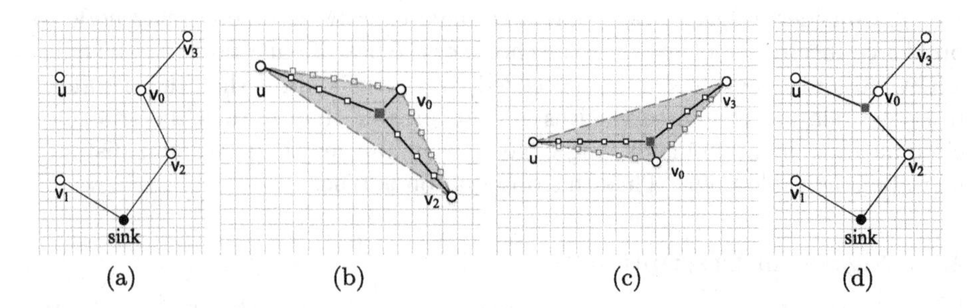

Fig. 4. An example of the FBS algorithm, squares in (b) and (c) represent relays.

Consider node u that will be added into G in Fig. 4. In Fig. 4(a), suppose node v_0 is physically nearest to u and it provides a path from u to sink s_0 that satisfies the hop constraint. Then we obtain two candidate triangles $\triangle_1$ and $\triangle_2$ shown in Fig. 4(b) and (c), both of which are formed with nodes $\{u, v_0, v_2\}$ and nodes $\{u, v_0, v_3\}$, respectively. Algorithm 1 scans the grid cells that partially or fully locate within the two triangles, and then calculates n_1^r and n_2^r. Here we use n_j^r to denote the number of relays that we can save if the Steiner tree is built on triangle $\triangle_j$. Algorithm 1 chooses the triangle achieving maximum n^r. If ties occur, it chooses the triangle that minimizes $h(u, s_0)$ and then deploys a relay node at the Steiner point to build a triangle Steiner tree. Here, we suppose triangle $\triangle_1$ has the maximum n^r and then is chosen; the final topology is shown in Fig. 4(d), which includes a relay node to construct a connected network.

By Observation 3 we know that the circumcenter of triangle $\triangle$ and the midpoint of $\triangle$'s long edge are special points that can achieve the best Steiner point with considerable probability. So Algorithm 1 also examines the two special points in line 3, when it searches the grid cells. Noticeably, deploying relays at Steiner points affect the path formation of other nodes and then their hop count. In Fig. 4(d), for instance, the inclusion of a relay node increases the hop count of node v_3. Therefore, when searching the grid cells to find the best Steiner point in line 3, Algorithm 1 needs to guarantee that besides node u, every node in V_G satisfies the hop constraint.

4.2.2 Relay Placement with Hop Constraint

The relay placement with hop constraint algorithm first initializes graph $G(V_G, E_G)$ by setting $V_G = \{s_0\}$ and $E_G = \emptyset$, and then puts all sensor nodes into set S. RPC iteratively connects each sensor to the sink node. The choice of which node will be added into G is determined by a Prime's-like heuristic, meanwhile any edges examined satisfy the hop constraint. Once RPC figures out the node to be added into G, it will invoke algorithm FBS to complete the relay placement process. The RPC algorithm is shown as follows:

Algorithm 2. Relay Placement with hop Constraint (RPC)

Input: locations of all the sensors and sink s_0, hop constraint k
Output: a connected Graph $G(V_G, E_G)$ with k-constrained hops
1: Initialize $G(V_G, E_G)$ with $V_G = \{s_0\}$ and $E_G = \emptyset$
2: Put all the sensors into a set S
3: Compute $L(s_i, s_0)$ for each sensor $s_i \in S$
4: **while** $S \neq \emptyset$ **do**
5: Of all edges $\{(u, v)|u \in S, v \in V_G\}$, find a shortest edge (u_0, v_0) that satisfies the hop constraint: $h(v_0, s_0) + \lceil L(u_0, v_0)/\rho \rceil \leq k \times \lceil L(u_0, s_0)/\rho \rceil$.
6: Run FBS to add node u_0 into $G(V_G, E_G)$
7: Remove node u_0 from S
8: **end while**
9: **return** $G(V_G, E_G)$

Theorem 2. *The network topology generated by Algorithm 2 is feasible for the HARP problem.*

Proof. In Algorithm 2, sensor nodes are added into G one by one, until all of them are connected with the sink. So Algorithm 2 is sure to yield a connected topology. The connectivity establishment for each sensor is executed by Algorithm 1, which ensures that the hop constraint of every node can be met when the grid cells are searched. Thus the theorem holds. □

Theorem 3. *The time complexity of Algorithm 2 is* $\mathcal{O}\left(n^3 + \frac{\mathcal{A}}{d^2} \times n\right)$, *where* $\mathcal{A}$ *denotes the area of the network region, and* d*, the side length of the grid cell.*

Proof. Line 5 of Algorithm 2 takes time of $\mathcal{O}(n^2)$; and the time cost of Algorithm 1 is mainly determined by the number of grid cells to be searched. The area of a grid cell of side length d is equal to d^2, and then the number of grid cells is bounded by $\frac{\mathcal{A}}{d^2}$. Therefore, the total time complexity of Algorithm 2 is $\mathcal{O}\left(n \times \left(n^2 + \frac{\mathcal{A}}{d^2}\right)\right)$. This theorem holds. □

5 Simulation

5.1 Experimental Settings and Metrics

In our simulations, we consider two cases: fixed network area with different network scales, and fixed sensor density with random deployments. In the first case, the network area is a 40×40 square, within which sensor nodes were randomly deployed. In the second case, we set the sensor density with 0.1 and 0.2, and also, the sensor nodes were randomly deployed. In all experiments ρ was set to 1 and the sink node located at the left bottom corner of the network area. We set the hop constraint k to 2. In Algorithm 1, the side length of each grid cell was set to 0.1. Each data point was the average of 40 repeated experiments with random seeds. We use three metrics to evaluate the performance of our work by comparing it with the Direct algorithm and STP-MSP [6] that is a 3-approximation algorithm. As a baseline, the Direct algorithm places relay nodes only on the straight line that connects the isolated sensor node with the sink, as Fig. 1(c) does. The three metrics are (1) the number of relays used, (2) the average hop count of the resulted paths, and (3) the maximum hop count of the resulted paths.

5.2 Results and Analyses

Figure 5 shows the resulted topologies of the STP-MSP and our designs when 200 sensor nodes were randomly deployed in a 40×40 square area. Note here that (1) STP-MSP runs without considering any hop constraint, and (2) in Fig. 5, we only show the relay nodes locating at the Steiner points and omit other relays that are evenly deployed on each edge. From Fig. 5 we can see that the topology returned by STP-MSP has much longer paths than that by the proposed RPC. Later figures will plot detailed comparisons.

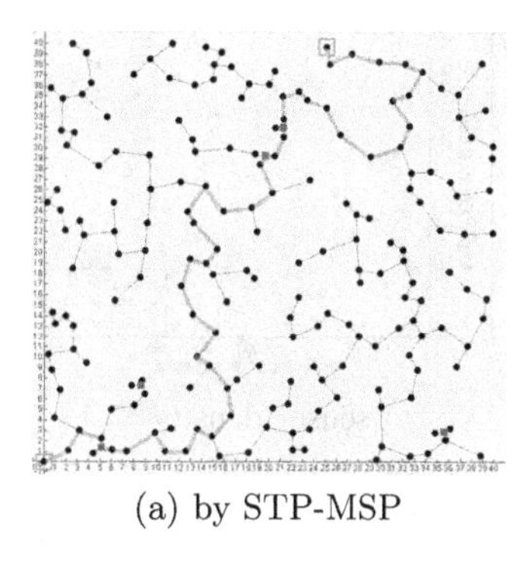

(a) by STP-MSP

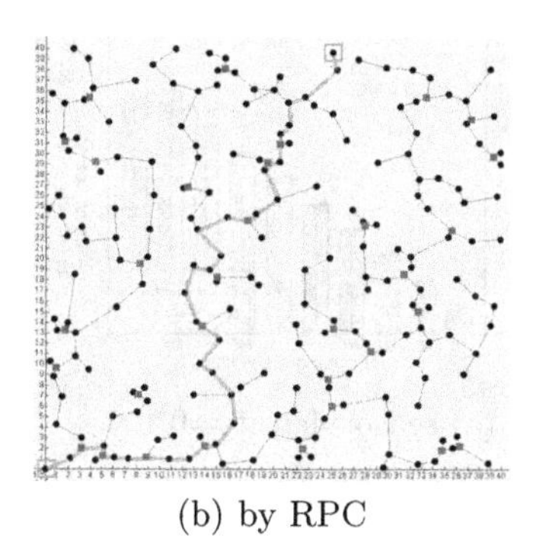

(b) by RPC

Fig. 5. Comparison of two resulted topologies with $n = 200$. Black dots and red squares represent the sensor nodes and the relay nodes placed at the Steiner points, respectively. (Color figure online)

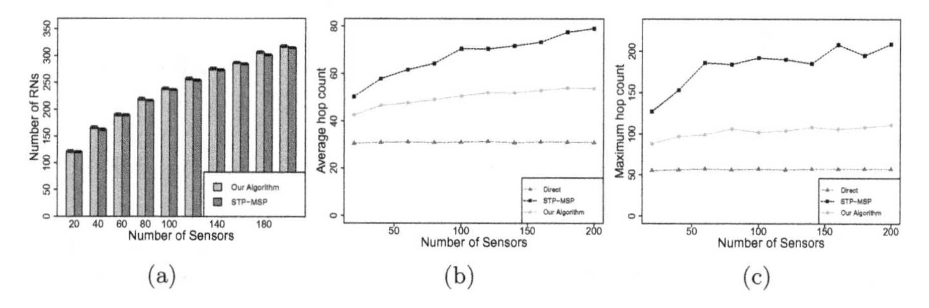

(a) (b) (c)

Fig. 6. Performance comparison under different network scales

Figure 6(a) compares the two algorithms (STP-MSP and RCP) with respect to the number of relays. The relay nodes needed by the Direct algorithm are far more than those of STP-MSP or RCP; so, we do not plot the number of relay nodes under Direct in Fig. 6(a), in order to make a clear comparison between STP-MSP and RCP. We can see that the relay nodes (RNs) used by our algorithm are slightly more (0.3–2.2%) than those of STP-MSP. Especially, as the network scale increases, the extra relay nodes of our algorithm over STP-MSP keeps relatively stable in amount. And then we think the proposed RPC can produce acceptable performance in terms of the number of relay nodes. Figure 6(b) and (c) shows the comparison of two algorithms in terms of average hop count and maximum hop count. In Fig. 6(b) and (c), we can see that the proposed RPC achieves shorter paths in comparison with STP-MSP, and the average and the maximum hop counts of RPC grow slowly with the number of sensor nodes increasing. Figure 6(b) and (c) collectively indicate that generally, the network created by RPC is more reliable than that by STP-MSP because of shorter delivery paths. In addition, compared to the Direct algorithm, the number of hops for each sensor node under our algorithm always satisfies the hop constraint ($k = 2$).

Figure 7 compares the performance of the three algorithms under two different densities of sensor deployment. It can be seen in Fig. 7(a) and (d) that

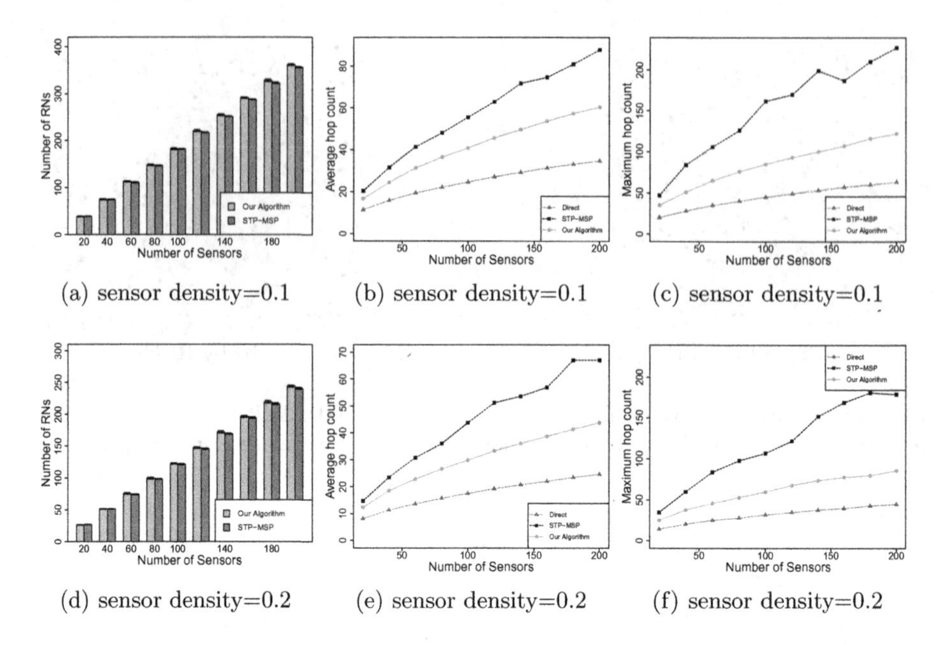

(a) sensor density=0.1 (b) sensor density=0.1 (c) sensor density=0.1

(d) sensor density=0.2 (e) sensor density=0.2 (f) sensor density=0.2

Fig. 7. Performance comparison with different densities of sensor deployment

for both algorithms, the number of relays needed increases in an approximately-linear manner, as the network scale increases, and that the relays needed by the proposed RPC and STP-MSP are almost identical in number. In the case with sparse sensor deployment (e.g., $n = 20$), specially, RPC needs relays 1.3% and 2.4% less than STP-MSP under the deployment densities of 0.1 and 0.2, respectively.

Figure 7(b) and (c) show the path performances of the three algorithms under the sensor deployment of density 0.1. We find that the average and the maximum hops both remain rising under the three algorithms, with the number of sensors increasing. But, our algorithm always keeps the hop constraint satisfied, while achieving better path performance than STP-MSP with almost the same amount of relays. Another advantage of RCP over STP-MSP is that RCP can achieve lower growth rates of the average and the maximum hops than STP-MSP. Combined with Fig. 6, we can draw a conclusion that the connected network topologies returned by our algorithm can achieve more acceptable network performances than the ratio-3 STP-MSP, regardless of the sensor deployment density and the network scale.

6 Conclusions

In this paper, we have presented the design (algorithm RPC) of approximately solving the HARP problem in wireless sensor networks. Based on greedy heuristics, RPC iteratively chooses the closest node that satisfies the hop constraint and adds it to the connected graph. RPC employs a locally-optimal algorithm

(FBS) which builds triangle Steiner trees in each RPC iteration and finds out the best Steiner points for the relay placement. We have also validated the performance of the proposed design and compared it with two baselines. The simulation results show the efficacy of our work. In the future, we will extend our designs to support fault tolerance and pursue approximation algorithms with proven competitiveness.

References

1. Younis, M., Akkaya, K.: Strategies and techniques for node placement in wireless sensor networks: a survey. Ad Hoc Netw. **6**(4), 621–655 (2008)
2. Manshahia, M.S.: Wireless sensor networks: a survey. Int. J. Sci. Eng. Res. **7**(4), 710–716 (2016)
3. Akyildiz, I.F., Su, W., Sankarasubramaniam, Y., Cayirci, E.: A survey on sensor networks. IEEE Commun. Mag. **40**(8), 102–114 (2002)
4. Potdar, V., Sharif, A., Chang, E.: Wireless sensor networks: a survey. International Conference on Advanced Information NETWORKING and Applications Workshops, pp. 636–641 (2009)
5. Ding, X., Sun, G., Yang, G., Shang, X.: Link investigation of IEEE 802.15.4 wireless sensor networks in forests. Sensors **16**(7), 987 (2016)
6. Cheng, X., Du, D.Z., Wang, L., Xu, B.: Relay sensor placement in wireless sensor networks. Wireless Netw. **14**(3), 347–355 (2008)
7. Lee, S., Younis, M.: Optimized relay node placement for connecting disjoint wireless sensor networks. Comput. Netw. **56**(12), 2788–2804 (2012)
8. Xu, X., Liang, W.: Placing optimal number of sinks in sensor networks for network lifetime maximization. In: IEEE International Conference on Communications, pp. 1–6 (2011)
9. Ma, C., Liang, W., Zheng, M., Sharif, H.: A connectivity-aware approximation algorithm for relay node placement in wireless sensor networks. IEEE Sens. J. **16**(2), 515–528 (2015)
10. Bhuiyan, M.Z.A., Cao, J., Wang, G.: Deploying wireless sensor networks with fault tolerance for structural health monitoring. IEEE Trans. Comput. **64**(2), 382–395 (2015)
11. Lee, S., Younis, M., Lee, M.: Connectivity restoration in a partitioned wireless sensor network with assured fault tolerance. Ad Hoc Netw. **24**(PA), 1–19 (2015)
12. Abbasi, A.A., Younis, M.F., Baroudi, U.A.: Recovering from a node failure in wireless sensor-actor networks with minimal topology changes. IEEE Trans. Veh. Technol. **62**(1), 256–271 (2013)
13. Ma, C., Wei, L., Meng, Z.: Delay constrained relay node placement in two-tiered wireless sensor networks: a set-covering-based algorithm. J. Netw. Comput. Appl. **93**, 76–90 (2017)
14. Senel, F., Younis, M.: Novel relay node placement algorithms for establishing connected topologies. J. Netw. Comput. Appl. **70**, 114–130 (2016)
15. Ma, C., Liang, W., Zheng, M.: Set covering-based approximation algorithm for delay constrained relay node placement in wireless sensor networks. Comput. Sci. **36**(8), 1–5 (2015)
16. Ma, C., Wei, L., Meng, Z.: Delay constrained relay node placement in wireless sensor networks: a subtree-and-mergence-based approach. Mob. Netw. Appl. 1–13 (2017). https://link.springer.com/journal/11036/onlineFirst/page/8
17. Lin, G.H., Xue, G.: Steiner tree problem with minimum number of steiner points and bounded edge-length. Inf. Process. Lett. **69**(2), 53–57 (1999)

Degrading Detection Performance of Wireless IDSs Through Poisoning Feature Selection

Yifan Dong[1], Peidong Zhu[1,2](✉), Qiang Liu[1](✉), Yingwen Chen[1], and Peng Xun[1]

[1] National University of Defense Technology, Changsha 410073, China
{dongyifan16,pdzhu,qiangliu06,xunpeng12}@nudt.edu.cn, csywchen@gmail.com
[2] Changsha University, Changsha 410022, China

Abstract. Machine learning algorithms have been increasingly adopted in Intrusion Detection Systems (IDSs) and achieved demonstrable results, but few studies have considered intrinsic vulnerabilities of these algorithms in adversarial environment. In our work, we adopt poisoning attack to influence the accuracy of wireless IDSs that adopt feature selection algorithms. Specifically, we adopt the gradient poisoning method to generate adversarial examples which induce classifier to select a feature subset to make the classification error rate biggest. We consider the box-constrained problem and use Lagrange multiplier and backtracking line search to find the feasible gradient. To evaluate our method, we experimentally demonstrate that our attack method can influence machine learning, including filter and embedded feature selection algorithms using three benchmark network public datasets and a wireless sensor network dataset, i.e., KDD99, NSL-KDD, Kyoto 2006+ and WSN-DS. Our results manifest that gradient poisoning method causes a significant drop in the classification accuracy of IDSs about 20%.

Keywords: Gradient poisoning · IDS · Feature selection
Adversarial examples

1 Introduction

With the rapid development of the Internet, network security has gained more and more attention. IDSs designed for wireless and wired networks makes up for the weakness of firewalls and provides effective intrusion detection measures to protect the system. However, the traditional IDSs usually need to manually construct detection rules, which is not only a time-consuming and laborious work but also greatly reduces the timeliness of rule's updates. To solve these problems, machine learning is taken into consideration. But, the data of network traffic is usually huge in size, which makes a great challenge to IDSs. Irrelevant and redundant features in these big data would slow the progress of classification

S. Chellappan et al. (Eds.): WASA 2018, LNCS 10874, pp. 90–102, 2018.
https://doi.org/10.1007/978-3-319-94268-1_8

down and increase the difficulty of making a correct classification. To solve these problems, feature selection algorithm as an important component of machine learning is designed to select a proper features subset by removing noise in data.

Although these IDSs using feature selection have made some achievements, the security of these algorithms is not well considered. These intrinsic vulnerabilities in these algorithms are potential dangers in the adversarial environment. Attackers usually modify the feature vectors that the classification model identifies as malicious samples, then produce a new feature vector that is classified as benign [1,2]. These attack methods usually modify the feature vectors, but modifying the features of the data requires a priori knowledge of the relevant data, which is difficult to achieve in real-world attacks.

Therefore, we develop an attack method that does not require modification of features. To verify the performance of wireless IDSs can be degraded through the use of machine learning security issues, we proposed this attack method based on gradient poisoning. We first extract the objective function of filter and embedded feature selection algorithms. We further use gradient optimization theory to find the adversarial objective functions of filter and embedded feature selection algorithms respectively. After that, we calculate the gradient direction forward to the biggest classification error. Then we use the backtracking line search method to update the gradient step size. We generate adversarial samples using the gradient direction and step size and inject them into the training set. Our experiments demonstrate that the detection rate of IDSs can be significantly reduced on four public network datasets including a wireless sensor dataset.

The contributions of this paper are summarized as follows: (1) An IDS attack method based on gradient poisoning is proposed. The attack can influence IDSs by lowering the detection rate. (2) The proposed method is based on the optimization theory to model the poisoning process of feature selection algorithm, which can be used to attack various mainstream feature selection algorithms, such as LASSO [3], Ridge regression [4], Elastic Net [5] and an intrusion detection method named Flexible Mutual Information Feature Selection (FMIFS) [6].

The rest of this paper is organized as follows. Section 2 introduces the related work. Section 3 briefly describes the adversary model and our assumptions about the attack method. Section 4 demonstrates the effectiveness of the attack method against feature selection algorithms and Sect. 5 introduces the experiments we did. In the end, we draw a conclusion and discuss future work in Sect. 6.

2 Related Work

2.1 Feature Selection Algorithms

Depending on whether extra learning algorithms are needed or not in the process of feature selection, and how to participate, feature selection algorithms can be divided into the wrapper, filter, and embedded feature selection algorithms.

With the fast increase of data dimensions, feature selection has become a part of IDSs as pre-processing step to improve the performance [7]. The authors in [6] adopted filter feature selection methods to select optimal feature for better classification in IDSs and performed well. In [8], the authors use Markov blanket model and decision tree analysis for selecting a feature to construct a hybrid architecture. Their empirical results indicate that selecting a significant input feature subset is important to design a lightweight, efficient and effective IDS. In [9], the authors develop an IDS based on flexible neural tree (FNT). This IDS add a pre-process step based on feature selection to improve detection performance. In this way, FNT only select 4 features from KDD99 dataset but the detection accuracy is up to 99.19%. Similarly, in [10], the authors also use the same dataset, reducing the feature set from 41 to 6 dimensions and improving the classification 1%. In our study, we concentrate on the filter and embedded class algorithms based IDSs.

2.2 Security Threats Towards Machine Learning

Since most algorithms did not consider possible attacks at the beginning of design, they have poor performance in the adversarial environment. In the existing research, attacks are divided into different types. Barreno et al. [11] originally proposed a taxonomy of attacks against machine learning algorithms. According to the impact of attacks on the classifier, attacks are divided into the causative attack and exploratory attack. Causative attacks change the geometric characteristics of a dataset. Attackers misleader classifiers by changing the training or test data to make these data have different distributions. Exploratory attacks are opposite to causative attacks. Exploratory attacks don't affect the training data and the classifier's learning process, they mainly through exploratory methods to get the information of the classifier and modify the data in the test data based on this information.

In general, causative attacks can cause great and long-term damage to the system because they change the distribution of the training data and mislead the classifier learning. Hence, we consider the causative attacks in our study. The training data used in training phase play a vital role in obtaining an excellent training model. Therefore, in order to reduce the performance of machine learning models, training data is frequently targeted by many attackers [12]. Poisoning attack is a typical causative attack that compromises availability and integrity of a model. In [13], the authors propose an attack scenario, a Naive Bayes was attacked by injecting craft malicious data in spam detection system. In [14], the authors attack a linear kernel SVM by poisoning attack in pdf detection system. Poisoning attack has already produced effects on other machine learning algorithms, but few to feature selection algorithms.

3 Adversarial Model and Assumptions

In this section, we present our adversary model and proper hypotheses for feature selection algorithms in IDS. Following the framework proposed by [15], our adversary model and hypotheses include four dimensions, i.e., adversarial goal, knowledge, capability of manipulating data and strategy.

1. **Adversarial Goal.** Adversarial goal is defined according to the violation of the system and the attack specificity. According to [11], Security violations include integrity, availability or privacy violation. In feature selection algorithms, we regard integrity violation as attacks which just modify the feature sets sightly. For availability violation, if attacker forces to select a feature subset which makes the biggest generalization error, availability is violated. Privacy violated means that the attacker can use reverse engineering to get information about model or users who use the system. In our poisoning attack, integrity and availability violation are mainly considered. Therefore, we assume that adversarial goal is to reduce the accuracy of classification by influencing the selection of feature subset.
2. **Adversarial Knowledge.** Adversarial knowledge can be considered from the concrete composition of the classifier, whether the attackers know the training data, the feature representation or the feature selection algorithms. Different attackers have different adversarial knowledge level. According to different levels of knowledge, we can divide attackers into complete knowledge attackers and limited knowledge attackers. In our work, in order to be realistic, we assume attackers have limited knowledge.
3. **Adversarial Capability.** Adversarial capability is whether attackers can get and modify the training data. Because the IDSs are updated every a fixed interval by using network data collected on the internet, attackers can obtain the network data and poison these algorithms at this moment. Therefore, we assume attackers cannot modify the original data and only can inject craft adversarial samples.
4. **Adversarial Strategy.** Following the methods in [16], we make our strategy to reach the adversarial goal with the attacker's knowledge level and the capability. We express our adversarial strategy as:

$$\begin{aligned} \max \quad & G(D+P|\theta) \\ s.t. \quad & \varphi(P) < \lambda\varphi(D) \quad 0 < \lambda < 1. \end{aligned} \tag{1}$$

where D is the original training data and P is the poison data produced by our attack algorithms. The adversarial model $\theta = (K, C, M)$, K is knowledge, C denotes capability and M is the method used in IDSs. The $G(D + P|\theta)$ denotes adversarial goal as the objective function, which represents the effect of the poison data P. The $\varphi(P)$ denotes the number of poison samples in P and $\varphi(D)$ denotes the number of training samples in D.

4 The Proposed Feature Selection Poisoning Method Using Gradient Optimization

4.1 Overall Framework of Poisoning Machine Learning Based IDSs

We present our methods to attack the feature selection algorithms used in IDSs. Our attack method is based on [16], but the method they adopted only against embedded feature selection algorithms based on spam filtering and malware in pdf detection system. Therefore, the method we improved is able to attack more feature selection algorithms. The framework of our attack method is illustrated in Fig. 1. Our framework is mainly divided into three components. The two components on the left side are based on the optimization theory to generate the adversarial objective function of embedded and filter respectively. The third component on the right side is based on the adversarial objective function to calculate the gradient to generate adversarial samples.

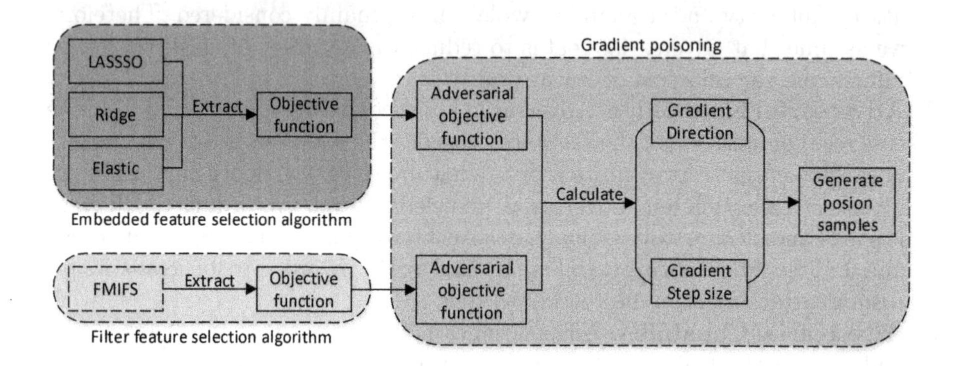

Fig. 1. The framework of the attack method

4.2 Formulation of Feature Selection Poisoning

Filter Feature Selection Poisoning. In this work, we choose to use the FMIFS proposed by [6] as an example, describing our attack method against the filter. FMIFS is a new variation of mutual information based filter feature selection. This new approach is designed to maximize the amount of information carried by a feature and minimizing the average of feature redundancy simultaneously. The objective function is defined as follows.

$$G_{MI} = \underset{f_i}{arg\,max}(I(C, f_i) - \frac{1}{|S|}\sum_{f_s} MR) \tag{2}$$

$$MR = \frac{I(f_i, f_s)}{I(C, f_i)} \tag{3}$$

Where Mutual Information (MI) is the information contained in a random variable about another random variable. $I(C, f_i)$ is the amount of information carried by a feature f_i about class C. MR represents the dependence of f_i and f_s about C. If $G_{MI} < 0$, feature selection algorithm will select a feature subset which has little MI. That is, the feature f_i is redundant about C. If $G_{MI} = 0$, the feature f_i cannot offer any information about classification, f_i is irrelevant for the classifier. If $G_{MI} > 0$, the feature f_i is relevant for classifier. Therefore, this feature selection algorithm will select features which let $G_{MI} > 0$. On the contrary, attacker needs to generate the poison samples by the following formula.

$$G(D+P|\theta) = \underset{f_i}{arg\,min}(I(C,f_i) - \frac{1}{|S|}\sum_{f_s} MR) \tag{4}$$

Embedded Feature Selection Poisoning. The embedded feature selection algorithms mainly include LASSO, Ridge regression and Elastic Net. These three algorithms use the same linear function $l(x) = a^T x + b$ and the loss function $L(y, l(x))$ is defined as follows.

$$L(y, l(x)) = \frac{1}{2}(l(x) - y))^2 \tag{5}$$

In general, the requirements for the choice of three algorithms can be represented as minimizing the following objective function.

$$\omega^* = \underset{\omega}{arg\,min} \sum_i L(y_i, l(x_i, \omega)) + \lambda \Omega(\omega) \tag{6}$$

Where λ is hyper-parameters to balance the loss and the regularization function $\Omega(\omega)$ of these two items. In addition to guaranteeing the minimum training error, we also hope to minimize the model test error, so we need to add the regularization function $\Omega(\omega)$ to constrain our model to be as simple as possible. For LASSO, the $\Omega(\omega) = ||\omega||_1$. For Ridge, $\Omega(\omega) = \frac{1}{2}||\omega||_2^2$. And for Elastic, $\Omega(\omega) = \eta||\omega||_1 + (1-\eta)\frac{1}{2}||\omega||_2^2$, $0 < \eta < 1$.

Under these definitions, based on the assumptions made by the attack model in Sect. 3, we can determine the attack objective function of the algorithm for embedded feature selection as:

$$G(D+P|\theta) = \underset{\omega^*}{max} \sum_p L(y_p, l(x_p, \omega)) + \lambda \Omega(\omega) \tag{7}$$

where the $l(x_i, \omega)$ is learned from raw data by minimizing ω^* in Eq. (6). We need to find the attack point x_p to poison the algorithm. Hence, we adopt the methods in Sect. 4.3.

4.3 Adversarial Sample Generation Using Gradient Optimization

For filter (embedded) feature selection algorithms used in IDSs, we use a gradient descent (ascent) method to find samples that minimize (maximize) its objective function and generate a large number of poisoned samples on the basis of this.

Take the embedded feature selection algorithm as an example, as described in Eq. (7), attacker's goal is determined by the function $l(x_p, \omega)$. Therefore, we iteratively calculate gradient direction and step size to generate the poison samples. The first step we calculate gradient is calculating the gradient direction by taking the derivative of Eq. (7) with respect to x_p. Considering Eq. (1) as a constrained optimization problem, we use the method named Lagrange multiplier to solve this problem. The second step is determining the gradient step size. We adopt the method named backtracking line search for tuning step size. Then we use the direction and step size for generating a poison sample satisfied attacker's objective function. The new sample is then added to the training set to be re-trained to obtain new model parameters. The gradient is calculated again to generate poisoned samples. This method is iterated until convergence. Our attack method is shown in Algorithm 1. We can calculate the complexity of our algorithm by follows:

$$\lim_{n,m \to \infty} n(m + (n + l)) \approx mn^2 \tag{8}$$

Where m is the complexity of the machine learning algorithm chosen in *step 3* and l is the complexity in *steps 5 to 11*. Thus, our algorithmic complexity is $O(mn^2)$ when n approaches infinity.

Algorithm 1. Gradient attack

Input: The training dataset D, the num of attack samples n, initial step size s and direction d, ε is a small positive constant.

Output: Poison data $D \cup P$

```
1  P = ∅;
2  for i ← 1 to n do
3      Model ← learned on dataset D ∪ P ;
4      for j in D ∪ P do
5          d ← calculate gradient ∇G(D + P) ;
6          x_{i+1} = x_i + ds;
7          P = P ∪ x_i;
8          s ← backtracking line search;
9          if G(D + P) − G(D + (P − 1)) ⩽ ε then
10             return D ∪ P;
11         end
12     end
13 end
14 return D ∪ P ;
```

5 Performance of Evaluation

5.1 Experimental Setup

In our experiments, we use three real public datasets including KDD99, NSL-KDD, kyoto2006+ [17] and a simulation dataset WSN-DS [18] to validate the effectiveness of our attack method with limited knowledge described in Sect. 3. The KDD99 dataset recorded network connectivity and system audit data for nine weeks, including four types of anomalies. However, the KDD99 dataset has been found to be flawed, so we use NSL-KDD and Kyoto 2006+ datasets on this basis. The former is a dataset that improves upon some of the problems with KDD99, the later is traffic data that continual collected by Kyoto University from honeypots since 2006. For details, we choose the "kddcup.data_10_percent corrected", "NSL-KDDTrain+" and the datasets of the days 200908 from Kyoto 2006+ as our experiments datasets. In particular, except for these three regular network datasets, we conducted experiments specifically on WSN-DS dataset collected from wireless sensor network.

We perform ten independent experiments on each dataset and each experiment randomly select 15,000 samples. Then we randomly selected 7,000 of them as training sets and the rest as test sets. To simulate the limit knowledge of attackers, we assume attackers only reach training data when IDSs are updated but attackers don't know the meaning of these data.

First, we make dataset available through data preprocessing as follows.

1. **Data Transforming.** KDD99, NSL-KDD and Kyoto2006+ contain both numeric and character data (i.e. TCP, UDP, ICMP, HTTP, FTP and so on), but the data feature selection algorithms required are numeric data, so we need to convert the character data to numeric.
2. **Data Normalization.** Data normalization is an essential step in data preprocessing when transforming is done. Normalization is to map the different range values to the same fixed range, Commonly $[0, 1]$.
3. **Label Binarization.** Because these four datasets have multi-type abnormal samples (e.g. Dos, Probe, U2R and so on), for convenience, we regard the label of normal data as 1 and abnormal data as -1.

And then, for FMIFS, we use Flexible Liner Correlation Coefficient based Feature Selection (FLCFS) [6] to calculate $I(C, f_i)$. FLCFS improved insensitivity to the non-linear correlation of Linear Correlation Coefficient (LCC) and LCC is one of the most effective methods to evaluate MI. For embedded, we use LassoCV class in sklearn tools. The LassoCV class is the first choice for Lasso regression and it uses cross-validation to help us select a suitable hyperparameter α, eliminating the hassle of filtering out α by our own. LassoCV is a must when faced with finding the key features in a bunch of high-dimensional features and LassoCV also works well for sparse linear relationships. Similar, we adopt RidgeCV and ElasticCV in sklearn tools. In poison stage, we use our method to generate the poison data P and then put P in the new training data to wait for the next training process. We evaluate our result by reporting the changing trend of the classification error rate and consistency.

5.2 Evaluation Metrics

To demonstrate the effectiveness of our attack method, we choose classification error rate and feature subset consistency as evaluation metrics.

1. **Classification Error Rate.** Classification error rate reflects the accuracy of IDSs detection. The higher the error rate, indicating that the lower the accuracy of the IDSs detection classification. The Classification error rate is defined as follows.

$$Clf\_err = \frac{FP + FN}{TP + TN + FP + FN} \tag{9}$$

where FP (False Positive) means the numbers of abnormal data but classified as normal. FN (False Negative) means the numbers of normal data but classified as abnormal. Hence we regard both FP and FN as mis-classification, TP (True Positive) and TN (True Negative) as proper classification.

2. **Consistency.** The feature subset consistency is defined as the distinction between the subset with attack and without attack. We adopt the Kuncheva's Consistency [19] to evaluate the stability of subsets. Kuncheva's Consistency can be described as follows.

$$C(A, B) = \frac{rd - k^2}{k(n-k)} \tag{10}$$

There are two subsets $A, B \subset X$, let $|A| = |B| = k$, where $0 < k < |X| = n$, and $r = |A \cap B|, -1 \leq C(A, B) \leq 1$. When $C > 0$, it means the two subsets are similar. When $C = 0$, it means the two subsets are independent. When $C < 0$, it means the two subsets are negatively correlated.

5.3 Results and Analysis

The results of ten independent replicate experiments are shown as follows. All the values in these curves are averaged by each experiment. In Fig. 2(a–d), the y-coordinate indicates the classification error rate, the x-coordinate indicates the percent of poison samples in the training data. We can see that all the methods show a lower error rate when there is no attack (i.e. at 0% poison samples), that is, the classification accuracy is very high. As the proportion of poison sample increasing, the error rate of classification increases continuously. When we just inject 15% samples, the classification error rate increased around 20%. We can see that the LASSO is the most affected, followed by ridge and elastic, the least affected is FMIFS. To better determine whether the impact on classification is due to the selection of feature subsets, we show the tend of consistency with the number of poisonings increasing in Fig. 2(e–h). We can see that our attack methods can affect those four algorithms effectively in every dataset. These feature subsets selected from these algorithms changed a lot after the attack. Even the subset of features before and after the attack showed a negative correlation.

In addition, in order to show the changes of the feature subset after the attack more clearly, we performed statistical analysis on the feature subset selected from each experiment. As shown in Table 1 for the KDD99 dataset when poisoning 15% samples, according to the top ten features selected by different feature

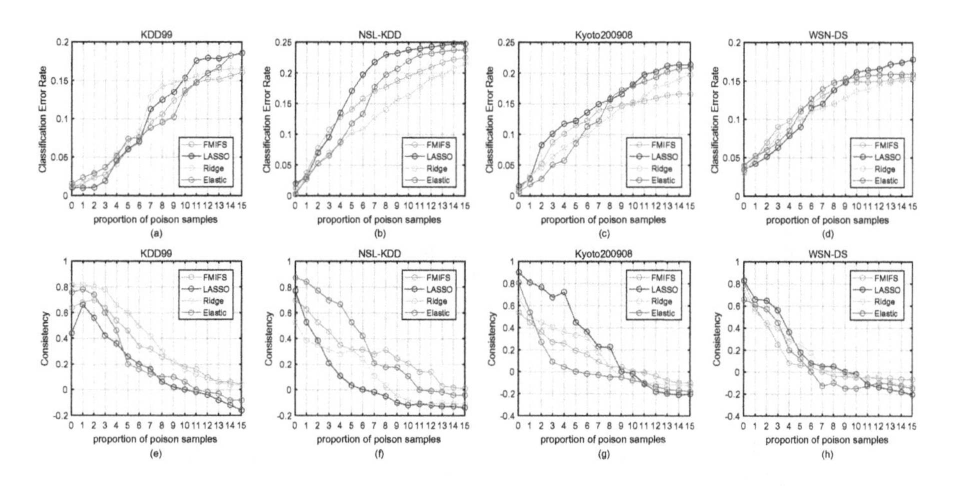

Fig. 2. The classification error rate and consistency under attacks

Table 1. Feature subsets before and after attack on KDD99

Algorithm	Before attack	After attack	Changes
FMIFS	f3, f5, f8, f12, f17, f19, f22, f23, f31, f33	f2, f5, f8, f11, f18, f19, f22, f23, f31, f33	30%
LASSO	f1, f2, f12, f13, f17, f22, f28, f31, f34, f37	f1, f3, f11, f13, f17, f25, f27, f33, f34, f37	50%
Ridge	f4, f7, f11, f21, f24, f29, f31, f32, f33, f37	f4, f7, f13, f21, f25, f29, f30, f32, f33, f38	40%
Elastic	f3, f5, f9, f12, f18, f19, f30, f32, f33, f38	f3, f5, f8, f12, f18, f19, f25, f32, f33, f38	50%

Table 2. Feature subsets before and after attack on NSL-KDD

Algorithm	Before attack	After attack	Changes
FMIFS	f3, f5, f6, f12, f23, f24, f29, f31, f32, f37	f3, f6, f11, f23, f24, f29, f30, f32, f34, f37	40%
LASSO	f1, f8, f9, f10, f11, f12, f19, f23, f31, f32	f1, f8, f9, f11, f16, f23, f27, f31, f33, f40	60%
Ridge	f1, f5, f8, f9, f10, f12, f17, f29, f31, f32	f1, f5, f8, f11, f13, f17, f27, f31, f32, f38	50%
Elastic	f5, f9, f10, f12, f17, f23, f26, f29, f31, f32	f2, f5, f10, f19, f26, f29, f30, f34, f38, f40	50%

selection algorithms, it can be seen that the feature subset selected by LASSO changes biggest, and the last column is the proportion of features subset changes. We can see that the degree which the features subsets are affected by the attack corresponds exactly to the classification error rate affected by the attack. Similar, as shown in Tables 2, 3 and 4, we can see the same trend.

Table 3. Feature subsets before and after attack on Kyoto 200908

Algorithm	Before attack	After attack	Changes
FMIFS	f2, f5, f8, f9, f12, f17, f18, f19, f22, f23	f2, f6, f8, f9, f15, f17, f18, f20, f21, f23	40%
LASSO	f2, f6, f8, f9, f11, f15, f17, f19, f20, f22	f2, f5, f8, f9, f13, f14, f15, f18, f20, f21	50%
Ridge	f1, f4, f5, f7, f10, f12, f15, f17, f19, f20	f2, f3, f5, f7, f10, f12, f15, f19, f21, f22	40%
Elastic	f2, f5, f6, f8, f10, f12, f16, f17, f19, f21	f2, f5, f7, f8, f11, f12, f16, f18, f20, f21	40%

Table 4. Feature subsets before and after attack on WSN-DS

Algorithm	Before attack	After attack	Changes
FMIFS	f1, f3, f4, f5, f7, f10, f13, f14, f16, f17	f2, f3, f4, f6, f7, f10, f13, f14, f15, f17	30%
LASSO	f2, f5, f7, f8, f12, f14, f15, f16, f17, f18	f1, f5, f6, f8, f11, f13, f14, f16, f17, f18	40%
Ridge	f1, f4, f5, f7, f9, f11, f12, f14, f17, f18	f1, f2, f5, f7, f10, f11, f12, f14, f16, f18	30%
Elastic	f3, f4, f5, f7, f8, f10, f11, f13, f15, f18	f1, f4, f5, f7, f9, f10, f11, f14, f15, f17	40%

6 Conclusions

Machine learning have been increasingly adopted in wireless IDSs to improve detection performance. Feature selection as an important component in machine learning is confronted with a broad and growing range of security threats. In this paper, we adopt a gradient poisoning attack method to generate poisoning samples to induce classifier to select a fake feature subset by searching the biggest classification error point. We make an explicit assumption about the attackers' goals, knowledge, capability and strategy. Through our experiments, they validate that the attack method is effective for the embedded and filter feature selection algorithms based on assumptions. We conduct experiments using three regular network-based public datasets and a wireless sensor network dataset, demonstrating that the detection rate of IDSs can be significantly reduced around 20% for the poison samples. For future work, we plan to focus on the expansion of existing attack methods and the robustness of machine learning algorithms.

Acknowledgment. This work is supported by the National Nature Science Foundation of China under Grant Nos. 61572514 and 61702539.

References

1. Grosse, K., Papernot, N., Manoharan, P., Backes, M., McDaniel, P.: Adversarial perturbations against deep neural networks for malware classification. arXiv preprint arXiv:1606.04435 (2016)
2. Xu, W., Qi, Y., Evans, D.: Automatically evading classifiers. In: Proceedings of the 2016 Network and Distributed Systems Symposium (2016)
3. Tibshirani, R.: Regression shrinkage and selection via the lasso. J. R. Stat. Soc. Ser. B (Methodol.) **58**, 267–288 (1996)
4. Hoerl, A.E., Kennard, R.W.: Ridge regression: biased estimation for nonorthogonal problems. Technometrics **12**(1), 55–67 (1970)
5. Zou, H., Hastie, T.: Regularization and variable selection via the elastic net. J. R. Stat. Soc.: Ser. B (Stat. Methodol.) **67**(2), 301–320 (2005)
6. Ambusaidi, M.A., He, X., Nanda, P., Tan, Z.: Building an intrusion detection system using a filter-based feature selection algorithm. IEEE Trans. Comput. **65**(10), 2986–2998 (2016)
7. Abraham, A., Jain, R., Thomas, J., Han, S.Y.: D-SCIDS: distributed soft computing intrusion detection system. J. Netw. Comput. Appl. **30**(1), 81–98 (2007)
8. Chebrolu, S., Abraham, A., Thomas, J.P.: Feature deduction and ensemble design of intrusion detection systems. Comput. Secur. **24**(4), 295–307 (2005)
9. Chen, Y., Abraham, A., Yang, B.: Feature selection and classification using flexible neural tree. Neurocomputing **70**(1–3), 305–313 (2006)
10. Mukkamala, S., Sung, A.H.: Significant feature selection using computational intelligent techniques for intrusion detection. In: Bandyopadhyay, S., Maulik, U., Holder, L.B., Cook, D.J. (eds.) Advanced Methods for Knowledge Discovery from Complex Data, pp. 285–306. Springer, London (2005). https://doi.org/10.1007/1-84628-284-5_11
11. Barreno, M., Nelson, B., Sears, R., Joseph, A.D., Tygar, J.D.: Can machine learning be secure? In: Proceedings of the 2006 ACM Symposium on Information, Computer and Communications Security, pp. 16–25. ACM (2006)
12. Liu, Q., Li, P., Zhao, W., Cai, W., Yu, S.: A survey on security threats and defensive techniques of machine learning: a data driven view. IEEE Access **99**, 1 (2018)
13. Wittel, G.L., Wu, S.F.: On attacking statistical spam filters. In: CEAS (2004)
14. Šrndic, N., Laskov, P.: Detection of malicious pdf files based on hierarchical document structure. In: Proceedings of the 20th Annual Network & Distributed System Security Symposium, pp. 1–16 (2013)
15. Biggio, B., Fumera, G., Roli, F.: Security evaluation of pattern classifiers under attack. IEEE Trans. Knowl. Data Eng. **26**(4), 984–996 (2014)
16. Xiao, H., Biggio, B., Brown, G., Fumera, G., Eckert, C., Roli, F.: Is feature selection secure against training data poisoning? In: International Conference on Machine Learning, pp. 1689–1698 (2015)

17. Song, J., Takakura, H., Okabe, Y., Eto, M., Inoue, D., Nakao, K.: Statistical analysis of honeypot data and building of Kyoto 2006+ dataset for NIDS evaluation. In: Proceedings of the First Workshop on Building Analysis Datasets and Gathering Experience Returns for Security, pp. 29–36. ACM (2011)
18. Almomani, I., Al-Kasasbeh, B., Al-Akhras, M.: WSN-DS: a dataset for intrusion detection systems in wireless sensor networks. J. Sens. **2016**(2), 1–16 (2016)
19. Kuncheva, L.I.: A stability index for feature selection. In: Artificial Intelligence and applications, pp. 421–427 (2007)

A Cooperative Jamming Based Secure Uplink Transmission Scheme for Heterogeneous Networks Supporting D2D Communications

Jingjing Fan[1], Yan Huo[1(✉)], Xin Fan[1], Chunqiang Hu[2], and Guanlin Jing[1]

[1] School of Electronics and Information Engineering, Beijing Jiaotong University, Beijing, China
yhuo@bjtu.edu.cn

[2] School of Big Data and Software Engineering, Chongqing University, Chongqing, China

Abstract. A heterogeneous network supporting D2D communications is a typical framework to cope with data sharing in a cyber physical system. In the network, there exists more interference than traditional cellular networks, which promotes interference cancellation (IC) technologies. Yet, how IC can improve security performance of D2D-enabled networks is still unknown. In addition, how to establish a secure physical transmission link for users in these networks is of great significance. In this paper, with IC or not, we first derive the expressions of the secrecy outage probability in D2D-enabled heterogeneous networks, respectively. Then, according to the transmission constraint and the security constraint, we present two cooperative jamming schemes based on devices in D2D networks for achieving secure uplink of cellular users. In these schemes, we formulate the above secure challenge as an optimization problem with or without fully IC and provide the corresponding optimal solutions to improve the secrecy performance of cellular uplink. Finally, numerical simulation demonstrates that the secure performance of cellular uplink is remarkably enhanced using our cooperative jamming schemes.

Keywords: Heterogeneous networks · D2D communications
Secure transmission · Cooperative jamming

1 Introduction

A variety of mobile smart devices such as smartphones, tablets and ultrabooks are gaining popularity in ever-evolving wireless communications technologies. Human demands for wireless services [1] over these smart devices [2] have not only satisfied with the traditional voice, text and other basic business. To support novel intelligent and convenient services to access to the Internet at anytime and anywhere, e.g., vehicular communications [3], wireless sensor networks [4], webcasting and real-time online gaming [5], device-to-device (D2D) networking

S. Chellappan et al. (Eds.): WASA 2018, LNCS 10874, pp. 103–114, 2018.
https://doi.org/10.1007/978-3-319-94268-1_9

for Internet of Things (IoT) has been presented as a promising technology for future wireless communications [6].

As a complement and extension of backbone networks, D2D communications can enable smart devices to communicate with each other directly without accessing the network infrastructure [7,8]. So far, many existing works on D2D mainly focus on radio resource management of stand alone [9,10] and network-assisted one [11] as well as interference management for either in-band or out-of-band case. These works consider the interference generated by devices as an obstacle to cellular communications. They employ power control to mitigate interference [12–14], allocate orthogonal spectrum to avoid interference [15,16], and design advanced coding schemes to cancel interference [17–19].

Apart from the above issues, secure data sharing between D2D users is more vulnerable due to the openness nature of broadcast communications [20]. Attacks on privacy, confidentiality and integrity may result in disclosure and unavailability of legitimate data. Although cryptography based approaches are introduced directly to cope with these threats [21,22], they may bring a new issue, i.e., low hardware resources may be not enough to support complex cryptographic operations. As a result, some researchers intend to employ cooperative jamming based physical layer security to study D2D secure issues.

The existing works on the design of secure mechanisms in D2D communications starts from different perspectives. From the perspective of system power, first, Zhang et al. introduce the secrecy capacity into a power control strategy to improve network secrecy performance with low computational complexity in [23]. In [24], a social relationship based jammer selection scheme is studied for D2D overlay to achieve secure cellular links according to pre-optimized power allocation results. Second, some researchers suggest exploiting spectrum resource to reward cooperative jammers. For example, the authors in [25] design a merge-and-split scheme using a max-coalition order and provide a coalitional game based cooperation mechanism with efficient resource block (RB) allocation. Also, the authors of [26] formulate a secrecy outage probability maximum problem being subject to RB allocation constraint. However, there is no work to explore secure transmission schemes for cellular uplink supporting D2D communications. As the D2D technology becomes pervasive in future IoT applications, it is important to study the secure data transmission under heterogeneous networks.

Considering mutual interference between cellular users and D2D nodes, we present two optimal secure solutions for cellular uplink based on the D2D communications. To be specific, we first derive the mathematical expressions of secrecy outage probability for the scenario with or without interference cancellation (IC). According to the theoretical analysis, we present different optimal solutions for these two scenarios. Our rigorous analyses demonstrate that the proposed solutions can achieve secure transmission for the cellular uplink.

The rest of the paper is organized as follows. In Sect. 2, we present the system model and necessary preliminaries. Next, we formulate the secrecy outage probability of the system and propose an optimal problem and provide two optimal solutions for the scenario with or without IC in Sect. 3. Finally, numerical simulations and a conclusion are drawn in Sects. 4 and 5, respectively.

2 System Model and Preliminary

2.1 Network Model

We consider a heterogeneous network supporting underlay device-to-device (D2D) communication, depicted in Fig. 1, which is composed of a macro cellular and several smart devices. A macro base station with IC capability, abbreviated as the IC-based MBS, is located at the center of the macro cellular. The IC-based MBS can provide primary data transmission for legitimate (cellular) users uniformly located within the communication coverage radius R. To achieve zero interference between legitimate users, we assume that all legitimate users are allocated with orthogonal frequency bands.

For the underlay density D2D communication, numerous single-antenna devices are scattered within the cellular network. A D2D transmitter intends to communicate with its corresponding receiver through sharing with cellular spectrum without interfering with legitimate users. We assume that the positions of these devices that share spectrum with the i-th cellular user CU_i follow a homogeneous Poisson point process (PPP) $\mathbf{\Phi}_{c,d}$ with intensity $\lambda_{c,d}$. Taking into account stationary devices, the distance between both sides of the underlay D2D communication is fixed and isotropic.

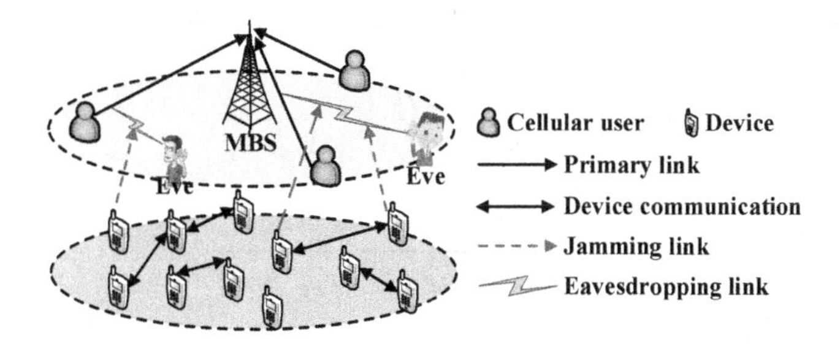

Fig. 1. A heterogeneous network model supporting D2D.

This heterogeneous network may be subject to eavesdropping threats due to randomly dynamic sharing spectrum. Some single-antenna malicious nodes (eavesdroppers) hidden in the network, i.e., Eve in Fig. 1, may arbitrarily wiretap legitimate signals sent from cellular users to the IC-based MBS. These malicious nodes only passively wiretap rather than actively attack (e.g., transmitting modified signals or interference) legitimate users because these active behaviours may be easily detected and captured by legitimate nodes. Note that the passive Eves are not collusive and also obey the distribution of the homogeneous PPP $\mathbf{\Phi}_e$ with intensity λ_e.

2.2 Transmission Model

In this subsection, we provide transmission model based on the above network to characterize received signals. The received signals at the Mth IC-based MBS from the cth cellular user (CU_c) and the eth Eve (Eve_e) can be expressed as

$$y_c = h_{cM} r_{cM}^{-\frac{\alpha}{2}} s_c + \sum_{k \in \Phi_{c,d}} h_{kM} r_{kM}^{-\frac{\alpha}{2}} s_k + n_c, \tag{1}$$

$$y_e = h_{ce} r_{ce}^{-\frac{\alpha}{2}} s_c + \sum_{k \in \Phi_{c,d}} h_{ke} r_{ke}^{-\frac{\alpha}{2}} s_k + n_e, \tag{2}$$

where s_c and s_k are the transmitted signals of CU_c and the kth D2D transmitter (D_k). Thus, the first term on the right-hand side of (1) and (2) are the received legitimate information sent by CU_c, respectively, while other terms of these two equations refer to the interference and noise. In particular, n_c and n_e are the additive white Gaussian noise (AWGN) over each transmission link, which are defined as complex normal distributions $\mathcal{CN}(0, \sigma_j^2), j \in \{c, e\}$.

In the transmission model, we introduce the small-scale quasi-static Rayleigh fading into the typical large-scale fading to characterize the practical channel fading of our wireless network model. As a result, all channel state coefficients ($h_{uv}, u, v \in \{c, k, e, M\}$) in above equations are defined as the small-scale fading distributed as $\mathcal{CN}(0,1)$. In addition, $r_{uv}^{-\alpha}$ represents the standard path loss for the large-scale fading, where r_{uv} is the distance from transmitter u to receiver v and α is the path loss factor.

According to the signal transmission model, we define the corresponding received signal-to-interference-plus-noise-ratio (SINR) as follows. First, the SINR to receive CU_c's signals at the Mth MBS and Eve_e can be expressed as

$$\gamma_c = \frac{p_c \left|h_{cM}\right|^2 r_{cM}^{-\alpha}}{\sum_{k \in \Phi_{c,d}} p_d \left|h_{kM}\right|^2 r_{kM}^{-\alpha} \Delta_k + \sigma_c^2}, \tag{3}$$

$$\gamma_e = \frac{p_c \left|h_{ce}\right|^2 r_{ce}^{-\alpha}}{\sum_{k \in \Phi_{c,d}} p_d \left|h_{ke}\right|^2 r_{ke}^{-\alpha} + \sigma_e^2}, \tag{4}$$

where $p_c = |s_c|^2$ and $p_d = |s_k|^2$ refer to the transmission power of CU_c and D_k.

Here, we assume that the base station in our model can adopt unconditional interference cancellation [27] to facilely cancel the signals that are stronger than a predefined threshold τ. Based on this, we employ an indicator function, Δ_k, to denote the kth device that sends its signals to interfere with CU_c, i.e.,

$$\Delta_k = \begin{cases} 0, \text{ if } p_d \left|h_{kM}\right|^2 r_{kM}^{-\alpha} > \tau \\ 1, \text{ else} \end{cases}.$$

3 Problem Formulation and Two Optimal Solutions

In this section, we first analyze the secrecy outage probability of CU_c. Then, we propose two optimization formulations for these users in our heterogeneous networks. Finally, we present two optimal solutions for the scenario with or without IC, respectively.

3.1 Secrecy Outage Probability (SOP) Analysis

The secrecy outage probability of CU_c can be expressed as

$$\begin{aligned} P_{\mathrm{SOP}} &\triangleq \Pr(\log_2(1+\gamma_c) - \log_2(1+\max\gamma_e) < Q) \\ &= \int_0^\infty \int_0^{\mathrm{Th}(x)} f_{\gamma_c}(y) g(x) \mathrm{d}y \mathrm{d}x \\ &= \int_0^\infty (F_{\gamma_c}(\mathrm{Th}(x)) - F_{\gamma_c}(0)) g(x) \mathrm{d}x \\ &= 1 - \int_0^\infty \Pr(\gamma_c > \mathrm{Th}(x)) \mathrm{d}G(x), \end{aligned} \tag{5}$$

where $f_{\gamma_c}(\cdot)$ and $g(\cdot)$ denote as the probability density functions (PDF) of CU_c and the most detrimental Eve, while $F_{\gamma_c}(\cdot)$ represents the cumulative density function (CDF) of CU_c. In addition, $\mathrm{Th}(x) \triangleq 2^Q(1+x) - 1$ is a threshold of secure transmission interruption. According to different capabilities of interference processing, here, we introduce two lemmas to analyze the scenarios with or without IC of the base station.

Lemma 1. *When it is difficult to achieve IC because of computational complexity,* (5) *can be further derived as follows.*

$$P_{\mathrm{SOP}}^{(\mathrm{N})} = 1 - \frac{2B}{\alpha A \lambda_{c,d}^2} \int_0^\infty \frac{1 - e^{-A\lambda_{c,d}\mathrm{Th}(x)^{\frac{2}{\alpha}}}}{\mathrm{Th}(x)^{\frac{2}{\alpha}}} \cdot e^{-\frac{B}{\lambda_{c,d}} x^{-\frac{2}{\alpha}}} x^{-1-\frac{2}{\alpha}} \mathrm{d}x, \tag{6}$$

where $A = \pi(\frac{p_d}{p_c})^{\frac{2}{\alpha}} \Gamma(1+\frac{2}{\alpha})\Gamma(1-\frac{2}{\alpha})R^2, B = \frac{\pi\lambda_e R^2}{A}$, *and* $\Gamma(\cdot)$ *is the gamma function.*

Proof. See Appendix A.

In particular, if we define the secrecy outage probability as zero (i.e., no communication interruption, $Q = 0$), then (6) can be simplified as below.

$$P_{\mathrm{SOP}}^{(\mathrm{N},0)} = 1 - \frac{1}{\lambda_e \pi R^2} + 2K_2(2\sqrt{\lambda_e \pi R^2}). \tag{7}$$

where $K_2(\cdot)$ represents the modified Bessel function of the second kind.

Lemma 2. *If the base station can easily cancel interference from other users,* (5) *can be rewritten as below.*

$$P_{\mathrm{SOP}}^{(\mathrm{I})} = 1 - \frac{2B}{\alpha\lambda_{c,d}} \int_0^\infty P e^{-\frac{B}{\lambda_{c,d}} x^{-\frac{2}{\alpha}}} x^{-1-\frac{2}{\alpha}} \mathrm{d}x. \tag{8}$$

Proof. See Appendix B.

3.2 Secure Transmission Schemes for CUs' Uplink

In order to achieve secure transmission for cellular users, we assume CU_c adopts the Wyner's encoding scheme. Thus, CU_c needs to satisfy a successful connection without secrecy outage in its uplink, i.e.,

- **Transmission Constraint (TC):** the secrecy rate $\log(1+\gamma_c)$ of CU_c needs to be higher than the target transmit rate denoted by R_t.
- **Security Constraint (SC):** the achievable minimum secrecy rate $\log(1+\gamma_c)-\log(1+\max\gamma_e)$ needs to be higher than the target secure transmit rate denoted by R_s.

Note that when TC is satisfied, SC can be transformed into $\log(1+\max\gamma_e) < R_t - R_s$. As a result, we can formulate an optimization problem to maximize secure transmission probability $P_{\mathrm{STP}} \triangleq P_{\mathrm{TC}}P_{\mathrm{SC}}$ of CU_c's uplink, i.e.,

$$\mathbf{P1}: \quad \max_{\lambda_{c,d}} P_{\mathrm{TC}}P_{\mathrm{SC}}, \tag{9}$$

where $P_{\mathrm{TC}} = \Pr(\log_2(1+\gamma_c) > R_t)$ and $P_{\mathrm{SC}} = \Pr(\log_2(1+\max\gamma_e) < R_t - R_s)$.

Lemma 3. *Considering fixed transmit power of CU_c and D_k, the secure transmission probability without IC capability, which is a unimodal function of $\lambda_{c,d}$ with a maximum value, can be further derived as $P_{\mathrm{STP}}^{(\mathrm{N})}$, i.e.,*

$$P_{\mathrm{STP}}^{(\mathrm{N})} = \frac{1-e^{-\hat{A}\lambda_{c,d}}}{\hat{A}\lambda_{c,d}} \cdot e^{-\frac{\hat{B}}{\lambda_{c,d}}}, \tag{10}$$

where $\hat{A} = (2^{R_t}-1)^{\frac{2}{\alpha}}A$, $\hat{B} = \frac{B}{(2^{R_t-R_s}-1)^{\frac{2}{\alpha}}}$.

Proof. See Appendix C.

Therefore, we can easily compute the optimal solution of **P1** by the Bisection method or the Golden-section search algorithm. Then, we present the solution for the scenario with IC by Lemma 4.

Lemma 4. *The secure transmission probability can be rewritten as an monotonically increasing function, $P_{\mathrm{STP}}^{(\mathrm{I})}$, for the scenario with the capability of IC.*

$$P_{\mathrm{STP}}^{(\mathrm{I})} = \frac{2\gamma(\frac{2}{\alpha}, \frac{\sigma_c^2 R^\alpha (2^{R_t}-1)}{p_c})}{\alpha R^2 (\frac{\sigma_c^2}{p_c}(2^{R_t}-1))^{\frac{2}{\alpha}}} \cdot e^{-\frac{\hat{B}}{\lambda_{c,d}}}, \tag{11}$$

where $\gamma(\cdot)$ represents the lower incomplete gamma function.

Proof. For the scenario with IC, $\mathrm{P}_{\mathrm{TC}}^{(\mathrm{I})}$ can be derived as below.

$$\mathrm{P}_{\mathrm{TC}}^{(\mathrm{I})} = P(|h_{cM}|^2 > \frac{r_{cM}^\alpha (2^{R_t}-1)\sigma_c^2}{p_c}) = \frac{2\gamma(\frac{2}{\alpha}, \frac{(2^{R_t}-1)\sigma_c^2 R^\alpha}{p_c})}{\alpha R^2 (\frac{(2^{R_t}-1)\sigma_c^2}{p_c})^{\frac{2}{\alpha}}}.$$

And $\mathrm{P}_{\mathrm{SC}}^{(\mathrm{I})}$ is still equal to $e^{-\frac{\hat{\beta}}{\lambda_{c,d}}}$. According to the definition of P_{STP}, (11) holds. Clearly, $P_{\mathrm{STP}}^{(\mathrm{I})}$ should be an monotonically increasing function of $\lambda_{c,d}$. This means that the larger $\lambda_{c,d}$ is, the greater interference with the eavesdropper caused by D2D users. As a result, we can analyze the optimization problem **P1** from the change of $P_{\mathrm{SC}}^{(\mathrm{I})}$. And this completes the proof of Lemma 4.

4 Numerical Simulation

In this section, we analyze the secure transmission probability of the scenarios with and without IC. Here, we assume that the communication radius of cellular users is 100 m, the eavesdroppers' intensity λ_e is 0.00001 m^{-2}, and the path loss factor α is 4. Besides, the power of AWGN is set to $\sigma_c^2 = 10^{-10}$ W. In all simulations, the extreme point $\lambda_{c,d}^*$ of $\mathrm{P}_{\mathrm{STP}}^{(\mathrm{N})}$ is obtained by the Bisection method.

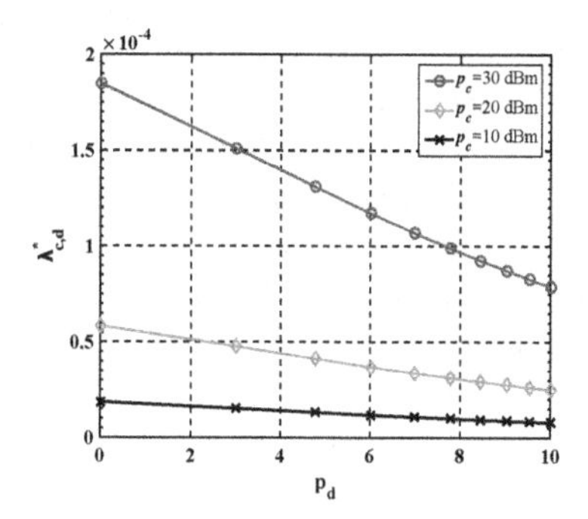

Fig. 2. $\lambda_{c,d}^*$ *v.s.* p_d.

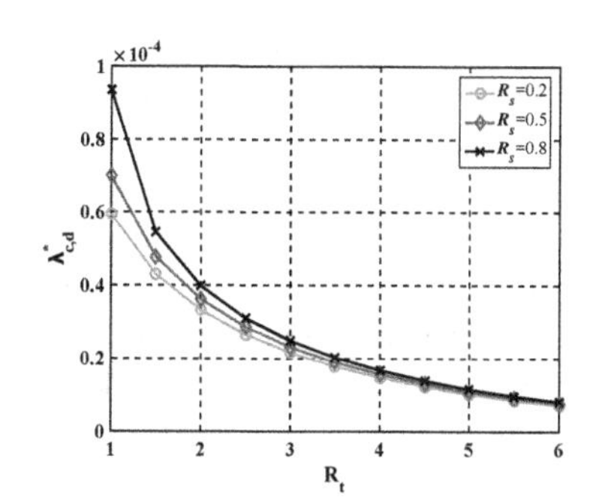

Fig. 3. $\lambda_{c,d}^*$ *v.s.* R_t.

In Fig. 2, we first analyze the impact of p_d on $\lambda_{c,d}^*$ for different transmit power of the cellular user with $R_t = 3$ bps/Hz and $R_s = 1$ bps/Hz. Obviously, a larger p_d can lead to a smaller $\lambda_{c,d}^*$. The reason is that a larger transmit power of D2D users may reduce the required number of D2D users to achieve the maximum $\mathrm{P}_{\mathrm{STP}}^{(\mathrm{N})}$. In addition, since the required number of D2D users grows with the growth of the cellular user' transmit power, $\lambda_{c,d}^*$ should increase with the increasing p_c.

Next, Fig. 3 analyzes the relationship between $\lambda_{c,d}^*$ and R_t for different R_s with $p_c = 20$ dBm and $p_d = 10$ dBm. It can be seen that $\lambda_{c,d}^*$ decreases with the increasing target secure transmit rate R_t. This is because that the restriction of the SINR at the cellular user is loosen so that the required number of D2D users is smaller. Moreover, the most detrimental Eve' SINR is limited to be smaller as the target secure transmit rate R_s increases from 0.2 to 0.8. As a result, we need more D2D users (i.e., the increasing $\lambda_{c,d}^*$) to ensure secure transmission.

Third, Fig. 4 shows the impact of $\lambda_{c,d}$ on the secure transmission probability with $p_c = 20$ dBm, $p_d = 10$ dBm, $R_s = 2.1$ bps/Hz, $R_t = 4$ bps/Hz. As we can see, for the scenario with IC, the secure transmission probability increases with the increasing of $\lambda_{c,d}$. Yet, the secure transmission probability is a unimodal function of $\lambda_{c,d}$ without IC, as proved in the Lemma 3.

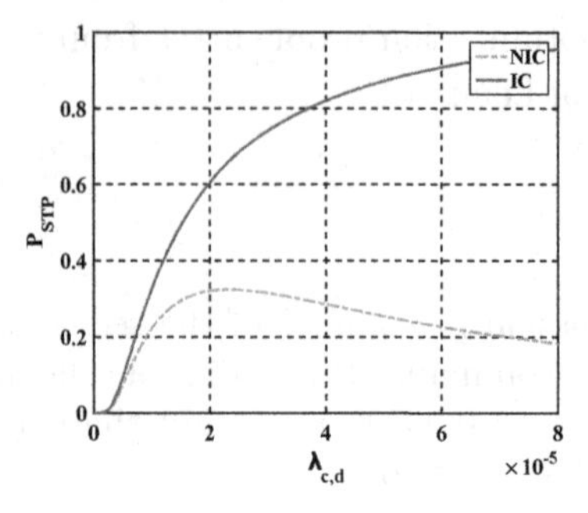

Fig. 4. $\mathrm{P_{STP}}$ *v.s.* $\lambda_{c,d}$.

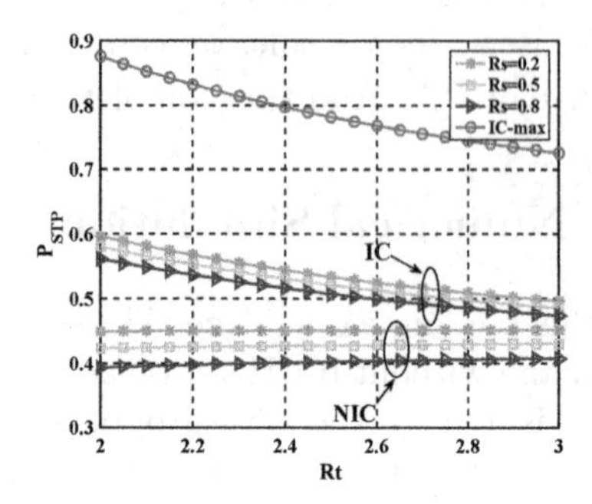

Fig. 5. $\mathrm{P_{STP}}$ *v.s.* R_t.

Finally, Fig. 5 shows $\mathrm{P_{STP}}$ versus R_t for different R_s with $p_c = 20$ dBm and $p_d = 10$ dBm. Note that "IC-max" represents the scenario with $\lambda_{c,d} \to \infty$. $\lambda^*_{c,d}$ used by the curves of IC are the same as the responding NIC. Obviously, under the scenario with IC, $\mathrm{P_{STP}}$ may decrease along with increasing R_t, which means that the data transmission rate and the security issue cannot be guaranteed at the same time. In addition, without IC, $\mathrm{P_{STP}}$ little changes with the increasing of R_t. However, under the scenario with or without IC, we can see that $\mathrm{P_{STP}}$ decreases as R_s increases regardless of whether IC is enabled or not. The reason is that a smaller P_{SC} caused by a larger R_s may lead to a smaller $\mathrm{P_{STP}}$.

5 Conclusion

In this paper, we study a secure physical transmission link for cellular users in a heterogeneous network. Considering the capability of IC for a base station, we derive the expressions of secrecy outage probability for the cellular uplink. Next, we formulate an optimization problem based on the secure transmission probability and present corresponding optimal solutions to ensure secure cellular uplink with or without IC. Numerical simulation results demonstrate that our solutions can provide secure uplink for cellular users.

Acknowledgments. This work was supported by the National Natural Science Foundation of China (Grant No. 61471028, 61572070, and 61702062), and the Fundamental Research Funds for the Central Universities (Grant No. 2017JBM004 and 2016JBZ003).

Appendix A: Proof of Lemma 1

To proof Lemma 1, we first discuss $F_{\gamma_c}(T) \triangleq 1 - \Pr^{(\mathrm{N})}(\gamma_c > T)$ based on the CDF definition. Here, $\Pr^{(\mathrm{N})}(\gamma_c > T)$ represents the coverage probability of the cellular link without IC capability, which is

$$\begin{aligned}\Pr^{(\mathrm{N})}(\gamma_c > T) &= P\left(|h_{cM}|^2 > \frac{r_{cM}^{\alpha}T}{p_c}\sum_{k\in\Phi_{c,d}} p_d\,|h_{kM}|^2\, r_{kM}^{-\alpha}\right)\\ &= \int_0^R \frac{2r_{cM}}{R^2}\mathcal{L}^{(\mathrm{N})}\left(\frac{r_{cM}^{\alpha}T}{p_c}\right)\mathrm{d}r_{cM},\end{aligned} \tag{12}$$

where $\mathcal{L}^{(\mathrm{N})}(s) = E_{\Phi_{c,d}}\left[e^{-s\sum_{k\in\Phi_{c,d}} p_d|h_{kM}|^2 r_{kM}^{-\alpha}}\right]$ denotes the Laplace transform of s for the scenario without IC capability. Specifically,

$$\begin{aligned}\mathcal{L}^{(\mathrm{N})}(s) &\overset{(a)}{=} \exp\left\{-\lambda_{c,d}\int_{R^2}\left(1-E_h\left[e^{-sp_d|h_{kM}|^2 r_{kM}^{-\alpha}}\right]\right)\mathrm{d}r_{kM}\right\}\\ &\overset{(b)}{=} \exp\left\{-2\pi\lambda_{c,d}\int_0^{\infty}\frac{sp_d r_{kM}}{sp_d + r_{kM}^{\alpha}}\mathrm{d}r_{kM}\right\}\\ &= \exp\left\{-\pi\lambda_{c,d}(sp_d)^{\frac{2}{\alpha}}\Gamma(1+\frac{2}{\alpha})\Gamma(1-\frac{2}{\alpha})\right\},\end{aligned} \tag{13}$$

where (a) follows the PDF of PPP, and (b) holds because of the definition of a surface integral. Accordingly, (12) can be rewritten as follows.

$$\Pr^{(\mathrm{N})}(\gamma_c > T) = \frac{1-e^{-\pi\lambda_{c,d}(\frac{Tp_d}{p_c})^{\frac{2}{\alpha}}\Gamma(1+\frac{2}{\alpha})\Gamma(1-\frac{2}{\alpha})R^2}}{\pi\lambda_{c,d}(\frac{Tp_d}{p_c})^{\frac{2}{\alpha}}\Gamma(1+\frac{2}{\alpha})\Gamma(1-\frac{2}{\alpha})R^2}. \tag{14}$$

Next, we can derive the PDF of the most detrimental Eve, $g(\cdot)$, based on the partial differential of the corresponding CDF, $G(\cdot)$. Assuming M as the best SINR of eavesdroppers, $G(M) \triangleq \Pr(\max\gamma_e < M)$ can be calculated as follows.

$$\begin{aligned}G(M) &= E\left[\prod_{\Phi_e}\left(1-e^{-\frac{r_{ce}^{\alpha}M}{p_c}\sum_{k\in\Phi_{c,d}} p_d|h_{ke}|^2 r_{ke}^{-\alpha}}\right)\right]\\ &\overset{(a)}{=} \exp\left\{-2\pi\lambda_e\int_0^{\infty} r_{ce}\mathcal{L}^{(\mathrm{N})}\left(\frac{r_{ce}^{\alpha}M}{p_c}\right)\mathrm{d}r_{ce}\right\}\\ &\overset{(c)}{=} \exp\left\{-\frac{\lambda_e}{\lambda_{c,d}(\frac{Mp_d}{p_c})^{\frac{2}{\alpha}}\Gamma(1+\frac{2}{\alpha})\Gamma(1-\frac{2}{\alpha})}\right\},\end{aligned} \tag{15}$$

where (c) holds by substituting $\mathcal{L}^{(\mathrm{N})}(\cdot)$.

As a result, the secrecy outage probability of CU_c can be expressed as

$$\begin{aligned}P_{\mathrm{SOP}}^{(\mathrm{N})} &= 1-\int_0^{\infty}\Pr^{(\mathrm{N})}(\gamma_c > \mathrm{Th}(\max\gamma_e))\mathrm{d}G(M)\\ &= 1-\frac{2B}{\alpha A\lambda_{c,d}^2}\int_0^{\infty}\frac{1-e^{-A\lambda_{c,d}\mathrm{Th}(x)^{\frac{2}{\alpha}}}}{\mathrm{Th}(x)^{\frac{2}{\alpha}}}\cdot e^{-\frac{B}{\lambda_{c,d}}x^{-\frac{2}{\alpha}}}x^{-1-\frac{2}{\alpha}}\mathrm{d}x,\end{aligned} \tag{16}$$

where $A = \pi(\frac{p_d}{p_c})^{\frac{2}{\alpha}}\Gamma(1+\frac{2}{\alpha})\Gamma(1-\frac{2}{\alpha})R^2$ and $B = \frac{\pi\lambda_e R^2}{A}$. And this completes the proof of Lemma 1.

Appendix B: Proof of Lemma 2

Similar to the proof of Lemma 1, we first provide the coverage probability of cellular links with IC capability follows [27].

$$\begin{aligned}\Pr^{(\mathrm{I})}(\gamma_c > T) &= P(|h_{cM}|^2 > \frac{r_{cM}^{\alpha}T}{p_c}\sum_{k\in\Phi_{c,d}} p_d |h_{kM}|^2 r_{kM}^{-\alpha}\Delta_k) \\ &= \int_0^R \frac{2r_{cM}}{R^2}\mathcal{L}^{(\mathrm{I})}(\frac{r_{cM}^{\alpha}T}{p_c})\mathrm{d}r_{cM},\end{aligned} \tag{17}$$

where $\mathcal{L}^{(\mathrm{I})}(s) = e^{-\lambda_{c,d}2\pi(\phi_1(s)+\phi_2(s)-\phi_3)}$ denotes the Laplace transform for the IC-enabled scenario. Here, $\phi_1(s) = \frac{1}{2}(sp_d)^{\frac{2}{\alpha}}\Gamma(1+\frac{2}{\alpha})\Gamma(1-\frac{2}{\alpha}), \phi_2(s) = \frac{1}{\alpha}(sp_d)^{\frac{2}{\alpha}}\Gamma(1+\frac{2}{\alpha})\Gamma(-\frac{2}{\alpha}, s\tau)$, and $\phi_3 = \frac{1}{\alpha}(\tau^{-1}p_d)^{\frac{2}{\alpha}}\Gamma(\frac{2}{\alpha})$.

As a result, the secrecy outage probability of CU_c can be derived as follows,

$$\begin{aligned}P_{\mathrm{SOP}}^{(\mathrm{I})} &= 1 - \int_0^{\infty} \Pr^{(\mathrm{I})}(\gamma_c > \mathrm{Th}(\max \gamma_e))\mathrm{d}G(M) \\ &= 1 - \frac{2B}{\alpha\lambda_{c,d}}\int_0^{\infty} P e^{-\frac{B}{\lambda_{c,d}}x^{-\frac{2}{\alpha}}} x^{-1-\frac{2}{\alpha}}\mathrm{d}x,\end{aligned} \tag{18}$$

where $P = \Pr^{(\mathrm{I})}(\gamma_c > \mathrm{Th}(x))$. And this completes the proof of Lemma 2.

Appendix C: Proof of Lemma 3

Assuming $T = 2^{R_t} - 1$ and $M = 2^{R_t - R_s} - 1$ in (14) and (15), then the transmission constraint and the security constraint can be rewritten as $P_{\mathrm{TC}} = \frac{1-e^{-\hat{A}\lambda_{c,d}}}{\hat{A}\lambda_{c,d}}$ and $P_{\mathrm{SC}} = e^{-\frac{\hat{B}}{\lambda_{c,d}}}$. Then, an extremum of P_{STP} can be computed by its first derivative of $\lambda_{c,d}$, i.e.,

$$\lambda_{c,d}^* \triangleq \arg\left\{\frac{\mathrm{d}P_{\mathrm{STP}}^{(\mathrm{N})}}{\mathrm{d}\lambda_{c,d}} = 0\right\} = \arg\left\{f(\lambda_{c,d}) = 0\right\}, \tag{19}$$

where $f(\lambda_{c,d}) = e^{\hat{A}\lambda_{c,d}} - \frac{\hat{A}\lambda_{c,d}^2}{\lambda_{c,d}-\hat{B}} - 1$. Then, We analyze the existence of the extremum and explain that the extremum is the maximum value.

Existence: Considering the properties of continuous function, we analyse two cases of $f(\lambda_{c,d})$. Because $\frac{\mathrm{d}f(\lambda_{c,d})}{\mathrm{d}\lambda_{c,d}} = \hat{A}\left[e^{\hat{A}\lambda_{c,d}} - 1 + (\frac{\hat{B}}{\lambda_{c,d}-\hat{B}})^2\right] > 0$ holds and $f(\lambda_{c,d})$ is continuous, $f(\lambda_{c,d})$ is larger than zero with $\lambda_{c,d} \in (0, \hat{B})$. Similarly, there must exists only one solution $\lambda_{c,d}^*$ to ensure $f(\lambda_{c,d}) = 0$ with $\lambda_{c,d} \in (\hat{B}, +\infty)$.

The Maximum Value: It's easy to know that $\frac{dP_{\mathrm{STP}}^{(\mathrm{N})}}{d\lambda_{c,d}} > 0$ with $\lambda_{c,d} \in (0, \hat{B}) \cap (\hat{B}, \lambda_{c,d}^{*})$. For the full range of $\lambda_{c,d}$, we still need to consider $\frac{dP_{\mathrm{STP}}^{(\mathrm{N})}(\hat{B})}{d\lambda_{c,d}} = \frac{\hat{A}}{\hat{B}} e^{-1-\hat{A}\hat{B}} > 0$. Similarly, $\frac{dP_{\mathrm{STP}}^{(\mathrm{N})}}{d\lambda_{c,d}} < 0$ with $\lambda_{c,d} \in (\lambda_{c,d}^{*}, +\infty)$.

According to the above analysis of the extremum of P_{STP}, we know that $P_{\mathrm{STP}}^{(\mathrm{N})}$ is a unimodal function of $\lambda_{c,d}$. And this completes the proof of Lemma 3.

References

1. Zheng, X., Cai, Z., Li, J., Gao, H.: A study on application-aware scheduling in wireless networks. IEEE Trans. Mob. Comput. **16**(7), 1787–1801 (2017)
2. Cheng, S., Cai, Z., Li, J., Fang, X.: Drawing dominant dataset from big sensory data in wireless sensor networks. In: 2015 IEEE Conference on Computer Communications (INFOCOM), pp. 531–539, April 2015
3. Cai, Z., Chen, Z.-Z., Lin, G.: A 3.4713-approximation algorithm for the capacitated multicast tree routing problem. Theor. Comput. Sci. **410**(52), 5415–5424 (2009)
4. Li, J., Cheng, S., Cai, Z., Yu, J., Wang, C., Li, Y.: Approximate holistic aggregation in wireless sensor networks. ACM Trans. Sen. Netw. **13**(2), 11:1–11:24 (2017)
5. Cheng, S., Cai, Z., Li, J., Gao, H.: Extracting kernel dataset from big sensory data in wireless sensor networks. IEEE Trans. Knowl. Data Eng. **29**(4), 813–827 (2017)
6. Huo, Y., Hu, C., Qi, X., Jing, T.: LoDPD: a location difference-based proximity detection protocol for fog computing. IEEE Internet Things J. **4**(5), 1117–1124 (2017)
7. Asadi, A., Wang, Q., Mancuso, V.: A survey on device-to-device communication in cellular networks. IEEE Commun. Surv. Tutor. **16**(4), 1801–1819 (2014)
8. Huo, Y., Tian, Y., Ma, L., Cheng, X., Jing, T.: Jamming strategies for physical layer security. IEEE Wirel. Commun. **25**(1), 148–153 (2018)
9. He, Z., Cai, Z., Cheng, S., Wang, X.: Approximate aggregation for tracking quantiles and range countings in wireless sensor networks. Theor. Comput. Sci. **607**(P3), 381–390 (2015)
10. Cai, Z., Lin, G., Xue, G.: Improved approximation algorithms for the capacitated multicast routing problem. In: Wang, L. (ed.) COCOON 2005. LNCS, vol. 3595, pp. 136–145. Springer, Heidelberg (2005). https://doi.org/10.1007/11533719_16
11. Cheng, S., Cai, Z., Li, J.: Curve query processing in wireless sensor networks. IEEE Trans. Veh. Technol. **64**(11), 5198–5209 (2015)
12. Lee, N., Lin, X., Andrews, J.G., Heath, R.W.: Power control for D2D underlaid cellular networks: modeling, algorithms, and analysis. IEEE J. Sel. Areas Commun. **33**(1), 1–13 (2015)
13. Huang, Y., Nasir, A.A., Durrani, S., Zhou, X.: Mode selection, resource allocation, and power control for D2D-enabled two-tier cellular network. IEEE Trans. Commun. **64**(8), 3534–3547 (2016)
14. Memmi, A., Rezki, Z., Alouini, M.S.: Power control for D2D underlay cellular networks with channel uncertainty. IEEE Trans. Wirel. Commun. **16**(2), 1330–1343 (2017)
15. Lin, X., Andrews, J.G., Ghosh, A.: Spectrum sharing for device-to-device communication in cellular networks. IEEE Trans. Wirel. Commun. **13**(12), 6727–6740 (2014)

16. Ma, C., Li, Y., Yu, H., Gan, X., Wang, X., Ren, Y., Xu, J.J.: Cooperative spectrum sharing in D2D-enabled cellular networks. IEEE Trans. Commun. **64**(10), 4394–4408 (2016)
17. Tanbourgi, R., Jakel, H., Jondral, F.K.: Cooperative interference cancellation using device-to-device communications. IEEE Commun. Mag. **52**(6), 118–124 (2014)
18. Ni, Y., Jin, S., Xu, W., Wang, Y., Matthaiou, M., Zhu, H.: Beamforming and interference cancellation for D2D communication underlaying cellular networks. IEEE Trans. Commun. **64**(2), 832–846 (2016)
19. Lv, S., Xing, C., Zhang, Z., Long, K.: Guard zone based interference management for D2D-aided underlaying cellular networks. IEEE Trans. Veh. Technol. **66**(6), 5466–5471 (2017)
20. Haus, M., Waqas, M., Ding, A.Y., Li, Y., Tarkoma, S., Ott, J.: Security and privacy in device-to-device (D2D) communication: a review. IEEE Commun. Surv. Tutor. **19**(2), 1054–1079 (2017)
21. Barki, A., Bouabdallah, A., Gharout, S., Traor, J.: M2M security: challenges and solutions. IEEE Commun. Surv. Tutor. **18**(2), 1241–1254 (2016)
22. Huang, L., Fan, X., Huo, Y., Hu, C., Tian, Y., Qian, J.: A novel cooperative jamming scheme for wireless social networks without known CSI. IEEE Access **5**, 26476–26486 (2017)
23. Zhang, R., Cheng, X., Yang, L.: Joint power and access control for physical layer security in D2D communications underlaying cellular networks. In: 2016 IEEE International Conference on Communications (ICC), pp. 1–6, May 2016
24. Wang, L., Wu, H., Liu, L., Song, M., Cheng, Y.: Secrecy-oriented partner selection based on social trust in device-to-device communications. In: 2015 IEEE International Conference on Communications (ICC), pp. 7275–7279, June 2015
25. Zhang, R., Cheng, X., Yang, L.: Cooperation via spectrum sharing for physical layer security in device-to-device communications underlaying cellular networks. IEEE Trans. Wirel. Commun. **15**(8), 5651–5663 (2016)
26. Alavi, F., Yamchi, N.M., Javan, M.R., Cumanan, K.: Limited feedback scheme for device-to-device communications in 5G cellular networks with reliability and cellular secrecy outage constraints. IEEE Trans. Veh. Technol. **66**(9), 8072–8085 (2017)
27. Ma, C., Wu, W., Cui, Y., Wang, X.: On the performance of interference cancelation in D2D-enabled cellular networks. Wirel. Commun. Mob. Comput. **16**(16), 2619–2635 (2016)

Delay-Constrained Throughput Maximization in UAV-Assisted VANETs

Xiying Fan[1,2], Chuanhe Huang[1,2(✉)], Xi Chen[1,2], Shaojie Wen[1,2], and Bin Fu[3]

[1] Computer School of Wuhan University, Wuhan 430072, China
huangch@whu.edu.cn

[2] Collaborative Innovation Center of Geospatial Technology, Wuhan University, Wuhan 430072, China

[3] Department of Computer Science, The University of Texas Rio Grande Valley, Edinburg, TX 78539, USA
bin.fu@utrgv.edu

Abstract. Efficient data dissemination in vehicular ad hoc networks (VANETs) is a challenging issue due to the high mobility of vehicles. We consider a novel mobile relaying technique by employing unmanned aerial vehicles (UAVs) to assist VANETs when the communication infrastructure is not available or network connectivity is poor. A throughput maximization problem with delay constraint is formulated to achieve high network throughput and guarantee real-time data dissemination. We reduce the graph knapsack problem to the throughput maximization problem, which is proved NP-hard. A polynomial time approximation scheme is proposed to solve the problem. Theoretical analysis that includes time complexity and approximation ratio of the algorithm is presented. The evaluation demonstrates the effectiveness of the proposed algorithm.

Keywords: VANETs · Data dissemination · UAVs
Throughput maximization · Polynomial time approximation scheme

1 Introduction

Vehicular ad hoc networks (VANETs), which aim to improve transportation safety and enable data services for in-vehicle consumption, have attracted a lot of interest [1]. Interconnected by means of vehicle-to-vehicle (V2V) communications, VANETs can exchange information about the states of vehicles and roads, as well as provide the infotainment services [2]. However, the high mobility and inherent intermittent connectivity of VANETs bring a variety of challenges, which could lead to transmission failures and affect the reliability of data transmission and routing. To deal with the issues, unmanned aerial vehicles (UAVs) can be applied to cooperate with the existing VANETs. In the areas

C. Huang—This work is supported by the National Science Foundation of China (No. 61772385, No. 61373040, No. 61572370), by National Science Foundation Early Career Award 0845376 and Bensten Fellowship of the University of Texas - Rio Grande Valley.

S. Chellappan et al. (Eds.): WASA 2018, LNCS 10874, pp. 115–126, 2018.
https://doi.org/10.1007/978-3-319-94268-1_10

where the infrastructures are difficult or too costly to install and maintain to provide ideal network coverage, UAVs can serve as a viable option, as they can collect information from an area of interest, and transmit the information to ground VANETs [3]. They can also act as relays to ground networks when direct multi-hop V2V communications are not available. In this work, we consider the cooperation of an ad hoc network of flying UAVs with existing VANETs through heterogeneous communications.

Consider the scenario where vehicles in an area of interest are demanding the same content, and a source vehicle carries the corresponding information. To complete the transmission, the data can be either transmitted over V2V links or Air-to-Ground (A2G) communication links. Aiming to improve the performance of data dissemination, we formulate the throughput maximization problem by optimizing the transmission rate of the links over which the data is transmitted.

We reduce the well-known 0/1 knapsack problem to the throughput maximization problem. Due to the property of transmission in VANETs, the problem can be regarded as the graph knapsack problem which is one of the classical NP-complete problems [4]. Then, we propose a polynomial time approximation algorithm for the graph knapsack problem based on the idea of the approximation scheme for subset sum problem [5]. Thus the throughput maximization problem can be solved and the end-to-end path with maximum throughput is obtained.

The remainder of the paper is organized as follows. Section 2 overviews the related work. Section 3 describes the network architecture and problem formulation. Section 4 proposes a polynomial time approximation scheme for the formulated problem. Section 5 presents the performance evaluation. Finally, Sect. 6 concludes the paper.

2 Related Work

Extensive research has been done to study how to develop efficient and reliable data dissemination algorithms for VANETs [6].

Tan et al. [7] proposed an analytical model to characterize the downlink average throughput and distribution achieved for each vehicle during the sojourn time by a Markov reward model. Zhang et al. [8] developed an analytical model for data delivery in real-time traffic and delay-tolerant traffic, in which the theoretical per-vehicle throughput was derived. Bharati and Zhuang [9] introduced a node cooperation based makeup strategy, then formulated an optimization problem and proposed a channel prediction scheme to select the best helper nodes. Wang et al. [10] developed a mathematical model to analyze delivery delay in a sparse highway scenario with intermittent connectivity for RSU deployment and analyzed the impact of the deployment distance between two neighbor RSUs. Zhu et al. [11] considered the critical problem of base stations for maximizing delay-constrained coverage of an urban area achieved by the vehicular network. Zhang et al. [12] presented the information propagation process in VANETs and categorized vehicles into traffic streams according to different speeds. Lin and

Rubin [13] proposed a Vehicular Backbone Network (VBN) protocol for dynamically synthesizing a networking scheme that supported message flows from a RSU to vehicles traveling along a highway. As the first to study high performance reliable transmission for bulk or stream-like data in DTNs [14], Zeng et al. proposed a dynamic segmented network coding scheme to efficiently exploit the transmission opportunity and adopted a dynamic segment size control mechanism. Xing et al. [15] studied the problem of multimedia transmission scheduling among multiple vehicles and then proposed a utility model to map the network throughput. Xing et al. [16] investigated a multimedia dissemination problem for large-scale VANETs, which took consideration the tradeoff of the delivery delay, the QoS of delivered data, and the storage cost in dropbox.

To improve the performance of VANETs in harsh environments, Zhou et al. [17] proposed multi-UAV-assisted VANETs in which UAV aerial networks were incorporated into ground VANETs and discussed the challenges. Oubbati et al. [18] studied how UAVs operating in ad hoc mode could cooperate with the ground vehicular networks and proposed a UAV-assisted routing protocol to assist data dissemination based on previous work [19]. Mozaffari et al. [20] analyzed the deployment of an UAV as a flying base station to provide the fly wireless communications to a given geographical area and derived an analytical framework for the coverage and rate analysis for the device-to-device communication network. Mozaffari et al. [21] investigated the optimal 3D deployment of multiple UAVs in order to maximize the downlink coverage performance with a minimum transmit power. Orfanus et al. [22] proposed to use of the self-organizing paradigm to design efficient UAV relay networks, to provide robust and connections to the devices on the military field.

However, most of these existing works do not consider the optimization of network throughput with delay constraint in UAV-assisted VANETs, which motivates our study.

3 Network Architecture and Problem Formulation

In this section, we first describe the network architecture, then formulate the throughput maximization problem.

3.1 Cooperative Network Architecture

To improve the reliability and efficiency of data dissemination in VANETs, we employ UAVs to form a cooperative air-to-ground network.

Most urban applications that use UAVs like small Quad-Copters do not fly at high altitudes, such as in [23]. Therefore, we assume that UAVs have a low and constant altitude during the flight in order to communicate with vehicles. IEEE 802.11p wireless interfaces with a large transmission range (i.e., up to 1000 m [24]) are assumed to be used by UAVs. In addition, assume that the selected urban area has a sufficient number of UAVs so that at each moment, the vehicles on the ground are covered by at least one UAV. Vehicles and UAVs are equipped with GPS and digital maps to obtain their current geographical positions.

Figure 1 depicts the cooperative network architecture of the UAV-assisted VANETs, which is composed of the UAV network and the ground VANETs. This scenario includes the communication links between UAVs and vehicles, which are the A2G links, and the traditional V2V links. It is a hybrid mode in which the network can use both A2G communications and V2V communications to facilitate data dissemination in VANETs.

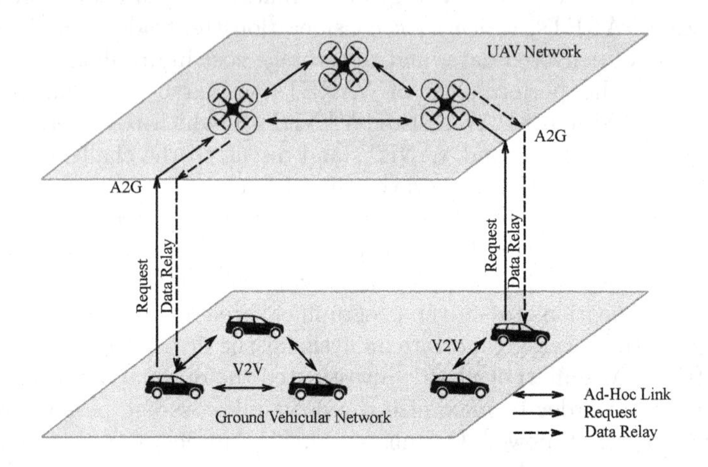

Fig. 1. Cooperative air-to-ground network

The network can be modeled as an edge-weighted graph $G(V;E)$, where V is a set of vehicular nodes and UAV nodes and E is a set of links. The weight of each edge is represented by the transmission condition of the corresponding link. We use tuple (w_e, d_e) to represent the weight of edge e, where w_e and d_e indicate the weighted transmission rate and transmission delay of e, respectively. (w_l, d_l) indicates the weight of link l.

3.2 Problem Formulation

Assume source vehicle s carries a packet, which needs to be transmitted to a number of vehicles in a specific area. A variety of end-to-end paths from vehicle s to vehicles in V may exist. The paths may only exist among vehicles over V2V links or they may contain a hybrid of A2G and V2V links. For simplicity, the A2G and V2V links are considered as common links hereinafter, the differences of the links are reflected by their weights.

As the network throughput can be mapped by transmission rate of the end-to-end path, in this study, we discuss how to optimize the transmission rate of each individual path to achieve the maximum throughput. To guarantee the real-time transmission, the end-to-end delay is limited to a predefined threshold.

We use $f(r_l) = \log r_l$ as a function of the transmission rate to depict the throughput as the logarithmic utility function $\log r_l$ can better reflect the transmission rate on a path and guarantee the maximum transmission rate. $f(r_l)$ is a continuous convex function, where l denotes a link on path p and r_l denotes transmission rate of link l.

The throughput maximization problem can be formulated as below:

$$\max \sum\nolimits_{l \in L} f(r_l) \cdot x_l$$
$$\text{s.t.} \sum\nolimits_{l \in L} d_l \cdot x_l \leq \delta$$

where $r_l \in (1, c_l)$, c_l indicates the maximum capacity of link l, d_l is the transmission delay of link l, δ is the predefined delay threshold. If link l is selected, x_l is equal to 1, otherwise, x_l is 0. Transmission delay d_l can be calculated by the following equation according to the channel model [25]: $d_l = \frac{K}{c_l - r_l}$, where K denotes the size of the packet.

4 Algorithm Design

We reduce the graph knapsack problem to the throughput maximization problem. Then we propose a polynomial time approximation algorithm to solve the graph knapsack problem.

From $d_l = \frac{K}{c_l - r_l}$, we can see that transmission delay d_l will increase when r_l increases. Since it is complex to obtain the optimal values, we calculate the approximate values for r_l, d_l. According to the equation, the range of d_l is $(K/(c_l - 1), \delta]$ and the range of r_l is $(1, c_l)$. Let r_l increase $1 + \epsilon$ each time until it reaches the largest value, where $\epsilon > 0$ is the approximation parameter. Then we get a list of values for r_l, represented as $\{1, 1+\epsilon, (1+\epsilon)^2, \cdots, (1+\epsilon)^t\}$, where $t < \log_{1+\epsilon} c_l$. Thus we have t pairs of (r_l, d_l). After $f(r_l)$, t pairs of corresponding values (w_l, d_l) are generated. The approximation ratio to obtain the approximate optimal value of r_l is $1 + \epsilon$.

As there may exist more than one path from source node s to node i, node i could have different pairs of weighted values (W_i, D_i) which indicates the total transmission rate and delay from source node s to i by adding the corresponding (w_e, d_e). Let Y denote the set of values. Assume Y is sorted into monotonically increasing order of W_i. We design a procedure *trim*() (see Algorithm 1) to remove unnecessary values of node i. The procedure scans the elements of Y in monotonically increasing order. An element is appended onto the returned list Y' only if it is the first element of Y or if it cannot be represented by the most recent values placed into Y'. The output of the procedure *trim*() described as Algorithm 1 is a trimmed, sorted list.

Given the trim procedure, we can construct our polynomial time approximation scheme for the graph knapsack problem, which is described as Algorithm 2. The approximation procedure takes as input a set $Q = \{(W_1, D_1), (W_2, D_2), \cdots, (W_n, D_n)\}$ (in arbitrary order), the delay threshold δ, and an approximation parameter ϵ. Algorithm 2 calls Algorithm 1 to trim the input list.

An approximate solution within a $1+\epsilon$ factor of the optimal solution will be returned by the scheme.

Algorithm 1. Trim (Y, δ, ϵ)

Require: Y: a list of $(W_i, D_i), \forall i \in V$;
  δ: the predefined delay threshold;
  ϵ: a real number;
Ensure: Y': a trimmed list of Y
1: let m be the length of list Y
2: $Y' = \{(W_i, D_i)\}$;
3: $last = (W_i, D_i)$;
4: **for** $j = [2, m]$ **do**
5:   **If** $D_i > \delta$
6:     Remove the tuple from list Y;
7:   **else**
8:     **If** $last_w \cdot (1+\epsilon/2n) < W_i$;
9:       append (W_i, D_i) onto the end of S';
10:       $last = (W_i, D_i)$;
11:     **else**
12:       **If** $last_d > D_i$
13:       remove $last$ from Y' and append (W_i, D_i) onto the end of Y';
14: **end for**
15: return Y'.

We have Lemma 1 to prove that the proposed scheme runs in polynomial time. Meanwhile, Theorem 1 is derived to show that there is a polynomial time approximation algorithm for the graph knapsack problem.

Lemma 1. *The algorithm for the 0/1 graph knapsack problem runs in* $O(\frac{3n^2 m \ln W^*}{\epsilon})$ *time, where* m *and* n *denote the number of edges and vertexes of graph* G*, respectively.*

Proof. We will show that the running time of the proposed scheme is polynomial in both $1/\epsilon$ and the size of the input. The first part of the algorithm runs in time $O(nm)$, since the initialization in line 1 takes $\Theta(n)$ time, each of the $|V|-1$ passes over the edges takes $\Theta(m)$ time, where $|V| = n, |E| = m$.

Now we analyze the running time of trim process. Assume W^* is the optimal weighted transmission rate of link l and $y.W < y.W^*$. After trimming, successive elements y and y' of Y have the relationship $y'.W/y.W > 1+\epsilon/2n$, that is, they differ by a factor of at $1+\epsilon/2n$. Thus, each list contains possibly the value 1 and up to $\lfloor \log_{1+\epsilon/2n} W^* \rfloor$ values. We have, the number of elements in each list Y is at most

---

Algorithm 2. Polynomial Time Approximation Scheme

Require: Q: a list of (W_u, D_u) for node u, every $u \in G.V$;
  δ: the predefined delay threshold;
Ensure: approximation solution P
1: INITIALIZE $G(V, E)$, set up the value of (w_e, d_e) for corresponding link;
2: Let $Y_u = \emptyset$ for every $u \in G.V$;
3: **for** $i = 1$ to $|G.V|$ - 1 **do**
4:   **for** each edge $e = (u, v) \in G.E$ **do**
5:     **for** each $(W_u, D_u) \in Y_u$ **do**
6:       Calculate $(W_v, D_v) = (W_u, D_u) + (w_e, d_e)$;
7:       Add (W_v, D_v) to Y_v;
8:     **end for**
9:     Trim(Y_v);
10:   **end for**
11: **end for**
12: Return P, which contains the set of links selected.

$$\begin{aligned}\lfloor \log_{1+\epsilon/2n} W^* \rfloor + 1 &= \frac{\ln W^*}{\ln(1+\epsilon/2n)} + 1 \\ &\leq \frac{2n(1+\epsilon/2n)\ln W^*}{\epsilon} + 1 \\ &< \frac{3n \ln W^*}{\epsilon} + 1\end{aligned}$$

In summary, the overall running time of the algorithm is $O(\frac{3n^2 m \ln W^*}{\epsilon})$. This bound of running time is polynomial in the size of the input n and $1/\epsilon$.

Theorem 1. *The approximation algorithm for 0/1 graph knapsack problem has an approximation ratio* $1 + \epsilon$*, and runs in* $O(\frac{3n^2 m \ln W^*}{\epsilon})$ *time.*

Proof. Let P^* denote the optimal solution of the problem. It is easily seen that $P \leq P^*$.

After trimming, successive tuples y and y' of Y' must have the relationship $y'.W/y.W > 1+\epsilon/2n$, where n indicates the number of nodes in G. Scan all the edges, find the path from source node s to a destination node u, let $v_1 v_2 \cdots v_t$ denote the path, $s = v_1, u = v_t$. From the proposed polynomial time approximation scheme, we know that a trim process is executed at the receiver node of each edge and there exists $(1+\epsilon/2n)$ factor approximation.

As to v_i, there are $i-1$ edges between s to v_i, therefore the approximation ratio should be $(1+\epsilon/2n)^{i-1}$. Then, v_i reaches v_{i+1} through edge $v_i v_{i+1}$, the approximation ratio at v_i after trim process should be $(1+\epsilon/2n)^{i-1} \cdot (1+\epsilon/2n)$, which is equal to $(1+\epsilon/2n)^i$. Since there are totally n nodes, the number of edges of the path is at most $n-1$. From the induction of the above procedure, an overall approximation ratio can be expressed as $(1+\epsilon/2n)^{n-1}$, which can be presented as below.

$$P/P^* \leq (1+\epsilon/2n)^{n-1}$$

Now, we need to show that $P/P^* \leq 1+\epsilon$. We do so by showing that

$$(1+\epsilon/2n)^{n-1} < 1+\epsilon.$$

Since $\lim_{n\to\infty}(1+x/n)^n = e^x$, we have $\lim_{n\to\infty}(1+\epsilon/2n)^{n-1} = e^{\epsilon/2}$. Since $\frac{d}{dn}(1+\epsilon/2n)^{n-1} > 0$, function $(1+\epsilon/2n)^{n-1}$ is monotonically increasing, which means the function increases with n as it approaches the limit $e^{\epsilon/2}$. So we have

$$(1+\epsilon/2n)^{n-1} \leq e^{\epsilon/2} \leq 1+\epsilon/2+(\epsilon/2)^2 \leq 1+\epsilon$$

Combine with $P/P^* \leq (1+\epsilon/2n)^{n-1}$, we have

$$P/P^* \leq (1+\epsilon),$$

the analysis of the approximation ratio completes.

5 Performance Evaluation

In this section, we first introduce the simulation settings, and then present the simulation results.

5.1 Simulation Settings

To evaluate the performance of the proposed algorithm, we implement our proposed scheme and the compared routing protocols. In the simulations, the following default settings are used.

We construct a simulation area, size of 3000 m × 4000 m. The mobility of vehicles are generated using VanetMobiSim mobility generator [26]. The speed of vehicles varies from 0 to 50 km/h. We deploy 10 UAVs and assume that by cooperating they can form a full coverage of the simulated area. The speed of UAVs varies from 0 to 60 km/h and the UAVs maintain a constant altitude that does not exceed 200 m during the flight. We use random walk mobility model for UAVs.

Extensive simulations are conducted to thoroughly investigate the efficiency of our proposed scheme in aspect of delivery ratio, delivery delay and throughput when the number of vehicles varies. We also compare the proposed scheme with two other state-of-the-art routing schemes which are VBN [13] and UVAR [18].

5.2 Simulation Results

(1) Impact of number of vehicles on delivery ratio

Figure 2 compares the delivery ratio under different number of vehicles for the compared schemes. As shown in the figure, the evaluated schemes achieve higher delivery ratio when there are more vehicles in the network. Besides, it can be seen that our proposed scheme and UVAR achieve better delivery ratio, due

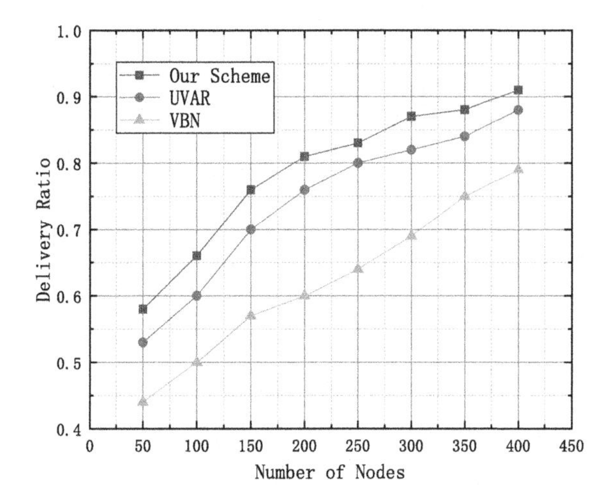

Fig. 2. Delivery ratio

to the advantage of the applied UAVs that can maintain a strong connectivity and guarantee a significant accuracy of the routing path selection. VBN mainly chooses the delivery paths based on cooperation among RSUs and vehicles, which cannot be accurate all the time and may lead to the selection of inappropriate paths, resulting in lower delivery ratios.

(2) Impact of number of vehicles on delivery delay

Figure 3 describes the impact of number of vehicles on data delivery delay. As the proposed approximation scheme executes selecting operations on each node, the

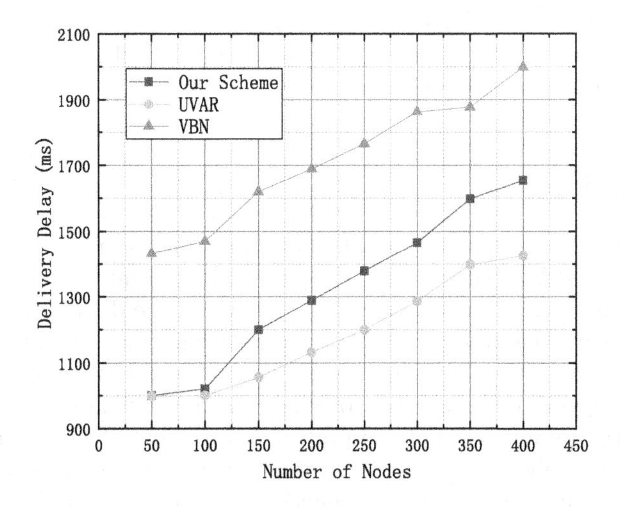

Fig. 3. Delivery delay

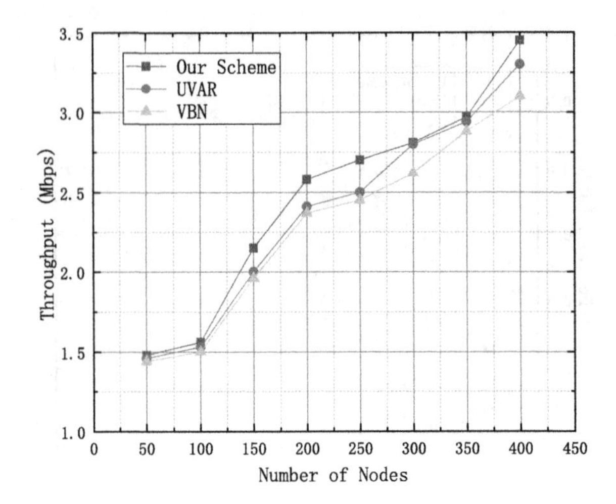

Fig. 4. Throughput

time our proposed scheme consumes is slightly higher than UVAR to complete the transmissions. VBN achieves the highest delay because it performs badly when the network connectivity is not good or when it is hard to find a relay node. It is also easy to notice that if traffic starts increasing, all the compared schemes obtains higher dissemination delay.

(3) Impact of number of vehicles on throughput

Figure 4 shows how the network throughput changes with the number of vehicle nodes. Our proposed scheme outperforms the other two routing protocols and has the highest network throughput. When number of vehicles is 50, the throughput of our scheme is 1.48 Mbps, higher than that of UVAR and VBN. The figure also shows that all the compared protocols achieve higher network throughput as the vehicle density level increases. As the number of vehicles increases to 400, the corresponding throughput of the three schemes increases to 3.45, 3.3 and 3.1 Mbps.

6 Conclusion

In this paper, we investigate efficient data dissemination in cooperative UAV-assisted VANETs. We formulate a throughput maximization problem by optimizing the transmission rate of the links over which the data is transmitted, to find the best delivery strategy for data dissemination. Then we reduce the graph knapsack problem to the throughput maximization problem and propose a polynomial time approximation scheme to achieve an approximate solution. Finally, we evaluate the performance of the proposed algorithm. The evaluation shows that the proposed algorithm is superior to other approaches. In the future work, more complex situations with UAV-vehicle communications and simulations will be considered.

References

1. Karagiannis, G., Altintas, O., Ekici, E., Heijenk, G., Jarupan, B., Lin, K., Weil, T.: Vehicular networking: a survey and tutorial on requirements, architectures, challenges, standards and solutions. IEEE Commun. Surv. Tutor. **13**(4), 584–616 (2011)
2. Cunha, F., Villas, L., Boukerche, A., et al.: Data communication in VANETs: protocols, applications and challenges. Ad Hoc Netw. **44**, 90–103 (2016)
3. Zhang, N., Zhang, S., Yang, P., et al.: Software defined space-air-ground integrated vehicular networks: challenges and solutions. IEEE Commun. Mag. **55**(7), 101–109 (2017)
4. Karp, M.R.: Reducibility Among Combinatorial Problems, pp. 85–103. Springer, Boston (1972). https://doi.org/10.1007/978-1-4684-2001-2_9
5. Cormen, H.T., Leiserson, E.C., Rivest, L.R., Stein, C.: Introduction to Algorithms, 3rd edn, pp. 1128–1133. The MIT Press, Cambridge (2009)
6. Benamar, N., Singh, K.D., Benamar, M., et al.: Routing protocols in vehicular delay tolerant networks: a comprehensive survey. Comput. Commun. **48**(8), 141–158 (2014)
7. Tan, W.L., Lau, W.C., Yue, O.C., et al.: Analytical models and performance evaluation of drive-thru internet systems. IEEE J. Sel. Areas Commun. **29**(1), 207–222 (2011)
8. Zhang, B., Jia, X., Yang, K., et al.: Design of analytical model and algorithm for optimal roadside AP placement in VANETs. IEEE Trans. Veh. Technol. **65**(9), 7708–7718 (2016)
9. Bharati, S., Zhuang, W.: CRB: cooperative relay broadcasting for safety applications in vehicular networks. IEEE Trans. Veh. Technol. **65**(12), 9542–9553 (2016)
10. Wang, Y., Zheng, J., Mitton, N.: Delivery delay analysis for roadside unit deployment in vehicular ad hoc networks with intermittent connectivity. IEEE Trans. Veh. Technol. **65**(10), 8591–8602 (2016)
11. Zhu, Y., Bao, Y., Li, B.: On maximizing delay-constrained coverage of urban vehicular networks. IEEE J. Sel. Areas Commun. **30**(4), 804–817 (2012)
12. Zhang, Z., Mao, G., Anderson, B.D.O.: Stochastic characterization of information propagation process in vehicular ad hoc networks. IEEE Trans. Intell. Transp. Syst. **15**(1), 122–135 (2014)
13. Lin, Y., Rubin, I.: Throughput maximization under guaranteed dissemination coverage for VANET systems. In: Information Theory and Applications Workshop (ITA), pp. 313–318 (2015)
14. Zeng, D., Guo, S., Hu, J.: Reliable bulk-data dissemination in delay tolerant networks. IEEE Trans. Parallel Distrib. Syst. **25**(8), 2180–2189 (2014)
15. Xing, M., He, J., Cai, L.: Maximum-utility scheduling for multimedia transmission in drive-thru internet. IEEE Trans. Veh. Technol. **65**(4), 2649–2658 (2016)
16. Xing, M., He, J., Cai, L.: Utility maximization for multimedia data dissemination in large-scale VANETs. IEEE Trans. Mob. Comput. **16**(4), 1188–1198 (2017)
17. Zhou, Y., Cheng, N., Lu, N., Shen, X.S.: Multi-UAV-aided networks: aerial-ground cooperative vehicular networking architecture. IEEE Veh. Technol. Mag. **10**(4), 36–44 (2015)
18. Oubbati, O.S., Lakas, A., Lagraa, N., et al.: UVAR: an intersection UAV-assisted VANET routing protocol. In: Wireless Communications and Networking Conference. IEEE (2016)

19. Oubbati, O.S., Lakas, A., Zhou, F., Gunes, M., Lagraa, N., Yagoubi, M.: Intelligent UAV-assisted routing protocol for urban VANETs. Comput. Commun. **107**, 93–111 (2017)
20. Mozaffari, M., Saad, W., Bennis, M., Debbah, M.: Unmanned aerial vehicle with underlaid device-to-device communications: performance and tradeoffs. IEEE Trans. Wirel. Commun. **15**(6), 3949–3963 (2016)
21. Mozaffari, M., Saad, W., Bennis, M., Debbah, M.: Efficient deployment of multiple unmanned aerial vehicles for optimal wireless coverage. IEEE Commun. Lett. **20**(8), 1647–1650 (2016)
22. Orfanus, D., de Freitas, P.E., Eliassen, F.: Self-organization as a supporting paradigm for military UAV relay networks. IEEE Commun. Lett. **20**(4), 804–807 (2016)
23. Gupta, L., Jain, R., Vaszkun, G.: Survey of important issues in UAV communication networks. IEEE Commun. Surv. Tutor. **18**(2), 1123–1152 (2016)
24. CAMP Vehicle Safety Communications Consortium: Vehicle Safety Communications Project: Task 3 Final Report: Identify Intelligent Vehicle Safety Applications Enabled by DSRC. National Highway Traffic Safety Administration, US Department of Transportation, Washington, D.C. (2005)
25. Guo, S., Dang, C., Yang, Y.: Joint optimal data rate and power allocation in lossy mobile ad hoc networks with delay-constrained traffics. IEEE Trans. Comput. **64**(3), 747–762 (2015)
26. Haerri, J., Fiore, M., Filali, F.: Vehicular mobility simulation with VanetMobiSim. Simulation **87**(4), 275–300 (2011)

Smart Device Fingerprinting Based on Webpage Loading

Peng Fang, Liusheng Huang(✉), Hongli Xu, and Qijian He

School of Computer Science and Technology,
University of Science and Technology of China, Hefei, China
{fape,cnstrong}@mail.ustc.edu.cn, {lshuang,xuhongli}@ustc.edu.cn

Abstract. Detecting devices connected to a network has become of serious importance for the network. Different devices differ in CPU scheduler, screen resolution and clock frequency, resulting in different performances when loading the same webpage. In this paper, we present a content-agnostic device identification method, a technique which decomposes webpage loading time and loads as the features to identify physical devices. This proposed method can deal with various types of devices such as mobiles, laptops, and other smart devices. We conduct experiments to evaluate the performance of the proposed method with real-world traffic. The experiment results demonstrate that the proposed method can accurately identify the types of devices from encrypted traffic and the recognition rate can reach 98.4%. To demonstrate the scalability of the method, we heuristically applied it to website identification and found that it has better effects than existing methods.

Keywords: Fingerprint · TCP · Page loading · Wireless · Website

1 Introduction

The network security evaluation and protection system which is represented by the vulnerability detection has been widely used. Device identification is the foremost step to launch network security attacks. For users, how to avoid being identified is the key to network security protection. Users must have a clear understanding of how their devices can be identified by different skills. Once attackers know the type of the target device, they can explore known vulnerabilities within applications and services running on this device, and then carry out the attack on this device. The most valuable information in the device identification process is the network traffic. More and more traffic is encrypted to protect the confidentiality of the communications [5]. However, most existing methods have the problem of feature redundancy and low recognition accuracy when dealing with the smart device identification of encrypted traffic.

There are now some powerful techniques for operating system fingerprinting [4,9,11–13], we push those techniques further and introduce the notion of smart device fingerprinting. Different devices may have different CPU schedulers,

S. Chellappan et al. (Eds.): WASA 2018, LNCS 10874, pp. 127–139, 2018.
https://doi.org/10.1007/978-3-319-94268-1_11

screen resolutions, operating systems and clock frequencies, resulting in different webpage loading schemes, including file downloading order and the caching strategy *etc.* We can employ the timing traces generated by the page loading process as features to fingerprint the devices. Due to the hardware differences between devices, the throughputs of different types of devices in loading the same webpage must be different. The identification process requires different devices to access the same page, is this condition too harsh?

As we know, more and more free WiFi covers corners around us. In general, the free WiFi requires the device to be authenticated. Authentication refers to the process of automatically opening a webpage through the default browser and requiring the user to register. The observer can recognize devices by using timing traces generated by accessing an automatic webpage. The main challenge in smart devices fingerprinting with the decomposed page loading process comes from the fact that the page loading traffic contains much noise. How to filter the captured traffic is a foremost problem. Due to the different web loading process need to download different files, so we heuristically employ the above method to identify websites, and the recognition rate can reach 95.9%. The main contributions of this paper are summarized as follows:

(1) We propose a content-agnostic smart device fingerprinting method, which extracts features by breaking down the webpage loading process. We denoise the raw data by means of TCP group reporting method. By cleaning the data, the correctness of the devices recognition is greatly improved.
(2) We propose two different feature extraction methods. The corresponding device recognition methods are called OSDF and DSDF, respectively. The principles of these two methods are also illustrated in this paper. These two methods can recognize both the smart devices and the websites.
(3) We conduct experiments to evaluate the performance of our proposed method with real-world traffic. The experiment results demonstrate that the smart devices can be identified with a high degree of correctness. This method can also determine the webpage the user is accessing.

2 Related Work

Our device fingerprinting technique is built on the basis of the operating system fingerprinting. There are now a number of powerful techniques for operating system fingerprinting. According to the way to get information, the operating system identification methods can be divided into active and passive recognition. As the passive operating system fingerprinting does not need to interact with the target host, we only introduce passive operating system fingerprinting below. The passive network analysis tools such as p0f, Ettercap, Satori and NetworkMiner [7,15,16,21], developed as parts of the Honeynet Project [20], fingerprinting the operating systems by checking TCP signatures. The biggest drawback of these tools is the need for real-time updates. We conduct experiments in the campus network using p0f for operating system identification. The recognition rate is less than 20%. Operating system versions are constantly evolving, in particular, new operating systems such as Win 8 and Win 10 have caused these tools to be unable to identify innovative operating systems effectively. So the new operating system fingerprinting methods have sprung up quickly. Many of them aim at machine learning methods. They combine existing tools and machine learning methods to identify operating systems. The correctness of recognition is much higher than that of a single tool. The most representative of the machine learning methods are Naive Bayes, Neural Networks, Decision Trees, *etc.* [1,2,14,18,19]. The above method is based on the protocol stack or traffic content to identify the operating systems, the drawback of that method is that they can not deal with the encrypted traffic, as a consequence, those methods are useless for device fingerprinting.

Kohno *et al.* proposed a remote physical device fingerprinting method which uses the clock skew as the features to identify the remote devices [9]. They employ three time factors, the first one is the time in seconds at which the measurer observed the packets, and the second one is the timestamp contained within the packet, the third one is the clock frequency. The above three factors are combined to calculate the clock skew. Their method was applied to a number of different practically useful goals, ranging from remotely distinguishing between virtual honeynets and real networks to counting the number of hosts behind a NAT. But the drawback of their method is that active fingerprinting method is needed when the timestamp is not contained in the packet by default, and the method cannot deal with the encrypted traffic.

To overcome the drawback of above method, Radhakrishnan *et al.* [17] given a passive physical device fingerprinting method-GTID. Due to different devices have different physical implementations, they employ the distribution of the inter-arrival time to fingerprint the devices. But their method has not referred to the types of the traffic. Actually, different types of traffic will cause different fingerprinting results, for example, the traffic generated by video watching and by webpage loading may cause different fingerprinting results. As the user will automatically open a webpage for identity authentication when connecting to a free WiFi, this paper employs the fact that different devices have different performances while loading the same webpage to fingerprint devices.

3 The Framework of Fingerprinting

3.1 Page Loading Scheme

If a client wants to access to an URL, the first step is to enter the correct domain name to the address bar. A request is made when a link is clicked. The browser will get the IP address of the requested page by submitting the DNS query to the superior. Then the exact IP address will be returned if there is no error. The client establishes a connection with the server through three handshakes. The browser then sends an HTTP request, the server responds to a permanent redirect which will cause the browser to track the redirect accordingly. Next, the server handles the request and returns an HTTP response. After that, the browser will display HTML content and send requests for resources embedded in HTML. Finally, the browser downloads the resources (including CSS and JS files) and renders synchronously.

The whole process will be different because of the differences between the devices. Webpage loading time can be decomposed into five parts: idle, scripting, rendering, painting and loading. An idle period refers to a period with no network activity. Different operating systems may have different idle time. The scripting represents the time of the JavaScript execution. Rendering includes the time spent on the layout, recalculation and the update of layer trees. In this step, servers can tell whether the user is on a desktop or mobile device from the user-agent. Once servers identify the device, different layouts will be designed for devices. Painting refers to the time that the contents are drawn to the screen, such as the decoding of the image. Loading refers to the time of parsing HTML. There are some differences in the above five items among various devices due to device memory, screen resolution, and processor diversity. This is the theoretical basis for identifying devices by page loading.

In order to further illustrate the feasibility of the method, we conduct experiments on the URL: www.taobao.com. What we need to notice is that the website was selected randomly from China's top 10 websites in 2016, which is not unique to our experiments. Three experiments were carried out on three devices to test the page loading scheme of different devices accessing the same webpage under the same network environment condition. The devices include iPhone 6 (iOS 10), iPad A1489 (iOS 9) and Lenovo Desktop (Windows 7 home). In order to reduce the impact of browsers, chrome is employed in three devices. The experiment process used Google's web developer tools to calculate the execution time and the number of requests for each part. The results are shown in Fig. 1. Figure 1(a) refers to the execution time of page loading. Lenovo Desktop has the longest scripting time while iPhone 6 has the shortest. Figure 1(b) represents the number of requests for different items. We can see that iPad has the largest image requests and HTML requests. iPhone 6 has the least number of requests in all projects. Different devices have different execution time and the different number of requests in all five parts. In the next section, we will show how JavaScript performs on different devices.

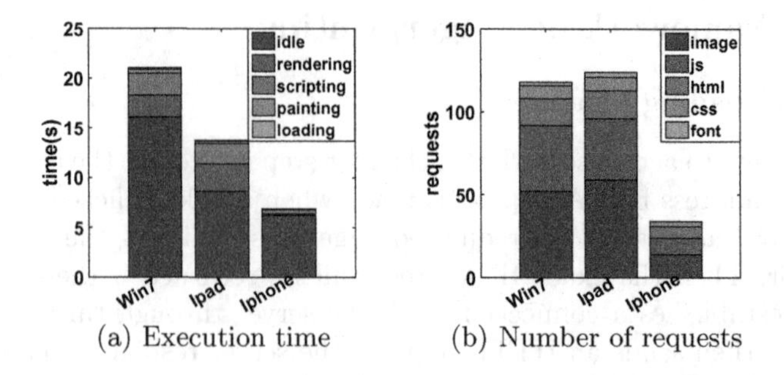

(a) Execution time (b) Number of requests

Fig. 1. The effect of three different devices on webpage loading. (a) refers to the execution time of page loading. (b) represents the number of requests.

Table 1. The execution time of JavaScript.

OS	Times (ms)									
	1	2	3	4	5	6	7	8	9	10
iPhone 6	373.8	370.4	378.4	367.2	378.0	367.9	383.4	360.6	365.8	363.1
iPad A1395	1818.6	1586.0	1578.6	1582.6	1582.4	1587.9	1603.7	1586.9	1586.5	1583.5
Lenovo Desktop	187.3	188.4	186.1	185.1	181.7	189.7	188.8	182.8	181.8	189.4
Lenovo Y430p	196.7	184.6	198.7	197.5	201.8	196.3	195.6	196.9	196.6	193.4
Kali	233.0	246.5	236.4	240.5	241.6	242.4	245.6	242.9	242.4	242.3

3.2 JavaScript Performance

JavaScript is a scripting language that has been widely used in web application development and is often used to add a variety of dynamic features to webpages to provide users with smoother and more enjoyable browsing. JavaScript is usually embedded in the HTML to achieve their functions. Given the limited processing power on devices, HTML rendering and JavaScript may become the bottleneck of the page loading. Serval factors jointly connect with the speed of the page processing, such as memory, screen resolution, OS(operating system, OS) and the browser.

The client browser interprets and executes code with JavaScripts. Scripts are downloaded from the server to the client and then carried out on the client. The process does not take up server-side resources. The client shared the task of the server through client script which significantly reducing the pressure on the server. Different devices have different JavaScript performances which also make it possible for us to identify the types of the devices. When the browser encounters the first script tag, the browser will execute the codes located between the script tags. The browser reads the contents of the code segment and parses it. Finally, the browser executes the script code. The global variables and functions defined by each script can be called by a script that is executed later.

To show the differences between different devices in JavaScript performance, we conduct experiments on different devices with the same browser-chrome. We use **SunSpider** [8] to test the performance of JavaScript of different devices with the same browser. SunSpider is a JavaScript benchmark. This benchmark tests the core JavaScript language only, not the DOM (Document Object Model, DOM) or other browser APIs (Application Programming Interface, API). It is designed to compare different versions of the same browser, as well as different browsers. We conducted 10 experiments on each device and found out the JavaScript execution time needed for different devices using the same browser. The devices include iPhone 6 (IOS 10.3.1), iPad A1395 (IOS 9.3.5), Lenovo Desktop (Win 7), Lenovo Y430p (Win 8) and Lenovo Desktop (Kali Linux 2.0). In order to make it easy to express, we remember Lenovo Desktop (Kali Linux 2.0) as Kali or Kali2. The test results are presented in Table 1. We can find that different devices have the different execution time of JavaScript. Different devices must have different throughput pattern during page loading. We will show how to use throughput as recognition features in the next section.

3.3 Feature Extraction

Different devices have different webpage loading manners. Based on this fact, we decompose the throughput generated during the page loading as the features to fingerprint smart devices. The specific implementation is as follows. We start from the DNS first frame query data and reassemble the IP fragments according to IP address and port number. We need to note that our scheme works for ICMP, UDP, and TCP protocols, but for each protocol in our tests, we only focused on one protocol without combining any protocols. Since almost all of the protocols involved in the process of loading webpage are TCP protocols, we focus on TCP connection only. A TCP connection includes three steps that are the handshake, data transmission, and the connection closure. We categorize the TCP data according to the connection. We only select the data associated with the requested URL. In order to select the right IP addresses, we need to do pre-investigation. Considering the question of whether the packets are ordered or not, we propose two methods for feature extraction.

Ordered Packet Feature Extraction. We set N different time bins to allow the packets to fall into the corresponding bins in order. Let t denotes the width of the time bin. Let s_i represent the number of bytes falling into the ith time bin and n_i denote the number of packets falling into ith time bin. For example, if we set the width of the time bin $t = 1s$, then the packets received during the $0 \sim 1s$ will be assigned to the first time bin. $S_o = ((s_1, n_1), (s_2, n_2), \ldots, (s_N, n_N))$ is the input features derived from the decomposition of the page load. To demonstrate the unique characteristics of each device, we illustrate the throughput distribution graphs for different devices accessing the same URL: http://wifi.360.cn/. The configuration information of these devices is displayed in Table 2. The compared results are presented in Fig. 2, where the width of the time bin $t = 0.1s$. We can find that different devices have different traffic distributions. Each device has the unique characteristic of periodicity. Lenovo Desktop and Laptop have the largest throughput than others while Kali has the smallest throughput.

Disordered Packet Feature Extraction. We set N different kinds of volume bins according to the size of the packets, and each type of bin has two different directions, upload, and download. Let w denotes the width of the volume bin. Let n_{i1} denote the number of upstream packets falling into the ith type of volume bin and n_{i2} represents the number of downstream packets falling into the ith type of volume bin. Thus $S_d = (n_{11}, n_{21}, \cdots, n_{N1}, n_{12}, n_{22}, \cdots, n_{N2})$ is the training features. To demonstrate the unique characteristics of each device, we illustrate the distribution graphs of volume bins for different devices accessing the same URL: http://wifi.360.cn/, where $w = 100$ *bytes*. The compared results are presented in Fig. 3. It is clear that different devices have different data packet size distribution rules when accessing the same webpage. Although the distribution graphs of iPhone6 and iPad A1395 are similar, the number of packets in their corresponding bins is different. Those discrepancies can assist us in fingerprinting physical smart devices.

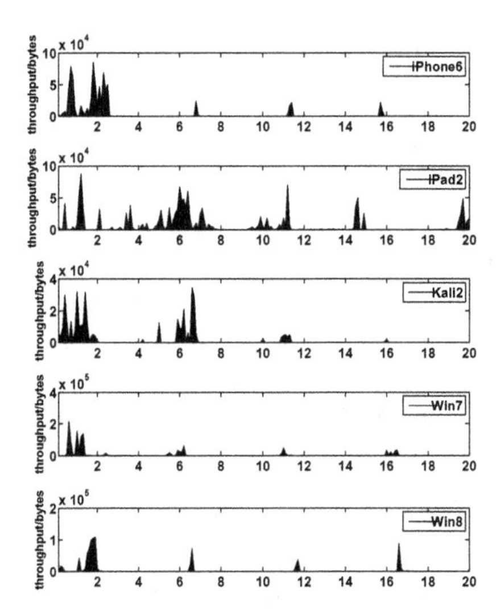

Fig. 2. The throughput of different devices. The time interval for each I/O graph is 0.1s.

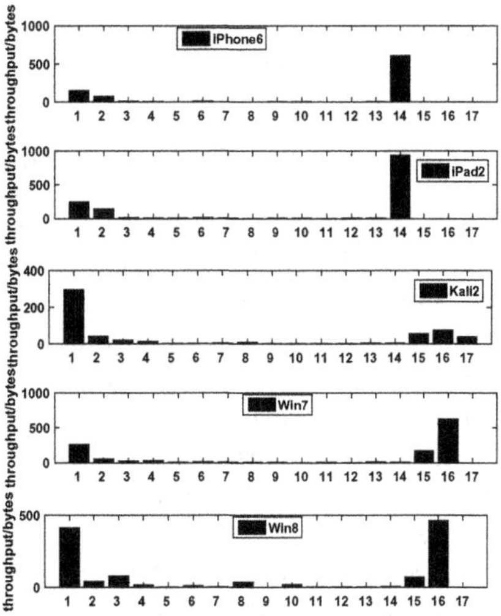

Fig. 3. Distribution of the number of packets of different sizes. The unit of the x-axis is 0.1 Kbytes.

Table 2. Specifications of devices used.

Device	Detail		
	OS	CPU	RAM
Iphone 6	iOS 10	1.4 GHz 2-core A8	1G
Ipad A1395	iOS 9	1 GHz 2-core A5	512M
Lenovo Desktop	Win 7	2.5 GHz 4-core i5	4G
Lenovo Y430p	Win 8	2.5 GHz 8-core i7	8G
Lenovo Desktop	Kali Linux 2.0	2.0 GHz 2-core	2G
ipad A1489	iOS 9	1.3 GHz 2-core A7	1G

4 Evaluation

4.1 Experiment Setup

When we connect to a free WiFi, the browser will automatically jump to a specific page. For example, welcome to connect 360 free WiFi pages. In order to make the website more illustrative, we selected an automatic connection site: http://wifi.360.cn/. If clients want to connect 360 free WiFi, they are required to access the same URL. The server can determine the types of devices based on the traffic generated by the clients accessing the webpage. We built a free 360 WiFi hotspot with a laptop, and each device can connect to our hotspot. We use WireShark to collect the traffic generated during the connection. In order to get enough traffic, each device visits the site 50 times. The details of the devices are presented in Table 2.

4.2 Content Agnostic Device Fingerprinting

In this section, we connect all the test devices to a same 360 hotspot and capture the packets with WireShark on the server side. When the device is connected, the device opens the default browser to access the same URL: http://wifi.360.cn/. The default browser is the chrome browser. We classify the captured traffic according to the connection and count the S_o and S_d. We compare three different machine learning methods: Naive Bayes, Decision Tree and Random Forest, and find that Random Forest has the best performance on our experiments. Next, we only use Random Forest as our machine learning method. We divide the data set into two randomly uniform parts, one of which is a training set and the other is a test set. According to the difference of the feature extraction process, we remember the two methods of this paper respectively as Orderly Smart Device Fingerprinting (**OSDF**) and Disorderly Smart Device Fingerprinting (**DSDF**).

Evaluation Metrics. We use three measures to quantify the detection accuracy: (i) *precision*, *i.e.*, the proportion that is currently divided into a class of devices that are properly classified. (ii) *recall*, *i.e.*, traffic from a given device are correctly detected by our scheme and (iii) $F - measure = \frac{2}{1/precision + 1/recall}$. For all three metrics, larger values indicate higher accuracy.

4.3 Compared with Existing Method

In this section, we compare our proposed method OSDF and DSDF with GTID [17]. GTID is a technique which relies on the distribution of the inter-arrival time (IAT). IAT measures the delay ($\triangle_t$) between successive packets and characterizes the traffic rate. The IAT feature vector is defined as:

$$f = \{\triangle_{t_1}, \triangle_{t_2}, \cdots, \triangle_{t_i}\} \quad (1)$$

where $\triangle_{t_i}$ is the inter-arrival time between packet i and $i-1$. Distributions capture the frequency density of events over discrete intervals of time. They define frequency count as a vector that holds the number of IAT values falling within each of the N equally spaced time bins. Then they regard the frequency count of N bins as the features to train the model.

We employ the same data set as shown in Sect. 4.1. The length of the bin is $3s$. N is selected according to the URL (http://wifi.360.cn/), that is, to ensure that the data transmission process is over. We need to note that, when generating features from the same captured traffic, different methods may have the different number of features. *i.e.*, the number of features of OSDF may be smaller than DSDF because of the width of the bin. We compare the three different methods on the test dataset, and the compared results are shown in Fig. 4.

The average F-measure of DSDF is 98.4% while GTID and OSDF are 83.9% and 90.4% respectively. Accordingly to Fig. 4, we can find that our method DSDF has the best performance over all the devices, and OSDF is in the second place. All these three methods can accurately identify the traffic generated by iPhone6, which indicates that iPhone6 has distinctive features different from other devices. Different devices have their distinct characteristics when accessing the same webpage, which easily reveals the users' privacy. This is exactly what we are going to do next, that is, to protect users' privacy through differential privacy technology. To demonstrate the scalability of the DSDF and OSDF, in the next section, we will employ these two methods to fingerprint the websites.

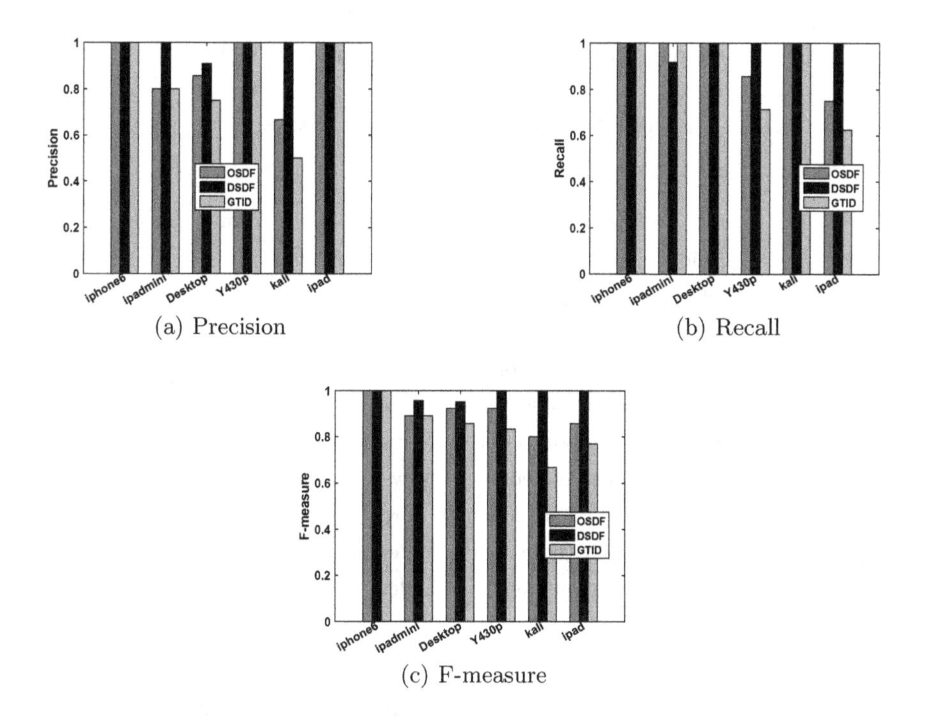

(a) Precision

(b) Recall

(c) F-measure

Fig. 4. The compared results of different devices. (a) Precision. (b) Recall. (c) F-measure.

5 Content Agnostic Website Traffic Fingerprinting

Website traffic fingerprinting technology can be used by legitimate network monitors to manage the security and health of the network. For example, in a Cross Site Scripting attack, when a user accesses a web server infected with a Cross Site Scripting malicious attack code, the malicious code is automatically downloaded and steals the user's information. By identifying the website visited by the victim, it is possible to block the access to the site to reduce the loss of the user as soon as possible. On the other hand, website traffic fingerprinting can also be abused by malicious users to track the internet behavior of legitimate users, eavesdropping the user's web browsing interest, habits, and other personal information, resulting in the disclosure of user's privacy.

Traffic analysis is the process of monitoring the properties and behavior of network traffic rather than its content [3]. In general, the content of the traffic refers to the application data transmitted in the network communication. The traffic attribute and the behavior are called the side channel information of the traffic, such as the time of the packet, the number of packets and the length of the packet. Website fingerprinting is a technique for identifying websites accessed by client users through traffic analysis technology. Although users use privacy enhancements to hide the traffic content contained in the website traffic, we can also identify the source of the traffic using traffic analysis technology.

Table 3. 2016 China's top ten websites.

Site	Detail		
	Address	Describe	Rank
Baidu	www.baidu.com	Search Engines	1
QQ	www.qq.com	News	2
Taobao	www.taobao.com	Shopping	3
Sina	www.sina.com.cn	News	4
Youku	www.youku.com	Video	5
Sohu	www.sohu.com	News	6
163	www.163.com	News	7
Hao123	www.hao123.com	Search Engines	8
Tudou	www.tudou.com	Video	9
PPS	www.pps.tv	Video	10

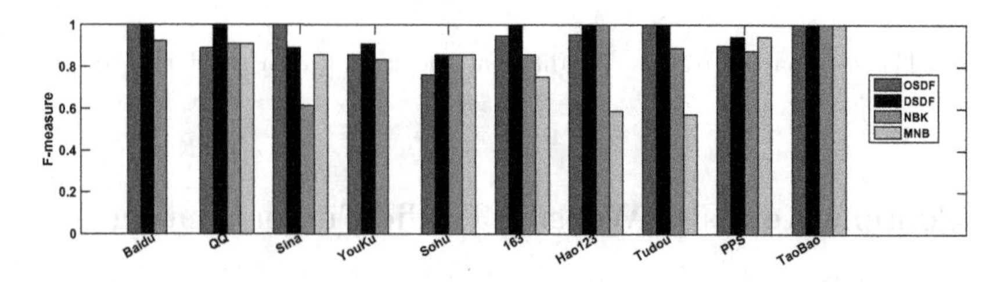

Fig. 5. The F-measure of fingerprinting different websites. The x-axis represents different websites and y-axis refers to the F-measure.

When a user visits a typical webpage, he needs to download several files, including HTML files for the webpage, images included in the page, and the referenced stylesheets. For example, if users visited Taobao's webpage at the website: www.taobao.com, they would download about sixty separate files. Each of these sixty files has a specific file size. The requested files become the fingerprint of the website. Different websites have a different number of requests and different size of files. We use these subtle features to identify the websites. We conduct experiments on 2016 China's top ten websites, and the websites are shown in Table 3. We access 10 websites and each for 10 times and use WireShark to capture the traffic. We denoise the captured traffic and perform feature extraction using the method presented in Sect. 3.3. We compare OSDF and DSDF with NBK [10] and MNB [6] method. NBK used the lengths of incoming and outgoing TCP/IP packets and discarded the order, and used Naive Bayes with normal kernel density estimation enabled. MNB can be described as an application of known text mining techniques to website fingerprinting. As before, the order of packets was discarded, and only lengths and frequencies were used. They applied a set of well-known text mining transformations to optimize their classifier-Multinomial Naive Bayes. The compared results are shown in Fig. 5.

We can find that DSDF has the best performance than other methods. Although OSDF has a low F-measure on some websites, the overall recognition accuracy rate is still high. The average F-measures of DSDF and OSDF are 95.9% and 93.1% respectively, while NBK and MNB are only 87.2% and 60.4%. The proposed method not only can quickly identify the devices in the network but also can effectively prevent the browsing of illegal, malicious websites. This is very important for the security management of wireless networks.

6 Conclusion

Different devices have different performances in webpage loading scheme. By decomposing the webpage loading process, we propose a content-agnostic device recognition method. This method uses the throughput, packet order, and the number of packets generated by the webpage loading process to establish a fingerprint for each device. To illustrate the effectiveness of the method, we identify devices in a real 360 free WiFi environment (http://wifi.360.cn/) and find that the average F-measure of device recognition can reach 98.4%. In order to show the scalability of the proposed method, we heuristically employ the above method to identify the websites. We conducted experiments on China's top ten websites and found that the average F-measure can reach 95.9%. This phenomenon indicates that our method can not only monitor the health of the network but also can be used to analyze users' online habits under legal conditions, which has potential commercial value.

Acknowledgments. This paper is supported in part by NSFC under Grant 61472383, Grant U1709217, Grant 61728207, and Grant 61472385, and in part by the Natural Science Foundation of Jiangsu Province in China under Grant BK20161257.

References

1. Al-Shehari, T., Shahzad, F.: Improving operating system fingerprinting using machine learning techniques. Int. J. Comput. Theory Eng. **6**(1), 57 (2014)
2. Beverly, R.: A robust classifier for passive TCP/IP fingerprinting. In: Barakat, C., Pratt, I. (eds.) PAM 2004. LNCS, vol. 3015, pp. 158–167. Springer, Heidelberg (2004). https://doi.org/10.1007/978-3-540-24668-8_16
3. Bruce, S.: Applied Cryptography, 2nd edn. Wiley, Hoboken (1996)
4. Chen, Y.C., Liao, Y., Baldi, M., Lee, S.J., Qiu, L.: OS fingerprinting and tethering detection in mobile networks. In: Proceedings of the 2014 Conference on Internet Measurement Conference, pp. 173–180. ACM (2014)
5. Gayle, D.: This is a secure line: the groundbreaking encryption app that will scrample your calls and messages (2013)
6. Herrmann, D., Wendolsky, R., Federrath, H.: Website fingerprinting: attacking popular privacy enhancing technologies with the multinomial naïve-bayes classifier. In: Proceedings of the 2009 ACM Workshop on Cloud Computing Security, pp. 31–42. ACM (2009)
7. Hjelmvik, E.: Passive network security analysis with networkminer. In: (IN)Secure, no. 18, pp. 1–100 (2008)
8. JetStream: Sunspider 1.0.2 javascript benchmark. https://webkit.org/perf/sunspider/sunspider.html Accessed 8 May 2017
9. Kohno, T., Broido, A., Claffy, K.C.: Remote physical device fingerprinting. IEEE Trans. Dependable Secure Comput. **2**(2), 93–108 (2005)
10. Liberatore, M., Levine, B.N.: Inferring the source of encrypted HTTP connections. In: Proceedings of the 13th ACM Conference on Computer and Communications Security, pp. 255–263. ACM (2006)
11. Lippmann, R., Fried, D., Piwowarski, K., Streilein, W.: Passive operating system identification from TCP/IP packet headers. In: Workshop on Data Mining for Computer Security, p. 40. Citeseer (2003)
12. Matoušek, P., Ryšavý, O., Grégr, M., Vymlátil, M.: Towards identification of operating systems from the internet traffic: IPFIX monitoring with fingerprinting and clustering. In: 2014 5th International Conference on Data Communication Networking (DCNET), pp. 1–7. IEEE (2014)
13. Matsunaka, T., Yamada, A., Kubota, A.: Passive OS fingerprinting by DNS traffic analysis. In: 2013 IEEE 27th International Conference on Advanced Information Networking and Applications (AINA), pp. 243–250. IEEE (2013)
14. Medeiros, J.P.S., Brito, A.M., Pires, P.S.M.: A data mining based analysis of Nmap operating system fingerprint database. In: Herrero, Á., Gastaldo, P., Zunino, R., Corchado, E. (eds.) CISIS 2009. AINSC, vol. 63, pp. 1–8. Springer, Heidelberg (2009). https://doi.org/10.1007/978-3-642-04091-7_1
15. Miłós, G., Murray, D.G., Hand, S., Fetterman, M.A.: Satori: enlightened page sharing. In: Proceedings of the 2009 Conference on USENIX Annual Technical Conference, p. 1 (2009)
16. Ornaghi, A., Valleri, M.: Ettercap. https://www.ettercap-project.org/. Accessed 28 Apr 2017
17. Radhakrishnan, S.V., Uluagac, A.S., Beyah, R.: GTID: a technique for physical device and device type fingerprinting. IEEE Trans. Dependable Secur. Comput. **12**(5), 519–532 (2015)
18. Sarraute, C., Burroni, J.: Using neural networks to improve classical operating system fingerprinting techniques. arXiv preprint arXiv:1006.1918 (2010)

19. Schwartzenberg, J.: Using machine learning techniques for advanced passive operating system fingerprinting. Master's thesis, University of Twente (2010)
20. Spitzner, L.: Know your enemy: passive fingerprinting. In: World Wide Web, March 2002
21. Zalewski, M.: p0f: passive OS fingerprinting tool. http://lcamtuf.coredump.cx/p0f.shtml. Accessed 6 May 2017

Trajectory Prediction for Ocean Vessels Base on K-order Multivariate Markov Chain

Shuai Guo, Chao Liu(✉), Zhongwen Guo, Yuan Feng, Feng Hong, and Haiguang Huang

Department of Computer Science and Engineering, Ocean University of China, Qingdao 266100, China
liuchao@ouc.edu.cn

Abstract. Trajectory prediction is a key problem in MDTN (Mobile Delay Tolerant Network). Because Vessel's moving pattern is in free space and easily influenced by the fish moratorium, tide, weather, etc., it brings new challenges in free-space vessel trajectory prediction. In addition, the trajectory characteristics of a vessel are different from that on land, causing traditional trajectory prediction method can't be directly used in ocean domain. To solve the problem above, we propose a novel trajectory prediction algorithm for ocean vessel called TPOV. We utilize k-order multivariate Markov Chain and multiple sailing related parameters to build state-transition matrixes. Through simulations and experiments on two-year trajectory data of two thousand vessels, we provide quantitative analysis of the proposed strategy. The results show that TPOV has high precision prediction with a minor error.

Keywords: Vessel trajectory prediction · Entropy analysis
Marine IoT · Ocean MDTN · K-order Markov Chain

1 Introduction

With the rapid development of IoT technology, everything in the world is going to interconnect via IoT [1]. According to Cisco's study [2], there will be 25 billion devices connected to the Internet by 2015 and 50 billion by 2020, including vessels and many other marine devices. Marine IoT has become a hot spot in the domain of marine technology, which shows huge potential research value for marine resources exploitation, environmental protection and so on. In particular scenarios, researchers apply marine IoT technologies to manage the whole ocean. For example, ocean observing system plays an important role in monitoring the ocean and forecast natural disasters. GOOS (Global Ocean Observing System) and U.S. IOOS (Integrated Ocean Observing System) are both integrated systems for observations, modeling, and analysis of heterogeneous marine

S. Guo and C. Liu—Equal Contributor.

S. Chellappan et al. (Eds.): WASA 2018, LNCS 10874, pp. 140–150, 2018.
https://doi.org/10.1007/978-3-319-94268-1_12

data. DARPA (Defense Advanced Research Project Agency) also unveils plans to create "Ocean of Things". However, there are several flaws existing in ocean data recollection and transmission: (1) limited cover of cellular signal, (2) high cost of satellite data transmission, (3) high demand for infrastructures such as submarine optical fiber cable.

Researchers are dedicated to building an efficient MDTN (Mobile Delay Tolerant Network) for ocean communication under a complex and changeable environment with few wireless infrastructures [3,4]. If trajectories can be known or predicted in advance, the delivery ratio and efficiency would be further improved [5,6]. However, vessel's mobility pattern is difficult to discover due to several reasons. Firstly, vessel's mobility could be easily limited by the fish moratorium, tide, weather and so on. Secondly, there is no central navigation device to get the destination ahead of schedule. Thirdly, there is no support of road topology (such as crossroad turning possibility, shortest path, stay point and so on), which makes vessel's trajectories hard to predict. Many trajectory prediction algorithms have been designed and applied [7–11]. However, some of them are designed for human, vehicle or river vessels, which cannot be directly adopted by ocean vessel trajectory prediction, and others are designed for short-term prediction within a small range or general mobility pattern.

In this paper, we propose a novel trajectory prediction for ocean vessel called TPOV and our main contributions of this paper are as follows.

(1) We use entropy to analyze the influence of different factors and orders on vessel trajectory uncertainty. The average entropy results can be directly used by trajectory prediction algorithm.
(2) Based on entropy results, we use location, speed, and direction to build transition matrix and 3-order Markov Chain, which can predict the trajectory for 4 h accurately.
(3) We conduct extensive trajectory-driven experiments, and the experimental results show TPOV has high precision prediction with a minor error.

The remaining of this paper is organized as follows: Sect. 2 introduces some related work about trajectory prediction. Section 3 details the design of proposed algorithm, which can predict vessel trajectory by analyzing raw trajectory data. Section 4 demonstrates the utility and efficiency of our algorithm with a set of comprehensive experiments on real datasets from the security and emergency center of Wenzhou oceanic and fishery administration in China. Finally, Sect. 5 gives a brief conclusion and possible direction of future work.

2 Related Work

In this section, we introduce some research related to our paper. Many researchers are working on trajectory prediction with high accuracy and feasibility for humans and vehicles. Song et al. research on 50000 user's base station records and use entropy to analyze them. He utilizes entropy results and Fano inequality to predict the transition pattern among base stations. The average prediction accuracy is 93% [12]. The mentioned methods did not consider other factors (such as time, weather). Zhu et al. [5] use 2-order Markov chain to predict taxies in Shenzhen and Shanghai, and verify their methods in VANETs field. But they do not consider other factors either.

Other researchers are devoted to fulfilling vessel trajectory prediction with high accuracy and feasibility. Li et al. [13] present a multi-step clustering method to find the regular vessel routes and detect abnormal trajectories, but fail to finds vessel's individuation. By using historical AIS (Automatic Identification System) data, a novel data-driven approach [14] is proposed to predict vessel's next position and time, performing well for medium time horizons ranging up to about 30 min. However, the algorithm is sensitive to the choice of certain decision factors. Perera et al. [7,8] use an artificial neural network for detecting and tracking multiple vessels and then propose an extended Kalman filter to predict vessel trajectory. To realize trajectory prediction for ocean vessels, we need to consider factors (such as location, direction, speed and so on) and utilize trajectory big data to build a personalized model for each vessel.

3 Algorithm Design

We propose a novel algorithm for predicting vessel trajectory by utilizing K-order multivariate Markov Chain. The architecture of TPOV is shown in Fig. 1, which contains four steps:

Step 1. Divide a given sea area into a grid whose cell has no overlap with each other. Each cell represents a small sea area. Instead of predicting a possible location where a vessel is, we aim at getting the cell which it may be in.

Step 2. In terms of conditional and marginal entropy, Choose vessel location, direction, and speed as key factors to analyze spatial-temporal regularity of vessel mobility.

Step 3. Calculate transition probability matrix and predict vessel trajectory by using K-order Markov Chain.

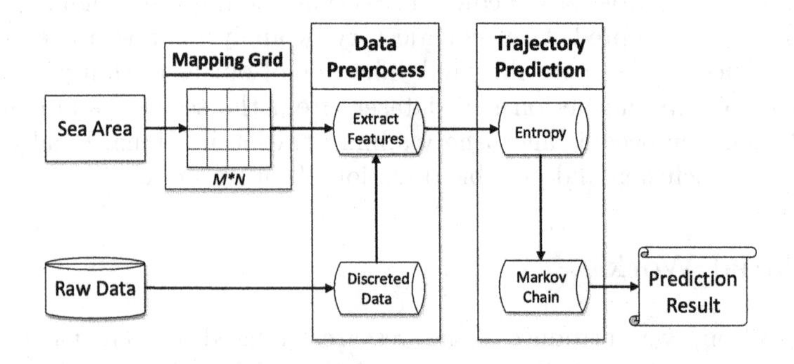

Fig. 1. Architecture of TPOV

3.1 Mapping Grid

Based on actual demand, we choose a certain sea area and separate it into an $M \times N$ grid (see Fig. 2) whose cell has no overlap with each other. Each cell represents a small area of the sea. The whole sea area S is defined as follows where s_i represents a cell:

$$S = \{s_1, s_2, \ldots, s_{(M \times N)} | s_i \cap s_j = \emptyset\} \tag{1}$$

Grid Size: Choose two points $A(x_1, y_1)$ and $B(x_2, y_2)$ in the same diagonal to determine the grid size where the vertical direction is parallel to North-South and the horizontal direction is parallel to East-West. The value of x represents longitude and y represents latitude.

Cell Size: Consider the size of each cell as a threshold θ_l, which can be calculated in a different approach. In this paper, based on vessel's average speed and sampling interval, we use $\theta_l = v \times t$, where v represents speed and t represents sampling interval, shown in Algorithm 1.

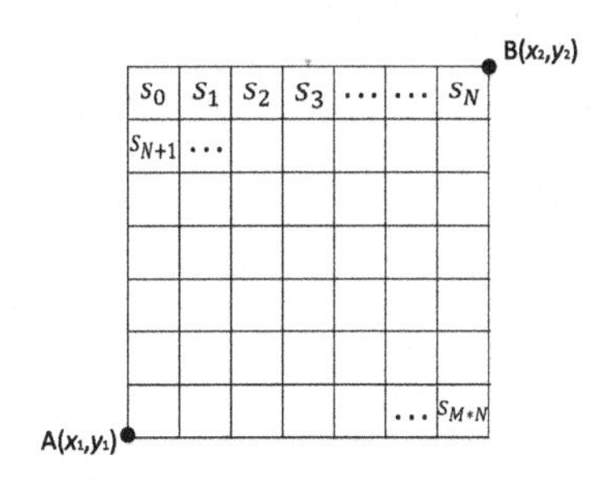

Fig. 2. Mapping grid of sea area

Algorithm 1. Mapping Grid

Input: Sea area S with point $A(x_1, y_1)$ and point $B(x_2, y_2)$
Vessel speed v and data sampling interval t
Raw trajectory data D
Output: Mapping Grid
1 Get the size of $S \leftarrow$ Gauss Map$(x_2 - x_1, y_2 - y_1)$;
2 Calculate average Speed and Sampling interval of each vessel from D;
3 Threshold $\theta_l = v \times t$;
4 Building a $M \times N$ grid S where $M = (y_2 - y_1)/\theta_l$, $N = (x_2 - x_1)/\theta_l$;
5 $S = \{s_1, s_2, \ldots, s_{(M \times N)} | s_i \cap s_j = \emptyset\}$;
6 return Mapping grid ;

3.2 Data Preprocessing

Each cell s_i represents a small part of the sea area S. Once vessel moves into s_i, its mobility can be influenced by the natural feature of this area or vessel event, which will be easily reflected by many factors such as speed and direction. We take the two factors above with location into consideration, define vessel trajectory characteristic λ as Eq. 2:

$$\lambda = \{Location,\ Speed,\ Direction\} \tag{2}$$

However, it is difficult to leverage Markov Chain to analyze these characteristics because all of them are continuous factors. Thus, we discrete those characteristics by several thresholds defined in Table 1 and preprocess raw trajectory data. Based on preprocessed data, we can also find the distribution of all the characteristics, dig correlations between each of them and consider these correlations as an important result, which can be used to predict vessel trajectory. The whole process is detailed in Algorithm 2.

Table 1. Definition of thresholds for data discretization

Factor	Threshold (discrete level)
Location	$\theta_l = v \times t.$
Speed	θ_s = speed/n
Direction	$\theta_d = 360°/n$

Algorithm 2. Data Preprocessing

```
   Input: Raw trajectory data RTD
          Thresholds: θl, θs, θd
          Grid of sea area S = {s1, s2, ..., s(M×N) | si ∩ sj = ∅}
   Output: Preprocessed data PD
 1 RTD = {t1, t2, ..., tk}, k ∈ VesselIDset;
 2 foreach k ∈ VesselIDset do
 3     sort tk by time stamp;
 4     foreach row in tk do
 5         if Gauss Map(row.loation) ∈ si then
 6             row.location ← Index of si ;
 7         end
 8         row.speed ← row.speed/θs ;
 9         row.direction ← row.direction/θd ;
10     end
11 end
12 return Preprocessed data PD;
```

3.3 Entropy Analysis

According to information theory, entropy analysis is a mathematical measure of the degree of disorder, or more precisely unpredictability in a dataset, with greater randomness implying high entropy and greater predictability implying lower entropy. Moreover, conditional entropy represents the probability of prediction ahead of knowing the previous status. We leverage entropy analysis to choose the order of Markov Chain and the key factors for vessel trajectory prediction as the following steps:

Step 1. Based on the preprocessed data from Algorithm 2, we assume $vessel_i$ has a trace t_i denoted in Eq. 3 where λ is defined in Eq. 2. M represents the number of time slots.

$$t_i = \{\lambda_0, \lambda_1, \ldots, \lambda_{(M-1)}\} \tag{3}$$

Step 2. For all the λ_j of t_i, we count how many times λ_j occurs and denote it as m_{λ_j}, where $0 \leq j \leq (L \times S \times D - 1)$.

- $L = \text{Max}(\lambda_j.\text{location})$
- $S = \text{Max}(\lambda_j.\text{speed})$
- $D = \text{Max}(\lambda_j.\text{direction})$

Step 3. In terms of frequency (m_{λ_j}/M), we calculate the marginal entropy of t_i (see Eq. 4) and extend t_i to a sequence of two tuples ${t_i}^1 = \{(\lambda_0, \lambda_1), (\lambda_1, \lambda_2), \ldots, (\lambda_{M-2}, \lambda_{M-1})\}$, count how many times $(\lambda_\phi, \lambda_\xi)$ occurs in ${t_i}^1$ and denote it as $m_{\phi,\xi}$, finally get joint $H({t_i}^1, t_i)$ entropy as Eq. 5.

$$H(t_i) = \sum\nolimits_{j=0}^{L \times S \times D-1} \frac{m_{\lambda_j}}{M} \times \log_2 \frac{1}{m_{\lambda_j}/M} \tag{4}$$

$$H(t_i^1|t_i) = \sum\nolimits_{\forall 0 \leq \phi,\xi \leq (L \times S \times D-1)} \frac{m_{\phi,\xi}}{M-1} \times \log_2 \frac{1}{m_{\phi,\xi}/(M-1)} \tag{5}$$

Step 4. Calculate conditional entropy of ${t_i}^1$ by utilizing Eq. 6.

$$H(t_i^1|t_i) = H(t_i^1, t_i) - H(t_i) \tag{6}$$

Step 5. Based on Step 1–Step 4, keep calculating the conditional entropy of ${t_i}^k$ and eventually the conditional entropy of ${t_i}^k$ is written as Eq. 7.

$$H(t_i^k|t_i t_i^1 \ldots t_i^{k-1}) = H(t_i^k, t_i t_i^1 \ldots t_i^{k-1}) - H(t_i^k, t_i t_i^1 \ldots t_i^{k-2}) - \ldots H(t_i) \tag{7}$$

In order to choose proper factors of vessel trajectory and orders of Markov Chain, we implement several experiments under each possible circumstances, shown in Figs. 3 and 4 (L represents *Location*, S represents *Speed*, D represents *Direction*), where a higher level of conditional entropy represents less uncertainty of vessel trajectory prediction. Figure 3 depicts that 3-order Markov performs better than lower-order Markov no matter what factors are utilized. In addition, with the increasing number of factors, the level of conditional entropy between 2-order Markov and 3-order Markov becomes closer but both two are getting further from 1-order Markov. Figure 4 reveals the influence of choosing proper factors on vessel trajectory prediction. It is obvious that leveraging (location, speed, direction) factors with 3-order Markov will be the best choice. In Fig. 4(a), little difference appears when using different factors and based on Fig. 4(b) and (c), (location, direction) works better than (location, speed) with higher-order Markov. In conclusion, we could get more precise prediction result by using higher-order Markov with multiple factors. However, it is unrealistic due to the consequent increase of temporal and spatial complexity for our algorithm.

Algorithm 3. Conditional Entropy Calculation

Input: Preprocessed data PD
Output: Conditional entropy
1 Group PD by same vessel ID;
2 PD $= \{t_1, t_2, \ldots, t_i\}, i \in VesselIDset$;
3 **foreach** $i \in$ *Vessel IDset* **do**
4 sort trace t_i by time stamp;
5 $t_i = \{\lambda_0, \lambda_1, \ldots, \lambda_{M-1}\}$;
6 **foreach** $\lambda \in t_i$ **do**
7 $m \leftarrow$ total appearance times of λ;
8 **end**
9 **for** j=0 *to* $L \times S \times D - 1$ **do**
10 $H(t_i) = H(t_i) + (\frac{m_j}{M} \times \log_2 \frac{1}{m_j/M})$;
11 **end**
12 extend to 1-order Markov;
13 $t_i{}^1 = \{(\lambda_0, \lambda_1), (\lambda_1, \lambda_2), \ldots, (\lambda_{M-2}, \lambda_{M-1})\}$;
14 repeat step 6–8;
15 **for** k=0 *to* $C^2_{L \times S \times D} - 1$ **do**
16 $H(t_i^1, t) = H(t_i^1, t_i) + (\frac{m_{\phi,\xi}}{M-1} \times \log_2 \frac{1}{m_{\phi,\xi}/(M-1)})$;
17 **end**
18 $H(t_i^1 | t_i) = H(t_i^1, t_i) - H(t_i)$;
19 extend step 6–18 to K-order Markov;
20 $H(t_i^k | t_i t_i^1 \ldots t_i^{k-1}) = H(t_i^k, t_i t_i^1 \ldots t_i^{k-1}) - H(t_i^k, t_i t_i^1 \ldots t_i^{k-2}) - \ldots H(t_i)$;
21 **end**
22 return Conditional entropy;

3.4 Vessel Trajectory Prediction

Assuming a K-order Markov Chain, we calculate the transition probability matrix P of each vessel under different time divisions. P represents vessels moving pattern under different conditions, which is denoted as follows:

$$P = \begin{bmatrix} p_{c^1, e^1} & \cdots & p_{c^1, e^{L \times S \times D}} \\ \vdots & \ddots & \vdots \\ p_{c^{(L \times S \times D)^k}, e^1} & \cdots & p_{c^{(L \times S \times D)^k}, e^{L \times S \times D}} \end{bmatrix} \quad (8)$$

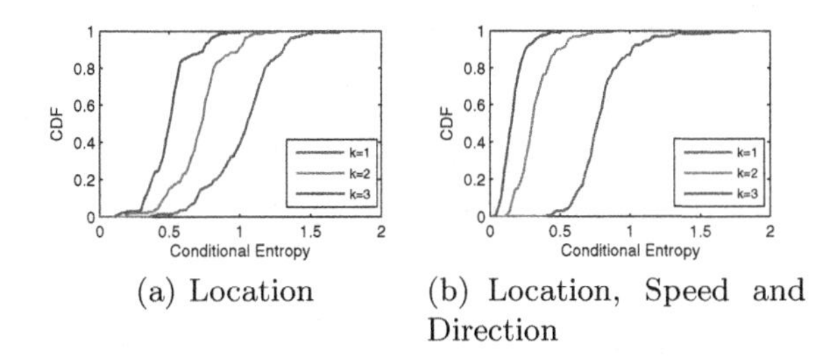

(a) Location (b) Location, Speed and Direction

Fig. 3. Conditional entropy comparison between different k with fixed factors

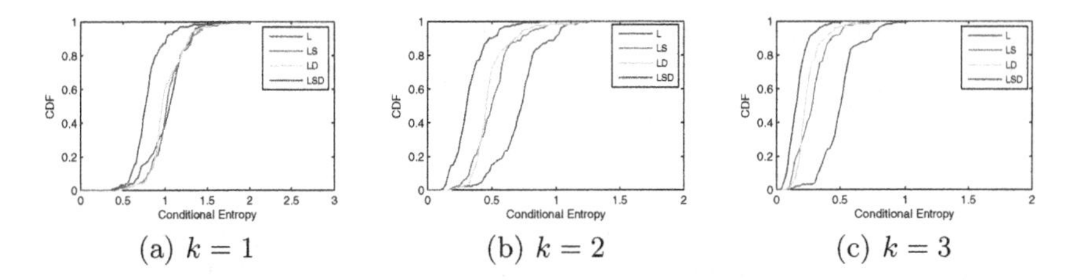

(a) $k = 1$ (b) $k = 2$ (c) $k = 3$

Fig. 4. Conditional entropy comparison between different factors with fixed k

c represents vessel current k state, e represents the next state. $p_{c^i,e^j} = M(e^j|c^i)/M(c^i)$, $M(c^i)$ represents the frequency of state c^i, $M(e^j|c^i)$ represents the frequency of current k state c^i with next state e^j.

4 Experiment Analysis and Evaluation

4.1 Experimental Trajectory Data

The experiment data in this paper are collected by the security and emergency center of Wenzhou oceanic and fishery administration in China. In order to demonstrate the utility and efficiency of TPOV, we extract more than 300 GB BDS (BeiDou Navigation Satellite System) records from January 1, 2016 to December 31, 2017, including 2116 fishing vessels. In our dataset, vessel trajectory is represented by a sequence of time slots, each of which contains the information of vessel ID, GPS_time, latitude, longitude, speed and direction (see Table 2). This trajectory dataset can be used in many other research domains, such as vessel mobility pattern mining, ocean MDTN, and ocean observation.

4.2 Mapping Grid

We choose an area of the East Sea of China. In terms of Beijing Geodetic Coordinate System (BJZ54), we leverage Gauss Map to transform GPS geodetic coordinates into Gauss geodetic coordinates and then build our own relative coordinates by utilizing point A(25.616141°N, 120.692423°E) in the bottom-left corner as the coordinate origin and point B(30.377660°N, 125.894656°E) to

Table 2. The format of trajectory data

Vessel ID	GPS_time (Linux)	Latitude	Longitude	Speed	Direction
29229	1420041972	28.8401154	122.316762	7.1	174
65644	1420041786	29.9443883	123.483638	1.1	48
... ...	... ...	... ...	... ...	... ...	... ...

determine the size of chosen sea area, choose vertical direction as x axis pointing the North and horizontal direction as y axis pointing the East. Finally, based on vessel average speed and sampling interval, we take vessel trajectory prediction error into consideration and set the cell length $\theta_l = 1$ km.

4.3 Vessel Trajectory Prediction

Based on entropy analysis result from Sect. 3.3, we choose (location, speed, direction) as key factors to continue our experiment and set thresholds. First of all, Markov Chain utilizes previous states with maximum probability to predict the next state, which makes a prediction at each time very important. However, maximum probability method can't guarantee the right prediction result. As shown in Fig. 5, there are still 17% of error ratio when using 1-order Markov. But with the higher order of Markov, error ratio decrease by 50% from 1-order Markov to 2-order Markov and even more than 70% to 3-order Markov.

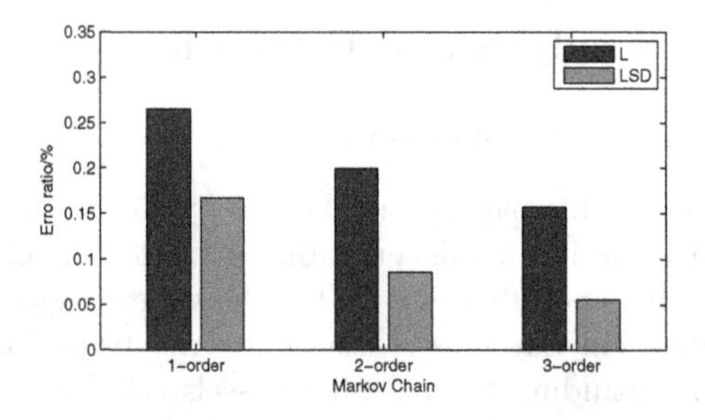

Fig. 5. Error ratio of Markov choosing maximum probability state at each time

In the end, we choose 3 vessels and compare their original trajectory with predicted trajectory (see Fig. 6). As described in Sect. 3.1, we leverage the center of the belonging cell as the predicted location, which can be optimized by considering the direction and the optimized results are shown in Fig. 7.

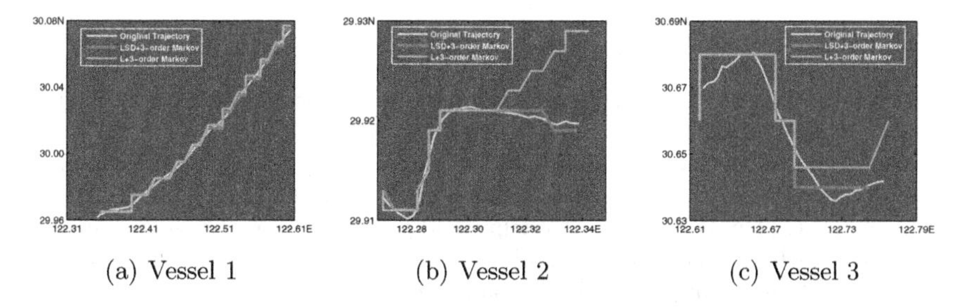

(a) Vessel 1 (b) Vessel 2 (c) Vessel 3

Fig. 6. Comparison between original trajectory and predicted trajectory under 3-order Markov with different factors

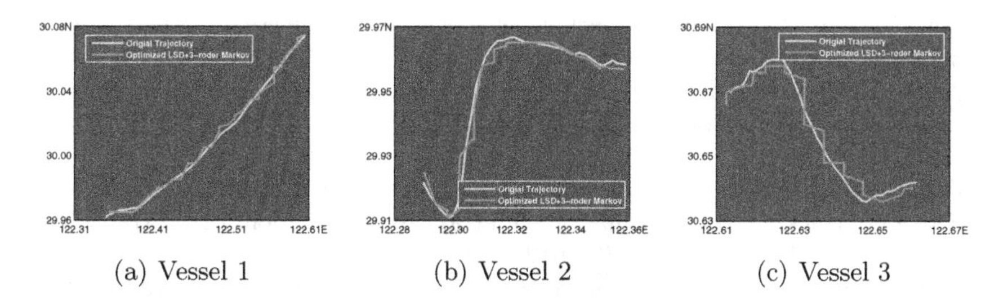

(a) Vessel 1 (b) Vessel 2 (c) Vessel 3

Fig. 7. Comparison between original trajectory and predicted trajectory under 3-order Markov with factor Location after optimized

5 Conclusion and Future Work

In this paper, we propose a novel trajectory prediction for ocean vessel called TPOV by utilizing K-order multivariate Markov Chain. Firstly, divide a given sea area into a grid whose cell has no overlap with each other. Secondly, considering many factors like weather, tide and fish moratorium, choose vessel's location, direction, and speed as key factors to analyze spatial-temporal regularity of vessel mobility. In terms of entropy analysis, we choose proper factors and order of Markov for our algorithm. Finally, calculate transition probability matrix and predict vessel trajectory by using K-order Markov method. For future work, we will implement more experiments to find better thresholds automatically and analyze statistics distributions of the preprocessed data. Furthermore, we will apply TPOV to solve ocean MDTN problems.

Acknowledgements. S. Guo and C. Liu contributed equally to this work and should be regarded as co-first authors. This work was supported by the National Key R&D Program 2016YFC1401900, the China Postdoctoral Science Foundation 2017M620293, the Fundamental Research Funds for the Central Universities 201713016, Qingdao National Labor for Marine Science and Technology Open Research Project QNLM2016ORP0405, and Natural Science Foundation of Shandong No. ZR2018BF006.

References

1. Internet of Things (IoT): A vision, architectural elements, and future directions. Future Gener. Comput. Syst. **29**(7), 1645–1660 (2013)
2. Evans, D.: The Internet of Things: how the next evolution of the internet is changing everything. Cisco Internet Bus. Solutions Group (IBSG) **1**, 1–11 (2011)
3. Shi, Y., Li, H., Du, W.C., et al.: Modeling and performance analysis of marine DTN networks with nodes-cluster in an ad hoc sub-net. In: International Conference on Computer Engineering and Information Systems (2016)
4. Liu, C., Guo, Z., Hong, F., et al.: DCEP: data collection strategy with the estimated paths in ocean delay tolerant network. Int. J. Distrib. Sens. Netw. **2014**(1), 155–184 (2014)
5. Zhu, Y., Wu, Y., Li, B.: Trajectory improves data delivery in urban vehicular networks. IEEE Trans. Parallel Distrib. Syst. **25**(4), 1089–1100 (2014)
6. Jeong, J., Guo, S., Gu, Y., et al.: Trajectory-based data forwarding for light-traffic vehicular ad hoc networks. IEEE Trans. Parallel Distrib. Syst. **22**(5), 743–757 (2011)
7. Perera, L.P., Oliveira, P., Soares, C.G.: Maritime traffic monitoring based on vessel detection, tracking, state estimation, and trajectory prediction. IEEE Trans. Intell. Transp. Syst. **13**(3), 1188–1200 (2012)
8. Perera, L.P., Soares, C.G.: Ocean vessel trajectory estimation and prediction based on extended Kalman filter. In: International Conference on Adaptive and Self-Adaptive Systems and Applications, pp. 14–20 (2010)
9. Yuan, C., Li, D., Xi, Y.: Campus trajectory forecast based on human activity cycle and Markov method. In: IEEE International Conference on Cyber Technology in Automation, Control, and Intelligent Systems, pp. 941–946. IEEE (2015)
10. Wang, B., Hu, Y., Shou, G., Guo, Z.: Trajectory prediction in campus based on Markov chains. In: Wang, Y., Yu, G., Zhang, Y., Han, Z., Wang, G. (eds.) BigCom 2016. LNCS, vol. 9784, pp. 145–154. Springer, Cham (2016). https://doi.org/10.1007/978-3-319-42553-5_13
11. Asahara, A., Maruyama, K., Sato, A., et al.: Pedestrian-movement prediction based on mixed Markov-chain model. In: ACM Sigspatial International Symposium on Advances in Geographic Information Systems, ACM-GIS 2011, Proceedings, 1–4 November 2011, Chicago, IL, USA, pp. 25–33. DBLP (2011)
12. Song, C., Qu, Z., Blumm, N., et al.: Limits of predictability in human mobility. Science **327**(5968), 1018–1021 (2010)
13. Li, H., Liu, J., Liu, R.W., et al.: A dimensionality reduction-based multi-step clustering method for robust vessel trajectory analysis. Sensors **17**(8), 1792 (2017)
14. Hexeberg, S., Flaten, A.L., Eriksen, B.O.H., et al.: AIS-based vessel trajectory prediction. In: International Conference on Information Fusion, pp. 1–8 (2017)

Experimental Study on Deployment of Mobile Edge Computing to Improve Wireless Video Streaming Quality

Li-Tse Hsieh[1], Hang Liu[1(✉)], Cheng-Yu Cheng[1], Xavier de Foy[2], and Robert Gazda[3]

[1] The Catholic University of America, Washington, DC 20064, USA
liuh@cua.edu
[2] InterDigital Canada, Ltée, Montreal, QC H3A 3G4, Canada
[3] InterDigital Communications, Inc., Conshohocken, PA 19428, USA

Abstract. Due to the rapid increase in wireless network traffic, especially video traffic, innovative network architectures and algorithms need be developed to reduce congestion and improve the quality of service (QoS). Multi-access edge computing or mobile edge computing (MEC) is a new paradigm that integrates computing and storage capabilities at the edge of the wireless network. In this paper, we design and implement a wireless access network-aware video streaming system based on the MEC concept, called Edge-Controlled Adaptive Streaming (ECAS). ECAS employs in-network video bitrate adaptation to improve data delivery efficiency. In our design, a MEC function intercepts HTTP requests from the client, and a rate adaptation mechanism is employed to decide the best video representation for the client. Updated HTTP requests, with modified video rates based on the adaptation decision, are forwarded to the video server. We also design a resource allocation and streaming bitrate adaptation algorithm to achieve the overall optimization of multiple video streams, with fairness, subject to wireless transmission capacity constraint. This network-assisted adaptive streaming approach allows more accurate estimation of wireless network states and can enhance the QoS of multiple video streams. A prototype is implemented to prove the concept, and the experimental results demonstrate that the proposed scheme significantly improves video streaming performance compared to the de facto standard video streaming technique, Dynamic Adaptive Streaming over HTTP (DASH).

Keywords: Multi-access Edge Computing (MEC) · Mobile edge computing
Wireless access network · Software Defined Network (SDN)
Dynamic Adaptive Streaming over HTTP (DASH) · In-network process

1 Introduction

Multimedia content, especially video traffic, has been growing at a rapid pace. It is anticipated that global mobile video traffic will increase 11-fold from 2016 to 2021, reaching 23 exabytes per month and accounting for 75% of total mobile data traffic by 2021 [1]. Innovative network architectures and protocols are needed to improve the

S. Chellappan et al. (Eds.): WASA 2018, LNCS 10874, pp. 151–163, 2018.
https://doi.org/10.1007/978-3-319-94268-1_13

efficiency and quality of service (QoS) of data delivery. The trend of combining advanced information technology and communications is creating unprecedented opportunity for innovation when designing network-centric services. The rapid growth in cloud computing is an outcome of such integration. However, the traditional cloud computing model that depends on centralized data centers has many limitations. For example, a mobile device may experience varying wireless connectivity due to fading, mobility, interference, and wireless traffic load, it is difficult to efficiently adapt traffic (video coding rate) to respond to quick changes in wireless network conditions using far-away data centers. To address these limitations, a new model, multi-access or mobile edge computing (MEC) [2–6] has recently gained attention in both academia and industry, which integrates computing and storage capabilities at the edge of the network, near mobile users, to provide real-time radio-aware services and enhance user experience. MEC can be deployed to process data in the wireless access networks, e.g. adapt video rates to changing network conditions, to improve the QoS of video streaming.

Dynamic Adaptive Streaming over HTTP (DASH) [7, 8] is the de facto standard video streaming technique and has been widely deployed by many video service providers such as YouTube, Netflix, etc. DASH enables streaming media content to the clients from conventional web servers using HTTP and TCP protocols, and it employs a client-based adaptive bitrate streaming strategy to adapt to the changing network conditions. The content, e.g. a movie or the live broadcast of a sports event is broken into a sequence of small HTTP-based file segments, each containing a short interval of content. Each of the content segments is encoded to a variety of alternative representations at different qualities and bitrates. While the current content segment is being played by a DASH client, the client automatically selects the next segment to download and play from alternative representations (e.g. resolution, bitrate, etc.) based on current network conditions, device capability, and user preference. The DASH-enabled client needs to read a Media Presentation Description (MPD) file before requesting the content segments, which describes the information of different representations of segments including timing, URL, video resolution, and encoded bitrates. It runs an adaptation algorithm that estimates the available bandwidth and selects the next segment with the bitrate and quality that can be downloaded in time for playback without causing stalls or re-buffering events under the current network conditions and is able to be handled by the device.

Such a client-based approach has several drawbacks. First, it is challenging for a client to obtain an accurate estimate of the available bandwidth in the network, especially in dynamic wireless network environments due to time-varying fading, shadowing, and interference. Existing DASH streaming rate adaptation schemes based on playback buffer occupancy and probing often cause picture freezing, loading interruptions, and long wait times. Second, each client performs an independent estimation and rate adaptation without information of the other users sharing the same network so that it cannot adjust its adaptation strategy cooperatively with the other users while it encounters network bandwidth fluctuations. This makes it difficult to achieve overall system optimization and fairness. Although TCP can achieve a long-term average throughput fairness, there is unfairness of available bandwidth among the clients during a time scale of streaming multiple content segments and instability in the content delivery arises. In addition, the abrupt changes in TCP window size due to packet loss cause unnecessary fluctuations of video quality. Finally, client-based rate adaptation cannot immediately detect the changes in network conditions, which leads to buffer underrun and video stalls.

In this paper, we design and implement a wireless-aware video streaming system based on the MEC concept, called Edge-Controlled Adaptive Streaming (ECAS), which employs in-network video bitrate adaptation to enhance the QoS of multiple video streams. In our design, in-network video processing and wireless transmission are integrated under a unified software-defined networking (SDN) platform to provide more flexible programmability. Instead of performing resource-consuming video transcoding, the MEC function intercepts the HTTP requests from the clients, and a rate adaptation algorithm is employed to decide the best video representations for the clients. The requests with modified video rates based on the adaptation decision are forwarded to the video server. We also design a resource allocation and streaming bitrate adaptation algorithm to achieve the best quality of service for multiple video streams with fairness, subject to wireless transmission capacity constraints. This network-assisted approach eliminates the limitations of the client-side adaptation in DASH because the MEC function has the best knowledge about the available network resources and can achieve the overall optimization of multiple video streams. Further, it is compatible with standard DASH without changes in clients and video servers. A prototype is implemented to prove the concept and the experimental results demonstrate that the proposed scheme significantly enhances video streaming quality compared to the de facto standard DASH streaming technique.

The remainder of the paper is organized as follows: Sect. 2 describes the system design and discusses the function of each component. Section 3 presents the resource allocation and streaming bitrate adaptation algorithm. Section 4 describes the prototype implementation. Section 5 demonstrates the experimental results and, finally, Sect. 6 concludes the paper.

2 System Design

2.1 System Architecture

Our Edge-Controlled Adaptive Streaming (ECAS) system design takes on a network-centric approach, instead of a client-side approach. Network operators may deploy dedicated micro-clouds or cloudlets in the wireless access networks or integrate the computation and storage hardware with base stations (BSs), access points (APs), backhaul switches, and routers, which can perform in-network customized computations to support various applications with wireless network awareness and low latency at the edge.

Figure 1 illustrates the system diagram of ECAS. Mobile users connect to the wireless access network through WiFi or cellular radios. There is no requirement to change the clients and video servers. The clients who attempt to access the videos send standard HTTP requests to the video server. The AP/BS is enhanced to report measured wireless network parameters, such as the link data rate of each mobile user and current bandwidth usage to the ECAS controller. The AP/BS also receives configuration and resource allocation policy commands from the controller. A SDN switch will route all the HTTP requests from the clients to the MEC server. The MEC server will adjust the requested video quality/bitrate and modify the requests according to the instruction of the Resource Optimizer that runs a resource allocation and bitrate adaptation algorithm to determine the best quality video representations for the clients based on the available wireless network bandwidth and the resources demanded by the clients. It minimizes the local network congestion and serves a stable streaming video flow to each user fairly to achieve the overall optimization of multiple video streams. The modified requests are forwarded to the video server through the switch. The video server responds with the corresponding video segments to the clients. This in-network processing is transparent to the clients.

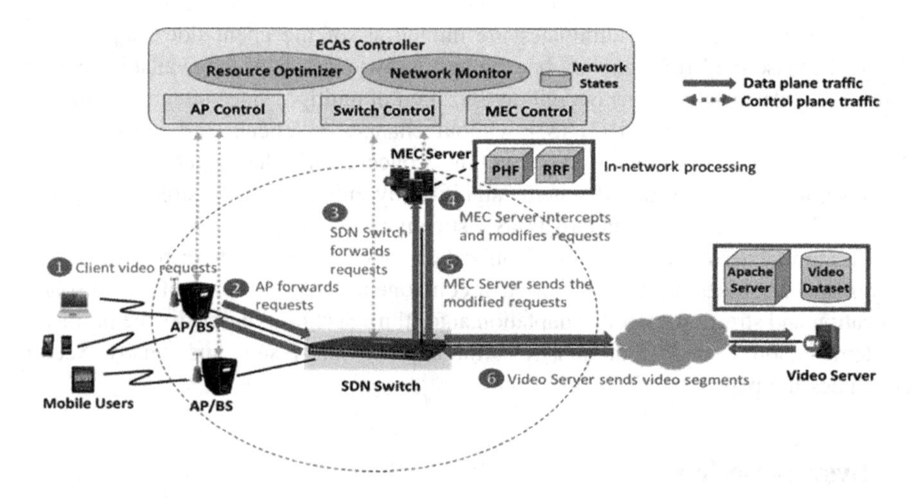

Fig. 1. ECAS system architecture.

In our design, wireless transmission, traffic routing, and in-network processing are managed by an ECAS controller. Extending the SDN concept [9–11], the controller takes a unified, network-wide view to generate configurations and policy rules for traffic forwarding as well as in-network services. The bandwidth, computation and storage resources can be abstracted and controlled. The network devices, such as routers, switches, APs/BSs, and MEC servers, have separate control and data interfaces. The control interfaces are used to communicate with the controller, reporting resource usage and service status, and receiving commands from the controller. The controller also has an open interface for external applications that allows third-party application/content providers to lease resources and deploy their own services. In this way, the data plane and control plane are decoupled, and in-network services can be flexibly programmed with software-based implementations on general-purpose hardware platforms. Such a smart computation-capable and programmable access network can provide data forwarding and in-network processing to optimize video delivery.

2.2 Key Elements

Enabling dynamic adaptation requires available wireless network resources to be monitored, traffic to be correctly steered, adaptation decisions to be made, in-network processing to be performed, integrity of content and protocol continuity to be maintained post adaptation, and flexible service interfaces to invoke these services.

Access Point: To support in-network processing and software-defined networking, the AP is enhanced with two modules, AP agent and wireless network monitor. The AP agent is responsible for wireless connection control and configuration functions. It interacts with the ECAS controller and enforces the wireless resource allocation, e.g. the maximum channel time of a user flow, based on the instruction from the controller. To reduce the controller load, the MAC function is split between the controller and the AP. The controller provides the high-level policy such as the maximum share of channel time that can be used to transmit a data flow. The AP will implement the schedule to transmit data using standard wireless protocols such as IEEE 802.11 based on the resource allocation policy. The wireless network monitor measures and reports user and network conditions to the controller, for instance, the number of the users, their channel conditions and data transmission rates.

SDN Switch: Standard SDN switches are used to support data forwarding with flexible programmability. The controller configures the rules in the SDN switch's forwarding table in order to forward the HTTP requests from the clients to the MEC server for further deep packet inspection and in-network processing, and the processed HTTP video requests are sent to the video server.

MEC Server: The MEC server acts as the gateway for the client communication with the video server, and it executes two main functions, packet holding function (PHF) and request rewriting function (RRF). The former intercepts and inspects the HTTP header. It redirects the HTTP video request packets to the RRF and forwards non-video HTTP packets to the SDN switch. The RRF maintains a video adaptation table that contains the representation and bitrate information assigned to each video flow. It uses the video adaptation table to adjust the video bitrates, i.e. the corresponding video representation in the requests from the clients and forward the modified requests to the video server. The video adaptation table can be set and updated by the resource optimizer in the controller through a control interface. The request rewriting function also forwards the HTTP requests to the ECAS controller through the control interfaces. The resource optimizer at the controller will then know the video segment representation and corresponding bitrate requested by the client. It can make wireless resource allocation and streaming bitrate adaptation decision and update the video adaptation table based on the resources demanded by the clients and the resources available in the wireless access network. This design clearly decouples the data plane and control pane, and provides flexible programmability, for example, different resource allocation and bitrate adaptation algorithms can be used.

ECAS Controller: The ECAS controller will monitor and manage all the wireless bandwidth and computing resources in the wireless access network. It includes three main functions, network monitor, resource optimizer, and SDN controller. The network monitor collects and abstracts the network resource and traffic information, including wireless channel conditions and data rate to each client, requested video rates, etc. through the control interfaces with the other network elements. The resource optimizer runs an algorithm to allocate network resources and determine video bitrates and quality for each stream based on the network and traffic demand information obtained from the network monitor. The SDN controller in ECAS provides network control functions and is enhanced from the standard SDN controller. It can control and configure not only the forwarding tables in the switches but also the MAC policy in the AP and video rate adaptation table rules in the MEC server. Given the capability of modern computing devices, as well as the potential scale of a wireless access network, the ECAS controller can be hosted in the MEC server or a separate compute cluster to address single-point failure and scalability concerns. Further, the system can be extended to multiple APs and the functions such as inter-AP communication and user handover are supported through the ECAS controller.

3 Resource Allocation and Rate Adaptation

As discussed above, the network-centric approach of our ECAS system design overcomes the limitations of DASH client-based streaming bitrate adaptation, which shifts the task of resource allocation and streaming adaptation to the component having the best knowledge about the network conditions and achieves overall system optimization while maintaining compatibility with existing HTTP infrastructure such as video web servers and DASH clients.

The flexible and programmable ECAS architecture allows employing various resource allocation and streaming bitrate adaptation algorithms. In this section, we present a simple algorithm that allocates wireless network resources and determines the bitrate for each stream to optimize overall performance with fairness when delivering multiple video streams to multiple clients that experience variable wireless channel quality and heterogeneous user demands.

Consider N video streams are sent to the clients. Adaptive modulation and forward error correction schemes are typically used in data transmissions by the AP [12, 13]. The link transmission rate for a user depends on its signal-to-noise ratio (SNR). For example, for 802.11 g [14], the user data rate will be 6, 9, 12, 18, 24, 36, 48, or

54 Mbps, dependent upon the channel SNR value. Assume that the data transmission rate of a video stream i to a client is r_i. Let α_i $(0 \leq \alpha_i \leq 1)$ denote the channel utilization, i.e., the share of time that the AP uses the wireless channel to transmit video stream i. Then the throughput of stream i is $T_i = \alpha_i r_i$. The data transmission should meet the wireless channel utilization constraint, that is,

$$\sum\nolimits_{i=1}^{N} \alpha_i \leq 1. \quad (1)$$

The client requested rate for video stream i is S_i^d, and the video rate after in-network adaptation is S_i^t. Then we have

$$S_i^t < S_i^d \quad \textit{if}\; x_i = 1\,(\text{video rate modified}), \text{and} \quad (2)$$

$$S_i^t = S_i^d \quad \textit{if}\; x_i = 0\,(\text{video rate not modified}) \quad (3)$$

where x_i is a variable to indicate whether the requested streaming bitrate is modified by the in-network adaptation function or not. Let δ denote the protocol overhead, which includes the video segment header, the lower layer headers and overhead to transmit video data, then, the required bandwidth to transmit the video stream is

$$T_i = \alpha_i r_i = \delta S_i^t. \quad (4)$$

As shown in Algorithm 1, if there is enough wireless bandwidth to transmit all N video streams at the quality each client desires or requests, the HTTP requests from the clients will be forwarded to the video server with no in-network rate adjustment and the clients will receive the video with their requested video bitrates and quality. If there is not enough wireless bandwidth to transmit the N video streams at the quality that the users requested, the algorithm will lower the video quality of a single stream at a time by one level, starting from the stream with the worst wireless link data rate because it requires the most channel time to transmit a given amount of data. The algorithm checks whether there is enough wireless bandwidth to transmit all the video streams each time a video stream rate is reduced. If not, it will lower the stream with the next worse link data rate by one level. The process continues. Once the quality of all the streams has been lowered by one level, if the wireless bandwidth is still insufficient to transmit the video streams, the algorithm begins another round, starting from the stream with the lowest link data rate and following with streams in the order of increasing link rate. If the resolution of all the streams are lowered to the minimum available quality but they still cannot be supported, some streams will be dropped, starting from the one with the worst link rate. This process continues until the quality of the video streams that can be delivered with the available wireless bandwidth are discovered.

Algorithm 1: Wireless Resource Allocation and Streaming Bitrate Adaptation

% N: The total number of video streams
% S_i^d: Desired video rate, i.e., the rate requested by the client for video stream i
%S_i^t: Data rate after wireless resource allocation and bitrate adaptation (*min*: the data rate required to provide the minimum video quality; 0: video stream i is dropped).
%r_i: Link data rate between AP and user i
% δ: Protocol and header overhead
%α_i: Channel utilization of stream i
% x_i: Video rate modified ($x_i = 1$) or not ($x_i = 0$)

1 *for* $i = 1{:}N$ % *Initialization*
2 $\alpha_i = \delta\, S_i^d / r_i;\ S_i^t = S_i^d; x_i = 0$
3 *end*
4 *If* $\sum_{i=1}^{N} \alpha_i \le 1$ *then*
5 *break;* % *no rate adjustment*
6 *else*
7 *sort the users' link data rate, i.e.,* $r_i \le r_j,\ if\ i < j;$
8 $i = 1;$
9 *while (1)*
10 *if* $S_i^t \neq min$ *then*
11 *reduce stream i bitrate by one level,*
12 *stream i bitrate adjusted and* $x_i = 1$
13 *else if* $(S_j^t == \min for\ j \ge i)\ \&\&\ ((S_j^t == 0\ for\ j < i) || i == 1))$
14 *stream i dropped,* $S_i^t = 0$ *and* $x_i = 0$
15 *end*
16 $\alpha_i = \delta\, S_i^t / r_i;$
17 *If* $\sum_{i=1}^{N} \alpha_i \le 1$ *then*
18 *break;*
19 *else* $i++;$
20 *if* $i > N$ *then* $i = 1;$
21 *end*
22 *end*

This algorithm attempts to fairly distribute the available wireless bandwidth to the users. It does not reduce any stream bitrate by two levels until all other streams have been lowered by one level. The algorithm's primary goal is to ensure as many users as possible to receive video, and its secondary goal is to provide the best video quality possible within the wireless bandwidth constraint.

4 Prototype Implementation

The ECAS system includes: (a) new modules in the APs that implement the wireless channel monitoring functions and SDN control interfaces, (b) a MEC server that implements the functions to process the HTTP requests from the clients and adjust the video bitrates in the HTTP requests along with the interfaces for configuring and running in-network processing, (c) a SDN switch that forwards HTTP requests to the MEC server for in-network processing and sends the modified HTTP requests to the video server based on the rules set up by the controller through the control interface, (d) a controller that implements the functions to collect network information, allocates resources, and controls the policies for wireless transmission, traffic routing, and in-network streaming bitrate adaptation. In this section, we will present our prototype implementation.

The AP is a Soft AP [15] running on an Ubuntu v14.04.5 host with Intel Core i5-4590 3.3 GHz CPU and 8G memory. A wireless network monitor module is implemented which records the information of associated clients and their RSSI and link data rates. The wireless network monitor periodically reports the network status to the ECAS controller through the AP agent. The AP agent implements the configuration, reporting and control interfaces and is integrated with Soft AP manager. The Soft AP manager fully implements IEEE 802.11 access point management and authentication, as well as DNSMasq that acts as DNS and DHCP server.

A Pica8 P-3297 48 x 1G SDN switch [16] is used to connect the AP and MEC server. It runs PicOS, an open network OS with OpenFlow 1.4 [17] and Open vSwitch (OVS) 2.0 [18] to support standard layer 2/3 protocols. The MEC server is hosted on a computer with Intel Core i5-4590 3.3 GHz CPU, 16 GB memory and Ubuntu v14.04.5. The packet holding function and packet rewriting function are implemented based on Squid v3.3 [19] transparent proxy server. The clients will not notice that the video requests were processed while the MEC server intercepts and modifies the HTTP requests. The MEC server has two network interfaces, the inbound interface used to intercept the clients' original HTTP requests, and the outbound interface used to send out the modified HTTP requests to the video server.

The ECAS controller is also hosted on a computer with Intel Core i5-4590 3.3 GHz CPU, 16 GB memory and Ubuntu v14.04.5. It co-locates a SDN Controller that runs Floodlight and controls the Pica8 switch. The network monitor, resource optimizer, and interfaces to control the AP and MEC server are implemented by Python. The network monitor module keeps listening on the TCP port reserved for communication with the AP and waiting for the periodical client reports. The reports are processed and then stored in a network status file. The resource optimizer reads the network status file and runs the algorithm to allocate the wireless resource and determine the bitrate for each stream. The control interfaces are implemented for communications between the resource optimizer module and the MEC server. We use VLC 3.0 DASH clients [20] running on Mac laptops that are connected to AP through built-in WiFi interface. The video server is Apache version 2.0 [21] hosted on a machine with Intel Core i5-4590 CPU, 16 GB memory and Ubuntu 14.04.5 version.

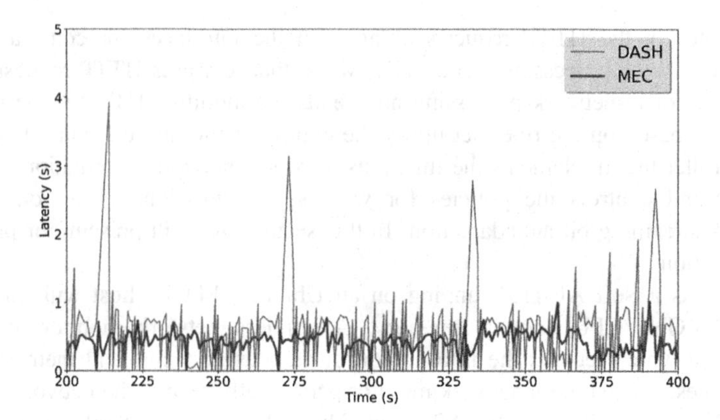

Fig. 2. End-to-end response time of MEC-based in-network streaming bitrate adaptation and client-based DASH. (Color figure online)

5 Experimental Evaluation

We used the above prototype to evaluate the MEC-based in-network rate adaptation service for a DASH video streaming application. DASH video segments generated from a movie trailer are used in the experiments. Six lengths (1, 2, 4, 6, 10, and 15 s) of segments are available and each of the segments is encoded at 15 to 20 representations with bitrates from 50 Kbps to 8000 Kbps giving different video resolutions and qualities.

We first compare the performance of MEC-based in-network rate adaptation and client-based DASH. In the test, three users simultaneously access video. For DASH, we use the default rate adaptation algorithm in VLC v3.0. For in-network rate adaptation, the MEC server obtains the video segment representation information by downloading the MPD files from the video server. It uses the above resource allocation and bitrate adaptation algorithm to calculate and distribute the network resource and determines the video bitrates and the corresponding video segment to be requested based on the measured bandwidth constraints.

We tested and compared the end-to-end response time. The end-to-end response time is defined as the duration from the time when the client sends a request to the time when the client receives the complete video segment. It is an important metric to measure the quality of DASH video streaming. When the segment arrival time is later than the expected playback time due to the large end-to-end response time, the video playback will stall. Although video stalling and picture freezing can be compensated by a larger initial playback delay and video buffer, the users will have to wait for a longer time to start video playing. We simulated the wireless network dynamics by changing the allowed bandwidth on the AP wireless interface connecting to the three clients with the Soft AP manager. The Soft AP manager provides an interface to manage total network bandwidth, and the wireless network monitor at the AP will report the network status changes to the ECAS controller. The wireless bandwidth is first designated at 9000 kbps for 30 s, and then lowered to 1800 kbps for 30 s to simulate the bad channel conditions that users may encounter. The bandwidth was changed to 7500 kbps for another 30 s. After that, the bandwidth was restored to 9000 kbps. This cycle was repeated. Figure 2 shows the results of playing a 450-second-long video. The red line represents the DASH scenario and the blue line represents the MEC-based in-network video adaptation. It is evident from the large peaks of the red line in Fig. 2 that the response time of DASH increases significantly when encountering large fluctuations in the network bandwidth. The increase in response time is due to the long time it takes for the DASH client to determine the change in network status and adjust its requested video bitrate. In contrast, with in-network video adaptation, the network monitor collects wireless network and traffic information in real time and can provide more accurate and faster rate adaptation. The resource optimizer lowers the bitrates of multiple video streams fairly as the available wireless transmission capacity decreases, which reduces the peak end-to-end latency and jitter. It mitigates the problems in client-based DASH and optimizes overall system QoS.

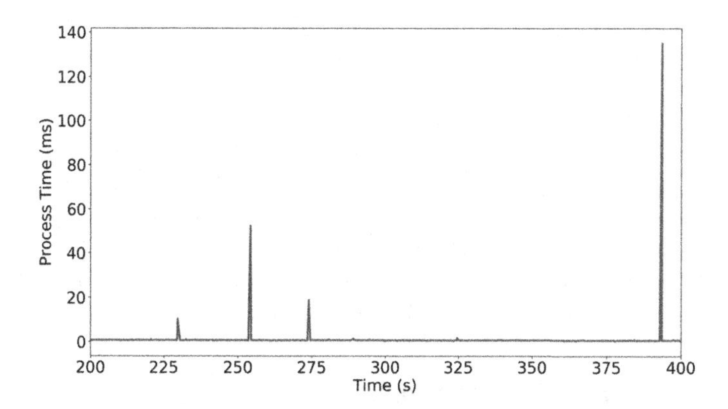

Fig. 3. Process time in the MEC server.

Figure 3 shows the processing latency in the MEC server when three users watch video under varying network conditions. The MEC processing latency is the total duration from the time when the HTTP request arrives at the inbound interface of the MEC server to the time the request is sent out at the outbound interface, including HTTP request interception, buffering, processing and rewriting time. The average latency is 1 ms. There are the peaks in Fig. 3 when the wireless network bandwidth fluctuates. This is because the resource optimizer needs to change the video bitrate adaptation table and the HTTP requests need to be modified to adapt to the new network conditions. The peak process time is less than 140 ms which shows a good latency performance.

Figure 4 shows the end-to-end response time when the clients request video segments with different bitrates and the MEC in-network streaming bitrate adaptation scheme is used. Note that if there is enough wireless bandwidth to transmit the video streams at the quality requested by the clients, no in-network bitrate adjustment is performed at the MEC server, and the HTTP requests from the clients are forwarded to the video server. The clients receive the video with their requested video quality and bitrates. If there is not enough wireless bandwidth to transmit the video streams at the quality that the users requested, the MEC server performs in-network rate adaptation as described before. As expected, when the clients request higher quality of video with higher bitrates and larger segment sizes (the green line), the responses time is increased compared to the case lower bitrate and quality video is requested (the blue and red lines).

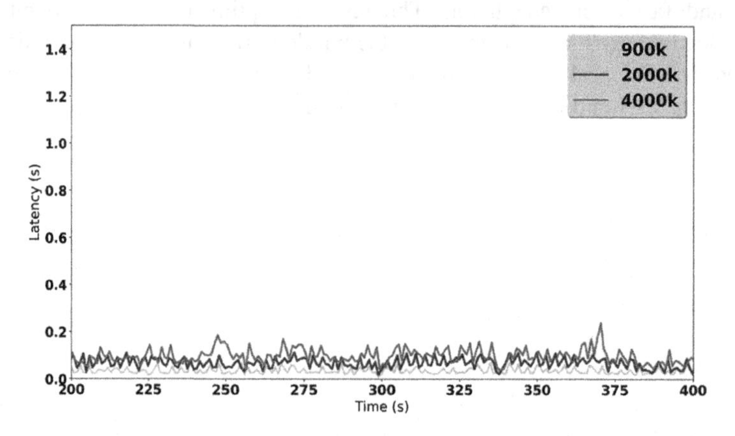

Fig. 4. End-to-end response time with different requested video encoding rates for the MEC-based in-network streaming bitrate adaptation scheme. (Color figure online)

6 Conclusions and Future Work

In this paper, we have designed ECAS, an in-network bitrate adaptation system for dynamic adaptive video streaming based on the MEC concept. We also implemented a prototype that integrates a MEC server, a SDN switch, an enhanced AP, video clients and a video server with an ECAS controller that provides a unified transmission and in-network processing control and orchestrates resource allocation, wireless transmission, traffic routing, and steaming bitrate adaptation. A resource allocation and in-network bitrate adaptation algorithm is designed to achieve the overall QoS optimization for multiple streams based on wireless network conditions and user traffic demands. The experimental results show that the proposed ECAS system significantly reduces the end-to-end latency and jitter of multiple video streams under dynamic network conditions compared to the de facto standard video streaming technique, DASH. As part of our future research, we will conduct more experiments on larger networks and investigate scalability issues.

References

1. Cisco Visual Network Index: Global Mobile Traffic Forecast Update 2016–2021 (2017)
2. Liu, H., Eldarrat, F., Alqahtani, H., Reznik, A., de Foy, X., Zhang, Y.: Mobile edge cloud system: architectures, challenges, and approaches. IEEE Syst. J. **99**, 1–14 (2017)
3. Patel, M., Naughton, B., Chan, C., Sprecher, N., Abeta, S., Neal, A.: Mobile-edge computing introductory technical white paper. Mobile-Edge Computing (MEC) Industry Initiative (2014)
4. ETSI Industry Specification Group (ISG) for Multi-access Edge Computing (MEC). http://www.etsi.org/technologies-clusters/technologies/multi-access-edge-computing
5. Tran, T.X., Hajisami, A., Pandey, P., Pompili, D.: Collaborative mobile edge computing in 5g networks: new paradigms, scenarios, and challenges. IEEE Commun. Mag. **55**(4), 54–61 (2017)
6. Ahmed, A., Ahmed, E.: A survey on mobile edge computing. In: Proceedings of International Conference on Intelligent Systems and Control, pp. 1–8 (2016)
7. ISO/IEC DIS 23009-1.2: Dynamic adaptive streaming over HTTP (DASH)
8. ETSI 3GPP TS 26.247: Transparent end-to-end packet-switched streaming service (PSS); Progressive Download and Dynamic Adaptive Streaming over HTTP
9. Banikazemi, M., et al.: Meridian: an SDN platform for cloud network services. IEEE Commun. Mag. **13**(2), 120–127 (2013)
10. Jain, S., et al.: B4: experience with a globally-deployed software defined WAN. In: Proceedings of Sigcomm, Hong Kong (2013)
11. Gudipati, A., Perry, D., Li, E., Katti, S.: SoftRAN: software defined radio access networks. In: Proceedings of ACM HotSDN (2013)
12. Biaz, S., Wu, S.: Rate adaptation algorithms for IEEE 802.11 networks: a survey and comparison. In: Proceedings of IEEE Symposium on Computers and Communications (2008)
13. Pavon, J., Choi, S.: Link adaptation strategy for IEEE 802.11 WLAN via received signal strength measurement. In: Proceedings of IEE ICC (2003)
14. IEEE 802.11 Wireless LAN Medium Access Control (MAC) and Physical Layer (PHY) Specifications (2012)
15. Soft AP. https://github.com/oblique/create_ap
16. Pica8 P-3297 datasheet. http://www.pica8.com/documents/
17. McKeown, N., et al.: OpenFlow: enabling innovation in campus networks. ACM SIGCOMM Comput. Commun. Rev. **38**(2), 69–74 (2008)
18. OVS configuration guide. http://www.pica8.com/wp-content/uploads/2015/09/v2.9/html/ovs-configuration-guide/
19. Squid proxy server. http://www.squid-cache.org/
20. VLC media player. https://nightlies.videolan.org/
21. Apache HTTP server. https://httpd.apache.org/

An Efficient Privacy-Preserving Data Aggregation Scheme for IoT

Chunqiang Hu[1,2(✉)], Jin Luo[1], Yuwen Pu[1], Jiguo Yu[3], Ruifeng Zhao[4], Hongyu Huang[5], and Tao Xiang[2,5]

[1] School of Big Data and Software Engineering, Chongqing University, Chongqing, China
{chu,thoreluo,yw.pu}@cqu.edu.cn

[2] Key Laboratory of Dependable Service Computing in Cyber Physical Society, Ministry of Education, Chongqing University, Chongqing, China
txiang@cqu.edu.cn

[3] School of Information Science and Engineering, Qufu Normal University, Jining, China
jiguoyu@sina.com

[4] Electric Power Dispatching and Control Center of Guangdong Power Grid Co. Ltd., Guangzhou, China
ruifzhao@126.com

[5] College of Computer Science, Chongqing University, Chongqing, China
hyhuang@cqu.edu.cn

Abstract. Internet of Things (IoT) provides the most flexibility and convenience in our various daily applications as the IoT devices can improve efficiency, accuracy and economic benefit in addition to reduced human intervention. However, security and privacy challenges are also arising in IoT. To address this challenge, in this paper, we present a privacy-preserving data aggregation scheme for IoT to preserve privacy of customers. In our scheme, the IoT devices slice their actual data, keep one piece to themselves, and send the remaining pieces to other group devices via symmetric key. Then each IoT device adds the received slices and the held piece together to get immediate result, which should be sent to the server after the computation. Finally, analysis shows that our scheme can guarantee the integrity, confidentiality of IoT device's data, and can resist external attack, internal attack and collusion attack and so on.

1 Introduction

The development of Internet of Things (IoT) has changed our lifestyle greatly by providing the most flexibility in our various daily applications, which include smart healthcare [13], smart grid [1,2,5,15], smart home [29], smart city [10], smart shopping [23], social network [4,9,18] and etc. Although the IoT applications have their characteristics in their own fields, they are essentially the same, i.e., any IoT application is formed by a number of IoT devices, which can not only collect real time data but also exchange the data to achieve better

S. Chellappan et al. (Eds.): WASA 2018, LNCS 10874, pp. 164–176, 2018.
https://doi.org/10.1007/978-3-319-94268-1_14

decisions. For example, in smart grid application, smart meters are deployed in houses. Each smart meter can record users' usage data in real-time and report them to an operation center in certain frequency, e.g. every 20 min, which can help the operation center get nearly real-time situational awareness, manage power distribution and adjust electricity price during peak periods.

Clearly, IoT will provide great benefits to us. However, it will also bring us many security and privacy issues [14,20,21,32]. An attacker can analyze the behaviors and habits of people by eavesdropping the transmissions between IoT devices and server or compromising the server [17,19,25]. According to the real-time usage data, attackers can monitor a person without any advanced tools [6,11,24]. For example, a burglar can know a person's lifestyle by monitoring his electrical consumption in smart grid applications [12,16]. Another instance is that an attacker can get the related records of a patient via analyzing the real-time records in smart healthcare application [17]. Therefore, it is a great challenge to protect users' privacy in the IoT applications. There are same issues in other IoT applications such as smart home, smart city and so on. To address the above challenge, several privacy preserving data aggregation schemes have been proposed [28,31]. However, most of them either focus on protecting the privacy of one single side only, namely the IoT devices, or leak intermediate results during communication to potential privacy attackers, or cannot resist collusion attacks. For example, the collaborating IoT devices could get the actual data of one IoT device in [6]. Individual's privacy can be possibly at risk via exposing the intermediate information.

In this paper, we propose an efficient privacy-preserving data aggregation scheme for IoT. In our scheme, we divide data collected by IoT devices into many parts randomly instead of adding noise signal from trusted third party (TTP). Then, each IoT device preserves one part (which is only known by himself) and exchanges the rest parts with others. By this way, all the collected data have been hidden before reported. Data Aggregator aggregates the received data without knowing each individual's actual data.

The remainder of this paper is organized as follows. In Sect. 2, we introduce the related work. In Sect. 3, we present our system model, security requirements, and the design goal. In Sect. 4, we introduce the Paillier cryptosystem. Then, we present our scheme in Sect. 5, which is followed by security analysis, performance evaluation in Sects. 6 and 7, respectively. Finally, we draw our conclusions in Sect. 8.

2 Related Works

Many existing privacy preserving data aggregation schemes for IoT applications have been developed. In this section, we mainly summarize state-of-the-art privacy preserving data aggregation schemes as follows.

Fan *et al.*'s scheme presents a secure data aggregation scheme by introducing an offline trusted third party, which can resist internal attackers [7]. However, it cannot meet data integrity requirement [3]. Thai *et al.*'s scheme preserves the privacy of usage data by adding the jamming signals, which are designed to be cancelled at Aggregator [30]. However, the data may be manipulated by a malicious attacker during communication. Saputro *et al.*'s scheme prevents usage data from revealing by using homomorphic encryption [27]. It can permit Aggregator to aggregate the usage data without decrypting. However, the scheme cannot resist internal attack. To overcome the issues, Li *et al.*'s scheme [22] proposes an efficient privacy-preserving demand aggregation (EPPDR) scheme by using homomorphic encryption to preserve users' privacy data. Although EPPDR has improved efficiency in terms of computation and communication overheads, it still brings much computation pressure and storage pressure to smart meters.

In summary, existing privacy preserving data aggregation scheme for IoT application can not provide privacy protection as either the private data of the user or the intermediate information may be disclosed. Moreover, most of them cannot resist collusion attacks. In this paper, we propose an efficient and practical data aggregation scheme in which collected data are confused to preserve users' privacy.

3 System Model, Security Requirements, and Design Goal

In this section, we formalize the system model, security requirements, and identify our design goals.

3.1 System Model

In our system model, we mainly focus on how to report the collected data of IoT devices to the server in a privacy-preserving way. We consider a two-level gateway topology in IoT as shown in Fig. 1, which includes three types of entities: The Server, Aggregator, and Devices. We assume that the Server covers m Aggregators, and that each Aggregator covers n IoT devices.

1. **Server:** The server provides space for IoT devices to store the collected data that can be retrieved by the customers. It is a trustable and powerful entity which can collect and process data.
2. **Aggregator:** The Aggregator is an honest but curious entity. Its responsibility is aggregating the collected data of IoT devices into an integrated one, then transmitting the aggregation result to the server.
3. **Device:** Every IoT Device, namely, a sensor, a smart meter, or an RFID reader, collects data, and preprocess them. For the sake of simplicity, the IoT devices will be abbreviated as Device. We assume that the Devices have the some computational and storage capability. The devices are honest but curious.

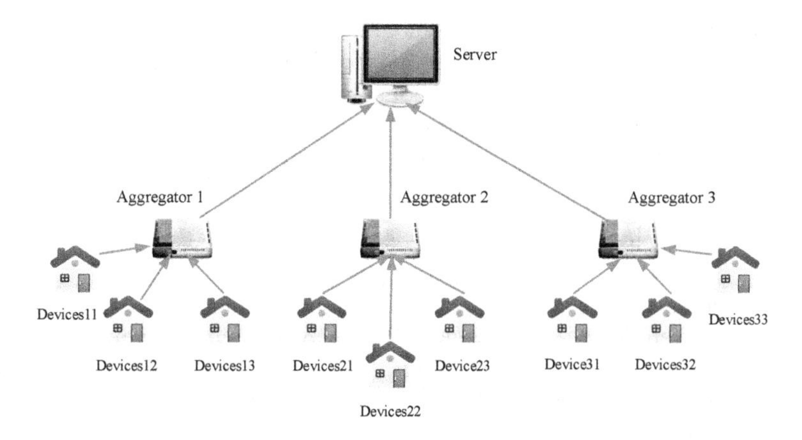

Fig. 1. Data aggregation model in IoT

3.2 Attacker Model and Security Requirements

In our attack model, we consider the following three attacks in IoT.

- **External Attack:** An adversary may compromise privacy of users by eavesdropping or modifying information transmitted from the Device to Aggregator on communication channels and also try to compromise the Aggregator to get all the privacy information of users.
- **Internal Attack:** Aggregator may be curious about device's collected data, which may lead to revealing the users' privacy information.
- **Collusion Attack:** The devices want to infer other devices' private via collusion activity.

In our system, the following security requirements should be achieved:

- **Privacy Preservation:** An adversary cannot acquire devices' data during system communications and operations. Even if several devices collude with each other, they cannot infer other devices' privacy data.
- **Authentication:** Aggregator should guarantee that the received data are valid and derived from legal entities.
- **Data Integrity:** When an adversary forges or modifies a report, the malicious operations should be detected.

3.3 Design Goal

According to the system model and security requirements, our design goal concentrates on proposing a secure, efficient, flexible and privacy-preserving data aggregation scheme. Specifically, the following design goals are to be achieved.

- **Security:** The proposed scheme should meet all the security requirements as above.

- **Efficiency:** The proposed scheme should consider communication-efficiency. In other words, the system should support real-time transmission of data from hundreds and thousands of users.
- **Flexibility:** The proposed scheme support "plug and play". Besides, it should be convenient for system to add a new device in a residential area.

4 Preliminaries

Homomorphic encryption [8] allows certain computation over encrypted data. Paillier cryptosystem [26] is a popular Homomorphic encryption scheme that provides fast encryption and decryption, which is a probabilistic asymmetric algorithm based on the decisional composite residuosity problem. It is adopted by the secure scalar product, which has been widely used in privacy preserving data mining. The Paillier cryptosystem is briefly introduced as follows:

Key Generation

- Choose two large prime numbers p and q randomly and independently of each other such that $\gcd(pq, (p-1)(q-1)) = 1$. This property is assured if both primes are of equal length.
- Compute $n = pq$ and $\lambda = lcm(p-1, q-1)$.
- Choose random integer g where $g \in \mathbb{Z}^*_{n^2}$.
- Ensure n divides the order of g by checking the existence of the following modular multiplicative inverse: $\mu = (\mathbf{L}(g^\lambda \, mod \, n^2))^{-1} \, mod \, n$, where function $\mathbf{L}$ is defined as $\mathbf{L}(x) = \frac{x-1}{n}$.
- The public key is (n, g).
- The private key is (λ, μ).

Encryption. Given a plaintext m where $0 \leq m < n$, select random r where $0 \leq r < n$ and calculate the ciphertext as: $c = g^m \cdot r^n \, mod \, n^2$.

Decryption. Given a ciphertext c, calculate the plaintext as: $m = \mathbf{L}(c^\lambda \, mod \, n^2) \cdot \mu \, mod \, n$.

Homomorphic Addition of Ciphertexts

- We assume that there are two message m_1, m_2. We can encrypt them with the public key independently and obtain ciphertexts c_1, c_2, which are denoted as following: $c_1 = g_1^m \cdot r_1^n \, mod \, n^2$, $c_2 = g_2^m \cdot r_2^n \, mod \, n^2$.
- We can calculate the product of c_1 and c_2, and obtain the result $E(m_1) \cdot E(m_2) = c_1 \cdot c_2 = (g_1^m \cdot r_1^n \, mod \, n^2) \cdot (g_2^m \cdot r_2^n \, mod \, n^2) = g^{m_1+m_2} \cdot (r_1 r_2)^n \, mod \, n^2 = E(m_1 + m_2)$. Hence, the sum of plaintext can be calculated from multiplication of the ciphertext.

5 Our Scheme

In this section, we present a privacy-preserving data aggregation scheme. We assume that the IoT devices have some computing power and storage. All IoT devices in the same residential area are regarded as one group. Each Aggregator manages a group of IoT Devices. The devices has a common secret key which is used to encrypt the data for system communication. They also have a private parameter τ. Additionally, each device has a unique identification ID which is only known by himself and Aggregator. We propose to assist devices' secure communication with Advanced Encryption Standard (AES) symmetrical encryption. Moreover, we provide a One-Time Pad for the devices to enhance security. For secure communication between devices and their corresponding Aggregator, we utilize homomorphic encryption scheme to protect private information. In the proposed schemes, the Aggregator can obtain the data aggregation result in a residential area without knowing actual data of each device. The curious and collusive devices cannot infer other devices' private usage data either. The scheme consists of the following five stages: (i) Data division: (ii) encryption and exchange, (iii) decryption and confusion, (iv) encryption and report, (vi) authentication and aggregation.

Data Division. We assume a topical residential area which comprises an Aggregator connected with a large number of residential users' devices $U = \{U_1, U_2, \ldots, U_n\}$. The devices collect data $M = \{M_1, M_2, \ldots, M_n\}$ respectively in a certain period. We divide the data $M_i(i \in (1, 2, \ldots, n))$ into n parts $W_{ij}(i \in (1, 2, \ldots, n), j \in (1, 2, \ldots, n))$ randomly. Namely,

$$\begin{cases} M_1 = \sum_{j=1}^{n} W_{1j}, \\ M_2 = \sum_{j=1}^{n} W_{2j}, \\ \cdots, \\ M_n = \sum_{j=1}^{n} W_{nj} \end{cases} \tag{1}$$

Encryption and Exchange. Each device only preserves one part W_{ii} and transmits the rest parts $\{W_{i1}, W_{i2}, \ldots, W_{in}(i \neq n)\}$ to other $n-1$ devices respectively. To provide an One-Time Pad, we assume that initial key is K_1, the subsequent secret keys K_n are shown as following:

$$\begin{cases} K_2 = H(\tau || K_1) \\ K_3 = H(\tau || K_2) \\ \cdots, \\ K_n = H(\tau || K_{n-1}) \end{cases} \tag{2}$$

We assume that One device Dv transmits $n-1$ part of data W_{ij} to other devices via the key, it can do a hash operation on the W_{ij} and real time T denoted as $ch = H(W_{ij} || T)$ and encrypt W_{ij} and ch with K_i denoted as $c = E_{K_i}(W_{ij} || T || ch)$. Then, D_v will transmit the ciphertext c to other devices.

Decryption and Confusion. When receiving ciphertext c, other devices will decrypt c with K_i denoted as $p = D_{K_i}(c)$ and obtain the plaintext W_{ij} and hash value ch. Then, the devices who receive the data can verify whether the message has been manipulated or replayed by doing a hash operation on W_{ij} and T and matching the result with ch. If the verification succeeds, W_{ij} is correct. Otherwise, W_{ij} will be discarded. When all devices finish this process, the data $M_i'(i \in (1, 2, \ldots, n))$ of each device is the sum of its preserved slice W_{ii} and received $n - 1$ slices $\{W_{1i}, W_{2i}, \ldots, W_{ni}(i \neq n)\}$ from other devices. All data of users are hidden via the operations as above. The confused usage data of n devices are shown as following:

$$\begin{cases} M_1' = W_{11} + \sum_{i=1}^{n} W_{i1}, & (i \neq 1) \\ M_2' = W_{22} + \sum_{i=1}^{n} W_{i2}, & (i \neq 2) \\ \ldots, \\ M_n' = W_{nn} + \sum_{i=1}^{n} W_{in} & (i \neq n) \end{cases} \tag{3}$$

Encryption and Report. After Devices' exchanging partial data with each other, the actual data has been blinded. Each device preserves the public key pk of the server and encrypts the blinded data with pk. All blinded data $M_i'(i \in (1, 2, \ldots, n)$ are encrypted, which is denoted as $C_i = E_{pk}(M_i')(i \in (1, 2, \ldots, n)$. In order to resist replay, impersonation and manipulation attack, devices also compute hash value of identification ID, real time T and ciphertext C_i denoted as $h = H(ID||T||C_i)$(H is a hash function, ID is user's unique identification, T is the real time, C_i is the preceding ciphertext). Finally, devices report ciphertext C_i and hash value h to Aggregator.

Authentication and Aggregation. When the Aggregator obtains the ciphertext C_i and the hash value h from the devices, it will verifies whether the message is from valid user and whether the message has been modified at first. Therefore, the Aggregator has a hash operation on ID, T, C_i and match them with h. If that succeeds, Aggregator will accept the ciphertext C_i. Finally, it gets a final result $C = \prod_{i=1}^{n} C_i$ and transmits that to the Server. When receiving C, the Server decrypts it with the private key sk and obtain the total data Tol of a residential area, which can be denoted as $Tol = D_{sk}(C)$. During the procedure, both Server and Aggregator do not know the actual data of each device.

6 Security Analysis

In this section, we analyze the security properties of the proposed schemes. In particular, our analysis focuses on how the schemes can resist various attacks and achieve the individual user's privacy preservation.

6.1 Resistance to Eavesdropping Attack

Theorem 1. *An adversary cannot obtain users' private usage data by eavesdropping the encrypted data during transmitting.*

Proof. Even if an adversary obtains the ciphertext of users' data, it cannot decrypt the ciphertext to reveal users' data without private key. Because all users' message is encrypted with symmetrical encryption or asymmetric encryption before transmitting, adversary cannot decrypt it by brute-force with a non-negligible probability.

6.2 Resistance to Replay Attack

Theorem 2. *If an adversary reports the same message to Aggregator or devices, it can be detected.*

Proof. We assume that an adversary A transmits message M to Aggregator or devices. M contains the hash value of real time T. When the Aggregator or devices receives M, which will be verified. If succeed, it indicates the message is not replayed. Otherwise, the Aggregator or devices will find out that there is a replay attack.

6.3 Resistance to Manipulation Attack

Theorem 3. *If an adversary manipulates the message between two IoT devices during communication, it can be detected.*

Proof. We assume that an IoT device A reports M to another IoT device B. M is the ciphertext $E(M_i||H(M_i||T))$ (M_i is the transmitted plaintext data). When B receiving M, it will decrypt M to obtain the hash value $H(M_i||T)$ and the transmitted data M_i. Then, B has a hash operation on M_i and the real time T, and verifies whether it matches the result with preceding hash value $H(M_i||T)$ or not. If the verification is passed, it indicates that the message M has not been manipulated.

Theorem 4. *If an adversary manipulates the message between IoT devices and Aggregator, it can be detected.*

Proof. We assume that an IoT device A reports M to Aggregator. M contains the hash value $H(ID||T||C)$ (H is a hash function, ID is A's unique identification, T is the real time, C is ciphertext of the blinded data) and the ciphertext C. When receiving M, Aggregator do a hash operation on A's ID, T, C and verifies whether it matches the result $H(ID, T, C)$. If the verification fails, it's indicated that the message has been manipulated.

6.4 Resistance to Impersonation Attack

Theorem 5. *If an adversary masquerades as another valid device reporting data to Aggregator it can be detected.*

Proof. If an adversary wants to masquerade as another valid device B and reports message to Aggregator, the adversary must be able to obtain the ID of B. However, the ID of B is only known by B and Aggregator. Thus, Aggregator can verify whether it is from a valid device by checking the hash value containing ID.

6.5 Resistance to Internal Attack

Theorem 6. *We assume that Aggregator is an internal attacker which is curious about all devices' privacy information. It still cannot obtain the actual data of all devices.*

Proof. We assume that Iot devices are $U = \{U_1, U_2, \dots, U_n\}$ and their reporting message are $E = \{E_1, E_2, \dots, E_n\}$ respectively. E is the ciphertext of blended data which are encrypted with the public key of Server. Aggregator does not have the corresponding private key to decrypt the ciphertext.

6.6 Resistance to Colluding Attack

Theorem 7. *The proposed scheme can resist collusion attack among the customers.*

Proof. We assume that there are n users, whose collected data is $M = \{M_1, M_2, \dots, M_n\}$ respectively. After dividing $M(i \in (1, 2, \dots, n))$ into n parts, users preserve one part and exchange the rest of $n-1$ parts with each. Assume that $n-1$ colluding users want to infer another user(A)'s information. Nevertheless, they only know $W_{ij}(i \neq j)$ which A has exchanged with them, but they don't know W_{ii} which is held by A himself. Moreover, they also cannot obtain the value of M_i', because M_i' has been encrypted before transmitting to Aggregator. Therefore, the colluding users cannot reveal other users' privacy information.

7 Performance Evaluation

In this section, we will evaluate the computational cost of the proposed schemes.

Computational cost of modular exponentiation and multiplication operations is much higher than that of hash functions and addition operations; so we will ignore the cost of hash operations and addition operations, and only focus on the computational cost incurred by encryption and decryption operations in this study.

We assume a topical group which comprises an Aggregator connected with a large number of users $U = \{U_1, U_2, \dots, U_n\}$. Note that C_e is the computational

Table 1. The operations of Aggregator, one device.

	Single IoT device	Aggregator
Data division	-	-
Encryption and exchange	$(n-1)A_e$	-
Decryption and confusion	$(n-1)A_d$	-
Encryption and report	$2C_e + C_m + C_o$	-
Authentication and aggregation	-	$(n-1)C_m$
Total cost	$(n-1)(A_e + A_d) + 2C_e + C_m + C_o$	

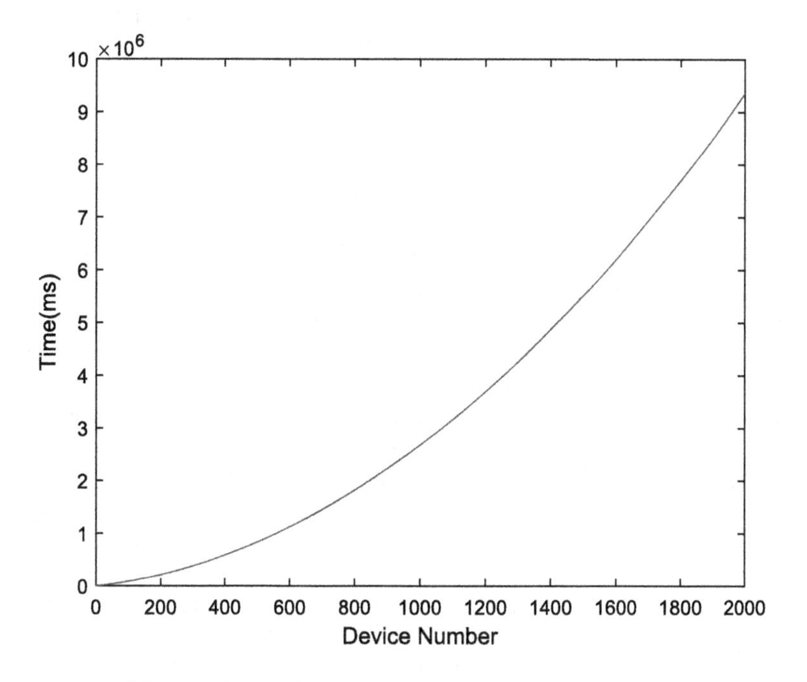

Fig. 2. The total communication costs in our scheme

cost of an exponentiation operation; C_m is the computational cost of a multiplication operation; C_o is the computational cost of a modulo operation, and A_e and A_d are respectively the computational cost of an AES encryption and an AES decryption.

For one device, no computation is needed for the following two phases: Data Division, Authentication and Aggregation. The Encryption and Exchange phase involves $n-1$ AES encryption operations, thus the cost is $(n-1)A_e$. The Decryption and Confusion phase involves $n-1$ AES decryption operations and a series of negligible addition operations, thus the cost is $(n-1)A_d$. The Encryption and Report phase involves two exponentiation operations, one multiplication operation and one modulo operation, thus the cost is $2C_e + C_m + C_o$. Hence, the total computational cost is $(n-1)(A_e + A_d) + 2C_e + C_m + C_o$.

Moving on to the Aggregator. No computation is needed for the former four phases. The Authentication and Aggregation phase involves $n-1$ multiplication operations, thus the total computational cost is $(n-1)C_m$.

Table 1 respectively summaries the computational complexities of the IoT device and Aggregator in each phase.

We conduct the experiments running in Python on a 2.60 GHz-processor 8 G-memory computing machine to study the operation costs. The result of experiments indicates that both an AES encryption operation and an AES decryption operation with 256-bit key costs less than 1 ms. When the Paillier key is 256-bit, an encryption operation almost costs 680 ms and an decryption operation costs 2 ms. So the total computational cost can be denoted as $S1 = n(n-1) \times 1 + n(n-1) \times 1 + n \times 680$, which is shown in Fig. 2.

8 Conclusion

In this paper, an efficient and reliable data aggregation schemes is proposed for IoT. It supports "plug and play" and preserves users' privacy data by confusing their data before they reported. For communication, we also use different encryption methods for different situations to guarantee confidentiality and integrity of usage data. Our scheme can reduce computational cost and improve communication efficiency. Besides, we have provided security analysis to demonstrate that our scheme can resist internal attack, external attack and collusion attack. For future work, we will try to find a new way to reduce data-exchange frequency when confusing users' data between IoT devices. Meanwhile, we also plan to further improve the scheme and deploy it in real-world IoT applications.

Acknowledgments. Thank all the reviewers for their helpful comments. This project was partial supported by the National Key R&D Program of China (No. 2017YFB0802000), the National Natural Science Foundation of China (Grant No. 61702062, 61672118, 61672321, 61373027), Science and Technology Project of Guangdong Power Grid Co. Ltd. (GDKJXM20180250), and Chongqing Research Program of Basic Research and Frontier Technology (No. cstc2014jcyjA40030).

References

1. Alrawais, A., Alhothaily, A., Hu, C., Cheng, X.: Fog computing for the internet of things: security and privacy issues. IEEE Internet Comput. **21**(2), 34–42 (2017)
2. Bao, H., Lu, R.: A new differentially private data aggregation with fault tolerance for smart grid communications. IEEE Internet Things J. **2**(3), 248–258 (2015)
3. Bao, H., Lu, R.: Comment on privacy-enhanced data aggregation scheme against internal attackers in smart grid. IEEE Trans. Ind. Inform. **12**(1), 2–5 (2016)
4. Cai, Z., He, Z., Guan, X., Li, Y.: Collective data-sanitization for preventing sensitive information inference attacks in social networks. IEEE Trans. Dependable Secur. Comput., 1 (2017)

5. Cai, Z., Zheng, X.: A private and efficient mechanism for data uploading in smart cyber-physical systems. IEEE Trans. Netw. Sci. Eng., 1 (2018)
6. Erkin, Z., Troncoso-Pastoriza, J.R., Lagendijk, R.L., Perez-Gonzalez, F.: Privacy-preserving data aggregation in smart metering systems: an overview. IEEE Signal Process. Mag. **30**(2), 75–86 (2013)
7. Fan, C.I., Huang, S.Y., Lai, Y.L.: Privacy-enhanced data aggregation scheme against internal attackers in smart grid. IEEE Trans. Ind. Inform. **10**(1), 666–675 (2014)
8. Fontaine, C., Galand, F.: A survey of homomorphic encryption for nonspecialists. EURASIP J. Inf. Secur. **2007**(1), 1–10 (2007)
9. He, Z., Cai, Z., Yu, J.: Latent-data privacy preserving with customized data utility for social network data. IEEE Trans. Veh. Technol. **67**(1), 665–673 (2018)
10. Hu, C., Cheng, X., Yu, J., Tian, Z., Lv, W., Chen, X.: Achieving privacy preservation and billing via delayed information release. IEEE Trans. Comput. (2018, submitted)
11. Hu, C., Cheng, X., Zhang, F., Wu, D., Liao, X., Chen, D.: OPFKA: secure and efficient ordered-physiological-feature-based key agreement for wireless body area networks. In: 2013 Proceedings of IEEE INFOCOM, pp. 2274–2282. IEEE (2013)
12. Hu, C., Huo, Y., Ma, L., Liu, H., Deng, S., Feng, L.: An attribute-based secure and scalable scheme for data communications in smart grids. In: Ma, L., Khreishah, A., Zhang, Y., Yan, M. (eds.) WASA 2017. LNCS, vol. 10251, pp. 469–482. Springer, Cham (2017). https://doi.org/10.1007/978-3-319-60033-8_41
13. Hu, C., Li, H., Huo, Y., Xiang, T., Liao, X.: Secure and efficient data communication protocol for wireless body area networks. IEEE Trans. Multi-Scale Comput. Syst. **2**(2), 94–107 (2016)
14. Hu, C., Li, W., Cheng, X., Yu, J., Wang, S., Rongfang, B.: A secure and verifiable access control scheme for big data storage in clouds. IEEE Trans. Big Data, 1 (2017)
15. Hu, C., Liu, H., Ma, L., Huo, Y., Alrawais, A., Li, X., Li, H., Xiong, Q.: A secure and scalable data communication scheme in smart grids. Wirel. Commun. Mob. Comput. **2018**, 1–17 (2018)
16. Hu, C., Yu, J., Cheng, X., Tian, Z., Akkaya, K., Sun, L.: CP_ABSC: an attribute-based signcryption scheme to secure multicast communications in smart grids. Math. Found. Comput. **1**(1), 77–100 (2018)
17. Hu, C., Zhang, N., Li, H., Cheng, X., Liao, X.: Body area network security: a fuzzy attribute-based signcryption scheme. IEEE J. Sel. Areas Commun. **31**(9), 37–46 (2013)
18. Huang, L., Fan, X., Huo, Y., Hu, C., Tian, Y., Qian, J.: A novel cooperative jamming scheme for wireless social networks without known CSI. IEEE Access **5**, 26476–26486 (2017)
19. Huo, Y., Dong, W., Qian, J., Jing, T.: Coalition game-based secure and effective clustering communication in vehicular cyber-physical system (VCPS). Sensors **17**(3), 475 (2017)
20. Huo, Y., Hu, C., Qi, X., Jing, T.: LoDPD: a location difference-based proximity detection protocol for fog computing. IEEE Internet Things J. **4**(5), 1117–1124 (2017)
21. Huo, Y., Tian, Y., Ma, L., Cheng, X., Jing, T.: Jamming strategies for physical layer security. IEEE Wirel. Commun. **25**(1), 148–153 (2018)
22. Li, H., Lin, X., Yang, H., Liang, X., Lu, R., Shen, X.: EPPDR: an efficient privacy-preserving demand response scheme with adaptive key evolution in smart grid. IEEE Trans. Parallel Distrib. Syst. **25**(8), 2053–2064 (2014)

23. Li, R., Song, T., Capurso, N., Yu, J., Couture, J., Cheng, X.: IoT applications on secure smart shopping system. IEEE Internet Things J. **4**(6), 1945–1954 (2017)
24. Liang, Y., Cai, Z., Yu, J., Han, Q., Li, Y.: Deep learning based inference of private information using embedded sensors in smart devices. IEEE Netw. (2018, to appear)
25. Lu, Y., Zhao, Z., Zhang, B., Ma, L., Huo, Y., Jing, G.: A context-aware budget-constrained targeted advertising system for vehicular networks. IEEE Access **6**, 8704–8713 (2018)
26. Paillier, P.: Public-key cryptosystems based on composite degree residuosity classes. In: Stern, J. (ed.) EUROCRYPT 1999. LNCS, vol. 1592, pp. 223–238. Springer, Heidelberg (1999). https://doi.org/10.1007/3-540-48910-X_16
27. Saputro, N., Akkaya, K.: Performance evaluation of smart grid data aggregation via homomorphic encryption. In: 2012 IEEE Wireless Communications and Networking Conference (WCNC), pp. 2945–2950. IEEE (2012)
28. Shim, K.A., Park, C.M.: A secure data aggregation scheme based on appropriate cryptographic primitives in heterogeneous wireless sensor networks. IEEE Trans. Parallel Distrib. Syst. **26**(8), 2128–2139 (2015)
29. Song, T., Li, R., Mei, B., Yu, J., Xing, X., Cheng, X.: A privacy preserving communication protocol for IoT applications in smart homes. IEEE Internet Things J. **4**(6), 1844–1852 (2017)
30. Thai, C.D.T., Lee, J., Ryu, J.Y., Quek, T.Q.: Privacy preservation with channel-based jamming for data aggregation in smart grids. In: 2017 IEEE International Conference on Communications (ICC), pp. 1–6. IEEE (2017)
31. Xing, K., Hu, C., Yu, J., Cheng, X., Zhang, F.: Mutual privacy preserving k-means clustering in social participatory sensing. IEEE Trans. Ind. Inform. **13**(4), 2066–2076 (2017)
32. Zheng, X., Cai, Z., Li, Y.: Data linkage in smart IoT systems: a consideration from privacy perspective. IEEE Commun. Mag. (2018, to appear)

Improving Security and Stability of AODV with Fuzzy Neural Network in VANET

Baohua Huang[1(✉)], Jiawei Mo[1], and Xiaolu Cheng[2]

[1] School of Computer and Electronic Information, Guangxi University, Nanning 530004, Guangxi, China
bhhuang66@gxu.edu.cn, 327998377@qq.com

[2] Department of Computer Science, Virginia Commonwealth University, Richmond, VA 23220, USA
chengx3@vcu.edu

Abstract. To improve security and stability of AODV in VANET (Vehicle Ad hoc Network), a secure and stable AODV, named GSS-AODV is proposed. GSS-AODV uses a fuzzy neural network to compute the node information in routing activities. The stability of nodes is computed to evaluate links. The link stability and the number of hops are considered in a balanced way, so a stable path with fewer hops is selected. GSS-AODV uses trust value of the node to evaluate the node security. The evaluation balances node security with network environment and node utilization to prevent malicious node attack. In routing maintenance, GSS-AODV uses genetic simulated annealing algorithm to optimize the parameters of the fuzzy neural network in real time to ensure that the calculated stability and trust value of node match the actual situation.

Keywords: VANET · Node security · Link stability · Fuzzy neural network AODV

1 Introduction

AODV (Ad hoc On-demand Distance Vector Routing) is a typical on-demand routing protocol widely used in VANET and plays an important role in the development of VANET [1]. For VANET, the most basic requirement is that the designed routing protocol is efficient, secure, and stable in unattended, harsh environments. It is very important to propose a stable routing algorithm with security policy.

At present for this problem, many domestic and foreign literature on the AODV protocol has been studied and improved. The literature [2] proposes an improved TAODV routing protocol based on trust mechanism to determine whether the node is a malicious node by comparing the trust value of the node. The literature [3] uses a fixed time window to judge whether the node is selfish or not, and there is a delay in judging the behavior of the node. Although the protocol in the literature [4] can detect changes in node behavior, there is a problem of insufficient evidence in calculating the trust value. The literature [5] put forward that the TARF routing protocol uses a neighbor

S. Chellappan et al. (Eds.): WASA 2018, LNCS 10874, pp. 177–188, 2018.
https://doi.org/10.1007/978-3-319-94268-1_15

table to record the trust degree and energy consumption of each neighbor node and is used to prevent attacks based on routing location. However, routing protocol increases routing load when broadcasting energy control packets. The literature [6] based on the AODV routing protocol, they use the public key to encrypt and identify IP addresses. Encryption technology will increase many communication, computation and memory costs in the key distribution process. The literature [7] proposed CBM-AODV, which combines the success rate and the link quality, improves the path stability in the routing part and can effectively prevent the link failure. The literature [8] proposed the LLA method to find a stable communication path, which focuses on the improvement of multi-hop paths and the improvement of link stability, which makes the routing meet the requirements of Quality of Service (QoS) and provide real-time security information services. The literature [9] proposed a reactive routing protocol AODVCS based on the biologically inspired cuckoo search algorithm. The protocol uses the Cuckoo Search Algorithm (CSA) to determine the shortest path between two nodes and adds a route trust prediction mechanism. While ensuring the complete routing security and reliability of the routing protocol packet delivery rate, end-to-end delay, and other performance.

This paper uses fuzzy neural networks to solve the stability and security of routing protocols, the node trust value and node stability are obtained through fuzzy calculation. On this basis, a stable AODV protocol with security policy is proposed based on the original AODV protocol. The node trust value and the node stability are obtained through the fuzzy neural network when the route is initiated according to various influencing factors of the vehicle and finally applied to the routine activities of the protocol. Protocols improve link stability, save routing repair costs, reduce the impact of malicious nodes, and improve network security. The simulation results show that compared with the AODV protocol, this protocol can improve network performance.

2 Improving AODV Protocol Based on Fuzzy Neural Network

2.1 Arithmetic Statement

Stability and security throughout the judgment of AODV improve every part of the program. Its success is directly related to the smooth operation of the entire routing protocols. In this paper, GASA-FNN is used to improve the scheme based on Genetic Simulated Annealing Algorithm (GASA). Firstly, security impact factors and stability impact factors of the extraction nodes are taken as the node's security measures and stability measures, and they are normalized. Then, fuzzy neural networks are used to carry out fuzzy calculations on safety metrics and stability metrics, and genetic simulated annealing is used to optimize parameters in the calculation process. Finally, node stability and node trust values are obtained and discriminated in the routing process. The algorithm structure is shown in Fig. 1.

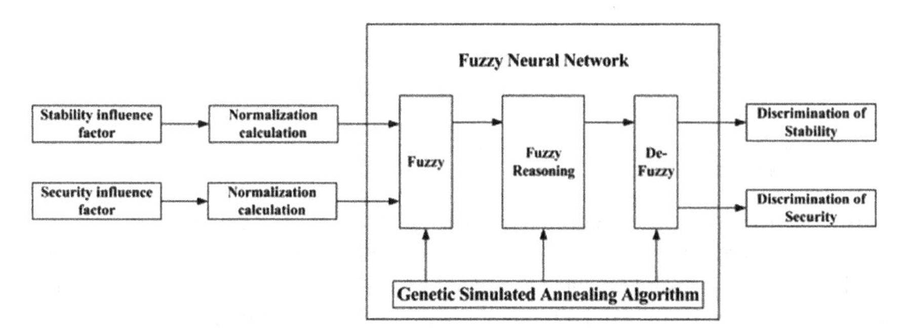

Fig. 1. GSS-AODV structure

2.2 Node Stability Metrics

Based on the classical AODV protocol algorithm, the relative velocity u and the relative distance d between nodes are extracted, and the node load q is the key factor to measure the node stability. After the above factors are normalized and comprehensively processed, the node stability can be obtained through fuzzy processing to be used in the improved GSS-AODV algorithm.

Set N_i neighbor nodes j of node i to form Φ_i collection.

If the velocity vectors of node i and its neighbor node j are respectively summed, the normalized relative velocity is defined as formula 1. Where, $j = 1, 2, \ldots, N_i$; $u_{\max}$ is the maximum relative rate between vehicles, such as within the city speed limit 60 km/h, the maximum relative rate of 120 km/h.

$$u_{ij}^{\text{normal}} = \frac{\left|\vec{\boldsymbol{u}}_i - \vec{\boldsymbol{u}}_j\right|}{u_{\max}} \tag{1}$$

Suppose the coordinates of node i and neighbor node j are (x_i, y_i), (x_j, y_j), then the relative distance between nodes is defined as formula 2.

$$d_{ij} = \sqrt{\left(x_i - x_j\right)^2 + \left(y_i - y_j\right)^2} \tag{2}$$

Then, the normalized relative distance is defined as formula 3, where $d_{\max}$ is the maximum communication distance, taking 250 m.

$$d_{ij}^{\text{normal}} = \frac{d_{ij}}{d_{\max}} \tag{3}$$

Suppose the load of neighbor j: q_i, that is, the number of packets stored in the cache queue of this node. Let $q_{\max}$ represents the total length of the queue cached by the node, that is, the maximum number of packets allowed to be stored. The normalized load is defined as formula 4.

$$q_{ij}^{\text{normal}} = \frac{q_{ij}}{q_{\max}} \tag{4}$$

2.3 Node Security Metrics

VANET internal nodes common several attacks are: random data packet loss in the process of forwarding; packet tampering and forgery; entice the surrounding nodes to send data packets to malicious nodes to launch black hole attacks. By analyzing these attacks, we can see that when the internal nodes are attacked, the repetition rate of data packets may be too large. When there are attacks such as black hole attacks and selective forwarding, there will be an abnormal number of packets sent [10]. The neighbor table corresponding to a node has a certain correlation, and the neighbor table between normal nodes should be repeated to some extent. Therefore, the packet content repetition rate, the number of packets [11], and the relevance of surrounding nodes can be used as detection factors for malicious nodes for the improved AODV algorithm. Based on the classical AODV protocol algorithm, the detection factors are extracted, normalized, and integrated, and then the node trust value is obtained through the fuzzy processing.

Set N_i neighbor nodes j of node i to form Φ_i collection.

As shown in the formula 5, $S_{ij}(t)$ represents the normalized packet repetition rate, $T_{ij}(t)$ represents the packet transmission factor, and $U_{ij}(t)$ represents the normalized node similarity. Among them, $p_{ij}(t)$ is the number of packets at the t moment, $sp_{ij}(t)$ is the number of repeated packets, and the $\Delta P(t)$ is the expected value of the number of packets. $U_{ij}(t)$ is measured by Adamic-Adar [12] indexes. $N(i)$ is the neighbor set of nodes i, c is the common neighbor of two nodes, $\log_k(c)$ is the logarithm of node degree.

$$\left\{ \begin{array}{l} S_{i,j}(t) = \dfrac{p_{i,j}(t) - sp_{i,j}(t)}{P_{i,j}(t)} \\ T_{i,j}(t) = \dfrac{|p_{i,j}(t) - \Delta p(t)|}{p_{i,j}(t)} \\ U_{i,j}(t) = \dfrac{\sum\limits_{c \in N_{i,j}(t)} \frac{1}{\log_k(c)}}{\sum\limits_{c \in N_{i,j}(t)} 1} \end{array} \right\} \tag{5}$$

2.4 Fuzzy Neural Network

In this paper, a multi-input and single-output neural network is used, and the structure is shown in Fig. 2.

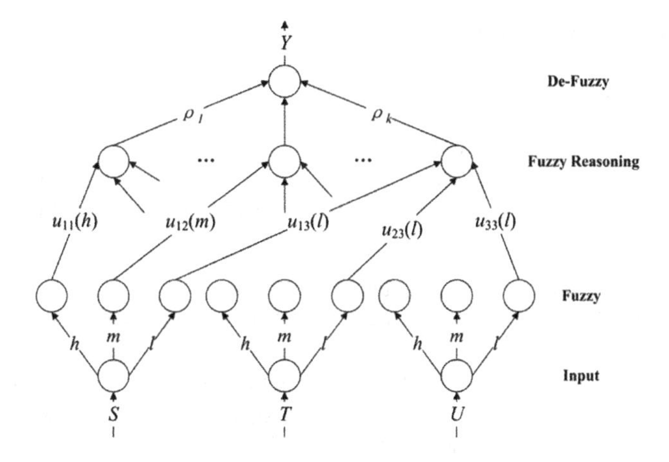

Fig. 2. Fuzzy neural network structure

The first layer is the input layer, which is responsible for passing the input variables to the second layer. The input value is the exact value and the number of nodes is the number of input variables. This layer has three neuron nodes, also known as three variables, *S*, *T*, and *U*. The second layer is the fuzzy layer, which is mainly to blur the input values. The *S*, *T*, and *U* are converted into three fuzzy subsets {high, middle and low}, which can be represented as {h, m, l}, and there are 9 nodes. U_{ij} means *j*-th membership fuzzy subset of variables *i*. This article uses the Gaussian function. The third layer is the fuzzy rules reasoning layer. Each node of this layer corresponds to a fuzzy rule, which is connected to the fuzzy subset of every variable in the second layer, and there are 27 nodes, which correspond to 27 rules of the inference. The fourth layer is the de-fuzzy layer. The fuzzy value of the fuzzy inference is converted into an exact value, and the gravity method is used to blur it, and the output value of the neural network is obtained. ρ_k is the connection weight of the third and fourth layers. *Y* is the deterministic solution of the problem and the node trust value.

2.5 Optimizing the Fuzzy Neural Network with Genetic Simulation Algorithm

In this paper, genetic algorithm as the main body, from a group of randomly generated initial population for the global optimal search process, first through the selection, crossover, mutation and other genetic operations to produce a new group of individuals, and then independent of the resulting each individual simulated annealing, and evaluation of new fitness groups, for individual selection, copying and other operations, the final result of its generation as individuals in the population. Run the process iteratively until some termination condition is satisfied. In summary, there is a thought that the simulated annealing algorithm is dissolved in the running of the genetic algorithm, which not only has the advantages of the genetic algorithm and the simulated annealing algorithm but also overcomes the corresponding deficiencies [13].

3 GSS-AODV Protocol Description

3.1 Routing Initiation

When a source node needs to communicate with a destination node, it first performs a route initiation process and broadcasts the RREQ packet to its neighboring neighbors.

The neighbor nodes that receive the RREQ package perform the following operations in turn:

1. Check for loop. If a loop discards the RREP package.
2. Check for duplicate RREQ packages. If a RREQ package is repeated, discard the RREQ package.
3. Check whether the reverse route has been established with the source node. If the new route has a higher quality or similar quality and is newer than the original route, update the last-hop route. Otherwise, a reverse path should be established with the source node to generate a route to the previous hop. The stability of the RREQ packet encapsulation is summed by the length of the route as the link quality of the current path.
4. Set a stable threshold with an initial value of 0.5 in the neighbor node. During a period, the neighbor nodes will use the node stability of the neighbor nodes stored in the neighbor table to calculate the average node stability and update the stability threshold. When the node stability is greater than 0.5 or the stability threshold, the RREQ packet is forwarded. This ensures that stable nodes with stability greater than 0.5 are always able to participate in forwarding. When the node is in the unstable state, the stability threshold can be relatively stable to participate in forwarding, improve node utilization, ensure link quality, and avoid the entire network forwarding of RREQ messages.
5. Neighbor nodes accumulate the stability and store them in the RREQ packet to continue broadcasting to the neighbor nodes. When the neighbor node is a relay node, different from the AODV protocol, the stability of the node still needs to be judged in this case, which helps the source node to consider the link quality of the entire path comprehensively. When node stability of the relay node meets the condition, the stability in the RREQ packet is summed up and divided by the current hop count. And is summed and averaged with the link quality of the path from the relay node to the destination node to obtain the link quality of the source node to the destination node. After being written into the RREP packet, the link quality is sent along the reverse path to the source node.
6. Considering neighbor node trust value, firstly calculates the average values of trust. All the neighbors trust in the trust value is less than 0.5 and less than the average value of nodes, are marked as in a state of distrust of neighbor node. They do not participate in the current node routing process.
7. The process ends after the destination node is reached.

3.2 Routing Choice

After the RREQ packet is continuously forwarded by the node, the node finally sends the packet to the destination node. After the destination node receives the RREQ packet, the node performs a routing process.

1. When the destination node receives the RREQ packet, it first waits for a route discovery period and continuously receives the RREQ packet before the waiting time.
2. After the waiting time is over, the destination node calculates the link quality according to the cumulative sum of node stability and the number of hops contained in the RREQ packet using formula 6, to evaluate the link stability:

$$LQ_m = \frac{St_m}{N} \tag{6}$$

   Where LQ_m represents the link quality of the m-th link, St_m represents the sum of node stability of the m-th link, and N is the number of link hops. The higher the node stability of the path is, the higher the number of hops and the smaller the number of hops is, the better the link quality is and the higher the link stability is. On the other hand, the stability of the link is easily broken.
3. If the RREQ packet is received during the process of receiving a higher quality or newer path than the link in the routing table, the corresponding path is replaced, and the routing table is updated.
4. Upon receiving the RREP packet, the source node first checks whether the reverse route is established with the destination node and then determines whether the route needs to be updated by comparing the quality of the link. This ensures that the reverse path is always stable, then selects the most stable link for data transfer.

3.3 Routing Maintenance

Through periodically sends the HELLO message to maintain a connection between neighbor nodes, the improved routing protocol for stability calculation of encapsulation of the neighbor node in the HELLO message neighbor list, the node that receives the HELLO message performs a route maintenance procedure.

When a node first receives a HELLO message from a neighbor node, it first adds a neighbor to the neighbor table and then uses the fuzzy neural network to calculate the node trust level of the corresponding node. And stores the repetition rate, the number of packets, the relevance of surrounding nodes, and the average trust values of all the current neighbors for the packet contents used in the current calculation as the training data of the simulated genetic annealing algorithm in the neighbor table. Each time a HELLO message is received from a neighbor node, the node first reads the node information encapsulated in the message, and then uses the fuzzy neural network to update the node trust value and then prolongs the lifetime of the corresponding neighbor node. From time to time, the node checks whether the survival time is less than the current time. If the neighbor nodes are lost, the node uses the simulated genetic annealing algorithm to optimize the parameters of the fuzzy neural network.

Each time a HELLO message is received from a neighbor node, the node first uses the node information encapsulated in the message, and then uses the fuzzy neural network to update the node stability and the node trust value, and then prolongs the lifetime of the corresponding neighbor node in the neighbor table. From time to time, the node checks whether the survival time of all the nodes is less than the current time and considers that the neighboring node is lost. The improved routing protocol, when the node is lost, the time difference t_d between the lost time and the first received HELLO message is the actual link time with the neighboring node. Calculate the node actually stable St_r by formula 7, where MT represents the average node link duration.

$$St_r = \frac{-1}{\left(\frac{t_d}{MT} + 1\right)} + 1 \tag{7}$$

This function has a stability of 0.5 at $t_d = MT$ and a positive limit of 1. Then St_r is used as the output of training data of genetic simulated annealing algorithm, and the corresponding training data of the neighbor table is input as training data to optimize the parameters of the fuzzy neural network. Finally, the lost neighbor node is deleted from the neighbor table. To ensure that the actual stability of the node always meets the current motion environment of the node, the node calculates the average link duration of all the neighboring nodes in the time and updates the MT from time to time.

4 Performance Simulation and Analysis

NS2 (Network Simulator Version 2) is used to simulate GSS-AODV and the original AODV protocol, and the random movement model of nodes is generated with it.

4.1 Stability Experimental Results

Setting the vehicle speed between 20 and 120 m/s, the routing performance of GSS-AODV and AODV shows from Figs. 3, 4, 5 and 6.

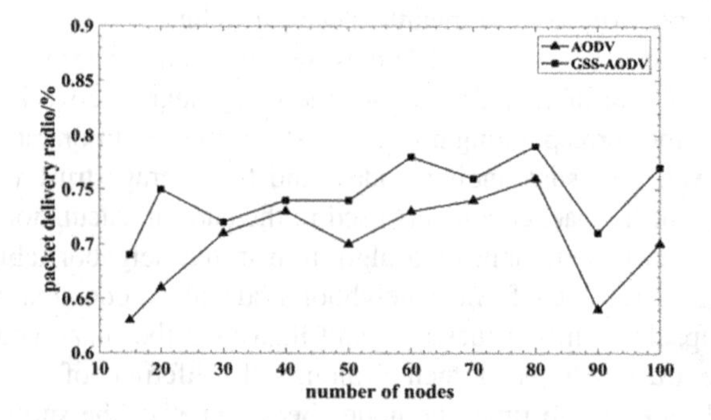

Fig. 3. Packet delivery ratio vs. number of nodes

Figure 3 shows that GSS-AODV finally obtains the link quality by calculating the node stability, which is used to evaluate the link stability. Therefore, the stable link is always selected in the route initiation and selection part to avoid data packet loss. In addition, GSS-AODV can dynamically adjust the parameters used by the fuzzy neural network according to different environments to improve the accuracy of selecting a stable link. Therefore, when the number of nodes is the same, the packet delivery rate of GSS-AODV protocol is always higher than that of AODV protocol, and the relative stability is small.

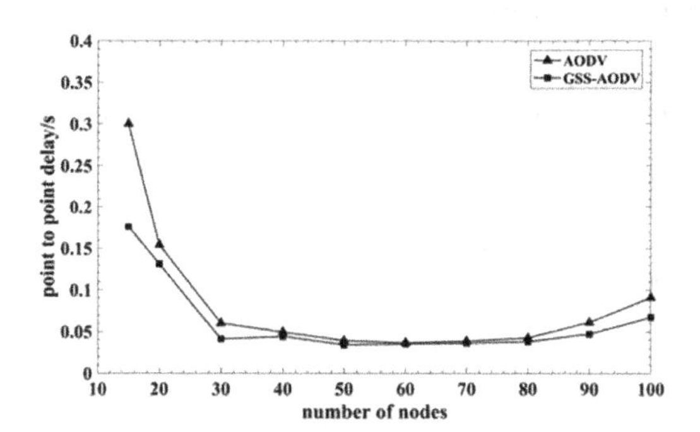

Fig. 4. Point to point delay vs. number of nodes

Figure 4 shows that because the GSS-AODV protocol considers the node load when selecting the link, the node load is taken as the factor to calculate the node stability. Therefore, when the number of nodes is small, GSS-AODV can select a stable and low-load link for data transmission, reduce the queuing time, and reduce the end-to-end delay. When the number of nodes continues to increase, the number of optional paths increases, the load on the nodes generally decreases, and the influence of the node load on the link stability decreases. GSS-AODV will optimize the algorithm parameters through simulated genetic annealing to reduce the weight of the node load in the calculation stability of the fuzzy neural network.

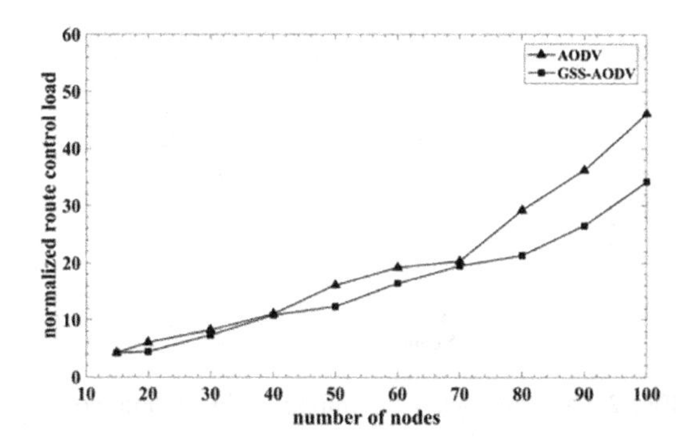

Fig. 5. Normalized route control load vs. number of nodes

Figure 5 shows that as the number of nodes increases, the control overhead of the AODV and GSS-AODV routing protocols also increases. However, the GSS-AODV protocol determines whether to forward the message according to the node stability decision result in the route initiation part, which limits the flooding of the RREQ message in the network. And GSS-AODV always selects the stable link to transmit data, the link is not easily broken, the number of route repairs is reduced, and the source node does not need to frequently initiate the route, thereby reducing the number of RREQ transmissions.

4.2 Security Experimental Results

Setting the number of vehicles to 100, the routing performance of the GSS-AODV and AODV under various number of black hole nodes shows from Figs. 6, 7 and 8.

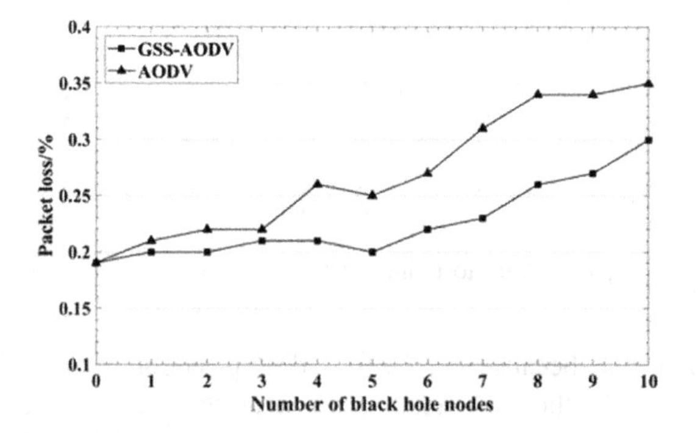

Fig. 6. Packet loss rate vs. the number of black hole nodes

Figure 6 shows that with the increase of the number of black hole nodes, more REEQ packets are phagocytic, and the loss rate of both protocols is on the rise.

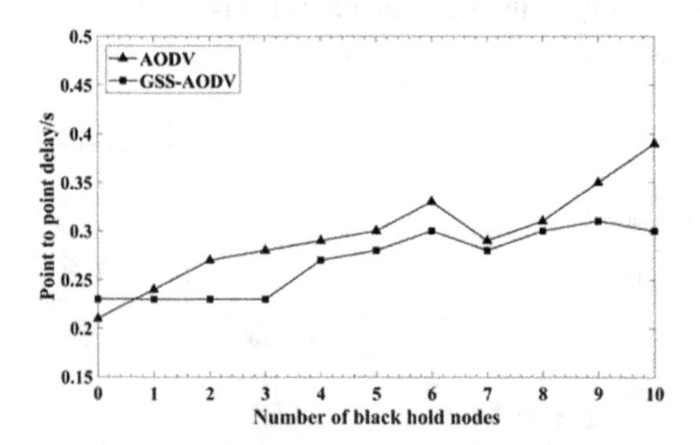

Fig. 7. Average delay vs. number of black hole nodes

Figure 7 shows that in the environment with fewer black hole nodes, the GSS-AODV protocol preferentially selects the nodes with higher trust values to participate in the routing process, which causes certain network delay. However, routing protocol optimizes the parameters according to the specific conditions and avoids prolonged delay, so that the average delay does not show a large gap. With the increase of black hole nodes, the GSS-AODV protocol always selects the nodes with higher trust values to participate in the routing process and reduces the probability of routing requests being swallowed by the attacking nodes.

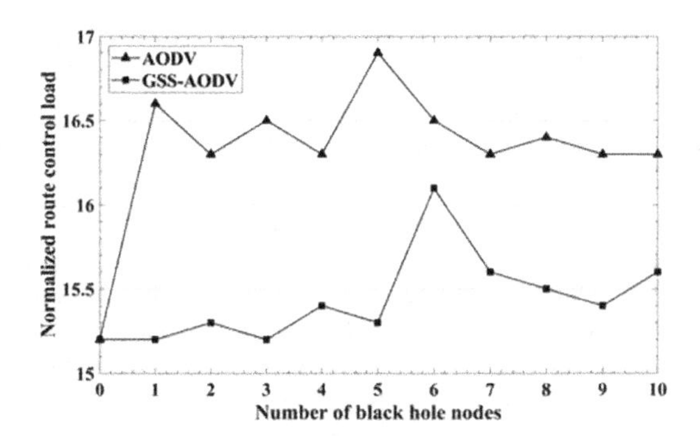

Fig. 8. Routing cost vs. number of black hole nodes

Figure 8 shows that when the number of black hole nodes increases, both the probability of losing RREQ and the number of control messages between nodes increase. Therefore, the routing overhead increases. However, since the GSS-AODV controls the forwarding of the RREQ through the node trust value in the routing initiation part and uses the genetic simulated annealing algorithm to control the parameters, the weight of the node similarity in the fuzzy neural network is adjusted according to the condition of the black hole node. In different environments, GSS-AODV can select the security node to participate in the routing process to reduce the routing overhead required for initiating the route initiation due to the black hole node attacks.

5 Conclusion

Stability and security are both hot issues in VANET research. This paper presents a stable AODV routing algorithm, named GSS-AODV, based on the fuzzy neural network. In GSS-AODV, the node stability and route length are considered in equilibrium, and the parameters are adjusted by genetic simulated annealing algorithm under different practical conditions to ensure that the calculated node stability is in accordance with the actual situation. The proposed algorithm takes into consideration of multiple attack models and adjusts the parameters through genetic simulated annealing algorithm in different practical environments to improve the accuracy of node trust value.

Experimental results show that GSS-AODV can choose the route with safer nodes, reduce the packet loss rate, reduce the average end-to-end delay, and normalize routing overhead.

Acknowledgements. This work was supported by National Natural Science Foundation of China under Grant No. 61262072.

References

1. Perkins, C., Belding-Royer, E., Das, S.: Request for comments: ad hoc on-demand distance vector (AODV) routing. Exp. Internet Soc. **6**(7), 90 (2003)
2. Jain, A., Prajapati, U., Chouhan, P.: Trust based mechanism with AODV protocol for prevention of black-hole attack in MANET scenario. In: Colossal Data Analysis and Networking (CDAN), pp. 1–5. IEEE (2016)
3. Hiroki, U., Sonoko, T., Hiroshi, S.: An effective secure routing protocol considering trust in mobile ad hoc networks. IPSJ J. **55**, 649–658 (2014)
4. Umeda, S., Takeda, S., Shigeno, H.: Trust evaluation method adapted to node behavior for secure routing in mobile ad hoc networks. In: Eighth International Conference on Mobile Computing and Ubiquitous Networking, pp. 143–148. IEEE, Melbourne (2015)
5. Zhan, G., Shi, W., Deng, J.: Design and implementation of TARF: a trust-aware routing framework for WSNs. IEEE Trans. Dependable Secur. Comput. **9**(2), 184–197 (2012)
6. Li, H., Singhal, M.: A secure routing protocol for wireless ad hoc networks. In: Hawaii International Conference on System Sciences, p. 225a. IEEE, Hawaii (2006)
7. Rak, J.: Providing differentiated levels of service availability in VANET communications. IEEE Commun. Lett. **17**(7), 1380–1383 (2013)
8. Rak, J.: LLA. A new anypath routing scheme providing long path lifetime in VANETs. IEEE Commun. Lett. **18**(2), 281–284 (2014)
9. Kout, A., Labed, S., Chikhi, S., et al.: AODVCS, a new bio-inspired routing protocol based on cuckoo search algorithm for mobile ad hoc networks. Wirel. Netw. **9**, 1–11 (2017)
10. Singh, B., Srikanth, D., Kumar, C.R.S.: Mitigating effects of black hole attack in mobile ad-hoc NETworks: military perspective. In: IEEE International Conference on Engineering and Technology, pp. 21–22. IEEE, Hammamet (2016)
11. Zhu, Y.-S., Dou, G.-Q.: A dimensional trust-based security data aggregation method in wireless sensor networks. J. Wuhan Univ. (Nat. Sci. Ed.) **59**(2), 193–197 (2013)
12. Adamic, L.A., Adar, E.: Friends and neighbors on the web. Soc. Netw. **25**(3), 211–230 (2003)
13. Xiao, W., Dong, H., Xin, Q.-L., et al.: Synthesis of large-scale multistream heat exchanger networks based on stream pseudo temperature. Chin. J. Chem. Eng. **14**(5), 574–583 (2006)

Exploiting Aerial Heterogeneous Network for Implementing Wireless Flight Recorder

Shanshan Huang[1], Zhen Wang[1], Chi Zhang[1(✉)], and Yuguang Fang[2]

[1] University of Science and Technology of China, Hefei 230026, China
{hss0201,wang1992}@mail.ustc.edu.cn, chizhang@ustc.edu.cn
[2] University of Florida, Gainesville, FL 32603, USA
Fang@ece.ufl.edu

Abstract. Flight recorder (or black box) equipped in a commercial aircraft plays a crucial role in investigating aviation accidents and incidents. While intended to be indestructible, the flight recorder can still be damaged in extreme conditions such as explosions, crashes, etc., or may become lost in the case of a crash in an inaccessible geographic location. In this paper, we propose a wireless flight recorder which can transmit the recorded operating data of a plane while in flight to a ground center through an aerial network with the aid of satellites. We utilize satellite links to provide widely available control channels and inter-aircraft wireless communications as a high-speed dynamic data plane in a software defined network framework, and carry out a network-wide global optimization for flight recorder data streaming in real time. By viewing the flight recorder data streaming task as a traffic engineering problem (in particular, as a multi-commodity flow problem), we aim at maximizing total streaming throughput and minimizing the usage of expensive and scarce satellite channels. We validate the feasibility and efficacy of proposed wireless flight recorder scheme by experimenting with a realistic setting for aerial heterogeneous networks and the real-time flights location information published by the International Air Transport Association (IATA).

Keywords: Aviation safety · Flight recorder
Aerial heterogeneous network · Multi-commodity flow problem

1 Introduction

Flight recorder (FR) [1] is an electronic device placed aboard an aircraft and commonly known as black box. It is used as the primary method of collecting flight data concerning the state of the aircraft. These data are acquired by sensors distributed throughout the fuselage of the aircraft and transmitted to the black box via the data bus. FR is composed of flight data recorder (FDR) and cockpit voice recorder (CVR) as shown in Fig. 1. The recorded data in FR can be used by

S. Chellappan et al. (Eds.): WASA 2018, LNCS 10874, pp. 189–200, 2018.
https://doi.org/10.1007/978-3-319-94268-1_16

the National Transportation Safety Board (NTSB) for investigating an aviation disaster, as well as for analyzing aircraft safety and engine performance. Due to the importance of FR, it is designed to survive the tremendous forces experienced during and after an air crash. However, not all black box can survive a crash. For example, there are 239 death in the MH370 flight incident [2] in 2014. Although lots of manpower and resources are invested, the flight recorder has not been found until now and the cause of this disaster is still unclear. Therefore, the survivability of FR is an urgent issue to be solved.

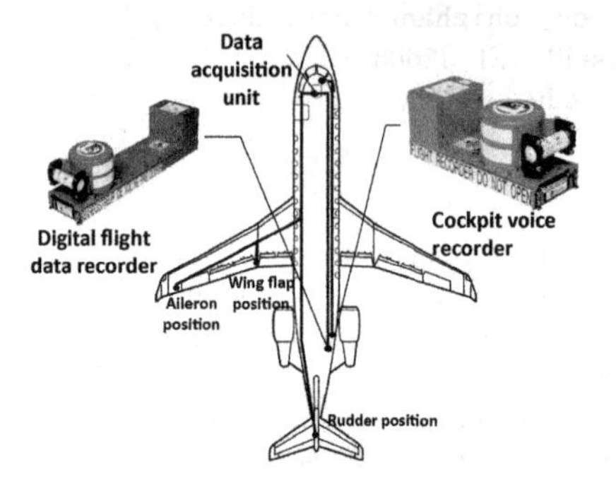

Fig. 1. Flight recorders.

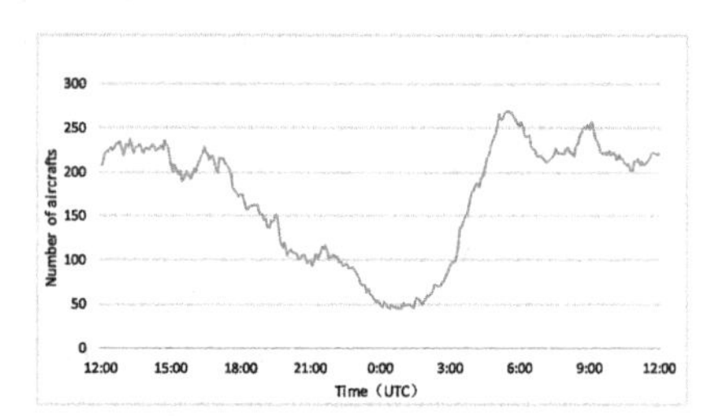

Fig. 2. The variation curve of the number of the aircrafts in North Atlantic aviation spaces.

According to the statistical records [3] in recent years, if we could know what happened aboard the aircraft, the disaster might be avoided. A good example is the Germanwings flight 4U 9525 disaster [4] in 2015. The first officer locked the door and let the captain be outside the cockpit. He then drove the aircraft and resulted in the crash. If we had known this in real time, we might unlock the door in a remote method, and the disaster might not happen. To this end, we propose to transmit the black box's data to a ground center. These information can be used to monitor the aircraft in real time and check the performance of the engine to reduce the accident rate.

It is possible to use one or several of the four currently available radio systems for transmitting data to ground. They are High Frequency (HF), Ultra High Frequency (UHF), Very High Frequency (VHF), and Satellite communication. The HF provides good signal coverage, but it is highly susceptible to atmospheric disturbances and relatively low in bandwidth. VHF and UHF can provide reliable communication when they are properly implemented. They are both light-of-sight propagation and the transmission range is about 200 km. Consequently, we can use VHF and UHF for air-to-air communication. Satellite communication has a worldwide coverage and the bandwidth meets our requirement, but the cost of satellite channel is high. Thus in our system, we only use it as an alternate when there is no neighbor aircrafts or ground stations receiving the FR's data.

According to Frank Dorank, a senior engineer for L3 Communications, the data rate required is 120 Kbps for cockpit voice and 3 Kbps for flight data, or a total of 123 Kbps [1]. The bandwidth offered by UHF and VHF can fulfil requirements of CVR and FDR's data transmission. The flight data contains the flight status which is more important than the cockpit voice. The latter is usually used for investigation after crash. Therefore, in our proposed scheme the flight data is transmitted through hop-by-hop mode to the ground center, and the cockpit voice is transmitted only one hop to nearby aircrafts and picked up after the aircraft landed.

The North Atlantic is the busiest oceanic airspace in the world. We use this scenario in our simulation, and the real flight data is from the IATA airline schedule database [5]. Figure 2 shows the variation curve of the number of aircrafts over this area on Dec 20, 2016. When the number of aircrafts is large enough to form a path from the aircraft to the ground base station. Then the packet will transmit through the airborne network. Otherwise, satellite will be used instead. That is, the airborne network and satellite are used alternately to transmit the black box's data to the ground center.

Ankan and Chougule [6] propose to transmit the FDR's data via XBEE-RF module to the ground station. But this approach only covers limited parameters in FDR, such as engine temperature, fuel level, speed, and location etc. Nagrath et al. [7] propose to make FR work as a DTN (Delay/Distruption Tolerate Network) device. When the analyzer program senses sudden abnormality in an airplane, it generates alarm to alert components of black box to start functioning as a DTN device and transmitting the data to remote DTN devices through long range wireless communication radio. But the flight may abnormally crash when the determination that the flight is not normal is made. In [8], the flight data in the black box is transmitted to the ground center through radio system when the aircraft is over the land, but satellite over seas. However, the cost of equipment and bandwidth of satellite communication are high.

In this paper, a hybrid approach for transmitting the data in FR is proposed in order to reduce the cost of transmitting data, where the aerial network as a primary approach and the satellite communication as a secondary way when there has no neighbor aircrafts or ground station around. The rest of this paper is organized as follows. System architecture is introduced in Sect. 2. Problem formulation is describe in Sect. 3. Simulation is provided in Sect. 4. Finally, the conclusion in presented in Sect. 5.

2 Wireless Flight Recorder System Architecture

As mentioned in Sect. 1, we use the ground base stations and satellites as infrastructures to receive black boxes' data. The dynamic nature of aerial networks often results in poor network performance because the links between nodes are constantly being formed and broken. Aircraft nodes need to learn knowledge of network changes instantly and use the new topology information to form a new routing solution. However, this mechanism is time-consuming and does not allow for fine-grained control over time.

Software-defined network (SDN) [9] provides a manageable, adaptable, and cost-effective framework for aerial network [10]. It decouples the control plane from data plane and then yields a centralized view of the network control plane. The controller can use all the available information to make optimized network decision. We also use SDN and ground infrastructures with the aid of satellite to form a centralized control plane and make the ground controller have the global view of the whole network. Our system architecture is illustrated in Fig. 3. which is composed of three components, the ground control system, the aerial system, and the satellite system.

2.1 Ground Control System

The ground control system includes base stations and a centralized controller. The centralized controller possesses all the information about the aerial network and takes the following responsibility.

(1) Topology management: the centralized controller contains a topology change detector to monitor the aerial network. Once it finds the topology has changed due to the mobility of aircrafts, it informs the controller to take actions.
(2) Network optimization: the controller has the information about the whole network, such as aircrafts positions, link delay and other useful metrics. With these information, the controller can calculate the routing path for each aircraft and send out the route table for intermediate nodes. Then the controller decides whether to apply the solution to the aerial network or not as the cost of control packet is increased. The details is described in Sect. 3.

2.2 Aerial System

The aerial system consists of every commercial aircraft which has a semi-stationary trajectory assigned by the International Civil Aviation Organization (ICAO). Each aircraft is a data source since the black box records data incessantly. This system is similar to the data plane in SDN architecture. Aircrafts transmit data according to the routing table which is calculated by the ground controller. Due to the mobility of aircrafts, the topology is time-dynamic. Thus the aircraft need to check whether the next hop in the routing table is in the transmission range. If the next hop is connected, then it will transmit the data to the next hop. Otherwise, it will send the data through the satellite system.

2.3 Satellite System

The satellite system contains all the available satellites. Satellite communication provides worldwide coverage and has good signal acquisition. The reliability is high during transmitting, and each aircraft equips the mobile transceiver system. But the cost of bandwidth and equipment is high. Thus satellite system would be used in control and data plane as a supplement to aerial system.

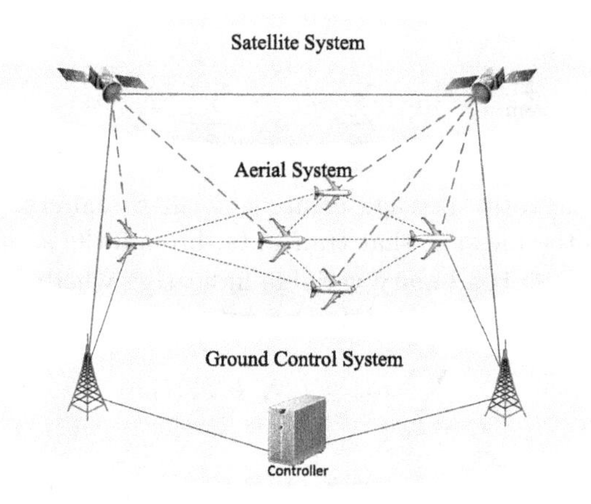

Fig. 3. Schematic diagram of wireless flight recorder system.

(1) Transmit control packet: when the aircraft is flying over seas or remote areas where no infrastructure has been deployed, the controller sends routing table through satellite. Otherwise, the routing table is sent by base station.
(2) Transmit black box data: as we mentioned before, the aircraft will check the next hop before transmitting. If the next hop is not in the transmission range, it will send data through satellite.

3 Wireless Flight Recorder Data Streaming Optimization

The centralized controller has a global view of the aerial network, such as the aircrafts' positions, data distribution, and wireless link states. We use these information to formulate a linear programming problem and obtain the optimal inter-aircraft routes by solving it. The centralized controller then sends the calculated routes to the aircrafts through the control channels. As mentioned before, satellites can provides alternate ways to transmit recorded data from the aircrafts, but the costs of satellite channels are very high. Thus, in this section we tend to minimize the usage of satellites and try to increase the utilization of inter-aircraft communications to support flight recorder data streaming tasks.

3.1 Problem Formulation

We abstract the network into an undirected graph. Due to the mobility of nodes, we model the network with a dynamic graph. At time t, the network is defined as $G(t) = (V(t), E(t))(t \in R)$, where $V(t)$ is the set of plane nodes in the network, and $E(t) = \{(u, v) : u, v \in V(t)\}$ is the set of all edges connecting pairs of nodes at time t. Let $P_v(t)$ be the set of paths that flight v can follow to transmit the data to the ground station at time t. m_v is the message created by aircraft v.

The objective function is

$$Z_{min} = min(\sum_{v} m_v - \sum_{p \in P_v(t)} x_{vp} m_v) \tag{1}$$

where $\sum_v m_v$ is the total message created by all the aircrafts at time t and $\sum_{p \in P_v(t)} x_{vp} m_v$ is the message that transmits through the aerial network. Here $x_{vp} \in \{0, 1\}, \forall p \in P_v(t)$ is a binary variable indicating whether path p is chosen for aircraft v to transmit data at time t or not with the following constraint,

$$\sum_{p \in P_v} x_{vp} \leq 1, \forall p \in P_v(t) \tag{2}$$

which ensures that each aircraft can be allocated at most one path to transmit data.

$$\sum_{v} \sum_{p \in P_v(t) | e \in p} m_v x_{vp} \leq c_e, \forall e \in E(t) \tag{3}$$

$$x_{vp} \sum_{e \in p} w_e \leq x_{vp} W, \forall v, \forall p \in P_v(t) \tag{4}$$

where w_e and c_e are the propagation delay and capacity of link e, and W is the end-to-end delay constrain of message m_v. Inequalities (3) and (4) show the capacity and delay constrain respectively.

The above problem is a 0–1 programming problem which is NP-hard. In order to solve the problem, we consider the column generation algorithm [11] with the combination of simplex method. This algorithm is usually used to solve large scale linear programming problems. It considers a small scale variables at each step and iterates to solve the entire problem. At each iteration, the solution of the (1)–(4) is a feasible set of paths for data transmitting to the ground station. We has $Z(t) \geq Z_{min}$ at each iteration. Define the optimal gap between current $Z(t)$ and the optimal Z_{min} as follows,

$$g(t) = Z(t) - Z_{min} \tag{5}$$

where $Z(t)$ continues to approach Z_{min} after each iteration, and $g(t)$ means the extra amount of data that need to use satellite to transmit in contrast with the optimal one decreasing after each iteration. Define the evolution of $g(t)$ as follows,

$$g(t+1) = (1-k)g(t) \tag{6}$$

where $k \in (0, 1)$ is a constant that represents the relation of next optimal gap and the current one. In order to verify the decreasing process of $g(t)$, we using CG algorithm solver in a static network topology with 20 nodes. The result is presented in Fig. 4. It gives the change of $g(t)$ with the number of iterations.

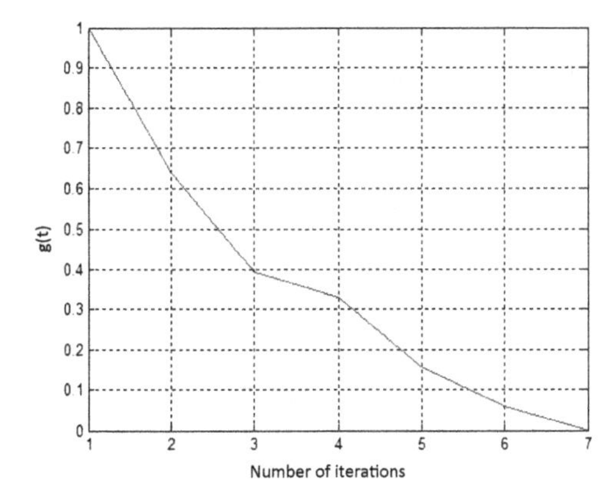

Fig. 4. Changes of $g(t)$ with the number of iterations.

3.2 Dynamic Network Optimization

The topology of aerial network changes with time due to the mobility of aircraft. In order to improve the stability of the network, aircrafts current positions and direction information can be used to predict the lifetime of the links between the nodes. We define threshold T_L. When the link's lifetime is longer than T_L, then we think the link is stable and can be added to the set $E(t)$. Otherwise, the link is not added to $E(t)$.

We have a topology change listener to monitor the topology. Once it finds the topology has changed, it will inform the controller to take actions. The controller will apply the new topology to solve the problem (1)–(4) using CG algorithm. But the optimal solution needs several iterations, especially in large scale. In this iterating period, the mobility of aircraft may result in a new topology, and there may be a new change before we get the optimal solution of (1)–(4). If we want to make the network always work on the optimal solution, we then need to apply the result of each iteration. But this will increase the control traffic in satellite channels. Hence we form another problem which decides whether or not to apply the solution at each iteration.

3.3 Route Reconfiguration Policy

In this section, we formulate a control problem which identifies whether or not to apply the solution of the solver at each slot. Let $\varepsilon(t)$ be the variable that the controller decides whether the network working on a new routing mechanism or not. $\varepsilon(t) = 1$ indicates that the controller applies the solution at time t, otherwise $\varepsilon(t) = 0$. Define the extra usage of satellite as $\sigma(t)$ which equals the difference between the current usage of satellites of the network and the output of the CG solver at slot t. The evolution of $\sigma(t)$ shows as follows,

$$\sigma(t) = (\sigma(t-1) + kg(t-1))(1 - \varepsilon(t)) \tag{7}$$

If $\varepsilon(t) = 1$, then $\sigma(t) = 0$, which means the network applies the solution of the solver at slot t. If $\varepsilon(t) = 0$, the $\sigma(t)$ is increased by $kg(t-1)$, which is the improvement of the satellite usage computed by the solver not applied to the network.

The objective is to formulate a control policy which selects the $\varepsilon(t)$ at each slot and minimizes the average surcharge usage of satellites. The constraint is the average number of update the network equals the number of we select the $\varepsilon(t) = 1$. Now, we formalize a stochastic optimization problem as follows,

$$min \lim_{T\to\infty} sup \frac{1}{T} \sum_{t=0}^{T-1} E[\sigma(t)] \tag{8}$$

$$s.t. \lim_{T\to\infty} sup \frac{1}{T} \sum_{t=0}^{T-1} E[\varepsilon(t)] \leq f_n \tag{9}$$

where $0 < f_n \leq 1$ is the frequency of updating routing table at each slot.

In order to make the above problem easy to solve, we apply the solution of the solver after each change of the network topology. It means that the slot after each jump of $g(t)$, $\varepsilon(t) = 1$. Define a renewal frame is the continues slots between two jump of $g(t)$ which we apply the solution of the solver. Let t_n be the n_{th} update after the n_{th} jump and $s_n = t_{n+1} - t_n$ be the number of slots in n_{th} frame.

Our goal is to get the control sequence by solving the above problem. We propose the solution based on the theory of drift-plus-penalty (DPP) algorithm which is proposed by Neely [12]. It indicates that a way to solve the stochastic optimization problem is the use of policies that greedily balance the penalty in a time slot with the instability of virtual queues. In this paper, we can define a virtual queue in order to track the cost of each update of routing table. The virtual queue is

$$Q(t_{n+1}) = \left[Q(t_n) - s_n f_n + \sum_{t=t_n+1}^{t_n+s_n} \varepsilon(t) \right]^+ \tag{10}$$

where $\sum_{t=t_n+1}^{t_n+s_n} \varepsilon(t)$ is the number of update in the last renewal frame and $s_n f_n$ is the max number of update in this frame. The mechanism of virtual queue is to introduce a cost which in our system is the cost of satellite. We track the virtual queue at the end of each renewal frame. The stability of virtual queue can guarantee that the system is feasible and satisfies (7).

In order to investigate the real-time stability of $Q(t_n)$. We define the quadratic Lyapunov drift as

$$\Delta(s, Q(t_n)) = E[Q^2(t_n + s_n) - Q^2(t_n) | \sigma(t_n) = s, Q(t_n)]$$

For a given policy, $\Delta(s, Q(t_n))$ presents the changes of $Q(t_n)$ in each slot. Any policy that can make the $(s, Q(t_n))$ has a bounded value as $\Delta(s, Q(t_n))$

can let the virtual queue $Q(t_n)$ be a mean rate stable. And these policies are all feasible to solve our problem.

Our primary objective is to minimize the average of the extra usage of satellites which is proposed in (8). Thus we combine the above Lyapunov drift and the objective (8) in each renewal frame as

$$Combine(s, Q(t_n)) = \Delta(s, Q(t_n)) + VE\left\{\sum_{t=t_n+1}^{t_n+s_n} \sigma(t)\right\} \tag{11}$$

where V is a constant which weights the two goals. Minimizing the above $Combine$ problem is a balance of stability and usage of satellites. We can extend the $Combine$ problem and get the upper bound. It can be optimized by a method of solving the following optimization problem

$$OPT(t_n; V) = minE\left\{\sum_{t=t_n+1}^{t_n+s_n} V\sigma(t) + Q(t_n)\varepsilon(t)\right\} \tag{12}$$

The optimization problem is to get a feasible control sequence $\varepsilon(t_n+1), \ldots, \varepsilon(t_n+s_n)$ in n-th frame which balances the extra usage $V\sigma(t)$ and the cost of update $Q(t_n)\varepsilon(t)$.

Algorithm 1. Network Decision Algorithm

Input: V

Output: Control sequence $\mathcal{S}$

1 $t_0 \leftarrow 0$

2 $Q(t_0) = V$

3 The slot after each jump of $g(t)$, let $g(t_n) = 1$;

4 Update virtual queue

$$Q(t_{n+1}) = \left[Q(t_n) - s_n f_n + \sum_{t=t_n+1}^{t_n+s_n} \varepsilon(t)\right]^+$$

5 Use standard dynamic programming algorithm to solve (13)

$$minE\left\{\sum_{t=t_n+1}^{t_n+s_n} V\sigma(t) + Q(t_n)u(t)\right\}$$

6 Get the output of the dynamic programming algorithm

$$\mathcal{S} = \varepsilon(t_n+1), \varepsilon(t_n+2), ...$$

7 **return** $\mathcal{S}$

We now give the update policy. It minimizes (12) in each renewal frame. The steps are given as follows. Applying the solution of the solver after each jump of $g(t)$ which means selects $\varepsilon(t) = 1$ after each change of the network and this can be seen as the beginning of a frame. Then update the virtual queue which monitors the evolution of the network state. At the beginning of each slot, it calls a dynamic programming algorithm which approximately solves (12). The return of this algorithm is a infinite control sequence $\varepsilon(t_n + 1), \varepsilon(t_n + 2), \ldots$. And the control algorithm uses this sequence as the decision of control problem in the n_{th} frame until the end of this frame and drops the remain subsequence and starts over. The algorithm can be seen as Algorithm 1. Due to the limits of pages, we do not talk about the detail of the dynamic programming algorithm here.

4 Simulation Results

In this section, we present the simulation results to verify the feasibility of the proposed wireless flight recorder system and assess the performance of recorded data streaming strategy in a realistic aeronautical scenario. We generate flight trajectories of aircrafts according to the airline flight schedule database published by the International Air Transport Association (IATA) and simulate in a 6-h time window, from 18:00 UTC to 24:00 UTC. The aerial network is formed in a rectangle area (from 45°N to 65°N latitude and 5°W to 60°W longitude), which means we only consider the aircrafts flying in this area. In the simulation we also assume that the aircrafts flying in this area have the same wireless communication range of 200 km for aircraft to ground station communications and 200 km for inter-aircraft communications.

All aircrafts generate streaming traffics with the same data rate of 3 bps for the FDR and 120 kbps for the CVR. The simulated scenario consists of six ground base stations as shown in Fig. 5.

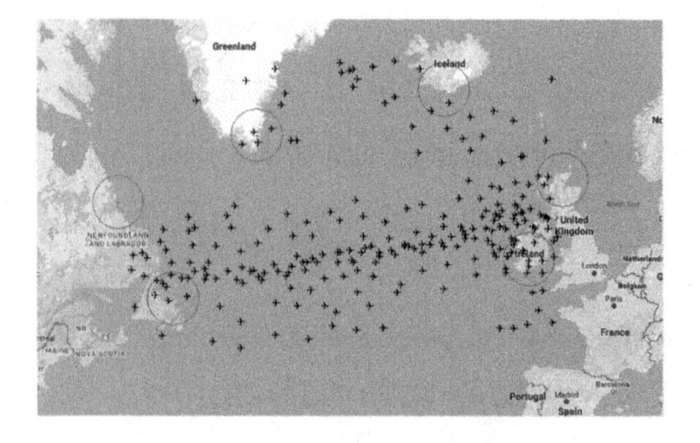

Fig. 5. Positions of the ground stations.

4.1 End-to-End Delay

Figure 6 shows the end to end delay of the packet which transmits from the aircraft to the ground station. We recorded each packet's delay time during the simulation time, and calculate the frequency of the packet's delay time. As it can be seen, most packet's end to end delay is below 250 ms, only a few around 500 ms. The reason is that there maybe no neighbors or ground stations around. Thus the aircraft needs to transmit the packet through satellite. The 500 ms is a two-way end to end propagation delay for a geostationary satellite link.

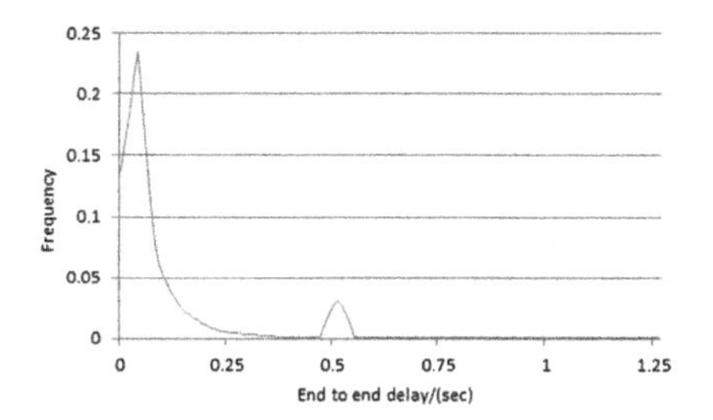

Fig. 6. End-to-end delay distribution.

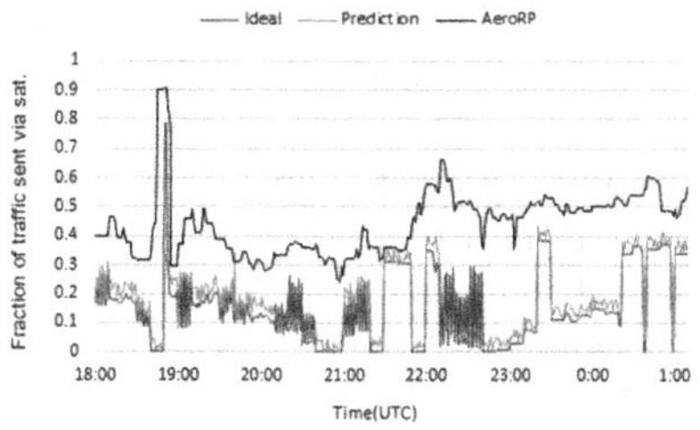

Fig. 7. Fraction of recorded data streaming sent via satellite.

4.2 Usage of Satellites

Figure 7 gives the ratio of the usage of satellites. We define the usage of satellites as

$$f = \frac{n}{m} \cdot 100\%$$

where m is the total number of packets that created by the aircrafts in every 5 min and n is the packets that transmit through satellite. We compare 3 situations in this figure. They are the idea situation, the prediction situation and the AeroRP situation.

(1) The ideal situation: in this situation, the time spent on transmitting routing table is not considered. If two nodes are in the transmission range, there is a link between them.
(2) The prediction situation: in this situation, we use the aircrafts' position, direction and speed to predict the live time of the link between two nodes. We define a live time threshold T_L, if a link's live time is longer than T_L, then the link can be used for data transmission, otherwise can't.
(3) The AeroRP situation: AeroRP is a routing protocol used for highly dynamic aerial network, the more detail can be found in [13].

From the result, it can be found that the prediction situation's usage of satellites is a little higher than the usage of ideal's and both of them are lower than the AeroRP situation. But in the real situation, we need to think about the time consumption of the transmission of the routing table, and the live time of the link needs to be considered.

5 Conclusion

In this paper, we propose to transmit the black box's data to the ground center through aerial network with the combination of satellite communication. The centralized controller has a holistic view of the whole network. We use this information to formulate an optimize problem and solve it using column generation algorithm. Due to the time-dynamic of the aerial network, another stochastic optimization problem is formulated to give the control sequence to decide whether to update the network in each slot. The simulation results show that our system is a feasible way to transmit the data in the black box to the ground center. In the future, we will try to transmit the video in the FR and try to reduce the usage of satellites in a better way.

Acknowledgment. This work was supported by the National Key Research and Development Program of China under grant 2017YFB0802202 and by the Natural Science Foundation of China (NSFC) under grants 61702474 and 91638301. The work of Y. Fang was partially supported by US National Science Foundation under grants CNS-1409797 and CNS-1343356.

References

1. Schoberg, P.R.: Secure ground-based remote recording and archiving of aircraft "Black Box" data. Naval Postgraduate School, Monterey (2003)
2. BBC News: Missing Malaysia plane: what we know, January 2015. http://www.bbc.com/news/world-asia-26503141
3. PlaneCrashInfo: Accident Database. http://www.planecrashinfo.com/database.htm
4. BBC News: Germanwings crash: what happened in the final 30 minutes, March 2017. http://www.bbc.com/news/world-europe-32072218
5. IATACodes: IATA Codes Database. http://iatacodes.org/
6. Ankan, A., Chougule, S.B.: Wireless flight data recorder (FDR) for airplanes. Adv. Mater. Res. **433**, 6663–6668 (2012)
7. Nagrath, P., Aneja, S., Purohit, G.N.: BlackBox as a DTN device. Int. J. Next-Gener. Comput. **6**, 57–65 (2015)
8. Shaji, N.S., Subbulakshmi, T.C.: Black box on earth-flight data recording at ground server stations. In: 2013 Fifth International Conference on Advanced Computing (ICoAC), pp. 400–404. IEEE (2013)
9. Open Networking Foundation: Software-Defined Networking (SDN) Definition. https://www.opennetworking.org/sdn-definition/
10. Iqbal, H., Ma, J., Stranc, K., et al.: A software-defined networking architecture for aerial network optimization. In: 2016 IEEE NetSoft Conference and Workshops (NetSoft), pp. 151–155 (2016)
11. Lbbecke, M.E.: Column generation. In: Wiley Encyclopedia of Operations Research and Management Science (2011)
12. Neely, M.J.: Stochastic network optimization with application to communication and queueing systems. Synth. Lect. Commun. Netw. **3**(1), 1–211 (2010)
13. Peters, K., Jabbar, A., Cetinkaya, E.K., et al.: A geographical routing protocol for highly-dynamic aeronautical networks. In: 2011 IEEE Wireless Communications and Networking Conference (WCNC), pp. 492–497 (2011)

TOA Estimation Algorithm Based on Noncoherent Detection in IR-UWB System

Yuanfa Ji[1,2], Jianguo Song[1(✉)], Xiyan Sun[1,2], Suqing Yan[1,2], and Zhengquan Yang[1]

[1] School of Information and Communication, Guilin University of Electronic Technology, Guilin 541004, China
jiyuanfa@163.com, 1638240755@qq.com, sunxiyanl@163.com
[2] Guangxi Key Laboratory of Precision Navigation Technology and Application, Guilin 541004, China

Abstract. In impulse radio-ultra wideband (IR-UWB) indoor positioning algorithm, ranging accuracy has always been the focus of the research. A threshold comparison (TC) algorithm based on the mean to minimum energy sample ratio (MMR) is proposed which dynamically set the normalization threshold based on the non-coherent detection energy sampling sequence. The new normalization threshold is defined, the corresponding function between MMR and the optimal normalization threshold is observed and the mathematical function model is established. According to a single sample value of the MMR, the energy block sequence number of the direct path (DP) is captured to realize the time estimation of arrival (TOA). The simulation results show that, compared with the classical TOA estimation algorithms, the MMR-TC algorithm has the best accuracy in all signal-to-noise ratio range under CM1 channel, and has the smallest error in CM2 channel.

Keywords: Impulse radio-ultra wideband (IR-UWB) · Mean to minimum energy sample ratio (MMR) · Normalized threshold · Direct path (DP) · Arrival time estimation (TOA)

1 Introduction

Nowadays, the global positioning system (GPS) technology has reached a perfect performance in outdoor applications, but its limitations are fully exposed in the indoor environment. IR-UWB uses microsecond or even nanosecond narrow pulse to modulate the information, with good time resolution and multipath discrimination ability. What's more, UWB can support large-capacity communication, which is very suitable for high-precision positioning applications, making up the shortages of GPS [1].

The National Natural Science Fundation of China (61362005, 61561016, 11603041) Innovation Project of Guet Graduate Education (2018YJCX19, 2018YJCX22) Guangxi information science experiment center funded project (桂科 AC16380014, 桂科 AA17202048, 桂科 AA17202033). The project of Design of pseudolite indoor positioning system design and its localization algorithm (ky2016YB164) in basic ability Promotion of young and middle-aged teachers in Universities of Guangxi province.

S. Chellappan et al. (Eds.): WASA 2018, LNCS 10874, pp. 201–210, 2018.
https://doi.org/10.1007/978-3-319-94268-1_17

In IR-UWB systems, the performance of TOA estimation directly influences the accuracy of the subsequent localization algorithms. The signal's direction-of-arrival (DOA) can be captured through coprime array [2, 3], when used in UWB indoor positioning system, the DOA of target node can be locked accurately in line of sight (LOS) environment, while the DP signal is attenuated severely and the detection error of DOA estimation increases in complex non-line of sight (NLOS) environment. The distance between the receiving and transmitting nodes can be calculated in literature [4], using the received signal strength (RSS) and angle of arrival (AOA), but there is serious multipath phenomenon in the indoor transmission of IR-UWB signals, so this scheme is not applicable in IR-UWB indoor positioning system.

Therefore, there are two types of way to process the received detection signal, that is, coherent solver and non-coherent solver. In Ref. [5], TOA estimation based on matched filter is proposed, which belongs to coherent solver and can achieve high precision of ranging. However, it is difficult for receiver to generate accurate matching template, and the operation is too complicated to realize in engineering. Based on the above, various TOA estimation algorithms based on non-coherent detection of energy have been proposed. Reference [6, 7] refers to three classical TOA estimation algorithms: Maximum Energy Selection (MES), Threshold Comparison (TC) and Maximum Energy Selection-Search Back (MES-SB). The MES algorithm chooses the largest energy block as the threshold. However, in the complex multipath environment, the strong path (SP) is always not DP, which leads to large TOA estimation error. Fix-TC does not have a fixed threshold to achieve good ranging accuracy in all SNR ranges. MES-SB algorithm is a further improvement on MES algorithm. However, due to the limitation of MES algorithm, accuracy is difficult to improve.

In the Ref. [8], a TOA estimation algorithm which combines coherence estimation and non-coherent estimation is proposed. Reference [9–11] proposed an algorithm in which the normalized threshold is set dynamically based on the ratio of energy sampling sequence. The accuracy of the above method is better than that of traditional Algorithms. In order to further improve the range precision of IR-UWB positioning system, based on the two-step TOA estimation method of energy detection, a TOA estimation algorithm based on MMR to set normalized threshold is proposed in this paper. By using CM1 and CM2 channel models recommended by IEEE802.15.4a [12], the performance of the traditional TOA estimation algorithm has been greatly improved, and the feasibility of MMR-TC algorithm has been verified. The MMR-TC algorithm lays a theoretical foundation for indoor high-precision positioning.

2 Signal Model of Receiving System

2.1 Expression of Receiving Signal

The multipath signal received by the IR-UWB system can be expressed as:

$$r(t) = \sum_{j=-\infty}^{\infty} k_j w_p(t - jT_f - c_j T_c) + n(t) \tag{1}$$

Where j and T_f are the sequence number and duration of the frame respectively, T_c is the duration of the chip, so the number of chips per frame is $N_c = T_f/T_c$, $k_j \in \{\pm 1\}$ is the random polarity code, $c_j \in \{0, 1, \ldots, N_c - 1\}$ is the pseudo-random sequence of different users and $n(t)$ is additive white Gaussian noise with zero-mean and variance σ^2. $w_p(t)$ is the waveform of a single pulse which goes through the multipath channel and reaches the receiving end, expressed as:

$$w_p(t) = \sqrt{\frac{E_b}{N_s}} \sum_{i=1}^{L} a_i w(t - \tau_i) \tag{2}$$

Where $w(t)$ is a single-diameter pulse waveform with unit energy; a_i and τ_i are the gain and arrival time of each single diameter; the first arrival of the single diameter is DP; the delay τ_1 is also the estimated arrival time of τ_{TOA}; E_b is symbolic energy, N_s is the number of required pulses to transmit a single symbol.

In order to keep the generality, k_j is set as 1, and the received signal is considered as frame-synchronized [13], that is, $\tau_{TOA} < T_f$. What's more, inter-frame interference (IFI) can be completely avoided, which means the condition $T_f > T_{CIR} + C_{\max} T_c$ is satisfied, where T_{CIR} is valid for channel impulse response Duration, $C_{\max}$ is the maximum value of the jump sequence. Then a high accuracy estimate of τ_{TOA} can be completed by detecting the precise position of the energy block where the DP is located.

2.2 Non-coherent Detection and TOA Estimation

The two steps of the TOA estimation method are as follows: Firstly, the non-coherent detection is used to estimate the energy block where the DP is located, and secondly, coherent detection is used to estimate the TOA [8]. The basic flow of two-step incoherent TOA algorithm based on energy detection is shown in Fig. 1 [14].

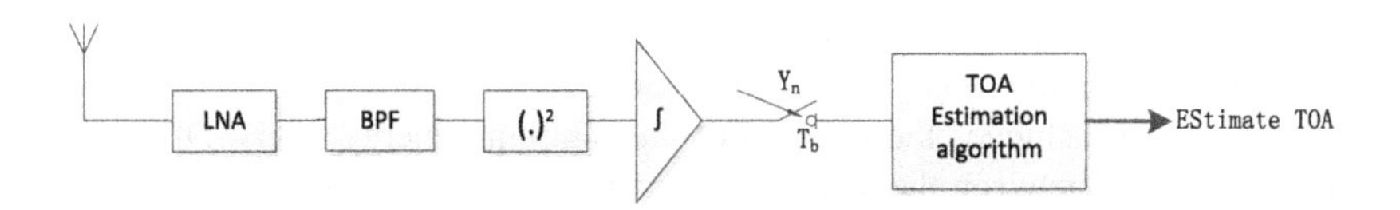

Fig. 1. Non-coherent TOA estimation based on energy detection

Once multipath signal arrives to the receiver, it orderly goes through low noise amplifier (LNA), bandpass filter (BPF) and square integrator. After that, we can obtain the energy sampling sequence of the signal Y_n. The integration period is T_b, and the number of energy blocks is $N_b = \lceil T_f/T_b \rceil$. If T_b is too long, which means the time resolution is low, the estimation error of DP position will increase. Conversely, if T_b is too short, although the DP position accuracy can be improved in a certain range, the comparison times of corresponding threshold and storage space will increase as N_b increases., All things considered, the value of T_b in 1–4 ns will be more reasonable. The TOA estimation algorithm determines the total non-coherent TOA estimation performance. The energy sampling sequence Y_n is then used for the specific TOA estimation algorithm to detect the energy block of DP and complete the TOA estimation based on energy detection. In order to make the TOA estimation more accurate, the energy blocks of multiple frames will be averaged.

3 Optimal Normalized Threshold Algorithm Based on MMR

3.1 The Proposed MMR Algorithm

In the energy-based detection algorithm, the normalized threshold setting is the main factor affecting the performance of the TOA estimation algorithm. How to set the threshold becomes the key to accurately extract the energy block. In this paper, an algorithm based on the ratio of the mean to minimum values of the energy sampling sequence is proposed. The MMR is defined as follows:

$$r = 10\log 10(\frac{mean\{Y_n\}}{\min\{Y_n\}}) \tag{3}$$

The normalization threshold K_{opt} is defined as:

$$K_{opt} = \frac{K - (mean\{Y_n\} + \min\{Y_n\})/2}{\max\{Y_n\} - (mean\{Y_n\} + \min\{Y_n\})/2} \tag{4}$$

In above equation, $\max\{Y_n\}$, $\min\{Y_n\}$ are the maximum and minimum values of the energy sampling sequence, respectively, and $mean\{Y_n\}$ is the mean of all energy sampling sequences. When the sampling frequency is large enough, the integral cycle T_b is set reasonably, and $\max\{Y_n\}$ is in the superimposed region of the useful signal and noise, $\min\{Y_n\}$ must be in the pure noise region, and the r value must reflect the certain SNR information. The TOA estimation algorithm in Fig. 1, as the research point of this paper, achieves the position of the sampling energy of the DP by firstly finding the function between the MMR value and the optimal normalization threshold, then setting the threshold dynamically.

3.2 Simulation Environment

MATLAB simulation environment settings are as follows: The channel are CM1 and CM2 channel that IEEE802.15.4a recommended, choose Gaussian secondary pulse as a signal modulation pulse, the duration of pulse is 1 ns, pulse forming factor is 0.45 ns; 1 side of the sender sent 1bit information, Each bit information occupies only one pulse number; system sampling frequency is 40 GHz; energy coherent integration period T_b

are set to 1 ns and 4 ns respectively; SNR value is {8, 10, 12, 14, 16, 18, 20, 22, 24, 26, 28, 30, 32} dB. By observing the power delay profile of impulse response on CM1 channel and CM2 channel [15], we can find that for more than 99% of the energy, the CM1 channel falls at 120 ns, and the CM2 channel is focused on less than 150 ns. It can be seen that, as the frame period is set long enough, inter-frame interference can be avoided. Therefore, frame period is set to 200 ns. Each simulation is set up with a real TOA value that random evenly distribute in; each channel simulation number is set to 2000.

3.3 The Relationship Between MMR and the Optimal Normalization Threshold

In this section, the relationship between MMR and signal-to-noise ratio is studied. Figure 2 shows the relationship between MMR and SNR for CM1 and CM2 channels. It can be seen that when the signal-to-noise ratio is greater than 12 dB, the MMR value increases with the increase of SNR. The integral period T_b is similar inversely proportional to the MMR value. In the same energy sampling period, the MMR values are different, but the trend is more regular and can reflect the channel feature in a certain degree.

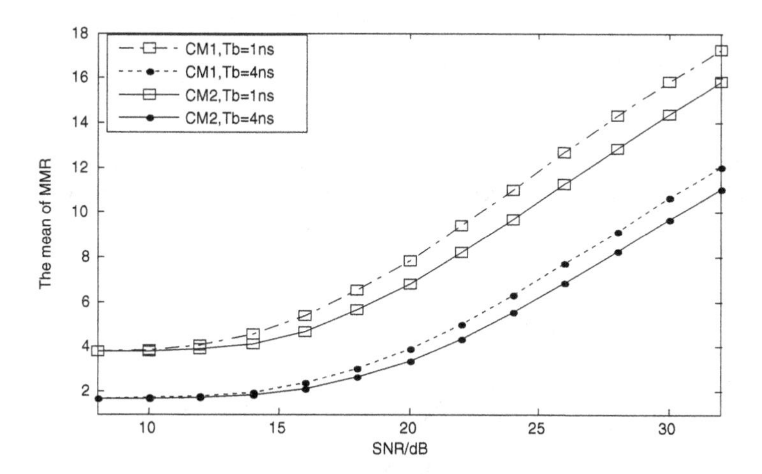

Fig. 2. The relationship between the statistical mean of MMR and the signal-to-noise ratio

The Mean Absolute Error (MAE) is used as an indicator to measure the TOA estimation algorithm. The MAE is defined as follows:

$$MAE = \frac{1}{N}\sum_{n=1}^{N} |t_n - \dot{t}_n| \quad (5)$$

Where t_n is true TOA, $\dot{t}$ is estimated TOA.

Figure 3 shows MAE with respect to the normalized threshold in different MMR value, with CM1 and CM2 channel and T_b fixed. Under the condition of a determining MMR value, the corresponding optimal normalization threshold, represented by K_{opt}, is obtained when the MAE reaches the minimum.

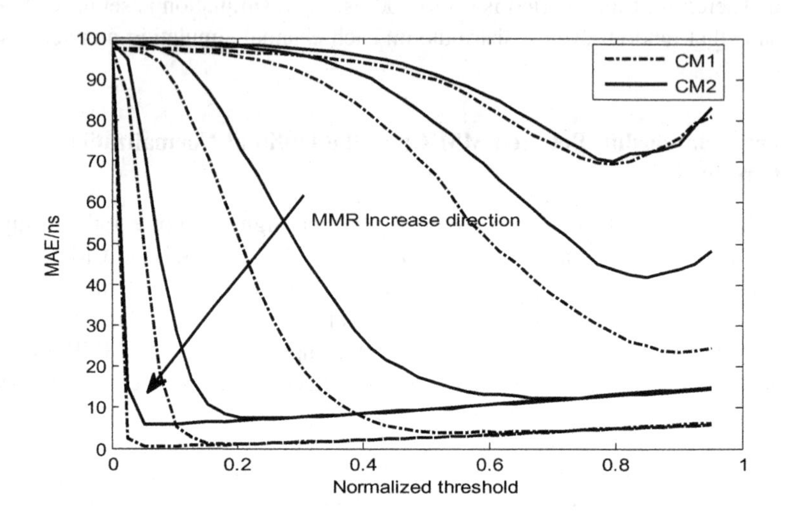

Fig. 3. The relationship between the estimation of MAE by TOA under different MMR with normalized threshold K_{opt}

Through a large number of experiments, we obtained the relation function between MMR and the optimal normalized threshold. Figure 4 shows the simulation results.

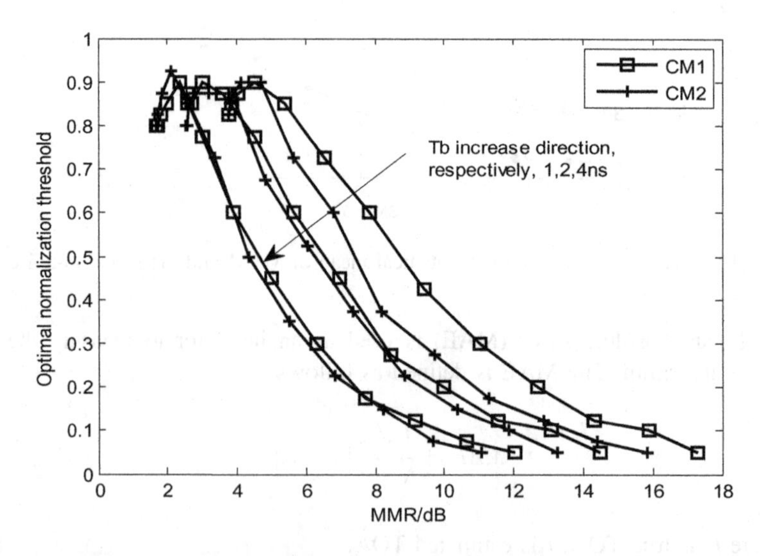

Fig. 4. Relationship and fitting between the optimal normalized threshold and MMR

It can be seen from Fig. 4, the discrete normalized threshold of the CM1 and CM2 channels are basically the same, and as the energy sampling period changes, the shape of the curve remains almost unchanged but just moves to the left or right. A large number of functions are selected to fit the relationship between the optimal normalized threshold and MMR, resulting in that Eq. (6) fits best. Furthermore, it is well matched with T_b under different channels.

$$K_{opt} = a * \exp(b * r) + c * \exp(d * r) \tag{6}$$

The parameters a, b, c and d are set according to the specific channel and the energy sampling period. Tables 1 and 2 show the parameters of CM1 and CM2, respectively. The coefficients are satisfied by 95% of the confidence interval.

Table 1. CM1 channel under different parameters of the specific parameters of the fitting

Parameter	1 ns	2 ns	4 ns
a	5.094	3.105	2.151
b	−0.2532	−0.2769	−0.3156
c	−9.7	−6.705	−9.239
d	−0.5712	−0.8938	−1.764

Table 2. CM2 channel under different parameters of the specific parameters of the fitting

Parameter	1 ns	2 ns	4 ns
a	3.333	3.089	1.852
b	−0.2596	−0.2918	−0.3036
c	−191.7	−12	−66.55
d	−1.621	−1.135	−3.147

According to the above fitting methods, we can obtain the function between the optimal normalized threshold and MMR under CM1 and CM2 mode according to the specific value of T_b.

3.4 MMR-TC Algorithm Specific Process

The TOA estimation algorithm with adjustable/adaptive normalized threshold based on MMR is as follows:

(1) Calculate the energy sampling sequence Y_n, substitute Y_n into Eq. (3) to calculate the MMR value;

(2) According to the specific channel, the energy integration cycle check table selects fitting parameters and MMR value, substitute them into Eq. (6) to calculate the optimal normalization threshold K_{opt} corresponding to the current MMR value;
(3) Substitute K_{opt} into the formula (4) to obtain the corresponding threshold, the calculation method as:

$$K = K_{opt} \ * \ \max\{Y_n\} + (1 - K_{opt}) \ * \ (mean(Y_n) + \min(Y_n))/2 \tag{7}$$

To prevent the threshold from exceeding the permitted scope, correct K.

$$K = \begin{cases} \min\{Y_n\}, & K < \min\{Y_n\} \\ \max\{Y_n\}, & K > \max\{Y_n\} \\ K, & others \end{cases} \tag{8}$$

(4) Substitute K of step 3 into Eq. (9) to estimate the TOA value.

$$T_{toa} = [\min\{n | Y_n > K\} - 0.5]T_b \tag{9}$$

4 Simulation Results and Analysis

The simulation environment is given in Sect. 3.2, and the MMR-TC algorithm proposed in this paper is compared with the MMR-TC (WU) algorithm proposed by Wu [9] and three classical non-coherent TOA estimation algorithms. Parameter setting, the fixed threshold Fix-TC algorithm is 0.4, the backtracking of MES-SB algorithm is set to 40 ns; energy sampling period is 1 ns, then introducing a real TOA random distribution in, finally we can get average value of 2000 independent simulation under CM1 and CM2 channel, the results are shown in Fig. 5.

It can be seen from the simulation results that: compared with other TOA estimation algorithms, the proposed algorithm is superior in MAE under all signal-to-noise ratios in CM1. In CM2 channel mode, though the MMR-TC algorithm is slightly worse than MES-SB algorithm between 18 dB and 24 dB, its overall advantage in distance error performance is obvious. Therefore, the results verify the feasibility and effectivity of the MMR-TC algorithm.

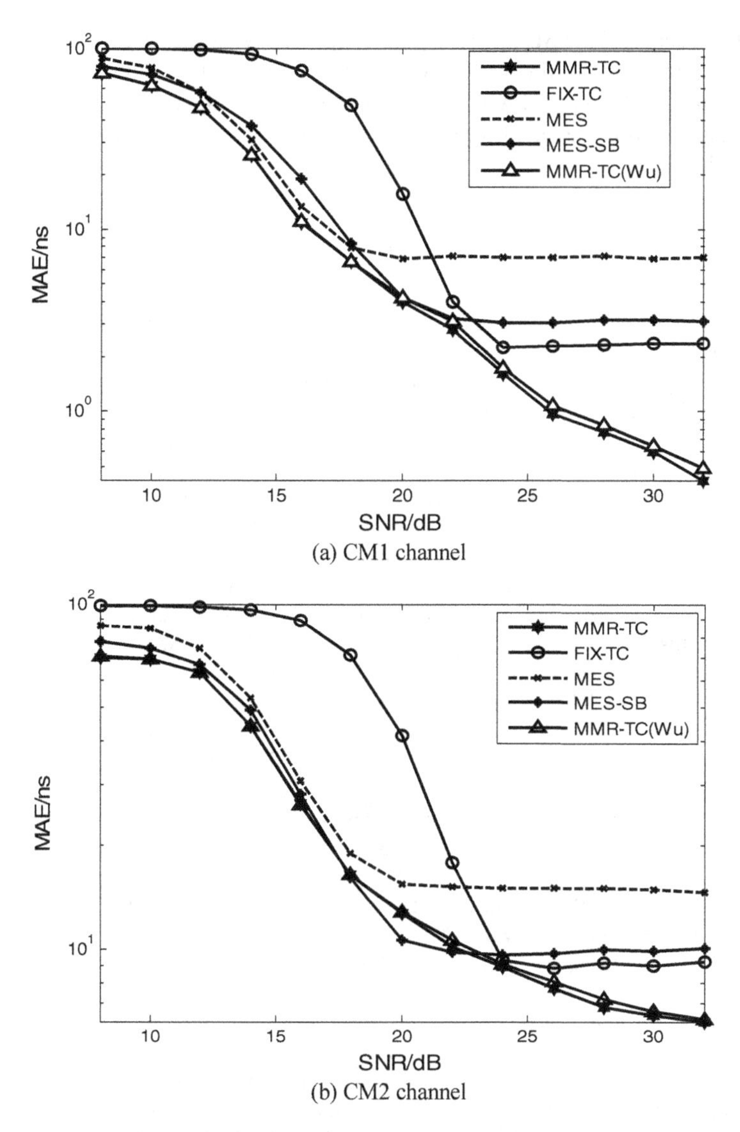

(a) CM1 channel

(b) CM2 channel

Fig. 5. Comparing the performance of MMR-TC algorithm and traditional TOA estimation algorithm

5 Conclusion

Distance measurement accuracy is an important index for measuring the indoor positioning performance of IR-UWB. In the TOA estimation algorithm based on energy detection, no classical localization algorithm has an advantage in all SNR. In general, UWB signal has a high SNR at the receiving. Therefore, it is necessary to improve the error performance of the TOA estimation algorithm in the high SNR segment to obtain high range accuracy. The MMR-TC algorithm presented in this paper is based on the normalized threshold of MMR. In the calculation of sample under the premise of sample calculation, compared with other TOA estimation algorithms, the MAE of the proposed algorithm is the smallest, which reflects the superiority and effectivity of the proposed algorithm.

References

1. Djeddou, M., Zeher, H., Nekachtali, Y., Drouiche, K.: TOA estimation technique for IR-UWB based on homogeneity test. ETRI J. **35**(5), 757–766 (2013)
2. Shi, Z., Zhou, C., Gu, Y., et al.: Source estimation using coprime array: a sparse reconstruction perspective. IEEE Sens. J. **17**(3), 755–765 (2017)
3. Zhou, C., Gu, Y., Zhang, Y.D., et al.: Compressive sensing-based coprime array direction-of-arrival estimation. IET Commun. **11**(11), 1719–1724 (2017)
4. Khan, M.W., Salman, N., Kemp, A.H.: Optimised hybrid localisation with cooperation in wireless sensor networks. IET Signal Process. **11**(3), 341–348 (2017)
5. Wu, S., Zhang, Q., Zhang, N.: TOA estimation based on match-filtering detection for UWB wireless sensor networks. J. Softw. **20**(11), 3010–3022 (2009)
6. Guvenc, I., Sahinoglu, Z.: Threshold-based TOA estimation for impulse radio UWB systems. Mitsubishi Electric Research Laboratories, Cambridge, MA, USA (2005)
7. Guvenc, I., Sahinoglu, Z.: Multiscale energy products for TOA estimation in IR-UWB systems. In: IEEE Global Telecommunications Conference, GLOBECOM 2005 (2005)
8. Gezici, S., Sahinoglu, Z., Molisch, A.F., Kobayashi, H., Poor, H.V.: Two-step time of arrival estimation for pulse-based ultra-wideband systems. EURASIP J. Adv. Signal Process. **2008** (1), 529134 (2008)
9. Wu, S., Zhang, Q., Zhang, G.: Novel threshold-based TOA estimation algorithm for IR-UWB systems. J. Commun. **29**(7), 7–13 (2008)
10. Zhai, S., Qian, Z., Wang, X., Sun, F.: TOA estimation based on normalized threshold for IR-UWB systems. J. Beijing Univ. Posts Telecommun. **38**(4), 19–23 (2015)
11. Luo, S., Li, Q., Ding, G., Wang, Y.G.: TOA estimation based on energy detection for IR-UWB system. Comput. Technol. Dev. **26**(11), 134–138 (2016)
12. Molisch, A.F., Balakrishnan, K., Cassioli, D., et al.: IEEE 802.15.4a Channel Model Summary Report [DB/ OL]. http://www.ieee802.org/15/pub/2004. Accessed 10 July 2008
13. Tian, Z., Giannakis, G.B.: A GLRT approach to data-aided timing acquisition in UWB radios-Part I: algorithms. IEEE Trans. Wirel. Commun. **4**, 2956–2967 (2005)
14. Xia, J., Zheng, L., Wang, M.: High-accuracy double-threshold TOA estimation in impulse-radio ultra wideband systems. Comput. Eng. Appl. **49**(10), 211–215 (2013)
15. Molisch, A.F.: IEEE802.15.4a channel model-final report. IEEE P802.15-04/662r0-SG4a **15** (12), 911–913 (2004, 2005)

Max-Min Fairness Scheme in Wireless Powered Communication Networks with Multi-user Cooperation

Ming Lei, Xingjun Zhang(✉), and Bocheng Yu

Xi'an Jiaotong University, Xi'an, China
xjzhang@xjtu.edu.cn

Abstract. In wireless powered communication networks (WPCNs), different channel states of energy harvesting and information transfer result in the large gap of throughput performance between user nodes. In this paper, the multi-user cooperation was used to improve the throughput fairness in the WPCN. We discuss one important fairness problem called the max-min throughput problem with resource allocation. The problem is modeled as a mixed-integer nonlinear programming. Although it is hard to be solved by existing methods, we use a branch and bound method based on the piece-wise linear method to obtain the lower bound. The simulation results that the proposed method achieves an average of 97% performance gain when path loss component $\alpha = 3$ with random progress method.

Keywords: WPCN · Multi-user cooperation · Max-min throughput
Resource allocation

1 Introduction

Harvesting energy from wireless power transfer (WPT) provides a promising solution to combat the power limitation in Wireless Sensor Networks. Based on the WPT, the wireless powered communication network (WPCN) can be used to prolong the operating lifetimes and provide a stable energy to its user nodes (UNs), such as sensor nodes [1]. Each UN in the WPCN harvests energy from an energy transmitter (ET) in the downlink, and uses the energy to transmit information in the uplink.

In WPCNs, the "near-far" effect leads to throughput performance unfairness [2]. These are widely researched about improving fairness to combat the "near-far" effect in the WPCN. In the duration of harvesting energy, it has been proposed that employing multi-antenna ET to perform the WPT with the energy beamforming [3]. In [4], the weighted sum-rate maximization problem was considered to incorporate fairness based on the energy beamforming. Due to the complexity and high cost of multi-antenna ET deployment, some solutions are researched by resource allocation to balance throughput among the UNs in the single-antenna ET case. Based on the same system model in [2], the authors considered the circuit power dissipation, which was not negligible especially for the WPT. They also proposed a low-complexity fixed point iteration algorithm for the max-min throughput problem in [5].

S. Chellappan et al. (Eds.): WASA 2018, LNCS 10874, pp. 211–222, 2018.
https://doi.org/10.1007/978-3-319-94268-1_18

Because cooperation is an efficient method to improve fairness in the WPCN, energy or transmission cooperation has been studied widely in [6–9]. In [10], a "harvest-then-cooperate" protocol was proposed, in which the UNs and their dedicated relay nodes harvest energy, and then cooperatively work to transmit the information of UNs to enhance fairness. Based on the user and relay cooperation, the relay selection problem was studied in [11]. The authors maximized the capacity under an energy transfer constraint. In [12], they proposed a Nash bargaining approach to optimal information transmission efficiency of source-destination pairs with multiple source-destination pairs and one relay. However, setting dedicated relay nodes adds additional costs and increases deployment complexity. User cooperation is another method to improve fairness performance. User cooperation utilized the UNs with harvested more energy to relay the information of UNs which have less energy or poorer transmission channel states [13]. In [14], they presented a new pricing strategy to motivate one UN with more energy to sell its excess energy to help another UN with less energy complete a UL information transfer. The "relay" UN placement and the UN's communication mode selection problem were also discussed. In [15], the paper considered that two communication groups cooperated with each other via the WPT and time sharing to fulfill their expected information delivering and achieve "win-win" collaboration. In [16], the letter focused on fairness-aware power and time allocation in the WPCN under the "harvest-then-transmit" protocol, where downlink wireless energy transfer was implemented at first and then uplink wireless information transfer took place in a spectrum-sharing fashion. They aimed to achieve the rate fairness of all UNs under three fairness criteria named max–min, proportional, and harmonic fairness. But they considered the single transmission method without user cooperation. In [13], they tested the user cooperation between two UNs and proved that it utilized to improve the throughput fairness in the simple two-hop WPCN. In the user cooperation, the forwarding UNs must sacrifice harvested energy and transmission time to improve the throughput of the UNs with less energy. It is very important for improving the throughput fairness to select reasonable forwarding path and balance the UNs' the amount of data transferred and harvested energy. However, it is neglected in the current literature.

In this paper, we study the max-min throughput (MMT) problem in WPCNs. We consider the transmission of UNs by multi-user cooperation without considering deploying relays among UNs. Each UN harvests energy from a hybrid access point (HAP) and is controlled by the HAP. The HAP ether provides the energy in the downlink or receives the information of the UNs. We formulate the problem subject to time allocation, energy harvesting, and traffic conservation constraints. The model of the problem is a mixed-integer nonlinear programming (MINLP), which can not be solved by existing methods directly. We propose a branch and bound method based on the piece-wise linear (PWL) method, where power control, time allocation and link scheduling are jointly optimized. The numerical results will be given to present the performance of the proposed method compared with some traditional methods.

2 System Model and Problem Formulation

We consider a WPCN that consisting of a HAP denoted by H and a set of UNs $\mathbb{N} = \{1, 2, \ldots, N\}$ without fixed energy sources in Fig. 1. According to the "harvest-then-transmit" protocol, the HAP provides wireless power to all UNs in the downlink, and then all UNs use the harvested energy to transmit information. Due to the "near-far" effect, UNs close to the HAP obtain more energy than those far from the HAP. More importantly, these "far" UNs consume more energy to transmit information due to the poorer channel quality. Therefore, we consider the multi-user transmission method to help the "far" UNs transmit information.

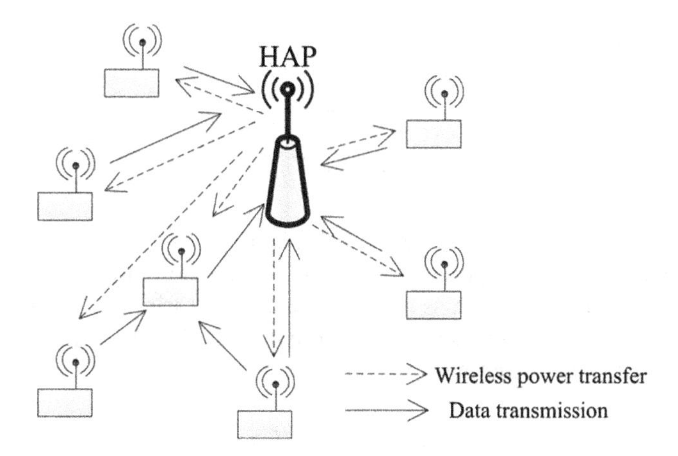

Fig. 1. Model of the multi-hop wireless powered communication network.

We denote the set of all UNs and HAP to V and use L_{ij} to denote the link formed between node i and node j. The uplink and downlink power gains of link L_{ij} are denoted by g_{ij} and h_{ij}, respectively. In a time block, the UNs harvest energy from the HAP in the duration of downlink, which is denoted by t_0. We use P_0 to denote the transmit power of the HAP. The available harvesting energy of UN i is given by the following [2].

$$E_i = \zeta P_0 h_{Hi} t_0 \tag{1}$$

where ζ represents the power conversion efficiency, generally is 50%–70%. Each UN starts to transmit the information through its output links after it gets the harvested energy. Receiving information and transmitting energy simultaneously will cause self-interference at the HAP. Thus, the HAP is only taken as a sink in the uplink and does not provide wireless power.

In the transmission uplink, only if the channel quality of link L_{ij} is better than that of link L_{iH}, the link may be active for forwarding information. Otherwise, the link L_{ij} is close and the UN i transmits information to the HAP directly. We call those links, which may be active, as the feasible links. The set of all feasible links is denoted by L. We introduce the binary variable U_{ij} to indicate whether the feasible link L_{ij} is active in the uplink, $L_{ij} \in L$. If the link L_{ij} is active, then $U_{ij} = 1$. Otherwise $U_{ij} = 0$. When the link L_{ij} is active, its reverse link must be silent in the uplink. Thus, we have

$$1 \geq U_{ji} + U_{ij}, \forall L_{ij} \in L. \tag{2}$$

Let t_{ij} denote the transmission time of link L_{ij}. For the convenience of discussion, we normalize a length of one time block into unit time. We can express the time constraint as follows:

$$\sum\nolimits_{L_{ij} \in L} t_{ij} + t_0 = 1, \tag{3}$$

$$t_{ij} \leq U_{ij}, \forall L_{ij} \in L. \tag{4}$$

Let P_i denote the transmission power of UN i, which takes the value in $[0, P_{max}]$. At the step of the transmission phase, the energy consumption also includes non-ideal circuit energy consumption, such as sampling, signal processing and so on. We use P_c to denote the additional circuit loss energy consumption [17]. The transmission energy consumption of UN i is expressed by

$$E_{ti} = \sum\nolimits_{L_{ij} \in L} (P_c + P_i) t_{ij}, \forall i \in \mathbb{N}. \tag{5}$$

Using TDMA method for transmission to avoid co-channel interference, that is, only one link can be active for data transmission at the same time. Therefore, the quality of each active link only depends on the corresponding signal-noise ratio (SNR) value. Let SNR_{ij} denote the SNR of link L_{ij}. Once the transmission node i of the link L_{ij} determines the level of transmission power, the link capacity C_{ij} can be calculated by

$$C_{ij} = W * log(1 + SNR_{ij}) \tag{6}$$

where $SNR_{ij} = \frac{P_i g_{ij}}{W \eta}$, W is the bandwidth, η is the noise power, log is the function of logarithm (base 2). Let r_{ij} indicate the amount of information through the link L_{ij} in the time block. Then, in the time t_{ij}, the amount of transmission information r_{ij} must satisfy the following capacity constraint:

$$r_{ij} \leq C_{ij} t_{ij}, \forall L_{ij} \in L. \tag{7}$$

When the receiving node j receives the data through the link L_{ij}, it needs to consume a certain amount of energy to the energy dissipation of the radio, which is caused by running the receiver circuitry and denoted by E_{elec} nJ/bit. The consume energy of UN i for receiving information is expressed as [18].

$$E_{ri} = E_{elec} \sum\nolimits_{L_{ui} \in L} r_{ui}, \forall i \in \mathbb{N}. \tag{8}$$

By the inequalities (1), (5), and (8), the energy constraint of UN i can be expressed as

$$E_{ri} + E_{ti} \leq E_i. \tag{9}$$

The data generated and forwarded by each UN needs to be transmitted through its multiple output links. We use f_i to represent the amount of data generated by UN i. Then, the UN i needs to satisfy the traffic constraint as follows:

$$\sum\nolimits_{L_{ij} \in L} r_{ij} - \sum\nolimits_{L_{ui} \in L} r_{ui} \geq f_i, \forall i \in \mathbb{N}. \tag{10}$$

Before the end of the time block, the data of each UN needs to be fully imported into the HAP. Therefore, The HAP needs to satisfy the following traffic conservation constraint:

$$\sum\nolimits_{i \in V/H} f_i = f_H. \tag{11}$$

Constraints (7) and (10) ensure the reliability of each active link. If (7) and (10) hold, then (11) is satisfied. Therefore, the constraint (11) can be relaxed. Let f denote the minimum value among the amount of transmission information of UNs in the time block. Then, the transmission data of each UN is no less than the values in the time block, that is

$$f_i \geq f, \forall i \in \mathbb{N}. \tag{12}$$

The max-min throughput problem is an important issue for optimizing throughput fairness. In this paper, we mainly consider the resource allocation in the max-min problem. With (12), the object function of max-min problem can be transformed to maximize the amount of data value f. The max-min problem is modeled as the MMT, which is

$$\begin{aligned} \text{MMT}: \quad & \max \quad f \\ & \text{s.t.}\ (1) - (10), (12) \\ & U_{ij} \in \{0, 1\}, r_{ij} \geq 0, f \geq 0, P_i \geq 0, t_0 \geq 0, t_{ij} \geq 0 \end{aligned}$$

The MMT is a MINLP model. Such programming problems are generally NP-hard problems [19]. The existing methods can not be used to solve the problem directly. It can be seen from the programming model that the main challenges for solving the model are as follows: (1) the multiplication of nonlinear *log* function and linear variable appears on the right side of the constraint (7) in the MMT; (2) there are some bi-linear product, such as $P_i t_{ij}$. In the next section, we propose converting the model to the mixed-integer linear programming (MILP) model by using the PWL method.

3 Branch and Bound Method Based on the PWL Method for the Max-Min Problem

In the MMT, the constraint (7) contains the *log* function, which leads to the main difficulty to solve the problem. The proposed PWL method is used to linearize the *log* function term [20]. The PWL method's idea is to approximate the *log* curve (base e) by a set of line segments and guarantees the gap between the piece-wise function and the *log* function (base e) denoted *ln* function to be less than a threshold. We denote the threshold to γ. For the sake of discussion, the following constraint is used to exchange the constraint (6), which is

$$C_{ij} = \frac{W}{ln2} ln\left(1 + SNR_{ij}\right) \tag{13}$$

Since the *ln* function of the corresponding active links L_{ij} is the function of the SNR_{ij} variable, we segment the interval of variable SNR_{ij}, which is $\left[0, P_{max} g_{ij}/W\eta\right]$, into multiple segments. Let D_{ij} denote the number of segments in the interval $\left[0, P_{max} g_{ij}/W\eta\right]$. The qth segment interval is expressed as $\left(\left(SNR_{ij}^q\right)_L, \left(SNR_{ij}^q\right)_U\right), q \in D_{ij}$. The slope of the qth piece-wise segment corresponding to SNR_{ij} is denoted by v_{ij}^q and expressed as:

$$v_{ij}^q = \frac{\ln\left(1 + \left(SNR_{ij}^q\right)_U\right) - ln\left(1 + \left(SNR_{ij}^q\right)_L\right)}{\left(SNR_{ij}^q\right)_U - \left(SNR_{ij}^q\right)_L} \tag{14}$$

Next, we determine the value of D_{ij} by the PWL method. So that the gap between the piece-wise function and the *log* function is less than the threshold γ. The process of the PWL method is shown in Algorithm 1.

Algorithm 1: PWL method

Input: $\gamma, P_{max}, g_{ij}, W, \eta$

1. Initialization: $q = 0,\ D_{ij} = 0,\ \left(SNR_{ij}^{q}\right)_L = 0$.
2. Use Newton method to solve the following equation as follows:
$-ln\left(v_{ij}^{q}\right) + v_{ij}^{q}\left(1 + \left(SNR_{ij}^{q}\right)_L\right) - 1 - ln\left(1 + \left(SNR_{ij}^{q}\right)_L\right) = \gamma.$
Determine the slope v_{ij}^{q}.
3. Use Newton method to solve the equation (14) to obtain $\left(SNR_{ij}^{q}\right)_U$. If $\left(SNR_{ij}^{q}\right)_U \geq P_{max}g_{ij}/W\eta$, then stop computer and set $\left(SNR_{ij}^{q}\right)_U$ and D_{ij} to $P_{max}g_{ij}/W\eta$ and q. Otherwise continue to step 4.
4. Use the equation $\left(SNR_{ij}^{q}\right)_U = \left(SNR_{ij}^{q+1}\right)_L$ to obtain the value of $\left(SNR_{ij}^{q+1}\right)_L$. Set $q{=}q + 1$, return to step 2.

Output: $\left(SNR_{ij}^{q}\right)_L$, $\left(SNR_{ij}^{q}\right)_U$ and D_{ij}

By Algorithm 1, the *log* function of variable SNR_{ij} can be transferred to the linear piece-wise one. The value of the linear piece-wise function is the lower bound of that of the *log* function. We reconvert constraint (7) with the *log* function to the following group of constraints:

$$r_{ij} \leq \frac{Wt_{ij}}{ln2}\left[v_{ij}^{q}\left(\frac{P_i g_{ij}}{W\eta} - \left(SNR_{ij}^{q}\right)_L\right) + ln\left(1 + \left(SNR_{ij}^{q}\right)_L\right)\right], \forall L_{ij} \in L, q \in D_{ij} \quad (15)$$

By replacing constraint (7) in the MMT with constraint (15), all terms are linear in the MMT except for the term $P_i t_{ij}$. A new variable α_{ij} is used to represent the bi-linear product term. The MMT can be transformed into the MMT1 as follows:

$$\begin{aligned}
\text{MMT1:}\quad & \max\ f \\
\text{s.t.}\ & f_i \geq f, \forall i \in \mathbb{N} \\
& 1 \geq U_{ji} + U_{ij}, \forall L_{ij} \in L \\
& \textstyle\sum_{L_{ij}\in L} t_{ij} + t_0 = 1 \\
& t_{ij} \leq U_{ij}, \forall L_{ij} \in L \\
& \textstyle\sum_{L_{ij}\in L} (P_c t_{ij} + \alpha_{ij}) + E_{elec}\sum_{L_{ui}\in L} r_{ui} \leq \zeta P_0 h_{Hi} t_0 \\
& \textstyle\sum_{L_{ij}\in L} r_{ij} - \sum_{L_{ui}\in L} r_{ui} \geq f_i, i \in V/H
\end{aligned}$$

$$r_{ij} \leq \frac{W}{ln2}\left[v_{ij}^{q}\left(\frac{\alpha_{ij} g_{ij}}{W\eta} - \left(SNR_{ij}^{q}\right)_L t_{ij}\right) + t_{ij} ln\left(1 + \left(SNR_{ij}^{q}\right)_L\right)\right], \forall L_{ij} \in L, q \in D_{ij}$$

$$U_{ij} \in \{0,1\}, r_{ij} \geq 0, f \geq 0, \alpha_{ij} \geq 0, t_0 \geq 0, t_{ij} \geq 0$$

The new model eliminates the variable P_i and is a MILP. It can be solved easily by existing optimization software with the branch and bound method, such as Gurobi and CPLEX. We now give the relationship between the solution of the MMT and that of the MMT1.

Lemma 1: The optimal solution of the MMT1 is a feasible solution of the MMT.

Proof: The only difference between the MMT and the MMT1 is the capacity constraint. The capacity constraint in the MMT1 is obtained by linearizing the right term of the corresponding constraint in the MMT. For the given values P_i and t_{ij}, the value of the right term of the capacity constraint (15) in the MMT1 is no more than that of the capacity constraint (7) in the MMT. The solution space of variable r_{ij} in the MMT1 is covered by that in the MMT. Because of the same solution spaces of other variables, the optimal solution in the MMT1 must be a feasible solution to the MMT. ■

We see from Lemma 1 that the optimal solution in the MMT1 provides a lower bound for the MMT.

4 Simulation Results

4.1 Simulation Setup

In this section, we simulate the network throughput performance used our proposed multi-user cooperation method. We consider three traditional transmission methods as performance benchmarks: One is Random Progress method (RA), in which data of UNs is routed with equal probability through a feasible link. The second one is greedy method (GR), in which the UNs transmit information by a feasible link which receiver node is closest to them. The third method is single-hop transmission method without user cooperation, in which each UN transmits its information to HAP directly. We use "MMT1" to denote the lower bound of optimal solution, which is obtained by solving MMT1 by our proposed method.

In the WPCN, the bandwidth is set to 1 MHz and the noise power $\eta = -90$ dBm. We assume the channel short-term fading is Rayleigh distributed. The channel power gains $g_{ij} = h_{ij} = 10^{-3}\rho_{ij}^2 d_{ij}^{-\alpha}$, $\forall i,j \in \{1,2,\ldots,N\}$ [2], ρ_{ij}^2 follows the Rayleigh distribution, d_{ij} denotes the distance between UN i and UN j, and α is the path loss. We assume a 30 dB average signal attenuation at the 1 m reference distance. The number of UNs is set to $N = 20$. We set $\gamma = 10^{-6}$. All of the methods are implemented by C++ plus GPLEX and operated on a Sony EA37EC notebook with an Intel Core i3 CPU (2.4 GHz) and 4 GB RAM. In each setup, we run 100 instances.

4.2 Simulation Results

Figure 2 displays the average throughput for various values of transmission power of the HAP with $\alpha = 3$. The optimized throughput performance tends to increase with P_0. This appearance is predictable, because larger P_0 gives more energy in the duration for harvested energy and more time for transmission. It is clear from the figures that the throughput used our proposed method or single-hop transmission method is more than that used GR or RA. The poor performance of the traditional methods can be attributed to some main reasons. First, multiple UNs may select a same "relay" UN for information forwarding. With the limited energy, the "relay" UN can become a bottleneck. Second, using the GR methods selects the closest "relay" UN for help forward information. However, the channel may be poor between the "relay" UN and the HAP. The harvested energy of the UN is smaller due to the poor channel. It leads to less information be forwarded by the "relay" UN and decrease the throughput performance. The throughput performance of our proposed method and that of single-hop transmission method are similar as P_0 is small. Our multi-hop transmission method becomes more conducive to throughput performance as increasing the intensity of harvested energy. This is because more energy can be harvested by UNs and used for forwarding information as P_0 become larger.

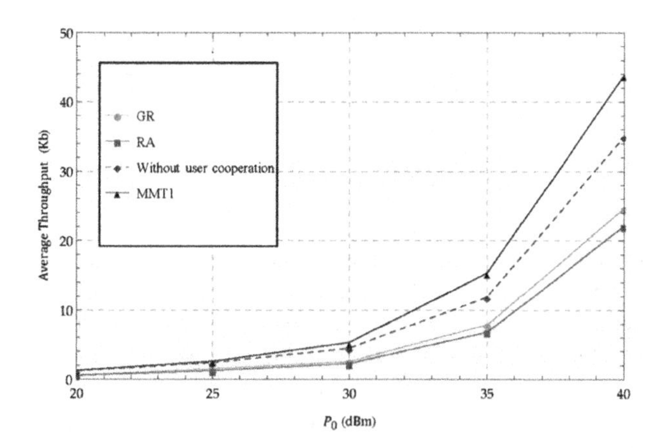

Fig. 2. Change in the transmission power of the HAP

In the second simulation study, the average throughput performance, under individual path loss, is evaluated using different methods with $P_0 = 30$ dBm in Fig. 3. As the path loss decreases, both channels for harvested energy and transmission improve. When the path loss is large, the performance of our proposed method is similar to that of single-hop transmission method without cooperation, but far better than that of the GR or RA method. This is because, as the channel states of the links is poor, each UN can scarcely provide the energy for forwarding. The traditional methods of forcing some nodes to become "relay" nodes can cause bottlenecks in throughput performance for these "relay" nodes. When the path loss is small, the throughput performance gap between the proposed method and the single-hop transmission method increases. This is because the better channels promote UNs to harvest energy and transmit information. The UNs begins to dominate more energy for forwarding information.

We further show the effect of the additional circuit loss energy consumption for transmission in Fig. 4. As the energy consumption increases, the performances of all methods decrease almost linearly with very small slopes because the additional energy consumption has a very low duty cycle in terms of transmission energy consumption.

In the transmission process, most of the energy is provided to UN i to transmit signals, that is $P_c \ll P_i, i \in \mathbb{N}$. The average throughput performance when using our proposed method is superior to those of the other methods.

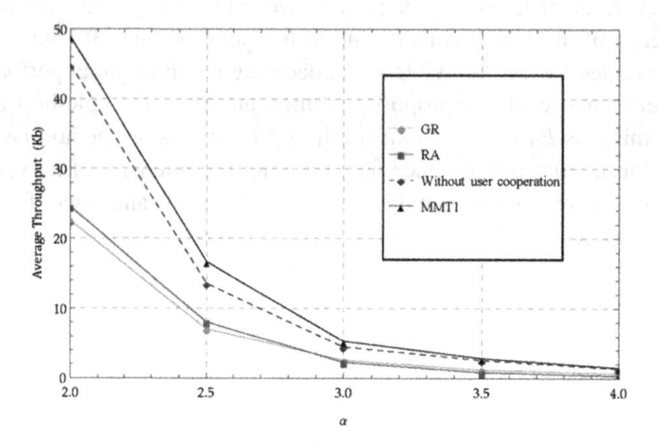

Fig. 3. Change in the path loss

We show the effect of the energy dissipation of the radio caused by running the receiver circuitry with $P_0 = 30$ dBm and $\alpha = 3$ in Fig. 5. Because the single-hop transmission method does not have the information forwarding, its throughput performance is not affected as this parameter value change. We can observe that the throughput performance of both tradition methods and our proposed method as the energy dissipation E_{elec} is larger. This is because UN needs to allocate more energy to cope with the energy cost of receiving the forwarding information as the energy dissipation is increased. As the cost of energy increases, the amount of forward information becomes less and less at each UN. The throughput performance of our proposed method will be closer to that of the single-hop transmission method.

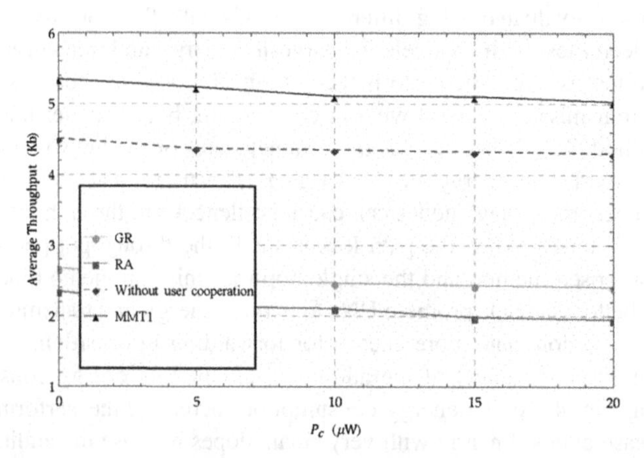

Fig. 4. Change in the circuit loss energy consumption for transmission.

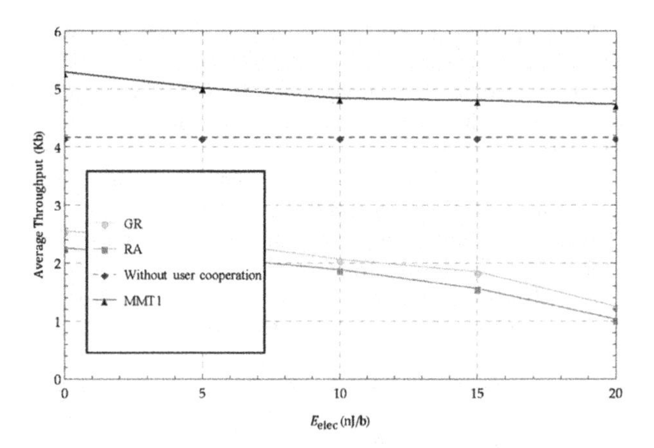

Fig. 5. Change in the amount of the receiver circuitry

5 Conclusion

In this paper, we present the time allocation and power control scheme with multiple user nodes in WPCNs. Since previous works mostly focus on single hop in WPCNs without user cooperation or fixed user nodes pairs with user cooperation, the schemes they used could not be directly used in a multiple user nodes transmission scenario. We derive a piece-wise linear based method to achieve the maximum minimal throughput and we also derive a lower bound for the max-min throughput problem. Simulation results show the proposed method is effective when compared with the random progress method and greedy method.

Acknowledgments. This work was supported by the project DETERMINE funded under the Marie Curie IRSES Actions of the European Union Seventh Framework Program (EU-FP7 Contract No. 318906), and the NSFC project (Grant No. 61572394).

References

1. Lin, S., Zhang, J.: ATPC: adaptive transmission power control for wireless sensor networks. ACM Trans. Sens. Netw. **12**(1), 223–236 (2016)
2. Ju, H., Zhang, R.: Throughput maximization in wireless powered communication networks. IEEE Trans. Wirel. Commun. **13**(1), 418–428 (2014)
3. Huang, W., Chen, H., Li, Y., Vucetic, B.: On the performance of multi-antenna wireless-powered communications with energy beamforming. IEEE Trans. Veh. Technol. **65**(3), 1801–1808 (2015)
4. Lee, H., Lee, K.J., Kong, H.B., Lee, I.: Sum rate maximization for multi-user MIMO wireless powered communication networks. IEEE Trans. Veh. Technol. **65**(11), 9420–9424 (2016)
5. Sun, Q., Zhu, G., Shen, C., Li, X., Zhong, Z.: A low-complexity algorithm for throughput maximization in wireless powered communication networks. In: IEEE Proceedings of IGBSG (2014)

6. Gurakan, B., Ozel, O., Yang, J., Ulukus, S.: Energy cooperation in energy harvesting communications. IEEE Trans. Commun. **61**(12), 4884–4898 (2012)
7. Nasir, A.A., Zhou, X., Durrani, S., Kennedy, R.A.: Wireless-powered relays in cooperative communications: time-switching relaying protocols and throughput analysis. IEEE Trans. Commun. **63**(5), 1607–1622 (2013)
8. Zhou, Z., Peng, M., Zhao, Z., Li, Y.: Joint power splitting and antenna selection in energy harvesting relay channels. IEEE Signal Process. Lett. **22**(7), 823–827 (2014)
9. Atapattu, S., Evans, J.: Optimal energy harvesting protocols for wireless relay networks. IEEE Trans. Wirel. Commun. **15**(8), 5789–5803 (2016)
10. Chen, H., Li, Y., Rebelatto, J.L., Ucha-Filho, B.F., Vucetic, B.: Harvest-then-cooperate: wireless-powered cooperative communications. IEEE Trans. Signal Process. **63**(7), 1700–1711 (2015)
11. Michalopoulos, D.S., Suraweera, H.A., Schober, R.: Relay selection for simultaneous information transmission and wireless energy transfer: a tradeoff perspective. IEEE J. Sel. Areas Commun. **33**(8), 1578–1594 (2013)
12. Zheng, Z., Song, L., Niyato, D., Han, Z.: Resource allocation in wireless powered relay networks: a bargaining game approach. IEEE Trans. Veh. Technol. **66**(2), 6310–6323 (2016)
13. Ju, H., Zhang, R.: User cooperation in wireless powered communication networks. In: IEEE Proceedings of GLOBECOM (2014)
14. Chen, H., Xiao, L., Yang, D., Zhang, T., Cuthbert, L.: User cooperation in wireless powered communication networks with a pricing mechanism. IEEE Access **5**, 16895–16903 (2017)
15. Xiong, K., Chen, C., Qu, G., Fan, P., Letaief, K.: Group cooperation with optimal resource allocation in wireless powered communication networks. IEEE Trans. Wirel. Commun. **16**(2), 3840–3853 (2017)
16. Guo, C., Liao, B., Huang, L., Li, Q., Lin, X.: Convexity of fairness-aware resource allocation in wireless powered communication networks. IEEE Commun. Lett. **20**(3), 474–477 (2016)
17. Pejoski, S., Hadzi-Velkov, Z., Duong, T.: Wireless powered communication networks with non-ideal circuit power consumption. IEEE Commun. Lett. **21**(6), 1429–1432 (2017)
18. Wang, C.F., Shih, J.D., Pan, B.H., Wu, T.Y.: A network lifetime enhancement method for sink relocation and its analysis in wireless sensor networks. IEEE Sens. J. **14**(6), 1932–1943 (2014)
19. Garey, M.R., Johnson, D.S.: Computer and Intractability: a Guide to the Theory of NP-Completeness, pp. 245–248. W. H. Freeman and Company, San Francisco (1979)
20. Shi, Y., Hou, Y., Kompella, S., Sherali, H.: Maximizing capacity in multi-hop cognitive radio networks under the SINR model. IEEE Trans. Mob. Comput. **10**(7), 954–967 (2011)

Cancer-Drug Interaction Network Construction and Drug Target Prediction Based on Multi-source Data

Chuyang Li[1], Guangzhi Zhang[1], Rongfang Bie[1], Hao Wu[1](✉), Yuqi Yang[1], Jiguo Yu[2], and Xianlin Ma[3]

[1] College of Information Science and Technology, Beijing Normal University, Beijing 100875, China
inslawliet@163.com, zgz_bnu@mail.bnu.edu.cn, {rfbie,wuhao}@bnu.edu.cn, 634542994@qq.com

[2] School of Computer Science, Qufu Normal University, Rizhao 276826, Shandong, China
jiguoyu@sina.com

[3] Beijing Rehabilitation Hospital of Capital Medical University, Beijing 100144, China
mxl1011@126.com

Abstract. With the finish of the human genome sequencing and the great progress in molecular biology like proteomics, many established authoritative international biomedical databases are completing continually in recent years. With these opening databases, all kinds of biological molecular networks can be constructed for potential disease gene detection and drug target prediction through network-based approaches. However, most methods do the drug target prediction along with data from only a single source, which have many limitations and tendencies. In this paper, we use multi-source data integrate with datasets from Uniprot, HGNC, COSMIC and DrugBank to do the anti-cancer drug target prediction more comprehensively. We construct Drug-Target network (DT network), Cancer-Gene network (CG network) and Cancer-Drug Interaction network (CDI network) based on the multi-source data we integrate, and do visualizations of the three networks in Cytoscape. In addition, we make an anti-cancer drug target prediction with the method of Random Walks on graphs, one of the most efficient method in biological molecular network analysis by now. Potential anti-cancer drug targets are predicted by calculating the correlation strengths between known cancer gene products and other proteins in CDI network with PersonalRank algorithm. Analysis of the prediction results shows that the potential anti-cancer drug targets we predicted are highly related with cancers both topologically and bio-functionally, which verifies the rationality and availability our method.

Keywords: PPI networks · Random Walks · Anti-cancer drug target prediction

S. Chellappan et al. (Eds.): WASA 2018, LNCS 10874, pp. 223–235, 2018.
https://doi.org/10.1007/978-3-319-94268-1_19

1 Introduction

With the finish of human genome sequencing and further proteomics research, drug development is more focused on the discovery of specific molecular drug targets. However, the potential complexity of the human interactome causes two major challenges in drug development: (1) a disease phenotype is rarely a consequence of an abnormality in a single effector gene product, but reflects various pathobiological processes that interact in a complex network; (2) drugs whose efficacy was predicted by specific target-binding experiments may work inefficiently or cause side effects for interacting with other non-disease cellular components in vivo. Therefore, it is difficult to simulate the pathological and pharmacological reactions in vivo through the traditional medical and biological experiments, leading to inefficient development of drug discovery. Compared with experimental methods, the advent of bioinformatics and network medicine provide a possible approach for the prediction of drug targets. Biomedical networks (like protein-protein interaction/interactome network [1, 2], drug-target networks [3], networks of gene data [4, 5] and so on) can well simulate the complex inter- and intracellular molecular interactions in vivo, which providing network-based comprehension of the molecular pathways of disease and aid drug relocations and discovery [6].

In the development of drug discovery, the prediction of drug targets for complex diseases such as cancers remains a research hotspot. Cancers seriously threaten the health and social development of human beings. The researches on the relationship between cancers and drug targets as well as the studied of anti-cancer drugs have always been the focus and difficulty for researchers. Based on the background above, in this paper we integrate authoritative biomedical datasets to construct a Cancer-Drug Interaction network and further predict new anti-cancer drug targets, which can not only help to accelerate the development of cancer drugs and bring hopes to cancer patients, but also provide a reference method for drug target prediction of other complex diseases.

2 Related Works

2.1 Drug Target Prediction

Currently, the bioinformatics methods for drug target prediction are still in its infancy. According to various data and biomedical theories or hypotheses, there are mainly four types of computing methods proposed for drug target prediction: (1) prediction methods based on molecular docking technology [7, 8]; (2) prediction methods based on gene sequencing [9, 10]; (3) prediction methods based on drug similarity [11, 12]; (4) prediction methods based on multi-source data integration [13]. The first three types of methods all do the drug target prediction along with data from a single source, and in some cases, molecular structure information is needed to analyze properties of drugs and proteins, which is identified and reconstructed with the assistance of some advanced image processing methods [14, 15] on object recognition, image composition, and 3D reconstruction. However, there are many limitations and tendencies for the molecular structure data is insufficient and inaccurate. In contrast, the methods based on multi-source data integration are more comprehensive and reliable, and the data required for analysis is more complete and easier to get, while how to integrate multi-source data effectively remains a big challenge [16].

In this paper we integrate datasets from Uniprot, HGNC, COSMIC and DrugBank four databases as a multi-data foundation to build Drug-Target network (DT network for short) and Cancer-Gene network (CG network for short), and construct Cancer-Drug Interaction network (CDI network for short) based on the Protein-Protein Interaction network (PPI network for short). Besides, we make visualizations of the three networks using *Cytoscape* software.

2.2 Random Walks Algorithms on Graph

As biomedical networks have many properties similar with networks in many other types (like social networks, wireless networks, webpage networks et al.), some network based methods of networks in other kinds [17, 18] can be referenced. And there are three mature types of network based methods to make various kinds of medical prediction at present: (1) linkage methods; (2) disease module-based methods; (3) diffusion-based methods. Barabási et al. [19] summarized and compared these three types of methods of network medicine in predicting disease gene expression based on the same dataset in 2011. As the conclusion, they obtained that Random Walks [20, 21], one diffusion-based method, had the best performance in potential disease gene prediction. As Random Walk models can well simulate the process of random diffusion in complex networks, algorithms of Random Walks on graphs have also been widely used in the analysis of biomolecular networks in recent years. Two classic algorithms of Random Walks on graphs are introduced below.

PageRank. PageRank is an algorithm used to rank webpages or websites, based on the interlinkage relationship among webpages (or webgraph in other words) [22, 23]. Let $G = (V, E)$ be a webgraph with N webpages, in which each node correspond to only one webpage and the directed edge between two webpages means we can link from one webpage to visit another. Consider a random walk on G starting at any node, and any step has an average probability *alpha* of continually jumping to neighbors (or called a damping factor *alpha*) and 1-*alpha* of stopping on the contrary. $PR(i)$ is the PageRank value of a webpage i, dependent on the PageRank values for each page j contained in the set $in(i)$ (the set containing all webpages linking to webpage i), divided by $|out(j)|$, the number of links from page j, as Eq. (1) displayed.

$$PR(i) = \frac{1 - alpha}{N} + alpha \sum_{j \in in(i)} \frac{PR(i)}{|out(j)|} \tag{1}$$

In PageRank, $PR(i)$ measures the relative importance of webpage i within all N webpages in G. we can rank websites by the PR values of each webpages, and know which websites are highly-visited while which are unpopular.

PersonalRank. PersonalRank is an algorithm based on PageRank, but not limitedly used in webgraph. Different from the PageRank algorithm, each random walk in PersonalRank always starts from a fixed node *root* in the graph or network, and the final PersonalRank value of node *u*, marked as *PR*(*u*), is calculated as Eq. (2). The damping factor *alpha* in Eq. (2) means the probability of continually moving the next step, similar to the *alpha* in PageRank [24].

$$PR(u) = (1 - alpha)r_u + alpha \sum_{i \in in(u)} \frac{PR(i)}{|out(i)|}, \; r_u = \begin{cases} 1, \; u = root \\ 0, \; u \neq root \end{cases} \tag{2}$$

In PersonalRank, *PR*(*u*) measures the relative correlation strength of node *u* to the start node *root* within the graph or network. We can choose nodes of interest as node *roots*, and rank by the *PR* values of other nodes to find which nodes are tightly correlated or have significant impacts to the nodes we are especially concerned.

Let $G = (V, E)$ be the graph representing the network, where V is the set of N nodes and E is the set of edges between node pairs. Set **A** as the column normalized adjacency matrix of G. Define $p_s(u)$ as the correlation strength of a node u to a start node s (marked as a node *root* in Eq. (2)), which is equal to the PersonalRank value $PR(u)$ in a random walk starting at node s. Let $\vec{\mathbf{p}}_s(V) = (p_s(1), p_s(2), \ldots, p_s(u), \ldots, p_s(N))^T$ be the correlation strength vector to the start node s, which can be computed by the iterative matrix algorithm of PersonalRank shown as follow in Fig. 1:

Input:	the column normalized adjacency matrix **A**;
	the start node *s*;
	the damping factor *alpha*;
Output:	the correlation strength vector $\vec{\mathbf{p}}_s(V)$
Initialization:	$\vec{\mathbf{p}}_s(V)^0 = \vec{\mathbf{r}}_s(V)$
	$\vec{\mathbf{r}}_s(V) = (p(1), \cdots, p(s-1), p(s), p(s+1), \cdots, p(n))^T = (0, \cdots, 0, 1, 0, \cdots, 0)^T$
While:	$\vec{\mathbf{p}}_s(V)^t$ has not converged
	$\mathbf{p}_s(V)^{t+1} = (1 - alpha) \times \vec{\mathbf{r}}_s(V) + alpha \times \mathbf{A}\vec{\mathbf{p}}_s(V)^t$

Fig. 1. The iterative matrix algorithm of PersonalRank. The convergence check requires the L_1-norm between consecutive $\vec{\mathbf{p}}_s(V)$ s to be less than a small threshold, e.g., 10^{-10}.

In this paper we use the iterative matrix algorithm of PersonalRank in CDI network we build to rank and discover highly-cancer-related proteins, and further make the drug prediction.

3 Cancer-Drug Interaction Network Construction

3.1 Multi-source Data Integration

We integrate datasets from Uniprot, HGNC, COSMIC and DrugBank four databases as a multi-data foundation to construct DT network, CG network and CDI network.

Uniprot. The protein data is from dataset *Reviewed swiss-prot* from the Uniprot Knowledgebase (UniprotKB, http://www.uniprot.org/uniprot/), with 20198 manually annotated and reviewed records of human proteins. According to the records, a total of 56223 pairs of protein-protein interaction are screened out, with 13783 proteins involved (including 7561 manually annotated and reviewed proteins and 6222 automatically annotated and not reviewed proteins).

HGNC. The data of genes is from HGNC (HUGO Gene Nomenclature Committee, https://www.genenames.org/), including a total of 45088 records of genes.

COSMIC. The data of cancer genes is from *cancer gene census* dataset of COSMIC (the catalogue of somatic mutations in cancer, http://cancer.sanger.ac.uk/cosmic), with 616 records of current released cancer genes involved.

DrugBank. The data of drugs and target proteins is from *approved drug target links* dataset of DrugBank (https://www.drugbank.ca/), with 7189 pairs of approved drug-target interactions involved (including 1670 FDA-approved drugs and 1798 target proteins).

According to the relationships among the four datasets shown as Fig. 2, there are several entry IDs or common fields to related the four datasets, so we can clearly integrate all the data of cancers, disease genes, drugs, target proteins and protein-protein interactions together by joining or text alignment, and only 1%–5% of the whole data are unavailable caused by the error records.

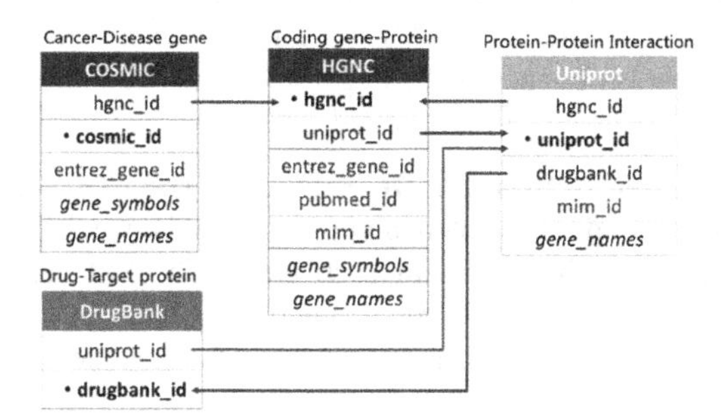

Fig. 2. Relationships of data from Uniprot, HGNC, COSMIC and DrugBank. According to the attributes of the four datasets, we can integrate all the data through their primary keys (shown as black bolded), entry IDs of each other (shown as black), entry IDs of other foreign databases (shown as grey), as well as some common fields (shown as purple italic). (Color figure online)

3.2 Network Construction and Visualization

Drug-Target Network (DT Network). With *approved drug target* dataset of DrugBank, the DT network can be visualized directly in *Cytoscape*. Set names of drugs as the source nodes and names of target proteins as the target nodes to obtain the visualization result of Fig. 3 (including 1669 drugs, 1797 target proteins and 7189 links).

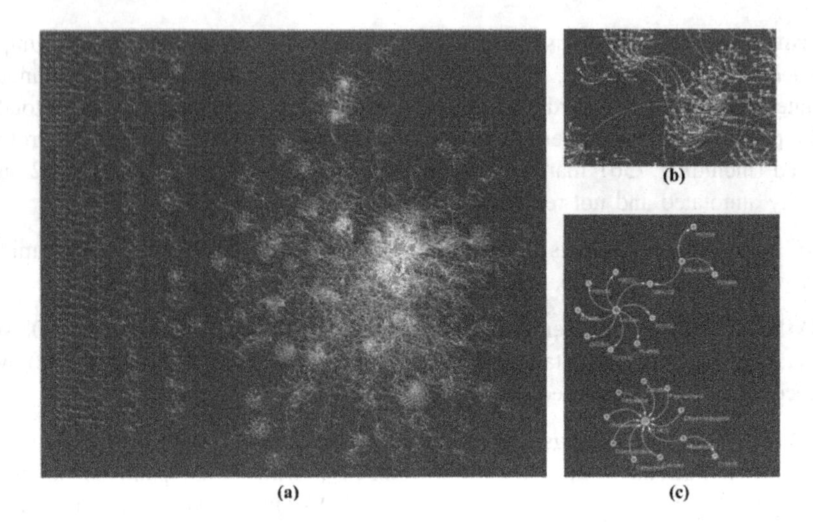

Fig. 3. DT network. (a) is the full visualization result of DT network. (b) and (c) are enlarged parts. Blue hexagons and pink circles correspond to drugs and target proteins, respectively. A link is placed between a drug node and a target node if the protein is a known target of that drug. The area of the drug (protein) node is proportional to the number of targets that the drug has (the number of drugs targeting the protein). (Color figure online)

Cancer-Gene Network (CG Network). The CG network is built and visualized in *Cytoscape* with integrated data of COSMIC and HGNC. Set names of cancers as the source nodes and symbols of genes as the target nodes to obtain the visualization result of Fig. 4 (including 75 cancers, 84 genes and 92 links).

Cancer-Drug Interactive Network (CDI Network). The CDI network is based on PPI network, in which a node is a protein and a link is set between two nodes if the two corresponding proteins can interact with each other. The PPI network can be constructed according to the records of protein interacting pairs in *Reviewed swiss-prot* dataset of UniprotKB. Then with the multi-source data integrated from the four databases, we can combine information of cancers, drugs together with proteins by marking both cancer gene products and drug target proteins in PPI network, and construct our CDI network visualized as shown in Fig. 5 (including a total of 13783 protein nodes and 56223 links, with 445 known cancer gene products proteins, 667 approved drug target proteins and 77 both the products and the targets as well).

The CDI network has more abundant information than PPI network, which can better aid the drug target prediction furthermore. The grey protein nodes in Fig. 5 as the majority has no record shows their relationships with any cancer or drug currently, whose impacts in cancers wait to be discovered through network analysis and prediction. In what followed, we refer to the proteins corresponding to the grey nodes in CDI network as candidate proteins.

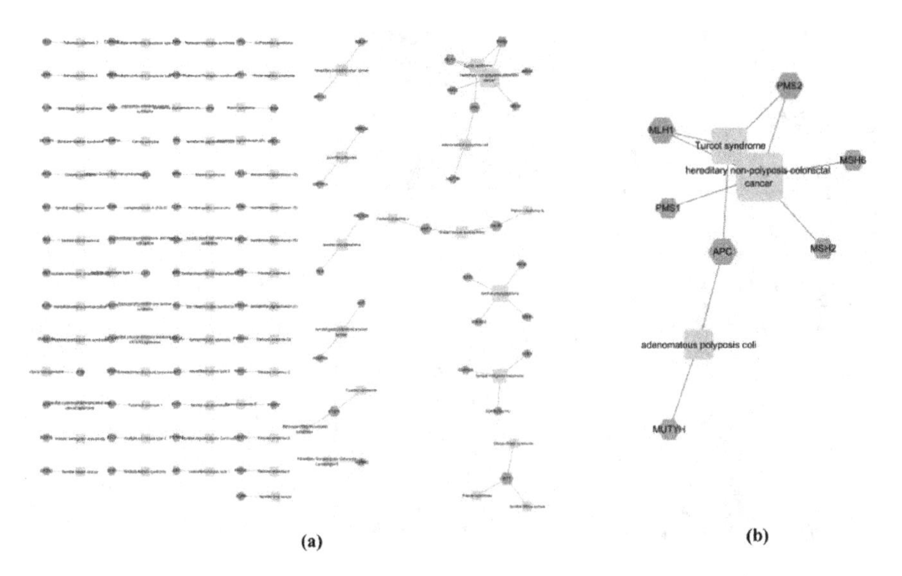

Fig. 4. CG network. (a) is the full visualization result of CG network while (b) is an enlarged part. Blue hexagons and pink rectangles correspond to cancers and genes, respectively. A link is placed between a cancer node and a gene node if the gene is an oncogene of the cancer. The area of the node is proportional to its degree. (Color figure online)

4 Anti-cancer Drug Target Prediction

Researches shows that modern anti-cancer drugs are rationally designed to alter or suppress aberrant oncogene activity, and the relations between cancer genes and the targets proteins of anti-cancer drugs are more direct than other kinds of diseases [25]. To ensure the high efficiency of anti-cancer drugs, the drug target proteins are preferably closely related to the protein products of cancer genes, that is, effective drug target proteins tend to highly correlated with cancer gene products in the PPI network. So it can be assumed that in our CDI network, if a protein is highly correlated to a known cancer gene product, the protein tends to have a significant impact on the cancer. Further, if a protein is tightly associated with many known cancer gene products, it can be a key protein that significantly affects cancers, which is likely to be a potential target protein for anti-cancer drugs. Therefore, by using the PersonalRank algorithm mentioned in Sect. 2.2 in the CDI network, we take each known cancer gene product as the *root* node in Eq. (2) to find out their highly-correlated proteins, among which the most common ones are predicted as our potential anti-cancer drug targets.

We use the iterative matrix algorithm of PersonalRank displayed in Fig. 1 in CDI network to rank and discover highly-cancer-related proteins, and further make the drug prediction. Let $G = (V, E)$ be the graph representing the CDI network, where V is the set of N protein nodes ($N = 13783$) and E is the set of undirected edges between protein interacting pairs, and the do programming and computing process of the algorithm with *Matlab*. Set the nodes correspond to 445 known cancer gene products as the start node s, respectively, and do the iteration. We can get the correlation strengths of all the

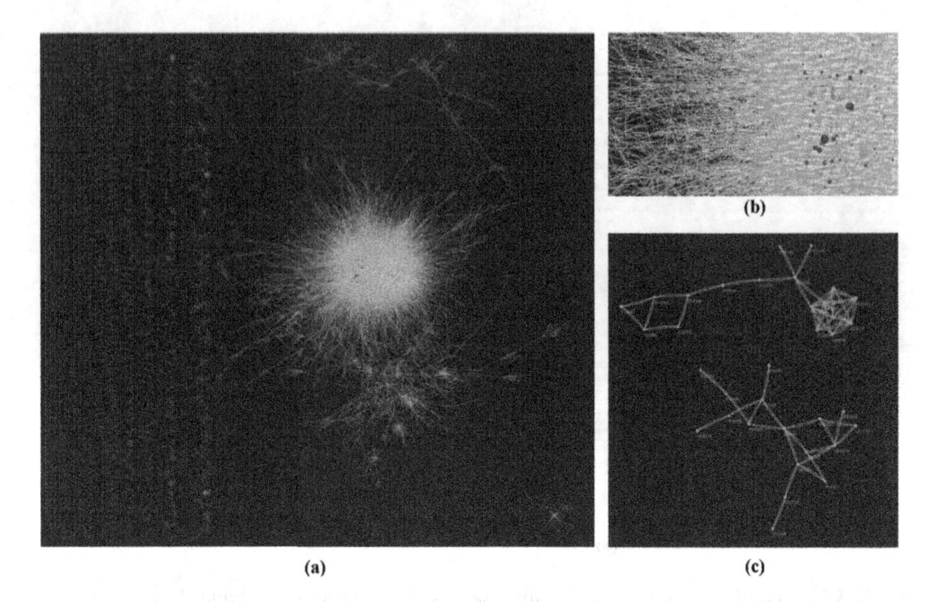

Fig. 5. CDI network. (a) is the full visualization result of CDI network while (b) and (c) are two enlarged parts. Each node is a protein and each pair of linked nodes correspond to a pairs of protein interaction. The color of nodes, displayed more clearly in (b), codes the properties of the proteins: pink nodes and blue nodes are cancer gene products and drug target proteins, respectively; deep blue nodes are both while grey nodes are not both. The area of the node is proportional to its degree. (Color figure online)

13783 proteins (including 445 known cancer gene products themselves and other 13338 candidate proteins) to each known cancer gene product in CDI network, through elements in converged $\vec{\mathbf{p}}_s(V)$ of each start node *s*.

4.1 Candidate Cancer-Related Proteins Prediction

Let *alpha* = 0.3 and $\vec{\mathbf{p}}_s(V)$ is converged at iterations $t = 50$ with the L_1-norm between consecutive $\vec{\mathbf{p}}_s(V)$ s to be less than 10^{-10}. We set a candidate proteins *u* with a correlation strength $p_s(u) > 0.001$ judged to be 'correlated with the known cancer gene product corresponding to the node *s*', while $p_s(u) < 0.001$ considered as 'unrelated to the known cancer gene product corresponding to the node *s*'. From the total of 13338 candidate proteins, 5682 candidate proteins are screened as correlated with at least one known cancer gene product.

Define S_c as a potential anti-cancer drug target score, which is equal to the number of known cancer gene products associated with each candidate proteins, and we obtain the statistics of Table 1.

Table 1. Statistical table of the number of candidate proteins (ranked by S_c).

Score (S_c)	The number of candidate proteins (unreviewed + reviewed)	The number of reviewed candidate proteins
≥ 5	1289	1144
≥ 10	463	426
≥ 20	155	149
≥ 50	36	35
≥ 100	6	5
≥ 150	3	3

Prediction Results. Excluding unreviewed proteins, the candidate proteins with top 10 S_c are selected as the potential anti-cancer drug target prediction results (see Table 2).

Table 2. Results of anti-cancer drug target prediction (proteins with top 10 S_c).

Uniprot ID	Protein name	Coding-gene name	Score (S_c)
Q08379	Golgin subfamily A member 2	GOLGA2	184
Q6A162	Keratin, type I cytoskeletal 40	KRT40, KA36	182
P62993	Growth factor receptor-bound protein 2	GRB2, ASH	150
Q7Z3S9	Notch homolog 2 N-terminal-like protein	NOTCH2 NL, N2N	108
Q9BRK4	Leucine zipper putative tumor suppressor 2	LZTS2, KIAA1813, LAPSER1	102
P16333	Cytoplasmic protein NCK1	NCK1, NCK	99
Q12933	TNF receptor-associated factor 2	TRAF2	95
Q15323	Keratin, type I cuticular Ha1	KRT31	94
Q5JR59	Microtubule-associated tumor suppressor candidate 2	MTUS2	94
Q96EB6	NAD-dependent protein deacetylase sirtuin-1	SIRT1	90

Result Analysis. Pan concluded in the book *The Molecular Biology of Gene and Diseases* that mutations that can induce tumor formation involves at least one cell growth-related regulatory genes, divided into 7 categories [25]:

(1) Growth factors, such as basic fibroblast growth factor (bFGF) and epidermal growth factor (EGF);
(2) Growth factor receptors such as epidermal growth factor receptor (EGFR);
(3) Proteins related to information transmission, such as RAS, GSP, RAF1 and SRC;
(4) Intracellular nuclear oncogene products, such as Myc and so on;
(5) Genes related to cell cycle control, such as CDK gene mutation;
(6) Genes related to apoptosis control, such as nuclear Bcl gene mutation;
(7) Genes involved in several major types of DNA damage repair genes.

According to the description about the biological functions of the 10 potential anti-cancer drug target proteins obtained from UniprotKB (summarized as Table 3, the complete descriptions can be searched at http://www.uniprot.org/uniprot/), comparing with the 7 categories of cancer inducements mentioned above, it is revealed that in addition to Keratin, type I cytoskeletal 40 (Q6A162) and Keratin, type I cuticular Ha1 (Q15323) which two hair-related proteins can not be analyzed due to too little functional description, the remaining 8 predicted anti-cancer drug targets are all linked to the above-mentioned cancerogenic categories (see Table 3).

The analysis of the prediction results obtained from multi-source data and network analysis shows a high correlation between our predicted anti-cancer drug targets and known cancer gene products as well as cancers, not only topologically but also bio-functionally. It confirms the correctness of the prediction results and the effectiveness of our method.

Table 3. Comparison between bio-functions of predicted targets and cancerogenic categories.

Uniprot ID	Summary of biological functions	Related category indexes
Q08379	Involved in cell division, related to cell cycle control gene CDK1	(5)
Q6A162	May play a role in late hair differentiation	–
P62993	Related to cell surface growth factor, EGF and RAS signaling, involved in apoptosis control	(1), (2), (3), (6)
Q7Z3S9	Involved in the Notch signaling pathway and neutrophil differentiation.	(3)
Q9BRK4	Required for cytokinesis, involved in the cell cycle and Wnt signaling	(3), (5)
P16333	Related to growth factors and RAS signaling, involved in DNA damage	(1), (3), (7)
Q12933	Related to the immune process, involved in cell survival and apoptosis	(6)
Q15323	Type I (acidic) hair/microfibrillar keratin	–
Q5JR59	Related to tubulin and microtubules	(3)
Q96EB6	Involved in transcriptional regulation, cell cycle, DNA damage response, metabolism, apoptosis and autophagy and many other aspects	(5), (6), (7)

4.2 Result Improvement with Information of Approved Drug Targets

In Sect. 4.2 we do anti-cancer drug target prediction only using information about known cancer genes. The problem is that the 10 predicted drug targets we get are quite likely to be highly associated with cancers, but they are possibly related to other non-cancerous pharmacological processes, which means, the anti-cancer drugs targeting the 10 predicted drug targets might have side effects and disrupt other drug therapies. As a cancer drug target, it is required to be 'highly correlated' with cancers and at the same time, as far as possible 'unrelated' with other pharmacological effects. Therefore, we use the information of approved drug targets in our CDI network, and calculate the correlation strengths of 10 predicted drug target to the approved drug targets (*alpha* = 0.3, iterations t = 70 when it is converged) with the same threshold of 0.001 to determine 'correlated' and 'unrelated', and then calculate the potential drug side effects score S_d (the number of approved drug targets correlated with each predicted drug targets, similar with S_c) to rescreen better potential anti-cancer drug targets.

Improved Prediction Results and Analysis. As low S_d of a protein means slight side effects while high S_d means serious side effects, we sort the10 predicted drug target proteins by S_d in ascending order and select the first three (shown as bolded in Table 4 as the improved prediction results.

Table 4. Improved results of anti-cancer drug target prediction (ordered by S_d).

Uniprot ID	Protein name	Score (S_d)
Q96EB6	**NAD-dependent protein deacetylase sirtuin-1**	**60**
Q12933	**TNF receptor-associated factor 2**	**105**
Q5JR59	**Microtubule-associated tumor suppressor candidate 2**	**114**
Q9BRK4	Leucine zipper putative tumor suppressor 2	128
P16333	Cytoplasmic protein NCK1	133
Q15323	Keratin, type I cuticular Ha1	143
Q7Z3S9	Notch homolog 2 N-terminal-like protein	162
P62993	Growth factor receptor-bound protein 2	176
Q08379	Golgin subfamily A member 2	210
Q6A162	Keratin, type I cytoskeletal 40	213

Analysis of Improved Results. We do a further analysis of the bio-functions of the 3 improved prediction results as follow:

(1) NAD-dependent protein deacetylase sirtuin-1 inhibits transcription by promoting DNA methylation, which is a phenomenon that has also been observed in cells of patients with sporadic breast cancer, colon cancer, and soft tissue tumors [25];
(2) TNF receptor-associated factor 2 is involved in the immune process and plays an important role in the control of cell survival and apoptosis. From the naming of this protein, it can be seen that it is related to tumor necrosis factor receptor (TNF receptor);
(3) Microtubule-associated tumor suppressor candidate 2 is a microtubule-associated candidate tumor suppressor protein that is associated with tubulin and binds microtubules. Microtubules are an important therapeutic target in tumor cells, and until the advent of targeted therapy microtubules were the only alternative to DNA as a therapeutic target in cancer [26]. In recent years, more and more discoveries have been made in the treatment of cancer through drugs that act on microtubules, which confirms this protein is also closely related to cancers.

Obviously, these three proteins are functionally more focused on tumor or cancer pathology and oncology compared to the remaining 7 proteins, and therefore may have less side effects as anti-cancer drug targets, which can better help to achieve effective cancer targeted therapies.

5 Conclusion

This paper addresses the problems about network construction based on multi-source data and drug target prediction with method of Random Walk on graph. we construct DT network, CG network and CDI network using multi-source biomedical data integrated from Uniprot, HGNC, COSMIC and DrugBank, and visualize the three networks with *Cytoscape*. Then we predict anti-cancer drug targets in CDI network using PersonalRank algorithm. The analysis of prediction results proves the rationality and effectiveness of the method, which lays the foundation for future research.

References

1. Rual, J.F., Venkatesan, K., Hao, T.: Towards a proteome-scale map of the human protein-protein interaction network. Nature **437**(7062), 1173–1178 (2005)
2. Guney, E., Menche, J., Vidal, M.: Network-based in silico drug efficacy screening. Nat. Commun. **7**(10331), 1–13 (2016)
3. Mestres, J., Gregoripuigjané, E., Valverde, S.: The topology of drug–target interaction networks: implicit dependence on drug properties and target families. Mol. BioSyst. **5**(9), 1051–1057 (2009)
4. Mehmood, R., Elashram, S., Bie, R.: Clustering by fast search and merge of local density peaks for gene expression microarray data. Sci. Rep. **7**, 45602 (2017)
5. Futreal, P.A., Coin, L., Marshall, M.: A census of human cancer genes. Nat. Rev. Cancer **4** (3), 177–183 (2004)
6. Goh, K.I., Cusick, M.E., Valle, D.: The human disease network. Proc. Natl. Acad. Sci. **104** (21), 8685–8690 (2007)
7. Keiser, M.J., Setola, V., Irwin, J.J.: Predicting new molecular targets for known drugs. Nature **462**(7270), 175–181 (2009)
8. Cheng, A.C., Coleman, R.G., Smyth, K.T.: Structure-based maximal affinity model predicts small-molecule druggability. Nat. Biotechnol. **25**(1), 71–75 (2007)
9. Li, Q., Lai, L.: Prediction of potential drug targets based on simple sequence properties. BMC Bioinform. **8**(1), 353 (2007)
10. Wang, Y., Zhao, X.M., Chen, L.: Gene function prediction using labeled and unlabeled data. BMC Bioinform. **9**(1), 1–14 (2008)
11. Campillos, M., Kuhn, M., Gavin, A.C.: Drug target identification using side-effect similarity. Science **321**(5886), 263–266 (2008)
12. Tatonetti, N.P., Liu, T., Altman, R.B.: Predicting drug side-effects by chemical systems biology. Genome Biol. **10**(9), 238 (2009)
13. Zhao, X.M., Iskar, M.: Prediction of drug combinations by integrating molecular and pharmacological data. PLoS Comput. Biol. **7**(12), e1002323 (2011)
14. Wu, H., Li, Y., Miao, Z.: Creative and high-quality image composition based on a new criterion. J. Vis. Commun. Image Represent. **38**(C), 100–114 (2016)

15. Wu, H., Li, Y., Miao, Z.: A new sampling algorithm for high-quality image matting. J. Vis. Commun. Image Represent. **38**(C), 573–581 (2016)
16. Wang, Y.: Repositioning drugs based on molecular network. Shanghai University, Shanghai (2015)
17. Yu, J., Chen, Y., Ma, L.: On connected target k-coverage in heterogeneous wireless sensor networks. Sensors **16**(1), 104 (2015)
18. Zhang, X., Yu, J., Li, W.: Localized algorithms for Yao graph-based spanner construction in wireless networks under SINR. IEEE/ACM Trans. Netw. **99**, 1–14 (2017)
19. Barabási, A., Gulbahce, N., Loscalzo, J.: Network medicine: a network-based approach to human disease. Nat. Rev. Genet. **12**(1), 56–68 (2011)
20. Can, T., Singh, A.K.: Analysis of protein-protein interaction networks using random walks. In: BIOKDD 2005, pp. 61–68. ACM, DBLP, Chicago (2005)
21. Lovász, L., Lov, L., Erdos, O.P.: Random walks on graphs: a survey. Combinatorics **8**(4), 1–46 (1993)
22. Li, Z., Yang, W., Xie, Z.: Research on PageRank algorithm. Comput. Sci. **38**(10A), 185–188 (2011)
23. Shui, C., Chen, T., Li, H.: Survey on automatic network layouts based on force-directed model. Comput. Eng. Sci. **37**(3), 457–465 (2015)
24. CSDN blog. http://blog.csdn.net/harryhuang1990/article/details/10048383. Accessed 18 Aug 2013
25. Pan, X.: The Molecular Biology of Gene and Diseases, 1st edn. Chemical Industry Press, Beijing (2014)
26. Dumontet, C., Jordan, M.A.: Microtubule-binding agents: a dynamic field of cancer therapeutics. Nat. Rev. Drug Discov. **9**(10), 790–803 (2010)

Enabling Efficient and Fine-Grained DNA Similarity Search with Access Control over Encrypted Cloud Data

Hongwei Li[1,2](✉), Guowen Xu[1], Qiang Tang[3], Xiaodong Lin[4], and Xuemin (Sherman) Shen[5]

[1] School of Computer Science and Engineering, University of Electronic Science and Technology of China, Chengdu, China
hongweili@uestc.edu.cn
[2] Science and Technology on Communication Security Laboratory, Chengdu, China
[3] New Jersey Institute of Technology, Newark, NJ, USA
[4] Department of Physics and Computer Science, Faculty of Science, Wilfrid Laurier University, Waterloo, Canada
[5] Department of Electrical and Computer Engineering, University of Waterloo, Waterloo, Ontario N2L 3G1, Canada

Abstract. DNA similarity search has proven to be an essential demand in human genomic researches. Since DNA sequences contain many sensitive personal information, the acquisition and dissemination of DNA data have been tightly controlled and restricted by authorities. Although the problem of private DNA similarity query has been an active research issue, the latest research findings are still inadequate in terms of security, functionality and efficiency. In this paper, we propose an Efficient DNA Similarity Search scheme (EDSS) which can achieve fine-grained query and data access control over encrypted cloud data. Our original contributions are fourfold. First, we creatively put forward a private edit distance approximation algorithm to realize the efficient and high accurate DNA similarity query. Second, we classify the whole DNA sequences and design a multiple genes search strategy to achieve complicated logic query such as mixed "AND" and "NO" operations on genes. Third, the proposed scheme can also efficiently support data access control by employing a novel polynomial based design. Finally, security analysis and extensive experiments demonstrate the high security and efficiency of EDSS compared with existing schemes.

Keywords: DNA similarity search · Fine-grained query
Access control · Privacy-preserving · Cloud computing

1 Introduction

DNA similarity search has proven to be an important basic demand in many human genomic researches, such as similar patient query [1,2], personalized

S. Chellappan et al. (Eds.): WASA 2018, LNCS 10874, pp. 236–248, 2018.
https://doi.org/10.1007/978-3-319-94268-1_20

medicine performs diagnoses [3,4] and genome sequencing [5,6]. For instance, genes BRCA1 and BRCA2 have been widely recognized as closely related to the high incidence of breast cancer, if we take DNA similarity query among the breast cancer patients ahead, early prevention and treatment can drastically degrade the morbidity of individuals possessing analogous genes with patients. However, the size of DNA data is always massive and the computation cost of DNA similarity query over them is often immense. To address this challenge, a promising and attractive way is to outsource the DNA data to public cloud, such as DNAnexus, Google Genomics[1], etc.

However, although the formidable storage and computing capabilities of the cloud server can significantly reduce user's storage cost and overwhelmingly improve the search efficiency over raw data, a widespread privacy concern is that the DNA data may be exposed to unauthorized third parties, especially the cloud server, which has full access rights to the user's DNA data. Under this circumstance, the user's sensitive data such as DNA sequences, genetic information, and even the health status can be easily accessed by the cloud server motivated with malicious intentions or economic temptations. Therefore, it is urgent to design a privacy-preserving policy, which can ensure the confidentiality of user's DNA data before being outsourced to the public cloud while still guarantee the similarity search functionality.

To tackle the above problems, a routine method is to encrypt the DNA data before outsourcing them to the cloud server. However, if we use the traditional encryption methods (e.g., AES [7]) directly to encrypt user's DNA data, data availability and operability will be significantly reduced [8]. Besides, from a practical point of view, the entire encryption process will be also deemed as meaningless [9–11]. Most of existing methods [2,3,6] are interactive protocols, and these protocols always assume the authorized party is remaining online and need multiple interactions during the entire querying process. For example, Wang et al [3] proposed a similar patient query model called GENSETS which can achieve unprecedentedly high efficiency and precision of DNA similarity query. However, the shortcoming of multiple interactions between numerous parties will profoundly limit the scalability of GENSETS. In addition, data access control and the function of fine-grained query such as mixed "AND" and "NO" operations on genes are also not taken into account in their paper. Gilad Asharov et al [2] also designed an efficient approximation algorithm to compute the edit-distance between sequences, even with high divergence among individual genomes. Unfortunately, similar to [3], the whole scheme is also devised from the perspective of efficiency. In order to solve the problem of high communication overhead, Wang et al [4] proposed a DNA sequence pattern matching scheme which only requires one round of communication compared with previous protocols. However, due to the complex encryption method based on bilinear group, their model has been criticized for its inefficiency and is also hard to apply to realistic scenes.

[1] https://cloud.google.com/genomics/.

In this paper, we propose an efficient DNA similarity search scheme which can enable fine-grained query and data access control over encrypted cloud data. Our original contributions can be summarized in four aspects as follows:

- We creatively put forward a private edit distance approximation algorithm which can realize the conversion of edit distance computation problem to the set difference size approximation problem. It will significantly reduce the number of elements to be matched in the querying process.
- We classify the whole DNA sequences according to the types of genes and design a multiple genes search strategy. In this way, the complicated logic query such as mixed "AND"and "NO" operations on genes can be achieved.
- We employ a novel polynomial based design to support efficient data access control, where the data owner can define who can access a specific DNA sequence by assigning different access roles to each user.
- We analyze the security of our proposed scheme in terms of both index and trapdoor privacy. Besides, extensive experiments demonstrate the high efficiency of our proposal compared with existing schemes.

The remainder of this paper is organized as follows. In Sect. 2, we outline the system model, threat model and security requirements. Then, we describe the approximate algorithm for editing distance in Sect. 3. In Sect. 4, we discuss our EDSS in details. Later, we carry out the security analysis and performance evaluation in Sects. 5 and 6, respectively. Finally, Sect. 7 concludes the paper.

2 System Model, Threat Model and Security Requirements

2.1 System Model

As shown in Fig. 1, our system model has three entities, i.e., data owner, the cloud server and search user. The primary tasks of data owner are to encrypt raw DNA sequences and index before outsourcing them, and the tasks of executing user's requests will be assigned to the cloud server. When an authorized search user receives the secret key from the data owner, target encrypted request (i.e., trapdoor) will be sent to the cloud server. After querying, the cloud server returns the corresponding query results to authorized search user.

2.2 Threat Model and Security Requirements

The cloud server is generally considered as "honest-but-curious", which means that the cloud server will execute search orders in strict accordance with prior procedures on the one hand. However, it may be "curious" to spy and infer the contents of user's data by its powerful computational capacity. Here we consider two threat models on the basis of the information available to the cloud server.

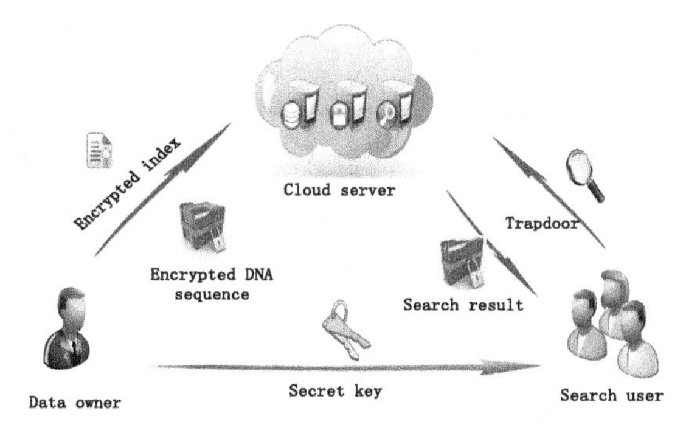

Fig. 1. System model

- **Known Ciphertext Model:** In this model, the cloud server knows nothing but the encrypted DNA sequence collection and index collection, which are both outsourced to the cloud server.
- **Known Background Model:** In this strong model, the cloud server has more information besides the known ciphertext model. In other words, extensive data analysis of a known database being similar to the target database have been invested prior by the cloud server.

Based on the threat models above, we define the security requirements as follows:

- *Confidentiality of data owner's DNA sequences:* User's DNA data are usually outsourced to the cloud server before being utilized. Taking privacy issues into account, these data should be encrypted ahead and not leaked to anyone except the data owner and authorized search users throughout the querying process.
- *Privacy protection of index and trapdoor:* The trapdoor and index are generated according to corresponding sets of gene editing operations. In order to prevent the cloud server from utilizing the relationships among index, trapdoor and encrypted DNA sequences to threaten the data privacy of data owner, the privacy of trapdoor and index should be protected well.
- *Unlinkability of trapdoors:* Unlinkability of trapdoors, means that if a DNA sequence is queried frequently, its trapdoors should be generated totally different. Because of the randomness of query, a search user may query one DNA sequence many times, and he will send a trapdoor generated by the same set of gene editing operations continually. If trapdoors are generated deterministically, the cloud server can utilize the relationships among trapdoors to further infer the contents of user's DNA data. Thus, for same query request, trapdoors should be generated randomly, rather than deterministic.

3 Private Edit Distance Approximation

The private edit distance approximation algorithm will be introduced in this section, which also plays a cornerstone role in our whole paper.

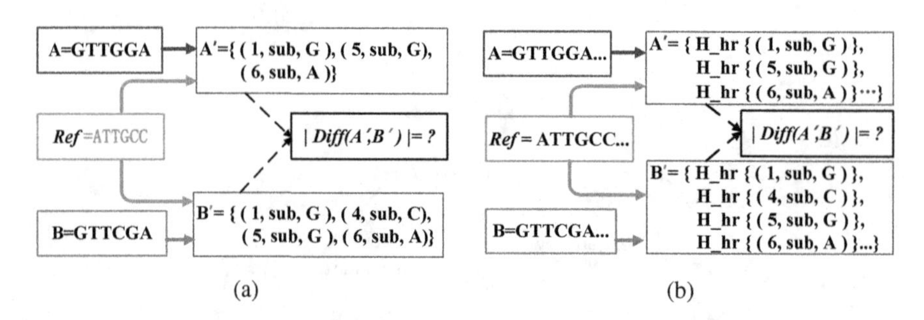

Fig. 2. Set differences between two sets of gene editing operations. (a) Under the plaintext environment. (b) Under the encrypted environment.

As shown in Fig. 2(a), when in the plaintext environment, given $Reference=ATTG\ CC$, $A = GTTGGA$, $B = GTTCGA$, A and B will be converted into $A' = \{(1, sub, G), (5, sub, G), (6, sub, A)\}$, $B' = \{(1, sub, G), (4, sub, C), (5, sub, G), (6, sub, A)\}$ utilizing the dynamic programming algorithms proposed in literature [2]. Accordingly, the set differences between A' and B' can be calculated as $|Diff(A', B')| = |A' - B'| + |B' - A'| = 1$, which is exactly equal to the edit distance between A and B. Therefore, in this way, the edit distance computation problem can be converted into the problem of solving set difference approximately. Similarly, as shown in Fig. 2(b), assuming there has two DNA sequences described as $A = (a_1 \cdots a_t)$ and $B = (b_1 \cdots b_t)$, given a reference $R = (r_1 r_2 \cdots r_t)$, A, B and R will be classified as $A' = (A_1, A_2, A_i \cdots, A_s)$, $B' = (B_1, B_2, B_i \cdots, B_s)$ and $R' = (R_1, R_2, R_i \cdots, R_s)$ respectively, where A_i, B_i and R_i denote the corresponding DNA fragments of one specific gene. Since we utilize the dynamic programming algorithms shown in [2], A' and B' will be further converted into the sets of gene editing operations A'' and B'', where we assume $A'' = (A_1^{'}, A_2^{'}, A_i^{'} \cdots A_s^{'})$ and $B'' = (B_1^{'}, B_2^{'}, B_i^{'} \cdots B_s^{'})$. Then, data owner selects a hash function $H_{hr} : F \longrightarrow \{-1, 1\}$ to convert all elements of sets into $\{-1, 1\}$, where hr is a secret key and F denotes the universe of all set elements. Because of the randomness of hash functions, for each element t in the set of gene editing operations, we have $E[H_{hr}(t)] = 0$, $E[H_{hr}^2(t)] = 1$, $\underset{t_1 \neq t_2}{E}[H_{hr}(t_1)H_{hr}(t_2)] = E[H_{hr}(t_1)] \cdot E[H_{hr}(t_2)] = 0$. Further, following results can be derived:

$$
\begin{aligned}
E[d_{A''}^2] &= E\left[\left(\sum_{i=1}^{s}\sum_{t\in A_i^{'}} H_{hr}(t)\right)^2\right] = E\left[2\sum_{i=1}^{s}\sum_{t\in A_i^{'}\&t_1\neq t_2} H_{hr}(t_1)H_{hr}(t_2)\right] + \\
& E\left[\sum_{i=1}^{s}\sum_{t\in A_i^{'}} H_{hr}^2(t)\right] + E\left[\sum_{t_k\in A_k^{'}\&t_j\in A_j^{'}\&k\neq j} H_{hr}(t_k)H_{hr}(t_j)\right] \\
&= \sum_{i=1}^{s} E\left[\sum_{t\in A_i^{'}} H_{hr}^2(t)\right] = \sum_{i=1}^{s} |A_i^{'}|
\end{aligned}
\tag{1}
$$

where $|A_i^{'}|$ denotes the number of elements in $A_i^{'}$. Similarly, $E[d_{B''}^2] = \sum_{i=1}^{s} |B_i^{'}|$. Known

$$d_{A''} - d_{B''} = \sum_{i=1}^{s} \sum_{t \in A_i^{'}} H_{hr}(t) - \sum_{i=1}^{s} \sum_{t \in B_i^{'}} H_{hr}(t)$$
$$= \sum_{i=1}^{s} \left(\sum_{t \in A_i^{'} - B_i^{'}} H_{hr}(t) - \sum_{t \in B_i^{'} - A_i^{'}} H_{hr}(t) \right) \quad (2)$$

We have

$$E\left[(d_{A''} - d_{B''})^2\right] = E\left[\sum_{i=1}^{s} \sum_{t \in A_i^{'} - B_i^{'}} H_{hr}^2(t)\right] + E\left[\sum_{i=1}^{s} \sum_{t \in B_i^{'} - A_i^{'}} H_{hr}^2(t)\right]$$
$$= \sum_{i=1}^{s} \left(|A_i^{'} - B_i^{'}| + |B_i^{'} - A_i^{'}|\right) = \sum_{i=1}^{s} |Diff(A_i^{'}, B_i^{'})| \quad (3)$$

Therefore, according to the Eq. 3, the edit distance computation problem of two DNA sequences can be successfully transformed into the set intersection approximation problem between two sets of gene editing operations under the encrypted environment. However, this edit distance approximation algorithm also suffers from errors, for example, if $Reference = ATTGCCCGA, A = GTTGGATAA, B = GTTCGATGA$, we can calculate the set differences between A' and B' as $|Diff(A', B')| = |A' - B'| + |B' - A'| = 4$, which is not equal to the actual edit distance (the actual value is 2). Fortunately, it has been proved that the error rate of the conversion from edit distance computation problem to the set difference size approximation problem are vary small in the literature [3], here we will not explain the details, interested readers can refer more details to literature [2,3].

4 Proposed Scheme

In this section, we propose an efficient DNA similarity search scheme which can achieve fine-grained query and data access control under the encrypted DNA data.

Algorithm 1. EDSS Model

Input: $A_i = (A_{i1}, \cdots A_{im}, \cdots A_{is}; t_{is}, \cdots t_{is+Y}; t_{is+Y+1}, \cdots t_{is+Y+U})$, where A_{im} denotes the set of gene editing operations for data owner i's m-th DNA fragment, $t_{is+y} (y = 0, 1, \cdots Y)$ indicates the corresponding coefficient of $f_i(x)$, $t_{is+Y+u} (u = 1, \cdots U)$ represents the dummy noises added by data owner. $Q = (Q_1, \cdots, Q_s; q^0, \cdots$
$q^Y; q_1, \cdots q_U)$, where Q_m $(m = 1, 2 \cdots s)$ denotes the set of gene editing operations for search user's m-th DNA fragment, $q (q \neq 0, q \in \widetilde{Y})$ represents the role of current search user, and $q_u (u = 1, 2, \cdots U)$ denotes the dummy noises added by search user.

Output: $\widehat{D_i} = \frac{\sum_{j=1}^{k} D_{ji}}{k}$
1: **for** $j = 1$ to $j = k$ **do**
2: data owner $i(i = 1, 2, \cdots N)$ chooses a pseudorandom permutation function π_j with a secret key $S_{kj} \leftarrow \{0,1\}^K$, where $\pi_j : \{0,1\}^K \times \{0,1\}^{2s+Y+U+1} \rightarrow \{0,1\}^{2s+Y+U+1}$, and a hash function $H_{r_j} : F \longrightarrow \{-1, 1\}$, where r_j is a secret key exploited to convert all elements in A_i into $\{-1, 1\}$.
3: each element t in the set of A_{im} will be calculated as $d_{A_{jim}} = \sum_{t \in A_{im}} H_{r_j}(t)$. Then, let $A_{ji} = (\underbrace{d_{A_{ji1}}, -\frac{1}{2}(d_{A_{ji1}})^2, \cdots d_{A_{jis}}, -\frac{1}{2}(d_{A_{jis}})^2}_{2s}; \underbrace{t_{is+0}, t_{is+1}, \cdots t_{is+Y}}_{Y+1};$ $\underbrace{t_{is+Y+1}, t_{is+Y+2}, \cdots t_{is+Y+U}}_{U})$, and the location of each element in the set A_{ji} will be changed as $A'_{ji}(\pi_{S_{kj}}(m)) = A_{ji}(m)$ utilizing the pseudorandom permutation function π_j .
4: **for** $y = 1$ to $y = 2s + Y + U + 1$ **do**
5: data owner randomly chooses a $(2s+Y+U+1)$ dimensional binary vector S_j and two $(2s+Y+U+1) \times (2s+Y+U+1)$ invertible matrices M_{1j}, M_{2j} to blind the A'_{ji}, where if $S_j[y] = 0$, $A'_{ji}(y)$ will be split as $D'_{A_{ji}}[y] = D''_{A_{ji}}[y] = A'_{ji}(y)$, otherwise, $D'_{A_{ji}}[y] + D''_{A_{ji}}[y] = A'_{ji}(y)$. Then, the encrypted index can be described as $I_{A_{ji}} = \{M_{1j}D'_{A_{ji}}, M_{2j}D''_{A_{ji}}\}$.
6: **end for**
7: data owner sends the encrypted index $I_{A_{ji}}$ to the cloud server.
8: **end for**
9: assuming the set that the search user wants to query is $Q = (Q_1, \cdots, Q_m, \cdots, Q_s; q^0, \cdots q^Y; q_{Y+1}, \cdots q_{Y+U})$.
10: **for** $j = 1$ to $j = k$ **do**
11: for every element t in the set of $Q_m(m = 1, \cdots s)$, if $Q_m \in \mathbf{Q}_G$, then $d_{Q_{jm}} = (0, 0)$; otherwise, $d_{Q_{jm}} = (\sum_{t \in Q_m} H_{r_j}(t), 1)$. Besides, let $Q_j = (\underbrace{d_{Q_{j1}}, d_{Q_{j2}}, \cdots d_{Q_{js}},}_{2s}$ $\underbrace{q^0, q^1, \cdots q^Y,}_{Y+1} \underbrace{q_{Y+1}, q_{Y+2}, \cdots q_{Y+U}}_{U})$, for every element in Q_j, we further calculate $Q'_j(\pi_{S_{kj}}(m)) = Q_j(m)$.
12: **for** $y = 1$ to $y = 2s + Y + U + 1$ **do**
13: if $S_j[y] = 0$, $D'_{Q_j}(y) + D''_{Q_j}(y) = Q'_j(y)$; otherwise, $D'_{Q_j}(y) = D''_{Q_j}(y) = Q'_j(y)$.
14: **end for**
15: $T_{Q_j} = \{M_{1j}^{-1}D'_{Q_j}, M_{2j}^{-1}D''_{Q_j}\}$
16: search user submits the encrypted trapdoor T_{Q_j} to the cloud server.
17: **end for**
18: when the cloud server receives $I_{A_{ji}}, T_{Q_j}(i = 1, 2, \cdots N, j = 1, 2, \cdots k)$ from the data owner and search user, respectively,
19: **for** $j = 1$ to $j = k$ **do**
20: the cloud server executes $D_{ji} = I_{A_{ji}} \cdot T_{Q_j}$ for every set $A_i(i = 1, \cdots N)$, and if$|D_{ji}| - D < 0$, save D_{ji}; otherwise, discard D_{ji}.
21: **end for**
22: return $\widehat{D_i} = \frac{\sum_{j=1}^{k} D_{ji}}{k}$ to authorized search user.

The EDSS model is shown in Algorithm 1. Specifically, we utilize following methods to realize the functions of fine-grained query and data access control.

4.1 Achieving Fine-Grain Query

In our EDSS, the whole DNA sequences will be classified by their corresponding genes before being outsourced to the cloud server, which will greatly facilitate the fine-grain query over encrypted cloud data. More concretely, we assume set $\boldsymbol{Q_G}$ (which is a subset in the types of all genes) contains those genes that the search user does not want to query. As shown in steps 10–17 in Algorithm 1, $d_{Q_{jm}}$ will be set as $d_{Q_{jm}} = (0,0)$ if $Q_m \in \boldsymbol{Q_G}$, otherwise, $d_{Q_{jm}} = (\sum_{t \in Q_m} H_{rj}(t), 1)$. In this way, when the cloud server executes the operations (i.e., $D_{ji} = I_{A_{ji}} \cdot T_{Q_j}$) over encrypted DNA sequences, every gene belongs to $\boldsymbol{Q_G}$ will be filtered due to the fact that $d_{Q_{jm}} \times (d_{A_{jim}}, -\frac{1}{2}(d_{A_{jim}})^2) = 0$, where both Q_{jm} and A_{jim} belong to $\boldsymbol{Q_G}$. Thus, the complicated logic query such as mixed "AND" and "NO" operations on genes can be achieved.

4.2 Data Access Control

Different users may have different access authorities to DNA sequences, we assume there are Y independent user roles, $\widetilde{Y} = (a_1, a_2, \cdots, a_Y)$, let

$$f_i(x) = \prod_{select\ a_{it} \in \widetilde{Y}} (x - a_{it}) = \sum_{y=0}^{y=Y} t_{iy} x^y \tag{4}$$

where $f_i(x)$ denotes the role polynomial of data owner i, and we use δ_i to represent the degree of $f_i(x)$, which satisfies that when $\delta_i < Y, t_{iy} = 0 (\delta_i < y \leq Y)$. Besides, we suppose user's roles $\widetilde{Y}$ that meet following conditions:

$$\begin{aligned} & a_1 > D > \max\left\{\left|\sum_{m=1}^{m=s}(d_{A_{jim}}, -\frac{1}{2}(d_{A_{jim}})^2) \cdot d_{Q_{jm}}\right|\right\} \\ & (i = 1, \cdots N, j = 1, \cdots k, m = 1, \cdots s) \\ & \sum_{y=1}^{y=g-1} a_y + 2D < a_g, (g = 2, 3, \cdots Y) \end{aligned} \tag{5}$$

Next, we will introduce our EDSS model how to achieve the data access control under the encrypted environment.

As illustrated in Algorithm 1, known

$$\begin{aligned} I_{A_{ji}} \cdot T_{Q_j} &= \{M_{1j} D'_{A_{ji}}, M_{2j} D''_{A_{ji}}\} \cdot \{M_{1j}^{-1} D'_{Q_j}, M_{2j}^{-1} D''_{Q_j}\} \\ &= \left(D'_{A_{ji}} \cdot D'_{Q_j} + D''_{A_{ji}} \cdot D''_{Q_j}\right) \\ &= \sum_{m=1}^{m=2s} A_{ji}(m) \cdot Q_j(m) + \sum_{y=0}^{y=Y} t_{is+y} q^y + \sum_{j=1}^{j=U} t_{is+Y+j} \cdot q_{Y+j} \end{aligned}$$

Case 1: if $q \neq 0$ and $q \in f_i(x)$, we have $\sum_{y=0}^{y=Y} t_{is+y}q^y = 0$. Then we assume both t_{is+Y+j} and $q_{Y+j} (i = 1, 2 \cdots, N, j = 1, 2, \cdots U)$ obey a same normal distribution with mean and variance as $(0, \varepsilon^2)$, where ε is a small positive number. Therefore,

$$E\left[\sum_{j=1}^{j=U} t_{is+Y+j} \cdot q_{Y+j}\right] = \left[\sum_{j=1}^{j=U} E(t_{is+Y+j}) \cdot E(q_{Y+j})\right] = 0$$

Thus, $E[I_{A_{ji}} \cdot T_{Q_j}]$ can be simplified as

$$E\left[I_{A_{ji}} \cdot T_{Q_j}\right] = E\left[\sum_{m=1}^{m=s} \left((d_{A_{jim}}, -\frac{1}{2}(d_{A_{jim}})^2) \cdot d_{Q_{jm}}\right)\right]$$

As discussed in Sect. 4.1, each DNA fragment belonging to $\boldsymbol{Q_G}$ will be filtered thanks to $d_{Q_{jg}} \times (d_{A_{jig}}, -\frac{1}{2}(d_{A_{jig}})^2) = 0$, where both Q_{jg} and A_{jig} belong to $\boldsymbol{Q_G}$. Besides, $|I_{A_{ji}} \cdot T_{Q_j}| - D < 0$ because of $D > \max\left\{\left|\sum_{m=1}^{m=s}(d_{A_{jim}}, -\frac{1}{2}(d_{A_{jim}})^2) \cdot d_{Q_{jm}}\right|\right\}$, which satisfies the condition of steps 19–21 in EDSS model.

Correctness: For arbitrary A_i, A_j and Q meeting the conditions of **Case 1**, it is easy to deduce that $E\left[I_{A_i} \cdot T_Q - I_{A_j} \cdot T_Q\right] = \sum_{k=1}^{k=s} (|Diff(A_{jk}, Q_k)| - |Diff(A_{ik}, Q_k)|)$. Therefore, if $I_{A_i} \cdot T_Q - I_{A_j} \cdot T_Q > 0 \Rightarrow \sum_{k=1}^{k=s} |Diff(A_{jk}, Q_k)| > \sum_{k=1}^{k=s} |Diff(A_{ik}, Q_k)|$, it shows that A_i is closer to Q.

Case 2: if $q \neq 0$ and $q \notin f_i(x)$, we know that

$$\begin{aligned} I_{A_{ji}} \cdot T_{Q_j} &= \{M_{1j}D'_{A_{ji}}, M_{2j}D''_{A_{ji}}\} \cdot \{M_{1j}^{-1}D'_{Q_j}, M_{2j}^{-1}D''_{Q_j}\} = D'_{A_{ji}} \cdot D'_{Q_j} + D''_{A_{ji}} \cdot D''_{Q_j} \\ &\approx \sum_{m=1}^{m=s}(d_{A_{jim}}, -\frac{1}{2}(d_{A_{jim}})^2) \cdot d_{Q_{jm}} + \sum_{y=0}^{y=Y} t_{is+y}q^y \end{aligned}$$

Due to the condition $a_1 > D > \max\left\{\left|\sum_{m=1}^{m=s}(d_{A_{jim}}, -\frac{1}{2}(d_{A_{jim}})^2) \cdot d_{Q_{jm}}\right|\right\}$, we have $\left|\sum_{y=0}^{y=Y} t_{is+y}q^y\right| \geq a_2 - a_1 > 2D$. Thus

$$\left|I_{A_{ji}} \cdot T_{Q_j}\right| > 2D - \left|\sum_{m=1}^{m=s}(d_{A_{jim}}, -\frac{1}{2}(d_{A_{jim}})^2) \cdot d_{Q_{jm}}\right| - D > 0$$

Therefore, if $q \neq 0$ and $q \notin f_i(x)$, the similarity scores between Q and A_i will be discarded since query Q is against the conditions shown in steps 19–21 of Algorithm 1. Overall, our EDSS model can achieve both the function of fine-grained query and data access control under the encrypted DNA data.

5 Security Analysis

As shown in Sect. 2.2, the main security requirements in EDSS are confidentiality of data owner's DNA sequence data, privacy protection of index and trapdoor, and the unlinkability of trapdoors. Therefore, we mainly analyze these security requirements in details and other security features are not the discussed fields in this paper.

5.1 Confidentiality of Data Owner's DNA Sequences

In EDSS, the plaintexts of data owner's DNA sequences are encrypted by traditional symmetric encryption technologies such as AES before being outsourced to the cloud server, and the secret key is assigned to authorized search user through a secure channel. Thus, it is impossible that if an attacker wants to spy the contents of DNA sequences maliciously without the secret key. Besides, the high security of AES has been proved in literature [7], therefore, the confidentiality of data owner's DNA sequences can be protected well.

5.2 Privacy Protection of Index and Trapdoor

The index $I_{A_{ji}}$ and trapdoor T_{Q_j} are generated through following phases. Firstly, the set of gene editing operations A_i and Q will be compressed into A_{ji} and Q_j by hash function H_{rj}, respectively. Then, pseudorandom permutation function π_j is employed to further change the location of each element in the set A_{ji} and Q_j. Finally, A'_{ji} and Q'_j will be encrypted utilizing secure KNN computation [12] with the symmetric key $\{S_j, M_{1j}, M_{2j}\}$. Due to the one-way properties of hash function H_{rj}, the confidentiality of secret key $\{S_{kj}, r_j, S_j, M_{1j}, M_{2j}\}$, and the high security of secure KNN computation [12], the privacy protection of index and trapdoor can be achieved.

5.3 Unlinkability of Trapdoors

We will prove that our scheme can achieve the unlinkability of trapdoors in the stronger threaten model, i.e., Known Background Model.

As discussed in Sect. 4, the form of trapdoor is $T_{Q_j} = \{M_{1j}^{-1}D^{'}_{Q_j}, M_{2j}^{-1}D^{''}_{Q_j}\}$. Specifically, each element in set Q_j will be mapped into $\{-1, 1\}$ by hash function H_{rj} firstly. For convenience, we assume there are 2^l possible mapped results for all of the elements in Q_j. In addition, the pseudorandom permutation function π has a mapped relation $\pi : \{0,1\}^K \times \{0,1\}^{2s+Y+U+1} \rightarrow \{0,1\}^{2s+Y+U+1}$, consequently, there are also $(2s+Y+U+1)!$ possible replaced positions for each request Q_j. On the other hand, we add U dummy noises to each query request Q_j randomly. Without loss of generality, we assume the size of each dummy noise are η_s, hence, there are $U2^{\eta_s}$ possible values for all of the dummy noises. In addition, we use a $(2s+Y+U+1)$ dimensional binary vector S_j to split the vector Q'_j, if $S_j[y] = 0$, as shown in Algorithm 1, the $Q'_j(y)$ will be split into two random values. Assuming the number of '0' in S_j are μ, and the size of each dimension in $D^{'}_{Q_j}$ and $D^{''}_{Q_j}$ are η_q bits, there are $2^{\eta_q \mu}$ possible values for split process. Note that η_s, μ and η_q are independent of each other, therefore, for two trapdoors, we can calculate the probability of them that are identical as follows:

$$P_1 = \frac{1}{2^l \cdot (2s+Y+U+1)! \cdot U2^{\eta_s} \cdot (2^{\eta_q})^{\mu}} = \frac{1}{(2s+Y+U+1)! \cdot U2^{l+\eta_s+\mu\eta_q}} \quad (6)$$

Therefore, if we select the lagrer η_s, μ and η_q, we can ensure the high security. For example, if we select a 1024-bit t, thus the probability $P_1 < 1/2^{1024}$. As a result, the probability that two trapdoors are the same is negligible.

6 Performance Evaluation

In this section, we will evaluate the performance of our proposed schemes by comparing with existing proposals [2–4]. Here we select data in batches from a real-world dataset GenBank [13], and implement our simulation experiments with an Intel Core i7 3.5 GHz system.

6.1 Functionality

As shown in Table 1, both literatures [2–4] and our EDSS model can achieve the DNA similarity search under the encrypted DNA data. However, since [3] mainly focuses on efficiency issues, it cannot support the functions of fine-grained query and data access control. Besides, because of a mutual protocol used in scheme [3], it needs k times communication rounds throughout the whole querying process, where k is a positive integer selected by search user. Similar to [3], literature [2] is also devising the whole scheme from the perspective of efficiency. In addition, due to the protocol of oblivious transfer that they adopted, it also needs 3 times communication rounds to achieve whole querying process. On the other hand, despite the scheme [4] only requires one round of communication compared with previous protocols, however, because of the complex encryption method based on bilinear group, their work is criticized for its defects in performance. What's more, both the functions of fine-grained query and data access control are also not supported in their protocol.

Table 1. Comparison of functionalities

	[3]	[2]	[4]	EDSS
DNA similarity search	√	√	√	√
Fine-grained query	×	×	×	√
Access control	×	×	×	√
Communication rounds	k	3	1	1

6.2 Computation Overhead of Similarity Query

According to the Fig. 3(a) and (b), we can see that the running time of EDSS to query the encrypted cloud data are less than [3], but not [2]. However, compared with [2], the functions of fine-grained query and data access control are supported in our EDSS. Besides, since the hash function, pseudorandom permutation function and the secure KNN computation are exploited to encrypt the raw DNA data, our EDSS shows more remarkable privacy protection performance. Therefore, EDSS possesses higher performance in terms of functionality, efficiency and security compared with existing models.

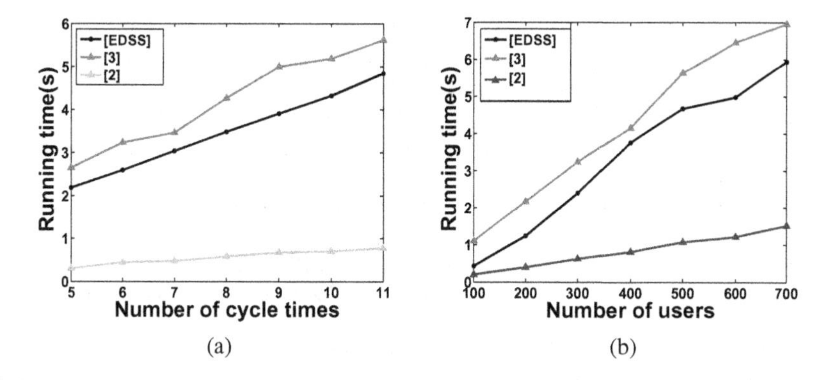

Fig. 3. Time for query. (a) For the different number of cycle times with the same number of users, $N = 500$. (b) For the different number of users with the same number of cycle times $k = 10$.

7 Conclusion

In this paper, we have proposed an efficient DNA similarity search scheme (EDSS) which can achieve fine-grained query and data access control over encrypted cloud data. Specifically, we set up a subset (which contains those types of genes that the search user does not want to query) of the whole DNA sequences beforehand, which will be filtered utilizing a multiple genes search strategy in the querying process. Thus, the complicated logic query such as mixed "AND" and "NO" operations on genes can be achieved. Then, we employ a novel polynomial based design to support efficient data access control, where the data owner can define who can access specific DNA sequences by verifying whether the user's role is a root of corresponding polynomials. Finally, security analysis and extensive experiments demonstrate the high security and efficiency of EDSS compared with existing schemes. For the future work, we intend to extend our proposal to support dynamic update of DNA data meanwhile guarantee the forward and backward privacy in the update process, and further improve the accuracy, security and performance of proposed scheme.

Acknowledgment. This work is supported by the National Key R&D Program of China under Grants 2017YFB0802300 and 2017YFB0802000, the National Natural Science Foundation of China under Grants 61772121, 61728102, and 61472065, the Fundamental Research Funds for Chinese Central Universities under Grant ZYGX2015J056.

References

1. Ebadollahi, S., Sun, J., Gotz, D., Hu, J., Sow, D., Neti, C.: Predicting patient's trajectory of physiological data using temporal trends in similar patients: a system for near-term prognostics. In: proceedings of AMIA Annual Symposium, pp. 192–196 (2010)
2. Asharov, G., Halevi, S., Lindell, Y., Rabin, T.: Privacy-preserving search of similar patients in genomic data. IACR Cryptology ePrint Archive, pp. 1–27 (2017)

3. Wang, X.S., Huang, Y., Zhao, Y., Tang, H., Wang, X., Bu, D.: Efficient genome-wide, privacy-preserving similar patient query based on private edit distance. In: Proceedings of ACM CCS, pp. 492–503 (2015)
4. Wang, B., Song, W., Lou, W., Hou, Y.T.: Privacy-preserving pattern matching over encrypted genetic data in cloud computing. In: Proceedings of IEEE INFOCOM, pp. 1–9 (2017)
5. Watson, M.: Illuminating the future of DNA sequencing. Genome Biol. **15**(2), 108–110 (2014)
6. Wang, R., Wang, X.F., Li, Z., Tang, H., Reiter, M.K., Dong, Z.: Privacy-preserving genomic computation through program specialization. In: Proceedings of ACM CCS, pp. 338–347 (2009)
7. Hamalainen, P., Alho, T., Hannikainen, M., Hamalainen, T.D.: Design and implementation of low-area and low-power AES encryption hardware core. In: EUROMICRO Conference on Digital System Design, pp. 577–583 (2006)
8. Li, H., Liu, D., Dai, Y., Luan, T.H., Shen, X.: Enabling efficient multi-keyword ranked search over encrypted cloud data through blind storage. IEEE Trans. Emerg. Top. Comput. **3**(1), 127–138 (2015)
9. Li, H., Yang, Y., Luan, T.H., Liang, X., Zhou, L., Shen, X.S.: Enabling fine-grained multi-keyword search supporting classified sub-dictionaries over encrypted cloud data. IEEE Trans. Dependable Secure Comput. **13**(3), 312–325 (2016)
10. Li, H., Lin, X., Yang, H., Liang, X., Lu, R., Shen, X.: EPPDR: an efficient privacy-preserving demand response scheme with adaptive key evolution in smart grid. IEEE Trans. Parallel Distrib. Syst. **25**(8), 2053–2064 (2014)
11. Xu, G., Li, H., Tan, C., Liu, D., Dai, Y., Yang, K.: Achieving efficient and privacy-preserving truth discovery in crowd sensing systems. Comput. Secur. **69**, 114–126 (2017)
12. Wong, W.K., Cheung, D.W.l., Kao, B., Mamoulis, N.: Secure kNN computation on encrypted databases. In: Proceedings of ACM SIGMOD, pp. 139–152 (2009)
13. GenBank Database (2017). https://www.ncbi.nlm.nih.gov/genbank/

Sampling Based δ-Approximate Data Aggregation in Sensor Equipped IoT Networks

Ji Li[1], Madhuri Siddula[1], Xiuzhen Cheng[2], Wei Cheng[3], Zhi Tian[4], and Yingshu Li[1](✉)

[1] Department of Computer Science, Georgia State University, Atlanta, GA 30303, USA
{jli30,msiddula1}@student.gsu.edu, yili@gsu.edu

[2] Department of Computer Science, The George Washington University, Washington, D.C. 20052, USA
cheng@gwu.edu

[3] Department of Computer Science, Virginia Commonwealth University, Richmond, VA 23284, USA
wcheng3@vcu.edu

[4] Department of Electrical and Computer Engineering, George Mason University, Fairfax, VA 22030, USA
ztian1@gmu.edu

Abstract. Emerging needs in data sensing applications result in the usage of IoT networks. These networks are widely deployed and exploited for various efficient data transfer. Wireless sensors can be incorporated in IoT networks to reduce the deployment costs and maintenance costs. One of the critical problems in sensor equipped IoT devices is to design an energy efficient data aggregation method that processes the maximum value query and distinct set query. Therefore, in this paper, we propose two approximate algorithms to process the maximum queries and distinct-set queries in wireless sensor networks. These two algorithms are based on uniform sampling. Solid theoretical proofs are offered which can make sure the proposed algorithms can return correct query results with a given probability. Simulation results show that both δ-approximate maximum value and δ-approximate distinct set algorithms perform significantly better than a simple distributed algorithm in terms of energy consumption.

1 Introduction

With the ever-increasing population, problems of everyday sustainability have become onerous. According to the survey by United Nations, 54% of the world's current population lives in urban areas. It is also expected that the percentage increases to 66 by 2050. With this escalation in urban population, new challenges have emerged. These challenges include the constant power supply, public safety, disaster prediction, and traffic maintenance. Smart City (SC) has become

S. Chellappan et al. (Eds.): WASA 2018, LNCS 10874, pp. 249–260, 2018.
https://doi.org/10.1007/978-3-319-94268-1_21

inevitable to address these challenges. Many cities like New York, Detroit, Singapore, and London are working towards smart city development. These cities have adopted for various technologies like smart parking services, intelligent street light systems, sensors to redirect traffic, and water conservation. Although these applications are designed for urban living, they should also be incorporated in rural areas so that more resources could be preserved for future generations. SC networks should collect data from all over the city to provide better information. So, to collect such well spread data, they exploit various sensor equipped devices in the city to collect data and interpret information at the city level.

In today's world, all the devices from smart home devices to intelligent transport systems are well connected to the Internet. Such a network with well-connected devices is called Internet-Of-Things (IoT). As the network grows, our need for intelligent devices develop and so does the necessity to sense various activities for the convenient living of people in the cities. Some of the applications like transportation, healthcare, and seamless internet connection widely use IoT technologies. The main aim in the IoT is to reduce cost and provide faster access to the data [12,16,21]. But the primary challenge is that the deployment of IoT network is expensive as it requires a large number of sensing devices. Additionally, it is also important to collect data autonomously and provide intelligent methods that address the issues of dynamic traffic, accommodating new services, channel conditions, and ever-increasing user requirements.

Sensors are the building blocks for many IoT devices. Utilizing sensors as the communication media helps us resolve many of the problems discussed earlier. Previous researchers have studied the issues of routing, data management, link scheduling, coverage and topology control in networks that include sensors for communication [3–7,20,22,23]. Although using sensors reduces the communication cost, it raises the issue of processing cost. Sensors collect data for a more extended period over vast networks. Therefore, we end up with massive data being processed at a single sensor node and thus increasing the processing cost. To address this issue, we need data aggregation at the sensor level. When data is aggregated, we only send the aggregated partial data into the network. This further raises the energy consumption issue as the aggregation costs much energy and the sensors are not equipped with a huge amount of power supply. According to [18], cost of transmitting one bit of data using wireless link is equivalent to the cost of executing 1000 instructions. So, reducing the data transmission is the major way to decrease the energy consumption. Hence, it is critical to design energy efficient data aggregation models for the sensor equipped IoT networks.

There are two kinds of aggregation queries: maximum query and distinct set query. The maximum query is to calculate the maximum of all the readings in the sensory data while the distinct set query is to calculate the unique values in the sensory data. Both the queries are essential for a given network. For example, while monitoring pollution, maximum query results in the most polluted area along with its values. Similarly, distinct set query shows the pollution levels in all the regions. Hence, the energy efficient data aggregation model should accommodate both queries in its development.

In practice, exact query results are not always necessary while approximate query results may be acceptable for conservation [9]. Therefore, in this paper, we propose an algorithm to process δ-approximate maximum queries and δ-distinct-set queries in IoTs. This algorithm is based on uniform sampling. Proposed algorithm returns the exact query results with probability not less than $1-\delta$ where the value of δ can be arbitrarily small.

The rest of the paper is organized as follows. Section 2 defines the problem. Section 3 provides the mathematical proof for the δ-approximate aggregation algorithms. Section 4 explains the proposed δ-approximate aggregation algorithms. Section 5 shows the simulation results and the related works is discussed in Sect. 6. Section 7 concludes the paper.

2 Problem Definition

Let us assume that we have a sensor equipped IoT network with n sensor nodes and s_{ti} is the sensory value of node i at time t. $S_t = \{s_{t1}, s_{t2}, \ldots, s_{tn}\}$ is used to denote the set of all the sensory data in the network at time t. We use $Dis(S_t) = \{s^d_{t1}, s^d_{t2}, \ldots, s^d_{t|Dis(S_t)|}\}$ to denote the distinct set of S_t, which contains the distinct values in S_t. For example, if we have $S_t = \{s_{t1}, s_{t2}, s_{t3}, s_{t4}, s_{t5}\}$ and $s_{t1} = 1, s_{t2} = 1, s_{t3} = 2, s_{t4} = 3, s_{t5} = 3$; then the $Dis(S_t) = \{1, 2, 3\}$. In this paper, we assume that the data is distributed randomly in the network while the spatial and temporal correlation of the sensory data is ignored.

In this paper, we focus on two aggregation operations on S_t, which are *max* and *distinct set*. The definition of the maximum value and distinct-set are as follows:

1. The exact maximum value denoted by $Max(S_t)$ satisfies $Max(S_t) = \max\{s_{ti} \in S_t | 1 \leq i \leq n\}$.
2. The exact distinct-set of S_t denoted by $Dis(S_t)$ satisfies that $\forall s \in S_t, \exists s^d \in Dis(S_t), s = s^d$ and $\forall s^d_x,\ s^d_y \in Dis(S_t), x \neq y \Rightarrow s^d_x \neq s^d_y$.

A naive method that solves the max and distinct set aggregation problems has three main steps.

1. Organize all the nodes in the network into an aggregation tree. The sink node broadcasts the aggregation operation in the network.
2. All the nodes in the network submit their sensory data to the sink node along the aggregation tree.
3. The intermediate nodes in the aggregation tree aggregate the partial results during the data transmission.

However, the above method will lead to an immense communication cost and computation cost for calculating exact aggregation result. Therefore, we propose a δ-approximate result for the above two aggregation operations. Let I_t and $\widehat{I_t}$ are the exact aggregation result and approximate aggregation result of S_t at time t respectively. The definition of the δ-estimator is as follows.

Definition 1 (δ-estimator). *For any δ $(0 \le \delta \le 1)$, $\widehat{I_t}$ is called the δ-estimator of I_t if* $\Pr(\widehat{I_t} \neq I_t) \le \delta$,

According to Definition 1, the problem of computing δ-approximate maximum value and δ-approximate distinct-set is defined as follows.

Input: (1) A sensor equipped IoT network with n nodes; (2) The sensory data set S_t; (3) Aggregation operator $Agg \in \{Max, DistinctSet\}$ and δ $(0 \le \delta \le 1)$.

Output: δ-approximate aggregation result of Agg.

3 Preliminaries

Let $u_1, u_2, ..., u_m$ denote m simple random samplings with replacement from S_t, $U(m) = \{u_1, u_2, ..., u_m\}$ is used to denote a uniform sample of S_t with sample size m, then we have the following conclusions.

1. u_i and u_j are independent of each other for all $1 \le i \neq j \le m$.
2. $\Pr(u_i = s_{tj}) = \frac{1}{n}$ for any $1 \le i \le m$, $1 \le j \le n$.

Based on the above conclusions, we have the following theorem.

Lemma 1. *For any given value $x \in Dis(S_t)$, we have*

$$\Pr(x \notin U(m)) = (1 - \frac{n_x}{n})^m$$

where n_x is the number of appearance of value x in S_t.

Proof: $\Pr(x \notin U(m)) = \Pr(u_1 \neq x \wedge u_2 \neq x \wedge \ldots u_m \neq x)$. Since all the samples $u_1, u_2, \ldots, u_m$ are independent with each other, we have

$$\Pr(x \notin U(m)) = \prod_{i=1}^{m} \Pr(u_i \neq x) = (\Pr(u_1 \neq x))^m.$$

Moreover, we also have

$$\Pr(u_1 \neq x) = 1 - \Pr(u_1 = x) = 1 - \frac{n_x}{n}.$$

Then this lemma is proved. □

To obtain the δ-approximate maximum value, we need the mathematical estimator. Let $\widehat{Max(S_t)}_u$ denote the uniform sampling based estimator of exact value $Max(S_t)$. Then $\widehat{Max(S_t)}_u$ is defined as

$$\widehat{Max(S_t)}_u = Max(U(m)) = \max\{u_i \in U(m) | 1 \le i \le m\}.$$

Based on Lemma 1, we have the following theorem.

Theorem 1. *$\widehat{Max(S_t)}_u$ is a δ-estimator of $Max(S_t)$ if*

$$m \geq \frac{\ln \delta}{\ln(1 - \frac{n_{min}}{n})}$$

where n_{min} is the number of appearances for the least appearing data.

Proof: Based on the condition, we have

$$m \ln(1 - \frac{n_{min}}{n}) \leq \ln \delta$$

$$(1 - \frac{n_{min}}{n})^m \leq \delta.$$

According to Lemma 1, we have

$$\Pr(Max(S_t) \notin U(m)) = (1 - \frac{n_{Max(S_t)}}{n})^m$$

where $n_{Max(S_t)}$ is the number of times the maximum value appears in S_t. Since $n_{Max(S_t)} \geq n_{min}$, we have

$$\Pr(Max(S_t) \notin U(m)) \leq (1 - \frac{n_{min}}{n})^m \leq \delta.$$

Then this theorem is proved. □

Let $\widehat{Dis(S_t)}_u$ denote the uniform sampling based estimator of exact result $Dis(S_t)$. Then $\widehat{Dis(S_t)}_u$ is defined as

$$\widehat{Dis(S_t)}_u = Dis(U(m)).$$

Based on Lemma 1, we have the following theorem.

Theorem 2. *$\widehat{Dis(S_t)}_u$ is a δ-estimator of $Dis(S_t)$ if*

$$m \geq \frac{\ln(1 - (1 - \delta)^{n_{min}/n})}{\ln(1 - \frac{n_{min}}{n})}$$

where n_{min} is the number of appearances for the least appearing data.

Proof: Based on the condition, we have

$$(1 - \frac{n_{min}}{n})^m \leq 1 - (1 - \delta)^{n_{min}/n}$$

$$(1 - (1 - \frac{n_{min}}{n})^m)^{n/n_{min}} \geq 1 - \delta$$

$$1 - \prod_{i=1}^{|Dis(S_t)|} (1 - (1 - \frac{n_{min}}{n})^m) \leq \delta$$

Let $n_{s_{ti}^d}$ denote the number of appearances for $s_{t_i}^d$, then we have

$$1 - \prod_{i=1}^{|Dis(S_t)|} (1 - (1 - \frac{n_{s_{ti}^d}}{n})^m) \leq \delta$$

since $n_{min} \leq n_{s_{ti}^d}$. Moreover, according to Lemma 1, we have

$$1 - \prod_{i=1}^{|Dis(S_t)|} (1 - \Pr(s_{ti}^d \notin U(m))) \leq \delta$$

$$1 - \prod_{i=1}^{|Dis(S_t)|} \Pr(s_{ti}^d \in U(m)) \leq \delta$$

$$1 - \Pr(\widehat{Dis(S_t)}_u = Dis(S_t)) \leq \delta$$

$$\Pr(\widehat{Dis(S_t)}_u \neq Dis(S_t)) \leq \delta$$

Then this theorem is proved. □

4 δ-Approximate Aggregation Algorithm

The theorems in Sect. 3 show how to calculate the required sampling size and sampling probability according to a given δ. However, we still have the following problems to be solved.

1. How does the sink node broadcast the sampling information in the whole network.
2. How to sample the sensory data from the entire network.
3. How to transmit and aggregate the partial aggregation results.

When the sample size m is calculated using the theorems in Sect. 3, there is a simple method to sample the sensory data.

1. The sink generates m random numbers from the set $\{1, 2, 3, \ldots, n\}$ and broadcasts them in the whole network.
2. The sensor node whose id is one of the m randomly selected ids sends its sensory data to the sink node.

However, the above algorithm has a huge energy cost during the first step since a significant amount of sampling information needs to be transmitted. To further reduce the energy cost, we divide the whole network into k disjoint clusters C_1, C_2, ... , C_k. Each cluster randomly selects one of its node as the cluster head. By using the method, proposed in [15], all the cluster heads in the network are organized as a minimum hop-count spanning tree rooted at the sink node. We then adopt the uniform sampling algorithm proposed by [8], described as follows.

1. The sink generates a series of random numbers Y_i with the probability $\Pr(Y_i = l) = \frac{|C_l|}{n} (1 \le i \le m)$,
2. Let m_l be the sample size of C_l. Then m_l is calculated by $m_l = |\{Y_i | Y_i = l\}|$.
3. The sink node sends the sample size $\{m_l \mid 1 \le l \le k\}$ to each cluster head. Each cluster head samples the sensory data in the cluster using the above naive sampling algorithm.

When the cluster head of the l-th cluster receives all the sampled sensory data, $U(ml)$, it calculates the partial aggregation result $R(U(m_l))$ according to aggregation operation Agg by using the following method.

$$R(U(m_l)) = \begin{cases} Max(U(m_l)) & \text{if } Agg = Max \\ Dis(U(m_l)) & \text{elsewhere} \end{cases}$$

The above process is explained in Algorithm 1. Then the partial aggregation result $R(U(m_l))$ is transmitted along the spanning tree to the sink node. To further reduce the transmission cost, the intermediate nodes in the spanning tree aggregate the received partial result while transmitting the data. The above process is explained in Algorithm 2.

Algorithm 1. Uniform Sampling Based Aggregation Algorithm

Input: δ, aggregation operator $Agg \in \{Max, DistinctSet\}$
Output: δ-approximate aggregation results

1: **if** $Agg = Max$ **then**
2: $\quad m = \lceil \frac{\ln \delta}{\ln(1-\frac{n_{min}}{n})} \rceil$
3: **else**
4: $\quad m = \lceil \frac{\ln(1-(1-\delta)^{n_{min}/n})}{\ln(1-\frac{n_{min}}{n})} \rceil$
5: **end if**
6: generate Y_i following $\Pr(Y_i = l) = \frac{|C_l|}{n}$,
7: $m_l = |\{Y_i \mid Y_i = l\}|$ $(1 \le i \le m, 1 \le l \le k)$, the sink sends m_l to each cluster head by multi-hop communication
8: **for** each cluster head of the clusters C_l $(1 \le l \le k)$ **do**
9: $\quad$ generates random numbers $k_1, k_2, \ldots, k_{m_l}$ then broadcast inside the cluster
10: **end for**
11: **for** each cluster member of C_l $(1 \le l \le k)$ **do**
12: $\quad$ send sensory value to cluster head if its $id \in \{k_1, k_2, \ldots, k_{ml}\}$
13: **end for**
14: **for** each cluster head of the clusters C_l $(1 \le l \le k)$ **do**
15: $\quad$ receive sample data $U(m_l)$ and calculate partial result $R(U(m_l))$
16: **end for**

According to the analysis in Sect. 3, for the sample size m, we have

$$m = \begin{cases} \lceil \frac{\ln \delta}{\ln(1-\frac{n_{min}}{n})} \rceil & \text{if } Agg = Max \\ \lceil \frac{\ln(1-(1-\delta)^{n_{min}/n})}{\ln(1-\frac{n_{min}}{n})} \rceil & \text{if } Agg = Dis \end{cases}$$

Algorithm 2. Partial Data Aggregation Algorithm

```
1: for each node j in the spanning tree do
2:   if j is the leaf node then
3:     Send R_j to its parent node
4:   else
5:     Receive partial results R_j1, R_j2, ..., R_jc from its children
6:     if Agg = Max then
7:       R_j = max(R_j1, R_j2, ..., R_jc)
8:     else
9:       R_j = ∪_{i=1}^{c} R_ji
10:    end if
11:    if j is the sink node then
12:      return R_j
13:    else
14:      Send R_j to its parent node
15:    end if
16:  end if
17: end for
```

Therefore, we have

$$m = \begin{cases} O(\ln \frac{1}{\delta}) & \text{if } Agg = Max \\ O(\ln(\frac{1}{1-(1-\delta)^{n_{min}/n}})) & \text{if } Agg = Dis \end{cases}$$

In practice, $|R_j|$ can be regarded as a constant. According to [8], the communication cost and the energy cost of the uniform sampling based δ-approximate aggregation algorithm is $O(\ln \frac{1}{\delta})$ if $Agg = Max$, while the cost is $O(\ln(\frac{1}{1-(1-\delta)^{n_{min}/n}}))$ if $Agg = Dis$.

5 Simulation Results

To evaluate the proposed algorithms, we have simulated a network with 1000 nodes. All the nodes are randomly distributed in a rectangular region of size 300 m × 300 m, and the sink is in the center of the region. The following strategy is used to define the clusters.

1. Divide the whole region into 10 × 10 grids.
2. Group the nodes in the same gird into the same cluster.
3. Randomly chose the cluster head among the nodes of the same grid.

For each node, the energy cost to send and receive one byte is set as 0.0144 mJ and 0.0057 mJ [10]. The communication range of each sensor node is set to be $30\sqrt{2}$m [1]. This kind of simulation setting can make every sensor node communicate with its cluster head by a one-hop message.

The first group of simulations is about the relationship between δ and the sample size. The results are presented in Fig. 1. These results show that the

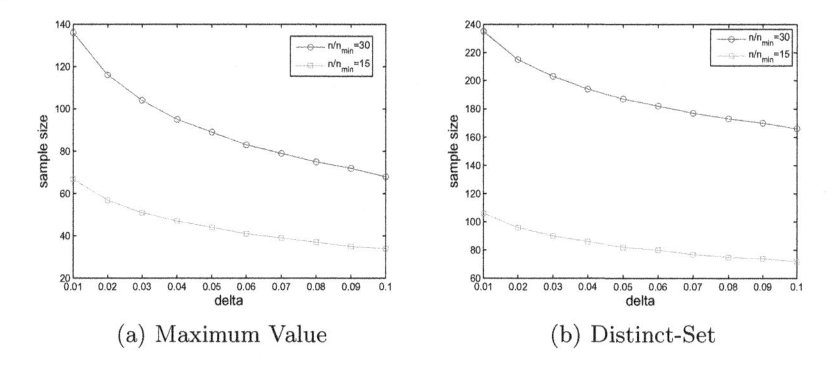

(a) Maximum Value (b) Distinct-Set

Fig. 1. The relationship between δ and the sample size.

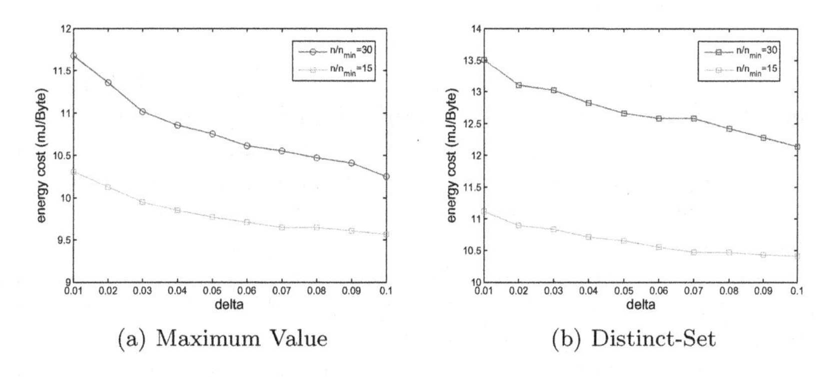

(a) Maximum Value (b) Distinct-Set

Fig. 2. The relationship between δ and the energy cost for the uniform sampling based aggregation algorithm.

sample size increases with the decline of δ. Moreover, the sample sizes are much smaller than the size of the network. For example, when $\delta = 0.01$, the sample size is about 67 for deriving δ-approximate maximum value. If $\frac{n}{n_{min}} = 15$, which indicates that we just need to sample 6.7% sensory data from the network to guarantee that the estimated maximum value being equal to the actual maximum value with the probability greater than 99%. Therefore, our uniform sampling based algorithm saves a tremendous amount of energy as it only needs a little amount of sensory data to be sampled and transmitted in the network. Moreover, we can see that in the same condition, the required sample size for the distinct-set aggregation is greater than that of the maximum value aggregation since the distinct-set aggregation needs to make sure all the distinct values are being sampled.

The second group of simulations is about the relationship between δ and the energy cost. The results are shown in Fig. 2. These results indicate that the energy cost increases with the decline of δ. We can also see that for the same condition, the energy cost for the distinct-set aggregation is higher than that of the maximum value aggregation, as the distinct-set aggregation has a greater sample size.

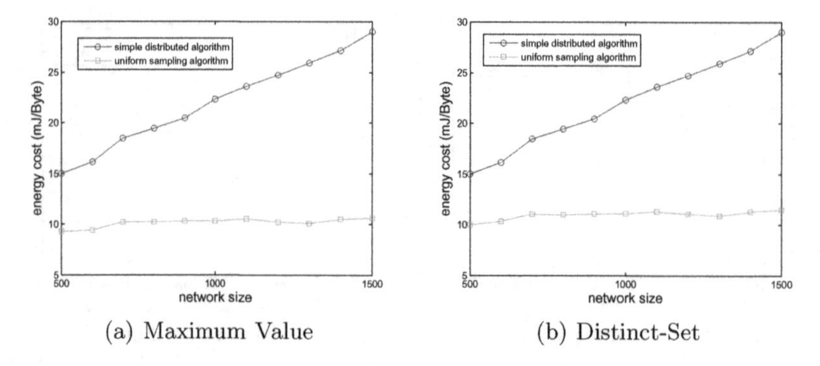

(a) Maximum Value

(b) Distinct-Set

Fig. 3. Energy cost comparison between the uniform sampling based aggregation algorithm and the simple distributed algorithm.

The third group of simulations is to compare the energy cost between the uniform sampling based aggregation algorithm and the simple distributed algorithm. The simple distributed algorithm is to collect all the raw sensory data and to aggregate the partial results during the transmission, which can always return accurate aggregation results. For the uniform sampling based aggregation algorithm, we set $\delta = 0.1$, $\frac{n}{n_{min}} = 15$, and the network size varies from 500 to 1500. The results are listed in Fig. 3. We can see that for all the proposed algorithms, the energy cost increases with the increase of the network size. Moreover, for the same network size, the energy cost of the uniform sampling based aggregation algorithm is much lower than that of the naive distributed algorithm. These results indicate that the uniform sampling based aggregation algorithm performs better in terms of energy consumption. It is also to be observed that with an increase in the network size, the energy cost of the naive distributed algorithm proliferates, while the energy cost of the uniform sampling based aggregation algorithm almost remains the same. The above phenomenon indicates that the uniform sampling based aggregation algorithm has even better performance when the network size is large.

6 Related Works

The sampling technique has been widely used in many fields, such as quantile calculation, data collection and top-k query. For example, [13,14] introduce approximate algorithms to calculate the quantiles in wireless sensor networks. This algorithm reduces energy cost by using the sampling technique. By using the sampling technique, [11] develops ASAP, an adaptive sampling approach to energy-efficient periodic data collection in sensor networks, whose basic idea is to use a dynamically changing node set as samplers. [19] uses samples of past sensory data to formulate the problem of optimizing approximate top-k queries under an energy constraint. However, all the above techniques cannot be used in our problem directly since the above operations differ a lot with the maximum query and distinct-set query.

The distinct-count query in wireless sensor networks has been widely studied in many existing works, such as [2,17]. [17] proposes an algorithm to calculate approximate distinct-count based on approximate frequency query results. [2] also proposes an algorithm to compute the approximate distinct-count. However, this algorithm is centralized and not appropriate for large scale wireless sensors. Moreover, all the above works are for the distinct-count query, which can only reflect the size of the distinct set instead of all the content of the distinct set. Therefore, the above works still cannot be used in our problem directly.

7 Conclusions

In this paper, the δ-approximate algorithms for the maximum value and distinct-set aggregation operations in sensor equipped IoT networks are proposed. These algorithms are based on the uniform sampling. Mathematical proofs have been made for better understanding of these algorithms. Additionally, we have also proposed mathematical estimators for the two algorithms. Moreover, we have derived the values for the sample size and the sample probability which satisfies the specified failure probability requirements of the final result. Finally, a uniform sampling based algorithm is provided.

Experiments are conducted for various delta values and the network sizes. The results are then compared between the naive method and the proposed algorithms. The simulation results indicate that the proposed algorithms have high performance with respect to the energy cost.

Acknowledgment. This work is partly supported by the NSF under grant No. 1741277, No. 1741279, No. 1741287, No. 1741338 and the National Natural Science Foundation of China under Grant NO. 61632010, 61502116, U1509216, 61370217.

References

1. Anastasi, G., Falchi, A., Passarella, A., Conti, M., Gregori, E.: Performance measurements of motes sensor networks. In: Proceedings of the 7th ACM International Symposium on Modeling, Analysis and Simulation of Wireless and Mobile Systems, pp. 174–181. ACM (2004)
2. Beyer, K., Haas, P.J., Reinwald, B., Sismanis, Y., Gemulla, R.: On synopses for distinct-value estimation under multiset operations. In: Proceedings of the 2007 ACM SIGMOD International Conference on Management of Data, pp. 199–210. ACM (2007)
3. Cai, Z., Chen, Z.Z., Lin, G.: A 3.4713-approximation algorithm for the capacitated multicast tree routing problem. Theor. Comput. Sci. **410**(52), 5415–5424 (2009)
4. Cai, Z., Lin, G., Xue, G.: Improved approximation algorithms for the capacitated multicast routing problem. In: Wang, L. (ed.) COCOON 2005. LNCS, vol. 3595, pp. 136–145. Springer, Heidelberg (2005). https://doi.org/10.1007/11533719_16
5. Cheng, S., Cai, Z., Li, J.: Curve query processing in wireless sensor networks. IEEE Trans. Veh. Technol. **64**(11), 5198–5209 (2015)

6. Cheng, S., Cai, Z., Li, J., Fang, X.: Drawing dominant dataset from big sensory data in wireless sensor networks. In: The 34th Annual IEEE International Conference on Computer Communications (INFOCOM 2015), pp. 531–539, April 2015
7. Cheng, S., Cai, Z., Li, J., Gao, H.: Extracting kernel dataset from big sensory data in wireless sensor networks. IEEE Trans. Knowl. Data Eng. **29**(4), 813–827 (2017)
8. Cheng, S., Li, J.: Sampling based (epsilon, delta)-approximate aggregation algorithm in sensor networks. In: 29th IEEE International Conference on Distributed Computing Systems, ICDCS 2009, pp. 273–280. IEEE (2009)
9. Considine, J., Li, F., Kollios, G., Byers, J.: Approximate aggregation techniques for sensor databases, pp. 449–460 (2004)
10. Crossbrow Inc.: MPR-Mote Processor Radio Board User's Manual
11. Gedik, B., Liu, L., Philip, S.Y.: ASAP: an adaptive sampling approach to data collection in sensor networks. IEEE Trans. Parallel Distrib. Syst. **18**(12), 1766–1783 (2007)
12. Han, M., Li, L., Xie, Y., Wang, J., Duan, Z., Li, J., Yan, M.: Cognitive approach for location privacy protection. IEEE Access **6**, 13466–13477 (2018)
13. He, Z., Cai, Z., Cheng, S., Wang, X.: Approximate aggregation for tracking quantiles and range countings in wireless sensor networks. Theor. Comput. Sci. **607**, 381–390 (2015)
14. Huang, Z., Wang, L., Yi, K., Liu, Y.: Sampling based algorithms for quantile computation in sensor networks. In: Proceedings of the 2011 ACM SIGMOD International Conference on Management of data, pp. 745–756. ACM (2011)
15. Lachowski, R., Pellenz, M.E., Penna, M.C., Jamhour, E., Souza, R.D.: An efficient distributed algorithm for constructing spanning trees in wireless sensor networks. Sensors **15**(1), 1518–1536 (2015)
16. Li, J., Cai, Z., Wang, J., Han, M., Li, Y.: Truthful incentive mechanisms for geographical position conflicting mobile crowdsensing systems. IEEE Trans. Comput. Soc. Syst. (2018)
17. Li, J., Cheng, S., Cai, Z., Yu, J., Wang, C., Li, Y.: Approximate holistic aggregation in wireless sensor networks. ACM Trans. Sens. Netw. (TOSN) **13**(2), 11 (2017)
18. Li, J., Li, J.: Data sampling control, compression and query in sensor networks. Int. J. Sens. Netw. **2**(1–2), 53–61 (2007)
19. Silberstein, A.S., Braynard, R., Ellis, C., Munagala, K., Yang, J.: A sampling-based approach to optimizing top-k queries in sensor networks. In: Proceedings of the 22nd International Conference on Data Engineering, ICDE 2006, p. 68. IEEE (2006)
20. Wang, Y.: Topology control for wireless sensor networks. In: Li, Y., Thai, M.T., Wu, W. (eds.) Wireless Sensor Networks and Applications, pp. 113–147. Springer, Heidelberg (2008). https://doi.org/10.1007/978-0-387-49592-7_5
21. Yan, M., Han, M., Ai, C., Cai, Z., Li, Y.: Data aggregation scheduling in probabilistic wireless networks with cognitive radio capability. In: 2016 IEEE Global Communications Conference (GLOBECOM), pp. 1–6. IEEE (2016)
22. Yu, J., Huang, B., Cheng, X., Atiquzzaman, M.: Shortest link scheduling algorithms in wireless networks under the SINR model. IEEE Trans. Veh. Technol. **66**(3), 2643–2657 (2017)
23. Yu, J., Wan, S., Cheng, X., Yu, D.: Coverage contribution area based k-coverage for wireless sensor networks. IEEE Trans. Veh. Technol. **66**(9), 8510–8523 (2017)

Poisoning Machine Learning Based Wireless IDSs via Stealing Learning Model

Pan Li, Wentao Zhao(✉), Qiang Liu(✉), Xiao Liu, and Linyuan Yu

National University of Defense Technology, Changsha 410073, Hunan, China
{wtzhao,qiangliu06}@nudt.edu.cn

Abstract. Recently, machine learning-based wireless intrusion detection systems (IDSs) have been demonstrated to have high detection accuracy in malicious traffic detection. However, many researchers argue that a variety of attacks are significantly challenging the security of machine learning techniques themselves. In this paper, we study two different types of security threats which can effectively degrade the performance of machine learning based wireless IDSs. First, we propose an Adaptive SMOTE (A-SMOTE) algorithm which can adaptively generate new training data points based on few existing ones with labels. Then, we introduce a stealing model attack by training a substitute model using deep neural networks (DNNs) based on the augmented training data in order to imitate the machine learning model embedded in targeted systems. After that, we present a novel poisoning strategy to attack against the substitute machine learning model, resulting in a set of adversarial samples that can be used to degrade the performance of targeted systems. Experiments on three real data sets collected from wired and wireless networks have demonstrated that the proposed stealing model and poisoning attacks can effectively degrade the performance of IDSs using different machine learning algorithms.

1 Introduction

Nowadays, various network intrusion behaviors significantly challenge the security and the availability of underlying networks, especially wireless networks. Hence, the defending technology against these malicious behaviors is emerging as an important research topic in information and communication technology (ICT) academia and industry. To improve the capability and the performance of intrusion detection systems (IDSs), introducing machine learning into intrusion detection has been attracting more attentions from researchers along with the advance of artificial intelligence. Typically, the machine learning techniques used in IDSs include Native Bayes (NB), Support Vector Machine (SVM), Decision Tree (DT), Logistic Regression (LR) and deep learning [1].

Although having achieved excellent performance, the machine learning-based IDSs are suffering many security threats in recent research [2,3]. These threats can be classified into two categories [4], i.e., causative attack and exploratory

S. Chellappan et al. (Eds.): WASA 2018, LNCS 10874, pp. 261–273, 2018.
https://doi.org/10.1007/978-3-319-94268-1_22

attack. Specifically, causative attack, also termed as poisoning attack, occurs in the training phase of learning models. The adversary always uses two strategies to launch poisoning attack. One is modifying the features or labels of initial training data [5], another one is injecting adversarial samples to the initial training data during the process of retraining models [2]. Both the two strategies will seriously affect the performance of learning models, even cause the learning models totally unavailable. On the other hand, exploratory attack, containing spoofing attack and inverse attack, exploits the vulnerabilities of training model to destroy the classifiers without affecting training phase of learning models. Specifically, spoofing attack can confuse the classifiers by crafting adversarial samples. For example, the adversary can generate adversarial samples to evade the detection of classifiers such as spam filtering [6] and malware detection system [7]. As for inversion attack, it mainly exploits the API information to destroy the privacy security of models, containing stealing the targeted models [8] and inferring the membership information of training data [9]. In this paper, because the targeted models are machine learning-based IDSs, which need collecting new training data to retrain the learning model periodically, we mainly study the poisoning attack with injecting adversarial samples. Meanwhile, to make the attack practically, we also study the stealing model attack before the poisoning attack.

Previous works regarding poisoning attack are mainly aiming at specific algorithms such as SVMs [10] and principal component analysis (PCA) [11], which limit the transferability of poisoning attack and need the adversary obtaining the algorithm information of targeted models. Besides, the attack assumptions are too strong to achieve in reality, for example, the adversary launching label contamination attack (LCA) [5] needs to access the training data and modify their labels. Another poisoning attack, called batch-EPD boundary pattern poisoning (BEBP) [12], does not need the adversary to modify the labels but also needs to access the whole training data. More important, previous works towards the security of IDSs mainly concentrate on poisoning and evasion attack. However, researchers seldom focus on the model and data privacy security of IDSs.

In this paper, we propose a poisoning strategy via stealing learning model, which can threat the security and the availability of IDSs. Specifically, the main contributions are as follows:

1. We propose an improved Synthetic Minority Oversampling Technique (SMOTE), called Adaptive SMOTE (A-SMOTE), which can adaptively generate new training data points based on few existing ones with labels.
2. Based on the above augmented training data, we imitate the targeted models by training substitute models using deep neural networks (DNNs). According to the experimental results, the decision outputs of the substitute models are almost the same as those of the targeted ones.
3. We propose a novel poisoning strategy to craft high quality adversarial samples using the generated substitute models. The notable advantages of the proposed poisoning strategy include two aspects: (1) it considers the targeted IDSs as black-box systems and does not need to access their learning models directly, and (2) the resulting adversarial samples generated from the substitute models can be used to attack against the targeted ones.

2 Adversary Model and Attack Assumptions

Generally, the precondition of attack needs reasonable attack assumptions, which are named as adversary model in adversarial machine learning [4]. Specifically, we introduce the adversary model in this paper as follows.

- **Adversary Goal.** Adversary goal is the final effect that the adversary wants to achieve. In this paper, the adversary has two goals. One goal is stealing the machine learning models of targeted IDSs, the other is poisoning against the targeted IDSs.
- **Adversary Knowledge.** Before launching attack, adversary needs some information related to the targeted IDSs, the information is named adversary knowledge. In previous works, the adversary needs to know the whole training data, or the algorithms, even the concrete parameters of targeted models. In this paper [12,13], the adversary just needs to know $\sigma\%$ of training data.
- **Adversary Capability.** Besides the adversary knowledge, the adversary should have ability to launch attack. In this paper, for the stealing model attack, the adversary just needs polling the classifiers with input data and get corresponding returned labels. While for the poisoning attack, the adversary can not modify the features or labels of training data and is able to inject adversarial samples during retraining the learning models. Meanwhile, the injecting data amount is limited to $\eta\%$ of training data at each time.

Based on above attack assumption, we detail the process of the poisoning method via stealing learning model in Sects. 3 and 4.

3 Stealing Model Attack by Training Substitute Model

In this section, we present the details of the proposed stealing model attack. The general process of training the substitute model is as follows.

- First, the adversary obtains the label L_{tr0} corresponding to the known part of training data X_{tr0} by enquiring the targeted IDSs.
- Second, the adversary generates synthetic data X_{syn} by augmenting X_{tr0} with A-SMOTE.
- Third, the adversary can obtain the returned labels L_{syn} by inputting the augmented data X_{syn} to the targeted IDSs.
- Finally, the adversary trains the substitute models using the augmented data $X_{tr0} \cup X_{syn}$ and their corresponding outputting labels $L_{tr0} \cup L_{syn}$.

3.1 Data Augmentation by Adaptive SMOTE

Owing to only knowing few training data of the targeted IDSs, the adversary should augment more data for two reasons. (a) Training substitute models needs enough training data to approximate the targeted models. (b) The further attacks, such as poisoning and evasion attack based on the substitute models, also need enough training data to better craft adversarial samples with high quality.

In this paper, we propose an Adaptive SMOTE (A-SMOTE) to generate synthetic data. SMOTE is a popular oversampling technique used to solve the unbalanced data problems [14]. The basic theory of SMOTE is:

$$\boldsymbol{x}_{new} = \boldsymbol{x} + rand(0,1) \times (\widetilde{\boldsymbol{x}} - \boldsymbol{x}), \quad (1)$$

where $\boldsymbol{x}_{new}$ represents the produced synthetic data, $\boldsymbol{x}$ refers to the initial sample, and $\widetilde{\boldsymbol{x}}$ is one sample randomly selected from the k-nearest neighbors of $\boldsymbol{x}$.

However, owing to just considering the distance relationship among the samples with same class label, the synthetic data generated by SMOTE are limited to the connecting lines of known data. Instead, A-SMOTE can generate more valid synthetic data. The formulation of A-SMOTE is:

$$\boldsymbol{x}_{new} = \boldsymbol{x} + rand(0,1) \times \boldsymbol{T}(\widetilde{\boldsymbol{x}} - \boldsymbol{x}), \quad (2)$$

where $\boldsymbol{T}$ is a vector which can adaptively adjust the active degree of pushing vector $(\widetilde{\boldsymbol{x}} - \boldsymbol{x})$. The details of A-SMOTE are shown in Algorithm 1, where X and Y represent the initial data and corresponding labels, respectively. L means the targeted class of synthetic data, and k refers to the number of nearest neighbors. Besides, X_{syn} and X_L represent the synthetic samples generated by A-SMOTE and the samples with the label L in X, respectively. N refers to the number of synthetic data generated from each sample point in X_L. In the calculating process of $\boldsymbol{T}$, $\boldsymbol{\tau}$ is a vector with the same dimensionality as the sample $\boldsymbol{x}$, and every variable of $\boldsymbol{\tau}$ is randomly generated according to uniform distribution from 0 to 1. Moreover, α is varying from p, which represents the percent of the same class to $\boldsymbol{x}$ among its k-nearest neighbors. Actually, p controls the varying of $\boldsymbol{T}$, which further affects the direction of generating synthetic data. Obviously, the higher proportion p will cause more active pushing vector $\boldsymbol{T}(\widetilde{\boldsymbol{x}} - \boldsymbol{x})$.

Algorithm 1. Adaptive SMOTE

1: **Input**: X, Y, L, N, k, α
2: **Output**: Synthetic samples X_{syn}
3: **Initialize**: $X_{syn} = \emptyset$
4: Select the X_L samples from X
5: **for** $i = 0$ to $size(X_L)$ **do**
6: Compute k nearest neighbors X_{ne} of $\boldsymbol{x}$, where $\boldsymbol{x} = X_L^i$
7: Select the targeted data X_{ne} with $y = L$ from k nearest neighbors
8: **for** $j = 0$ to $N - 1$ **do**
9: Random select $\widetilde{\boldsymbol{x}}$ from X_{ne}
10: Random generate a vector $\boldsymbol{\tau} \sim rand(0,1)$ with the same dimension to $\boldsymbol{x}$
11: $\boldsymbol{T} = \alpha \times \frac{size(X_{ne})-1}{k} \times \boldsymbol{\tau} + \boldsymbol{1}$
12: Calculate $\boldsymbol{x}_{new}$ by E.q(2), $X_{syn} = X_{syn} \cup \boldsymbol{x}_{new}$
13: **end for**
14: **end for**

As shown in Fig. 1, the k-nearest neighbors of a sample data point $\boldsymbol{x}1$ all have the same class labels as that of $\boldsymbol{x}1$, so the pushing vectors generated by A-SMOTE will be active. In Fig. 1, the pushing vectors of $\boldsymbol{x}1$ generated by SMOTE and A-SMOTE are illustrated by green lines in the left figure and green sectors in the right one, respectively. Regarding the sample data point $\boldsymbol{x}2$, there are only two sample data points with the same label as that of $\boldsymbol{x}2$ among its k-nearest neighbors. Hence, its pushing vectors generated by A-SMOTE will be less active. Accordingly, the pushing vectors of $\boldsymbol{x}2$ generated by SMOTE and A-SMOTE are illustrated by black lines in the left figure and the right one of Fig. 1, respectively.

3.2 Training Substitute Model Using DNNs

The targeted IDSs can be treated as black-box models because their concrete algorithms or architectures are unknown to the adversary. So it is hard to inverse the specific algorithms even parameters of the targeted IDSs, while training a substitute model can be a good choice to steal the targeted model.

In this paper, we use DNNs as substitute models, and the general function of DNNs model F with n layers can be expressed as follow:

$$F(\boldsymbol{x}) = f_n(\theta_n; f_{n-1}(\theta_{n-1}; \dots f_2(\theta_2; f_1(\theta_1; \boldsymbol{x}))), \tag{3}$$

where f_n and θ_n are the function and weights parameters of nth layer. As the substitute models, DNNs have the two advantages. (a) With enough layers and parameters, DNNs have strong representational abilities so as to better approximate the targeted models. (b) The back-propagation algorithm, one of the significant character of DNNs, can successively propagate error gradients with respect to network parameters from the networks output layer to its input layer.

4 Poisoning Data Crafting

Based on the substitute models, the further attacks, such as poisoning and evasion attack, are easier to achieve. In this paper, we mainly study the poisoning attack based on the substitute models, which has two advantages compared with previous works. (a) With the substitute models, there is no need to continually access the targeted IDSs, which may induce that the adversary is suspected by IDSs. (b) The adversary can choose better adversarial samples by simulating the poisoning attack on the substitute models.

Poisoning Attack with Boundary Pattern Data. In [12], the author proposed a poisoning attack method named batch-EPD boundary pattern poisoning (BEBP), which uses the boundary pattern data as potential poisoning data. The boundary pattern data satisfy two characters: (a) They are closed to the discriminant plane between normal and abnormal data. (b) They are classified as normal data even though they very close to abnormal data. Specifically, the

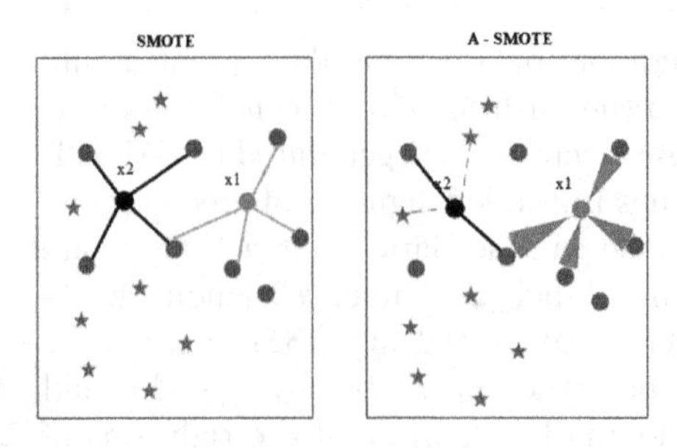

Fig. 1. Synthetic data areas generated by SMOTE (left) and A-SMOTE (right)

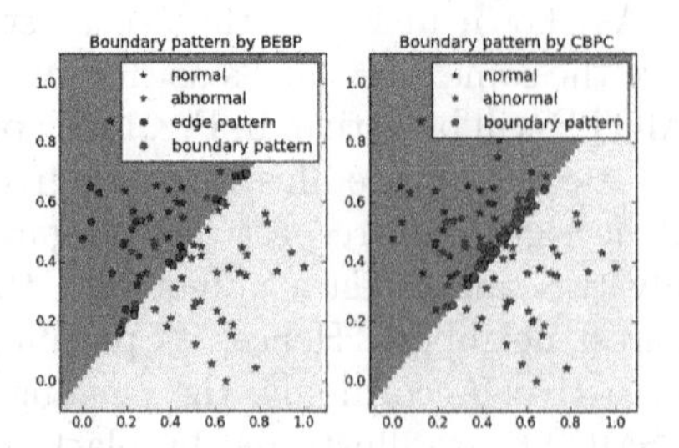

Fig. 2. Comparative results of boundary pattern data generated by BEBP (left) and CBPC (right)

adversarial samples $\mathcal{D}_a$ derived from boundary pattern data can be formulated as:

$$\mathcal{D}_a = \{\mathbf{x}_a | d(\mathbf{x}_a, \mathbf{x}_b) < \varepsilon, f(\mathbf{x}_b) = 0\}, \tag{4}$$

where $d(\mathbf{x}_a, \mathbf{x}_b)$ is the Euclidean distance between two vectors, and ε denotes the chosen threshold between an adversarial sample $\mathbf{x}_a \in \mathcal{D}_a$ and the boundary.

Poisoning Data Crafting with Centre-Drifting Boundary Pattern Crafting. BEBP method [12] has been demonstrated achieving pretty poisoning effect. However, for large scale network datasets, it is time consuming to calculate their edge pattern data and corresponding normal vectors. Besides, only using edge pattern data will waste the inner data information, especially the number of training data is very limited. So here we propose a simple and fast boundary pattern data crafting method named Centre-drifting Boundary Pattern Crafting (CBPC), the comparative result on a synthetic data set between BEBP and CBPC is shown in Fig. 2.

Algorithm 2. Centre-drifting Boundary Pattern Crafting(CBPC)

1: **Input**: $\mathbf{x}_n$, $\mathcal{N}$, target learning model M, m, λ
2: **Output**: A boundary pattern $\mathcal{D}_a$ generated from $\mathbf{x}_n$
3: Initialize $\lambda_0 = \lambda$, $\mathbf{x}_a^0 = \mathbf{x}_n$, $\mathcal{D}_a = \emptyset$;
4: **for** $i = 0, \cdots, m-1$ **do**
5:   **if** $f_M(\mathbf{x}_a^i) == N$ **then**
6:     **if** $(\exists \mathbf{x}_b, f_M(\mathbf{x}_b) = 0$ **and** $d(\mathbf{x}_a^i, \mathbf{x}_b) < \varepsilon)$ **then**
7:       $\mathcal{D}_{bp} = \mathcal{D}_{bp} \bigcup \{\mathbf{x}_a^i\}$;
8:     **end if**
9:     $\mathbf{x}_a^{i+1} = \mathbf{x}_a^i + \lambda_i \cdot \mathcal{N}$; $\lambda_{i+1} = \lambda_i$;
10:   **else**
11:     $\mathbf{x}_a^{i+1} = \mathbf{x}_a^i - \lambda_i \cdot \mathcal{N}$; $\lambda_{i+1} = \lambda_i/3$;
12:   **end if**
13: **end for**

In CBPC, adversarial sample $\mathcal{D}_a$ is generated by pushing the normal training data $\mathcal{X}_N$ towards the abnormal data $\mathcal{X}_A$. As is shown in Eq. (5), the pushing vector $\mathcal{N}$ can be obtained by calculating the vector between the two class centres.

$$\mathcal{N} = \frac{1}{M}\sum_{i=1}^{M} x_{Ai} - \frac{1}{N}\sum_{j=1}^{N} x_{Nj}, \tag{5}$$

where x_{Ni} and x_{Aj} is the ith normal data and jth abnormal data in training data, respectively. Specifically, we perform the following two operations based on $\mathcal{X}_N$ and $\mathcal{N}$: Firstly, selecting a normal data $\mathbf{x}_n \in \mathcal{X}_n$. Then, pushing $\mathbf{x}_n$ outwards along the direction of $\mathcal{N}$ until the generated data points are near to the discriminant plane of classifiers. The data shifting is formally defined by

$$\mathbf{x}_a^i = \mathbf{x}_a^{i-1} + k_{i-1} \cdot \lambda_{i-1} \cdot \mathcal{N}, \tag{6}$$

where

$$\begin{cases} k_i = 1, \lambda_i = \lambda_{i-1}, & \text{if } f(\mathbf{x}_a^{i-1}) \text{ is normal} \\ k_i = -1, \lambda_i = \lambda_{i-1}/3, & \text{otherwise} \end{cases} \tag{7}$$

The pseudo code of the CBPC is shown in Algorithm 2, where m is the maximal number of iterations, λ means the initial shifting step size, $\mathbf{x}_n$ and $\mathcal{N}$ represent one normal sample from initial training data and corresponding centre shifting direction, respectively. In particular, we first shift $\mathbf{x}_n$ outwards along $\mathcal{N}$ according to Eqs. (6) and (7), where $\mathbf{x}_a^i$ and λ_i determine the generated adversarial sample and the shifting step size in the ith iteration. Note that $\mathbf{x}_a^0 = \mathbf{x}_n$. Furthermore, the output of a target learning model M ($f_M(\mathbf{x})$) with respect to an input sample $\mathbf{x}$ falls into $\{N, A\}$ representing *Normal* and *Abnormal*, respectively. Finally, we select valid adversarial samples (i.e. boundary pattern points) according to the Eq. (4). For simplicity, ε is set to λ.

Poisoning Attack with Various Poisoning Strategies. Combining the above stealing model and poisoning attack method, we propose a novel poisoning strategy. Specifically, we first generate synthetic data X_{syn} by A-SMOTE based on X_{tr0}. Then, we train the substitute model M_{sub} on $X_{all}(X_{syn} \cup X_{tr0})$ using DNNs. After that, we use CBPC to craft adversarial samples $\mathcal{D}_a$ by simulating the poisoning attack against M_{sub}. Finally, we use $\mathcal{D}_a$ to launch poisoning attack against the targeted IDSs. For simplicity, we name this poisoning strategy as A-SMOTE&CBPC. Similarly, we can obtain the various poisoning strategies, i.e., X_{tr0}&CBPC and SMOTE&CBPC represent the strategies which poison the targeted models using adversarial samples generated by simulating poisoning attack with CBPC method against M_{sub} trained on X_{tr0} and $X_{tr0} \cup X_{syn}^1$ (generated by SMOTE), respectively. A-SMOTE&BEBP denotes as the strategy which poisons the targeted models using adversarial samples generated by simulating poisoning attack with BEBP method against M_{sub} trained on X_{all}.

5 Experiments and Results

In this section, we first examine the imitation performance of the substitute models trained on training data generated by different data augmentation methods.

Then, we compare the poisoning effect of various poisoning strategies. All experiments are carried out on three network intrusion detection datasets, and all targeted models are realized by the official classifiers in *sklearn*[1]. Specifically, the targeted models include NB-Gaussian, LR, SVM with a sigmoid kernel (SVM-sigmoid), SVM with a linear kernel (SVM-linear) and SVM with a radial basis function kernel (SVM-RBF).

5.1 Experimental Settings

Data Set. To demonstrate the performance of the proposed attack method, we choose two wired network data sets, i.e., *NSL-KDD*[2], *Kyoto 2006+*[3] and one wireless network data set *WSN-DS*. *NSL-KDD* is a revised version of *KDD-CUP99*, and it contains five categories of samples (one normal and four abnormal). Moreover, each sample has 41 features. *Kyoto 2006+* proposed in [15] is another recognized data set for performance evaluation. The data set has been collected from honeypots and regular servers that are deployed at the Kyoto University since 2006. Moreover, *Kyoto 2006+* contains three types of samples, i.e., normal, known attack and unknown one, and each sample has 24 features. Apart from these two widely used wired network data sets, we also evaluate on one wireless network data set *WSN-DS* [16], which has 18 features and five categories of samples (one normal and four abnormal).

Data Preprocessing. Considering that the goal of poisoning attacks is to reduce the performance of IDSs detecting abnormal behaviors, we treat all samples with abnormal labels in each data set as a whole regardless of their specific types of attacks. Similar to general IDSs [17], we preprocess and perform data normalization with respect to all samples such that each feature value is normalized into a range of $[0, 1]$.

Referring to [17], we adopt 6472 samples as training data and 6820 samples as evaluating data that are randomly selected from the $KDDTrain+$ of *NSL-KDD* data set. Similarly, we randomly select 13292 samples from the traffic data collected during 27–31, August 2009 regarding the *Kyoto 2006+* data set and the same number of data from *WSN-DS* data set. In order to minimize the fluctuation of experimental results brought by random data sampling, we independently rerun the stealing model and poisoning attacks 10 times.

Performance Metric. To evaluate the performance of stealing model and poisoning attack, we first introduce two common metrics in machine learning, i.e, accuracy ($ACC = \frac{TP+TN}{TP+TN+FN+FP}$) and detecting rate ($DR = \frac{TP}{TP+FN}$), where true positive ($TP$) is the number of truly abnormal samples that are classified as abnormal ones by IDSs, true negative (TN) means the number of

[1] http://scikit-learn.org.
[2] http://www.unb.ca/cic/datasets/nsl.html.
[3] http://www.takakura.com/kyoto_data/.

truly normal samples that are treated as normal ones, false positive (FP) refers to the number of truly normal samples classified as abnormal ones, and false negative (FN) represents the number of truly abnormal samples classified as normal ones.

For the proposed poisoning attack, we adopt ACC and DR to evaluate the performance reduction of machine learning-based IDSs. While for the stealing model attack, we refer to [18] and adopt label matched rate (LMA) as a metric. The calculation of LMA is similar to ACC, the only difference between them is that LMA adopts the labels of targeted model outputting as ground truth.

5.2 Experimental Results and Analysis

Training Substitute Model. According to the adversarial model, the adversary can access σ% training data, here we set $\sigma <= 5$. For 6372 training data, the number of initial training data (X_{tr0}) known by adversary is set as 300.

The parameters in SMOTE and A-SMOTE are set as follow: $N = 4, k = 5, \alpha = 0.5$. Besides, the substitute models are trained using DNNs with two hidden layers whose number of nodes are 60 and 20 respectively. The LMA of substitute models trained with different training data are shown in Table 1, where X_{tr0} is the substitute model trained on known training data X_{tr0}, X_{tr0}+SMOTE and X_{tr0}+A-SMOTE represent the substitute models trained on X_{tr0} merging the synthetic data generated by SMOTE and A-SMOTE, respectively. From Table 1 we can see that the DNNs can effectively approximate the targeted models. Besides, X_{tr0}+SMOTE and X_{tr0}+A-SMOTE can obtain higher imitation performance. Especially, X_{tr0}+A-SMOTE performs slight better than the other two on *Kyoto 2006+* and *WSN-DS*.

Table 1. The LMA of the substitute models trained using various training data on NSL-KDD, Kyoto 2006+ and WSN-DS. Boldface means the best one.

Data set	Training data	NB	LR	SVM-sigmoid	SVM-linear	SVM-RBF
NSL KDD	X_{tr0}	96.058%	96.595%	97.668%	**97.796%**	96.244%
	X_{tr0}+SMOTE	**96.483%**	97.182%	**97.811%**	97.682%	97.033%
	X_{tr0}+A-SMOTE	96.435%	**97.395%**	97.729%	97.730%	**97.050%**
Kyoto 2006+	X_{tr0}	91.752%	95.298%	96.355%	95.444%	94.507%
	X_{tr0}+SMOTE	**96.000%**	**97.601%**	97.284%	97.609%	97.136%
	X_{tr0}+A-SMOTE	94.934%	96.890%	**97.497%**	**97.628%**	**97.170%**
WSN DS	X_{tr0}	99.909%	99.099%	99.651%	99.460%	99.103%
	X_{tr0}+SMOTE	99.893%	**99.260%**	99.650%	99.481%	99.203%
	X_{tr0}+A-SMOTE	**99.909%**	99.169%	**99.691%**	**99.518%**	**99.212%**

Experimental Results of Various Poisoning Strategies. For the poisoning attack, we set the poisoning proportion $\eta = 7\%$ referring to [12]. In order to demonstrate the effect of proposed A-SMOTE&CBPC, we set two groups of experiments from following aspects: (a) Comparing the poisoning strategies based on different training data and corresponding substitute models, we evaluate the effect of adversarial samples crafted by simulating poisoning attack using *CBPC* against the substitute models trained on various training data. Specifically, we examine three poisoning strategies, i.e., X_{tr0}&CBPC, SMOTE&CBPC, A-SMOTE&CBPC. (b) Comparing the poisoning strategies using different poisoning crafting methods, we evaluate the poisoning effect of adversarial samples crafted by simulating poisoning attack using different methods against the substitute models trained on the same training data X_{all}. Specifically, we examine two poisoning strategies, i.e., A-SMOTE&BEBP and A-SMOTE&CBPC. Besides, referring to the *BoilFrog* attack [13] and *Chronic Poisoning* [12], we simulate 7 rounds poisoning attack with all above poisoning strategies.

Table 2. Comparative results of *ACC* on NSL-KDD, Kyoto2006+ and WSN-DS after 7 rounds poisoning attack using various poisoning strategies. Boldface means little difference with the best one.

Data set	Training data	NB	LR	SVM-sigmoid	SVM-RBF	SVM-linear
NSL KDD	BASIC	87.534%	94.708%	94.246%	95.122%	95.255%
	X_{tr0}&CBPC	81.500%	89.811%	87.828%	94.940%	91.956%
	SMOTE&CBPC	86.002%	88.428%	87.832%	94.076%	92.074%
	A-SMOTE&BEBP	84.634%	89.357%	88.993%	**91.557%**	90.536%
	A-SMOTE&CBPC	**76.079%**	**85.552%**	**82.414%**	94.467%	**87.780%**
Kyoto 2006+	BASIC	95.521%	98.336%	97.169%	98.217%	98.969%
	X_{tr0}&CBPC	75.249%	88.163%	86.974%	89.679%	87.754%
	SMOTE&CBPC	72.157%	**68.237%**	**62.968%**	**87.751%**	72.529%
	A-SMOTE&BEBP	**67.336%**	86.931%	88.556%	**87.611%**	88.282%
	A-SMOTE&CBPC	71.252%	**68.230%**	65.840%	**87.477%**	**69.355%**
WSN DS	BASIC	97.648%	97.669%	97.729%	97.679%	97.814%
	X_{tr0}&CBPC	97.436%	95.823%	92.911%	95.898%	96.558%
	SMOTE&CBPC	**96.967%**	**85.485%**	**75.285%**	**90.702%**	**93.028%**
	A-SMOTE&BEBP	97.657%	97.516%	97.729%	97.679%	97.814%
	A-SMOTE&CBPC	**96.892%**	88.117%	**76.082%**	**90.713%**	**94.270%**

The comparative results are shown in Table 2 and Fig. 3, which represent the performance of different classifiers after 7 rounds poisoning attack using various attack strategies. Specifically, BASIC represents the performance of classifiers without being poisoned, X_{tr0}&CBPC, SMOTE&CBPC, A-SMOTE&CBPC

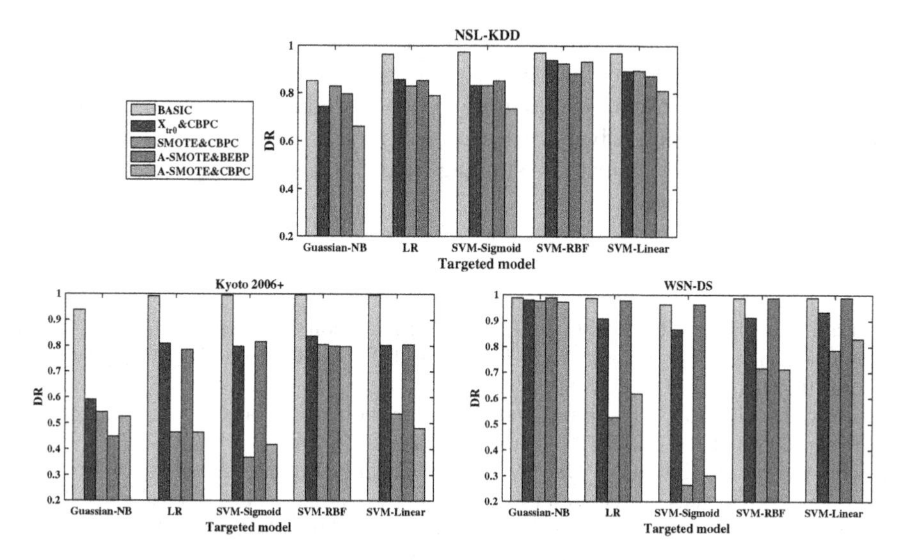

Fig. 3. Comparative results of DR on NSL-KDD, Kyoto 2006+ and WSN-DS after 7 rounds poisoning attack using various poisoning strategies

and A-SMOTE&BEBP refer to the various poisoning strategies mentioned in Sect. 4.2.

As shown in Table 2 and Fig. 3, A-SMOTE&CBPC is the best poisoning strategy among them. Specifically, from the view of using the same poisoning method but different substitute models trained on various training data, i.e., A-SMOTE&CBPC, SMOTE&CBPC, and X_{tr0}&CBPC, we can find that using the substitute models trained on data generated by A-SMOTE can achieve the best poisoning effect both on NSL-KDD and Kyoto 2006+. While for data set WSN-DS, the proposed attack strategies can obtain comparative poisoning effect to the best one. On the other hand, comparing different poisoning methods using the same substitute models, i.e., A-SMOTE&CBPC and A-SMOTE&BEBP, it is obvious that the CBPC can obtain more effective adversarial samples than BEBP, which demonstrated the superiority of proposed poisoning method.

6 Conclusions and Future Work

In this paper, we have proposed a novel poisoning strategy which can effectively poison machine learning-based IDSs via stealing targeted models. Specifically, we first propose an improved synthetic data generating method named A-SMOTE to provide proper data for training substitute models using DNNs. Then, we introduce an effective poisoning method, called CBPC, which can exploit the information of augmented training data to craft poisoning samples. After that, we present a poisoning strategy, which can poison the targeted models using poisoning samples generated by simulating poisoning attack against the substitute models. Experiments on three real data sets demonstrate the effectiveness of the proposed stealing model method and poisoning strategy.

In future, it is worthwhile to do more in-depth studies based on the substitute model, such as crafting adversarial sample to evade the detection. Moreover, research on defending against the stealing model and poisoning attack will be a meaningful work as well.

References

1. Tsai, C.F., Hsu, Y.F., Lin, C.Y., Lin, W.Y.: Intrusion detection by machine learning: a review. Expert Syst. Appl. Int. J. **36**(10), 11994–12000 (2009). https://doi.org/10.1016/j.eswa.2009.05.029
2. Kloft, M., Laskov, P.: Online anomaly detection under adversarial impact. In: Proceedings of the AISTATS 2010, pp. 405–412 (2010)
3. Liu, Q., Li, P., Zhao, W., Cai, W., Yu, S., Leung, V.C.M.: A survey on security threats and defensive techniques of machine learning: a data driven view. IEEE Access **6**, 12103–12117 (2018). https://doi.org/10.1109/ACCESS.2018.2805680
4. Barreno, M., Nelson, B., Sears, R., Joseph, A.D., Tygar, J.D.: Can machine learning be secure? In: Proceedings of the ASIACCS 2006, pp. 16–25. ACM (2006). https://doi.org/10.1145/1128817.1128824
5. Zhao, M., An, B., Gao, W., Zhang, T.: Efficient label contamination attacks against black-box learning models. In: Proceedings of the IJCAI 2017, pp. 3945–3951 (2017). https://doi.org/10.24963/ijcai.2017/551
6. Wittel, G.L., Wu, S.F.: On attacking statistical spam filters. In: Proceedings of the CEAS 2004 (2004). http://www.ceas.cc/papers-2004/170.pdf
7. Hu, W., Tan, Y.: Generating adversarial malware examples for black-box attacks based on GAN (2017). https://arxiv.org/abs/1702.05983
8. Tramèr, F., Zhang, F., Juels, A., Reiter, M.K., Ristenpart, T.: Stealing machine learning models via prediction APIs, pp. 601–618 (2016)
9. Shokri, R., Stronati, M., Song, C., Shmatikov, V.: Membership inference attacks against machine learning models. In: Proceedings of the Symposium on Security and Privacy 2017, pp. 3–18 (2017). https://doi.org/10.1109/SP.2017.41
10. Biggio, B., Nelson, B., Laskov, P.: Poisoning attacks against support vector machines. In: Proceedings of the ICML 2012, pp. 1467–1474 (2012)
11. Biggio, B., Fumera, G., Roli, F., Didaci, L.: Poisoning adaptive biometric systems. In: Gimel'farb, G., et al. (eds.) SSPR/SPR 2012. LNCS, vol. 7626, pp. 417–425. Springer, Heidelberg (2012). https://doi.org/10.1007/978-3-642-34166-3_46
12. Li, P., Liu, Q., Zhao, W., Wang, D., Wang, S.: BEBP: an poisoning method against machine learning based IDSs (2018). https://arxiv.org/abs/1803.03965
13. Rubinstein, B.I., Nelson, B., Huang, L., Joseph, A.D., Lau, S., Rao, S., Taft, N., Tygar, J.D.: Antidote: understanding and defending against poisoning of anomaly detectors. In: Proceedings of the IMC 2009, pp. 1–14. ACM, New York (2009). https://doi.org/10.1145/1644893.1644895
14. Chawla, N.V., Bowyer, K.W., Hall, L.O., Kegelmeyer, W.P.: Smote: synthetic minority over-sampling technique. J. Artif. Intell. Res. **16**(1), 321–357 (2002). https://doi.org/10.1613/jair.953
15. Song, J., Takakura, H., Okabe, Y., Eto, M., Inoue, D., Nakao, K.: Statistical analysis of honeypot data and building of Kyoto 2006+ dataset for NIDs evaluation. In: Proceedings of the BADGERS 2011, pp. 29–36. ACM, New York (2011). https://doi.org/10.1145/1978672.1978676

16. Almomani, I., Al-Kasasbeh, B., Al-Akhras, M.: WSN-DS: a dataset for intrusion detection systems in wireless sensor networks. J. Sens. **2016**(2), 1–16 (2016). https://doi.org/10.1155/2016/4731953
17. Ambusaidi, M.A., He, X., Nanda, P., Tan, Z.: Building an intrusion detection system using a filter-based feature selection algorithm. IEEE Trans. Comput. **65**(10), 2986–2998 (2016). https://doi.org/10.1109/TrustCom.2014.15
18. Papernot, N., McDaniel, P., Goodfellow, I., Jha, S., Celik, Z.B., Swami, A.: Practical black-box attacks against machine learning. In: Proceedings of the ASIACCS 2017, pp. 506–519. ACM, New York (2017). https://doi.org/10.1145/3052973.3053009

Signal-Selective Time Difference of Arrival Estimation Based on Generalized Cyclic Correntropy in Impulsive Noise Environments

Sen Li(✉), Bin Lin, Yabo Ding, and Rongxi He

Department of Information Science and Technology,
Dalian Maritime University, Dalian 116026, China
{listen, binlin, ybd, hrx}@dlmu.edu.cn

Abstract. This paper addresses the issue of time difference of arrival (TDOA) estimation of cyclostationary signal under impulsive noise environments modeled by α-stable distribution. Since α-stable distribution has not finite second-order statistics, the conventional cyclic correlation based signal-selective TDOA estimation algorithm does not work effectively. To resolve this problem, we define the generalized cyclic correntropy (GCCE) which is a robust cyclic correlation and can be reviewed as an extension of the generalized correntropy for cyclostationary signal. A robust signal-selective TDOA estimation algorithm based on GCCE is proposed. The computer simulation results demonstrate that the proposed algorithm outperforms the conventional cyclic correlation and the fractional lower order cyclic correlation based algorithms.

Keywords: TDOA estimation · Impulsive noise
Fractional lower order cyclic correlation · Generalized cyclic correntropy

1 Introduction

Time difference of arrival (TDOA) estimation is an important issue in signal processing. It is widely used in areas such as communications, radar, sonar and biomedical applications [1, 2]. Conventional time delay estimation methods are based on the stationary signal model, such as generalized cross-correlation method, generalized phase spectrum method and adaptive time delay estimation method [3]. Most of the artificial signals used in communications, telemetry, and sonar systems are nonstationary signal and have cyclostationarity properties [4]. In these systems, the received source signals are often mixed with statistically related interferences which overlaps with the frequency spectrum of the source signals. This situation causes the performance degradation of traditional TDOA estimation methods. In response to this problem, Gardner and Chen [5, 6] used second order cyclic correlation to improve the

This work was supported in part by the National Natural Science Foundation of China under Grants 61301228, 61371091 and the Fundamental Research Funds for the Central Universities under Grant 3132016331 and 3132016318.

S. Chellappan et al. (Eds.): WASA 2018, LNCS 10874, pp. 274–283, 2018.
https://doi.org/10.1007/978-3-319-94268-1_23

traditional generalized cross-correlation method and achieved remarkable results. Zhang [7] proposed an adaptive TDOA estimation method based on the cyclostationarity feature. In order to solve the limitations of TDOA estimation method based on second order cyclic correlation, a joint TDOA and Doppler estimation method based on cyclic ambiguity function is proposed to eliminate the influence of Doppler frequency difference on TDOA estimation [8]. In general, the interferences and signals have different cycle frequencies, and noise does not have cyclostationarity characteristics. Therefore, the above mentioned estimation algorithms based on the cyclostationarity characteristic of the signals has stronger anti-interference ability than the traditional methods.

One common assumption made by the above methods is that the ambient noise is assumed to be Gaussian distributed. However, in many real world applications the noise often exhibits non-Gaussian properties and sometimes is accompanied by strong impulsiveness [9]. For example, natural sources such as atmospheric noise resulting from thunder storms, car ignitions, microwave ovens and other types of man-made signal sources generally result in aggregating noises that may produce high amplitudes during small time intervals. To address this type of noise the α-stable distribution was proposed as a better and suitable noise model [10]. It has been also shown to have potential in characterizing various impulsive noises via selecting different values of the parameter α.

Since α-stable distribution has no finite second order statistics (SOS), the SOS-and HOS-based estimation methods are generally not applicable. Therefore, to address this issue the fractional lower order statistics (FLOS) was recently proposed such as the fractional lower order correlation (FLOC) [11] which can be regarded as equivalents of the correlation under SOS. However, FLOS requires a priori knowledge of the α-stable distribution, which is difficult to estimate in some practical applications. To solve this problem, a new robust correlation referred to as the correntropy was defined in [12], and was further expanded to the generalized correntropy by extending the Gaussian kernel function to the generalized Gaussian kernel function [13].

To explore the cyclostationarity of the signal in the impulse noise environment, *p*th-order cyclic correlation (PCC) and fractional lower order cyclic correlation (FLOCC) were defined and used for TDOA estimation [14, 15] and localization [16, 18] in impulsive noise. PCC and FLOCC can be considered as the fractional lower order cyclic statistics (FLOCS) which also need to pre-estimate the characteristic exponent α before use them. Based on the concept of correntropy, the authors of Ref. [19–21] defined cyclic correntropy (CCE) and cyclic correntropy spectrum (CCES) which were applied to solve the frequency estimation and spectrum sensing problem in impulsive noise environment.

In this paper, the generalized cyclic correntropy (GCCE) is defined in order to apply generalized correntropy for cyclostationary signals, and the CCE is a special case of GCCE. Then a new TDOA estimation method based on GCCE is developed. Computer simulation experiments are presented to illustrate the performance superiority of the proposed methods over the traditional second cyclic correlation (CC) and FLOCC based method.

2 α-Stable Distribution

This section describes a noise model specified by α-stable distribution with its characteristic function given by

$$\phi(t) = e^{\{jat - \gamma|t|^{\alpha}[1 + j\beta \mathrm{sgn}(t)\varpi(t,\alpha)]\}} \tag{1}$$

where γ and a are the dispersion and location parameters respectively and $\varpi(t, \alpha)$ is defined by

$$\varpi(t, a) = \begin{cases} \tan\frac{\pi\alpha}{2}, & \text{if } \alpha \neq 1 \\ \frac{2}{\pi}\log|t|, & \text{if } \alpha = 1 \end{cases} \tag{2}$$

the sign function $\mathrm{sgn}(t)$ is given as

$$\mathrm{sgn}(t) = \begin{cases} t/|t|, & \text{if } t \neq 0 \\ 0, \text{ if } t = 0 \end{cases} \tag{3}$$

In particular, α $(0 < \alpha \leq 2)$ is the characteristic exponent that measures the thickness of the tails of the distribution where the smaller α is, the thicker its tails. β is the symmetry parameter, if $\beta = 0$, the distribution in which case the observation is referred to as the symmetry α-stable (SαS) distribution. When $\alpha = 2$ and $\beta = 0$, the α-stable distribution becomes a Gaussian distribution. An important difference between the Gaussian and the α-stable distribution is that the former has only first two moments while the latter does not have any statistics when the moments of order is greater than or equal to α.

3 Generalized Cyclic Correntropy

To make use of the cyclostationarity of signal in impulsive noise environment, the fraction lower order cyclic statistics PCC and FLOCC were defined in [14–18]. Although the PCC and FLOCC are robust in α-stable impulsive noise, they require a priori knowledge of the α-stable distribution, which is difficult to estimate in some practical applications. This leads to the need for a better statistics which can deal with cyclostationary signals and the α-stable distributed noise simultaneously. In order to deal with this problem, cyclic correntropy was proposed and used on frequency estimation [19], spectrum sensing [20] and DOA estimation [21]. The cyclic correntropy can be regarded as a kernel method which transforms data from input data space to high dimensional feature space by using the Gaussian kernel function and then computes the cyclic correlation of the transformed data. Although the Gaussian kernel function has good smoothing characteristic and is strictly positive definite, it is not always the best choice. In this section, the generalized Gaussian kernel function was proposed to use and a new robust cyclic correlation named as generalized cyclic correntropy (GCCE) is proposed to extend the fundamental definitions and applications of cyclic correlation-like methodologies when it is hard to obtain the priori knowledge of α-stable distributed noise.

Given a stochastic process $x(t)$, the generalized correlation function called generalized correntropy is defined as:

$$V_x(\tau) = E[G_{\mu,\upsilon}(x(t), x(t-\tau))] \tag{4}$$

where $E[\cdot]$ is the expectation operator and $G_{\mu,\upsilon}(\cdot)$ is the generalized Gaussian function given by

$$\begin{aligned} G_{\mu,\upsilon}(x(t), x(t-\tau)) &= \frac{\mu}{2\upsilon\Gamma(1/\mu)}\exp(-\left|\frac{x(t) - x(t-\tau)}{\upsilon}\right|^{\mu}) \\ &= \gamma_{\mu,\upsilon}\exp(-\lambda|x(t) - x(t-\tau)|^{\mu}) \end{aligned} \tag{5}$$

where $\Gamma(\bullet)$ is the gamma function, $\mu > 0$ is the shape parameter, $\upsilon > 0$ is the scale parameter, $\lambda = 1/\upsilon^{\mu}$ is the kernel parameter, and $\gamma_{\mu,\upsilon} = \mu/(2\upsilon\Gamma(1/\mu))$ is the normalization constant.

If the generalized correntropy $V_x(\tau)$ is periodic and the period is T, it can be expanded by Fourier series

$$V_x(\tau) = \sum_{\varepsilon} V_x^{\varepsilon}(\tau)e^{j2\pi\varepsilon t} \tag{6}$$

where $\varepsilon = n/T$, n is an integer and ε is considered as the cyclic frequency. Define the Fourier coefficients $V_x^{\varepsilon}(\tau)$ as the generalized cyclic correntropy (GCCE) which can be given by

$$V_x^{\varepsilon}(\tau) = \frac{1}{T}\int_{-T/2}^{T/2} V_x(\tau)e^{-j2\pi\varepsilon t}dt \tag{7}$$

Employing (4) into (7), GCCE could be further expressed as

$$\begin{aligned} V_x^{\varepsilon}(\tau) &= \lim_{T\to\infty}\frac{1}{T}\int_{-T/2}^{T/2} G_{\mu,\upsilon}(x(t), x(t-\tau))e^{-j2\pi\varepsilon t}dt \\ &= \left\langle G_{\mu,\upsilon}(x(t), x(t-\tau))e^{-j2\pi\varepsilon t}\right\rangle_t \end{aligned} \tag{8}$$

In particular, when cyclic frequency $\varepsilon = 0$, $V_x^{\varepsilon}(\tau)$ becomes the generalized correntropy. Similar to that correntropy is a special case of generalized correntropy, CCE is a special case of GCCE with shape parameter fixed to $\mu = 2$.

Property 1: If the GCCE of the signal $x(t)$ is $V_x^{\varepsilon}(\tau)$, then the GCCE of the signal $x'(t) = x(t-D)$ is $V_{x'}^{\varepsilon}(\tau) = V_x^{\varepsilon}(\tau)e^{-j2\pi\varepsilon D}$.

Proof: Set $v = t - D$, then $t = v + D$

$$\begin{aligned} V_{x'}^{\varepsilon}(\tau) &= \left\langle G_{\mu,\upsilon}(x'(t), x'(t-\tau))e^{-j2\pi\varepsilon t}\right\rangle_t \\ &= \left\langle \gamma_{\mu,\upsilon}\exp(-\lambda|x(t-D) - x'(t-D-\tau)|^{\mu})e^{-j2\pi\varepsilon t}\right\rangle_t \\ &= \left\langle \gamma_{\mu,\upsilon}\exp(-\lambda|x(v) - x'(v-\tau)|^{\mu})e^{-j2\pi\varepsilon(v+D)}\right\rangle_{v+D} \\ &= \left\langle \gamma_{\mu,\upsilon}\exp(-\lambda|x(v) - x'(v-\tau)|^{\mu})e^{-j2\pi\varepsilon v}\right\rangle_v e^{-j2\pi\varepsilon D} \\ &= V_x^{\varepsilon}(\tau)e^{-j2\pi\varepsilon D} \end{aligned}$$

Property 2: If the GCCE of the signal $x(t)$ is $V_x^{\varepsilon}(\tau)$, set $z(t) = x(t - D)$, then the cross GCCE of $x(t)$ and $z(t)$ is

$$V_{zx}^{\varepsilon}(\tau) = \left\langle G_{\mu,\upsilon}(z(t), x(t-\tau))e^{-j2\pi\varepsilon t}\right\rangle_t = V_x^{\varepsilon}(\tau - D)e^{-j2\pi\varepsilon D}$$

Proof: Set $v = t - D$, then $t = v + D$

$$\begin{aligned} V_{zx}^{\varepsilon}(\tau) &= \left\langle G_{\mu,\upsilon}(z(t), x(t-\tau))e^{-j2\pi\varepsilon t}\right\rangle_t \\ &= \left\langle \gamma_{\mu,\upsilon}\exp(-\lambda|x(t-D) - x(t-\tau)|^{\mu})e^{-j2\pi\varepsilon t}\right\rangle_t \\ &= \left\langle \gamma_{\mu,\upsilon}\exp(-\lambda|x(v) - x(v+D-\tau)|^{\mu})e^{-j2\pi\varepsilon(v+D)}\right\rangle_{v+D} \\ &= \left\langle \gamma_{\mu,\upsilon}\exp(-\lambda|x(v) - x(v-(\tau-D))|^{\mu})e^{-j2\pi\varepsilon v}\right\rangle_v e^{-j2\pi\varepsilon D} \\ &= V_x^{\varepsilon}(\tau - D)e^{-j2\pi\varepsilon D} \end{aligned}$$

4 TDOA Estimation Algorithm Based on GCCE

In general, the signal model for TDOA estimation can be expressed as follows

$$x(t) = s(t) + w(t) + n_1(t) \tag{9}$$

$$y(t) = s(t - D_1) + w(t - D_2) + n_2(t) \tag{10}$$

where $x(t)$ and $y(t)$ are the received signals of two independent sensors, $s(t)$ is the periodic signal of interest (SOI), $w(t)$ denotes the periodic interference signal with different cycle frequencies from $s(t)$, $n_1(t)$ and $n_2(t)$ are additive noise that do not have cyclostationarity characteristics, D_1 is the TDOA to be estimated, D_2 is the TDOA of the periodic interference signal. For the convenience of analyzing the problem, it is assumed that SOI, interference and additive noise are statistically independent from each other.

When $n_1(t)$ and $n_2(t)$ are additive Gaussian noises, the cyclic cross-correlation of $x(t)$ and $y(t)$ at the cycle frequency ε_0 which is the cycle frequency of SOI is:

$$R_{xy}^{\varepsilon_0}(\tau) = \left\langle x(t)y^*(t+\tau)e^{-j2\pi\varepsilon_0 t}\right\rangle_t = R_s^{\varepsilon_0}(\tau - D_1)e^{-j\pi\varepsilon_0 D_1} \tag{11}$$

The peak of $\left|R_s^{\varepsilon_0}(\tau)\right|$ does not occur at $\tau = 0$ for all the cycle frequencies but $\left|R_s^{\varepsilon_0}(\tau)\right|$ is an even function. Take BPSK signal as the example, the peak of $\left|R_s^{\varepsilon_0}(\tau)\right|$ occurs at $\tau = 0$ when the cyclic frequency ε_0 is equal to twice the carrier frequency but $\left|R_s^{\varepsilon_0}(\tau)\right|$ has a pair of peaks straddling at $\tau = 0$ when the cyclic frequency ε_0 is equal to the baud rate, so the TDOA can be estimated by the peak point or the midpoint of the pair of peaks according to the selected cyclic frequency. This method is referenced as second order cyclic correlation (CC) based TDOA estimation method [5, 6].

When $n_1(t)$ and $n_2(t)$ are additive α-stable noise, the above process fails because $R_{xy}^{\varepsilon_0}(\tau)$ are not finite values due to the effect of noise [14]. To address this problem, a FLOCC based TDOA estimation method was proposed in [15] by using FLOCC to instead of CC. However, FLOCC based method requires a priori knowledge of α-stable distribution. However, the priori knowledge is difficult to obtain in some practical applications, therefore in this paper we replace CC with CCE and GCCE, the proposed methods are referenced as CCE or GCCE based TDOA estimation algorithm.

5 Simulation Results

In this section, a series of numerical experiments under different conditions are conducted to compare the estimation performance of our proposed CCE and GCCE based algorithms with that of the CC and FLOCC based algorithms. In the experiments, the SOI signal is BPSK signal with the carrier frequency $f_{c_1} = 0.25f_s$, the baud rate is $f_{d_1} = 0.0625f_s$ and the TDOA $D_1 = 100T_s$. The interference signals are BPSK signal with carrier frequency $f_{c_2} = 0.2f_s$, baud rate $f_{d_2} = 0.025f_s$ and the TDOA $D_2 = 50T_s$. The sampling frequency is $f_s = 10^7$ Hz and the sample number is 10000. The signal to interference ratio (SIR) is set to be 0 dB. As the characteristic of the α-stable distribution makes the use of the standard SNR meaningless, a new SNR measure, generalized signal-to-noise ratio (GSNR) is defined as $GSNR = 10\log_{10}\left(\sigma_s^2/\gamma\right)$, where σ_s is the variance of the signal. The fraction lower order parameter for FLOCC based algorithm is $p = (\alpha - 0.1)/2$, the core length of the CCE based algorithm is $\sigma = 1$ [22], the shape and scale parameter of GCCE based algorithm is $\mu = 5$, $\lambda = 0.01$.

Figure 1 shows the normalized estimation results of CC, FLOCC, CCE and GCCE based TDOA estimation algorithms when the cycle frequency is $\varepsilon_0 = f_{d_1}$ in the α-stable impulsive noise with the characteristic exponent $\alpha = 1.3$ and GSNR = -2 dB. It can be seen that there are symmetrical peaks on both sides of the TDOA point $\tau = 100$ for the CCE and GCCE based algorithms, symmetric peaks also appear in the FLOCC based method, but are not obvious. Due to the impact of impulse noise, more peaks appeared in the CC based algorithm resulting in the TDOA cannot be estimated correctly.

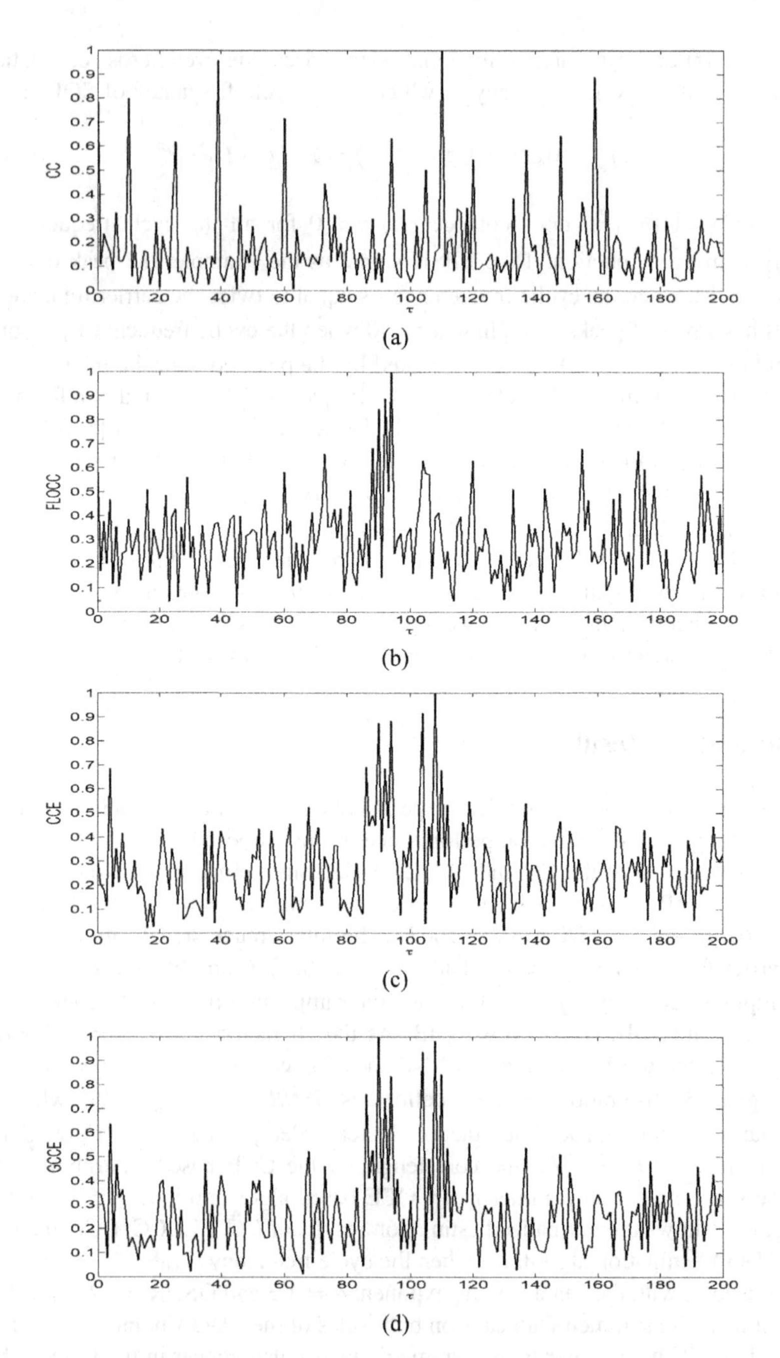

Fig. 1. Normalized TDOA estimation result (a) CC based method (b) FLOCC based method (c) CCE based method (d) GCCE based method

Define the MSE of the TDOA estimation is

$$MSE = \frac{\sum_{i=1}^{N} \left(D_1 - \widehat{D}_1(i) \right)^2}{N} \tag{12}$$

Where $\widehat{D}_1(i)$ is the ith estimation of the TDOA D_1, N is the number of simulations. In the following two simulation experiments, N is set to be 1000. Fixed the characteristic exponent of the α-stable impulsive noise at $\alpha = 1.3$, the MSE performance curves at different GSNRs of the CC, FLOCC, CCE and GCCE based algorithms are given in Fig. 2. The results demonstrate that the, CCE and GCCE based algorithms are much superior to the CC and FLOCC based algorithms, especially in the case of lower GSNR.

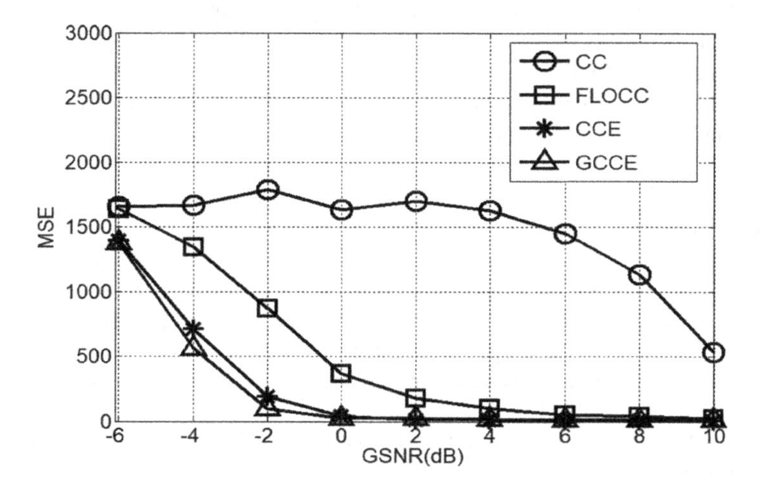

Fig. 2. MSE vs. GSNR

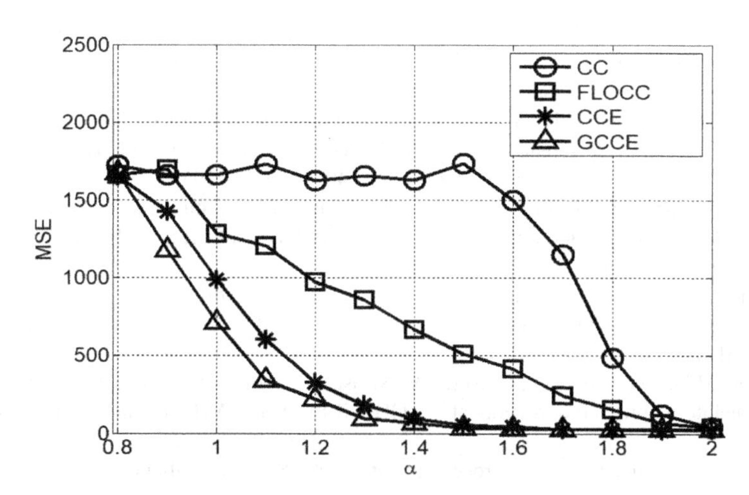

Fig. 3. MSE vs. characteristic exponent

Figure 3 plots the performance of the four algorithms varying with different values of the characteristic exponent of the α-stable impulsive noise when GSNR $= -2\,\text{dB}$. It is demonstrated that the impulsive noise inhibition ability of the CCE and GCCE based algorithms is better than that of the FLOCC based algorithm. In particular, the performance of the GCCE based algorithm is effective than the CCE based algorithm because the shape parameter of the GCCE is adjustable, while the shape parameter of the CCE is fixed to 2.

6 Conclusion

In this paper, we study the TDOA estimation problem of cyclostationary signals in impulsive noise environment. Since conventional cyclic correlation does not make sense in the impulse noise environment, a robust cyclic correlation GCCE is defined and the CCE is a special case of it. Two new signal-selective TDOA estimation algorithms based on CCE and GCCE are proposed. Simulation results show that the proposed algorithms outperform the conventional CC and FLOCC based methods, especially in low GSNR and strong impulsive noise environments.

References

1. Yoon, J.Y., Kim, J.W., Lee, W.Y., Eom, D.S.: A TDOA-based localization using precise time-synchronization, In: Proceedings of the 14th International Conference on Advanced Communication Technology, pp. 1266–1271, February 2012
2. Ni, H., Re, G., Chang, Y.: A TDOA location scheme in OFDA based WMANs. IEEE Trans. Consum. Electron. **54**(3), 1017–1021 (2008)
3. Wang, H.Y., Qiu, T.S.: Adaptive noise cancellation and time delay estimation. Dalian University of Technology Press, Dalian (1999). (In Chinese)
4. Gardner, W.A., Napolitano, A., Paura, L.: Cyclostationarity: half a century of research. Signal Process. **86**(4), 639–697 (2006)
5. Gardner, W.A., Chen, C.K.: Signal selective time-difference of arrival estimation for passive location of man-made signal sources in highly corruptive environments, part I: theory and method. IEEE Trans. Signal Process. **40**(5), 1168–1184 (1992)
6. Gardner, W.A., Chen, C.K.: Signal selective time-difference of arrival estimation for passive location of man-made signal sources in highly corruptive environments, part II: algorithms and performance. IEEE Trans. Signal Process. **40**(5), 1185–1197 (1992)
7. Zhang, Y., Wang, C.M., Collins, L.M.: Adaptive time delay estimation method with signal selectivity. In: IEEE International Conference on Acoustics, Speech and Signal Processing, vol. 2, pp. 1477–1480 (2002)
8. Huang, Z.T., Zhou, Y.Y., Jiang, W.L., Lu, Q.Z.: Joint estimation of Doppler and time difference of arrival exploiting cyclostationary property. IEE Proc. - Radar Sonar Navig. **149** (4), 161–165 (2002)
9. Button, M.D., Gardiner, J.G., Glover, I.A.: Measurement of the impulsive noise environment for satellite-mobile radio systems at 1.5 GHz. IEEE Trans. Veh. Technol. **51**(3), 551–560 (2002)
10. Nikias, C.L., Shao, M.: Signal Processing with Alpha-Stable Distributions and Applications. Wiely, New York (1995)

11. Li, S., He, R.X., Lin, B., Sun, F.: DOA estimation based on sparse representation of the fractional lower order statistics in impulsive noise. IEEE/CAA J. Autom. Sin. **5**(4), 860–868 (2016). https://doi.org/10.1109/JAS.2016.7510187
12. Liu, W.F., Pokharel, P.P., Principe, C.: Correntropy: properties and applications in non-Gaussian signal processing. IEEE Trans. Signal Process. **55**(11), 5286–5298 (2007)
13. Chen, B.D., Xing, L., Zhao, H.Q., Zheng, N.N., Principe, J.C.: Generalized correntropy for robust adaptive filtering. IEEE Trans. Signal Process. **64**(3), 3376–3387 (2016)
14. Guo, Y.: The study on novel time delay estimation methods based on stable distribution. Ph.d. dissertation. Dalian University of Technology (2009)
15. Liu, Y., Qiu, T.S., Sheng, H.: Time-difference-of-arrival estimation algorithms for cyclostationary signals in impulsive noise. Signal Process. **92**(9), 2238–2247 (2012)
16. You, G.H., Qiu, T.S., Yang, J.: A novel DOA estimation algorithm of cyclostationary signal based on UCA in impulsive noise. AEUE-Int. J. Electron. Commun. **67**(6), 491–499 (2013)
17. You, G.H., Qiu, T.S., Zhu, Y.: A novel extended fractional lower order cyclic MUSIC algorithm in impulsive noise. ICIC Express Lett. **6**(9), 2371–2376 (2012)
18. You, G.H., Qiu, T.S., Song, A.M.: Novel direction findings for cyclostationary signals in impulsive noise environments. Circuits Syst. Signal Process. **32**(6), 2939–2956 (2013)
19. Luan, S.Y., Qiu, S.T., Zhu, J., Yu, L.: Cyclic correntropy and its spectrum in frequency estimation in the presence of impulsive noise. Signal Process. **120**, 503–508 (2016)
20. Fontes, A.I.R., Rego, J.B.A., Martins, A.M., Silveira, L.F., Principe, J.C.: Cyclostationary correntropy: definition and applications. Expert Syst. Appl. **69**, 110–117 (2017)
21. Li, S., Chen, X.J., He, R.X.: Robust cyclic MUSIC algorithm for finding directions in impulsive noise environment. Int. J. Antennas Propag. Article ID 9038341 (2017)
22. Song, A.M.: New algorithms for time delay estimation and beamforming under stable distribution noise. Ph.d. dissertation. Dalian University of Technology (2015)

iKey: An Intelligent Key System Based on Efficient Inclination Angle Sensing Techniques

Ke Lin[1], Jinbao Wang[2(✉)], Jianzhong Li[1], Siyao Cheng[1], and Hong Gao[1]

[1] School of Computer Science and Technology,
Harbin Institute of Technology, Harbin, China
linkehit@163.com, {lijzh,csy,honggao}@hit.edu.cn

[2] The Academy of Fundamental and Interdisciplinary Sciences,
Harbin Institute of Technology, Harbin, China
wangjinbao@hit.edu.cn

Abstract. The elderly may have different aspects of inconvenience in their daily life. Among them, many old people have trouble remembering things even just happened hours ago. They often forget whether they have locked the door while leaving so that they may have to return and check. Such situation also happens to many younger people that do not concentrate their mind while locking the door. In this paper, an intelligent key system, iKey, is proposed to solve such problem. It can be deployed on an existing key to detect user's locking actions and store locking status in the form of time. Related hardware architecture and working process are proposed. The sensing module based on inclination angle sensors is designed to reduce the amount of data generated. Furthermore, efficient locking detection algorithms are proposed accordingly. Such system and techniques can also be applied in knobs or rotating handles of machines and facilities to detect illegal operations and to avoid user's forgetting to operate them.

Keywords: Ubiquitous computing · Cyber-physical system
Human activity recognition

1 Introduction

Recent years, the aging of the population has become growing concerned in many countries. According to researches, the number and ratio of aged people are increasing dramatically all over the world [1]. However, the elderly may have many aspects of inconvenience in their daily life for the degeneration of their body functions. For example, they may have trouble keeping balance while walking, hearing things clearly, and calling the emergency when they are in danger. Hence, efforts to improve the life quality of the aged people by researchers in all fields are urgently demanded.

S. Chellappan et al. (Eds.): WASA 2018, LNCS 10874, pp. 284–295, 2018.
https://doi.org/10.1007/978-3-319-94268-1_24

Among the problems the old people are faced with, hypomnesia is particularly common. People with hypomnesia are hard to remember things even just happened several hours ago. We find a very common situation in the daily life of the elderly who are easy to forget things. When they leave their home for a while, they are often hard to remember whether they have locked the door. In fact, such circumstance not only happens to the old people, but also to some younger people when they are quite busy or very tired. They often leave their home or office while thinking about some other things instead of concentrating their mind on locking the door. As a result, they may decide to return and check, which leads to the waste of time and resources.

A possible solution is to design network-based smart home systems that are mounted on the door to detect the status of the lock. When the user wants to check whether the door is locked, he may connect to the system through the mobile phone. However, such systems have some drawbacks. Firstly, smart home systems can be very expensive. Secondly, such systems often require complicated configurations which are not suitable for the elderly. Thirdly, even though plenty of efforts have been made to improve the security and reliability of the network-based smart systems [2–5], they still have the risk of privacy leakage [6,7]. If they were invaded by hackers, they would threat the safety of users.

Therefore, designing an easily-deployable and low-cost system to solve the above mentioned problem is still eagerly demanded. The system should be cheap in price, convenient to use, and require least configurations. Moreover, it had better not replace anything existing, and make least modification to the locks and doors.

To satisfy above requests, we decide to attach an intelligent embedded system to the existing keys. It recognizes the user's actions while holding the key. If a locking action is detected, related locking status will be updated or stored in the system. The user can ensure whether he has locked the door by checking it.

The first task of such system is the recognition of user's actions while using the key. A straightforward idea is to take advantage of the inertial sensors such as 3-axis accelerometers. Such MEMSs (micro-electro-mechanical systems) have been widely used in motion recognitions [8,9]. Their sensing values vary with the moving status of the objects they are attached to. With a proper sampling rate, we can obtain time series related to the object's moving process. In other motion recognition scenarios [10], instance-based strategies are primarily used for mining the time series of motions or human activities. Such strategies firstly collect training sets of sensory data. Then, algorithms based on Dynamic Time Warping [11] and K Nearest Neighbors [12,13] can be used for classification.

However, we don't consider such sensors and algorithms are quite suitable for our task for the following reasons. Firstly, the amount of data generated by 3-axis accelerometers or other inertial sensors every second can be very large. Besides, the type of data is float numbers sampled from analog signals. To process such large amount of float numbers at a time can be an extremely heavy burden to a low-cost embedded system. Secondly, instead of mining patterns from the training set, the KNN-based methods simply store the training set and require all of them while classifying. Such character is a challenge to the embedded systems with limited storage resources. Thirdly, the communication between the micro controlling unit and the sensors like accelerometers often requires SPI or IIC bus, or at least a multi-channel analog-to-digital converter. Frequent bus or ADC operations are very resource and energy consuming.

With above considerations, we propose the intelligent key system, iKey, based on multiple inclination angle sensors. Such design leads to much fewer data generation. It also avoids the using of buses and ADCs, making it more energy and resource efficient. In addition, we propose a compressing-based strategy for locking actions detection. It both minimizes the training set and reduces the data processed while pattern recognition. All above features make iKey a lightweight and efficient system for the task.

Besides an intelligent key, the system and techniques we propose can also be applied in the knobs or rotating handles on all kinds of machines and facilities. The users can check whether the knob or handle is operated at a certain time. The user's forgetting to operate them and illegal operations can be detected with the help of our system.

The main contributions of the paper are listed as follows:

1. A prototype of an easy-deployable intelligent key system based on inclination angle sensing techniques, iKey, is proposed. To the best of our knowledge, it is the first intelligent key system based on inclination angle sensing.
2. Corresponding compressing-based locking detection algorithms are designed to detect the locking actions with high efficiency.
3. Experiments are carried out to study the features of the proposed algorithms and verify the effectiveness of the iKey system.

The remaining parts of this paper are organized as follows. Section 2 shows the overview of the system. Section 3 gives the working process of the system. Section 4 proposes the design of the sensing module. Section 5 provides the locking detection algorithms. Section 6 presents the experimental results. Finally, Sect. 7 concludes the whole paper.

2 System Overview

2.1 The Deployment of iKey

Figure 1 presents how the iKey is deployed on common keys. Figure 1(a) shows how the system is integrated on a newly-manufactured key. On one side, there's a four-digit screen to show the time of user's last locking action or the time interval from the last locking to now. On the other side, a touching sensor is deployed to be used as a switch. In order to simplify the user interactions, this switch is used for both showing the time and starting the detecting process. As can be seen from the figure, if the user locks the door with such a key, his hand will automatically touch the touching sensor. An additional step to turn it on while locking is not required. All the other circuits are embedded inside the key

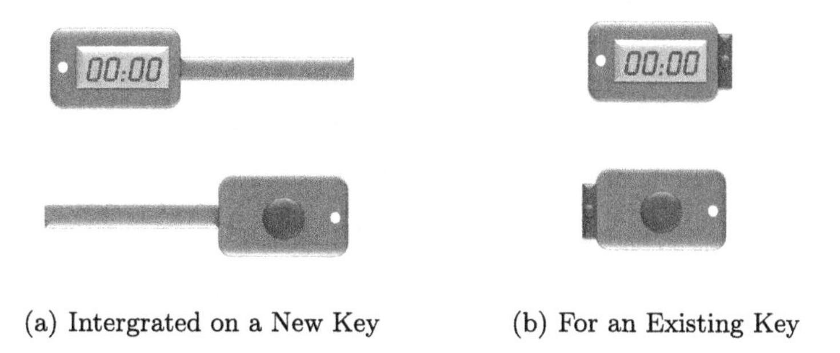

(a) Intergrated on a New Key (b) For an Existing Key

Fig. 1. The deployment of iKey

handle. Figure 1(b) shows the design for an existing key. A cavity is set in the handle with two metal clips and screws. An existing key can be placed into the cavity and clamped by the clips and screws so that it's not essential for the user to replace his original keys.

2.2 Application Scenario

The application scenario of the iKey system can be described as follows. After leaving his home or office for a while, the user may suddenly realize that he is not quite certain about whether he locked the door while leaving. Consequently, he may decide to return. With the help of iKey system, the user can just pick it out and touch the touching sensor. The screen will show the time of his last locking (or the time interval from now to his last locking). If such time is around the time of his leaving (or the time interval is about the time he has been out), it means he must have locked the door. Otherwise, he had better return to lock the door carefully.

2.3 Hardware Architecture

The hardware architecture of the system is illustrated by Fig. 2. As shown in the figure, it includes the power module, the sensing module, the touching sensor, the micro controlling unit (MCU), the real time clock chip (RTC), and the LCD screen with related driving chips. Besides, a debugging interface is preserved for developers to connect a larger debugging screen or use serial ports.

The power module provides the power supply for all the other modules. The voltage and capacity of it can be chosen according to the type of the MCU. The sensing module is a vital component. It is used for sensing the user's action while holding the key and sending the related signals to the MCU. The detailed design of this module will be discussed in Sect. 4.

The MCU is the core of the whole system. It manages the duty cycle of the system, receives signals from the sensors, communicates with the RTC to fetch and store the time, and displays information to the LCD screen. Besides, all the data processing tasks and detection algorithms are run on the MCU.

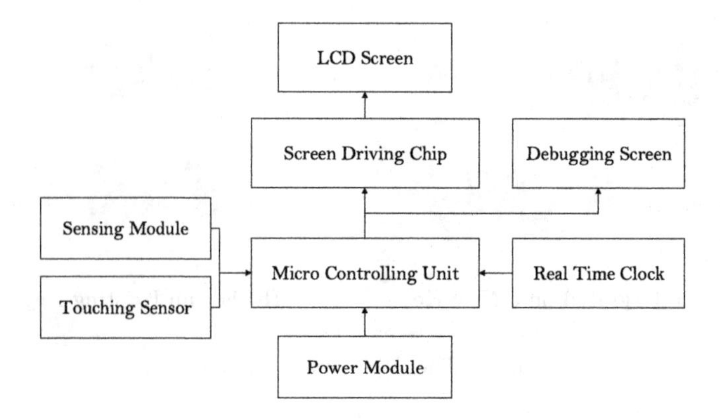

Fig. 2. The hardware architecture

Since iKey is a low-duty-cycle system, the MCU is in low-power-consumption mode for most time. Hence, a real time clock is needed to maintain the time synchronization. The RTC is always awake even though the MCU is asleep. As far as we are concerned, the power consumption of an RTC is so low that a single battery can guarantee its working for a long period of time [14].

The touching sensor is connected to an interrupt interface of the MCU with waking up function. It can be regraded as the power switch. When it is touched by hand, the MCU wakes up to its duty. The LCD screen is used for showing information to the user. In order to reduce the pins occupied by the screen and the processing burden of the MCU, a screen-driving chip is deployed to manage the displaying-related tasks.

3 The Working Process

We design two working modes for the iKey system, the "Locking Time Mode" and the "Time Interval Mode". We name them "mode L" and "mode I" respectively.

In mode L, the screen shows the time of user's last locking action when the user switch it on. The working process of this mode is listed as follows:

Step 1. When the system initializes, it starts the low-power-consumption mode. The inner clock of the MCU is stopped while the RTC remains its working status. The interrupt controller disables all the interrupts except the waking-up interrupts. The screen is also turned off to save energy.

Step 2. When the touching sensor is touched, it sends an interrupt to the waking-up interface. The MCU wakes up to fetch the time of the user's last locking from its inner storage and show it on the LCD screen. If no such time was recorded before, the screen displays some error messages, such as "EE:EE".

Step 3. Then, the system starts to collect and process the data from the sensing module to detect the locking actions of the user.

Step 4. If the MCU detects a locking action, it will instantly read the current physical time from the RTC and store it in the inner storage. Then it returns to the low-power-consumption mode.

Step 5. If an unlocking action is detected, the system will stop its detecting process and return to the low-power-consumption mode.

Step 6. If the MCU detects no locking action for a long period of time, which exceeds the awake-time threshold, it will return to sleep without any changes.

We add an additional step to detect the unlocking actions because some of the locks are just locked or unlocked by one round of rotation. When the key is pulled out after unlocking, it also has to be rotated back. Such rotating back process may affect the recognition of locking actions. To overcome this problem, the detecting process ends immediately it detects an unlocking action.

However the above working process has a disadvantage. In that process, the RTC works as a time clock runs synchronously with the physical time. When it is used over long time, the time of RTC may drift and become inaccurate. We have to add additional buttons or involve other devices to adjust the time.

To solve the problem, we propose mode I. In this mode, the RTC works only as a timer. It starts (or restarts) to count time from zero when a locking action is detected. When the user switches the system on, the screen shows the time interval from now to the last locking action which is right the current reading of RTC. In mode I, the **Step 2** and **Step 4** above can be modified as follows:

Step 2. When the touching sensor is touched by the user, it sends an interrupt to the waking-up interface. The MCU wakes up to fetch the reading of RTC, and then displays it on the LCD screen.

Step 4. If the MCU detects a locking action, it will clear the existing time of the RTC and restart it from "00:00". Then the MCU returns to the low-power-consumption mode.

The advantages of mode I are obvious. Firstly, the time counted by the RTC becomes very limited for frequent restarting. Thus, the time drift error won't accumulate to an unbearable value. Then, the time setting of iKey when manufacturing is not essential. Moreover, even the whole system runs out of the battery, the time adjustment is not required because the RTC works only as a timer. Finally, the MCU doesn't have to record any time-related information.

4 Design of the Sensing Module

Due to the disadvantages of inertial sensors mentioned in Sect. 1, we need alternative sensors with fewer data generation. The solution we find is the inclination angle sensors (shown in Fig. 3(a)). Inclination angle sensors are firstly designed for detecting the falling down of some large mechanical facilities. Each inclination angle sensor has its own target angle range related to its direction of placement. When the sensor is falling down to its target angle range, the output voltage changes. Considering that locking actions are a set of rotations, we can combine multiple inclination angle sensors to detect the rotating status of the key. To take three sensors as an example, the structure of the multiple sensors can be designed as Fig. 3(b) or Fig. 3(c) shows. The blue symbols in the figures show the direction of the axis of the key rotating (towards inside the paper).

However, inclination angle sensors are also sensitive to vibrations during the rotations. The locking actions are typical quick rotating process with vibrations. Thus, the output signals are not stable when the sensor is towards its target angle range. Instead, they change in the form of continuous square waves (Sample outputs of our experiments can be found in Sect. 6). With above consideration, we set the communication between the MCU and the sensors through the interrupts in order to reduce the computing burden of the MCU. The sensors are connected to the interrupt interfaces of the MCU with proper circuits.

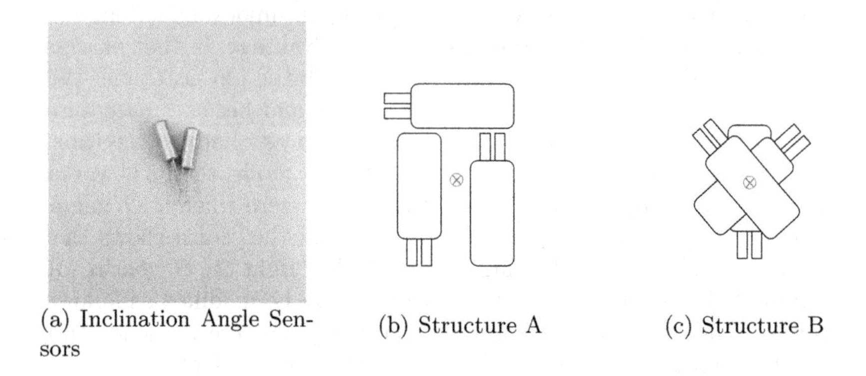

(a) Inclination Angle Sensors (b) Structure A (c) Structure B

Fig. 3. Inclination angle sensors and their combinations (Color figure online)

5 The Locking Detection

With the design of the sensing module mentioned in last section, the signals the MCU receives are a series of interrupt requests from the sensors. We can number the signals from the n-th sensor with n. The aim of the locking detection is to recognize the locking actions from the data stream of the interrupt requests from all the sensors.

The overall idea of our algorithm is to compress the data stream of interrupt requests with a sliding window. However, due to the unstableness of the inclination sensor signals, the data stream of the interrupt requests can be noisy. Therefore, we discuss some methods to filter the noisy sliding windows at first.

We define the fluctuation factor of a sliding window w as $f(w)$. Specially, $f(w) = 0$ if $len(w) \leq 1$. When $len(w) \geq 2$, $f(w)$ can be calculated by Eq. 1. The $len(w)$ represents the count of elements in w.

$$f(w) = (\sum_{k=1}^{len(w)-1} g(k))/(len(w)-1) \quad where\ g(k) = \begin{cases} 1 & if\ w[k] \neq w[k-1] \\ 0 & otherwise \end{cases} \tag{1}$$

We also define $d(w) = Count(i) - Count(j)$ to evaluate the distribution of the requests in a sliding window, where i is the request with the largest count in the sliding window w, and j is the request with second largest count in w.

We set thresholds δ_f for $f(w)$ and δ_d for $d(w)$, for example, $\delta_f = (len(w) - 2)/(len(w) - 1)$ and $\delta_d = 1$. If $f(w)$ exceeds δ_f, it means there is obvious fluctuation in the sliding window (E.g. a sliding window like [ababab]). If $d(w)$ is too small to reach δ_d, it means there are two major requests that have large count in a window (E.g. a sliding window like [aaabbb]). It's hard to identify which sensor is mainly activated at that moment. We decide to discard such unreliable windows to avoid their interference. Similarly, we can also discard the windows where the value of $Count(i)$ is smaller than a certain threshold δ_i, for example, half the size of a sliding window.

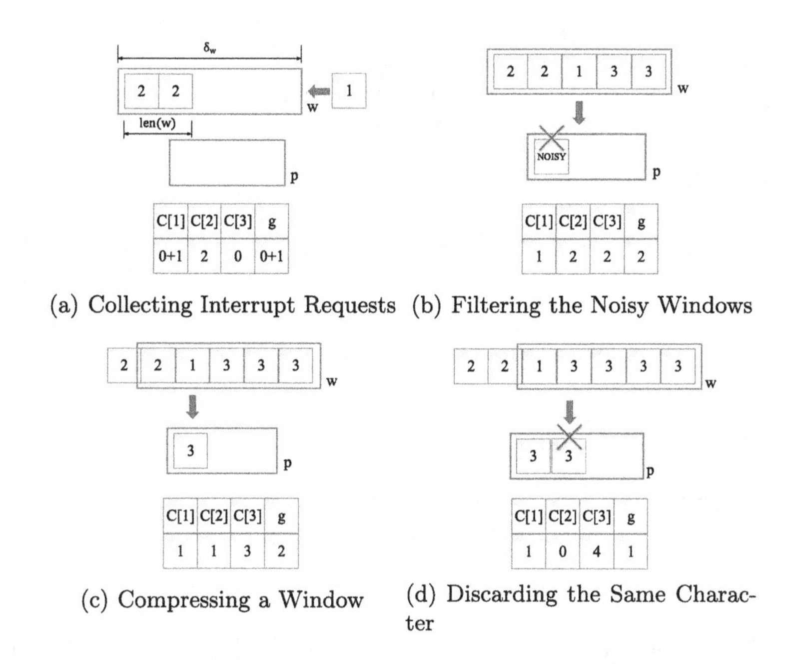

(a) Collecting Interrupt Requests (b) Filtering the Noisy Windows

(c) Compressing a Window (d) Discarding the Same Character

Fig. 4. Examples of different operations in compressing

Our compressing strategy can be defined by Algorithm 1. Firstly, the data structures needed in the following steps are initialized (line 1–4). w is the sliding window. p is the compression result of the stream (we call it a compressed stream). C storages the count for each requests in the sliding window ($C[r] = Count(r)$). g is used for calculating the $f(w)$ defined by Eq. 1. Then, it begins to wait for interrupt requests from the sensing module unless the awake time t reaches threshold δ_{time} (line 5). When a request r is received, it is appended into the window. The statistics of window w are updated accordingly (line 6–10). An example of the operation is presented in Fig. 4(a).

Algorithm 1. The Locking Detection Algorithm

```
  Input: the number of sensors N, the window size δ_w (the threshold of len(w)),
         the awake time thresehold δ_time, δ_f, δ_d, δ_i, Pattern Set P
  Output: The classification result of the action
1 w ← an empty deque
2 C ← an array of N zeros
3 p ← an empty string
4 g ← 0
5 while t < δ_time do
6   if an interrupt request r is received then
7     if len(w) ≠ 0 and r ≠ w[len(w) − 1] then
8       g = g + 1
9     C[r] ← C[r] + 1
10    w.append(r)
11    if len(w) = δ_w then
12      i ← the request # with the largest count
13      j ← the request # with the second largest count
14      if C[i] − C[j] < δ_d or C[i] < δ_i or g/(len(w) − 1) > δ_f then
15        continue
16      if len(p) = 0 or i ≠ p[len(p) − 1] then
17        p.append(i)
18        if P_lock.find(p) ≠ null then
19          Return "lock"
20        if P_unlock.find(p) ≠ null then
21          Return "unlock"
22      if w[0] ≠ w[1] then
23        g = g − 1
24      C[w[0]] ← C[w[0]] − 1
25      remove the first element w[0] from w
26 Return "noise"
```

If w is filled up, the algorithm starts to compress it (line 11). At first, the requests with the largest and the second largest count are found out (line 12–13). Then, the sliding window are filtered with the method discussed above (as shown in Fig. 4(b))(line 14–15). After that, we know that when one of the sensors is rotated to its target angle, the MCU receives a series of same interrupt requests from it. To minimize the data, we only have to preserve one of them. Therefore, the algorithm compresses w to a single character i, which appears the most times in w (i is already obtained in line 12). The other types of requests with smaller counts are treated as noises and eliminated. Considering that long continuous sequence of same interrupt requests only indicates a same angle the key rotates to, when neighboring sliding windows are compressed to the same character, we only have to keep one of them. Hence, only when i is different from the last character in p, it is appended to p (as presented in Fig. 4(c)) (line 16–17). Otherwise, i is discarded (as shown in Fig. 4(d)).

Afterwards, every time when p is updated, the algorithm attempts to match p with the patterns in the pattern set. If p is matched with a "locking" or "unlocking" pattern, the algorithm returns that result and stops the detecting process (line 18–21). The patterns are obtained from plenty of labeled locking and unlocking actions with the same compressing strategy. The manufacturer of keys can do plenty of locking and unlocking actions and compress the collected data stream to patterns. Our experiments proves that both the length of patterns and the size of the pattern set are small. If all the matching processes fail, the first element will be removed from the window to make it not full again. Meanwhile, related statistics are updated accordingly (line 22–25). Then, the system loops to wait for another interrupt request. If no locking action is detected till the time is up, the algorithm returns "noise" (line 26).

6 Experimental Results

We implement a prototype system for the iKey. We choose the low-power-consumption MCU with multiple interrupt interfaces, STC15W56S4. The screen is a 4-digit LCD screen, and it's driving chip for displaying is HT1621B. That chip can convert the 12-line displaying interface to a 3-line serial interface. The real time clock we use is a DS1302 RTC. The sensing module consists of three SW520D inclination angle sensors. Figure 5 shows an example of output signal from a single sensor when it rotates to its target angle.

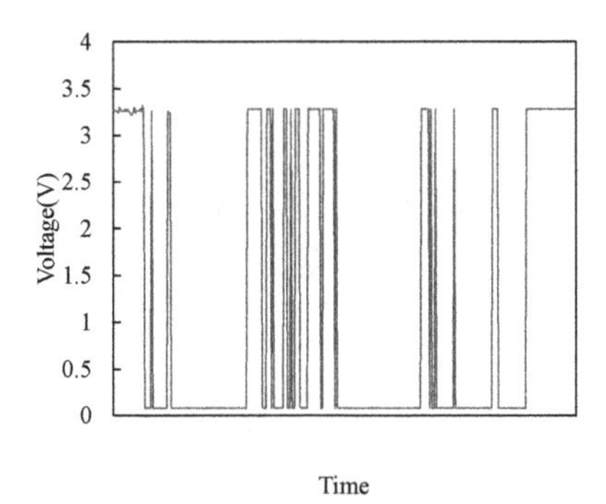

Fig. 5. The original signal from a single sensor

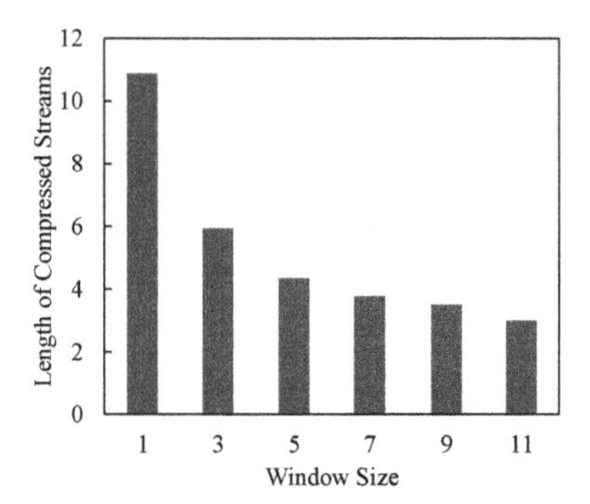

Fig. 6. The average length of the compressed streams

Figure 6 illustrates how the average length of the compressed streams varies with the sliding window size δ_w. As shown in the figure, the length of compressed streams decreases with the increasing of the window size. However, too small δ_w makes the length of compressed streams too long, which affects the noise removal and the efficiency of the matching process. Conversely, if δ_w is too big, some critical features will be lost for the too short compressed streams generated. The optimized value of δ_w is related to the physical structure and deployment of the system, so it should be assigned by experiments when manufacturing. We set a moderate value $\delta_w = 5$ in the our following experiments. Meanwhile, it can also be inferred from the figure that the compressed streams are short with a proper δ_w, which improves the efficiency of the system.

We also carry out experiments to study the distribution of the compressed streams of locking actions. As Fig. 7 shows, 75.8% of the locking actions can be compressed to a single pattern, which means about 75% accuracy can be reached in detecting those locking actions with only one pattern. If a second pattern is added, 15.5% more actions can be further detected. The else covers only about 7%. The distribution approximately satisfies the Zipf's law [15]. Such result also indicates that the size of the pattern set is so small that the matching process is not resource-consuming.

Figure 8 indicates the experimental results on the accuracy of iKey prototype system. The first two bars show the accuracy of the locking and unlocking detection. An overall accuracy of 95% is reached. Due to the design of our system, the system is in low-power-consumption mode during most time. Therefore, most non-locking actions such as shaking can hardly affect the its usage. But we still want to evaluate the robustness of iKey in case of the touching sensor is pressed by mistake. So we also carry out experiments on shaking it while keeping the iKey awake. The last three bars present the accuracy of classifying shaking up and down, back and forth, left and right as noises. In our experiments, all the shaking actions mentioned above are correctly classified as noises.

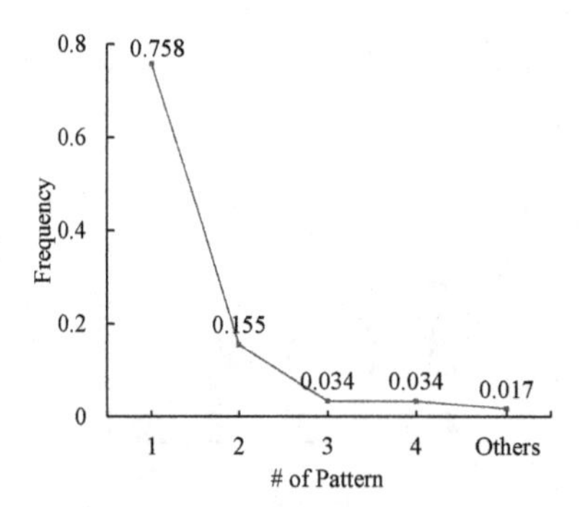

Fig. 7. The distribution of the compressed streams

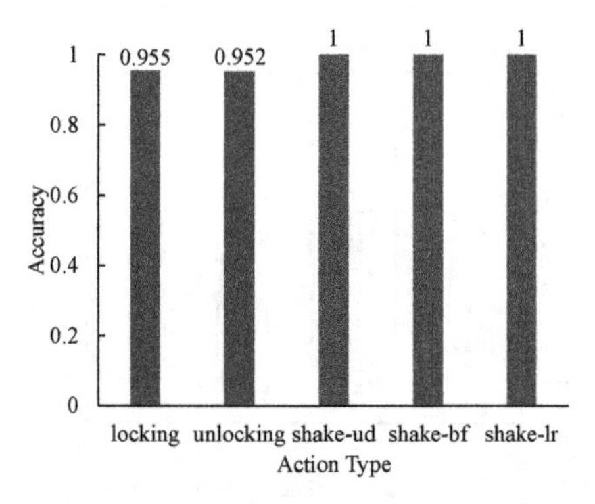

Fig. 8. The accuracy of different actions

7 Conclusion

This paper studies the intelligent key system based on inclination angle sensing techniques. At first, we propose the prototype of a locking detection intelligent key system, iKey. Related hardware architecture and working processes are given accordingly. Then we discuss the design of the sensing module, providing the structure design of the multiple sensors. Thirdly, algorithms are designed for the locking detection based on data from the sensors. Eventually, experiments are carried out to evaluate the effectiveness and features of our proposed system and algorithms.

Acknowledgments. This work is supported in part by the National Natural Science Foundation of China under Grant No. 61632010, 61502116, 61370217, and U1509216.

References

1. Lutz, W., Sanderson, W., Scherbov, S.: The coming acceleration of global population aging. Nature **451**(7179), 716–719 (2008)
2. Cai, Z., Zheng, X.: A private and efficient mechanism for data uploading in smart cyber-physical systems. Trans. Netw. Sci. Eng. (TNSE) (2018)
3. Liang, Y., Cai, Z., Yu, J., Han, Q., Li, Y.: Deep learning based inference of private information using embedded sensors in smart devices. IEEE Netw. Mag. (2018)
4. Zhang, L., Cai, Z., Wang, X.: FakeMask: a novel privacy preserving approach for smartphones. IEEE Trans. Netw. Serv. Manag. **13**(2), 335–348 (2016)
5. Zheng, X., Cai, Z., Li, Y.: Data linkage in smart IoT systems: a consideration from privacy perspective. IEEE Commun. Mag. (2018)
6. Sanchez, I., Satta, R., Fovino, I.N., Baldini, G., Steri, G., Shaw, D., Ciardulli, A.: Privacy leakages in smart home wireless technologies. In: Proceedings of International Carnahan Conference on Security Technology, pp. 1–6. IEEE (2014)
7. Duan, Z., Li, W., Cai, Z.: Distributed auctions for task assignment and scheduling in mobile crowdsensing systems. In: Proceedings of International Conference on Distributed Computing Systems, pp. 635–644. IEEE (2017)
8. Liu, J., Zhong, L., Wickramasuriya, J., Vasudevan, V.: uWave: accelerometer-based personalized gesture recognition and its applications. Pervasive Mob. Comput. **5**(6), 657–675 (2009)
9. Zhang, X., Chen, X., Li, Y., Lantz, V., Wang, K., Yang, J.: A framework for hand gesture recognition based on accelerometer and EMG sensors. IEEE Trans. Syst. Man Cybern. Part A Syst. Hum. **41**(6), 1064–1076 (2011)
10. Lin, K., Cheng, S., Li, Y., Li, J., Gao, H., Wang, H.: SHMDRS: a smartphone-based human motion detection and response system. In: Yang, Q., Yu, W., Challal, Y. (eds.) WASA 2016. LNCS, vol. 9798, pp. 174–185. Springer, Cham (2016). https://doi.org/10.1007/978-3-319-42836-9_16
11. Keogh, E.J., Pazzani, M.J.: Scaling up dynamic time warping for data mining applications. In: Proceedings of International Conference on Knowledge Discovery and Data Mining, pp. 285–289. ACM (2000)
12. Zhang, S., Li, X., Zong, M., Zhu, X., Wang, R.: Efficient kNN classification with different numbers of nearest neighbors. IEEE Trans. Neural Netw. Learn. Syst. **29**(5), 1774–1785 (2017)
13. Song, G., Rochas, J., Beze, L., Huet, F., Magoules, F.: K nearest neighbour joins for big data on mapreduce: a theoretical and experimental analysis. IEEE Trans. Knowl. Data Eng. **28**(9), 2376–2392 (2016)
14. Maxim Integrated. https://para.maximintegrated.com/en/results.mvp?fam=rtc&tree=master
15. Powers, D.M.W.: Applications and explanations of Zipf's law. In: Advances in Neural Information Processing Systems, vol. 5, no. 4, pp. 595–599 (1998)

Massive MIMO Power Allocation in Millimeter Wave Networks

Hang Liu[1(✉)], Chinh Tran[1], Jan Lasota[1], Son Dinh[1], Xianfu Chen[2], and Feng Ouyang[3]

[1] The Catholic University of America, Washington, DC, USA
liuh@cua.edu
[2] VTT Technical Research Centre of Finland, Oulu, Finland
[3] Johns Hopkins Applied Physics Laboratory, Laurel, MD, USA

Abstract. Massive multiple-input multiple-output (MMIMO) is a key technology for 5G mobile communication systems, which enables to simultaneously form and transmit multiple directional signal beams to multiple mobile terminals (MTs) on the same frequency channel with high array beamforming gains and throughput. One of the challenges in MMMIO beamforming is how to allocate the transmit power to multiple beams sent from a MMIMO base station to multiple MTs and schedule data transmissions, given heterogeneous traffic and channel conditions of multiple MTs. Furthermore, the statistics of users' packet arrivals and channel states may not be known a priori and vary over time. In this paper, we propose a framework to optimize MMIMO beam power allocation and transmission scheduling in millimeter wave networks with time-varying traffic and channel conditions. The optimization problem is formulated as a Markov decision process (MDP) with the objective to minimize the overall queueing delay of multiple MTs by taking their heterogeneous and dynamic traffic and channel states into account. An online reinforcement learning scheme is designed which allows achieving the long-term optimal system performance with no requirement for a priori knowledge of user traffic statistics and wireless network states. Evaluation results show that our proposed scheme outperforms the state-of-the-art baselines.

Keywords: Massive MIMO · Millimeter wave communications
Power allocation · Machine learning · Markov decision process
Dynamic networks

1 Introduction

Massive multiple-input multiple-output (MMIMO) and millimeter wave (mmWave) communications are the cornerstones for 5G mobile systems to achieve tens of Gbps data rates to meet dramatic increases in wireless traffic demands. Conventional radio communication spectrum below 10 GHz is becoming overcrowded and cannot keep up with the bandwidth requirement for rapid traffic growth [1, 2]. The wireless industry is looking to mmWave bands (25–100 GHz) for 5G communications [3–5] where orders of magnitude more spectrum is available. However, shifting mobile communication services into the mmWave spectrum has its own challenges due to unfavorable channel

S. Chellappan et al. (Eds.): WASA 2018, LNCS 10874, pp. 296–307, 2018.
https://doi.org/10.1007/978-3-319-94268-1_25

characteristics of mmWave frequency such as large pathloss, which reduces service coverage and impairs communication performance [6, 7]. On the other hand, short wavelength of mmWave frequency allows implementing antenna arrays with a large number of elements in a reasonable size, which can form highly directional beams to combat the large propagation loss and increase throughput. Massive MIMO is thus considered as a key technology that will enables 5G communications in mmWave bands and achieve the needed increase in network capacity [8, 9].

Massive MIMO is different from conventional MIMO in terms of signal processing principles and hardware architecture. In massive MIMO systems, the number of antennas (hundreds) at the base station (BS) is much greater than the number of active mobile terminals (several or tens) [10, 11]. Advanced signal processing techniques can be used to leverage the large number of antennas, and concurrently generate multiple directional signal beams, termed beamforming. Each of the beams focuses a great amount of radiated signal energy on an intended mobile terminal (MT). Beamforming enables the BS to transmit/receive multiple signal beams simultaneously to/from multiple MTs on the same frequency channel with high array beamforming gains. The total throughput increases with the number of antennas at the BS.

One of the challenges in beamforming is how to allocate the transmit power to multiple beams, which determines the signal amplitude and throughput of each beam. Conventionally, beamforming algorithms at the physical layer aim to maximize the total throughput of multiple data beams [12, 13], given an allowed value of the total transmit power. However, MTs may have heterogeneous user traffic demands and experience different channel conditions. Beam transmit power allocation should take these factors into consideration. In addition, the statistics of users' packet arrivals and channel states may not be known a priori and vary over time. The power allocation policy should not be "one-shot" optimization, but it should achieve the optimal overall system performance in a long timescale. For example, one user has a good channel now, but may get a bad channel in the next time slot, it may be better to allocate more power to this user to transmit more data to it in the current time slot. Furthermore, latency is another key performance indicator because 5G will support new ultra-low latency communication applications, e.g. augmented reality (AR), connected autonomous vehicles, and real-time industrial control.

In this paper, we propose an optimization framework that allocates the transmit power to the multiple beams sent from a MMIMO base station to multiple MTs and schedules data transmissions in mmWave networks with time-varying traffic and channel conditions. The MMIMO power allocation and transmission scheduling optimization problem is formulated as a Markov decision process (MDP) with the objective to minimize the queueing delay of the MTs by taking into consideration their heterogeneous and dynamic traffic and channel states. An online machine learning (ML) scheme is designed which allows achieving the long-term optimal system performance with no requirement for a priori knowledge of user traffic statistics and wireless network states. Evaluation results show that our proposed scheme outperforms the baseline schemes.

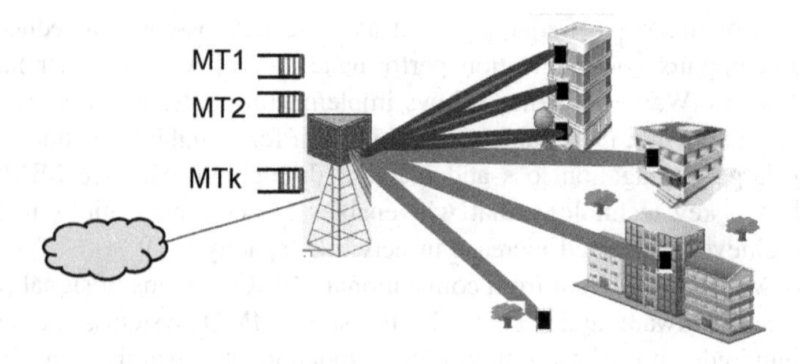

Fig. 1. Massive MIMO mmWave cellular network.

2 System Model

As illustrated in Fig. 1, we investigate a mmWave cellular network in which the BS is equipped with massive MIMO antennas and simultaneously transmits multiple signal beams to multiple MTs with heterogeneous and dynamic traffic and channel states. We concentrate on the downlink transmissions in this paper, nevertheless, the analysis can be extended to the uplink transmissions in a reverse way with the MMIMO BS receiving multiple data streams from the MTs.

We assume that the BS is equipped with M antennas and there are a set $\mathcal{K} = \{1, \ldots, K\}$ of single-antenna MTs in the network. The BS simultaneously transmits K data symbols, $\mathbf{s} = [s_1, s_2, \ldots, s_k, \ldots, s_K]^T$, on a mmWave channel, one symbol to a MT. Let y_k be the signal received by MT k. The signal vector $\mathbf{y} = [y_1, y_2, \ldots, y_k, \ldots, y_K]^T$ received by the MTs can be expressed by [14, 15]

$$\mathbf{y} = \mathbf{HWs} + \mathbf{n}, \tag{1}$$

where $\mathbf{W}$ is the $M \times K$ massive MIMO beamforming matrix, $\mathbf{H}$ is the $K \times M$ channel matrix, and $\mathbf{n}$ is noise and interference. The rows of the channel matrix, $\mathbf{H} = [\mathbf{h}_1, \ldots, \mathbf{h_k}, \ldots, \mathbf{h_K}]^T$, represent the channels or spatial signatures associated with each MT. Massive MIMO is based on the fact that for large M, simple receiver architectures yield near-optimal performance. In particular, the channels and the beams for MTs in distinct locations theoretically become orthogonal as the size of the array grows [16, 17]. Simple conjugate beamforming, also known as matched filter precoding with $\mathbf{W} = \mathbf{H}^H$, can asymptotically eliminate inter-user interference and suppress noise for large M, provided that the channel state information (CSI) for each MT can be acquired, that is, $\mathbf{W} = [\mathbf{w}_1, \ldots, \mathbf{w_k}, \ldots, \mathbf{w_K}]$ and $\mathbf{w_k} = \mathbf{h}_k^H$. The signal received by MT k is then

$$y_k = \mathbf{h}_k \mathbf{h}_k^H s_k + n_k, \tag{2}$$

The mmWave channel vector with a linear antenna array can be modelled as [18, 19],

$$\mathbf{h_k} = \frac{\beta_k}{1+b_k^e}[1, e^{-j\pi\varphi_k}, \ldots, e^{-j\pi(M-1)\varphi_k}], \tag{3}$$

where b_k is the distance between the BS and MT k, e is the pathloss exponent, φ_k is the normalized direction, and β_k is the fading attenuation coefficient. The distribution of β_k is not known a priori. Let $\theta_k = |\mathbf{h}_k \mathbf{h}_k^H|^2$ denote the channel state of MT k. The achievable data rate between the BS and MT k can then be expressed as [14, 20],

$$R_k = B \log_2 \left(1 + \frac{\alpha_k \mathcal{P}}{\sigma^2} \theta_k \right), \tag{4}$$

where $\mathcal{P}$ is the total transmitted power, B is the channel bandwidth, and σ^2 is the noise power. α_k, $k \in \mathcal{K}$ are the power allocation coefficients that are subject to the constraint,

$$\sum\nolimits_{k=1}^{K} \alpha_k \leq 1. \tag{5}$$

Let $\rho = \mathcal{P}/\sigma^2$ be the ratio between the transmit power and the noise power, that is, the transmit signal to noise ratio (SNR). We can see from Eq. (4) that the achievable data rate to MT k depends on the transmit power allocated to its signal beam, the distance between the BS and MT k, and the fading attenuation coefficient of its channel vector.

We consider that the time horizon is discretized into scheduling slots, each of which is of a fixed duration δ. Let $\theta_k^t \in \Theta$ denote the channel state of MT k at slot t, where Θ is the set of possible channel states for a MT. The channel coherence time is typically much longer than a slot duration in practical cellular scenarios. We thus assume a block-fading model for the channels of the MTs [21, 22], with which the state of a channel experienced by a MT is identical during the period of each scheduling slot. Then θ_k^t is quasi-static within each slot, but it changes between different slots. We can model the channel state transition across the time horizon as a finite-state Markov chain. However, the state transition probability is not known a priori. The achievable data rate R_k^t between the BS and MK k at time slot t can be obtained from (4). We can observe that given θ_k^t, the data rate to MT k in slot t, R_k^t is a strictly monotonically increasing function of the transmit power $\alpha_k^t \mathcal{P}$. The BS will allocate the transmit power of MMIMO beams and schedule data transmissions to the MTs slot by slot.

Let A_k^t be the number of new packets randomly arrived at the BS in time slot t for MT k. Assume the packet size is L. The packet arrival process is assumed to be independent among the MTs and independently and identically distributed (i.i.d.) across scheduling slots. However, the distribution of A_k^t is also not known a priori. There is a downlink transmission queue maintained for each MT at the BS. The arrived packets get queued until transmission. Let Q_k^t denote the queue length in the number of packets for MT k at the BS at the beginning of scheduling slot t. The queue evolution of MT k can be expressed as

$$Q_k^{t+1} = Q_k^t + A_k^t - C_k^t, \tag{6}$$

where C_k^t is the number of packets that are transmitted from the BS to MT k in time slot t, which is,

$$C_k^t = \min\left\{Q_k^t, \frac{\delta R_k^t}{L}\right\}. \tag{7}$$

3 Problem Formulation

The network state at each time slot t can be represented by $\boldsymbol{x}^t = \{\boldsymbol{x}_k^t : k \in \mathcal{K}\}$, where $\boldsymbol{x}_k^t = (Q_k^t, \theta_k^t) \in \chi$ denotes the state for a MT k that encapsulates the queue state and local channel state. We want to design a power allocation control policy $\boldsymbol{\alpha} = \{\alpha_k, k \in \mathcal{K}\}$. At the beginning of slot t, based on the observation of the network state $\boldsymbol{x}^t$, the BS strategically decides the power allocation to transmit K signal beams to K MTs according to the control policy, that is, $\boldsymbol{\alpha}(\boldsymbol{x}^t) = \{\alpha_k(\boldsymbol{x}^t), k \in \mathcal{K}\}$. The power allocation $\alpha_k(\boldsymbol{x}^t)$ to a beam k determines the amount of data sent to MT k in time slot t based on Eq. (4).

Stochastic user data arrivals and dynamic network states present challenges and make naïve one-shot system optimization schemes unstable and unable to achieve the optimal performance on a long timescale. Therefore, we want to develop an optimization framework that minimizes the expected long-term delay of multiple MTs. Let $d(\boldsymbol{x}^t, \boldsymbol{\alpha}(\boldsymbol{x}^t)) = \sum_{k \in \mathcal{K}} d_k(\boldsymbol{x}_k^t, \alpha_k(\boldsymbol{x}^t)/K$ denote the average delay of K MTs at time slot t with $d_k(\boldsymbol{x}_k^t, \alpha_k(\boldsymbol{x}^t)$ being the delay of MT k, which is a function of the network state and power allocation performed by the BS. The expected long-term delay of multiple MTs depends on the power allocation policy, which can be expressed as

$$D(\boldsymbol{\alpha}) = E_{\boldsymbol{\alpha}}\left[\lim_{T\to\infty}\frac{1}{T}\sum\nolimits_{t=1}^{T} d(\boldsymbol{x}^t, \boldsymbol{\alpha}(\boldsymbol{x}^t)\right] = E_{\boldsymbol{\alpha}}\left[\lim_{T\to\infty}\frac{1}{TK}\sum\nolimits_{t=1}^{T}\sum\nolimits_{k\in\mathcal{K}} d_k\left(\boldsymbol{x}_k^t, \alpha_k(\boldsymbol{x}^t)\right)\right] \tag{8}$$

The expectation is over the stochastic network states and the BS's decision of the transmit power for each of the beams $k \in \mathcal{K}$ induced by the power allocation control police $\boldsymbol{\alpha}$. In other words, the expected long-term delay depends on the network state and the control policy. The long-term average delay of a stable queue is equal to the long-term average queue length divided by the average packet arrival rate according to Little's law [23]. Therefore, the long-term average delay is,

$$D(\boldsymbol{\alpha}) = E_{\boldsymbol{\alpha}}\left[\lim_{T\to\infty}\frac{1}{TK}\sum\nolimits_{t=1}^{T}\sum\nolimits_{k\in\mathcal{K}}\frac{Q_k\left(\boldsymbol{x}_k^t, \alpha_k(\boldsymbol{x}^t)\right)}{\lambda_k}\right], \tag{9}$$

where λ_k is the average data arrival rate of MT k. The objective is to design an optimal power allocation control policy $\boldsymbol{\alpha}^*$ that minimizes the expected long-term delay, which can be formulated as

$$\boldsymbol{\alpha}^* = \underset{\boldsymbol{\alpha}}{\operatorname{argmin}}\, E_{\boldsymbol{\alpha}}\left[\lim_{T\to\infty}\frac{1}{TK}\sum\nolimits_{t=1}^{T}\sum\nolimits_{k\in\mathcal{K}}\frac{Q_k\left(\boldsymbol{x}_k^t, \alpha_k(\boldsymbol{x}^t)\right)}{\lambda_k}\right],$$
$$\text{s.t.} \sum\nolimits_{k\in\mathcal{K}}\alpha_k \leq 1. \tag{10}$$

In addition, the power allocation policy $\boldsymbol{\alpha} = \{\alpha_k : k \in \mathcal{K}\}$ induces a probability distribution over a sequence of network states $\boldsymbol{x}^t$ along a series of time slots, which can be modeled by a controlled Markov chain with the following state transition probability,

$$\Pr\{\boldsymbol{x}^{t+1}|\boldsymbol{x}^t, \boldsymbol{\alpha}(\boldsymbol{x}^t)\} = \prod\nolimits_{k\in\mathcal{K}}\Pr\{Q_k^{t+1}|Q_k^t, \boldsymbol{\alpha}(\boldsymbol{x}^t)\} \times \prod\nolimits_{k\in\mathcal{K}}\Pr\{\theta_k^{t+1}|\theta_k^t\}. \tag{11}$$

The optimization problem in (10) is in general a Markov decision process (MDP) problem with the average cost criterion.

Let $V_{\boldsymbol{\alpha}}(\boldsymbol{x})$ be the value function for a network state $\boldsymbol{x}$ under a control policy $\boldsymbol{\alpha}$. The optimal state value function, which is given by $V(\boldsymbol{x}) = V_{\boldsymbol{\alpha}*}(\boldsymbol{x}), \forall\boldsymbol{x}$, can be achieved by solving a Bellman's optimality equation as follows [24],

$$V(\boldsymbol{\chi}) = \min_{\boldsymbol{\alpha}}\{d(\boldsymbol{\chi}, \boldsymbol{\alpha}(\boldsymbol{x})) + \sum\nolimits_{\boldsymbol{\chi}'} P_r(\boldsymbol{\chi}'|\boldsymbol{\chi}, \boldsymbol{\alpha}(\boldsymbol{x}))V(\boldsymbol{\chi}')\} - v, \tag{12}$$

where $v = D(\boldsymbol{\alpha}^*)$ is the optimal expected long-term delay of multiple MTs, $\boldsymbol{\chi}' = \{(Q_k', \theta_k') : k \in \mathcal{K}\}$ represents the subsequent network state, and $d(\boldsymbol{\chi}, \boldsymbol{\alpha}(\boldsymbol{x})) = \sum_{k\in K} Q_k/\lambda_k K$.

Solving (12) is generally a very challenging problem. The conventional methods to solve the MDP problem are based on value and policy iteration [24, 25]. However, these methods require for a complete knowledge of the network state transition probabilities and the packet arrival statistics which are not known a priori in our scenario. In addition, the size of the network state space is in the order of $\prod_{k\in\mathcal{K}}(|Q_k| \times |\theta_k|)$, where $|\,.\,|$ means the cardinality of the set. The network state space grows exponentially as the number of MTs increases. The conventional methods suffer from exponential computation complexity due to the huge network state space even with a reasonable number of MTs.

4 Power Allocation Optimization Algorithm

In this section, we focus on developing a practically efficient scheme with low-complexity for achieving a near optimal power allocation and transmission policy with no requirement for a priori knowledge of statistical network state transition and packet arrival information by exploring several techniques including online reinforcement learning, state approximation, and decomposition. First, we define a post-decision state for each scheduling slot [26]. The packet arrivals are independent of the power allocation decision. At the current scheduling slot, the post-decision state is defined as $\tilde{\mathbf{x}} = \{(\tilde{Q}_k, \tilde{\theta}_k) : k \in \mathcal{K}\}$, where $\tilde{Q}_k = Q_k - C_k$ and $\tilde{\theta}_k = \theta_k$, $\forall k \in \mathcal{K}$. The optimal state-value function satisfying (12) can hence be expressed by,

$$V(\boldsymbol{\chi}) = \min_{\boldsymbol{\alpha}}\{d(\boldsymbol{\chi}, \boldsymbol{\alpha}(\boldsymbol{x})) + \tilde{V}(\tilde{\boldsymbol{\chi}})\}, \tag{13}$$

where $\tilde{V}(\tilde{\boldsymbol{\chi}})$ is the optimal post-decision state-value function that satisfies the Bellman's optimality equation,

$$\tilde{V}(\tilde{\boldsymbol{\chi}}) = \sum\nolimits_{\boldsymbol{\chi}'} \Pr(\boldsymbol{\chi}'|\tilde{\boldsymbol{\chi}}) V(\boldsymbol{\chi}') - v. \tag{14}$$

The optimal state-value function can be obtained from the optimal post-decision state-value function by performing minimization over all feasible power allocation decision. The optimal power allocation policy is thus expressed as

$$\boldsymbol{\alpha}^*(\boldsymbol{x}) = \underset{\boldsymbol{\alpha}}{\operatorname{argmin}}\left(d(\boldsymbol{\chi}, \boldsymbol{\alpha}(\boldsymbol{x})) + \tilde{V}(\tilde{\boldsymbol{\chi}})\right). \tag{15}$$

The channel states and packet arrival statistics for MTs are independent each other. We can hence decompose the optimal post-decision state-value function. Mathematically,

$$\tilde{V}(\tilde{\boldsymbol{\chi}}) = \sum\nolimits_{k \in K} \tilde{V}_k\left(\tilde{Q}_k, \tilde{\theta}_k\right). \tag{16}$$

Given the optimal power allocation control policy $\boldsymbol{\alpha}^* = \left\{\alpha_k^* : k \in \mathcal{K}\right\}$, the post-decision state-value function for MT $k \in \mathcal{K}$, $\tilde{V}_k\left(\tilde{Q}_k, \tilde{\theta}_k\right)$ satisfies,

$$\tilde{V}_k\left(\tilde{Q}_k, \tilde{\theta}_k\right) = \sum\nolimits_{\theta_k', A_k} P_r\left(\theta_k' | \tilde{\theta}_k\right) P_r(A_k) V_k(Q_k', \theta_k') - v_k, \tag{17}$$

where v_k is the optimal long-term average delay for MT k. The optimal per-MT state-value function can be expressed as,

$$V_k(Q_k', \theta_k') = \frac{Q_k'}{K\lambda_k} + \tilde{V}_k\left(\tilde{Q}_k', \tilde{\theta}_k'\right), \tag{18}$$

where $\tilde{Q}_k'$ and $\tilde{\theta}_k'$ are the post-decision queue and channel states of MT k at the subsequent scheduling slot.

Through the proposed linear decomposition of the post-decision state-value function, the problem to solve a complex post-decision Bellman's optimality Eq. (14) is broken into much simpler MDPs. The linear decomposition approach is a special case of the feature-based decomposition method [27] but provides a near-optimal approximation of the state-value functions [27, 28]. Additionally, in order to deploy a power allocation policy based on the network state $\boldsymbol{x} = \{\boldsymbol{x}_k : k \in \mathcal{K}\}$ with $\boldsymbol{x}_k = (Q_k, \theta_k)$, the base station has to keep the state-value function with $\prod_{k \in \mathcal{K}} (|Q_k| \times |\theta_k|)$ values. Using the linear decomposition, only $K \times |Q_k| \times |\theta_k|$ values need to be stored.

The channel states during the next scheduling slot and the number of packet arrivals at the end of the current slot are unavailable beforehand. Instead of directly calculating the post-decision state value functions in (17), we propose an online machine learning algorithm to approach $\tilde{V}_k\left(\tilde{Q}_k, \tilde{\theta}_k\right)$, $\forall k \in \mathcal{K}$. Based on the observations of the network state $\boldsymbol{x}_k^t = \left(Q_k^t, \theta_k^t\right)$, the number of packets departures C_k^t, the number of packet arrivals A_k^t at the current scheduling slot t and the resulting network state $\boldsymbol{x}_k^{t+1} = \left(Q_k^{t+1}, \theta_k^{t+1}\right)$ at the next slot $t+1$, the post-decision state value function can be updated on the fly,

$$\tilde{V}_k^{t+1}\left(\tilde{Q}_k^t, \tilde{\theta}_k^t\right) = (1-\varepsilon^t)\tilde{V}_k^t\left(\tilde{Q}_k^t, \tilde{\theta}_k^t\right) + \varepsilon^t\left(V_k^t(Q_k^{t+1}, \theta_k^{t+1}) - \tilde{V}_k^t\left(\tilde{Q}_k^{(r)}, \tilde{\theta}_k^{(r)}\right)\right), \quad (19)$$

where $\varepsilon^t \in [0, 1)$ is the learning rate, $\tilde{Q}_k^{(r)} \text{and} \tilde{\theta}_k^{(r)}$ are the reference queue and channel states for MT k, respectively. The power allocation of the MMIMO beams, $\boldsymbol{\alpha}^t = \left\{\alpha_k^t, k \in \mathcal{K}\right\}$ at scheduling slot t is determined as,

$$\begin{aligned} \boldsymbol{\alpha}^t &= \underset{\boldsymbol{\alpha}}{\operatorname{argmin}} \sum\nolimits_{k \in K} \left(\frac{\tilde{Q}_k^t}{\lambda_k K} + \tilde{V}_k^t\left(\tilde{Q}_k^t, \tilde{\theta}_k^t\right)\right), \\ \text{s.t.} & \sum\nolimits_{k \in K} \alpha_k^t \leq 1. \end{aligned} \quad (20)$$

The online machine learning algorithm for estimating the optimal post-decision state value function and determining the optimal power allocation policy is summarized in Algorithm 1.

Algorithm 1. Online Learning Algorithm for Optimal Post-Decision State Value Function

1. **Initialize** the post-decision state value function $\tilde{V}_k^t(\tilde{Q}_k, \tilde{\theta}_k)$, $\forall \tilde{Q}_k, \forall \tilde{\theta}_k$, and $\forall k \in \mathcal{K}$ for $t = 1$.
2. **Repeat**
3. At the beginning of scheduling slot t, the BS observes the network state $\boldsymbol{x}^t = \{\boldsymbol{x}_k^t : k \in \mathcal{K}\}$ with $\boldsymbol{x}_k^t = (Q_k^t, \theta_k^t)$, and determines the transmit power allocation of MMIMO beams $\boldsymbol{\alpha}^t = \{\alpha_k^t, k \in \mathcal{K}\}$ according to (20).
4. After transmitting the packets for the MTs using the allocated power, the BS observes the post-decision state, $\tilde{\boldsymbol{x}}^t = \{\tilde{\boldsymbol{x}}_k^t : k \in \mathcal{K}\}$, where $\tilde{\boldsymbol{x}}_k^t = \left(\tilde{Q}_k^t, \tilde{\theta}_k^t\right)$ with $\tilde{Q}_k^t = Q_k^t - C_k^t$ and $\tilde{\theta}_k^t = \theta_k^t$.
5. With A_k^t, $\forall k \in \mathcal{K}$ new packets arrived at the end of slot t, the network state transits to $\boldsymbol{x}^{t+1} = \{\boldsymbol{x}_k^{t+1} : k \in \mathcal{K}\}$ with $\boldsymbol{x}_k^{t+1} = (\tilde{Q}_k^t + A_k^t, \theta_k^{t+1})$ at the following scheduling slot $t+1$
6. The BS calculates $V_k^t(Q_k^{t+1}, \theta_k^{t+1})$, $\forall k \in \mathcal{K}$ according to (18), and updates the post-decision state value function $\tilde{V}_k^{t+1}(\tilde{Q}_k^t, \tilde{\theta}_k^t)$ according to (19).
7. The scheduling slot index is updated by $t \leftarrow t+1$.
8. **Until** a predefined stopping condition is satisfied.

5 Evaluation Results

In this section, we evaluate the performance of the proposed machine learning-based MMIMO beam power allocation scheme. For the purpose of performance comparison, the following three baseline schemes are also simulated.

(1) Baseline 1: The BS allocates equal transmit power to the MMIMO beams;
(2) Baseline 2: The BS allocates the power to the MMIMO beams for maximizing the total data rate at each time slot t without considering the queue states of the MTs, that is, $\max_{\alpha} \sum_{k \in \mathcal{K}} R_k^t$.
(3) Baseline 3: At each time slot t, the BS allocates the power to the MMIMO beams for minimizing the total queue of the MTs at the end of time slot t, that is, $\min_{\alpha} \sum_{k \in \mathcal{K}} (Q_k^t - C_k^t)$.

We simulated multiple network scenarios with different number of MTs, MT locations, traffic, transmit power, and other system parameters. Due to the page limit, we present the results for a typical setting to compare the performance of the proposed MMIMO beam allocation algorithm to that of the baselines and gain the insights. In the simulations, the mmWave channel between the BS and a MT is modeled with three states characterizing the line-of-sight (LOS), the non-LOS (NLOS), and the blocking. The channel states $\theta_k^t, \forall k \in \mathcal{K}$ of different MTs are independent and evolve according to a Markov chain model [29]. The study of the system performance with a more accurate mmWave channel model is part of our future work. We assume that the packet arrivals for a MT $k \in \mathcal{K}$ follows a Poisson arrival process with an average arrival rate of λ_k packets per second. We choose the learning rate as $\varepsilon^t = \varepsilon_0/(\log t + 1)$ with $\varepsilon_0 = 0.6$.

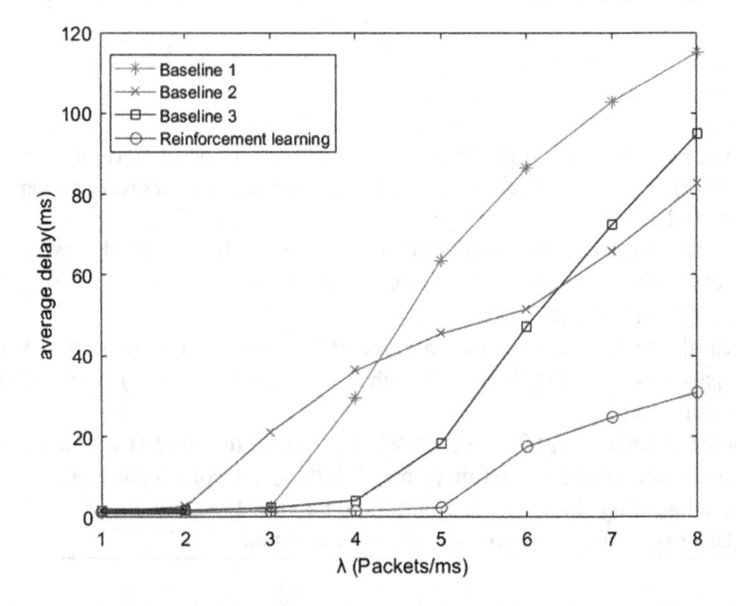

Fig. 2. Average delay versus average packet arrival rate for the proposed reinforcement learning-based scheme and baselines.

Figure 2 shows the average delay of three MTs as the total packet arrival rates changes. The three MTs locate in a cell with 43 m, 35 m, and 20 m from the MMIMO BS, and their mean Poisson packet arrival rates are λ_1, $\lambda_2 = 3\lambda_1$, and $\lambda_3 = 6\lambda_1$, respectively. Other system parameters are set as follows, the total transmit power is 1 W, the channel bandwidth is 500 MHz, the pathloss exponent e is 3, the transmit signal to noise ratio (SNR) ρ is 60 dB, the number of antennas at BS is 128, the slot duration is 1 ms, and the packet size is 1000 bytes. The curves indicate that our proposed reinforcement learning based scheme outperforms all the three baseline schemes. The reason is that with the baseline schemes, the BS makes shortsighted power allocation decisions. Using our proposed scheme, the BS not only consider the current transmission performance but also takes into account the performance in the future when making the power allocation to the MMIMO beams and determining the number of packets to send to multiple MTs.

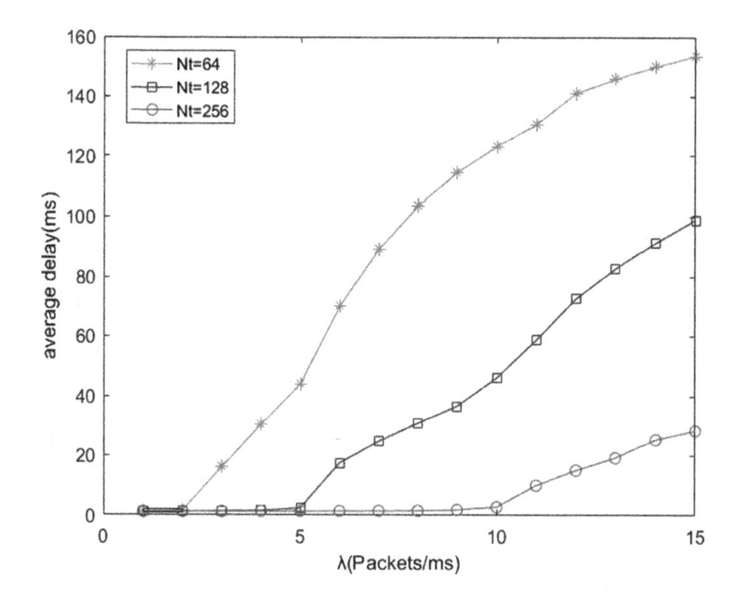

Fig. 3. Average delay versus average packet arrival rate for different number of antennas at the MMIMO BS.

Figure 3 shows the effects of the number of antennas at the MMIMO base station as the proposed reinforcement learning-based MMIMO beam power allocation scheme is used. When the number of antennas is larger, the antenna gains will be higher, so is the achievable data rate. Thus, the average delay can be lower.

6 Conclusions

This paper proposes a framework to optimize multi-beam power allocation in massive MIMO mmWave networks with time-varying traffic and channel conditions. More particularly, the optimization problem is formulated as a Markov decision process (MDP) with the objective to minimize the overall queueing delay of multiple MTs by taking into consideration of their heterogeneous and dynamic traffic and channel states. By decomposing the post-decision state value function, a low-complexity online reinforcement learning scheme is developed to achieve the optimal power allocation policy. The proposed scheme does not need a priori knowledge of wireless channel state transition probabilities and user packet arrival statistics. The evaluation results show that our proposed scheme outperforms the baseline schemes.

References

1. The Spectrum Crunch. https://www.nist.gov/ctl/spectrum-crunch
2. Assessment of the Global Mobile Broadband Deployments and Forecasts for International Mobile Telecommunications, ITU-R. Technical report M.2243 (2015)
3. Rappaport, T., Sun, S., Mayzus, R., Zhao, H., Azar, Y., Wang, K., Wong, G.N., Schulz, J. K., Samimi, M., Gutierrez, F.: Millimeter wave mobile communications for 5G cellular: it will work. IEEE Access. **1**, 335–349 (2013)
4. Bleicher, A.: Millimeter waves may be the future of 5G phones. IEEE Spectr. **50**(6), 11–12 (2013)
5. Boccardi, F., Heath, R., Lozano, A., Marzetta, T., Popovski, P.: Five disruptive technology directions for 5G. IEEE Commun. Mag. **52**(2), 74–80 (2014)
6. Bai, T., Desai, V., Heath Jr., R.W.: Millimeter wave cellular channel models for system evaluation. In: International Conference on Computing, Networking and Communication (2014)
7. Rappaport, T., Gutierrez, F., Ben-Dor, E., Murdock, J., Qiao, Y., Tamir, J.: Broadband millimeter-wave propagation measurements and models using adaptive-beam antennas for outdoor urban cellular communications. IEEE Trans. Antennas Propag. **61**(4), 1850–1859 (2013)
8. Larsson, E.G., Edfors, O., Tufvesson, F., Marzetta, T.: Massive MIMO for next generation wireless systems. IEEE Commun. Mag. **52**(2), 186–195 (2014)
9. Roh, W., Seol, J.-Y., Park, J., Lee, B., Lee, J., Kim, Y., Cho, J., Cheun, K., Aryanfar, F.: Millimeter-wave beamforming as an enabling technology for 5G cellular communications: theoretical feasibility and prototype results. IEEE Commun. Mag. **52**(2), 106–113 (2014)
10. Rusek, F., Persson, D., Lau, B.K., Larsson, E., Marzetta, T., Edfors, O., Tufvesson, F.: Scaling up MIMO: opportunities and challenges with very large arrays. IEEE Signal Process. Mag. **30**(1), 40–60 (2013)
11. Hoydis, J., ten Brink, S., Debbah, M.: Massive MIMO in the UL/DL of cellular networks: how many antennas do we need: IEEE. J. Sel. Areas Commun. **31**(2), 160–171 (2013)
12. Kim, T., et al.: Tens of Gbps support with mmWave beamforming systems for next generation communications. In: IEEE Globecom (2013)
13. Yin, B., et al.: High-throughput beamforming receiver for millimeter wave mobile communication. In: IEEE Globecom (2013)

14. Biglieri, E., et al.: MIMO Wireless Communications. Cambridge University Press, Cambridge (2007)
15. Ngo, H., Larsson, E., Marzetta, T.: Energy and spectral efficiency of very large multiuser MIMO systems. IEEE Trans. Commun. **61**, 1436–1449 (2013)
16. Marzetta, T.: Noncooperative cellular wireless with unlimited numbers of base station antennas. IEEE Trans. Wirel. Commun. **9**(11), 3590–3600 (2010)
17. Lu, L., Li, G., Swindlehurst, A., Ashikhmin, A., Zhang, R.: An overview of massive MIMO: benefits and challenges. IEEE J. Sel. Top. Signal Process. **8**(5), 742–758 (2014)
18. Ding, Z., Dai, L., Schober, R., Poor, V.: NOMA meets finite resolution analog beamforming in massive MIMO and millimeter-wave networks. IEEE Wirel. Commun. Lett. **21**(8), 1879–1882 (2017)
19. Bjornson, E., Sanguinetti, L., Hoydis, J., Debbah, M.: Designing multi-user MIMO for energy efficiency: when is massive MIMO the answer. In: IEEE Wireless Communication and Networking Conference (2014)
20. Yang, H., Marzetta, T.: Performance of conjugate and zero-forcing beamforming in large-scale antenna systems. IEEE J. Sel. Areas Commun. **31**(2), 172–179 (2013)
21. Yang, W., Durisi, G., Koch, T., Polyanskiy, Y.: Block-fading channels at finite block-length. In: Proceedings 10th International Symposium on Wireless Communication Systems, Ilmenau, Germany (2013)
22. Lau, V., Liu, Y., Chen, T.: On the design of MIMO block-fading channels with feedback-link capacity constraint. IEEE Trans. Commun. **52**(1), 62–70 (2004)
23. Bertsekas, D.P., Gallager, R.G.: Data Networks. Prentice Hall, Upper Saddle River (1987)
24. Puterman, M.L.: Markov Decision Processes. Wiley, Hoboken (1994)
25. Burnetas, A., Katehakis, M.: Optimal adaptive policies for markov decision processes. Math. Oper. Res. **22**(1), 222–255 (1997)
26. Chen, X., Liu, P., Liu, H., Wu, C., Ji, Y.: Multipath transmission scheduling in millimeter wave cloud radio access networks. In: Proceedings of IEEE ICC 2018, Kansas City, MO (2018)
27. Tsitsiklis, J.N., Van Roy, B.: Feature-based methods for large scale dynamic programming. Mach. Learn. **22**(1), 59–94 (1996)
28. Sutton, R.S., Barto, A.G.: Reinforcement Learning: An Introduction. MIT Press, Cambridge (1998)
29. Mezzavilla, M., Goyal, S., Panwar, S., Rangan, S., Zorzi, M.: An MDP model for optimal handover decisions in mmWave cellular networks. In: EuCNC, Athens, Greece (2016)

A Detection-Resistant Covert Timing Channel Based on Geometric Huffman Coding

Jianhua Liu, Wei Yang(✉), Liusheng Huang, and Wuji Chen

School of Computer Science and Technology,
University of Science and Technology of China, Hefei 230027, China
qubit@ustc.edu.cn

Abstract. Network covert timing channel is a communication mechanism that transfers secret messages by modulating the timing characteristics of network traffic. It is targeted for secret information transmission on networks which can ensure security and confidentiality. However, most proposed covert timing channels can be detected by several detection methods such as regularity testing, distribution shape testing, entropy-based testing and recent machine learning based methods. In this paper, we design and implement a novel covert timing channel by leveraging Geometric Huffman Coding (GHC) to realize covert and overt channel matching. In network experiments and simulations, it is demonstrated that the proposed channel is undetectable against not only the traditional detection methods but also the latest machine learning based methods. Meanwhile, it maintains a reasonable transmission capacity of 2.25 bits/packet much higher than binary channels.

Keywords: Covert timing channels · Detection-resistant · Security

1 Introduction

Covert channel was originally defined by Lampson in 1973 as channels "not intended for information transfer at all, such as the service program's effect on system load" [9]. A covert channel aims to conceal the very existence of secret communication to any third party or eavesdropper which is essential to secure data transmission. Nowadays, it is widely used for secret information transmission on networks.

Two general categories of covert channels have existed since the pioneering work of Lampson: covert storage channels (CSCs) and covert timing channels (CTCs). In a CSC, one side directly or indirectly writes to a particular storage location and another side reads from that location. For example, the sender transmits secret data to the receiver by modifying unused/free bits in the packet header [12,13,19]. However, most CSCs prove to be detected by checking specific fields [6]. On the other hand, a CTC usually embeds secret messages into the

S. Chellappan et al. (Eds.): WASA 2018, LNCS 10874, pp. 308–320, 2018.
https://doi.org/10.1007/978-3-319-94268-1_26

timing characteristics of network traffic such as modulating the message into inter-packet delays (IPDs). Similarly, several methods have been proposed to disrupt or detect CTCs. For disruption, network jammers are used by modifying all traffic passing through it regardless of whether it is covert or not [8]. However, it will unfairly punish innocent traffic as well which leads to waste of computing resources. For detection, it primarily leverages statistical tests to distinguish covert traffic from legitimate traffic. Up to now, the known detection methods can be categorized into four broad classes: shape testing, regularity testing, entropy-based testing and the latest machine learning based method. Most proposed CTCs suffer from weak non-detectability against above detection methods. Although recently some CTCs are trying to evade some detections but they either fail one specific testing such as entropy-based testing or have a low transmission capacity such as binary channels. The main contributions of this paper can be summarized as follows:

- We systematically design a novel CTC that is resistant to the latest detection methods including entropy based tests. Our method utilizes Geometric Huffman Coding (GHC) to achieve covert and overt channel matching by minimizing the divergence between the covert traffic distribution and legitimate traffic distribution.
- Compared to binary channels, our CTC encodes 2-bit binary code or 3-bit binary code in one interval according to a specific prefix-free matcher which results in a higher transmission capacity.
- Our approach is validated by testing the non-detectability of our CTC using a large set of HTTP traces collected on our institute network and the results are compared with three other CTCs. The experimental results show that our method outperforms several other CTCs in non-detectability against the latest detection methods.

2 Related Work and Motivation

In this section, we firstly present a brief introduction of several network traffic based CTCs. Then, we discuss the existing detection methods for CTCs and their performance. After that, we give a brief introduction of GHC and our motivation.

2.1 Covert Timing Channels

To date, many methods designing and constructing CTCs have been proposed.

In [6], the authors first designed and implemented an IP timing channel based on a simple *on-off switching* technique. The two parties agree on an interval time and the reception and absence of a packet within the interval represent bit '1' and bit '0' respectively. However, strict synchronization mechanism is required between two parties and the capacity is relatively low since each packet signals only one bit.

In [16], the authors proposed a covert timing channel called JitterBug by hiding covert bits in the form of jitter in normal keystroke traffic. JitterBug operates by creating delays in keypresses to change inter-packet arrival time of keystrokes. It works even in the absence of synchronization since it is based on IPDs.

The authors proposed a covert timing channel, called *Time Replay channel* to evade detection in [5]. The IPDs of the covert channel are set to mimic a previously recorded overt traffic which is divided into two bins separated by the media. A random delay value from the higher bin is selected to embed '1' or from the lower bin to embed '0'. However, the undetectability of this channel is not experimentally confirmed, let alone against all known detection methods.

Compared to binary channels, the authors improved the covert transmission capacity by leveraging a *L-bits to n-packets* scheme in [15]. This method works by mapping binary strings of length L to multiple packet transmission times of size n. It is able to achieve a rate of 5 bits/sec for telnet traffic.

Liu *et al.* fulfilled undetectability by using a model-based modulation scheme that approximated any legitimate traffic distribution in [11]. However, the security of this CTC is only experimentally verified against regularity and shape test without considering entropy-based tests.

A detection-resistant covert timing channel based on IPD shaping, named Liquid, was proposed in [18]. According to the paper, half of IPDs are used for shaping and half for transmitting hidden message. Thus, Liquid only achieves the transmission capacity of 1.4 bits/s which is quite low.

Another covert timing channel aims to evade detection by leveraging distribution matching in [10]. It also suffers from the same disadvantage of low transmission capacity since it is just a binary covert channel.

2.2 Detection Tests

Detection methods can be broadly categorized into three classes: shape tests, regularity tests and entropy-based tests. Shape tests describe the first-order statistics such as mean, variance, and distribution including Kolomorov-Smirnov (KS) Test and Kullback-Leibler (KL) Divergence Test. Regularity tests represent the second or higher-order statistics such as correlations in the data. As for entropy-based tests, they provide a measure of uncertainty in a process. To the best of our knowledge, the latest detection method against CTCs proposed in [17] is a Support Vector Machine-Based framework which utilizes machine learning method to distinguish covert traffic samples from overt traffic samples.

(1) Kolomorov-Smirnov Test

The KS test is a nonparametric method which is widely used to determine whether one sample is drawn from the reference distribution or two samples are drawn from the same distribution since it is sensitive to differences in both location and shape of the empirical cumulative distribution functions of the two samples. It is leveraged as a method to distinguish covert traffic from legitimate traffic in [1]. The test score is defined as:

$$KSD = \max_{x} |F_1(x) - F_2(x)| \tag{1}$$

In our test, $F_1(x)$ and $F_2(x)$ are two empirical cumulative probability distributions of the IPDs of the two traffic samples being compared.

(2) Kullback-Leibler Divergence Test

In [1,2], KL divergence measure was utilizedd as a detection metric for CTCs. It is a measure of how one probability distribution diverges from a second expected probability distribution. The KL divergence from Q to P is often denoted as $D_{KL}(P||Q)$:

$$D_{KL}(P||Q) = \sum_{i} P(i) \log \frac{P(i)}{Q(i)} \tag{2}$$

In our test, $P(i)$ and $Q(i)$ are two probability mass functions of a covert traffic sample and an overt traffic sample respectively.

(3) Regularity Test

The regularity measures the variation of a process within the stream itself. A method of detecting covert timing channels was investigated in [6] by Cabuk *et al.* In detail, the traffic is separated into non-overlapping windows of size w and for each window, the standard deviation is computed. Then, the regularity is obtained as the standard deviation of all pairwise differences.

$$regularity = STDEV(\frac{|\sigma_i - \sigma_j|}{\sigma_i}), i < j, \forall i, j \tag{3}$$

In our test, to calculate regularity for each sample of 2000 IPDs, it is divided into 20 windows with size 100.

(4) Entropy-based Test

Gianvecchio *et al.* proposed a new entropy-based approach to detect various covert timing channels in [7]. The entropy presents a measure of uncertainty in a process. It is defined as:

$$H(X_1, \ldots, X_m) = - \sum_{X_1, \ldots, X_m} P(X_1, \ldots, X_m) \log P(X_1, \ldots, X_m) \tag{4}$$

Due to the limited sample size, the entropy rate is estimated with the corrected conditional entropy, which was first used on biological processes [14]. According to the paper, the corrected conditional entropy is defined as:

$$CCE(X_m|X_1, \ldots, X_{m-1}) = H(X_m|X_1, \ldots, X_{m-1}) + perc(X_m)EN(X_1) \tag{5}$$

Here, $H(X_m|X_1,...,X_{m-1})$ is the conditional entropy which is defined as:

$$H(X_m|X_1,...,X_{m-1}) = H(X_1, X_2,...,X_m) - H(X_1, X_2,...,X_{m-1}) \quad (6)$$

and $perc(X_m)$ represents the percentage of unique patterns of length m, $EN(X_1)$ is the first order entropy with m fixed at 1. In our test, the probability density functions are substituted with empirical probability density functions based on the method of histograms. The data is binned into 5 equiprobable bins recommended by the authors in [7].

(5) SVM-based Test

The latest detection method against CTCs is a Support Vector Machine-Based framework. It comprehensively exploits four characteristics to distinguish covert traffic samples from overt traffic samples which are the KS test score, regularity score, entropy score and corrected conditional entropy score which are explained above.

2.3 GHC and Our Motivation

Geometric Huffman Coding was first proposed in [3] to realize matching dyadic distributions to channels by minimizing the KL distance between a dyadic probability mass functions (PMF) and a weighted version of the capacity achieving PMF. A non-negative vector is defined as a vector with non-negative real entries and at least one positive entry. The algorithm aims to approximate x by a dyadic PMF by solving the matching problem

$$\min_{dyadic\ d} D(d||x) \quad (7)$$

In this equation, $D(d||x)$ is the Kullback-Leibler divergence from x to d. The GHC is similar to Huffman Coding which should be familiar to readers but with different updating rules [4]. Suppose in some step, x has n entries and is sorted in descending order. The two updating rules are as follows:

- Rule 1: Updating x

  The two smallest entries x_n and x_{n-1} are replaced by $x^{'}$ with the following rule.

$$x^{'} = \begin{cases} x_{n-1} & if\ x_{n-1} \geq 4x_n \\ 2\sqrt{x_{n-1}x_n} & if\ x_{n-1} < 4x_n \end{cases}$$

- Rule 2: Updating the binary tree

  The entries 1 to n correspond to n root nodes of n trees.
  If $x_{n-1} \geq 4x_n$: remove the whole tree that ends at node n and associate probability zero with the leaves.
  If $x_{n-1} < 4x_n$: join node n and node $n-1$ in a parent node.

Finally, $d = GHC(x)$ where d is the dyadic PMF that is induced by the prefix free tree constructed by GHC.

Inspired by this, we leverage GHC to achieve covert and overt channel matching by minimizing the KL divergence from overt traffic sample to covert traffic based on a binning strategy. Thus, the proposed CTC can evade KL test and entropy-based test. In addition, the modulated IPDs can be changed dynamically by PRNG (pseudorandom number generator) to deal with regularity test.

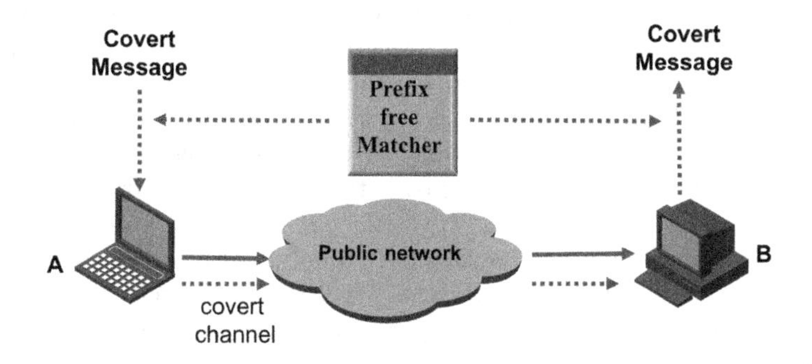

Fig. 1. Covert timing channel prototype

3 Covert Timing Channel Design

In this section, we firstly describe the prototype of our CTC. Then, the construction of the essential prefix-free matcher is demonstrated. After that, we introduce the encoding and decoding process in detail.

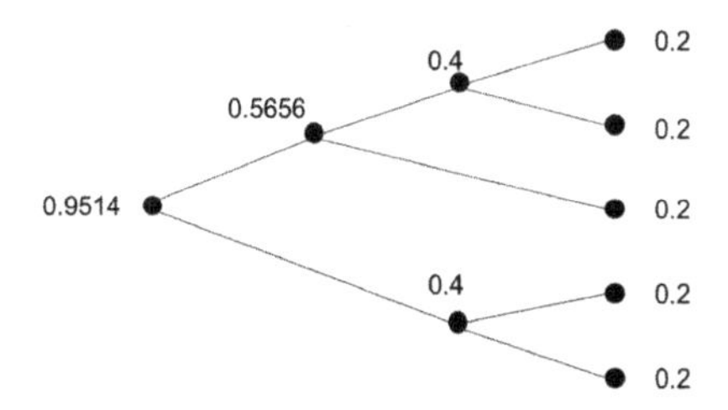

Fig. 2. Construction of the prefix free tree

Prefix-free Matcher		
code	#bin	IPD range
00	1	$F^{-1}(0.0)\sim F^{-1}(0.2)$
01	2	$F^{-1}(0.2)\sim F^{-1}(0.4)$
10	3	$F^{-1}(0.4)\sim F^{-1}(0.6)$
110	4	$F^{-1}(0.6)\sim F^{-1}(0.8)$
111	5	$F^{-1}(0.8)\sim F^{-1}(1.0)$

Fig. 3. Prefix free matcher

3.1 The Prototype of Our CTC

As illustrated in Fig. 1, the sender and receiver reside on two distant computers which are the host A and B respectively. The covert messages to be transmitted are binary strings. The sender A encodes covert message by parsing the stream by a prefix-free matcher and maps it into intervals, then controls the sending time accordingly. On the other hand, the receiver B obtains the intervals and decodes the message according to the matcher. Worthy to note, the two covert parties determine the prefix-free matcher by observing legitimate traffic distributions before beginning secret communication.

3.2 Construction of Prefix-Free Matcher

The prefix-free matcher is determined by two covert communication parties before transmission by observing and leveraging legitimate traffic distribution. According to [7], first of all, we partition the observed legitimate traffic sample into 5 equiprobable bins. Thus, the proportion of each bin is 0.2 which means the target PMF is $x = (0.2, 0.2, 0.2, 0.2, 0.2)^T$. Figure 2 illustrates the process of using GHC to construct a dyadic approximation of the PMF d from x. Therefore, the induced dyadic PMF is $d = (2^{-2}, 2^{-2}, 2^{-2}, 2^{-3}, 2^{-3})$.

Obviously, the cumulative distribution probability ranges of the 5 bins are: $[0, 0.2]$, $(0.2, 0.4]$, $(0.4, 0.6]$, $(0.6, 0.8]$ and $[0.8, 1.0]$ respectively. Given a cumulative distribution probability p, the corresponding target IPD value can be calculated:

$$IPD_{value} = F^{-1}(p) \tag{8}$$

where F^{-1} is the inverse cumulative distribution function of an observed legitimate traffic sample. Combined with the prefix free tree constructed above, the prefix-free matcher is shown in Fig. 3. Each prefix code corresponds to a code sample, namely the bin number. For each bin, the IPD range can be calculated using the inverse cumulative distribution function. Therefore, each code sample can be mapped to any random value within its corresponding IPD range.

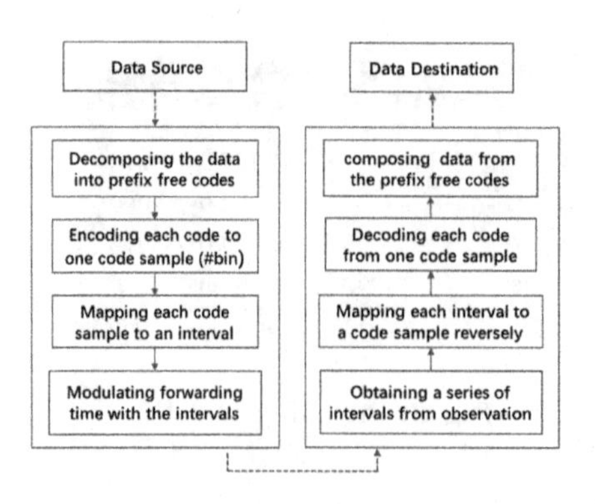

Fig. 4. Encoding/decoding process

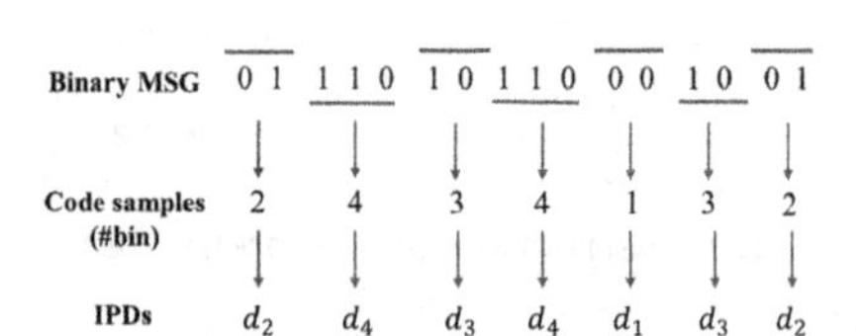

Fig. 5. An example of the encoding process

3.3 Encoding/Decoding Mechanism

Figure 4 shows the information transmission process which consists of encoding and decoding process. Suppose the sender wishes to transmit a binary message $\{b_1, b_2, b_3, \dots\}$. First of all, the bit stream is parsed by the prefix-free matcher to produce code symbols (#bin) $\{s_1, s_2, s_3, \dots\}$. Then, the code symbols are modulated into IPDs $\{d_1, d_2, d_3, \dots\}$ by mapping each code symbol to a random IPD within the corresponding range by PRNG. During the mapping process, PRNG is leveraged so that the modulated IPDs can be changed dynamically within specific ranges in order to evade producing regular patterns.

For example, the binary stream is set to be 0111010110001001. As illustrated in Fig. 5, firstly by parsing the stream through the prefix matcher, the code symbols (#bin) are obtained as $\{2, 4, 3, 4, 1, 3, 2\}$. Next, by using PRNG, the modulated IPDs are acquired by mapping the symbols to random values within the corresponding IPD ranges. The modulated IPD d_k for symbol k is:

$$d_k = RAND(F^{-1}(0.2 * (k-1)), F^{-1}(0.2 * k)) \tag{9}$$

where $RAND$ is the function to generate a random number within a specific range by PRNG.

To be thorough, there is one possible problem when parsing the binary stream. There may be no matched prefix code when the ending of a stream is only one bit or two bits such as '11'. To solve this problem, we simply abandon the ending one bit or two bits which is acceptable since the missing rate is quite low.

The receiver observes the intervals and decodes the message by performing the reverse operation of encoding process. Firstly, the symbol codes are obtained by determining which ranges the IPDs are within according to the prefix-free matcher. Then, each symbol code is reversely mapped to one prefix code and the secret message is acquired by combining all achieved prefix free codes in order.

4 Experimental Results

In this section, we validate the effectiveness of our proposed method through a series of experiments and compare the detection resistance of our channel with on-off switching CTC, L-bits-to-N-packets CTC and Time-relay CTC. In addition, the hidden transmission capacity of our channel is demonstrated.

4.1 Data Selection

HTTP packet is the most popular allowable one in networks and HTTP traffic is very suitable for being taken as covert traffic. In our experiment, we collected packets from a gateway of our institute which was recorded using Wireshark and relevant HTTP packets are filtered out. This is our overt traffic for our proposal and other three CTCs.

For on-off CTC, three different interval values 40 ms, 60 ms, 80 ms are chosen according to [6]. After every 100 packets, the three transmission intervals are cycled to avoid creating regular patterns on the traffic.

For Time-Replay CTC, the observed legitimate HTTP traffic is sorted and divided into two bins. A random delay value from the higher bin is selected to embed '1' or from the lower bin to embed '0'.

For L-bits-to-N-Packets CTC, we choose an 8-bits to 3-packets scheme for simplicity and efficiency. In addition, we select the minimum delay to be 50 ms and the scalar to be 10 ms recommended by the authors in [15]. Therefore, each of the 8-bit binary strings is mapped to a sequence of packet inter-transmission times (T_1, T_2, T_3) and T_i only takes values from the set $E = \{T : 50 + k * 10, 0 \leq k \leq 10\}$.

Generally, for evaluating the detection resistance ability of our method, we collect five datasets, including:

- Legitimate HTTP: 200,000 HTTP packets
- on-off CTC: 200,000 HTTP packets
- Time-Replay CTC: 200,000 HTTP packets
- L-bits-to-N-packets CTC: 200,000 HTTP packets
- Our CTC: 200,000 HTTP packets.

4.2 Detection Resistance

In our experiments, we run detection tests on samples of covert CTCs and legitimate traffic in order to validate the effectiveness of our proposed approach. There are totally 100 samples for each CTC and overt traffic. Each traffic sample consists of 2000 IPDs.

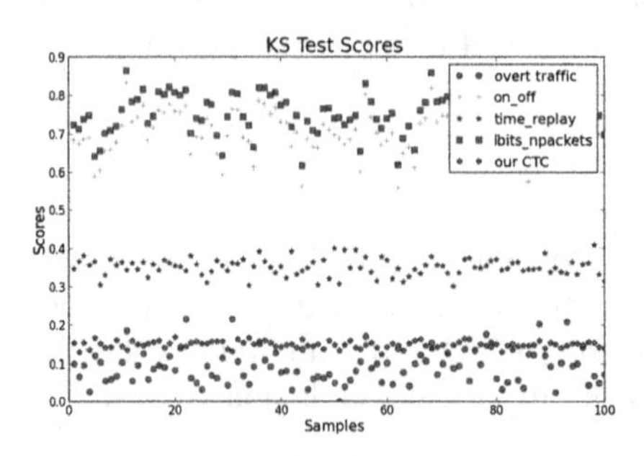

Fig. 6. KS test scores for CTCs

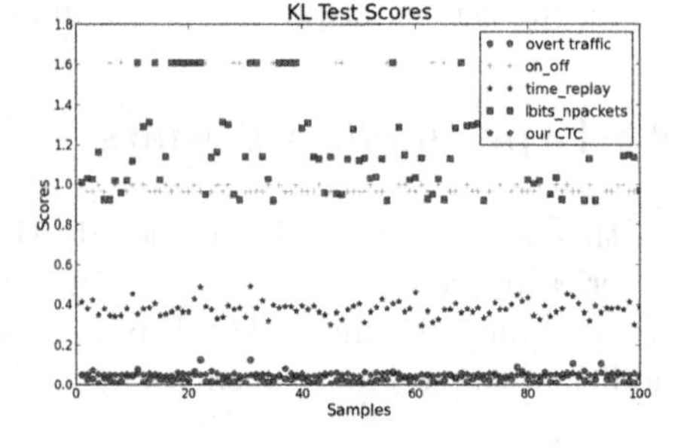

Fig. 7. KL test scores for CTCs

Figure 6 shows that the KS test score for our CTC is relatively quite small and closest to the legitimate traffic samples while the scores for the other three CTCs are remarkably higher. Since the KS test score measures the distance from the covert traffic sample and the legitimate traffic sample, the smaller the test score is, the closer a covert sample is to the normal behavior. In other words, a large test score means that a sample does not fit the normal behavior, indicating the possible existence of a covert timing channel. Accordingly, we can draw the conclusion that our CTC is closest to the legitimate traffic and can evade KS test successfully because of presenting quite normal behavior.

In Fig. 7, we can immediately see that our CTC has the lowest divergence from the legitimate traffic which is quite close to zero and has very low variance. Meanwhile, the KL test scores for on-off CTC and L-bits-to-N-packets CTC are remarkably higher and vary greatly among samples. For time-replay CTC, the scores are lower than on-off and L-bits-to-N-packets CTC but still too high to evade detection since KL divergence is a dissimilarity measure used to compare IPD distributions of the overt traffic and the covert traffic. However, by leveraging Geometric Huffman Coding, our method successfully minimizes the KL divergence.

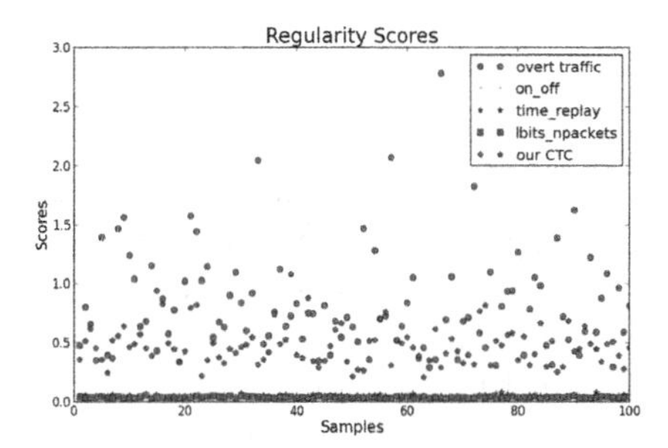

Fig. 8. Regularity scores for all channels

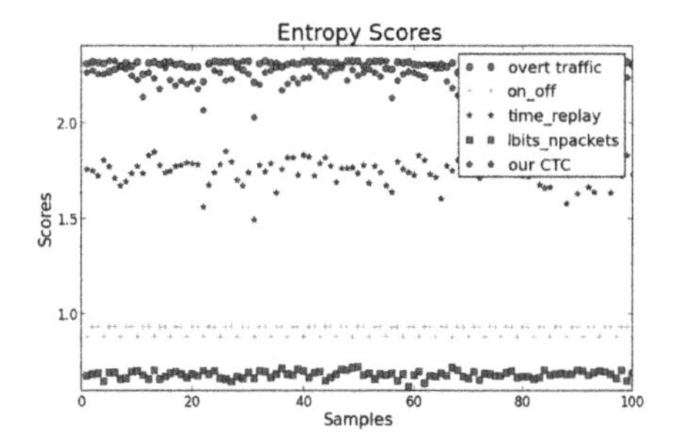

Fig. 9. Entropy scores for all channels

The regularity test is used to detect covert timing channels based on the fact that low regularity score corresponds to a highly regular sample, indicating the possible occurrence of a covert timing channel. As shown in Fig. 8, the covert traffic generated by our method has higher regularity score and higher variance, while the other three CTCs have much lower scores compared to the legitimate traffic. It can bc drawn from the figure that the threshold '0.2' can be used to distinguish the on-off CTC, time-replay CTC and L-bits-to-N-Packets CTC from the legitimate traffic while it is apparent that our CTC is resistant against the regularity test.

According to the entropy-based detection method proposed in [7], the phenomenon that the entropy is low suggests a possible covert timing channel because it means the sample does not uniformly fit the appropriate distribution. In addition, if the corrected conditional entropy test score is lower or higher than the cutoff scores, there may be a covert channel since low corrected conditional entropy test score means the sample is highly regular while a high conditional entropy test score which is near the first-order entropy shows that the sample is in lack of correlations.

As we can see from Figs. 9 and 10, the entropy and corrected conditional entropy of our CTC and the overt traffic are very close and high which is in

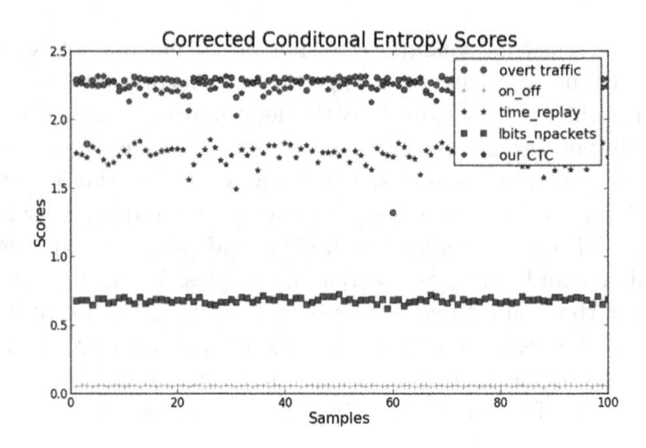

Fig. 10. Corrected conditional entropy for overt traffic, our CTC and three other CTCs

accord with the fact that the legitimate traffic is irregular and lack of correlations. For time-replay CTC, the scores are lower than overt traffic but higher than the other two CTCs since it tries to reduce regularity and correlations by selecting random IPDs within two bins separated by the media. For on-off CTC and L-bits-to-N-packets CTC, they fail to avoid creating regular patterns, thus should have low quite scores exactly as shown in the picture. Therefore, our CTC is closest to legitimate traffic in entropy and corrected conditional entropy test, thus certainly can evade the entropy-based detection.

As for the latest detection method which is a SVM-based framework, the four characteristics which are four kinds of scores used for distinguishing covert traffic from overt traffic are of no meaning since the above results shows that our CTC is much close to legitimate traffic for every score. Therefore, it can be deduced that our CTC is detection-resistant to this latest SVM-based method.

4.3 Channel Capacity

As demonstrated in Sect. 3, each prefix-free code is mapped to one IPD which means each packet can transmit 2 bits or 3 bits. The three prefix free codes '00', '01' and '10' are of length 2 while the other two '110' and '111' are of length 3. According to the probability of each code, it can be drawn that each packet can transmit 2.25 bits on average. For on-off and time-replay CTC, each packet can only transfer one binary bit. Therefore, the transmission capacity of our method is up to one time higher than on-off and time-replay CTC and similar to the throughput of L-bits-to-N-packets CTC with L fixed at 8 and N fixed at 3.

In general, our method is undetectable against the entropy-based detection and other known tests while maintaining reasonable transmission capacity higher than binary channels.

5 Conclusion and Future Work

In this paper, we design and implement a novel covert timing channel by leveraging Geometric Huffman Coding to realize covert and overt channel matching. The divergence between the covert traffic distribution and legitimate traffic distribution is minimized and regular patterns are evaded to reduce regularity.

We evaluate the detection resistance of our method compared to other three CTCs. Experimental results show that our proposal outperforms other methods significantly and proves to be undetectable against the distribution shape testing, regularity testing and entropy-based detection method. In addition, it can be obviously deduced that our CTC is detection-resistant to the latest SVM-based detection method. Furthermore, we demonstrate that our CTC can achieve a reasonable transmission capacity of 2.25 *bits/packet* which is higher compared to other binary CTCs. Generally speaking, our method not only resists all current known detection methods experimentally but also maintains a reasonable transmission capacity higher than binary CTCs.

In future, we plan to continue our work in several meaningful ways. Firstly, we will explore methods to improve the transmission capacity of our CTC such as transmitting over multiple channels. In addition, we plan to improve the robustness of our channel so as to deal with network jitter by leveraging some methods such as redundancy.

Acknowledgments. This work was supported by the National Natural Science Foundation of China (Nos. 61572456, 61672487), the Anhui Province Guidance Funds for Quantum Communication and Quantum Computers and the Natural Science Foundation of Jiangsu Province of China (No. BK20151241).

References

1. Archibald, R., Ghosal, D.: A covert timing channel based on fountain codes. In: 2012 IEEE 11th International Conference on Trust, Security and Privacy in Computing and Communications (TrustCom), pp. 970–977. IEEE (2012)
2. Archibald, R., Ghosal, D.: A comparative analysis of detection metrics for covert timing channels. Comput. Secur. **45**, 284–292 (2014)
3. Bocherer, G., Mathar, R.: Matching dyadic distributions to channels. In: 2011 Data Compression Conference (DCC), pp. 23–32. IEEE (2011)
4. Böcherer, G.: Capacity-achieving probabilistic shaping for noisy and noiseless channels. Ph.D. thesis, Hochschulbibliothek der Rheinisch-Westfälischen Technischen Hochschule Aachen (2012)
5. Cabuk, S.: Network covert channels: design, analysis, detection, and elimination (2006)
6. Cabuk, S., Brodley, C.E., Shields, C.: IP covert timing channels: design and detection. In: Proceedings of the 11th ACM Conference on Computer and Communications Security, pp. 178–187. ACM (2004)
7. Gianvecchio, S., Wang, H.: Detecting covert timing channels: an entropy-based approach. In: Proceedings of the 14th ACM Conference on Computer and Communications Security, pp. 307–316. ACM (2007)

8. Kang, M.H., Moskowitz, I.S., Chincheck, S.: The pump: a decade of covert fun. In: 21st Annual Computer Security Applications Conference, pp. 7–pp. IEEE (2005)
9. Lampson, B.W.: A note on the confinement problem. Commun. ACM **16**(10), 613–615 (1973)
10. Liu, G., Zhai, J., Dai, Y.: Network covert timing channel with distribution matching. Telecommun. Syst. **49**(2), 199–205 (2012)
11. Liu, Y., Ghosal, D., Armknecht, F., Sadeghi, A.-R., Schulz, S., Katzenbeisser, S.: Hide and seek in time—robust covert timing channels. In: Backes, M., Ning, P. (eds.) ESORICS 2009. LNCS, vol. 5789, pp. 120–135. Springer, Heidelberg (2009). https://doi.org/10.1007/978-3-642-04444-1_8
12. Lou, J., Zhang, M., Fu, P.: Design of network covert transmission scheme based on TCP. Netinfo. Secur. 34–39 (2016)
13. Martins, D., Guyennet, H.: Attacks with steganography in PHY and MAC layers of 802.15.4 protocol. In: Fifth International Conference on Systems and Networks Communications, pp. 31–36 (2010)
14. Porta, A., Baselli, G., Liberati, D., Montano, N., Cogliati, C., Gnecchi-Ruscone, T., Malliani, A., Cerutti, S.: Measuring regularity by means of a corrected conditional entropy in sympathetic outflow. Biol. Cybern. **78**(1), 71–78 (1998)
15. Sellke, S.H., Wang, C.-C., Bagchi, S., Shroff, N.: TCP/IP timing channels: theory to implementation. In: 2009 IEEE INFOCOM, pp. 2204–2212. IEEE (2009)
16. Shah, G., Molina, A., Blaze, M.: Keyboards and covert channels. In: Conference on Usenix Security Symposium, p. 5 (2006)
17. Shrestha, P.L., Hempel, M., Rezaei, F., Sharif, H.: A support vector machine-based framework for detection of covert timing channels. IEEE Trans. Dependable Secur. Comput. **13**(2), 274–283 (2016)
18. Walls, R.J., Kothari, K., Wright, M.: Liquid: a detection-resistant covert timing channel based on IPD shaping. Comput. Netw. **55**(6), 1217–1228 (2011)
19. Zander, S., Armitage, G., Branch, P.: A survey of covert channels and countermeasures in computer network protocols. IEEE Commun. Surv. Tutor. **9**(3), 44–57 (2007)

Spark: A Smart Parking Lot Monitoring System

Blake Lucas and Liran Ma(✉)

Texas Christian University, Fort Worth, TX 76129, USA
{b.g.lucas,l.ma}@tcu.edu

Abstract. Parking on a college campus is understood to be a challenge for commuters. With a rising matriculation rate in the United States, the task of finding parking on an expansive campus grows even more daunting. However, the rising prominence of the Internet of Things has initiated a paradigm shift in data-analysis computing. The point of data collection is often outlier locations, removed from existing infrastructure, and parking lots are no exception. Using proximity sensors, solar power, and cellular communication, we can create such an IoT system to monitor parking lot in- and outflows. The parking data collected can be analyzed to create a smarter, more efficient parking experience.

Keywords: Data analytics · Internet of Things (IoT)
Outlier data collection · Proximity sensing · Self-contained systems

1 Introduction

In the United States, the matriculation of students—ages 18 to 24—to four-year institutions has increased from roughly 17% in 1973 to almost 30% in 2015 [11]. The increase in students seeking higher education, while beneficial, puts a strain on available institution resources. While not echoed universally, campuses are not expanding at the same rate their student populations are, and this disparity is most obvious in campus parking. With an ever-growing population to educate, most degree-granting institution infrastructure spending has remained fixed around 6% of the allocated campus budget from 1999–2015, albeit decreasing slightly [11]. Of the campus expansion occurring parking assuredly does not reap the benefits before education and dormitory facilities. Another solution must be organized in order to facilitate dispersion of the parking gridlock tangling campus commuters.

Spending on the Internet of Things (IoT) reached $624 billion dollars worldwide, according to the International Data Corporation, and is expected to increase further, to $1 trillion in 2020 [5]. The widespread investment in IoT and related applications has created a paradigm shift in data analytics: data collection occurs at "outlier points," or locations removed from existing infrastructure [7,19]. Parking lots are no exception. Hence, we can utilized the ever-expanding IoT ecosystem to create a system that will monitor parking lot occupancy in the form of automobile in- and outflow. Once these outlier collection points are

S. Chellappan et al. (Eds.): WASA 2018, LNCS 10874, pp. 321–332, 2018.
https://doi.org/10.1007/978-3-319-94268-1_27

functional, we can utilize a data analytics engine to process the collected parking lot traffic data and provide commuters information about parking beyond merely the number of spots remaining. Time remaining until selected parking lots are full or a suggested time of departure are two examples of insights that could ease parking strain.

This concept is not novel. Grodi *et al.*, reference a similar system infrastructure in [3], using a triangulation method of Zigbee-connected sensors that feed into a central database of parking data, accessible to commuters by a variety of mediums: smartphone application, website, etc. Hong *et al.*, in [4], discuss the possibility of using queue theory, jointly allowing parking garage operators to simulate parking flow given a variety of conditions while presenting commuters with availability. Commercial solutions are presently available for parking garages [13,16]. However, there are limitations associated with each of these system variants that prohibit adaptation to an outlier location.

The first challenge of any outlier data collection system is electrical power. Unlike a dead wristwatch, a dead sensor is never right. Regardless of calibration and system accuracy, reliable power must be provided to every sensor at all times. In parking garage solutions, [4,13,16], power is provided to the structure. In a lot solution, however, each module would need to generate its own power. In such a vast triangulation sensor network as described in [3], that increases cost significantly. Additionally, if the system cannot be operated in an event-driven nature, isolated power stations—batteries or solar panels—are difficult to implement.

Network connectivity is another hurdle to such IoT systems. Access points must be provided, and in either a parking lot or garage, this poses a problem. In parking lots, existing WiFi or other communication networks are often not available, and, since an access point cannot be interrupt-driven, powering an access point is again subject to the difficulties that befall a polling-loop system. While most parking garage systems do not advertise data outside of the physical infrastructure (light fixtures above spaces and counting screens, as displayed in [13,16]), providing a singular access point through multiple layers of concrete is nevertheless a difficult task, and outside the scope of this research; therefore, our design must be able to handle both wired and wireless connections.

Additionally, an effective solution cannot require infrastructure modification—an expensive, time-consuming process. As most IoT implementations are monitoring addendums to existing systems, the hardware selected should maintain a small material footprint and be capable of existing within various environments without substantive modification or costly, intricate installation. We attempt to address these shortcomings by designing, implementing, and evaluating such a parking lot system named *Spark*, for *S*mart *Park*ing.

Having established Spark's design requirements, the remainder of this paper is organized as follows: Sect. 2 presents theory and related applications adapted for this development. An overview of the Spark system is presented in Sect. 3. Section 4 describes the technical details of the developed system. Finally, we evaluate the existing design, noting its limitations and required future work, in Sect. 5 and conclude the discussion of Spark in Sect. 6. Data regarding sensor calibration is presented in Appendix A.

2 Related Work

Martinez *et al.*, define several methods to model "energetically autonomous sensors", emphasizing two stages of autonomous power: "harvesting" and "buffering" energy until it is necessary to consume it [8]. In the collection (harvesting) phase discussed in [8], Mondal and Paily present a method to manage solar panel power supplies for IoT applications in [10]. The implementation developed by Mondal and Paily increases efficiency as input photo-voltaic cell voltage increases [10, Fig. 14]. While their design is not directly employed, we draw on their charge-pump design [10, Fig. 5] in our implementation of a power station, presented in Sect. 3. A holistic IoT power design is presented and analyzed by Shafiee *et al.*, in [15]. However, Shafiee *et al.*, focus on the creation of a SoC, or system-on-a-chip; as such, the integrated circuit design concepts presented are not directly applicable to our discrete device centered (DDC) design. We employ DDC methodology in favor of creating printed circuit board deliverables, which are more cost-effective for our relatively small-product-volume solution and can be directly implemented.

Various IoT frameworks—defined for our scope as server platforms capable of recording, analyzing, and presenting outlier-collected data—are demonstrated in [6,18]. Additionally, the concepts presented in [1] are important in maintain the efficiency and privacy of our network transmissions. Laubhan *et al.*, in [6], present a mixed wired and wireless sensor network, which lends itself to adoption in a parking garage for our system. Yelmarthi *et al.*, propose multilayer, multi-sensor implementations through the use of Raspberry Pi's [18]. We seek to create a standalone solution with much less power consumption and a smaller material footprint.

3 System Design

Our proposed parking lot monitoring method consists of four discrete modules: *Sentinel*, *Scout*, *Power*, and *Uplink*. Figure 1 shows the positioning of the modules on a basic parking lot[1]. The Sentinels count the number of cars entering and leaving the lot, while the Scouts monitor the status of individual spaces. Communication is accomplished via the Uplink, a custom implementation of a WiFi access point. Technical specifics will be presented in Sect. 4. The Spark design is entirely modular, meaning custom implementations do not require much re-engineering or cost.

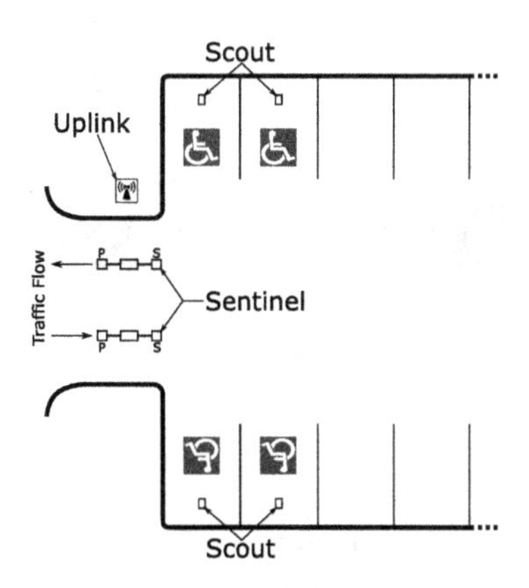

Fig. 1. An example deployment layout.

[1] The *Power* modules are not shown in Fig. 1 to avoid clutter.

It is worth noting that the concept of an IoT system supported by solar panels and cellular modules is not inherently unique. However, we seek to adapt these standard IoT principles to develop a system capable of operating away from existing infrastructure. Such a parking lot monitoring system is not commercially available anywhere.

3.1 Sentinel

The Sentinel module consists of two proximity sensors separated by roughly 0.6 m, joined to a host processor. Each proximity sensor is configured to interrupt when it detects a car covering it (a state referred to as a sensor "closing") and again when the car has passed (referred to as a sensor "opening").

The sensors are designated as either the Primary or Secondary (P or S), with the only distinction being the P sensor should always be placed at the mouth of the parking lot entryway (see Fig. 1). The Sentinel modules determine the direction of the cars, entering or leaving, by tracking the order in which the sensor interrupts are received, as demonstrated in Fig. 2. If the P sensor interrupts first and is still closed when the S sensor closes, the lot occupancy count is incremented. If the S sensor interrupts first, and the P sensor closes while S is still closed, the lot occupancy count is decremented. If either sensor closes while the other sensor is open, the count is not changed. **Both sensors must be closed for a count change to occur,** as shown in Fig. 2(c). This precaution prevents accidental counting and makes intentionally interfering with the module's operation more difficult.

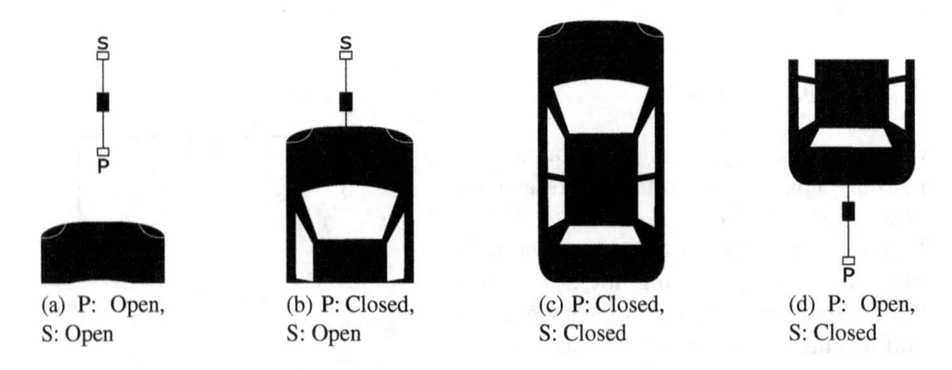

(a) P: Open, S: Open (b) P: Closed, S: Open (c) P: Closed, S: Closed (d) P: Open, S: Closed

Fig. 2. The operation of sentinel as a car enters.

Data transmission is accomplished by mounting a connection PCB on top of the Sentinel's PCB. Four lines—power, ground, data (SDA), and clock (SCL)—are provided on the Sentinel for this purpose, with the SDA and SCL lines forming an I^2C bus (see Fig. 3). When the lot occupancy count should be changed, the Sentinel sends a corresponding I^2C command via this bus. Our design uses WiFi for communication, but any protocol could be employed—CAN bus, Ethernet, etc—as long as the attached board that implements the protocol understands the Sentinel's I^2C message schema.

3.2 Scout

The Scout module monitors the status of a single spot and is most useful in parking lots with a mixture of reserved and general-availability spaces, as shown in Fig. 1. There is no way for a Sentinel to differentiate a car that will occupy a handicapped space from one that will occupy a non-reserved space, thus, Scout modules can be placed on the reserved spaces to monitor their statuses independently of generally available parking. The addition of Scouts allows the end-user application to present an accurate number of available regular spaces and available reserved ones.

The Scout module utilizes a proximity sensor that closes when a car pulls into a space and opens when the car leaves. Communication is accomplished in a similar way to the Sentinel (discussed further in Sect. 4). These modules could be used to monitor the occupancy count of an entire parking lot; such a system would result in a more accurate count that need not be kept synchronized but would increase power requirements and cost.

3.3 Power

As briefly discussed in Sect. 1, power is one of the greatest obstacles of an outlier IoT system. Spark provides two alternatives: the *Direct Solar* and *Rotating Battery* methods. Both alternatives are analyzed more in Sect. 5, but it is important to note that, despite their differences and relative advantages, both designs have uses in the Spark system.

Direct Solar. This method uses a solar panel/battery pair. The solar panel powers the attached load directly while trickle charging the battery. If the solar panel is not capable of providing enough energy, the charging stops and the battery is connected to the load instead.

Rotating Battery. The largest difference in the Rotating Battery method is that the solar panel is never directly connected to the output; while one battery provides power to the module, the solar panel recharges the other. A nano-power comparator determines which battery has a higher potential—using the solar panel voltage for reference—and connects that battery to the output.

When the solar panel voltage is lower than the highest battery voltage, the comparator is useless, so passive switching is accomplished by diode-bias balancing. Whenever there is a potential difference between the batteries equal to the threshold voltage of a single diode, the battery with the higher potential will drive the load entirely. This method is included only as a safeguard, to ensure that power will always be available for the device. If the batteries are switching passively, no solar charging has taken place for a significant length of time, which—barring inclement weather—is indicative of a possible need for maintenance, or even component failure.

3.4 Uplink

If communication infrastructure is inaccessible, an independent access point needs to be provided to allow the Sentinel or Scout modules to transmit data to our cloud platform. Since cellular modules, especially 4G or LTE varieties, are too expensive to be viable built-on options, Spark Uplink serves as a simplified WiFi[2] Hotspot. Uplink employs an LTE modem in tandem with a WiFi controller acting as an access point, available to nearby Spark devices. The modem is turned off until a request is sent to the WiFi controller. To limit per-transmission bandwidth, the WiFi controller keeps an active count of net activity for a few seconds during and after the modem's activation. **Only one transmission is sent per period of activity**, as opposed to sending multiple (possibly offsetting) transmission packets in rapid succession. Such an implementation of a Hotspot allows us to conserve power, bandwidth, and, ultimately, cost.

4 Implementation

Each module in the Spark design is PCB-based. We will demonstrate the process by which we designed and created these PCB's in this section.

4.1 Sentinel

A Sentinel module consists of three components: the main processor called the Handler, the connection, and the Sensors.

Handler. The microcontroller unit (MCU) utilized for the Handler is the PIC16F18325 by Microchip. This MCU was selected because it possesses two Master Synchronous Serial Ports (MSSP's). The MSSP's can function as separate I^2C busses. Thus, two identical sensors can be connected to one MCU without need for address translation. The Sensor modules are connected to the Handler's PCB via RJ-45 connectors mounted under the board (see Figs. 3, 4). The bus to which the P sensor is connected is called the "primary bus," and, by convention, this bus also manages the Connection module.

Fig. 3. Sentinel prototype.

[2] As discussed previously, our design uses WiFi as the primary communication method. Neither WiFi nor Spark Uplink is required for operation if existing frameworks can be used or custom hardware can be provided; we assume neither to present a complete, self-contained solution.

Connection. For data transmission, we employ an ESP-WROOM-02 WiFi module for its availability, price ($2.70), and programmability (via the Arduino IDE). The drawbacks of this module are its high power consumption and limited external interrupt capability. Limited interrupts prevent the WROOM-02 from acting as an I^2C slave, meaning for it to communicate with the Handler, we must employ another PIC16F18325 MCU to act as a surrogate, or buffer, between the Handler and WROOM-02.

Due to its high power consumption, the WROOM-02 spends most of its time in "Deep Sleep" mode, where a pulse on the RESET pin is the only way to reactivate the module. Two pins, CTS (Clear to Send) and DTR (Data Ready), are connected between the surrogate MCU and the WROOM-02. The active-low CTS pin indicates whether it is safe to reset the WROOM-02, and the DTR pin is the RESET line of the WROOM-02. When data is received from the Handler, the surrogate MCU checks the status of CTS. If CTS is low, the surrogate pulses DTR to wake the WROOM-02. If CTS is high, the surrogate waits until the WROOM-02 requests I^2C data or until CTS goes low. The use of a CTS pin prevents unnecessary resetting of the ESP module, especially critical if an Over-the-Air (OTA) update is occurring.

Sensors. The proximity sensor employed for this solution is the VCNL4200 by Vishay [14]. Each sensor was placed on a circuit board designed to fit atop an RJ-45 Ethernet connector.

The VNCL4200 requires a calibration resistance to set the current through the infrared-emitting diode (IRED). We have chosen a resistance of 68 Ω. The characteristic curves for the VCNL4200 and further discussion of this choice are available in Appendix A.

Fig. 4. Sensor module.

4.2 Scout

Scouts utilize the same electrical hardware as the Sentinel. To preserve modularity, a fundamental concept of Spark's design, Scout includes the ability to stack a Connection module, identical to Sentinel. While this does increase cost slightly, failure to implement a stackable Connection in the Scout design means that we have not adequately accounted for both wired and wireless connections.

The same proximity sensor modules used in Sentinel are used in Scout, extended from the board via an RJ-45 connector. This allows the module to be offset from the parking space and decreases production cost by maintaining modularity.

4.3 Power

Functional schematics for each design—Direct Solar & Rotating Battery—are shown in Fig. 5. CMOS pairs determine which source is linked to the output;

the selected transistors are IRF9510PBF & IRF510PBF Power P- and N-MOSFET's, respectively. These transistors were selected because of their high forward-current rating and similar, ≥1 S forward transconductance. TO-220-3 packages are used in each design. All power connections use JST RCY connectors for flexibility in PCB housing design. SB130-B Schottky diodes were used for their low threshold voltage and high forward current rating.

Battery recharging is accomplished using the MCP73833 IC. Our battery charging circuit is adapted from [9]. The battery charging circuit acts as a comparator in the Direct Solar method (Fig. 5(a)): when it is active, the solar panel can supply enough energy to the system, and the output voltage activates the NMOS connecting the solar panel to the load. When the battery charger is off, the load switches to the battery.

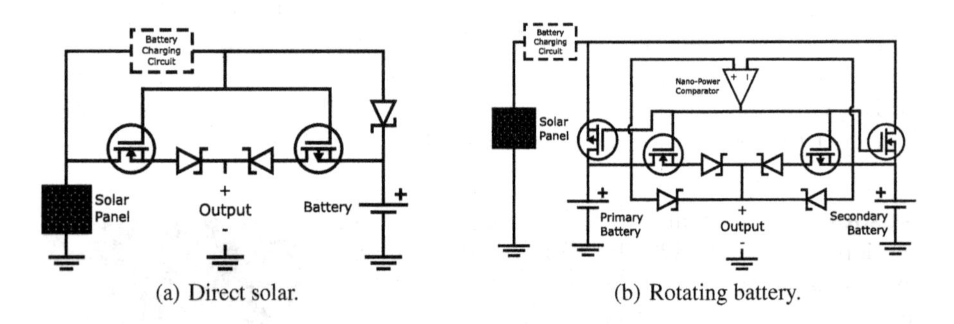

(a) Direct solar. (b) Rotating battery.

Fig. 5. Functional power schematics.

4.4 Uplink

Uplink pairs an ESP-WROOM-02 with a u-blox SARA-R4 LTE Cat-M1 Modem. The PCB antenna on the WROOM-02 may suffer crosstalk with the LTE Modem even if a u.FL antenna is used [12], so we must take precautions to isolate their signals. Additional signal filtering may also be needed. Since the SARA-R4 operates on LTE Bands 2, 4, 5, and 12 [17], all of which fall below the WiFi standard of 2.4 GHz, filtering could ease the effects of crosstalk, if observed.

Since the WROOM-02 is acting as an access point, we cannot run it in Deep Sleep mode. It is possible, though, to deactivate the R4 Modem until it is needed. The WROOM-02 can also act as a WiFi Server, therefore no complicated routing firmware needs to be written; the WROOM-02 can act purely as a middleman between Spark modules and cloud platform.

The WROOM-02 is used again for Uplink in spite of its high current requirement because it does not require a host microcontroller. However, it may ultimately be advantageous to sacrifice module cost for energy conservation.

5 Evaluation

The two most effective ways to evaluate our design are performance and cost, specifically in terms of energy consumption per device and cost per spot monitored. Note that, at the time of this writing, only the Sentinel prototype is available for testing.

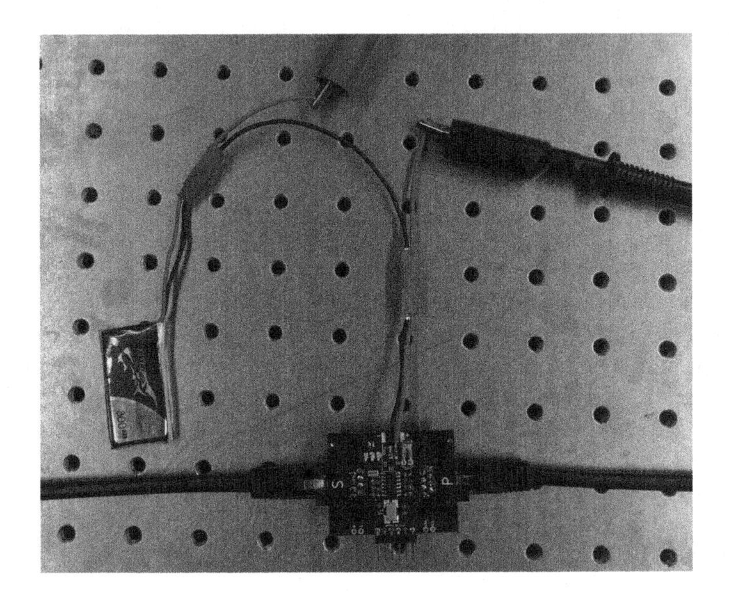

Fig. 6. Sentinel power measurement setup.

5.1 Power

The Sentinel, without a Connection module (as shown in Fig. 6) consumes 3.55 mA (11.7 mW) awake and 2.30 mA (7.59 mW) in sleep mode. Using an 800 mAh battery, this energy consumption translates to over 12.5 days of operation[3] with no solar power bolstering. This typical case analysis assumes a 70:30 idle/active ratio.

5.2 Cost

Cost analysis is most informative when interpreted as the total cost to the institution to implement our system. Currently, this summary focuses solely on the cost of the electrical components, not of mechanical housings or fasteners.

[3] Our communication module consumes 74.6 mA of current awake, and less than 1 mA in Deep Sleep mode. Since it is only active for 5–10 s at a time, its effect on current consumption is minimized and is not included in the four day estimate.

Table 1. Basic parking lot implementation.

Module	Quantity	Price ($)
Sentinel	2	24.66
Scout	0	16.56
Uplink	1	32.73
Rotating battery	3	17.39
TOTAL	6	134.22

Table 2. Complex parking lot implementation.

Module	Quantity	Price ($)
Sentinel	3	24.66
Scout	15	16.56
Uplink	1	32.73
Direct solar	15	13.34
Rotating battery	3	17.39
TOTAL	37	607.38

Per-Lot

For a basic parking lot with no reserved spaces, Spark implementation costs are shown in Table 1. However, as the lot begins to increase in complexity, cost begins to increase, as demonstrated in Table 2.

While these numbers may seem large, consider that, in the absence of reserved spaces, cost is independent of size. Since Spark tracks cars in and out, instead of just which spaces are occupied, large lots do not cost proportionately more than smaller ones. Parking garages can be covered easily, even at individual floor levels. A Basic lot of 200 spaces costs less than $1.00 per spot. A Complex lot of 200 spaces costs roughly $3.00 per spot. For context, Vanderbilt University charges $600 per year per student for parking. Harvard and Pennsylvania University charge over $2,000 [2].

Comparative

Other parking solutions comparable to ours do exist; some are discussed in Sect. 1. However, these products do not publicize their cost, and require customers to schedule installation with a verified vendor [13,16]. Conversely, our system is simple enough to be sold as a standalone product, requiring no complex installation procedures.

6 Conclusion

This paper presents the design, implementation, and evaluation of a self-contained, low-cost parking lot monitoring system that is applicable to any facility. Coupled with the Internet of Things, such a system can be employed to enhance the parking experience of any densely populated institution, such as college campuses. While our implementation presented is in its infancy, the concepts and methods presented herein provide a comprehensive overview of an autonomous, IoT-capable sensor network, which is widely desirable beyond merely monitoring a parking lot.

A Sensor Calibration

The VCNL4200 sensor does not provide a proximity reading in distance units. Rather, it provides a unitless, 16-bit figure. We need to be able to attach a distance meaning to this figure in order to determine whether or not the reading corresponds to the height of a car's chassis.

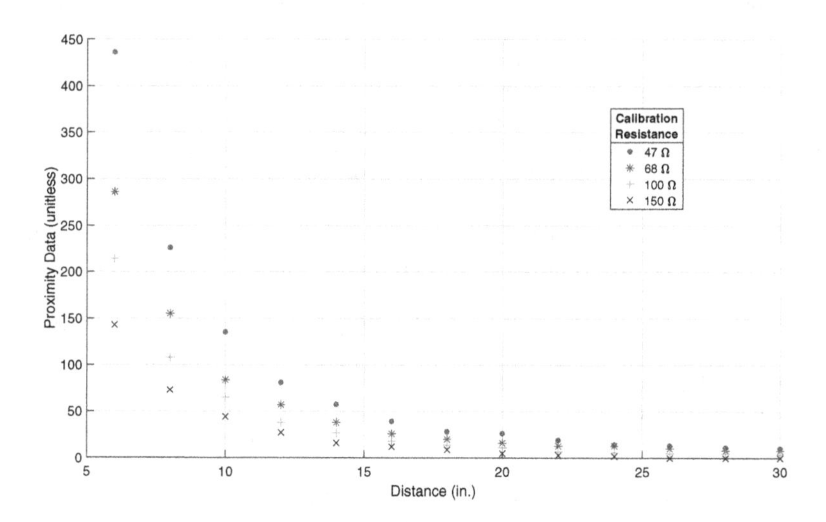

Fig. 7. Proximity data vs. distance (in.) for VCNL4200.

In order to effectively configure our sensor to detect a car, we need enough current through the IRED to overcome the lack of reflectivity of a car's underside but not so much that we drain our energy reserves. We used a piece of plain cardboard with medium reflectivity, placed at successive 2 in. intervals along a tape measure, to construct characteristic curves for the VCNL4200 under our operating conditions (1/320 duty cycle, 250 measurements/s, 200 mA IRED, 16-bit output). These conditions can be changed to reduce energy consumption or improve range.

The curves recorded in Fig. 7 are clearly exponential but also display quasi-linear behavior beyond a certain point. We moniker this value the "Point of Consistency," because results measured at and after this distance are repeatable regardless of precision. At closer distances, two to four inches especially, the output values vary too widely to be determinant. At around 14 in., however, the results were consistent within ± 2 (for the resistances shown above). Our target measurement height is 8 in., but designing for a factor of safety of 2 requires we target 16 in. The 68 Ω resistor displays linear behavior at 16 inches and beyond, thus we select that resistance value.

References

1. Cai, Z., Zheng, X.: A private and efficient mechanism for data uploading in smart cyber-physical systems. Trans. on Netw. Sci. Eng. April 2018
2. Friedman, S., Hamburger, J.: See how Vanderbilt's parking costs compare to other top universities. Vanderbilt Hustler, August 2016
3. Grodi, R., Rawat, D.B., Rios-Gutierrez, F.: Smart parking: parking occupancy monitoring and visualization system for smart cities. In: IEEE SoutheastCon (2016)
4. Hong, T.P., Soh, A.C., Jaafar, H., Ishak, A.J.: Real-time monitoring system for parking space management services. In: IEEE Conference on Systems, Process & Control, pp. 149–153, December 2013
5. International Data Corporation: IDC Forecasts Worldwide Spending on the Internet of Things to Reach $772 Billion in 2018, December 2017. https://www.idc.com/getdoc.jsp?containerId=prUS43295217
6. Laubhan, K., Talaat, K., Riehl, S., Aman, M.S., Abdelgawad, A., Yelamarthi, K.: A low-power IoT framework: from sensors to the cloud. In: ICM 2016, pp. 648–652 (2016)
7. Liang, Y., Cai, Z., Yu, J., Han, Q., Li, Y.: Deep learning based inference of private information using embedded sensors in smart devices. IEEE Netw. (2018)
8. Martinez, B., Montón, M., Vilajosana, I., Prades, J.D.: The power of models: modeling power consumption for IoT devices. IEEE Sens. J. **15**(10), 5777–5789 (2015)
9. Microchip Technology Inc.: MCP7383X Li-Ion System Power Path Management Reference Design (2008)
10. Mondal, S., Paily, R.: Efficient solar power management system for self-powered IoT node. IEEE Trans. Circuits Syst. I Regul. Pap. **64**(9), 2359–2369 (2017)
11. National Center for Education Statistics: Digest of Education Statistics (2016). https://nces.ed.gov/programs/digest/d16/
12. Pattnayak, T., Thanikachalam, G.: Antenna design and RF layout guidelines. Cypress, G edn
13. Q-Free: Single Space Monitoring. https://www.q-free.com/solution/single-space-monitoring/
14. Schaar, R.: Designing the VCNL4200 into an application. Vishay, Semiconductors, December 2017
15. Shafiee, N., Tewari, S., Calhoun, B., Shrivastava, A.: Infrastructure circuits for lifetime improvement of ultra-low power IoT devices. IEEE Trans. Circuits Syst. I Regul. Pap. **64**(9), 2598–2610 (2017)
16. SignalTech: RedStorm Parking Guidance System. https://www.signal-tech.com/products/parking/redstorm_parking_guidance_system
17. u-blox: SARA-R4 Series: System Integration Manual, 7 edn. January 2018
18. Yelmarthi, K., Abdelgawad, A., Khattab, A.: An architectural framework for low-power IoT applications. In: ICM 2016, pp. 373–376 (2016)
19. Zheng, X., Cai, Z., Li, Y.: Data linkage in smart IoT systems: a consideration from privacy perspective. IEEE Wirel. Commun. (2018)

A Hybrid Model Based on Multi-dimensional Features for Insider Threat Detection

Bin Lv[1,2], Dan Wang[1,2], Yan Wang[1](✉), Qiujian Lv[1,2], and Dan Lu[1,2]

[1] Institute of Information Engineering, Chinese Academy of Sciences, Beijing 100093, China
{lvbin,wangdan3,wangyan,lvqiujian,ludan}@iie.ac.com
[2] School of Cyber Security, University of Chinese Academy of Sciences, Beijing 100049, China

Abstract. Insider threats have shown their power by hugely affecting national security, financial stability, and the privacy of many people. A number of techniques have been proposed to detect insider threats by comparing behaviors among different individuals or by comparing the behaviors across different time periods of the same individual. However, both of them always fail to identify the certain kinds of inside threats due to the fact that the behaviors of insider threats are complex and diverse. To deal with this issue, this paper focuses on constructing a hybrid model to detect insider threats based on multi-dimensional features. First, an Across-Domain Anomaly Detection (ADAD) model is proposed to identify anomalous behaviors that deviate from the behaviors of their peers based on the isolation Forest algorithm. Second, an Across-Time Anomaly Detection (ATAD) model is proposed to measure the degree of unusual changes of a user's behavior based on an improved Markov model. What's more, we propose a hybrid model to integrate the evidence from the above two models ADAD and ATAD. To evaluate the performance of the proposed models comprehensively, we implement a series of experiments with the 17-month data. The experimental results show that the ADAD and ATAD models are robust and the hybrid model can outperform the two separated models obviously.

Keywords: Insider threat detection · Information fusion
Hybrid model · Isolation Forest · Markov model

1 Introduction

Insider threats are threats with malicious intent directed towards organizations by people internal to the organization [1], which include physical sabotage activities, theft of confidential data, business secrets, and fraud. Financial loss and reputation damage caused by insider threat far outweighs that caused by external attacks. According to one of the latest articles from CSO magazine [2] about

S. Chellappan et al. (Eds.): WASA 2018, LNCS 10874, pp. 333–344, 2018.
https://doi.org/10.1007/978-3-319-94268-1_28

external and internal attacks, whilst it took about 50 days to fix a data breach caused by an internal attack, it only took 2 to 5 days in the case of external attacks. Nowadays, researchers have proposed different models to prevent or detect the presence of attacks [3,5]. These methods can be categorised to two types, data-driven detection model [3,4] and behavior driven detection model [5,6].

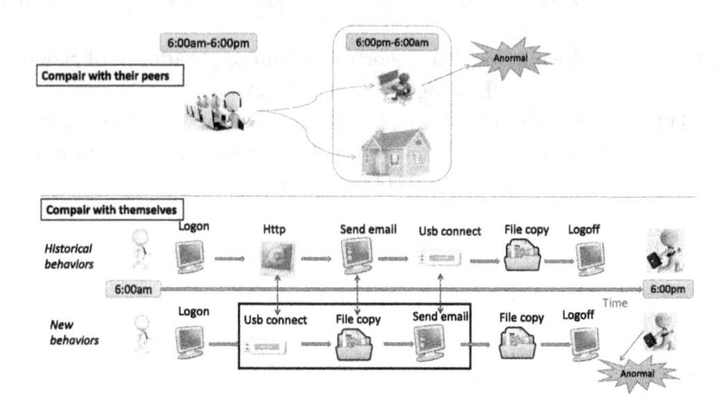

Fig. 1. Example of insider threat.

(1) The first model aims to find a normal portrait in all users data that is used as a reference in the insider threats detection [7]. The upper subfigure of Fig. 1. After getting off work, the general behavior is going home to have a rest. By contrast, few of users secretly deal with the company's confidential documents in the workplace when others are sunk in sleep at midnight. These behaviors are more likely to be an insider threat. However, merely comparing the behaviors among peers, this kind of models fails to detect a malicious insider who tries to behave like a normal user to cover up his evil.
(2) The second model regards the abnormal changes in the behavior as the basis for insider threat detection by comparing behaviors of themselves in different time periods. The lower subfigure of Fig. 1 shows an example to depict a malicious insiders detection based on the behavior-driven method [8]. Concretely, the normal behaviors of the user is defined as click on the browser after booting, view the email, and then reply over a long period. In this case, the behaviors that connects the mobile device after booting, and has a series of operations on the file-copy to steal the company's confidential documents would be regarded as insider threats since they can't match the normal behaviors. It is worth mentioning that this kind of insider threat can get rid of the detection of the first model where time period is used as judging criteria, but it still can be detected successfully here by the second model. However, the model is unable to recognize situations where an user systematically attacks an organization over a long-term period [3].

Hence, the detection of insider threat cannot be defined as a data or behavior driven problem independently [3]. It is necessary to define this problem as a combination of data and behavior driven problem, based on which a new model is to be proposed.

In this paper, we propose a hybrid model that combines a data-driven model with a behavior driven model to detect insider threat which is more robust and accurate in practice. First, the multi-dimensional features extracted from data collected from an enterprise network is formatted and fed separately into the two separate models. Second, each model generates an abnormal score to represent the degree of users' unusual behaviors. Finally, the abnormal scores of two models are fused to be the final abnormal score of each user, and an user is identified to be an insider threat if the anormal score exceeds the threshold. After a wide range experiments, it is verified that the hybrid model can detect insider threats in a more robust and accurate manner. Overall, the contributions of this paper can be summarized as follows:

1. We apply the Isolation Forest to detect behavioral inconsistencies among the behaviors of users. In this model, temporal features are extracted from multi-source data and combined comprehensively to construct the user's portrait.
2. We propose an improved Markov model to identify users' unusual changes. The proposed model considers all historical behaviors of users as the contextual information. Compared with the existing detection methods of Markov model [7], our proposed approach is more effective.
3. We propose a hybrid model based on multi-dimensional features for insider threat detection. The model is able to determine malicious insiders that act inconsistent with their peers and have unusual changes compared with their historic behaviors. By using this model, we obtain an accuracy of 95% in number of experiments, which is of great significance in industry and scientific research.

The remainder of this article is structured as follows. Section 2 presents the related work in the line of detection of insider threat. Then, Sect. 3 presents the details of our approach. Next, in Sect. 4, we detail our implementation of the models and analyze the results of experiments. Section 5 concludes this paper and presents the limitation of our work as well.

2 Related Work

The topic of insider threat has been paid much attention in the literature recently. Researchers have proposed different models aiming at preventing or detecting the insider threat attacks [10,11]. To elicit the state of art, the work presented here is focuses on the approaches of detecting insider threat based on data-driven methods and behavior-driven methods, respectively.

With regard to the data-driven methods, Mathew et al. detected inside threat on account of user access patterns [12], Eberle et al. used social graphs to detect the abnormal behaviors [13]. More recently, Eldardiry et al. have also proposed a system for managing insider attacks by comparing users' behaviors based on peer baselines [14]. Goldsmith applied a layered architecture by fusing across multiple levels information to detect anomalies from heterogeneous data [15]. Hoda et al. detected peer groups of users and modeled user behavior with respect to these peer groups [12]. Subsequently, they detected insider activity by identifying users who deviated from their peers. There are also other approaches based on data-driven to detect abnormal behaviors [8,9]. However, they did not factor in the changes of user behaviors over time. We note that while a common activity is not suspicious, a rare change of the order common activity can be.

There are several literatures based on behavior-driven models [7,10,11]. Rashid took the change of the user's behavior over time to detect the anomaly and achieved some results [5]. Bishop examined the application of process modeling and subsequent analyses to the insider problem [10]. However, these behavior-driven models just work out based on unusual changes of user behaviors, and they will miss recognizing situations where an user systematically attacks an organization over an extended time-framework.

3 Key Methodologies

We propose a hybrid model based on multi-dimensional features to detect insider threat that deviates from the portraits of normal users as well as who tries to behave like a normal user to cover up his evil. The structure of our proposed model is illustrated in Fig. 2. The hybrid model consists of two components that are named Across-Domain Anomaly Detection (ADAD) and Across-Time Anomaly Detection (ATAD). We get abnormal scores from the two components, then fuse them as the basis for insider threat detection. In this section, we describe the dataset used in this paper at first. Then we present our ADAD model and ATAD model in detail. Finally, we introduce the fusion method to combine ADAD and ATAD.

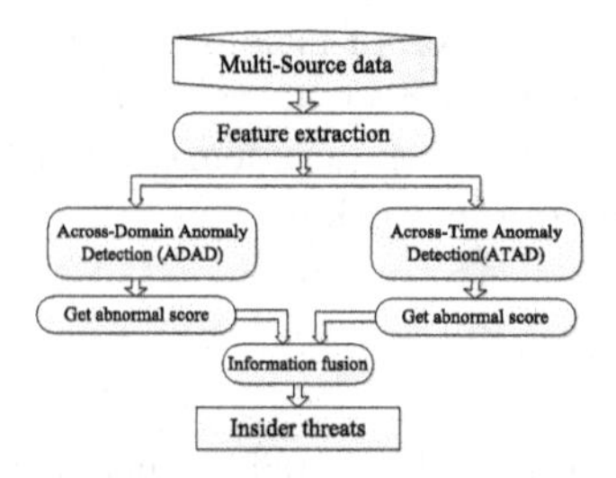

Fig. 2. Anomaly detection framework.

3.1 The Data Set

Due to the lack of availability of proper insider threat datasets, we utilize the insider threat dataset published by CERT Carnegie Mellon University for this research [5]. The 17-month period dataset "R4.2.tar.bz" is chosen for this analysis. This dataset consists of six types of data records (HTTP, logon, device, file,

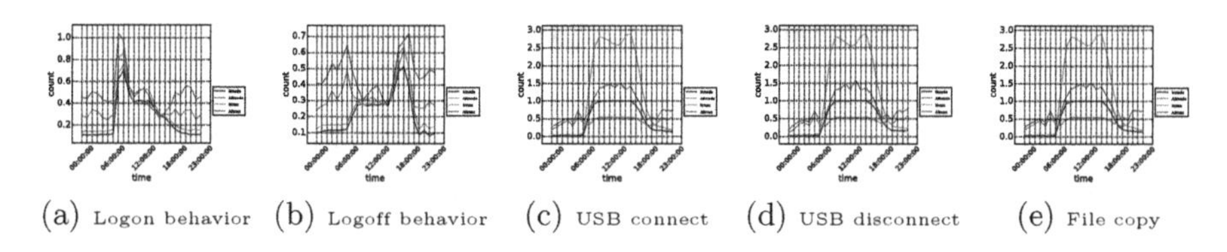

(a) Logon behavior (b) Logoff behavior (c) USB connect (d) USB disconnect (e) File copy

Fig. 3. Users' behaviors. "Nmode/Nmax" presents the mode/max number of normal users behaviors, and the "ABmode/ABmax" presents the mode/max number of abnormal users behaviors.

email and psychometric) of 1000 employees over a 17 months period. All HTTP records contain user, PC, URL and web page content with time stamps. The file "Logon.csv" consists of user logon/logoff activities with the corresponding PC with timestamps. The data file "device.cs" indicates "insert/remove" actions of the relevant user, PC, and timestamp. Details of file copies are stored in file "file.csv" with date, user, PC, filename, and content. Note that the Dataset contains the ground truth for each user (when they are acting maliciously or not), which allows us to monitor the success or failure of our experiment.

3.2 Across-Domain Anomaly Detection (ADAD)

This framework will utilize multi-domain information inputs, such as logon records and operation on file to identify abnormal users, who behave differently from their peers. We first extract temporal features from multi-domain data, upon which insider threats are detected.

Feature Extraction. *Logon/Logoff Behaviors.* As most disgruntled insiders tend to commit malicious logon or logoff activities after hours, these behaviors are used to identify malicious insiders.
Removable Media Usage. Removable media is one of the most popular methods used in theft of Intellectual Property. The use of removable media can be an excellent information source for identifying suspicious events.
File Copy Behaviors. File copying is an easy method to steal confidential evidence from organizations. So the number of occurrences of this behavior can give some useful information to detect insider threats.

To profile users' behaviors, we extract several temporal features to present the occurrences of different behaviors. For each user, we first count the times of each behavior for every hour, and then calculate the average of the maximum counts and mode counts of each behavior. The maximum counts indicate the change range of user behaviors, and the mode counts present the general behavior of most users. In order to find out whether the number of occurrences of these behaviors is different between normal users and abnormal users, Fig. 3 investigates their distributions. We find that there is a big difference in behavior between normal users and abnormal users at different times in a day. Compared

Table 1. Selected feature set.

Module	Features (per 6 h)	The number of features
Logon events	Max/Mode logon counts per 6 h	4
Logoff events	Max/Mode logoff counts per 6 h	4
Removable media	Max/Mode USB connection counts per 6 h	4
	Max/Mode USB disconnection counts per 6 h	4
File copy events	Max/Mode filecopy counts per 6 h	4

with normal users, these abnormal users have more frequent operations in midnight. Therefore, we divide the 24 h of a day into 4 time segments: (0:00–06:00), (00:60–12:00), (12:00–18:00), and (18:00–24:00). The maximum or mode counts of user behaviors with regard to the five domains are calculated for the 4 segments respectively. Table 1 is an illustration of the feature set.

In addition, in order to select domain features that have big impacts on users' behavior, we apply the Principal Component Analysis (PCA) [16] to allocate each feature with a score value. The features with high scores are used as the input to the model.

Anomaly Detection. Due to the complex background of insider threat problem, it is hard to pinpoint an user as a malicious insider. This section focuses on implementing an anomaly detection algorithm based on the features identified in the last subsection. The anomaly detection algorithm adopted in this analysis is the "Isolation Forest (IForest)" algorithm [17], which stands out in effectively separating anomalous events from the rest of the instances. In the end, the IForest gives an anormal score R_{ADAD} for each user and outputs the users who may be an insider threat explicitly.

3.3 Across-Time Anomaly Detection (ATAD)

In this section, an improved model (IM) based on Markov is proposed to detect insider threats. We first construct a model to represent the regular behaviors of an user. Then, we compare the temporal behavior in the recent past with the regular behaviors to detect unusual changes of his behaviors.

Model Constructing. Markov model (MM) [18] is a powerful tool to model temporal sequence information. It has been widely used in temporal pattern recognition problems (e.g. speech recognition, bioinformatics, gesture recognition), as well as in the area of intrusion detection [18]. When constructing an improved model (IM), we take users behaviors as a temporal sequence for MM.

Users have different behaviors on computers every day. The historical behaviors of an user can be represented as a sequence of observations $B = (a_1, a_2, \ldots, a_n)$, where a_i is the behavior that he/she is served at time i. The *order-k* (or "*O(k)*") Markov assumes that the behavior can be predict from the current *context*; that is, the sequence of the k most recent symbols in the behavior history $(a_{n-k+1}, \ldots, a_n)$. Moreover, the underlying Markov model represents each state as a context, and transitions represent the possible behaviors that follow that context.

Let sub-string $B(i,j) = a_i a_{i+1} \ldots a_j$ for any $1 \leq i \leq j \leq n$. We think of the user's behaviors as a random variable X. Let *X(i,j)* be a string $X_i X_{i+1} \ldots X_j$ representing the sequence of random variates $X_i, X_{i+1}, \ldots X_j$ for any $1 \leq i \leq j \leq n$. Define the context $c = B(n-k+1, n)$. Let A be the set of all possible behaviors. The Markov assumption is that X behaves as follows, for all $a \in A$ and $i \in \{1, 2, \ldots, n\}$:

$$\begin{aligned} P(X_{n+1} &= a|X(1,n) = B) \\ &= P(X_{n+1} = a|X(n-k+1,n) = c) \\ &= P(X_{i+k+1} = a|X(i+1,i+k) = c), \end{aligned} \tag{1}$$

where $P(X_i = a_i|\ldots)$ denotes the probability that X_i takes the value a_i. The first two lines indicate the assumption that the probability depends only on the context of the k most recent behaviors. The latter two lines indicate the assumption of a stationary distribution; that is, the probability is the same anywhere the context is the same.

These probabilities can be represented by a *transition probabilities matrix M*. Both the rows and columns of M are indexed by *length-k* strings from A^k so that

$$P(X_{n+1} = a|X(1,n) = B(1,n)) = M(s,s'), \tag{2}$$

where $s = B(n-k+1,n)$ is the current context and $s' = B(n-k+2,n)$ is the next context.

In order to obtain *M*, we generate an estimate $\hat{P}$ from the current history B using the current context c. The probability for the next transition symbol to be a is

$$P(a) = \hat{P}(X_{n+1} = a|B) = \frac{N(ca,B)}{N(c,B)}, \tag{3}$$

where *N(s', s)* denotes the number of times the substring *s'* occurs in the string *s*.

The estimate predicts the symbol $a \in A$ with the maximum probability $\hat{P}(X_{n+1} = a|B)$; that is, the symbol that most frequently followed the current context c in prior occurrences in the history. For the user's behavior, the current behavior is closely related to the previous one, but there is not much contact with the behavior before the previous one. We also found that the M matrix obtained when $k > 1$ is sparser than the M matrix obtained when $k = 1$ through experiments. So, in this paper, we use $k = 1$ to obtain the M matrix.

Anomaly Detection. We define the temporal behaviors in the recent past by opening up D continuous observation windows with the length of N. The users behaviors in the *i-th* observation window is defined as $B_i = a_{i1}, a_{i2}, \ldots, a_{iN}$, where N denotes the *N-th* behavior in the window i. Given the M matrix produced in the first step, the anormal score R_{ATAD} of the user is calculated as

$$R_{ATAD} = \frac{\sum_{d=1}^{D} \prod_{n=1}^{N} P(a_{dn})}{D}, \tag{4}$$

After obtaining the anormal score for all users, we set a threshold T carefully which is used as a critical parameter of our model to distinguish the anomalous users. If the anormal score of the user is below the threshold, he/she is identified as anomalous. About the value of T, we can save the anormal scores generated by IM algorithm and test different values of T to choose a proper one. Besides, one can also imagine a human security analyst increasing T from 0 in order to get the value greater than the scores of most anomalous instances.

3.4 Information Fusion

In this section, our goal is to combine anormal scores that have been generated from the two models proposed above and achieve a higher detection accuracy. A method based on weight fusion is developed to combine the anomaly sores derived from ADAD and ATAD. The combined anormal score $R_{combine}$ is computed as

$$R_{combine} = W * R_{ATAD} + (1 - W) * R_{ADAD}, \tag{5}$$

where R_{ADAD} (R_{ATAD}) denotes the anormal score of ADAD (ATAD), and W denotes the weight of R_{ATAD}. We then set a threshold $T_{combine}$, which is used to distinguish anomalous users. If the anormal score of the user is below the threshold, the user is regarded as the anomalous one.

4 Experimental Results

This section will provide a comprehensive evaluation on our proposed models. First, we assess the performance of ADAD, and then discuss the results of the ATAD. Finally, we provide a comparative assessment of our proposed hybrid model by comparing with the ADAD and ATAD.

4.1 Across-Domain Anomaly Detection (ADAD)

The ADAD model identifies the anomalies by using the IForest. We use *accuracy, precision* and *recall* to evaluate the performance of this model. *Precision* is the fraction of the data entries labeled malicious that are truly malicious; *recall* is the fraction of malicious entries that are classified correctly; *accuracy* is the fraction of all entries that are classified correctly [19].

Table 2. The experimental results of ADAD.

Domain	Features	Accuracy	Precision	Recall
Logon	MAX	89%	32%	47%
	MODE	87%	22%	34%
Logoff	MAX	88%	23%	34%
	MODE	85%	14%	21%
USB connect	MAX	78%	65%	27%
	MODE	79%	70%	28%
USB disconnect	MAX	80%	73%	30%
	MODE	79%	69%	28%
Filecopy	MAX	80%	60%	28%
	MODE	77%	44%	20%
All domains	MAX	82%	68%	34%
	MODE	79%	52%	31%
All domains combined with PCA method		**87%**	**79%**	**35%**

Table 2 shows the results of our models ADAD with different parameters. We find that USB connection and disconnection are easy to detect. While logon and file domains are weak in detecting insider threats. It appears that normal users and abnormal users show great variations in their behavior of device usages, but are more uniform in the logon and file behaviors. Moreover, the mode counts are more effective than maximum counts. The result indicates that the mode represents a more significant difference between the normal and the abnormal. It is obvious that some features interfere with the ability to correctly identify insider threats. The model can achieve a higher detection accuracy via decomposing the features to a 2-D space by *PCA*, which denoises the data at the same time.

4.2 Across-Time Anomaly Detection (ATAD)

In this section, we first introduce the performance metrics for evaluating the detection model and then provide a comparative assessment of our proposed models with existing MM detection methods as before.

For ATAD, the threshold T is essential to classify users as anomalous or not. So the metrics Receiver Operating Characteristic curves (or *ROC* curves) is applied to evaluate the models [5].

Comparison Methods. MM is an extremely powerful tool to model temporal sequence information. It was also used in the general area of intrusion detection by some notable works [18]. The IM is compared with the existing detection model Markov.

Comparison Between IM and MM. 60 days from the 16th month and the 17th month of dataset R4.2.tar.bz have been used for ATAD. Specifically, the first 53 days in the 60 days are treated as the training set to build regular behaviors model and the rest are used as the testing set. Choosing 53 days as the training set can avoid the user's regular behaviors changing, which has an effect on building the normal model. The testing set is composed of records of a week, as behaviors of a week can build a complete work cycle profile for an user. IM mode sets the length of the observation window N as a day. The results of the IM model and MM model are shown in Fig. 4. It can be seen that the IM model is more effective than MM from the results. For MM, if an user has one unusual change during this week, the change will have an significant influence on the detection result. However, this effect can be avoided by IM. When scoring the user's behavior, IM uses the week's average anormal score to evaluate an user's behavior which can reduce the impact of accidental unusual change on detecting insider threats. Hence, it detects insider threats more robustly.

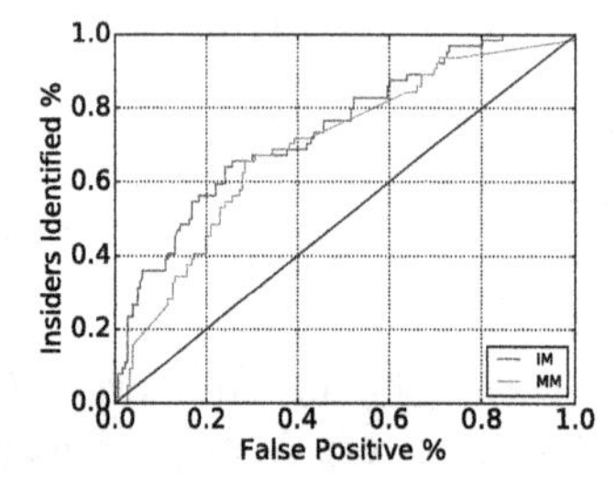

Fig. 4. ROC Curve showing the differences between MM algorithm and IM algorithm.

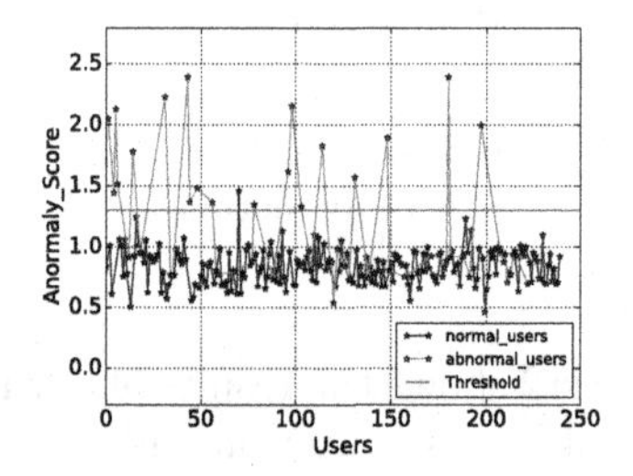

Fig. 5. Abnormal score distribution. (Color figure online)

4.3 Information Fusion

For the fusion method, the most important issue is to determine the weight of each component. So we let W, the weight of ATAD, increase from 0 to 1 with increment of 0.1. Then we use the metrics *accuracy, precision* and *recall* to quantitatively evaluate the models.

We compare the fusion method with the ADAD and the ATAD, and the results of detecting insider threats are shown in Table 3. To remove amplitude variation and only focus on the underlying distribution shape on data, the scores are normalized before the weight fusion. When $W = 0$, it is the result of individual ADAD method. When $W = 1$, it is the result of individual ATAD method. We can see the ADAD is better than the ATAD, and fusing two model achieves better performance. When W varies between 0.1 and 0.3, the precision of the hybrid model achieves highest. Since the ADAD model is a data-driven model, it fails to detect a malicious insider who tries to behave like a normal user to cover up his evil. However, the ATAD can make up for this deficiency by comparing behaviors of users in different time periods. Therefore, the hybrid model which combines the individual ATAD and ADAD scores can outperform each of them.

Table 3. The result of the information fusion experiment.

W	0 (ADAD)	0.1	0.2	0.3	0.4	0.5	0.6	0.7	0.8	0.9	1 (ATAD)
Precision	79.17%	**90%**	**94.44%**	**95%**	74.73%	90%	85%	82.35%	66.67%	61.9%	60%
Recall	35.19	**35.18%**	**31.48%**	**35.19%**	33.33%	33.33%	31.48%	25.92%	25.92%	24.07%	27.78%

To get a better visualization of the results based on our approaches, we mark the positive samples with the red and negative samples with blue. When W is 0.3, we chose a threshold of 1.3 to determine insider threats. Figure 5 is an indication of how the anormal scores are distributed when W is 0.3. The purple color horizontal line is equivalent to the threshold. Users whose anormal scores exceed the threshold can be considered as the anomalous. In fact, the model has some limitations. Our current procedure for determining whether the user is anomalous or not is to compare the anormal score with a threshold. This requires us to manually set a threshold value, which is a hyperparameter of our mode that significantly affects the results. We will explore how to set this parameter in the future study.

5 Conclusion

In this paper, we proposed a hybrid model that combined a data-driven model with a behavior-driven model to detect insider threat in a more robust and accurate manner. First, the multi-dimensional features extracted from data collected from the enterprise network is formatted and fed separately into two separate models. Second, each model generates an abnormal score to represent the degree of users' unusual behaviors. Finally, the abnormal scores of two models were fused as the final abnormal scores for each user, and an user was detected as an insider threat if the anormal score exceeded the threshold. After a wide range of experiments, it is verified that the hybrid model can detect insider threats with a high accuracy of 95%, which is of great significance in industry and scientific research. In the future work, we will take the user's job role [15] into account to improve the efficiency of our method for anomaly detection.

Acknowledgment. This work was supported by the National Natural Science Foundation of China (No. 61372062).

References

1. Gavai, G., et al.: Detecting insider threat from enterprise social and online activity data. In: ACM CCS International Workshop on Managing Insider Security Threats, pp. 13–20. ACM (2015)
2. By the numbers: cyber attack costs compared (2016). http://www.csoonline.com/article/3074826/security/bythe-numbers-cyber-attack-costs-compared.html. Accessed 31 May 2016

3. Cappelli, D., Moore, A., Trzeciak, R.: The CERT Guide to Insider Threats: How to Prevent, Detect, and Respond to Information Technology Crimes (Theft, Sabotage, Fraud). Addison-Wesley Professional, Boston (2012)
4. Young, W.T., et al.: Use of domain knowledge to detect insider threats in computer activities (2013)
5. Rashid, T., Agrafiotis, I., Nurse, J.R.C.: A new take on detecting insider threats: exploring the use of hidden Markov models. In: International Workshop, pp. 47–56 (2016)
6. Eldardiry, H., Sricharan, K., Liu, J., et al.: Multi-source fusion for anomaly detection: using across-domain and across-time peer-group consistency checks. Comput. Inform. **31**(3), 575–606 (2014)
7. Winston, W.L.: Operations Research: Applications and Algorithms. Duxbury Press, Belmont (1994)
8. Zeadally, S., et al.: Detecting insider threats: solutions and trends. Inf. Secur. J. Glob. Perspect. **21**(4), 183–192 (2012)
9. Gamachchi, A., Sun, L., Boztas, S.: A graph based framework for malicious insider threat detection. In: Hawaii International Conference on System Sciences (2017)
10. Legg, P.A., et al.: Towards a conceptual model and reasoning structure for insider threat detection. J. Wirel. Mob. Netw. Ubiquit. Comput. Dependable Appl. **4**(4), 20–37 (2013)
11. Bishop, M., et al.: Insider threat detection by process analysis. In: Proceedings of IEEE SPW, pp. 251–264 (2014)
12. Mathew, S., Petropoulos, M., Ngo, H.Q., Upadhyaya, S.: A data-centric approach to insider attack detection in database systems. In: Jha, S., Sommer, R., Kreibich, C. (eds.) RAID 2010. LNCS, vol. 6307, pp. 382–401. Springer, Heidelberg (2010). https://doi.org/10.1007/978-3-642-15512-3_20
13. Eberle, W., Graves, J., Holder, L.: Insider threat detection using a graph-based approach. J. Appl. Secur. Res. **6**(1), 32–81 (2010)
14. Eldardiry, H., et al.: Multi-domain information fusion for insider threat detection. In: Proceedings of IEEE SPW, pp. 45–51, May 2013
15. Legg, P.A., Buckley, O., Goldsmith, M., et al.: Automated insider threat detection system using user and role-based profile assessment. IEEE Syst. J. **11**(2), 503–512 (2017)
16. Jolliffe, I.: Principal Component Analysis. Wiley, Hoboken (2005)
17. Liu, F.T., Ting, K.M., Zhou, Z.H.: Isolation forest. In: 2008 Eighth IEEE International Conference on Data Mining, pp. 413–422, December 2008
18. Ye, N.: A Markov chain model of temporal behavior for anomaly detection, pp. 171–174 (2000)
19. Ko, L.L., et al.: Insider threat detection and its future directions. Int. J. Secur. Netw. **12**(3), 168 (2017)

Throughput Analysis for Energy Harvesting Cognitive Radio Networks with Unslotted Users

Honghao Ma, Tao Jing, Fan Zhang, Xin Fan, Yanfei Lu, and Yan Huo(✉)

School of Electronics and Information Engineering, Beijing Jiaotong University, Beijing, China
yhuo@bjtu.edu.cn

Abstract. Considering a cognitive radio network (CRN) with the energy harvesting (EH) capability, we design a sensing-based flexible timeslot structure for a secondary transmitter (ST). This structure focuses on an unslotted transmission mode between two primary users (PUs). In this structure, the ST can decide whether to transmit data or to harvest energy based on the sensing results. Aiming to maximize the long-term average achievable throughput of the secondary system, we study an optimal policy, including the optimal energy harvesting time as well as the optimal transmit power. To reduce the computational complexity, we also derive an effective suboptimal policy by maximizing the upper bound on the throughput. Finally, simulation results demonstrate that the proposed flexible timeslot structure outperforms the conventional fixed timeslot structure in terms of average achievable throughput.

Keywords: Cognitive radio network · Energy harvesting
Unslotted primary users · Average available throughput

1 Introduction

A cognitive radio network (CRN) with energy harvesting (EH) is expected as a promising solution for green communications [1,2]. Different from energy-efficient protocol designs [3–5], the EH technology may increase the battery life of wireless devices by replenishing energy from various energy sources [6,7], while CRNs can improve the spectral efficiency by opportunistic access schemes. In view of the inherent "harvesting-sensing-throughput" tradeoff in EH CRNs [8,9], it is crucial for an EH secondary user (SU) to effectively utilize energy (i.e., charging or discharging) to improve system performance and spectral efficiency [10,11].

Existing studies focused on the optimal energy management and spectrum sensing policies in EH CRNs with time-slotted PUs. In this scenario, the channel of a PU remains idle or busy invariably in one slot while changes with a certain probability in another. Also, SUs need to synchronize with the PUs' slot to achieve cooperative communications. In particular, the authors presented a saving-sensing-transmitting timeslot division strategy for an EH secondary

S. Chellappan et al. (Eds.): WASA 2018, LNCS 10874, pp. 345–356, 2018.
https://doi.org/10.1007/978-3-319-94268-1_29

transmitter (ST) to maximize its expected achievable throughput in [8]. In [12] the authors analyzed impacts of sensing and access probabilities as well as the energy queue capacity on the maximum achievable throughput in a multi-user EH CRN. The authors of [13] investigated the optimal energy-efficient resource allocation schemes for the EH CRNs. Moreover, information-energy cooperative strategies in [14,15] are investigated to further improve the energy efficiency.

Yet, PUs may send signals in an unslotted manner during actual transmission [16,17]. PUs and SUs may be hard to be synchronized because of incompatible communication protocols. As a result, it is urgent to study novel spectrum access strategies and/or power allocation schemes for SUs with unslotted primary systems. To the best of our knowledge, little has been done to improve the achievable throughput of an SU in the unslotted EH CRNs. Our exploration of this uncharted area requires addressing the following challenging questions: (i) How to formulate a primary traffic model in the unslotted scenario? (ii) How to design an unslotted data transmit policy of an SU with energy constraint?

Considering these questions, the authors formulated the duration of either an idle state or a busy one as an exponential distributed random variable in [16–18]. They calculated the prior probability of channel being idle using the mean durations of these two states. Similarly, [19,20] employed the channel's state transition matrix to derive stationary probabilities of channel states. Then, several solutions were proposed for the conventional CRNs without EH. The authors first derived the optimal frame duration and transmit power for energy-unconstrained SUs in [17]. Then, they studied the optimal power control policy in [16] to maximize SUs' energy efficiency in the presence of unslotted PUs and sensing errors. In [19], the authors designed two data transmit policies with idle and busy sensing results to fully utilize unslotted channels. Also, the authors of [20] elaborated an optimal dual sensing-interval policy to maximize ST's spectrum utilization to achieve opportunistic energy harvesting for primary signals.

By adopting the same assumption of [20], we present a novel sensing-based flexible timeslot structure for the unslotted EH CRN in this article. Instead of achieving SU's optimal spectrum utilization, we intend to maximize its achievable throughput. Moreover, we assume that an ST may harvest energy from the ambient environment. After that, it employs a part of the stored energy to sense the primary channel. If the channel is sensed as idle, the ST will send data via the channel; otherwise, it should continue to harvest energy then re-sense the channel after energy harvesting. The main contributions are as follows:

- Considering an EH CRN with an unslotted primary system, we propose a novel sensing-based flexible timeslot structure for the EH ST to maximize its long-term average achievable throughput.
- For the throughput maximization problem, we employ a differential evolution (DE) algorithm to derive the optimal policy, including the optimal harvesting time and the optimal transmit power.
- To reduce the computational complexity, we also derive an effective suboptimal policy by maximizing the upper bound on the achievable throughput.

The remainder of this paper is organized as follows. An EH CRN architecture including primary and secondary systems is described in Sect. 2. In Sect. 3, we formulate and solve the achievable throughput maximization problem of the secondary system. Section 4 demonstrates the throughput performance of our flexible timeslot structure. Finally, we conclude this paper in Sect. 5.

2 System Model

Figure 1 is an EH CRN with Rayleigh fading channels. Channel power gains $|h_{pp}|^2$, $|h_{sp}|^2$ and $|h_{ss}|^2$ of PT-PR, ST-PR and ST-SR links are exponentially distributed variables with unit mean. Moreover, noise power is assumed to be N_0 for all users.

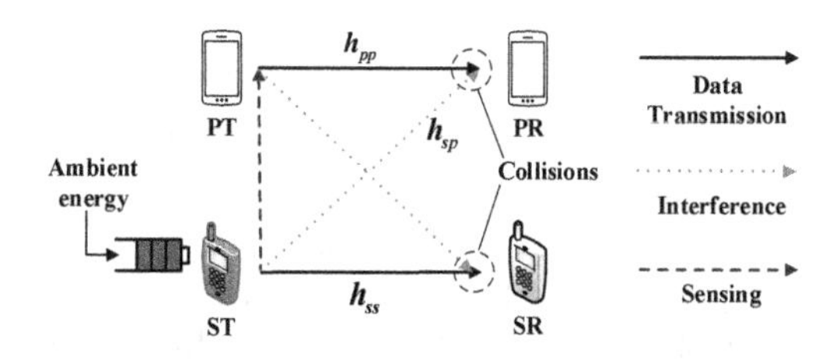

Fig. 1. System model of an EH CRN.

2.1 Primary System Model

A PT sends signals to its PR in an unslotted mode[1] with a fixed transmit power P_p. The corresponding channel state alternatively transfers between busy and idle with random durations. Similar to [17,19,20], busy and idle durations can be modeled as independent exponentially distributed random variables with mean $E_1 = \frac{1}{\lambda_1}$ and $E_0 = \frac{1}{\lambda_0}$.[2] Thus, the busy and idle probabilities are defined as

$$p_1 = \Pr\{S(t) = 1\} = \frac{\lambda_0}{\lambda_0 + \lambda_1}, \text{and } p_0 = \Pr\{S(t) = 0\} = \frac{\lambda_1}{\lambda_0 + \lambda_1}. \tag{1}$$

where $S(t) = 1$ (or $S(t) = 0$) indicates the channel is busy (or idle) at time t. And the primary communication is successful only when the received SINR at the PR is higher than a predefined threshold β [21].

[1] In this paper, we consider a single-user unslotted primary system without sensing ability, so ST's transmit may not prevent the PT from reactivation.

[2] Similar to [19], λ_1 and λ_0 can be known at an ST by probing the channel in a specified learning period.

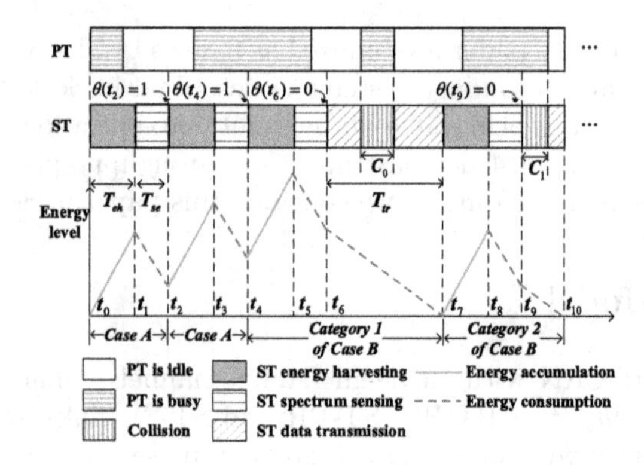

Fig. 2. The sensing-based flexible timeslot structure.

2.2 Secondary System Model

In the secondary system, an EH-enabled ST completely depends on the energy harvested from the ambient environment to communicate with its energy-unconstrained secondary receiver (SR) when the channel is idle. Here, we propose a novel sensing-based flexible timeslot structure for ST to realize effective transmissions, shown as Fig. 2. This structure includes three processes, i.e., energy harvesting, spectrum sensing, and data transmission, with durations of T_{eh}, T_{se}, and T_{tr}, respectively. Due to hardware duplex limitations [8,22], we assume that the ST can perform only one process at any time.

Energy Harvesting. In this process, the ST harvests energy from the ambient environment. The energy flows follow an i.i.d random process with mean P_{eh}, thus the average harvested energy is $e_h = P_{eh}T_{eh}$ during T_{eh}. These energy is stored in the battery and the battery capacity is assumed to be infinite. Note that the energy loss caused by harvesting is negligible in this paper.

Spectrum Sensing. After energy harvesting, the ST carries out spectrum sensing with an energy detector. Since T_{se} is much smaller than E_1 and E_0, it is reasonable to think that the channel state remains unchanged during T_{se} [19]. If the channel is sensed as idle, the ST transmits; otherwise, it converts to energy harvesting process immediately. The sensing accuracy is measured by the detection probability P_d and the false alarm probability P_f. For a target detection probability P_d^*, the false alarm probability P_f can be written as

$$P_f = Q\left(\sqrt{2\gamma_{ST}+1}\cdot Q^{-1}(P_d^*) + \gamma_{ST}\sqrt{T_{se}\cdot f_s}\right), \tag{2}$$

where $Q(x) = \frac{1}{\sqrt{2\pi}}\int_x^\infty e^{-t^2/2}\mathrm{d}t$ is a Q-function, f_s is the ST's sampling frequency to primary signals and γ_{ST} is the received SNR of primary signals at the ST.

For analysis simplicity, we assume the energy consumption for spectrum sensing e_s is proportional to T_{se}, i.e., $e_s = P_{se}T_{se}$, where P_{se} is the power consumption per unit of sensing time.

Data Transmission. It is assumed that the ST has perfect information of h_{ss} but only knows the distributions of h_{pp} and h_{sp}. When the channel is sensed as idle, the ST will exhaust all its residual energy E_r to transmit [8]. The transmit power, P_{tr}, is consistent during transmission and the transmit time $T_{tr} = \frac{E_r}{P_{tr}}$ varies with different E_r.

3 Problem Formulation

In this section, we first formulate the long-term achievable throughput of the secondary system, then derive a differential evolution based optimal policy to maximize this throughput. Finally, we further develop a suboptimal policy to reduce the computational complexity by maximizing the upper bound on the average achievable throughput.

3.1 Formulation of the Average Achievable Throughput

According to whether it transmits data after spectrum sensing, we classify ST's successive operations into *Case A* and *Case B*, as shown in Fig. 2.

Case A: The ST harvests energy and senses the channel, but not transmits data. *Case A* happens when the channel is sensed as busy (no matter what the exact channel state is) with the probability of $P_A = p_1 P_d^* + p_0 P_f$, where $p_1 P_d^*$ is the probability that the channel is correctly sensed as busy and $p_0 P_f$ is the probability that ST wrongly senses the idle channel as busy.

Case B: After the channel is sensed as idle, the ST performs a data transmission. Here, *Case B* can be further divided into *Scenario 1* and *Scenario 2* according to the real channel state:

1. *Scenario 1*: It happens when the channel is really idle and no false alarm is generated with the probability of $P_{B1} = p_0(1 - P_f)$.
2. *Scenario 2*: It happens when the channel is actually busy but wrongly sensed as idle by the ST with the probability of $P_{B2} = p_1(1 - P_d^*)$.

In an unslotted primary system, PUs can start or stop transmitting at any time. Thus in *Scenario 1*, the PT might occupy the channel when ST is transmitting. This inevitably leads to interference between two systems, i.e., collisions. ST's maximum instantaneous transmit rate under the idle channel is $r(P_{tr}) = \log_2(1 + \alpha P_{tr})$, where $\alpha = \frac{|h_{ss}|^2}{N_0}$. While a collision happens, secondary transmission fails and $r(P_{tr})$ reduces to zero for P_p is much higher than P_{tr}. And the outage probability of PUs is

$$P_{out} = \Pr[\frac{|h_{pp}|^2 P_p}{|h_{sp}|^2 P_{tr} + N_0} < \beta] = 1 - \frac{P_p e^{-\frac{\beta N_0}{P_p}}}{\beta P_{tr} + P_p}. \quad (3)$$

According to [20,23], given $S(t)=0$, the channel's expected idle time during $[t, t+T_{tr}]$ is a function of P_{tr}, i.e.,

$$T_0^0(P_{tr}) = \frac{E_r}{P_{tr}} - C_0(P_{tr}), \tag{4}$$

where

$$C_0(P_{tr}) = \frac{E_r}{P_{tr}} p_1 + \frac{e^{\left(\frac{-(\lambda_1+\lambda_0)E_r}{P_{tr}}\right)} - 1}{\lambda_1+\lambda_0} p_1 \tag{5}$$

is the average collision time and has been illustrated in Fig. 2.

For *Scenario 2*, the ST might have opportunities to enable successful transmission after the current primary transmission is finished. And the expected idle time during $[t, t+T_{tr}]$ provided that $S(t)=1$ is

$$T_0^1(P_{tr}) = \frac{E_r}{P_{tr}} - C_1(P_{tr}), \tag{6}$$

where

$$C_1(P_{tr}) = \frac{E_r}{P_{tr}} p_1 - \frac{e^{\left(\frac{-(\lambda_1+\lambda_0)E_r}{P_{tr}}\right)} - 1}{\lambda_1+\lambda_0} p_0. \tag{7}$$

The ST may go through *Case A* k times before going through *Scenario 1* or *Scenario 2* of *Case B* with the probability $P_{B1}{P_A}^k$ or $P_{B2}{P_A}^k$, respectively. And the residual energy E_r for each condition is $E_r(T_{eh}) = (1+k)(T_{eh}P_{eh} - T_{se}P_{se})$. Therefore the average achievable throughput of the secondary system for a long-term period can be written as

$$R(T_{eh}, P_{tr}) = \sum_{k=0}^{\infty} \frac{r(P_{tr})\Delta(P_{tr})P_{tr}}{T(T_{eh})P_{tr} + E_r(T_{eh})} \cdot {P_A}^k, \tag{8}$$

where $\Delta(P_{tr}) = P_{B1}T_0^0(P_{tr}) + P_{B2}T_0^1(P_{tr})$ is ST's average effective transmit time, and $T(T_{eh}) = (1+k)(T_{eh}+T_{se})$.

3.2 Average Achievable Throughput Maximization

We formulate the maximization problem of the long-term average achievable throughput as

$$\begin{aligned}
\mathbf{P1:} \quad & \max_{T_{eh}, P_{tr}} \quad R(T_{eh}, P_{tr}) && \text{(9a)}\\
& s.t. \quad P_{tr} \geq 0 && \text{(9b)}\\
& \qquad e_h \geq e_s && \text{(9c)}\\
& \qquad P_{out} \leq P_{out}^{max} && \text{(9d)}\\
& \qquad 0 \leq T_{eh} \leq T_{eh}^U. && \text{(9e)}
\end{aligned}$$

In **P1**, (9c) refers to the average energy causality constraint. P_{out}^{max} in (9d) is the outage probability threshold of the primary system. And T_{eh}^U in (9e) is the maximum allowed energy harvesting time.

Substituting the expressions of e_h and e_s into (9c), we obtain

$$T_{eh} \geq T_{eh}^L, \text{ where } T_{eh}^L = \frac{P_{se}T_{se}}{P_{eh}}. \tag{10}$$

And according to (9d), we can get

$$P_{tr} \leq P_{tr}^U, \text{ where } P_{tr}^U = \frac{1}{\beta}\left(\frac{P_p e^{-\frac{\beta N_0}{P_p}}}{1 - P_{out}^{max}} - P_p\right). \tag{11}$$

Then **P1** can be rewritten as

$$\begin{aligned} \textbf{P2:} \quad & \max_{T_{eh}, P_{tr}} \quad R(T_{eh}, P_{tr}) \\ & \quad s.t. \quad T_{eh}^L \leq T_{eh} \leq T_{eh}^U \\ & \qquad\quad\; 0 \leq P_{tr} \leq P_{tr}^U. \end{aligned} \tag{12}$$

Here, **P2** is a non-trivial problem. We employ the differential evolution (DE) algorithm [24] to solve it. The optimal policy derivation is detailed in Algorithm 1, where the population size N_p, mutation factor F, crossover constant C_r and the maximum number of generation G_{max} are set to be 30, 0.85, 0.7 and 100, respectively. The total time complexity of Algorithm 1 is $O(2N_pG_{max})$.

Algorithm 1. The DE-based Optimal Policy Derivation.

1: Set $G = 0$;
2: Randomly select $(T_{eh}(i), P_{tr}(i))$ from $T_{eh} \in [T_{eh}^L, T_{eh}^U]$ and $P_{tr} \in [0, P_{tr}^U]$;
3: **while** $G < G_{max}$ **do**
4: $\quad G = G + 1$;
5: $\quad$ **for** $i \in \{1, 2, ..., N_p\}$ **do**
6: $\quad\quad$ Randomly pick a, b and $c \in \{1, ..., N_p\} - \{i\}$;
7: $\quad\quad D = (T_{eh}(b), P_{tr}(b)) - (T_{eh}(c), P_{tr}(c))$;
8: $\quad\quad (T_{eh}^{var}, P_{tr}^{var}) \leftarrow (T_{eh}(a), P_{tr}(a)) + F \cdot D$;
9: $\quad\quad$ Randomly pick k_1 and $k_2 \in [0, 1]$;
10: $\quad\quad$ **if** $k_1 > Cr$ **then**
11: $\quad\quad\quad T_{eh}^{var} \leftarrow T_{eh}(i)$;
12: $\quad\quad$ **end if**
13: $\quad\quad$ **if** $k_2 > Cr$ **then**
14: $\quad\quad\quad P_{tr}^{var} \leftarrow P_{tr}(i)$;
15: $\quad\quad$ **end if**
16: $\quad\quad$ **if** $R(T_{eh}^{var}, P_{tr}^{var}) > R(T_{eh}(i), P_{tr}(i))$ **then**
17: $\quad\quad\quad (T_{eh}(i), P_{tr}(i)) \leftarrow (T_{eh}^{var}, P_{tr}^{var})$;
18: $\quad\quad$ **end if**
19: $\quad$ **end for**
20: **end while**
21: **return** $(T_{eh}^*, P_{tr}^*) \leftarrow \text{argmax}\{R(T_{eh}(i), P_{tr}(i)), i = 1, ..., N_p\}$;

3.3 Suboptimal Solution Derivation

Since the optimal policy of Algorithm 1 has a relatively high computational complexity, we next propose a suboptimal but effective policy as below.

Theorem 1. *The average achievable throughput of the secondary system $R(T_{eh}, P_{tr})$ is upper bounded by*

$$R_U(T_{eh}, P_{tr}) = \frac{p_0(1-P_f)r(P_{tr})\phi_1(T_{eh})}{\phi_1(T_{eh})P_{tr} + \phi_2(T_{eh})} \cdot \frac{1}{1-P_A}, \tag{13}$$

where $\phi_1(T_{eh}) = T_{eh}P_{eh} - T_{se}P_{se}$, $\phi_2(T_{eh}) = T_{eh} + T_{se}$.

Proof. Since $Q(x)$ is a monotonously decreasing function, $P_f \leq P_d^*$ holds according to (2). Then we have

$$\Delta(P_{tr}) \leq (1-P_f)(p_0 T_0^0(P_{tr}) + P_1 T_0^1(P_{tr})). \tag{14}$$

Substituting (4) and (6) into (14), we can further derive that $\Delta(P_{tr}) \leq \frac{p_0(1-P_f)E_r}{P_{tr}}$. Additionally, $\sum_{k=0}^{\infty} P_A^k = \frac{1}{1-P_A}$. Therefore,

$$R(T_{eh}, P_{tr}) \leq \frac{p_0(1-P_f)r(P_{tr})\phi_1(T_{eh})}{\phi_1(T_{eh})P_{tr} + \phi_2(T_{eh})} \cdot \frac{1}{1-P_A}. \tag{15}$$

This completes the proof of Theorem 1.

Then we discard the constant components in $R_U(T_{eh}, P_{tr})$ and maximize it under the same constraints with **P2**, i.e.,

$$\begin{aligned} \textbf{P3:} \quad & \max_{T_{eh}, P_{tr}} \quad \frac{r(P_{tr})\phi_1(T_{eh})}{\phi_1(T_{eh})P_{tr} + \phi_2(T_{eh})} \\ & \quad s.t. \quad T_{eh}^L \leq T_{eh} \leq T_{eh}^U \\ & \qquad\qquad 0 \leq P_{tr} \leq P_{tr}^U. \end{aligned} \tag{16}$$

To solve **P3**, we present the following Theorem 2.

Theorem 2. *For **P3**, the optimal energy harvesting time $T_{eh}^{'}$ is T_{eh}^U and the optimal transmit time $P_{tr}^{'}$ is given as*

$$P_{tr}^{'} = \min\{P_{tr}^U, \frac{\varphi}{\alpha \cdot W\left(\frac{\varphi}{e}\right)} - \frac{1}{\alpha}\}, \tag{17}$$

where $W(\cdot)$ refers to the Lambert W function and $\varphi = \frac{\alpha(T_{eh}^U P_{eh} - T_{se}P_{se})}{T_{eh}^U + T_{se}} - 1$.

Proof. We first define a function

$$f(T_{eh}, P_{tr}) \triangleq \frac{r(P_{tr})\phi_1(T_{eh})}{\phi_1(T_{eh})P_{tr} + \phi_2(T_{eh})}. \tag{18}$$

It can be easily proved that the first order partial derivative of $f(T_{eh}, P_{tr})$ with respect to T_{eh} is positive for any $T_{eh} \in [T_{eh}^L, T_{eh}^U]$. Therefore, $f(T_{eh}, P_{tr})$ is a monotonically increasing function for T_{eh} and the optimal harvesting time $T_{eh}^{'}$ of **P3** is obtained at T_{eh}^U.

Substituting $T_{eh}^{'} = T_{eh}^U$ into (18), we have $f(P_{tr}) = \frac{r(P_{tr})\phi_3}{\phi_4 P_{tr}+\phi_3}$, where $\phi_3 = T_{eh}^U P_{eh} - T_{se}P_{se}$ and $\phi_4 = T_{eh}^U + T_{se}$. We denote the stationary point of $f(P_{tr})$ by P_{tr}^s and it can be derived as

$$\left.\frac{\partial f(P_{tr})}{\partial P_{tr}}\right|_{P_{tr}=P_{tr}^s} = 0 \Rightarrow \alpha\phi_4 P_{tr} + \phi_3 - \phi_4(1+\alpha P_{tr})ln(1+\alpha P_{tr}) = 0$$
$$\Rightarrow P_{tr}^s = \frac{\varphi}{\alpha \cdot W\left(\frac{\varphi}{e}\right)} - \frac{1}{\alpha}, \tag{19}$$

where $W(\cdot)$ refers to the Lambert W function and $\varphi = \frac{\alpha\phi_4}{\phi_3} - 1$. Since $\frac{\partial f(P_{tr})}{\partial P_{tr}} \geq 0$ when $0 \leq P_{tr} \leq P_{tr}^s$ and $\frac{\partial f(P_{tr})}{\partial P_{tr}} \leq 0$ when $P_{tr} \geq P_{tr}^s$, the optimal transmit power of **P3** is $P_{tr}^{'} = \min\{P_{tr}^U, P_{tr}^s\}$. This completes the proof of Theorem 2.

Now, we can derive $(T_{eh}^{'}, P_{tr}^{'})$ as a suboptimal policy to maximize the average achievable throughput by using Theorem 2. Next, we will evaluate its performance in Sect. 4.

4 Simulation Results

In this section, simulation results are presented to evaluate the performance of our proposed sensing-based flexible timeslot structure. Unless mentioned explicitly, simulation parameters (mainly referred to [21,25]) are set as Table 1.

Table 1. Simulation parameters

Notations	Meanings	Values
β	SINR threshold of the PR	5
P_{out}^{max}	Outage probability threshold of PUs	0.1
P_p	PT's transmit power	50 mW
P_{se}	ST's sensing power	110 mW
T_{se}	ST's sensing time	1 ms
f_s	ST's sampling frequency	1 MHz
γ_{ST}	SINR of primary signal at the ST	−10 dB
N_0	Noise power	−40 dBm

Firstly, we present the impacts of average idle duration E_0 and average busy duration E_1 on the maximum average achievable throughput of the optimal policy, as shown in Fig. 3. The target detection probability P_d^* and P_{eh} are set to be 0.9 and 20 mW, respectively. As we can see, the maximum average

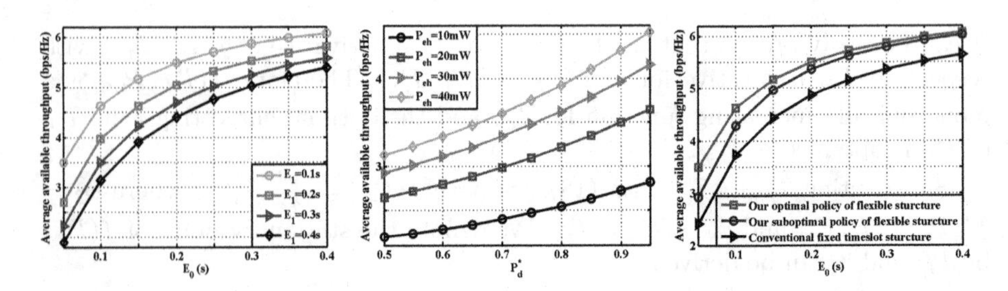

Fig. 3. R versus E_0.

Fig. 4. R versus P_d.

Fig. 5. Throughput.

achievable throughput increases with E_0 for a given E_1. The reasons come from two aspects. One is that the increase of E_0 results in a higher p_0 and a lower P_A when E_1 is kept constant. And the decreased P_A provides the ST with more spectrum access opportunities. The other is that the increase of E_0 prolongs ST's average effective transmit time and finally leads to a throughput improvement. In addition, it can be seen that the maximum average achievable throughput raises up with a slow-down tendency, due to p_0 converges to 1 as E_0 enlarges. These also explain the throughput decline resulting from the increase of E_1 for a fixed E_0. In the case of the same p_0 (i.e., $E_0 = 0.1$ s with $E_1 = 0.2$ s and $E_0 = 0.2$ s with $E_1 = 0.4$ s), the throughput is absolutely dominated by E_0 thus a higher E_0 gains a better throughput performance.

Next, we analyze the impact of the target detection probability P_d^* on the maximum average achievable throughput of the optimal policy in Fig. 4. Here, E_0 and E_1 are set to be 0.05 s and 0.1 s, respectively. As shown in Fig. 4, the maximum average achievable throughput grows significantly with the growth of P_d^*. Since P_A monotonically increases with P_d^*, the ST has fewer spectrum access opportunities. To take full advantages of these rare opportunities, the ST may reduce T_{eh} to sense the spectrum more frequently and then improve P_{tr} to acquire a higher instantaneous transmit rate. The decrease of T_{eh} and the increase of P_{tr} jointly result in a higher maximum average achievable throughput.

4.1 Impacts of Key Parameters on the Optimal Policy

Finally, we discuss the impact of P_{eh} on the throughput performance. From Fig. 4 we can observe that a higher average achievable throughput can be achieved by employing a higher P_{eh}. But the higher P_{eh} is, the smaller the throughput gain is obtained. This implies that blindly promoting energy harvesting performance is not always desirable. The reason is that the increment of harvesting rate could not lead to a proportional throughput gain as we expected.

4.2 Analysis of the Average Achievable Throughput

In this subsection, we compare the maximum average achievable throughput of the optimal policy and suboptimal policy of our proposed flexible timeslot structure in Fig. 5. A conventional fixed timeslot structure proposed in [8] is introduced as our reference. Note the difference of throughput between the optimal and suboptimal policies is quite small, which indicates that our suboptimal policy can provide a proper approximation to the optimal policy. In addition, the average available throughput grows gradually when the growth of E_0. The reason is that the larger E_0 can provide more spectrum access opportunities for an ST. Accordingly, the results in Fig. 5 illustrate that our policies (including the optimal policy and suboptimal policy) outperform the conventional one both in achievable throughput. This demonstrates that our proposed flexible structure has more freedom to harvest and sense than a fixed timeslot structure.

5 Conclusion

In this paper, we propose a flexible timeslot structure for an EH CRN. The ST with the energy harvesting capability in this structure can share the same spectrum with an unslotted primary system without degrading primary transmission. To achieve this goal, we formulate an optimal policy derivation problem to maximize the long-term average achievable throughput of the secondary system under energy causality constraint and the SINR requirement of the primary system. Then, we design a DE-based optimal policy derivation to find the optimal solution and further provide a relative suboptimal policy to reduce the computational complexity. Numerical results demonstrate that our flexible timeslot structure is superior to the conventional fixed one. These results also indicate that it is necessary to design this flexible structure for the secondary system when PUs communicate with each other in the unslotted mode.

Acknowledgments. This work was supported by the National Natural Science Foundation of China (Grant No. 61471028, 61571010, and 61572070), and the Fundamental Research Funds for the Central Universities (Grant No. 2017JBM004 and 2016JBZ003).

References

1. Ren, J., Hu, J., Zhang, D., Guo, H., Zhang, Y., Shen, X.: RF energy harvesting and transfer in cognitive radio sensor networks: opportunities and challenges. IEEE Commun. Mag. **56**(1), 104–110 (2018)
2. Ahmed, M.E., Kim, D.I., Kim, J.Y., Shin, Y.: Energy-arrival-aware detection threshold in wireless-powered cognitive radio networks. IEEE Trans. Veh. Technol. **66**(10), 9201–9213 (2017)
3. Cheng, S., Cai, Z., Li, J., Gao, H.: Extracting kernel dataset from big sensory data in wireless sensor networks. IEEE Trans. Knowl. Data Eng. **29**(4), 813–827 (2017)
4. Li, J., Cheng, S., Li, Y., Cai, Z.: Approximate holistic aggregation in wireless sensor networks. In: IEEE International Conference on Distributed Computing Systems, p. 11 (2015)
5. Cheng, S., Cai, Z., Li, J., Fang, X.: Drawing dominant dataset from big sensory data in wireless sensor networks. In: Computer Communications, pp. 531–539 (2015)
6. Shi, T., Cheng, S., Cai, Z., Li, J.: Adaptive connected dominating set discovering algorithm in energy-harvest sensor networks. In: IEEE INFOCOM 2016 - The IEEE International Conference on Computer Communications, pp. 1–9 (2016)

7. Shi, T., Cheng, S., Cai, Z., Li, Y., Li, J.: Exploring connected dominating sets in energy harvest networks. IEEE/ACM Trans. Netw. **25**(3), 1803–1817 (2017)
8. Yin, S., Qu, Z., Li, S.: Achievable throughput optimization in energy harvesting cognitive radio systems. IEEE J. Sel. Areas Commun. **33**(3), 407–422 (2015)
9. Hu, C., Li, H., Huo, Y., Xiang, T., Liao, X.: Secure and efficient data communication protocol for wireless body area networks. IEEE Trans. Multi-Scale Comput. Syst. **2**(2), 94–107 (2016)
10. Zhang, F., Jing, T., Huo, Y., Jiang, K.: Outage probability minimization for energy harvesting cognitive radio sensor networks. Sensors **17**(2), 224 (2017)
11. Hu, C., Li, R., Mei, B., Li, W., Alrawais, A., Bie, R.: Privacy-preserving combinatorial auction without an auctioneer. EURASIP J. Wirel. Commun. Netw. **2018**(1), 38 (2018)
12. Yun, H.B., Baek, J.W.: Achievable throughput analysis of opportunistic spectrum access in cognitive radio networks with energy harvesting. IEEE Trans. Commun. **64**(4), 1399–1410 (2016)
13. Yadav, R., Singh, K., Gupta, A., Kumar, A.: Optimal energy-efficient resource allocation in energy harvesting cognitive radio networks with spectrum sensing. In: Vehicular Technology Conference, pp. 1–5 (2017)
14. Xu, B., Chen, Y., Carrin, J.R., Zhang, T.: Resource allocation in energy-cooperation enabled two-tier NOMA hetnets toward green 5G. IEEE J. Sel. Areas Commun. **35**(12), 2758–2770 (2017)
15. Zhang, R., Chen, H., Yeoh, P.L., Li, Y., Vucetic, B.: Full-duplex cooperative cognitive radio networks with wireless energy harvesting. In: IEEE International Conference on Communications (2017)
16. Ozcan, G., Gursoy, M.C., Tang, J.: Spectral and energy efficiency in cognitive radio systems with unslotted primary users and sensing uncertainty. IEEE Trans. Commun. **65**(10), 4138–4151 (2017)
17. Ozcan, G., Gursoy, M.C., Tang, J.: Power control for cognitive radio systems with unslotted primary users under sensing uncertainty. In: IEEE International Conference on Communications, pp. 1428–1433 (2015)
18. Zhang, F., Jing, T., Huo, Y., Jiang, K.: Throughput optimization for energy harvesting cognitive radio networks with save-then-transmit protocol. Comput. J. **60**(6), 911–924 (2017)
19. Messina, A.: Power and transmission duration control in un-slotted cognitive radio networks. In: Computer Applications and Information Systems, pp. 1–6 (2014)
20. Pratibha, P., Li, K.H., Teh, K.C.: Optimal spectrum access and energy supply for cognitive radio systems with opportunistic RF energy harvesting. IEEE Trans. Veh. Technol. **66**(8), 7114–7122 (2017)
21. Che, Y.L., Duan, L., Zhang, R.: Spatial throughput maximization of wireless powered communication networks. IEEE J. Sel. Areas Commun. **33**(8), 1534–1548 (2014)
22. Luo, S., Rui, Z., Teng, J.L.: Optimal save-then-transmit protocol for energy harvesting wireless transmitters. IEEE Trans. Wirel. Commun. **12**(3), 1196–1207 (2013)
23. Mehanna, O., Sultan, A.: Inter-sensing time optimization in cognitive radio networks. Comput. Sci. **72**, 5533 (2010)
24. Yin, S., Zhang, E., Yin, L., Li, S.: Saving-sensing-throughput tradeoff in cognitive radio systems with wireless energy harvesting. In: 2013 IEEE Global Communications Conference (GLOBECOM), pp. 1032–1037, December 2013
25. Park, S., Kim, H., Hong, D.: Cognitive radio networks with energy harvesting. IEEE Trans. Wirel. Commun. **12**(3), 1386–1397 (2013)

A Crowdsourcing-Based Wi-Fi Fingerprinting Mechanism Using Un-supervised Learning

Xiaoguang Niu[1,2(✉)], Chun Zhang[1], Ankang Wang[1], Jingbin Liu[2,3], and Zhen Wang[4]

[1] School of Computer Science, Wuhan University, Wuhan, China
{xgniu, czhc, ak_wang}@whu.edu.cn
[2] Collaborative Innovation Center of Geospatial Technology, Wuhan, China
[3] State Key Laboratory of Information Engineering in Surveying, Mapping and Remote Sensing, Wuhan University, Wuhan, China
jingbin.liu@whu.edu.cn
[4] Troops 62115 PLA, Beijing, China
471135881@qq.com

Abstract. In recent years, the Wi-Fi fingerprint-based indoor localization methods are widely applied to more and more ubiquitous applications. One of the key concerns is how to efficiently collect Wi-Fi fingerprint to reflect the harsh indoor environmental dynamics. However, continuous Wi-Fi fingerprinting confronts a contradiction: consumption in fingerprint collection and the real-time accuracy of fingerprint. We find that location fingerprint variations are related to crowd spatial distribution, and the distributions often varies periodically. Based on these observations, this paper proposes a crowdsourcing-based Wi-Fi fingerprinting mechanism using un-supervised learning, which exploit the historical data similar to the current fingerprint with particle filter method to enrich the data updating location fingerprint and generated updated location fingerprint with Gaussian process regression. Experimental results show that in our experimental environment, compared with the location fingerprints which are updated with only current data, the mean square error of the updated location fingerprints is reduced significantly.

Keywords: Wi-Fi fingerprinting · Indoor localization · Crowdsourcing Un-supervised learning

1 Introduction

In recent years, the Wi-Fi fingerprint-based localization methods are widely applied. In such methods, the accuracy of localization greatly depends on how well the fingerprints fit into the groundtruth at the time. In online localization, the crowd spatial distributions, the locations of the objects and the changes of APs make updating location fingerprint very important. However, periodical complete location fingerprinting can take a lot of manpower, material resources and time. A common solution is updating location fingerprint with crowdsourcing data. However, in the practical application of simple crowdsourcing, there is often an insufficient amount of data, and the historical data are ignored.

S. Chellappan et al. (Eds.): WASA 2018, LNCS 10874, pp. 357–373, 2018.
https://doi.org/10.1007/978-3-319-94268-1_30

By observing and analyzing experimental data, we have the following three findings: (1) When the environment does not change drastically, the groundtruth fingerprints will change cyclically with the cyclical changes of the crowd spatial distribution, (2) For a certain area, the variation of crowd spatial distributions are regular and those distributions can be divided into different and a limited number of categories, (3) Location fingerprints corresponding to same or similar crowd spatial distribution are similar in some aspects.

Based on these observations, we propose a crowdsourcing-based Wi-Fi fingerprinting mechanism using un-Supervised Learning. In the mechanism, current data and historical data similar to the current are used to update location fingerprints dynamically. First, location, time and statistical features of location fingerprint sequences (LFS) are extracted. Then historical LFS from crowdsourcing are divided into multiple categories called scenes using density-based spatial clustering of applications with noise. Data in the same scene are corresponding to the same or similar crowd spatial distribution. While crowdsourcing, particle filter and hidden Markov model (HMM) is used to match the current data with those existed scenes. Finally, new location fingerprints are generated with the current and the matched historical data by Gaussian process regression.

The major contributions of this paper consist of the following aspects:

(1) The paper proposed a crowdsourcing mechanism to update location fingerprints in Wi-Fi-based indoor localization, making it less necessary to fingerprint again completely. A lot of time and resources will be saved.
(2) The mechanism enriched data which can be used to update current location fingerprints without extra manpower and devices. These means the location fingerprints in use will fit the groundtruth better, and the accuracy of localization will improve.

2 Related Work

Recently, the proposed Wi-Fi-based indoor localization can be divided into two types: The Radio-Propagation-Based and the Location-Fingerprint-Based. The Radio-Propagation-Based methods [1, 2] utilize the knowledge of radio propagation to locate the users. However, there are some drawbacks in this type of method: difficulties of handling multipath effect and the high cost of getting or setting APs' locations. Therefore, the Location-Fingerprint-Based methods are more widely studied and applied. In RADAR [3], users' locations are estimated by the nearest-neighbor algorithm. Afterwards, some methods calculate RSSs' distribution in location fingerprints and estimate users' locations by probabilistic method [4–6]. For example, the Horus uses a Bayesian Network to locate users.

The location fingerprint databases play a significant role in fingerprint-based method, so is the update of location fingerprints. The most direct method is to collect all location fingerprints at set intervals. It undoubtedly cost high amount of resources in this method. Methods in [7, 8] generate signal map with the help of mobile devices. Work in [9, 10] presents SLAM-based methods to construct location fingerprints.

These methods utilize inertial sensors or Depth cameras to locate. Some methods based on transfer learning [11, 12] are proposed to reduce the fingerprinting cost. For example, in [12], unlabeled and labeled data from different floors are embedded into the same latent space to propagate labels. The method in [11] updates non-reference points' location fingerprints by utilizing the correlations of location fingerprints between reference point and non-reference points.

Recently, methods [13–18] using crowdsourcing data are paid more attention to. The methods in [13, 14] build location fingerprint databases with user's uploaded data set. The LAAFU [15] utilizes data from crowdsourcing to detect altered APs and filter them out before localization. The location fingerprints can be adaptively updated by non-parametric Gaussian process regression [19] in LAAFU. The work in [17] present a location-estimation approach based on manifold co-regularization, which is a machine learning technique for building a mapping function between data.

A problem for crowdsourcing is that sometimes the amount of the data is not enough as expected. By analyzing and utilizing men's influences on RSSI, our method enriching the crowdsourcing data with historical data collected when there was a similar crowd spatial distribution in the same area.

3 Overview

In this phase, we introduce the overview of the proposed mechanism. Figure 1 shows the architecture.

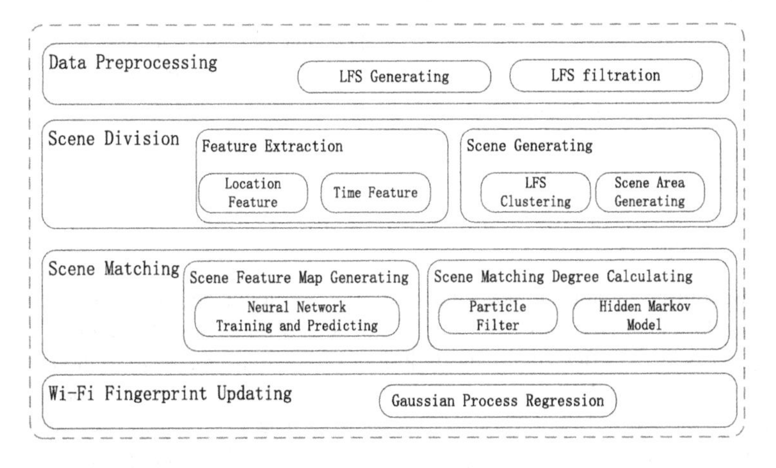

Fig. 1. System architecture

3.1 Data Preprocessing

The proposed mechanism serves a converged multi-source indoor localization project. We obtained an initial location fingerprint database, which was collected in no-man's environments. In online localization phase, we received data used to update location fingerprints by crowdsourcing. These data were organized as location fingerprint sequences. Then we selected LFSs which were collected by gatherers standing still or walking straight, which could be judged by pedestrian dead reckoning (PDR). These selected LFSs were used in latter work.

3.2 Scene Division

In this phase, we divided historical LFSs into different scenes. A one-to-many relationship exists between a kind of crowd location distribution in an enclosed area and scenes. In other words, RSSIs of LFSs from a scene are collected under conditions with same or similar crowd spacial distributions.

Firstly, we extracted features of historical LFSs. The features consist of spatial features, time features and RSSI statistical features. Secondly, we clustered these historical LFSs to generate scenes. Thirdly, we find the area each scene covers.

3.3 Scene Matching

After scene division, we had a number of scenes. In scene match, we found which scenes matched the circumstance in recent times. Firstly, we generated several feature maps for each scene. Secondly, we extracted the features of recently LFSs and match each LFSs with a scene by particle filter and hidden Markov model, then calculate the matching degree for every scene. Thirdly, we selected the most likely scenes which did not overlap with each other from the matched scenes.

3.4 New Location Fingerprints Generating

In this phase, we utilized the latest and matched historical data to update the location fingerprints in area where data were dense. Gaussian process regression was used to generate RSSI for each AP in the area.

4 Design

4.1 Data Preprocessing

The crowdsourcing data include acquisition time, localization results of different method including PDR and device information. A location fingerprint sequence (LFS) consists of a series of nodes including crowdsourcing data. A node is corresponding to an upload. Data in the same LFS were collected by the same participants in a continuous period of time. Therefore, the nodes in a LFS were close in space and orderly in time. We selected those LFSs corresponding to standing still or walking straight for the next work. The reason is that we need to exclude the interference of the carrier of the collecting device.

4.2 Scene Division

LFS Feature Extraction. After observing and analyzing the location fingerprints in our laboratory for four weeks, we found that the crowd spatial distribution had a great influence over the location fingerprints and crowdsourcing data.

Some important regularities can be used to enrich data updating location fingerprints with historical crowdsourcing data. For those LFSs corresponding to same or similar crowd spatial distribution in an enclosed area, some features of them are similar. Meanwhile, the features corresponding to different crowd spatial distributions are often not close to each other. These regularities apply to the following three kinds of features of LFSs corresponding to standing still: (1) RSSI statistical features, (2) Time features, (3) Spatial features.

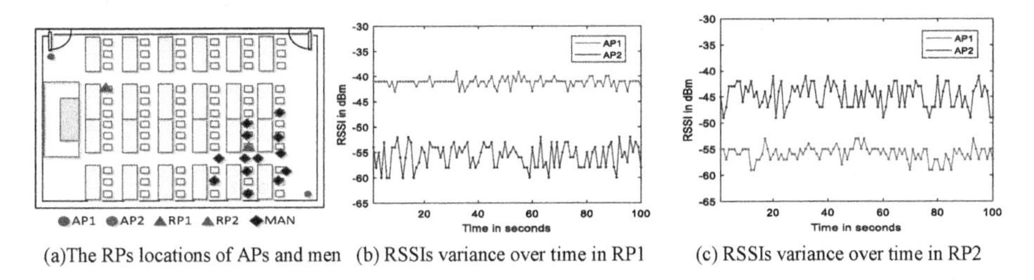

(a)The RPs locations of APs and men (b) RSSIs variance over time in RP1 (c) RSSIs variance over time in RP2

Fig. 2. The RPs, APs and crowds' locations and RSSIs in two RPs

Firstly, we illustrate why we used the RSSI statistical features and how we extracted them. In our experiments, we found that the statistical features is related to crowd spatial distribution. The statistical features include average, maximum and standard deviation of RSSI. We call the locations where location fingerprints are collected reference points (RP). Figure 2 shows the RSSIs variances over time in two RPs. the RPs and the crowds in a classroom.

In Fig. 2, there was no man between AP1 and RP1. The AP1's RSSI in RP1 is more stable than the others. RP2 was surrounded by 13 standing people, which led to an instability of the RSSIs in RP2. Under the same crowd spatial distributions, the RSSIs of different APs in the same position were affected differently. The standard deviation of an AP's RSSI was higher when there were crowd between the collecting device and the AP. And the same AP's RSSI in different RPs were affected differently, too. The distances between men around, AP and the collecting device also matters.

We denote the RSSI statistical feature of a LFS as F^S, a 3M-dimensional vector. f_i^s represents the ith elements in F^S:

$$f_i^s = \begin{cases} r_i^{avg} & (1 \leq i \leq M) \\ r_{i-M}^{std} & (M+1 \leq i \leq 2M) \\ r_{i-2M}^{\max} & (2M+1 \leq i \leq 3M) \end{cases} \tag{1}$$

where

$$r_i^{avg} = \alpha \cdot \frac{1}{S} \sum_{j=1}^{S} r_{ij}^s \tag{2}$$

$$r_i^{std} = \beta \cdot \frac{1}{S-1}\left(\sum_{j=1}^{S}\left(r_{ij} - R_{ij}^{avg}\right)^2\right)^{\frac{1}{2}} \quad (3)$$

$$r_i^{\max} = \gamma \cdot \max\left\{r_{ij}^{s}\right\}_{j=1}^{S} \quad (4)$$

In Eqs. (1), (2), (3) and (4), M is the number of APs, r_{ij} is the ith AP's RSSI of the jth node in a LFS. r_i^{avg}, r_i^{std}, and r_i^{max} represent the ith AP's RSSI average, standard deviation and maximum in this LFS. α, β, and γ are the coefficients affecting the importance of the corresponding statistical features.

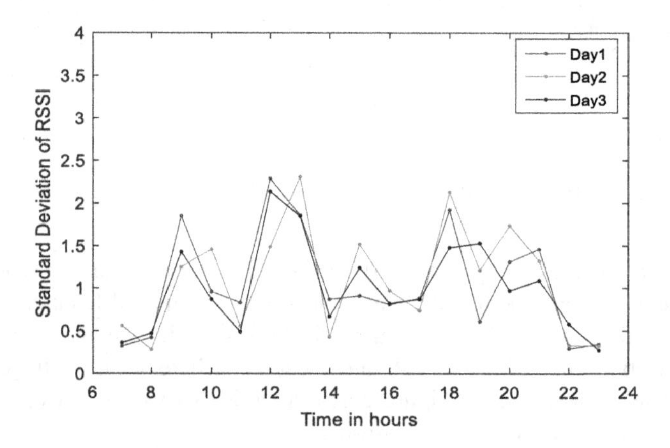

Fig. 3. RSSIs' standard deviation over time

Secondly, we introduce the time features. As mentioned, crowd spatial distributions have influence over location fingerprints. In other words, if the environment and APs do not change drastically, cyclical human activities may lead to cyclical changes of location fingerprints in this area. Our experimental data show that the changes of location fingerprints is cyclical. Figure 3 shows the hourly standard deviation of a AP's RSSIs in the RP near the door of our laboratory from 7 to 23 in three days.

Figure 3 shows the standard deviations of RSSIs of some stable APs in a day. The standard deviations are higher at 9, 12 and 21 o'clock, which are the times when we enter or leave the lab. These means human's cyclical activities have cyclical influences on RSSI. Based on this observation, we extract the cyclical time feature of a LFS as:

$$F^c = (\sin(T^c), \cos(T^c)) \quad (5)$$

where

$$T^c = \frac{1}{S}\sum_{i=1}^{S} t_i^c \quad (6)$$

Where t_i^c is the ith node's acquisition time counting from 0 o'clock that day. For those LFSs with similar acquisition time, their cyclical time features are close. In addition, we realized that the new LFS is more important than the old ones. Therefore, we extract the time difference feature:

$$F^D = 1 - e^{-\alpha T^D} \tag{7}$$

In Eq. (7), T^D denotes the time in days that has passed since the LFS was collected and the coefficient α affects the importance of newer data. The closer the value of this feature to 1, the newer the data. This feature can be used to determine how important a LFS is and whether it is going to be used.

Thirdly, we illustrate spatial features. RSSIs in adjacent locations are close. The crowd's influence on RSSI on the distant area is unstable or not obvious. In addition, Human activities far away from each other are with lower correlation, which means correlation of distant location fingerprints' change is not high enough.

We represent the spatial features of a LFS with a 2D vector:

$$F^{Loc} = (x_l, y_l) \tag{8}$$

where

$$x_l = \frac{1}{N_p} \sum_{i=1}^{N_p} x_i \tag{9}$$

$$y_l = \frac{1}{N_p} \sum_{i=1}^{N_p} y_i \tag{10}$$

In Eqs. (9) and (10), x_i is X coordinates of the ith node in the LFS and y_i has a similar definition. N_P is the number of nodes in the LFS. x_l and y_l are the average coordinates of all node in a LFS.

Scene Generating. In this section, we divide historical LFSs into different scene by density-based clustering algorithm. The LFSs in the same scene are similar in these three kind of features. Firstly, we cluster historical LFSs with the location and cyclical time features. The distance between the ith LFS and the jth LFS is denoted as:

$$D_{ST} = \left(\alpha\left|F_i^C - F_j^C\right|^2 + \gamma\left|F_i^{Loc} - F_j^{Loc}\right|^2\right)^{\frac{1}{2}} \tag{11}$$

In Eq. (11), α and γ are the weights of cyclical time features and spatial features. LFSs in the same cluster are close in location and time.

Secondly, for every cluster, we cluster LFBs in it with RSSI statistical feature. The distance of the ith and the jth LFSs is denoted as:

$$D_{RS} = \left| F_i^S - F_j^S \right| \tag{12}$$

Where F_i^s is the RSSI statistical feature of the *i*th LFSs.

Thirdly, we will find the covered area for each scene. The location fingerprints in a scene's area will be updated with LFSs in this scene. We divide the enclosed area into girds with same shape and size. A RP only locates in the center of a grid. For each scene, we count the numbers of nodes of LFSs corresponding to standing still inside for every grid. If the amount is enough, the grid and the RP inside this grid are covered by the scene. All covered grids constitute scene A's area. Figure 4 shows the covered RPs of scene A.

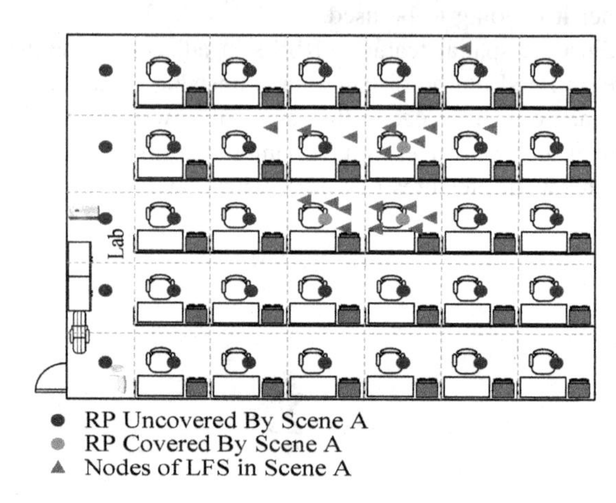

Fig. 4. RPs covered by scene A's

4.3 Scene Matching

Scene Feature Map Generating. There are three kinds of scene feature maps. They are average RSSI feature map, maximal RSSI feature map and RSSI standard deviation feature map. We utilizing LFSs corresponding to standing still to generate the three feature maps for each AP for each scene. For each AP, three fully connected neural networks were trained separately to fit these three kinds of feature map. The inputs of a neural network are the coordinates of LFSs belonging to the scene and the outside initial location fingerprints which are closest to the edge of the scene's area. We use a N_i-by-2 matrix L to denote the inputs:

$$L = \begin{vmatrix} x_1 & y_1 \\ x & y_2 \\ \vdots & \vdots \\ x_{N_i} & y_{N_i} \end{vmatrix} \tag{13}$$

where N_i is the number of input samples and the *i*th row of L denotes the coordinates of the *i*th sample. The outputs of the neural network is a Ni-by-1 matrix, denoted as O.

Each row of the outputs denotes the scaled value in the corresponding position on the feature map.

The cost function of the neural network is

$$J(W,b,L,O) = \frac{1}{2}\left\|C \cdot (h_{W,b}(L) - O)\right\| + \lambda\|W\|_2 \tag{14}$$

where W and b were the weights and biases vectors of the neural network. $h_{W,b}(L)$ is the outputs of the neural network. λ is a L2 regularization coefficient. C is a N-by-N diagonal matrix with diagonal elements $\{c_i\}_{i=1}^{N}$ defined in Eq. (15):

$$c_i = \frac{1 - F_i^{TD}}{\max\{1 - F_j^{TD}\}_{j=1}^{N}} \tag{15}$$

where F_i^{TD} is the time difference feature of the LFS's corresponding to the ith input sample. We minimize the cost function in the training step by gradient descent method.

Scene Matching Degree Calculating. We calculate the matching degree between recent LFSs and scenes by particle filter. Firstly, we extract the features of recent LFSs. Secondly, we select scene with similar time features then search for the most matched part in these scene's feature map for every LFS by particle filter. Thirdly, we match every recent LFS with a scene by hidden Markov model (HMM). The way we extract recent LFSs' features has been described in Sect. 4.2. We will illustrate the second and third step in detail.

For each LFS, we use particle filter to search for a most matched scene. We firstly illustrate how to generate a particle. The state of the ith particle consists of a scene ID denoted as I_{si}^{p} and the coordinates denoted as a 2-dimension vector $L_i^p = \left(L_i^{px}, L_i^{py}\right)$. We use l_i^s to denote the number of LFSs in the ith scene. We assume that the total number of particles is N_p, the total number of LFS is N_{LFS}, and the total number of scene is N_S In the initialization stage, the number of the particles we put in the ith scene is denoted as

$$n_i = \left\lfloor \frac{N_P \cdot l_i^s}{\sum_{j=1}^{N_{LFS}} l_j^s} \right\rfloor \tag{16}$$

If we calculate the number of particles for a scene according to Eq. (16), the total number of particles may be less than N_p. We put the remaining particles into the scene with maximal number of LFSs to solve this problem. Then we judge whether a particle's coordinates are in the area of the scene it belongs. If it is not, the weight of the particle will be set to 0. Otherwise, we extract the particle's features. We put the coordinates of each particle into all the features of his scenario and combine all the obtained values as a feature vector which has a same structure as LFS's. We denote the Euclidean distance between the particle's features and the LFS's as d_i^{ps}. Then we calculate the weight of the ith particle:

$$w_i^p = \frac{1}{1 + d_i^{ps}} \tag{17}$$

In the resampling stage, we discard the particles of which weight is less than 80% particles'. The particles of which weight is larger than 90% particles' will be replicated. The we add Gaussian noise, of which mean is 0 and standard deviation is 0.5, to the particle's coordinates before next iteration. The loop will stop when the number of iteration is high enough or all particle's weights is 0. For every LFS-scene pair, we selected the largest weight among all particles in this scene as the similarity. The similarity between the ith scene and the jth LFS are denoted as $\mathrm{m}_{i,j}$.

In our method, a hidden state corresponds to a scene. An observation corresponds to a LFS. We use the similarity between a LFS and a scene as the emission probability. The transition probability matrix is denoted as:

$$P^T = \begin{vmatrix} p_{1,1}^T & \cdots & p_{1,N_S}^T \\ \vdots & \ddots & \vdots \\ p_{N_S,1}^T & \cdots & p_{N_S,N_S}^T \end{vmatrix} \tag{18}$$

where $p_{i,j}^T$ is the transition probability from the ith scene to the jth scene. The transition probability matrix will be updated at set intervals. It is important to note that these scenes are generated in scene division and N_S may change every time the matrix is reconstructed. In our mechanism, we calculate the transition probability as follows.

Firstly, we divide LFSs in a scene into several sub-scenes with time difference feature. Therefore, LFSs in a sub-scene were collected at a close time. We give an example to explain the difference between scenes and sub-scenes. We generated a scene consisting of LFSs which had been collected from 11:37 to 12:06. Then this scene was divided into several sub-scenes. The LFSs in sub-scene 1 were collected from 11:37 to 11:52 in 23/9/2017. The LFSs in sub-scene 2 were from 11:54 to 12:06 in 24/9/2017. The LFSs in sub-scene 3 were from 11:41 to 11:49 in 26/9/2017.

After sub-scene division, we sort these sub-scenes by collection time. For the ith scene, we denote the set of the sub-scenes in it as $\mathrm{S}_{\mathrm{i}}^{\mathrm{sub}}$. And we denote the set of the sub-scenes appearing within 5 min after those sub-scenes in ith scene disappear as S_i^{ASub}. The element $s_{a,b}^{Sub}$ in S_i^{ASub}. denotes the bth sub-scene in the ath scene. The transition probability from the ith scene to the jth scene is

$$p_{i,j}^T = \frac{1 + \left|\left\{s_{j,b}^{Sub} \in S_i^{ASub}\right\}\right|}{N_S + |S_i^{Sub}|} \tag{19}$$

The higher more frequently the jth scene appears after the ith scene disappear, the higher the transition probability from the ith scene to the jth scene.

We use a matrix M denote the matching degrees between LFSs and scenes:

$$M = \begin{vmatrix} m_{1,1} & \cdots & m_{1,N_S} \\ \vdots & \ddots & \vdots \\ m_{N_{LFS},1} & \cdots & m_{N_{LFS},N_S} \end{vmatrix} \quad (20)$$

where

$$m_{i,j} = (p_{k,j}^T + \theta)/(1 + \theta) \cdot q_{i,j} \quad (21)$$

In Eq. (21), $m_{i,j}$ denotes the matching degree between the ith LFS and the jth scene. k is the sequence number of the scene which the ith LFS belongs to. $p_{k,j}^T$ is the transition probability from the kth scene to the jth scene. θ is a coefficient affecting the importance of $p_{k,j}^T$. Then we calculate the total matching degree for each scene. The ith scene's matching degree is denoted as

$$m_i^s = \sum_{j=1}^{N_{LFS}} m_{j,i} \quad (22)$$

Nonoverlapping Scene Selection. In this phase, we select scenes without overlap area. The nonoverlapping scenes selection algorithm is described in Algorithm 1.

```
Algorithm 1: Nonoverlapping Scenes Selection
Input:
Candidate Scene Set Sc.
Ss ←empty set
while Sc is not empty
  select scene with maximal matching degree in Sc
  if scene does not overlap with all scenes in Ss
    Ss ← Ss + scene
  end if
  Sc ← Sc - scene
end while
end
Output: selection set of nonoverlapping scenes ←Ss
```

4.4 New Location Fingerprints Generating

In this phase, we utilize the matched historical data and the latest data to update location fingerprints in use. Due to the propagation characteristics of Wi-Fi signals and the complex indoor environment, the spatial distribution of RSSI often presents a complex nonlinearity. Therefore, we utilize Gaussian Process Regression to predict RSSIs in those RPs covered by matched scenes for each AP.

The training set consist of nodes of LFSs in matched scenes. We use acquisition time and locations of nodes as the inputs of the regression. These are based on the following two considerations: (1) The difference between RSSIs in RPs of which locations are closer is often smaller, (2) Older data have less influences than newer data. The covariance function of Gaussian Process Regression is defined as:

$$k(v, v^*) = \sigma_1^2 \exp\left[\frac{-\|v - v^*\|}{2l_1^2}\right] + \sigma_2^2 \exp\left[\frac{-\|v - v^*\|}{2l_2^2}\right] + \sigma_n^2 \delta(v, v^*) \tag{23}$$

where v and v* are two points' 3-D input vectors consisting of coordinates and acquisition time in days. The first term of Eq. (23) is used to reflect the general changing trend of RSSIs over space and time. The second term is used to reflect small fluctuation, such as fluctuation over acquisition time. l_1, l_2, σ_1, σ_2 and σ_n are hyper parameters. Let the number of training samples be N_T and the number of test samples be N_P. And let the first N_T samples be training samples. We define the covariance matrix as

$$M = \begin{vmatrix} M_T & M_{TP} \\ M_{TP}^{\mathrm{T}} & M_{PP} \end{vmatrix} \tag{24}$$

where M_T is the covariance matrix of training samples, M_{PP} is the covariance matrix of test samples and M_{TP} is defined as

$$M_{TP} = \begin{vmatrix} k(v_1, v_{1+N_T}) & \cdots & k(v_1, v_{N_P+N_T}) \\ \vdots & \ddots & \vdots \\ k(v_{N_T}, v_{1+N_T}) & \cdots & k(v_{N_T}, v_{N_P+N_T}) \end{vmatrix} \tag{25}$$

where v_i denotes the input vector of the ith sample. The output O_P is a N_P-D column vector of which the ith element denote the corresponding AP's RSSI of the ith test sample. O_P is calculated as follow:

$$O_P = M_{TP}^{\mathrm{T}} \cdot M_T^{-1} \cdot O_T \tag{26}$$

where O_T is a N_T-D vector, of which the ith element denote the corresponding AP's RSSI of the ith training sample.

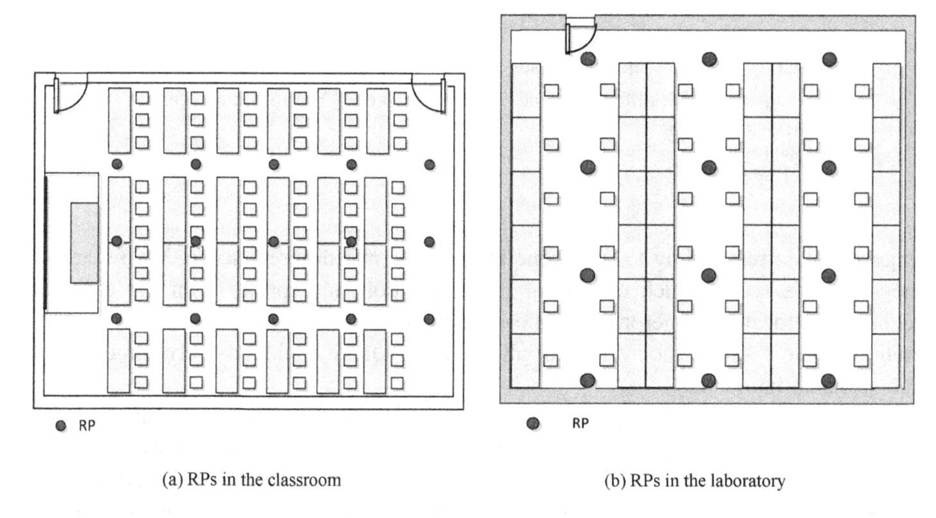

(a) RPs in the classroom

(b) RPs in the laboratory

Fig. 5. Layouts of experimental areas and RPs

5 Evaluation

5.1 Experimental Setup

In order to test the correctness and effectiveness of our method, we built an experimental environment at our laboratory and a classroom in Wuhan University. The shapes and sizes of the two areas are different. Their layouts and the RPs are shown in Fig. 5. We placed 30 AP randomly in and around the laboratory before the experiment at a height of 1 m, and the locations of the APs did not change during the experiment.

We used the RSSIs of these APs to generate the location fingerprints. Our collection devices in the laboratory were the raspberry pies installed Wi-Fi module. Each device was attached to a pole at a height of 1 m. We place these devices in the RPs to collect the location fingerprints continuously. We used Xiaomi 5 mobile phone to crowdsourcing. They were bundled on the left shoulder of the crowdsourcing participants. In our laboratory, we chose 10 volunteers to take part in the crowdsourcing. In the experiment, participants were required to keep their daily habits as much as possible. Before the experiment, we first collected the location fingerprints in the no-man laboratory as the initial position fingerprints. During the experiment, the position fingerprints we collected would be used as groundtruth for judging the proposed mechanism. The experiment in the classroom was similar to the one in our laboratory. The difference was that the total number of participants in classroom was 23. These experiments lasted for 4 weeks.

5.2 Scene Division

In this phase, we evaluate the accuracy of scene division. Scene division aims to divide LFSs with similar groundtruth location fingerprints into the same cluster. Therefore, we calculated the mean square deviations between groundtruth RSSIs of location

Table 1. I

Area	Mean square deviations of RSSIs		
	With different scene	With same scene	With no man scene
Lab	3.42	2.31	2.79
Classroom	5.86	3.85	5.67

fingerprints corresponding to sub-scene in the same and different scene. We separated the no man's scene which consist of initial location fingerprints from the different scenes. The location fingerprints in the whole enclosed areas were taken into consideration. Table 1 shows the average of all the mean square deviations between different sub-scene combinations.

According to Table 1, there are more difference between the location fingerprints from different scenes than those from the same scene. And the differences between location fingerprints of sub-scene from the same scenes are less than the difference between this sub-scene's and the no man's scenes. These means scene division makes sense.

5.3 Scene Matching

Scene Matching aims to select one or a set of scenes, of which the corresponding groundtruth location fingerprints are closest to latest ones. We define the RSSI mean square error between the generated location fingerprints corresponding to the selected scene and latest groundtruth location fingerprints as RSSI error, the lower is better. In order to verify the correctness of each step, we compared the performance of the following three matching strategies: (1) always using the initial location fingerprints, (2) only considering the similarities between latest LFSs and the candidate scenes, (3) considering the latest matched scenes on the basis of (2).

Although we update location fingerprints in areas covered by matched scenes, we aim to reduce the RSSI errors in the whole area. Therefore, we calculated the RSSI errors of the three strategies for every RP in the whole area. Figure 6 shows the CDF of RSSI errors. We found that there was no significant difference between the number of RPs with low RSSI error in strategy (1) and (2), while strategy (3)'s was obvious less than theirs. With RSSI errors increasing, number of RPs corresponding to strategy (1) became larger than strategy (2)'s.

Figure 7 shows the RSSI errors corresponding to selected scenes with different matching degree. For ease of comparison, in strategy (1), the corresponding RSSI error of all matching degrees are the RSSI mean square deviation between the initial and the latest location fingerprints. Because the calculation methods of matching degree in strategy (2) and (3) are different, each strategies' matching degrees are zoomed to [0.1, 1]. Then we divided those scenes into ten groups according to matching degree. We found that, in strategy (2) and (3), the RSSI errors dropped with matching degree increasing and the RSSI errors of scenes with high matching degree are lower than errors of strategy (1). These mean the matching degree can reflect the similarity between latest and historical data. And this step can find historical data which can be used to reduce the RSSI errors of latest location fingerprints. In addition, the changing trend of RSSI errors in strategy (3) was more stable than strategy (2). Meanwhile, in strategy (3), the RSSI errors corresponding to the highest matching degree is lower than those in strategy (2). These means considering the latest matched scenes makes sense.

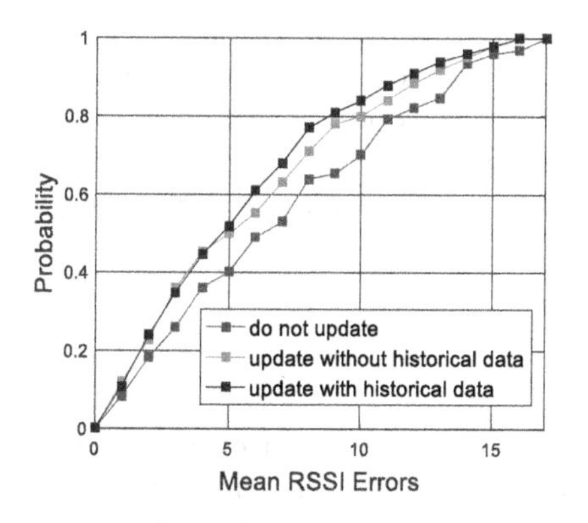

Fig. 6. CDF of RSSI errors

Fig. 7. RSSI errors of different matching degrees

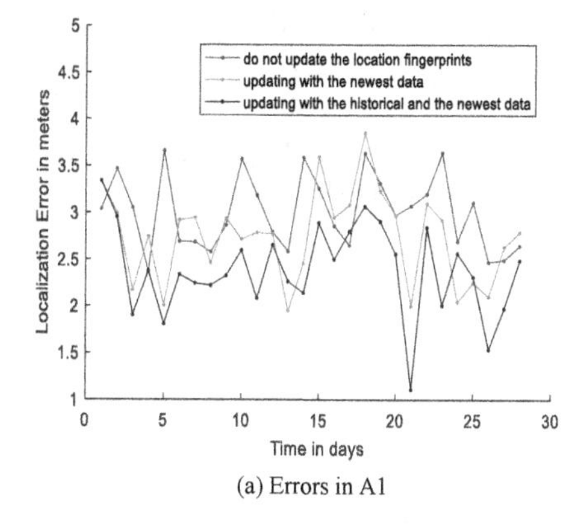

(a) Errors in A1

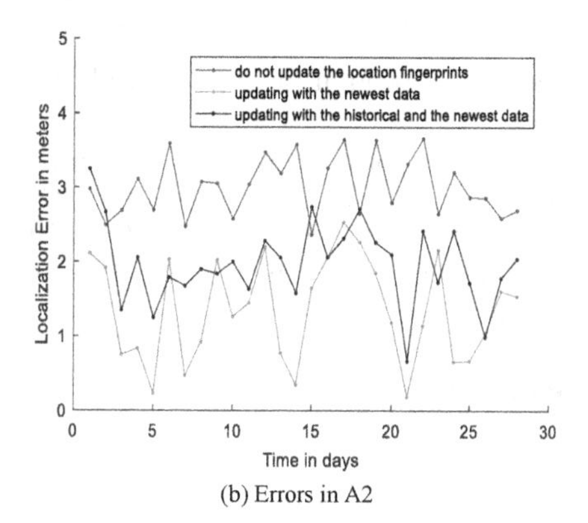

(b) Errors in A2

Fig. 8. Localization errors over time of three strategies

5.4 Location Fingerprint Updating

We compared the following three kinds of strategies: (1) updating location fingerprints with latest and historical data. (2) update fingerprints with only latest data. (3) do not update fingerprints. Then We analyzed the errors in following two ranges: (1) the matched scenes' area. (2) area where the newest data were dense. We call the former A1 and the latter A2.

Figure 8 shows the mean localization error in meters of WKNN per day using different strategies in our laboratory and the classroom. By using historical data, errors in A1 decreased. However, in A2, errors were lower when we did not use historical data. There might be two reasons for this: (1) the matched historical data might not be accurate as the latest data, (2) while using the historical data, those fingerprints outside A2 might be updated, which means these fingerprints might be more similar to those in A2. This might cause interference to the positioning. We will study this in future work.

When we update location fingerprints with the latest and historical data, compared with using only latest data, the RSSI error is reduced by 3% – 23% under different scenes and by an average of 9.3% in our experiment. We can find that in most cases, updating location fingerprints reduces the deviation between estimated location fingerprints and the groundtruth, and improves the accuracy of localization by 9.2% in area with enough matched historical data in our experiments compared with strategy not using historical data.

6 Conclusion

This paper presents an adaptive Wi-Fi fingerprinting mechanism for dynamic indoor environment. Compared with the existing mechanisms, our mechanism takes the influence of the crowd around on location fingerprints into consideration. We cluster the historical data collected in environments with similar distribution of crowd's location. Then, we calculate the similarity between latest and historical crowdsourcing data, and enriches data which can be used to update latest location fingerprints without extra consume. The experimental results reveal that the enriched data can help to update the location fingerprints with higher accuracy and in wider area.

Acknowledgement. This work was partially supported by the National Key Research Development Program of China (2016YFB0502201), the National Natural Science Foundation of China NSFC (U1636101, 61572370) and CERNET Next Generation Internet's Technology Innovation Project (NGII20160324, NGII20170633).

References

1. Priyantha, N., Chakraborty, A., Balakrishnan, H.: The Cricket location-support system. In: Proceedings of ACM MobiCom 2000, pp. 32–43 (2000)
2. Savvides, A., Han, C., Strivastava, M.B.: Dynamic fine-grained localization in Ad-Hoc networks of sensors. In: Proceedings of ACM MobiCom 2001, pp. 166–179 (2001)
3. Bahl, P., Padmanabhan, V.N.: RADAR: an in-building RF-based user location and tracking system. In: Proceedings of IEEE INFOCOM 2000, pp. 775–784 (2000)
4. Youssef, M., Agrawala, A.: The Horus WLAN location determination system. Wirel. Netw. **14**(3), 357–374 (2008)
5. Mirowski, P., Milioris, D., Whiting, P., Ho, T.K.: Probabilistic radio-frequency fingerprinting and localization on the run. Bell Labs Tech. J. **18**(4), 111–133 (2014)
6. Kerckhofs, G., Schrooten, J., Van Cleynenbreugel, T., Lomov, S.V., Wevers, M.: Empirical evaluation of signal-strength fingerprint positioning in wireless LANs. In: Proceedings of MSWIM 2010, pp. 5–13 (2010)
7. Ji, Y., Pandey, S., Agrawal, P.: ARIADNE: a dynamic indoor signal map construction and localization system. In: Proceedings of ACM MobiSys 2006, pp. 151–164 (2006)

8. Wu, C., Yang, Z., Liu, Y., Xi, W.: WILL: wireless indoor localization without site survey. In: Proceedings of IEEE INFOCOM 2012, pp. 64–72 (2012)
9. Huang, J., Millman, D., Quigley, M., Stavens, D., Thrun, S., Aggarwal, A.: Efficient, generalized indoor WiFi GraphSLAM. In: Proceedings of IEEE ICRA 2011, pp. 1038–1043 (2011)
10. Mirowski, P., Palaniappan, R., Ho, T.K.: Depth camera SLAM on a low-cost WiFi mapping robot. In: Proceedings of IEEE TePRA 2012, pp. 1–6 (2012)
11. Zheng, V.W., Xiang, E.W., Yang, Q., Shen, D.: Transferring localization models over time. In: Proceedings of AAAI 2008, pp. 1421–1426 (2008)
12. Wang, H.Y., Zheng, V.W., Zhao, J., Yang, Q.: Indoor localization in multi-floor environments with reduced effort. In: Proceedings of PerCom 2010, pp. 244–252 (2010)
13. Yang, S., Dessai, P., Verma, M., Gerla, M.: FreeLoc: calibration-free crowdsourced indoor localization. In: Proceedings of IEEE INFOCOM 2013, pp. 2481–2489 (2013)
14. Park, J.G., Charrow, B., Curtis, D., Battat, J., Minkov, E., Hicks, J., et al.: Growing an organic indoor location system. In: Proceedings of ACM MobiSys 2010, pp. 271–284 (2010)
15. He, S., Lin, W., Chan, S.H.G.: Indoor localization and automatic fingerprint update with altered AP signals. IEEE Trans. Mob. Comput. **16**(7), 1897–1910 (2017)
16. Zhuang, Y., Syed, Z., Georgy, J., El-Sheimy, N.: Autonomous smartphone-based WiFi positioning system by using access points localization and crowdsourcing. Pervasive Mob. Comput. **18**(C), 118–136 (2015)
17. Pan, S.J., Kwok, J.T., Yang, Q., Pan, J.J.: Adaptive localization in a dynamic WiFi environment through multi-view learning. In: Proceedings of AAAI 2007, pp. 1108–1113 (2007)
18. Wu, C., Yang, Z., Xiao, C.: Automatic radio map adaptation for indoor localization using smartphones. IEEE Trans. Mob. Comput. **17**(3), 517–528 (2018)
19. Mchutchon, A., Rasmussen, C.E.: Gaussian process training with input noise. In: Proceedings of NIPS 2011, pp. 1341–1349 (2011)

Secure and Efficient Outsourcing of Large-Scale Matrix Inverse Computation

Shiran Pan[1,2,3], Qiongxiao Wang[1,2], Fangyu Zheng[1,2](✉), and Jiankuo Dong[1,2,3]

[1] State Key Laboratory of Information Security, Institute of Information Engineering, Chinese Academy of Sciences, Beijing, China
{panshiran,wangqiongxiao,zhengfangyu,dongjiankuo}@iie.ac.cn

[2] Data Assurance and Communication Security Research Center of Chinese Academy of Sciences, Beijing, China

[3] School of Cyber Security, University of Chinese Academy of Sciences, Beijing, China

Abstract. Matrix inverse computation (MIC) is one of the fundamental mathematical tasks in linear algebra, and finds applications in many areas of science and engineering. In practice, MIC tasks often involve large-scale matrices and impose prohibitive computation costs on resource-constrained users. As cloud computing gains much momentum, a resource-constrained client can choose to outsource the large-scale MIC task to a powerful but untrustworthy cloud. As the input of and the solution to the MIC task usually contain the client's private information, appropriated mechanisms should be placed for privacy concerns. In this paper, we employ certain matrix transformations and construct an outsourcing scheme known as SEMIC, which can solve the MIC task in a masked yet verifiable manner. Thorough theoretical analysis shows that SEMIC is correct, verifiable, and privacy-preserving. Extensive experimental results demonstrate that SEMIC significantly reduces the computation costs of the client. Compared with the most related work, our solution offers enhanced privacy protection without impairing the efficiency.

Keywords: Matrix inverse computation · Cloud computing
Verifiable outsourcing · Privacy preserving

1 Introduction

Matrix inverse computation (MIC) is one of the most basic algebraic task in scientific and engineering applications (e.g., mobile localization [1] and MIMO wireless communication [2]). Nowadays, real-world applications often involve MIC tasks with large-scale input matrices, which often incur prohibitive computation costs. For example, a modern laptop computer may spend more than 1 min to compute the inverse of a $2{,}000 \times 2{,}000$ matrix, while a weaker device (e.g., a

S. Chellappan et al. (Eds.): WASA 2018, LNCS 10874, pp. 374–386, 2018.
https://doi.org/10.1007/978-3-319-94268-1_31

smart phone) may spend even much longer time. Consequently, in this paper we consider the cloud computing paradigm in which a resource-constrained client outsources such computation intensive tasks to a powerful cloud, which aids the client to compute the inverses of large-scale matrices. By such outsourcing, the client can release itself from heavy computation burden and enjoy the cloud's nearly unlimited computing resources in a pay-per-use manner.

As the cloud might be untrustworthy [3], the combination of MIC and cloud computing also arouses some privacy concerns. First, the cloud may save its computing resources and provide invalid computation results, hoping not to be detected by the client [4,5]. Second, the data processed (e.g., the input matrix) and generated (e.g., the inverse matrix) during the computation in the cloud often contains the client's sensitive information, which may be exploited by the cloud for various purposes [4,6]. Therefore, it is of significant importance for the client to deploy appropriate mechanisms to verify the results by the cloud and protect sensitive information.

In the literature, there have been many secure outsourcing schemes on scientific calculations related to matrix transformations (such as [4–7], just to name a few). The research effort most related to ours are the MIC outsourcing schemes proposed in [8,9]. In Lei et al.'s solution [8], the client employs random sparse matrices of special structure to transform the input matrix into a masked form, which is sent to the cloud; then, the client receives an intermediate result from the cloud, based on which it restores and verifies the inverse of the input matrix. As a following study, Hu et al. [9] employs chaotic systems to generate the secret key sets (which can reduce the key storage space) and additionally proposes a new verification algorithm to check the computation result by the cloud (which costs relatively less time compared to [8]). Though the solution in [9] is more efficient, we find that the new verification algorithm in [9] is not rigorous and an invalid result may happen to pass the verification (we will be more specific in Sect. 5.1). Furthermore, we observe that, for efficiency reasons both schemes [8,9] employ extremely sparse matrices for masking the input matrix before outsourcing. However, their masking techniques [8,9] only address the privacy of non-zero elements in the input matrix, but would leak the number of the zero elements, which also contains some privacy information in many scenarios [4,9,10] (e.g., the input matrix is a 0/1 matrix [10] or an extremely sparse one [4]). Thus, these two schemes are actually not secure enough (the privacy breach will be detailed in Sect. 5.1). In this paper, we tackle the problem of secure and efficient MIC outsourcing, and construct a novel scheme for enhanced privacy assurance. Our technical contributions can be summarized as follows:

(i) We formally define the problem of secure outsourcing of matrix inverse computation (MIC). Specifically, we propose system/threat models for privacy-preserving MIC outsourcing, formulate the basic components of the outsourcing scheme, and identify the design goals.
(ii) We elaborately construct a scheme called SEMIC for secure and efficient outsourcing of MIC. Thorough theoretical analysis shows that our proposal is correct, verifiable, and privacy-preserving. Compared with related schemes

[8,9], our proposal offers enhanced privacy protection, i.e., additionally hiding the number and positions of zero elements in the input matrix.

(iii) We implement and evaluate SEMIC, and compared it with two most related schemes [8,9] in terms of efficiency. Extensive experimental results demonstrate that SEMIC is comparable with both schemes, and can benefit a client with significantly reduced computation costs.

This paper is organized as follows. Mathematical preliminaries are presented in Sect. 2. The problem statement regarding securely outsourcing MIC tasks is in Sect. 3. Section 4 presents our proposal. Theoretical analysis is conducted in Sect. 5, while experimental evaluation is in Sect. 6. Section 7 concludes the paper.

2 Preliminaries

2.1 Matrix-Matrix and Matrix-Vector Multiplications

For matrix $\boldsymbol{M} \in \mathbb{R}^{l\times m}$ and matrix $\boldsymbol{N} \in \mathbb{R}^{m\times n}$, their product is $\boldsymbol{MN} \in \mathbb{R}^{l\times n}$. The computation complexity of such matrix-matrix multiplication is $\mathcal{O}(lmn)$.

When $n = 1$, $\boldsymbol{N} \in \mathbb{R}^{m\times n}$ is reduced to a column vector $\boldsymbol{n} \in \mathbb{R}^{m\times 1}$. The product of matrix $\boldsymbol{M}$ and vector $\boldsymbol{n}$ is also a column vector $\boldsymbol{Mn} \in \mathbb{R}^{l\times 1}$. The computation complexity of such matrix-vector multiplication is $\mathcal{O}(lm)$.

2.2 Privacy-Preserving Matrix Transformation

Definition 1 (Computational Indistinguishability [6]**).** *Let $\boldsymbol{X} = (x_{i,j}) \in \mathbb{R}^{m\times n}$ be a random matrix with entries in its j-th column following a uniform distribution over the interval $[-R_j, R_j]$ $(j \in [1,n])$. Given a matrix $\boldsymbol{Y} = (y_{i,j}) \in \mathbb{R}^{m\times n}$, $\boldsymbol{X}$ and $\boldsymbol{Y}$ are computationally indistinguishable if for any probabilistic polynomial-time distinguisher $D(\cdot)$, there exists a function $nelg(\cdot)$ negligible in the security parameter λ satisfying*

$$|\Pr(D(x_{i,j}) = 1) - \Pr(D(y_{i,j}) = 1)| \leq nelg(\lambda), \ \forall i \in [1,m], \ j \in [1,n],$$

where $D(\cdot)$ outputs 1 when it identifies that the input is sampled from a uniform distribution over $[-R_j, R_j]$ or 0 otherwise.

Based on Definition 1, Salinas et al. [6] proposes a privacy-preserving matrix transformation scheme, which can hide the value of each element in a given matrix in the sense of computational indistinguishability. Consider a matrix $\boldsymbol{A} = (a_{i,j}) \in \mathbb{R}^{m\times n}$ with each entry being in the range $[-2^l, 2^l]$ $(l > 0)$. To protect the privacy of (i.e., to hide) $\boldsymbol{A}$, we first generate a matrix $\boldsymbol{P} = \boldsymbol{uv}^{\mathrm{T}} \in \mathbb{R}^{m\times n}$, where each entry of the vector $\boldsymbol{u} \in \mathbb{R}^{m\times 1}$ follows a uniform distribution over the interval $(-2^p, 2^p)$ $(p > 0)$ and each entry v_j of $\boldsymbol{v} \in \mathbb{R}^{n\times 1}$ is an arbitrary positive number sampled from $[2^l, 2^{l+q}]$ $(q > 0)$. Then, we can hide the value of each element in $\boldsymbol{A}$ by applying a matrix addition as $\boldsymbol{A}' = \boldsymbol{A} + \boldsymbol{P}$, where $p_{i,j}$ and $a'_{i,j} = a_{i,j} + p_{i,j}$ denote the elements on the i-th row and j-th column

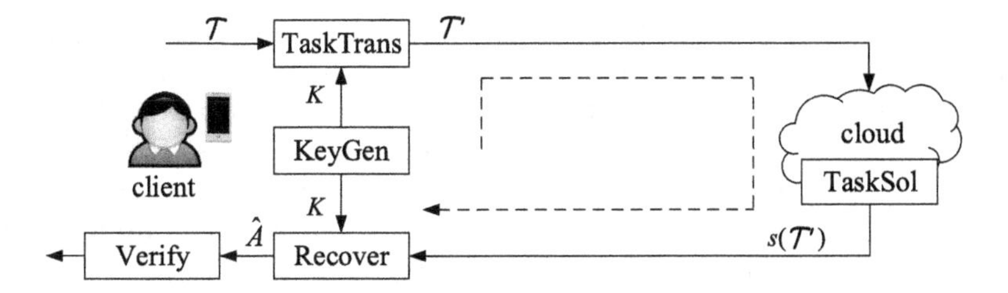

Fig. 1. Architecture of secure outsourcing of matrix inverse computation (MIC). It involves two participants: the client and the cloud. The client first generates its secret key K and uses it to transform the original MIC task $\mathcal{T}$ (i.e., computing $f(\boldsymbol{A})$) into a masked and outsourced computing task $\mathcal{T}'$. Subsequently, it receives the intermediate result $s(\mathcal{T}')$ by the cloud and then restores $\hat{\boldsymbol{A}}$, which is expected to be the solution to $\mathcal{T}$. Last, the client checks the correctness of $\hat{\boldsymbol{A}}$ to decide whether or not to accept it.

of $\boldsymbol{P}$ and $\boldsymbol{A}'$, respectively. Let $\boldsymbol{K} \in \mathbb{R}^{m\times n}$ be a random matrix with entries in its j-th column following the uniform distribution over $[-2^p v_j, 2^p v_j]$. According to Theorem 1 in [6], the elements in $\boldsymbol{K}$ and those in $\boldsymbol{A}'$ are computationally indistinguishable following Definition 1.

3 Problem Formulation

3.1 System Model

Denote the MIC function as $f(\boldsymbol{A})$, which takes as input a non-singular matrix $\boldsymbol{A} \in \mathbb{R}^{m\times m}$ and outputs its inverse $\boldsymbol{A}^{-1} \in \mathbb{R}^{m\times m}$. Formally, $f(\boldsymbol{A}) = \boldsymbol{A}^{-1}$. Generally, employing textbook algorithms (e.g., Gauss-Jordan elimination and LU decomposition [11, Sect. 2.3–2.4]) for computing $f(A)$ incur $\mathcal{O}(m^\rho)$ costs, which can be reduced to $\mathcal{O}(m^{2.373})$ ($2.373 \leq \rho \leq 3$) with a more fast algorithm [12]. In this paper, we consider an outsourcing framework illustrated in Fig. 1, in which a resource-constrained client turns to a powerful but untrustworthy cloud to securely solve an MIC task $\mathcal{T}$ (i.e., to compute $f(\boldsymbol{A})$). The solid lines indicate certain data flows while the dashed line indicates the workflow.

In this architecture, an MIC outsourcing scheme consists of 5 algorithms (**KeyGen**, **TaskTrans**, **TaskSol**, **Recover**, **Verify**). As in Fig. 1, only **TaskSol** is conducted by the cloud alone. First, the client invokes **KeyGen** to generate its secret key K, and transforms the original MIC task $\mathcal{T}$ into a masked and outsourced form $\mathcal{T}'$ by invoking **TaskTrans**. Subsequently, the cloud executes **TaskSol** and returns the client the solution to $\mathcal{T}'$. Based on this intermediate result, the client then invokes **Recover** to derive $\hat{\boldsymbol{A}}$, which is expected to be the solution to $\mathcal{T}$ (i.e., $\boldsymbol{A}^{-1}$). Last, the client calls **Verify** to check the correctness of the restored solution. For privacy concerns, the secret key K is employed for transforming the task and recovering the final solution.

However, as the client is assumed to be resource-constrained, even transforming $\mathcal{T}$ into $\mathcal{T}'$ and recovering $\hat{\boldsymbol{A}}$ from $s(\mathcal{T}')$ may be too heavyweight (e.g., incurring larger than $\mathcal{O}(m^3)$ costs). In our case study, the computation complexities of task transformation and solution recovery are on par with locally solving $\mathcal{T}$. Thus, the client needs the cloud's assistance for generating the masked task and restoring the solution, which means the **TaskGen** and **Recover** modules are both interactive. All examinations on the correctness of such "cloud-aided task transformation" and "cloud-aided solution recovery" are left to the **Verify** module (we will be more specific in Sect. 4).

3.2 Threat Model

Following the most related work [8,9], we consider a deceptive and curious cloud, which is known as the "fully malicious" adversary. Specifically, the cloud may return invalid computation results (for saving its computing resources) while hoping not to be detected by the client. In addition, based on what it is allowed to know, the cloud may also try to infer the private information associated with the original MIC task. This information refers to the input matrix (including the number and positions of zero elements) and the output computation result (i.e., the inverse of the input matrix).

3.3 Formalized Components

1. **KeyGen**$(\lambda) \mapsto K$: On input the security parameter λ, the client executes this probabilistic key generation algorithm to yield the secret key K.
2. **TaskTrans**$(\mathcal{T}, K) \mapsto \mathcal{T}'$: The client executes this deterministic task transformation algorithm to convert with K the original task $\mathcal{T}$ to a masked form $\mathcal{T}'$, which is outsourced to the cloud.
3. **TaskSol**$(\mathcal{T}') \mapsto s(\mathcal{T}')$: The cloud invokes this deterministic task solving algorithm to compute an intermediate result $s(\mathcal{T}')$, which is expected to be the solution to the masked task $\mathcal{T}'$ and sent to the client.
4. **Recover**$(s(\mathcal{T}'), K) \mapsto \hat{\boldsymbol{A}}$: The client executes this deterministic recovery algorithm to restore $\hat{\boldsymbol{A}}$ from $s(\mathcal{T}')$ with K. The restored result $\hat{\boldsymbol{A}}$ is expected to be the solution to the original task $\mathcal{T}$.
5. **Verify**$(\mathcal{T}, \hat{\boldsymbol{A}}) \mapsto$ `true`/`false`: The client invokes this probabilistic verification algorithm to check whether $\hat{\boldsymbol{A}}$ is the valid solution to $\mathcal{T}$. If so, the client outputs `true` and accepts $\hat{\boldsymbol{A}}$; otherwise, it outputs `false` indicating an error.

3.4 Design Goals

We identify three design goals for MIC outsourcing schemes.

Definition 2 (Correctness). *An MIC outsourcing scheme is said to be correct if for the $s(\mathcal{T}')$ that is returned by a cloud following the designed algorithms, then the $\hat{\boldsymbol{A}}$ output by* **Recover** *is the valid solution to $\mathcal{T}$, i.e., $\hat{\boldsymbol{A}} = f(\boldsymbol{A})$.*

Definition 3 (Verifiability [7]). *An MIC outsourcing scheme is said to be verifiable if for any intermediate result $s(\mathcal{T}')$ that is returned by the cloud but not the valid solution to $\mathcal{T}'$, there exists a function $nelg(\cdot)$ negligible in the security parameter λ satisfying* $\Pr(\textbf{Verify}(\mathcal{T}, \hat{\boldsymbol{A}})) \mapsto \texttt{true}) \leq nelg(\lambda)$.

Definition 4 (Privacy preserving). *An MIC outsourcing scheme is said to be privacy-preserving if the cloud cannot infer any of the input matrix (including the number and positions of zero elements) and the output computation result (i.e., the inverse of the input matrix).*

4 Proposed Scheme

In this section, we present a secure and efficient MIC outsourcing scheme called SEMIC to compute the inverse of the matrix $\boldsymbol{A} \in \mathbb{R}^{m \times m}$ ($m \gg 1$). The basic idea is as follows. The client converts the input matrix $\boldsymbol{A}$ to a masked form $\boldsymbol{A}'$, and transforms the original task into an outsourced task, which is to compute the inverse of $\boldsymbol{A}'$. Based on the computation result by the cloud, the client then derives and verifies the solution to the original MIC task, i.e., $\boldsymbol{A}$'s inverse matrix.

Following the formal definitions presented in Sect. 3.3, SEMIC comprises 5 algorithms (**KeyGen**, **TaskTrans**, **TaskSolve**, **Recover**, **Verify**). Next, one by one we elaborate on the components of SEMIC, some of which are relatively simple but others are a bit complex. For example, the **TaskTrans** and **Recover** modules are both interactive.

4.1 KeyGen(λ) $\mapsto \boldsymbol{K} = \boldsymbol{R}$

On input the security parameter λ, the client randomly generates an invertible dense matrix $\boldsymbol{R} \in \mathbb{R}^{m \times m}$. According to Levy-Desplanques Theorem ([13, Sect. 5.6]), a strictly diagonally dominant matrix is definitely invertible. Therefore, the client could choose as its secret key a row diagonally dominant matrix $\boldsymbol{R} = (r_{i,j}) \in \mathbb{R}^{m \times m}$ satisfying $\sum_{j \neq i} |r_{i,j}| < |r_{i,i}|$ for $i \in [1, m]$.

4.2 TaskTrans($\boldsymbol{A}, \boldsymbol{K}$) $\mapsto \boldsymbol{A}'$

To protect the input matrix $\boldsymbol{A} \in \mathbb{R}^{m \times m}$, the client intends to mask it via matrix multiplication with $\boldsymbol{R}$ to generate $\boldsymbol{A}' = \boldsymbol{RA}$. Since $\boldsymbol{R}$ is a random dense matrix, such mask can hide the value and position of each element in $\boldsymbol{A}$ (including zero elements). However, locally calculating $\boldsymbol{RA}$ incurs as large as $\mathcal{O}(m^3)$ costs (cf. Sect. 2.1), which is on par with that of directly computing $f(\boldsymbol{A})$ ($\mathcal{O}(m^\rho)$, cf. Sect. 3.1). Hence, the client needs the cloud's assistance for such mask.

For efficiently computing $\boldsymbol{A}' = \boldsymbol{RA}$, we construct **TaskTrans** illustrated in Table 1, which employs the privacy-preserving matrix transformation scheme in Sect. 2.2 as a building block and involves one-round client-cloud interaction. In step 1 of **TaskTrans** the client should appropriately select $(\boldsymbol{u}_i, \boldsymbol{v}_i)$ ($i = 1, 2$) to ensure $\boldsymbol{R}$ and $\boldsymbol{A}$ are masked in the sense of computational indistinguishability (cf. Definition 1). Now we explain why computing $\boldsymbol{RA}$ with our **TaskGen** imposes lower than $\mathcal{O}(m^3)$ costs. The point is to reduce matrix-matrix multiplication to a series of matrix-vector multiplications. Take $\boldsymbol{P}_1\boldsymbol{Y}$ in step 5 for example. This is actually calculated as $\boldsymbol{P}_1\boldsymbol{Y} = \boldsymbol{u}_1\boldsymbol{v}_1^{\mathrm{T}}\boldsymbol{Y} = \boldsymbol{u}_1(\boldsymbol{v}_1^{\mathrm{T}}\boldsymbol{Y})$, where the computation costs for calculating $\boldsymbol{v}_1^{\mathrm{T}}\boldsymbol{Y}$ and $\boldsymbol{u}_1(\boldsymbol{v}_1^{\mathrm{T}}\boldsymbol{Y})$ are both $\mathcal{O}(m^2)$. That is, the total cost for

Table 1. A one-round interactive **TaskGen**, via which the client can efficiently compute $\boldsymbol{A}' = \boldsymbol{R}\boldsymbol{A}$ with the aid of the cloud. The total computation complexity of the client's operations is only $\mathcal{O}(m^2)$.

client		cloud
1. appropriately select $\boldsymbol{u}_1, \boldsymbol{u}_2, \boldsymbol{v}_1, \boldsymbol{v}_2 \in \mathbb{R}^{m\times 1}$; $\boldsymbol{P}_1 = \boldsymbol{u}_1\boldsymbol{v}_1^{\mathrm{T}}$, $\boldsymbol{P}_2 = \boldsymbol{u}_2\boldsymbol{v}_2^{\mathrm{T}}$; $\boldsymbol{X} = \boldsymbol{R} + \boldsymbol{P}_1, \boldsymbol{Y} = \boldsymbol{A} + \boldsymbol{P}_2$.		
	2. $\boldsymbol{X}, \boldsymbol{Y}$ $\longrightarrow$	
		3. $\boldsymbol{S} = \boldsymbol{X}\boldsymbol{Y}$.
	4. $\boldsymbol{S}$ $\longleftarrow$	
5. $\boldsymbol{A}' = \boldsymbol{S} - (\boldsymbol{R}\boldsymbol{u}_2)\boldsymbol{v}_2^{\mathrm{T}} - \boldsymbol{u}_1(\boldsymbol{v}_1^{\mathrm{T}}\boldsymbol{Y})$.		

Table 2. A one-round interactive **Recover**, via which the client can efficiently restore $\hat{\boldsymbol{A}}$ with the aid of the cloud, where $\hat{\boldsymbol{A}}$ is expected to be the inverse of $\boldsymbol{A}$. The total computation complexity of the client's operations is $\mathcal{O}(m^2)$.

client		cloud
1. appropriately select $\boldsymbol{u}_3, \boldsymbol{v}_3 \in \mathbb{R}^{m\times 1}$; $\boldsymbol{P}_3 = \boldsymbol{u}_3\boldsymbol{v}_3^{\mathrm{T}}$, $\boldsymbol{Z} = \boldsymbol{R} + \boldsymbol{P}_3$.		
	2. $\boldsymbol{Z}$ $\longrightarrow$	
		3. $\boldsymbol{T} = \hat{\boldsymbol{A}}'\boldsymbol{Z}$.
	4. $\boldsymbol{T}$ $\longleftarrow$	
5. $\hat{\boldsymbol{A}} = \boldsymbol{T} - (\hat{\boldsymbol{A}}'\boldsymbol{u}_3)\boldsymbol{v}_3^{\mathrm{T}}$.		

such a divide-and-conquer approach to computing $\boldsymbol{P}_1\boldsymbol{Y}$ is reduced from $\mathcal{O}(m^3)$ to $\mathcal{O}(m^2)$. Lemma 1 tells the correctness of our protocol.

Lemma 1. *If the cloud honestly executes* ***TaskTrans****, then* $\boldsymbol{A}' = \boldsymbol{R}\boldsymbol{A}$.

Proof. If the cloud follows the protocol, then $\boldsymbol{S} = \boldsymbol{X}\boldsymbol{Y}$.
$\therefore\ \boldsymbol{A}' = \boldsymbol{X}\boldsymbol{Y} - (\boldsymbol{R}\boldsymbol{u}_2)\boldsymbol{v}_2^{\mathrm{T}} - \boldsymbol{u}_1(\boldsymbol{u}_2^{\mathrm{T}}\boldsymbol{Y}) = (\boldsymbol{R} + \boldsymbol{P}_1)\boldsymbol{Y} - \boldsymbol{R}\boldsymbol{P}_2 - \boldsymbol{P}_1\boldsymbol{Y} = \boldsymbol{R}(\boldsymbol{A} + \boldsymbol{P}_2) - \boldsymbol{R}\boldsymbol{P}_2 = \boldsymbol{R}\boldsymbol{A}$. □

4.3 TaskSol($\boldsymbol{A}'$) $\mapsto \hat{\boldsymbol{A}}'$

Given the masked matrix $\boldsymbol{A}'$, the cloud computes the inverse of $\boldsymbol{A}'$ and then returns the computing result $\hat{\boldsymbol{A}}'$ to the client. For the computation, the cloud can choose from several feasible approaches, such as Gaussian-Jordan elimination and LU decomposition [11, Sect. 2.3–2.4].

4.4 Recover($\hat{\boldsymbol{A}}', K$) $\mapsto \hat{\boldsymbol{A}}$

Upon receiving $\hat{\boldsymbol{A}}'$, the client invokes **Recover** to use K to restore $\hat{\boldsymbol{A}}$, which is expected to be the inverse of $\boldsymbol{A}$. If the cloud follows the **TaskTrans** and **TaskSol**

Algorithm 1. Verification of the restored solution to the original task

Input: original matrix $\boldsymbol{A}$ and restored solution $\hat{\boldsymbol{A}}$
Output: `true`/`false`
1: **for** $i = 1$ to l ($\ll m$) **do**
2: select a random $(m \times 1)$-dimensional 0/1 vector $\boldsymbol{\alpha}$, compute $\boldsymbol{p} = \hat{\boldsymbol{A}}(\boldsymbol{A}\boldsymbol{\alpha}) - \boldsymbol{I}\boldsymbol{\alpha}$
3: **if** $\boldsymbol{p} \neq (0, 0, \cdots, 0)^{\mathrm{T}}$ **then** return `false`
4: **end for**
5: return `true`

modules, then $\hat{\boldsymbol{A}}' = (\boldsymbol{A}')^{-1} = (\boldsymbol{R}\boldsymbol{A})^{-1} = \boldsymbol{A}^{-1}\boldsymbol{R}^{-1}$. Thus, for recovering $\boldsymbol{A}^{-1}$, the client needs to compute $\hat{\boldsymbol{A}}'\boldsymbol{R}$, which also imposes $\mathcal{O}(m^3)$ costs. This indicates the clients has to turn to the cloud for such matrix multiplication. Thus, similar to **TaskTrans**, the **Recover** module is also interactive, which is illustrated in Table 2. Notice that in step 1 the client should appropriately select $(\boldsymbol{u}_3, \boldsymbol{v}_3)$ to ensure $\boldsymbol{R}$ is masked in the sense of computational indistinguishability (cf. Definition 1). The cost of the client for computing $\hat{\boldsymbol{A}}'\boldsymbol{R}$ is only $\mathcal{O}(m^2)$. The complexity analysis is similar to that of **TaskTrans** and omitted here due to space limitation.

4.5 Verify$(\boldsymbol{A}, \hat{\boldsymbol{A}}) \mapsto$ `true/false`

To detect possible cheating behaviors by the cloud, the client should check the correctness of the "cloud-aided problem transformation" as well as the "cloud-aided solution recovery." Fortunately, in SEMIC it suffices to verify whether the restored $\hat{\boldsymbol{A}}$ is the inverse of $\boldsymbol{A}$, i.e., whether $\hat{\boldsymbol{A}}\boldsymbol{A} = \boldsymbol{I}$.

Since the computation costs for calculating $\hat{\boldsymbol{A}}\boldsymbol{A}$ are as high as $\mathcal{O}(m^3)$, we do not really perform this matrix multiplication for a deterministic verification, which would negate the benefit of cloud computing. Instead, following [8] we employ Freivalds' technique [14] to design a probabilistic but efficient verification procedure shown in Algorithm 1, the computation complexity of which is only $\mathcal{O}(lm^2)$ thanks to the matrix-vector multiplications. The client accepts $\hat{\boldsymbol{A}}$ if Algorithm 1 returns `true`, or rejects $\hat{\boldsymbol{A}}$ otherwise. In the next section we prove the effectiveness of the verification algorithm, which can balance between security and efficiency. For example, $l = 80$ ($\ll m$) seems enough for achieving almost perfect verifiability.

5 Analytic Evaluation

5.1 Security

Theorem 1. SEMIC *correctly solves the MIC task* $\mathcal{T}$.

Proof. Following Definition 2, we next prove that if the cloud follows the designed protocol, then the $\hat{\boldsymbol{A}}$ output by **Recover** is the valid solution to $\mathcal{T}$, i.e., $\hat{\boldsymbol{A}} = f(\boldsymbol{A}) = \boldsymbol{A}^{-1}$ or equivalently, $\hat{\boldsymbol{A}}\boldsymbol{A} = \boldsymbol{I}$.

If the cloud is honest, **TaskTrans** outputs $\boldsymbol{A}' = \boldsymbol{R}\boldsymbol{A}$ (following Lemma 1) and the $\hat{\boldsymbol{A}}'$ output by **TaskSol** is the inverse of $\boldsymbol{A}'$, i.e., $\hat{\boldsymbol{A}}' = (\boldsymbol{A}')^{-1} = \boldsymbol{A}^{-1}\boldsymbol{R}^{-1}$. In **Recover**, the cloud returns $\boldsymbol{T} = \hat{\boldsymbol{A}}'\boldsymbol{Z} = \boldsymbol{A}^{-1}\boldsymbol{R}^{-1}(\boldsymbol{R}+\boldsymbol{P}_3) = \boldsymbol{A}^{-1} + \hat{\boldsymbol{A}}'\boldsymbol{P}_3$ to the client, which calculates $\hat{\boldsymbol{A}} = \boldsymbol{T} - (\hat{\boldsymbol{A}}'\boldsymbol{u}_3)\boldsymbol{v}_3^{\mathrm{T}} = (\boldsymbol{A}^{-1} + \hat{\boldsymbol{A}}'\boldsymbol{P}_3) - \hat{\boldsymbol{A}}'\boldsymbol{P}_3 = \boldsymbol{A}^{-1}$.

That is, SEMIC correctly solves the MIC task $\mathcal{T}$. □

Theorem 2. SEMIC *is a verifiable MIC outsourcing scheme.*

In SEMIC for any $\hat{\boldsymbol{A}}$ that is not $\boldsymbol{A}$'s inverse, $\Pr(\mathbf{Verify}(\boldsymbol{A}, \hat{\boldsymbol{A}}) \mapsto \mathtt{true}) \leq 2^{-l}$, where l is the number of check rounds in Algorithm 1. The proof is quite similar to that of Theorem 4 in [8]. We omit the proof here due to the space limitation.

Remark 1 (Possible verification failure in [9]). In Hu et al.'s scheme [9], for verifying whether the restored $\hat{\boldsymbol{A}}$ is the inverse of the input matrix $\boldsymbol{A}$, the client generates a random $(1 \times m)$-dimensional vector $\boldsymbol{\gamma} = [\gamma_1, \cdots, \gamma_m]$ and accepts $\hat{\boldsymbol{A}}$ if $(\boldsymbol{\gamma}\hat{\boldsymbol{A}})(\boldsymbol{A}\boldsymbol{\gamma}^{\mathrm{T}}) = \sum_{i=1}^{m} \gamma_i^2$. However, we observe that it may be possible for a matrix $\hat{\boldsymbol{A}}$ that is not the inverse of $\boldsymbol{A}$ but happens to pass the verification. We next present a simple example to illustrate the risk of the "false positive." Assume $\boldsymbol{A} = \begin{bmatrix} 1 & 0 \\ 0 & 2 \end{bmatrix}$ and the restored matrix $\hat{\boldsymbol{A}} = \begin{bmatrix} 4 & 0 \\ 0 & -1 \end{bmatrix}$ (which is obviously not the inverse of $\boldsymbol{A}$). When choosing $\boldsymbol{\gamma} = [1, 1]$ for verification, the client computes $(\boldsymbol{\gamma}\hat{\boldsymbol{A}})(\boldsymbol{A}\boldsymbol{\gamma}^{\mathrm{T}}) = [1, 2] \times [4, -1]^{\mathrm{T}} = 2 = 1^2 + 1^2$, which means the invalid $\hat{\boldsymbol{A}}$ passes the verification by accident. The explanation is that $(\boldsymbol{\gamma}\hat{\boldsymbol{A}})(\boldsymbol{A}\boldsymbol{\gamma}^{\mathrm{T}}) = \sum_{i=1}^{m} \gamma_i^2$ is only a necessary condition for $\hat{\boldsymbol{A}} = \boldsymbol{A}^{-1}$ but not a sufficient condition. Even though the probability of such a "false positive" is small, the verification algorithm in [9] is essentially not a deterministic one. However, we fail to find any estimation on the probability of verification failure in [9], and we conjecture that [9] actually sacrifices security for efficiency.

Theorem 3. SEMIC *is privacy-preserving against the fully malicious adversary.*

Proof. In SEMIC, the cloud knows $(\boldsymbol{X}, \boldsymbol{Y})$ (from **TaskTrans**), $\boldsymbol{A}'$ (from **TaskSol**), and $\boldsymbol{Z}$ (from **Recover**). Next, we prove that the cloud cannot infer the input matrix $\boldsymbol{A}$ (including the number and the positions of zero elements) nor $\boldsymbol{A}$'s inverse matrix based on its knowledge.

According to step 1 of **TaskTrans**, $\boldsymbol{R}$ and $\boldsymbol{A}$ are masked in the sense of computational indistinguishability (cf. Definition 1) to generate $\boldsymbol{X}$ and $\boldsymbol{Y}$, respectively. Following Sect. 2.2, the cloud cannot distinguish them from random matrices. Thus, the cloud is unable to derive any element in $\boldsymbol{R}$ and $\boldsymbol{A}$ from their masked forms. Similarly, in the **Recover** module $\boldsymbol{R}$ is masked in the sense of computational indistinguishability, so the cloud also cannot derive the value of any element in $\boldsymbol{R}$ from $\boldsymbol{Z}$. This proves the privacy of the secret key $\boldsymbol{R}$. Due to the space limitation, here we omit the proof on the "computational indistinguishability," and one can refer to Theorem 1 in [6] for details. Without knowing $\boldsymbol{R}$, the cloud cannot recover the input matrix $\boldsymbol{A}$ from decrypting $\boldsymbol{A}' = \boldsymbol{R}\boldsymbol{A}$ nor recover its inverse $\boldsymbol{A}^{-1}$ from decrypting $\hat{\boldsymbol{A}}' = \boldsymbol{A}^{-1}\boldsymbol{R}^{-1}$. It should be noted that since $\boldsymbol{R}$ is a random dense matrix, the matrix product $\boldsymbol{A}' = \boldsymbol{R}\boldsymbol{A}$ can hide the value and position of each element in $\boldsymbol{A}$ (including zero elements). In addition, even though the cloud knows $\hat{\boldsymbol{A}}'$ and $\boldsymbol{T} = \boldsymbol{A}^{-1} + \hat{\boldsymbol{A}}'\boldsymbol{P}_3$, it still cannot obtain $\boldsymbol{A}^{-1}$ (and thus $\boldsymbol{A}$) since $\boldsymbol{P}_3$ is randomly generated by the client.

To sum up, the cloud cannot derive any element in $\boldsymbol{A}$ and its inverse $\boldsymbol{A}^{-1}$ based on its knowledge. This proves the privacy preserving of SEMIC. □

Remark 2 (Privacy breaches in [8,9]). In the most related solutions [8,9], the input matrix $\boldsymbol{A}$ is encrypted with two extremely sparse matrices $(\boldsymbol{M}, \boldsymbol{N})$ as $\boldsymbol{A}' = \boldsymbol{M}\boldsymbol{A}\boldsymbol{N}$. However, since $\boldsymbol{M}$ $(\boldsymbol{N})$ has only one non-zero elements in each row/column, the values of zero elements in $\boldsymbol{A}$ are not changed after such matrix encryption (though their positions are rearranged). One can refer to [8,9] for details. That is, the encrypted matrix $\boldsymbol{A}'$ reveals the number of zero elements in $\boldsymbol{A}$, which may cause the privacy leakage (cf. Sect. 1). Different from both schemes [8,9], SEMIC employs random dense matrix for masking $\boldsymbol{A}$, and thus offers enhanced privacy protection (i.e., additionally preserving the number privacy of zero elements in $\boldsymbol{A}$).

5.2 Complexity

In **TaskTrans**, the client needs to spend $\mathcal{O}(m^2)$ costs generating $\boldsymbol{P}_1$ and $\boldsymbol{P}_2$; then, computing $(\boldsymbol{R}\boldsymbol{u}_2)\boldsymbol{v}_2^{\mathrm{T}}$ and $\boldsymbol{u}_1(\boldsymbol{v}_1^{\mathrm{T}}\boldsymbol{Y})$ takes the client $\mathcal{O}(m^2)$ costs. In **Recover**, the client first spends $\mathcal{O}(m^2)$ costs generating $\boldsymbol{P}_3$ and then $\mathcal{O}(m^2)$ costs computing $(\hat{\boldsymbol{A}}'\boldsymbol{u}_3)\boldsymbol{v}_3^{\mathrm{T}}$. In **Verify**, since l (e.g., 80) is far smaller than m, the client only needs to spend $\mathcal{O}(m^2)$ costs checking whether $\hat{\boldsymbol{A}}(\boldsymbol{A}\boldsymbol{\alpha}) = \boldsymbol{0}$.

To sum up, the overall computation complexity is $\mathcal{O}(m^2)$, which is lower than that of locally solving the MIC task ($\mathcal{O}(m^\rho)$, $2.373 \leq \rho \leq 3$, cf. Sect. 3.1). As to be seen in the next section, the improvement of computation complexity from $\mathcal{O}(m^\rho)$ to $\mathcal{O}(m^2)$ is significant when m is large.

6 Experimental Evaluation

In this section, we evaluate SEMIC with large-scale matrices with dimensions $m \times m$ ranging from 500×500 to $3{,}000 \times 3{,}000$; any element in these matrices is randomly, uniformly, and independently sampled from the same interval $[-1024.0, 1024.0]$. All algorithms are implemented with MATLAB. We implement the client on a laptop computer with an Intel i5-5200U processor running at 2.20 GHz and 8 GB memory, and emulate the cloud on a desktop workstation with an Intel i7-6700 processor running at 3.40 GHz and 8 GB memory. Following [8], we employ a textbook algorithm (herein LU decomposition method [11, Sect. 2.4]) for computing the matrix inversion.

6.1 Efficiency of SEMIC

To evaluate the overhead for computing the inverse of matrices with various dimensions, we adopt the processing time as the performance metric and test two scenarios for comparison. In the first scenario, the client deploys SEMIC, and the processing time on both the client and the cloud sides is measured. However, we exclude the **KeyGen** algorithm (which is fairly lightweight, as specified in Sect. 4.1) for measurement because it can be conducted offline. Specifically, $K = \boldsymbol{R}$ and is independent of $\boldsymbol{A}$ and can be generated by the client beforehand (as long as m is determined). For our interactive **TaskGen** and **Recover**, time costs on the client and the cloud sides are measured, respectively. For **Verify**, the parameter l (the number of check rounds in Algorithm 1) is set to 80, which assures almost perfect verifiability following Theorem 2. In the second scenario, the client does not turn to the cloud. It directly adopts the LU decomposition method to compute the matrix inversion "locally."

Table 3 illustrates the client's processing time (averaged from repeated but independent experiments) in the two scenarios. In all experiments, the client's processing time in the second scenario (i.e., local computing) is significantly larger than the total time in the first scenario (i.e., privacy-preserving cloud computing). To make this clear, we adopt their ratio as the indicator for cost savings. We observes that this indicator increases as the dimensions increase, which means the larger the input matrix is the more cost savings are obtained.

Table 3. The client's processing time (in seconds) for computing the inverses of large-scale matrices with different dimensions. It takes $t_{\mathsf{SEMIC.client}}$ in all when employing SEMIC (which is the sum of client's online costs of **TaskTrans**, **Recover**, and **Verify**), or t_{direct} when locally adopting LU decomposition for computing the matrix inversion.

m	t_{SEMIC}				t_{direct}	$\frac{t_{\mathrm{direct}}}{t_{\mathsf{SEMIC.client}}}$
	TaskTrans	**Recover**	**Verify**	Client (Σ)		
500	0.013	0.005	0.036	0.054	1.123	20.796
1,000	0.042	0.017	0.168	0.227	9.145	40.286
1,500	0.098	0.036	0.323	0.457	33.572	73.462
2,000	0.237	0.059	0.483	0.779	74.857	96.094
2,500	0.264	0.101	0.769	1.134	159.618	140.757
3,000	0.411	0.152	1.129	1.693	307.505	181.633

Table 4. Processing time (in seconds) of the client by deploying SEMIC and the schemes in [8,9] for computing the inverses of large-scale matrices with different dimensions.

m	500	1,000	1,500	2,000	2,500	3,000
Lei et al.'s [8]	0.085	0.328	0.763	1.364	2.198	3.227
Hu et al.'s [9]	0.023	0.091	0.303	0.564	0.753	1.049
SEMIC	0.054	0.227	0.457	0.779	1.134	1.693

6.2 Performance Comparison

In this subsection, we are interested in the performance comparison between SEMIC and two related schemes proposed in [8,9]. In doing so, however, we fail to find any implementation and experimental evaluation in [9]. Hence, we take the trouble to implement each component of the scheme in [9]. For a fair comparison, all three secure outsourcing schemes ([8,9], and SEMIC) are implemented on the same laptop/workstation configuration specified in the previous subsection. Again we adopt the total processing time on the client side (as in Table 4) to measure the performance. In favor of Lei et al.'s scheme [8], we deliberately optimize it by excluding measuring certain operations that can be conducted offline. The results are shown in Table 4, where the dimensions of the input matrix increase from 500×500 to $3{,}000 \times 3{,}000$. Generally, the experimental results demonstrate that SEMIC is comparable to the other two schemes in terms of efficiency. On the other hand, as shown in Sect. 5.1, SEMIC actually offers enhanced privacy protection compared with the other two schemes. Thus, we conclude that SEMIC achieves a balance between security and efficiency.

7 Conclusion

In this paper, we have presented a new secure and efficient outsourcing scheme known as SEMIC for solving large-scale MIC tasks, which offers enhanced privacy protection compared with the state of the art. Theoretical analysis shows that SEMIC guarantees the privacy of both the input matrix of and the solution to the MIC task. We have also conducted extensive experiments to evaluate our proposal. The results show that SEMIC brings significant cost savings to the client for solving large-scale MIC tasks. In addition, performance comparison of SEMIC and related schemes from the state of the art shows that SEMIC maintains high efficiency.

Acknowledgment. The authors would like to thank the anonymous reviewers for their valuable comments. This work was partially supported by National Key R&D Program of China under Grant No. 2017YFC0822704 and 2017YFB0802103.

References

1. Cheung, K.W., So, H.C., Ma, W.-K., Chan, Y.T.: Least squares algorithms for time-of-arrival-based mobile location. IEEE Trans. Signal Process. **52**, 1121–1130 (2004)
2. Prabhu, H., Rodrigues, J., Edfors, O., Rusek F.: Approximative matrix inverse computations for very-large MIMO and applications to linear pre-coding systems. In: 2013 IEEE Wireless Communications and Networking Conference, WCNC 2013, pp. 2710–2715 (2013)
3. Ren, K., Wang, C., Wang, Q.: Security challenges for the public cloud. IEEE Internet Comput. **16**, 69–73 (2012)
4. Chen, X., Huang, X., Li, J., Ma, J., Lou, W., Wong, D.S.: New algorithms for secure outsourcing of large-scale systems of linear equations. IEEE Trans. Inf. Forensics Secur. **10**, 69–78 (2015)
5. Wang, C., Ren, K., Wang, J., Wang, Q.: Harnessing the cloud for securely outsourcing large-scale systems of linear equations. IEEE Trans. Parallel Distrib. Syst. **24**, 1172–1181 (2013)

6. Salinas, S., Luo, C., Chen, X., Li, P.: Efficient secure outsourcing of large-scale linear systems of equations. In: 2015 IEEE Conference on Computer Communications, INFOCOM 2015, pp. 1035–1043. IEEE (2015)
7. Pan, S., Zheng, F., Zhu, W.-T., Wang, Q.: Harnessing the cloud for secure and efficient outsourcing of non-negative matrix factorization. In: the 6th IEEE Conference on Communications and Network Security, CNS 2018. IEEE (2018)
8. Lei, X., Liao, X., Huang, T., Li, H., Hu, C.: Outsourcing large matrix inversion computation to a public cloud. IEEE Trans. Cloud Comput. **1**, 78–87 (2013)
9. Hu, C., Alhothaily, A., Alrawais, A., Cheng, X., Sturtivant, C., Liu, H.: A secure and verifiable outsourcing scheme for matrix inverse computation. In: 2017 IEEE Conference on Computer Communications, INFOCOM 2017, p. 9. IEEE (2017)
10. Yu, Y., Luo, Y., Wang, D., Fu, S., Xu, M.: Efficient, secure and non-iterative outsourcing of large-scale systems of linear equations. In: 2016 IEEE International Conference on Communications, ICC 2016, p. 6. IEEE (2016)
11. Strang, G.: Introduction to Linear Algebra. Wellesley-Cambridge Press, Wellesley (2003)
12. Strassen, V.: Gaussian elimination is not optimal. Numer. Math. **13**, 354–356 (1969)
13. Horn, R.A., Johnson, C.R.: Matrix Analysis. Cambridge University Press, Cambridge (2013)
14. Motwani, R., Raghavan, P.: Randomized Algorithms. Cambridge University Press, Cambridge (1995)

Household Electrical Load Scheduling Algorithms with Renewable Energy

Zhengrui Qin[1(✉)] and Qun Li[2]

[1] School of Computer Science and Information Systems, Northwest Missouri State University, Maryville, MO 64468, USA
zqin@nwmissouri.edu
[2] Department of Computer Science, College of William and Mary, Williamsburg, VA 23185, USA
liqun@cs.wm.edu

Abstract. Efficient household electrical load scheduling benefits not only individual customers by reducing electricity cost but also the society by reducing the peak electricity demand and saving natural resources. In this paper, we aim to design efficient load scheduling algorithms for a household considering both real-time pricing policies and renewable energy sources. We prove that household load scheduling problem is NP-hard. To solve this problem, we propose several algorithms for different scenarios. The algorithms are lightweight and optimal or quasi-optimal, and they are evaluated through simulations.

1 Introduction

With the development of information and communication technologies, many smart meters have recently been deployed in the power grid, and more will be deployed in the near future. With these smart meters, time-varying pricing policy, which encourages customers to use power wisely, becomes practical. With time-varying pricing, consumers are motivated to shift their high-load appliances to off-peak periods, in which unit price is usually low. Furthermore, some houses may be equipped with a renewable energy system, such as a photovoltaic panel and/or a wind turbine, which also drives customers to seek smart load scheduling such that they can benefit more from renewable energy. In a word, efficient household electrical load scheduling benefits not only individual customers by reducing electricity cost but also society by reducing the peak electricity demand and saving natural resources.

In this paper, we formulate a general household load scheduling problem that considers both time-varying price and renewable energy sources. We show that finding an optimal solution is NP-hard. Then, we propose several load scheduling algorithms for different household appliances, with or without renewable energy. Different from previous work, our goal in this paper is to design generic lightweight and efficient algorithms such that our solution can handle all types of scheduling scenarios. We group household appliances into three categories:

S. Chellappan et al. (Eds.): WASA 2018, LNCS 10874, pp. 387–399, 2018.
https://doi.org/10.1007/978-3-319-94268-1_32

(1) schedulable and uninterruptable appliances; (2) schedulable and interruptable appliances; and (3) real-time appliances. For one appliance alone in each category, we propose optimal solutions. For multiple appliances of different categories, we propose efficient heuristic algorithms. Our main contributions in this paper are as follows:

- We formulate a generic household load scheduling problem which includes real-time pricing, with and without renewable energy.
- We design optimal algorithms to schedule any individual appliance. Especially, for electric water heater (EWH)-like appliances, we design an optimal algorithm using dynamic programming.
- We propose heuristic algorithms for multiple commonly used appliances.
- We evaluate our dynamic programming and heuristic algorithms via simulation studies. The results validate the efficiency and performance of our proposed algorithms.

2 Related Work

Time-varying price policies have been proposed since the last century. Examples are real-time pricing, time-of-use pricing, critical-peak pricing, and so on [1,2]. Given time-varying price, power cost saving via load scheduling has been extensively studied, such as [3–8]. In [5], the authors investigated residential load control in presence of a real-time pricing combined with inclining block rates. In [4], the authors discussed the load scheduling in households, buildings and warehouses, but they only considered uninterruptable appliances. In [3] and [7], the authors modeled the energy consumption scheduling with a carefully selected utility function, and they formulated to maximize the utility function minus electricity cost; however they consider a house as a whole and ignore the detailed appliance scheduling. In [6], Du and Lu have investigated the electrical load scheduling problem for a specific appliance, EWH. They solved the minimization problem with a multiloop heuristic algorithm, which, however, may not be optimal. In [8], Lu *et al.* have studied the scheduling for another specific appliance, HVAC (heating, ventilation and cooling).

Different from previous work, we tackle the household electrical load scheduling problem in a general sense by considering all types of appliances and renewable energy, and propose lightweight and efficient algorithms to solve it.

3 Problem Formulation

We group household appliances into three categories, based on their usage characteristics: (1) A_1, appliances that are schedulable but not interruptable, such as washer and dryer; (2) A_2, appliances that are both schedulable and interruptable, such as air conditioning unit (AC), and EWH; (3) A_3, real-time appliances, such as TV and microwave. Besides various appliances, a house may also have a renewable energy system, such as a photovoltaic system and/or a wind turbine system, which can store the renewable energy in a battery. We assume that the capacity of the battery that stores renewable energy is B kwh, and it can only be discharged up till a fraction c of its capacity, such as $c = 20\%$. We denote the renewable energy generating rate by G_t and the battery's discharge rate by F_t[1].

3.1 Constraints

For each appliance a, it usually has a preferred power range, denoted by $[r_a^{min}, r_a^{max}]$. For each schedulable appliances, there is a preferred time window, denoted by $[t_a^\alpha, t_a^\omega]$. Let $R_{a,t}$ denote the power of appliance a at time t, and let C_a denote the duration of a cycle for an appliance in A_1. Then general constraints for appliances in three categories are as follows:

$\forall a \in A_1$:

$$\begin{cases} r_a^{min} \le R_{a,t} \le r_a^{max}, & t \in [t_a, t_a + C_a] \subset [t_a^\alpha, t_a^\omega]; \\ R_{a,t} = 0, & \text{otherwise.} \end{cases} \tag{1}$$

$\forall a \in A_2$:

$$\begin{cases} r_a^{min} \le R_{a,t} \le r_a^{max} \text{ or } 0, & t \in [t_a^\alpha, t_a^\omega]; \\ R_{a,t} = 0, & t \notin [t_a^\alpha, t_a^\omega]. \end{cases} \tag{2}$$

$\forall a \in A_3$:

$$\begin{cases} r_a^{min} \le R_{a,t} \le r_a^{max}, & \text{if } a \text{ is on;} \\ R_{a,t} = 0, & \text{otherwise.} \end{cases} \tag{3}$$

Many appliances may have other specific constraints. For instance, the temperature of the water in an EWH cannot be too high or too low. We describe these specific constraints in a general form:

$$Specific\ Constraint(a), \forall a \in A. \tag{4}$$

Each house has a circuit breaker that automatically protects the household electricity system from overloading or short circuit. Thus, each house is constrained by the maximum power, denoted by R_{max}:

$$\sum_{a \in A} R_{a,t} - F_t \le R_{max} \tag{5}$$

Battery constraint:

$$cB \le B_0 + \sum_{t=0}^{i} (G_t - F_t)\delta t \le B \tag{6}$$

where B_0 is the energy already stored at the beginning of the first time slot.

[1] $B = G_t = F_t = 0$ if there is no renewable energy system.

3.2 Formulation

The ultimate goal for household load scheduling is to minimize the total electricity cost while fulfilling user's satisfaction. We assume the scheduling time domain is $[0, H]$, which is divided into m time slots with length of $\delta t = \frac{H}{m}$. The formulation is as follows.

$$Min: \sum_{t=1}^{m} (\sum_{a \in A} R_{a,t} - F_t) * \delta t * P_t + \sum_{t=0}^{H} \sum_{a \in A} D_{a,t} \quad (7)$$

Subject to: Constraints (1), (2), (3), (4), (5), and (6).

The first term in the objective function is the total electricity cost, and the second term is the penalty imposed by delayed services or unfulfilled services, where $D_{a,t}$ is the penalty of appliance a at time t.

4 Scheduling Algorithms

In this section, we will first show in a theorem that the problem formulated in Sect. 3 is NP-hard. Then we will design lightweight algorithms to solve the problem in different scenarios.

Theorem 1: The household load scheduling problem formulated in Sect. 3 is NP-hard.

Proof: We prove the theorem by the method of restriction, in which we show that a special case of the problem (less complicated than the original one) is NP-hard.

The special case is designed as follows. The renewable energy generating rate is G_0 all the time, and the discharge rate is F_0 all the time, where $F_0 \leq G_0$. Furthermore, there are only schedulable and uninterruptable appliances, which have the same usage duration, H/m, and the same scheduling window, $[0, H]$, where m is an integer. Finally, the power of each appliance is less than F_0. Since renewable energy is free, the cost minimization should first check whether renewable energy alone can satisfy the power demand. This is equivalent to packing all the jobs into m bins, with height of F_0 and width of H/m. Since all the usage durations are H/m, the solution with jobs scheduled across two bins is not better than the one without jobs across two bins. Therefore, it is equivalent to asking, whether we can pack all the appliances into less than or equal to m bins. This is obviously a bin packing problem, which is NP-hard. ∎

Within a household, constraint Eq. (5) seldom takes effect; that is, the total consumption of a household usually does not surpass R_{max}. The circuit breaker seldom triggers unless there is a short circuit. Thus, we can schedule appliances individually in most cases, since they affect each other only when constraint Eq. (5) comes into effect. With this insight, we propose algorithms for individual appliances in each category, and ignore the second term in Eq. (7).

4.1 Algorithms for Different Scenarios

As the renewable energy is storable, the energy in the battery is non-descending if no renewable energy is used. We denote the amount of renewable energy available at time slot i, $i \in [1, m]$, by $RE_0(i)$, if no renewable energy is used.

Algorithm for One Appliance in A_1. The basic idea is to reserve as much renewable energy as possible at high-price periods for *each possible time window*. For each possible time window, we sort the price in descending order. For the highest-price time slot, we reserve as much renewable energy as needed and update renewable energy available for each time slot in this time window. Then we continue with the second-highest-price slot, and so on, until no more renewable energy can be scheduled.

Algorithm 1. Optimal scheduling for a washer alone.

Input: t_a^α, t_a^β, $C_a = i\delta t$.
Output: the start time t_0.

1 $t_0 = t_a^\alpha$;
2 $cost$ = a big number;
3 **for** $j \leftarrow t_a^\alpha$ **to** $t_a^\beta - i$ **do**
4 $\quad P_{temp,1:i} = sort(P_{j:j+i-1})$ // in descending order;
5 $\quad c = 0$;
6 $\quad RE = RE_0$;
7 $\quad$ **for** $k \leftarrow 1$ **to** i **do**
8 $\quad\quad t_* = \{t | t \in [j, j+i-1] \ \& \ P_t = P_{temp,k}\}$;
9 $\quad\quad r = min\{RE(t_*), R_a * \delta t\}$;
10 $\quad\quad RE = REupdate(t_*, RE, r)$;
11 $\quad\quad c = c + (Ra * \delta t - r) * P_{t_*}$;
12 $\quad$ **if** $c < cost$ **then**
13 $\quad\quad cost = c$;
14 $\quad\quad t_0 = j$;

The function $REupdate(t_*, RE, r)$ in Algorithm 1 is to update the available renewable energy at each time slot after r amount of renewable energy is reserved at time slot t_*. $REupdate(t_*, Re, r)$ returns an array with m elements. The function is as follows.

Algorithm for One Appliance in A_2. Here we take an EWH as a representative of schedulable and interruptable appliances. For an EWH alone, the formulation for time domain $[0, H]$ is as follows.

Min: $\sum_{t=0}^{H} R_{a,t} * \delta t * P_t$
Subject to:

$$\theta_t = f(R_{a,t}, \theta_{am}, \theta_{t-1}, \delta t) \quad (8)$$

Function REupdate(i,RE,r)

```
1 for k ← i to,m do
2   RE(k) = RE(k) − r;
3 for k ← i − 1 to 1 do
4   RE(k) = min{RE(k), RE(i)};
5 return RE.
```

$$\theta_{low} \leq \theta_t \leq \theta_{high} \tag{9}$$

where f is the hot water thermal dynamics function, θ_{am} is the ambient temperature (the temperature of cold feed water), θ_t is the hot water temperature at time slot t, and θ_{low} and θ_{high} are the lower and upper limits of the comfort temperature band respectively. Here the penalty of dissatisfaction is not considered in the objective function, since the comfort temperature band, i.e., Eq. (9) is enforced.

EWH has been specifically studied by Du and Lu [6]. They have formulated the scheduling of EWH in a similar way, and solved it by a heuristic algorithm, which can achieve good performance but is not optimal. In the following, we aim to find the optimal solution through dynamic programming.

We equally divide the time domain $[t_0, H]$, such as a day, into m time slots, and divide the comfort temperature range $[\theta_{low}, \theta_{high}]$ into n domains. Then a grid mesh is formed in which x-axis is time t, y-axis is temperature θ, and each grid point (t_i, θ_j) means that hot water's temperature at time t_i is θ_j, $\forall 1 \leq i \leq m$ and $1 \leq j \leq n$, as shown in Fig. 1. The basic idea of our algorithm is to find the best scheduling for the transition from the starting point (t_0, θ_0) to another point (t_i, θ_j), where $t_0 < t_i$. The transition must go from the left side to the right side. For instance, Fig. 1 shows a transition from point P_1 to point P_2. We also need schedule the amount of renewable energy consumed during each transition. Similar to the temperature domain, we divide $RE_0(m)$ equally into l portions, and only an integer number of portions can be used during each time slot.

We use a function $Tr(t_0, \theta_0, t_i, \theta_j, r_i)$ to denote the cost of the optimal scheduling for the transition from the starting point (t_0, θ_0) to another point (t_i, θ_j), where r_i is the amount of renewable energy consumed so far at the point (t_i, θ_j). For the scheduling time domain $[t_0, H]$, the least cost is $Tr(t_0, \theta_0, H, \theta_H, r_m)$, where θ_H is the temperature of hot water at the end of the time domain. The dynamic programming method is shown in Algorithm 2.

Algorithm 2 outputs the least cost c_{min}. Once obtaining c_{min}, we can trace back to get the optimal path. Note that in the dynamic programming, the term $Tr(t_{i-1}, \theta_{j'}, t_i, \theta j, r)$ has to be calculated. The term represents the optimal cost from one point at time t_{i-1} (the beginning of time slot i) to another point at its next time t_i (the end of time slot i). We will calculate this term using an equivalent thermal parameter model [9,10]. According to this model, the temperature changing rate is non-linear with respect to time. The higher the initial temperature of EWH, the more difficult to increase the water temperature;

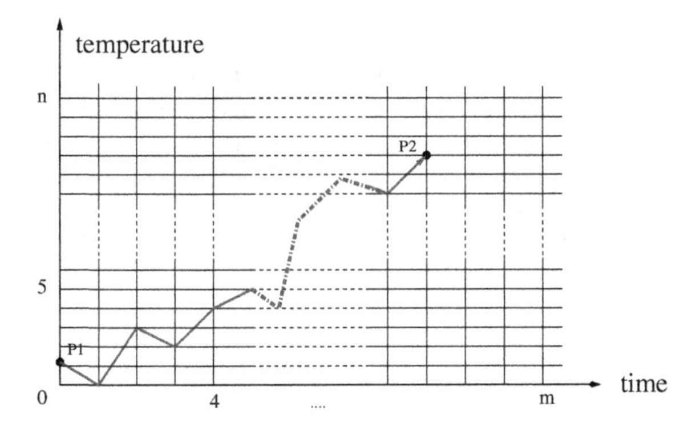

Fig. 1. A transition from point P_1 to point P_2. The time domain is divided into m slots and the temperature domain is divided into n slots. A valid transition should go from left to right.

Algorithm 2. Optimal heating scheduling for EWH for storable renewable energy.

Input: H, n, m, l, t_0, θ_0.
Output: The least cost, c_{min}.

1 $t_i = t_0 + i * H/m, \forall\, i \in [1, m]$;
2 $\theta_j = \theta_0 + j * (\theta_{high} - \theta_{low})/n, \forall\, j \in [0, n]$;
3 $\delta r = R(m)/l, \forall k \in [0, l]$;
4 **for** $i \leftarrow 2$ **to** m **do**
5   **for** $j \leftarrow 0$ **to** n **do**
6   **for** $k \leftarrow 1$ **to** $\lfloor R(i)/\delta r \rfloor$ **do**
7             $r_i = \delta r * k$;
8             $Tr(t_0, \theta_0, t_i, \theta_j, r_i) = min_{\forall j' \in [1,n], \forall r \in \{0,\ldots,r_i\}} \{Tr(t_0, \theta_0, t_{i-1}, \theta_{j'}, r_i - r) + Tr(t_{i-1}, \theta_{j'}, t_i, \theta_j, r)\}$;

9 $c_{min} = min_{\forall r \in [0,R]} \{Tr(t_0, \theta_0, t_m, \theta_0, r)\}$. // Find out the least cost;

the higher the initial temperature of EWH, the faster the water temperature decreases. We need to consider in total three types of transitions: (1) from (t_1, θ_1) to (t_2, θ_1), (2) from (t_1, θ_1) to (t_2, θ_2), where $\theta_1 < \theta_2$, and (3) from (t_1, θ_1) to (t_2, θ_2), where $\theta_1 > \theta_2$.

We assume that we have flat electricity prices within each time slot, which is usually the case in practice. For the first type of transition, the best scheduling should avoid high temperature during the transition, since high temperature is slow to reach and fast to vanish. Therefore, the best scheduling is switching on and off EWH alternatively, heating it for a moment and then letting it cool down to θ_1, and so on. For the second type of transition, the best scheduling should avoid temperature higher than θ_2. The last switch-on should directly increase the temperature to θ_2 without any cooling down. So the best scheduling has two sub-transitions: from (t_1, θ_1) to (t_x, θ_1) and from (t_x, θ_1) to (t_2, θ_2). The first sub-transition uses the same strategy as in the first type of transition, and the second sub-transition does not have any cooling down. Similar to the second type of transition, the third one is also split into two sub-transitions: from (t_1, θ_1) to (t_x, θ_2) and from (t_x, θ_2) to (t_2, θ_2). The first sub-transition does not have any heating, and the second sub-transition uses the same strategy as in the first type of transition.

Algorithm for One Appliance in A_3. In this case, the algorithm is very simple. The optimal solution is to assign as much renewable energy as needed to the time slots in price descending order.

Algorithm for All Appliances. As proved in the beginning of this section, this problem is NP-hard. In previous subsections, we propose algorithms for different types of appliances individually. Here we consider all types of appliances together. For a household, constraint (5) usually does not take effect; that is, the consumption of a household does not surpass R_{max}. We propose a heuristic algorithm, which schedules appliances according to their priorities. We assume the real-time appliances have the highest priority, followed by schedulable but uninterruptable appliances, then by the schedulable and interruptable appliances. We refer to this algorithm still as Algorithm MA.

4.2 Complexity and Performance

Our goal in this paper is to provide lightweight algorithms for an NP-hard problem—the household load scheduling problem. All algorithms proposed in this section are indeed lightweight. We list the complexity of all algorithms in Table 1, which are all polynomial.

Table 1. The complexity of proposed algorithms.

Scenarios	One in A_1	One in A_2	One in A_3	Multiple appliances
Complexity	$O(m)$	$O(mn^2l^2)$	$O(1)$	$O(mn^2l^2)$

As to performance, algorithms for each individual appliance are optimal or quasi-optimal (in the dynamic program, we use discretized temperature and renewable energy values), no matter whether there is renewable energy. As to the heuristic algorithms for multiple appliances of different types with renewable energy, they are not optimal. We cannot even derive a theoretical approximation bound for them, since no such bound exists as shown in the next subsection.

4.3 Discussion

Here we show that with renewable energy, no α-approximation algorithm exists.

Suppose we are given n appliances, each with a duration of C_j, during which it has a power of R_j. To simplify things, we assume all appliances are uninterruptable, and that the appliances can be scheduled at any time from 0 to T. The cost of one unit of energy at time t is P_t. The goal is to assign each appliance a time interval of length C_j such that the total cost of the energy consumed is minimized.

Claim 1: There exists no α-approximation algorithm for the household scheduling problem for any α.

Proof: We prove it by reduction. The reduction is from 2-partition. Given n items with sizes $a_1, \ldots, a_n$, can they be partitioned into two subsets of equal size? We set $RE_i = \frac{1}{2}\sum_{j=1}^{n} a_j$ for $i = 1, 2$, and $RE_i = 0$ otherwise, and create n jobs with $R_j = a_j$ and $C_j = 1$. Then a schedule of 0 cost corresponds to a partition, and any α-approximation must also be a schedule of 0 cost (to get any approximation guarantees, one has to exclude the possibility of a 0 cost schedule). ■

5 Evaluation

In this section, we will evaluate our algorithms for different appliances under different scenarios via simulations. We mainly evaluate our dynamic programming based algorithm (referred to as *Price-Driven Dynamic Programming algorithm* (PDDP)) for schedulable and interruptable appliances, and compare the cost with those obtained from existing schemes. We also evaluate the approximation ratio for the proposed heuristic algorithms under different scenarios.

5.1 Performance of PDDP Algorithm for an EWH

Here we will evaluate our algorithm, PDDP, for EWH with simulation. In [6], the authors compared their algorithm (we call it *Du & Lu's algorithm*) with the non-price-sensitive algorithm and the transactive control algorithm [11], and concluded that their algorithm outperforms the other two. Hereby, we will compare PDDP algorithm with Du & Lu's algorithm, the best of the three algorithms evaluated in [6].

We evaluate our PDDP algorithm with Ameren's residential real time pricing program [12]. Ameren's real time pricing program is an electric supply rate option in which customers pay electricity supply rates that vary by the hour. With this program, hourly prices for the next day are set the night before and can be communicated to customers so that they can determine the best time of day to use major appliances. Figure 2 shows the real-time price for Nov 3 (Sunday), Nov 4 (Monday), and Nov 7 (Thursday), 2013.

Furthermore, according to the past usage profile [13], we can forecast the daily hot water usage for a household. Based on [13], we can obtain hourly fraction of daily hot water draw. Suppose a typical household has 5 persons who consume about 180 gallons hot water daily. Thus, we obtain the hot water daily usage rate for a typical house as shown in Fig. 3.

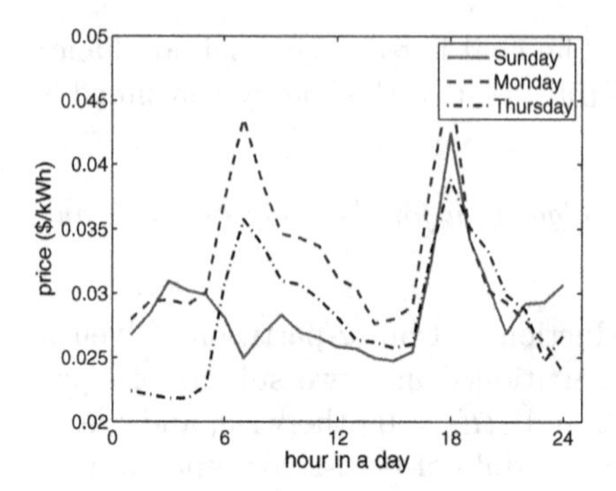

Fig. 2. Daily price.

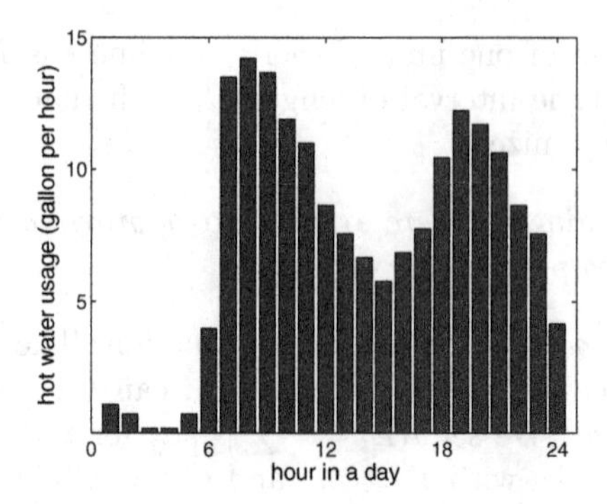

Fig. 3. Daily water usage forecast.

We simulate PDDP algorithm and Du & Lu's algorithm for a whole day. At the beginning, the temperature in the water tank is at the lower limit of the temperature band; at the end, the temperature goes back its original value. Figure 4 shows the EWH temperature profile using water usage in Fig. 3 and real price on Monday in Fig. 2. As we can see, Du & Lu's algorithm has high-temperature intervals, while PDDP only has medium-temperature intervals. Furthermore, the locations of intervals above the lower temperature bound are different. The high-temperature intervals of Du & Lu's algorithm reside in low-price periods, while those of PDDP are not necessarily in low-price periods. These differences contribute to 4 cent saving of electricity. Of course, the savings are different for different days due to different daily real time price. We conduct the simulation for the whole November, 2013, by using the water usage in Fig. 3 and Ameren's real time price. Figure 5 shows the daily cost for both algorithms (upper panel) and the daily saving (bottom panel). The daily saving can be as large as 24 cents and as small as 0.5 cents as well, but never be negative. Therefore, PDDP always outperforms Du & Lu's algorithm. On average, the daily cost for EWH is $\$0.913 \pm 0.085$, and that for Lu and Du's algorithm is $\$0.968 \pm 0.084$. The daily saving is $0.055, accounting for about 6% saving; and the monthly saving is about $1.65.

5.2 Approximation Ratios of Algorithm for Multiple Appliances

Since Algorithm MA, algorithm for multiple appliances, is heuristic, we evaluate its performance using the approximation ratio, denoted by ar, which is defined by the following equation:

$$ar = \frac{\text{the cost from } Alg.MA}{\text{the optimal cost}} \tag{10}$$

We set up two simulations. The first one has three appliances, one of each type; namely a stove, a washer, and an EWH. Note that Algorithm MA schedules appliances one by one based on their priorities. By running Algorithm MA, we can get the approximate cost. We also get the optimal solution by a variation

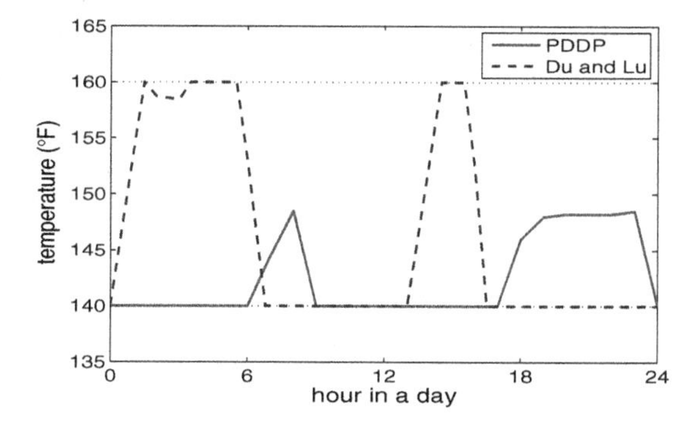

Fig. 4. Hot water temperature (°F) profiles for *PDDP algorithm* and *Du & Lu's algorithm.*

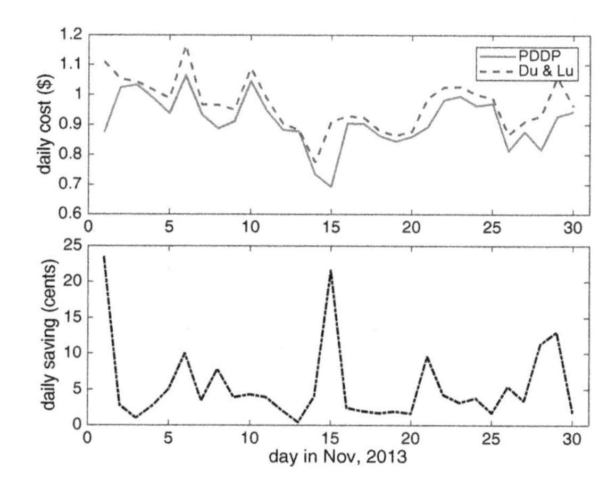

Fig. 5. Daily hot water cost comparison in Nov, 2013, for a typical household.

of our dynamic programing. Then we can calculate ar according to Eq. (10). The second simulation has an EWH, an AC, a washer and some real-time appliances, by considering that a household usually has two main schedulable and interruptable appliances, i.e., AC and EWH. In this simulation, we obtain the optimal cost through a brute-force search with dynamic programming. In both simulations, for a washer, we randomly pick a scheduling time window $[t_a^\alpha, t_a^\beta]$; for an EWH, we still use daily hot water profile in [13], with some random noise; for an AC, we set temperature range as [72,75]° F.

For the renewable energy, we consider the solar power generated by a photovoltaics (PV) array. We utilize *PVWatts Calculator* at National Renewable Energy Laboratory [14] to calculate the hourly solar power. In our simulation, the PV array is fixed on the roof of a household at Maryville, Missouri, with a size of 25 m^2, 20° of the tilt angle and 180° of the azimuth angle. In each round, we randomly pick one day from the whole year.

We run the simulations multiple rounds and then calculate the maximum ratio and the average ratio. The results of the simulations are listed in Table 2.

Table 2. The approximation ratios of algorithms for multiple appliances using Alg-MA.

Simulation #	Mean	Maximum	# of rounds
1st	1.49	2.57	2000
2nd	1.14	2.06	2000

As we can see, the heuristic algorithm does not work very well. The reason is that the schedule of one appliance affects that of another. Scheduling of each appliance only considers minimizing its own cost, and it is likely that shifting some renewable energy from one appliance to another achieves smaller overall total cost.

6 Conclusion

In this paper, we formulate a generic household load scheduling problem which considers both real-time pricing and renewable energy sources. We prove that the household load scheduling problem is NP-hard. We divide household appliances into three categories, and propose algorithms for individual appliances from each category and for multiple appliances from all categories. Specifically, for schedulable and interruptable appliances, such as EWH and AC, we propose a dynamic programming algorithm, which guarantees an optimal or quasi-optimal solution. For scenarios with multiple appliances from all three categories, we propose a heuristic algorithm which schedules appliances in each category in different priority order. We evaluate our algorithms with simulations, validating their efficiency and performance.

References

1. Baughman, M., Siddiqi, S.: Real-time pricing of reactive power: theory and case study results. IEEE Trans. Power Syst. **6**(1), 23–29 (1991)
2. Herter, K.: Residential implementation of critical-peak pricing of electricity. Energy Policy **35**(4), 2121–2130 (2007)
3. Samadi, P., Schober, R., Wong, V.: Optimal energy consumption scheduling using mechanism design for the future smart grid. In: Second IEEE International Conference on Smart Grid Communications (2011)
4. Goudarzi, H., Hatami, S., Pedram, M.: Demand-side load scheduling incentivized by dynamic energy prices. In: Second IEEE International Conference on Smart Grid Communications (2011)
5. Mohsenian-Rad, A., Leon-Garcia, A.: Optimal residential load control with price prediction in real-time electricity pricing environments. IEEE Trans. Smart Grid **1**(2), 120–133 (2010)

6. Du, P., Lu, N.: Appliance commitment for household load scheduling. IEEE Trans. Smart Grid **2**(2), 411–419 (2011)
7. Samadi, P., Mohsenian-Rad, A., Schober, R., Wong, V., Jatskevich, J.: Optimal real-time pricing algorithm based on utility maximization for smart grid. In: First IEEE International Conference on Smart Grid Communications, pp. 415–420 (2010)
8. Lu, J., Sookoor, T., Srinivasan, V., Gao, G., Holben, B., Stankovic, J., Field, E., Whitehouse, K.: The smart thermostat: using occupancy sensors to save energy in homes. In: Proceedings of the 8th ACM Conference on Embedded Networked Sensor Systems (2010)
9. Taylor, Z., Pratt, R.: The effects of model simplifications on equivalent thermal parameters calculated from hourly building performance data. In: Proceedings of the ACEEE Summer Study on Energy Efficiency in Buildings, pp. 10–268 (1988)
10. Pratt, R., Taylor, Z.: Development and testing of an equivalent thermal parameter model of commercial buildings from time-series end-use data. In: Pacific Northwest National Laboratory (PNNL), Richland, WA, Technical report, April 1994
11. Katipamula, S., Chassin, D., Hatley, D., Pratt, R., Hammerstrom, D.: Transactive controls: market-based gridwisetm controls for building systems. Technical report, Pacific Northwest National Laboratory (PNNL), Richland, WA (2006)
12. Ameren real time pricing program. https://www.ameren.com
13. Fairey, P., Parker, D.: A review of hot water draw profiles used in performance analysis of residential domestic hot water systems. Technical report, Florida Solar Energy Center, vol. 15 (2004)
14. National Renewable Energy Laboratory, "PVWatts Calculator". http://pvwatts.nrel.gov/index.php

An Empirical Study of OAuth-Based SSO System on Web

Kaili Qiu[1,2], Qixu Liu[1](✉), Jingqiang Liu[1], Lei Yu[1], and Yiwen Wang[1]

[1] School of Cyber Security, University of Chinese Academy of Sciences, Beijing 100049, China
liuqixu@ucas.ac.cn

[2] School of Computer Science and Technology, Tianjin University, Tianjin 300350, China

Abstract. More and more websites use OAuth 2.0 protocol to provide SSO services to ease password management for users. Although OAuth 2.0 has been implemented carefully by following many guidelines, still some parts have been ignored. In this paper, we discover a new attack mode for hijacking the account in the OAuth-based SSO system. We conduct an empirical study for the proposed attack on top 500 Chinese websites of Alexa supporting SSO services by 6 IdPs. Our results uncover four vulnerabilities that allow attackers hijack the victim's account without knowing the user's username and password. Closer examination reveals that 68.67%, 12.87%, 68.67% and 59.66% of the websites are vulnerable to the four vulnerabilities respectively and 45.49% of the websites can be conducted proposed complete attack. To defend this attack, we provide developers simple practical recommendations to the critical vulnerable nodes.

Keywords: SSO system · OAuth 2.0 · Account hijacking

1 Introduction

Many websites such as Sina Weibo provide the web applications a service by API to let user login the web applications with the social accounts. By using OAuth 2.0, these web applications can ease the password management for their users, as well as promoted by the well-known social websites. In addition, it reduces the cost of user experience when users first visit a new web application. Due to its enhancement in sharing with other websites and convenience provided to users, it has been widely implemented in the real world. What followed is the increasing risk of the third-party account hijacking. Although some mainstream social networking websites have a very high level of security for their user accounts, there are still some channels through which the resources of accounts can be accessed without the username and password. When OAuth 2.0 used as a Single-Sign-On(SSO) scheme, it has been deployed commercially for authentication and authorization by many websites or applications. Consequently, it

S. Chellappan et al. (Eds.): WASA 2018, LNCS 10874, pp. 400–411, 2018.
https://doi.org/10.1007/978-3-319-94268-1_33

is very imperative to conduct a study on the vulnerabilities in the SSO system based on OAuth 2.0 protocol.

The security of OAuth 2.0 protocol has been studied by several researches [1–3] using formal methods. The results of those analyses suggest that the protocol is secure. However, the implementation of SSO system based on OAuth 2.0 in practice is different. Li et al. [4] discovered the CSRF attack to hijack victim's account by embedding a malicious link which hosts on malicious website and can initiate requests into the webpage. Wang et al. [5] conducted their study on three platforms and investigated the verification in impersonation attack. Sun and Beznosov [6] also studied the impersonation attack without limitation in use of credentials. Although they proposed some means like correctly usage of state parameter to defend CSRF attack, enhancement in detection of malicious RP or limitation in usage of credentials to defend impersonation attack, there are still some ignored vulnerabilities which can be utilized to carry out the account hijacking attack.

To ensure the security of OAuth 2.0 protocol in SSO system, we propose an attacker-guided analysis methodology to discover the ignored vulnerabilities. The security we analyzed in this paper is evaluated in three aspects: (a) whether the *redirect_uri* can be modified; (b) whether the RP can be embedded into external resources; (c) whether the IdP verifies the binding between the authorization code and the current user.

Our study was conducted on the top 500 Chinese websites of Alexa, including 6 IdPs. It covers four vulnerabilities we discovered in this paper. The results reveal that: (1) OAuth-based SSO services are implemented widely nowadays. (2) The OAuth-based SSO system are suffering from many vulnerabilities. (3) 59.66% of the websites we have examined don't verify the binding of the current user and authorization code. (4) the IdPs of Tencent QQ and Douban is better than others in limiting the *redirect_uri*.

The remainder of this paper is organized as follows. Section 2 provides background about the OAuth 2.0 protocal and the researches of predecessors on the security of OAuth 2.0. Section 3 summaries the account hijacking attacks in the SSO system of OAuth 2.0 and describes the new attack mode for hijacking account. In Sect. 4, we describe the approach and adversary model we use in this paper and give the detail description of the analytical methodology. Section 5 presents the empirical experiments and the analyses of the results. In Sect. 6 we propose some mitigations against the account hijacking attack. Section 7 concludes this paper.

2 Background and Related Work

2.1 OAuth 2.0 Used for SSO

OAuth 2.0 authorization framework [7] aims to offer an open and standardized protocol for authorization. It makes it possible for users to allow the third-party applications limited access to their protected data without exposing their own credentials to the third-party application. With OAuth 2.0 being one of the most popular frameworks, authorization have found widespread adoption in the web over the last years and it seems like will be implemented more in the future. There are four roles defined in this protocol: *resource owner*, *resource server*, *client*, and *authorization server*. Resource owner is an entity capable of granting access to the protected resource and it always refers to users in practice. Resource server is the server hosting the protected resources. Authorization server is the server issuing access token to the client after successfully authenticating the resource owner and obtaining authorization code. In reality, these two servers are in a common server. We call it IdP (identity provider) in the rest of the paper. Client is an application making requests for protected resource on behalf of the user with his authorization. It is called RP (Relying Party) in the rest.

In order to meet the needs of different conditions, four authorization modes have beed proposed. They are Authorization Code mode, Implicit mode, Resource Owner Password Credentials and Client Credentials. Due to the widespread use in authorization code in website, we only analyze the work flow of login with the third-party based on Authorization code mode. Its works roughly as follows, see Fig. 1.

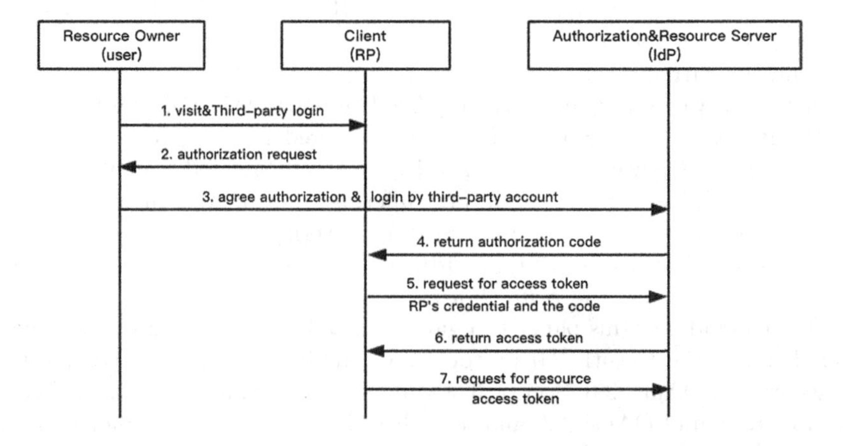

Fig. 1. The work flow of the authorization code mode in SSO.

When a user visits a RP and chooses to login with Third-party account, RP redirects the user to IdP which storing the Thrid-party accounts. Then the RP requests the user for authorization. The user provides his own credential (username and password) to the IdP to approve authorization. After the IdP checking the credential successfully, the IdP will redirect the user back to the RP with an authorization code. Subsequently, the RP provides the code with the client id and client secret (if it is required) to request for the access token. Once receiving the access token, the RP will be able to use the access token to access the user's resources which held at the IdP.

2.2 Related Work

Many researches have been done on the security of OAuth 2.0 protocol itself. IETF has presented threat models in RFC6819 [8] for OAuth 2.0 and proposed that the authorization code flow has been proved to be secure as long as the TLS is properly used. Pai et al. [1] confirmed a security issue described in the OAuth 2.0 using Alloy Framework. And Chari et al. [2] analyzed OAuth 2.0 in the Universal Composability Security Framework [9], and the result shows that the protocol is secure if all endpoints from IdP and RP are SSL-protected. However, all these work is based on abstract models of OAuth 2.0 and delicate implementation details are ignored.

To understand the implementation security of the OAuth 2.0 protocol in the real world, many researchers have studied the SSO system deployed with OAuth 2.0 in web, android and ios platform, mainly including CSRF attack and Impersonation attack to hijack the account. The researches of vulnerabilities in mobile platform were conducted by Wang et al. [11] and Chen et al. [12,13]. Li and Mitchell [14] investigated the CSRF attack with no state parameter implemented or incorrectly implemented. The CSRF attack with state parameter through embedding malicious link hosted on malicious websites, XSS or Sql Injection is studied by many researchers [5,6,15–17]. Hu et al. [18] presented the Impersonation attack for APP and investigated its impact on 12 major OSN providers. Our work focuses on the Impersonation attack for hijacking the account instead of the Application. Wang et al. [5] discovered the impersonation attack for account hijacking by building malicious RPs to gather the user's account information and the access token is always the credential they used. And Sun et al. [6] completed the impersonation attack by observing the unencrypted communications. However, we discover a new attack mode for account hijacking against the *redirect_uri* with the limitation that the credentials can only be used once. Prior to this, this kind of vulnerability had been submitted to CNVD (CHINA NATIONAL VULNERABILITY DATABASE) [19] and CNVD has shown the analysis of this attack. And we eavesdrop the authorization code through Referer instead of doing a lot of work to build a malicious website.

3 Account Hijacking Attack Against OAuth 2.0

3.1 Already Existing Attacks for Account Hijacking

Many Online Social Networks are using OAuth 2.0 protocol to grant access to API endpoints nowadays. It provides many potential vulnerabilities to carry out the account hijacking attack. There are two major attacks for hijacking account.

Cross Site Request Forgery Attack. In this attack, a malicious website causes the browser to initiate a request of the attacker to the target site. Then the target site will execute actions without the involvement of the user. The attacker causes the browser to send the target site a request containing the attacker's own authorization code or access token to associate the attacker's IdP account with the victim's current account. The OAuth 2.0 specification recommends the developers using the state parameter correctly to defend this attack due to that it will verify the source of the request.

Impersonation Attack. An impersonation attack is conducted by acquiring the SSO credentials (token, code, user profile etc.). Then the credentials are used to impersonate the victim to login successfully. There are many ways to obtain the credentials, such like that the attacker can build a malicious RP to lure the victim to login in the RP so that he can gather the user's information which can be used to replay in impersonating the victim on the RP. However, the credentials should not limited in one-time use in the attack they proposed.

3.2 New Attack Mode for Account Hijacking

As we can see, these existing ways of hijacking user's account are difficult to carry out due to various mitigations proposed by the researches. We describe a new way of account hijacking attack through eavesdropping the authorization code which is similar to the impersonation attack, but breaking the limitation that the code can only be used once by modifying the *redirect_uri*.

This attack focuses on the authorization code mode of the four grant authorization modes. The attacker logins the RP on behalf of the victim by using the victim's authorization code eavesdropped by the attacker. Consequently, the attacker can impersonate the victim to request for the access token. Moreover, he can login the websites as victim's account and access to the other websites information bound to the account, viewing sensitive information or performing authorization operations. Even worse is that the attacker can use the victim account for illegal information dissemination, fraud and other illegal activities.

In order to avoid replay attack, the OAuth 2.0 protocol specifies that the authorization code only can be used one time. To complete attack, three conditions should be satisfied: (1) The authorization code can be eavesdropped in some means. Through investigation, we find that the code is as a parameter of the uri in the web application. When the IdP sends the authorization code to the user, it will redirect the user to the RP and the authorization code is adhere to the uri as a parameter in the RP's browser. It is very easy to cause the leakage of the code to some extent. We use two methods to steal the authorization code in later experiments: embedding external resources or observing http traffic between RP and IdP. (2) The authorization code stolen could not be used. In order to obtain an unused authorization code, the attacker can forge an authorization link by modifying the *redirect_uri*. When the victim clicks the forged authorization link and logins, the attacker can acquire unused code. (3) There is no binding of current user and authorization code. In this way the attacker can login the victim's account successfully without the other information but the authorization code.

The work flow of the complete attacks are divided into two small classes according to the different ways in eavesdropping the authorization code. We show the flow of this attack roughly in Fig. 2.

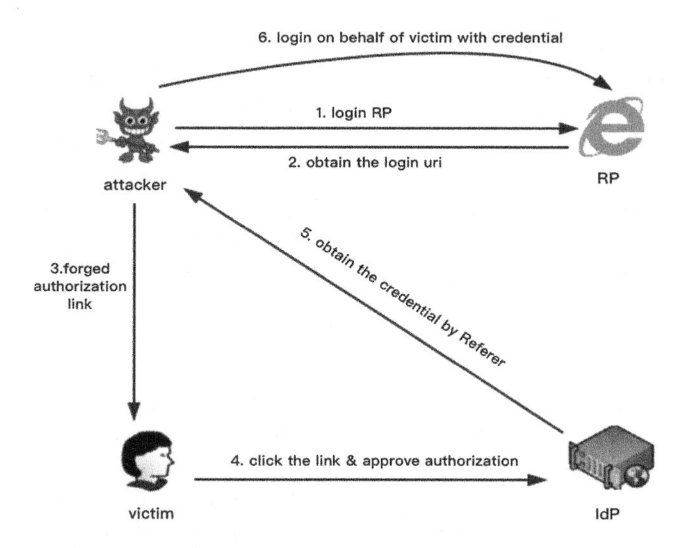

Fig. 2. The flow of the attack.

1. The attacker chooses to login the RP with third-party.
2. The attacker obtains the normal authorization link that login by the third-party account.
3. The attacker sends the forged authorization link which is modified.
4. The victim clicks the link and he is redirected to the IdP with approval of authorization to the RP.
5. The attacker obtains the victim's credential.
6. The attacker uses the victim's credential to login the website on behalf of the victim.

4 Our Approach

Before we proposed the adversary model, we make some assumptions: (1) The user's browser is not compromised. (2) The RP and the IdP are benign. (3) The attackers are capable of eavesdropping the authorization code transferred between the RP and the IdP based on HTTP protocol. Due to the mature in the technique of ARP spoofing, we make this assumption to facilitate our studies. Trabelsi [20] conducted the lab exercises about performing Denial of Service(DoS) and Man-in-the-Middle(MiM) attack using ARP cache poisoning. Bull et al. [21] demonstrated ARP poisoning attack across every major hypervisor platform.

4.1 Adversary Model

Our adversary model aims at impersonating the victim on the RP website by eavesdropping the user's authorization code. Due to the different way to filch the authorization code, there are two adversary models in our work.

(a) The first type of adversary has the capabilities of a web attacker, he can share malicious links or post comments which contain malicious content in the benign RP. The malicious content might trigger the web browser to send http/https request to the attacker's server so that the attacker obtain the authorization code through the HTTP Referer.
(b) The second type of adversary can intercept and replay unencrypted network traffic between the RP and the IdP. When the adversary and the RP are in a local area network, the adversary can launch arp spoofing.

4.2 Methodology

Facing the challenges when conducting a security analysis of real-world OAuth-based SSO on websites, many researchers [6,10] employed a methodology that they treat the IdPs and RPs as a black box and focused on the traffic during the OAuth-based SSO transaction. Apart from focusing on the traffic between the RPs and the IdPs, we also concern about the modifiable of the *redirect_uri* and the embeddedness of the RP website. Here, our methodology consists of four aspects, corresponding four kinds of vulnerabilities:

- Modification of the *redirect_uri*. For a higher rate of success, we use the homologous uri replace the former *redirect_uri*. The subpages of the target website are a good choice for us. The way of modifying the *redirect_uri* is replacing with the homologous webpage an embedded external resource. Or we examine whether the *redirect_uri* can be replaced by another uri based on HTTP protocol so as to make the code unused.
- Embedment the RP website. Through posting the comment in the RP website with an embedded external resource which the address of the resource is belong to us, we can obtain the authorization code by utilizing the HTTP Referer. HTTP Referer is an HTTP header field that identifies the address of the webpage (i.e. the URI or IRI) that linked to the resource being requested [22]. When a user visits a webpage with an external resource embedded, two requests launched. The first one request for the webpage and the other one request for the resource. As a result, the browser will send the request that includes address of the webpage to the attacker. Hence, the attacker will obtain the authorization code in such a way of embedding external resource into the target website.
- Interception of the traffic between the RPs and the IdPs. We use the ARP spoofing to intercept the traffic when we are in the same local area network with the RP. When our MAC address is associate with another host's IP address such as the default gateway of the local area network, any traffic meant for that IP address will be send to the us instead. Hence, we will obtain the response and request, and then we transmit the response to the victim. As the transmission is in plain text based on HTTP protocol, the attacker can eavesdrop the authorization code.
- Verification of the binding of the authorization code and the current user. In most situations, the attacker can impersonate the victim because of lacking

sufficient verification. In this step, we register the RP's account and login the RP with IdP's account. We obtain our unused authorization code by Burpsuite tool. Then, we login the RP in a new browser (no any information about the user information) only with the authorization code.

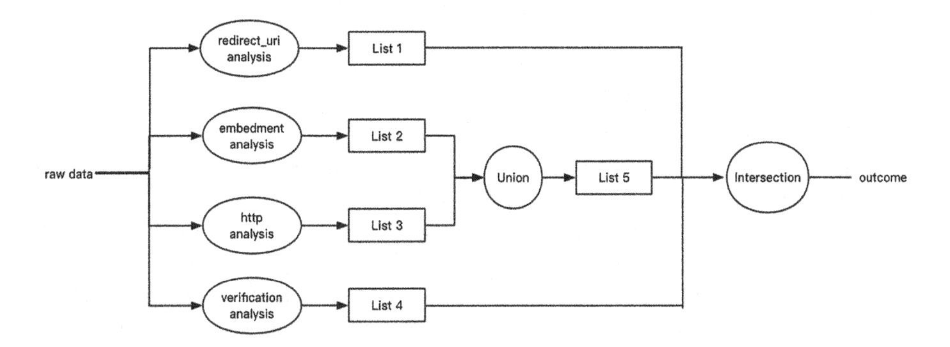

Fig. 3. The shematic of methodology.

The schematic of the methodology is show in Fig. 3. First, we conducted the above four analyses on the websites to obtain the four files: (1) Redir file. The list of the websites which the *redirect_uri* can be modified. (2) Embed file. The list of the websites which the webpages can be embedded external resources. (3) HTTP file. The list of the websites which the communication with the IdP is based on HTTP protocol. (4) Unverify file. The list of websites which do not implement the verification of the current user and the authorization code. Secondly, we take the union of the Embed file and the HTTP file and regard them as the Code file. The Code file means the websites in this file can be eavesdropped authorization code through intercepting the unencrypted traffic or HTTP Referer of embedded external resource. Eventually, we obtain the final file, containing the websites which can be carried out the complete attack by taking the intersection of Redir file, Code file and Unverify file.

5 Experiments and Analysis

According to the above methodology, we examined top 500 Chinese websites of Alexa, of which there are 233 websites implementing the OAuth-based SSO services. Here we regard the websites of the same domain as one. And we select 6 widely used IdPs for our studies, using IdP1, IdP2, IdP3, IdP4, IdP5 and IdP6 to identify them respectively. To begin with the studies, we register the test accounts for each IdPs and each RPs.

Firstly, we examined the 233 website for the four vulnerabilities mentioned above respectively. Besides, we count the number of the website that can be

Table 1. The percentage of the vulnerabilities. **v1**: the vulnerability of modification of uri. **v2**: the vulnerability of embedding external resource. **v3**: the vulnerability of interception of the traffic between the RPs and the IdPs based on HTTP. **v4**: the vulnerability of no verification between the current user and the authorization code.

Type	v1	v2	v3	v4	Attack
Percentage	68.67%	12.87%	68.67%	59.66%	45.49%

carried out the complete attack we proposed according to outcome file. The result is shown in the Table 1.

From the outcome and the process of the analysis, we find that more than half of the websites are vulnerable to the modification of *redirect_uri*. And the authorization code can be transmitted on the HTTP channel as long as the *redirect_uri* is revisable. The developer does not notice that the restriction of the *redirect_uri* is significant to the security and we conduct a detailed analysis of it subsequently. The v4 is relatively less than the former, but still serious in practice. 59.66% of the websites do not concern about the verification, it often causes the impersonation attack. Embedding an external resource successfully is most in the circumstance that the website owns some plates that we can submit our resources. It always happens in a small part of websites in top 500. According to our methods, we discover a total of 106 websites that exist the attack we proposed, including 22 websites which can be attacked by embedding the external resources.

By analyzing the 6 major IdPs, we discover the security in modifying the *redirect_uri* which is registered in different IdPs. The different IdPs show different security levels with the developers' implementation methods. The result is show in Table 2.

Table 2. The popularity and security in different IdPs

IdPs	Popularity		Alterability in URI	
	N	%	N	%
IdP1	211	90.56%	0	0
IdP2	146	62.67%	38	26.03%
IdP3	192	82.40%	143	74.47%
IdP4	10	4.29%	5	50.00%
IdP5	13	5.58%	9	69.23%
IdP6	9	3.86%	0	0

From the 233 websites we examined, the mostly used IdPs are IdP1, IdP3 and IdP2 and they respectively account for 90.56%, 82.40%, and 62.67% in all websites which provide SSO services. In our experiment, there is no one website that implementing the SSO by IdP1 and IdP6 can be modified the *redirect_uri*. And according to the result, we can know that the IdP3 is more vulnerable than the others. Although IdP4 and IdP5 have a high rate of alterability, we can only come to this conclusion that IdP4, IdP5 exist the **v1** attack due to that the test sample is few.

6 Mitigation

OAuth 2.0 protocol have been implemented widely in the real-world, and it seems like the increasing numbers of websites will deploy OAuth 2.0 for SSO services in their existing systems. And it is a significant danger. Below we make some recommendations in the implementation about the threat critical nodes we mentioned above. These suggestions all provided for the developers of the client application.

- Confine *Redirect_uri*. To prevent the attacker to modify the *redirect_uri* and forged the user authorization link, the developer of the client application should use whitelist technology just like specificing one or several *redirect_uri* instead of just setting the *redirect_uri* to homologous.
- Use HTTPS Protocol in the Communication. It protects the privacy and integrity of the data transferred against eavesdropping the authorization code. RFC6819 also recommends HTTPS to secure.
- Limitation in External Resources. In the comment area of a website, there should be a higher limit to the insertion of external resources.
- Use verification mechanism between the code and the current user. HTTP cookie is a choice in logining the sites with the authorization code. Cookie is a piece of data sent from the website and stored on the user's computer by the use's web browser while the user is browsing. If the mechanism of the login using authorization code include the cookie when verifying one's identity, this attack will be greatly mitigated due to the difficulties in obtain the victim's cookie.

7 Conclusion

This paper discovers a new attack mode for account hijacking in OAuth-based SSO system on web. The experiment results show that 45.49% of the websites are vulnerable to carry out the new account hijacking attack. The RP websites have different extent to the various vulnerabilities, 68.67%, 12.87%, 68.67% and 59.66% of the v1, v2, v3, v4 in the websites we examined. These vulnerabilities are not only vulnerable in the attack we proposed, but also a threat to other attack like impersonate attack. Developers should pay more attention to the application when it is developed. To defend against this attack, some simple recommendations have been proposed to mitigate this attack and secure the OAuth-based SSO system.

Acknowledgments. The authors would like to thank the anonymous reviewers for their helpful comments for improving this paper. This work is supported by the National Key R&D Program of China (2016YFB0801604).

References

1. Pai, S., Sharma, Y., Kumar, S., Pai, R.M., Singh, S.: Formal verification of OAuth 2.0 using alloy framework. In: Communication Systems and Network Technologies, pp. 655–659. IEEE, Jammu (2011)
2. Chari, S., Jutla, C., Roy, A.: Universally composable security analysis of OAuth v2.0. Report 2011/526 (2011)
3. Fett, D., Küsters, R., Schmitz, G.: A comprehensive formal security analysis of OAuth 2.0. In: Proceedings of the 2016 ACM SIGSAC Conference on Computer and Communications Security, pp. 1204–1215. ACM, Vienna (2016)
4. Li, W., Mitchell, C.J., Chen, T.: Mitigation CSRF Attacks on OAuth 2.0 and OpenID Connect. https://arxiv.org/abs/1801.07983
5. Wang, H., Zhang, Y., Li, J., Gu, D.: The Achilles' heel of OAuth: a multi-platform study of OAuth-based authentication. In: Proceedings of the 32nd Annual Conference on Computer Security Application, pp. 167–176. ACM, Los Angeles (2016)
6. Sun, S.T., Beznosov, K.: The devil is the implementation details: an empirical analysis of OAuth SSO system. In: Proceedings of the 2012 ACM Conference on Computer and Communications Security, pp. 378–390. ACM, Raleigh (2012)
7. The OAuth 2.0 Authorization Framework. https://tools.ietf.org/html/rfc6749
8. OAuth 2.0 Threat Model and Security Considerations. https://tools.ietf.org/html/rfc6819
9. Canetti, R.: Universally composable security: a new paradigm for cryptographic protocols. In: Proceedings 42nd IEEE Symposium on Foundations of Computer Science, pp. 136–145. IEEE, Washington, D.C. (2001)
10. Wang, R., Chen, S., Wang, X.: Signing me onto your accounts through Facebook and Google: a traffic-guide security study of commercially deployed single-sign-on services. In: Proceedings of the 2012 IEEE Symposium on Security and Privacy, pp. 365–379. IEEE, Washington, D.C. (2012)
11. Wang, H., Zhang, Y., Li, J., Liu, H., Yang, W., Li, B., Gu, D.: Vulnerability assessment of OAuth implementations in android application. In: Proceedings of the 31st Annual Computer Security Applications Conference, pp. 61–70. ACM, Los Angeles (2015)
12. Shehab, M., Mohsen, F.: Securing OAuth implementations in smart phones. In: Proceedings of the 4th ACM Conference on Data and Application Security and Privacy, pp. 167–170. ACM, San Antonio (2014)
13. Chen, E.Y., Pei, Y., Chen, S., Tian, Y., Kotcher, R., Tague, P.: OAuth demystified for mobile application developers. In: Proceedings of the 2014 ACM SIGSAC Conference on Computer and Communications Security, pp. 892–903. ACM, Arizona (2014)
14. Li, W., Mitchell, C.J.: Security issues in OAuth 2.0 SSO implementations. In: Chow, S.S.M., Camenisch, J., Hui, L.C.K., Yiu, S.M. (eds.) ISC 2014. LNCS, vol. 8783, pp. 529–541. Springer, Cham (2014). https://doi.org/10.1007/978-3-319-13257-0_34
15. Slack, Q., Frostig, R.: OAuth 2.0 implicit grant flow analysis using Murphi. http://www.stanford.edu/class/cs259/WWW11/

16. Dill, D.L.: The Mur ϕ verification system. In: Alur, R., Henzinger, T.A. (eds.) CAV 1996. LNCS, vol. 1102, pp. 390–393. Springer, Heidelberg (1996). https://doi.org/10.1007/3-540-61474-5_86
17. Bansal, C., Bhargavan, K., Delignat-Lavaud, A., Maffeis, S.: Discovering concrete attacks on website authorization by formal analysis. J. Comput. **22**, 601–657 (2014)
18. Hu, P., Yang, R., Li, Y., Lau, W.C.: Application impersonation: problems of OAuth and API design in online social networks. In: Proceedings of the Second ACM Conference on Online Social Networks, Dublin, pp. 271–278 (2014)
19. CNVD-2018-01622. http://www.cnvd.org.cn/webinfo/show/4397
20. Trabelsi, Z.: Hands-on lab exercises implementation of DoS and MiM attacks using ARP cache poisoning. In: Proceedings of the 2011 Information Security Curriculum Development Conference, pp. 74–83. ACM, Kennesaw (2011)
21. Bull, R., Matthews, J.N., Trumbull, K.A.: VLAN hopping, ARP poisoning and man-in-the-middle attacks in virtualized environments. https://ronnybull.com/assets/docs/bullrl_defcon24_slides.pdf
22. HTTP_referer. https://en.wikipedia.org/wiki/HTTP_referer

Turning Legacy IR Devices into Smart IoT Devices

Chuta Sano[1], Chao Gao[1(✉)], Zupei Li[1], Zhen Ling[2], and Xinwen Fu[3]

[1] University of Massachusetts Lowell, Lowell, MA 01854, USA
{schuta, cgao, zlil}@cs.uml.edu
[2] Southeast University, Nanjing, China
zhenling@seu.edu.cn
[3] University of Central Florida, Orlando, FL 32816, USA
xinwenfu@cs.ucf.edu

Abstract. In this paper, we introduce a low-cost setup to convert an infrared (IR) controllable device to a smart IoT device. We design and implement a circuit board containing an IR sensor and multiple IR LEDs in parallel that transmit up to around *seventeen meters*. The board has an interface that can be connected to a Raspberry Pi and other similar devices. Our IR signal learning tool can record IR signals of an IR remote while our IR replay tool can replay the recorded signals to control the corresponding IR device. Therefore, a smartphone can be used to remotely control IR devices through an MQTT broker on a cloud server since the Raspberry Pi can be connected to the cloud. We also introduce the security implications of infrared communication using our setup and demonstrate attack scenarios. For example, a drone armed with our device can remotely turn off a TV - https://youtu.be/rPbzPbWrbf8 or http://v.youku.com/v_show/id_XMzQ0Njc5MzM3Ng.

Keywords: Internet of Things · Infrared · Smart home · Drone

1 Introduction

The Internet of Things (IoT) is a world-wide network of uniquely addressable interconnected objects [1, 2]. Many household devices including TVs, fans, air conditioners and toys have IR receivers and are controlled through IR remote. We may want to connect these legacy IR devices to the Internet and control them remotely while maintaining an acceptable cost level.

In this paper, we aim to design a cheap bridge between IR devices and Internet. In other words, we want to implement a smart IR remote with the Internet capability. To this end, we design a custom IR signal recording and replay circuit board, which can be connected to a Raspberry Pi and similar low-cost single board computers. For example, the new Raspberry Pi Zero W with the WiFi capability costs only $10 dollars. Without loss of generality, we will use the popular Raspberry Pi as an example in this paper to demonstrate how our IR board is used. We implement an IR recording tool that can record IR signals from an IR remote of any IR device. Our IR replay tool can replay the signal and control the IR device. Therefore, the Raspberry Pi can be connected to the

S. Chellappan et al. (Eds.): WASA 2018, LNCS 10874, pp. 412–424, 2018.
https://doi.org/10.1007/978-3-319-94268-1_34

Internet and a smartphone can be used to remotely control the IR device through the Raspberry Pi. The smartphone and Raspberry Pi can be interconnected with a IoT broker on the cloud such as Amazon EC2.

The contributions of this paper are summarized as follows. We introduce a low-cost and extendable model to transform IR remote controllable devices into smart IoT devices. To the best of our knowledge, this work is the first to fully address IR playback through both hardware and software. An IR transceiver module is made for the Raspberry Pi along with software. We also introduce the security implications of infrared communication using our setup and demonstrate attack scenarios. For example, we demonstrate that a remote-controlled drone armed with our device can turn off/on a TV. Please refer to the YouTube video https://youtu.be/rPbzPbWrbf8 or YouKu video http://v.youku.com/v_show/id_XMzQ0Njc5MzM3Ng.

The rest of this paper is organized as follows. We introduce background knowledge including infrared communications, Raspberry Pi and Message Queuing Telemetry Transport (MQTT) in Sect. 2. The hardware and software of the IR board is elaborated in Sect. 3. We discuss the security implications of the device in Sect. 4 and evaluate the board in Sect. 5. Related work is introduced in Sect. 6 and we conclude the paper in Sect. 7.

2 Background

2.1 Infrared Communications

Infrared communications start with the sender rapidly turning pulses into a series of 940 nm wavelength electromagnetic waves, usually via LEDs, which the receiver decodes into bits based on a scheme. The sender modulates the pulses at usually 38 kHz and encodes the signal to reduce noise and jitter, such as from the sun, while increasing accuracy of transmission.

Figure 1 shows the three well known forms of encoding a signal: *pulse distance*, *pulse length* and *bi-phase*. In this subsection, encoder refers to the sender, decoder refers to the receiver, "on" refers to the infrared source (e.g., an LED) being on, and "off" refers to the infrared source being off.

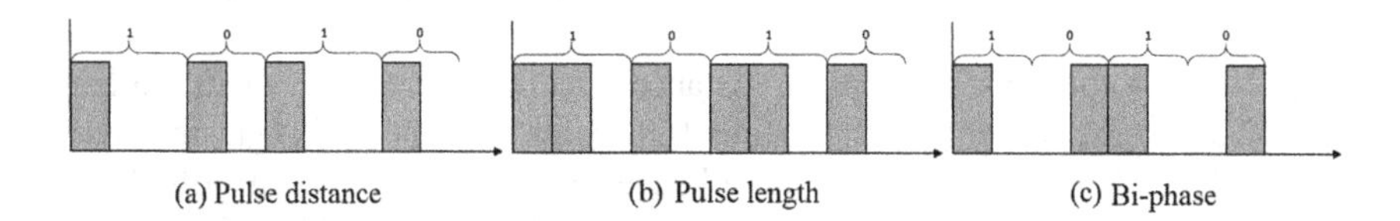

(a) Pulse distance (b) Pulse length (c) Bi-phase

Fig. 1. IR signal encoding

The pulse distance encoding scheme takes a constant on-length and two different off-lengths for a 1 and a 0. It encodes a binary 1 with an *on* for on-length microseconds and then an *off* for off-length-1 ms and likewise, a binary 0 with an on for on-length microseconds and then an off for off-length-0 ms. Note that the on duration is fixed for both binary 1 and 0 and the off duration determines whether it is interpreted as a 1 or 0.

The pulse length encoding scheme, similar to pulse distance, takes two different on-lengths for a 1 and a 0 and an off-length; it encodes a binary 1 by a sequence of on for on-length-1 ms then an off for off-length and likewise, a binary 0 with an on for on-length-0 ms and then an off for off-length microseconds. In this case, the off duration for both binary 1 and 0 is fixed and the on duration determines whether the signal is detected as a 1 or a 0.

The bi-phase encoding scheme takes a timeslot, also known as time window, and encodes a binary 1 by a sequence of *on* → *off* and a binary 0 by a sequence of off → on. The timeslot determines how long each off and on would be; it is a constant that must be exchanged to both the sender and the receiver beforehand. Unfortunately, bi-phase encoding may potentially cause decoding to be flipped; if the decoder does not detect the start of a bi-phase and misses an odd multiple of timeslots, it can incorrectly detect a binary 1 as a 0 and vice versa. Figure 2 is an example of bi-phase decoding flip. The sequence 10110 is encoded but because the receiver misses the first on pulse, it incorrectly decodes the sequence as 01001. Therefore, in practice, a known header is implemented to guarantee that the decoder and encoder are in sync, for example with the RC5 protocol.

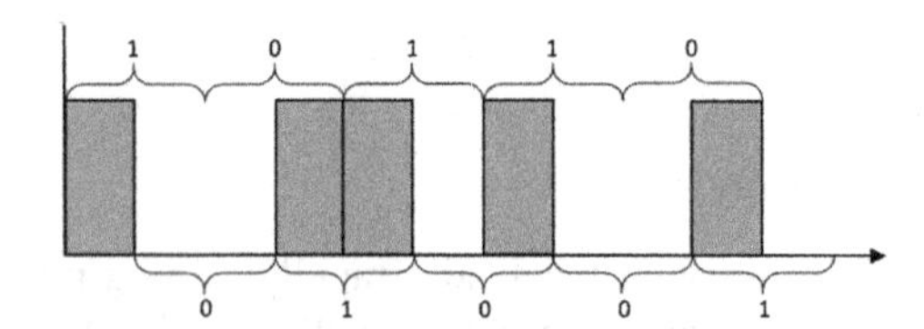

Fig. 2. Bi-phase decoding flip

Many manufacturers follow a variation of the RC5 or the NEC protocols, which are both modulated at 38 kHz. The RC5 protocol uses the bi-phase encoding scheme with the time slot being 1.78 ms and consists of a 2-bit header "11", an alternating bit, 5 address bits, and 6 command bits. The alternating bit alternates on retransmission within a button press, for example if a button is continuously pressed [3]. RC5 is most commonly used by American and European manufactured audio and video equipment such as speakers.

The NEC protocol uses the pulse distance encoding scheme with its on-length being 562.5 ms, the off-length for 0 being 1687.5 ms, and the off-length for 1 being 562.5 ms. The NEC protocol starts with a header of 9 ms on and 4.5 ms off. Then, it sends 8 address bits followed by its inverted form and then 8 data bits followed by its inverted form for a total of 32 bits. For example, the full body of a command with an address of 01001000 and data of 00000001 will be 01001000, 10110111 (inverted address), 00000001, 11111110 (inverted data). Finally, it sends an additional 562.5 ms of on as a footer. In cases of a repeat, for example when the volume button is continuously pressed, it waits for approximately 40 ms, sends a 9 ms on, a 2.25 ms off, and finally a 562.5 ms on [4].

A problem with emulating infrared communications on an OS like Linux is that the protocols are sensitive to time, and using the system clock as a form of time tracking is not necessarily precise enough. Furthermore, since processes share time with other processes, the infrared communication process cannot run accurately enough.

2.2 Raspberry Pi and PiGPIO Library

Raspberry Pi is a lightweight computer that runs on an ARM CPU. Although various OSes are compatible with Raspberry Pi, in this paper, all Raspberry Pis use Raspbian, which is a Debian-based Linux OS. In this work, Raspberry Pi 2 and 3 were used, but below we establish common features between other models to emphasize the point that any Raspberry Pi and other similar devices can be used.

The PiGPIO library is a GPIO interface written in C that can handle time sensitive GPIO tasks by running a helper daemon. Its primary use in this project was to generated time accurate pulse waves. Using traditional file I/O methods to send pulses was not precise enough due to various delays introduced by the overhead (e.g., I/O and time-sharing with other processes).

2.3 MQTT

Message Queue Telemetry Transport (MQTT) is a popular protocol to implement IoT communications due to its lightweight and simple nature. MQTT is a topic-based publish/subscribe messaging system. A topic is a unique string that serves as the identifier for a type of message. A publisher is any client that sends messages, which contain the topic and the payload, whereas a subscriber is any client that listens for incoming messages from some topic. A client or node is any system that connects to a broker and publishes and/or subscribes to topics.

Mosquitto [5] is an open source implementation of MQTT 3.1 including MQTT over TLS. Mosquitto provides a broker executable along with publish and subscription tools. Paho-mqtt is a library for MQTT that provides APIs to subscribe and publish to an MQTT broker.

3 IR Transceiver and Smart IR Devices

In this section, we first introduce the hardware and software of our long-range IR transceiver. We then briefly discuss how to convert legacy IR devices into smarts one with our IR transceiver so that we can control it from the Internet anywhere.

3.1 Hardware

Figure 3 illustrates the schematic of our IR transceiver board, Fig. 4 shows the PCB design and Table 1 lists its parts. A typical usage setup will consist of a Raspberry Pi and a transceiver module. The transceiver uses four GPIO interface to connect the Raspberry Pi. It receives power from two GPIO pins, a 5 V pin, and a ground pin from the Raspberry Pi. The sender GPIO pin (J-1) receives signals from the Raspberry Pi to turning on and off the LEDs on and off respectively and the timing is controlled by the Raspberry Pi. The receiver pin (J-2) of the transceiver connects to the Raspberry Pi which records the signals.

The transmitter portion of the transceiver has three important improvements in design compared with related work [6]. First, PNP (Q1) transistors are used to pull a 5 V power pin instead of directly drawing from the GPIO pin which only offers 3.3 V power. This increases the intensity of the light, leading to higher range of IR transmission. Second, power supply is routed through a 220 uF capacitor (C2) and a 0.1 uF capacitor (C1). Because the infrared protocols are fired in short bursts, the 220 uF capacitor can store charge while the LED is off to increase stability in cases where the Raspberry Pi fails to transmit steady current, and the 0.1 uF capacitor removes small electric noise. Finally, four IR LEDs (D1–D4) are placed in parallel to maximize coverage. the two outer LEDs are wide and short-ranged, whereas the two inner LEDs are narrow and long-ranged. The GPIO command is passed through the PNP transistor to each NPN transistor (Q2–Q5), which draws power from the 5 V pin in parallel. This design allows a range of approximately 10.0 m compared to a trivial GPIO-resistor-LED design which reaches approximately 2.0 m.

Table 1. Parts list

ID on PCB design	Part name
C1	Ceramic 0.1 uF capacitor (COM-08375)
C2	220 uF capacitor with 5 V + rating
D1, D3	Wide IR LED (IR333C/H0/L10)
D2, D4	Narrow IR LED (IR333-A)
Q1	PNP transistor (PN2907)
Q2, Q3, Q4, Q5	NPN transistor (PN2222)
TSSP58038	38 K IR receiver module
R1	1 KΩ 1/4 W 5% resistor

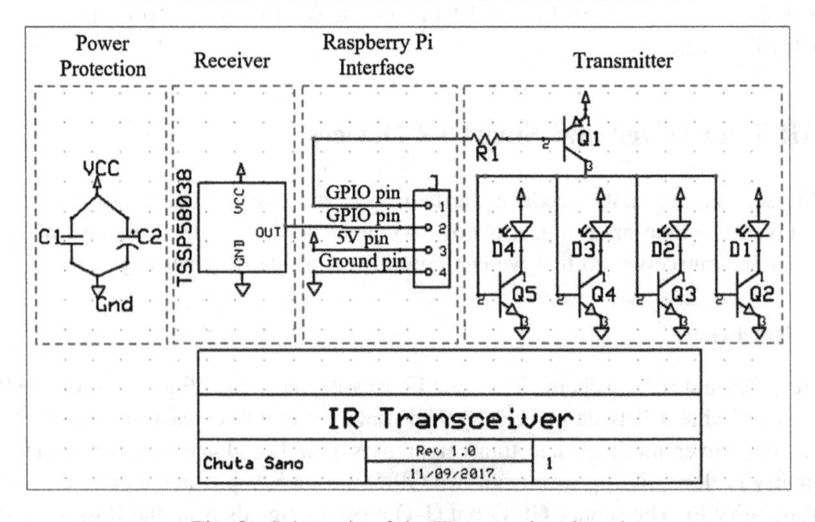

Fig. 3. Schematic of the IR transceiver board

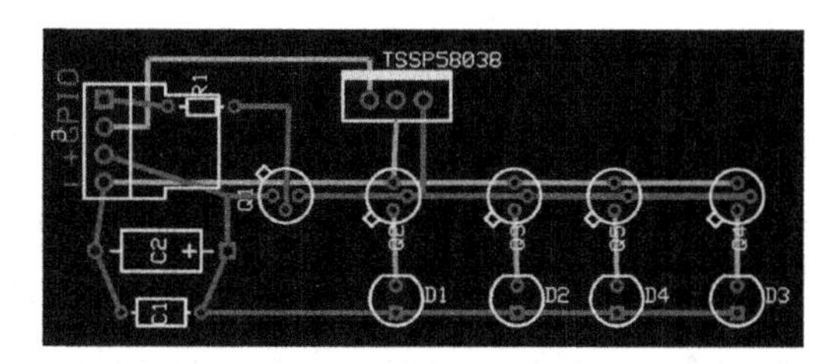

Fig. 4. PCB design of the IR transceiver board

The receiver module uses an integrated 38 kHz module of 940 nm peak wavelength, which can automatically filter out infrared light that is not modulated at 38 kHz. This decision is made for higher stability and distance compared to a generic light sensor and is justified because consumer IR protocols mostly use 38 kHz modulation [7].

3.2 Software

Separate software to interface with the sender and the receiver is designed along with a lightweight library that abstractifies much of the low-level intricacies in infrared communication.

The receiver software, written in C++, records the durations of low and high inputs from the receiver and records them into a raw format consisting of positive numbers representing the duration in microseconds to turn on the LED and negative numbers representing the duration in microseconds to turn off the LED. A user can associate a name to a recording. The recording is saved to "ircodes.txt" by default but can be optionally specified in the receiver software's first argument. The saved recording consists of a text file with two lines; the first line contains "name: <name>" where <name> refers to the chosen name in the prompt from the receiver. The second line consists of a list of integers delimited by space. A positive integer refers to turning on the LED for that many microseconds, and a negative integer refers to turning off the LED for that many microseconds. Additional recordings can be concatenated into the same file.

The sender software uses the PiGPIO library for nanosecond level accuracy in GPIO control, which is necessary to correctly replay signals. The software takes the name of the command as the first argument, matches it to the line reading "name: <name>" and as described previously, replays the list of numbers accordingly. The sender software attempts to read from "ircodes.txt" by default but the file name can be optionally specified in the second argument.

We have also written an extensive C++ library that abstractifies the sending of IR signals using our long-range IR transceiver. The sender library implements the previously mentioned raw format scheme, pulse-distance, pulse-length, and bi-phase encodings, and the NEC and RC5 protocols.

3.3 Smart IR Devices

Using the IR transceiver hardware and software, we can now connect our long-range IR transceiver to a Raspberry Pi and control an IR device by recording and replaying its IR signals. Since a Raspberry Pi has the Internet capability, we can connect the Raspberry Pi to the Internet and control the IR device from anywhere.

We give an example setup of controlling an IR device from the Internet. We set up a Mosquitto broker on an Amazon EC2 server so that we can publish messages through the smartphone. A Raspberry Pi can subscribe to a topic and send the appropriate signal upon receival of a message to the connected IR device.

4 Security Implications

In this section, we discuss the threats of the long-range IR transceiver that may be used to attack IR controllable devices of various kinds.

4.1 Replay Attack

Similar devices (from the same brands) do not change their IR codes on a per device basis, so following a simple replay procedure like introduced in the previous sections can allow anyone to be able to control the same devices. TVBGone follows a similar approach where the authors pre-record various brands of TV on/off toggle signals and replay it. With TVBGone, replaying all its code take approximately 2 min, which is certainly a very reasonable time frame to attack any TV [6]. Our long-range IR transceiver is controlled by a Raspberry Pi and can be easily customized to attack any IR controllable device.

4.2 Brute-Force Attack

The brute-force attack traverses all possible bits of a given protocol. For NEC, the set of all possible commands are based on unique address and command bits, both being 8 bits long, for a total of 16 bits or 2 bytes. For RC5, there are 5 address bits and 6 command bits for a total of 11 bits of entropy. A simple ascending approach where the address bit and command bit are considered one number and incremented was implemented. For example, the brute-force algorithm for NEC would start at 0x0000 (address = 0x00 and command = 0x00), and then try 0x0001 (address = 0x00 and command = 0x01), and so on until 0xFFFF. Generally, small items and products made in Asia use the NEC protocol.

As an example of the brute-force attack, a remote-controlled light strip, "BINZET 5 M 50 LEDs 3AA Battery Operated Copper Wire String Light LED Fairy Light LED Starry Light Cool White Festival Accent Light with Remote Control," was attacked via NEC brute-force, the objective being to turn on the light strip. The starting bits for the address and command bits were set to 0x00 and 0x00 respectively; the light strip was brute-force attacked within milliseconds, which was expected because the light strip's actual command was NEC with address 0x00 and command 0x02.

The simple incremental approach that was implemented as mentioned above traverses through all possible commands in one minute and twenty seconds per address. Therefore, this approach is estimated to take up to five hours and forty minutes in the worst case. However, in a fully optimized scenario, which can be implemented by creating a kernel module that accepts queueing of pulses to send (meaning subsequent commands are sent without delay), each command takes 67.5 ms to send, and there are 65536 distinct commands; the worst-case scenario can be shorted to just over one hour. Due to lack of feedback (the automated system cannot detect whether the right code was sent or not), the worst case must be considered over the average case.

An RC5 brute-force attack can be implemented the same way; the RC5 protocol, which only has eleven varying bits and takes 24.9 ms to send, can be completely brute-force attacked within fifty-one seconds.

An interesting observation was that the light strip accepted three command signals within the address 0x00. The timings at which the incorrect commands were accepted were varied and unpredictable; the reason for this unpredictable behavior remains unclear and may be a hardware issue on the lightstrip.

4.3 Drone Attack

One difficulty of attacking IR devices is the attacking IR transceiver has to be close to the target IR devices. To circumvent accessibility issues, we can put the Raspberry Pi and our IR transceiver on a drone. The Raspberry Pi can be connected to the Internet so that we can control the IR transceiver from anywhere.

Figure 5 shows an example of drone attack with a DJI Phantom 2. To minimize weight and therefore maximize flight duration, the on-board camera was removed, and a Raspberry Pi 2 with Anker PowerCore + Mini 3350 mAh were mounted. The battery powers the Raspberry Pi 2. The drone flew outside a second-floor conference room. This attack shows a method for attackers to partially circumvent the line of sight requirement; if a line of sight exists from outside through a window, then a drone can be leveraged to provide line of sight for the infrared setup.

Fig. 5. Drone attack against a TV through a window

5 Evaluation

In this section we introduce the device setup in terms of hardware and software. We estimate the cost of the entire setup, evaluate our hardware, software, as well as the attacks on infrared communication, and then discuss limitations and potential improvements to the system.

5.1 IR Device Setup

Figure 6 shows the smart IR device test setup. On the left side is an IR controlled LED (wrapped around a tree-like artifact) with its IR receiver. On the right side, the IR transceiver circuit board is connected to the Raspberry Pi Zero W [8] through GPIO pins. The four connected pins from bottom to top on the IR transceiver are connected to GPIO 22 (sender pin), GPIO 23 (receiver pin), a 5 V power pin, and a ground pin on Raspberry Pi Zero W, respectively. The Raspberry Pi Zero W is connected to a monitor, a power supply and a keyboard and mouse dongle through a USB OTG cable.

5.2 Cost

The total cost of the IR transceiver module is approximately $5.00 due to the variability of PCB printing (for example building one module is very expensive due to PCB printing). This experiment was done on the Raspberry Pi Zero W, which costs only $10.00. The software is computationally trivial, more than enough for a Raspberry Pi Zero W. Therefore, the total cost of the entire setup can be minimized to approximately less than $15.00 with a Raspberry Pi Zero W.

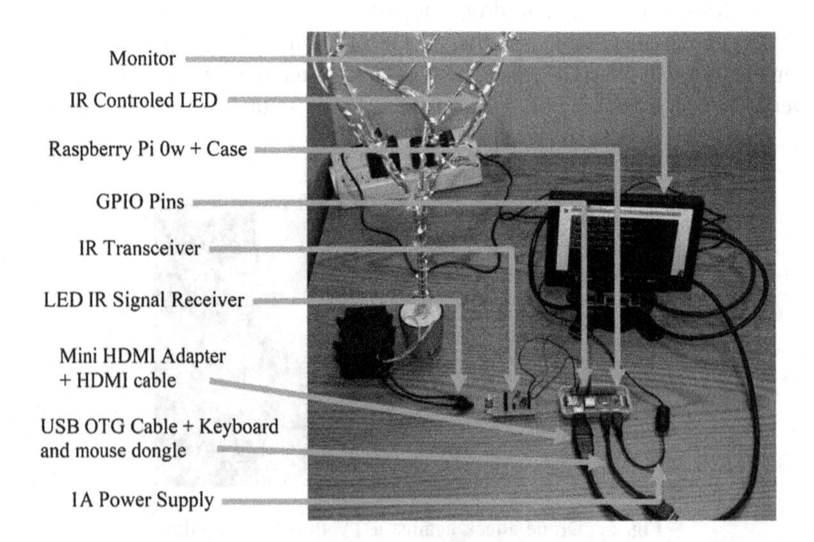

Fig. 6. IR device test setup

5.3 Range of Coverage

The range of the signals depends on multiple factors of the environment. High reflectivity of walls, narrow space, and minimal external noise (e.g., minimizing lights from sun or incandescent light bulbs) all increase the coverage and range of the signal. A similar approach to Acoustic Theory can be taken to fully describe IR transmission in rooms. Another huge factor is the sensitivity and noise tolerance of the IR sensor in the device: a higher quality IR sensor (higher sensitivity and better noise filtering) will also increase the range of the signal.

Table 2. Maximum distance for given situations

Situation	Maximum distance (m)
Inside, a building corridor	38.00
Inside, 13 × 13 m room with no pockets	17.70
Outside, through a window, receiver inside a room	13.00
Outside, receiver facing the sun	1.77
Outside, receiver facing away from the sun	4.26
Outside, in shade	5.83

Table 2 illustrates the maximum range of the IR transceiver module for indoor and outdoor use. The maximum range was determined by finding the highest distance between our transceiver and the aforementioned lightstrip module such that a signal was responsive ten out of ten times. As expected, the distance was maximized inside where there are likely minimal noise. Moreover, windows can decrease the range, mainly because of its reflectivity. Therefore, to perform a drone attack presented in Sect. 4.3, the drone should be at most 13 m away from the target device, and in most cases the drone can be very close (within 0.5 m) to the window, meaning target devices that are at most 12.5 m away from windows are vulnerable. All three outside cases show that the bounce of the light is very important for high distance. Furthermore, the experiment showed that the receiver tends to be impacted by the sun more than the sender; the sender side is not affected as much as shown by its maximum distance being similar to the Outside, in shade case.

The software was mainly developed as a usability improvement to the Linux Infrared Remote Control (LIRC) package. Although LIRC is a powerful tool, its usability is lacking for simple signal replaying purposes, where attempting to learn a signal not only failed at very noticeable rates but also took approximately thirty seconds. The entire recording process of our software is very quick and does not fail. Furthermore, the software implements various APIs to abstractify GPIO interaction, simplifying any generic infrared communication needs.

5.4 Security Analysis

Although it is clear that infrared communication has no security features and therefore suffers from replay attacks and brute-force attacks, lack of accessibility and impact are major problems that need to be addressed to make attacking infrared worthwhile.

Lack of accessibility comes from two factors: distance and blockage. Distance of infrared light is limited because as previously described, there are too many sources of potential interference, meaning infrared light becomes increasingly unstable with longer range. Blockage comes from various sources that block infrared light, such as walls. A possible workaround using drones was proposed.

The drone attack demonstrated that any IR devices with line of sight, for example through windows, to target devices can be exploited. To a determined attacker, using a similar attack alleviates some accessibility concerns. A potential issue with drone attack is its motor noise may attract the attention of the target. A drone's endurance (flight time per charge) may also affect the effectiveness of the attack.

We now discuss the impact of the attack. Turning off someone's TV is not as significant of an issue as stealing someone's credit card information. However, IR controllable air conditioners, fans, and thermal control units, which are popular in Asian countries, are certainly devices that need to be kept secure. Exploiting those devices to arbitrarily change room temperatures can cause health risks but also a non-trivial amount of monetary damage. Even controlling a TV can cause significant harm if an attacker were to turn it on and maximize the volume at night.

5.5 Limitations

Because of the nature of the infrared spectrum, it behaves almost identical to the visible spectrum. We essentially need a line of sight between the transceiver and the device to perform any IR communication. We worked on addressing this issue by maximizing both the distance and coverage of the IR module. At shorter distances, line of sight is not necessary because the strong and spread signal will bounce off other surfaces and reach the destination. However, this issue implies that multiple setups may be needed depending on the layout of the house and where the legacy IR devices are located.

Another problem is that there is lack of feedback due to the one-way communication of infrared controlled devices. For example, users receive feedback from turning on their TV by visually seeing the television turn on; they know to press the power button again if they visually see that the TV does not turn on in a reasonable timeframe. However, turning on the TV through Internet means that a user cannot get feedback about the TV's status unless the user is already within reasonable range from the TV, which would make the added connectivity meaningless. Therefore, extra sensors, such as camera or mic must be added, and those sensor inputs must be trained on a per device basis to fully consider the legacy device as IoT. Nevertheless, in an indoor scenario, interference is rare because of the high intensity of the emitted lights from the proposed design, meaning an infrared equivalent of packet loss is rarely an issue assuming the placement of the transceiver module is not too far, so users can assume with high confidence that any sent signal will be received.

6 Related Works

Overall, we were not able to find any published works that fully covered both the hardware and software aspects of IR playback introduced in this paper.

The Linux Infrared Remote Control (LIRC) is a software package for Linux that abstractifies much of the low-level details in infrared communication. It runs a helper daemon on the kernel level, which allows precise timing when sending infrared commands [9]. Although LIRC is very powerful, its recording process is very tedious, taking close to a minute to learn simple commands and sometimes even failing. Our software is a usability improvement to the LIRC.

TVBGone is a lightweight module that sends pre-trained TV on/off signals. It is controlled by an IC chip, and precise timing is obtained through a ceramic resonator. Its circuit design contained many clever tricks to increase stability and range in the transmission, which this project heavily drew upon (see Sect. 2.1). Its purpose is to turn off any TV, and it does so by hardcoding pre-recorded TV power toggle signals in the IC chip [6]. Therefore, it is different in design from our work which aims to record and playback any signal.

IrSlinger is a simple infrared sender which uses the PiGPIO daemon for precise timing. It uses a PN2222 transistor to route 5 V current to the LED instead of using the 3.3 V from GPIO [10]. However, it is potentially unstable because of the possibly unstable current from Raspberry Pi, and its coverage is not too large because it only uses one LED. Furthermore, the software does not fully automate playback, and users need to manually record and hardcode recorded values into a program.

Arduino Universal Remote autodetects NEC, RC5, and RC6 to handle for their repeat codes. It can only record and play back one signal. In our work, repeat codes were not handled at all because through testing we were not able to find any devices that were not responsive to sending the same code.

IRRemberizer uses a IR photo transistor to record signals instead of a IR transceiver, which allows IRRememberizer to record modulations of 30 k to 60 kHz at the cost of increased unreliability and therefore decreased range due to the photo resistor more likely to be affected by interferences from external infrared sources [11]. In this paper, a 38 kHz IR receiver module was used instead, which allowed higher stability and range in the recording process.

7 Conclusion

In this paper, we introduce a cost-effective method to transform legacy IR controlled devices into Internet connected smart IoT devices through a Raspberry Pi with an IR transceiver module. The cost of the IR transceiver is around $5.00. Our IR transceiver design achieves long-range coverage. We also introduce an extensive IR communication library to show security flaws in common IR protocols. A drone armed with our device may pose severe threats against IR controllable devices of various kinds.

Acknowledgments. This work was supported in part by US NSF grants 1461060, 1642124, and 1547428, by National Science Foundation of China under grants 61502100 and 61532013, by Jiangsu Provincial Natural Science Foundation of China under Grant BK20150637, by Ant Financial Research Fund. Any opinions, findings, conclusions, and recommendations in this paper are those of the authors and do not necessarily reflect the views of the funding agencies.

References

1. Xu, G., Yu, W., Griffith, D., Golmie, N., Moulema, P.: Toward integrating distributed energy resources and storage devices in smart grid. IEEE Internet Things J. **4**, 192–204 (2017)
2. Lin, J., Yu, W., Zhang, N., Yang, X., Zhang, H., Zhao, W.: A survey on internet of things: architecture, enabling technologies, security and privacy, and applications. IEEE Internet Things J., Enabling Technologies (2017)
3. Altium Limited: Philips RC5 Infrared Transmission Protocol, 13 Sept 2017. http://techdocs.altium.com/display/FPGA/Philips+RC5+Infrared+Transmission+Protocol
4. Altium Limited: NEC Infrared Transmission Protocol, 13 Sept 2017. http://techdocs.altium.com/display/FPGA/NEC+Infrared+Transmission+Protocol
5. Mosquitto (2018). https://mosquitto.org/
6. Adafruit Industries: TVBGone, 4 Jan 2018. https://cdn-learn.adafruit.com/downloads/pdf/tv-b-gone-kit.pdf
7. Gotschlich, M.: Remote Controls – Radio Frequency or Infrared White Paper, Infineon Technologies AG (2010). https://www.infineon.com/dgdl/RF2ir+WhitePaper+V1.0.pdf?fileId=db3a30432b57a660012b5c16272c2e81
8. Raspberry Pi Zero W (2018). https://www.raspberrypi.org/products/raspberry-pi-zero-w/
9. Christoph Bartelmus: Linux Infrared remote control (LIRC), 26 May 2016. http://www.lirc.org/
10. Schwind, B.: Sending Infrared Commands From a Raspberry Pi Without LIRC, 29 May 2016. http://blog.bschwind.com/2016/05/29/sending-infrared-commands-from-a-raspberry-pi-without-lirc/
11. Sensacell: IR Rememberizer- IR Remote Control Recorder/Player, 6 Aug 2014. https://forum.allaboutcircuits.com/blog/ir-rememberizer-ir-remote-control-recorder-player.648/

Solving Data Trading Dilemma with Asymmetric Incomplete Information Using Zero-Determinant Strategy

Korn Sooksatra[1], Wei Li[1(✉)], Bo Mei[2], Arwa Alrawais[3], Shengling Wang[4], and Jiguo Yu[5(✉)]

[1] Georgia State University, Atlanta, GA, USA
ksooksatra1@student.gsu.edu, wli28@gsu.edu
[2] Texas Christian University, Fort Worth, TX, USA
b.mei@tcu.edu
[3] The George Washington University, Washington, DC, USA
alrawais@gwmail.gwu.edu
[4] Beijing Normal University, Beijing, China
wangshengling@bnu.edu.cn
[5] Qufu Normal University, Qufu, Shangdong, China
jiguoyu@sina.com

Abstract. Trading data between user and service provider is a promising and efficient method to promote information exchange, service quality improvement, and development of emerging applications, benefiting individual and society. Meanwhile, data resale (i.e., data secondary use) is one of the most critical privacy issues hindering the ongoing process of data trading, which, unfortunately, is ignored in many of the existing privacy-preserving schemes. In this paper, we tackle the issue of data resale from a special angle, i.e., promoting cooperation between user and service provider to prevent data secondary use. For this purpose, we design a novel game-theoretical algorithm, in which user can unilaterally persuade service provider to cooperate in data trading, achieving a "win-win" situation. Besides, we validate our proposed algorithm performance through in-depth theoretical analysis and comprehensive simulations.

1 Introduction

The explosive progress of emerging applications, such as big data, Internet of Things (IoT), and smart city, greatly facilitates the ubiquitous data generation/collection via smart devices across every field in real life. Everyday, 2.5×10^{18} bytes of data are created [7]. With such a tremendous amount of data, individuals can enjoy various personalized services offered by service providers; in return, by collecting data from individuals, service providers can further improve their service quality as well as develop new products. For example, when a service provider intends to develop a mobile diagnosis app, he could pay for mobile users in exchange of the desired test data. Actually, to efficiently share or exchange

S. Chellappan et al. (Eds.): WASA 2018, LNCS 10874, pp. 425–437, 2018.
https://doi.org/10.1007/978-3-319-94268-1_35

information, data is able to be traded in online data markets, such as Microsoft Azure [12]. Companies, organizations, and government agencies are also investigating ways to make their data tradable for profit, e.g., the National Environment Agency (NEA) of Singapore has made a purchase guide for selling Singapore's climate data online [17]. *Data trading will definitely become an inevitable trend, benefiting individuals and society.*

Meanwhile, it is worth mentioning that a large amount of data has been available but not effectively shared/traded due to people's struggle between data trading and data privacy. On one hand, individuals are willing to trade their personal data with service providers for services or rewards (e.g., payment, gift card, and coupon) while caring about their data privacy [18]. Particularly, *data secondary use* by a third party and marketing is one major type of privacy concerns [9]. Moreover, "89% of users avoid companies that do not protect their privacy" according to the 2016 TRUSTe/NCSA Consumer Privacy Infographic [18]. On the other hand, personal data might be shared or sold by the service providers [1], and data prices in black market are attractive [24]. More specifically, 55% of iOS applications and 59.7% Android applications surreptitiously leak users' personal data [21]. *The contradiction between the user's need for privacy protection and the service provider's pursuit for profit makes data trading in a dilemma position.*

To preserve data privacy, a number of schemes have been proposed in literature [2–6,8,10,11,13,16,19,20,23]; in particular, [8,13,16,19,20] tackle privacy issues by exploiting game theory to balance the trade-off between benefit and privacy cost. However, when applying such schemes for data trading, data usability (e.g., the quality and the accuracy of data) is affected more or less, therefore reducing service provider's benefit. For instance, the user's actions, such as stopping an online transaction, withholding personal information, and unsubscribing a service, have significantly negative impacts on data usability and service provider's benefit. Besides, in many practical scenarios (e.g., healthcare), re-selling personal data is prohibited and preserving data privacy cannot effectively prevent data resale from service provider.

Inspired by the aforementioned observations, in this paper, we raise a new challenging problem: *whether and how a scheme fairly benefits both user and service provider such that accurate data can be collected from user while data resale can be prevented at service provider side, achieving a "win-win" situation?* Since this problem has never been discussed in literature, we have to deal with the challenges coming from game formulation, solution analysis, and performance evaluation. In a nutshell, our goal is to develop a game-theoretical scheme such that the user accurately submits data while the service provider does not resell the user's data to others. More specifically, we first model our problems as an iterated data trading dilemma game with considering two scenarios: *"data for service"* and *"data for monetary reward"*. Then, a novel algorithm is proposed by utilizing Zero-Determinant Strategy [15], in which the user can successfully convince the service provider to cooperate through unilaterally controlling the service provider's expected payoff. Finally, our proposed algorithm is evaluated via profound theoretical analysis and intensive simulations in terms of cooperation probabilities and average payoffs. Our contributions are summarized below.

- To the best of our knowledge, this the first work to focus on the problem of data trading dilemma, thus filling the blank in literature.
- We formulate an iterated data trading dilemma game with considering two practical scenarios, including data for service and data for monetary reward.
- In the proposed algorithm, the cooperation of the two players can be achieved by unilaterally setting the service provider's expected payoff at the user side.
- We rigorously prove the two players' cooperation probabilities and achievable maximum payoffs.
- The simulation results validate the effectiveness of our proposed algorithm in terms of cooperation probability and average payoff.

The rest of this paper is organized as follows. Our proposed game models and algorithm is detailed in Sects. 2 and 3, respectively. After presenting simulations in Sect. 4, we conclude this paper and brief our future work in Sect. 5.

2 Game Formulation

In this paper, we focus on the two trading scenarios: (i) *"data for service"* where the user trades data for requested services from the service provider; and (ii) *"data for monetary reward"* where the user supplies data to the service provider for monetary rewards, such as money, gift card, and coupon. In such scenarios, both the user and the service provider have their own concerns. On one hand, the user is willing to share personal data for services or monetary rewards while caring about data privacy; and particularly, the user would consider the worst case where the service provider resells all collected data. On the other hand, the service provider offers data-based services via collecting the user's data or pay monetary reward to the user for incentivizing data supply, but he may resell the user's data to a third party (e.g., an advertisement company) for extra profits without the user's permission. The trading process could happen in a multi-round manner. For instances, the user is a faithful user to her favorite application/service (e.g., Google Map) and uses it frequently; and in mobile crowdsensing systems, the user equipped with smart devices is active in contributing sensory data. The multi-round interactions between the user and the service provider can be modeled as an **iterated data trading dilemma game**, which is detailed in the following.

2.1 Data for Service

Firstly, in the scenario of *"data for service"*, the user submits her data to the service provider who in return, provides the user with the requested data-based service in a round-by-round manner. At the end of each round, the user receives the requested service if she gives the service provider her personal information. During each round, the service provider may resell the user's private information. We assume that the user will be able to aware whether the service provider leak her information to someone else or not after passing a number of rounds. For example, the user can find privacy leakage via getting spam emails, junk advertisements and is able to realize which companies are the cause of privacy leakage. When being aware of privacy leakage, the user cannot do anything to prevent that leakage, but can take strategic actions to reduce privacy loss in the future, such as sending partial personal data, adding dummy data into true data, and withholding data.

In the iterated data trading dilemma game, the user and the service provider are either *cooperative* or *defective*. If the user accurately gives the required data to the service provider, the user is cooperative; otherwise, she is defective. On the other hand, if the service provider does not sell the user's data to anyone else, he is cooperative; otherwise, he is defective. Accordingly, in each round, there are four possible situations of the interactions between the two players, which are summarized in Table 1.

Table 1. Payoffs regarding user's data release strategy.

User	Service provider	
	Cooperation	Defection
Cooperation	R_u, R_s	$R_u - m, R_s + a$
Defection	$R_u + n - w$, $R_s - b$	$R_u - m + n - w$ $R_s + a - b$

Table 2. Payoffs regarding user's data pricing strategy.

User	Service provider	
	Cooperation	Defection
Cooperation	R_u, R_s	$R_u - m, R_s + a$
Defection	$R_u + d, R_s - c$	$R_u - m + d$, $R_s + a - c$

Our analysis for these four situations is addressed as follows. **Situation 1:** Both the two players are cooperative. In this situation, the user can obtain a payoff, denoted by R_u, from exchanging personal data for the requested service; and the service provider can receive a payoff, denoted by R_s, from offering service and collecting the user's data. **Situation 2:** The user is cooperative but the service provider is defective. Under this situation, the service provider's payoff is increased to $R_s + a$ where a is the revenue from selling the user's data to a third party, while the user's payoff is reduced to $R_u - m$ where m is the monetarily equivalent privacy loss incurred by data resale at the third party. **Situation 3:** The user defects but the service provider cooperates. This situation indicates that the user reports inaccurate data or partial information to the service provider who keeps the user's data confidential. As a result, the user's payoff changes to $R_u + n - w$ where n is the benefit from preserving personal data and w is the benefit loss caused by a decrease in the service quality. On the other hand, the service provider's payoff is reduced to $R_s - b$ where b is the benefit loss yielded by the user's inaccurate/partial information. **Situation 4:** Both of them defect each other. Based on the analysis in situations 1 to 3, the user's payoff is $R_u - m + n - w$ and the service provider's payoff is $R_s + a - b$.

2.2 Data for Monetary Reward

Secondly, we consider the scenario of *"data for monetary reward"*, in which the user trades personal data with the service provider who pays monetary rewards for compensating the user's data generation/collection cost. Similar to the analysis in Sect. 2.1, the user and the service provider have two states, i.e., cooperative and defective. In this scenario, besides reporting inaccurate data and withholding partial information, the user could also adaptively adjust the price for her data.

The actions and the payoffs of the two players are analyzed in the following. **Situation 1:** When the two players cooperate with each other, they can respectively gain payoffs R_u and R_s where R_u is the benefit from trading personal data with the service provider and R_s is the revenue of purchasing the user's data. **Situation 2:** If the user is cooperative but the service provider chooses to defect, the service provide can get an increased payoff $R_s + a$ while the user's payoff is reduced to $R_u - m$. **Situation 3:** If the user defects but the service provider cooperates, the payoffs can be calculated according to two cases. (i) In reality, the service provider usually determines the payment according to data quality, thus when using data release strategy, the user's payoff is $R_u + n - w'$ where w' represents the benefit loss incurred by reducing data quality and the service provider's payoff is decreased to $R_s - b$. (ii) Intuitively, the user could ask for a higher price to compensate the loss of privacy leakage from the service provider to a third party even though data resale cannot be controlled. So, we have $R_u + d$ for the user where d indicates the benefit from asking a higher price for personal data and $R_s - c$ for the service provider in which c denotes the revenue loss due to the increase of data price. **Situation 4:** When both of them are defective, we have two cases: (i) their payoffs respectively are $R_u - m + n - w'$ and $R_s + a - b$ when the user adjusts data release strategy; or (ii) the payoffs respectively are $R_u - m + d$ and $R_s + a - c$ when the user changes pricing strategy.

From the aforementioned analysis, we observe that under these two scenarios, the two players' payoffs can be calculated in the same way when the user adopts data release strategy. Thus, for simplicity, we assume $w = w'$, which indeed does not affect the performance of our proposed game model and algorithm. To sum up, the two players' payoffs in the scenario of *"data for monetary reward"* can be computed via Tables 1 and 2.

2.3 Iterated Data Trading Dilemma Game

The dilemma of data trading lies in two aspects: (i) the user should carefully determine *whether to accurately release data to the service provider at the risk of data resale*; and (ii) the service provider needs to decide *whether to resell the user's data for extra profits when facing a privacy-aware user*. In this paper, we mainly study the cases where the data trading dilemma appears. This is because both the two players do not have incentive to cooperate with each other if such dilemma does not exist.

Notice that the user cannot learn the exact price paid from a third party to the service provider but can estimate it through prior knowledge, as data prices in black market can be obtained from public [24]. That is, the user can only know that the value of a is drawn from a certain distribution $\mathcal{D}$ within a certain range. Specifically, when the user adopts data release strategy, a is in range $[1, b-1]$

because: (i) a should be larger than 0 as otherwise the service provider will have no incentive to sell the information; and (ii) a should not be smaller than b as otherwise the service provider does not have incentive to cooperate. With the similar analysis, we have $a \in [1, c-1]$ when the user chooses pricing strategy.

Theorem 1. *The interactions between the user and the service provider can be modeled as an iterated data trading dilemma game with the user employing data release strategy if* $1 \leq a \leq b-1, a < m, b > n-w$, *and* $w < n < w+m$.

Proof. According to Prisoners' Dilemma [14], there are two conditions for being modeled as data trading dilemma in our problem. **Condition 1:** Nash equilibrium is the mutual defection state. **Condition 2:** The mutual cooperation is the global best outcome.

To meet Condition 1, Eqs. (1) and (2) should hold; that is, we must have $w < n < w+m$ and $a < b$.

$$R_u + n - w > R_u > R_u - m + n - w > R_u - m. \tag{1}$$

$$R_s + a > R_s > R_s + a - b > R_s - b. \tag{2}$$

From Condition 2, we should simultaneously satisfy Eqs. (3), (4), and (5), and obtain $a < m$ and $b > n - w$.

$$R_u + R_s > (R_u - m) + (R_s + a). \tag{3}$$

$$R_u + R_s > (R_u + n - w) + (R_s - b). \tag{4}$$

$$R_u + R_s > (R_u - m + n - w) + (R_s + a - b). \tag{5}$$

Then, to be modeled as iterated data trading dilemma, Eqs. (6) and (7) should be satisfied. As a result, there must have $a < m$ and $b > n - w$.

$$2R_u > (R_u + n - w) + (R_u - m). \tag{6}$$

$$2R_s > (R_s + a) + (R_s - b). \tag{7}$$

Moreover, based on our aforementioned analysis, we have $1 \leq a \leq b-1$.

Therefore, with $1 \leq a \leq b-1, a < m, b > n-w$, and $w < n < w+m$, the interactions between the user and the service provider can be formulated to be an iterated data trading dilemma game.

Theorem 2. *The interactions between the user and the service provider can be modeled as an iterated data trading dilemma game with the user utilizing pricing strategy when* $1 \leq a \leq c-1, a < m, d < c$, *and* $d < m$.

The proof process of Theorem 2 is similar to that of Theorem 1, which is omitted due to page limit.

Remarks: According to Theorems 1 and 2, one can see that the interactions between the user and the service provider can be modeled to be an iterated data trading dilemma game regardless of the types of the user's strategy (i.e., data release and pricing strategies). In other words, no matter which type of strategy the user takes, the essential properties of the corresponding iterated data trading dilemma games are the same. Therefore, in the following part of this paper, we only focus on the scenario of *"data for service"*.

2.4 Player's Strategy

In such iterated data trading dilemma game, the service provider and the user aim to maximize their received payoffs via strategically taking *cooperation* or *defection* actions. It is worth noting that the two players are *asymmetric* with respect to their concerns, dilemmas, and abilities to gather trade information.

As a powerful role in the data trading market, the service provider can learn the trade information regarding himself and the user. For simplicity, we assume that the service providers in the market are the same type, and we will extend our work to a more practical and complicated scenario where different types of service providers co-exist. Notice that a service provider may lack sufficient knowledge about other service providers. Thus, to well survive among in the market while maximizing payoff with incomplete information, a service provider could adapt action over the course of repeated plays of a game in an evolutionary manner regardless the strategies or payoffs of his opponents [14,15]; that is, in this paper, the service provider is supposed to be an evolutionary player.

Compared with the service provider, the user is at a weak position without knowing the exact revenue gained by reselling her data to a third party. In addition, since the service provider is an evolutionary player, the user's strategy cannot yield any impact on the service provider's strategy [14,15]. Nevertheless, the user can persuade the service provider to play a desired strategy by unilaterally affecting the service provider's payoff, which will be introduced in next section.

3 Algorithm Design

To effectively solve the iterated data trading dilemma, it is desired to promote cooperation between the user and the service provider in the data trading market. For this purpose, we propose a novel algorithm for the user to convince the service provider to cooperate, in which Zero-Determinant Strategy [15] is utilized. The main idea of the proposed algorithm is to control the service provider's expected payoff by exploiting the relationship between his expected payoff and the user's cooperation probability, thus building up cooperation between the two players.

3.1 Zero-Determinant Strategy

Zero-Determinant Strategy [15] is employed to set the service provider's expected payoff and determine the user's cooperation probability at each iteration t.

Stage 1: Control Service Provider's Expected Payoff

Denoted by $\mathbf{p}_z^t = [p_{z1}^t, p_{z2}^t, p_{z3}^t, p_{z4}^t]$ the vector of cooperation probability of the user at iteration t, in which there are four elements:

- The first one, p_{z1}^t, is the cooperation probability at iteration t when both the user and the service provider are cooperative at iteration $t-1$.
- The second one, p_{z2}^t, is the cooperation probability at iteration t when the user cooperates but the service provider defects at iteration $t-1$.

- The third one, p_{z3}^t, is the cooperation probability at iteration t when the user is defective but the service provider is cooperative at iteration $t-1$.
- The last one, p_{z4}^t, is the cooperation probability at iteration t when both the user and the service provider defects at iteration $t-1$.

When using Zero-Determinant Strategy, the relationship between the expected payoffs of the two players can be expressed as follows [15],

$$\alpha E_{zu}^t + \beta E_{zs}^t + \gamma = \frac{D(\mathbf{q}_z^t, \mathbf{p}_z^t, \alpha \mathbf{S}_u + \beta \mathbf{S}_s + \gamma \mathbf{1})}{D(\mathbf{q}_z^t, \mathbf{p}_z^t, \mathbf{1})}, \tag{8}$$

where $\mathbf{q}_z^t = [q_{z1}^t, q_{z2}^t, q_{z3}^t, q_{z4}^t]$ is the cooperation probability vector of the service provider at iteration t, $\mathbf{S}_u$ is the payoff vector of the user, $\mathbf{S}_s$ is the payoff vector of the service provider, $\mathbf{1}$ is the vector with all components 1, and α, β, and γ are system parameters. In particular, $D(\mathbf{q}_z^t, \mathbf{p}_z^t, \mathbf{1})$ can be computed by

$$D(\mathbf{q}_z^t, \mathbf{p}_z^t, \mathbf{f}) = \begin{pmatrix} -1 + q_{z1}^t p_{z1}^t & -1 + q_{z1}^t & -1 + p_{z1}^t & f_1 \\ q_{z2}^t p_{z3}^t & -1 + q_{z2}^t & p_{z3}^t & f_2 \\ q_{z3}^t p_{z2}^t & q_{z3}^t & -1 + p_{z2}^t & f_3 \\ q_{z4}^t p_{z4}^t & q_{z4}^t & p_{z4}^t & f_4 \end{pmatrix}.$$

In Eq. (8), it is possible for the user to unilaterally make the determinant in the numerator vanish. Specifically speaking, if the user's cooperation probabilities satisfy $\mathbf{p}_z^t = \alpha \mathbf{S}_u + \beta \mathbf{S}_s + \gamma \mathbf{1}$, the following linear relationship between E_{zu}^t and E_{zs}^t can be found, i.e.,

$$\alpha E_{zu}^t + \beta E_{zs}^t + \gamma = 0. \tag{9}$$

Equation (9) indicates that with $\mathbf{p}_z^t = \alpha \mathbf{S}_u + \beta \mathbf{S}_s + \gamma \mathbf{1}$, the user is able to unilaterally control the service provider's expected payoff by setting $\alpha = 0$ regardless the service provider's strategy. Moreover, p_{z2}^t and p_{z3}^t can be determined in terms of p_{z1}^t and p_{z4}^t, i.e.,

$$p_{z2}^t = \frac{b p_{z1}^t - E_a(1 + p_{z4}^t)}{b - E_a} \text{ and } p_{z3}^t = \frac{E_a(1 - p_{z1}^t) - b p_{z4}^t}{b - E_a}.$$

Thus, the service provider's expected payoff at iteration t can be computed as:

$$E_{zs}^t = \frac{(1 - p_{z1}^t)(R_s + E_a - b) + R_s p_{z4}^t}{1 - p_{z1}^t + p_{z4}^t}. \tag{10}$$

In Eq. (10), the maximum value (denoted by max^Z) of E_{zs}^t is R_s, and the minimum value (denoted by min^Z) of E_{zs}^t is $R_s + E_a - b$. Therefore, the user can control the service provider's expected payoff according to p_{z1}^t and p_{z4}^t where p_{z1}^t can be approximated as $p_{z1}^t = \frac{E_{zs}^t - (R_s + E_a - b)}{R_s - (R_s + E_a - b)}$. Then, by substituting the approximate expression of p_{z1}^t into Eq. (10), the user can determine p_{z4}^t with a certain value of E_{zs}^t.

Stage 2: Convince Service Provider's Cooperation

As analyzed in Stage 1, if the service provider cooperates, he would gain an increase in his expected payoff E_{zs}^t; otherwise, he would suffer a decrease in his expected payoff E_{zs}^t. After determining E_{zs}^t, the user can compute obtain the vector $\mathbf{p}_z^t$. A vector $\mathbf{P}_s = [P_{cc}, P_{cd}, P_{dc}, P_{dd}]$ is utilized to predict the service provider's action and is calculated through the historical statistics in the first t_p ($<t$) iterations. With $\mathbf{P}_s$, we can carry out the following ways to set E_{zs}^t.

Case 1: the service provider cooperates at iteration $t-1$. If $P_{cc} > P_{cd}$, set $E_{zs}^t = E_{zs}^{t-1} + \frac{max^Z - E_{zs}^{t-1}}{2}$; otherwise, set $E_{zs}^t = E_{zs}^{t-1} - \frac{E_{zs}^{t-1} - min^Z}{2}$.

Case 2: the service provider defects at iteration $t-1$. If $P_{dc} > P_{dd}$, set $E_{zs}^t = E_{zs}^{t-1} + \frac{max^Z - E_{zs}^{t-1}}{2}$; otherwise, set $E_{zs}^t = E_{zs}^{t-1} - \frac{E_{zs}^{t-1} - min^Z}{2}$.

Thus, given E_{zs}^t at iteration t, the user can obtain the cooperation probability vector $\mathbf{p}_z^t$ based on the computation in Stage 1.

3.2 Theoretical Analysis

From an evolutionary point of view, the winner is the one who continues longer and the loser is the one who gives away first, which provides us with a long-term vision to analyze the two players' strategies.

Theorem 3. *In the long term, the service provider's cooperation probability reaches to 1 in the proposed algorithm, indicating that cooperation is his best strategy to maximize payoff.*

Proof. As an evolutionary player, the service provider [22] can update his cooperation probability, represented by q^t, at iteration t by Eq. (11).

$$q^{t+1} = q^t(\frac{E_c^t}{E^t}), \tag{11}$$

where E_c^t and E^t respectively denote the service provider's expected payoff with cooperation and the service provider's expected payoff at iteration t. Moreover, we have: (i) the service provider's expected payoff when he cooperates is $E_c^t = p^t R_s + (1-p^t)(R_s - b) = R_s - b + p^t b$; (ii) the service provider's expected payoff when he defects is $E_d^t = R_s + E_a - b + p^t b$; and (iii) the service provider's expected payoff is $E^t = q^t E_c^t + (1-q^t)E_d^t$.

According to Eq. (11), it can be noticed that q^t can increase when $E_c^t > E^t = q^t E_c^t + (1-q^t)E_d^t$, which further implies that q^t increases if $E_c^t > E_d^t$.

In our proposed algorithm, there are two situations to be considered:

(1) When the service provider keeps the user's data secret, he obtains reward. Thus, his expected payoff increases from E^t to E^{t+1}, which means $E^{t+1} > E^t$. Although q^{t+1} is unknown, it can be statistically determined as $q^{t+1} = \frac{q^t t+1}{t+1} = q^t + \frac{1-q^t}{t+1}$. Accordingly, we have

$$E_c^{t+1} = \frac{E^{t+1} - (1-q^{t+1})E_d^{t+1}}{q^{t+1}} = \frac{E^{t+1} - (1-(q^t + \frac{1-q^t}{t+1}))E_d^{t+1}}{q^t + \frac{1-q^t}{t+1}}.$$

As t grows up to a large number, we have $\lim_{t\to\infty} E_c^{t+1} = \frac{E^{t+1}-(1-q^t)E_d^{t+1}}{q^t}$. Thus, with $E_d^{t+1} = E_d^t$, we can obtain

$$E_c^{t+1} \approx \frac{E^{t+1} - (1-q^t)E_d^{t+1}}{q^t} > \frac{E^t - (1-q^t)E_d^{t+1}}{q^t} = E_c^t.$$

(2) When the service provider tends to sell the user's data to a third party, he is punished by decreasing his expected payoff from E^t to E^{t+1}, i.e., $E^{t+1} < E^t$. In this case, q^t is updated as $q^{t+1} = \frac{q^t t}{t+1} = q^t - \frac{q^t}{t+1}$ and E_d^{t+1} is computed as

$$E_d^{t+1} = \frac{E^{t+1} - q^{t+1}E_c^{t+1}}{1-q^{t+1}} = \frac{E^{t+1} - (q^t - \frac{q^t}{t+1})E_c^{t+1}}{1-(q^t - \frac{q^t}{t+1})}.$$

Therefore, there is $\lim_{t\to\infty} E_d^{t+1} = \frac{E^{t+1}-q^t E_c^{t+1}}{1-q^t}$, indicating that

$$E_d^{t+1} \approx \frac{E^{t+1} - q^t E_c^{t+1}}{1-q^t} < \frac{E^t - q^t E_c^{t+1}}{1-q^t} = E_d^t,$$

where the last equality holds because $E^t = E_c^{t+1}$.

In conclusion, in the proposed algorithm, E_c^t increases when E^t decreases. As a result, q^t will always increase at every iteration and eventually reach 1.

Theorem 4. *In the long term, the user's cooperation probability reaches to 1 in the proposed algorithm, which means cooperation is her best strategy for maximizing payoff.*

Proof. Theorem 3 shows that the service provider always adapts cooperation when $q^t = 1$. Thus, the user will increase E_{zs}^t after q^t reaches 1. In addition, notice that p_{z1}^t is an increasing function of E_{zs}^t and that $p_{z1}^t = 1$ when E_{zs}^t is equal to the maximum value R_s. In other words, as q^t goes to 1 over time, E_{zs}^t reaches to R_s and p_{z1}^t increases to 1.

As a result, the user's cooperation probability can finally increase to 1 in the proposed algorithm.

Theorem 5. *In our iterated data trading dilemma game, the maximum achievable payoffs of the service provider and the user are R_s and R_u, respectively.*

Proof. From Table 1, and Theorems 3 and 4, this theorem can be directly proved.

Remarks: Theorem 3 through Theorem 5 can theoretically prove that in the data trading dilemma, the two players can cooperate with each other and receive their maximum payoffs, achieving a "win-win" situation. Actually, we can obtain the same conclusion when applying our proposed algorithm into the scenario of *"data for monetary reward"*.

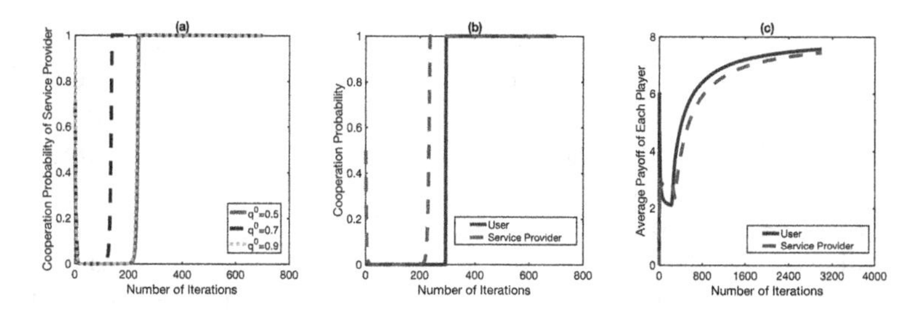

Fig. 1. Results of zero-determinant strategy-based algorithm.

4 Simulation

In this section, we evaluate the performance of our proposed algorithm in terms of *cooperation probability of each player* and *average payoff of each player*. The parameters are set as follows: $R_u = 8, R_s = 8, n = 5, w = 3, m = 8, b = 8$, and $a = 3$. The user computes the expected value of a with the knowledge that a uniformly and randomly distributes in range $[1, b-1]$. The prediction probability vector $\mathbf{P}_s$ is obtained in the first $t_p = 100$ iterations.

Since the properties of the iterated data trading dilemma game under the scenarios of data for service and data for monetary reward are the same, the performance of the proposed algorithm under the two scenarios are also the same. For example, with $R_u = 8, R_s = 8, d = 2, m = 8, c = 8$ and $a = 3$, we obtain the same results in the scenario of data for monetary reward. Due to page limit, we only present the simulation results in the scenario of data for service. The cooperation probability and the average payoff output by the algorithm, Zero-Determinant Strategy, are reported in Fig. 1.

Figure 1(a) shows the service provider's cooperation probability over iterations, in which the service provider's cooperation probability eventually converges 1 with different initial cooperation probabilities. The service provider's cooperation probability drops at first because the expected payoff of defection is still greater than the expected payoff of cooperation. Nevertheless, when the expected payoff of cooperation is greater than that of defection (see the concave part of cooperation probability in Fig. 1(a)), the cooperation probability increases until reaching 1. These results indicate that our proposed algorithm can offer incentive to the service provider for cooperation, which is consistent with Theorem 3.

The cooperation probabilities of the service provider and the user are compared in Fig. 1(b). At the beginning, their cooperation probabilities are reduced, for which the reason lies in two aspects: (i) the user reduces the cooperation probability to decrease the expected payoff of defection of the service provider; and (ii) the service provider tries to decrease the cooperation probability because the expected payoff of defection is not small enough. Also, the time the user needs to increase her cooperation probability to 1 is longer than the time the service provider uses to enlarge his cooperation probability to 1. This is because the user needs a time period to compute the prediction probability vector $\mathbf{P}_s$ for confirming that the service provider will tend to be cooperative in next iteration. After such a period, the user continues to increase the service provider's expected payoff, further promoting cooperation. Eventually, the two players definitely take cooperation action, which confirms the effectiveness of our proposed algorithm and validates our theoretical analysis.

According to Fig. 1(c), the service provider' average payoff is greater than the user's at the very beginning, because defection is the best choice for the service provider during such a time period. However, as the number of iteration increases, the user collects sufficient information to motivate the service provider to join cooperation by increasing the expected payoff of cooperation and reducing the expected payoff of defection. As a result, the service provider becomes cooperative for a higher payoff. Besides, the two players' payoffs do not exceed their achievable maximum values ($R_s = R_u = 8$ in simulations), which has been proved in Theorem 5.

From the simulation results and the above analysis, we can draw the following conclusions: (i) the proposed algorithm can effectively promote cooperation between the user and the service provider, preventing data resale; (ii) each player can obtain a high payoff that is close to the achievable maximum value, achieving a "win-win" situation; and (iii) the simulation results are consistent with our theoretical analysis.

5 Conclusion

In this paper, to prevent data resale for a service provider, we formulate an iterated data trading dilemma game between the user and the service provider. To solve such data trading dilemma, we propose a scheme that possesses three main novelties: (i) the user can unilaterally control the service provider's expected payoff regardless the service provider's strategy; (ii) the user can unilaterally promote cooperation with the service provider; (iii) both two players can receive their maximum payoffs via cooperation.

References

1. Armstrong, S.: What happens to data gathered by health and wellness apps? Br. Med. J. (Online) **353** (2016)
2. Cai, Z., He, Z., Guan, X., Li, Y.: Collective data-sanitization for preventing sensitive information inference attacks in social networks. IEEE Trans. Dependable Secure Comput. (2018, to appear)
3. Cai, Z., Zheng, X.: A private and efficient mechanism for data uploading in smart cyber-physical systems. Trans. Netw. Sci. Eng. (2018, to appear)
4. Han, M., Li, L., Xie, Y., Wang, J., Duan, Z., Yan, M.: Cognitive approach for location privacy protection. IEEE Access (2018, to appear)
5. Hu, C., Li, R., Li, W., Yu, J., Cheng, X.: Efficient privacy-preserving dot-product computation in mobile computing. In: ACM Workshop on Privacy-Aware Mobile Computing (PAMCO), July 2016

6. Hu, C., Li, W., Cheng, X., Yu, J.: A secure and verifiable secret sharing scheme for big data storage. IEEE Trans. Big Data (2018, to appear)
7. IBM. https://www.ibm.com/analytics/us/en/technology/big-data/
8. Jin, R., He, X., Dai, H.: On the tradeoff between privacy and utility in collaborative intrusion detection systems-a game theoretical approach. In: Proceedings of the Hot Topics in Science of Security: Symposium and Bootcamp, pp. 45–51 (2017)
9. Krasnova, H., Günther, O., Spiekermann, S., Koroleva, K.: Privacy concerns and identity in online social networks. Identity Inf. Soc. **2**(1), 39–63 (2009)
10. Li, W., Larson, M., Hu, C., Li, R., Cheng, X., Bie, R.: Secure multi-unit sealed first-price auction mechanisms. Secur. Commun. Netw. **9**(16), 3833–3843 (2016)
11. Liang, Y., Cai, Z., Yu, J., Li, Y.: Deep learning based inference of private information using embedded sensors in smart devices. IEEE Mag. (2018, to appear)
12. Microsoft: Microsoft Azure: cloud computing platform and services. https://azure.microsoft.com/en-us/
13. Njilla, L.Y., Pissinou, N., Makki, K.: Game theoretic modeling of security and trust relationship in cyberspace. Int. J. Commun. Syst. **29**(9), 1500–1512 (2016)
14. Osborne, M.J.: An Introduction to Game Theory, 1st edn. Oxford University Press, Oxford (2003)
15. Press, W.H., Dyson, F.J.: Iterated Prisoner's Dilemma contains strategies that dominate any evolutionary opponent. Proc. NAS **109**(26), 10409–10413 (2012)
16. Shokri, R., Theodorakopoulos, G., Troncoso, C.: Privacy games along location traces: a game-theoretic framework for optimizing location privacy. ACM TOPS **19**(4), 11 (2016)
17. National Environment Agency of Singapore: Singapore's climate information and data. http://www.nea.gov.sg/weather-climate/climate/singapore's-climate-information-data
18. TrustArc: 2016 TRUSTe/NCSA consumer privacy infographic - US Edition. https://www.truste.com/resources/privacy-research/ncsa-consumer-privacy-index-us/
19. Wang, W., Zhang, Q.: A stochastic game for privacy preserving context sensing on mobile phone. In: IEEE INFOCOM, Toronto, Canada, April 2014
20. Wang, W., Ying, L., Zhang, J.: A game-theoretic approach to quality control for collecting privacy-preserving data. In: IEEE Allerton. Allerton Park and Conference Center, USA, September 2015
21. Wang, X., Yang, Y., Tang, C., Zeng, Y., He, J.: DroidContext: identifying malicious mobile privacy leak using context. In: Proceedings of the IEEE Trustcom/BigDataSE/ISPA, Tianjin, China, August 2016
22. Weibull, J.W.: Evolutionary Game Theory. MIT Press, Cambridge (1997)
23. Zheng, X., Cai, Z., Li, J., Gao, H.: Location-privacy-aware review publication mechanism for local business service systems. In: IEEE INFOCOM, Atlanta, USA, May 2017
24. Computer Science Zone: Security and the internet of things. http://www.computersciencezone.org/security-internet-of-things/

Data Uploading Mechanism for Internet of Things with Energy Harvesting

Gaofei Sun[1,2], Xiaoshuang Xing[1(✉)], and Xiangping Qin[1]

[1] School of Computer Science and Engineering, Changshu Institute of Technology, Suzhou, China
{gfsun,xing,xpqin}@cslg.edu.cn

[2] Provincial Key Laboratory for Computer Information Processing Technology, Soochow University, Suzhou, China

Abstract. To facilitate uploading of sensing data with a tremendous and still growing number of devices in Internet of Things (IoT), is one of the most pressing tasks today. It is not sensible to equip each IoT device with cellular or other wide range access technologies, thus the network access management and sensing data fusion are necessary to solve the paradox of spectrum drain and endless user experience. In this framework, we considered a practical scenario where exist heterogeneous IoT devices with energy harvesting and limited short range access technologies, e.g. WiFi, BLE4.0, and their data can be uploaded through an access point (AP) in given area. The AP needs to schedule the data uploading of heterogeneous IoT devices, and conservatively satisfy the quality of experience (QoE) requirements, e.g. delay, sensing interval, data rate. First, we modeled the energy harvesting and sensing data of IoT devices by Markov chain and probability transfer matrix, and derived the expression of urgency function which can clearly distinguish the urgency of data transmission among devices. Secondly, an auction-based IoT devices data uploading mechanism is proposed, which satisfies the expected economic robustness with low communication overhead. Finally, we performed extensive simulations to verify the proposed data uploading scheme and algorithm. The simulation results indicate that the proposed scheme works well.

1 Introduction

Internet of Things (IoT) has become one of the most prosperous paradigms and key components in future 5G network [1]. Just like the cellular system which connects humans on this planet, the IoT would connect objects, devices, sensors etc. all over the world, and eventually achieve the inter-operations between devices and humans. Moreover, IoT integrates with heterogeneous network systems and has stimulated a large amount of novel interdisciplinary concepts such as Cloud of Things, Web of Things and Social Internet of Things [2]. However, it is not an easy task to connect these IoT devices due to following aspects: First, the estimated quantity of IoT devices up to 2020 is about 50 billion, this would

S. Chellappan et al. (Eds.): WASA 2018, LNCS 10874, pp. 438–449, 2018.
https://doi.org/10.1007/978-3-319-94268-1_36

bring big challenges to spectrum scarcity. Secondly, IoT devices are an extension of human's sense organ, thus can be deployed in harsh environment where humans barely go to. The devices may work without wireless network coverage or cable connection, which bring a practical trouble for Internet connections, or cause privacy issue [3–5]. Thirdly, the IoT devices are sensible to energy consumption on sensing or data transmission due to the limited battery capacity or performance of energy harvesting [6,7].

Fortunately, the international organizations, IEEE, 3GPP, ETSI, and industries, Huawei, Microsoft, LORA have already proposed special wireless access technologies for IoT devices, e.g. 802.15.4, LTE-M, NB-IoT, M2M, LPWA, supporting low power devices and also mission-critical applications. In [8], Lauridsen *et al.* compared the LTE-M and NB-IoT in rural area by extensive experiments, and derived the tradeoff in outdoor and indoor devices. This study indicates that it is not enough to satisfy the requirement of IoT devices by single access technology. In [9], Niyato *et al.* proposed a wireless gateway for energy harvesting IoT sensors which acts as a caching to avoid activating sensors frequently. It is sensible to implement a central node for data caching of IoT devices, and also brings down the requirement on wireless access to backbone networks. In [10], Boualouache *et al.* proposed a BLE based data collection prototype using well-know open technologies, Arduino and Andriod, and achieves the energy efficiency of the designed system. In [11], Chan *et al.* proposed the Hint protocol, which is slotted and schedule-oriented for uploading IoT device's data with multiple blocks, and can significantly reduce overhead in transmission and contention. Considering these former work, we derive the following conclusions: First, any single IoT wireless access technology can not solve the data transmission of various IoT devices, e.g. indoor or outdoor, energy harvesting or power supply; Secondly, gateway IoT devices are needed for data transmission scheduling or fusion, caching.

In this framework, we considered the scenario where several IoT devices are equipped with short range wireless transmission technologies, such as WiFi, BLE4.0, or cognitive access [12,13], and a center gateway device is deployed to act as access point (AP) to the Internet. The main contributions are listed as follows:

- We considered a practical scenario with various IoT devices in a given area, e.g. different sensing periods, data frame length, energy states. The urgent function for these devices are derived, and a novel data uploading mechanism was proposed to satisfy the transmission requirements of IoT devices appropriately.
- We incorporated the depth k into the urgency function when devices have not data to transmit in future periods, which further improves the data transmission efficiency.
- We proposed the auction-based data uploading mechanism by incorporating the auction mechanism. Also, the economic robustness are confirmed. Simulation results indicate that our proposed mechanism work well for practical usages.

The rest of this paper is organized as follows. In Sects. 2 and 3, we describe the system model and formulate the urgency function of IoT devices. In Sect. 4, we formulate the data uploading as an optimization problem, and propose an auction based mechanism and also verifies its economic robustness. Further, the practical issues are analyzed in Sect. 5. In Sect. 6, the simulation results are provided. Finally, we conclude this work in Sect. 7.

2 System Model

Considering a scenario that several IoT devices located in a given area, collecting data of environment monitoring. Each device is equipped with an energy-harvesting component, the energy can be collected from the solar, or wireless radiation, etc. Moreover, they are also equipped with limited data transmission ability, that is short range communication technology, such as WiFi, BLE, due to the cost and geo-location constraints of deployment. Thus, these IoT devices can transmit their data through an AP in this given area as depicted in Fig. 1.

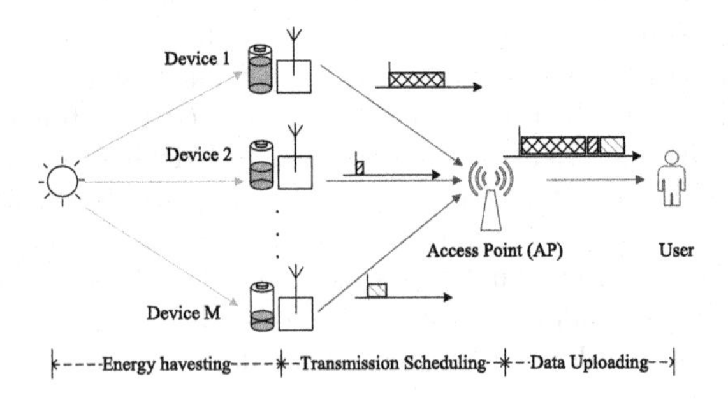

Fig. 1. System model.

To be practical, we consider that these IoT devices are heterogeneous, functioned for different purposes, such as image acquisition, temperature monitoring, or status observation, etc. For simplicity, we set up following assumptions: First, The AP approves the IoT device for data transmission once a time, and the approved IoT device can use the whole bandwidth alone. Thus, the difference on sensing data between various IoT devices is how much time they need to transmit; Secondly, IoT devices are allowed to transmit in a slotted manner, and there exists a common control (CC) channel between AP and IoT devices for control message exchanging. As we depict in Figs. 2 and 3 devices transmit data through AP with different transmission period and data length. By carefully scheduling, each data frame can be transmit within the current data frame, unless its data unchanged and don't need to be transmitted.

Suppose there are M IoT devices, denoted as $\mathcal{M} = \{1, \cdots, i, \cdots, M\}$. We character IoT device i by following factors: the data uploading period T_i, which means the device should sense the required object within T_i slots; the data length

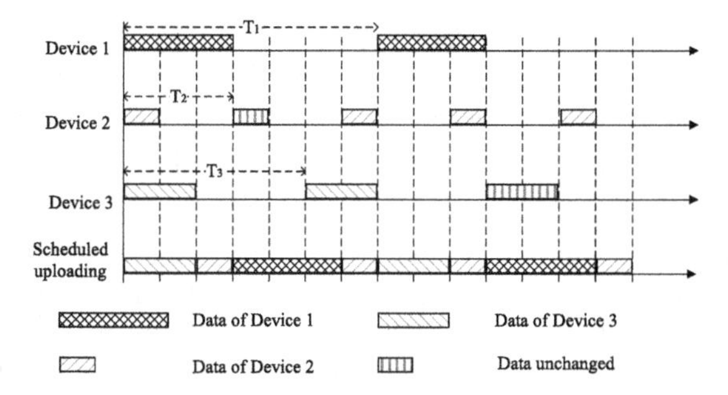

Fig. 2. Frame structure.

s_i, which means the required slots to transmit its sensing data; energy state e_i, which represents the current battery energy. For simplicity, we assume that the number of energy state is identical for all IoT devices, thus divide the battery power into $K+1$ levels (including the empty state). Once the device acquire one unit power then the current power level increases by 1, otherwise, by sensing and transmission, the power level decreases by several steps, according to data length. Further, we introduce a matrix for describing the energy harvesting of IoT device i as Ξ_i in a $(K+1) \times (K+1)$ matrix

$$\Xi_i = \begin{bmatrix} \xi_{0,0}^i & \xi_{0,1}^i & \cdots & \xi_{0,K}^i \\ \xi_{1,0}^i & \xi_{1,1}^i & \cdots & \xi_{1,K}^i \\ \vdots & \vdots & \ddots & \vdots \\ \xi_{K,0}^i & \xi_{K,1}^i & \cdots & \xi_{K,K}^i \end{bmatrix} \quad (1)$$

Here, ξ_{k_1,k_2}^i means the probability that the energy state of device i changes from state k_1 to state k_2. For example, $\xi_{0,0}^i$ means the probability that device i do not acquire energy and remain in empty energy, $\xi_{0,k}^i$ means the probability that device i has acquired k unit energy from state 0. Note that, Ξ_i only represents the energy harvesting ability of device i, but not represents the energy consumption of data sensing and transmission. Hence, the element $\xi_{k_1,k_2}^i, k_1 > k_2$ has zero probability.

We denote the energy threshold of device i as e_i^{req}, device i can perform the data sensing when $e_i \geq e_i^{req}$, and its energy state turns to $\bar{e}_i = e_i - e_i^{req}$. We assume that the data transmission needs a small part of energy compared to the energy consumption of sensing, and thus we omit the consideration on energy requirement for data transmission, the device i can transmit data with $e_i = 0$.

We further classify the sensing data into several classes, and denote variable $d_1, \cdots, d_{L_i}$ represented the sensing data states of device i, and L_i denotes the number of data states. Note that, we omit the accurate value of data, and take the data transition as a Markov chain with finite states. Hence, we introduce matrix for describing the data transition of IoT device i as

$$\Gamma_i = \begin{bmatrix} \gamma^i_{1,1} & \gamma^i_{1,2} & \cdots & \gamma^i_{1,L_i} \\ \gamma^i_{2,1} & \gamma^i_{2,2} & \cdots & \gamma^i_{2,L_i} \\ \vdots & \vdots & \ddots & \vdots \\ \gamma^i_{L_i,1} & \gamma^i_{L_i,2} & \cdots & \gamma^i_{L_i,L_i} \end{bmatrix} \tag{2}$$

Here, $\gamma^i_{m,n}$ represents the probability of transferring from state m to state n. Note that, the dimension of Γ_i is different referring to specified sensing object.

Based on descriptions of IoT devices above, we need to figure out that, whether IoT device i need to transmit its sensing data at a given moment. The condition of sensing data transmission are listed as follows:

- The current energy state satisfies $e_i \geq e_i^{req}$, then device i can perform the data sensing and acquire fresh data. Otherwise, no data is acquired and transmitted.
- The sensing data has changed in current period referring to the last period. Otherwise, the IoT device would not need to send data for saving energy.

3 Urgency Function of IoT Devices

It is apparently that not all M IoT devices need to transmit their sensing data at the same time, thus we omit the trivial case when only one IoT device need to transmit its sensing data at a time. Hence, we need to sort the sequence of transmission requirements of IoT devices. We denote the urgency function of IoT device i as $G_i(T_i, n_i, \{d_i^{-1}, d_i, d_i^1, d_i^k, \cdots\})$, which is a function of sensing data in last period d_i^{-1}, current period d_i, and also in next k period d_i^k. Here, n_i denotes the remain slots until the next data sensing period comes. Note that, sensing data $\{d_i^k\}$ related to the energy state of device i, thus $d_i = \phi$ when the current energy state $e_i = 0$, and $d_i^1 = \phi$ when IoT device i can not obtain energy during the current period. We derive the urgency function of IoT device i as follows:

- The last period sensing data d_i^{-1} was sent successfully, and also data is unchanged in current period, that is $d_i^{-1} = d_i$. In this case, device i need not to participate in transmission competitions, thus denote $G_i = 0$.
- The sensing data satisfies $d_i \neq d_i^{-1}$, thus IoT device i needs to transmit. The urgency function of device i depends on the remaining slots for transmitting its sensing data. We analyze the situation in following cases:
  1. device i also needs to transmit new data d_i^1 in next period, thus we have $e_i^1 \geq e_i^{req}$ and $d_i^1 \neq d_i$. Hence, we have $G_i(n_i) = P(e_i^1 \geq e_i^{req}, d_i^1 \neq d_i)$.
  2. device i needs not to transmit in next period but to transmit in the period following the next. Hence, we have $G_i(n_i + T_i) = P(e_i^1 < e_i^{req} \text{ or } d_i^1 = d_i) \cdot P(e_i^2 \geq e_i^{req}, d_i^2 \neq d_i)$.

  Similar to the derivation of $G_i(n_i), G_i(n_i + T_i)$ above, we can derive $G_i(n_i + k \cdot T_i), \forall k = 2, \cdots$.
- The last period sensing data d_i^{-1} is not transmitted successfully, and the data in current period unchanged or energy is below threshold, this case is the same as the case above.

- The last period sensing data d_i^{-1} is not transmitted successfully, and also $d_i^{-1} \neq d_i$, thus data d_i^{-1} is discarded, and the data d_i in current period needs to be sent. Hence, this case is also the same as the second case above.

Next, we need to calculate the expression for $G_i(n_i + k \cdot T_i)$. We begin with $G_i(n_i) = P(e_i^1 \geq e_i^{req}, d_i^1 \neq d_i)$. As data transition and energy transition are independent, we have $G_i(n_i) = P(e_i^1 \geq e_i^{req}) \cdot P(d_i^1 \neq d_i) = P(e_i^1 \geq e_i^{req}) \cdot \sum_{j \neq d_i} \gamma_{d_i,j}^i$. The energy state e_i^1 can be divided into two cases: First, $\bar{e}_i \geq e_i^{req}$, thus no matter whether device i obtain energy outside the sensing can be performed with probability 1; Second, $\bar{e}_i < e_i^{req}$, whether device i performs the data sensing depending on obtaining energy to achieve the minimum energy e_i^{req}, the probability can be calculated as $\sum_{j=e_i^{req}}^{K} \xi_{\bar{e}_i,j}^i$. Based on analysis above, we derive the expression of $G_i(n_i)$ as in Eq. (3).

$$G_i(n_i) = \begin{cases} \sum\limits_{j \neq d_i} \gamma_{d_i,j}^i, & \bar{e}_i \geq e_i^{req}, \\ \sum\limits_{j=e_i^{req}}^{K} \xi_{\bar{e}_i,j}^i \cdot \sum\limits_{j \neq d_i} \gamma_{d_i,j}^i, & \bar{e}_i < e_i^{req}. \end{cases} \tag{3}$$

$$G_i(n_i + T_i) = \begin{cases} \gamma_{d_i,d_i} \cdot \sum\limits_{j \neq d_i} \gamma_{d_i,j}^i, & \bar{e}_i \geq 2e_i^{req}, \\ \gamma_{d_i,d_i} \cdot \sum\limits_{j \neq d_i} \gamma_{d_i,j}^i \cdot \sum\limits_{j=e_i^{req}}^{K} \xi_{\bar{e}_i - e_i^{req},j}^i, & e_i^{req} \leq \bar{e}_i < 2e_i^{req}, \\ \sum\limits_{j \neq d_i} \gamma_{d_i,j}^i \cdot \sum\limits_{j < e_i^{req}} \sum\limits_{e_i^{req} \leq k \leq K} \xi_{\bar{e}_i,j}^i \xi_{j,k}^i + & \\ \gamma_{d_i,d_i} \cdot \sum\limits_{j \neq d_i} \gamma_{d_i,j}^i \cdot \sum\limits_{j \geq e_i^{req}} \sum\limits_{e_i^{req} \leq k \leq K} \xi_{\bar{e}_i,j}^i \xi_{j-e_i^{req},k}^i, & \bar{e}_i < e_i^{req}. \end{cases} \tag{4}$$

Similarly, we derive the expression of $G_i(n_i + T_i)$ as in Eq. (4). Hence, we derive the urgency function of device i as

$$G_i = \sum_k \Gamma(n_i + k \cdot T_i) G_i(n_i + k \cdot T_i), \tag{5}$$

where $\Gamma(\cdot)$ is a decreasing function of $n_i + k \cdot T_i$.

4 Proposed Data Upload Mechanism

In this section, we need to analyze the data uploading scheduling problem facing by AP. All the information that the AP may know is the urgency function we derived in section above. Besides, devices have different length of data to be transmit, e.g. device i has a data length s_i, this should be considered in data uploading mechanism design. Hence, we need to ensure the most urgency device to transmit its data among M devices by examining variables (G_i^t, s_i). It is apparently that this scenario similar to the economic model as auction in monopoly market. Moreover, we incorporate auction mechanism by these considerations: First, The communication overhead and cost should be minimized. By using auction, the devices who need to transmit sensing data only send their bids to AP, and the AP decides which device can transmit in idle slot. Thus, the CC channel can be easily constructed. Secondly, as the AP only receives the bids of devices, it needs to make sure that devices send the actual bids reflecting their actual transmission requirements. Therefore, the auction mechanism is suitable for ensuring truthfulness of devices.

First, we denote the utility function of device i as

$$u_i = G_i \cdot f(s_i, T_i), \tag{6}$$

where $f(\cdot)$ is a function of data length s_i and period T_i but irrelevant to special devices. We conduct the second price auction as follows:

The device i's equilibrium bid strategy in a second-price auction is

$$b_i = u_i, \tag{7}$$

and the winning device's payment is $p_k = \max_{i \neq k} u_i$ when device k is the winner. Hence, the winning device k's bid satisfied $b_k \geq b_i, \forall i \in 1, \cdots, M$, and its payoff equals to $u_k - \max_{i \neq k} u_i$.

It is essentially crucial to ensure that the proposed mechanism reserves all the economic robustness [14], as listed below:

1. *Truthfulness:* also called *strategy proofness.* An auction is *truthfulness* if for each device $i \in [1, M]$, it cannot improve its own utility by bidding higher or lower than its true transmission requirement.
2. *Individual Rationality:* the data uploading auction is *individual rationality* if no winning device is paid less than its ask and no winning device pays more than its bid, i.e., the utilities of devices are no less than 0.
3. *Ex-post Budget Balance:* the data uploading auction is *ex-post budget balanced* if the AP or auctioneer's profit $\Phi_{AP} \geq 0$. The profit Φ_{AP} is defined as the difference between the revenue collected from devices and the cost of providing data uploading from backbone network.
4. *Incentive:* the devices fare best when they truthfully reveal any private information to the auctioneer and trust in that the auctioneer will schedule the transmission efficiently and fairly.

It is apparently that the applied secondary-price auction satisfies these economic robustness, thus ensures the devices submit the truthful transmission requirement. We summarize the data uploading mechanism as in Algorithm 1.

Algorithm 1. data uploading process of proposed mechanism

Initialization stage

1: The IoT devices broadcast their data transmission requests to AP through CC channel.
2: The AP decides whether devices' requests can be approved according to equation (14).

Data uploading stage

3: **while** the current slot is idle, set as $T = 0$ **do**
4: The AP informs IoT devices for bids submission.
5: The IoT devices compute their bids by urgency function G_i and $f(s_i, T_i)$ according to equation (7), then submit bid b_i to AP for devices $i \in M$.
6: The AP selects the highest bid among devices, and the winning device begins its data transmission.
7: After the data uploading finished, set $T = T + \bar{s_i}$, where $\bar{s_i}$ is the data length of winning device.
8: **end**

5 Practical Issues

Along with the data uploading mechanism described above, there still leave several issues to be solved when it applies to practical usage.

- The number of served devices M. The parameter M is crucial to the data uploading, otherwise collision can be induced by too many devices. The AP should carefully ensure the number of devices approved in data uploading auction process. As we described device i by transmission period T_i, data length s_i, energy state transition matrix E_i and data transition matrix D_i, we will derive the transmission constraints on devices.
  Considering the time period $T_L \gg T_i, \forall i \in M$, and let $T_L = c_i \cdot T_i, \forall i \in M$ by carefully selecting T_L, where c_i is the number of data sensing period T_i. Hence, for device i, the slots needed for possible data transmission in T_L is denoted as $c_i \cdot s_i \cdot P_i^S$, where P_i^S is the probability for data transmission. The probability P_i^S is related to two things: the energy state and the data state, thus we have $P_i^S = P(e_i \geq e_i^{req}, d_i \neq d_i^{-1})$. By independence of these two variables, we have

$$P_i^S = P(e_i \geq e_i^{req}) \cdot P(d_i \neq d_i^{-1}) \tag{8}$$

  We begin with deriving $P(d_i \neq d_i^{-1})$. By solving $\pi_i^d \Gamma_i = \pi_i^d$ we can derive the stationary state distribution π_i^d of device i. Hence, we have

$$P(d_i \neq d_i^{-1}) = \sum_{i=1}^{L_i} \pi_i^d \sum_{j \neq i} \gamma_{ij}^i \tag{9}$$

  Moreover, the energy state transition is related to the data sensing process. That is, when energy state satisfies $e_i \geq e_i^{req}$, the data sensing needs to be performed, and then the current energy state becomes $e_i - e_i^{req}$. For facilitating, we denote the matrix H as

$$H_i = \begin{bmatrix} 1\,0 \cdots\cdots 0 \\ 0\,1 \cdots\cdots 0 \\ 0\,0\;1\;\cdots 0 \\ \vdots\;\vdots\;\;\vdots\;\;\ddots\;\vdots \\ 0\,0\;1\;\;1\;0 \end{bmatrix} \tag{10}$$

For all column $k \geq e_i^{req}$ of H_i, we have $H_{k,k}^i = 0, H_{k,k-e_i^{req}}^i = 1$. By considering the energy consumption of data sensing, we derive the energy transition as $\Xi_i \cdot H_i$. Finally, we derive the stationary state distribution π_i^e as a solution of $\pi_i^e \Xi_i \cdot H_i = \pi_i^e$. Hence, we have

$$P(e_i \geq e_i^{req}) = \sum_{i,\forall e_i \geq e_i^{req}}^{K} \pi_i^e \tag{11}$$

Summarizing the equations above, M devices can upload their data through AP without collision must satisfied:

$$\sum_{i=1}^{M} c_i s_i P(e_i \geq e_i^{req}) \cdot P(d_i \neq d_i^{-1}) \leq L \tag{12}$$

that is

$$\sum_{i=1}^{M} \frac{s_i \cdot \sum_{i,\forall e_i \geq e_i^{req}}^{K} \pi_i^e \cdot \sum_{i=1}^{L_i} \pi_i^d \sum_{j \neq i} \gamma_{ij}^i}{T_i} \leq 1 \tag{13}$$

- The expressions of $\Gamma(\cdot), f(\cdot)$. As we denoted before, $\Gamma(\cdot)$ is a decreasing function referring to remaining slots for transmission, thus it is better to have a fast decreasing curve with relatively small slots, and a flat floor curve with relatively large slots, such as x^{-1}. function $f(\cdot)$ used for balancing the transmission between long period and short period, and also long sensing data and short sensing data. We will set this function as T_i/s_i in simulations.
- The depth k when calculating $G_i(n_i + kT_i)$. By setting $k > 1$, users may transmit their sensing data in more than 1 period, thus the successful transmission probabilities may increase. However, this parameter needs to tradeoff between the enhanced performance and increased complexity.

Table 1. Parameter of different classes

Device classes	Period (slots)	Data (slots)	Probability for transmission
Class A	10	4	0.35
Class B	6	2	0.42
Class C	3	1	0.23

6 Simulation Results

In this section, we verify the performance of our data uploading mechanism. Considering that there are 9 users with 1 AP for data uploading in given area. Further, we denotes these users as 3 classes, each with different sensing data length and transmission period, which depicted by heterogeneous energy harvesting matrices and data transition matrices. Given these matrices, we generate users' data sequences and derive their data transmission probabilities as in Table 1.

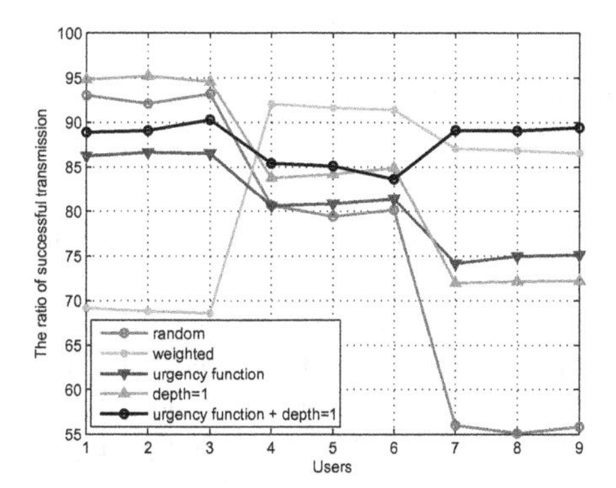

Fig. 3. Successful transmission of 9 devices refer to different G_i. (Color figure online)

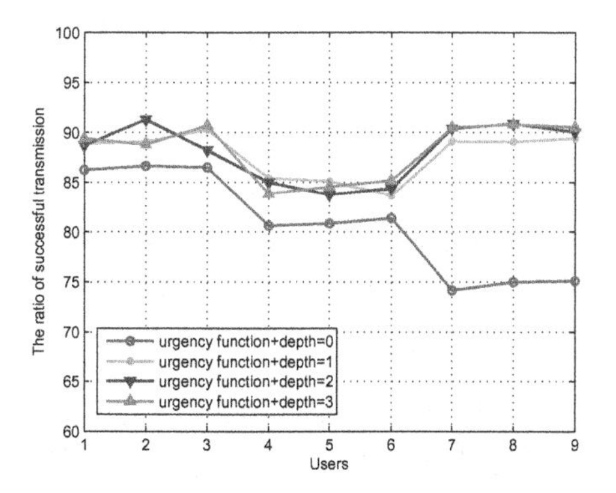

Fig. 4. Successful transmission of 9 devices refer to different k.

In Fig. 3, we compared the performance with different urgency functions. Note that, we take random selection among 9 devices for transmission as a benchmark scheme for comparison, which depicted as red curve. We observe that the class C devices with short period time have the lowest successful transmission rate, about 55%. If we take the simple method, by using the weighted vector, multiply the function $f(\cdot)$ only, as depicted in green curve, the ratio of class C devices increases a lot by severe damaging the class A devices. Further more, we perform the simulations under the urgency function but depth $k = 0$, and depth $k = 1$ only, and also the urgency function with depth $k = 1$. The results indicate that by using urgency function and setting depth $k = 1$ achieve better performance and balance among devices, with ratio above 85% for each class devices.

In Fig. 4, we compare the performance with different depth k. As expected, by considering the future data sensing and energy harvesting state, all devices increase their successful data transmission. However, further increments on depth k can not bring obvious improvement in performance, but increase the computation complexity. It is apparent that the optimal depth is $k = 1$ under this simulation scenario.

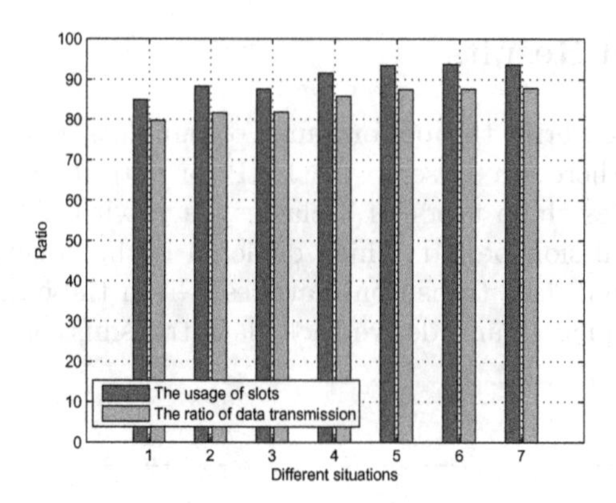

Fig. 5. The efficiency of slot usage and data uploading.

In Fig. 5, we compared the efficiency of slot usage and data transmission under different urgency functions, where 1–7 in x-axis represent cases listed as random selection, weighted, $G(\cdot)$ only, and urgency function with $k = 0, 1, 2, 3$. We observed that the proposed mechanism not only increase the successful data transmission, from 85% to above 90%, and also increase the ratio of slot usage. However, by different sensing period constraints, the slot usage can not increase even by using higher depth k.

7 Conclusion

In this paper, we have analyzed the sensing data uploading of heterogeneous IoT devices with energy harvesting located in given area. By modeling the energy harvesting and data sensing process as Markov chains, we derived the urgency function which represents the urgency of data transmission among devices. Further, an auction based device selection process is proposed in order to decrease the communication overhead, and also ensure the truthfulness on urgency computation of devices. Finally, by extensive simulations, we verify the performance of our proposed mechanism. Moreover, it is still an open question to analyze the data uploading under multiple AP scenario, and leave as our future work.

Acknowledgment. The authors would like to thank the support from the Natural Science Foundation of China (61602062, 61702056), Educational Commission of JiangSu Province (17KJB520001), the Natural Science Foundation of JiangSu Province (BK20160410), the Provincial Key Laboratory for Computer Information Processing Technology, Soochow University (KJS1521).

References

1. Palattella, M.R., Dohler, M., Grieco, A., Rizzo, G., Torsner, J., Engel, T., Ladid, L.: Internet of Things in the 5G era: enablers, architecture, and business models. IEEE J. Sel. Areas Commun. **34**(3), 510–527 (2016)
2. Xu, K., Qu, Y., Yang, K.: A tutorial on the Internet of Things: from a heterogeneous network integration perspective. IEEE Netw. **30**(2), 102–108 (2016)
3. Liang, Y., Cai, Z., Yu, J., Han, Q., Li, Y.: Deep learning based inference of private information using embedded sensors in smart devices. IEEE Netw. Mag. (2018)
4. Zheng, X., Cai, Z., Li, Y.: Data linkage in smart IoT systems: a consideration from privacy perspective. IEEE Commun. Mag. (2018)
5. Cai, Z., Zheng, X.: A private and efficient mechanism for data uploading in smart cyber-physical systems. Trans. Netw. Sci. Eng. (TNSE) (2018)
6. Shi, T., Cheng, S., Cai, Z., Li, Y., Li, J.: Exploring connected dominating sets in energy harvest networks. IEEE/ACM Trans. Netw. **25**(3), 1803–1817 (2017)
7. Chen, Q., Gao, H., Cai, Z., Cheng, L., Li, J.: Energy-collision aware data aggregation scheduling for energy harvesting sensor networks. In: IEEE INFOCOM, Las Vegas, Nevada (2018)
8. Lauridsen, M., Kovacs, I.Z., Mogensen, P., Sorensen, M., Holst, S.: Coverage and capacity analysis of LTE-M and NB-IoT in a rural area. In: IEEE 84th Vehicular Technology Conference (VTC-Fall), Montreal, QC (2016)
9. Niyato, D., Kim, D.I., Wang, P., Song, L.: A novel caching mechanism for Internet of Things (IoT) sensing service with energy harvesting. In: IEEE International Conference on Communications (ICC), Kuala Lumpur (2016)
10. Boualouache, A.E., Nouali, O., Moussaoui, S., Derder, A.: A BLE-based data collection system for IoT. In: 2015 First International Conference on New Technologies of Information and Communication (NTIC), Mila (2015)
11. Chan, T.Y., Ren, Y., Tseng, Y.C., Chen, J.C.: eHint: an efficient protocol for uploading small-size IoT data. In: IEEE Wireless Communications and Networking Conference. (WCNC), San Francisco, CA (2017)
12. Gao, L., Xu, Y., Wang, X.: MAP: multiauctioneer progressive auction for dynamic spectrum access. IEEE Trans. Mob. Comput. **10**(8), 1144–1161 (2011)
13. Gao, L., Wang, X., Xu, Y., Zhang, Q.: Spectrum trading in cognitive radio networks: a contract-theoretic modeling approach. IEEE J. Sel. Areas Commun. **29**(4), 843–855 (2011)
14. Luong, N.C., Hoang, D.T., Wang, P., Niyato, D., Kim, D.I., Han, Z.: Data collection and wireless communication in Internet of Things (IoT) using economic analysis and pricing models: a survey. IEEE Commun. Surv. Tutor. **18**(4), 2546–2590 (2016)

Enabling ZigBee Link Performance Robust Under Cross-Technology Interference

Yingxiao Sun[1], Zhenquan Qin[1(✉)], Junyu Hu[1], Lei Wang[1], Jiaxin Du[1], and Yan Ren[2]

[1] Key Laboratory for Ubiquitous Network and Service Software of Liaoning Province, School of Software, Dalian University of Technology, Dalian, China
yingxiao.sun1@gmail.com, hujunyu1222@gmail.com, dujoyce1023@gmail.com
{qzq,lei.wang}@dlut.edu.cn

[2] College of Computer Science and Engineering, Xinjiang University of Finance and Economics, Urumqi, China
yhat@sohu.com

Abstract. The unlicensed ISM spectrum is becoming increasingly populated by ubiquitous wireless networks, such coexistence can inevitably incur the cross-technology interference (CTI) for the scarcity of spectrum resources. Existing approaches for defending against such interferences often modify hardware condition, which can be ineffective when encountering with the massive deployment of legacy ZigBee devices and uncertain WiFi APs. In this paper, we build an Adaptive Transmission Estimation (ATE) model to mitigate the negative effect on ZigBee side based on a channel quality quantification metric called Channel Idle Degree (CID) and logistic regression, which can indicate ZigBee devices to adjust packet rate efficiently. Particularly, our mechanism that is a lightweight, and software-level approach without any modification on the hardware of devices, can be easily implemented on the off-the-shelf ZigBee devices. Extensive experimental results show that, under the coverage of different WiFi traffic, our lightweight mechanism can achieve 4× and 1.5× performance gains over WISE and CII. Particularly, when WiFi traffic is 4 Mbps, we can still obtain over 90%. Furthermore, the energy consumption efficiency can be reduced markedly.

Keywords: Cross-technology interference · ZigBee · WiFi

1 Introduction

With the exponential growth in the development of wireless networks operating in the unlicensed spectrum, these co-located heterogeneous wireless networks, however, usually use different technologies and follow different communication primitives, generating RF cross-technology interference (CTI) inevitably. Among

S. Chellappan et al. (Eds.): WASA 2018, LNCS 10874, pp. 450–461, 2018.
https://doi.org/10.1007/978-3-319-94268-1_37

these wireless networks, the growing adoption of ZigBee technology for the implementation of monitoring and control systems applications with different requirements in timeliness, energy-efficiency and reliability. Since most of ZigBee applications above are performance-sensitive and the inherent open characteristic of wireless channel, ZigBee network is more vulnerable, therefore the transmission efficiency and communication reliability is of great importance.

The wireless data traffic congesting the spectrum rises numerous communication challenges, of which Cross Technology Interference (CTI) is a typical communication problem, which is highly uncertain and raises the need for agile methods that assess the channel condition and take steps to maximize transmission success. Because many of the cross technology networks are expected to overlap with each other in both temporal and spatial domain, this inevitably determines the performance of co-located networks if it is not properly mitigated. We address a key question related to the trend of heterogeneous networks coexistence: *which factors should the interference mitigation mechanisms satisfy to be applied to the legacy ZigBee networks such as wireless sensor networks?* We strive for a mitigation approaches in which there is a tradeoff between lightweight software-based level and performance enhancement. The contribution of this paper are summarized as follows:

1. We conduct theoretical analysis of idle duration of WiFi traffic captured in the office building with complicated WiFi Aps coverage. We reveal that there are abundant white spaces during WiFi time domain large enough for ZigBee device to transmit packet.
2. We propose the Channel Idle Degree (CID) to estimate the channel idle state. CID is calculated by background RSSI despite the limited ability of ZigBee nodes and PRR of current channel.
3. Based on CID and LR, we build our ATE model, which can get a appropriate trade between link performance and delay sensitivity. ATE predicts the channel state based on the just acquired CID value and decision boundary, and then instructs ZigBee device to adjust packet rate intelligently.
4. Our method focuses on ZigBee side without any modification to WiFi devices, which introduces negligible performance degradation to WiFi and can be applied to massive deployment of legacy ZigBee network.

The remainder of this paper is organized as follows. Section 2 introduces the key components in ATE. Section 3 demonstrates the deployment and analysis of experiment. Section 4 reviews some existing works. Followed by the conclusions we make in Sect. 5.

2 Adaptive Transmission Estimation Model

2.1 Channel Idle Degree

We argue that there exists abundant idle duration of WiFi channel which can be leveraged by ZigBee devices to coexist with WiFi. Figure 1 illustrates the

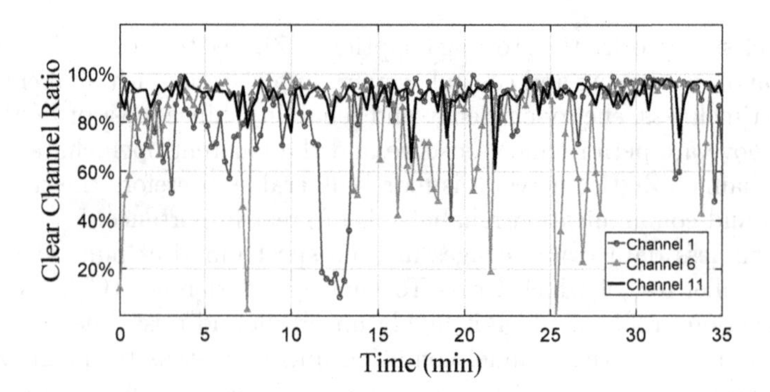

Fig. 1. Clear channel time

channel utilization trace captured in a real-life WiFi network consisting of 2 FAT APs and more than 20 fixed users. It should be noted that we choose the 3 orthogonal WiFi channels (1, 6, 11) here, because these channels cover the whole shared frequency band. It is clear that the three channels are free in most of time. The main reason is that WiFi traffic is highly burst leaving abundant idle durations between 802.11 frames.

To distinguish between a channel where the traffic is burst with large inactive periods and a channel that has very high frequency periodic traffic with the same energy profile. We present channel idle degree (CID) to utilize the idle duration in WiFi channels. CID can help ZigBee side build a prediction model to grasp the real-time change in shared channel so as to organize the transmission.

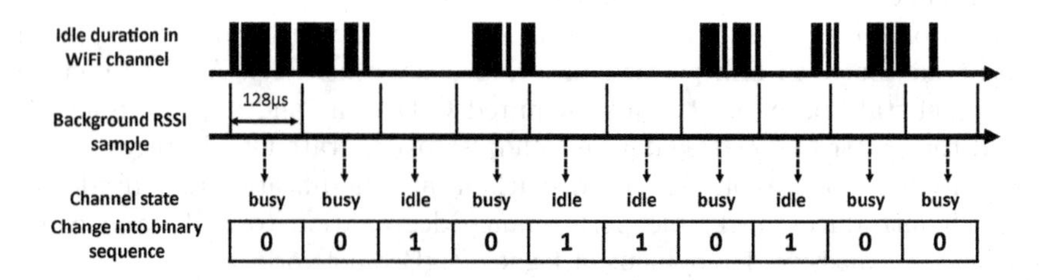

Fig. 2. Binary sequence derived from channel state.

Definition of CID. Considering the RSSI in some channel is periodically sampled, suppose the acceptable noise and interference threshold is defined as R_T, therefore the channel can be considered idle if measured $RSSI \leq R_T$. Then, L denotes the total sample times in a constant period. To discretize the consecutive background RSSI, the binary sequence R is used to indicate RSSI changes on the channel. Each sampled RSSI can be binarized as R_i:

$$R_{i_{1\leq i\leq L}} = \begin{cases} 0 & RSSI > R_T, \\ 1 & otherwise. \end{cases} \quad (1)$$

Figure 2 is a simple conversion example that we convert the channel state into binary sequence according to the background RSSI. As we discussed, not only the proportion of the idle time but also the character of the channel vacancy during a sampling window should be reflected in the metric. Therefore, we want to rank a channel with larger vacancies higher even if the sum of the idle time might be the same. Hence, we define CID as:

$$CID = \frac{1}{L}\sum_{j=1}^{k} v_j^{(1+\beta)} \quad (2)$$

where k denotes the total number of the vacancies, $\beta > 0$ is the bias, and each vacancy consists of consecutive R_i that equals 1, v_j denotes the length of jth vacancy. Actually, in each L-samples sampling window, $0 \leq k \leq \frac{L}{2}$ and $v_j > 0$. Our metric is based on the fundamental inequality that $a^2 + b^2 \leq (a+b)^2$ when $a \geq 0, b \geq 0$, where the larger value indicates a better channel state.

CID calculated using Eq. (2) takes values between 0 and x^{β}, which varies in a large range that depends on L and β. In order to lower bound data and compare among CID values merely computed with different parameter values, we normalize the CID as:

$$CID = \frac{1}{L}(\sum_{j=1}^{k} v_j^{(1+\beta)})^{\frac{1}{(1+\beta)}} \quad (3)$$

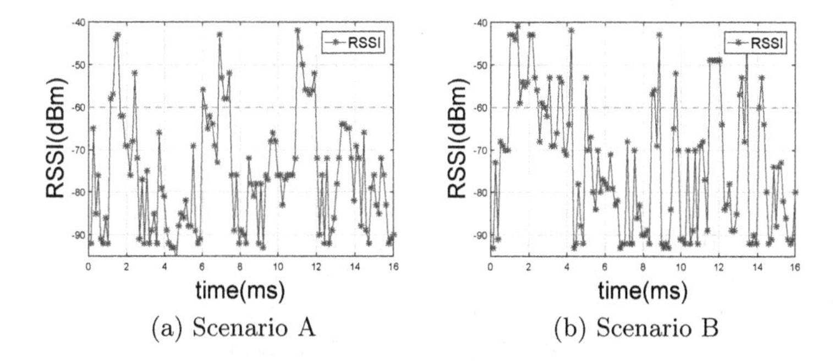

(a) Scenario A (b) Scenario B

Fig. 3. Background RSSI of the channel 12

In Eq. (3), the range of CID is [0, 1], regardless of L and β values. As the CID values approach to 1, channel state presents a better tendency, which means that the ZigBee transmission have a high probability of success. Figure 3 illustrates the vacancies distribution of channel in two scenarios with a similar sum of idle durations calculated with a R_T. Since 802.11g channel 1 covers the 802.15.4

channel 11, 12, 13 and 14, so we pick the data of 802.15.4 channel 12 here to avoid the effect of WiFi power imbalance (power density at the center of the bandwidth is higher than that on the border). However, as we argued earlier, a consecutive vacancy is more desirable than several short durations. When L is 120, the CID value in scenario A and scenario B is 0.4316 and 0.7754 respectively with the same 80% channel idle ratio, and it therefore highlights the differences in terms of quality between the two channels.

Sampling Window Size. How to select a compromised sampling window size for obtaining a meaningful CID value is a key problem. Short sampling frequency may lead to the uncertainty about the near future state of channel. While sampling too long may result in missing the dynamics. However, choosing an appropriate window size is almost impossible as a result of dynamic networks. An optimum sampling window size relatively is required to ensure the channel quality stable. Hence we hand-pick the sample window size according to the statistical of experiment results.

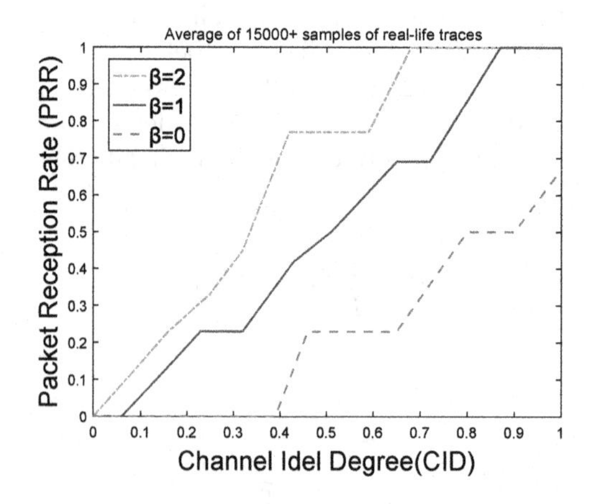

Fig. 4. PRR and CID compulated according to Eq. (3).

Correlation with PRR. For the case with concrete CID, we introduce Packet Reception Rate (PRR) that is a widely used reliability metric to evaluate it. Figure 4 shows the correlation between CID and PRR. we count CID for β values 0, 1, 2. Note that β is an integer here for the limited computation ability of sensors. When $\beta = 0$, CID equals to an average energy metric as mentioned before, when CID values increase up to 0.4, there is almost no package received yet. In this case, CID values vary faster than PRR, which also indicates that average energy metric cannot cover well enough the complexity of channel state. On the other hand, higher β values indicate longer sampling window size required, which lose the prediction effectiveness of channel state. As shown in Fig. 4, the curves help us pick up an appropriate β value. Similar to the sampling window size, finding an optimum β value is non-trivial.

2.2 Logistic Regression

Here we adopt the logistic regression (LR) to obtain the corresponding CID threshold. The reasons why we choose LR are as follows: (i) Firstly, LR happens to be an efficient classification algorithm compared with several other common algorithms (SVM, Naive Bayesian), LR can achieve 99.2593% detection accuracy over other algorithms. (ii) Secondly, the low complexity of calculation make it a feasible way faced with the limited resource constrains.

Denote y as the Bernoulli random variable representing whether the packet is successfully received or not, where 1 and 0 denotes reception success and failure respectively. Then, x is defined as a 4×1 feature vector $[x_0, x_1, x_2, x_3]^T$, where x_0 is fixed to 1 (intercept term), x_1 is the R_T we have defined in the Eq. (1), x_2 represents the CID value, and Vector x_3 is the average of the last several RSSI values in a sampling window. Particularly, x_3 here reflects the effect of latest channel state on prediction result. We give vector x as follows:

$$x = \begin{bmatrix} x_0 \\ x_1 \\ x_2 \\ x_3 \end{bmatrix} = \begin{bmatrix} 1 \\ R_T \\ CID \\ avgR \end{bmatrix}$$

where is an indispensable parameter, Note that the numerical value of R_T is closely related to the packet loss. According to Logistic regression, the probability of a next packet received or loss to be delivered can be given as:

$$\begin{aligned} P(y = 1|x) &= h_\theta(x) = \frac{1}{1 + \mathrm{e}^{-\theta^{\mathrm{T}} x}} \\ P(y = 0|x) &= 1 - h_\theta(x) = \frac{\mathrm{e}^{-\theta^{\mathrm{T}} x}}{1 + \mathrm{e}^{-\theta^{\mathrm{T}} x}} \end{aligned} \tag{4}$$

$h_\theta(x)$ is a sigmoid function and $0 \leq h_\theta(x) \leq 1$. $\theta = [\theta_0, \theta_1, \theta_2]$ indicates the parameters need to be adjusted. In addition, let $[x^i, y^i]$ denotes the ith sample of the total m samples, where $1 \leq i \leq m$. For each sample, we get the cost function as:

$$Cost(h_\theta(x), y) = -y \log(h_\theta(x)) - (1 - y) \log(1 - h_\theta(x)) \tag{5}$$

Following the cost function of one sample, the logistic regression cost function of the training set is:

$$\begin{aligned} J(\theta) &= \frac{1}{m} \sum_{i=1}^{m} Cost(h_\theta(x^{(i)}), y^{(i)}) \\ &= -\frac{1}{m} \left[\sum_{i=1}^{m} y^i \log h_\theta(x^{(i)}) \right. \\ &\qquad \left. + (1 - y^{(i)}) \log(1 - h_\theta(x^{(i)})) \right] \end{aligned} \tag{6}$$

The parameter θ should meet the goal of minimizing the $J(\theta)$. We use the gradient descent to solve for the minimum $J(\theta)$, where for each element of vector θ:

$$\theta_j := \theta_j - \alpha \frac{\partial}{\partial \theta_j} J(\theta) \tag{7}$$

where α is a constant that related to the converge speed. We consider that the model have converged when the difference between two iterations is very small.

2.3 ATE Model Building

Collect Training Samples. Note that we collect the CID before each ZigBee packet transmit under the real-life WiFi traffic with different R_T using 16 TelosB motes for almost 4 h in an office building. The nodes will continue sending the packets whatever the calculated CID is with the L setting to 20. Therefore, we can gain the CID of all the ZigBee packets successfully received or lost.

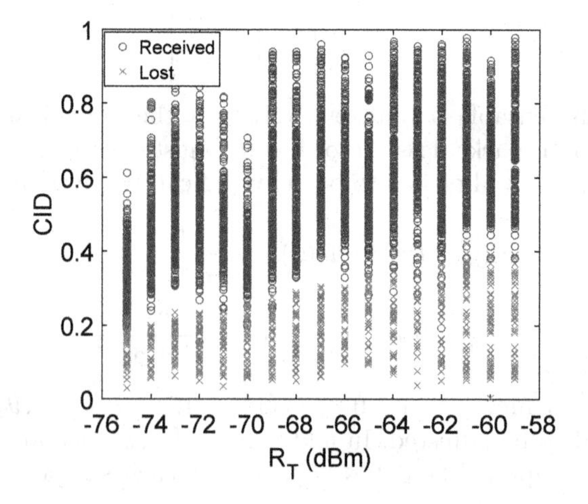

Fig. 5. Distribution of all packets' CID

Train the CID Threshold. In Fig. 5, the positive and the negative samples are clearly separated. Here we use the decision boundary to find the optimal CID threshold of each N_T. In particular, $\theta^{\mathrm{T}} x = 0$ in Eq. (4) is the decision boundary function and we can optimize the parameters θ by minimizing the cost function $J(\theta)$ using the gradient descent method. Therefore, we choose the most appropriate CID threshold of different N_T to predict the channel idle state so that the ZigBee packet can leverage to harvest better performance.

The Workflow of Model. We give the workflow of ATE as follows: **(i)** Whether a packet is successfully received or not, we record its CID value to build a feature vector. **(ii)** We leverage the logistic regression to train collected samples. And then the trained model can output a decision boundary. **(iii)** To make a new prediction, the feature vector x consisting of CID and R_T will be input in the model. According to the decision boundary $h_\theta(x)$ shown in Algorithm 1 and CID calculated after a sampling window period, the ZigBee node will decide to omit the prediction process and increase packet rate directly or wait for another sample window.

Algorithm 1. Decision to adjust packet rate

Input:

L, R_T,

Set $R = 0$; $CID = 0$; $avgR = 0$; $fSize = 8$

Begin:

1: **for all** $RSSI$ value in one sample window(S) **do**
2: **if** $RSSI \leq R_T$ **then**
3: $Temp \leftarrow 1$
4: **else**
5: $Temp \leftarrow 0$
6: **end if**
7: $R \leftarrow (R << 1) + Temp$
8: **end for**
9: **for all** v_j in R **do**
10: $CID \leftarrow CID + v_j^2$
11: **end for**
12: Normalize the CID:
13: $CID \leftarrow \frac{1}{L}(CID)^{\frac{1}{2}}$
14: Calculate the sum of latest fSize RSSI values in S
15: **if** $Reads \geq size(S) - fSize + 1$ **then**
16: $avgR \leftarrow avgR + RSSI$
17: **end if**
18: $avgR \leftarrow (avgR >> 3)$
19: Use CID, R_T and $avgR$ to build the feature vector:
20: $x = [1, R_T, CID, avgR]$
21: **if** $h_\theta(x) \geq 0.5$ **then**
22: Jump over the prediction process
23: Increase the packet rate
24: **else**
25: Wait for next sample window
26: **end if**

3 Experiment and Evaluation

3.1 Experiment Deployment

To compare our scheme with WISE [2] and CII [1]. We run the CII, WISE and our scheme on three pairs of TelosB motes simultaneously under the same WiFi interference pattern. Especially, to avoid the internal interference between three pairs of nodes, we set APE, CII, and WISE on the 802.15.4 channel 12, 13 and 14, respectively. And we set the laptop called A with the D-ITG WiFi traffic generator installed to work on the 802.11g channel 1. This channel covers the 802.15.4 channel 11, 12, 13 and 14 so that the source of interference is unique.

We use another laptop B, which also installs D-ITG [3] to communicate with the laptop A to build the WiFi interference environment. The receiver of each pair of TelosB motes is connected to the laptop C. The senders are bound together and we place them 5 m, 10 m, 15 m far from laptop A. On the receiver side, both three receivers collect the successfully arrived packets and send them to the laptop for the sake of later analysis.

3.2 Performance Evaluation

To evaluate the performance of our scheme, we conduct a contrast experiments with WISE and CII under different WiFi traffic intensity.

Throughput. Figure 6(a) and (b) show the throughput of ZigBee under different WiFi traffic interferences with 3 schemes respectively. When WiFi traffic is larger than 2Mbps, both CII and WISE especially WISE drop linearly. Since the CII scheme once detects the channel state is busy, it will suspend its transmission, so the throughput of CII can lower than WISE sometimes, while WISE always transmit packet in a default packet rate if there is a packet to deliver without consideration of current channel state, as a result, most of its packet are lost due to interference. Our scheme is based on the CID threshold to decide whether to increase the ZigBee packet rate or not under the current channel state. Extensive experiments show that with the increasing communication distance to 10 m, ATE still shows better performance under the same occasion.

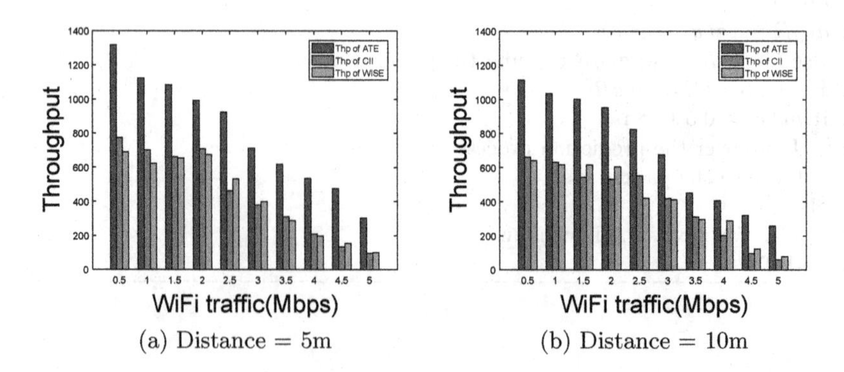

(a) Distance = 5m (b) Distance = 10m

Fig. 6. Throughput under different WiFi interferences

Packet Reception Rate. Figure 7(a) and (b) shows the PRR of ZigBee nodes under different WiFi throughput. We can clearly see that as the throughput of WiFi increase, the PRR of WISE have the sharply reduction, while the reduction

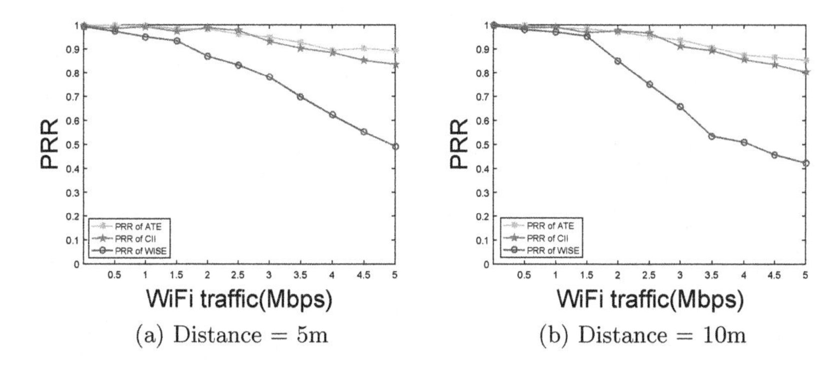

(a) Distance = 5m (b) Distance = 10m

Fig. 7. PRR under different WiFi interferences

of PRR in our scheme is very slow even when the WiFi traffic increases to $4Mbps$, the PRR is near 90% better than WISE obviously, while our PRR nearly coincide with CII, since CII will suspend its transmission when detecting the collision hazard, therefore its PRR is high inevitably. In addition, our scheme can predict channel state as well as a high probability to avoid the WiFi interference, our scheme has a very high robustness.

Energy Consumption. We suppose that the receiver side has enough energy, so we discuss the energy consumption effectiveness of the coexisted system mainly focusing on the transmitter side. We define the energy consumption effectiveness of the given system and observe its values of different schemes,

$$\eta = \frac{E_{Receive} \times Count_R}{E_{Deliver} \times Count_D + E_{Trans} + E_{other}}$$

where $E_{Deliver}$ and $E_{Receive}$ denote the energy consumption of delivering or receiving a packet, $Count_D$ and $Count_R$ denote the original delivery packet number and received packet number respectively. E_{trans} is the energy consumption of transmission process, and E_{other} is a fixed energy constant.

As shown in Fig. 8, our scheme has much higher throughput and PRR compared with WISE. It is obvious that WISE has transmitted more packets and lost the most of them under heavy intensity of WiFi interference. That is to say our scheme have saved most of the energy that should be used for the lost packets. And only when the node has a packet to transmit, it will keep the radio module on to sample the channel background RSSI. Therefore, under the cross-technology interference environment, the ZigBee node utilizing our scheme will consume less energy.

4 Related Work

The broadcast nature of wireless medium makes it inherently vulnerable to interference from heterogeneous networks overlapping in the same band. such cross-technology interference may introduce diverse temporal and spatial resolution on

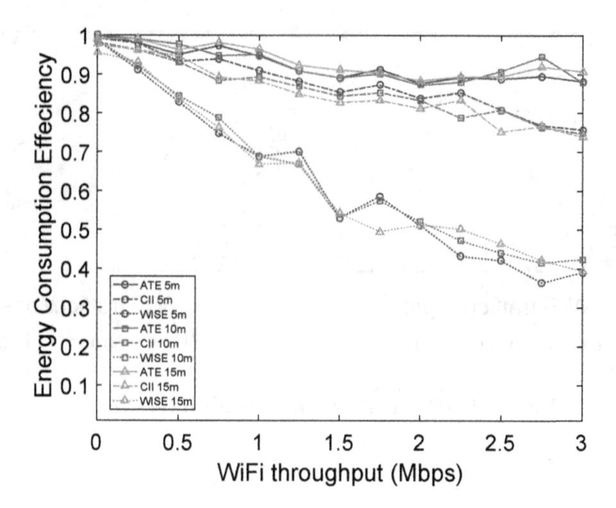

Fig. 8. Energy consumption under different WiFi throughput

link or network communication performance of heterogeneous wireless networks [4]. In order to accurately estimate the corruption of co-existed networks, many existing mechanisms present the in-packet RSSI-based sampling techniques [4,5]. However, when encountering with highly noisy environment, the mechanism will suffer inevitable errors, and hence it degrade the accuracy of channel state estimation.

Consider asymmetric interference, [8] proposed a scheme to control WiFi activities so as to enhance ZigBee performance, however, which will degrade WiFi performance. Some researchers designed a special hybrid device to enhance the visibility of low-power ZigBee as well as to fairly complete the channel access with WiFi devices therein. [6] proposed a such device called signaler, which could be senses by WiFi devices through continuously sending CBT due to its high transmission power. [7] introduced a dedicated node to monitor the channel quality. However, the coordination between the hybrid device and other device is complex and unreliable, and such a method will be ineffective when facing with massive deployment of legacy ZigBee network.

Alternatively, to enhance the coexistence in temporal domain, WISE [2] harnesses the statistical model of white space, the size of ZigBee packets are reduced to an appropriate value according to the idle duration, which will increase the communication overhead as well as decrease the transmission efficiency. Besides, WISE will suspend the transmission during each WiFi burst, resulting in poor performance for delay-sensitive applications. [1] built a channel idle state prediction model with CII, which can effectively quantify the white space. Unfortunately, it has the same delay problem.

5 Conclusion

Since cross-technology interference may lead to severe performance degradation of sensitive ZigBee networks, especially under high density WiFi interference. In order to improve the ZigBee link performance under the interference of co-located WiFi APs, in this paper, we propose a new metric called channel idle degree (CID) to quantify current channel quality. Based on CID and logistic regression, we build ATE to instruct ZigBee nodes to adjust packet rate. Our mechanism introduces negligible performance degradation to WiFi and can be applied to massive deployment of legacy ZigBee network. Extensive experimental results show that, under the coverage of different WiFi traffic, our mechanism can achieve 4× and 1.5× performance gains over WISE and CII. Particularly, when WiFi traffic is up to 4Mbps, We still attain over 90% PRR. Furthermore, the ZigBee energy-cost can be also reduced markedly.

Acknowledgments. This work was supported in part by the Fundamental Research Funds for the Central University under Grant DUT17LAB16 and Grant DUT2017TB02, and in part by the Tianjin Key Laboratory of Advanced Networking, School of Computer Science and Technology, Tianjin University, Tianjin 300350, China.

References

1. Hu, J., Qin, Z., Sun, Y.: CII: a light-weight mechanism for ZigBee performance assurance under WiFi interference. In: Proceeding of IEEE ICCCN (2016)
2. Huang, J., Xing, G., Zhou, G., Zhou, R.: Beyond co-existence: exploiting WiFi white space for Zigbee performance assurance. In: Proceeding of IEEE ICNP (2010)
3. A tool for the generation of realistic network workload for emerging networking scenarios. https://doi.org/10.1016/j.comnet.2012.02.019
4. Wang, S., Yin, Z., Song, M., He, T.: Achieving spectrum efficient communication under cross-technology interference (2017)
5. Hu, J., Qin, Z., Sun, Y.: CARE: corruption-aware retransmission with adaptive coding for the low-power wireless. In: IEEE International Conference on Network Protocols (2016)
6. Zhang, X., Shin, K., Kang, G.: Enabling coexistence of heterogeneous wireless systems: case for ZigBee and WiFi. In: Proceeding of ACM MobiHoc (2011)
7. Gomes, R., Rocha, G., Lima F., Abel C.: Distributed approach for channel quality estimation using dedicated nodes in industrial WSN. In: Proceeding of IEEE PIMRC (2014)
8. Kim, Y., Lee, S., Lee, S.: Coexistence of ZigBee-based WBAN and WiFi for health telemonitoring systems. IEEE J. Biomed. Health Inform. **20**, 222–230 (2016)

Quadrant-Based Weighted Centroid Algorithm for Localization in Underground Mines

Nazish Tahir[1], Md. Monjurul Karim[2], Kashif Sharif[2(✉)], Fan Li[2(✉)], and Nadeem Ahmed[1]

[1] School of Electrical Engineering and Computer Science, National University of Sciences and Technology, Islamabad, Pakistan
{12msitntahir,nadeem.ahmed}@seecs.edu.pk

[2] School of Computer Science and Technology, Beijing Institute of Technology, Beijing, China
{mkarim,kashif,fli}@bit.edu.cn

Abstract. Location sensing in wireless sensor networks (WSNs) is a critical problem when it comes to rescue operation in underground mines. Most of the existing research on node localization uses traditional centroid algorithm-based approach. However, such approaches have higher localization error, which leads to inaccurate node precision. This paper proposes a novel quadrant-based solution on weighted centroid algorithm that uses received signal strength indicator for range calculation and distance improvement by incorporating alternating path loss factor according to the mine environment. It also makes use of four beacon nodes instead of traditional three with weights applied to reflect the impact of each node for the centroid position. The weight factor applied is the inverse of the distance estimated. Simulation results show higher localization accuracy and precision as compared to traditional weighted centroid algorithms.

Keywords: Localization · Centroid algorithm · RSSI
Sensor networks

1 Introduction

There has been an increasing demand for miner's safety that requires development of location sensing infrastructure [1]. WSNs are a potential solution that can provide vital localization information [2]. However, there is a need to develop fast and reliable localization algorithm that can work with a WSN based sensing infrastructure. Localization algorithms rely on the existing technologies, such as RFID, Wi-Fi, GSM, and the likes. However, these technologies have their

The work of F. Li was supported by the National Natural Science Foundation of China (NSFC) under Grant 61772077, and Grant 61370192.

S. Chellappan et al. (Eds.): WASA 2018, LNCS 10874, pp. 462–472, 2018.
https://doi.org/10.1007/978-3-319-94268-1_38

limitations. For example, in the case of RFID, the card reading time causes delays if applied to multiple personnel. Besides, the localization accuracy is low and requires an additional RFID antenna in the existing WSN motes to build a workable algorithm [3]. Similarly, in case of Wi-Fi, the signal attenuation is very high in an underground environment. Therefore, the signals can never be relied on to be accurate in case of a mining incidents. There is also high computational cost involved while working with Wi-Fi signals, which is undesirable for wireless motes working in the underground environment [4]. Weak signal strength of GSM/4G technologies results in a significant reduction of localization accuracy for a cellular-based localization algorithm [5]. ZigBee is one technology that is more suited for underground environments due to its inherent characteristics [6]. ZigBee based WSNs have recently emerged as a flexible power capacity system that offers low energy consumption, small size, reliable performance in underground environments, and ease of use and deployment.

Many localization solutions based on WSNs already exist. Smart Dust [7], Pico Radio [8], and SCADDS [9], all address node localization with energy management in ZigBee. Works have been done in node localization at the application layer [10] as well. Therefore, ZigBee based WSN can be considered as a potential solution to the miner's localization issues [2]. Moreover the use of location information is not limited to WSNs only [11–14], and hence any improvement in localization algorithms benefits a larger application domain. In this paper, we introduce a novel localization algorithm that makes use of Received Signal Strength Indicator (RSSI) ranging and alternating path loss factor to correct and calculate possible distances between the nodes. We use four, instead of three of these nodes to compute the centroid position on a dedicated ZigBee infrastructure. Additionally, applied weights are used to estimate distances between the selected nodes and the unlocalized node.

The remainder of this paper is organized as follows. In Sect. 2, we discuss the related work. Section 3 presents the proposed quadrant-based weighted centroid algorithm. Section 4 presents performance evaluation, and Sect. 5 concludes the paper.

2 Related Work

A number of approaches have been introduced to address the localization-related fundamental issues and to improve the location precision by combining the ZigBee technology with RSSI based location algorithms. Some of the traditional proposed localization systems with minor alteration to WSNs are Centroid systems [15], APIT (Approximate Point in Triangulation) [16], and DV-Hop [17]. Their principal focus is on the improvement in location precision, reduction of the localization error and enhancement of the cost efficiency.

Jian and He-ping [18] propose a novel algorithm for wireless sensor networks based in a coal mine environment, that uses RSSI algorithm for distance measurements and weighted centroid algorithm for node localization. The improved algorithm uses multi-hop transmission and power control devices. That means if any beacon node is damaged in the network, the beacon node next to it will increase its transmission power to connect with the unknown node to participate in the localization process. The authors also propose that since the underground environment is harsh and the electromagnetic wave propagation becomes problematic in such a medium, the wireless signals face scattering attenuation. A polarization factor is added to the RSSI values to counter the vertical and horizontal polarization effect on electromagnetic waves. The unknown node archives the average RSSI received from each beacon node and selects three best RSSI values, which have the highest values received from the network. These values are converted into distance by formulas that include propagation loss in the underground mines. After that, weights are applied to improve the results. The least localization error recorded is 4.87 m from this improved version, which still needs considerable improvement, given the mine environment.

Fan et al. [19] propose a new scheme of weighted centroid algorithm designed for improved RSSI ranging. This algorithm uses the distance as well as RSSI values and corrects the distance between the beacon nodes. The RSSI and weights applied are reciprocal of the distance calculated. Additionally, this scheme uses the Free-Space model for the simulation environment to apply the traditional principles of weighted centroid algorithm. The RSSI ranging is accomplished by using the signal strength measurements and the distance between the beacon nodes, the coordinates of which are known. The results indicate that centroid algorithm with RSSI ranging shows better positioning and less localization error as compared to the traditional centroid algorithm. The results are improved by 10% as compared to traditional methods.

Another simple scheme has been proposed by Xie et al. [20], where none of the complexities related to path loss and distance between the anchor nodes are reflected. The scheme only relies on the received RSSI information to localize the unknown node. The authors establish the relationship between the received power at the unknown node from an anchor node and the transmission power of the two nodes placed one meter apart. By using the equation from path loss in a Free Space model, the authors calculate distance. After the distance calculation, normalized weights are assigned to the least distant nodes. The average localization error is improved as compared to the traditional weighted centroid localization algorithm. The least error recorded by is 4.32 m while the improved version had the least localization error of 3.75 m. The scheme also claims to be faster and efficient than traditional localization methods as it does not calculate β, the path loss factor, and the distances between the beacon nodes. Hence, this can preserve a significant amount of time and energy. However, this scheme can reasonably fail in the underground environment as path loss factor cannot be ignored in such harsh settings and avoiding the calculation of path loss can result in erroneous measurements. Although all the schemes mentioned above claim to perform better than the traditional weighted centroid algorithm, they can still undergo significant improvement to enhance localization accuracy.

3 Quadrant-Based Weighted Centroid Algorithm

In this article, we propose an improved localization scheme based on the weighted centroid algorithm. The design includes a set of static beacon nodes whose coordinates are fixed at the time of deployment. The deployed system is used to track an un-localized node carried by a mobile target. We have designed our network in a way that the beacon nodes are programmed to send beacon frames at a regular interval in the network. The un-localized nodes receive periodic beacon frames transmitted by the beacon nodes. These beacon frames contain information, such as beacon node ID and coordinates (X, Y) of each sender. The RSSI values are calculated at the receiving node. These values represent the signal strength in dBm.

As explained above, weighted centroid algorithm coupled with RSSI distance measurement provides promising results in an underground environment. We propose a localization algorithm that takes two essential aspects of the node localization in underground mines into account, namely; the path loss factor and an improved weighted centroid algorithm, both of which work together to give improved localization results as compared to other techniques. Our proposed scheme comprises of four phases: (a) path loss calculation, (b) distance estimation, (c) position estimation, and (d) quadrant calculation. We will give a brief overview on each of them in the following sections.

3.1 Pathloss Calculation

The path loss calculation phase involves finding the path loss in the vicinity of each beacon node. The existing literature on radio propagation model in different environments [21,22] suggests that a non-isotropic path loss exists because of the variation in the propagation medium and direction. Neglecting the path loss factor while considering underground environments can lead to highly erroneous results. The feasibility of any localization algorithm largely depends on the correct estimation of this constant, which varies based on the environment. Most of the models assume the path loss to be between 2–4 when considering the underground environment. However, such assumptions do not always hold true due to the varying and harsh underground environment. Therefore, to analyze the radio propagation pattern, the irregularity of the path loss must be defined.

A series of calibrations performed in [23] show that keeping the path loss factor constant to compute distance despite the alternating factors in the underground environment exhibits certain drawbacks. This verifies that various mediums, such as walls, obstacles, glass, etc. affect the signal attenuation differently. Therefore, using a uniform signal propagation constant, neglects the interference properties of these materials. So calculating a single path loss factor for all beacon nodes may lead to high errors in the localization process.

In the proposed scheme, when an un-localized node receives the beacon frames at predefined intervals, it calculates the RSSI values. This RSSI value is used to approximate the distance from each beacon node by using the path loss factor. Any error in path loss factor estimation will result in erroneous distance estimation. This first phase aims to improve the estimation process of path loss factor.

Beacon nodes are already aware of their coordinates as they are fixed at the time of deployment. The fundamental principle is that beacon nodes can exchange beacon messages with each other and calculate RSSI of the received beacon messages. As the actual distance between the respective beacon nodes is already known, RSSI can be easily related to the actual distances to calculate realistic path loss factor involved in the beacon message transaction. Each beacon node can calculate and advertise this path loss factor in its future beacon messages. The unknown node will receive the path loss factor from the beacon messages and can use it to improve the distance calculation based on RSSI. When the beacon frames are exchanged, beacon nodes calculate their relative distances compared to other beacon nodes by geometric line formula.

This distance is stored at each beacon node such that every beacon node in the network is aware of its neighbor's coordinates and their respective distances from each other. After determining each other's distances, beacon nodes calculate the path loss factor by the measured RSSI values and the relative distances by reversing the linear RSSI equation.

$$n_i = -\left(\frac{RSSI - A}{10\log_{10}d_i}\right) \tag{1}$$

The value of A is an absolute value calibrated by determining the signal strength between two nodes set at 1 m apart. In this phase, the path loss factor is calculated to be used in the distance estimation. After path loss calculation by each beacon node, we proceed to the distance estimation phase.

3.2 Distance Estimation

Since each beacon node in the network has estimated the path loss factor, it will propagate this information to other nodes within the beacon frame advertisement. The un-localized node can now receive beacon frames containing the information of their transmitted beacon node IDs, coordinates, and the path loss factor along with the RSSI values calculated upon the arrival of each beacon. The un-localized node upon receiving this information stores this information and averages the multiple path loss values received from various beacon nodes to obtain an average path loss factor value in the current environment.

Once the average path loss factor value is available, an un-localized node is capable of converting its RSSI measurements into the distance through the use of this average path loss factor. However, the principal challenge in using raw RSSI values is that it is prone to high sensitivity of its environmental variations. The fluctuating nature of RSSI measurement limits the accuracy of the distance estimation. Even when a mobile node is not changing its position and the whole network is static, RSSI values vary over time for the same distances. Thus, to reduce this complex wavering of the signals when the unknown node is mobile,

RSSI smoothing is required. To smooth the RSSI values while the user moves arbitrarily in the network, we calculate the simple moving average of measured RSSI values for each beacon node as

$$RSSI_{avg(i)} = \frac{RSSI_{avg(i)} + RSSI_{new(i)}}{2} \tag{2}$$

where $RSSI_{avg}$ is the previously averaged and $RSSI_{new}$ is the newly received $RSSI$ from beacon node i. Averaged $RSSI$ values obtained using Eq. (2) are then converted into distance with estimated path loss factor derived from the phase 1 of our algorithm. The un-localized node now has calculated the distance to each beacon node from which it has received a series of beacon messages. The information of neighbor beacon nodes is stored at the un-localized node according to the calculated distance in ascending order.

3.3 Position Estimation

We select at least three reference nodes to determine the position of the unknown node. These three nodes are the three least distant beacon nodes from the un-localized node. A traditional weighted centroid algorithm is applied, which calculates the position of the ascending node based on range measurements from the three beacon nodes received at the same time. The algorithm requires the coordinates of these three reference nodes $(X_i,\ Y_i)$, the distances d_i between the un-localized node and the beacon nodes, which has already been calculated in the last phase and sorted according to the distance.

Now weights are applied based on the distance from each beacon node. From the weighted centroid principle, we have

$$X_{est} = \left(\sum_{i=1}^{n} \frac{X_i/d_i}{1/d_i}\right), Y_{est} = \left(\sum_{i=1}^{n} \frac{Y_i/d_i}{1/d_i}\right) \tag{3}$$

where X_i, Y_i are the x-coordinator, y-coordinator of beacon nodes and d_i is their respective distances. The weighing factor used is

$$w_i = \left(\frac{1}{d_i}\right) \tag{4}$$

In other words, we can say that the closer the distance, the greater the value of the weight. In this way, a node closest to the un-localized node will have more influence on the position estimate than a node that is farther away. After this phase has completed, we have an estimated position of the un-localized node.

3.4 Quadrants Estimation

Once the estimated position is calculated, the partially-localized node will use this position to divide its surrounding area and nodes into four quadrants. Each quadrant will be a mathematically calculated area based on the coarse calculation of X and Y for the unknown node. These quadrants are calculated by the unknown node through sorting the X and Y of the neighboring beacon nodes into four squares from four directions. For this purpose, the unknown node will run a query on its existing beacon information to separate $X_{beacons} > X_{est}$ and $Y_{beacons} > Y_{est}$into first quadrant, $X_{beacons} > X_{est}$and $Y_{beacons} < Y_{est}$into second quadrant, $X_{beacons} < X_{est}$ and $Y_{beacons} < Y_{est}$into third quadrant, and $X_{beacons} < X_{est}$ and $Y_{beacons} > Y_{est}$into fourth quadrant. After this division, the unknown node will select the least distant node from each quadrant, so that now it has four nodes, one from each quadrant to perform the weighted centroid algorithm. The four nodes selected will undergo the position estimation again in which the weighted centroid algorithm will be applied to get a new estimated position.

The idea behind the division of the wireless sensor network into four quadrants is that the un-localized node should be localized from four directions in a 2D tunnel plane. A node which is going to be located through the information received from all directions will achieve better accuracy in localization as compared to a node using three least distant nodes as is the usual practice of a typical weighted centroid localization algorithm.

4 Performance Evaluation

The proposed algorithm has been simulated in Network Simulator(NS) 2.34 at the lower stack MAC level. Transmission power is set to be 0.008 W. The calibration constant A is set at -22.628 dBm, which is an absolute value obtained when the nodes are set at 1 m apart. Results are compared in two different scenarios. The first scenario includes the physical dimension of the outdoor environment, i.e., $100\,\text{m} \times 100\,\text{m}$. A maximum total of 40 reference nodes are scattered or randomly placed in the square test area. During the simulation, a user carrying an unknown node walks from an initial point with coordinate (10, 10) to a finish point with coordinate (60, 60). This movement is directed diagonally across the plane of the simulation. The speed of the movement is set at 3.5 m/s. The movement is tracked at every second interval of the simulation. The second scenario sets up the simulation area of $5\,\text{m} \times 100\,\text{m}$, which is the closest possible dimension of a mine tunnel as shown in Fig. 1. A total number of 40 nodes are placed at an equal distance of 5 m apart around the walls of the tunnel. The coordinates of these nodes are fixed and known by the system. The unknown node will start moving from its source point (2, 10) to its destination point (2, 80). The movement is set at the speed of 3.5 m/s.

First, a round of traditional WCL (Weighted Centroid Localization) algorithm is applied to the simulation area followed by the quadrant-based weighted centroid algorithm. Figure 2 shows the comparison of position estimation for both traditional weighted centroid algorithm and the quadrant-based weighted centroid algorithm for an open area. The error is recorded as soon as the simulation starts with the unknown node moving diagonally across the simulated area.

The localization error over the course of time is observed to be lesser for the quad-based weighted centroid algorithm with the maximum error observed of

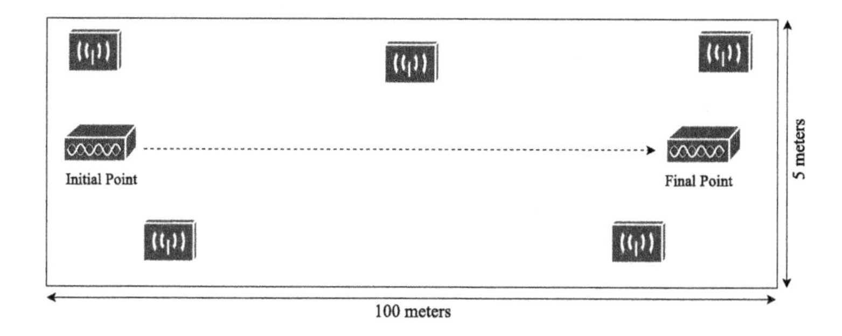

Fig. 1. Topology of 5 m × 100 m mine tunnel with 40 nodes placed 5 m apart.

3.62 m and the minimum of 0.39 m as shown in Fig. 2. Around 50% improvement in localization accuracy is observed as compared to the traditional weighted centroid algorithm. RSSI refinement through the use of path loss factor improves the position accuracy of the mobile node.

Similarly, simulations were performed with a setup of 5 m × 100 m, which closely depicts a mine tunnel dimensions. The error is recorded as soon as the simulation starts with the unknown node moving straight over the simulated area which depicts a miner walking straight inside a mine tunnel.

It is evident that, in Fig. 3, the proposed quadrant-based weighted centroid algorithm outperforms the traditional weighted centroid algorithm. The localization error is much reduced now with around 1.25 m to be maximum and 0.07 m to be minimum. Around 70% improvement in localization accuracy is observed as compared to the traditional weighted centroid algorithm. This accuracy is a significant improvement as the localization error has dramatically reduced while keeping beacon nodes orientation and the estimated path loss factor into account.

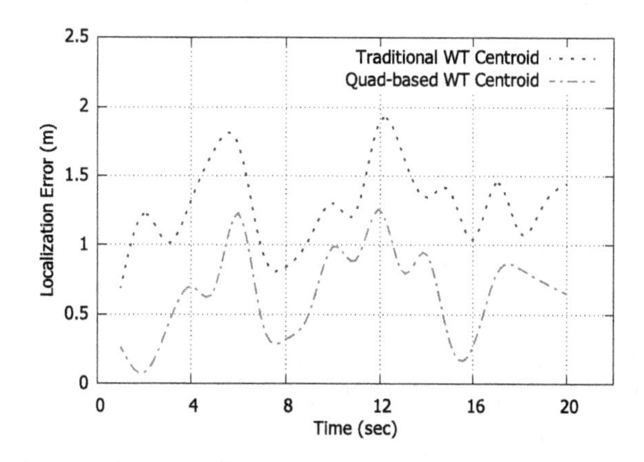

Fig. 2. Comparison of traditional weighted centroid algorithm with Quadrant-based weighted centroid localization error in 100 m × 100 m area.

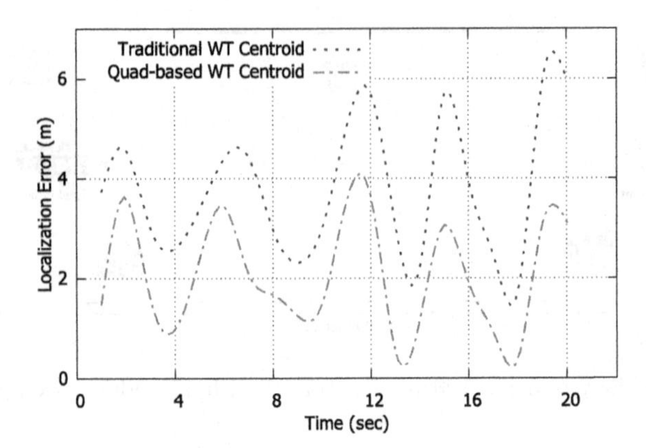

Fig. 3. Comparison of localization error of traditional weighted centroid algorithm with Quadrant-based weighted centroid algorithm in a mine tunnel.

It can be seen that results obtained in the mine tunnel are better than that of the open area. This is because the mine tunnels are narrow and close spaces in a width of 5 m where a miner would always be in proximity of the nodes placed in the tunnel. The path loss factor observed in a mine tunnel is less than 1.94 m on average. On the other hand, in an open area where the nodes are randomly deployed, a person will sometimes be surrounded by closely spaced nodes and, in other instances, he will have the beacon nodes placed far apart from him. This difference in the topology creates the differences in the RSSI values received. The path loss estimation in an open area is observed to be around 1.85 on average. However, the division of the area under consideration into quadrants still significantly improves the results.

In comparison with other RSSI-based location algorithms, our proposed system exhibits resemblance of calibration of the propagation constant, but the alternating path loss calculation is not suggested in other schemes. The accuracy of the localization is immensely improved as compared to other proposed systems by using the path loss and the quadrant division by our algorithm.

5 Conclusion

We have proposed a novel localization algorithm that makes use of RSSI ranging and alternating path loss factor to correct and calculate possible distances between the existing nodes. Besides, this method enables extended functionality by using four instead of three of these nodes to compute the centroid position on a dedicated ZigBee infrastructure. It has been observed that the localization error can be reduced by performing distance correction through the use of alternating path loss factor and dividing the area into four quadrants to pin down the un-localized node. The results obtained by our proposed algorithm exceed in accuracy as compared to the results obtained by the traditional weighted centroid algorithm. The accuracy has been improved up to 50% in open space and 70%

in the underground mine tunnels. The results can be further improved by implementing the algorithm in real underground environment, and some extensions can be applied to the proposed quadrant-based weighted centroid algorithm to improve the localization accuracy further. In our proposed algorithm, the path losses advertised by beacons are the *inward* path losses. However, this may be different from the path losses in other directions. Therefore, in future, more work will be carried out on calculating the accurate path loss factor. The impact of using various propagation models can also be applied so that more accuracy can be achieved if real-time RSSI values of underground mines are used in the simulation. We plan to implement our proposed algorithm on real sensor network hardware to ascertain its performance in real mine conditions.

References

1. Tian, F., Dong, Y., Sun, E., Wang, C.: Nodes localization algorithm for linear wireless sensor networks in underground coal mine based on RSSI-similarity degree. In: 2011 7th International Conference on Wireless Communications, Networking and Mobile Computing, pp. 1–4, September 2011
2. Forooshani, A.E., Bashir, S., Michelson, D.G., Noghanian, S.: A survey of wireless communications and propagation modeling in underground mines. IEEE Commun. Surv. Tutor. **15**(4), 1524–1545 (2013)
3. Want, R.: An introduction to RFID technology. IEEE Pervasive Comput. **5**(1), 25–33 (2006)
4. Liu, J.: Survey of wireless based indoor localization technologies. Department of Science & Engineering, Washington University (2014)
5. Fatima, B., Shah, M.A.: Self organization based energy management techniques in mobile complex networks: a review. Complex Adapt. Syst. Model. **3**(1), 2 (2015)
6. Biddut, M.J.H., Islam, N., Karim, M.M., Miah, M.B.A.: An analysis of QoS in ZigBee network based on deviated node priority. JECE **2016** (2016)
7. Warneke, B., Last, M., Liebowitz, B., Pister, K.S.J.: Smart dust: communicating with a cubic-millimeter computer. Computer **34**(1), 44–51 (2001)
8. Rabaey, J.M., Ammer, M.J., da Silva, J.L., Patel, D., Roundy, S.: PicoRadio supports ad hoc ultra-low power wireless networking. Computer **33**(7), 42–48 (2000)
9. Estrin, D., Heidemarm, J., Govindan, R.: Scalable coordination architectures for deeply distributed systems (SCADDS), a DARPA sponsored research project. Technical report, Information Science Institute, University of southern California (2014)
10. Akyildiz, I., Su, W., Sankarasubramaniam, Y., Cayirci, E.: Wireless sensor networks: a survey. Comput. Netw. **38**(4), 393–422 (2002)
11. Wang, Y., Cai, Z., Tong, X., Gao, Y., Yin, G.: Truthful incentive mechanism with location privacy-preserving for mobile crowdsourcing systems. Comput. Netw. **135**, 32–43 (2018)
12. Wang, J., Cai, Z., Li, Y., Yang, D., Li, J., Gao, H.: Protecting query privacy with differentially private k-anonymity in location-based services. Pers. Ubiquit. Comput. (2018)
13. Zheng, X., Cai, Z., Li, J., Gao, H.: Location-privacy-aware review publication mechanism for local business service systems. In: IEEE Conference on Computer Communications, IEEE INFOCOM 2017, pp. 1–9, May 2017

14. Li, J., Cai, Z., Yan, M., Li, Y.: Using crowdsourced data in location-based social networks to explore influence maximization. In: The 35th Annual IEEE International Conference on Computer Communications, IEEE INFOCOM 2016, pp. 1–9, April 2016
15. He, T., Huang, C., Blum, B.M., Stankovic, J.A., Abdelzaher, T.: Range-free localization schemes for large scale sensor networks. In: Proceedings of the 9th Annual International Conference on Mobile Computing and Networking, MobiCom 2003, pp. 81–95. ACM, New York (2003)
16. Dalce, R., Val, T., Van den Bossche, A.: Comparison of indoor localization systems based on wireless communications. Wirel. Eng. Technol. **2**(04), 240 (2011)
17. Huang, Q., Selvakennedy, S.: A range-free localization algorithm for wireless sensor networks. In: 2006 IEEE 63rd Vehicular Technology Conference, vol. 1, pp. 349–353, May 2006
18. Jian, L., He-ping, L.: A new weighted centroid localization algorithm in coal Mine wireless sensor networks. In: 2011 3rd International Conference on Computer Research and Development, vol. 3, pp. 106–109, March 2011
19. Fan, H., He, G., Tao, S., Xu, H.: Weighted centroid localization algorithm based on improved RSSI ranging. In: Proceedings 2013 International Conference on Mechatronic Sciences, Electric Engineering and Computer (MEC), pp. 544–547, December 2013
20. Xie, S., Hu, Y., Wang, Y.: Weighted centroid localization for wireless sensor networks. In: 2014 IEEE International Conference on Consumer Electronics - China, pp. 1–4, April 2014
21. Scott, T., Wu, K., Hoffman, D.: Radio propagation patterns in wireless sensor networks: new experimental results. In: Proceedings of the 2006 International Conference on Wireless Communications and Mobile Computing, IWCMC 2006, pp. 857–862. ACM, New York (2006)
22. Biaz, S., Yiming, J., Qi, B., Wu, S.: Dynamic signal strength estimates for indoor wireless communications. In: Proceedings of the 2005 International Conference on Wireless Communications, Networking and Mobile Computing, vol. 1, pp. 602–605, September 2005
23. Lau, E.E.L., Chung, W.Y.: Enhanced RSSI-based real-time user location tracking system for indoor and outdoor environments. In: Proceedings of the 2007 International Conference on Convergence Information Technology, ICCIT 2007, pp. 1213–1218. IEEE Computer Society, Washington (2007)

Exploration of Human Activities Using Sensing Data via Deep Embedded Determination

Yiqi Wang[1], En Zhu[1](✉), Qiang Liu[1], Yingwen Chen[1], and Jianping Yin[2](✉)

[1] College of Computer, National University of Defense Technology, Changsha 410073, China
enzhu@nudt.edu.cn
[2] Dongguan University of Technology, Dongguan, China
jpyin@dgut.edu.cn

Abstract. Clustering analysis is one of promising techniques of uncovering different types of human activities from a set of ubiquitous sensing data in an unsupervised manner. Previous work proposes deep clustering to learn feature representations that favor clustering tasks. However, these algorithms assume that the number of clusters is known *a priori*, which is often impractical in the real world. Determining the number of clusters from high dimensional data is challenging. On the other hand, the lack of the number of clusters make it difficult to extract low dimensional features appropriate for clustering. In this paper, we propose Deep Embedding Determination (DED), a method that can determine the number of clusters and extract appropriate features for the high dimensional real data. Our experimental evaluation on different datasets shows the effectiveness of DED, and the excellent performance of DED in exploring the human activities using sensing data.

Keywords: Human activity analysis · Deep clustering
Determination of cluster number · Sensing data

1 Introduction

Nowadays, ubiquitous sensing is an emerging research direction in the field of wireless sensor networks. The goal of ubiquitous sensing is to establish a distributed and interconnected computing infrastructure that transparently supports human activities. Specifically, the ubiquitous sensing uses powerful perception capability provided by sensing techniques to implement ubiquitous computing infrastructure, where a variety of sensors interconnect with each other to form a sensor network and generate a high volume of multimodal sensor data flows. Due to the prevailing deployment of wearable devices and mobile terminals, people are able to analyze their sensing data and recognize human daily activities using the ubiquitous sensing technology.

S. Chellappan et al. (Eds.): WASA 2018, LNCS 10874, pp. 473–484, 2018.
https://doi.org/10.1007/978-3-319-94268-1_39

Clustering analysis is one of promising techniques to recognize human activities from ubiquitous sensing data. Basically, clustering is a fundamental research topic in machine learning, which aims to reveal class structure in a data set by partitioning it in an unsupervised manner. In the case of human activity recognition, people can utilize clustering analysis to uncover different types of human activities from a set of sensing data in an unsupervised manner. The clustering analysis has been extensively studied in various applications; however, the performance of existing methods is negatively affected by the "curse of dimensionality" and the lack of priori knowledge (i.e., the number of clusters). To deal with high dimensional data, deep clustering algorithms [1–4] have been proposed. However, these algorithms all need to know the number of clusters. The number of clusters K is significant to clustering problems and obtaining a good estimation of K is not trivial, though. Prior work [5–9] in determining the number of clusters suffers from the "curse of dimensionality".

In this paper, we revisit cluster analysis and ask: Can we use a data driven approach to solve jointly for the unknown number of clusters and feature space? To answer the question, we present Deep Embedding Determination (DED), an effective algorithm for predicting the number of clusters in the high-dimensional dataset. DED exploits the advantages of both feature representation learning techniques and density-based clustering algorithms. It learns appropriate embedded features (DED features) using a deep convolutional autoencoder (CAE) and t-SNE visualization techniques, and then predicts the number of clusters by using a new density-based clustering algorithm. Combining the representation learning capability of CAE and the pairwise distance modeling capability of t-SNE, DED reduces the dimension of input data without losing much similarity information. The DED features preserve pairwise distances in 2 dimensions that are suitable for the subsequent density-based clustering algorithm. At last DED outputs the number of clusters and gives robust statistics that indicate a good number of clusters from multiple runs.

Our experiments show significant improvements over state-of-the-art clustering methods in the determination of the number of clusters on different datasets. We evaluate DED on the MNIST [10], HASY [11], USPS [12], and UCI HAR [13] datasets and comparing the clustering analysis on the UCI HAR dataset on different feature spaces (input, CAE, t-SNE and DED feature spaces). The DED features outperform other features in terms of K estimation by a significant margin. In addition, our experiments show that DED is very robust with respect to the choices of hyperparameters, given that robustness is a significantly important property for unsupervised learning problem [1]. The contributions of our work are summarized as below:

- DED is the first work to determine the number of clusters in high dimensional datasets, which enables other deep clustering methods to train specific models and to improve the clustering results of existing clustering algorithms.
- A real-world application of DED to human activity analysis using sensing data demonstrate that DED is very effective to find out implicit valued

information, e.g, the number of different types of activities, in an unsupervised manner.
- DED provides a better embedded feature space for the further clustering analysis of unknown datesets.

2 Related Work

As an essential topic in machine learning, clustering has been widely studied from different perspectives. Traditional clustering methods such as k-means [14] and Gaussian Mixture Models (GMMs) [15] are effective for a wide range of problems, but they cannot deal with high dimensional input [20]. To tackle this problem, deep clustering methods [1,2,4,16,17] have been developed to achieve feature transformation via deep neural networks. These algorithms achieve high clustering accuracy, but all of them require the number of clusters as input.

Estimating the optimal number of clusters is an important but challenging problem. Most approaches set an evaluation metric and optimize it as a function of the number of clusters, such as the silhouette statistic [5] and the gap statistic [6]. Density-based clustering algorithms group data points based on region density and do not require the number of cluster, which provide an alternative method to determine the number of clusters. Density-based clustering algorithms include DBSCAN [7] and mean shift [8]. DBSCAN takes separated regions of high density as different clusters and discards points of region of low density as noises. mean shift maps the input data into a vector space and determines clusters by finding data points of high probability density. A novel density-based clustering algorithm (CFSFDP) which is based on the density peaks has been proposed [9]. It characterizes cluster centers by finding points surrounded by neighbors with lower density and relatively far from other points of high density. Yet most of the methods mentioned above suffer from the "curse of dimensionality".

Feature transformation from high-dimensional space to low-dimensional space is an important issue in clustering, and it can be done by applying dimensionality reduction techniques like Principle Component Analysis (PCA) and autoencoder (AE). An autoencoder is an unsupervised model consisting of the encoder and the decoder. The decoder uses the output of encoder to reconstruct the input of encoder. The aim of AE is to learn representative features of low dimension via minimizing reconstruction loss. Convolutional autoencoder (CAE) [18] is the combination of convolutional neural network and autoencoder, which preserves spatial locality by sharing weights among different locations of the input. T-SNE [19] is also an effective dimensionality reduction technique, which minimizes the Kullback-Leibler (KL) divergence between a data distribution and an embedded distribution and reveal the data structure.

DED combines the virtues of the CAE and the t-SNE technique to extract low-dimensional features that favor the K estimation. Empirically, algorithms that determine the number of clusters tend to suffer from the "curse of dimensionality". DED preserves the salient structure of the input data and models the pairwise distances in 2-dimensional embedded features.

3 Deep Embedding Determination

Given a set of points $\{x_i\}_{i=1}^n \subset X$ where the number of clusters is not known *a priori*, we want to predict the number of clusters of this set. First we use a nonlinear mapping F_w to extract the latent features $\{z_i\}_{i=1}^n \subset Z$, which are both low dimensional and appropriate for clustering. Next we take advantage of a clustering algorithm which can make the best use of the embedded features. Deep neural network is a good candidate for the nonlinear mapping, given their theoretical function approximation properties [24] and feature learning capabilities [25]. However, due to the lack of explicit objective function towards clustering, DNN-based unsupervised learning techniques do not perform well on clustering problems. To solve this, previous work [1–4] has proposed clustering loss in different ways but they all require K as input. Most of density-based clustering algorithms do not require K, but they are sensitive to the dimension of data and even fail with feature representations extracted by deep autoencoder.

The network structure is demonstrated in Fig. 1. There are two essential components in our method: the feature extraction model and the density-based clustering algorithm. The feature extraction model aims to get the low dimensional features which are appropriate for clustering. The density clustering algorithm is used to estimate the number of clusters on the latent features. We introduce these two components in the following subsections.

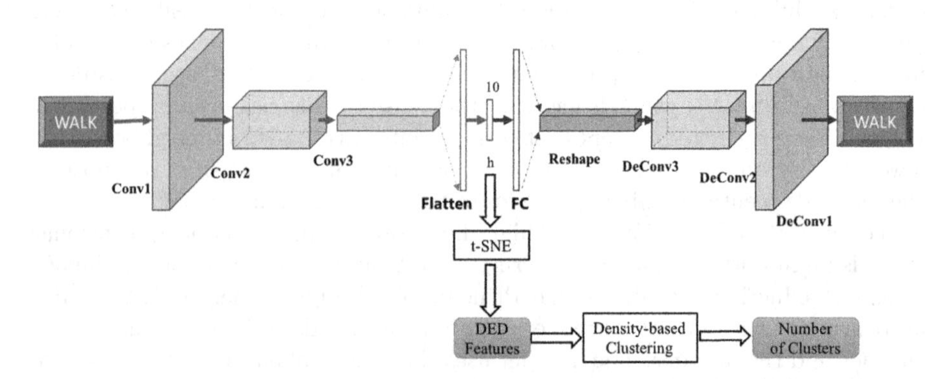

Fig. 1. Network Structure of DED: in the first step, a deep convolutional autoencoder is trained to extract 10-dimensional features with reasonable reconstruction loss; in the second step, t-SNE algorithm is used to define a non-linear mapping from feature space M to a 2-dimensional feature space Z; at last, a density-based clustering algorithm returns the number of clusters with the input of 2-dimensional features.

3.1 Feature Representation for Clustering

We propose to extract features to be suitable for estimating the number of clusters in an unsupervised manner in two steps. In the first step, a CAE is trained to extract 10-dimensional features. The autoencoder preserves intrinsic local structure of data in 10-dimensional feature space M. However, in order to better fit the subsequent density-based clustering method, the features need to be further reduced to lower dimension. In the second step, DED uses t-SNE algorithm which defines a non-linear mapping from feature space M to a 2-dimensional feature space Z. It minimizes the mismatch between feature space M and Z in terms of pairwise distances. We find that the feature space Z obtained by these two successive steps are more suitable for the task of estimating the number of clusters. Empirically, the density-based clustering algorithm does not perform well with high dimensional data. Low dimensional features (e.g. 2-dimensional) extracted directly by CAE or t-SNE suffer from either significant information loss or sensitiveness to high dimensional noise.

The deep convolutional autoencoder (CAE) learns representations in an unsupervised manner and preserves intrinsic local structure in data. The reconstruction loss is used to train the network and the goal is to let latent code represent input data well. Intuitively, the capability of feature representation decreases as the dimension of latent feature decreases. Thus, to obtain appropriate feature representations with reasonable reconstruction loss, the dimension of latent feature space is set to 10. Note that the features extracted by CAE are locations of data points in feature space M, while only pairwise distance is needed by the following density-based clustering algorithm.

As the density-based clustering algorithm favors low-dimensional features with pairwise distance, t-SNE algorithm is used to further reduce the dimension of feature space M while maintaining the pairwise distance as accurately as possible. T-SNE algorithm minimizes the mismatch between feature space M and Z in terms of pairwise distance. The objective function is given by

$$C = KL(P||Q) = \sum_i \sum_j p_{ij} log \frac{p_{ij}}{q_{ij}}. \tag{1}$$

where KL is KullbackLeibler divergence that measures the non-symmetric difference between two probability distributions P and Q. p_{ij} represents the pairwise similarity between point x_i and x_j in feature space M and q_{ij} represents the pairwise similarity between point y_i and y_j in low-dimensional feature space Z. The similarity is computed by a Student-t distribution and the distributions q_{ij} and p_{ij} are defined as

$$q_{ij} = \frac{exp(-||y_i - y_j||^2)}{exp(-\sum_{k \neq l} ||y_k - y_l||^2)}. \tag{2}$$

$$p_{ij} = \frac{exp(-||x_i - x_j||^2/2\sigma^2)}{exp(-\sum_{k \neq l} ||x_k - x_l||^2/2\sigma^2)} \tag{3}$$

Normally t-SNE is not applied directly on high-dimensional data due to its sensitiveness to noise. Features extracted by CAE are suppressed without severely distorting the pairwise distance, thus serve as a proper input for t-SNE. The dimension of feature space Z is set to 2. So far, the feature space Z has the following properties: (1) extremely low dimension (2) containing no extra information but pairwise distance. These two properties are both essential to the subsequent density-based clustering algorithm.

3.2 Density-Based Clustering Algorithm

Given the 2-dimensional feature space Z with pairwise distance information, we still cannot overcome the problem that close clusters (e.g. "4" and "9" in MNIST) are mixed together, which leads to a poor clustering result. However, we observe that typically there are two distinguishable clusters with each consisting of comparable amounts of data points "4" and "9". These two clusters are still identifiable to the density-based clustering algorithm.

Thus, we propose to use the density-based clustering algorithm CFSFDP [9]. In this algorithm, Local density (ρ or rho) is one of the leading criteria. In replacement of the indicator function, we use the Gaussian kernel in density estimation to improve the robustness of this value. The density is defined by

$$\rho_i = \sum_j exp(-\frac{||xi - xj||^2}{2\sigma^2}). \tag{4}$$

where x_i is an embedded 2-dimensional data point and σ is the cutoff distance. The algorithm is sensitive only to the relative magnitude of density in different points, indicating that the results of the analysis are relatively robust with respect to the cutoff distance. We have also tried to use the mean distance to nearest K neighbor as the cutoff distance. However, we have given up this attempt because a new hyperparameter has to be introduced.

The minimum distance with higher density (δ) is another important criterion, which is measured by computing the minimum distance between the point x_i and any other point with higher density:

$$\delta_i = min_{j:\rho_j > \rho_i}(d_{ij}). \tag{5}$$

where d_{ij} is the distance between x_i and x_j. For the point with highest density

$$\delta_i = max_j(d_{ij}) \tag{6}$$

our goal is to find points that have both high ρ and δ, which are cluster centers and thus give the number of clusters.

Algorithm 1 demonstrates the framework of DED. First, a convolutional autoencoder is trained to extract 10-dimensional features from the input data. Next, DED uses t-SNE technique to do further dimensionality reduction and to preserve pairwise similarity as much as possible. At last DED estimates the number of clusters via the density-based clustering algorithm. We suggest that users run DED for several times, and use the nearest integer of the mean of the estimations as the number of clusters. According to the experiments, 10 runs are usually enough for a credible estimation for the number of clusters.

Algorithm 1. Framework of Deep Embedding Determination.

Require:
Input data from feature space X
Cutoff distance σ (default: 0.50)
Density threshold for ρ (default: 3.50)
Distance threshold for δ (default: 0.10)
Number of iterations n (default: 10)
Ensure:
Number of clusters K
1: Extract 10-dimensional CAE feature in space M from feature space X using convolutional autoencoder;
2: Extract 2-dimensional DED feature in space Z from feature space M using t-SNE;

3: **for** $i = 1$ to n **do**
4: Calculate Density and minimum distance with higher density on DED features of subset
5: Select cluster centers based on the thresholds of ρ and δ, and count the number of clusters k
6: **end for**
7: Calculate the mean μ, variance and the number of correct predictions of k from n runs
8: **return** μ;

4 Experiments

We evaluate DED on five datasets and compare its ability in determining the right number of clusters against other clustering algorithms, including the silhouette [5], gap [6], DBSCAN [7], mean shift [8], and density-based clustering algorithm [9]. Unlike popular k-means based clustering algorithms, the algorithms we compare do not require the number of clusters K as input and are completely based on the distance matrix. Also, we have done clustering analysis on different feature spaces, and prove the advantage of the DED features.

4.1 Dataset Description

The following datasets are used in our experiments.

MNIST: The MNIST dataset consists of 70000 handwritten digits. The images are centered and of the size 28 by 28 pixels.

USPS: The USPS dataset contains 9298 gray-scale handwritten digit images with the size of 16×16 pixels. The features are floating point in [0, 2].

HASYv2: The HASY dataset of handwritten symbols contains 168,233 instances of more than 300 classes. Due to the limited complexity of our model, we do not scale to full HASYv2 dataset. We sample subsets of 8 and 12 classes, which we call HASY-8 and HASY-12, respectively.

HCI HAR: The UCI HAR dataset contains 10299 instances of six human activities. Each instance consists of 561 features.

The statistics of datasets are summarized in Table 1.

Table 1. Dataset description

Dataset	# Points	# Classes	Dimension
MNIST	70000	10	784
USPS	9298	10	256
HASY-8	8000	8	1024
HASY-12	11996	12	1024
HCI HAR	10299	6	561

4.2 Experimental Setup and Evaluation Metric

We run our algorithm 100 times on five datasets and report the average number of predictions (mean), the variance and the number of correct predictions (hit rate). We take the nearest integer of the mean as the output of our algorithm. The performance of DED will be given in Table 2.

Table 2. Performance of DED in terms of the average number of clusters, variance and the hit rate of the correct predictions among 100 runs. For DED, the correct numbers of clusters have the largest number of occurrences on all datasets.

Dataset	MNIST	USPS	HASY-8	HASY-12	HCI HAR
Ground truth	10	10	8	12	6
Mean	10.1	10.0	8.2	11.7	6.4
Hit rate	72/100	42/100	49/100	39/100	29/100
Variance	0.43	0.86	0.68	1.00	0.63

Parts of our model, such as the CAE network architecture and the output dimension of t-SNE, are universal to all machine learning problems. Thus, we mainly evaluate the effectiveness of three clustering-related hyperparameters, which are cutoff distance and thresholds for ρ and δ. The same set of hyperparameters are used across all datasets. Each time, we vary one hyperparameters over 6 possible choices. The range of hyperparameters and the range of predictions are given in Fig. 2.

For the comparison purpose, we run other cluster number determination algorithms on the MNIST dataset, the result is demonstrated in Table 3. Note the gap method and the sillhouette method require a range of number of clusters as input and in our experiments we set the number range as [2, 20].

In order to demonstrate the effectiveness of the proposed features, we extract different features via various method, and then take these features as input of CFSFDP to get the estimation of number of clusters of the UCI HAR dataset. Figure 3 illustrates the decision graphs of CFSFDP running on six different feature spaces, which consist of two 2-dimensional feature spaces extracted by the

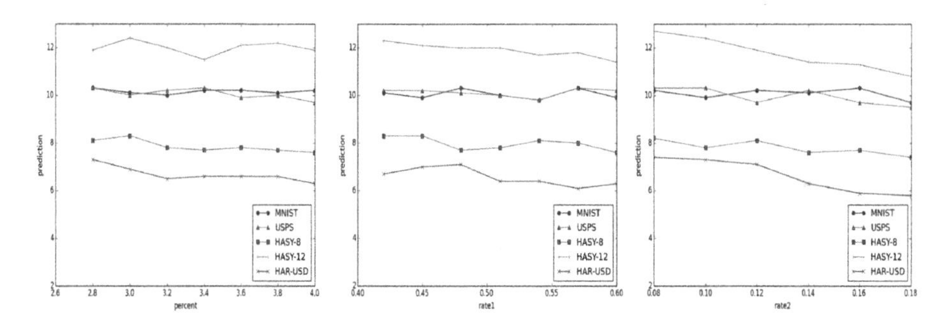

Fig. 2. Robustness analysis of three hyperparameters: we vary one parameter at a time with the other two fixed, and report the results of DED for five datasets.

Table 3. Comparison of average prediction of MNIST with different methods. "hit rate" represents the ratio of number of correct prediction and the number of all runs (i.e., 10).

Metric	Gap	Silhouette	DBSCAN	Mean shift	CFSFDP	DED
Mean	19.0 (±0)	2.0 (±0)	0 (±0)	1.0 (±0)	1.3*k* (±58.5)	10.1 (±0.3)
Hit rate	0/10	0/10	0/10	0/10	0/10	9/10

PCA technique and the tSNE algorithm respectively, three 10-dimensional feature spaces extracted by an autoencoder, the PCA technique and the proposed method respectively, and one raw feature space.

4.3 Implementation

For the sake of simplicity without loss of fairness, we apply the same deep convolutional autoencoder network on all datasets. Specifically, we set the encoder network structure to $conv^5_{32} \rightarrow conv^5_{64} \rightarrow conv^3_{128} \rightarrow FC_{10}$, where $conv^k_n$ denotes a convolutional layer with n filters and kernel size of k by k. A stride of length 2 is used as default. The decoder network is a mirror of the encoder. Except for input and output layers, all internal layers are activated by Relu nonlinearity function [23].

The CAE network is trained for 400 iterations and the minibatch size is set to 256. The Adam optimizer [22] with default learning rate is used. We set convergence threshold to $\delta = 0.1$ and the checkpoint interval to $T = 100$. All of the above parameters are shared across datasets and are set to achieve a reasonably good CAE reconstruction loss. Our implementation is based on Python and Keras [21]. Dataset specific tunings of these parameters definitely improve the performance on each dataset. But generally there is no validation set available in unsupervised learning scenario, so we avoid any unrealistic parameter tuning.

We set the number of components to 2 for the t-SNE algorithm. For the subsequent density-based clustering algorithm, the density is estimated with the

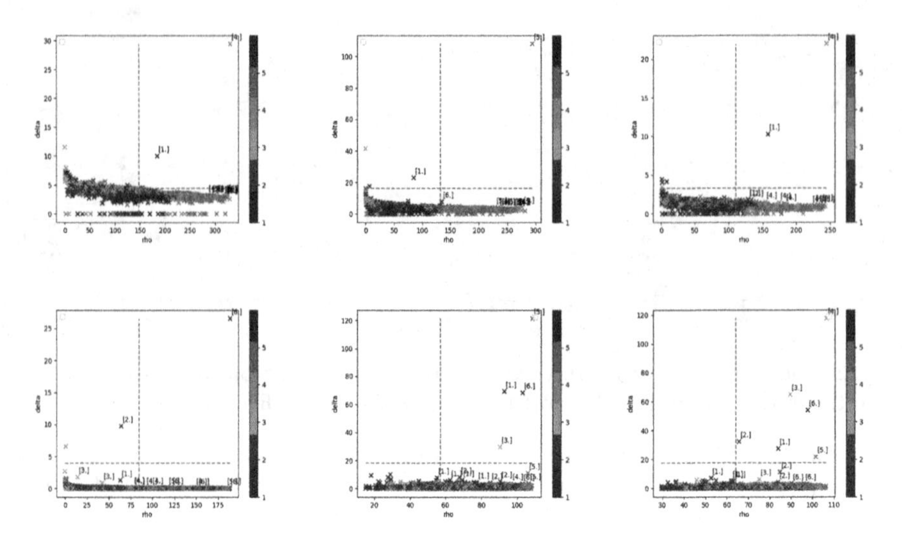

Fig. 3. Decision graphs of CFSFDP running on the raw feature, the 10-dim ae feature. The 10-dim pca feature, the 2-dim feature, the 2-dim tsne feature and the 10-dim DED feature spaces

cutoff distance percent as 4%. The density threshold and distance threshold are set to 0.55 and 0.18 respectively. All of the above hyperparameters are shared across datasets to achieve the best performance on Table 2.

4.4 Results

We show the effectiveness of DED on all datasets on Table 2. As we can see, the closest integer of the average number of clusters is the most stable statistics and always equal to the ground truth in our experiments. It also has the highest frequency of occurrence and accounts for 40–80% of total runs. Thus, we do not report the frequency of occurrence of other false predictions. The small variance indicates that the average is not likely to blur with neighboring integers. The robustness is significant to our clustering algorithm since cross validation is not feasible in real data for clustering analysis. In order to investigate the effect of hyperparameters, we vary one of three hyperparameters (cutoff distance σ, density threshold ρ and distance threshold δ) each time over 6 possible choices and report the mean statistic. Note that all hyperparameters are normalized values that range from 0 to 1. We investigate a broad range of each hyperparameter. As Fig. 2 shows, DED is robust with respect to these three hyperparameters on all datasets. Moreover, the optimal hyperparameter combination for each dataset varies slightly.

We report the performance of comparing methods on the MNIST dataset on Table 3. Note that a range of possible K value has to be provided to the silhouette and gap algorithms. As shown in Table 3, our method outperform all the other techniques to a great margin. The proposed feature space is better for the clustering analysis than other feature space, according to the Fig. 3. Only in the decision graph on the proposed features can we get the correct estimation of the number of clusters of the UCI HAR dataset, which means DED offers a good feature space for the further analysis of human activities in this dataset.

5 Conclusion and Future Work

This paper has introduced the Deep Embedding Determination (DED) algorithm, which automatically determines the number of clusters in large datasets of high dimension. DED first combines the virtues of the convolutional autoencoder and the t-SNE technique to extract low dimensional embedded features. Then it determines the number of clusters using an density-based clustering algorithm. Our experiments on different datasets validate the effectiveness of DED and show that our method outperforms other algorithms by a great margin in estimating the number of clusters. DED offers robustness with respect to hyper-parameter settings, which is important since cross-validation is impractical in real-world cluster analysis.

DED can be an important component complementary to many existing algorithms. When the number of clusters K is not given, DED can provide the number of clusters K of high dimensional datasets, which enables other deep clustering methods to train specific models. DED shows its great potential in the exploration of unknown real-world dataset. It not only offers a good estimation of the number of clusters, and provides good features for the further analysis. In the domain of human activities analysis, how to make use of these features is meaningful and remained to be explored.

Acknowledgments. This work was supported by the National Key R&D Program of China 2018YFB1003202 and the National Natural Science Foundation of China (Project no. 61773392, 61702539 and 61672528).

References

1. Xie, J., Girshick, R., Farhadi, A.: Unsupervised deep embedding for clustering analysis. In: International Conference on Machine Learning, pp. 478–487 (2016)
2. Guo, X., Gao, L., Liu, X., Yin, J.: Improved deep embedding clustering with local structure preservation. In: International Joint Conference and Artificial Intelligence (2017)
3. Jiang, Z., Zheng, Y., Tan, H., Tang, B., Zhou, H.: Variational deep embedding: an unsupervised and generative approach to clustering. In: International Joint Conference and Artificial Intelligence (2017)
4. Dizaji, K.G., Herandi, A., Huang, H.: Deep clustering via joint convolutional autoencoder embedding and relative entropy minimization. In: ICCV (2017)
5. Kaufman, L., Rousseeuw, P.: Finding Groups in Data: An Introduction to Cluster Analysis. Wiley, Hoboken (1990)
6. Tibshirani, R., Walther, G., Hastie, T.: Estimating the number of clusters in a data set via the gap statistic. J. R. Stat. Soc.: Ser. B (Stat. Methodol.) **63**, 411–423 (2001)

7. Ester, M., Kriegel, H.P., Sander, J., Xu, X., et al.: A density-based algorithm for discovering clusters in large spatial databases with noise. In: KDD, vol. 96, pp. 226–231 (1996)
8. Comaniciu, D., Meer, P.: Mean shift: a robust approach toward feature space analysis. IEEE Trans. Pattern Anal. Mach. Intell. **24**, 603–619 (2002)
9. Rodriguez, A., Laio, A.: Clustering by fast search and find of density peaks. Science **344**, 1492–1496 (2014)
10. LeCun, Y., Cortes, C., Burges, C.J.: MNIST handwritten digit database. AT&T Labs (2010)
11. Thoma, M.: The HASYv2 dataset. arXiv preprint arXiv:1701.08380 (2017)
12. Hull, J.J.: A database for handwritten text recognition research. IEEE Trans. Pattern Anal. Mach. Intell. **16**, 550–554 (1994)
13. Anguita, D., Ghio, A., Oneto, L., Parra, X., Reyes-Ortiz, J.L.: A public domain dataset for human activity recognition using smartphones. In: 21th European Symposium on Artificial Neural Networks, Computational Intelligence and Machine Learning, ESANN 2013 (2013)
14. MacQueen, J., et al.: Some methods for classification and analysis of multivariate observations. In: Proceedings of the Fifth Berkeley Symposium on Mathematical Statistics and Probability, vol. 1, pp. 281–297 (1967)
15. Bishop, C.M.: Pattern Recognition and Machine Learning. Springer, Heidelberg (2006)
16. Tian, F., Gao, B., Cui, Q., Chen, E., Liu, T.: Learning deep representations for graph clustering. In: AAAI, pp. 1293–1299 (2014)
17. Yang, J., Parikh, D., Batra, D.: Joint unsupervised learning of deep representations and image clusters. In: Proceedings of the IEEE Conference on Computer Vision and Pattern Recognition, pp. 5147–5156 (2016)
18. Masci, J., Meier, U., Cireşan, D., Schmidhuber, J.: Stacked convolutional auto-encoders for hierarchical feature extraction. In: Honkela, T., Duch, W., Girolami, M., Kaski, S. (eds.) ICANN 2011. LNCS, vol. 6791, pp. 52–59. Springer, Heidelberg (2011). https://doi.org/10.1007/978-3-642-21735-7_7
19. Maaten, L., Hinton, G.: Visualizing data using t-SNE. J. Mach. Learn. Res. **9**, 2579–2605 (2008)
20. Steinbach, M., Ertöz, L., Kumar, V.: The challenges of clustering high dimensional data. In: Wille, L.T. (ed.) New Directions in Statistical Physics, pp. 273–309. Springer, Heidelberg (2004). https://doi.org/10.1007/978-3-662-08968-2_16
21. Chollet, F., et al.: Keras: deep learning library for Theano and TensorFlow (2015). https://keras.io/k
22. Kingma, D., Ba, J.: Adam: a method for stochastic optimization. arXiv preprint arXiv:1412.6980 (2014)
23. Glorot, X., Bordes, A., Bengio, Y.: Deep sparse rectifier neural networks. In: Proceedings of the Fourteenth International Conference on Artificial Intelligence and Statistics, pp. 315–323 (2011)
24. Hornik, K.: Approximation capabilities of multilayer feedforward networks. In: Neural Networks, vol. 4, pp. 251–257. Elsevier (1991)
25. Bengio, Y., Courville, A., Vincent, P.: Representation learning: a review and new perspectives. IEEE Trans. Pattern Anal. Mach. Intell. **35**, 1798–1828 (2013)

Reinforcement Learning for a Novel Mobile Charging Strategy in Wireless Rechargeable Sensor Networks

Zhenchun Wei[1,2,3], Fei Liu[1], Zengwei Lyu[1(✉)], Xu Ding[1,2,3(✉)], Lei Shi[1,2,3], and Chengkai Xia[1]

[1] School of Computer Science and Information Engineering, Hefei University of Technology, No. 193 Tunxi Road, Hefei 230009, China
lvzengwei@mail.hfut.edu.cn, dingxu@bjmu.edu.cn
[2] Engineering Research Center of Safety Critical Industrial Measurement and Control Technology, Ministry of Education, Hefei, China
[3] Key Laboratory of Industry Safety and Emergency Technology, Hefei, Anhui, China

Abstract. The charging strategy for the mobile charger (MC) has been a hot research topic in wireless rechargeable sensor networks. We focus on the charging path for the MC, since the MC stops at each sensor node until the sensor node is fully charged. Most of the existing reports have designed optimization methods to obtain the charging path, with the target like minimizing the charging cost. However, the autonomous charging path planning for the MC in a changeable network is not taken into consideration. In this paper, Reinforcement Learning (RL) is introduced into the charging path planning for the MC in WRSNs. Considering the influences of the energy variation and the locations of the sensor nodes, a novel Charging Strategy in WRSNs based on RL (CSRL) is proposed so that the autonomy of the MC is improved. Simulation experiments show that CSRL can effectively prolong the lifetime of the network and improve the driving efficiency of the MC.

Keywords: Charging strategy · Wireless rechargeable sensor network
Reinforcement Learning

1 Introduction

In recent years, the wireless rechargeable sensor networks (WRSNs) has become a hot issue to solve the lifetime problem for the wireless sensor networks (WSNs) [1, 2]. Wireless chargers [3] are introduced into the WSNs. Most of the existing reports have designed optimization methods to obtain the charging path, with the target like minimizing the charging cost. However, with incomplete information, if the MC can autonomously plan and adjust the charging path, it will charge more efficiently in a changeable sensor network.

Reinforcement Learning (RL) [4] is a popular method of machine learning, which can be applied in many situations where the environment information is incomplete. The path planning makes the mobile robots autonomously learn to walk in the terrain

S. Chellappan et al. (Eds.): WASA 2018, LNCS 10874, pp. 485–496, 2018.
https://doi.org/10.1007/978-3-319-94268-1_40

with obstacles, which is an important branch of RL applications. There are few reports about the charging path planning in WRSNs based on RL. In this paper, the Markov Decision Process (MDP) is used to describe the charging planning process for the MC. To prolong the lifetime of the network and improve the driving efficiency, two main factors are considered. One is the energy distribution of the network caused by the different energy consumption of the sensor nodes. The other is the influence of sensor nodes' locations on the MC's driving efficiency. The main contributions of this paper are as follows.

1. It is the first time to propose a charging strategy in WRSNs from the respect of Reinforcement Learning. According to the current state of the network, the MC continuously explores and hunts for its charging target so that it can autonomously obtain a charging path in WRSNs.
2. This paper improves the exploration and exploitation of the state-action value matrix Q, and proposes the evaluation mechanism of charging strategy. Based on the evaluation value, the state-action value matrix Q generated by an exploration period can be properly exploited.
3. We demonstrate the effectiveness of CSRL proposed compared with the following two algorithms. One is the algorithm without the evaluation mechanism. The other is the greedy algorithm where the MC always chooses the lowest-energy sensor node to charge. The results indicate that CSRL outperforms the two algorithms.

2 Related Works

2.1 Charging Strategies in WRSNs

The research on the charging strategy for the MC in WRSNs is a hot issue. According to the mobility of wireless chargers, there are two methods for different applications, the deployment of static chargers and the charging path planning for mobile chargers. For example, the static chargers are used in the changeable environments such as the jungle or the other occasions where they are hard for chargers to move. However, the mobile chargers (MCs) are suitable for a WRSN distributed in a relatively flat terrain. In addition, the researchers have proposed the single MC charging strategy for the small-scale sensor networks, and the multi-MCs [5] charging strategy for the larger-scale sensor networks.

Most of the researches of charging path planning by optimization methods can improve some performance of the network to some extent. Xu et al. [6] proposed MMES-LME based on an improved MAX-MIN ants system, and then proposed an equalization strategy constrained by a mobile charger. Based on the Wireless Identification and Sensing Platform (WISP), Fu et al. [7] narrowed the search space to the smallest enclosed space (SES), and then obtained the charging path for RFID Reader by discretizing the charging power and clustering the anchors. Liang et al. [8] proposed two approximate algorithms to maximize the total charging rewards in two different scenarios. It is the first time that Rao et al. [9] considered the influence of the charging distance and charging angle on the charging efficiency. They obtained the Hamiltonian path by Concorde TSP solver to find the best charging position for each sensor node. All the above reports proved superiority of the proposed algorithms with full simulation experiments. However, few of them improve the autonomy of the MCs.

2.2 Reinforcement Learning for Path Planning

RL is an important branch of Learning Algorithm. Through the interaction and trial with the environment, the agent makes use of the reward signal of evaluation to optimize the planning. Introducing RL into the path planning for mobile robots, the robots, treated as the agents, has its own "brain" to plan paths in a changeable environment. According to most of the studies, the optimal path means a collision-free path meeting the pre-set conditions.

To adapt to the changeable network environment [10, 11] and improve the autonomous planning capability of the MC in WRSNs, this paper combines RL with charging strategy in WRSNs.

3 Network Model and Problem Description

The structure model of the WRSN in this paper is shown in Fig. 1. N sensor nodes, a charging service station (CS), a base station (BS) and a MC are deployed over a 2-D monitored area [12, 13]. The locations of each sensor node, the CS and the BS are fixed and known. The MC replenishes energy for itself and learns charging strategy at the CS.

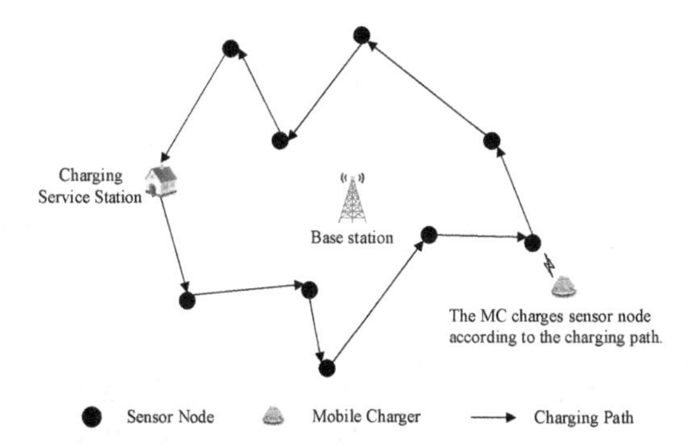

Fig. 1. Structure model of the WRSN with a MC

Definition 1. Charging Round: After learning the charging path planning, the MC sets off from the CS to charge all sensor nodes in the network, and then returns to the CS. This period is recorded as a charging round.

Definition 2. Vacation Period: The MC stays at the CS for resting and learning.

In this paper, the autonomous path planning in WRSNs is to prolong the lifetime of the network and improve the driving efficiency of the MC. It is assumed that the vacation time of the MC is certain, and the energy carried by the MC satisfies all the energy consumption in a charging round. Based on RL, an effective charging strategy for the MC is obtained while the MC is staying at the CS, where the network corresponds to the environment, and the MC corresponds to the agent in RL.

4 Learning Model Construction

In this paper, the charging strategy for the MC in WRSNs can be represented by a quadruple $<\mathcal{S}, \mathcal{A}, \mathcal{P}, \mathcal{R}>$. $\mathcal{S}$ is the state space, which represents the state set of the MC. $\mathcal{A}$ is the action space, which represents the actions set of the MC. $\mathcal{P}$ is the learning policy of the agent, which represents the exploration-exploitation policy of the charging path. $\mathcal{R}$ is the reward of the agent, which represents the reward generated by the MC's actions.

Definition 3. Time Step: At the time that the ME arrives at a sensor node.

At time step k, the Q-value $Q(s^k, a^k)$ of state-action-pair (SAP) (s^k, a^k) represents the value of action a^k when the MC is in state s^k.

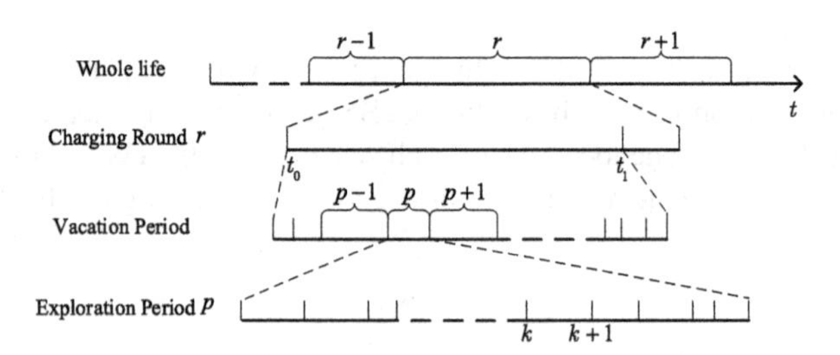

Fig. 2. Time-series plots of the charging strategy in WRSNs based on RL

As shown in Fig. 2, in each charging round, the MC learns the charging strategy in the vacation period before it leaves from the CS at time t_1. A vacation period contains several exploration periods. During each exploration period, the MC determines action a^k according to the state s^k at time step k. Through the exploration-exploitation policy, the MC chooses action a^k, receives reward r^{k+1}, and obtains a charging strategy at the end of this exploration period. Finally, an effective charging strategy for MC in WRSNs based on RL is achieved.

4.1 State Model

In this paper, the state space of the MC consists of its own state and the state of each sensor node in the network. The state space $\mathcal{S}$ is defined as a two-tuple.

$$\begin{aligned} \mathcal{S} &= \langle \mathcal{S}_{MC}, \mathcal{S}_{network} \rangle \\ \mathcal{S}_{MC} &= \{ s_{MC} | s_{MC} \in \mathcal{N} \} \\ \mathcal{S}_{network} &= \{ (\overrightarrow{e}, \overrightarrow{d}, \overrightarrow{visited}) \} \end{aligned} \tag{1}$$

In the Eq. 1, $\mathcal{N} = \{0, 1, \cdots, N, N+1\}$. $Node = \{n_1, n_2, \cdots, n_N\}$ represents the set of N sensor nodes. $\mathcal{S}_{MC}$ represents the set of the current location states of the MC. $s_{MC} = 0$ means the MC is at the CS while no sensor node has been visited. $s_{MC} = i(i \in \{1, 2, \cdots, N\})$ means the MC is at sensor node n_i. $s_{MC} = (N+1)$ means the MC is at the CS while all the sensor nodes have been visited. $\mathcal{S}_{network}$ represents the set of the current states of the sensor network, which consists of three parts, the current energy states of the network $\overrightarrow{e} = (e_0, e_1, \cdots, e_{N+1})$ as show as Eq. 2, the distance between the MC and the sensor nodes or the CS $\overrightarrow{d} = (d_0, d_1, \cdots, d_{N+1})$, and the flag $\overrightarrow{visited} = (visited_0, visited_1, \cdots, visited_{N+1})$ whether the sensor node is visited.

$$e_j = \begin{cases} E_j/E_{\max}, j \in \{1, 2, \cdots N\} \, and \, E_{\min} \leq E_j \leq E_{\max} \\ E_{CSmin}/E_{\max}, j = 0 \, and \, 0 < E_{CSmin} < E_{\min} \\ E_{CSmax}/E_{\max}, j = N+1 \, and \, E_{CSmax} > E_{\max} \end{cases} \tag{2}$$

$E_{\max}$ is denoted as the maximum capacity of the sensor node's battery, and sensor nodes should work above $E_{\min}$. E_j is the remaining energy of sensor node n_j. It is required that the MC must set out from the CS and return to the CS until all sensor nodes have been visited, so $0 < E_{\mathrm{CSmin}} < E_{\min}$ when $j = 0$, and $E_{\mathrm{CSmax}} > E_{\max}$ when $j = N+1$. If $j \in \{1, 2, \cdots N\}$, e_j means the current energy state of sensor node n_j, and d_j means the distance between the MC and sensor node n_j. If $i = 0 \, or \, (N+1)$, e_j means the "energy state" of the CS, and d_j means the distance between the MC and the CS.

To ensure that all sensor nodes are visited only once in each charging round, the flag $visited_j$ is set to 1 after sensor node n_j has been visited and charged by the MC.

Definition 4. Terminating State: The state at the end of a charging round is called terminating state. The state is the time when the MC returns to the CS at the time step. The time step is called terminating time step K. The terminating state is defined as $\mathcal{S}^K = \langle \mathcal{S}^K_{MC}, \mathcal{S}^K_{network} \rangle, s^K_{MC} = N+1$. At this moment, all the sensor nodes have been visited once. That is, $\overrightarrow{visited} = (1, 1, \cdots, 1)$.

4.2 Action Model

As for the MC, it is assumed that all sensor nodes are reachable in the network. That is, the MC can choose any sensor node whose flag is 0. In this paper, the action space is defined as a two-tuple.

$$\begin{aligned} &\mathcal{A} = \langle \mathcal{A}_{Move}, \mathcal{A}_{Charge} \rangle \\ &\mathcal{A}_{Move} = \{a_{Move} | a_{Move} \in \{1, 2, \cdots \mathrm{N}+1\}\} \\ &\mathcal{A}_{Charge} = \{a_{Charge} | a_{Charge} \in \{1, 2, \cdots \mathrm{N}+1\}\} \end{aligned} \tag{3}$$

In Eq. 3, We denote $a = a_{Move} = a_{Charge}$. The MC will return to the CS to prepare for the next charging round when $a = (N+1)$, and visit and fully charge sensor node n_i when $a = i(i \in \{1, 2, \cdots, N\})$.

4.3 Reward Model

According to the Sects. 4.1 and 4.2, the MC executes action a^k in state s^k, which means the MC visits and fully charges sensor node $n_i(i = a^k)$.Therefore, the MC's state $s_{MC}^{k+1} = a^k$ at time step $(k+1)$. In addition, the current energy states $\overrightarrow{e^{k+1}} = (e_0^{k+1}, e_1^{k+1}, \cdots, e_{N+1}^{k+1})$ and the distance states $\overrightarrow{d^{k+1}} = (d_0^{k+1}, d_1^{k+1}, \cdots, d_{N+1}^{k+1})$ are both updated. To calculate the $e_j^{k+1}(j \in \{1, 2, \cdots, N\})$ according to Eq. 2, the remaining energy E_j^{k+1} of sensor node n_j at time step $(k+1)$ is as Eq. 4.

$$E_j^{k+1} = \begin{cases} E_{\max}, j = s_{MC}^{k+1} \\ E_{\max} - p_j \Delta t^{k+1}, j = s_{MC}^k \\ E_j^k - p_j \Delta t^{k+1}, j \neq s_{MC}^k \, and \, j \neq s_{MC}^{k+1} \end{cases} \tag{4}$$

p_j is the energy consumption rate of sensor node n_j. Δt^{k+1} represents the period when the MC leaves from sensor node $n_i(i = s_{MC}^k)$ and then fully charges sensor node $n_i(i = s_{MC}^{k+1})$. When $j = s_{MC}^{k+1}$, the remaining energy of sensor node n_j is $E_{\max}$. When $j = s_{MC}^k$, the remaining energy of sensor node n_j is $\left(E_{\max} - p_i \Delta t^{k+1}\right)$ because it is fully charged at time step k. The remaining energy of the other sensor nodes $n_j(j \neq s_{MC}^k \, and \, j \neq s_{MC}^{k+1})$ is $\left(E_j^k - p_j \Delta t^{k+1}\right)$ because they have not been charged.

Δt^{k+1} consists of two parts as shown in Eq. 5, the driving time Δt_{dr}^{k+1} from sensor node $n_i(i = s_{MC}^k)$ to $n_i(i = s_{MC}^{k+1})$, and the charging time Δt_{ch}^{k+1} for sensor node $n_i(i = s_{MC}^{k+1})$.

$$\Delta t^{k+1} = \Delta t_{dr}^{k+1} + \Delta t_{ch}^{k+1} \tag{5}$$

$$\Delta t_{dr}^{k+1} = d_i^k / v, \quad i = s_{MC}^{k+1} \tag{6}$$

$$\Delta t_{ch}^{k+1} = \frac{E_{\max} - \left(E_i^k - p_i \Delta t_{dr}^{k+1}\right)}{u - p_i}, \quad i = s_{MC}^{k+1} \tag{7}$$

It is assumed that v is the driving speed and u is the charging power of the MC. $\left(E_i^k - p_i \Delta t_{dr}^{k+1}\right)$ represents the remaining energy of sensor node $n_i(i = s_{MC}^{k+1})$ at the time that the MC arriving at it.

The reward r^{k+1} of the MC can be calculated by the reward function $r^{k+1} = \rho(s^k, a^k, s^{k+1})$, while the MC executes action a^k in state s^k at the time step k, and the MC's state transfers to the next state s^{k+1}. In this part, the following two points should be taken into consideration.

1. To avoid the death of the network because of low-energy sensor node, the MC preferentially charges lower-energy sensor nodes as much as possible.
2. To improve the driving efficiency of the MC, the closer sensor nodes have higher precedence to be charged.

Therefore, the reward r^{k+1} is defined as follows.

$$\rho(s^k, a^k, s^{k+1}) = \alpha_1 \times \frac{1}{e_i^{k+1}} + \alpha_2 \times \frac{K}{d_i^k}, \quad i = s_{MC}^{k+1} \tag{8}$$

$\alpha_1(0 < \alpha_1 \leq 1)$ and $\alpha_2(0 < \alpha_2 \leq 1)$ respectively represent the proportions of the two factors in the reward function, the energy factor of the next-hop sensor node and the distance factor, where $\alpha_1 + \alpha_2 = 1$. K represents the unit distance value. e_j^{k+1} represents the energy state of sensor node n_i or the "energy state" of CS at the time step $(k+1)$, which can be calculated by Eq. 2.

5 CSRL: A Charging Strategy Based on RL

5.1 Exploration-Exploitation Policy

Simulated Annealing (SA) [14] is used to select the action. To achieve a superior charging strategy quickly, the greater the Q-value is, the more likely the action is to be selected, especially in the beginning inexperienced circumstances. According to the Metropolis Algorithm, the selecting probability is shown in Eq. 9.

$$\begin{aligned} p^k &= \exp[\frac{Q^k(s^k, a^k) - Q^k(s^k, (a^k)')}{T^k}] \\ T^k &= (\lambda)^k T_0 \end{aligned} \tag{9}$$

T_0 is the initial temperature, and $\lambda(0 < \lambda < 1)$ is the adjustment parameter.

In the learning period of the MC, the action in a state and its corresponding reward are uncertain. Therefore, the update of the Q-value is as Eq. 10.

$$Q(s^k, a^k) = Q(s^k, a^k) + \eta(r^{k+1} + \gamma \max_{a^{k+1}} Q(s^{k+1}, a^{k+1}) - Q(s^k, a^k)) \tag{10}$$

where η is the update factor, and γ $(0 < \gamma < 1)$ is the discount factor. r^{k+1} is the reward at time step $(k+1)$. According to the Sect. 4.3, the reward function is related to the energy and the distance of the target sensor node pointed by the action. Therefore, the reward function updates the corresponding Q-value, which measures the value when the action is performed in one state.

During the learning period of MC, the Q-value matrix is updated by Eq. 10 at the end of exploration period p, and the MC obtains a charging strategy π_p at the same time. However, the update of the Q-value matrix is limited because the update process is not able to evaluate the whole charging strategy π_p. If π_p makes the network's lifetime longer or the MC's charging efficiency better, we hope exploration period p have a greater influence on the update of the Q-value. Therefore, an evaluation mechanism is proposed for every period of the exploration. The evaluation indicator can be obtained at the end of exploration period p according to Eq. 11.

$$V_p = \beta_1 \times \frac{E_{ave}}{E_{\max}} + \beta_2 \times \frac{K}{d} \tag{11}$$

where $\beta_1 + \beta_2 = 1$. $\beta_1 (0 < \beta_1 \leq 1)$ and $\beta_2 (0 < \beta_2 \leq 1)$ represent the proportions of the energy state of the next-hop sensor node and the total driving distance d of the MC. K represents the unit distance value. E_{ave} represents the average of the remaining energy of all sensor nodes in the network.

The evaluation indicator is used to evaluate the charging strategy obtained by exploration period p, so as to evaluate the weight of the updated Q-value.

$$Q_p^k = Q_{p-1}^k + V_{p-1} \times \Delta Q_{p-1}^k \tag{12}$$

where $Q_p = 0$ if p is the first exploration period. Apart from this, the Q-value matrix in exploration period p is based on the former exploration period. In addition, the matrix generated in this exploration period is not directly updated into the matrix of the previous period, instead, updated to a new matrix ΔQ_p of the same order. At the end of exploration period p, evaluation indicator V_p and the new matrix ΔQ_p will be added to the final Q-value matrix Q_p.

5.2 Algorithm Description

As for the charging strategy based on RL proposed in this paper, the learning model of the MC is shown in Fig. 3. A charging path and a matrix ΔQ are generated in each exploration period. To avoid converging to the local optimum, the reasonable evaluation indicator V is designed. V and ΔQ are combined to get the initial matrix Q of the next exploration period. The MC obtains an effective charging strategy according to the state and the reward.

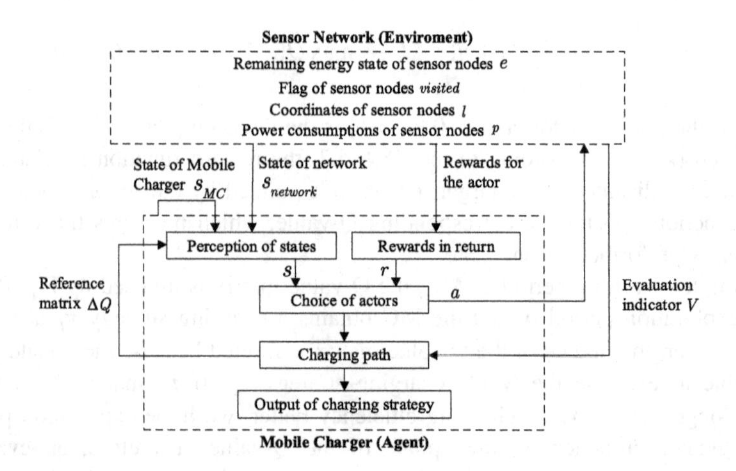

Fig. 3. Learning model of the MC

The algorithm of the charging strategy based on RL is described as follows.

CSRL Algorithm

//**input:** the remaining energy $e_i(t_0)(j \in \{1, 2, \cdots, N\})$ at time t_0, and the related parameters about the WRSNs
//**output:** the charging strategy for the MC
1. **initialization:** t_0, Δt, $maxepisode$;
2. Calculate the time MC leaves from the CS according to $t_1 = t_0 + \Delta t$;
3. $p \leftarrow 0$;
4. repeat
5. $p \leftarrow p + 1$;
6. Calculate the $\vec{e}$ according to Eq. 2, and record time t_1 as time step $k \leftarrow 0$;
7. Initialize state s^k, $s^k_{MC} \leftarrow 0$;
8. repeat
9. Execute action a^k obtained according to SA;
10. Calculate the new state s^k;
11. Calculate the reward r^{k+1} according to Eq. 8;
12. Calculate the matrix ΔQ^k_p according to Eq. 10;
13. $k \leftarrow k + 1$;
14. **until** $k = N + 1$
15. Calculate V_p according to Eq. 11;
16. Update Q_p according to Eq. 12;
17. **until** $p = maxepisode$
18. Obtain a charging path;
19. **if** time $t = t_1$ **do**
20. The MC sets off from the CS, and charges each sensor node;
21. end if

6 Simulation Analysis

Twenty sensor nodes are deployed over a 1000 m × 1000 m area randomly, in which a BS is located at (500 m, 500 m) and the CS (0 m, 0 m). The parameters are set as follows. $E_{max} = 10800$ J. $E_{min} = 540$ J. The energy consumption rate p_i is a random number varying from 0.05 W ~ 1.5 W. The driving speed $v = 10$ m/s. The charging power of the MC $U = 10$ W. Maximum exploration iterations $maxepisode = 100$. $\alpha_1 = \beta_1 = 0.6$, $\alpha_2 = \beta_2 = 0.4$, $\eta = 0.8$, $\gamma = 0.4$, $T_0 = 100$, $\lambda = 0.98$.

The charging strategy proposed in this paper is called CSRL, which is compared with the following two algorithms. (1) CSRL without the evaluation mechanism; (2) the greedy algorithm for charging the lowest-energy sensor node every time.

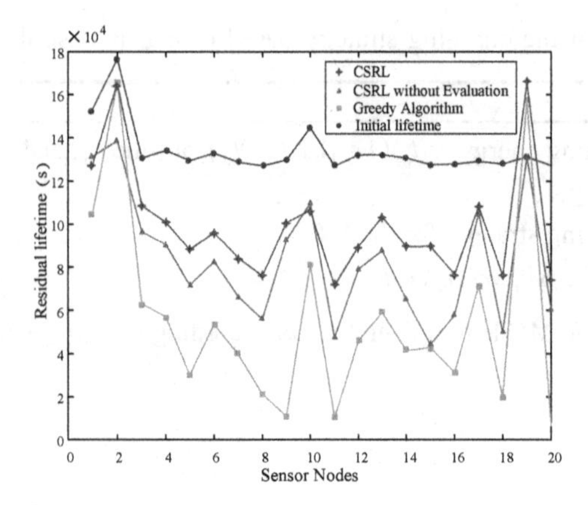

Fig. 4. Comparison of the remaining lifetime of 20 senor nodes of the three algorithm

The remaining lifetime of the twenty sensor nodes of these three algorithms is shown in Fig. 4. These three algorithms have run for 100 times independently. The purple dots in Fig. 4 represent the initial lifetime of the twenty sensor nodes. Obviously, the CSRL performs well than the others.

The mean, variance, minimum and maximum of the sensor nodes' remaining lifetimes are shown in Table 1. Compared with CSRL without the evaluation mechanism, the mean remaining lifetime obtained by CSRL is increased by 26.80% and variance is reduced by 22.58%. The minimum and maximum of remaining lifetimes in CSRL are both higher than the other two algorithms. Therefore, CSRL is effective in improving the lifetime of sensor nodes and the balance of the remaining lifetime of sensor nodes in the network.

Table 1. Experimental results of remaining lifetime of sensor nodes in the network

Algorithms	Mean (s)	Variance (s^2)	Minimum (s)	Maximum (s)
CSRL	9.7450×10^4	7.2398×10^8	6.9800×10^4	1.6550×10^5
CSRL without evaluation	7.7045×10^4	9.3524×10^8	3.5600×10^4	1.3880×10^5
Greedy algorithm	5.5040×10^4	1.8741×10^9	6.8100×10^3	1.6250×10^5

In this paper, the MC's driving efficiency is measured by the MC's total driving distance in a charging round. In a charging round, the MC's total driving distance for 100 experiments of the three algorithms are shown in Fig. 5. The red line shows the median of the results for each algorithm. The black horizontal lines at the top and bottom indicate the maximum and minimum of the results respectively. The red cross represents the outlier of the results. From the distribution in Fig. 5, CSRL performs better than the other two algorithms.

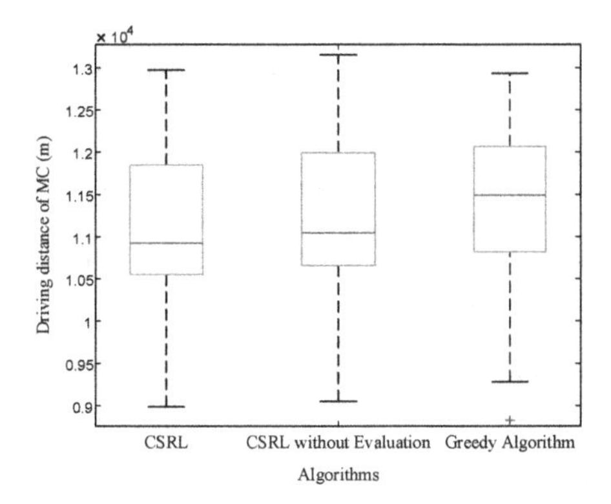

Fig. 5. Comparison of the MC's total distance in a charging round of the three algorithms (Color figure online)

In general, CSRL in this paper can prolong the remaining lifetime of the sensor nodes and reduce the total driving distance of the MC.

7 Conclusion

To our best knowledge, it is the first time to propose the charging strategy in WRSNs based on Reinforcement Learning. The energy factors and the location factors are considered to evaluate the charging path in each exploration, with the discretization of the state space and the action space. Simulation experiments show that the proposed algorithm is excellent in prolonging the lifetime of the network and improving the driving efficiency of the MC, compared with CSRL without the evaluation mechanism and the greedy algorithm. In the future work, the on-line charging strategy in WRSNs based on RL is considered, where the energy consumptions of sensor nodes are time-variant.

References

1. Chen, Q., Gao, H., Cai, Z., Cheng, L., Li, J.: Energy-collision aware data aggregation scheduling for energy harvesting sensor networks. In: International Conference on Computer Communications. IEEE, Honolulu (2018)
2. Shi, Y., Xie, L., Hou, Y.T., Sherali, H.D.: On renewable sensor networks with wireless energy transfer. In: The 30th IEEE International Conference on Computer Communications, pp. 1350–1358. IEEE, Shanghai (2011)
3. Kurs, A., Karalis, A., Moffatt, R., Joannopoulos, J.D., Fisher, P., Soljai, M.: Wireless power transfer via strongly coupled magnetic resonances. Science **317**(5834), 83–86 (2007)
4. Yau, K.L.A., Goh, H.G., Chieng, D., Kwong, K.H.: Application of reinforcement learning to wireless sensor networks: models and algorithms. Computing **97**(11), 1045–1075 (2015)

5. Liang, W., Xu, W., Ren, X., Jia, X., Lin, X.: Maintaining large-scale rechargeable sensor networks perpetually via multiple mobile charging vehicles. ACM Trans. Sens. Netw. **12**(2), 1–26 (2016)
6. Xu, J., Yuan, X., Wei, Z., Han, J., Shi, L., Lyu, Z.: A wireless sensor network recharging strategy by balancing lifespan of sensor nodes. In: Wireless Communications and Networking Conference, pp. 1–6. IEEE, San Francisco (2017)
7. Fu, L., Cheng, P., Gu, Y., Chen, J., He, T.: Optimal charging in wireless rechargeable sensor networks. IEEE Trans. Veh. Technol. **65**(1), 278–291 (2016)
8. Liang, W., Xu, Z., Xu, W., Shi, J., Mao, G., Das, S.K.: Approximation algorithms for charging reward maximization in rechargeable sensor networks via a mobile charger. IEEE/ACM Trans. Netw. **25**(5), 1–14 (2017)
9. Rao, X., Yang, P., Yan, Y., Zhou, H., Wu, X.: Optimal recharging with practical considerations in wireless rechargeable sensor network. IEEE Access **5**(99), 4401–4409 (2017)
10. Cai, Z., Zheng, X.: A private and efficient mechanism for data uploading in smart cyber-physical systems. IEEE Trans. Netw. Sci. Eng. **5**(1), 1–9 (2018)
11. Zheng, X., Cai, Z., Li, Y.: Data linkage in smart IoT systems: a consideration from privacy perspective. IEEE Commun. Mag. **10**(2), 12–20 (2018)
12. Liang, Y., Cai, Z., Yu, J., Han, Q., Li, Y.: Deep learning based inference of private information using embedded sensors in smart devices. IEEE Netw. Mag. **5**(8), 33–43 (2018)
13. Shi, T., Cheng, S., Cai, Z., Li, Y., Li, J.: Exploring connected dominating sets in energy harvest networks. IEEE/ACM Trans. Netw. **5**(12), 1–15 (2017)
14. Goudarzi, S., Wan, H.H., Anisi, M.H., Soleymani, S.A.: MDP-based network selection scheme by genetic algorithm and simulated annealing for vertical-handover in heterogeneous wireless networks. Wirel. Pers. Commun. **92**(2), 399–436 (2017)

A Multi-objective Algorithm for Joint Energy Replenishment and Data Collection in Wireless Rechargeable Sensor Networks

Zhenchun Wei[1,2,3], Liangliang Wang[1], Zengwei Lyu[1(✉)], Lei Shi[1,2,3(✉)], Meng Li[1], and Xing Wei[1,2,3]

[1] School of Computer Science and Information Engineering, Hefei University of Technology, Hefei, China
lvzengwei@mail.hfut.edu.cn, thunder10@163.com
[2] Engineering Research Center of Safety Critical Industrial Measurement and Control Technology, Ministry of Education, Hefei, China
[3] Key Laboratory of Industry Safety and Emergency Technology, Hefei, Anhui Province, China

Abstract. In the existing researches on the Wireless Rechargeable Sensor Networks (WRSNs), the charging path is scheduled firstly, and then the method of data collection is decided based on the path, which fails to ensure the high charging service quality and the performance of data collection. To solve this problem, a multi-objective path planning optimization model is proposed with the objectives of maximizing the remaining lifespan of sensor nodes and the amount of data collection. To deal with it, a Multi-Objective Discrete Fireworks Algorithm (MODFA) based on grid is proposed in this paper. Simulation results show that the algorithm proposed has better performance than NSGA-II, SPEA-II and MOEA/D in term of the diversity and convergence of Pareto front.

Keywords: Wireless Rechargeable Sensor Networks · Energy replenishment Data collection · Path planning · Multi-objective discrete fireworks algorithm

1 Introduction

The energy problem has been a key issue for Wireless Sensor Networks (WSNs). In recent years, with the development of wireless energy transmission technology, the appearance of WRSNs [1] using wireless charging technology and mobile wireless charger has solved this problem and provided good research prospects. As the data collection and transmission [2, 3] are the most important functions of WSNs, some researchers treated the Mobile Charger (MC) as a sort of mobile sink. Therefore, according to whether MCs are responsible for data collection, MCs can be divided into two categories. One is only responsible for charging, and the other is responsible for data collection and energy replenishment.

To solve the energy replenishment in WRSNs, Shi et al. [4] proved that maximizing the ratio of MC's vacation time is to minimize the traveling distance. Xie et al. [5] proposed that the MC can charge multiple sensor nodes in a limited range at the same time and jointly optimized the MC charging strategy. Reference [6] studied how to

S. Chellappan et al. (Eds.): WASA 2018, LNCS 10874, pp. 497–508, 2018.
https://doi.org/10.1007/978-3-319-94268-1_41

minimize the number of MCs to keep the network working when the MCs carry with limited energy.

The researches of joint energy replenishment and data collection in WRSNs, Guo et al. [7] scheduled the traveling path according to the remaining energy of sensor nodes and optimized the data generation rate of sensor nodes, link rate, and the sojourn time at the anchor point to maximize the network utility. Chen et al. [8] studied the data aggregation scheduling of energy-collision aware in the network. Shi et al. [9] studied how to find the largest number of connected dominating sets to ensure the extension of the network life. At the same time, it is proved that the problem is NP-Complete, and the four approximate algorithms are verified by algorithm simulation. The MC in [10] followed a predetermined path and aimed to reduce energy consumption by ensuring that sensor nodes can be replenished energy in time. References [11, 12] studied kernel dataset from big sensory data to improve network performance. Because of the high cost of multi-MC used in the past, Tan et al. [13] proposed a new MC scheduling algorithm based on periodic energy replenishment and data collection with the limited number of MC to maintain the normal operation of the network.

In the above researches that only considered energy replenishment, MCs were not considered for data collection to improve network performance. Most of the articles about the joint energy replenishment and data collection scheduled a traveling path and then optimized the data collection without joint consideration of energy replenishment and data collection. Therefore, in view of the above shortcomings, we not only consider the lifespan of sensor nodes in the network, but also consider the improvement of the MC performance in data collection.

Because the MC visits the sensor nodes only once, the different order of the visited sensor nodes will results in the different remaining energy and the amount of data collection, which cannot guarantee the remaining lifespan of sensor nodes and the amount of data collection in the network obtain maximum at the same time. The remaining lifespan of the sensor nodes and the amount of data collection have a conflicting relation, so this is a multi-objective optimization problem. In this paper, a Multi-Objective Discrete Firework Algorithm (MODFA) based on grid is proposed to solve this problem. The main contributions of this paper are summarized as follows:

- We propose a multi-objective MC path planning model that maximizes the average remaining lifespan and the amount of data collection of sensor nodes in the network.
- We firstly propose a MODFA to solve the path planning problem of MC, and use a grid-based partition to filter the Pareto optimal solution set. The simulation results show that the algorithm has better performance than NSGA-II, SPEA-II and MOEA/D.

The remaining of this paper is organized as follows. We introduce the network model in Sect. 2. We describe the multi-objective optimization problem model and two objective functions in Sect. 3. We present the specific algorithm in Sect. 4. Simulation results and analysis are given in Sect. 5. We conclude the work in Sect. 6.

2 Network Model

N sensor nodes are distributed over a 2-D monitored area in a WRSN. The specific positions of sensor nodes are known and fixed. There is a MC used for energy replenishment and data collection. A service station is used for energy replenishment and data uploading by MC. As shown in Fig. 1, there is no fixed sink in the network. All the sensor nodes are equipped with wireless rechargeable batteries which share the same capacity. The maximum capacity of each battery is denoted as E_{max}. To keep sensor nodes working properly, the minimum capacity of each battery is denoted as E_{min}. The MC with sufficient energy sets off from the service station, visits the network to replenish energy for sensor nodes and collect data simultaneously. When all the sensor nodes are visited, the MC will come back to the service station.

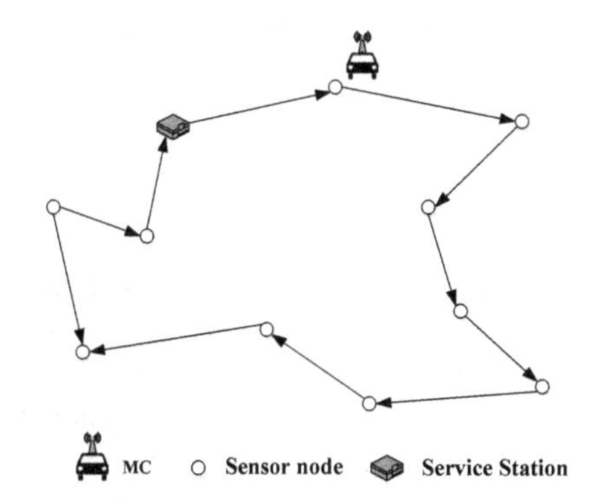

Fig. 1. A WRSN with a MC and a service station

3 Model Construction

3.1 Working Cycle

The path in which the MC visits in one cycle includes the service station and all sensor nodes. Denote Q as the traveling path, where π_0 is the service station and $\pi_i (1 \leq i \leq N)$ is the ith sensor node in the traveling path. So Q can be denoted as

$$Q = (\pi_0, \pi_1, \cdots, \pi_N, \pi_0). \tag{1}$$

The MC travels at a speed V in the network. The time it takes for the MC to visit between two adjacent sensor nodes or between the sensor nodes and the service station in a working cycle is denoted as

$$\tau_{ij}^{d} = \frac{d_{\pi_i \pi_j}}{V}, \tag{2}$$

where $d_{\pi_i \pi_j}$ represents the distance between two adjacent sensor nodes or sensor nodes and the service station in one cycle.

3.2 Strategy of Energy Replenishment and Data Collection

It is assumed that sensor node $i(1 \leq i \leq N)$ generates data at a constant speed of r_i with a constant energy consumption power p_i. Obviously, the remaining energy of sensor node i at time t is denoted as $e_i(t)$ which satisfies

$$E_{min} \leq e_i(t) \leq E_{max}. \tag{3}$$

Therefore, the remaining lifespan of sensor node i at time t is denoted as

$$L_i(t) = \frac{e_i(t) - E_{min}}{p_i}. \tag{4}$$

It is assumed that the MC arrives at sensor node i at time t_i, and the charging power is denoted as U. The charging time that the MC takes to replenish energy to E_{max} for sensor node i is denoted as

$$\tau_i^c = \frac{E_{max} - e_i(t_i)}{U - p_i}. \tag{5}$$

During the process of visiting sensor nodes, the MC not only replenishes energy for sensor nodes but also collects data generated by sensor nodes. When the MC is transmitting data with the sensor node, the transmission rate between the MC and the sensor nodes is denoted as r. It is supposed that the data amount of the sensor node i is denoted as $c_i(t)$ at time t and the $c_i(t)$ at any moment does not exceed the maximum data storage amount of the sensor node. When the MC collects all the data of the sensor nodes, the transmission time is denoted as

$$\tau_i^t = \frac{c_i(t_i)}{r - r_i}. \tag{6}$$

The sojourn time of the MC in each sensor node i, on the one hand, needs to consider the energy replenishment time of the sensor node. On the other hand, it needs to consider data transmission time of the sensor node. Therefore, we need to consider both the energy replenishment time and data transmission time. That is to say that the sojourn time of the MC at sensor node i is $\tau_i = max\{\tau_i^c, \tau_i^t\}$.

According to the sojourn time of the MC at each sensor node and the traveling time between two sensor nodes, we can get the time when the MC visits each sensor node i, shown as

$$t_i = \sum_{i=0}^{j-1} \frac{d_{\pi_i \pi_{i+1}}}{V} + \sum_{i=1}^{j-1} \tau_i. \tag{7}$$

After all the sensor nodes have been visited by the MC, the time when the MC starts from the service station and returns to the service station after visiting each sensor node is denoted as

$$T = \sum_{i=0}^{N-1} \frac{d_{\pi_i \pi_{i+1}}}{V} + \frac{d_{\pi_N \pi_0}}{V} + \sum_{i=1}^{N} \tau_i. \tag{8}$$

3.3 Average Remaining Lifespan and Data Collection

Average Remaining Lifespan. When the MC stays at sensor node i for time τ_i, and then the remaining lifespan of sensor node i is denoted as

$$L_i(t_i + \tau_i) = \frac{U\tau_i + e_i(t_i) - p_i\tau_i - E_{min}}{p_i}. \tag{9}$$

The remaining lifespan of other sensor nodes j in the network at time $(t_i + \tau_i)$ is denoted as

$$L_j(t_i + \tau_i) = \frac{e_j(t_i) - p_j\tau_i - E_{min}}{p_j}. \tag{10}$$

When the MC accesses the last sensor node π_N in the network, the average remaining lifespan of all the sensor nodes is denoted as

$$\overline{L(t_{\pi_N} + \tau_{\pi_N})} = \frac{1}{N} \sum_{i \in N} L_i(t_{\pi_N} + \tau_{\pi_N}). \tag{11}$$

Data Collection. When the MC collects data from sensor node i in the network, the amount of data collected by the MC from sensor node i depends on the sojourn time τ_i. Therefore, the amount of data collected by the MC at sensor node i is denoted as

$$C_i = \begin{cases} r\tau_i & (\tau_i^t \geq \tau_i) \\ r\tau_i^t + r_i(\tau_i - \tau_i^t) & (\tau_i^t < \tau_i) \end{cases} \tag{12}$$

When the MC visits all sensor nodes in the network, the total amount of data collection in one cycle is denoted as $C = \sum_{i \in N} C_i$.

In order to make the network lifespan as long as possible, we need to prolong the average remaining lifespan of the sensor nodes in the network. In view of the data collection, it is hoped that the MC can upload as much the amount of data as possible when the MC returns to the service station. As these two objectives have a conflicting relationship, the two objectives cannot reach the maximum at the same time. Therefore, the problem is formulated as

$$\begin{aligned} obj: &\quad \max\left\{\overline{L(t_{\pi_N} + \tau_{\pi_N})},\ \ C\right\} \\ s.t.: &\quad (3) \sim (7), (9) \sim (12), \end{aligned}$$

where π_i and τ_i are variable, p_i, U, V, r_i and r are constant.

4 Multi-objective Discrete Fireworks Algorithm Based on Grid

4.1 Basic Discrete Fireworks Algorithm

The fireworks algorithm firstly proposed by Tan et al. [14] is a heuristic algorithm based on the phenomenon of fireworks explosion. The fireworks algorithm simulates the principle of natural fireworks explosions. The algorithm will select a certain number of fireworks during each iteration, and generates sparks through the explosion operation to expand the local searching.

The discrete fireworks algorithm [15] is similar to the traditional fireworks algorithm [14]. There are still explosion, mutation and selection operations, but the mapping rules are removed. In discrete fireworks, 2-opt and 3-opt is used for explosion operations, and 2h-opt is used for mutation operations.

4.2 Multi-objective Discrete Fireworks Algorithm Based on Grid

Much of the researches on the fireworks algorithm mainly involves single objectives, and there are few researches on the multi-objective fireworks algorithm. In 2013, Zheng et al. [16] combined the fireworks algorithm with the differential evolution algorithm to solve the fertilization problem of oilseed crops. Liu et al. [17] firstly put the S-metric into the multi-objective fireworks algorithm to improve the performance of the algorithm. Multi-objective fireworks algorithms above are used to solve the continuous problems. Because the path planning of this paper belongs to the discrete problem. Therefore, we propose a multi-objective discrete fireworks algorithm based on the basic discrete fireworks algorithm to solve the problem.

Fitness Assignment Strategy. There are a variety of fitness assignment strategies for multi-objective optimization algorithms, and in the MODFA, we employ a strategy based on the Pareto strength used in SPEA-II [18] for an individual solution x_i in the fireworks population P. A strength value $S(x_i)$ is calculated according to the number of other individuals it dominates.

$$S(x_i) = \left|\left\{x_j \in P \middle| x_i \succ x_j\right\}\right| \tag{13}$$

where $\succ$ denotes the Pareto dominance relation. And the raw fitness value of x_i is determined by the strengths of its dominators:

$$R(x_i) = \sum_{(x_j \in P) \wedge (x_j \succ x_i)} S(x_j). \tag{14}$$

The raw fitness value assignment process provides a non-dominated ordering, but only non-dominated sorting is not enough. Density information must be referenced to distinguish fireworks individuals with the same raw fitness value, and the density is calculated as

$$D(x_i) = \frac{1}{\sigma_k(x_j) + 2}, \quad (15)$$

where $\sigma_k(x_j)$ is the distance from x_i to its k*th* nearest fireworks and k is set to equal to the square root of fireworks population size *NP*. Therefore, the fitness value is denoted as $F(x_i) = R(x_i) + D(x_i)$.

Calculation of the Number of Fireworks Explosion. In the algorithm presented in [14], the number of sparks for each firework x_i is respectively defined as follows,

$$s_i = s_m \bullet \frac{F_{\max} - F(x_i) + \varepsilon}{\sum_{i=1}^{p} (F_{\max} - F(x_i)) + \varepsilon}, \quad (16)$$

where s_m is the total number of exploded fireworks, $F_{\max}$ is the maximum fitness value of the fireworks population, $F(x_i)$ is the fitness value of the individual fireworks x_i, and ε is a small constant to avoid zero-division-error.

In order to prevent the number of explosions that some fireworks individuals produce too much or too few during the explosion, the upper and lower limits of the number of explosions per individual fireworks are limited. The number of sparks a for each firework is respectively defined as follows

$$S_i = \begin{cases} S_{\min} & if \quad S_i < S_{\min} \\ S_{\max} & else \quad if \quad S_i > S_{\max} \\ S_i & else \end{cases} . \quad (17)$$

Calculation of Non-dominated Solution Density Information. Due to the fact that there may be no dominance among fireworks individuals in archive, once the number of non-dominated solutions exceeds the size of the archive, it is difficult to select the non-dominated solution to maintain the size of the archive. Because diversity is one of the evaluation criteria of the algorithm, we use the density information of fireworks individuals as the partial order information for the non-dominated solutions.

The typical density estimation methods are mainly based on crowding distance method [19] and SPEA-II [18]. The main problems of these two algorithms are high computational complexity. In this paper, based on the adaptive mesh method in PAES [17], the basic idea is to space the objective into small areas by using the meshing method. As the density information of the fireworks, the algorithm adjusts the size of the grid adaptively as the algorithm evolves. The archive set is redistributed to determine the density information of each fireworks.

For the two-dimensional optimization of the objective space optimization problem, the algorithm is implemented as follows:

Algorithm 1. Density estimation based on grid

Input: Non-dominated solutions of iteration generations t

Output: Density information

1. Calculate the boundary of the objective space in evolution of t generations: $\left(min\ F_1^t, max\ F_1^t\right)$ and $\left(min\ F_2^t, max\ F_2^t\right)$.
2. Calculate the size of grid : $VF_1^t = \frac{max\ F_1^t - min\ F_1^t}{M}$, $VF_2^t = \frac{max\ F_2^t - min\ F_2^t}{M}$, where M is the number of grids set for a single objective dimension.
3. Traverse the each solution in the non-dominated solution, and calculate the grid number where it is located. For each solution, its grid number consists of two parts: $\left(Int(\frac{F_1^i - minF_1^t}{VF_1^t}) + 1, Int(\frac{F_2^i - minF_2^t}{VF_2^t}) + 1\right)$.
4. Calculate the number of non-dominated solutions in each grid.

Non-dominated Archive Maintenance. The select of an archive is to maintain the size of the archive when the number of non-dominated solutions exceeds the size of the archive to remove poor quality individuals from non-dominated solutions. We employ a strategy based on the method used in [20]. When the number of non-dominated solutions *PN* is larger than the size of archive *NA*, the fireworks individuals with small value of density information are deleted according to the grid information. For each grid with more than one individual, the number of individuals to be deleted according to the following formula:

$$GN_i = Int(\frac{PN - NA}{PN} \times Grid[i] + 0.8), \tag{18}$$

where $Grid[i]$ is the number of non-dominated solutions of grid i and $Int(\bullet)$ is the rounding function.

5 Experiments and Evaluation

5.1 Experimental Settings

Based on the network model, we randomly deploy four groups of network instances $(Net - i, i = 1, 2, 3, 4)$ in a sensor monitoring area 1000 m × 1000 m to analyze the algorithm. The number of sensor nodes in the $Net - 1$ and $Net - 2$ are 20 while the number of sensor nodes in $Net - 3$ and $Net - 4$ are 30. And the service station is located at the coordinate $(0, 0)$. Simulation parameters are shown as follows. E_{max}= 10.8K J, E_{min} = 540 J, V = 5 m/s, U = 10 W. The energy consumption power $p_i(W)$ is a random number varying from 0.01 to 0.2. The amount of data collection generated by the sensor node is randomly set between 1000 and 2000 and the unit is kbit/s. The data transmission rate between the MC and sensor nodes r is 200 kbit/s. The data generation of rate of the sensor node r_i(kbit/s) is a random number from 10 to 20.

5.2 Performance Metrics

In order to evaluate the quality of non-dominated solution sets found by the comparing algorithms, the hyper-volume (HV) performance measure proposed by Zitzler and Thiele [21] is employed in this paper. HV can estimate the diversity and convergence of Pareto front. The HV does not need to know the true Pareto front, and the larger the value, the better the algorithm performance. The HV performance measure can be mathematically defined as follows:

$$HV(PS, Z^*) = \lambda(U_{p \in PS}([f_1(p), Z_1^*] \times [f_2(p), Z_2^*] \times \cdots [f_m(p), Z_m^*])), \tag{19}$$

where PS is the Pareto front solutions, p is a Pareto front solution, λ is the Lebesgue measure, Z^* is a reference point and $f_i(\bullet)$ is the objective value of the i th objective.

5.3 Analysis of Simulation

In order to evaluate the performance of the proposed algorithm, under the same conditions, we also compares it with three classical algorithms, NSGA-II, SPEA-II and MOEA/D. The experimental data in Table 1 are statistic values over 30 independent runs.

Table 1. Performance comparison of four algorithms under four network instances

Network instances	Statistics	MODFA	NSGA-II	SPEA-II	MOEA/D
Net-1	Mean	**2.857**	2.714	2.790	2.475
	Median	**2.858**	2.709	2.795	2.498
	SD	**1.715**	3.380	3.421	15.768
	Max	**2.894**	2.788	2.848	2.823
	Min	**2.817**	2.659	2.717	2.246
Net-2	Mean	**1.983**	1.903	1.951	1.818
	Median	**1.981**	1.907	1.946	1.820
	SD	**0.699**	2.102	1.329	10.969
	Max	**1.995**	1.936	1.978	1.971
	Min	**1.967**	1.854	1.927	1.614
Net-3	Mean	**6.808**	6.345	6.565	5.618
	Median	**6.798**	6.354	6.563	5.513
	SD	**5.627**	12.034	9.282	39.377
	Max	**6.973**	6.549	6.756	6.627
	Min	**6.663**	6.106	6.378	5.073
Net-4	Mean	**8.141**	7.455	7.816	6.781
	Median	**8.136**	7.443	7.831	6.524
	SD	**5.763**	14.647	12.447	52.681
	Max	**8.251**	7.803	7.989	7.949
	Min	**8.023**	7.159	7.573	6.294

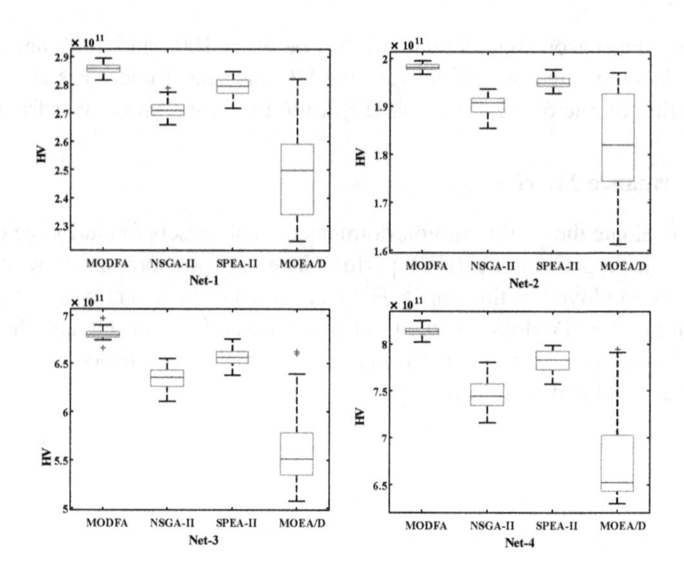

Fig. 2. HV box plots of four algorithms under four network instances

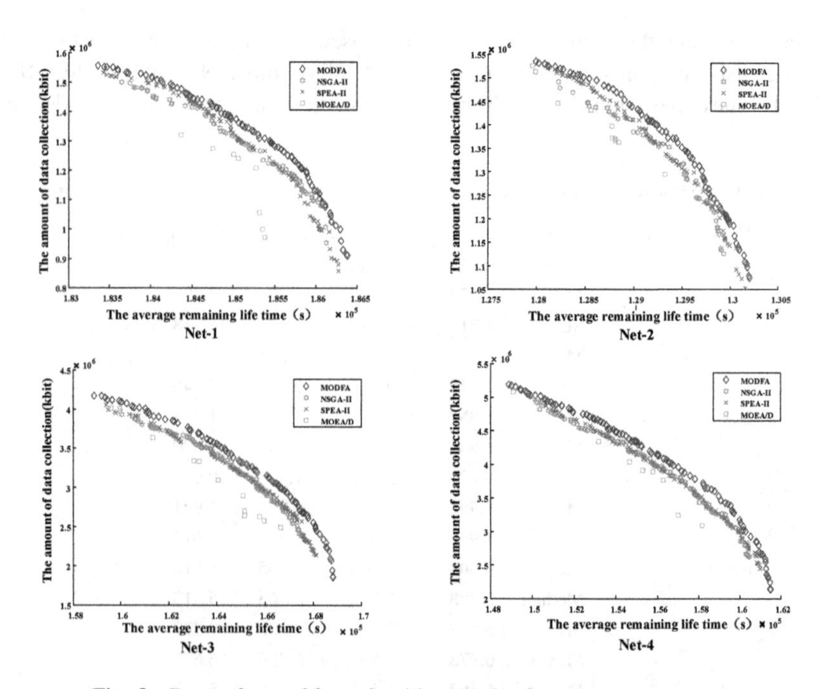

Fig. 3. Pareto front of four algorithms under four network instances

As shown in Table 1, for four kinds of network instances, the proposed MODFA algorithm has better performance than three classical algorithms, NSGA-II, SPEA-II and MOEA/D, under various statistic values. The reason why MODFA has good performance is due to the use of locally optimized 2-opt and 3-opt strategies, meanwhile, the filtering method based grid can reduce the complexity of the algorithm. In order to more vividly show the HV value of each algorithm after running, we shows the values through the box plots. The box plots as shown in Fig. 2.

As shown in Fig. 2, MODFA algorithm is obviously better than the other three algorithms in comparison. Finally, according to the size of the HV value, we select the largest group of data from the respective algorithm HV to display the Pareto front. As shown in Fig. 3, in the four instances, the Pareto front of the MODFA in this paper has better performance than the other three classic algorithms.

6 Conclusion

To solve the problem of joint energy replenishment and data collection in the WRSNs, we propose a multi-objective path planning model that maximizes the average remaining lifespan and data collection of the network to prolong lifespan of the network and improve performance of the network. For the problem model, we propose a multi-objective discrete fireworks algorithm based on grid, which has better performance than other algorithms in comparison.

References

1. Xie, L., Shi, Y., Hou, Y., Lou, A.: Wireless power transfer and applications to sensor networks. IEEE Wirel. Commun. **20**(4), 140–145 (2013)
2. Cheng, S., Cai, Z., Li, J.: Curve query processing in wireless sensor networks. IEEE Trans. Veh. Technol. **64**(11), 5198–5209 (2015)
3. Cai, Z., Zheng, X.: A private and efficient mechanism for data uploading in smart cyber-physical systems. IEEE Trans. Netw. Sci. Eng. **5**(1), 1–9 (2018)
4. Shi, Y., Xie, L., Hou, Y., Sherali, H.: On renewable sensor networks with wireless energy transfer. In: INFOCOM, pp. 1350–1358. IEEE, Shanghai (2011)
5. Xie, L., Shi, Y., Hou, Y., Lou, W., Sherali, H.: Multi-node wireless energy charging in sensor networks. IEEE/ACM Trans. Netw. **23**(2), 437–450 (2015)
6. Liang, W., Xu, W., Ren, X., Jia, X., Lin, X.: Maintaining sensor networks perpetually via wireless recharging mobile vehicles. In: Local Computer Networks, pp. 270–278. IEEE, Edmonton (2014)
7. Guo, S., Wang, C., Yang, Y.: Joint mobile data gathering and energy provisioning in wireless rechargeable sensor networks. IEEE Trans. Mobile Comput. **13**(12), 2836–2852 (2014)
8. Chen, Q., Gao, H., Cai, Z., Cheng, L., Li, J.: Energy-collision aware data aggregation scheduling for energy harvesting sensor networks. In: INFOCOM, pp. 1–9. IEEE, Honolulu (2018)
9. Shi, T., Cheng, S., Cai, Z., Li, Y., Li, J.: Exploring connected dominating sets in energy-harvest networks. IEEE/ACM Trans. Netw. **25**(3), 1803–1817 (2017)
10. Xie, L., Shi, Y., Hou, Y., Lou, W., Sherali, H.: A mobile platform for wireless charging and data collection in sensor networks. IEEE J. Sel. Areas Commun. **33**(8), 1521–1533 (2015)
11. Cheng, S., Cai, Z., Li, J., Fang, X.: Drawing dominant dataset from big sensory data in wireless sensor networks. In: INFOCOM, pp. 531–539. IEEE, Hong Kong (2015)

12. Cheng, S., Cai, Z., Li, J., Gao, H.: Extracting kernel dataset from big sensory data in wireless sensor networks. IEEE Trans. Knowl. Data Eng. **29**(4), 813–827 (2017)
13. Tan, D., Chu, S., Liu, B., Chen, C., Dang, H., Perumal, T.: Mobile charging and data gathering in multiple sink wireless sensor networks: how and why. In: International Conference on System Science and Engineering, pp. 550–553. IEEE, Ho Chi Minh City (2017)
14. Tan, Y., Zhu, Y.: Fireworks algorithm for optimization. In: Tan, Y., Shi, Y., Tan, K.C. (eds.) ICSI 2010. LNCS, vol. 6145, pp. 355–364. Springer, Heidelberg (2010). https://doi.org/10.1007/978-3-642-13495-1_44
15. Tan, Y.: Fireworks Algorithm: A Novel Swarm Intelligence Optimization Method. Springer, Berlin (2015). https://doi.org/10.1007/978-3-662-46353-6
16. Zheng, Y., Song, Q., Chen, S.: Multi-objective fireworks optimization for variable-rate fertilization in oil crop production. Appl. Soft Comput. **13**(11), 4253–4263 (2013)
17. Liu, L., Zheng, S., Tan, Y.: S-metric based multi-objective fireworks algorithm. In: Evolutionary Computation, pp. 1257–1264. IEEE, Sendai (2015)
18. Zitzler, E., et al.: SPEA2: improving the strength pareto evolutionary algorithm. Evol. Methods Design Optim. Control **7**(3), 1–21 (2001)
19. Deb, K., Pratap, A., Agarwal, S., Meyarivan, T.: A fast and elitist multi-objective genetic algorithm: NSGA-II. IEEE Trans. Evol. Comput. **6**(2), 182–197 (2002)
20. Yang, J., Zhou, J., Fang, R., Li, Y., Liu, L.: Multi-objective particle swarm optimization based on adaptive grid algorithms. J. Syst. Simul. **20**(21), 5843–5847 (2008)
21. Zitzler, E., Thiele, L.: Multi-objective evolutionary algorithms: a comparative case study and the strength Pareto approach. IEEE Trans. Evol. Comput. **3**(4), 257–271 (1999)

Privacy-Preserving Personal Sensitive Data in Crowdsourcing

Ke Xu, Kai Han(✉), Hang Ye, Feng Gao, and Chaoting Xu

School of Computer Science and Technology/Suzhou Institute for Advanced Study, University of Science and Technology of China, Hefei, People's Republic of China
{kexu,yehang,gf940312,xct1996}@mail.ustc.edu.cn, hankai@ustc.edu.cn

Abstract. Spatial crowdsourcing system refers to sending various location-based tasks to workers according to their positions, and workers need to physically move to specified locations to accomplish tasks. The workers are restricted to report their real-time sensitive position to the server so as to keep in coordination with the crowdsourcing server. Therefore, implementing crowdsourcing system while preserving the privacy of workers sensitive information is a key issue that needs to be tackled. We discard the assumption of a trustworthy third party cellular service provider (CSP), and further propose a local method to achieve acceptable results. A differential privacy model ensures rigorous privacy guarantee, and Laplace mechanism noise is introduced to preserve workers sensitive information. Finally, we verify the effectiveness and efficiency of the proposed methods through extensive experiments on real-world datasets.

Keywords: Crowdsourcing · Differential privacy
Sensitive information

1 Introduction

Nowadays, with the rapid proliferation of all kinds of smartphones and the convenience of mobile Internet, crowdsourcing has emerged as a significant computing technology which utilizes human intelligences. In particular, numerous crowdsourcing-based platforms, such as CrowdFlower [1], Gigwalk [15], Gmission [5] and etc., which leverages the wisdom of crowd to perform the specialized assignment appropriately and accurately. This new framework encourages active workers to participate in to perform specified tasks that are vicinity to the required locations. The crowd of workers have shift their conventional idea of data consumers to the role of gathering data to gain some deserved rewards (e.g., money, reputation). In crowdsourcing system, smartphone users are engaged to provide pervasive and inexpensive tasks of data collecting and computing eventually. The application of crowdsourcing has developed incredibly. It has been widely used in ride sharing, traffic or environment monitoring.

Specifically, the roles in the whole crowdsourcing system are categorized into three types: crowding platform (i.e., server), crowdworkers (i.e., workers), crowdsourcer (i.e., requester) [3]. The platform is responsible for distributing atomic

S. Chellappan et al. (Eds.): WASA 2018, LNCS 10874, pp. 509–520, 2018.
https://doi.org/10.1007/978-3-319-94268-1_42

tasks or viral tasks to workers and in charge of the data collecting job. The workers are the ones who are concentrating on finishing the small units of work in return for monetary payment. The responsibility of crowdsourcer is aiming to carrying out computationally hard tasks and divide them into several subtasks. In [16], the tasks on crowdsourcing platform can be published in two distinct modes: Worker Selected Tasks (WST) and Server Assigned Tasks (SAT). On the one hand, in WST mode [15], online workers are allowed to select arbitrary tasks delivered by the crowdsourcer in vicinity without the permission of crowdsoucing platform. On the other hand, in SAT mode [6], the workers are restricted to report their real-time position to the server so as to keep in coordination with the crowdsourcing server, then the server will decide how to allocate tasks reasonably in terms of some optimal functions. Meanwhile, in WST mode, workers are unnecessary to track the locations of workers while the crowdsourcing platform are never interrupted to follow the tracks of workers in SAT mode, so the issues on the privacy of these workers needs to be protected are rather hot problems.

Differential Privacy (DP) [11,25] is a relatively new notion of privacy, and also is one of the most popular privacy notations. It is actually implemented by noise mechanism which adds a random noise to the output data. With the privacy definition, To et al. [19] developed a new framework for protecting workers' locations by introducing the cellular service provider (CSP) [20] as a trustworthy third party. private spatial decomposition, which partition the geospatial data into smaller regions and obtain statistics on the points with each region, and designed to enhance the accuracy of the entire crowdsourcing system [7]. Fan et al. [12,23] divide the space into four equal subspaces using quadtree. Recursively partitioning on quadtree is highly efficient compared with partitioning of kd-tree [21,22]. [4,13] consider the privacy concerns that are hard to solve and propose a flexible optimization framework that can be adjusted to trade-offs with the joint efforts of platform and workers. As far as we know, there already have been some related works protecting workers location information on the crowdsourcing system, however, there are few works currently pay scalable attention to privacy-preserving of workers sensitive information, which discards the entirely reliable third party. It is a novel notion that proposed appropriately in accordance with the characteristics of reality, so achieving a desired privacy-preserving result is not a non-trivial problem.

In this paper, we formulate the privacy protection strategies without compromising the third party from a particular worker to a crowdsourcing platform as a non-trivial problem that follows two criteria: (1) efficiency of our proposed method, (2)utility of our method. Note that the aforementioned standards remains a secret for us, which motivates us to consider the fundamental factors in crowdsourcing system. We show that these two criteria can hardly be optimized simultaneously. We immediately divide the data publishing into three part: data preprocessing, information filtering and noise addition. In the first period, we generate a most suitable data structure to storage workers' sensitive information. Next, we filter out the insensitive information by exponent mechanism of differential privacy. Finally, we further add appropriate noises after information filtering to ensure the privacy leakage problem. In summary, the main contributions of our work are listed as follows:

- We identify the specific challenges of privacy-preserving in crowdsourcing system, and we further develop a model that illustrates this issue.
- We abandon the assumption that the third party CSP is rather convincing and adopt a local method to publish sensitive data sets.
- We conduct both extensive numerical evaluations and performance analysis to show the effectiveness and efficiency of our designed method using real-world datasets, and analyze the key factors associated with hierarchical method.

The paper is structured as follows. We present related preliminaries in Sect. 2. Next we develop our model to solve the problem in details in Sect. 3. We discuss the experimental results and analyze crucial factors in Sect. 4, respectively. Section 5 summarizes related work. Finally, we conclude the work in Sect. 6.

2 Preliminaries

Intuitively, Differential Privacy (DP)has grown as the standard in privacy protection, thanks to its strong mathematical guarantees rooted in related statistical analysis. DP ensures the attacker fail to deduce whether a particular individual in or not in the original data, thereby protecting the workers' privacy.

Definition 1 ((ε, δ)-differential privacy) [11]**:** Let D and D' be two neighboring datasets which differ on at most one record, denoted as $|D \Delta D'| = 1$, a randomized mechanism M: D→ R, Ω(M) be the set of all possible outputs of M in D and D', algorithm M gives (ε,δ)-differential privacy if:

$$\Pr[M(D) \in \Omega] \leq exp(\varepsilon) \times Pr[M(D') \in \Omega] + \delta \quad (1)$$

The parameter ε is called privacy budget, which controls the level of privacy guarantee. The smaller ε is, the higher security becomes. If $\delta = 0$, the randomized mechanism M gives ε -differential privacy by its strictest definition. Thus, (ε,δ)-differential privacy in some degree provide freedom to violate strict differential privacy for some low probability events.

Theorem 1 ***(Sequential Composition)*** [17]**:** Say we get a set of privacy algorithms M $= \{M_1, M_2, ..., M_m\}$. For each M_i satisfies a ε_i-differential privacy guarantee for the same dataset, M will provide $\sum_{i=1}^{m} \varepsilon_i$-differential privacy.

Sequential composition undertakes the privacy guarantee for a combination of the entire differential privacy process. When a set of randomized mechanisms have been conducted on the same dataset, the total privacy budget is the sum of all privacy budgets.

Theorem 2 ***(Parallel Composition)*** [18]: Say we get a set of privacy algorithms M $= \{M_1, M_2, \ldots, M_m\}$. For each M_i satisfies a ε_i-differential privacy guarantee on a disjoint subset of the whole dataset, M will provide max (ε_1,ε_2,. . . ,ε_m)-differential privacy.

Parallel composition corresponds to situation where a quantity of private mechanisms are applied to a disjoint dataset. Consequently, the privacy guarantee only depends on the largest privacy budget.

Definition 2 (Sensitivity) [10]**:** Given neighboring datasets D and D', for a query function f: D→ R, the sensitivity of f is defined as

$$\Delta f = \max_{(D,D')} |f(D) - f(D')|_1 \tag{2}$$

Sensitivity Δf is closely related to the query f. It is regarded as the maximal differential between the query results on neighboring datasets. Currently, two basic mechanisms are widely used to guarantee differential privacy: the Laplace mechanism and the Exponential mechanism.

Definition 3 (Laplace mechanism) [10]**:** Given dataset D and a function f: D→ R, Δ f is the sensitivity of f, representing the maximal value on the output of f when deleting any tuple in D. The randomized algorithm

$$M(D) = f(D) + Laplace(\frac{\Delta f}{\varepsilon}) \tag{3}$$

satisfies ε-differential privacy. We use Laplace(x) to represent the noise sampled from a Laplace distribution with a scaling of x.

Definition 4 (Exponential mechanism) [17]**:** Let (q, r) be a function of dataset D that measures the quality of output r $\in$ Range, Δ q represents the sensitivity of r. The exponential mechanism M satisfies ε-differential privacy if:

$$M(D) = (r : Pr[r \in Range] \propto exp(\frac{\varepsilon q(D,r)}{2\Delta q})) \tag{4}$$

For non-numeric queries, differential privacy utilizes the exponential mechanism to randomized the results.

3 Designed Model

We consider the problem of privacy-preserving spatial crowdsourcing task assignment in the SAT mode. The crux of our method is to how to choose data structure to storage workers' sensitive information and apply a differential privacy mechanism to each worker's location information. As mentioned in Sect. 1, data structure selection is a non-trivial step. Previous literature assumes that the third party Cell Service Provider (CSP) is completely convinced, but this may not be the case in real-world scenarios. Our proposed method adopts a novel model to protect workers' locations. In this way, the approach recommends tasks to each worker with a better success ration and a stronger privacy guarantee. Figure 1 shows the basic model of the proposed framework consisting of each worker's three components:

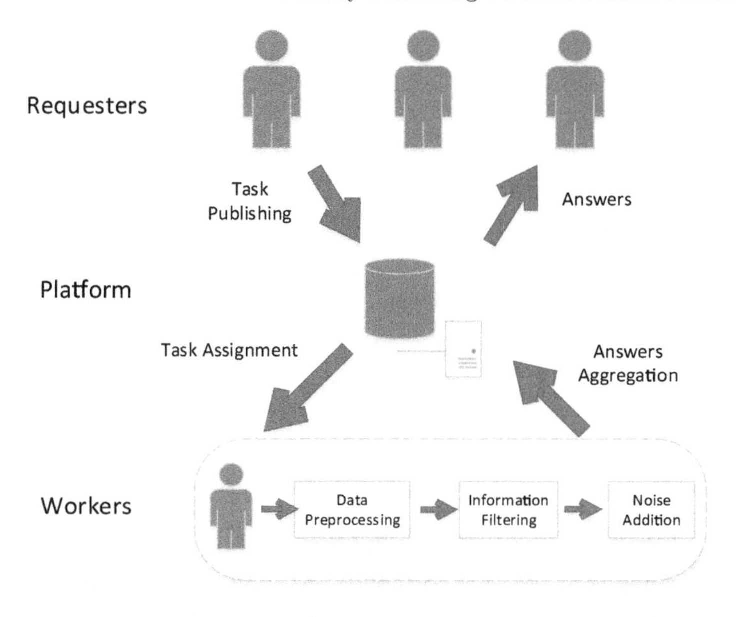

Fig. 1. Our system model for task recommendation in crowdsourcing systems

Data Preprocessing: In this component, each worker collects various statistics periodically in the background. After that, they will preprocess the detailed data for a short time. The sensitive information of each worker are not delivered to the third party, so this step can protect the private context information of participating workers well.

Information Filtering: In this component, based on the statistics preprocessing component, each worker then select the most sensitive location information. Note that workers are allowed to decide how much private information they are willing to share with others. Therefore, we set a constant threshold θ and eliminate those location information that are below this threshold. We may achieve an ideal result according the Exponent mechanism to complete sensitive information selection.

Noise Addition: In this component, based on the information filtering and Laplace mechanism, each worker then adds suitable noises to these sensitive information. It goes without saying that the noises are subjected to laplace distribution. After adding some noise in the original data, we achieve a series of brand new datasets.

3.1 Detailed Explanation

We just described the basic system model for task recommendation in Spatial Crowdsourcing system. Next we will represent a description of the details in these components and explain why we take these steps to tackle the barriers in the entire crowdsourcing systems.

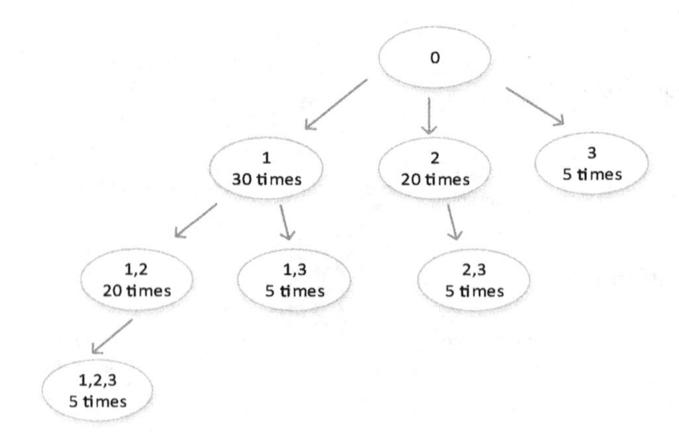

Fig. 2. Creating a Trie-Tree

Data Preprocessing: We aim to provide good efficiency, privacy and utility in our proposed framework. Each worker firstly achieve a transaction database D including the work's identity, the frequency of locations they visited during a long period and when the worker arrive at these locations. Apparently, there is no problem that we ignore some detailed location information such as the longitude and latitude of locations. Then we randomly select several items (e.g., I items) to represent the whole database D. After finishing that, it's vital for us to determine which data structure to storage our location information. We make a thorough decision to choose Trie-Tree to represent workers' context information. It's non-trivial for us to draw a solid conclusion that we select the data structure Trie-Tree. We summarize the reasons as follows: (1) It is most suitable to maintain the link between the location data and overwrite the original data set as well; (2) It reduces the number of noise addition in that we add noise into each node rather than each original data. It disturbs true visit frequency in each node, we are required to add noises only once into each node. In other words, all the original data sets in node are covered with some noise, it is unnecessary for us to add extra noises into each original data. Therefore, workers can protect his/her location privacy at a relatively low cost of utility. Figure 2 shows an example describing how we store workers' sensitive location information. We assume that the root node in level 0, it is clearly see that there are $\sum_{i=1}^{n} C_n^i = 2^i - 1$ nodes in the Trie-Tree and all the 1st-items in the level 1. Here, there are four different nodes in level 1 and total $2^3 - 1 = 7$ nodes in the tree. More specifically, we take node 1, 2 as an example, it denotes the combination of node 1 and node 2 in level 1, the constant value 20 means the least number of visiting times in the node. As we all know, node 1 is visited 30 times and node 2 is 20, so we achieve the minimum value 20 in node set 1, 2. Likewise, it is rather easy to get all the other nodes in our Trie-Tree.

Information Filtering: After representing detailed location data sets in Trie-Tree, we continue to finish sensitive location information selection based on Exponent mechanism. We firstly traverse the entire tree by level and eliminate those nodes in which the constant value is smaller than the specified threshold θ. By the way, we determine how much θ would be according the subsequent experiments. Furthermore, we naturally derive the nodes set S in which the visiting times of each node are above θ. Finally, we filter out n nodes in set S according to the exponent mechanism of differential privacy. The specific operations are as follows:

Step 1: Input the sensitive location information set S, then we successively take out each node in it and mark them in turn:

$$tag(S, s_i) = q(s_i) \tag{5}$$

where q(s_i) denotes the actual visiting frequency of node s_i.

Step 2: calculate the weight of each node:

$$s_i.w = exp(\frac{\varepsilon_1 \times tag(S, s_i)}{2\Delta tag}) \tag{6}$$

Where ε_1 means the privacy budget of exponent mechanism, the function Δtag means the sensitivity of node s_i.

Step 3: randomized figure out the top n nodes according the following equation, then we make up a new set C:

$$Pr(s_i) = \frac{s_i.w}{\sum_{i=1}^{n} s_i.w} \tag{7}$$

The goal we design the equation is to follow the exponent mechanism. s_i.w is derived from step 2.

In this section, we adopt exponent mechanism to finish information selection because of the merits of exponent mechanism. Not only we can evaluate the privacy protection by privacy budget, but it improves the efficiency of our proposed algorithm according to filtering out insensitive location information.

Noise Addition: We carry on adding suitable noises into elements in set C followed by information selection. That is to say, noises which are subjected to laplace distribution are appended into the top n nodes in C. Similarity, we use privacy budget ε_2 to evaluate privacy leakage in that the approach improves the efficiency of our algorithm and strengths the utility of data sets. The details are as follows:

$$q(c_i) = q(c_i) + laplace(\frac{\Delta q}{\varepsilon_2}) \tag{8}$$

where ε_2 means the privacy budget of laplace mechanism, Δq is the global sensitivity of function q. In this paper, both function tag and q mean the sensitive location visiting frequencies of each worker. In order to understand the laplace mechanism, we list the probability density function of laplace mechanism:

$$Pr(x, \lambda) = \frac{e^{-\frac{|x|}{\lambda}}}{\lambda} \tag{9}$$

where λ denotes that there is no correlation between noises and database, it merely concerned with sensitivity of function and privacy parameters. Here, x means the actual visiting frequencies.

We eventually formulate a completely new set E and then publish it to our crowdsourcing platform. At this point, We successfully complete the process of privacy preserving and deliver it to the unreliable third party.

4 Experimental Study

4.1 Experiments Setup

We use a real-world dataset: Gowalla. It contains the check-in history of users in allocation-based social network. It includes some detailed data such as the type of each user, the longitude and latitude of the location and the time when users visit the locations. For our experiments, we assume that these users are workers of the spatial crowdsourcing system, and their locations are those of the most recent check-in points. We transfer the original data into a database D in which it records the visiting frequencies of users in a month. The algorithms are implemented in java 8, and the experiments were performed on Intel(R) Core(TM) i7 2.40 GHz CPU and 8 GB main memory.

4.2 Experimental Results

We evaluate our proposed method from the following two aspects: *efficiency* and *utility*. When each worker receives a series of recommended tasks, he/she decides to choose the appropriate task to conduct. Workers may spend a lot of time finishing information filtering when the size of task set is substantial. Thus, the efficiency of information filtering is directly related to the Trie-Tree. The recommendation system should create Trie-Tree for a short time to ensure thee efficiency of information filtering. Furthermore, utility represents the accuracy and data protection level. From the perspective of the crowdsourcing platform, the utility is expected accuracy of data protection. Meanwhile, the utility is the degree of privacy preserving from the perspective of the workers. The utility for both stakeholders is closely related to the privacy protection level.

Efficiency. We analyze the efficiency of our method from timeliness of creating a Trie-Tree and extracting data from the tree. We vary the number of nodes from 8 to 512 to observe the situation of creating an entire Trie-Tree. t_{create} denotes how long we need to build a tree. As is shown in the following Table 1, We apparently see that building a tree is not a time-consuming process during the whole period of privacy protection in spite of the size of nodes set.

Table 1. Time to create Trie-Tree

Number of nodes	8	32	64	128	256	512
$T_{create}/\times 10^{-6}$s	9	21	33	59	114	146

Table 2. Time to extract sensitive information.

ε_1	0.01	0.02	0.10	0.40	0.90	1.30
n/s	3120	4382	41297	48633	52614	75621

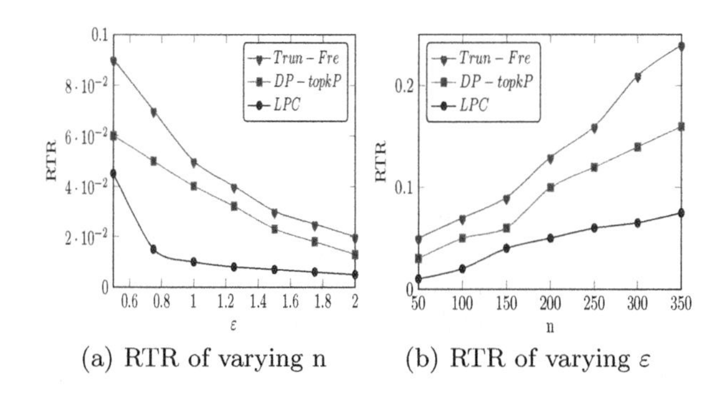

(a) RTR of varying n (b) RTR of varying ε

Fig. 3. RTR of varying n and ε.

From the Table 2, we also see that the efficiency of extracting data from tree is so high, because it depends on the essential structure and characteristics of Trie-Tree. In some degree, the time of extracting data grows longer as the privacy budget becomes bigger.

Utility. First of all, we use Local Protection in Crowdsourcing (LPC) to represent our method. Then we show the performance on ratio of rejecting true nodes (RTR), which significantly represents the utility of algorithms. Figure 3(a) plots the RTR and number of nodes with n ranging from 50 to 350. At the same time, we fix privacy level to 0.5. For each n, the x-axis represents the number of nodes in set, and the y-axis represents the ratio. The ratio increases as the size of nodes set grows bigger. LPC runs better than Trun-Fre [2] and DP-topkP [24]. This is reasonable because we add noises into sets rather than nodes merely.

Secondly, we analyze the ratio of RTR and the privacy protection level. Since noise needs to be added to provide (ε,δ)-differential privacy, the RTR is achieve at the cost of accuracy. Before the experiment is conducted, we strictly control the number of nodes to be 150. We illustrate the trade-off between the privacy parameter ε and RTR in Fig. 3(b). It shows the RTR and number of nodes with ε ranging from 0.5 to 2.0. For each ε, the x-axis represents the privacy budget in our algorithm, and the y-axis represents the ratio. Our method outperforms than Trun-Fre [2] and DP-topkP [24] as well.

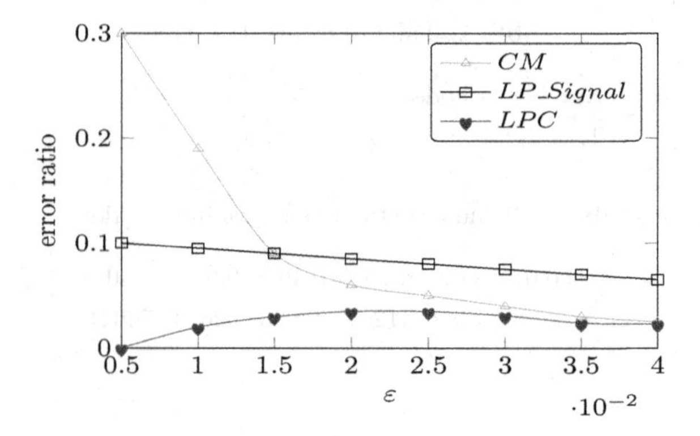

Fig. 4. Error of varying ε.

Lastly, we compare our novel method LPC with previous existing approaches CM [9] and LP_Signal [14]. Define F_{before} denotes the frequencies of locations before adding noises and F_{after} means frequencies after doing that. Thus error signifies the ratio of F_{before} and the differences between F_{after} and F_{before}:

$$Error = \frac{||F_{after} - F_{before}||_2}{||F_{before}||_2} \tag{10}$$

The experiment results are shown in the Fig. 4. It is clear for us to see that the impact of varying privacy levels on the errors. We can observe that the minimum error in our method compared with other existing algorithms. Moreover, whether the privacy budget ε is high or not, our method also achieve the steady and desired results. Since the error is inevitable with a small range, we conclude the expected errors of our privacy-preserving approach is almost close the optimal one.

5 Related Work

In this section, we review some previous work related to our problem in this literature. Differential privacy [8,11] is a strict privacy definition that is independent of prior knowledge. With the definition of privacy, To et al. [19] proposed a framework for protecting workers' locations by illustrating the cellular service provider (CSP) as a third party. The partition algorithm generates a private spatial decomposition (PSD) is widely conducted by the following works such as dividing grids [23], creating kd-trees [21,22] and quadtree [12]. However, the main shortcoming of these methods is that the privacy-preserving depends on the third party in some degree.

6 Conclusion

Privacy issues are increasingly becoming concerning with the popularity of spatial crowdsourcing. In this paper, we introduced a novel differentially private approach for spatial crowdsourcing, which enables the participation of various workers without compromising their sensitive information privacy. The magnitude of noise is minimized by fully utilizing the given privacy budget, which is crucial for efficiency and utility of our method. To ensure effective privacy protection, we select the Trie-Tree to storage workers' sensitive information rather than sending them to the completely trust third party. Because a trustworthy data collector is merely an assumption, which contradicts the reality and common sense. Comparisons between our method and existing approaches that privacy-preserving effects is dramatically enhanced by a series of experiments.

Acknowledgments. This work is partially supported by National Natural Science Foundation of China (NSFC) under Grant No. 61772491, No. 61472460, and Natural Science Foundation of Jiangsu Province under Grant No. BK20161256. Kai Han is the corresponding author.

References

1. CrowdFlower. http://crowdower.com
2. Bhaskar, R., Laxman, S., Smith, A., Thakurta, A.: Discovering frequent patterns in sensitive data. In: Proceedings of the 16th ACM SIGKDD International Conference on Knowledge Discovery and Data Mining, pp. 503–512. ACM (2010)
3. Cao, C.C., She, J., Tong, Y., Chen, L.: Whom to ask?: jury selection for decision making tasks on micro-blog services. Proc. VLDB Endow. **5**(11), 1495–1506 (2012)
4. Cao, C.C., Tong, Y., Chen, L., Jagadish, H.: WiseMarket: a new paradigm for managing wisdom of online social users. In: Proceedings of the 19th ACM SIGKDD International Conference on Knowledge Discovery and Data Mining, pp. 455–463. ACM (2013)
5. Chen, Z., Fu, R., Zhao, Z., Liu, Z., Xia, L., Chen, L., Cheng, P., Cao, C.C., Tong, Y., Zhang, C.J.: gMission: a general spatial crowdsourcing platform. Proc. VLDB Endow. **7**(13), 1629–1632 (2014)
6. Cheng, P., Lian, X., Chen, L., Han, J., Zhao, J.: Task assignment on multi-skill oriented spatial crowdsourcing. IEEE Trans. Knowl. Data Eng. **28**(8), 2201–2215 (2016)
7. Cormode, G., Procopiuc, C., Srivastava, D., Shen, E., Yu, T.: Differentially private spatial decompositions. In: 2012 IEEE 28th International Conference on Data Engineering (ICDE), pp. 20–31. IEEE (2012)
8. Dwork, C.: Differential privacy: a survey of results. In: Agrawal, M., Du, D., Duan, Z., Li, A. (eds.) TAMC 2008. LNCS, vol. 4978, pp. 1–19. Springer, Heidelberg (2008). https://doi.org/10.1007/978-3-540-79228-4_1
9. Dwork, C.: Differential privacy in new settings. In: Proceedings of the Twenty-First Annual ACM-SIAM Symposium on Discrete Algorithms, pp. 174–183. SIAM (2010)
10. Dwork, C.: A firm foundation for private data analysis. Commun. ACM **54**(1), 86–95 (2011)
11. Dwork, C., Kenthapadi, K., McSherry, F., Mironov, I., Naor, M.: Our data, ourselves: privacy via distributed noise generation. In: Vaudenay, S. (ed.) EUROCRYPT 2006. LNCS, vol. 4004, pp. 486–503. Springer, Heidelberg (2006). https://doi.org/10.1007/11761679_29

12. Fan, L., Xiong, L., Sunderam, V.: Differentially private multi-dimensional time series release for traffic monitoring. In: Wang, L., Shafiq, B. (eds.) DBSec 2013. LNCS, vol. 7964, pp. 33–48. Springer, Heidelberg (2013). https://doi.org/10.1007/978-3-642-39256-6_3
13. Gong, Y., Wei, L., Guo, Y., Zhang, C., Fang, Y.: Optimal task recommendation for mobile crowdsourcing with privacy control. IEEE Internet Things J. **3**(5), 745–756 (2016)
14. Jia, O., Jian, Y., Shaopeng, L., Yubao, L.: An effective differential privacy transaction data publication strategy. J. Comput. Res. Dev. **10**, 007 (2014)
15. Kazemi, L., Shahabi, C.: GeoCrowd: enabling query answering with spatial crowdsourcing. In: Proceedings of the 20th International Conference on Advances in Geographic Information Systems, pp. 189–198. ACM (2012)
16. Kazemi, L., Shahabi, C., Chen, L.: GeoTruCrowd: trustworthy query answering with spatial crowdsourcing. In: Proceedings of the 21st ACM SIGSPATIAL International Conference on Advances in Geographic Information Systems, pp. 314–323. ACM (2013)
17. McSherry, F., Talwar, K.: Mechanism design via differential privacy. In: 48th Annual IEEE Symposium on Foundations of Computer Science, FOCS 2007, pp. 94–103. IEEE (2007)
18. McSherry, F.D.: Privacy integrated queries: an extensible platform for privacy-preserving data analysis. In: Proceedings of the 2009 ACM SIGMOD International Conference on Management of Data, pp. 19–30. ACM (2009)
19. To, H., Ghinita, G., Shahabi, C.: A framework for protecting worker location privacy in spatial crowdsourcing. Proc. VLDB Endow. **7**(10), 919–930 (2014)
20. Wang, J., Liu, S., Li, Y., Cao, H., Liu, M.: Differentially private spatial decompositions for geospatial point data. China Commun. **13**(4), 97–107 (2016)
21. Xiao, Y., Gardner, J., Xiong, L.: DPCube: releasing differentially private data cubes for health information. In: 2012 IEEE 28th International Conference on Data Engineering (ICDE), pp. 1305–1308. IEEE (2012)
22. Xiao, Y., Xiong, L., Yuan, C.: Differentially private data release through multidimensional partitioning. In: Jonker, W., Petković, M. (eds.) SDM 2010. LNCS, vol. 6358, pp. 150–168. Springer, Heidelberg (2010). https://doi.org/10.1007/978-3-642-15546-8_11
23. Xiong, P., Zhang, L., Zhu, T.: Reward-based spatial crowdsourcing with differential privacy preservation. Enterp. Inf. Syst. **11**(10), 1500–1517 (2017)
24. Zhang, X., Wang, M., Meng, X.: An accurate method for mining top-k frequent pattern under differential privacy. J. Comput. Res. Dev. **51**(1), 104–114 (2014)
25. Zhu, T., Li, G., Zhou, W., Philip, S.Y.: Differentially private data publishing and analysis: a survey. IEEE Trans. Knowl. Data Eng. **29**(8), 1619–1638 (2017)

A First Step Towards Combating Fake News over Online Social Media

Kuai Xu[1](✉), Feng Wang[1], Haiyan Wang[1], and Bo Yang[2]

[1] Arizona State University, Glendale, USA
{kuai.xu,fwang25,haiyan.wang}@asu.edu
[2] Jiangxi University of Finance and Economics, Nanchang, China
jxncluoming@hotmail.com

Abstract. Fake news has recently leveraged the power and scale of online social media to effectively spread misinformation which not only erodes the trust of people on traditional presses and journalisms, but also manipulates the opinions and sentiments of the public. Detecting fake news is a daunting challenge due to subtle difference between real and fake news. As a first step of fighting with fake news, this paper characterizes hundreds of popular fake and real news measured by shares, reactions, and comments on Facebook from two perspectives: Web sites and content. Our site analysis reveals that the Web sites of the fake and real news publishers exhibit diverse registration behaviors and registration timing. In addition, fake news tends to disappear from the Web after a certain amount of time. The content characterizations on the fake and real news corpus suggest that simply applying term frequency - inverse document frequency (tf-idf) and Latent Dirichlet allocation (LDA) topic modeling is inefficient in detecting fake news, while exploring document similarity with the term and word vectors is a very promising direction for predicting fake and real news. To the best of our knowledge, this is the first effort to systematically study the Web sites and content characteristics of fake and real news, which will provide key insights for effectively detecting fake news on social media.

1 Introduction

The last decade has witnessed the rapid growth and success of online social networks, which has disrupted traditional media by fundamentally changing how, who, when, and where on the distribution of the latest news stories. Unlike traditional newspapers or magazines, anyone can spread any information at any time on many open and always-on social media platforms without real-world authentications and accountability, which has resulted in unprecedented circulation and spreadings of fake news, social spams, and misinformation [1–5].

Driven by the political or financial incentives, the creators of fake news generate and submit these well-crafted news stories on online social media, and subsequently recruit social bots or paid spammers to push the news to a certain popularity [6–8]. The recommendation and ranking algorithms on social media,

S. Chellappan et al. (Eds.): WASA 2018, LNCS 10874, pp. 521–531, 2018.
https://doi.org/10.1007/978-3-319-94268-1_43

if failed to immediately detect such fake news, likely surface such news to many other innocent users who are interested in the similar topics and content of the news, thus leading to a viral spreading process on social media. These rising social spams [9], click baits [10] and fake news [1], mixed with real news and credible content, create challenges and difficulties for regular Internet users to distinguish credible and fake content.

Towards effectively detecting, characterizing, and modeling Internet fake news on online social media [11], this paper proposes a new framework which systematically characterizes the Web sites and reputations of the publishers of the fake and real news articles, analyzes the similarity and dissimilarity of the fake and real news on the most important terms of the news articles via tf-idf and LDA topic modeling, as well as explores document similarity analysis via Jaccard similarity measures between fake, real and hybrid news articles.

The contributions of this paper are three-fold:

- We systematically characterize the Web sites and reputations of the publishers of the fake and real news articles on their registration patterns, Web site ages, and the probabilities of news disappearance from the Internet.
- We analyze the similarity and dissimilarity of the fake and real news on the most important terms of the news articles via term frequency - inverse document frequency (tf-idf) and Latent Dirichlet allocation (LDA) topic modeling.
- We explore document similarity between fake, real or hybrid news articles via Jaccard similarity to distinguish, classify and predict fake and real news.

The remainder of this paper is organized as follows. Section 2 describes the background of the fake news problem over online social media and describes data-sets used in this study. Section 3 characterizes the Web sites and reputations of the publishers of the fake and real news articles, while Sect. 4 focuses on analyzing the similarity and dissimilarity of the fake and real news on the most important terms of the news articles. In Sect. 5, we show the promising direction of leveraging document similarity to distinguish fake and real news by measuring their document similarity. Section 6 summarizes related work in detecting and analyzing fake news and highlights the difference between this effort with existing studies. Finally, Sect. 7 concludes this paper and outlines our future work.

2 Background and Data-Sets

As online social media such as Facebook and Twitter continue to play a central role in disseminate news articles to billions of Internet users, fake and real news share the same distribution channels and diffusion networks. The creators of fake news, motivated by a variety of reasons including financial benefits and political campaigns, are very innovative in writing the news stories and attractive titles that convince thousands of regular people to read, like, comment, forward. Such high engagement in a short time period can make the news go viral with little challenges or doubts on authenticity, verification or fact checking.

In this paper we explore the research data shared from a recent study in [12]. The data consists three data-sets, each of which includes hundreds of fake and real news stories over a 3-month time-span from dozens of fake news sites as well as well-respected major news outlets including New York Times, Washington Post, NBC News, USA Today, and Wall Street Journal. These three data-sets are referred to as *dataset 1*, *dataset 2*, and *dataset 3* throughout the rest of this paper. For each fake or real news article, the data includes the tile of the story, the Web URL of the news story, the publisher of the news and the total engagement, measured by the total number of shares, likes, comments, and other reactions of the news received on Facebook.

3 Characterizing Fake and Real News

In this section, we study a variety of subjective features on the publishers of real and fake news such as the registration behaviors of publishers' Web sites, the sites ages of the publishers, and the probability of the news disappearance on the Internet.

3.1 Web Site Registration Behavior of the Publishers

The real or fake news publishers typically have to go through the domain registration process, which allows anonymous domain registrar to serve as a proxy for publishers who prefer to hind their identities. If a publisher chooses to remain anonymous, the Internet whois database will show the proxy, e.g., Domains By Proxy, LLC as the registration organization. Most popular and well known newspaper typically choose to use the real organization name during the registration process. For example, the registration organization for wsj.com is Dow Jones & Company, Inc, which owns Wall Street Journal newspaper.

Our findings show that the majority of the fake news publishers register their Web sites via proxy services to remain anonymous, while all the real news publisher use their real identifies during the domain registration process. As shown in Table 1, over 78% of the domains publishing fake news are registered via proxy services to hide their true identities of the domain owners, while less than 2% of the domains publishing real news are registered in such a fashion. Thus we believe such patterns can become a powerful feature for machine learning models to distinguish fake and real news.

Table 1. Domain registration with proxy service for hiding domain owners' identify

Category	Dataset 1	Dataset 2	Dataset 3	Average
Fake news	90%	65%	80%	78.3%
Real news	5%	0%	0%	1.7%

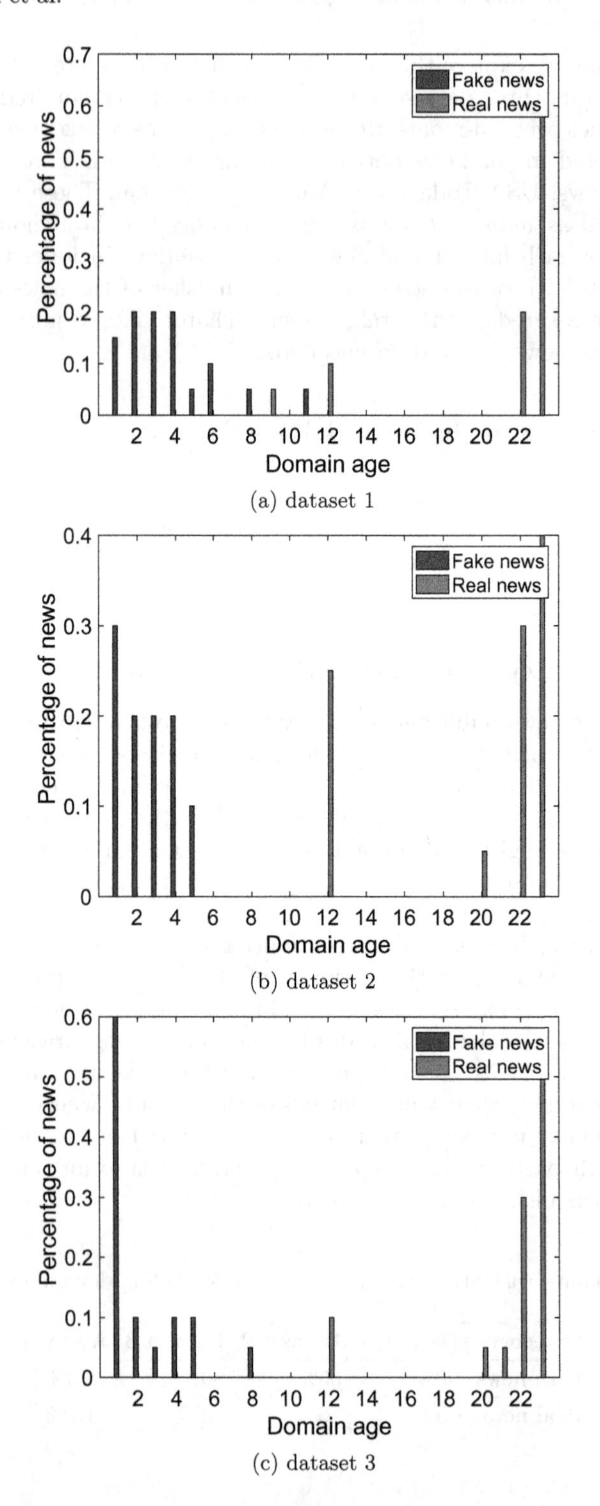

(a) dataset 1

(b) dataset 2

(c) dataset 3

Fig. 1. The Web site age distribution of the fake news publishers vs. real news publishers

3.2 Internet Site Ages of the Publishers

Beside the domain registration behavior, we also study the ages of the domains for the fake and real news in three data-sets. For each data-set, we characterize the domain age distribution for the fake and real news, respectively. As illustrated in Fig. 1, all data-sets exhibit consistent observations which reveal the very short domain ages for fake news, and the long domain ages for real news. This result is not surprising in that the credible newspapers registered their domains in early 1990s when Internet and Web start to attract attentions, while the fake news driven publishers often temporarily register the sites for the purpose of spreading fake news in a very short time of period.

3.3 Probability of News Disappearance

Credible news agency tends to maintain high quality sites that keep the published news for a long time. However, fake news sites often take the news offline after achieving the short-term goals of misleading the readers. Our analysis on the fake and real news corpus confirm such common practice.

Table 2. Page not found due to news disappearing.

Category	02/16–04/16	05/16–07/16	08/16–10/16	Average
Fake news	40%	70%	55%	55%
Real news	0%	0%	0%	0%

As shown in Table 2, the three data-sets of fake news corpus exhibit consistent news disappearing patterns, while the real news corpus has zero news that are taken offline. Thus we believe news disappearance could become a valuable feature for differentiating or modeling fake and real news.

In summary, our preliminary results on these popular fake and real news reveal substantial difference between fake and real news on the quality of the new pages, as well as the reputations of the publishing domains reflected by the domain ages as well as the interesting usage of the registration proxies.

4 Topics and Content of Fake and Real News

In this section, we first identify the most important topics of each fake or real news article via tf-idf analysis [13]. Subsequently, we explore the probabilistic LDA topic model to understand the difference or similarity of topics between labeled fake and real news.

Table 3. Top terms ranked by tf-idf values in fake, real and hybrid news corpus

Fake news corpus	Real news corpus	Hybrid news corpus
Violence	Trump	Comey
Trade	Nation	Transgender
Palin	Melania	Putin
Nuclear	Intelligence	Fraud
Mexico	FBI	Obama
Isis	Corrupt	Nuclear
Goods	Conway	Corrupt
Country	Conservative	Melania
Comey	Hillary	Isis
Canada	Wikileaks	Trump

4.1 Important Topics Identifications via tf-idf Analysis

tf-idf (term frequency - inverse document frequency) is a widely used statistical technique for extracting the most important term, t, or word, w, of a document in a document corpus, D. The tf-idf value of a term t, in the document d, tf-idf(t, d), is a product of the term frequency $tf(t,d)$ and the inverse document frequency $idf(t,D)$, i.e.,

$$tf\text{-}idf(t,d) = tf(t,d) * idf(t,D). \quad (1)$$

Table 3 shows that the most important terms extracted from fake, real and hybrid news corpus via tf-idf analysis are quite similar, thus relying on these terms alone is inefficient for detecting fake news.

4.2 Latent Dirichlet Allocation Topic Modeling

Topic models are widely used for understanding the content of documents based on word usage. In this paper, we explore Latent Dirichlet Allocation (LDA), a probabilistic topic model, to capture the topics of fake, real and hybrid news corpus respectively. The goal of LDA topic modeling on fake and real news is to understand the difference or similarity of topics between labeled fake and real news.

Tables 4, 5 and 6 illustrate the three topics with 5 most frequent terms for each corpus. As shown in these tables, the fake and real news share strong similarity in the overall topics, thus topic model alone is not an effective approach to detect or differentiate fake or real news in the real world.

Table 4. LDA topics for fake news corpus

Top 1	Top 2	Top 3
Trump	President	Candidate
Clinton	News	Black
Comey	State	Will
Hillary	American	One
Donald	Time	Said

Table 5. LDA topics for real news corpus

Top 1	Top 2	Top 3
Trump	Facebook	Republican
Clinton	Romney	Democratic
Donald	People	Authoritarian
President	Source	Politician
People	See	Party

Table 6. LDA topics for hybrid fake and real news corpus

Top 1	Top 2	Top 3
People	Trump	Trump
Authoritarian	Clinton	Donald
Politician	Republican	People
Party	Democratic	Make
American	President	Will

5 Document Similarity Analysis for News Predictions

As the LDA topics are inefficient to distinguish fake and real news, our followup analysis to explore document similarity between fake, real or hybrid news articles. First, we randomly divide the labeled fake and real news into training sets and test sets with a spit ratio of 67% for training and 33% for test.

For each fake or real news $\mathtt{n}$ in the test corpus, we measure the document similarity between $\mathtt{n}$ and every news in the fake news training set $\mathcal{F}$ and the real news training set $\mathcal{R}$. In particularly, we calculate Jaccard similarity $J(doc_1, doc_2)$, a widely used similarity measure between two documents doc_1 and doc_2 with the following equation

$$J(doc_1, doc_2) = \frac{doc_1 \cap doc_2}{doc_1 \cup doc_2}, \tag{2}$$

where doc_1 and doc_2 are represented with the vectors, typically sparse, of terms in the documents.

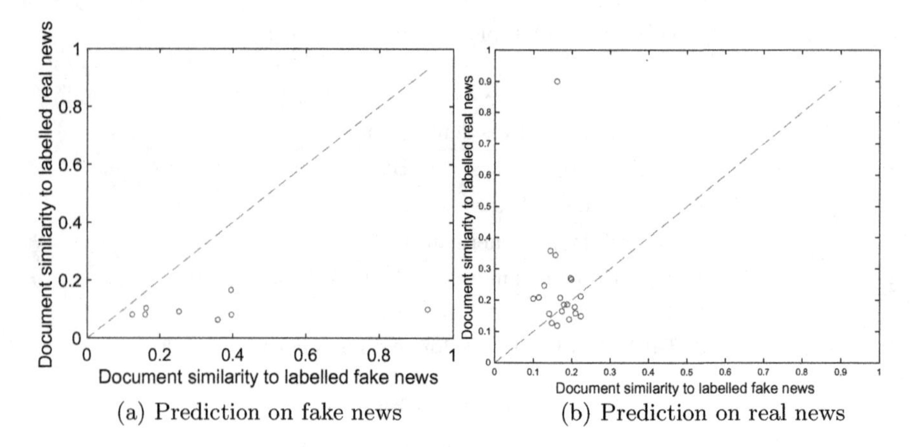

(a) Prediction on fake news (b) Prediction on real news

Fig. 2. The prediction on fake and real news based on labeled fake and real news corpus

Figure 2(a) shows the fake news in the test set have a much higher average document similarity with the news in the fake news training set $\mathcal{F}$ than with those in $\mathcal{R}$. However, Fig. 2(b) shows the real news in the test set have surprising similar document similarity with the news in the real news training set $\mathcal{R}$ and with those in $\mathcal{F}$. Thus as shown in Fig. 2(a), document similarity can potentially detect fake news. One of our future work is to systematically quantify the precision and recall of detecting both fake and real news in a large-scale news corpus.

In summary, our preliminary analysis on the topics and content of fake and real news reveals that it is very challenging to simply exploring the tf-idf and LDA topic modeling to effectively detecting fake news. However, our study also shows the promising aspect of leveraging document similarity to distinguish fake and real news by measuring the document similarity of the news under tests with the known fake and real news corpus.

6 Related Work

In recent years, several algorithms [1,2,8,14–20] have been proposed to detect the dissemination of information, misinformation or fake news. For example [2] exploits the diffusion patterns of information to automatically classify and detect misinformation, hoaxes or fake news, while [8] proposes linguistic approaches, network approaches, and a hybrid approach combining linguistic cues and network-based behavior insights for identifying fake news. In addition, [1] reviews the data mining literature on characterizing and detecting fake news on social media.

Similarly, [14] proposes a SVM-based algorithm for predicting misleading news with predictive features such as absurdity, grammar, punctuation, humor, and negative affect, and [15] uses logistic regression to distinguish credible news from fake news based on n-gram linguistic, embedding, capitalization, punctuation, pronoun use, sentiment polarity features. A recent effort in [16] formulates the fake news mitigation as the problem of optimal point process intervention in a network, and combines reinforcement learning with a point process network activity model for mitigating fake news in social networks.

In addition, [21] classifies the task of fake news detection into three different types: serious fabrications, large-scale hoaxes, and humorous fakes, and discusses the challenges of detecting each type of fake news. To address the lack of labeled data-sets for fake news detection, [22] introduces a real-world data-set consisting of 12,836 statements with real or fake labels. In [17], the authors locates the hidden paid posters who get paid for posting fake news via modeling the behavioral patterns of paid posters.

7 Conclusions and Future Work

As fake news and disinformation continue to grow in online social media, it becomes imperative to gain in-depth understanding on the characteristics of fake and real news articles for better detecting and filtering fake news. Towards effectively combating fake news, this paper characterizes hundreds of very popular fake and real news from a variety of perspectives including the domains and reputations of the news publishers, as well as the important terms of each news and their word embeddings. Our analysis shows that the fake and real news exhibit substantial differences on the reputations and domain characteristics of the news publishers. On the other hands, the difference on the topics and word embedding shows little or subtle difference between fake and real news. Our future work is centered on exploring the word2vec algorithm [23], a computationally-efficient predictive model based on neural networks for learning the representations of words in the high-dimensional vector space, to learn word embedding of the important words or terms discovered via the aforementioned tf-idf analysis. Rather than comparing the few important words of each new article, word2vec will allow us to compare the entire vector and embeddings of each word for broadly capturing the similarity and dissimilarity of the content in the fake or real news.

Acknowledgements. This work was supported in part by National Science Foundation Algorithms for Threat Detection (ATD) Program under the grant DMS #1737861.

References

1. Shu, K., Sliva, A., Wang, S., Tang, J., Liu, H.: Fake news detection on social media: a data mining perspective. ACM SIGKDD Explor. **19**(1), 22–36 (2017)
2. Tacchini, E., Ballarin, G., Della Vedova, M.L., Moret, S., de Alfaro, L.: Automated fake news detection in social networks. Technical report UCSC-SOE-17-05, School of Engineering, University of California, Santa Cruz (2017)
3. Mitchell Waldrop, M.: News feature: the genuine problem of fake news. Proc. Natl. Acad. Sci. **114**(48), 12631–12634 (2017)

4. He, Z., Cai, Z., Wang, X.: Modeling propagation dynamics and developing optimized countermeasures for rumor spreading in online social networks. In: Proceedings of IEEE International Conference on Distributed Computing Systems (ICDCS), June 2015
5. He, Z., Cai, Z., Yu, J., Wang, X., Sun, Y., Li, Y.: Cost-efficient strategies for restraining rumor spreading in mobile social networks. IEEE Trans. Veh. Technol. **66**(3), 2789–2800 (2017)
6. Shao, C., Ciampaglia, G.L., Varol, O., Flammini, A., Menczer, F.: The spread of fake news by social bots. arXiv.org, July 2017
7. Thorne, J., Chen, M., Myrianthous, G., Pu, J., Wang, X., Vlachos, A.: Fake news stance detection using stacked ensemble of classifiers. In: Proceedings of EMNLP Workshop: Natural Language Processing meets Journalism, September 2017
8. Conroy, N.J., Rubin, V.L., Chen, Y.: Automatic deception detection: methods for finding fake news. In: Proc. Assoc. Inf. Sci. Technol **52**(1) (2015)
9. Markines, B., Cattuto, C., Menczer, F.: Social spam detection. In: Proceedings of International Workshop on Adversarial Information Retrieval on the Web, April 2009
10. Chen, Y., Conroy, N.J., Rubin, V.L.: Misleading online content: recognizing clickbait as false news. In: Proceedings of ACM on Workshop on Multimodal Deception Detection, November 2015
11. Ruchansky, N., Seo, S., Liu, Y.: CSI: a hybrid deep model for fake news detection. In: Proceedings of ACM on Conference on Information and Knowledge Management, November 2017
12. This Analysis Shows How Viral Fake Election News Stories Outperformed Real News on Facebook. https://www.buzzfeed.com/craigsilverman/viral-fake-election-news-outperformed-real-news-on-facebook
13. Salton, G., McGill, M.J.: Introduction to Modern Information Retrieval. McGraw-Hill, New York City (1983)
14. Rubin, V.L., Conroy, N.J., Chen, Y., Cornwell, S.: Fake news or truth? Using satirical cues to detect potentially misleading news. In: Proceedings of Annual Conference of the North American Chapter of the Association for Computational Linguistics: Human Language Technologies, June 2016
15. Hardalov, M., Koychev, I., Nakov, P.: In search of credible news. In: Dichev, C., Agre, G. (eds.) AIMSA 2016. LNCS (LNAI), vol. 9883, pp. 172–180. Springer, Cham (2016). https://doi.org/10.1007/978-3-319-44748-3_17
16. Farajtabar, M., Yang, J., Ye, X., Xu, H., Trivedi, R., Khalil, E., Li, S., Song, L., Zha, H.: Fake news mitigation via point process based intervention. In: Proceedings of International Conference in Machine Learning (ICML) (2017)
17. Chen, C., Wu, K., Srinivasan, V., Zhang, X.: Battling the Internet water army: detection of hidden paid posters. In: Proceedings of IEEE/ACM International Conference on Advances in Social Networks Analysis and Mining (ASONAM), August 2013
18. Wang, F., Wang, H., Xu, K., Wu, J., Jia, X.: Characterizing information diffusion in online social networks with linear diffusive model. In: Proceedings of IEEE International Conference on Distributed Computing Systems (ICDCS), Philadelphia, PA, July 2013
19. Wang, F., Wang, H., Xu, K.: Diffusive logistic model towards predicting information diffusion in online social networks. In: Proceedings of IEEE ICDCS Workshop on Peer-to-Peer Computing and Online Social Networking (HOTPOST), Macao, China, June 2012

20. Dai, G., Ma, R., Wang, H., Wang, F., Xu, K.: Partial differential equations with robin boundary condition in online social networks. Discret. Contin. Dyn. Syst. - Ser. B (DCDS-B) **20**(6), 1609–1624 (2015)
21. Rubin, V.L., Chen, Y., Conroy, N.J.: Deception detection for news: three types of fakes. In: Proceedings of the Association for Information Science and Technology, February 2015
22. Wang, W.Y.: Liar, liar pants on fire: a new benchmark dataset for fake news detection. In: Proceedings of Annual Meeting of the Association for Computational Linguistics (ACL), July 2017
23. Mikolov, T., Chen, K., Corrado, G.S., Dean, J.: Efficient estimation of word representations in vector space. In: Proceedings of International Conference on Learning Representations (2013)

Predicting Smartphone App Usage with Recurrent Neural Networks

Shijian Xu[1], Wenzhong Li[1(✉)], Xiao Zhang[1], Songcheng Gao[1], Tong Zhan[1], Yongzhu Zhao[2], Wei-wei Zhu[3], and Tianzi Sun[4]

[1] State Key Laboratory for Novel Software Technology, Nanjing University, Nanjing, China
141220120@smail.nju.edu.cn, lwz@nju.edu.cn
[2] State Grid Shaanxi Electric Power Company, Xi'an, China
[3] State Grid Gansu Electric Power Company, Lanzhou, China
[4] State Grid Inner Mongolia East Electric Power Company, Hohhot, China

Abstract. Nowadays millions of apps are available and most of users install a lot of apps on their smartphones. It will cause some troubles in finding the specific apps promptly. By predicting the next app to be used in a short term and launching them as shortcuts can make the smartphone system more efficient and user-friendly. In this paper, we formulate the app usage prediction problem as a multi-label classification problem and propose a prediction model based on Long Short-term Memory (LSTM), which is an extension of the recurrent neural network (RNN). The proposed model explores the temporal-sequence dependency and contextual information as features for prediction. Extensive experiments based on real collected dataset show that the proposed model achieves better performance compared to the conventional approaches.

1 Introduction

Nowadays, smartphones are ubiquitous in people's daily life. Mobile applications are easily accessible and there are various types. This results in the rising number of apps installed in smartphones. According to the report of Statista[1], by March 2017, the number of apps available for download on Google Play is 2.8 million and that on Apple App Store is 2.2 million. The report conducted by App Annie[2] said that by 2017, the average number of apps installed on smartphones is about 80 to 90 and the average number of apps used per day is up to 10. This phenomena has caused some troubles for users to find the specific app they want to use. When a user is going to use some particular apps, they are not always available promptly and searching for the app can be time-consuming. So it is important to predict the next apps to be used.

[1] https://www.statista.com/statistics/276623/number-of-apps-available-in-leading-app-stores/.

[2] https://www.appannie.com/en/insights/market-data/global-consumer-app-usage-data/.

S. Chellappan et al. (Eds.): WASA 2018, LNCS 10874, pp. 532–544, 2018.
https://doi.org/10.1007/978-3-319-94268-1_44

Some efforts have been made to achieve prompt access to some specific apps. For example, some smartphone systems provide homescreen smart assistance that lets users open the apps they just used according to the chronological order. But this "prediction" is far from accurate. Some works proposed the usage of contextual information to predict apps. Verkasalo [1] found that the location and time of the day are the most important contextual features for app usage prediction. Shin et al. adopted a lot of context data collected from different kinds of sensors to enhance prediction result [2]. Apart from context analysis, machine learning approaches were widely used and many works used the naive Bayes classifers or hidden Markov model [6]. However, the existing approaches neglected the temporal-sequence dependency of app usage, which cannot capture the long-term patterns revealed by the historical information. In this paper, we propose the usage of recurrent neural network (RNN) model for app prediction, which can exploit the long-term temporal-sequence dependency to achieve better prediction precision.

Given the fact that people commonly use multiple apps at the same time (e.g., a user may shopping on Amazon while chatting with friends on WhatsApp to decide what to buy), we formulate the app prediction problem as a multi-label prediction problem: Given a list of apps installed on the smartphone and the historical records of when and where the apps were used, predict the apps to be used by the user in the next time interval. To solve the problem, we propose a prediction model based on Long Short-term Memory (LSTM), which is an enhancement of the RNN model. We implement the proposed model in Android smartphone platform and conduct extensive experiments based on real collected dataset. The results show that the proposed LSTM model outperform the baselines for app usage prediction.

The contributions of the paper are summarized below:

- We propose a novel LSTM based model for app usage prediction, which leverages both temporal-sequence dependency and contextual information to enhance the prediction precision.
- We show that app usage has strong temporal and spacial correlations, based on which we extract contextual features such as time and location to construct the prediction model.
- We implement the prediction model in Android platform and conduct experiments based on real collected dataset, which verifies the performance of the proposed approach.

2 Related Work

In this section, we briefly summarize related works on app usage prediction and temporal-sequence prediction.

2.1 App Usage Prediction

The problem of predicting smartphone app usage has been studied in many prior works. Most of them are based on contextual information for app prediction. As

Verkasalo [1] found, when and where the apps were used are the most important context information for prediction. Besides, other sensor contexts like the Wi-Fi signals, battery usage, etc., were also found useful for app prediction [2]. Parate et al. [3] proposed an app prediction algorithm called APPM that can work without requiring sensor context. Xu et al. [4] proposed a framework that jointly considered three key drivers of app usage: the community behavior, the contextual signals, and the user-specific preferences. They combined different kinds of sensor data and contextual features into a entirety which was called the App Bag and then computed the similarities between them for prediction.

Some works attempted to apply machine learning approaches for app prediction. Shin et al. [2] analyzed the context collected from users' phones and proposed a personalized naive Bayes model for each user. The prediction exploited the probabilistic models to infer the probabilities of apps with respect to a given context. Baeza-Yates et al. [6] proposed a method called PTAN (Parallel Tree Augmented Naive Bayesian Network) that combined the basic features collected from smartphone sensors and session features capturing the latent relations between app actions to predict the app usage.

Yan et al. [5] proposed a framework called FALCON to predict and prelaunch the apps. They found that the app usage can be a temporal burst. So it first predicted whether a burst of an app is underway. Then it checked whether the current location suggests the use of that app. Based on them, the app will be decided to be prelaunched or not.

However, the existing approaches failed to exploit the temporal-sequence information properly and none of them tried to jointly make use of historical data and context information for prediction, which is the focus of our work.

2.2 Temporal-Sequence Prediction

Various machine learning approaches have been applied to predict smartphone app usage, like naive Bayes classifiers [2], SVM, and so on. But few of them used neural networks. The neural network has a powerful representation ability and it has been successfully applied in a lot of areas, especially in computer vision [7,8] and natural language processing [9,11]. LSTM (Long Short-Term Memory) is a variant of recurrent neural network (RNN) which can capture the temporal dependency and alleviate the gradient vanishing problem. LSTM had been successfully applied in the areas of mobile computing for prediction. Li et al. [12] used LSTM for blood pressure prediction. They proposed a recurrent model with contextual layer which can cope with temporal-sequence data and contextual data properly. Wang et al. [13] proposed a spatiotemporal model based on LSTM for the prediction of traffic in cellular networks.

3 Data Collection and Observation

3.1 Data Collection

We develop a background Android app to collect the data of application usage and its context information via smartphone. The collected information includes the apps used (the package name), the timestamp, and the location of the smartphone. We make a coarse representation of the temporal feature by dividing a day into four time intervals, denoted by 0 (from 0 a.m. to 6 a.m.), 1 (from 6 a.m. to 12 a.m.), 2 (from 12 a.m to 18 p.m.) and 3 (from 18 p.m. to 24 p.m.). The detailed descriptions of the collected data are listed in Table 1.

Table 1. Data description

Data	Discription
App	The apps used, sorted by timestamp
Location	The altitude and latitude of the cellphone
Time	Time interval in a day when the apps used

We recruited 42 student volunteers to participate in the data collection. The students were asked to install the app in their Android smartphones as background service. These data were collected every five minutes and uploaded to the server everyday lasting for one month.

3.2 Data Observation

After data collection, we need to preprocess the dataset. We first remove the repeated data submissions due to the network delay and data cache. Besides, there exist some data submissions that are unqualified. Some volunteers diabled some sensors in their phones, like GPS, which caused the missing data. After removing these data, we finally have a dataset consists of 31 volunteers. After data preprocessing, we make some observations to the dataset.

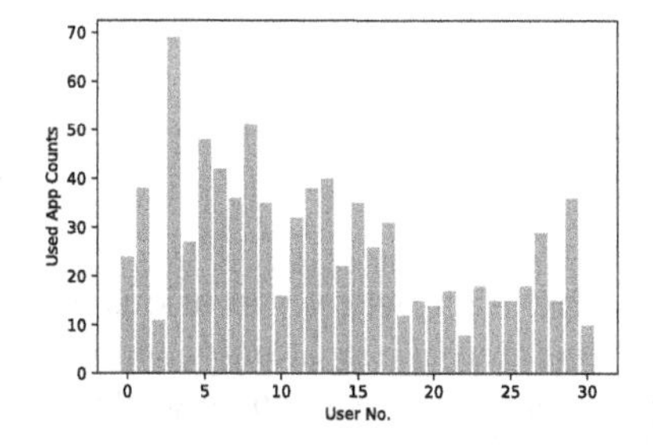

Fig. 1. The number of used apps.

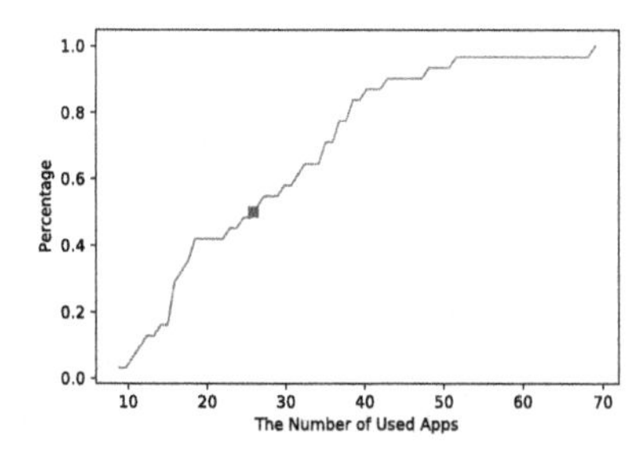

Fig. 2. CDF of the number of used apps. (Color figure online)

Number of APPs Used. We count the numbers of apps used of each user. The result is shown in Fig. 1. We also draw the CDF (cumulative distribution function) of the number of used apps, which result is shown in Fig. 2. As shown in the figures, most of users have the number of used apps between 15 to 40. The average value (shown as the green point in Fig. 2) is 27. These numbers are relatively large and the users will have difficulties in selecting the specific app they want to use from so many apps. Therefore, accurate prediction of the next used app can save a lot of troubles for smartphone users.

Intuitively, app usage could be related to the context. For example, some people like to read emails in the morning; some people like to chat with friend in the evening; some people tend to open a Map app when they are driving to work. App usage could be correlated to the time and location. We explore the temporal and spacial correlations below.

Temporal Correlation. Temporal correlation implies that people tend to use different kinds of apps in different time. To show the correlation, we randomly choose three users, and draw the frequency of using different apps in different time. The results are shown in Fig. 3. The warmer the color is, the more frequently the app is used.

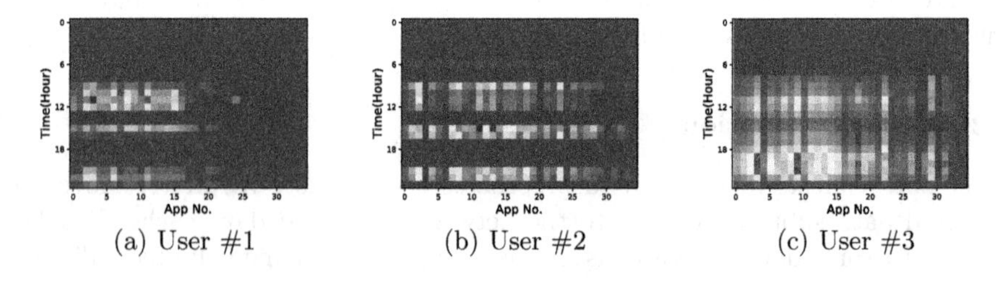

(a) User #1 (b) User #2 (c) User #3

Fig. 3. Temporal correlation (Color figure online)

As shown in the figure, it is clear that there is a pattern for users to use apps in different time slots. Especially we can see that people use fewer apps during 0 a.m. to 6 a.m. The three users have different patterns in app usage in different intervals. Some apps are more frequently used in some specific time intervals, which shows strong temporal correlations.

Spacial Correlation. We also analyze the correlation between locations and the use of apps. Intuitively, the usage of some apps could be location-dependent. For example, people will use more game apps when they are at home while use more social apps when they are at work. The analysis of spacial correlation of three random users is shown in Fig. 4.

As shown in the figure, the three users also have different patterns in app usage in different locations. Some apps are used more frequently in some specific locations, which shows strong spacial correlations.

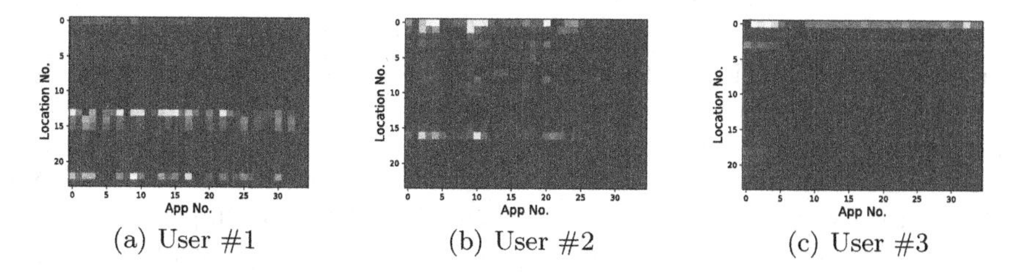

(a) User #1 (b) User #2 (c) User #3

Fig. 4. Spacial correlation

4 Problem Formulation

We define $\mathbf{x}^{(t)}$ as the input instance at timestamp t, where $\mathbf{x}^{(t)}$ is a vector that consists of several information: the app indicators (0–1 variables) that represent whether the app was used at t, and the contextual information at t (typically including the time and location of the cellphone usage). We define $\mathbf{y}^{(t)}$ as the label of the instance $\mathbf{x}^{(t)}$, where $\mathbf{y}^{(t)}$ is a vector that consists of the app indicators (0–1 variables) representing the apps that are actually used at the next timestamp. The app usage prediction problem is: find a mapping $f(\mathbf{x}^{(t)}) \rightarrow \hat{\mathbf{y}}^{(t)}$, so that the predicted label $\hat{\mathbf{y}}^{(t)}$ matches the actual label $\mathbf{y}^{(t)}$ as well as possible.

In the next section, we propose machine learning models to solve the problem.

5 Prediction Models

In this section, we introduce the models used for app usage prediction.

5.1 Recurrent Neural Network

Recurrent neural network (RNN) is a kind of artificial neural network. Unlike the standard neural network which assumes the training and test examples are independent, RNN can model input and/or output consisting of sequences of elements that are not independent. Further, recurrent neural networks can simultaneously model sequential and time dependencies on multiple scales. In structure, RNN includes the edges that span adjacent time steps, introducing a notation of time to the model [14]. The computation at each time step on the forward pass of RNN can be presented as:

$$\mathbf{h}^{(t)} = \sigma(\mathbf{W}^{hx}\mathbf{x}^{(t)} + \mathbf{W}^{hh}\mathbf{h}^{(t-1)} + \mathbf{b}_h) \tag{1}$$

$$\hat{\mathbf{y}}^{(t)} = softmax(\mathbf{W}^{yh}\mathbf{h}^{(t)} + \mathbf{b}_y) \tag{2}$$

Here the sigmoid[3] function $\sigma(\cdot)$ is defined as $\sigma(x) = 1/(1+e^{-x})$, which has a characteristic "S"-shaped curve and can squash the input into (0, 1) interval;

[3] https://en.wikipedia.org/wiki/Sigmoid_function.

the softmax[4] function $softmax(\cdot)$ is defined as $softmax(\mathbf{x})_j = e^{z_j} / \sum_{k=1}^{K} e^{z_k}$, which is a generalization of the logistic function that works on a vector; $\mathbf{h}^{(t)}$ is the result of sigmoid function at time step t; $\mathbf{W}^{hx}$ and $\mathbf{W}^{hh}$ are the weight matrices; $\mathbf{b}_h$ and $\mathbf{b}_y$ are biases.

However, because of the vanishing gradient problem, the traditional RNN can't capture very long historical dependency. Therefore an improvement version of RNN called Long Short-Term Memory (LSTM) is considered for better prediction performance.

Table 2. Description of notations

Notations	Description
t	The time step
$\mathbf{x}^{(t)}$	The input instance at time step t
$\mathbf{h}^{(t-1)}$	The output of sigmoid function at time step $t-1$
$\mathbf{y}^{(t)}$	The output at time step t
$\mathbf{b}_h, \mathbf{b}_y, \mathbf{b}_g, \mathbf{b}_i, \mathbf{b}_f, \mathbf{b}_o$	Bias
$\mathbf{i}, \mathbf{f}, \mathbf{o}$	Input gate, forget gate and output gate
$\mathbf{g}$	Input node
$\mathbf{s}$	Internal state
ϕ, σ	*tanh* function and *sigmoid* function
$\mathbf{W}$	Weight matrices

5.2 Long Short-Term Memory (LSTM)

In 1997, *Hochreiter* and *Schmidhuber* proposed the LSTM model in [15] in order to overcome the problem of vanishing gradients [10]. But nowadays, the LSTM is one of the most successful RNN structure. Unlike traditional RNN, LSTM replaces the activation function of the neurons to an intermediate type of storage via the memory cell, which is a composite unit that contains a node with a self-connected recurrent edge of fixed weight one, ensuring that the gradient can pass across many time steps without vanishing or exploding [14].

The following calculations describe the process of parameter update, which are performed at each time step.

$$\mathbf{g}^{(t)} = \phi(\mathbf{W}^{gx}\mathbf{x}^{(t)} + \mathbf{W}^{gh}\mathbf{h}^{(t-1)} + \mathbf{b}_g) \tag{3}$$

$$\mathbf{i}^{(t)} = \sigma(\mathbf{W}^{ix}\mathbf{x}^{(t)} + \mathbf{W}^{ih}\mathbf{h}^{(t-1)} + \mathbf{b}_i) \tag{4}$$

$$\mathbf{f}^{(t)} = \sigma(\mathbf{W}^{fx}\mathbf{x}^{(t)} + \mathbf{W}^{fh}\mathbf{h}^{(t-1)} + \mathbf{b}_f) \tag{5}$$

$$\mathbf{o}^{(t)} = \sigma(\mathbf{W}^{ox}\mathbf{x}^{(t)} + \mathbf{W}^{oh}\mathbf{h}^{(t-1)} + \mathbf{b}_o) \tag{6}$$

[4] https://en.wikipedia.org/wiki/Softmax_function.

$$\mathbf{s}^{(t)} = \mathbf{g}^{(t)} \odot \mathbf{i}^{(t)} + \mathbf{s}^{(t-1)} \odot \mathbf{f}^{(t)} \tag{7}$$

$$\mathbf{h}^{(t)} = \phi(\mathbf{s}^{(t)}) \odot \mathbf{o}^{(t)} \tag{8}$$

Here $\mathbf{g}, \mathbf{i}, \mathbf{f}$ and $\mathbf{o}$ are input node, input gate, forget gate and output gate. The input node takes activation with *tanh* function from the input layer $\mathbf{x^{(t)}}$ and the hidden layer $\mathbf{h^{(t-1)}}$. The **gate** is a sigmoid unit that likes the input node, takes activation from $\mathbf{x^{(t)}}$ and $\mathbf{h^{(t-1)}}$. They are called *gate* because if the output value of them is zero, then flow from the other node is cut off. In reverse, if the value is one, then all flow is passed through. The *tanh* function $\phi(\cdot)$[5] is defined as $tanh(x) = \frac{e^x - e^{-x}}{e^x + e^{-x}}$, which is another commonly used activation function in neural network. $\mathbf{W}^{ab}$ is the weight matrix between $\mathbf{a}$ and $\mathbf{b}$; $\odot$ is pointwise multiplication. The detailed description of these natations are listed in Table 2.

5.3 The Proposed Model

We adapt the LSTM model to fulfill the app prediction task as follows.

First, we use PCA to transform the features ($\mathbf{x}^{(t)}$) into fixed size input. PCA is the abbreviation of Principal Component Analysis, which was invented by Pearson in 1901 [17]. It is a commonly used method to reduce the dimension of data while keeping the components that have large contributions to the variance of the dataset. PCA can be done by eigenvalue decomposition of a data covariance (or correlation) matrix. In order to make sure that the lengths of features of all users are larger than the fixed size, we set the fixed input size to be as small as possible, which is 10 in the experiments.

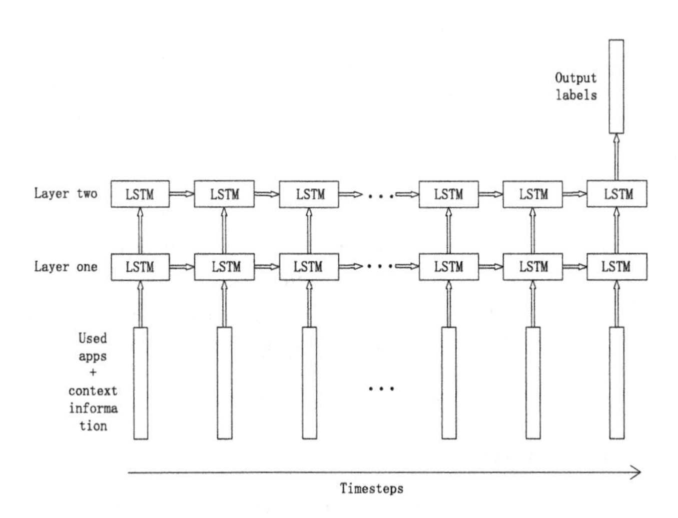

Fig. 5. The proposed LSTM model.

[5] https://en.wikipedia.org/wiki/Hyperbolic_function.

Second, we adopt a two layer LSTM model to predict the app usage in the next time slot. The structure of our model is shown in Fig. 5. As Fig. 5 shows, we use a two layer LSTM model. The input for LSTM should be a 3-dimensional tensor like $(samples, timesteps, input\_dim)$. Here, *samples* is the number of instances, *timesteps* is the length of history that LSTM needs, and *input_dim* is the dimension of each feature array. The first layer's input_dim is the same as the dimension of each feature array. The *output_dim* of the first layer is 50 and we use dropout layer after the first layer in order to prevent overfitting. Then the output will be inputed to the second layer, whose output_dim is 100. We also use dropout layer here. And finally, the output of layer two is inputed into the dense layer and the sigmoid activation layer to form the final result.

Last, we construct this model by Keras[6] with tensorflow as the backend. When compiling this model, we use MSE (Mean Squared Error) as the loss function and the optimization algorithm is $RMSprop$ (Root Mean Square Prop), which is an improved method of *AdaGrad* [19], proposed by Geoff Hinton [18].

6 Performance Evaluation

In this section, we conduct experiments on the collected dataset to evaluate the performance of the proposed approach.

6.1 Baseline Algorithms

We compare the proposed method with four traditional multi-label prediction methods.

- Binary Relevance (BR) [16]. Binary Relevance treats each dimension of the label as a separate single-class classification problem and applies basic classification algorithms such as Logistic Regression.
- Classifier Chains (CC) [16]. Classifier Chain treats each label as a part of a conditioned chain of single-class classification problems, which takes the label dependencies into consideration.
- Label Powerset (LP) [16]. Label Poweset treats each label combination as a separate class with one multi-class classification problem.
- Multi-Label k-NN (MLkNN) [16]. It is the generalization of kNN algorithm which makes classification decision based on the k nearest neighbours of the center point.

6.2 Evaluation Metrics

We adopt some commonly used metrics for performance comparison.

- **Accuracy.** It evaluates the proportion of correctly predicted labels to the total number of labels for each instance.

[6] https://keras.io.

- **Precision.** It evaluates the proportion of correctly predicted labels to the total number of predicted labels for each instance.
- **Recall.** It evaluates the proportion of correctly predicted labels to the total number of true labels for each instance.
- **F1-socre.** $F1$-score is the weighted harmonic average of the precision and recall.

The definitions of the performance metrics can be found in [16].

6.3 Numerical Results

For each user in the dataset, we train an individual prediction model using the proposed method, and the result are the average over all the users.

Comparison of Different Methods. We compare the proposed LSTM model with several baselines. Here we adopt a LSTM model without context, which means in this model the context information such as location and time are not used. The input of the LSTM model is a 0–1 array encoding the used apps. The results are shown in Table 3.

According to the table, the proposed LSTM model outperforms the baselines greatly in every performance metric, which proves that the histories of app usage have great effects on the prediction performance.

Table 3. Comparison of different methods.

	Accuracy	Precision	Recall	F_1-score
Binary relevance	0.4483	0.5728	0.6886	0.6254
Classifier chain	0.4550	0.5682	0.7130	0.6324
Label powerset	0.5067	0.6621	0.6814	0.6922
MLkNN	0.5982	0.7806	0.7419	0.7608
LSTM	**0.6596**	**0.7935**	**0.8079**	**0.8006**

Performance of LSTM with Different Context. We study the performance of LSTM with different contextual information by adding the combination of time and location to the inputs. Apart from LSTM without context, we have used three other LSTM models, of which the results are compared in Table 4.

According to the table, either combing with time or location can improve the performance of LSTM. The best performance achieves when LSTM is combined with both time and location information, which achieves the precision 0.80 in predicting app usage. It verifies that contextual information helps to enhance the performance of the proposed model.

Table 4. Comparison of different LSTM models.

	Accuracy	Precision	Recall	F_1-score
LSTM	0.66998	0.80137	0.81776	0.8095
LSTM+Time	0.67071	0.80314	0.81414	0.8086
LSTM+Location	0.66795	**0.80613**	0.80238	0.8042
LSTM+Time& Location	**0.67364**	0.80234	**0.81776**	**0.8099**

Hyperparameter Analysis. We study the performance with the choice of the hyperparameter of the LSTM model. The most important hyperparameter of the model is the timestep, which measures the length of historical records we use. The different length of history can apparently have great effect on the prediction performance.

In the experiments, we varies the length of timestep from 30 to 190. The performance results are shown in Fig. 6.

With the increasing of timestep, the accuracy, precision, recall and F-score zigzag go up and reach the peak. After that, they zigzag go down. The reason is explained below. If the length of historical records is too short, it may fail to reveal the APP usages. However, since the number of instances is fixed, if the size of the timestep is too large, it will produce fewer training samples, which may cause the model underfitting. Besides, it is observed that the prediction performance achieves the best when the length of timestep is around 120.

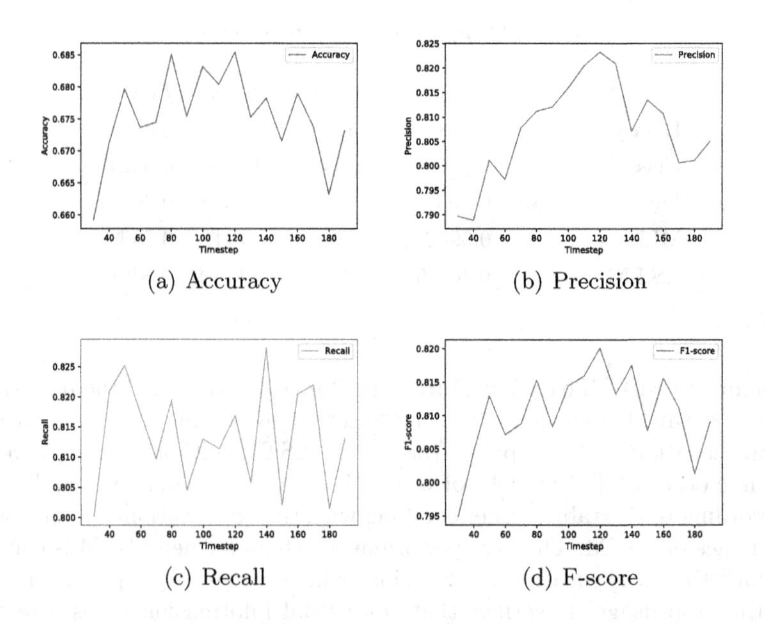

(a) Accuracy (b) Precision

(c) Recall (d) F-score

Fig. 6. Performance of the model under different hyperparameters.

7 Conclusion

Smartphone app usage prediction is important for saving people's time and improving user experience. In this paper, we propose a recurrent model that leverages the temporal-sequence information and contextual information to predict the smartphone's app usage pattern. We first analyze people's habit of using smartphones and formulate the app prediction problem as a multi-label classification problem. Then we make some observation on the collected dataset, which shows that app usage has a strong correlation with time and location. We extract temporal-sequence features and contextual features, which are used to developed a prediction model based on LSTM. We conduct experiments based on the collected dataset, which show that the LSTM model outperforms the baseline algorithms, and the contextual features can further improve the system performance, which achieves 0.80 precision in app usage prediction.

Acknowledgements. This work was partially supported by the National Key R&D Program of China (Grant No. 2017YFB1001801), the National Natural Science Foundation of China (Grant Nos. 61672278, 61373128, 61321491), the science and technology project from State Grid Corporation of China (Contract No. SGSNXT00YJJS1800031), the Collaborative Innovation Center of Novel Software Technology and Industrialization, and the Sino-German Institutes of Social Computing.

References

1. Verkasalo, H.: Contextual patterns in mobile service usage. Pers. Ubiquit. Comput. **13**(5), 331–342 (2009)
2. Shin, C., Hong, J.H., Dey, A.K.: Understanding and prediction of mobile application usage for smart phones. In: Proceedings of the 2012 ACM Conference on Ubiquitous Computing, pp. 173–182. ACM (2012)
3. Parate, A., Böhmer, M., Chu, D., Ganesan, D., Marlin, B.M.: Practical prediction and prefetch for faster access to applications on mobile phones. In: Proceedings of the 2013 ACM International Joint Conference on Pervasive and Ubiquitous Computing, pp. 275–284. ACM (2013)
4. Xu, Y., Lin, M., Lu, H., Cardone, G., Lane, N., Chen, Z., Campbell, A., Choudhury, T.: Preference, context and communities: a multi-faceted approach to predicting smartphone app usage patterns. In: Proceedings of the 2013 International Symposium on Wearable Computers, pp. 69–76. ACM (2013)
5. Yan, T., Chu, D., Ganesan, D., Kansal, A., Liu, J.: Fast app launching for mobile devices using predictive user context. In: Proceedings of the 10th International Conference on Mobile Systems, Applications, and Services, pp. 113–126. ACM (2012)
6. Baeza-Yates, R., Jiang, D., Silvestri, F., Harrison, B.: Predicting the next app that you are going to use. In: Proceedings of the Eighth ACM International Conference on Web Search and Data Mining, pp. 285–294. ACM (2015)
7. Donahue, J., Anne Hendricks, L., Guadarrama, S., Rohrbach, M., Venugopalan, S., Saenko, K., Darrell, T.: Long-term recurrent convolutional networks for visual recognition and description. In: Proceedings of the IEEE Conference on Computer Vision and Pattern Recognition, pp. 2625–2634 (2015)

8. Venugopalan, S., Xu, H., Donahue, J., Rohrbach, M., Mooney, R., Saenko, K.: Translating videos to natural language using deep recurrent neural networks. arXiv preprint arXiv:1412.4729 (2014)
9. Mikolov, T., Karafiát, M., Burget, L., Černocký, J., Khudanpur, S.: Recurrent neural network based language model. In: Proceedings of the Eleventh Annual Conference of the International Speech Communication Association (2010)
10. Bengio, Y., Simard, P., Frasconi, P.: Learning long-term dependencies with gradient descent is difficult. IEEE Trans. Neural Netw. **5**(2), 157–166 (1994)
11. Sutskever, I., Vinyals, O., Le, Q.V.: Sequence to sequence learning with neural networks. In: Advances in Neural Information Processing Systems, pp. 3104–3112 (2014)
12. Li, X., Wu, S., Wang, L.: Blood pressure prediction via recurrent models with contextual layer. In: Proceedings of the 26th International Conference on World Wide Web, pp. 685–693. International World Wide Web Conferences Steering Committee (2017)
13. Wang, J., Tang, J., Xu, Z., Wang, Y., Xue, G., Zhang, X., Yang, D.: Spatiotemporal modeling and prediction in cellular networks: a big data enabled deep learning approach. In: INFOCOM 2017-IEEE Conference on Computer Communications, pp. 1–9. IEEE (2017)
14. Lipton, Z.C., Berkowitz, J., Elkan, C.: A critical review of recurrent neural networks for sequence learning. arXiv preprint arXiv:1506.00019 (2015)
15. Hochreiter, S., Schmidhuber, J.: Long short-term memory. Neural Comput. **9**(8), 1735–1780 (1997)
16. Zhang, M.L., Zhou, Z.H.: A review on multi-label learning algorithms. IEEE Trans. Knowl. Data Eng. **26**(8), 1819–1837 (2014)
17. Pearson, K.: On lines and planes of closest fit to systems of points in space. Philos. Mag. **2**(11), 559–572 (1901). https://doi.org/10.1080/14786440109462720
18. Hinton, G.: Neural Networks for Machine Learning: Lecture 6a, Overview of Mini-Batch Gradient Descent (2016)
19. Duchi, J., Hazan, E., Singer, Y.: Adaptive subgradient methods for online learning and stochastic optimization. J. Mach. Learn. Res. **12**(Jul), 2121–2159 (2011)

A Self-organizing Base Station Sleeping Strategy in Small Cell Networks Using Local Stable Matching Games

Yiwei Xu[1](✉), Panlong Yang[2], Jian Gong[1], and Kan Niu[1]

[1] Electronic System Engineering Company of China, Beijing, China
elysian.hsu.7@gmail.com
[2] University of Science and Technology of China, Hefei, China
panlongyang@gmail.com

Abstract. A distributed small-base stations (s-BSs) sleeping approach is proposed for optimizing energy efficiency (EE) in small cell networks (SCNs). Different from the existing studies, the associating preferences of both s-BSs and user equipments (UEs) are considered to enhance the flexibility of optimization. The SCNs sleeping problems are modeled as matching markets. The s-BSs sleeping problem is disassembled into two subproblems: optimization of the associations and turning off of the superfluous s-BSs. Based on the matching game theory, we put forward a Matching Game Based UE Association Algorithm (MGBAA) in settling the first sub-problem and a local matching sleeping algorithm (LMSA) in solving the second. The proposed sleeping strategy optimizes the irrational associations as well as moderates the redundant s-BSs without whittling down network throughput. The simulation results verify that the suggested sleeping approach deployment scheme surpasses the existing cell planning scheme with the objective of the energy efficiency ratio (EER) under various scenarios.

Keywords: Sleeping strategy · Matching game · Small cell networks

1 Introduction

1.1 Backgrounds and Motivation

With regard to the future 5G era, it is predictable that millions more BSs with higher functionality and billions more smartphones and other devices with higher data rates will be associated [1]. The volatile growth in mobile traffic incurs extravagant energy consumption [2], which brings about running out of non-renewable energy resources and prompts harmful pollution to the environment [3]. To supply the energy consumption, mobile cellular network operators and devices spend more than $10 billion on electricity. This expenditure keeps on growing at a high rate [4]. Due to this, reducing energy consumption on cellular networks has drawn increasing attention in the recent past [5]. Consequently, the

S. Chellappan et al. (Eds.): WASA 2018, LNCS 10874, pp. 545–556, 2018.
https://doi.org/10.1007/978-3-319-94268-1_45

energy-friendly Base Stations (BSs) advancement is predictable to be a global architectural design for 5G [6]. In this way, the setback of energy efficiency optimization has become a crucial issue. To cope with the dilemma between the huge energy consumption and the network throughput, the sleeping strategy is indispensable to be conducted.

1.2 Related Works

Plenty of research has found out that BSs are typically underutilized, yet BSs with few or even without any communication event will typically consume more than 90% of their daily peak energy in the statistic part [7]. Conversely, since the electricity bill of the entire cellular networks primarily comes from energy consumption on BSs [8], BS sleeping-active deployment strategy has been established as one of the most effective energy efficient technologies [5]. Different from when in the active mode, during the sleeping mode, s-BSs only consume limited power and energy in sustaining the foundational functions so as to be woken up when topology changes as mentioned in [9]. Furthermore, for networks with multiple s-BSs, the optimization problem is complex to be resolved by centralized strategies, which may inevitably lead to an overload. Accordingly, to deal with the urgent dilemmas mentioned before, we will examine the sleeping strategy of BSs in small cell network under a self-organizing scheme [10].

1.3 Challenges and Contributions

There are three key challenges of the suggested matching game strategy. First, the BS sleeping strategies in previous studies [9,12–14] have not regarded the preference of both BSs and UEs, neither introduced the matching game theory for both-way selections between BSs and UEs. This work brings about matching games into self-organized sleeping strategy. Secondly, as the conventional matching approach is under static nodes and fixed preference lists [15–17], it is challenging to match the varying lists. Thirdly, the conventional matching reassigns all the disconnected UEs all over again after BS's turning off each time. Nevertheless, it certainly sets off heavy computational complication and consumes much energy.

The most important contributions of this paper can be summarized as follows:

- Establishing matching theory into sleeping strategy operation in SCNs, rather than random associations and cell planning scheme in [9] is an innovation. In multiple-SBs and multiple-UEs circumstances, a many-to-one matching market model is built up for initialization and updating of all connections. Varied from the conventional fixed one-off matching process, our process is circulatory.
- The exchange of information is extremely decreased in this local matching approach, where the decisions will only be made by the reconnecting UEs, whose former server s-BSs are turned off, so as to evade wasting of resources and complexity.

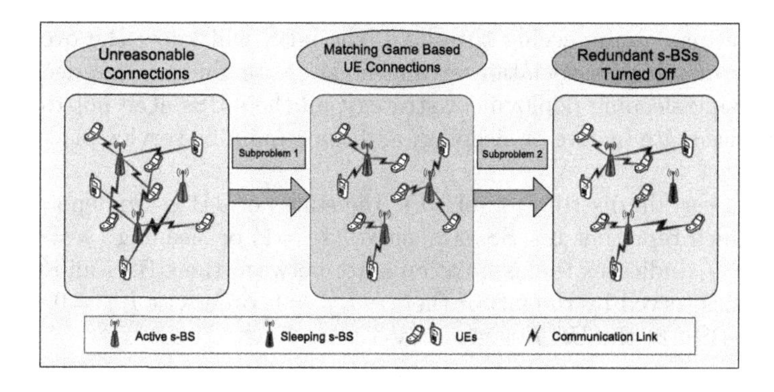

Fig. 1. The framework of the proposed sleeping strategy deployment process.

- A self-organizing BSs sleeping-active strategy utilizing Local Matching Sleeping Algorithm (LMSA) is put forward. The suggested algorithm directs optimizing sleeping strategy in converging to a two-sided stable matching. The results confirm that the suggested algorithm can advance EE without compromising the UEs' throughput.

The rest of this article is organized as follows. In Sect. 2, we present the system model and formulate the problem of BSs sleeping-active deployment based on matching game theory. In Sect. 3, we propose a BSs sleeping-active deployment algorithm using a local stable matching game to solve the assignment problem and prove that it can converge to a stable matching state. In Sect. 4, simulation results and discussion are presented. Finally, we make conclusions in Sect. 5.

2 System Model and Problem Formulation

2.1 System Model

An SCN is considered with N candidate s-BSs $\mathcal{N} = \{n_1, n_2, \ldots, n_N\}$ and M randomly distributed UEs $\mathcal{M} = \{m_1, m_2, \ldots, m_M\}$. These locations of candidate s-BSs are selected uniformly to meet the demand of random uniformly distributed UEs. To serve as much as possible UEs in this region, the locations of the candidate s-BSs can be arranged as dense as possible, according to the density of the UEs at a certain traffic circumstance. However, the more s-BSs we place, the more basic consumption will cost, and the deployment of s-BSs stays settled once the plan is decided. Therefore, it is necessary to alleviate the power and energy consumption through turning active s-BSs into sleeping mode.

We presume that every s-BS has two working modes, active mode and sleeping mode, between which they can switch freely. When the random distribution of the UEs changes, a portion of s-BSs may decide to switch to sleeping mode or active mode, and the network association topology between the s-BSs and the UEs may be rescheduled. Under a certain distribution of the UEs, the idle s-BSs tend to switch off. When the number of the UEs associated to one certain s-BS reaches the maximum serving threshold, the s-BS will know the overflow and trigger the matching association arrangement. Note that in this decentralized self-organizing sleeping deployment strategy, all the s-BSs need not to be aware of the service state (active or sleeping) and the remaining service places of other s-BSs.

Under a certain distribution of UEs, the states of s-BSs are denoted by $\mathbf{S} = [s_i]_{1\times N}$, which represent if s-BSs are active ($s_i = 1$) or sleeping ($s_i = 0$). While $L = [l_{i,k}]_{N\times M}$ indicates the association state between the s-BS i and the UE k, if the UE k is served by the s-BS i there is $l_{i,k} = 1$, otherwise $l_{i,k} = 0$. The SNR from the s-BS i to the UE k is given by

$$\gamma_{i,k} = \frac{p_{i,k} g_{i,k}}{\sigma}, \tag{1}$$

where $p_{i,k}$ denotes the transmission power allocated to UE k by s-BS i per channel, $g_{i,k}$ is the channel gain between s-BS i and UE k, and σ is the power of UE terminal noise. The channel gain is defined by $g_{i,k} = {d_{i,k}}^{-\alpha}$, where $d_{i,k}$ is the distance between s-BS i and UE k and α is the attenuation factor. And according to this definition, it is assumed that UE k can be connected with s-BS i when the SNR exceeds a certain threshold. Assume that every UE will be served only by one s-BS, that is $\sum_{j=1}^{N} l_{j,k} = 1$. And on the basis of the former research [9,12,14,19], the UE will mostly prefer the one offering the largest SNR if the largest one has enough serving places and is available for this UE. Thus, the Shannon capacity gained by UE k is obtained as $R_{i,k} = \log_2(1+\gamma_{i,k})$, where $\gamma_{i,k}$ is the SNR given in (1).

The communication cost is alleviated by switching off redundant s-BSs to sleeping mode, on the basis of ensuring all the UEs being well served. Although the s-BSs consume very limited power and energy during the sleeping mode, we can't turn the s-BSs off blindly, without considering the UEs' satisfaction. Therefore, to measure the effectiveness of the system, modified from [9], we define the energy efficiency ratio (EER) of the whole network system as (2) shown, which is designed as the ratio of the total capacity gained by all the UEs and the total energy consumption of the system:

$$\eta_{total} = \frac{\sum\limits_{i=1}^{N} s_i \sum\limits_{k=1}^{M} l_{i,k} R_{i,k}}{\sum\limits_{i=1}^{N} \left[s_i (\sum\limits_{k=1}^{M} p_{i,k} l_{i,k} + P_i^A) + (1 - s_i) P_i^S \right]}, \tag{2}$$

where P_i^A and P_i^S is the circuit power consumption of s-BS i when it is active or sleeping, respectively. The decision of the s-BSs is not only about switching on or off, but also about serving which UEs in its service area that will promote the system EER higher. As a result, the s-BSs will judge and make decisions about the active or sleeping mode by the current locations of UEs.

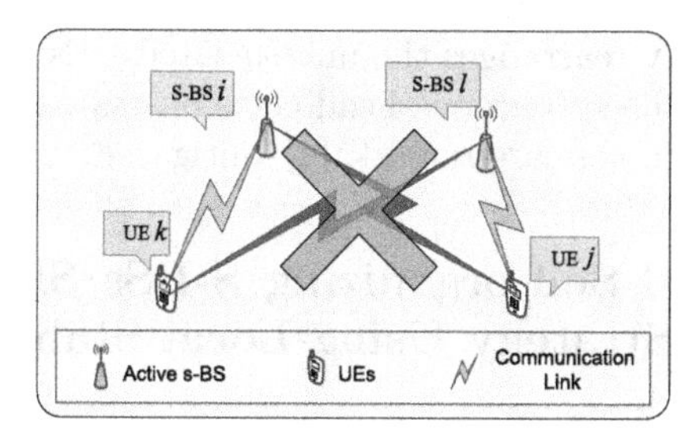

Fig. 2. The suboptimal blocking situation with the red paths and the optimal situation with the yellow paths. (Color figure online)

2.2 Problem Formulation

As depicted in Fig. 1, we disassemble the entire puzzle into two subproblems, the optimizing many-to-one association between the s-BSs and UEs, and the tentative closing procedure of redundant s-BSs according to the current UE distribution.

Subproblem 1: Matching Associations Between S-BSs and UEs. Illustrated in the Fig. 1 that the first objective is to determine the optimized connections between s-BSs and UEs. In the preliminary associations of the initialized networks, some unreasonable connections will block the performance of the following sleeping strategies. Specifically, some UEs may connect to their suboptimal s-BS instead of the available optimal one. For example in Fig. 2, the red paths show that the UE k is connecting to the farther s-BS l with lower SNR. However, the optimal matches are the green paths instead. We define the optimization problem as (3) described,

$$\begin{aligned} &\nexists \quad \mu\left(n_i, m_j\right) \quad \cup \quad \mu\left(n_l, m_k\right), \\ &s.t. \quad \gamma_{n_l,m_j} > \gamma_{n_i,m_j} \quad \cup \quad \gamma_{n_l,m_j} > \gamma_{n_l,m_k} \end{aligned} \tag{3}$$

where $\mu_{n,m}$ is defined as the association results that contain the connection between s-BS n and UE m, and $\gamma_{n,m}$ is the SNR between s-BS n and UE m, which will be the preference function. This means when γ_{n_l,m_j} is larger than γ_{n_i,m_j} and at the same time γ_{n_l,m_j} is larger than γ_{n_l,m_k}, then, the association pair can not be $\mu\left(n_i, m_j\right)$ and neither $\mu\left(n_l, m_k\right)$.

Subproblem 2: EE Promotion by S-BSs Sleeping. The second objective of this article is to find the s-BSs sleeping-active deployment to achieve the maximization of EER, which is presented as (4). Dealing with task $\mathcal{P}_2$, it is bold to promote this self-organizing sleeping-active deployment approach using the local matching game.

$$\mathcal{P}_2 : \max(\eta_{total}) \tag{4}$$

Different from the traditional matching game models, in which the system always reschedules all the UEs over again when one s-BSs is turned off, the proposed approach only rearranges the unassociated UEs served by the turned-off s-BS. In this way, the system overhead costs on rearrangement and decision making for updating process are remarkably mitigated.

3 The Proposed Self-organizing S-BSs Sleeping Deployment Strategy Using Local Stable Matching Game

3.1 Stable Matching Games

It is difficult to solve the first subproblem via classical optimization approaches. Furthermore, for a large-scale SCNs, it is more desirable using a decentralized, self-organizing method in which the s-BSs and the UEs can interact and make decisions based on their local information without relying on a centralized entity [10]. To this end, matching theory is a promising approach to address decentralized resource management in SCNs [15–17,21,22].

To tackle Subproblem 1, we build the problem as a many-to-one matching game model, where the s-BSs and the UEs are matched to each side as two market in (5). Suppose that both s-BSs' and UEs' preference relations are based on Eq. (1).

$$\mathcal{P}_1 : \mathcal{G}(\mathcal{N}, \mathcal{M}, \succ_n, \succ_m, \theta(\mathcal{M}), \theta(\mathcal{N})) \tag{5}$$

As mentioned earlier, s-BSs and UEs are two sides of preference relations, $\theta(\mathcal{M})$ is the preference lists of the UEs, $\theta(\mathcal{N})$ is the preference lists of s-BSs. The goal is to search the optimized matching solution as (3) mentioned along the process of switching off redundant s-BSs.

Based on the matching game theory and the traditional stable matching game [23,24], we propose a matching game based UE association algorithm (MGBAA) and a novel distributed self-organizing local matching sleeping algorithm (LMSA) of SCN for EER optimization. The algorithm flow chart of the total solution is generally given in Fig. 3, where Subproblem 1, used to be addressed by the traditional random association algorithm (TRAA), now is settled by the proposed MGBAA, instead. While Subproblem 2, used to be settled by traditional matching sleeping algorithm (TMSA), is now dealt by the proposed Algorithm 1 LMSA. MGBAA and LMSA will be presented elaborately in details in the following subsections.

3.2 The Framework of the Proposed Sleeping Strategy

As illustrated in the framework Fig. 1 and the algorithm flow diagram described in Fig. 3, the whole problem is divided into three stages and two subproblems, including the unreasonable connection stage, the optimized connection stage and the redundant s-BSs turned off stage. Subproblem 1 is optimized associations between s-BSs and UEs, and Subproblem 2 is EER promotion by s-BSs sleeping strategy.

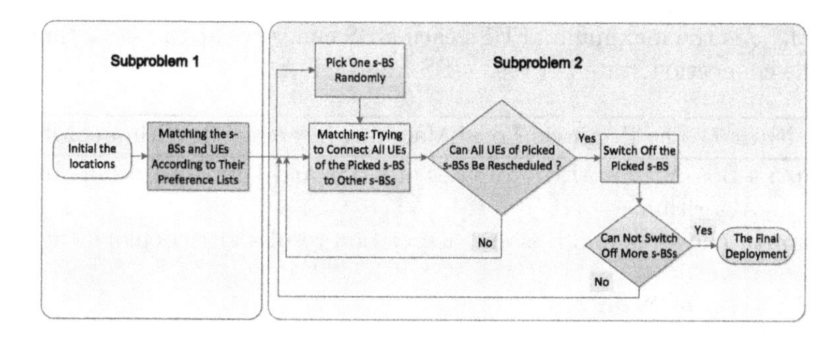

Fig. 3. The algorithm flow chart of the proposed sleep deployment strategy.

3.3 Subproblem 1: Matching Game Based UE Association Algorithm (MGBAA)

The MGBAA mainly consists of three stages: the generation of preference lists, matching evaluation, rearrangement of still-not-served UEs as following described. Instead of the randomly connection scheme in the traditional initializing connection algorithm proposed in [9], which is called traditional random association algorithm (TRAA) in the Sect. 4 for convenience, the proposed MGBAA comprehensively considers the preference of both the s-BSs and the UEs.

During the matching process, every UE reports its favor sequence list of s-BSs, according to the preference function (1), which is designed as $\theta(k) = \gamma_{i,k}$. Then, it will be fundamental to verdict whether the UE is in its service area also by (1). Meanwhile, the s-BS itself also needs to decide its preference list and whether it has extra places for the UEs proposed to connect. The association process will provide the premier connection results between the nodes of the s-BSs and the UEs.

3.4 Subproblem 2: The Proposed Local Matching Sleeping Algorithm (LMSA)

Assume one s-BS i is randomly chosen, preparing to be turned off to sleeping mode. Then, the UEs once served by s-BS i are rescheduled to other active s-BSs from its preference sequence set $\theta(m) = \gamma_{i,m}$ and build the new connection with the next optional s-BS p if the serving range condition are satisfied. The available place left of s-BS i is calculated with

$$P(i) = M_{\max} - \sum_{k=1}^{M} l_{i,k}, \tag{6}$$

where $M_{\max}$ is the maximum of UEs each s-BS can serve at the same time, and $l_{i,k}$ is the connection state between s-BS i and UE k.

Algorithm 1. The Proposed Local Matching Sleeping Algorithm (LMSA)

Input: s-BSs N,UEs M,coordinates of s-BSs and UEs. The results in Algorithm.

Output: The state of s-BSs and association results after deployment.

```
1  for iteration do
2      for i = 0 to N do
3          Randomly choose one s-BS i to turn off. Reschedule the UEs once
           served by the chosen s-BS i.;
4          for m ∈ {a, l_{i,a} = 1} do
5              Obtain the next option s-BS n of UE m from its preference
               sequence set θ(m) = γ_{i,m·} ;
6              if P(n) = M_max − Σ_{k=1}^{M} l_{n,k} > 0 then
7                  if γ_{n,m} > Γ then
8                      l_{i,m} = 0. l_{n,m} = 1.;
9                      break
10                 end
11             end
12             else
13                 if s(i) = 1 then
14                     Connect the top M_max UEs to the option s-BS n and
                       rearrange the last UE.;
15                     for p ∈ {b, l_{n,b} = 1} do
16                         Sort the γ_{n,p}, fetch the smallest UE q.
17                         l_{n,q} = 0.
18                     end
19                 end
20             end
21         end
22         if UEs m ∈ {a, l_{i,a} = 1} are successfully rearranged. then
23             S_i = 0.;
24             for m ∈ {a, l_{i,a} = 1} do
25                 l_{i,m} = 0.
26                 l_{n,m} = 1.
27             end
28         end
29         else
30             Cancel the off process and recover the settings.;
31         end
32     end
33 end
```

4 Simulation Results and Discussions

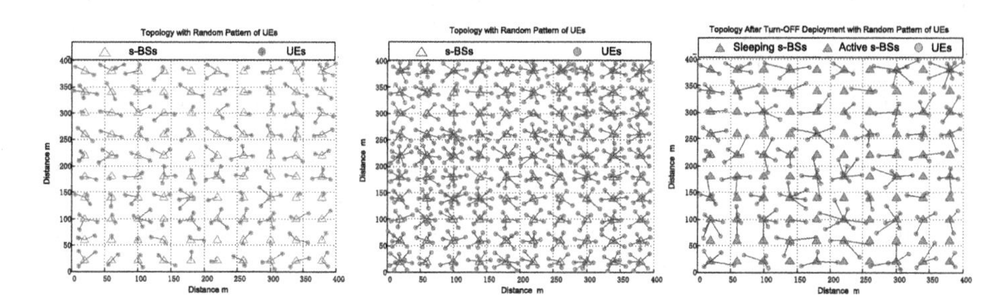

Fig. 4. The initial optimized associations with low-density UEs using the proposed MGBAA.

Fig. 5. The initial optimized associations with high-density UEs using the proposed MGBAA.

Fig. 6. The associations after sleeping strategy under low-density UEs using the proposed LMSA.

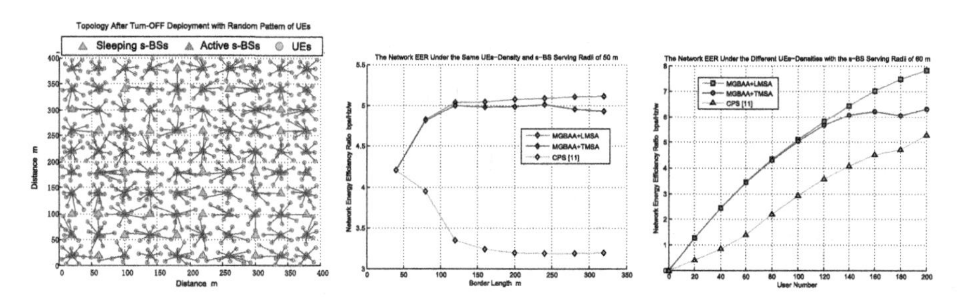

Fig. 7. The associations after sleeping strategy under low-density UEs using the proposed LMSA.

Fig. 8. The network EER under the same UEs-density with the s-BSs covering radii of 50 m.

Fig. 9. The network EER under the different UEs-density with the s-BSs covering radii of 60 m.

4.1 Simulation Scenario

In the serving area of 400 m $\times$ 400 m, the 100 candidate s-BSs' locations are uniformly distributed in two scenarios. One scenario is sparse with 200 randomly distributed UEs. The other scenario is dense with 600 stochastic scattered UEs. Partly referenced [9], the transmission power $p_{i,k}$ is settled to 16 dBm, the circuit power consumption of active s-BS is 39 dBm, and the circuit power consumption of sleeping s-BS is 35 dBm. The threshold of maximum number one s-BS can serve, is set to 10 UEs. And the serving radii of the s-BSs are set to 40 m.

4.2 The Association Results by the Proposed MGBAA

As illustrated in Figs. 4 and 5, the optimized association topology graphs under sparse UE distribution and dense UE distribution are shown, respectively. Before the sleeping-active deployment by turning off the superfluous s-BSs, the low-density distributed UEs are all associated and served by the corresponding s-BSs nearby as Fig. 4 depicted. However, it is very wasted for the active s-BSs to serve only a few UEs, far less than the threshold of s-BSs serving maximum. As a result, the EER of the system under this topology is fairly low. Even if the UE distribution is denser as Fig. 5 illustrated, a few less-loaded s-BSs can still be switch off by both traditional matching sleeping algorithm (TMSA) and the proposed local matching sleeping algorithm (LMSA), Algorithm 1.

4.3 The Sleeping Strategy Results by the Proposed LMSA

The sleeping deployment results after switching off the redundant s-BSs are illustrated in Figs. 6 and 7. After the active-sleeping deployment, the redundant s-BSs are switched off, while the necessary s-BSs are still active. It is obviously indicated by the sleeping deployment results that no matter the UE density is sparse or dense, there still some redundant s-BSs can be switched off.

4.4 Performance Comparison

The LMSA is compared with the cell planning scheme (CPS) proposed in [9].

Comparison Under the Same UE Density. First and foremost, the situation of the same UE density and the same s-BS density but different network scope is considered in this part. As illustrated in Fig. 8, as the area becoming larger, the EER has reached a steady trend. It is observed that the proposed LMSA has surpassed both the traditional TMSA and the CPS in [9].

Comparison Under the Different UE Density. Secondly, we consider the EER under the same serving area and the same s-BS density with the changing UEs' number in this part. As depicted in the following Fig. 9. After the turning off deployment, both LMSA and TMSA have a great promotion on EER. Even more, LMSA and the TMSA have all surpassed the cell planning scheme in [9]. Meanwhile, the EER results of the proposed local matching method LMSA have been on even terms with those of the TMSA, and even somehow surpassed the EER gained by the TMSA. And the good news is that the proposed LMSA needs less information exchange than the TMSA.

5 Conclusion

Based on many-to-one matching game theory, this paper proposed a novel local self-organizing sleeping deployment strategy. Different from the traditional random associating approaches and the cell planning scheme in [9], we considered the preference of both s-BSs and UEs in the association procedure. We disassembled the entire puzzle into two subproblems, the association process between s-BSs and UEs and the tentative sleeping procedure of the redundant s-BSs. We proposed a matching game based UE association algorithm (MGBAA) and the proposed local matching sleeping algorithm (LMSA). As the simulation results shown, the proposed sleeping strategy deployment scheme outperforms the existing cell planning scheme with the objective of the energy efficiency ratio (EER). However, in the coming future, the dynamicity of UEs patterns will be the next challenge for self-organizing sleeping deployment strategy, and the diversity of the traffic patterns are waiting for deeper exploration. Secondly, it would be innovative and challenging to re-investigate the sleeping deployment strategy when the BSs are equipped with sensor and energy harvesting ability [25–27].

Acknowledgment. This research is partially supported by National key research and development plan 2017YFB0801702, NSFC with No. 61625205, 61632010, 61772546, 61772488.

References

1. Chih-Lin, I., Rowell, C., Han, S., Xu, Z., Li, G., Pan, Z.: Toward green and soft: a 5G perspective. IEEE Commun. Mag. **52**(2), 66–73 (2014)
2. Marsan, M.A., Meo, M.: Network sharing and its energy benefits: a study of European mobile network operators. In: Proceedings of IEEE GLOBECOM, pp. 2561–2567 (2013)
3. Son, K., Kim, H., Yi, Y., Krishnamachari, B.: Base station operation and user association mechanisms for energy-delay tradeoffs in green cellular networks. IEEE J. Sel. Areas Commun. **29**(8), 1525–1536 (2011)
4. Wong, W., Yu, Y., Pang, A.: Decentralized energy-efficient base station operation for green cellular networks. In: IEEE GLOBECOM (2012)
5. Feng, D., Jiang, C., Lim, G., Cimini Jr., L.J., Feng, G., Li, G.Y.: A survey of energy-efficient wireless communications. IEEE Commun. Surv. Tutor. **15**(1), 167–178 (2013)
6. Andrews, J.G., Buzzi, S., Choi, W., Hanly, S.V., Lozano, A., Soong, A.C.K., Zhang, J.C.: What will 5G be? IEEE J. Sel. Areas Commun. **32**(6), 1065–1082 (2014)
7. Oh, E., Krishnamachari, B., Liu, X., Niu, Z.: Toward dynamic energy efficient operation of cellular network infrastructure. IEEE Commun. Mag. **49**(6), 56–61 (2011)
8. Auer, G., et al.: How much energy is needed to run a wireless network? IEEE Wirel. Commun. **18**(5), 40–49 (2011)
9. Zhou, L., Sheng, Z., Wei, L.: Green cell planning and deployment for small cell networks in smart cities. Ad Hoc Netw. **43**, 30–42 (2016)
10. Zhang, Z., Long, K., Wang, J.: Self-organization paradigms and optimization approaches for cognitive radio technologies: a survey. IEEE Wirel. Commun. **20**(2), 36–42 (2013)
11. Roth, A.E., Sotomayor, M.A.O.: Two-Sided Matching: A Study in Game-Theoretic Modeling and Analysis. Cambridge University Press, Cambridge (1992)
12. Cao, D., Zhou, S., Niu, Z.: Optimal combination of base station densities for energy-efficient two-tier heterogeneous cellular networks. IEEE Trans. Wirel. Commun. **12**(9), 4350–4362 (2013)

13. Tsilimantos, D., Gorce, J.M., Altman, E.: Stochastic analysis of energy savings with sleep mode in OFDMA wireless networks. In: IEEE INFOCOM, pp. 1097–1105, April 2013
14. Soh, Y.S., Quek, T.Q.S., Kountouris, M., Shin, H.: Energy efficient heterogeneous cellular networks. IEEE J. Sel. Areas Commun. **31**(5), 840–850 (2013)
15. Jorswieck, E.A.: Stable matchings for resource allocation in wireless networks. In: 17th International Conference on Digital Signal Processing (DSP), Corfu, Greece, pp. 1–8 (2011)
16. Pantisano, F., Bennis, M., Saad, W., Valentin, S., Debbah, M.: Matching with externalities for context-aware user-cell association in small cell networks. In: Global Communications Conference (GLOBECOM), GA, Atlanta, pp. 4483–4488 (2013)
17. Gu, Y., Saad, W., Bennis, M., Debbah, M., Han, Z.: Matching theory for future wireless networks: fundamentals and applications. IEEE Commun. Mag. **53**(5), 52–59 (2015)
18. Zhang, X., Zhou, S., Yan, Y., Xing, C., Wang, J.: Energy efficient sleep mode activation scheme for small cell networks. In: 2015 IEEE 82nd Vehicular Technology Conference (VTC Fall), Boston, MA, pp. 1–4 (2015)
19. Peng, J., Hong, P., Xue, K.: Energy-aware cellular deployment strategy under coverage performance constraints. IEEE Trans. Wirel. Commun. **14**(1), 69–80 (2015)
20. Xu, Y., Chen, J., Wu, D.: Toward 5G : a novel sleeping strategy for green distributed base stations in small cell networks. In: 12th International Conference on Mobile Ad Hoc and Sensor Networks (MSN) (2016)
21. Liu, D., Xu, Y., Xu, Y., Ding, C., Xu, K., Xu, Y.: Distributed satisfaction-aware relay assignment: a novel matching-game approach. Trans. Emerg. Telecommun. Technol. **27**(8), 1087–1096 (2016)
22. Liu, D., Xu, Y., Shen, L., Xu, Y.: Self-organizing multiuser matching in cellular networks: a score-based mutually beneficial approach. IET Commun. **10**, 1928–1937 (2016)
23. Roth, A.: Deferred acceptance algorithms: history, theory, practice, and open questions. Int. J. Game Theory **36**(3), 537–569 (2008)
24. An, C., Zhang, L., Liu, W.: A spectrum allocation algorithm based on matching game. In: 2009 5th International Conference on Wireless Communications, Networking and Mobile Computing, pp. 1–3 (2009)
25. Chen, Q., Gao, H., Cai, Z., Cheng, L., Li, J.: Energy-collision aware data aggregation scheduling for energy harvesting sensor networks. In: 37th Annual IEEE International Conference on Computer Communications (INFOCOM 2018) (2018)
26. Shi, T., Cheng, S., Cai, Z., Li, J.: Adaptive connected dominating set discovering algorithm in energy-harvest sensor networks. In: 35th Annual IEEE International Conference on Computer Communications (INFOCOM 2016) (2016)
27. Shi, T., Li, J., Gao, H., Cai, Z.: Coverage in battery-free wireless sensor networks. In: 37th Annual IEEE International Conference on Computer Communications (INFOCOM 2018) (2018)

Dealing with Dynamic-Scale of Events: Matrix Recovery Based Compressive Data Gathering for Sensor Networks

Zhonghu Xu, Shuo Zhang, Jing Xu, and Kai Xing(✉)

School of Computer Science, University of Science and Technology of China, Hefei 230026, Anhui, China
{xzhh,zshuo,jxu125}@mail.ustc.edu.cn, kxing@ustc.edu.cn

Abstract. The mass data produced in sensor networks has triggered a large variety of applications, e.g., smart city, environmental monitoring, etc. However, gathering such data from vast number of sensors throughout the network is a daunting and costly work. Previous works suffer from either a high communication overhead or a poor data recovery resulted in compressive sensing due to the high risk of sparsity violation. This paper introduces a new data gathering method to address two problems: one is how to compress and gather the large volume data effectively, the other is how to keep various time/space-scale event readings unaltered in the gathered data. According to state-of-the-art, either problem can be solved well but never both at the same time. This paper presents the first attempt to tackle with both problems simultaneously for sensor networks, from theoretical design to practical experiments with real data. Specifically, we take advantage of the redundancy and correlation of the sensor data cross time and spatial domain, and based on which we further introduce our low-rank matrix recovery design effectively recovering the gathered data. The experiment results with real sensor datasets indicate that the proposed method could recover the original data with event readings almost unaltered, and generally achieve SNR 10 times (10 db) better than typical compressive sensing method, while keeping the communication overhead as low as compressing sensing based data gathering method.

Keywords: Sensor data gathering · Data compression
Matrix recovery

1 Introduction

Sensor networks are envisioned as vastly distributed networks comprised of a number of wired or wireless sensor nodes, with each equipped with multiple sensing modules to monitor environmental measurands, e.g., air, temperature, light, sound, etc. The large volume of data generated from sensor networks attracts lots

S. Chellappan et al. (Eds.): WASA 2018, LNCS 10874, pp. 557–569, 2018.
https://doi.org/10.1007/978-3-319-94268-1_46

of research effort and has triggered a wide range of applications, such as smart city, transportation, infrastructure, agriculture, ocean, etc. Since each sensor is resource constrained with limited capacity in computation and storage, the mass data of the network is usually gathered from the sensor nodes to the data sink using a multi-hop communication way.

In sensor networks, such a way of transmitting data requires considerably effort of communication and storage. A traditional way of solving such a problem includes distributed source coding techniques [1,2], in-network collaborative wavelet transform [3], clustered data aggregation and compression [4,5] and so on. They are not robust enough to deal with event readings and have limited capacity in compression. In recent years, the combination of the compressed sensing theory with sensor networks [6] flushes. However, the compressive sensing method is based on constant sparsity. Such a situation hardly holds in real cases, and data with changing sparsity would impact the recovery quality significantly.

Another problem of compressive sensing based data gathering is the event data gathering and recovery. A common method to tackle this is that sensor readings d can be decomposed as $d_n + d_\alpha$ under the assumption that d_α is sparse in time domain. However, a change in environment may be recorded by a significant amount of sensors which would further make d_α not directly sparse in spatial domain. Besides,d_n and d_α are not necessarily sparse in the same domain. It is doubtful that they can be combined to create d and preserve ds sparsity cross time and space domain.

Inspired by the idea of taking the temporal correlations between historically reconstructed data into account, we consider the data sensed at each node and at each time interval, and construct them as a data matrix. Based on the correlation of the sensed data temporally and spatially, we could compress the data in the framework of matrix recovery, and address the event data gathering problem at the same time. Compared with the existent work of data gathering in sensor networks, our approach has the following contributions:

- According to state-of-the-art, either event fidelity or data compression and gathering problem can be solved well but never both at the same time in sensor networks. This paper presents the first attempt to tackle with both problems simultaneously in sensor networks with diverse time/space-scale events.
- The proposed study on a real sensor network observes that constant sparsity hardly holds in real cases, while low-rank property may be true, which may provide a fresh vision for research in both compressive sampling applications and sensor networks with diverse time/space-scale events. Moreover, we further generalize the low-rank based optimization problem to nuclear norm based optimization in our matrix recovery design, to make our approach general and robust.

- Theoretical analysis indicates that our matrix recovery based method is robust over diverse time/spacescale event readings, which is a significant improvement compared to existing compressive sensing based data gathering approaches.
- Extensive experimental study is conducted based on real sensor network datasets. The results show that the proposed method has the ability to keep event readings almost unaltered, while generally achieves SNR about 10 **times (10 db) better** than typical compressive sensing method.
- Our method is able to reduce global scale communication cost without introducing intensive computation or complicated transmissions at each sensor.

The following paper is organized as follows. Section 2 provides the preliminary knowledge of Matrix Recovery. Section 3 formulates the data gathering problem into the framework of Matrix Recovery, and introduces our data gathering and recovery method. In Sect. 4, communication overhead of the proposed method is analyzed with comparison to compressive sensing. Afterwards, Sect. 5 presents the experimental results with real sensor datasets from [7]. Finally, we conclude this paper in Sect. 6.

2 Preliminaries

In matrix recovery, we treat the original signal as $\boldsymbol{X}$, where $\boldsymbol{X}$ here is a matrix with a scale of MN. Different from compressive sensing, $\boldsymbol{X}$ is no longer to be sparse even in a proper basis. Let $rank(\boldsymbol{X}) = r$, in MR problem r is usually much smaller than $min\{M, N\}$.

Similar to CS, we hope to recover $\boldsymbol{X}$ from some linear combinations of $\boldsymbol{X}_{ij}$. And in theory, the number of the combinations needed is at most $cr(M+N)$, where c is a constant [8].

Let A be a linear map from R^{MN} space to R^p space, the combinations we get could be written as $A(\boldsymbol{X})$, and what we only need to solve is the following optimization problem:

$$\min_{\boldsymbol{X}} ||\boldsymbol{X}||_* \quad \text{s.t.}\, A\boldsymbol{X} = \boldsymbol{b} \tag{1}$$

where $||\cdot||_*$ is the nuclear norm (the sum of σ_{ii} in SVD decomposition) and $\boldsymbol{b}$ is the vector of these combinations we could get. When considering noise in the measurements, we could further modify the above problem in the following format:

$$\min_{\boldsymbol{X}} \quad \mu||\boldsymbol{X}||_* + ||A\boldsymbol{X} - b||_{L_2} \quad \text{s.t.}\, A(\boldsymbol{X}) = \boldsymbol{b} \tag{2}$$

Like CS method, which transform the l_0 norm (the number of nonzero entries) to the l_1 norm (the sum of absolute values of the entries), here the nuclear norm (the sum of the singular values) replace the rank (the number of the nonzero singular values) and thus the problem becomes a convex optimization problem, and make sure it could be solved if $p \leq Cn^{5/4}rlogn$ (where $n = max(M, N)$ and C is a constant) [8].

3 Matrix Recovery Based Data Gathering Methodology

3.1 Network Model and Design Principle

In this paper, we consider a static sensor network comprised of N resource-constrained sensors, in which a base station (BS) collects data from the network. The intuition of our design is taking both the temporal correlation and spatial correlation into consideration together. Let $\boldsymbol{X}$ be a $m \times n$ matrix, with the rows representing the time and the columns representing sensors (in the meaning of space domain). Suppose there are M time instances and N sensors. The readings of each node is defined as $\boldsymbol{X}_{ij}(1 \leq i \leq M, 1 \leq j \leq N$, where $\boldsymbol{X}_{ij}$ is the reading of jth sensor at the time instance i.

Let A be a linear map from R^{MN} space to R^p space, $A(\boldsymbol{X}) = \boldsymbol{\Phi} \cdot vec(\boldsymbol{X})$ where $\boldsymbol{\Phi}$ is a $p \times MN$ matrix and vec is a linear map to transform a matrix into a vector by overlaying one column on another. Specifically, the matrices Φ is a random matrix generated to satisfy the RIP condition [9]. Before deployment, each sensor is pre-installed the same pseudo-random number generator. Once a sensor produces a reading, it generates a random vector with a length of p using the random seed according to current time instance and its ID. Each element of the random vector is *i.i.d.* sampled from a Gaussian distribution with mean 0 and variance $1/p$. With this pseudo-random number generation method, the random vectors generated at each sensor could be reproduced in the sink by using the same generator.

The dimension of the range of map A, namely p, means that: to recover $\boldsymbol{X}$, we only need know p elements. The key of the recovery is solving the following optimization problem:

$$\min_{\boldsymbol{X} \in R^{M \times N}} \quad \frac{1}{2}||A(\boldsymbol{X}) - b||_F^2 + \mu||\boldsymbol{X}||_* \quad \text{s.t.}\, A\boldsymbol{X} = \boldsymbol{b} \tag{3}$$

where the first part of the objective function is designed for noise, and the second part for low rank. Typically, p need to be not less than $cr(3m + 3n - 5r)$ [8].

Remark 1. We assume sensors have erroneous measurements. These erroneous measurements usually occurs at sporadic points and do not impact the sparsity of the data in the network, thus abnormal reading recovery and detection could still work well in compressive sensing based data gathering. However, these abnormal readings are different from event data in that event readings are usually geo-concentrated and interrelated, and reported from a group of sensors in close proximity. Such event readings occurred in various time and space scale would lead to diverse sparsity of the data, which would further result in sparsity violation and thus lead to poor recovery in compressive sensing based data gathering methods.

Remark 2. If we consider the $NM-dim$ vectors at the same time, it may not be easy to find a good basis to make the N signal vectors sparse. Interestingly, the

$N \times M$dim data matrix is approximately low rank.[1] Therefore, Matrix Recovery (MR) method is sufficient to compress such data, and more importantly, the issue of diversescale event data gathering problem which cannot be well addressed by CS method could be tackled under this framework.

3.2 Data Gathering Method

In this section, we generalize the currently data gathering method from compressive sensing to the realm of matrix recovery. Such an extension has two-fold advantages: one is making use of the correlation pattern of the data in time domain, and considering the data structure in both space and time domain together; the other is the event data collection, which would mute the power of CS method, could be dealt with here by this new MR method.

During data gathering, sensor readings are merged while being relayed along the collection path, e.g., treetype or chain-type topology, to the sink. Suppose a leaf node s_j is to report its reading at time instance t_1, s_j first generates a random vector Φ_{1j} of length p at time instance t_1 (using both t_1 and its ID s_j as seed to generate a random vector). Then s_j calculates the vector $\boldsymbol{X}_{1j}\Phi_{ij}$. At time instance t_2, s_j generates another random vector Φ_{2j} and calculates $\boldsymbol{X}_{2j}\Phi_{2j}$, and add it to the vector $\boldsymbol{X}_{1j}\Phi_{ij}$. Till time instance t_M, s_j gets $\boldsymbol{X}_{Mj}\Phi_{Mj}$ and finally has the vector sum $S_j = \sum_{i=1}^{M} \boldsymbol{X}_{ij}\Phi_{ij}$.

As a result, each sensor s_j gets its vector sum S_j till time instance t_M. When sensor s_j relays the vector S_j to the next sensor s_i, s_i will add S_j with its vector sum S_i, then relay the vector sum of $S_i + S_j$ to the next sensor along the collection path and eventually to the sink. The sink finally gets $\sum_{i=1}^{M} \boldsymbol{X}_{ij}\Phi_{ij}$.

Remark 3. During data gathering, each node sends out only one message of length $cr(3M + 3N - 5R)$ along the collection path regardless of their hop distance to the sink (details are presented in Sect. 4).

To deal with event data, we make a little modification of the proposed matrix recovery based data gathering framework. At first, we recall that the row of the data matrix $\boldsymbol{X}$ represents the data acquired at one time for all sensors and each column of matrix $\boldsymbol{X}$ represents the data got from one sensor at different time instances.

Abnormal readings (including erroneous readings and diverse-scale event readings) could be classified into two categories. One is the internal error, which often occurs in a few sensors, caused by, for example, noise, systematic errors. The other category is the external event, which may take place due to environmental changes, like temperature and pressure. The former, internal error, is often sparse in spatial domain, while the latter is usually low rank in time domain

[1] Indeed, the matrix could be recovered by solving the nuclear-norm based MR optimization problem rather than the low rank based MR optimization problem, the details would be elaborated in Sect. 4.

leading to the data matrix remains low rank, which together may destroy the sparsity structure of the data.

Inspired by this, we decompose the data matrix into two parts, the normal one and the abnormal one: $\boldsymbol{X} = \boldsymbol{X}_n + \boldsymbol{X}_s$. Then we have

$$A\boldsymbol{X} = A(\boldsymbol{X}_n + \boldsymbol{X}_s) = A \cdot [I, I](\boldsymbol{X}_n, \boldsymbol{X}_s)^T = [A, A][\boldsymbol{X}_n, \boldsymbol{X}_s]^T \quad \text{s.t. } A(\boldsymbol{X}) = \boldsymbol{b} \tag{4}$$

According to Eq. 1, we consider $[A, A]$ as a new linear map and solve such a problem in the framework of the matrix recovery. That is, when we get an observed vector $y \in R^p$, we suppose the linear map is $[A, A]$ instead of A itself, and recover the matrix $\boldsymbol{X}^* \in R^{2M \times N}$.

3.3 Data Recovery Method

Note that Eqs. 3 and 4 are essentially same. We consider the general form of the following minimization problem:

$$\min_{\boldsymbol{X} \in R^{m \times n}} \quad \frac{1}{2}||A(\boldsymbol{X}) - b||_F^2 + \mu||\boldsymbol{X}||_* \quad \text{s.t. } A(\boldsymbol{X}) = \boldsymbol{b} \tag{5}$$

where $\mu > 0$ is a given parameter, and $A(x) = \Phi T(x)$, and $T(\cdot)$ is a operator to transform a matrix into a vector by overlaying one column of $\boldsymbol{x}$ on another. Φ is a $p \times MN$ random matrix.

We consider this problem instead of the original problem in Eq. 2, because Eq. 3 is the Lasso problem of Eq. 2, and in relaxed conditions, its solution is the solution of Eq. 2 [10].

We could transform the above in the following form:

$$\min_{\boldsymbol{X} \in R^{m \times n}} \quad F(x) \triangleq f(x) + P(x)$$

where $f(x) = \frac{1}{2}||A(\boldsymbol{X}) - b||_F^2$ and $P(x) = \mu||\boldsymbol{X}||_*$

In this form, the first part is differential and convex while the second part is convex but may not be differential. Through basic calculations in linear algebra, we get the expression that

$$\nabla f(\boldsymbol{X}) = A^*(A(\boldsymbol{X}) - b)$$

Here, A^* is the dual operator of A, and it is easy to see that $A^*(\boldsymbol{X}) = \Phi^T \boldsymbol{X}$. Thus,

$$\nabla f(\boldsymbol{X}) = A^*(A(\boldsymbol{X}) - b) = \Phi^T(\Phi^* T(X) - b)^*$$

Note that ∇f is linear, and thus it is Lipschitz continuous. So we are able to determine a positive constant L_f to satisfy the following in equation:

$$||\nabla f(\boldsymbol{X}) - \nabla f(\boldsymbol{Y})||_F \leq L_f||\boldsymbol{X} - \boldsymbol{Y}||_F$$

Lemma 1. *A rough estimation of L_f would be*

$$\sqrt{MN \cdot \min_i \{(\Phi^T\Phi)_i^2)\}},$$

where $(\Phi^T\Phi)_i^2)^2$ is the ith column of the matrix $\Phi^T\Phi$.

Proof. $||\nabla f(\boldsymbol{X}) - \nabla f(\boldsymbol{Y})||_F^2 = ||\Phi^T(\Phi^* T(\boldsymbol{X} - \boldsymbol{Y}))||_2^2$

if we set $\Phi^T\Phi = \begin{pmatrix} a_{11} & \cdots & a_{1,MN} \\ \vdots & & \vdots \\ a_{p1} & \cdots & a_{p,MN} \end{pmatrix}$, $T(X-Y) = \begin{pmatrix} x_{11} \\ \vdots \\ x_{MN} \end{pmatrix}$,

and $h = \max_i \{(\Phi^T\Phi)_i^2\}$,then

$$\begin{aligned} ||\Phi^T\Phi T(X-Y)||_2^2 &= \sum_{j=1}^{p}(\sum_{i=1}^{MN} a_{ji}x_i) \le h(x_1 + \ldots + x_{MN})^2 \\ &\le M \cdot N \cdot h(x_1 + \ldots + x_{MN})^2 \\ &= MNh||X-Y||_F^2 \end{aligned}$$

Thus $L_f \le \sqrt{MNh}$.

Remark 4. The lemma above is just a rough estimation of L_f. In experiments, we find that the L_f could be much smaller if the matrix is sampled from a Gaussian distribution, and a small L_f will help converging quickly.

Let's consider the following operator:

$$\begin{aligned} Q_\tau(X,Y) &\triangleq f(Y) + < \nabla f(Y), X-Y > + \frac{\tau}{2}||X-Y||_F^2 + P(X) \\ &= \frac{\tau}{2}||X-G||_F^2 + P(X) + f(Y) - \frac{1}{2\tau}||\nabla F(Y)||_F^2 \end{aligned}$$

where $\tau > 0$ is a given parameter. As the above function is a strong convex function of X, it has a unique minimizer. Let $S_\tau(G)$ be the minimizer of $\frac{\tau}{2}||X - G||_F^2 + P(X)$, and according to [11], we have $S_\tau(G) = U \cdot diag((\delta - \mu/\tau)_+) \cdot V_T$ where U, V, δ comes from the SVD decomposition of $G = U \cdot diag(\delta) \cdot V^T$. On the basis of the lemma above, we could get an estimation of the procedure and speed of convergence of data recovery.

According to accelerated proximal gradient (APG) design proposed in [8,11], Let $t_0 = t_1 = 1$ and $\tau_k = L_f$, and $\{X_k\}, \{Y_k\}, \{t_k\}$ be the sequence generated by APG. For $i = 1, 2, 3, \cdots$

- Step 1: Set $Y_k = X_k + \frac{t^{k-1}-1}{t^k}(X_k - X_{k-1})$
- Step 2: Set $G = Y_k - (\tau_k)^{-1}A^*(A(Y_k) - b)$
- Step 3: Set $X^{k+1} = S_{\tau_k}(G)$
- Step 4: Set $t_{k+1} = \frac{1+\sqrt{1+4(t_k)^2}}{2}$

Then we have $F(X_k) - F(X^*) \le \frac{2L_f||X^*-X_0||_F^2}{(k+1)^2}$. Thus, $F(X_k) - F(X^*) \le \varepsilon$ if $k \ge \sqrt{\frac{2L_f}{\varepsilon}}(||X_0||_F + \chi) - 1$, where χ has been defined in the lemma above.

Remark 5. From this lemma, we could find that the data recovery algorithm has a $O(\sqrt{L_f/\varepsilon})$ iteration complexity.

For the stopping condition of data recovery, we define $\delta(x)$ as $dist(0, \partial(f(x)) + \mu||X||_*)$, where $\delta(x)$ represents the speed of convergence of data recovery steps. It is natural to stop the process when $\delta(x)$ is small enough.

Note that it is difficult to compute $\delta(x)$, since $||X||_*)$ is not differential. But fortunately in APG designs given in [11] provide a good upper bound for $\delta(x)$.

They obverse that $\partial(\mu||X_{k+1}||_*) \geq \tau_k(G_k - X_{k+1})$ and in our method

$$\tau_k(G_k - X_{k+1}) = \tau_k(Y_k - X_{k+1}) - \nabla f(Y_k) = \tau_k(Y_k - X_{k+1}) - \Phi^T(\Phi \cdot vec(Y_k) - b)$$

Here the operator *vec* is defined as the same in the above, that is transforming a matrix into a vector by overlaying one column on another. So we set

$$\begin{aligned} S_{k+1} &\triangleq \tau_k(Y_k - X_{k+1}) + \nabla f(X_{k+1}) - \nabla f(Y_k) \\ &= \tau_k(Y_k - X_{k+1}) + A^*(A(X_{k+1}) - A(Y_k)) \\ &= \tau_k(Y_k - X_{k+1}) + \Phi^T(\Phi \cdot T(Y_k - X_{k+1})) \end{aligned}$$

Then we would find $S_{k+1} \in \partial(f(X_{k+1}) + \mu||X_{k+1}||_*)$. Thus, it is easy to find $\delta(X_{k+1}) \leq ||S_{k+1}||$.

In this way, we could give the following condition as the stopping condition, that is

$$\frac{||S_{k+1}||_F}{\tau_k \max\{1, ||X_{k+1}||_F\}} \leq Tol$$

where Tol is a moderately small tolerance.

Lemma 2. *For any $\mu > 0$, the optimal solution X^* of Eq. 3 is bounded according to* [8,11]. *And $||X||_F < \chi$ Where*

$$\chi = \begin{cases} min\{||b||_2^2/(2\mu), ||X_{LS}||_*\} & if\ A\ is\ surjective \\ ||b||_2^2/(2\mu) & Otherwise \end{cases}$$

with $X_LS = A*(AA^*)^{-1}$ On the basis of the lemma, we could get an estimation of the speed of convergence of this algorithm. And the conclusion is shown in the corollary below.

4 Theoretical Analysis

Before presenting our results in the framework of matrices recovery, we introduce a little bit more notation here to smooth the reading here. Let X_0 be an $M \times N$ matrix of rank r with singular value decomposition (SVD) $\boldsymbol{U\Sigma V^*}$. Without loss of generality, let us impose the conventions $M \leq N$, where Σ is $r \times r$, $\boldsymbol{U}$ is $M \times r$, $\boldsymbol{V}$ is $N \times r$.

In the low-rank matrix reconstruction problem, the subspace T is the set of matrices of the form $\boldsymbol{UY} + \boldsymbol{XV}$ where $\boldsymbol{X}$ and $\boldsymbol{Y}$ are arbitrary $M \times r$ and $N \times r$ matrices. The span of matrices of the form $\boldsymbol{UY}$ has dimension Mr, the span of $\boldsymbol{XV}$ has dimension Nr, and the intersection of these two spans has dimension r^2. Hence, we have $d_T = dim(T) = r(M + Nr)$. $T^{\perp}$ is the subspace of matrices spanned by the family ($\boldsymbol{xy}$), where $\boldsymbol{x}$ (respectively $\boldsymbol{y}$) is any vector orthogonal to $\boldsymbol{U}$ (respectively $\boldsymbol{V}$). The spectral norm denoted by $||\cdot||$ is dual to the nuclear norm. The subdifferential of the nuclear norm at $\boldsymbol{X}_0$ is given by

$$\partial||\boldsymbol{X}_0||_* = \{\boldsymbol{Z} : P_T(\boldsymbol{Z}) = \boldsymbol{U}\boldsymbol{V}^* and\, ||P_{T^\perp(\boldsymbol{Z})}|| \leq 1\}$$

Note that the Euclidean norm of $\boldsymbol{UV}$ is equal to $\sqrt{r}$.

Theorem 1. *Let X_0 be an arbitrary $M \times N$ rank-r-matrix and $||\cdot||$ be the matrix nuclear norm. For a Gaussian measurement map Φ with $m \leq c{\cdot}r(3M + 3N - 5r)$ for some $c > 1$, the recovery is exact with probability at least $1-2e^{(1-c)n/8}$, where $n = max(M, N)$.*

Gaussian measurement map Φ above e takes the form of a linear operator whose i^{th} component is given by $[\Phi(Z)]_i = \mathrm{tr}(\Phi_i \cdot Z)$.

Above, Φ_i is an $M \times N$ random matrix with i.i.d., zero-mean Gaussian entries with variance $1/p$. This is equivalent to defining Φ as an $p \times (MN)$ dimensional matrix acting on $vec(\boldsymbol{Z})$, the vector composed of the columns of Z stacked on top of one another. In this case, the dual multiplier is a matrix taking the form

$$Y = \Phi \cdot \Phi_T(\Phi \cdot \Phi_T) - 1(UV^*),$$

Here, Φ_T is the restriction of Φ to the subspace T. Concretely, one could define a basis for T and write out Φ_T as an $p \times d_T$ dimensional matrix.

According to this theorem, we could find that each sensor only needs to send a message with the length of $cr(3M + 3N - 5r)$ at the end of time M, and in that case, there will be an overwhelming probability that our method could recover the original matrix.

Moreover, this result is relatively close to the oracle result. It is proved that we need at least $p \leq (M + Nr)r$ measurements to recover matrices of rank r, by any method whatever. Namely if $p < (M + Nr)r$, we will always have two distinct matrices M and M_0 of rank at most r with the property $A(M) = A(M_0)$ no matter what A is. To see this, we may fix two matrices $U \in R^{M \times r}, V \in R^{N \times r}$ orthogonal columns, and consider the linear space of matrices of the form

$$T = \{UX^*YV^* : X \in R^{N \times r}, V \in R^{N*r}\}$$

The dimension of T is $r(M + Nr)$. Thus, if $p < (M + Nr)r$, there exists $M = UX^*YV^* = 0$ in T such that $\Phi(M) = 0$. This proves the claim since $\Phi(UX^*) = \Phi(YV^*)$ for two distinct matrices of rank at most r. Once again, in contrast to similar results in compressive sensing, the number of measurements required is within a constant of the theoretical lower limit—there is no extra log factor.

To illustrate this statement in details, the length of vector, in compressive sensing based data gathering method, sent by each sensor each time must be $O(\log N)$ respectively according to the recent results on low-rand data [Simple Bounds for Recovering Low-complexity Models] [12], and the total length of messages sent by all sensors during all M time instances will be $O(MN\log(N))$ in compressive sensing, which will be larger than the $O(rN(3M + 3N - 5r))$ of our method when M is larger than $O(N/\log(N))$. When self collection duration M reaches the same order of magnitude of N, even assuming that the sensor data fulfill the sparsity condition and CS methods could set the length of message sent by each sensor at $2K$–$4K$, the total communication overhead will be $O(K \times MN)$, which still has similar communication overhead compared with our method.

In the case with noise, in order to estimate the estimation error and its upper bound, we introduce the restricted isometry property (RIP):

Definition 1. *For each integer $r = 1, 2, \ldots, n$, the isometry constant δr of A is the smallest quantity such that*

$$(1 - \delta r)||X||_F^2 \leq ||A(X)||_2^2 \leq (1 + \delta r)||X||_F^2$$

holds for all matrices of rank at most r.

We say that A satisfies the RIP at rank r if δ_r is bounded by a sufficiently small constant between 0 and 1.

Theorem 2. *If the noise z satisfy that $||\Phi^*(z)|| \leq \varepsilon$ and $||\Phi T(z)||_\infty \leq \eta$, for some $\varepsilon \leq \eta$, if $\delta r < \frac{1}{3}$ with $r \leq 2$, suppose X^* is the solution of the recovery method, then $||X - X^*||_F \leq (\varepsilon + \eta)_+$*

Therefore, if the Φ is chosen properly, e.g., random matrix with *i.i.d.* zero-mean Gaussian entries with variance $1/p$, the error of our MR method could be controlled even under the noise setting.

In conclusion, the communication volume of each sensor node is $cr(3M + 3N - 5r)$, and the communication overhead of the sensors network, i.e. the total number of message sent will be $Ncr(3M + 3N - 5r)$. Comparing it with the original CS method, the corresponding communication volume at each sensor and total communication overhead in the network are $M \cdot (2cs \log N + s)$ and $MN \cdot (2cs \log N + s)$ respectively for low-rank data [12], which means our method outperforms the typical CS methods in terms of communication.

Remark 6. Remark: According to the analysis, the longer the sampling period at each sensor (namely the value of M), the better the communication performance the proposed MR method could achieve.

5 Exprimental Analysis

Our experiment is conducted with real sensor datasets [7], including both temperature and humidity data. In each data set, there are about 55 sensors continuously generating sensor readings. Specifically, we select a 115-h period for the experiments, and compare our method with compressed sensing based method.

5.1 Comparison Study on Temperature Data

As shown in Fig. 1(d), the temperature data recovered by our MR method is plotted in a contour map in 3 dimensions. Compared with the contour map of the original temperature data given in Fig. 1(a), It is easy to find that the two maps are almost the same, which indicates a high quality of data recovery. This observation is further confirmed in the SNR results computed for each nodes readings. As shown in Fig. 2(a), our MR method achieves about 20 db gain in the recovered data.

Compared with CS based results in Fig. 1(g), it is interesting to observe that although the CS method could recover the matrix to some degrees, the recovered data looks unstable and altered in multiple areas. It is clear that the exactness of recovery is much worse compared to our method. This observation is further confirmed in Fig. 2(a), the SNR results computed for each nodes readings. As shown in Fig. 2(a), the quality of the recovered data of our MR method is about 10 times (10 db) better than that of the CS based method.

5.2 Comparison Study on Humidity Data with Events

Monitoring emergency and distinguish sensors with problem is an important assignment for both our method and CS based data gathering methods. Here we propose the experiments for the data with event readings to study the performance of our MR method compared with the method of CS. The humidity data with small-range event and large-range event recovered by our MR method are plotted in 3D contour maps in Fig. 1(e) and (f) respectively. Compared with the contour map of the original humidity data given in Fig. 1(b) and (c), It is easy to find that the recovered maps by MR are almost the same as the original ones, including the small hill of event in the map Fig. 1(e) and the large-range event in the map Fig. 1(f), which indicates a high quality of data recovery. This observation is further confirmed in the SNR results computed for each nodes readings. As shown in Fig. 2(b) and (c), our MR method generally achieves about 20 db gain in the recovered data given either small-range event or large-range event.

Compared with CS based results in Fig. 1(h) and (i), it is clear that the CS based method could recover the data to some degrees. However, this method failed to recovered exactly and some areas are obviously altered. Moreover, both small-range event and large-range event are almost overwhelmed in the noise. It is obvious that the exactness of recovery is much worse compared to our method. This observation is further confirmed in Fig. 2(b) and (c), the SNR results computed for each nodes readings. As shown in Fig. 2(b) and (c), we can observe that the quality of the recovered data of our MR method is generally about 10 times (10 db) better than that of the CS based method in the data with large-range event.

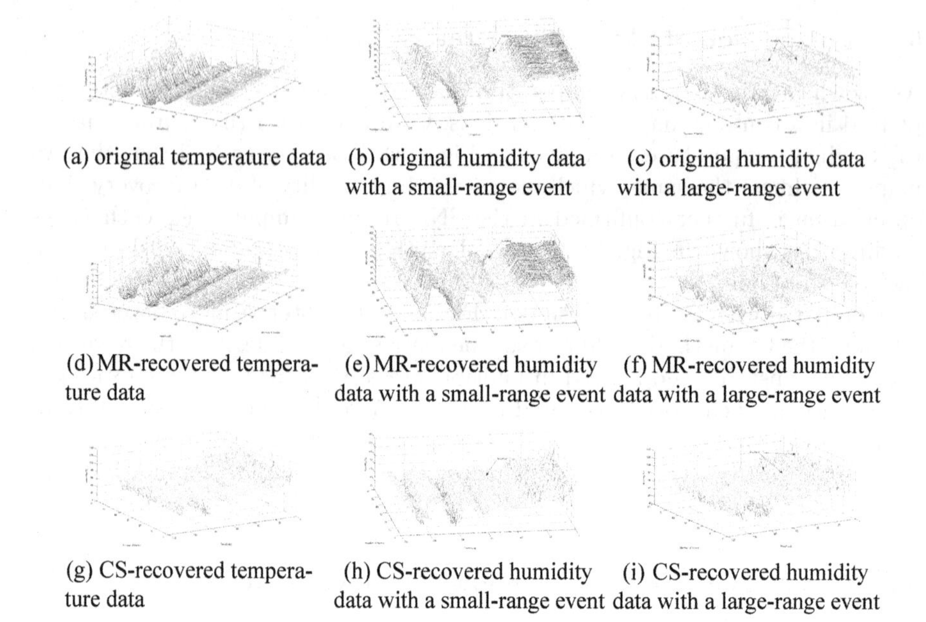

(a) original temperature data (b) original humidity data with a small-range event (c) original humidity data with a large-range event

(d) MR-recovered temperature data (e) MR-recovered humidity data with a small-range event (f) MR-recovered humidity data with a large-range event

(g) CS-recovered temperature data (h) CS-recovered humidity data with a small-range event (i) CS-recovered humidity data with a large-range event

Fig. 1. 3D contour map of original, MR-recovered and CS-recovered temperature and humidity data

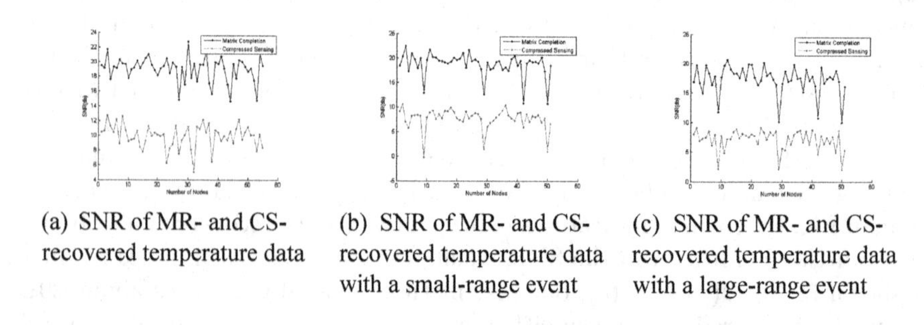

(a) SNR of MR- and CS-recovered temperature data (b) SNR of MR- and CS-recovered temperature data with a small-range event (c) SNR of MR- and CS-recovered temperature data with a large-range event

Fig. 2. SNR comparison of MR and CS methods among different datasets

6 Conclusion

In this paper, we have shown the power of matrix recovery in data compression, gathering and recovery in sensor networks. In particular, we have demonstrated that our MR method could solve both of the following questions: how to compress and gather the large volume data effectively, and how to keep various time/spacescale event readings unaltered in the gathered data. We also demonstrated via theoretical analysis that our method outperforms the original CS method in terms of communication overhead when dealing with low rank data. Finally, the experiments indicate that our MR method could be satisfyingly implemented in the realworld sensor networks, and has the ability to achieve much better data recovery quality. Our work provides a new angle of view in both compressive sampling applications and sensor networks with diverse time/spacescale events, and suggests a general design given the relaxation from low-rank based optimization to nuclear norm based optimization.

References

1. Tapparello, C., Simeone, O., Rossi, M.: Dynamic compression-transmission for energy-harvesting multihop networks with correlated sources. IEEE ACM Trans. Netw. **22**(6), 1729–1741 (2014)
2. Zahedi, A., Ostergaard, J., Jensen, S.H., Naylor, P., Bech, S.: Distributed remote vector Gaussian source coding for wireless acoustic sensor networks, pp. 263–272 (2014)
3. Orozco, A.L.S., Corripio, J.R., Hernandez-Castro, J.C.: Source identification for mobile devices, based on wavelet transforms combined with sensor imperfections. Computing **96**(9), 829–841 (2014)
4. Kasirajan, P., Larsen, C., Jagannathan, S.: A new data aggregation scheme via adaptive compression for wireless sensor networks. ACM Trans. Sens. Netw. **9**(1), 1–26 (2012)
5. Yang, G., Xiao, M., Zhang, S.: Data aggregation scheme based on compressed sensing in wireless sensor network. J. Netw. **8**(1), 556–561 (2013)
6. Luo, C., Wu, F., Sun, J., Chen, C.W.: Compressive data gathering for large-scale wireless sensor networks. In: International Conference on Mobile Computing and Networking, pp. 145–156 (2009)
7. Mao, X., Miao, X., He, Y., Li, X.Y., Liu, Y.: Citysee: Urban CO_2 monitoring with sensors. In: 2012 Proceedings IEEE INFOCOM, pp. 1611–1619 (2012)
8. Cands, E.J., Recht, B.: Exact matrix completion via convex optimization. Found. Comput. Math. **9**(6), 717 (2009)
9. Candes, E.J., Tao, T.: Near-optimal signal recovery from random projections: universal encoding strategies? IEEE Trans. Inf. Theory **52**(12), 5406–5425 (2006)
10. Richard, E., Savalle, P.A., Vayatis, N.: Estimation of simultaneously sparse and low rank matrices. Comput. Sci. (2012)
11. Toh, K.-C., Yun, S.: An accelerated proximal gradient algorithm for nuclear norm regularized least squares problems. Pac. J. Optim. **6**(3), 615–640 (2009)
12. Candes, E., Recht, B.: Simple bounds for low-complexity model reconstruction. Acta Botanica Gallica Bulletin De La Socit Botanique De France **156**(3), 477–486 (2011)

Proactive Caching for Transmission Performance in Cooperative Cognitive Radio Networks

Jiachen Yang[1], Huifang Xu[1], Bin Jiang[1](✉), Gan Zheng[2], and Houbing Song[3]

[1] Tianjin University, Tianjin, China
{yangjiachen,xhftju,jiangbin}@tju.edu.cn
[2] Loughborough University, Loughborough, UK
g.zheng@lboro.ac.uk
[3] Embry-Riddle Aeronautical University, Daytona Beach, USA
songh4@erau.edu

Abstract. This paper considers cooperation between the primary and secondary network via content caching. The core idea is that secondary basestation caches some popular primary files and thus directly send content to the close primary user, in return some spectrum for secondary basestation to serve their own user. We propose a cooperation scheme which is jointly completed with content caching and transmission schemes, to maximize SU's data transmission rates while PU's target rate is achieved. In addition, we formulate the optimal caching allocation into a concave problem in terms of the data transmission rates for any given power allocation. We then provide an effective bisection search algorithm. Simulation results indicate that significant performance gain for both systems over the cooperation without caching. It also shows that our proposed scheme can achieve larger rate region than traditional relay cooperation scheme.

Keywords: Content caching · Cooperative cognitive radio
Power allocation

1 Introduction

With the global mobile data services exponentially growing, the deployment of the next generation (5G) system will meet new challenges such as high data rates, mobility patterns and quality of experience (QOE). Recent studies have shown that mobile video streaming will generate more than 69 percent of mobile data traffic by the end of 2019, which will still grow rapidly [1,2]. Therefore, there needs more wireless spectrum resources to satisfy the high data traffic.

Cooperative cognitive radio networks (CCRNs) have been a promising solution to solve the spectrum scarcity problem of wireless networks. Zhang et al. [3] proposed multi-hop routing algorithms reducing the probability of spectrum handoff and rerouting upon PUs' arrival. The simultaneous consideration of both

S. Chellappan et al. (Eds.): WASA 2018, LNCS 10874, pp. 570–579, 2018.
https://doi.org/10.1007/978-3-319-94268-1_47

primary and secondary activities in the actual spectrum accessing in CRNs is taken into account in [4].

CCRN has attracted great attention in recent years, mainly from information theoretical viewpoint [5–8]. Considering the practical constraint of the limited energy of the SBS, the additional energy cooperation is proposed in [9,10], using energy harvesting or wireless energy transfer techniques [11,12]. It shows that joint cooperation creates stronger incentives for both systems to cooperate and substantially improves the system performance. However, most existing researches only consider the information and energy cooperation in CCRN ignore the time delay because the primary data need to be first fetched from the primary basestation (PBS) to the secondary basestation (SBS), and then delivered to the PU. Cai et al. [13] investigated a cross-layer routing method for a multi-channel multihop CRN to minimize the delay from the source to a common destination, and in [14] they proposed a cross-layer distributed opportunistic routing protocol, in which the spectrum sensing and the relay selection are jointly considered with the purpose of decreasing the delivery delay from source to destination. Motivated by this problem and the recent advances in wireless edge caching [15–18], we aim to exploit caching capability at the SBS which pre-fetches some popular files for the primary users (PUs), so that it does not need to get it from the PBS, thus eliminates the delay due to transmission from the PBS to the SBS and save more energy to serve secondary users (SUs). There are rich literatures on edge caching. In [15–17] joint caching placement has been investigated in femto base stations with limited cache storage. In [18], the caching scheme that stores the most popular contents (MPC) based on the zeroforcing beamforming (ZFBF) transmission to achieve cooperation gain is studied. However, there is little work employing the edge caching technique to improve the performance of cognitive radio networks, especially CCRN. Caching in cognitive networks is investigated in [19], where data retrieval probability is derived, but primary-secondary cooperation is not considered. In [20], cooperative caching in cognitive radio networks is studied and the cooperation caching scheme in cooperative cognitive networks are still not investigated.

In this paper, we propose the cooperation between the primary and secondary systems in CCRN via content caching at the SBS based on the transmission. We formulate a problem that jointly optimizes cache storage allocation of the primary and secondary systems and power allocation for SBS to maximize SU's transmission rate with the constraint on the minimum data transmission rate for PU. We then investigate the impact of cache allocation on the PU's content transmission rates and prove it is a convex optimization problem. We further propose an effective bisection algorithm which enables the SBS to find the optimal power and caching allocation.

2 System Model

Network Model. In this paper, we consider the cooperation between a primary system and a secondary system in a basic four-node cognitive radio network via

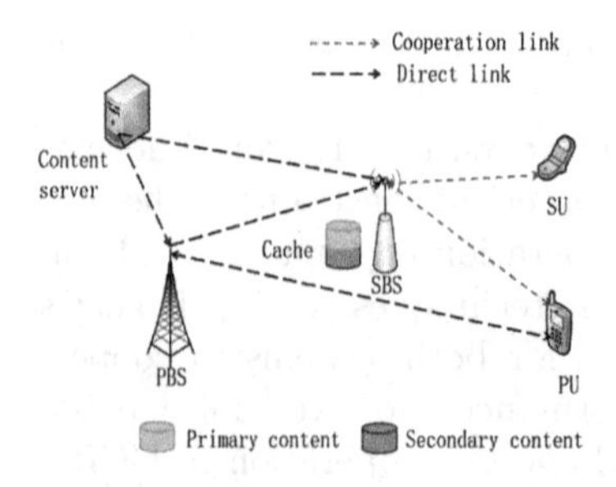

Fig. 1. The system model of CCRN.

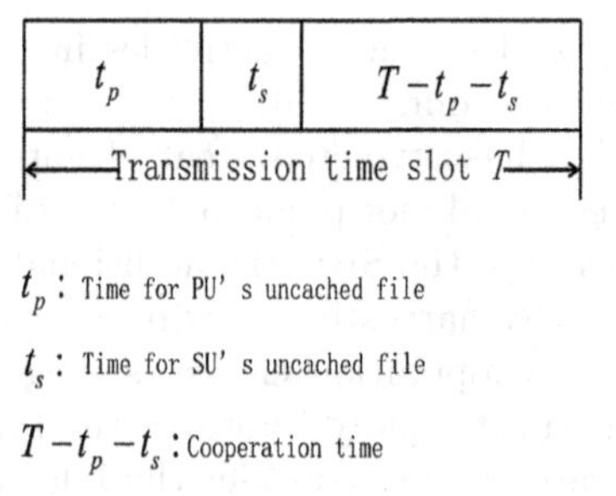

Fig. 2. Time allocation of a time slot.

content caching, as shown in Fig. 1. The primary system consists of a PBS and a PU, while the secondary system contains a SBS and a SU. All terminals have a single antenna except that the SBS has N_a antennas. The SBS is equipped with a limited cache capacity C to cache the primary and secondary contents. The SBS can simultaneously serve the primary and secondary users with appropriate power using ZFBF [18,22]. We assume that the direct transmission channel between the PBS and PU is poor, for instance, there is strong shadowing effect for the PU that lies at the cell edge, thus, it needs the secondary basestation to help relay the PU's data in order to satisfy the requirement of the primary user. We assume that SBS in the secondary system can sense the spectrum environment. When the PBS is not serving the primary user, the licensed spectrum belongs to the primary system can be shared to the secondary system. In addition, we consider all channels are quasi-static, thereby, the channels gain remain a constant value during a period of time.

Caching Model. We consider that the primary and the secondary system are different, so the PUs and the SUs are interested in different video contents. For instance, the users served by small cells like traditional videos, while users of the mactocell prefer to the modern videos. We denote the library of files requested by the PU and SU as $F_p \triangleq \{1,2,3\cdots M\}$ and $F_s \triangleq \{1,2,3\cdots N\}$, respectively. For simplicity, we assume that primary and secondary content have the same size. Without loss of generality, we suppose that the popularities of primary content and secondary content follow the Zipf law which is widely used in the literatures [18,21] and are given by $f_i^p = \frac{i^{-\gamma_p}}{\sum_{n=1}^{M} n^{-\gamma_p}}$ $(i \in F^p)$ and $f_j^s = \frac{j^{-\gamma_s}}{\sum_{n=1}^{N} n^{-\gamma_s}}$ $(j \in F^s)$, where γ_p and γ_s denote the primary and secondary file popularity, respectively. We also assume that both the primary and the secondary file popularities are descending in the index, i.e., $f_1^p \geq f_2^p \geq \cdots \geq f_M^p$ and $f_1^s \geq f_2^s \geq \cdots \geq f_N^s$, with $\sum_{i=1}^{M} f_i^p = 1$ and $\sum_{j=1}^{N} f_j^s = 1$, respectively. The primary and the secondary contents which are stored in the content server are obtained by the wireless backhaul. In this paper, we consider the secondary basestation can pre-cache some popular files of the primary and secondary systems in off-peak period to serve the PU and SU. In return, the SBS gains more transmission time to access of the primary spectrum to serve the primary and secondary requests, which leads to a win-win situation for both systems. The primary and secondary contents employ the MPC caching scheme according to the file popularity. We assume that the capacity C_o of the SBS's total cache capacity is used to the primary contents to serve the PU while the remaining capacity of $(C - C_o)$ is reserved to store its own content.

Transmission Model. We assume the total bandwidth that is licensed to the primary system is W and the time duration that a primary user is allowed to transmit a requested file over bandwidth W is T. A time slot T will be split into three parts based on the cached primary and secondary content, as shown in Fig. 2. The PBS and the SBS transmit with fixed power P_p and P_s, respectively. The SBS also splits its power to serve the PU and its own SU. Let x_p and x_s denote the signal of a requested file transmitted by the PBS and the SBS. The received signals at the PU from the PBS and the SBS are given by:

$$y_p = \sqrt{P_p h_p}|d_p|^{-\frac{\alpha}{2}} x_p + \eta_p, \tag{1}$$

$$y_{s,p} = \sqrt{\beta P_s h_{s,p}}|d_{s,p}|^{-\frac{\alpha}{2}} x_p + \eta_{s,p}, \tag{2}$$

and the received signal at the SU is

$$y_s = \sqrt{(1-\beta) P_s h_s}|d_s|^{-\frac{\alpha}{2}} x_s + \eta_s, \tag{3}$$

where h_p, d_p are channel fading coefficient and the distance from the PBS to the PU, respectively, $h_{s,p}$ and h_s are channel fading coefficients from the SBS to the PU respectively, $d_{s,p}$ and d_s denote the distances of the SBS to the PU and its own SU. The terms $\eta_p, \eta_{s,p}$ and η_s represent Gaussian noise distributed with zero mean at the PU and SU, respectively. β denotes the ratio of the SBS power allocated to serve the PU while the remaining portion of $(1-\beta)$ is reserved to serve the SU.

Aiming to achieve the cooperation gain, we consider the MPC caching strategy. Specifically, the SBS equipped with N_a antennas can use ZFBF approach to simultaneously serve the primary and secondary users over the licensed bandwidth [22]. Without loss of generality, the licensed bandwidth is normalized to 1 MHz. Based on the cached content of the PU and SU, we divide the transmission model into four types.

Type 1: When the primary and secondary files that are requested by the PU and SU are cached in the SBS, the SBS can simultaneously serve the PU and SU over the whole time T using ZFBF, the transmission data for PU and SU are respectively expressed as:

$$R_1^p = R_p \cdot \Pr(T\log_2(1 + \frac{\beta P_s |h_{sp}|^2 d_{sp}^{-\alpha}}{N_0}) \geq R_p), \tag{4a}$$

$$R_1^s = R_s \cdot \Pr(T\log_2(1 + \frac{(1-\beta) P_s |h_s|^2 d_s^{-\alpha}}{N_0} \geq R_s), \tag{4b}$$

where N_0 is noise power.

Type 2: When the primary content requested by the PU which is cached in the SBS in advance and the secondary file is not cached in the SBS. Thus, the SBS needs a period t_s $(0 \leq t_s \leq T)$ to obtain the secondary file from the content server, the remaining time $(T - t_s)$ is used to transmit the PU's and SU's data. The transmission data for PU and SU are respectively given by:

$$R_2^p = R_p \cdot \Pr((T - t_s)\log_2(1 + \frac{\beta P_s|h_{sp}|^2 d_{sp}^{-\alpha}}{N_0}) \geq R_p). \tag{5a}$$

$$R_2^s = R_s \cdot \Pr((T - t_s)\log_2(1 + \frac{(1-\beta)P_s|h_s|^2 d_s^{-\alpha}}{N_0} \geq R_s). \tag{5b}$$

Type 3: Similarly, when the primary content requested by the PU is uncached content, the SBS takes t_p $(0 \leq t_p \leq T)$ time to obtain the file. The requested secondary file has been cached in the SBS. Therefore, there is the remaining time $(T - t_p)$ used to transmit the PU's and SU's data. The transmission data for PU and SU are respectively given by:

$$R_3^p = R_p \cdot \Pr((T - t_p)\log_2(1 + \frac{\beta P_s|h_{sp}|^2 d_{sp}^{-\alpha}}{N_0}) \geq R_p). \tag{6a}$$

$$R_3^s = R_s \cdot \Pr((T - t_p)\log_2(1 + \frac{(1-\beta)P_s|h_s|^2 d_s^{-\alpha}}{N_0} \geq R_s). \tag{6b}$$

Type 4: Both of the requests from the PU and SU have not been cached at the SBS, thereby, the cooperation period $(T - t_s - t_p)$ is utilized to transmit the data of PU and SU by the SBS with different power allocation. The transmission data for PU and SU are respectively expressed as:

$$R_4^p = R_p \cdot \Pr((T - t_s - t_p)\log_2(1 + \frac{\beta P_s|h_{sp}|^2 d_{sp}^{-\alpha}}{N_0}) \geq R_p). \tag{7a}$$

$$R_4^s = R_s \cdot \Pr((T - t_s - t_p)\log_2(1 + \frac{(1-\beta)P_s|h_s|^2 d_s^{-\alpha}}{N_0} \geq R_s). \tag{7b}$$

3 Problem Formulation

In this paper, we aim to jointly optimize the caching capacity and power allocation strategies of the SBS so as to maximize the utility of the SBS, which is denoted as the total effective data transmission rates served by the SBS. Mathematically, based on the above model analysis, we can formulate the problem as:

$$\begin{aligned} \max_{C_o,\beta} \quad & R^S(C_o,\beta) \triangleq p_p \cdot p_s \cdot R_1^S + p_p \cdot R_2^S + p_s \cdot R_3^S + R_4^s \\ s.t. \quad & R^P(C_o,\beta) \triangleq (p_p \cdot p_s \cdot R_1^P + p_p \cdot R_2^P + p_s \cdot R_3^P + R_4^p) \geq R_{th} \\ & 0 \leq t_s \leq T, 0 \leq t_p \leq T, and\ 0 \leq t_s + t_p \leq T, \\ & 0 \leq C_o \leq C, 0 \leq \beta \leq 1, \end{aligned} \tag{8}$$

where $p_p \triangleq \sum_{i=1}^{C_o} f_i^p$ and $p_s \triangleq \sum_{j=1}^{C-C_o} f_j^s$ are the probabilities of the caching primary and secondary content, respectively. The $R_1^P \triangleq R_1^p + R_4^p - R_2^p - R_3^p$ and $R_1^S \triangleq R_1^s + R_4^s - R_2^s - R_3^s$, $R_2^P \triangleq R_2^p - R_4^p$ and $R_2^s \triangleq R_2^s - R_4^s$, $R_3^P \triangleq R_3^p - R_4^p$ and $R_3^S \triangleq R_3^s - R_4^s$ are the merging rates for the PU and SU, respectively.

There are multiple variables involved in the above problem, and we first study the property of caching placement C_o given power allocation β. It is easy to see that the cache allocation C_o is an integer variable and different values determine the amount of files of primary and secondary. For a given β, to find an optimal threshold C_o which satisfies PUs target transmission rate, one can exploit the exhaustive search with $O(C)$ computational complicity. However, there is not an efficient approach. To tackle this challenge, we introduce a continuous variable $q = C_o/C$, $(0 \leq q \leq 1)$. After we obtain an optimal q, then optimal value C_o can be approximated by:

$$C_o = \lceil qC \rceil, \tag{9}$$

where $\lceil \cdot \rceil$ is the ceiling function.

The cumulative distribution function (cdf) of the file popularity distribution of primary and secondary p_p and p_s following approximation of the sum of Zipf probabilities [21,23] is useful:

$$p_p = \sum_{i=1}^{C_o} f_i^p \approx \frac{C_o^{1-\gamma_p} - 1}{M^{1-\gamma_p} - 1}, \quad p_s = \sum_{j=1}^{C-C_o} f_j^s \approx \frac{(C - C_o)^{1-\gamma_s} - 1}{N^{1-\gamma_s} - 1} \tag{10}$$

Then, substitute (9) and (10) into $R^P(C_o, \beta)$, the effective data transmission of PU is written as:

$$\begin{aligned} R^P(q,\beta) \triangleq\ & \tfrac{(qC)^{1-\gamma_p}-1}{M^{1-\gamma_p}-1} \cdot \tfrac{((1-q)C)^{1-\gamma_s}-1}{N^{1-\gamma_s}-1} \cdot R_1^P + \tfrac{(qC)^{1-\gamma_p}-1}{M^{1-\gamma_p}-1} \cdot R_2^P \\ & + \tfrac{((1-q)C)^{1-\gamma_s}-1}{N^{1-\gamma_s}-1} \cdot R_3^P + R_4^p. \end{aligned} \tag{11}$$

The formula (11) is a complex polynomial about caching allocation q and power ratio β, thus, we further study its convexity below.

Theorem 1. *For any given power ratio β, $R^P(q)$ is a concave function about q.*

Proof: For any given power ratio β, all R_1^P, R_2^P, R_3^P and R_4^P are fixed value according to (4a), (5a), (6a) and (7a). For the considered system, all transmission rates are positive. Therefore, it is only a single variable function of content cache proportion q. We can briefly derive the second derivative of $R^P(q)$ which is respectively given by:

$$\begin{aligned} R^{P\prime}(q) \triangleq\ & (1-\gamma_p)\tfrac{C(qC)^{-\gamma_p}}{M^{1-\gamma_p}-1} \cdot \tfrac{((1-q)C)^{1-\gamma_s}-1}{N^{1-\gamma_s}-1} \cdot R_1^P \\ & - (1-\gamma_s)\tfrac{(qC)^{1-\gamma_p}-1}{M^{1-\gamma_p}-1} \cdot \tfrac{((1-q)C)^{-\gamma_s}}{N^{1-\gamma_s}-1} \cdot R_1^P \\ & + (1-\gamma_p)\tfrac{C(qC)^{-\gamma_p}}{M^{1-\gamma_p}-1} \cdot R_2^P - (1-\gamma_s)\tfrac{((1-q)C)^{-\gamma_s}}{N^{1-\gamma_s}-1} \cdot R_3^P \end{aligned} \tag{12}$$

$$\begin{aligned} R^{P''}(q) \triangleq & -C^2(1-\gamma_p)\gamma_p \frac{(qC)^{-\gamma_p-1}}{M^{1-\gamma_p}-1}\cdot\frac{((1-q)C)^{1-\gamma_s}-1}{N^{1-\gamma_s}-1}\cdot R_1^P \\ & -C^2(1-\gamma_p)(1-\gamma_s)\frac{(qC)^{-\gamma_p}}{M^{1-\gamma_p}-1}\cdot\frac{((1-q)C)^{-\gamma_s}}{N^{1-\gamma_s}-1}\cdot R_1^P \\ & -C^2(1-\gamma_s)(1-\gamma_p)\frac{(qC)^{-\gamma_p}}{M^{1-\gamma_p}-1}\cdot\frac{((1-q)C)^{-\gamma_s}}{N^{1-\gamma_s}-1}\cdot R_1^P \\ & -C^2(1-\gamma_s)\gamma_s\frac{(qC)^{1-\gamma_p}-1}{M^{1-\gamma_p}-1}\cdot\frac{((1-q)C)^{-\gamma_s-1}}{N^{1-\gamma_s}-1}\cdot R_1^P \\ & -C^2(1-\gamma_p)\gamma_p\frac{C(qC)^{-\gamma_p-1}}{M^{1-\gamma_p}-1}\cdot R_2^P - C^2(1-\gamma_s)\gamma_s\frac{((1-q)C)^{-\gamma_s-1}}{N^{1-\gamma_s}-1}\cdot R_3^P \end{aligned} \tag{13}$$

where $\frac{(1-\gamma_p)}{M^{1-\gamma_p}-1} \geq 0$ and $\frac{(1-\gamma_s)}{N^{1-\gamma_s}-1} \geq 0$ always hold true because $\gamma_p, \gamma_s \geq 0$. We also notice that $(qC)^{-\gamma_p} \geq 0$ and $((1-q)C)^{-\gamma_s} \geq 0$ due to $q \in [0,1]$. Hence, $R^{P''}(q) \leq 0$, $R^P(q)$ is concave, and this completes the proof.

Then, we study the properties of $R^P(q,\beta)$ about β.

Theorem 2. *For any given caching allocation q, $R^P(q,\beta)$ is increasing in regard to β, and the objective function $R^S(q,\beta)$ is decreasing in β according to equations* (4a)–(7a) *and* (4b)–(7b). *The character is straightforward and the proof is omitted.*

Algorithm 1. A Bisection Search Algorithm to Solve the Problem 8

1: Initialize upper and lower bounds β^U, β^L for β
2: Feasibility check: Given β^U, find the optimal q^* by solving the 12 to be zero because the function $R^P(q,\beta)$ is concave given β. If the result is $R^U(\beta^U) < R_{th}$. The problem is not feasible for the given β. Quit the algorithm. Otherwise, continue.
3: Given $\beta = (\beta^U + \beta^L)/2$, we find the optimal q^* via solving the 12 to be zero;
4: If $R^P(\beta) < R_{th}$, set $\beta^L = \beta$; otherwise set $\beta^U = \beta$;
5: Repeat the above step 3–4 until convergence, and return the optimal β^* and the corresponding the caching allocation q^*.

---

4 A Bisection Algorithm

This section presents an effective algorithm to cope with the complicated problem. Based on the above Theorems 1 and 2, we design the one-dimensional bisection search algorithm to solve problem (8), which maximizes the objective function of (8) while achieving the target data transmission of the primary network. The main idea is motivated by the basis bisection algorithm for solving the quasi-convex optimization problem. Generally speaking, we first consider the fact that the constraint $R^P(q,\beta) \geq R_{th}$ should hold with equality e.g., $R^P(q,\beta) = R_{th}$ to maximize the objective. Then, for a given power allocation β, we obtain the corresponding optimal caching allocation $\bar{q}$ by setting the (12) is zero, and the optimal q^* that maximize $R^P(q)$ is given by $q^* = \min(\bar{q}, 1)$. Next, combined with the optimal cache proportion, the minimum power portion β for satisfying the constraint $R^P(q,\beta) = R_{th}$ is obtained by the proposed bisection search algorithm. Finally, we get a joint optimal cache placement q and power ratio β for problem (8) via implementing our proposed scheme. The bisection algorithm is formally presented in Algorithm 1.

5 Simulation Results

In this section, simulation results are presented to estimate the performance of the proposed caching cooperation scheme in CCRN and the impact of system parameters. The simulation parameters are listed in Table 1 unless otherwise stated.

Table 1. Simulation parameters

Simulation parameters	Values
Primary content library, F_p	500
Secondary content library, F_s	300
Cache capacity of SBS, C	100
The number of antenna for SBS, N_a	2
Transmit power of PBS, P_p,	40 dBm
Transmit power of SBS, P_s,	20 dBm
Path loss exponent, α	3
Primary content popularity, γ_p	0.8
Secondary content popularity, γ_s	1.1
Noise power, N_0	−104 dBm
Distance of SBS to PU, d_{sp}	200 m
Distance of SBS to SU, d_s	300 m
Transmission time slot, T	10 s
Rate requirement of PU, R_p	2 Mbps
Rate requirement of SU, R_s	1 Mbps

In Fig. 3, we plot the PU's rates for the transmission of its own content as a function of caching portion q for three different values of the primary file popularity γ_p. We assume that the fixed power allocation β is 0.3. It is observed that the optimum caching ratio for the primary content is gradually reducing with γ_p increasing, similarly, the achievable transmission rate of the primary user increases with γ_p. This is because PU's most popular content is more concentrated when the file popularity is increasing, which contributes to requiring small caching storage to cache PU's most popular content, the remaining more cache capacity can be utilized for SU's content, which greatly improve the cooperation gains of the SBS.

Figure 4 illustrates the content transmission rates regions of both PU and SU for different cooperation schemes. It is seen that the proposed cooperation scheme outperforms the traditional rely cooperation scheme, which the achievable rate region of the proposed are greatly enlarged due to the content caching cooperation. This is because the SBS can directly transmit the cached content to the PU and SU, which saves more transmission time to access the licensed bandwidth for the SBS to achieve more cooperation gains.

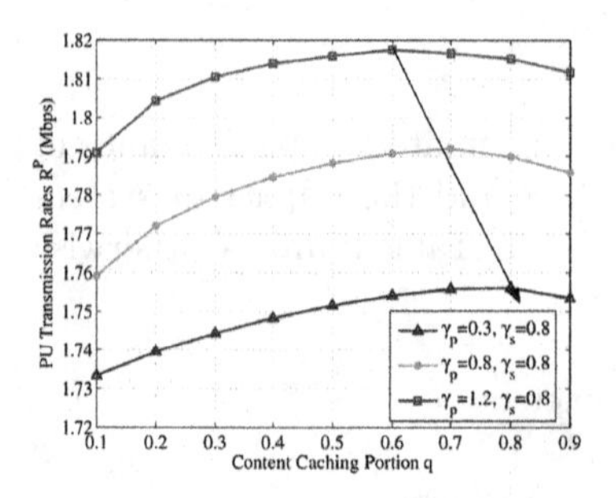

Fig. 3. The impact of PU transmission rates over Zipf parameters γ_p.

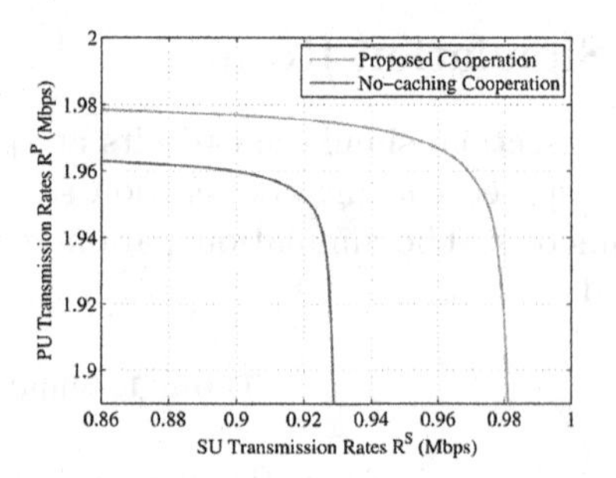

Fig. 4. SU and PU transmission rates regions of the different cooperation schemes.

6 Conclusion

In this paper, we proposed a content caching cooperation scheme for primary and secondary network in CCRN. we studied the problem of maximization of the SUs transmission rate subject to the PUs target rate. The jointly optimal cache allocation and power ratio were found by convex optimization together with a effective bisection search algorithm. Simulation results showed that the proposed cooperation outperforms traditional relay cooperation scheme.

Acknowledgments. This research is partially supported by National Natural Science Foundation of China (No. 61471260 and No. 61271324), and Natural Science Foundation 490 of Tianjin (No. 16JCYBJC16000).

References

1. Liu, D., Chen, B., Yang, C., Molisch, A.: Caching at the wireless edge: design aspects, challenges, and future directions. IEEE Commun. Mag. **54**(9), 22–28 (2016)
2. Tang, J., Quek, T.: The role of cloud computing in content-centric mobile networking. IEEE Commun. Mag. **54**, 52–59 (2016)
3. Zhang, L., Cai, Z., Li, P., Wang, L., Wang, X.: Spectrum-availability based routing for cognitive sensor networks. IEEE Access **5**, 4448–4457 (2017)
4. Lu, J., Cai, Z., Wang, X., Zhang, L., Li, P., He, Z.: User social activity-based routing for cognitive radio networks. Pers. Ubiquitous Comput. **13**, 1–17 (2018)
5. Jovicic, A., Viswanath, P.: Cognitive radio: an information-theoretic perspective. IEEE Trans. Inf. Theory **55**(9), 3945–3958 (2009)
6. Duan, Y., Liu, G., Cai, Z.: Opportunistic channel-hopping based effective rendezvous establishment in cognitive radio networks. In: Wang, X., Zheng, R., Jing, T., Xing, K. (eds.) WASA 2012. LNCS, vol. 7405, pp. 324–336. Springer, Heidelberg (2012). https://doi.org/10.1007/978-3-642-31869-6_28
7. Srinivasa, S., Jafar, S.A.: COGNITIVE RADIOS FOR DYNAMIC SPECTRUM ACCESS - the throughput potential of cognitive radio: a theoretical perspective. IEEE Commun. Mag. **45**(5), 73–79 (2007)

8. Cai, Z., Ji, S., He, J., Wei, L., Bourgeois, A.: Distributed and asynchronous data collection in cognitive radio networks with fairness consideration. IEEE Trans. Parallel Distrib. Syst. **25**(8), 2020–2029 (2014)
9. Krikidis, I., Laneman, J., Thompson, J., Mclaughlin, S.: Protocol design and throughput analysis for multi-user cognitive cooperative systems. IEEE Trans. Wirel. Commun. **8**(9), 4740–4751 (2009)
10. Zheng, G., Ho, Z., Jorswieck, E.A., Ottersten, B.: Information and energy cooperation in cognitive radio networks. IEEE Trans. Signal Process. **62**(9), 2290–2303 (2014)
11. Zhou, X., Zhang, R., Chin, K.H.: Wireless information and power transfer: architecture design and rate-energy tradeoff. IEEE Trans. Commun. **61**(11), 4754–4767 (2012)
12. Chen, J., Lv, L., Liu, Y., Kuo, Y., Ren, C.: Energy efficient relay selection and power allocation for cooperative cognitive radio networks. IET Commun. **9**(13), 1661–1668 (2015)
13. Cai, Z., Ji, S., He, J., Bourgeois, A.: Optimal distributed data collection for asynchronous cognitive radio networks. In: Proceedings of the IEEE 32nd International Conference on Distributed Computing Systems (ICDCS), pp. 245–254 (2016)
14. Cai, Z., Duan, Y., Bourgeois, A.: Delay efficient opportunistic routing in asynchronous multi-channel cognitive radio networks. J. Comb. Optim. **29**(4), 815–835 (2015)
15. Shanmugam, K., Golrezaei, N., Dimakis, A.G., Molisch, A.F., Caire, G.: Femtocaching: wireless content delivery through distributed caching helpers. IEEE Trans. Inf. Theory **59**(12), 8402–8413 (2013)
16. Maria, G., Jess, G., Javier, M., Deniz, G.: Wireless content caching for small cell and D2D networks. IEEE J. Sel. Areas Commun. **34**(5), 1222–1234 (2016)
17. Yang, C., Yao, Y., Chen, Z., Xia, B.: Analysis on cache-enabled wireless heterogeneous networks. IEEE Trans. Wirel. Commun. **15**(1), 131–145 (2016)
18. Weng, C., Psounis, K.: Distributed caching and small cell cooperation for fast content delivery. In: Proceedings of the ACM International Symposium, pp. 127–136 (2015)
19. Zhao, J., Gao, W., Wang, Y., Cao, G.: Delay-constrained caching in cognitive radio networks. In: IEEE INFOCOM, pp. 2094–2102 (2014)
20. Si, P., Yue, H., Zhang, Y., Fang, Y.: Spectrum management for proactive video caching in information-centric cognitive radio networks. IEEE J. Sel. Areas Commun. **34**(8), 2247–2259 (2016)
21. Taghizadeh, M., Micinski, K., Biswas, S., Ofria, C., Torng, E.: Distributed cooperative caching in social wireless networks. IEEE Trans. Mob. Comput. **12**(6), 1037–1053 (2013)
22. Wong, K., Pan, Z.: Array gain and diversity order of multiuser MISO antenna systems. Int. J. Wirel. Inf. Netw. **15**(2), 82–89 (2008)
23. Chen, Z., Lee, J., Quek, T.Q.S., Kountouris, M.: Cooperative caching and transmission design in cluster-centric small cell networks. IEEE Trans. Wirel. Commun. **16**(5), 3401–3415 (2017)

N-Guide: Achieving Efficient Named Data Transmission in Smart Buildings

Siyan Yao, Yuwei Xu(✉), Shuai Tong, Jianzhong Zhang, and Jingdong Xu

College of Computer and Control Engineering,
Nankai University, No. 38, Tongyan Road, Tianjin 300350, China
{siyan,nktongshuai}@mail.nankai.edu.cn,
{xuyw,zhangjz,xujd}@nankai.edu.cn

Abstract. Recently, Named Data Networking (NDN) has been proposed as a promising data delivery solution for smart building scenario because of its innovative properties such as name-based routing, in-network caching, and multi-source transmission. Due to the rapid growth in the number of devices and applications, it is difficult for NDN to meet the diverse transmission requirements only by leveraging the original forwarding strategy. To deal with this problem, we propose N-Guide, a novel protocol which not only supports all the transmission modes of applications but also implements the data aggregation in the scenario with large number of devices. Since our protocol directs the interest packets by the physical and application information in names just like a guide serving for the named data in smart buildings, we name it N-Guide. To verify the effectiveness of N-Guide, we make contrast simulations on ndnSIM. The results show that N-Guide outperforms two state-of-the-art NDN approaches.

Keywords: NDN · Smart buildings · Interest forwarding

1 Introduction

The global smart building market has continued to evolve with an increasing focus on creating a comfortable indoor environment with low energy consumption. Different from the buildings installed with some management systems, a smart building is able to connect all the devices and sensors by a variety of communication technologies, obtain massive data through the low-power IoT transmission, and make a deep analysis by leveraging artificial intelligence.

In smart buildings, efficient data transmission is extremely difficult due to the large number of devices, heterogeneous network environment and increasing application requirements. In early days, BACnet and LonWorks were widely used on point-to-point links or buses in the building environments [1], but they were designed for the local area network and could not provide access to external networks. With the popularity of Internet technology, it was feasible to encapsulate data into TCP or UDP packets and transmit them over IP network. In

S. Chellappan et al. (Eds.): WASA 2018, LNCS 10874, pp. 580–592, 2018.
https://doi.org/10.1007/978-3-319-94268-1_48

consideration of IPv4 address exhaustion, an IPv6-based solution, 6LoWPAN was improved for smart buildings [2]. However, it also faced the issues of complicated address configuration and performance degradation caused by middleware. Therefore, with the rapid development of IoT applications, it is necessary and urgent to put forward an efficient data transmission solution for smart buildings.

Recently, Named Data Networking (NDN) has been proposed with its innovative properties of name-based routing, in-network caching, and multi-source transmission. It provides a promising solution to deal with transmission problems in smart buildings [3]. First, NDN is a typical information-centric networking which focuses on data rather than host, so it avoids complicated address configuration. Second, the properties of multi-source transmission and in-network caching make it ideal for data collection [4]. Considering these advantages, some researchers have discussed how to deploy NDN framework in smart buildings. Although it is a potential alternative, there are still two key issues need to be solved. Firstly, the combination of hierarchical naming scheme and Longest Prefix Matching (LPM) is not flexible enough to meet the diverse transmission requirements. In smart buildings, the user may request data from a device or ask for the information of a specified area. If the naming scheme is not well designed, it is difficult to achieve efficient interest forwarding only by LPM. Secondly, there is no complete solution for data aggregation in smart buildings presently. How to determine the location and range of aggregation is still an important issue in the progress of data collection. Thus, it is difficult and challenging to design an excellent solution for named data transmission in smart buildings.

In this paper, we propose N-Guide for achieving the efficient named data transmission in smart buildings. The novelty of our work lies in three aspects.

- To cover all application types, we redesign a complete naming scheme which divides the whole namespace into physical and application spaces. For physical space, a 4-level structure is defined as the physical part of name to describe the topology of smart buildings. For application space, an application part is constructed by the information of objects, operation types and corresponding time. Besides, a bridge is built to connect the two parts so that every name applies as a whole for LPM (Subsect. 3.1).
- To accurately direct interests, we propose a 2-stage forwarding strategy which first transmits interest downstream to the level of room gateways, then disseminates it to the data providers by leveraging the application information. Because of the well-designed naming scheme, our strategy is not only able to transmit the single data, but also complete the task of data collection for the specified temporal and spatial ranges (Subsect. 3.2). Besides, the corresponding methods of Forwarding Information Base (FIB) and Pending Interest Table (PIT) managements are optimized to provide rapid data transmission (Subsect. 3.3).
- To reduce the communication overhead, the data aggregation function is supported by our well-designed naming scheme. In the hierarchical name, we can specify the location, operation type and range of aggregation so as to retrieve the final result by sending only one interest (Subsect. 3.4).

2 Background and Motivations

2.1 Smart Building Scenario

A smart building is a heterogeneous environment which is characterized by different types of devices, communication technologies, and service patterns. As shown in Fig. 1, its network topology is a tree. The root router is attached to a control center, which is responsible for receiving requests from applications and converting them to interests. These interests are first sent to the root router, then forwarded by the routers at each level. Finally, the gateways forward every interest to the network interfaces connecting to the devices and sensors. According to the entries in PITs, the data packets are returned back to the control center along the reverse path. In addition, the transmission requirements in smart building scenario can be summarized as following four modes:

- **1-1-1**: one interest is sent for one data packet from one provider, e.g. a user queries the brightness of a lamp remotely.
- **1-1-P**: one interest is sent for a data flow generated by one provider in period P, e.g. a user queries the power of sweeping robot once a day.
- **1-N-1**: one interest is sent for N data packets from N providers by one each, e.g. a user queries the temperature readings of all sensors in a room.
- **1-N-P**: one interest is sent for the data flows generated by N providers in period P, e.g. a user queries the temperature readings of all sensors in a room once an hour.

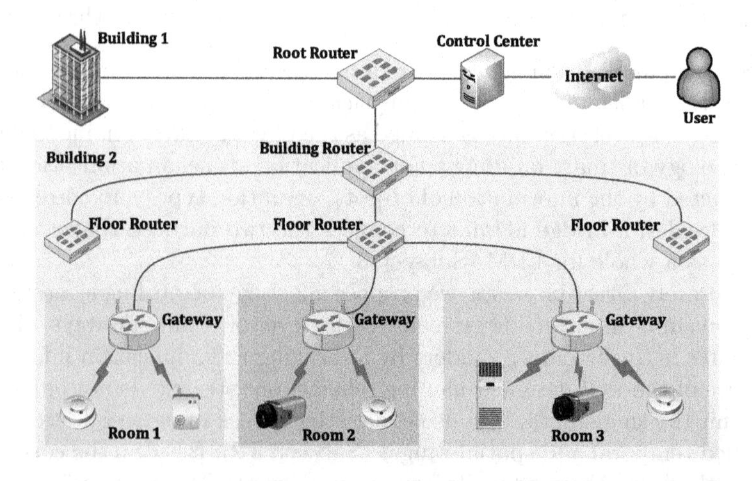

Fig. 1. Smart building topology.

2.2 Named Data Transmission in Smart Buildings

Although NDN is a promising solution, efficient named data transmission is also considered as a huge challenge in [5]. A cache-aware named-data forwarding method is proposed in [6], which requires the sensors to find whether there is a data copy in upstream neighbors before selecting the outgoing interfaces. However, this work only aims at WSN and cannot apply to the heterogeneous network environment of smart buildings. In [7,8], the building management system (BMS) is deployed with original NDN framework. They follow the hierarchical naming scheme and make forwarding decision by LPM. Since the different kinds of information are scrambled in a name, this method becomes inflexible as the number of applications increases. For example, one interest with name */Temperature/Building*1*/Floor*2*/Room*1 requests the temperature readings in room 1, and another with name */PM*2.5*/Building*1*/Floor-*2*/Room*1 requests the value of PM2.5. Both of them target to the same room, but we have to make forwarding decisions according to different entries in FIB. In [9], the NDN with flooding mode is implemented in a realistic scenario. In this mode, the NDN router broadcast the interests to all the possible interfaces, which introduces a lot of redundancy. At present, there is no solution achieving the efficient data transmission in smart building by leveraging the various information in names.

3 N-Guide

To achieve efficient named data transmission in smart buildings, we propose a novel protocol named N-Guide, which consists of 3 parts: a complete naming scheme covering all kinds of applications for smart buildings, a 2-stage forwarding strategy directing interests and data packets by leveraging the information of physical and application spaces, and a data aggregation method supporting multiple operation types in the specified temporal and spatial ranges.

3.1 Naming Scheme

In N-Guide, we propose a 2-space-based naming scheme, which can not only exquisitely construct a name by the information from physical and application spaces, but also make it still apply to the principle of LPM.

In consideration of the applications in smart buildings, the whole namespace is composed of two parts: physical space and application space. In the physical space, we can find the location information of named data, such as the numbers of room, floor and building. In the application space, we can know the providers, operation types and transmission requirements of named data. Although the information from two spaces is helpful for named data transmission, a well-designed naming scheme is necessary to make LPM work efficiently in interest forwarding. As shown in Fig. 2, all the elements of a name are classified into two parts mapping to the physical and application spaces. For each part, the elements are arranged hierarchically. It is worth mentioning that these two parts are not isolated from each other. The element */OBJ* is a bridge connecting them. It is considered as the lowest level in physical space, as well as the entity providing named data in application space. Therefore, each name in our scheme is treated as a whole when LPM is applied.

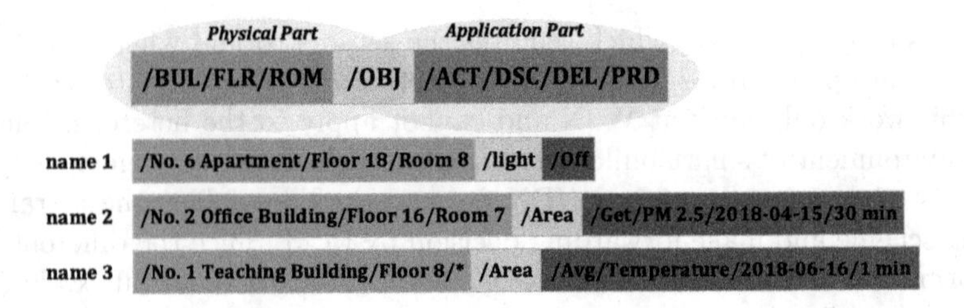

Fig. 2. Naming scheme and examples.

In view of the tree topology in smart building scenario, four levels are defined in the physical part. In Fig. 2, they are */BUL*, */FLR*, */ROM* and */OBJ*. First, the */BUL* represents the name or number of target building, especially when a controller is in charge of a group of buildings. Second, the */FLR* represents the floor number, including both the ground and underground parts. Third, the */ROM* means an enclosed area with fixed structure, such as office, conference room, hall and so on. Finally, */OBJ* is the lowest level of physical space. It could be a smart device or a sensor deployed in the room. In our scheme, these four elements construct a hierarchical physical part mapping to the tree topology. Additionally, some applications will not use all the levels of physical space. For example, a remote user wants to get the average temperature of *No.*1 teaching building. In this case, the asterisk $*$ is used as the placeholder in the levels of */FLR* and */ROM*, and the level of */OBJ* is set to */Area* to indicate that this request is target for an area not an object. Thus, the interest name in this example is expressed as */No.* 1 *Teaching Building/*/*/Area/Avg/Temperature*.

In application space, the name is designed as */OBJ/ACT/DSC/DEL/PRD*. As the bridge between two isolated spaces, the */OBJ* describes the object information, which could be the named data provider or the controlled device. The */ACT* specifies the operation type of current request. In N-Guide, it includes four basic and five aggregation types. For a variety of application requirements, the basic types are */On*, */Off*, */Set* and */Get*, while for the target of data aggregation, the types of */Avg*, */Max*, */Min*, */Count* and */CountIf* are supported. Besides, the */DSC* is set as an extension which helps to describe the data request in detail. Finally, two temporal elements are included in our naming scheme. The */DEL* identifies the moment when current interest expires, and the */PRD* describes the data transmission period at provider side.

In Fig. 2, we take three names as examples to demonstrate that our naming scheme meets all the transmission requirements in smart buildings. In the 1st example, an owner finds the light is still on after leaving room, so he sends a named command to turn it off. Since this is a real-time request, neither */DEL* nor */PRD* needs to be set. In the 2nd example, the air purification system needs to periodically obtain the readings of PM2.5 on the 1st basement, so it sends an interest to the building controller. As shown in Fig. 2, the transmission period is set to half an hour, and this interest will expire on a new day. In the last example, the central air conditioning system sends an interest to get the average temperature on the 8th floor every minute. Although its name is similar to the one in the 2nd example, it supports the basic data aggregation function.

3.2 Forwarding Strategy

To achieve the efficient named data transmission in smart buildings, the forwarding strategy should be able to (1) support all the transmission modes of applications, (2) direct the interests accurately to the providers, (3) return the data packets rapidly. In N-Guide, we propose a 2-stage forwarding strategy which can not only make full use of the beneficial information contained in the name, but also make all the forwarding decisions by LPM.

Generally in smart buildings, the objects described in interests are data providers, and the areas concerned by consumers are usually the places where the providers locate. Thus, the information of physical space is helpful for named data forwarding. Besides, the network topology of smart buildings is a tree covering four fixed levels, and remains unchanged for a long time. In view of above analysis, we divide the interest forwarding process into two stages.

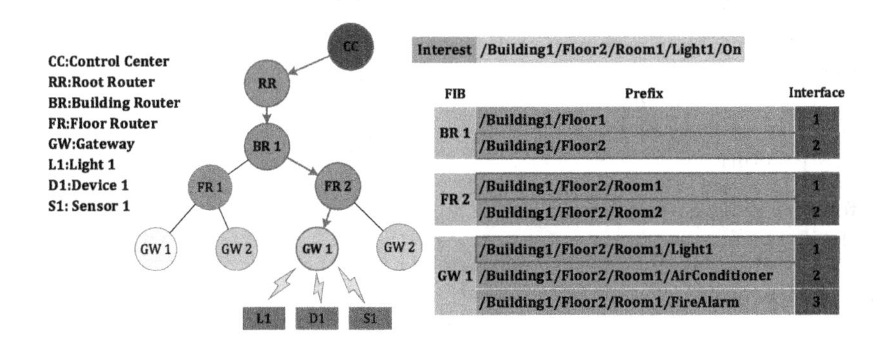

Fig. 3. 2-stage forwarding strategy.

First, we forward the interest according to the information described in the physical part. In the tree topology, we try to transmit the interest from the root router to the gateways of corresponding rooms. During the interest traveling downstream, the outgoing interface at each intermediate node is selected from FIB entries according to LPM. It is important to note that when the placeholder /* appears in one level, all the downstream interfaces should be selected for the interest forwarding. Take the names in Fig. 2 as examples, the first two interests are both transmitted to the gateways of target rooms, while the 3rd interest is disseminated to all the gateways on the 8th floor because of the /* in the room level. Once the interest reaches the gateways, the transmission progress enters the second stage, where the information in application part is used to direct the interest. Like the original NDN, the gateway determines the outgoing interfaces by looking up the historic entries in FIB. During this process, the principle of LPM is also adopted to match the interest with the FIB entries. If there is a successful matching entry, the interest is forwarded to the corresponding interfaces. Otherwise, it is sent to all the interfaces except the incoming one.

The progress of our 2-stage forwarding strategy is described in Algorithm 1. When a node ***N*** receives an interest ***I(n)*** from the incoming interface ***F***, it searches Content Store (CS) and PIT in turn for the matching data and historical entry. Only after confirming that it neither can provide the matching data nor has forwarded the same interest, ***N*** starts searching FIB to make forwarding decision. Different from the original NDN, an extra step is added before FIB searching in order to find whether the name ***n*** contains the placeholder /*. If a placeholder /* appears at the next level of ***N***'s corresponding position in ***n***, ***N*** directly sends the interest to all the downstream interfaces. In this way, all ***N***'s child nodes will receive the interest ***I(n)***. In addition, ***N*** only needs to look up the matching entry in FIB by LPM. If there is an optimal solution, ***I(n)*** is sent to the outgoing interfaces listed in the matching entry. Otherwise, all the possible downstream interfaces are used for forwarding ***I(n)***. It is important to

Algorithm 1. 2-stage forwarding algorithm

```
   Input: N: Node; F: Incoming Interface; L: Node Level; n: Interest Name;
          D(n): Data with name n;
 1 for each entry e ∈ CS do
 2     if e.name = n then
 3         send D(n) to F;
 4         return;
 5 for each entry e ∈ PIT do
 6     if e.name = n then
 7         e.IncomingInterfaceList.insert(F);
 8         return;
 9 max ← 0;
10 maxEntry ← null;
11 for each i ∈ [1,n.size] do
12     if n.element(i) = "*" and L = i-1 then
13         send D(n) to N.interfaces except F;
14         return;
15 for each entry e ∈ FIB do
16     j ← 0;
17     while j < min(n.size, e.name.size) do
18         if n.element(j) = e.name.element(j) then
19             j ← j+1;
20             continue;
21         else
22             break;
23     if j > max then
24         max ← j;
25         maxEntry ← e;
26 send D(n) to maxEntry.interfaces except F;
27 return;
```

note that although our naming scheme consists of physical and application parts, it still applies as a whole for LPM in the FIB searching progress.

In Fig. 3, the interest with name */Building*1*/Floor*2*/Room*1*/Light*1*/on* is transmitted by 2-stage forwarding strategy. Since there is no placeholder in the physical part, this interest travels directly to room1. In this progress, the intermediate routers only need to find out the optimal entry from their FIBs by LPM, and send this interest to the corresponding interfaces. After receiving the interest, the gateway also searches its own FIB. At this time, the matching progress leverages not only the physical part but also the application part.

3.3 FIB and PIT Managements

For efficient interest forwarding, FIB management is an important part of our N-Guide. Different from original NDN, N-Guide takes advantage of the location information. Since the elements in the physical part correspond to the fixed levels of tree topology, the FIB sizes of root router, building router and floor router in Fig. 3 are constant. Only several permanent entries need to set in the FIBs of those NDN routers, which specify the outgoing interfaces to their child nodes. For the gateway of each room, the name prefix no longer directly corresponds to the topological location of child nodes, so the application part is added to help the gateway make forwarding decision. At this time, LPM is implemented in the same way as the original NDN, and the FIB size becomes unfixed.

To support all the transmission modes in smart building scenario, an improvement is also made on PIT which provides the breadcrumbs for named data packets. Different from original NDN which deletes the PIT entry after receiving the matching data at the first time, N-Guide provides a novel PIT management method by setting reasonable lifetimes for different entries. For the four transmission modes mentioned in Subsect. 2.1, we design two different calculation methods for the PIT entry lifetime T_L. First, although the interest coverages are quite different, both of **1-1-1** and **1-N-1** require the providers to send back only one packet. Thus, the current node concerns about whether the data can be obtained before the application deadline t_d. Assuming that the node is allowed to send the interest k times before dropping the request, the PIT entry lifetime $T_L(i)$ for the ith ($1 \leq i \leq k$) interest forwarding is calculated according to the first equation of (1). To get the data as soon as possible, the binary exponential backoff is adopted for the ith interest transmission. In addition, t_c is the current moment, and Δt is the reserved time for returning the matching data back to consumer. Second, the modes of both **1-1-P** and **1-N-P** require that the node should obtain the data periodically. Thus, if the node does not receive the returning data, it can retransmit the interest $k-1$ times during one period of T. In the second equation of (1), the PIT entry lifetime $T_L(i)$ for the ith interest is calculated according to the period specified in the name by the application.

$$T_L(i) = \begin{cases} \left\lfloor \dfrac{t_d - t_c - \Delta t}{2^k - 1} \right\rfloor \cdot i & T = 0 \\ \left\lfloor \dfrac{T - \Delta t}{2^k - 1} \right\rfloor \cdot i & T > 0 \end{cases} \tag{1}$$

3.4 Data Aggregation

Since there are a huge number of devices and sensors in the smart buildings, data aggregation is necessary to reduce redundancy and network overhead. Thus, it is provided as an essential function in our N-Guide. On basis of the analysis in Subsect. 2.2, data aggregation applies to the three transmission modes of **1-1-P**, **1-N-1** and **1-N-P**. To realize this function, we need to solve three issues of (1) the physical range and aggregation location, (2) the temporal range and period, and (3) the aggregation type.

In our naming scheme, four fixed levels are defined in physical parts. If the placeholder /* is found in one level, it means that the data received by the nodes in this level should be aggregated. For the modes of **1-N-1** and **1-N-P**, if the placeholder appears, the element of */OBJ* should be assigned as */Area*, and the subsequent elements in physical part should also be filled by /*. Besides, the aggregation node is determined by the last non-placeholder element in physical part. Take the 3rd name in Fig. 2 for example, the NDN router at the 8th floor is responsible for aggregation of temperature readings. As for time dimension, the duration and period of aggregation are specified by the elements of */DEL* and */PRD* respectively. Finally, the aggregation types can be described by the element */ACT* such as */Avg*, */Max*, */Min*, */Count* and */CountIf*.

4 Simulation

4.1 Simulation Setup

To verify N-Guide, we run a series of simulations on ndnSIM. A standard tree topology is built in our simulation scenario. As shown in Fig. 4, the root is a NDN router attached by the control center, which receives all the interests and forwards them downstream. Since all the data transmissions occur in the same building, only one building node is set in the 2nd level. In Fig. 4, two floor nodes are set as the children of building, and each of them covers two rooms deployed with the gateways. Finally, multiple devices and sensors are deployed in each room. There are three kinds of communication technologies connecting them to the gateways, including ethernet, Wi-Fi, and IEEE 802.15.4.

4.2 Performance Comparison

In this subsection, we compare N-Guide with another two methods in four different transmission modes. Those two methods are respectively marked as O-NDN and B-NDN in the following simulations. O-NDN represents the original NDN [7], and B-NDN represents the approach adopting the broadcasting strategy [9].

First, the three methods are compared in the basic **1-1-1** mode. The interest flow with random names is generated by the control center. When the devices and sensors receive an interest, they look up their CSs by name, and send back the successful matching data. In Fig. 5, the success ratios of N-Guide and O-NDN are equal to 1. That means almost all the interests are satisfied. Since B-NDN

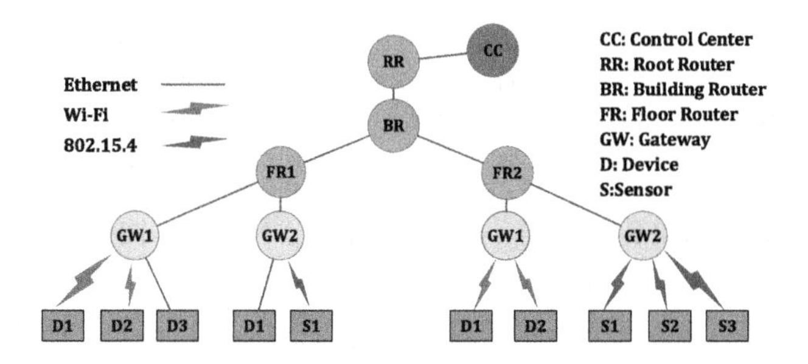

Fig. 4. Smart building simulation scenario.

forwards the interest to all the outgoing interfaces, it introduces a large number of redundant copies consuming network resources and even causing wireless transmission conflicts. As a result, its success ratio only reaches 84%. Since B-NDN forwards interest without FIB, we only compare the average FIB sizes of N-Guide and O-NDN. Figure 6 describes the average FIB sizes at the building router and floor router with respect to the device number. It is obvious that the FIB size of O-NDN grows linearly with the increment of device number, while the size of N-Guide remains unchanged. It is because that O-NDN creates the FIB entry according to the number of interests generated by applications, while N-Guide does this based on the fixed topology information. Besides, the coverage of building rooter is larger than floor router, so it needs to maintain more FIB entries to meet the transmission requirements for more devices. Figure 7 shows the PIT sizes of three methods with respect to the interest frequency and topology level. As the frequency grows from 10 packets per second (pps) to 1000 pps, the PIT sizes increase. Due to the redundancy, B-NDN introduces much more PIT entries than the other methods. Both N-Guide and O-NDN direct interests accurately, so their PIT sizes maintain in an acceptable range even though the frequency increases exponentially. In addition, as the topology level moves down, the interests are more likely to be satisfied with the named data, and the PIT entries are eliminated more rapidly. Thus, the PIT sizes of N-Guide and O-NDN decrease with the level decrement. However, B-NDN obtains the opposite trend. The broadcasting strategy introduces more and more unsatisfied PIT entries with the level decrement. As a result, the gateway gets the largest PIT size in B-NDN.

Second, we make performance comparison among the three methods in **1-1-P** mode. Different from O-NDN and B-NDN, N-Guide provides the PIT entry with a long lifetime. Thus, the consumer only needs to send one interest to retrieve the periodic data. Figure 9 shows the ratios of interest and data packet numbers with respect to the data delivery frequency. It is clear that the ratio of N-Guide decreases with the increment of data delivery frequency. Besides, since every data packet is got by sending one interest in O-NDN and B-NDN, their ratios are always equal to 1.

Third, we evaluate the performance of three methods in **1-N-1** mode. Figure 10 shows the ratios of interest and data packet numbers with respect to the device number. Similar to the situation in Fig. 9, the ratio of N-Guide decreases with the increment of device number, while the ratios of O-NDN and

B-NDN are the constant 1. In addition, the PIT sizes corresponding to different levels are described in Fig. 8 with the expansion of interest coverage. In Fig. 8, the x axis indicates three different coverages of interest, while y axis shows three topology levels. As shown in Fig. 8, N-Guide maintains a smaller PIT than the others in all cases. Even though the coverage of interest expands, the PIT size of N-Guide does not increase significantly like O-NDN and B-NDN. Therefore, N-Guide can obtain the best performance in the scenario with a large number of devices.

As **1-N-P** is the combination of **1-1-P** and **1-N-1**, the performance of three methods in **1-N-P** is similar to those two modes. Figure 11 shows the ratios of interest and data packet numbers. For N-Guide, its ratio decreases with the increment of device number and data delivery frequency. For O-NDN and B-NDN, their ratios are the constant 1, which are not plotted in Fig. 11. Finally, we verify the aggregation function of N-Guide. Figure 12 shows the ratios of data and interest packet numbers. It is clear that if we turn off the aggregation function, the ratio of N-Guide grows rapidly with the increment of device number and data delivery frequency. Once the aggregation is enabled, we can obtain the constant ratio equal to 1. At this time, every interest sent by consumer requires that the aggregation node only need to return a calculated result regardless of device number or data delivery frequency.

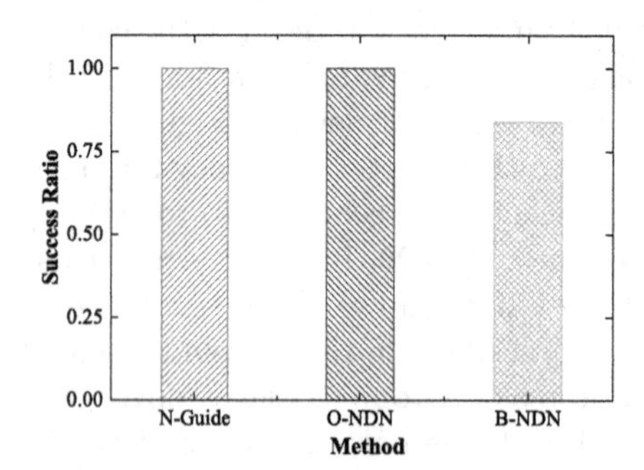

Fig. 5. Success ratio comparison (**1-1-1**).

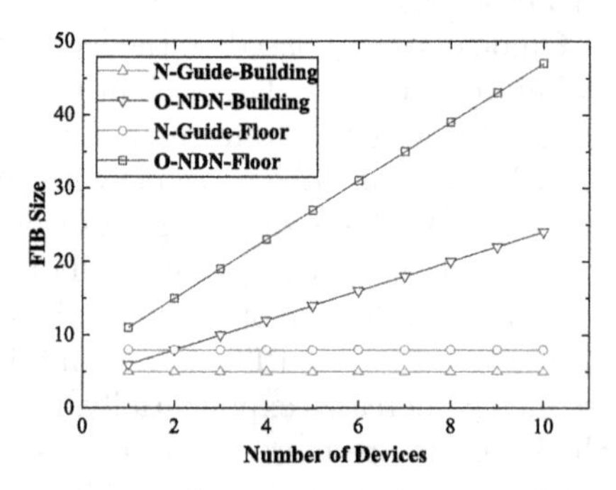

Fig. 6. FIB size comparison (**1-1-1**).

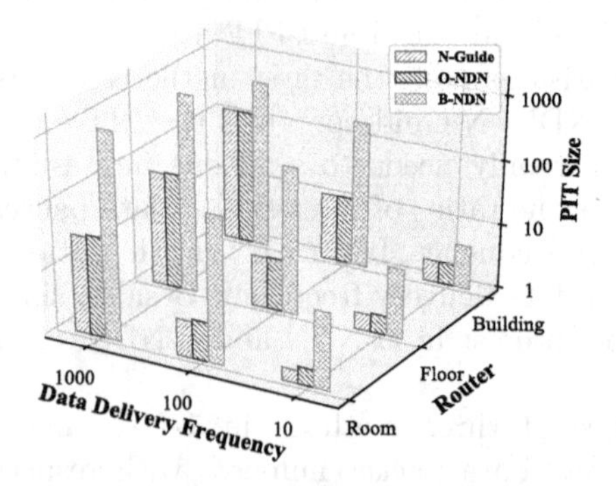

Fig. 7. PIT size comparison (**1-1-1**).

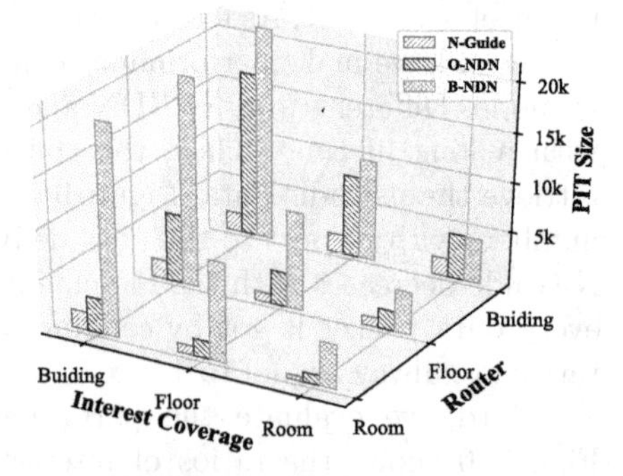

Fig. 8. PIT size comparison (**1-N-1**).

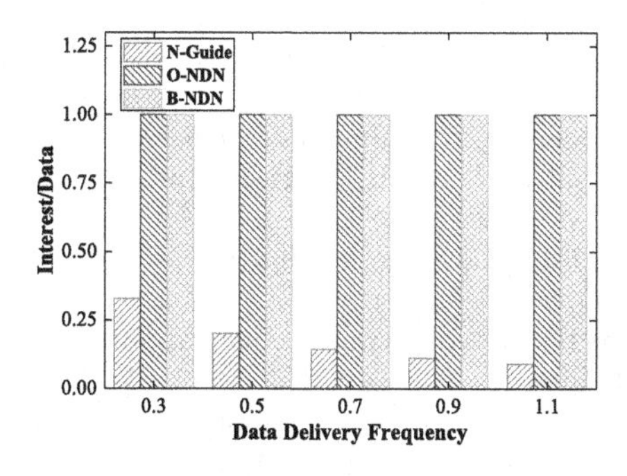

Fig. 9. Interest/data comparison (**1-1-P**).

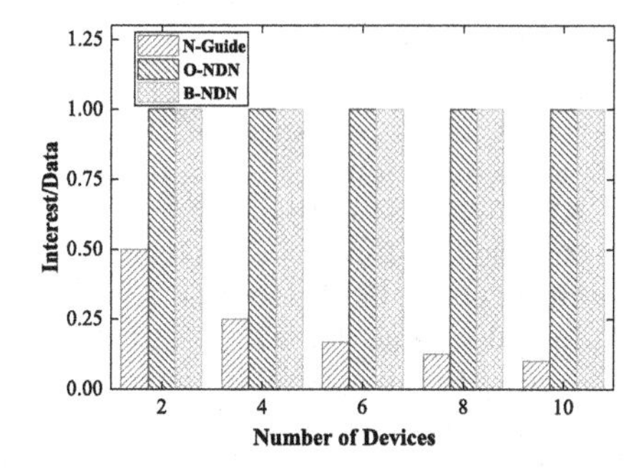

Fig. 10. Interest/data comparison (**1-N-1**).

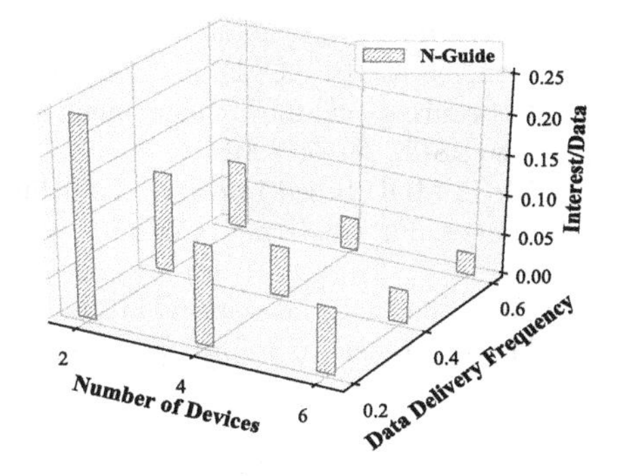

Fig. 11. Interest/data comparison (**1-N-P**).

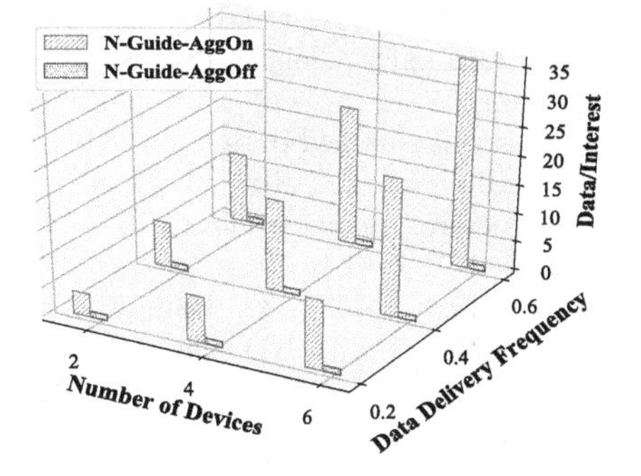

Fig. 12. Aggregation performance (**1-N-P**).

5 Conclusion

For the target of efficient named data transmission in smart buildings, we have proposed a novel protocol N-Guide in this paper. By leveraging the topology and application information in names, N-Guide is able to direct the interest accurately and retrieve named data rapidly. Besides, N-Guide also supports the aggregation function by setting three key elements in its naming scheme. To validate the performance of N-Guide, we have made the contrast simulations on ndnSIM. The results show that N-Guide obtains the same success ratio as original NDN but with smaller PIT and FIB sizes. Additionally, it reduces the network overhead by supporting three special transmission modes and aggregation function. In a word, N-Guide is an outstanding solution for named data transmission in smart buildings.

Acknowledgement. This work was supported by the National Natural Science Foundation of China (No. 61702288, 61702287), the Natural Science Foundation of Tianjin in China (No. 16JCQNJC00700), the Open Project Fund of Guangdong Key Laboratory of Big Data Analysis and Processing (No. 2017003), and the Fundamental Research Funds for the Central Universities (No. 7933).

References

1. Dietrich, D., Bruckner, D., Zucker, G., Palensky, P.: Communication and computation in buildings: a short introduction and overview. IEEE Trans. Ind. Electron. **57**, 3577–3584 (2010)
2. Zanella, A., Bui, N., Castellani, A., Vangelista, L., Zorzi, M.: Internet of things for smart cities. IEEE Internet Things J. **1**(1), 22–32 (2014)
3. Liu, B., Jiang, T., Wang, Z., et al.: Object-oriented network: a named-data architecture toward the future internet. IEEE Internet Things J. **4**(4), 957–967 (2017)
4. Shang, W., Bannis, A., Zhang, L., et al.: Named data networking of things (invited paper). In: 2016 IEEE First International Conference on Internet-of-Things Design and Implementation (IoTDI), pp. 117–128 (2016)
5. Amadeo, M., Campolo, C., Iera, A., Molinaro, A.: Named data networking for IoT: an architectural perspective. In: European Conference on Networks and Communications (EuCNC), pp. 1–5 (2014)
6. Zhang, Z., Ma, H., Liu, L.: Cache-aware named-data forwarding in internet of things. In: 2015 IEEE Global Communications Conference (GLOBECOM), pp. 1–6 (2015)
7. Shang, W., Ding, Q., Marianantoni, A., et al.: Securing building management systems using named data networking. IEEE Netw. **28**(3), 50–56 (2014)
8. Muralidharan, S., Roy, A., Saxena, N.: MDP-IoT: MDP based interest forwarding for heterogeneous traffic in IoT-NDN environment. Future Gener. Comput. Syst. **2**(4), 1–17 (2017)
9. Baccelli, E., Mehlis, C., Hahm, O., et al.: Information centric networking in the IoT: experiments with NDN in the wild. In: ACM Conference on Information-Centric Networking (ACM ICN), pp. 77–86 (2014)

Privacy-Preserving Task Assignment in Skill-Aware Spatial Crowdsourcing

Hang Ye, Kai Han(✉), Ke Xu, Feng Gao, and Chaoting Xu

School of Computer Science and Technology/Suzhou Institute for Advanced Study, University of Science and Technology of China, Hefei, People's Republic of China
{yehang,kexu,gf940312,xct1996}@mail.ustc.edu.cn, hankai@ustc.edu.cn

Abstract. Spatial Crowdsourcing (SC) is a emerging outsourcing platform that allocates spatio-temporal tasks to a set of workers. However, it usually demands workers to upload their privacy information to untrustworthy entities, which causes a privacy breach. In this paper, we consider a complex SC scenario, in which each worker has a skill, whereas each spatial task requires a set of skills. Under this scenario, we propose a novel framework that can protect the location and skill privacy of workers while providing effective task assignment. Experimental results on both real and synthetic data sets show the effectiveness and efficiency of our proposed framework.

Keywords: Spatial crowdsourcing · Privacy · Skill · Task assignment

1 Introduction

With the popularity of mobile Internet and smart devices, people can easily accomplish some location-based tasks that are adjacent to their current positions, such as environmental sensing, recording video, repairing houses, and preparing for parties at some locations. In that case, Spatial Crowdsourcing (SC) [8] is emerging as a transformative platform that distribute spatio-temporal tasks to a set of specified workers, who physically move to specified locations and complete this tasks. Lately, SC has applied in both academia and industry.

In the crowdsourcing platform, the requester upload tasks to a spatial crowdsourcing server (SC server) which matches the tasks with registered workers based on the location and other information. However, this process raises serious privacy disclosure as the SC server is not a trusted entity, thus revealing the workers' privacy. It is imperative to take corresponding measures to protect the privacy of the workers and then a large quantity of workers are willing to participate in spatial tasks eventually.

In reality, not all spatial tasks are as simple as taking pictures or videos, environmental sensing, or reporting temperatures, which can be easily accomplished by workers. We now consider the complex tasks which can only be completed by the skilled workers with specific expertise (e.g., painting walls or installing wire). For example, there is a complex task of renovating a house, requiring workers to

S. Chellappan et al. (Eds.): WASA 2018, LNCS 10874, pp. 593–605, 2018.
https://doi.org/10.1007/978-3-319-94268-1_49

painting the walls, fixing roofs and repairing the door. Motivated by the example, in this paper, we will formally define the complex spatial tasks assignment problem. We not only need to effectively assign tasks according to workers location and skill, but also protect location and skill of workers. It is essential to ensure location and skill privacy of workers as it may be utilized to infer sensitive individual information such as alternative lifestyles, political views, and work place. The hot issue about location privacy has been extensively studied in [2,3,9]. [9] designs some interfering locations without revealing the real location of each worker, then they send the wrapped and unrealistic location to service provider. In [2], k-anonymity focuses on the problem that the user's location cannot be distinguished from other k users. [3] makes a further improvement in security of their proposed model and develops an emerging mechanism of peer-to-peer anonymization. Aforementioned scheme about location privacy are not applicable to our scenario due to the complexity of the tasks.

In recent research, To et al. [14] proposed a framework that achieves the protection of workers location. However, it merely considers the location without paying attention to the skill situation. In the work of [13], they keep the premise that workers have to know the exact location of the requester in advance, then they adopt additive homomorphic encryption to protect the worker location privacy. Certainly, [13] creates a privacy threat because there is no trust relationship between requesters and workers.

In our paper, we propose a framework that preserves both location and skill privacy of workers. In our framework, the SC server only has access to sanitized data of workers according to differential privacy (DP) [6]. Every worker is subscribed to a cellular service provider (CSP) that can access to worker location and skill information, and provides the data to the SC servers in noise form according to DP. The CSP has a trust relationship with workers as it sings a contract with its subscribers, prescribing the clauses and conditions. To ensure the SC server assign tasks effectively using noise data, we adopt the Private Spatial Decomposition (PSD) approach proposed in [4], and further improve the decomposition method according to the skill information. Adding noise to the PSD also introduces another difficulty that fake points need to be created in the PSD. Thus, we adopt geocast [11] mechanism to disseminate task requests to workers to prevent the SC server from distinguishing these points.

Specifically, we make the following contributions.

(1) We identify the specify privacy challenges of task assignment in SC system, and we propose a novel framework to protect the location and skill privacy of workers while providing effective assignment of tasks and high quality of service, which can be a footstone for future studies in this area.
(2) We consider both skill and location information when assigning tasks in SC system, which allows SC to be applied in more complex task scenarios.
(3) We utilize a geocast mechanism when disseminating task requests to workers to conquer the restrictions imposed by DP, and the communication overhead during this process is considered in the geocast region construction.
(4) We conduct extensive experiments on real and synthetic data sets and show the efficiency and effectiveness of our proposed approaches.

We present the background in Sect. 2. Section 3 describes the problem formulation. Sections 4 and 5 show the proposed solution. Experimental results and related work are presented in Sects. 6 and 7, respectively. Finally, Sect. 8 concludes this paper.

2 Background

2.1 Differential Privacy

Definition 1. Given any two data sets D and $D^{'}$ which differ in only one element, for any set of outcomes S, a randomized algorithm M gives ϵ-differential privacy [6] if the probability distribution of the mechanism output on D and $D^{'}$ is bounded by

$$Pr\left[M\left(D\right) \in S\right] \leq exp\left(\epsilon\right) \cdot Pr\left[M\left(D^{'}\right) \in S\right] \tag{1}$$

Definition 2. Given any two datasets D and $D^{'}$ which differ in one element, the sensitivity of the released query set QS [6] is

$$\delta\left(QS\right) = \max_{D,D'} \|QS_D - QS_{D'}\|_1 \tag{2}$$

Given the sensitivity, a sufficient condition to achieve ϵ-DP is to add to each query result randomly distributed Laplace noise with mean $\delta(QS)/\epsilon$ [5].

Theorem 1 (Sequential composition). If M_i are a set of analyses, each providing ϵ_i-DP, then their sequential composition satisfies $(\sum \epsilon_i)$-DP [14].

Theorem 2 (Parallel composition). If M_i are a set of analyses, each providing ϵ_i-DP, then their parallel composition satisfies $\max(\epsilon_i)$-DP [14].

2.2 Private Spatial Decomposition (PSD)

The author in [4] first proposed the concept of PSD to construct a spatial dataset that achieves DP. A PSD is a spatial index where each index node points to a spatial region and the value of each node is the noise count of the data points enclosed by that nodes region. More information about PSD can be found in [4]. In our paper, we adapt the AG [12] approach to construct the PSD.

3 Problem Formulation

3.1 Problem Definition

In our paper, we consider a complex task scenario in SC. These tasks possess different spatial locations and cannot be accomplished by normal workers, but require the skilled workers with specific expertise. In this paper, we temporarily suppose that each worker has only one type of professional skill. In Sect. 5.3, we will further consider the worker's proficiency in the skill and represent it as a score. Later, we will divide this score into different levels, and the workers who are more proficient in skills will be divided into higher levels. Our objective is to protect the privacy of workers before they accept a task. Assume that $\Omega = \{s_1, s_2, \ldots, s_n\}$ is a universe of n skills, the definitions of task, worker and privacy are as follows:

Definition 3 (Complex Spatial Task). A task is defined as ($\{s_1{:}k_1, s_2{:}k_2, \ldots\}$, L, R), where the collection holds a series of pairs which consists of skills and their corresponding numbers of workers, the L is the location of the task, and the R is the reward given by the task.

Definition 4 (Skilled Worker). A worker is defined as (S, L), where S(S $\subseteq \Omega$) is the worker's skill, and L is the location of the worker.

Definition 5 (Privacy). We define privacy as all sensitive information for workers, which mainly includes worker identity, location and skill information.

3.2 System Model

Figure 1 shows the system model of our proposed privacy-preserving framework. There are mainly four parties in the system: the requesters, workers, the SC server, and the trusted third party CSP. Our system works as follows. First of all, workers send their locations and skills information to the CSP (Step 1), and then once the SC server receives the task request (Step 2), it will send the skill information required by the task to the CSP (Step 3) which then collects updates and releases a Skill-based Private Spatial Decomposition (S-PSD) according to skill information and the privacy budget ϵ agreed upon with workers (Step 4). The S-PSD is accessed by the SC server which then queries the S-PSD to decide a geocast region (GR) that contains workers in close proximity to the task and with the skills required for the task. Next, the SC server initiates a geocast communication [11] process to disseminate the task request to all workers with corresponding skills and within GR (Step 5). Note that it requires creation of fake locations and skills in the S-PSD when sanitizing a dataset based on DP. If the SC server is allowed to directly contact workers, it simple to identify these fake workers by trying to establish communication channel with workers, which would breach privacy of workers. Therefore, we use geocast in our framework to prevent the SC server from directly contacting workers.

After receiving the task request, a worker decides whether to perform this task. If the worker is willing to perform this task, he or she sends a confirmation message to the SC server (Step 6), otherwise do not respond.

3.3 Privacy Model and Assumptions

In the process of task assignment, all workers are candidates for real-time task allocation, thus locations and skills of all workers are monitored continuously.

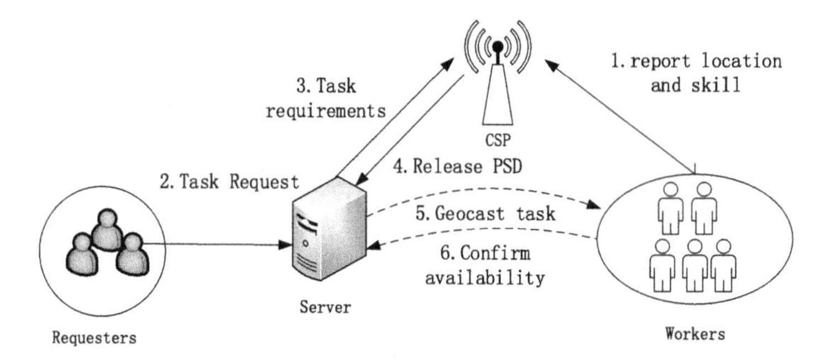

Fig. 1. Privacy framework for spatial crowdsourcing

Before workers consent, we need to protect the workers locations and skills from leaking to any non-trusted entity, including the SC server and task requesters. The SC server may be associate with diverse backgrounds, such as academic institutions and profit organizations, therefore workers cannot trust the SC server. On another hand, the CSP has already established a trust relationship with workers by signing service agreement, furthermore, they are reached the mutually-agreed rules about information disclosure. Therefore, reporting the information of workers to CSP does not result in additional information leaked.

Although the CSP is trusted by workers, it can not be used to complete the task allocation. Because task assignment involves a variety of issues such as managing profiles and interacting with various task requester, but the CSP has no expertise or financial interest to participate in such services. Thence, in our framework, the role of the CSP is to collect locations and skills from registered workers, add noise to it according to DP, and release the sanitized data to SC server for task assignment.

3.4 Performance Metrics

(1) **Assignment Success Rate (ASR).** Due to the uncertainty of S-PSD, the SC server may fail to find enough workers for a task. ASR measures the ratio of accepted tasks to all task requests. The task is accepted when the number of workers willing to perform the task meets the requirements of the task.
(2) **Completion Quality.** Even if a task is accepted by some workers, the completion quality of task may not meet the requirements. We can assign tasks to workers with high skill level to improve the completion quality.
(3) **System Overhead.** We use the average number of notified workers (ANW) [14] to quantify system overhead, which mainly includes communication overhead of disseminating task requests and computational overhead of algorithms that construct geocast region. ANW is the average number of notified workers per task request.

4 Constructing the S-PSD

In this section, we revise the state-of-the-art Adaptive Grid (AG) method proposed in [12] to construct S-PSD. The whole division process is divided into two steps. The first step, AG uniformly divided the location domain into $m_1 \times m_1$ cells. In order to minimize the error due to DP partition, the level-1 granularity is chosen as

$$m_1 = max\left(10, \left\lceil \frac{1}{4}\sqrt{\frac{N \times \epsilon}{k_1}} \right\rceil\right) \tag{3}$$

where N is the total number of workers with the skills required for the task, ϵ is the total privacy budget, and $k_1 = 10$ [12]. Next, the CSP issues a noise count query for each level-1 cell using part of the total privacy budget: $\epsilon_1 = \alpha \times \epsilon$, where $0 < \alpha < 1$. Specifically, in each level-1 cell, CSP counts the number of each type skill workers required by the task, and then adds random Laplace noise with mean $1/\epsilon_1$ to it. Since workers in different cells are disjoint, workers with different skills are disjoint as well, this stage provides ϵ_1-DP according to Laplace mechanism and Theorem 2.

The second step, each level-1 cell is divided into $m_2 \times m_2$ level-2 cells, where m_2 is chosen as

$$m_2 = \left\lceil \sqrt{\frac{N' \times \epsilon_2}{k_2}} \right\rceil \tag{4}$$

where $\epsilon_2 = \epsilon - \epsilon_1$, $N^{'}$ is the total noise count of workers in the level-1 cell, and k_2 is a constant and generally takes $\sqrt{2}$ [14]. Similar to the first step, CSP obtains the noise count of each kind skill workers in level-2 cells, but the mean of Laplace is $1/\epsilon_2$. Similarly, this stage provides ϵ_2-DP. Therefore, the data release of AG provides ϵ-DP according to Theorem 1. Finally, CSP releases these noise counts together with the structure of AG.

An example of an adaptive grid is shown in Fig. 2. We assume that the task requires two skills. Each level-1 cell provides the noise counts for two types of skill workers required by task. The level-2 cells provide noise counts like level-1 cells.

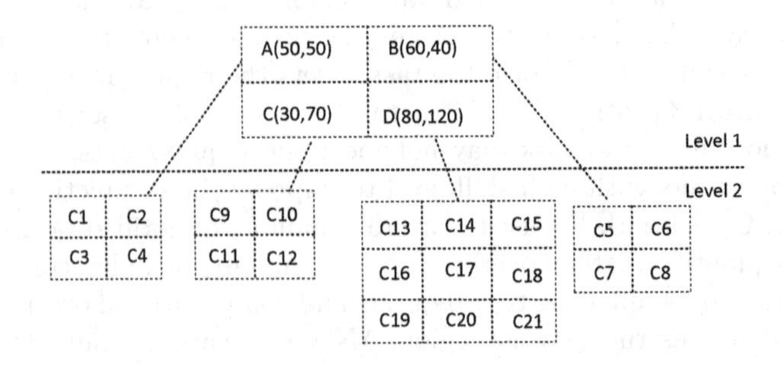

Fig. 2. Example of an adaptive grid

5 Task Assignment

After the S-PSD is released, the SC server queries it and constructs a geocast region GR. All workers with skills required by task in the GR will receive the task request.

5.1 Acceptance Rate Description

Intuitively, the probability of a worker accepting a task primarily depends on the total reward, r, total number workers of task required, k, and the distance, d. Let p denote the probability. We model p using the function that satisfies the following requirements. MTD is the maximum travel distance that a high percentage of workers are willing to travel [14], r_{min} is the minimum reward that most workers are willing to accept.

(1) p ∈ [0,1];
(2) If d > MTD or per capita reward r/k is less than r_{min}, then p = 0;
(3) When r/k is fixed, p will be decreased with the increasing of d;
(4) When d is fixed, p will be increased with the increasing of r/k.

According to the above features, we construct the acceptance probability model as follows:

$$p = \begin{cases} \frac{2}{\pi} arctan\left(\mu \frac{r}{kd}\right) & \text{if } d \leq MTD \text{ } and \text{ } \frac{r}{k} \geq r_{min} \\ 0 & \text{otherwise} \end{cases} \tag{5}$$

where μ is used for tuning the scale of p growth.

Next, we compute the probability AR that at least k workers accept the task in a given region, which is determined by each individual worker and the number of workers in the region. We first compute the overall acceptance rate of a task within a level-2 cell and then extend it to the entire geocast region.

We assume that the task requires n kinds of skills, and requires k_i workers for the i-th skill ($1 \leq i \leq n$), the number of the i-th skill workers is m_i in the level-2 cell. Workers are independent of each other, d is the average of the minimum distance and maximum distance of the task to the level-2 cell, so each worker in the level-2 cell has the same p. The acceptance rate AR in the level-2 cell is calculated as follows:

$$AR = \left[1 - \sum_{i=0}^{k_1 - 1} \binom{m_1}{i} p^i \left(1-p\right)^{m_1 - i}\right] \cdots \left[1 - \sum_{i=0}^{k_n - 1} \binom{m_n}{i} p^i \left(1-p\right)^{m_n - i}\right] \tag{6}$$

A GR may contains multiple level-2 cells, so we need to iteratively calculate the AR in the geocast region. Let $AR_x^{old}(j)$, $AR_x^{add}(j)$ and $AR_x^{new}(j)$ denote the probability that j workers with the x-th skill accept the task in the original, added, and new regions ($1 \leq x \leq n$), respectively. Then we have

$$AR_x^{new}(j) = \sum_{i=0}^{j} \left(AR_x^{old}(i) \times AR_x^{add}(j-i)\right) \tag{7}$$

Then the probability that at least k_x workers with the x-th skill accept the task in the GR is as follows:

$$AR_x = 1 - \sum_{j=0}^{k_x - 1} AR_x^{new}(j) \tag{8}$$

Therefore, the overall AR of the task in the geocast region is calculated as:

$$AR = \prod_{x=1}^{n} AR_x = \left[1 - \sum_{j=0}^{k_1 - 1} AR_1^{new}(j)\right] \cdots \left[1 - \sum_{j=0}^{k_n - 1} AR_n^{new}(j)\right] \tag{9}$$

5.2 Geocast Region Construction

Given a task, the SC server needs to construct a GR based on greedy algorithm, and then notify all workers in the GR. The pseudocode of the greedy algorithm is depicted in Algorithm 1. It first initializes the set CS with the level-2 cell that covers the task t, then loops through CS. At each loop, the algorithm selects the cell with the maximum increment of AR from CS and add it to GR. The algorithm ends until the AR of GR exceeds the threshold AR_{th} or CS is empty.

5.3 Completion Quality Improvement

Algorithm 1 only considers the acceptance rate of the task and ignores the completion quality of the task. To improve the completion quality of the task, we let the worker give a score of 0 to 1 when reporting the skill, indicating their proficiency. Then the score is divided into l level, the same level of workers for a group. Set aside the privacy budget ϵ_3 when constructing S-PSD, then count the number of workers within each score level in each level-2 cell, and add Laplace

Algorithm 1. Greedy Algorithm Based on AR

Input: task t, MTD, AR_{th}
Output: geocast region GR
1 R is the square of length 2×MTD centered at t
2 $AR(\cdot)$ is the acceptance rate of t in a region
3 CS={the level-2 cell that covers t}
4 GR=∅, $AR_{max} = 0$
5 **while** $AR_{max} < AR_{th}$ *and* $CS \neq \varnothing$ **do**
6 $\quad cell_{max} = argmax_{cell \in CS} AR(cell \bigcup GR)$
7 $\quad$ add $cell_{max}$ to GR
8 $\quad AR_{max} = AR(GR)$
9 $\quad neighbors = (\{cell_{max}'s\ neighbors\} - GR) \bigcap R$
10 $\quad CS = neighbors \bigcup (CS - \{cell_{max}\})$
11 return GR

noise with mean $2/\epsilon_3$ to them. Finally, these noise counts are also released with AG. The data release of AG still provides ϵ-DP according to Theorem 1.

The quality of the task completed by multiple workers takes the average of the following interval:

$$\varphi = \sum_{i=1}^{l} \frac{w_i}{w} r_i \tag{10}$$

where l is the number of levels, w is the total number of workers, w_i is the number of workers in level i, and r_i is the interval of level i. Now, we use the following formula to balance AR and quality when constructing GR,

$$U = \alpha AR + (1 - \alpha)\,\bar{\varphi} \tag{11}$$

where α is the balance factor between 0 and 1. We modify Algorithm 1 to choose new cells to add to GR based on U instead of AR until the acceptance rate and quality reach the corresponding threshold. We get Algorithm 2 from this.

5.4 Communication Cost

Propagation of a task request within the GR can be achieved in two ways. The CSP either sends a message to each worker in the region, or send the message to several workers and let the message transferred hop-by-hop within the GR. The former cost is proportional to ANW, which may be large. Therefore, it is better use hop-by-hop communication for geocasting. We added a skill condition for geocast communication, so geocast in this paper is slightly different from previous research [11]. Only the workers with skills required by the task and within the GR will be informed. In our paper, we use ANW to measure system overhead.

Algorithm 2. Greedy Algorithm Based on AR and Quality

Input: task t, MTD, AR_{th}, φ_{th}
Output: geocast region GR

1 R is the square of length 2×MTD centered at t
2 $AR(\cdot)$ is the acceptance rate of t in a region
3 CS={the level-2 cell that covers t}
4 GR=∅, $AR_{max} = 0$, $\varphi_{max} = 0$
5 **while** $(AR_{max} < AR_{th}$ or $\varphi_{max} < \varphi_{th})$ and $CS \neq \varnothing$ **do**
6 $\quad cell_{max} = argmax_{cell \in CS} U\,(cell \bigcup GR)$
7 $\quad$ add $cell_{max}$ to GR
8 $\quad AR_{max} = AR\,(GR)$
9 $\quad \varphi_{max} = \bar{\varphi}\,(GR)$
10 $\quad neighbors = (\{cell_{max}\text{'s } neighbors\} - GR) \bigcap R$
11 $\quad CS = neighbors \bigcup (CS - \{cell_{max}\})$
12 return GR

6 Experimental Evaluation

In this section, we evaluate the performance of our framework from ASR, completion quality and ANW by conducting experimentally with both real and synthetic datasets.

6.1 Experimental Methodology

We obtain the real data from Meetup data set, which was crawled from meetup.com. There are 5,153,886 users, 5,183,840 events, and 97,587 groups in Meetup. Each user has a location and a series of tags, each group also has a series of tags, and each event has a location and a group who created the event. The tags of the event are equivalent to the tags of the group who creates the event. We use the locations and tags of events to represent the locations and the required skills of tasks, and then randomly generate an integer value for each skill, indicating the number of workers with that skill required by the task. The reward r of the task is also randomly generated. In addition, we utilize the locations and tags of users to represent the locations and the skills of workers. Since we only consider the situation of single-skilled workers in this paper, we choose one of the most commonly used tag from the user's tags as the skill of workers. A score is generated for each worker's skill to measure the skill level of workers. Considering the limited travel distance of workers, we choose one city from meetup dataset to conduct the experiment.

For synthetic data, we uniformly generate locations of workers and tasks in a 2D data space $[0,1]^2$. Furthermore, we randomly associate one user in Meetup data set to a worker, and use tag of the user as his/her skill. Similarly, we randomly select an event and use tags of the event as the required skills of the task. We also randomly generate other needed data. We randomly select 1000 tasks for each experiment, and measure the performance of our method with respect to ASR, quality and ANW. All experimental results are run more than 10 times on random seeds.

6.2 Experimental Results

We use GA0 to represent the original AG PSD algorithm [12] without considering skills, use GA1 to represent our Algorithm 1, and use GA2 to represent our Algorithm 2. Figures 3a, b, and c are experimental results on a real data set, and the other three are the results on a synthetic data set. AR_{th} and φ_{th} are the thresholds for acceptance rate and completion quality, respectively. We can see from Fig. 3a and d that as AR_{th} increases, the ASR of GA0, GA1 and GA2 also grows. In addition, GA2's ASR is slightly smaller than that of GA1 because GA2 balances acceptance and quality. Compared with GA0, GA1 and GA2 can averagely increase ASR by 15% and 10%, respectively.

Figure 3b and e show the changes in the number of tasks that satisfy the quality threshold when varying φ_{th}. The number of tasks that reach the quality

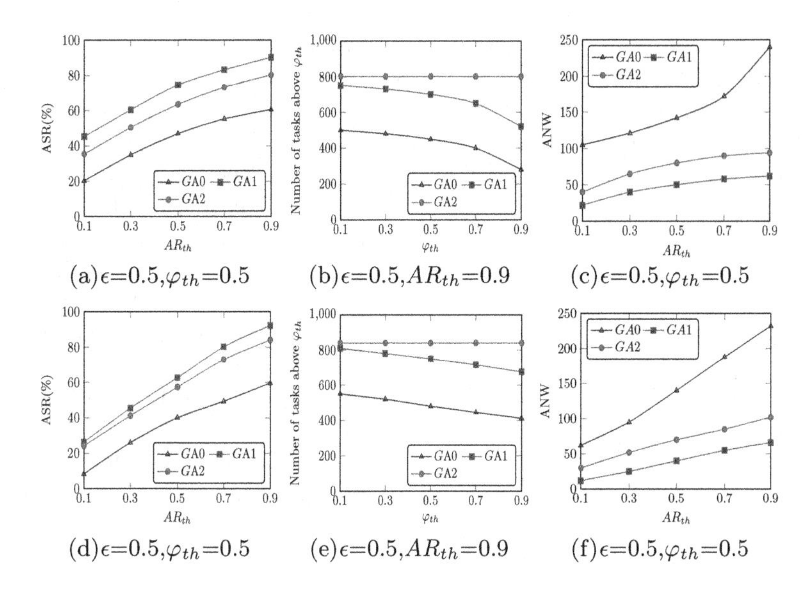

Fig. 3. Performance comparison

threshold in GA2 accounts for about 80% of the total and that in GA1 reaches about 50% of the total, but that in GA0 is below 50% of the total.

The changes in system overhead when varying AR_{th} are presented in Figs. 3c and f. As expected, for a higher AR_{th}, a larger GR should be selected, thus the overhead will increase. The ANW of GA0 is much higher than that of GA1 and GA2 since GA0 ignores the skill requirements of tasks. GA2 sacrificed part of the overhead to improve the completion quality of the task, so its ANW is slightly higher than that of GA1.

GA1 and GA2 need to traverse each level-2 cell at most and calculate the AR of the task on the GR, so their time complexity is linearly related to the number of level-2 cells and the time to compute AR. The above experimental results and analysis show that GA2 improves the task completion quality without significantly affecting the system performance, which proves the efficiency and effectiveness of our method.

7 Related Work

Crowdsourcing has been investigated in both the research community and the enterprise, whereas spatial crowdsourcing aroused concerns only in recent years. Previous works on privacy protection mainly concentrate on location privacy in location-based services. Until recent years, privacy protection in crowdsourcing began to be studied. The research in [10,13] resolve the issue of privacy-preserving in SC by using homomorphic encryption. The former proposes a secure task assignment protocol based on a semi-honest third party and additive homomorphic encryption. The latter introduces a dual-server design and a novel secure indexing technique called SKD-tree to increase system efficiency.

The work in [14] utilizes DP to protect the location privacy of workers and introduces a trusted entity CSP to provide worker data to SC server in noise form. The paper [7] not only provides privacy guarantees for workers but also further improves the completion quality of tasks by considering the reputation of workers. In the research of [15], the author proposes a differentially private method for reward-based SC, and constructs a novel contour plot with a DP guarantee. The paper [1] protects the skill privacy of workers by letting each worker perturb locally his or her profile and copes with the resulting perturbation by leveraging a taxonomy.

However, the aforementioned studies rarely consider protecting both location and skill information of workers at the same time, and based on this, perform effective task assignment. This paper not only solves the problem of privacy leakage, but also enables crowdsourcing to be applied in complex scenarios.

8 Conclusion

In this paper, we consider skills-based task scenarios in spatial crowdsourcing. We propose a privacy framework to protect the location and skill privacy of workers while providing effective assignment of tasks. Our experiments on both real and synthetic data sets demonstrated the effectiveness and efficiency of the proposed framework from task acceptance rate, completion quality and system overhead. As future work, we will extend our framework to solve the problem of privacy protection of multi-skilled workers.

Acknowledgements. This work is partially supported by National Natural Science Foundation of China (NSFC) under Grant No. 61772491, No. 61472460, and Natural Science Foundation of Jiangsu Province under Grant No. BK20161256. Kai Han is the corresponding author.

References

1. Béziaud, L., Allard, T., Gross-Amblard, D.: Lightweight privacy-preserving task assignment in skill-aware crowdsourcing. In: Benslimane, D., Damiani, E., Grosky, W.I., Hameurlain, A., Sheth, A., Wagner, R.R. (eds.) DEXA 2017. LNCS, vol. 10439, pp. 18–26. Springer, Cham (2017). https://doi.org/10.1007/978-3-319-64471-4_2
2. Chow, C.Y., Mokbel, M.F., Aref, W.G.: Casper*: query processing for location services without compromising privacy. ACM Trans. Database Syst. (TODS) **34**(4), 24 (2009)
3. Chow, C.Y., Mokbel, M.F., Liu, X.: Spatial cloaking for anonymous location-based services in mobile peer-to-peer environments. GeoInformatica **15**(2), 351–380 (2011)
4. Cormode, G., Procopiuc, C., Srivastava, D., Shen, E., Yu, T.: Differentially private spatial decompositions. In: 2012 IEEE 28th International Conference on Data Engineering (ICDE), pp. 20–31. IEEE (2012)

5. Dwork, C.: Differential privacy: a survey of results. In: Agrawal, M., Du, D., Duan, Z., Li, A. (eds.) TAMC 2008. LNCS, vol. 4978, pp. 1–19. Springer, Heidelberg (2008). https://doi.org/10.1007/978-3-540-79228-4_1
6. Dwork, C.: Differential privacy. In: van Tilborg, H.C.A., Jajodia, S. (eds.) Encyclopedia of Cryptography and Security, pp. 338–340. Springer, Boston (2011). https://doi.org/10.1007/978-1-4419-5906-5_752
7. Gong, Y., Zhang, C., Fang, Y., Sun, J.: Protecting location privacy for task allocation in ad hoc mobile cloud computing. IEEE Trans. Emerging Top. Comput. **6**, 110–121 (2015)
8. Kazemi, L., Shahabi, C.: GeoCrowd: enabling query answering with spatial crowdsourcing. In: Proceedings of the 20th International Conference on Advances in Geographic Information Systems, pp. 189–198. ACM (2012)
9. Kido, H., Yanagisawa, Y., Satoh, T.: An anonymous communication technique using dummies for location-based services. In: Proceedings of the International Conference on Pervasive Services, ICPS 2005, pp. 88–97. IEEE (2005)
10. Liu, B., Chen, L., Zhu, X., Zhang, Y., Zhang, C., Qiu, W.: Protecting location privacy in spatial crowdsourcing using encrypted data. In: EDBT, pp. 478–481 (2017)
11. Navas, J.C., Imielinski, T.: GeoCast—geographic addressing and routing. In: Proceedings of the 3rd Annual ACM/IEEE International Conference on Mobile Computing and Networking, pp. 66–76. ACM (1997)
12. Qardaji, W., Yang, W., Li, N.: Differentially private grids for geospatial data. In: 2013 IEEE 29th International Conference on Data Engineering (ICDE), pp. 757–768. IEEE (2013)
13. Shen, Y., Huang, L., Li, L., Lu, X., Wang, S., Yang, W.: Towards preserving worker location privacy in spatial crowdsourcing. In: 2015 IEEE Global Communications Conference (GLOBECOM), pp. 1–6. IEEE (2015)
14. To, H., Ghinita, G., Shahabi, C.: A framework for protecting worker location privacy in spatial crowdsourcing. Proc. VLDB Endow. **7**(10), 919–930 (2014)
15. Xiong, P., Zhang, L., Zhu, T.: Reward-based spatial crowdsourcing with differential privacy preservation. Enterp. Inf. Syst. **11**(10), 1500–1517 (2017)

HEVC Lossless Compression Coding Based on Hadamard Butterfly Transformation

Xi Yin[1], Weiqing Huang[2](✉), and Mohsen Guizani[3]

[1] Institute of Formation Engineering, Chinese Academy of Sciences, Beijing 100093, China

[2] University of Chinese Academy of Sciences, Beijing 100040, China
huangweiqing@iie.ac.cn

[3] Department of Electrical and Computer Engineering, University of Idaho, Idaho, USA

Abstract. Efficient video transmission has always been an important challenge in the field of computer networks. Elegant compression coding strategy will signicantly alleviate the heavy burden on network bandwidth. High Efficiency Video Coding (HEVC) is currently the newest video coding standard widely used in different fields, including lossy compression coding and lossless compression coding. In HEVC, each coding unit consists of a luma coding unit and a chromaticity coding unit. In this paper, a transformation based on Walsh-Hadamard butterfly transformation is applied to residuals of the luma coding block, and the element of the residual block is the difference between the frame brightness block and the reference pixel according to the prediction method. With the transformation proposed in this paper, the bits of residual coding can be effectively economized, and therefore, the bit rate of non-destructive coding transmission in frames can be reduced. Adaptively selecting between the proposed method and the direct residual transmission methods of HEVC itself effectively improves the performance of the coding. Experiment results show that compared with HM15.0, the bit rate of the videos is reduced for the five standard sequences.

Keywords: HEVC · Video coding · Hadamard butterfly transformation Non-destructive coding

1 Introduction

HD video is the primary form of information transmission nowadays, which has become much more important in the entertainment industry and information transmission. Structurally, a video is comprised of frame images, an HD image is about 3M, and an hour of uncompressed HD video requires about 300G storage, which is an enormous pressure to both the storage and the bandwidth of the network. Therefore,

This work was supported in part by the National Key Research and Development Program of China under Grant 2016YFB0801001 and Grant 2016YFB0801004.

S. Chellappan et al. (Eds.): WASA 2018, LNCS 10874, pp. 606–621, 2018.
https://doi.org/10.1007/978-3-319-94268-1_50

how to efficiently transmit video has always been an important issue in the field of computer networks. Many video coding techniques have been proposed in this area to alleviate the heavy burden on network bandwidth. Video compression technology is a natural choice for efficient video transmission. In many cases, it is acceptable to compress video into low-quality formats. However, there is still a very high demand of good video quality in the areas of medicine, astronomical observation, form detection and identification.

Video encoding usually includes both lossy encoding and lossless encoding. The lossy encoding usually has a high compression ratio, as the quality demands of video recovery is not very strict. On the contrary, the lossless encoding has a lower compression ratio due to the requirement of the decoded image needed to be the same as the original image. Various types of related, redundant information exist in the video data, and as a result, if encoding the video directly during the process of the video compression, the data that needs to be stored is large in volume. In video, the redundant information includes information entropy redundancy [9], time redundancy and space redundancy [2–4], which correspond to the source information transmission, time domain and spatial information. Information entropy redundancy means that there is a certain correlation in the information to be transmitted, which leads to an increase in the transmission burden when transmitting the same information. Partial information entropy redundancy can be eliminated effectively by entropy coding. In the same sub-image, the adjacent image content has a certain similarity; pixel values are closer, which cause spatial redundancy. The background area is relatively large in the same image. The time redundant information is based on the video's time domain, meaning, the contents of images have similarities between similar video images. The motion of the objects in the image is continuous, and the correlation of the data in similar images is called the time redundancy information. Variety of international video standards came into being to promote video conferencing, video real-time transmission, digital TV, and high-definition video storage transmissions, including MPEG-4 [3], H.264 [5, 6] and HEVC [8]. HEVC is the newest high efficiency video coding standard jointly developed by the two major standardization organizations ITU and ISO/IEC after H.264. Although H.264 video coding standard has achieved good results, there is still room for improvement in the compression ratio, as the field of expertise has higher requirements and is not meeting the requirements well. HEVC continues the main framework of H.264, improves some of the technologies, and leads to a significant improvement of the performance.

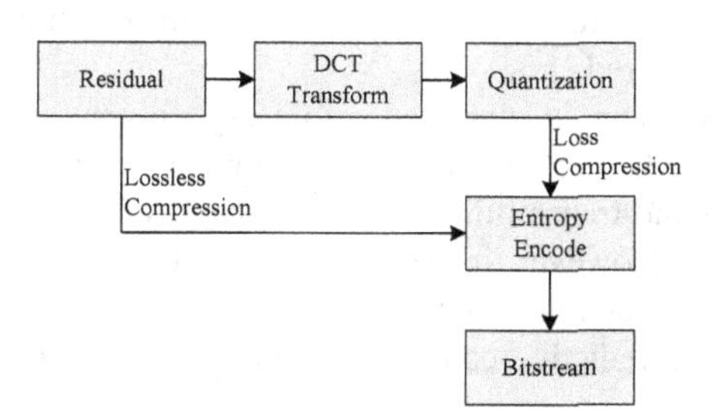

Fig. 1. Lossy and lossless compression coding framework of HEVC.

HEVC is a hybrid coding that combines predictive coding and transform coding, which is entropy coding after the transformation and quantization operations of a residual unit during coding is a normal lossy coding. Images obtained by lossy codes are not exactly same as the original images and do not affect the human visual sensory effects. In some applications that require strict image restoration, lossy coding does not meet the requirements, so lossless coding is indispensable. Due to the transformation coefficients after DCT are fractional, the coefficients are approximated as integers for convenience of transmission, and the high-frequency part is further weakened during quantization. Lossless coding in HEVC, similar to the coding structure of H.264, is skipped with energy lossy transformation and quantization steps, and direct entropy coding. As shown in Fig. 1, the encoding process of HEVC can clearly be understood [1].

Prediction plays a crucial role in video coding. If the residuals obtained after the prediction are relatively small, the number of bits to be encoded is relatively small, and the compression ratio is low. Otherwise, the compression ratio is high. It is very important to get a good prediction. In the process of prediction, the prediction mode plays a decisive role in the prediction of the object. The mode decision is decisive for the prediction of video compression coding. The process of transforming and quantifying the residual unit after prediction can make the energy concentrate in the low frequency domain and reduces the performance of transmission in high frequency regions. Hence, the bits of entropy coding can be reduced while maintaining the image quality. One of the research areas is how to compress a video image better and more effective and meet the needs of people for visual watching. On the other hand, the existence of lossless video compression is necessary, while there is a loss of residual energy in the encoding process when transforming and quantifying. This can result in a partial distortion of the decoded image. The compression ratio of the compressed video is not high while entropy coding is performed directly on the residuals obtained after the prediction. Therefore, how to improve the compression ratio of lossless coding and reduce the bit rate during coding is another research area.

2 Related Works

Video coding standards are based on the well-known Shannon Theory: the coding entity is a pixel or pixel block in the video sequence, and finally displayed in the form of a video image on an output device such as a display or a projector. Many video coding standards are put forward as the first generation of video coding technology. However, these standards are regardless of the video information, and the consumer's perception is not considered.

The video compression is classified into lossy compression and lossless compression according to whether the reconstructed image is exactly same as the original image after encoding and transmitting. In recent years, lossy pressure has been widely applied to most fields. However, in some specific fields, lossless codes are required. Common lossless compression coding includes dictionary coding, Huffman coding, arithmetic coding and predictive coding [7]. When compressing and encoding video images, the redundant information that needs to be considered usually includes time redundant information, spatial redundant information and chrominance redundant information. Time Redundancy and Spatial Redundancy are mentioned above, chroma redundancy information can be eliminated by converting to redundant information in other domains.

Coding generally include the following three methods: a method based on differential pulse code modulation (DPCM), a pixel-by-pixel prediction method, and a coding format based on improved information entropy. In the algorithms based on the residual differential coding (RDPCM), the image is first predicted based on the block and then processes the predicted residual block based on the DPCM method, such as the pixel-by-pixel prediction mode. Many other variations of this method are described in [17, 18]. Lee et al. [17] proposed the earliest H.264/AVC intra-frame lossless method. They first proposed spatial prediction based on blocks, and then performed pixel-by-pixel operations on residual cells, which are limited to horizontal or vertical prediction methods. In the horizontal prediction mode, the prediction of the current residual pixel value is performed by subtracting the left-side neighboring pixel value for each residual pixel. The residual values obtained in the vertical prediction mode are similar with those in the horizontal prediction mode. However, they did not describe how residuals are processed in other prediction directions. In [19–21], a prediction method is proposed based on H.264/AVC. In this, they applied the DPCM linear differential prediction coding instead of the original block-based prediction in H.264/AVC. As the reference pixel can be the reconstructed pixel value in the current block, and is no longer based on blocks, the residual value obtained after the prediction of the pixel value in the current block is smaller, the prediction accuracy is greatly improved, and the compression efficiency is greatly improved too. As the methods of selecting the optimal mode based on blocks, the optimal value of prediction cannot be obtained for each pixel in the blocks. When the image content is changed greatly, there will be obvious block effects. In [11, 12], authors adopted an enhanced intra-frame prediction mode based on pixels in the lossless intra-frame coding of H.264/AVC where they obtained a more accurate prediction value of the pixels.

The same RDPCM approach is applied to both intra and inter-coding [13, 14]. In the inter-frame coding, RDPCM is applied to vertical prediction, horizontal, or neither of them. It is necessary that setting a flag to mark each transformation unit whether or not to adopt the RDPCM, another flag needs to be set to mark the direction of the RDPCM. While in intra-frame coding, the only time the RDPCM is used without a flag is when the prediction mode is horizontal or vertical.

The pixel-wise prediction method is usually used in intra-frame lossless compression coding. As lossless coding skips the steps of transform, using pixel-based prediction instead of block-based prediction, it usually achieves very accurate prediction results. Many of these methods are described in [15–17]. Although these methods provide the best compression performance [18], these methods need to distinguish the attributes; they also lead to the disadvantage of the method: their pixel-based features are inconsistent with the block-based video codec system. In [15], the prediction mechanism without no pixel-wise is introduced.

In the lossy coding, the transform coefficients obtained from prediction residuals need to be entropy coded. Whereas in lossless coding, the prediction residuals need to be directly entropy coded. Considering the statistical differences between the transform coefficients and the prediction residuals after quantization, some of the algorithms for lossless coding that change at the entropy coding end predict the pixel values [19]. Yang and Faryar [20] proposed a prediction approach based on context, which uses a pixel-based unit to adaptively select intra-frame or inter-frame coding. This way, the prediction applied to the image can be added to the intra mode more easily. In [20], Brunello et al. proposed a compression scheme based on a run-off compensation, three-dimensional linear prediction and exponential Golomb entropy coding [21].

For the study of lossless compression coding algorithm in HEVC, Sanchez et al. [22] proposed relevant algorithms for HEVC. Since lossless compression skips the steps of transformation and quantization, and the statistical laws of transformed residuals and transformed residual coefficients vary widely, some relevant solutions are described in [23–25]. By improving the CABAC entropy coding to improve the compression efficiency of lossless coding, Gao et al. [26] modified the previous scheme and further improved the binarization. This applies to the Columbus coding method of the prediction residuals with a large amplitude. Tan et al. [27] proposed a novel method. By encoding the residual correctly (for which better effect can be obtained in terms of video compression performance), the method first predicts each row of pixels to derive a residual. Then, based on the residual at the beginning of the row, subtracts the residual at the beginning of the row from the other residuals in the row to obtain a new prediction residual.

Although various forms of video coding have been handled well in HEVC, they can be further improved in compression coding efficiency. In intra-frame lossless coding, in order to ensure that the image is not distorted, no information can be lost. In the transmission, the lossy transformation and quantization steps are directly skipped, and the residual is directly transmitted. In this paper, the corresponding transformation based on the Walsh-Hadamard butterfly transform is taken for the residuals transmitted directly. The Walsh-Hadamard butterfly transform is improved, which is applied to the residuals of the direct transmission without lossy coding. We test five standard test sequences of HEVC, and the experimental results show that the coding bit rate of CLASS A sequence is reduced by 3.11% on average, as well as reduced by 0.41% for CLASS B, increased by 0.55% for CLASS C, reduced by 1.13% for CLASS D, and increased by 1.91% for CLASS E. The coding efficiency cannot be improved for each test sequence by applying Walsh-Hadamard butterfly transformation to the residuals. Therefore, we count the number of blocks based on the residuals coded by Walsh-Hadamard coding, and set a flag to identify whether to transform the residual or not. We test five standard test sequences of HEVC with the improved method. The experimental results show that when the Walsh-Hadamard butterfly transform is performed on the residual block, the bit rates of the five test sequences are reduced, and the coding efficiency with the average increase of 0.92%.

3 The Proposed HEVC Lossless Compression Coding Hadamard Matrix Theory

3.1 Matrix Theory

Hadamard matrix is a square array with an element of ± 1 as an element and any two lines are orthogonal to each other. When $n \times n$ matrix A = (a_{ij}) is orthogonal, and $a_{ij} = \pm 1$:

$$\sum_{k=1}^{n} a_{ik}a_{jk} = \begin{cases} 0, i \neq j \\ n, i = j \end{cases} \tag{1}$$

The matrix A is called n order Hadamard matrix. It is obvious that if and only if $AA^T = nI_n, A = (a_{ij})_{n \times n}$ is Hadamard matrix, where A^T is the transposition of A, and it is also a $n \times n$ matrix.

3.2 Walsh-Hadamard Transformation

Walsh-Hadamard transformation and inverse transformation are defined as:

$$\begin{cases} f(t) = \sum\limits_{i=0}^{N-1} W(i)wal_h(i,t) \\ W(i) = \frac{1}{N}\sum\limits_{t=0}^{N-1} f(t)wal_h(i,t) \end{cases} \tag{2}$$

where wal_h (i, t) is a discrete Walsh function arranged according to Hadamard.

One dimensional Walsh-Hadamard transformation can be expressed by (3).

$$\begin{pmatrix} f(0) \\ f(1) \\ \vdots \\ f(N-1) \end{pmatrix} = H_N \begin{pmatrix} W(0) \\ W(1) \\ \vdots \\ W(N-1) \end{pmatrix} \tag{3}$$

The inverse transformation is as follows:

$$\begin{pmatrix} W(0) \\ W(1) \\ \vdots \\ W(N-1) \end{pmatrix} = \frac{1}{N} H_N \begin{pmatrix} f(0) \\ f(1) \\ \vdots \\ f(N-1) \end{pmatrix} \tag{4}$$

The two-dimensional Walsh-Hadamard transformation:

$$W = \frac{1}{N} H_N f(x, y) H_N \tag{5}$$

where H_N indicates a N order Hadamard matrix, and $f(x, y)$ is a N order square matrix. The inverse transformation is shown by (6).

$$f(x, y) = \frac{1}{N} H_N W H_N \tag{6}$$

The original matrix $f(x, y)$ is highly concentrated in a specific area after performing the orthogonal transformation, and the rest is zero. Therefore, the data compression in Walsh-Hadamard transform signal processing is very useful. As shown in Fig. 2.

Fig. 2. (left to right) Lena Image, two-dimension discrete Hadamard transform of Lena Image and two-dimension discrete Hadamard inverse transform of Lena Image.

As a discrete orthogonal transform, Hadamard transform can remove the correlation information in the spatial domain effectively. The Hadamard transform algorithm is relatively simple and the operation speed is fast. In the video image processing, Hadamard transformation has been widely used in radar, communication, image transmission, biomedical and other related fields. Though Hadamard transform has a fast butterfly transformation, it is only achieved by adding elements and adding operations. In some application directions, Hadamard transformation plays a very important role in related fields instead of DFT transformation.

3.3 Improved Hadamard Butterfly Transformation

In the process of the traditional n-order Hadamard butterfly transformation, we obtain the value of the U element and $u + N/2$ element of each line and each column by performing the addition and subtraction after $\log_2 N$ addition and subtraction transformation, respectively. The operation above is repeated recursively until a Hadamard butterfly transformation is completed. Taking the 8-order Hadamard butterfly transformation as an example, Fig. 3 shows the process of the traditional butterfly transformation.

We applied an improved butterfly transform of Hadamard. Each element of each row or column is obtained by making each of the U elements and $N - u - 1$ elements of sum and difference, taking the elements obtained as the inputs of the next

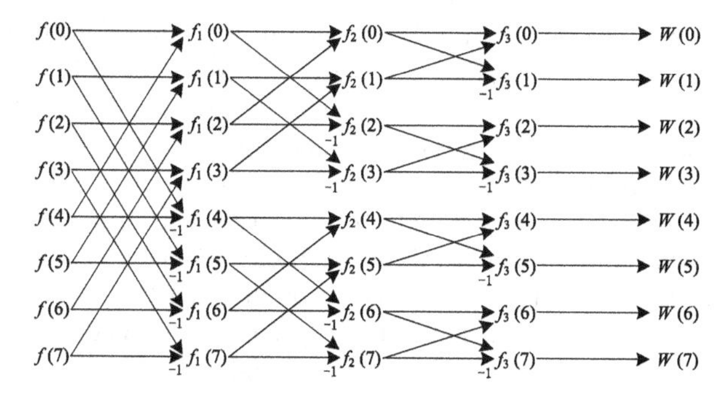

Fig. 3. Hadamard butterfly transformation algorithm.

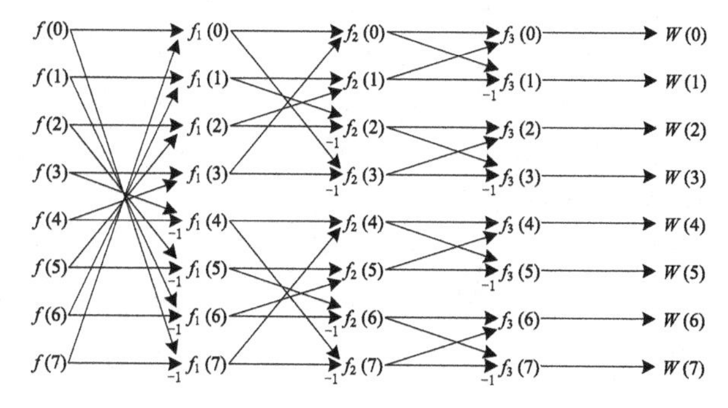

Fig. 4. Symmetric Hadamard butterfly transformation algorithm.

transformation until a butterfly transformation ends. Since the algorithm proposed in this paper has an operation of being divided by 2 in the process of Hadamard butterfly transformation in order to ensure that the number of transformation is integers in the coding process, the algorithm applied in this paper has the relation shown by (7).

$$\begin{cases} f_i(u) = f_{i-1}(u) + f_{i-1}(N-1-u) \\ f_i(N-1-u) = f_{i-1}(u) - f_{i-1}(N-1-u) \end{cases} \tag{7}$$

As shown in Fig. 4, take the 8-order Hadamard Butterfly transformation as an example. In the process of the first transformation, $f_1(0)$ is obtained by summing the 1st and 8th elements of an input row or column of elements. $f_1(7)$ is the difference between the 1st and 8th elements of a row or column of input. $f_1(1)$ is summed by the 2nd and 7th elements of the input row or column of elements. $f_1(6)$ is the difference between the elements of the 2nd and 7th of the input row or column of elements. $f_1(2)$ is obtained by summing the 3rd and 6th elements of an input row or column of elements. $f_1(5)$ is the difference between the 2nd and 6th elements of the input row or column of elements. $f_1(3)$ is the sum of the 4th and 5th elements of the input row or column. $f_1(4)$ is the difference between the 4th and 5th elements of the input row or column of elements. In the process of the next transformation, the increment in the group is $N/2$, and the butterfly transformation is performed in the group in a similar fashion as described above. Each time the transformation is divided into $N'/2$, where N is the number of fractions of the last transformation in the butterfly transformation. When the transformation is initiated, $N' = N$, we obtain the transformed row or column element values after $\log_2 N$ such transformations.

As shown in Fig. 5, the inverse transformation is a transformation that is grouped opposite of the Hadamard butterfly transformation. We obtain the element values for each row or column before the transformation by dividing the transformed element's value by N. Therefore, the Hadamard Butterfly shape transformation and its reversible transformation can be realized by an iterative method.

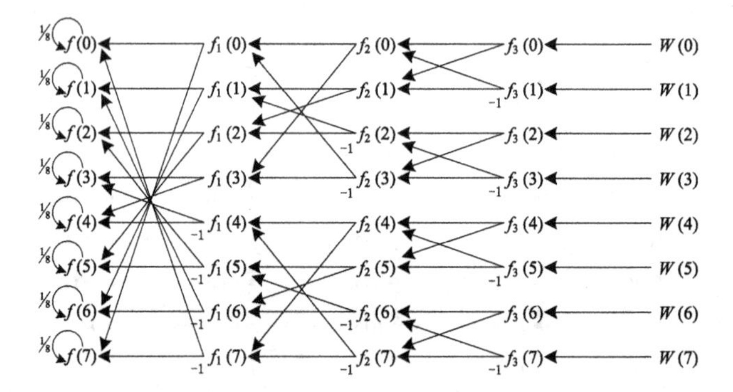

Fig. 5. Symmetrical Hadamard butterfly inverse transformation

3.4 HEVC Intra Frame Lossless Coding

As the elements in the Hadamard matrix are all +1 and −1, during the Hadamard transformation of residual blocks, the integer addition and subtraction operations on coefficients in residual blocks are performed. In this paper, the Hadamard butterfly transformation is applied to the luma residual blocks of different sizes, so as to achieve HEVC intra frame lossless coding and save the number of bits to code for the direct transmission residuals. The Hadamard butterfly transformation based on rows is implemented by luma residual block, and the transmission of bit stream is reduced by the relationship between the difference between transformation coefficients and the result parity. For example, two coefficients a and b, if $a + b$ is an even number, then a − b must be an even number, and vice versa. If $a + b$ is an odd number, then a − b must be an odd number. In this article, when a − b is an even or odd number, $a + b$ is expressed with $(a + b)/2$ or $(a + b{+}1)/2$. When decoding, we first determine the location of the two coefficients. If the position number is even, then we find the corresponding number and position, and multiply the value by 2 before the reassignment. If the number of the position is odd, then multiply the value by 2 and minus 1 before reassignment. This process is also performed $\log_2 N$ times, where N is the length or width of the luma residual block.

$$\begin{cases} f_i(u) = \frac{1}{2}f_i(u), if \quad f_i(N-1-u) \bmod 2 == 0, \\ f_i(u) = \frac{1}{2}[f_i(u)+1], if \quad f_i(N-1-u) \bmod 2! == 0. \end{cases} \tag{8}$$

where N represents the number of elements per group in the butterfly transformation process. In this paper $N = 2^n$, ($n \in [1, 2, 3, 4, 5]$), $f_i(u)$ represents the sum of the two representative symmetrical transform coefficients in the process of, $f_i(N-1-u)$ represents the difference between two numbers. The Inverse change is shown by (9).

$$\begin{cases} f_i(u) = 2f_i(u), if \quad f_i(N-1-u) \bmod 2 == 0, \\ f_i(u) = 2f_i(u) - 1, \; if \quad f_i(N-1-u) \bmod 2! == 0. \end{cases} \tag{9}$$

where N represents the number of elements in each group during the inverse transformation process. $N = 2^n$, ($n \in [1, 2, 3, 4, 5]$), f_i (u) represents the sum of two numbers in the symmetrical position, f_i (N− 1− u) represents the difference between two numbers.

3.5 Hadamard Butterfly Transformation for Residual Block

For saving the code stream for HEVC intra lossless encoding transmission, we apply an improved Hadamard butterfly transformation to the luma residual blocks which reduces the code stream effectively. In the decoding process, we can obtain the residual before the transformation by the inverse transformation.

We test five sequences of HEVC by Hadamard butterfly transforming the corresponding luma residuals, the experimental results obtained in HM15.0. As shown in Fig. 6, the average performance of Class C and Class E loses by 0.49% and 1.91%, the video sequence size of Class C and Class E is 832 * 480 and 1280 * 720. It is reasonable to infer that not all of the Hadamard butterfly transformations on sequences are promoted in coding performance.

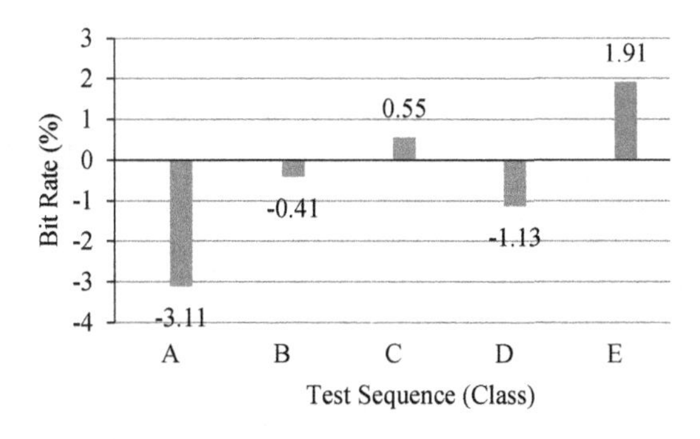

Fig. 6. Results of one dimensional row transformation (no sign bit).

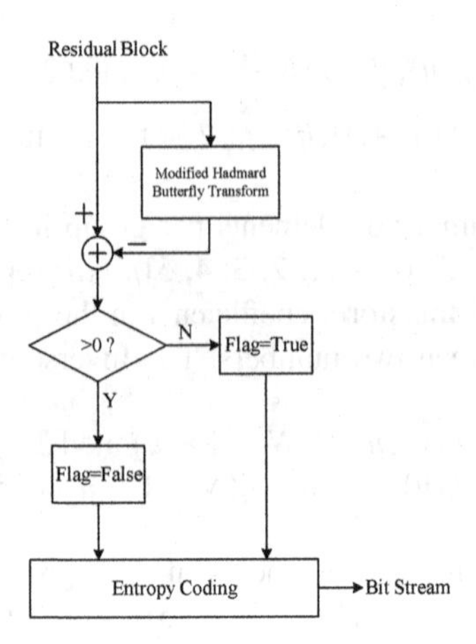

Fig. 7. Selection of two processing modes for residual unit.

It is known from the above that the Hadamard butterfly transformation can bit stream effectively. However, a small residual may become a large one after such a transformation, which leads to computer memory overflow problems. In view of the situation, we set a sign bit from the adaptive judgment to choose the improved Hadamard butterfly transformation for the luma residual block. As shown in Fig. 7, the residual blocks are directly transmitted, and the residual blocks are modified by Hadamard butterfly transform, the corresponding coded bits are calculated and then selected according to their bit rate. If the bit rate after the transformation is less than the bit rate of the direct transmission residuals, the flag bit is set to true which means that the Hadamard butterfly transformation is chosen to improve the residual error. Conversely, the flag bit is set to false, which means that the residual is not selected to transform, but directly to encode, to form a code stream after the transmission.

As shown by Fig. 8, we test five standard sequences of HEVC, the reference software is HM15.0, and each test sequence hasa performance improvement.

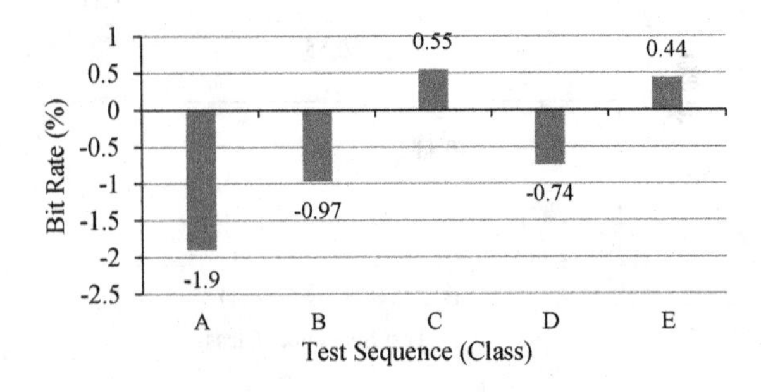

Fig. 8. Results of one dimensional row transformation (setting a symbol bit).

Similarly, we apply the improved Hadamard butterfly transform based on column to residual blocks, and a flag bit is set to select the Hadamard butterfly transform. As shown by Fig. 9, the performance of each test sequence has been improved for five standard test sequences of HEVC. However, the bit rate has decreased to some extent relative to one dimensional row transformation.

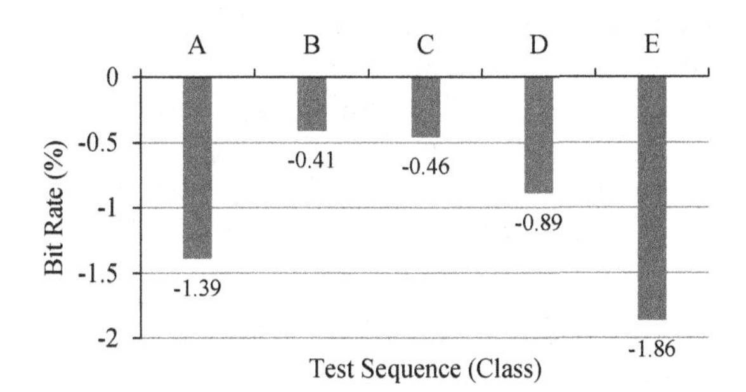

Fig. 9. One column transformation results (set a flag).

Similarly, we propose a two-dimensional improved Hadamard butterfly transform for luma residual blocks, and set flag bits to select Hadamard butterfly transform according to the size of the bit rate. Five standard test sequences of HEVC are conducted. As shown in Fig. 10, we apply the residual blocks to the butterfly transformation based on line then the two-dimensional Hadamard butterfly transform based on column; and the butterfly transformation based on column then the two-dimensional Hadamard butterfly transform based on line. The results of the two methods are basically the same, but the two-dimensional transform can save more bit rate than the one-dimensional transformation in the experimental results.

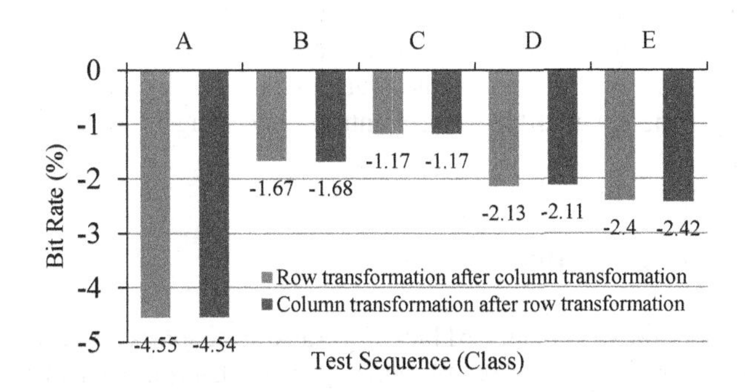

Fig. 10. Comparison of experimental results of two-dimensional transformation (setting a symbol bit)

3.6 Selection of Lossless Coding Transformation Mode

According to the experimental results of intra frame lossless coding, the bit rate of each test sequence is reduced after setting the flag bit when the improved Hadamard butterfly transform is performed on residual blocks based on row (or column, or row after row, or row before row). More than half of the residual blocks selected improved Hadamard butterfly transformation, the number of residual block selecting two-dimensional improved Hadamard butterfly transform is accounted for about 40% of the total number of blocks. This paper proposes three ways of transformation based on row, column, column and then row, and HEVC's direct transmission residuals. There are four ways to deal with residuals. For these four ways, two flag bits are set. Flag1 indicates whether Hadamard row transformation is performed for residual blocks, and flag2 indicates whether Hadamard row transformation is needed for residual blocks.

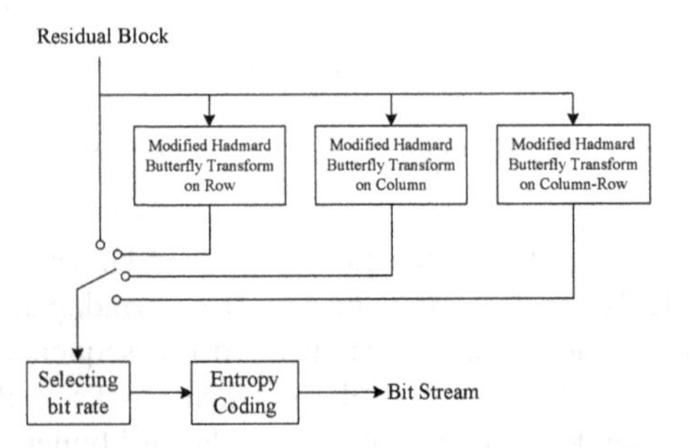

Fig. 11. Four types of residual block processing mode selection.

When flag1 = 0, flag2 = 0, it represents the lossless encoding of HEVC. When flag1 = 0, flag2 = 1, it represents the line transform of improved Hadamard butterfly transform corresponding to the residual block. When flag1 = 1, flag2 = 0, it represents the column transform corresponding to residual block. When flag1 = 1, flag2 = 1, it represents the two-dimensional modified Hadamard butterfly transformation for the residual block. As shown in Fig. 11, the four modes are marked, and then we transfer the bit stream of the residual block after entropy encoding of transmission by the bit rate selection.

4 Experimental Results

In this paper, an improved Hadamard butterfly transformation is proposed, in which the adaptive selection of the size of the bit rate is used to determine which transformation mode, the residual block, column or two-dimensional Hadamard butterfly transform. In the experiments, the method of changing the residual block is chosen by setting the flag bit. In the process of encoding and transmission after each residual unit is transformed, it is estimated whether the bit rate of the coded residual block after different Hadamard Butterfly transforms is the best choice dealing with residual blocks, then the entropy is

Table 1. Resolution ratio, categories and length of test sequence

Resolution ratio		Categories	Length
2560 × 1600	CLASS A	Traffic	150
		PeopleOnStreet	150
1920 × 1080	CLASS B	Kimono	240
		ParkScene	240
		Cactus	500
		BasketballDrive	500
		BQTerrace	600
832 × 480	CLASS C	BasketballDrill	500
		BQMall	600
		PartyScene	500
		RaceHorsesC	300
416 × 240	CLASS D	BasketballPass	500
		BQSquare	600
		BlowingBubbls	500
		RaceHorses	300
1280 × 720	CLASS E	FourPeople	600
		Johnny	600
		KristenAndSara	600

Table 2. Experiment results of different transform models

	Categories	Bit rate	Average
CLASS A	Traffic	−4.16%	−4.56%
	PeopleOnStreet	−4.96%	
CLASS B	Kimono	−1.98%	−1.92%
	ParkScene	−3.00%	
	Cactus	−1.22%	
	BasketballDrive	−1.58%	
	BQTerrace	−1.81%	
CLASS C	BasketballDrill	−0.77%	−1.56%
	BQMall	−1.86%	
	PartyScene	−1.26%	
	RaceHorsesC	−2.37%	
CLASS D	BasketballPass	−4.65%	−2.52%
	BQSquare	−0.80%	
	BlowingBubbls	−1.20%	
	RaceHorses	−3.42%	
CLASS E	FourPeople	−3.23%	−2.88%
	Johnny	−2.91%	
	KristenAndSara	−2.50%	−2.69%

encoded by selecting the corresponding Hadamard butterfly transformation to transform the residual. The Hadamard butterfly transform is adopted in the adaptive selection, which saves the bit rate transmission and improves the performance of HEVC lossless coding.

This algorithm is implemented in the HM15.0. We use the full I-frame detection to measure the lossy performance by the bit rate. Tables 1 and 2 show that after adding the flag bit to select four transformation modes, more bit rate is saved than the situation of selecting one transformation mode. The average performance is improved by 2.69%.

5 Conclusion

In this paper, we propose a butterfly transformation algorithm of an improved Hadamard matrix, and apply it to the intra luma block HEVC lossless coding. The proposed algorithm applies the relationship between the parity of sum and difference, expects the reduction of transformed element value, and reduces the bits of residual coding, which improves the HEVC lossless intra coding compression ratio. For ensuring the residuals being encoded in the way of saving coding bits most, we mark different ways of transformation as two flags. The encoder selects the optimal transformation mode by the bit rate of each two flags, and transmits the symbol bits which represents the transformation to the decoder. The decoder receives the signal and decodes the residual in the same way. The algorithm reduces the bit rate of HEVC CLASS A sequence by an average of 4.56%; the bit rate of CLASS B HEVC sequence is reduced by an average of 1.92%; the bit rate of CLASS C HEVC sequence is reduced by an average of 1.56%; the bit rate of CLASS D HEVC sequence is reduced by an average of 2.52%; the bit rate of CLASS E HEVC sequence is reduced by an average of 2.88%, the average performance is improved by 2.69%.

References

1. Wei, S.T., Kuo, P.C., Li, B.D., Yang, J.F.: Efficient residual coding algorithm based on Hadamard transform in lossless H. 264/AVC. IET Image Process. **8**(4), 194–198 (2013)
2. Heindel, A., Wige, E., Kaup, A.: Low-complexity enhancement layer compression for scalable lossless video coding based on HEVC. IEEE Trans. Circuits Syst. Video Technol. **27**(8), 1749–1760 (2017)
3. Zhao, P., Liu, Y., Liu, J., Yao, R.: Perceptual rate-distortion optimization for H.264/AVC video coding from both signal and vision perspectives. Multimed. Tools Appl. **75**(5), 2781–2800 (2016)
4. Sbiaa, F., Kotel, S., Zeghid, M., et al: A selective encryption scheme with multiple security levels for the H.264/AVC video coding standard. In: Proceedings of the IEEE International Conference on Computer and Information Technology, pp. 391–398. IEEE (2016)
5. Sullivan, G.J., Ohm, J., Han, W.J., Wiegand, T.: Overview of the high efficiency video coding (HEVC) standard. IEEE Trans. Circuits Syst. Video Technol. **22**(12), 1649–1668 (2012)
6. Pitchaipillai, P., Eswaramoorthy, K.: H.264/MPEG-4 AVC video streaming evaluation of LR-EE-AOMDV protocol in MANET. J. Comput. Inf. Technol. **25**(1), 15–29 (2017)

7. Hong, S.W., Kwak, J.H., Lee, Y.L.: Cross residual transform for lossless intra-coding for HEVC. Sig. Process. Image Commun. **28**(10), 1335–1341 (2013)
8. Dey, B., Kundu, M.K.: Enhanced macroblock features for dynamic background modeling in H.264/AVC video encoded at low-bitrate. IEEE Trans. Circuits Syst. Video Technol. **28**(3), 616–625 (2016)
9. Niu, K., Yang, X., Zhang, Y.: A novel video reversible data hiding algorithm using motion vector for H.264/AVC. Tsinghua Sci. Technol. **22**(5), 489–498 (2017)
10. Gaj, S., Patel, A.S., Sur, A.: Object based watermarking for H.264/AVC video resistant to RST attacks. Multimed. Tools Appl. **75**(6), 3053–3080 (2016)
11. Song, L., Luo, Z., Xiong, C.: Improving lossless intra coding of H.264/AVC by pixel-wise spatial interleave prediction. IEEE Trans. Circuits Syst. Video Technol. **21**(12), 1924–1928 (2011)
12. Wang, L.L., Siu, W.C.: Improved lossless coding algorithm in H.264/AVC based on hierarchical intra prediction. In: Proceedings of the IEEE International Conference on Image Processing, pp. 2009–2012 (2011)
13. Flynn, D., Marpe, D., Naccari, M., et al.: Overview of the range extensions for the HEVC standard: tools, profiles, and performance. IEEE Trans. Circuits Syst. Video Technol. **26**(1), 4–19 (2015)
14. Jeon, G., Kim, K., Jeong, J.: Improved residual DPCM for HEVC lossless coding. In: Proceedings of the IEEE Conference on Graphics, Patterns and Images, pp. 95–102 (2014)
15. Zhou, M., Gao, W., Jiang, M., Yu, H.: HEVC lossless coding and improvements. IEEE Trans. Circuits Syst. Video Technol. **22**(12), 1839–1843 (2012)
16. Kim, K., Jeon, G., Jeong, J.: Piecewise DC prediction in HEVC. Sig. Process. Image Commun. **29**(9), 945–950 (2014)
17. Lee, Y.L., Han, K.H., Sullivan, G.J.: Improved lossless intra coding for H.264/MPEG-4 AVC. IEEE Trans. Image Process. **15**, 2610–2615 (2006)
18. Alvar, S.R., Kamisli, F.: On lossless intra coding in HEVC with 3-tap filters. Sig. Process. Image Commun. **47**, 252–262 (2016)
19. Zhang, M.F., Jia, H.U., Zhang, L.M.: Lossless video compression based on the time-space adaptive prediction. Comput. Eng. Sci. **26**(10), 49–50 (2004)
20. Yang, K.H., Faryar, A.F.: A contex-based predictive coder for lossless and near-lossless compression of video. In: Proceedings of the International Conference on Image Processing, pp. 144–147 (2002)
21. Brunello, D., Calvagno, G., Mian, G.A., Rinaldo, R.: Lossless compression of video using temporal information. IEEE Trans. Image Process. **12**(2), 132–139 (2003)
22. Sanchez, V., Aulí-Llinàs, F., Serra-Sagristà, J.: DPCM-based edge prediction for lossless screen content coding in HEVC. IEEE J. Emerg. Sel. Top. Circuits Syst. **6**(4), 497–507 (2016)
23. Choi, J.A., Ho, Y.S.: Efficient residual data coding in CABAC for HEVC lossless video compression. SIViP **9**(5), 1055–1066 (2015)
24. Wige, E., Yammine, G., Amon, P., et al: Pixel-based averaging predictor for HEVC lossless coding. In: Proceedings of the IEEE International Conference on Image Processing, pp. 1806–1810. IEEE (2014)
25. Antony, A., Sreelekha, G.: HEVC-based lossless intra coding for efficient still image compression. Multimed. Tools Appl. **76**(2), 1–20 (2015)
26. Gao, W., Jiang, M., Yu, H.: Binarization scheme for intra prediction residuals and improved intra prediction in lossless coding in HEVC. US, US9277211 (2016)
27. Tan, Y.H., Chuohao, Y., Zhengguo, L.: Lossless coding with residual sample-based prediction. JCTVC-K0157, 10–19 (2012)

U-MEC: Energy-Efficient Mobile Edge Computing for IoT Applications in Ultra Dense Networks

Bowen Yu[1], Lingjun Pu[1,2](✉), Qinyi Xie[1], Jingdong Xu[1], and Jianzhong Zhang[1]

[1] College of Computer and Control Engineering, Nankai University, Tianjin, People's Republic of China
pulingjun@nankai.edu.cn

[2] Guangdong Key Laboratory of Big Data Analysis and Processing, Guangzhou, People's Republic of China

Abstract. With the development of the Internet of Things (IoT) technologies, many kinds of IoT devices as well as many colorful IoT applications are emerging and absorb great attention. In this paper, we propose U-MEC, a Mobile Edge Computing framework deployed in ultra dense networks, in order to narrow the resource gap between the constantly increasing demand of IoT applications and the restricted supply of IoT devices. To improve the energy efficiency of this framework, we present a comprehensive model and formulate a mixed integer programming problem to capture task offloading, user association and base station switching. We propose an online scheduling algorithm, which exploits current system information only by invoking Lyapunov optimization, Lagrange multiplier, and sub-gradient techniques. The simulation results show that our framework achieves a better performance in energy consumption compared with several benchmark schemes.

1 Introduction

Recent years have witnessed an increasing development of IoT technologies such as sensing, computation and communication capacity. In this context, many IoT applications such as vehicular crowdsensing, intelligent surveillance, and environmental monitoring are emerging and absorb great attention. In general, these IoT applications require IoT devices to periodically collect environmental data with fruitful sensors and onboard process them with various kinds of machine learning algorithms (e.g., vision processing) to generate corresponding knowledge. Nevertheless, as the big data era is evolving, the computation and energy capacity of IoT devices is feeble to satisfy the diverse requirements (e.g., delay and sensing density) of IoT applications.

Although mobile edge computing (MEC) is viewed as a very promising paradigm to cope with those challenges, which provides a significant amount of computation resources to complement IoT devices at the network edge [12], it

S. Chellappan et al. (Eds.): WASA 2018, LNCS 10874, pp. 622–634, 2018.
https://doi.org/10.1007/978-3-319-94268-1_51

suffers from the ever-increasing number of IoT devices due to the serve communication competitions at cellular access networks. As for this issue, mobile network operators advocate the principle of ultra dense networks (UDN), which provides sufficient cellular accesses by deploying a considerable number of base stations (BSs). However, small base stations (SBSs) such as picocells, microcells, and femtocells working together will lead to not only inevitable interference among them but also a large amount of energy consumption [4]. Therefore, how to integrate MEC with UDN to facilitate IoT applications while keeping the energy efficiency of UDN becomes a significant problem in the MEC research domain.

Minimizing energy consumption for mobile edge computing has drawn great attention from industry to academia. The studies in [1,14] focus on optimizing the energy consumption for multi-devices computing offloading. [7,9] study the problem of optimizing energy consumption or energy efficiency for mobile devices by scheduling computing offloading and edge cloud resources. However, these researches only consider the single BS scenario which will be inadequate for multi-BSs scenarios like UDN. [2,15] study energy efficient BS switching strategies for multi-BSs, while these strategies cannot be applied to MEC directly, since they lack of taking the QoS of tasks into consideration [3].

In this paper, we propose U-MEC, a Mobile Edge Computing framework deployed in ultra dense networks, which jointly captures task offloading, user association, and BS switching scheduling. Its goal is to minimize the energy consumption of IoT devices and BSs while meeting the QoS requirements of IoT applications. We take macro base station (MBS) as the centralized controller that is responsible to schedule (1) how many tasks should be executed locally in each device and how many tasks are offloaded to the edge cloud; (2) which BS each device should be connected to; (3) which BSs should be turned OFF to reduce energy costs. Different from previous works, we also capture the BS switching cost in our framework [15]. In order to obtain an energy efficient scheduling policy, U-MEC has to fulfill the following requirements.

- It should consider the overall energy consumption in terms of devices and base stations simultaneously.
- It should achieve energy minimization while keeping the QoS of applications.
- It should be scalable in order to apt to large-scale deployment of devices.

Based on these requirements, we provide a comprehensive framework model in Sect. 2, which includes device and cloudlet computing capacity, SBS sleeping and user association, application task, as well as task queueing and energy consumption. On this basis, we formulate a mixed integer programming problem in Sect. 3, which aims to minimize the time-average energy consumption of IoT devices and BSs while meeting the QoS requirements (e.g., latency) of the applications through the scheduling of task offloading, user association and BS switching decision. In Sect. 4, we invoke Lyapunov optimization, Lagrange multiplier and subgradient techniques to propose an energy efficient online algorithm for the formulated problem. We evaluate our algorithm with extensive simulations in Sect. 5 and conclude the paper in Sect. 6.

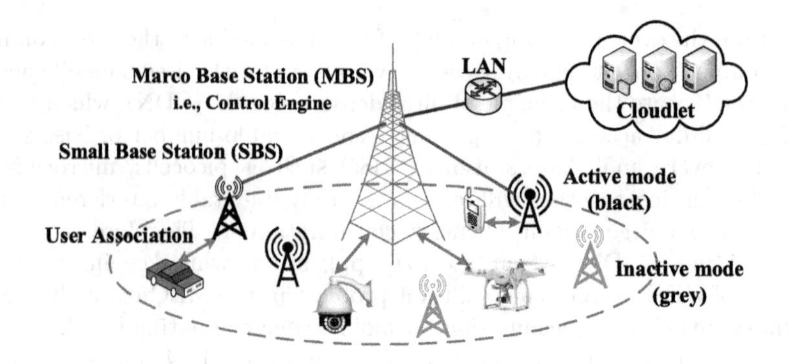

Fig. 1. An illustration of our framework.

2 System Model

As illustrated in Fig. 1, we consider a cloudlet-enabled UDN scenario, in which mobile network providers cooperate with individuals, enterprises or governments to build an IoT application platform. Briefly, they install a series of IoT applications in the cloudlet and deploy an application manager in the MBS to supervise IoT devices serving IoT applications. We consider that a set $M = \{1, 2, \ldots, m\}$ of IoT devices has registered on the platform, and some of them are periodically requested to run a suite of crowdsensing applications. In addition, we consider a set $N = \{0, 1, 2, \ldots, n\}$ of base stations. The index 0 represents the MBS and the rest ones represent SBSs. The MBS will keep in the active mode to guarantee the cellular coverage, and each SBS can be scheduled to be in the inactive mode for energy conservation. To proceed, we exploit a vehicular crowdsensing application to illustrate how U-MEC operates. In general, a vehicular crowdsensing application requires a crowd of vehicles to periodically sense data from immediate surroundings, in-situ process them such as data fusion and image/video processing, and transmit the processed results to the centralized application manager for post-processing. In U-MEC, the recruited vehicles can conduct the pre-processing tasks locally and then upload the results or can offload the sensed data to process in the cloudlet via the application manager control.

2.1 Device and Cloudlet Computing Capacity

We assume that the available computing capacity $s_i(t)$ in a time slot is indicated by the stabilized CPU working frequency together with the current CPU load, which can be estimated by the online frequency-history window method [6]. In addition, we assume that the available computing capacity of cloudlet for IoT applications is denoted by v, which can be a fixed value set by the application manager in the MBS.

2.2 SBS Sleeping and User Association

In general, each SBS can be in either active or inactive mode, and therefore we introduce a binary control variable $b_j(t) \in \{0,1\}$ to indicate whether a SBS $j \in N$ is active in a time slot. We assume the MBS and all the SBSs share the same block of spectrum, and leverage FDMA as the cellular access scheme. That is, the spectrum of a SBS j consists of h_j channels and each of its associated device is allocated with one channel. As our framework considers task offloading and processed result uploading, we only take device uplink capacity into account. In addition, we do not consider the interference among different IoT devices or simply capture it by introducing a maximum interference temperature value [10]. We also assume that each IoT device i transmits with a given power $q_i(t)$ in a time slot, and the SINR $\gamma_{ij}(t)$ can be measured via uplink training. In this context, for each device i associated with a base station j, its achievable uplink capacity can be written as $u'_{ij}(t) = a_{ij}(t)u_{ij}(t)$, $u_{ij}(t) = B\log(1+\gamma_{ij}(t))$ where $a_{ij}(t) \in \{0,1\}$ is a binary control variable to indicate whether a device i is associated with a base station j in a time slot, and B is the bandwidth of each channel. Intuitively, we have the following constraints:

$$a_{ij}(t) \le b_j(t), \quad (1)$$
$$\textstyle\sum_{j=0}^{n} a_{ij}(t) = 1, \quad (2)$$
$$\textstyle\sum_{i=1}^{m} a_{ij}(t) \le h_j, \quad (3)$$

where Constraint (3) is due to the fact that only the active base stations can serve IoT devices; Constraint (5) is due to the fact that each IoT device can establish one and only one cellular link; Constraint (6) is due to the number of IoT devices simultaneously served by a base station j can not exceed its total number of channels.

2.3 Application Task

For simplicity, we assume that all IoT devices will serve one IoT application[1]. In addition, we take the pre-processing of sensed data at the device side (e.g., one or several functional modules) as a whole task for ease of application management. In other words, similar to that considered in many existing works [7,8], our framework will concentrate on the task workload scheduling (i.e., sensed data offloading) for IoT applications. In this context, we denote by $\lambda_i(t)$ the amount of data sensed by device i in a time slot and denote by $\overline{\lambda}$ the average amount of data required to sense in each time slot, which is decided by the application manager. Then, we denote the processing density of the application task by ρ (i.e., the average amount of CPU cycles to process a bit of sensed data), and consider the processed result size equals to a fraction φ of the input sensed data.

[1] We will study the multi-application case in the future work.

2.4 Task Queueing and Energy Consumption

For each device i, we denote the task workload queue by $Q_i(t)$ indicating the amount of remaining task workloads required to be processed, and denote the task workload queue in the cloudlet by $L(t)$. In addition, we introduce a control variable $x_i(t)$ to indicate the number of task workloads processed locally and a control variable $y_i(t)$ to indicate the number of task workloads offloaded to the cloudlet for each IoT device i in each time slot. In this context, we can easily obtain the queue dynamics as follows:

$$\begin{aligned} Q_i(t+1) &= Q_i(t) + \lambda_i(t) - x_i(t) - y_i(t), \\ L(t+1) &= [L(t) + \textstyle\sum_{i=1}^{m} y_i(t) - \frac{v}{\rho}]^+, \end{aligned} \tag{4}$$

where $[x]^+ = \max(x, 0)$ and $\lambda_i(t)$ indicates the amount of data sensed by device i in a time slot[2]. Note that since the queue backlog cannot be negative and the number of processed task workloads cannot exceed the device processing capacity, we have the following constraint:

$$x_i(t) + y_i(t) \leq Q_i(t) + \lambda_i(t), \tag{5}$$

$$x_i(t) \leq \frac{s_i(t)}{\rho}, \tag{6}$$

$$y_i(t) + \varphi x_i(t) \leq \textstyle\sum_{j=0}^{n} u'_{ij}(t). \tag{7}$$

To proceed, the energy consumption of each device i can be represented as

$$E_i(t) = p_i(t)\frac{\rho x_i(t)}{s_i(t)} + \textstyle\sum_{j=0}^{n} q_i(t)\frac{\varphi x_i(t) + y_i(t)}{u'_{ij}(t)},$$

The first term indicates the energy consumption of local workload processing, where p_i is the CPU working power according to the stabilized CPU working frequency in a time slot [7], and the rest part is the time required for task workload processing. The second term indicates the energy consumption of processed result uploading and that of task workload offloading. The energy consumption of each base station consists of a baseline cost and a transmission cost [13]. The baseline cost occurs due to maintaining the BS in the active mode (e.g., used by the cooling system, power amplifier, and baseband units). The transmission cost is dependent on the traffic load (e.g., used by the radio frequency unit). The real system study in [13] shows that the baseline cost is the dominance of the total BS energy consumption. Therefore, the energy consumption of the MBS is denoted by e_0 and that of each SBS j is represented by $e_j b_j(t)$ in a time slot. Besides, we introduce BS switching cost which is the "penalty" due to switching base stations from sleep mode to active mode. As such, we denote it as $\delta_j[b_j(t) - b_j(t-1)]^+$, where δ_j indicates the energy consumption of signaling, user profile loading and state-migration processing. To sum up, the energy consumption of each SBS j can be represented as

$$E_j(t) = e_j b_j(t) + \delta_j [b_j(t) - b_j(t-1)]^+.$$

[2] $\overline{\lambda_i}$ indicates the average amount of data required to sense by device i in each time slot, which is decided by the application manager offline.

3 Problem Formulation

3.1 Latency Constraint

We consider that many IoT applications in practice are deadline-sensitive. That is, the incoming sensed data should be processed, and its processed results should be uploaded before a specific deadline d_{max}. To jointly capture the deadline-sensitive, large-scale (i.e., multiple devices) and long-term characteristics of IoT applications, we define the following application latency constraint based on the Little's law:

$$\frac{\frac{1}{T}\sum_{t=0}^{T-1}\{\sum_{i=1}^{m}Q_i(t)+L(t)\}}{m\overline{\lambda}} \leq d_{max}.$$

For convenience, we define $U(t)=\sum_{i=1}^{m}Q_i(t)+L(t)$. Then the latency constraint can be rewritten as

$$\frac{1}{T}\sum_{t=0}^{T-1}U(t) \leq m\overline{\lambda}d_{max}. \tag{8}$$

3.2 Energy-Minimizing Problem

In our framework, the objective is to design a joint control algorithm for task offloading, user association and BS switching in the application manager, so as to minimize the energy consumption of network-wide devices and SBSs for task processing, and meanwhile satisfying application latency requirement in a long-term perspective. To this end, we formulate the following optimization problem:

$$\begin{aligned} P_1: \min_{x,y,a,b} \quad & \lim_{T\to\infty}\sum_{t=0}^{T-1}\left[\sum_{i=1}^{m}E_i(t)+\sum_{j=1}^{n}E_j(t)\right], \\ \text{subject to} \quad & (1), (2), (3), (5), (6), (7), (8) \text{ and} \\ & x_i(t), y_i(t) \in R^+, \qquad (9) \\ & a_{ij}(t), b_j(t) \in \{0,1\}. \qquad (10) \end{aligned}$$

4 Algorithm Design

The main challenges of solving the problem P_1 lie in (1) the time-average objective function and constraints require future system information that is not easy to obtain, and (2) the problem has integer variables which makes the problem not be easy to tackle. To this end, we resort to Lyapunov optimization, Lagrange multiplier and subgradient techniques to design an efficient online algorithm.

4.1 Problem Transformation

To capture the time-average latency constraint (8), we introduce the following virtual queue $B(t)$.

$$B(t+1) = [B(t) + \sum_{i=1}^{m} Q_i(t) + L(t) - m\bar{\lambda}d_{max}]^+. \quad (11)$$

We specify the initial value of $B(t)$ is 0 when $t = 0$. If $B(t)$ is stable, then the constraint (8) will be satisfied. Let $\boldsymbol{\Theta}(t) = [\boldsymbol{Q}(t), L(t), B(t)]$ be the vector of all queues in our framework, then we define the Lyapunov function as

$$L(\boldsymbol{\Theta}(t)) = \tfrac{1}{2}\left\{\sum_{i=1}^{m} Q_i(t)^2 + L(t)^2 + B(t)^2\right\}.$$

The Lyapunov drift for one time slot is

$$\Delta(\boldsymbol{\Theta}(t)) = \mathbb{E}\left\{L(\boldsymbol{\Theta}(t+1)) - L(\boldsymbol{\Theta}(t))|\boldsymbol{\Theta}(t)\right\}.$$

We can minimize the time-average energy consumption and stabilize all queues in our framework for each time slot by minimizing the upper bound of

$$\Delta(\boldsymbol{\Theta}(t)) + V\left[\sum_{i=1}^{m} E_i(t) + \sum_{j=1}^{n} E_j(t)\right], \forall t, \quad (12)$$

where V is the tradeoff between the energy consumption and the stability of the queues. Substituting the queueing update rules (4) and (11) into Eq. (12), then we have the following inequality.

$$\begin{aligned} &\Delta(\boldsymbol{\Theta}(t)) + V\left[\sum_{i=1}^{m} E_i(t) + \sum_{j=1}^{n} E_j(t)\right] \\ \le &\sum_{i=1}^{m}\Big[\xi_i(t)x_i(t) + \psi_i(\boldsymbol{a}(t),t)(y_i(t) + \varphi x_i(t))\Big] + \sum_{j=1}^{n}\omega_j(t)b_j(t) + K + k, \end{aligned} \quad (13)$$

where

$$\begin{aligned} \xi_i(t) &= -Q_i(t) - \lambda_i(t) - B(t) + m\bar{\lambda}d_{max} + \frac{v}{\rho} + V\rho\frac{p_i(t)}{s_i(t)}, \\ \psi_i(\boldsymbol{a}(t),t) &= -Q_i(t) - \lambda_i(t) + L(t) - \frac{v}{\rho} + V\frac{q_i(t)}{\sum_{j=1}^{n} a_{ij}(t)u_{ij}(t)}, \\ \omega_j(t) &= V(e_j + \mathbb{1}_{\{b_j(t-1)=0\}}\Delta e_j), \\ K &= \frac{1}{2}\Big[m\lambda_{max}^2 + mx_{max}^2 + (m^2+m)y_{max}^2 + 2(\frac{v}{\rho})^2 + (m\bar{\lambda}d_{max})^2 \\ &\quad + 2mx_{max}y_{max} + m^2(\lambda_{max} - x_{max})^2\Big], \\ k &= \left[\sum_{i=1}^{m} Q_i(t)\lambda_i(t)\right] - L(t)\tfrac{v}{\rho} - B(t)(m\bar{\lambda}d_{max}) \\ &\quad - (\frac{v}{\rho} + m\bar{\lambda}d_{max})\sum_{i=0}^{m}\lambda_i(t) + \frac{v}{\rho}(m\bar{\lambda}d_{max} - B(t)). \end{aligned}$$

Note that $\lambda_{max} = \max(\lambda_i(t)|i \in M, t \in (0,1,2,\ldots,T))$, $x_{max} = \max(x_i(t)|i \in M, t \in (0,1,2,\ldots,T))$, $y_{max} = \max(y_i(t)|i \in M, t \in (0,1,2,\ldots,T))$, K is a constant, $\mathbb{1}_{\{b_j(t-1)=0\}} = 1$ when $b_j(t-1) = 0$ and $\mathbb{1}_{\{b_j(t-1)=0\}}$ is a constant when $b_j(t-1)$ is given.

Minimizing the right-hand-side of (13) is equivalent to

$$
\begin{aligned}
P_2: \min_{\boldsymbol{x},\boldsymbol{y},\boldsymbol{a},\boldsymbol{b}} \quad & R(t) = \sum_{i=1}^{m} \left[\xi_i(t)x_i(t) + \psi_i(\boldsymbol{a}(t),t)(y_i(t) + \varphi x_i(t))\right] \\
& + \sum_{j=1}^{n} \omega_j(t) b_j(t), \\
\text{subject to} \quad & (1), (2), (3), (5), (6), (7), (9), (10).
\end{aligned}
$$

4.2 Minimizing the Upper Bound

The problem P_2 is a quadratic mixed integer programming problem which is generally NP-hard. To solve this problem, we relax the 0–1 integer constraint (10) into $[0, 1]$. Then, P_2 is transformed into

$$
\begin{aligned}
P_3: \min_{\boldsymbol{x},\boldsymbol{y},\boldsymbol{a},\boldsymbol{b}} \quad & R(t) \\
\text{subject to} \quad & (1), (2), (3), (5), (6), (7), (9) \text{ and} \\
& 0 \leq a_{ij}(t), b_j(t) \leq 1. \qquad (14)
\end{aligned}
$$

The decision variable $a_{ij}(t)$ is coupled with $x_i(t)$, $y_i(t)$ and $b_j(t)$ not only in the objective function but also in the constraints (3) and (7), which makes P_3 be hard to solve directly. In order to develop an efficient algorithm, we decompose P_3 into two phases. We first find the optimal solution of $a_{ij}(t)$ by given a set of feasible $x_i(t)$, $y_i(t)$ and $b_j(t)$. Second, optimize $x_i(t)$, $y_i(t)$ and $b_j(t)$ through the value of $a_i(t)$ obtained in the first phase. We repeatedly loop the two phases until all the variables converge.

For given values of $\boldsymbol{x}$, $\boldsymbol{y}$ and $\boldsymbol{b}$, problem P_3 is transformed into

$$
\begin{aligned}
P_4: \min_{\boldsymbol{a}} \quad & R(t)|x_i(t), y_i(t), b_j(t) \\
\text{subject to} \quad & (1), (2), (3), (7), (14).
\end{aligned}
$$

We can initially set $x_i(t) = \min(Q_i(t) + \lambda_i(t), \frac{s_i(t)}{\rho})$, $y_i(t) = Q_i(t) + \lambda_i(t) - x_i(t)$, and all $b_j(t) = 1$. The problem P_4 is a convex optimization problem and can be solved in the dual domain. The dual problem of P_4 is given by

$$
\text{Dual} - P_4: \max_{\boldsymbol{\theta},\boldsymbol{\omega}} \quad f(\boldsymbol{\theta}, \boldsymbol{\omega})
$$

where $\boldsymbol{\theta}$ and $\boldsymbol{\omega}$ are the Lagrange multipliers for constraints (1), (7), and

$$
\begin{aligned}
f(\boldsymbol{\theta}, \boldsymbol{\omega}) = \min_{\boldsymbol{a}} \Big\{ & R(t) + \sum_{i=1}^{m} \sum_{j \in N} \theta_{ij}(a_{ij}(t) - b_j(t)) \\
& + \sum_{i=1}^{m} \omega_i \Big[y_i(t) + \varphi x_i(t) - \sum_{j=1}^{n} a_{ij}(t) u_{ij}(t)) \Big] \Big\}.
\end{aligned}
$$

The optimal solution of Dual$-P_4$ can be obtained by the subgradient method.

$$
\theta_{ij}^{iter+1} = \left[\theta_{ij}^{iter} - \tau(a_{ij}^{iter}(t) - b_j(t))\right]^{+}, \qquad (15)
$$

$$\omega_i^{iter+1} = \left[\omega_i^{iter} - \tau\left(y_i(t) + \varphi x_i(t) - \sum\nolimits_{j=1}^{n} a_{ij}^{iter}(t)u_{ij}(t)\right)\right]^+, \quad (16)$$

where τ is the step size of each iteration. Since P_4 is an LP, its constraints are linear and it satisfies the Slater condition, then strong duality holds. In other words, the duality gap between P_4 and its dual problem is zero.

Given $\boldsymbol{\theta}$, $\boldsymbol{\omega}$, $\boldsymbol{x}$, $\boldsymbol{y}$ and $\boldsymbol{b}$, $f(\boldsymbol{\theta}, \boldsymbol{\omega})$ is a standard LP to obtain $\boldsymbol{a}$.

$$\begin{aligned} P_5 : \quad & f(\boldsymbol{\theta}, \boldsymbol{\omega}) \\ \text{subject to} \quad & (2), (3), (14). \end{aligned}$$

Then we find the optimal $\boldsymbol{x}$, $\boldsymbol{y}$ when $\boldsymbol{a}$ and $\boldsymbol{b}$ are obtained. P_6 is a standard LP that can be solved by many mature algorithms (e.g., the simplex method).

$$\begin{aligned} P_6 : \min_{\boldsymbol{x}, \boldsymbol{y}} \quad & R(t)|a_{ij}(t), b_j(t) \\ \text{subject to} \quad & (5), (6), (7), (9). \end{aligned}$$

We find the optimal $\boldsymbol{b}$ by solving P_7 when $\boldsymbol{a}$, $\boldsymbol{x}$, and $\boldsymbol{y}$ are obtained.

$$\begin{aligned} P_7 : \min_{\boldsymbol{b}} \quad & R(t)|a_{ij}(t), x_i(t), y_i(t) \\ \text{subject to} \quad & (1), (14). \end{aligned}$$

Note that P_7 can be obtained by the following subgradient method.

$$b_j^{iter+1}(t) = [b_j^{iter} - \tau \sum\nolimits_{i=1}^{m} \theta_{ij}^{iter}]^+. \quad (17)$$

As described in Algorithm 1, we repeatedly solve the problem Dual$-P_4$ to P_7 until all the variables converge.

Algorithm 1. The centralized algorithm to get $\boldsymbol{a}$, $\boldsymbol{b}$, $\boldsymbol{x}$ and $\boldsymbol{y}$ for one time slot

---

Initialize $\boldsymbol{a}$, $\boldsymbol{b}$, $\boldsymbol{x}$, $\boldsymbol{y}$, $\boldsymbol{\theta}$ and $\boldsymbol{\omega}$;
while $\boldsymbol{b}$ does not converge; **do**
  while $\boldsymbol{\theta}$ and $\boldsymbol{\omega}$ do not converge; **do**
    Solve problem P_5 with a LP solver;
    Update $\boldsymbol{\theta}$ and $\boldsymbol{\omega}$ as in (15) and (16);
  end while
  Solve problem P_6 with a LP solver;
  Update $\boldsymbol{b}$ as in (17);
end while
Update $Q_i(t)$, $L(t)$ and $B(t)$ with Eqs. (4) and (11).

4.3 Algorithm Analysis

Now we explain the reasons why $a_{ij}(t)$ and $b_j(t)$ are integer solutions briefly. Let us rewrite P_5 in the following standard form.

$$\min_{a} \quad \mathbf{Ca}$$
$$\text{subject to} \quad \mathbf{Aa} \leq \mathbf{c}.$$

The constraint matrix $\mathbf{A}$ can be proved to have the totally unimodular feature. We should emphasize that, if the constraint matrix of an LP satisfies totally unimodularity, then it has all integral vertex solutions [11]. In this context, the variables $a_{ij}(t)$ in the optimal solution of the problem P_5 are either 0 or 1. According to the constraint (3), if any $a_{ij}(t) = 1$, then $b_j(t) = 1$. Therefore, the variables b_j in the optimal solutions of P_7 are also integers in $\{0, 1\}$.

We suppose that the average arrival task amount for each IoT device $(\overline{\lambda} + \varepsilon)$ is strictly within the system capacity region, and ε is a small positive value. Then we can have the bound of energy consumption and queues as follows.

$$\lim_{T\to\infty} \frac{1}{T} \sum_{t=0}^{T-1} \mathbb{E}\Big\{\sum_{i=1}^{m} Q_i(t) + L(t) + B(t)\Big\} \leq \frac{1}{\varepsilon}(K + V\sum_{i=1}^{m} J^*(\overline{\lambda} + \varepsilon)), \quad (18)$$

$$\lim_{T\to\infty} \frac{1}{T} \sum_{t=0}^{T-1} \mathbb{E}\Big\{\sum_{i=1}^{m} E_i(t) + \sum_{j=1}^{n} E_j(t)\Big\} \leq [\sum_{i=1}^{m} J^*(\overline{\lambda} + \varepsilon)] + \frac{K}{V}, \quad (19)$$

where $J^*(\overline{\lambda} + \varepsilon)$ is the offline minimum energy consumption for each IoT device.

5 Performance Evaluation

We use the opportunistic network simulator ONE [5] to evaluate our method with the virtual urban area scenario. We set the number of devices in the simulation

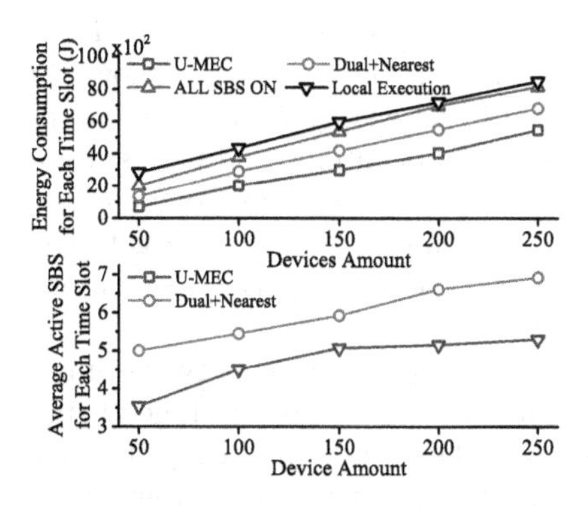

Fig. 2. Different strategies results.

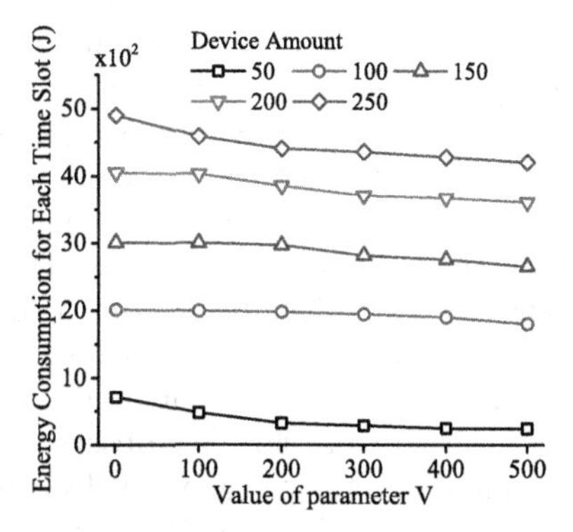

Fig. 3. Average energy consumption.

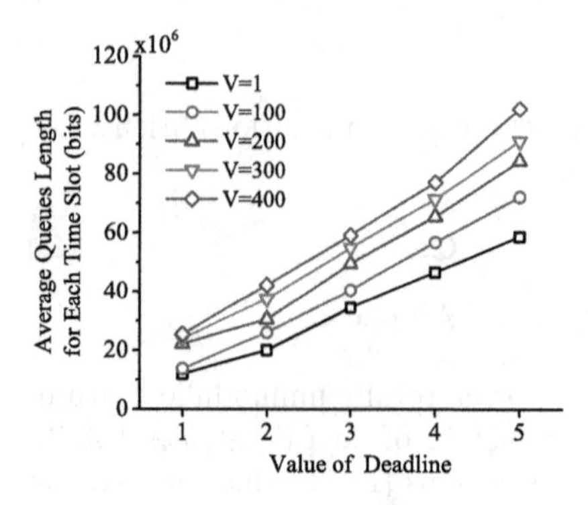

Fig. 4. Average queues lengths.

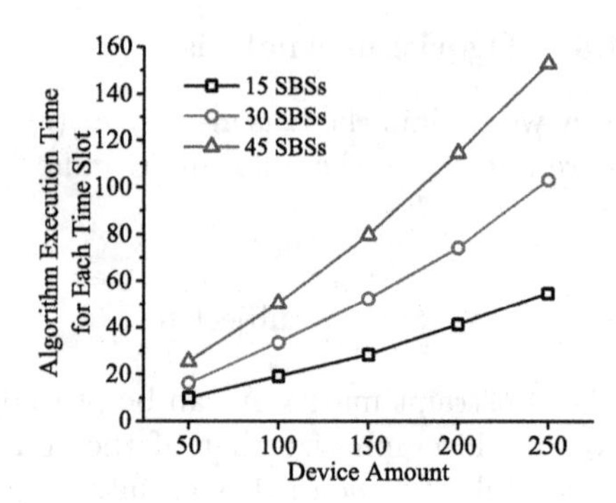

Fig. 5. Algorithm execution time.

from 50 to 250, and the device moves on the roads of the scene according to the working day movement model. We consider the stabilized CPU frequency of each IoT device is 1.0 GHz (i.e., 10^9 cycles/s), and the available computing capacity of it is the CPU cycles of its stabilized CPU frequency times a random fluctuation (i.e., from 0.5 to 1). We consider the CPU power and transmit power of each IoT device is 30 W and 5 W. As for the base stations, we set the MBS with the cloudlet in the center of the scene and set 30 SBSs randomly in the area. We set the number of channel for both MBS and SBSs to 20, which means each of base station can be associated with 20 devices simultaneously. We set the power consumption of the SBS to 100 Joule, and set the switching cost to 200 Joule. The transmission bandwidth between IoT device and base stations is 10 MHz. As for cloudlet, we consider that the computing capacity of it is 50×10^9 cycles. The average arrived task amount of each device is 10^6 bits per time slot. We set the computing density of the task to 1000 cycles/bit, the output radio to 0, and the maximum delay to 1 time slot.

We compare our algorithm with the following three strategies. **(1) DualControl + Nearest Association**. This method uses a strategy similar to DualControl [7] in the scheduling of task offloading and each IoT device connects to the nearest BS. **(2) ALL SBS ON**. This method uses DualControl for scheduling task offloading, and the SBSs are always active. **(3) Local Execution**. All of the IoT devices execute tasks locally. The results are shown in Fig. 2. We observe that our algorithm consumes less energy than other methods through joint scheduling task offloading, user associations, and base station switching. In terms of user association and base station switching, our method turns on 1 or 2 base stations less than the comparison method per time slot. The result shows it is meaningful to consider jointly task offloading, user associations and BS switching in the scheduling processing.

Figure 3 shows the impact of different values of V on average energy consumption for each time slot. We observe as the value of V increases, average energy consumption for each time slot decreases, sharing the same trend in Eq. (19). We evaluate the average lengths of queues for each time slot in our system under different values of V and deadline d_{max} in Fig. 4. As the value of d_{max} increases, the accumulated tasks in our system also increases, which coincides with constraint (8). Meanwhile as d_{max} increases, the amount of remaining tasks in our system also increases, which confirms with Eq. (18).

Then we evaluate the execution time of our algorithm with different numbers of devices and SBSs. Since our algorithm needs to loop through Dual$-P_4$ to problem P_7, as the number of devices and the number of base stations increases, the execution time of each time slot also increases. To lower the execution time, we can divide the devices in the scenario into independent subgraphs and execute the algorithms in parallel.

6 Conclusion

In this paper, we proposed U-MEC, a Mobile Edge Computing framework deployed in Ultra dense networks, which can increase the computing resources of IoT devices to satisfy the requirement of IoT applications. To achieve the energy efficiency of U-MEC, we presented a comprehensive model and formulated a mixed integer programming problem for scheduling the task offloading, user association and BS switching. We devised an online scheduling algorithm by invoking Lyapunov optimization, Lagrange multiplier and subgradient techniques. The simulation results indicated that our framework achieved a better performance in energy consumption than several benchmark schemes.

Acknowledgement. This work was supported by the National Natural Science Foundation of China (No. 61702288, 61702287), the Fundamental Research Funds for the Central Universities (No. 7933), the Natural Science Foundation of Tianjin in China (No. 16JCQNJC00700), and the Open Project Fund of Guangdong Key Laboratory of Big Data Analysis and Processing (No. 2017003).

References

1. Chen, X., Jiao, L., Li, W., Fu, X.: Efficient multi-user computation offloading for mobile-edge cloud computing. IEEE/ACM Trans. Netw. **24**, 2795–2808 (2016)
2. Feng, M., Mao, S., Jiang, T.: Boost: base station on-off switching strategy for energy efficient massive MIMO HetNets. In: IEEE INFOCOM (2016)
3. Feng, M., Mao, S., Jiang, T.: Base station on-off switching in 5G wireless networks: approaches and challenges. IEEE Wirel. Commun. **24**, 46–54 (2017)
4. Ge, X., Tu, S., Mao, G., Wang, C.X., Han, T.: 5G ultra-dense cellular networks. IEEE Wirel. Commun. **23**, 72–79 (2016)
5. Keränen, A., Ott, J., Kärkkäinen, T.: The one simulator for DTN protocol evaluation. In: ACM International Conference on Simulation Tools and Techniques (2009)
6. Kim, J.M., Kim, Y.G., Chung, S.W.: Stabilizing CPU frequency and voltage for temperature-aware DVFS in mobile devices. IEEE Trans. Comput. **64**, 286–292 (2015)
7. Kim, Y., Kwak, J., Chong, S.: Dual-side optimization for cost-delay tradeoff in mobile edge computing. IEEE Trans. Veh. Technol. **67**, 1765–1781 (2018)

8. Kwak, J., Kim, Y., Lee, J., Chong, S.: DREAM: dynamic resource and task allocation for energy minimization in mobile cloud systems. IEEE J. Sel. Areas Commun. **33**, 2510–2523 (2015)
9. Mao, Y., Zhang, J., Letaief, K.B.: Dynamic computation offloading for mobile-edge computing with energy harvesting devices. IEEE J. Sel. Areas Commun. **34**, 3590–3605 (2016)
10. Ng, D.W.K., Schober, R.: Resource allocation and scheduling in multi-cell OFDMA systems with decode-and-forward relaying. IEEE Trans. Wirel. Commun. **10**, 2246–2258 (2011)
11. Schrijver, A.: Theory of Linear and Integer Programming. Wiley, Hoboken (1998)
12. Wang, S., Zhang, X., Zhang, Y., Wang, L., et al.: A survey on mobile edge networks: convergence of computing, caching and communications. IEEE Access **5**, 6757–6779 (2017)
13. Yan, M., Chan, C.A., Li, W., et al.: Network energy consumption assessment of conventional mobile services and over-the-top instant messaging applications. IEEE J. Sel. Areas Commun. **34**, 3168–3180 (2016)
14. You, C., Huang, K., Chae, H., Kim, B.H.: Energy-efficient resource allocation for mobile-edge computation offloading. IEEE Trans. Wirel. Commun. **16**, 1397–1411 (2017)
15. Yu, N., Miao, Y., Mu, L., Du, H., Huang, H., Jia, X.: Minimizing energy cost by dynamic switching on/off base stations in cellular networks. IEEE Trans. Wirel. Commun. **15**, 7457–7469 (2016)

Localization of Thyroid Nodules in Ultrasonic Images

Ruiguo Yu[1], Kai Liu[1], Xi Wei[2], Jialin Zhu[2], Xuewei Li[1(✉)], Jianrong Wang[1], Xiang Ying[3], and Zhihui Yu[3]

[1] School of Computer Science and Technology, Tianjin University, Tianjin, China
{rgyu,kedixa,lixuewei,wjr}@tju.edu.cn

[2] Tianjin Medical University Cancer Institute and Hospital, Tianjin, China
weixi198204@126.com, sally2010zhu@126.com

[3] School of Computer Software, Tianjin University, Tianjin, China
{xiang.ying,yzh2012}@tju.edu.cn

Abstract. Thyroid nodule is a common clinical condition. Ultrasound is usually used to make a preliminary diagnosis, because it is convenient and cheap. Therefore, the study of thyroid ultrasound images of thyroid nodules has it's significance and value. This paper investigates the problem of locating thyroid nodules in ultrasound images by manual signs. The solution to this problem is divided into three parts: first, image processing processes the image preliminary and find the approximate location of the signs; then, sign recognition recognize the signs accurately using CNN models; finally, boundary adjustment is used for the final adjustment of the border. Experimental results show that the algorithm proposed in this paper can accurately locate the nodules in thyroid ultrasound images.

Keywords: Thyroid nodule · Ultrasonic image processing
Convolutional neural network

1 Introduction

Thyroid nodules are tumors locating in the thyroid gland, which is a common clinical condition. With the rapid development of human society, people's living pressure is getting bigger and bigger, and the prevalence of common diseases also increases year by year. Epidemiological studies have indicated that approximately 5% of women and 1% of men resident in iodine-sufficient areas have palpable thyroid nodules. However, by the age of 60 years about 50% of the general population is estimated to have at least one thyroid nodule [1]. Nodule incidence increases with age, and is increased in women, in people with iodine deficiency, and after radiation exposure [2]. There are about 90,000 new cases of thyroid cancer in China each year, and about 6,800 of them died of thyroid cancer [3]. Ultrasonography is the most accurate and cost-effective method for

S. Chellappan et al. (Eds.): WASA 2018, LNCS 10874, pp. 635–646, 2018.
https://doi.org/10.1007/978-3-319-94268-1_52

evaluating and observing thyroid nodules [2], which is cheap, convenient, non-invasive, non-radiation, and has many other advantages, so a large number of patients with thyroid nodules will use ultrasound examination first, and then a further diagnosis according to the doctor's advice will be done. Therefore, the study of thyroid nodules ultrasound image is of great significance.

Before the study of thyroid nodules ultrasound images, we must first screen a large number of ultrasound images, and mark the location of nodules in order to facilitate the follow-up treatment, which will take a lot of time and effort. This paper which combines image processing and convolutional neural network presents a method for locating thyroid nodules based on manual signs in ultrasound images that can analyze the ultrasound images quickly, and filter out well-tagged images, and then mark the specific location of thyroid nodules using a rectangular box in ultrasound images. This method saves a lot of time in data processing, laying a solid foundation for further processing.

2 Related Work

This paper studies the problem of target location in the image. In recent years, some articles use image processing methods to solve this problem, and some use deep convolutional neural networks which are good at dealing with images.

Gall and Lempitsky [4] proposed a method of detecting object categories on natural images They train a class-specific Hough forest that directly maps the image patch to the probabilistic about the possible location of the object. Rong et al. [5] proposed an improved Canny edge detection algorithm, which introduced gravitational field intensity to replace image gradient, and obtained the gravitational field intensity operator to keep more useful edge information. Hui [6] summarized a number of ways to detect manual signs in images, among of them, the localization algorithm of manual signs based on SUSAN algorithm and Hough transform has higher precision and identify speed. Ma [7] proposed a cross mark location method based on sub-pixel and template matching, which has higher recognition rate and identify speed.

Fang et al. [8] proposed a method of automatic mark detection of x-ray images under the framework of machine learning, they use the covariance-based descriptors to represent the features of marks in X-ray images, and utilize the cascade of LogitBoost classifiers based on covariance features to learn the marker detector. Girshick et al. [9] proposed a method of using CNN to locate the target candidate region in the image, and achieve a mean average precision of 53.3% on VOC 2012. Girshick [10] proposed a method named Fast R-CNN, the method first feed forward the whole image and then take samples on the convolutional feature based on the results of the candidate area generation algorithm, finally find the target position in each candidate area, which solve the problem of low efficiency of R-CNN. Ren et al. [11] generate candidate regions from the conv5 feature, and integrates candidate area networks into end-to-end training throughout the network, which further improve the efficiency and reached the realtime effect.

3 Thyroid Nodule Localization Algorithm

This section will introduce the algorithm proposed in this paper in detail, the algorithm is divided into three parts: image preprocessing, manual sign recognition and boundary adjustment. Image preprocessing will carry out preliminary processing and analysis of the original image to find the location where there may be manual signs; manual sign recognition will carefully distinguish the result of the previous step to determine the exact location of manual signs; due to the difference of machines, the complex background of the image, and the small scale of manual sign, there may be some errors and omissions in manual sign recognition part, border adjustment will comprehensively analyze these factors, and make sure the entire thyroid nodule is included in the rectangle given by the algorithm with minimum possible error. An example of ultrasound image is shown in Fig. 1.

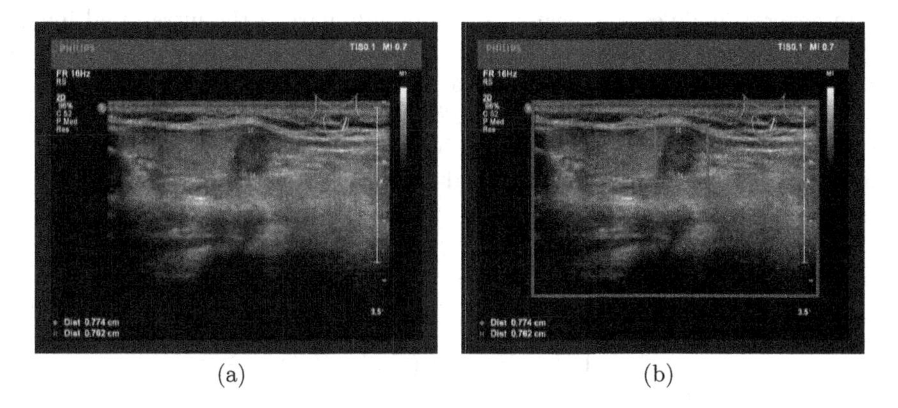

Fig. 1. (a) is an original thyroid ultrasound image, the blue rectangle in (b) is the area where we are looking for nodules, and the red rectangle is the location of the nodule. (Color figure online)

3.1 Image Preprocessing

The purpose of image preprocessing is to process the ultrasound image at the pixel level using image processing method, in order to remove extra information in the image, and find the positions that may be manual signs. In an ultrasound image, except for the ultrasound scanning result in the middle, there may be some extra marks around the image, such as machine type, diagnosis time, image scale, length and width of nodules, and other information due to the difference of machines. Therefore, these extra marks need to be removed first to prevent adverse effects in following processing.

In most cases, there are three channels of red, green, and blue when the images are saved to the computer, but the region of interest in the ultrasound image is generally gray-scale. So the ultrasound images are firstly converted to gray-scale, not only will it not have a negative impact, but it will also improve the efficiency of following processing. Suppose that the original image of the ultrasound is represented as A, the gray-scale image is represented as B, the pixel value of channel z, row x and column y is represented as $A(x, y, z)$, and $A(x, y, z) \in [0, 255], x \in [0, H), y \in [0, W), z \in \{r, g, b\}$, H represents the height of the image, W represents the width of the image. Gray-scale image can be expressed by Eq. 1 [12].

$$B(x, y) = w_r \times A(x, y, r) + w_g \times A(x, y, g) + w_b \times A(x, y, b) \tag{1}$$

In Eq. 1, w_r, w_g, w_b respectively represents the weight of red channel, green channel and blue channel to the gray value. Generally, w_r is 0.299, w_g is 0.587, and w_b is 0.114. Sometimes, in order to avoid floating-point operations, the weight is first multiplied by a power of 2 to get an approximate integer, and then we calculate the multiplication and addition in integer field, and finally divided by the power of 2 using bit operation.

In general, nodules are located in the middle part of the ultrasound image, but the extra marks around the image is nothing to do with the nodules. First of all, we remove the extra marks, and leave the real ultrasound information only. Using *top* to represent the top border, *bottom* for the bottom border, *left* for the left border and *right* for the right border, ϵ is an integer which means the threshold to distinguish the background and foreground.

$$f_1(x) = \begin{cases} H, & if \frac{1}{W} \sum_{y=0}^{W-1} B(x, y) > \epsilon \\ |x - \frac{H}{2}|, & if \frac{1}{W} \sum_{y=0}^{W-1} B(x, y) \le \epsilon \end{cases} \tag{2}$$

$$\begin{aligned} top &= \arg\min_x f_1(x), x \in (0, \frac{H}{2}) \\ bottom &= \arg\min_x f_1(x), x \in (\frac{H}{2}, H) \end{aligned} \tag{3}$$

$$f_2(y) = \begin{cases} W, & if \frac{1}{bottom - top + 1} \sum_{x=top}^{bottom} B(x, y) > \epsilon \\ |y - \frac{W}{2}|, & if \frac{1}{bottom - top + 1} \sum_{x=top}^{bottom} B(x, y) \le \epsilon \end{cases} \tag{4}$$

$$\begin{aligned} left &= \arg\min_y f_2(y), y \in (0, \frac{W}{2}) \\ right &= \arg\min_y f_2(y), y \in (\frac{W}{2}, W) \end{aligned} \tag{5}$$

For the convenience of description, the gray value of the part outside the rectangle defined by top, bottom, left, and right in the original image is set to 0, and the new image is represented as R. If the original image contains thyroid nodules, it must lie in image R. When the doctor is doing an ultrasound diagnosis of the patient, different signs will be marked on the left and right, top and bottom of the nodules. Our purpose is to find those locations that may be the manual signs in image R. These manual signs usually have higher brightness in ultrasound images, but it is difficult to locate these signs only by the brightness, because in addition to these markers, the ultrasound image itself will have some higher brightness areas. However, the brightness of ultrasound images generally changes gradually, but the brightness near the manual signs usually has a large change. For such regions with wide range and large change in brightness, using first-order differential and second-order differential will have good effects.

The derivatives of discrete functions can be defined by using differences, see Eq. 6.

$$\begin{aligned} \frac{\partial f}{\partial x} &= f'(x) = f(x+1) - f(x) \\ \frac{\partial^2 f}{\partial x^2} &= \frac{\partial f'(x)}{\partial x} = f'(x+1) - f'(x) \\ &= (f(x+2) - f(x+1)) - (f(x+1) - f(x)) \\ &= f(x+2) - 2f(x+1) + f(x) \end{aligned} \tag{6}$$

The above equation is expanded from $x+1$, and then we expand the second derivative of f from x, we can get Eq. 7, finally, the gradient of the binary discrete function can be present as Eq. 8.

$$\frac{\partial^2 f}{\partial x^2} = f''(x) = f(x+1) + f(x-1) - 2f(x) \tag{7}$$

$$\begin{aligned} \nabla^2 f(x,y) &= \frac{\partial^2 f}{\partial x^2} + \frac{\partial^2 f}{\partial y^2} \\ &= f(x+1,y) + f(x-1,y) + f(x,y+1) \\ &\quad + f(x,y-1) - 4f(x,y) \end{aligned} \tag{8}$$

The operator in Eq. 8 is Laplace operator L. Applying L to image R, we get the differential of image R, and represent the differential image as G. For the convenience of handling, binarization is applied to the differential image in Eq. 9, where η represents a threshold between 0 and 255.

$$\begin{aligned} C &= R \times L \\ G(x,y) &= \begin{cases} 255, & if \quad C(x,y) > \eta \\ 0, & if \quad C(x,y) \leq \eta \end{cases} \end{aligned} \tag{9}$$

In the binary image G, the point where the gray value is 255 is the point of interest. Generally, there may be many points with high gray scale value of 255 around the manual signs. Therefore, we use the search algorithm to deal with image G, put nearby bright points together, and then find the center of gravity of each group of points. Finally, in each center of gravity, there maybe a manual sign.

3.2 Manual Sign Recognition

Due to the large noise in the ultrasound image, an there are some other signs as well, the position of manual signs found by the above method may not be the real sign, it is also possible to find some wrong signs. Therefore, we need to find the positions of real signs from the positions above. In ultrasound images, a pair of '+' (plus sign) are commonly used to indicate the left and right border, and a pair of 'x' (multiplication sign) to indicate the top and bottom border of nodules. However, these signs usually have different sizes and shapes due to different machines types and different image sharpness, or even blends with its surroundings, making it difficult to be identified accurately using common methods. Convolutional neural network has a good effect in image classification [13–16], and in this paper, convolutional neural network is used to identify the real manual signs.

Convolution neural network (CNN) [17] is a kind of feed-forward neural network. In traditional neural networks, neurons are fully connected between two adjacent layers. Due to the large number of pixels contained in the image, the traditional fully connected neural network will greatly increase the number of connections if each pixel corresponds to a single neuron. Therefore, the computer can not process such data and the neural network can not be trained normally. CNNs connect contiguous two layers of neurons by convolution, and each neuron responds only to nearby neurons, and then make the image smaller through the pool layers to further reduce the number of connections and improve computing efficiency. CNNs are used in recognition of characters when they are proposed, and achieved good results. In recent years, it has been applied to the natural image classification, video classification, object recognition, object localization, machine translation, etc., and has been given play to its tremendous advantages.

This paper uses a multi-layer CNN, the input layer is a 19×19 image, and then it passes through a 3×3 convolution layer and a 2×2 pooling layer to reduce the size of the image and learn its shallow features, then learn the deep features through a layer of 5×5 convolution layer and a 2×2 pooling layer, finally there are two fully connected layers, with 256 neurons and 2 neurons in each layer, the last layer is the output layer of the neural network which is to indicate whether the input image is a manual sign or not. After each convolutional layer, a ReLU layer [18] is added as an activation function. The calculation of ReLU is simple, and ReLU are good at avoid gradient saturation problems and makes the gradient descent algorithm converge faster. The structure of the convolutional network is shown in Fig. 2. The cross entropy loss function is shown in Eq. 10.

$$Loss = \sum_i (y_i \log(\hat{y}_i) + (1 - y_i) \log(1 - \hat{y}_i)) \tag{10}$$

Deep learning requires a lot of training data, for small amount of data may not be able to learn the real knowledge. Data-enhanced methods are used to get more data if we lacking training data. For image data, new data is usually acquired by using a transform that does not change the image features. In this

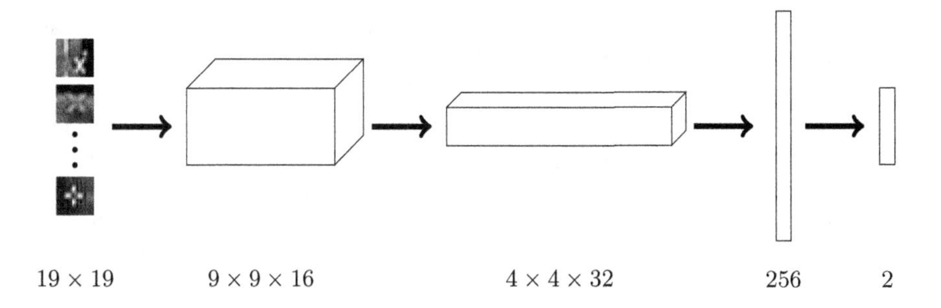

Fig. 2. Convolutional neural network structure

paper, the original images are rotated by 0°, 90°, 180° and 270°, respectively. The data of each rotation are inverted by the vertical central axis of the image, and eight images are obtained from one original image, greatly increasing the number of data. Through data enhancement, the trained model can see more data and learn more knowledge without influencing the image features. Later experiments prove that data enhancement does improve the recognition accuracy.

3.3 Boundary Adjustment

After finding the signs, it is easy to draw the bounding rectangle of the nodule based on the coordinates of signs, but the actual situation is often more complicated. For ease of expression, the signs that define the left and right boundaries are called plus signs, the signs that define the top and bottom boundaries are called multiplication signs, plus signs and multiplication signs are called manual signs. To label the location of the nodule, the algorithm requires a pair of plus sign and a pair of multiplication sign to determine the boundary. In practice, there may be fewer than four signs actually founded due to a recognition error in the algorithm for finding the signs or the doctor's forgetting to mark signs on some images, and there may also be some wrong locations found because the ultrasound image contains some symbols similar to the manual signs, making the algorithm get a mistake.

According to the actual situation, the plus sign in the ultrasound image will not be missed, and the correctness of the algorithm to identify the plus sign is also very high. Therefore, this paper assumes that the recognition algorithm will find at least two plus signs to determine the left and right border of the nodule, otherwise this algorithm can not give the location of nodules. Discuss the following situations.

1. More than two plus signs. This may be due to misidentifying a multiplication sign as a plus sign, or identifying a non-sign position as a plus sign. The CNN model gives the probability of each sign, we take the two plus signs with the highest probability in this situation.

2. No multiplication sign. Connect two plus signs into line segment, make a square centering on the midpoint of the line segment.
3. Only one multiplication sign. If the sign happens to be on either side of the line segment, then take the line as axis of symmetry, make the symmetry point of this sign as another multiplication sign. Otherwise, process it as case 2.
4. Two or more multiplication signs. Only keep those on both sides of the line segment, if there are more than one on each side, keep the nearest one.

In the above discussion, the position of sign generally refers to its geometric center position. When doctors mark the boundaries of nodules, they may mark at the inside of the nodules. In order to avoid the error caused by this problem, this paper extends the boundary of rectangular box outward by half the width of the signs.

The method proposed in this paper can be expressed as Fig. 3.

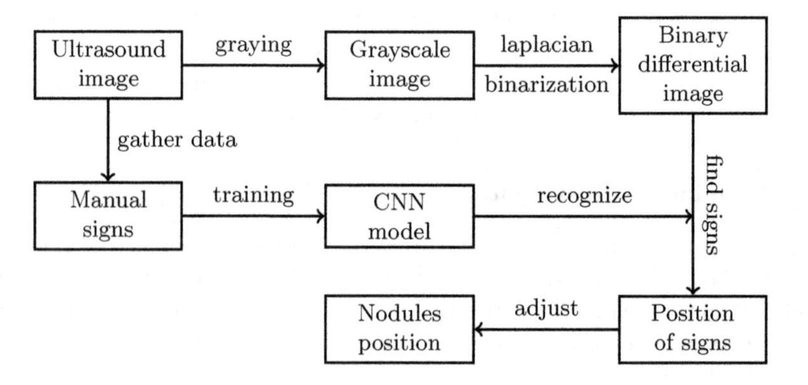

Fig. 3. Algorithm proposed in this paper

4 Experiment and Analysis

The experimental data in this paper is 500 real ultrasound diagnostic images in cooperation with Tianjin Medical University Cancer Institute and Hospital, including 280 malignant nodules and 220 benign nodules. In order to ensure that the privacy of patients is not infringed upon, detailed treatment has been done before the data is obtained to ensure that the experimental data are hidden from all sensitive information and only used for scientific research. There is only a set of signs that defines the location of the nodules marked by professional doctors on each image. In this paper, the part of neural network is implemented by Python language and TensorFlow library is used as the main tool, the rest are implemented by C++ language. All the code runs on a computer with Intel® Core™ i5-4590 CPU @ 3.3 GHz, 8.00 GB memory.

In the sign recognition experiment, we manually cropped 500 plus signs, 500 multiplication signs, and 500 non signs, amounting to 1500 data. Part of the data is shown in Fig. 4. Small amount of data may reduce the accuracy of the neural network, and data enhancement method is introduced in the paper, the training data is expanded by some transforms like flipping and rotating without changing the feature of signs, and at last we get the data which is 8 times the amount of the original data. This paper trained two neural network models to identify the plus sign and the multiplication sign. When we training to identify the model of the plus sign, the plus data is used as the positive sample and the remaining data as the negative sample. When training the model to identify the multiplication sign, we use multiplication data as the positive sample, and the rest data as the negative sample. In the training of neural networks, this paper randomly selects 70% data as the training set, the rest 30% as a test set.

Since these signs have different sizes and different shapes, all the cropped image signs are at the size of 19×19 making the training data have the same size and to ensuring that the features of signs are not lost.

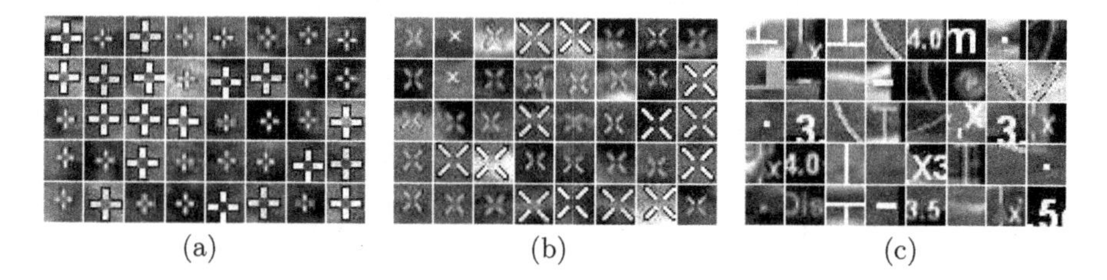

Fig. 4. (a) represents the plus signs, (b) represents the multiplication signs, (c) shows negative samples, the data in (c) are taken from ultrasound images where the brightness changes significantly or similar to the signs in (a) and (b).

The loss function is shown in Eq. 10, after training the models, the loss curve is shown in Fig. 5, in which the blue line (Loss 1) uses data with data enhancement, and the red line (Loss 2) uses data without data enhancement. It turns out that the use of data enhancement can reduce training errors. The accuracy curve is shown in Fig. 6, where the blue line (Accu 1) uses data with data enhancement, and the red line (Accu 2) uses data without data enhancement. It is obvious that data enhancement increases the accuracy of the algorithm.

Manual sign recognition is actually a classification problem, this paper compares the CNN method with other common classification algorithms, and the comparison results are shown in Table 1. It shows that CNN has better performance than other algorithms.

The CNN model can accurately and quickly determine whether a position has a plus sign or a multiplication sign, which will be combined with image preprocessing and boundary adjustment algorithm, and then we can achieve the

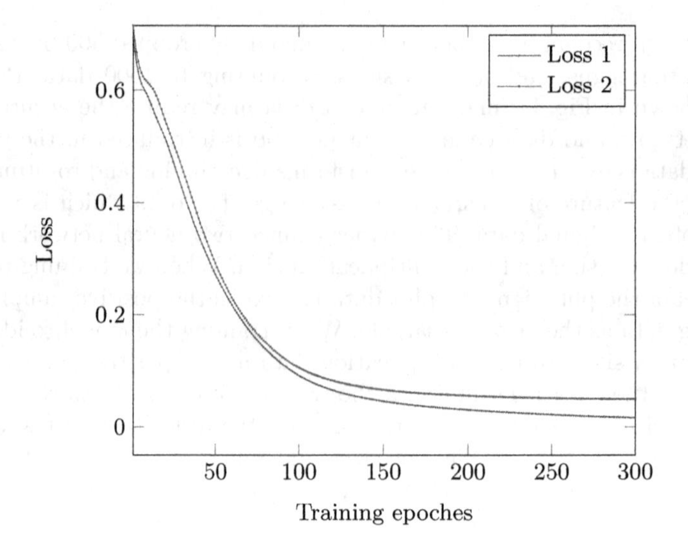

Fig. 5. Loss curve (Color figure online)

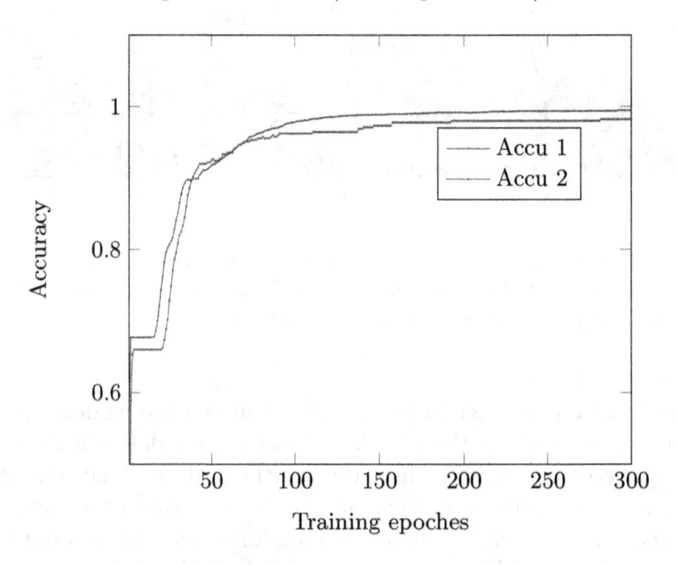

Fig. 6. Accuracy (Color figure online)

goal of this paper. The size of the ultrasound image is about 720 × 576, and the algorithm proposed in this paper only cost about 50 ms to deal with one image. Some of the results of the algorithm are shown in Fig. 7.

Table 1. Compare with other algorithms

Method	Accuracy	Precision	Recall	f1-score
CNN	99.47%	99.74%	98.62%	99.18%
KNN	97.86%	95.86%	97.59%	96.72%
LogisticRegression	92.11%	87.03%	88.82%	87.91%
Naive bayes	88.39%	85.78%	76.78%	81.03%
Random forest	91.92%	95.70%	78.50%	86.25%
Decision tree	94.11%	88.94%	93.38%	91.11%
SVM	94.06%	95.41%	85.73%	90.31%

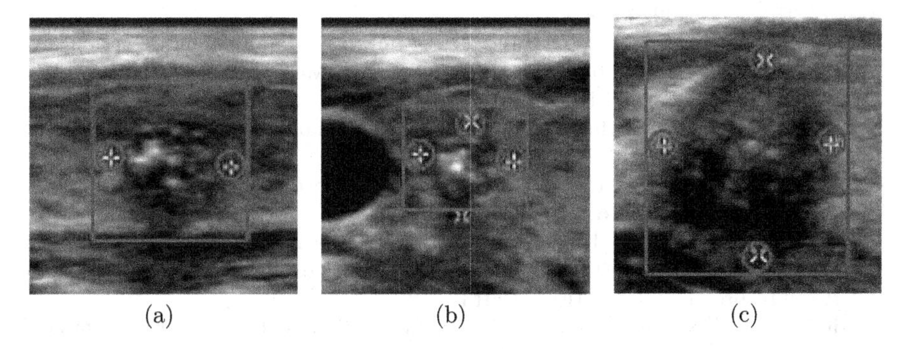

(a) (b) (c)

Fig. 7. Signs are surrounded by red circles and thyroid nodules are surrounded by red rectangles, a multiplication sign in (b) is lost by mistake. (Color figure online)

5 Conclusion

This paper presents an algorithm for locating thyroid nodules in thyroid nodules ultrasound images. First, preprocessing the image using image processing methods. Then CNN is used to train the model for accurately identifying the manual signs. Finally, we adjust the found boundary to ensure that the thyroid nodules in the ultrasound image are completely contained in the rectangular box given by the algorithm.

The algorithm proposed in this paper can accurately and quickly analyze and process a large number of thyroid nodules ultrasound images, screen out clearly marked ultrasound images and give the location of thyroid nodules, laying a solid foundation for further research. Experiments show that the proposed algorithm has a good performance in terms of correctness and operational efficiency.

References

1. Paschou, S.A., Vryonidou, A., Goulis, D.G.: Thyroid nodules: a guide to assessment, treatment and follow-up. Maturitas **96**, 1–9 (2017)
2. Colatrella, A., Loguercio, V., Mattei, L., Trappolini, M., Festa, C., Stoppo, M., Napoli, A.: Best practice & research clinical endocrinology & metabolism. Best Pract. Res. Clin. Endocrinol. Metab. **24**, 635–651 (2010)
3. Chen, W., Zheng, R., Baade, P.D., Zhang, S., Zeng, H., Bray, F., Jemal, A., Yu, X.Q., He, J.: Cancer statistics in China, 2015. CA Cancer J. Clin. **66**(2), 115–132 (2016)
4. Gall, J., Lempitsky, V.: Class-specific hough forests for object detection. In: Criminisi, A., Shotton, J. (eds.) Decision Forests for Computer Vision and Medical Image Analysis, pp. 143–157. Springer, Heidelberg (2013). https://doi.org/10.1007/978-1-4471-4929-3_11
5. Rong, W., Li, Z., Zhang, W., Sun, L.: An improved Canny edge detection algorithm. In: 2014 IEEE International Conference on Mechatronics and Automation (ICMA), pp. 577–582. IEEE (2014)
6. Hui, Z.: Summary of artificial landmarks and its methods application in close-range photogrammetry. Geosp. Inf. **7**(6), 30–32 (2009)
7. Ma, Y.: Application of the sub pixel and template matching in the corss mark location. Inf. Commun. **6**, 32–34 (2012)
8. Fang, F., Liu, Y., Yao, J., Li, Y., Xie, R.: Automatic marker detection from x-ray images. In: 2013 IEEE International Conference on Robotics and Biomimetics (ROBIO), pp. 1689–1694. IEEE (2013)
9. Girshick, R., Donahue, J., Darrell, T., Malik, J.: Rich feature hierarchies for accurate object detection and semantic segmentation. In: Proceedings of the IEEE Conference on Computer Vision and Pattern Recognition, pp. 580–587 (2014)
10. Girshick, R.: Fast R-CNN. In: IEEE International Conference on Computer Vision, pp. 1440–1448 (2015)
11. Ren, S., He, K., Girshick, R., Sun, J.: Faster R-CNN: Towards real-time object detection with region proposal networks. In: Advances in Neural Information Processing Systems, pp. 91–99 (2015)
12. Saravanan, C.: Color image to grayscale image conversion. In: 2010 Second International Conference on Computer Engineering and Applications (ICCEA), vol. 2, pp. 196–199. IEEE (2010)
13. Krizhevsky, A., Sutskever, I., Hinton, G.E.: Imagenet classification with deep convolutional neural networks. In: Advances in neural information processing systems, pp. 1097–1105 (2012)
14. Szegedy, C., Liu, W., Jia, Y., Sermanet, P., Reed, S., Anguelov, D., Erhan, D., Vanhoucke, V., Rabinovich, A., et al.: Going deeper with convolutions. In: CVPR (2015)
15. Simonyan, K., Zisserman, A.: Very deep convolutional networks for large-scale image recognition. arXiv preprint arXiv:1409.1556 (2014)
16. He, K., Zhang, X., Ren, S., Sun, J.: Deep residual learning for image recognition. In: Proceedings of the IEEE Conference on Computer Vision and Pattern Recognition, pp. 770–778 (2016)
17. LeCun, Y., Bottou, L., Bengio, Y., Haffner, P.: Gradient-based learning applied to document recognition. Proc. IEEE **86**(11), 2278–2324 (1998)
18. Nair, V., Hinton, G.E.: Rectified linear units improve restricted Boltzmann machines. In: Proceedings of the 27th International Conference on Machine Learning (ICML-10), pp. 807–814 (2010)

Cost Reduction for Micro-Grid Powered Data Center Networks with Energy Storage Devices

Guanglin Zhang(✉), Kaijiang Yi, Wenqian Zhang, and Demin Li

College of Information Science and Technology, Engineering Research Center of Digitized Textile and Apparel Technology, Ministry of Education, Donghua University, Shanghai 201620, China
glzhang@dhu.edu.cn

Abstract. The data center networks including multiple geo-distributed data centers begin to surge for meeting the ever-increasing Internet demand. To reduce the electricity bills in data centers, the current efforts center on developing sustainable data centers and improving energy efficiency. In this paper, we power data center with a microgrid which have conventional generators (CG), renewable energy sources (RES) and connected with electricity markets. To smooth the RES uncertainty, we integrate energy storage devices into microgrid. Besides, we consider any kind of energy can be charged into batteries or directly cover the demand of data centers. A stochastic program is formulated by integrating the geo-distributed load balancing, the energy management in microgrid and server configuration while guaranteeing the quality of service experience by end users. We design an online algorithm using Lyapunov optimization techniques without foreseeing any future information. The numerical evaluations based on real-world data sets corroborate the superior performance in reducing cost compared with previous works.

Keywords: Data center networks · Microgrid · Lyapunov
Energy storage devices · Renewable energy sources

1 Introduction

With the penetration in cloud computing, a growing number of data centers are built to meet the massive network requirements, such as data backup and transmission. Since single data center is hard to provide stable and reliable service, the current trend is to create data center networks including multiple data centers distributed in different areas. Even so, those companies that own data centers, such as Google, have to pay millions of dollars for just electricity cost.

Motivated by the great interests, both academia and industry make efforts to reduce the electricity cost caused by energy consumption. Most researchers focus their attention on the engineering techniques designs, such as energy-efficient chips, multicore servers, distributed power supply, virtualization and advanced

S. Chellappan et al. (Eds.): WASA 2018, LNCS 10874, pp. 647–659, 2018.
https://doi.org/10.1007/978-3-319-94268-1_53

cooling systems [1]. For instance, the thermal-aware allocation maximize the energy efficiency by keeping the temperature of all the IT operators below a certain threshold [2]. Proposed in [3], some policies aim at reducing data center operational cost by virtualization. Asghari put forward to optimize energy management in multi-core servers with speed scaling capability in [4].

Designing more optimal algorithm is another effective way to reduce the electricity cost. These referred works can be generalized into three different levels. First level focus on single server. For example, to reduce the cost on resource allocation, a joint optimization framework is proposed to scale dynamic voltage and frequency for individual cores [5]. The second level base on single data center. Specifically, [6] introduce a novel three-state model beyond turning on/off and improve the response time of service using the intermediate state. The last level is the interdata center level. Proposed in [7], a new virtualization scheme called ToU-aware Provisioning (ToUP) is designed for an interdata center networks. Another work combines multipath TCP and segment routing to save the cost by using a four-layer data center network simulated software defined network [8].

In addition, exploiting renewable energy is particularly important to prevent carbon emission. To smooth the uncertainty of RES caused by weather condition, there exist two feasible ways: one is to use scenario-based stochastic optimization approach based on renewable source samples from historical data; another is to use Lyapunov optimization [9]. In this paper, we consider the second one. The use of energy storage devices can smooth the randomness and instability by charging and discharging [10]. For purpose of power supply reliability and QoS, we assume each data center in data center network is powered by a micro-grid in case of an emergency and power shortage. Besides, the energy from RES, CG and electricity markets can be charged into storage devices as reserve power supply or directly cover the consumption in data centers. We investigate the problem of exploiting geo-distributed data center network powered with the micro-grid to minimize the average electricity cost and use Lyapunov optimization technique to solve it. Our contribution can be summarized as follows:

- We power the data center with a micro-grid which is composed of renewable energy source, conventional generators and batteries and use wholesale electricity markets as alternative energy. Each energy supply can be used to directly cover energy requirement in data center or charged into batteries.
- We formulate the problem as a stochastic program by dynamically determining interdata workload, server configuration and the energy allocation in micro-grid. Moreover, we design an online algorithm to minimize the electricity cost by adopting the Lyapunov optimization technique.
- The data sets used in this paper, such as real time electricity prices and wind power, are from real-world. Therefore, the results show the proposed algorithm indeed make sense in minimizing the electricity costs.

2 System Model

We consider a data center network include N geographically distributed data centers and K front-end proxy servers, which are respectively denoted by

$\mathcal{D} = \{D_1, D_2, \ldots, D_N\}$ and $\mathcal{S} = \{S_1, S_2, \ldots, S_K\}$. Besides, we assume the time is slotted into a discrete-time scheduling horizon $\mathcal{T} := \{1, \cdots, T\}$.

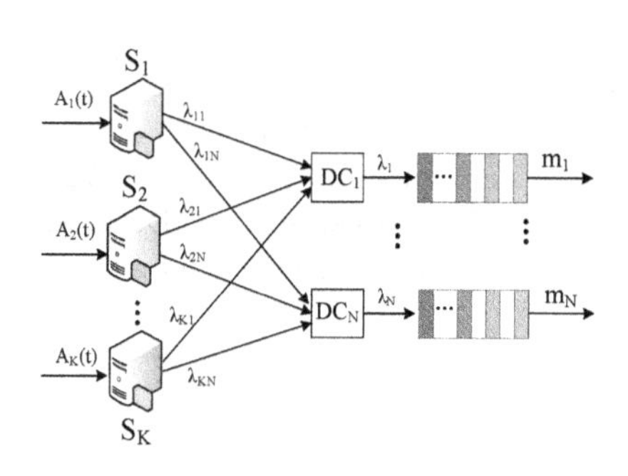

Fig. 1. The workload scheduling in DCs

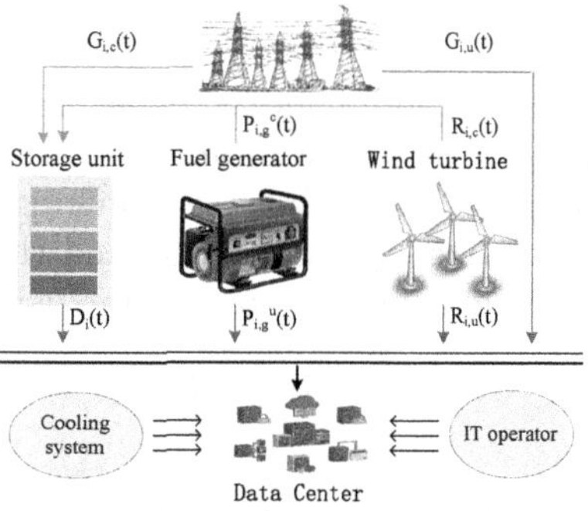

Fig. 2. A sustainable data center powered by micro-grid

2.1 The Workload and QoS Model

We first describe the workload allocation in the cloud network. Before processed by data centers, customer requests will arrive at front-end proxy servers S_j, and then be allocated to data center D_i. The average arrival rate of workload at S_j is denoted as $A_j(t)$, $j \in 1, \cdots, K$. The average rate of workload from S_j to D_i is denoted as $\lambda_{ji}(t), i \in \{1, 2, \cdots, N\}$ and it is positive. The workload allocation in data centers is shown in Fig. 1. We formulate the process as

$$\sum_{i=1}^{N} \lambda_{ji}(t) = A_j(t), \forall j = 1, \cdots, K. \tag{1}$$

Let $\lambda_i(t)$ represent average arrival rate of workloads at D_i, and we have

$$\lambda_i(t) = \sum_{j=1}^{K} \lambda_{ji}(t), \forall i = 1, \cdots, N. \tag{2}$$

According to the service level agreement (SLA) between the servers and customers, quality of service for customers need some paraments to measure. In this paper, we consider the average response time $d_i(t)$ as the parameter. Referred to [12], we use $M/m/n$ queue model to analyze the average response time. To meet the customer requirements, there need $m_i(t)$ active servers with average service rate μ_i to collaboratively work, where $m_i(t)$ is an integral variable, and

$$0 \leq m_i(t) \leq M_i, \forall i = 1, \cdots, N. \tag{3}$$

According to the queueing theory in [11], the number of active servers, service rate and traffic arrival rate must subject to the follow constraint,

$$d_i(t) \geq \frac{P_Q}{m_i(t)\mu_i - \lambda_i(t)}, \forall i = 1, \cdots, N, \tag{4}$$

where P_Q represents the queueing probability. In this paper, we assume $P_Q = 1$ because the servers are always busy if turned on, and assume each server consume energy H_i, when running at rate μ_i. The consumption include the cooling energy consumption and the servers computing consumption. Besides, the energy consumption must satisfy the follow constraints for each DC_i:

$$m_i(t)H_i \leq R_i^u(t) + G_i^u(t) + P_{i,g}^u(t) + D_i(t). \tag{5}$$

2.2 The Energy Supply Model

In this model, each data center is powered by a micro-grid as shown in Fig. 2. The micro-grid include renewable energy sources, energy storage units, conventional generators. Let $P_{i,g}(t)$, with upper bound $P_{i,g}^{\max}$, denote the energy output of conventional generator in D_i at slot t. The power can directly cover the server requirement or be charged into battery, with proportion of $P_{i,g}^u(t), P_{i,g}^c(t)$.

$$0 \leq P_{i,g}(t) \leq P_{i,g}^{\max}, \forall i = 1, \cdots, N, \tag{6}$$

$$P_{i,g}^u(t) + p_{i,g}^c(t) = P_{i,g}(t), \forall i = 1, \cdots, N. \tag{7}$$

In each period t, we denote the power generated from renewable energy as $R_i(t)$. The renewable energy is also used for charging the storage units or directly covering the energy consumption, with proportion of $R_i^c(t), R_i^u(t)$.

$$0 \leq R_i^c(t) + R_i^u(t) \leq R_i(t), \forall i = 1, \cdots, N. \tag{8}$$

If microgrids fail to meet energy requests, the electricity market is alternative. We denote the power from electric market as $G_i(t)$ with maximum $G_{i,\max}$. The same in above, the power can also be divided into two parts: $G_i^u(t)$ and $G_i^c(t)$.

$$0 \leq G_i(t) \leq G_{i,\max}, \forall i = 1, \cdots, N, \tag{9}$$

$$G_i^u(t) + G_i^c(t) = G_i(t), \forall i = 1, \cdots, N. \tag{10}$$

Since the conventional generators consume fossil fuel, which lead additional costs. We define fuel price as $w_i(t)$, which is specified in later part.

2.3 The Battery Model

As in [13], we can utilize the spatial variation of electricity price to minimize the cost by storing energy when the price is low for later shortage. Let $C_i(t)$ and $D_i(t)$ define the charged and discharged energy, respectively. If $C_i(t) \geq 0$,

$D_i(t) = 0$ and vice versa. The charged energy come from conventional generator, renewable energy source and electricity market,

$$C_i(t) = R_i^c(t) + G_i^c(t) + P_{i,g}^c(t), \forall i = 1, \cdots, N. \tag{11}$$

Denote battery level as $B_i(t)$. And assume that the energy leakage in battery is negligible and they operate independently of each other, we obtain the dynamics,

$$B_i(t+1) = B_i(t) + C_i(t) - D_i(t), \forall i = 1, \cdots, N. \tag{12}$$

Since physical limitations of batteries, the energy discharge rate (charge rate) has an upper bound $C_{i,\max}$ ($D_{i,\max}$) for each data center D_i. So we have

$$0 \le C_i(t) \le C_{i,\max}, \forall i = 1, \cdots, N, \tag{13}$$

$$0 \le D_i(t) \le D_{i,\max}, \forall i = 1, \cdots, N. \tag{14}$$

Besides, for the authenticity of the model, the battery energy level should be always positive and lower than the battery capacity, denoted as $B_{i,\max}$,

$$0 \le B_i(t) \le B_{i,\max}, \forall i = 1, \cdots, N. \tag{15}$$

Reformulating the constraints (13), (14) and (15), we get two equivalent constraints to limit the charge and discharge rate for each data center DC_i,

$$C_i(t) - D_i(t) \ge -\min\{D_{i,max}, B_{i,max}\}, \tag{16}$$

$$C_i(t) - D_i(t) \le \min\{C_{i,max}, B_{i,max} - B_i(t)\}. \tag{17}$$

In this model, we assume the service life of battery is long enough that we can ignore its single use price.

2.4 The Electricity Model

The electric power grid in US is organized into different reliability regions each of which has its own regional transmission organization (RTO) or independent system operator (ISO) to ensure the reliability of electricity supply [14]. Since the electricity price is spatial and temporal, we denote the time-varying electricity price at D_i as $P_i(t)$, which vary in $[P_{i,\min}, P_{i,\max}]$. The electricity prices in the cloud network is a vector, denoted as $\mathbf{P(t)} = (P_1(t), P_2(t), \cdots, P_N(t))$, and the electricity bought from the electricity market is also a vector, denoted as $\mathbf{G(t)} = (G_1(t), G_2(t), \cdots, G_N(t))$. And the fuel price in D_i is $w_i(t)$. From the above discussion, during the time slot t, the total electricity cost of N data centers is described as the follow:

$$f(t) = \sum_{i=1}^{N} \{P_i(t)G_i(t) + w_i(t)P_{i,g}(t)\}.$$

3 Optimization Problem and Solution

Our objective problem aims at minimizing the average-time expectation of electricity cost during all the time period. In this paper, we can achieve the objective by determining the follow control decisions: (1) the workload flow $\{\lambda_i(t)\}$; (2) the number of active servers $\{m_i(t)\}$; (3) the charge/discharge rate $\{C_i(t), D_i(t)\}$; (4) the energy in microgrid $\{R_i^c(t),\ R_i^u(t),\ G_i^c(t),\ G_i^u(t),\ P_{i,g}^c(t),\ P_{i,g}^u(t)\}$. Then we describe the stochastic program as problem **P1**:

$$\begin{aligned} \min \lim_{T\to\infty} \sup \frac{1}{T} \sum_{t=0}^{T-1} \sum_{i=1}^{N} \mathbb{E}\{G_i(t)P_i(t) + w_i(t)P_{i,g}(t)\} \\ s.t. \quad (1), (3)\text{–}(11), (15)\text{–}(17). \end{aligned} \tag{18}$$

Taking into account the simplicity of the solution, we need relax the target constraints. First we define the time-average expected usage of the battery as:

$$\overline{C_i} - \overline{D_i} = \lim_{T\to\infty} \sup \frac{1}{T} \sum_{t=0}^{T-1} \mathbb{E}\{C_i(t) - D_i(t)\}. \tag{19}$$

Based on the evolution of battery energy in (12), we move the $B_i(t)$ to the left and sum it over $t \in \{0, 1, \cdots, T-1\}$. After taking expectation of both sides, we obtain following equation about the time-average expected usage of the battery,

$$\mathbb{E}\{B_i(T)\} - \mathbb{E}\{B_i(0)\} = \sum_{t=0}^{T-1} \mathbb{E}\{C_i(t) - D_i(t)\}. \tag{20}$$

Since $0 \le B_i(t) \le B_{i,max}$ holds for each time slot, dividing the equation by T and letting $T \to \infty$ we have $\overline{C_i(t)} - \overline{D_i(t)} = 0$. Then we transform **P1** into **P2**:

$$\begin{aligned} \min \lim_{T\to\infty} \sup \frac{1}{T} \sum_{t=0}^{T-1} \sum_{i=1}^{N} \mathbb{E}\{G_i(t)P_i(t) + w_i(t)P_{i,g}(t)\}, \\ s.t. \quad (1), (3)\text{–}(11), (13), (14), \\ \overline{C_i} - \overline{D_i} = 0. \end{aligned} \tag{21}$$

The proposed algorithm is based on Lyapunov optimization techniques. In the algorithm, we use perturbed weights for determining the traffic distribution, server configuration and discharging/charging decisions. By legitimately revising the weight parameter V, the energy level in the battery will never overflow the feasible region regardless of whether the battery is charged or discharged. Before deducing the algorithm, we denote the perturbation parameter as θ_i, which will be specified later, and define the virtual energy queue as follows,

$$\widetilde{B}_i(t) \triangleq B_i(t) - \theta_i, \forall i = 1, \cdots, N. \tag{22}$$

Then we should define the modified Lyapunov as follow,

$$L(t) \triangleq \frac{1}{2}\widetilde{B}_i(t)^2, \forall i = 1, \cdots, N, \tag{23}$$

and define the one-period conditional Lyapunov drift as the following equation,

$$\triangle(t) \triangleq \mathbb{E}\{L(t+1) - L(t) \mid \widetilde{B}(t)\}, \forall i = 1, \cdots, N. \tag{24}$$

According to the Lyapunov optimization framework, we can obtain the following drift-plus-penalty term by adding the penalty function to the above one-period Lyapunov drift. Taking the evolution of battery into account, the drift-plus-penalty term subjects to the following inequality:

$$\begin{aligned}\triangle(t) + V\mathbb{E}\{f(t)\} \leq C + \mathbb{E}\{\sum_{i=1}^{N}[\widetilde{B}_i(t)(R_i^c(t) + P_{i,g}^c(t) + G_i^c(t) - D_i(t))] \mid \widetilde{B}(t)\} \\ + V\mathbb{E}\{\sum_{i=0}^{N}[G_i(t)P_i(t) + w_i(t)P_{i,g}(t)] \mid \widetilde{B}(t)\}.\end{aligned} \tag{25}$$

where $C = \sum_{i=1}^{N} \frac{\max\{C_{i,\max}^2, D_{i,\max}^2\}}{2}$. Then our algorithm is to determine the control decisions to minimize R.H.S of (25). So **P2** can be rewritten as **P3**:

$$\min \sum_{i=1}^{N}\{\widetilde{B}_i(t)[R_i^c(t) + P_{i,g}^c(t) + G_i^c(t) - D_i(t)] + V[G_i(t)P_i(t) + w_i(t)P_{i,g}(t)]\}, \tag{26}$$

subject to (1), (3)–(11), (13), (14) and $\overline{C_i} - \overline{D_i} = 0$.

The weight parameter V regulates the impact of the virtual battery on optimization result. A larger V leads the deviation of virtual battery to hardly affect on optimization result, and as a result, long-term target needs a greater number of time periods to mitigate the change. In other words, A smaller V indicates that the virtual battery has a higher impact on the optimization outcome.

As in [16], according to the KKT conditions, we obtain that, if the optimal solution exists, the arrival rate of workloads at data center i must satisfy constraint $\lambda_i^*(t) = m_i^*(t)\mu_i - \frac{1}{d_{i,max}}$. After summing (1) for $j = \{1, 2, \cdots, K\}$ and (2) for $i = \{1, 2, \cdots, N\}$, we have:

$$\sum_{i=1}^{N} m_i^*(t)\mu_i = \sum_{j=1}^{K} A_j(t) + \sum_{i=1}^{N} \frac{1}{d_{i,max}}. \tag{27}$$

Replacing the constraints (1), (2) with (27), **P3** can be rewritten as **P4**:

$$\begin{aligned}\min \quad & \sum_{i=1}^{N} \widetilde{B}_i(t)[P_{i,g}^c(t) + G_i^c(t) + R_i^c(t) - D_i(t)] + V[G_i(t)P_i(t) + w_i(t)P_{i,g}(t)] \\ s.t. \quad & (3), (5)\text{–}(11), (13), (14), (27).\end{aligned} \tag{28}$$

4 Algorithm Performance Analysis

In this section, we analyze thoroughly our algorithm in terms of the feasibility and performance. Before that, we define the perturbation parameter θ and the maximum of control parameter V_{max} as follow:

Definition 1. *For guaranteeing that the energy level in battery will never overflow the battery capacity, the parameter V has a maximum value. And suppose the maximum V and perturbation parameter θ_i are defined as follow:*

$$V_{max} \triangleq \min \frac{B_{i,max} - C_{i,max} - D_{i,min}}{P_{i,max}}, \tag{29}$$

$$\theta_i \geq VP_{i,\max} + D_{i,\max}. \tag{30}$$

When deduce the solution to **P3**, we get the following property.

Lemma 1. *To minimize* **P3**, *the optimal solution has the following property: Suppose $\widetilde{B}_i(t) \geq 0$, we always choose $C_i^*(t) - D_i^*(t) \leq 0$.*

4.1 Feasibility of the Algorithm

All control decisions we made satisfy the constraints in **P3**. If the energy level in battery will never over the capacity, we can guarantee all constraints in **P1** are also satisfied, which means our optimal decisions are feasible to the objective problem **P1**. About the energy level in battery, we have the following theorem.

Theorem 1. *Assume the initial energy in battery subject to the constraint, $B_{i,ini} \in [0, B_{i,max}]$. Under condition that $V \in (0, V_{max}]$, the algorithm can ensure that the battery energy level $B_i(t)$ always varies in the range $[0, B_{i,max}]$.*

Proof. As we know $B_{i,ini} \in [0, B_{i,max}]$, to verify $0 \leq B_i(t) \leq B_{i,\max}$, we just prove that $0 \leq B_i(t+1) \leq B_{i,\max}$ is satisfied by mathematical induction.

1. First we prove the right part of the equality, $B_i(t) \leq B_{i,\max}$: if $\widetilde{B}_i(t) \geq 0$, from lemma 2, we have $C_i(t) - D_i(t) \leq 0$. Then $B_i(t+1) = B_i(t) + C_i(t) - D_i(t) \leq B_i(t) \leq B_{i,\max}$ can be obtained; if $\widetilde{B}_i(t) \leq 0$, according to the definition of $V_{\max}$, we have $B_i(t) \leq VP_{i,\max} + D_{i,\max} \leq B_{i,\max}$. Then the right part of the inequality holds for each time slot.
2. Then we should prove $B_i(t) \geq 0$: if $\widetilde{B}_i(t) \geq 0$, we obtain $B_i(t+1) \geq VP_{i,\max} + D_{i,\max} + C_i(t) - D_{i,\max} \geq 0$; if $\widetilde{B}_i(t) \leq 0$, we divide the inequality into two parts: when $D_{i,\max} \leq B_i(t) \leq VP_{i,\max} + D_{i,\max}$, we obtain $B_i(t+1) = B_i(t) + C_i(t) - D_i(t) \geq B_i(t) - D_{i,\max} \geq 0$; if $0 \leq B_i(t) \leq D_{i,\max}$, we just prove the value of the objective when $D_i(t) = 0$ is always larger than the case of $D_i(t) \geq 0$, which indicates the battery do not discharge at this period. The process is formulated as,

$$\begin{aligned}&(B_i(t) - \theta)(P_{i,g}^c(t) + G_i^c(t) + R_i^c(t)) + VP_i(t)C_i(t)\\&\leq -D_i(t)(B_i(t) - \theta) - VP_i(t)D_i(t).\end{aligned} \tag{31}$$

After induction and integration to the above, we just prove $B_i(t) - VP_{i,\max} - D_{i,\max} + VP \leq 0$. Because $VP_i(t) - VP_{i,\max} \leq 0$ and $B_i(t) \leq D_{i,max}$, the above inequality is satisfied. Then $B_i(t+1) \geq 0$ holds for each period.

4.2 Asymptotic Optimality

In this subsection, we verify that the solution achieved under the proposed algorithm will progress the optimality as the parameter V increase in the range $(0, V_{\max})$. We demonstrate the asymptotic optimality by the following theorem.

Theorem 2. *If $A(t)$ and $C(t)$ are i.i.d. over slots, then the time-average expected electricity cost under our algorithm is within bound B/V of the optimal value:*

$$\limsup_{T\to\infty} \frac{1}{T} \sum_{t=0}^{T-1} \sum_{i=1}^{N} \mathbb{E}\{f^{'}(t)\} \leq Q^{OPT} + C/V, \tag{32}$$

where C is the constant given in (25) and $f^{'}(t)$ is the cost under our proposed algorithm, and Q^{OPT} represents the optimal solution to the original problem **P1**.

Proof. As described in Lyapunov optimization techniques, our algorithm is always trying to greedily minimize the R.H.S. of (25) at each period t under any possible feasible control policy including our proposed algorithm and other suitable policy. Therefore, by plugging our proposed policy in the right hand of the inequality (25), we get an inequality:

$$\triangle_V(t) \leq C + V\mathbb{E}\{\sum_{i=1}^{N}[f(t)] \mid \widetilde{B}_i(t)\} \leq C + VQ^{OPT}. \tag{33}$$

Then taking the expectation of both sides and summing over $t \in \{0, 1, 2, \ldots, T-1\}$, we obtain:

$$V\sum_{i=1}^{N}\sum_{t=0}^{T-1} \mathbb{E}\{f(t)\} \leq CT_{+}VTQ^{OPT} - \mathbb{E}\{L(T)\} + \mathbb{E}\{L(0)\}. \tag{34}$$

Since $\mathbb{E}\{L(0)\}$ are finite and $\mathbb{E}\{L(t)\} \geq 0$, we divide both side of (34) by T where $T \to \infty$. Then we can guarantee our algorithm performance by the follow:

$$\limsup_{T\to\infty} \frac{1}{T} \sum_{i=1}^{N} \sum_{t=0}^{T-1} \mathbb{E}\{f(t)\} \leq Q^{OPT} + C/V, \tag{35}$$

where Q^{OPT} is objective value under optimal situation and C is defined in (25).

5 Numerical Evaluation

5.1 Experimental Setup

In this section, results of simulation based on real-world data sets are presented to analyze the proposed algorithm performance. In order to better simulate the real power supply, we divide the time into a discrete sequence every five minutes according to the price change. For better display the illustration, we just consider four data centers and one front-proxy server. The arrival requests at proxy server exponentially distribute in $10^6 \sim 5 * 10^6$ with means of $1.1 * 10^6$ (requests/5-min), we show the requests for 10 days in Fig. 3(b).

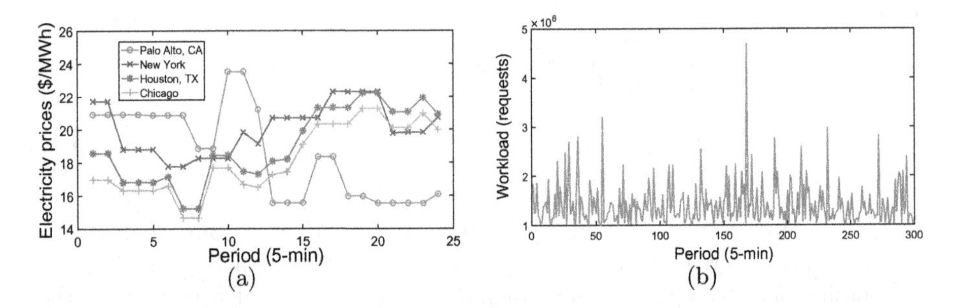

Fig. 3. (a) 5-min average real-time electricity prices. (b) 5-min average workload

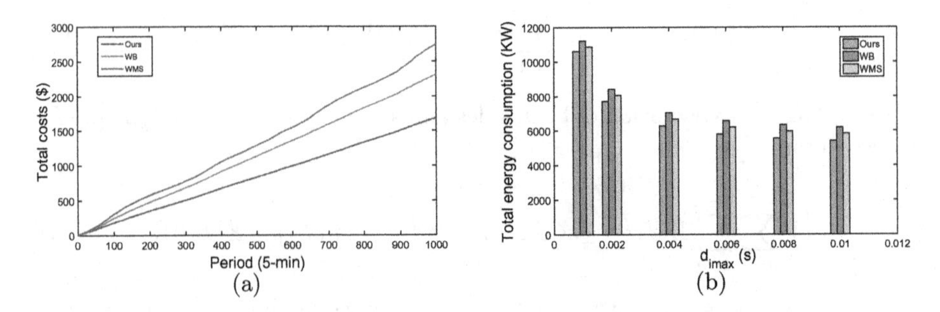

Fig. 4. (a) Total cost under three algorithm. (b) Energy consumption as $d_{i,\max}$ change.

The electricity prices come from the public government sources [14]. We adopt the 5-min locational marginal prices (LMP) in real-time electricity markets at four different cities: Chicago, New York, Palo Alto, CA and Houston, TX. The prices are from July 1 to July 5, 2016, including 1441 5-min time slots. We show the prices over two hours in Fig. 3(a). The fuel prices w_i in CG are set as $w_i(t) = (1/T)\sum_{t=1}^{T} P_i(t)$ [9]. We assume the maximum of power generated from CG is 20 KWh and assume the maximum charged (discharged) energy is 24 KWh. Besides, we get the wind power trace from [15].

We assume the servers in same data center are homogeneous. The data centers in above four cities are equipped with different servers respectively. By calculating the parameter in [12], all these servers consume power $H_i = 325\,\mathrm{W}$ per second, while their service rate are respectively $\mu_i = [2\ 2\ 1.5\ 2.5]$ (in unit of requests per second). And the number of servers in this four data centers can be presented as $M_i = [15000\ 10000\ 10000\ 10000]$.

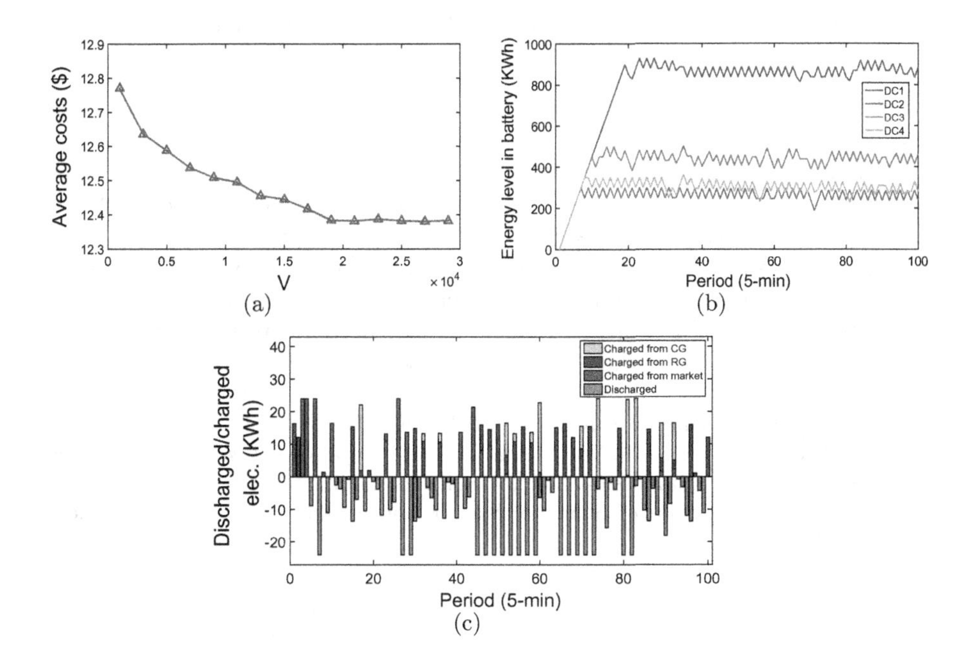

Fig. 5. (a) Average cost with the change of V. (b) Energy level in battery 1, 2. (c) The discharged and charged energy in data center 1.

5.2 Numerical Results

To analyze the performance of our algorithm, we compare it with two previous works [12,16]: (1) Without microgrid supply (WMS): in this scheme data centers dynamically route interdata traffic and manage energy, but energy supply just depends on electricity market; (2) Without battery (WB): here the system can manage the interdata workloads and servers, but not use storage devices to exploit the electricity prices variation.

When we choose $V = V_{\max}$, Fig. 4(a) depicts the comparison of the total electricity cost under above three schemes for 1000 time periods. Compared with WMS and WB, the cost under our algorithm decrease by 38.6% and 27.2%, respectively. As in Fig. 4(b), we increase the average response time $d_{i,\max}$, total energy consumption will decrease but under our algorithm the consumption is always less than others.

Besides, we adjust parameter V to obtain the average cost. The results, shown in Fig. 5(a), show that the average cost under proposed algorithm progress the optimality as V increase. Let $V = V_{\max}$, the energy trajectory in all batteries over 100 time slots, illustrated in Fig. 5(b), show battery level will never overflow the feasible region. Since the parameter in each data center is not same, the state of energy stable in different values. From the figure, the Theorem 2 is confirmed. To better observe the energy flow in microgrid, we show the charged (discharged) energy including energy charged from CG, RG and electricity market in Fig. 5(c).

6 Conclusion

By leveraging spatio-temporal electricity prices in markets and fuel prices, we design an online algorithm to manage the power in microgrid and workload allocation in data center network for reducing the energy cost. The proposed algorithm represents excellent performance in distributing more workload to data center where electricity price is lower. When we improve the parameter V, the cost approximate optimal value. Extensive numerical results test the effectiveness of the algorithm in saving the electricity cost without specific assumption.

Acknowledgment. This work is supported by the NSF of China under Grant No. 71171045, No. 61772130, and No. 61301118; the Innovation Program of Shanghai Municipal Education Commission under Grant No. 14YZ130; and the International S&T Cooperation Program of Shanghai Science and Technology Commission under Grant No. 15220710600.

References

1. Zeadally, S.: Energy-efficient networking: past, present, and future. J. Supercomput. **62**(3), 1093–1118 (2012)
2. Hameed, A., Khoshkbarforoushha, A., Ranjan, R., et al.: A survey and taxonomy on energy efficient resource allocation techniques for cloud computing systems. Computing **98**(7), 751–774 (2016)
3. Birke, R., Podzimek, A., Chen, L.Y., Smirni, E.: State-of-the-practice in data center virtualization: toward a better understanding of VM usage. In: 2013 43rd Annual IEEE/IFIP International Conference on Dependable Systems and Networks, DSN, Budapest, pp. 1–12 (2013). https://doi.org/10.1109/DSN.2013.6575350
4. Asghari, N.M., Mandjes, M., Walid, A.: Energy-efficient scheduling in multi-core servers. Comput. Netw. **59**(4), 33–43 (2014)
5. Wang, Y., Chen, S., Goudarzi, H., Pedram, M.: Resource allocation and consolidation in a multi-core server cluster using a Markov decision process model. In: International Symposium on Quality Electronic Design, ISQED, Santa Clara, CA, pp. 635–642 (2013)
6. Nguyen, B.M., Tran, D., Nguyen, Q.: A strategy for server management to improve cloud service QoS. In: 2015 IEEE/ACM 19th International Symposium on Distributed Simulation and Real Time Applications, DS-RT, Chengdu, pp. 120–127 (2015)

7. Kantarci, B., Mouftah, H.T.: Inter-data center network dimensioning under time-of-use pricing. IEEE Trans. Cloud Comput. **4**(4), 402–414 (2016)
8. Pang, J., Xu, G., Fu, X.: SDN-based data center networking with collaboration of multipath TCP and segment routing. IEEE Access **5**, 9764–9773 (2017)
9. Chen, T., Zhang, Y., Wang, X., Giannakis, G.B.: Robust workload and energy management for sustainable data centers. IEEE J. Sel. Areas Commun. **34**(3), 651–664 (2016)
10. Deng, W., Liu, F., Jin, H., Wu, C., Liu, X.: MultiGreen: cost-minimizing multi-source datacenter power supply with online control. In: Proceedings of ACM International Conference on Future Energy System, Berkeley, CA, pp. 149–160 (2013)
11. Guo, Y., Ding, Z., Fang, Y., Wu, D.: Cutting down electricity cost in internet data centers by using energy storage. In: 2011 IEEE Global Telecommunications Conference - GLOBECOM 2011, Houston, TX, USA, pp. 1–5 (2011)
12. Guo, Y., Fang, Y.: Electricity cost saving strategy in data centers by using energy storage. IEEE Trans. Parallel Distrib. Syst. **24**(6), 1149–1160 (2013)
13. Harsha, P., Dahleh, M.: Optimal management and sizing of energy storage under dynamic pricing for the efficient integration of renewable energy. IEEE Trans. Power Syst. **30**(3), 1164–1181 (2015)
14. Federal Energy Regulatory Commission. http://www.ferc.gov/
15. National Renewable Energy Laboratory. http://wind.nrel.gov
16. Rao, L., Liu, X., Ilic, M.D., Liu, J.: Distributed coordination of internet data centers under multiregional electricity markets. Proc. IEEE **100**(1), 269–282 (2012)

An Information Classification Collection Protocol for Large-Scale RFID System

Jumin Zhao[1(✉)], Haizhu Yang[1], Wenjing Li[2], Dengao Li[1], and Ruijuan Yan[1]

[1] Taiyuan University of Technology, Taiyuan, China
{zhaojumin,lidengao}@tyut.edu.cn, 530406653@qq.com, 1171373912@qq.com
[2] University of California, Santa Barbara, USA

Abstract. Sensor-enabled RFID tag can store information about the state of the corresponding objects or the surrounding environment, which has generated a lot of interests from industries. But in the process of communication, tags may need to transfer different types of information which causes different lengths of time slot. In the fixed time slot, less information transmission in large slot will reduce the time efficiency. In this paper, we study the problem on how to design efficient protocols to collect such sensor information from numerous tags in a large-scale RFID system, and propose an information classification collection protocol (ICCP). It classifies interrogated tags into multiple categories, then, orders each tag through hash function, finally, removes any wasteful slots when transmitting data to the reader. Extensive simulation results show that ICCP can significantly reduce the communication overhead and execution time of the protocol, compared with all the existing solution.

Keywords: Sensor-enabled RFID tag · Information collection
Multiple categories · Time slot

1 Introduction

RFID systems have been deployed for varieties of applications, in terms of large-scale information collection, the reader can quickly collect different types of information carried by the tag, which may be commodity attributes [1], and may be information about the environment or status of the objects monitored by the micro sensor [2]. Through the rapid collection of tag information, managers can be more convenient to understand the status of goods, and can quickly and accurately find out the problems with the products and its treatment [3, 4].

Apparently, how to efficiently collect multiple types of information is a critical problem for the large scale RFID systems. In this paper, we study how to design a fast and efficient protocol for collecting multiple types of tag information. The ideal information collection protocol (LB) is that all tags are sent back to the information in turn, and the reader can receive them successfully. The basic protocol (PLC) is that the reader sequentially polls all tags and collects the information carried by the tag one by one. In [5], Chen et al. designed two protocols, called Single-hash Information

S. Chellappan et al. (Eds.): WASA 2018, LNCS 10874, pp. 660–670, 2018.
https://doi.org/10.1007/978-3-319-94268-1_54

Collection protocol (SIC) and Multi-hash Information Collection protocol (MIC), to read sensor-produced data from a large number of tags with optimal execution time. In [6], Qiao et al. investigated the information collection problems from the aspect of energy efficiency. The Tag-Ordering Polling Protocol (TOP) and the enhanced version are proposed for a reader to collect sensor information from a subset of tags in the system with minimum energy consumption. In [7], Yue et al. proposed a time-efficient information collection protocol based on multi-reader RFID systems. However, they did not consider the reader to collect a variety of different types of tags information which will cause reading error or consume too long time at the same time and so on. If we can avoid redundant information transmission, shorten the vector length and reduce the number of collision slots in the vector, it will greatly shorten the time of information collection and improve the efficiency of protocol execution.

In this paper, we propose the ICCP protocol to collect various types of information of storing in tags in large-scale RFID systems. It is mainly divided into three phases, the first phase is classifying all tags according to the amount of information needed to transfer, the second phase is sorting each type of tag through a number of hash functions, the third phase is compressing the recovery vector sequence of the second phase, and the empty slot is eliminated. It reduces the execution time and improve the efficiency of system.

2 System Model and Problem Description

The problem is to design an efficient protocol that allows the reader to quickly and accurately collect information about all the tags in the recognition area. We denote all tags within the range of recognition as set M, it will be divided into n classes through the classifying phase, which $M = \{X_1, X_2, \ldots, X_n\}$. And we assume that the tag has enough energy to successfully receive commands from the reader and send its own information to the reader. In order to improve the accuracy of the protocol, information received by the reader is as accurate as possible, and it can be detected by the CRC check code.

In the C1G2 standard, reader broadcasts a 96 bit tag identifier that requires the slot length to be t_{id}. Tag transmits the same number of bits, and it required slot length is t_{inf}. It is important to note that the time taken by a reader to send an identifier to the tag is different from the time which is taken by tag to send an identifier to the reader. Because the communication frequency of the reader and the tag is generally different [8]. We assume that the number of all tags is m in the range of reader identification for the RFID system, the lower limit of total time is $m \times t_{inf}$ for all tags. A message needs to be repeatedly transmitted several times to be successfully received by the reader, due to signal interruption and interference during transmission, so we need to design a protocol that expects execution time closer to this lower limit.

3 Information Classification Collection Protocol (ICCP)

3.1 Classifying Phase

Each tag transfers a different amount of information to the reader, the amount of information that each tag transfers to the reader is different, which need different lengths of time slots. Usually, time slots are defined as the time required for the longest message that tags need to be transmitted it ensures the accuracy of the transmission, but it will increase the information processing time of the entire system. And transmit a small amount of information in large slot will result in waste of time. Therefore, it is very important to classify them according to the amount of the information, and it is possible to reduce transmission time to a large extent by allowing different information to be transmitted in different lengths of time slots.

The reader first broadcasts a request command with parameters <f, r>, where f is the number of slots in a virtual frame, r is a random seed and will change in every round. Note that the request command defines the communication between reader and tags, and the specific details, including the data transfer rate, encoding, etc. The reader can identify all of tags in the region through the hash function to get the bloom vector. The virtual frame will never be transmitted, it only serves as a vehicle to inform all tags that their classify number, which can save the energy of tags to a large extent and reduce the cost in the communication process. Upon receiving this request, each tag through H(id; r) mod (f + 1) label its classify number, where id is the part of tag's ID and H(·) is a hash function. In this phase, only single slot is useful for us, and the reader can forecast the result by using the same formula. If there are different types of tags get the same category code, the reader re-sends a new frame length f and random number until the different types of tags have different class codes.

Take Fig. 1 for example, there are five different types of tags. First, the reader broadcasts a command that carries of <f, r_1> to all tags. Then the tags of received the command calculate the mapping value through H (id, r_1) mod f, and the reader can also get the same result through the same calculation. These five categories of tags, c1, c2, c3, c4, c5, through calculate and get the values of 0, 3, 1, 3 and 2, the results as their category number, you can see that the class 2 and class 4 are not successfully distinguished, which requires the reader to reselect the random number and sent to the tag. After the second time calculated, the result of these five types of tags calculated is 2, 1, 4, 2 and 3, as the second category code. Finally, the category number of these five classes are "02", "31", "14", "32", "23", which have been successful to assigned their own category code.

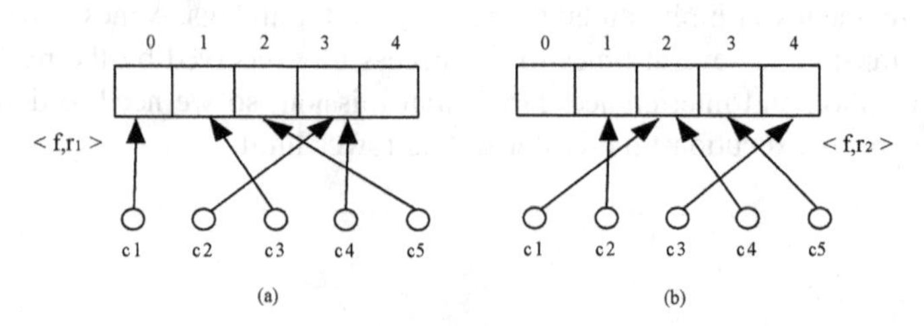

Fig. 1. Classifying phase

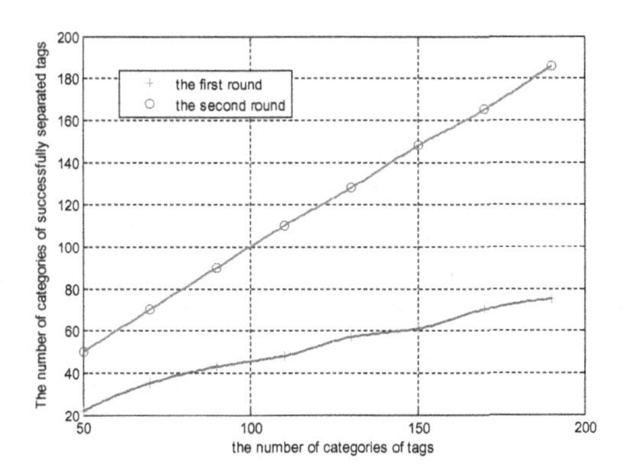

Fig. 2. Classifying phase simulation

In classifying phase, the reader only needs to send a command containing the random number r and the frame length f to all tags. The tags that have received this command will calculate the value through H (id, r_i) mod f as its class number. Not need to transmit information other than the command, so the execution time is only the reader to send the command time plus the calculation time of tags, it is very short. At the same time, because there is no need to transmit the bloom vector, there is no false positive probability problem at this stage, which greatly reduces the error. Through experimental simulation, as shown in Fig. 2, we can see that only about 37% of tag categories can be successfully separated after the first classification round, but after the second round, almost all tag categories will be successfully separated. And in general applications, two rounds are enough.

3.2 Ordering Phase

The Ordering phase consists of k rounds, each round involves one partition process and multiple hash calculations. In the partition process, the reader divides one group tags in X(i) into multiple partitions. First the reader broadcasts the category code for all types of tags, and the tags that receive the commands are activated when their category code successfully matches with the received, otherwise, the tags will remain silent. Then, the reader constructs an indicating vector V_J according to each partition W_j $(j \in [1, w])$. The length of the vector equals the number of partitions. And each tag will be assigned a partition index. The reader broadcasts a request with parameters <w, r_w>, where w is the total number of partitions and r_w is a random seed. Note that the number of partitions w can be calculated as $w = \left\lceil \frac{Xi}{g} \right\rceil$, where X_i is the cardinality of the one group and g is the number of tags in each partition.

An illustrative example is given in Fig. 3, in the first round, the reader broadcasts <w, r_w>, which makes all of tags in group X(i) divided into 9 partitions, then broadcasts the vector V_J to inform each tag what its slot interval. Then the reader uses a number of hash functions, expressed in h_i, $(1 < i < k)$, sorting all the tags of one partition to determine the specified single time slot. Considering an arbitrary time-slot, we call it a single slot if only one tag replies in this slot, or an empty slot if no tag replies in this slot, or a busy slot if one or more tags respond in this slot. And in this phase, the reader uses "0" to represent an empty slot with an idle channel and a busy slot with a busy channel. The length of a slot for a tag to transmit a one-bit short response is denoted as ts, usually equal to $\frac{t_{inf}}{96}$. Note that it can be set larger than the time of one-bit data transmission for better tolerance of clock drift in tags.

In the multiple hash calculation process, the reader broadcasts a request with parameter <g, r_{gm}>, where r_{gm} is the random seed in this process. All tags that receive this command will begin counting from the first hash function until the order of their responses is determined. If the result of the i-th hash function is calculated as n, then the tag will check whether the (n + 1) th bit of the received vector is i, if correct, it will use this slot position to reply information to the reader. If not, it continues to use the (i + 1) th hash function to calculate. After all m hash functions are used, the reader has a subset of tags that are assigned to slots, and it also knows which hash function each of these tags should use. The problem of communicating information to the tags will be addressed shortly.

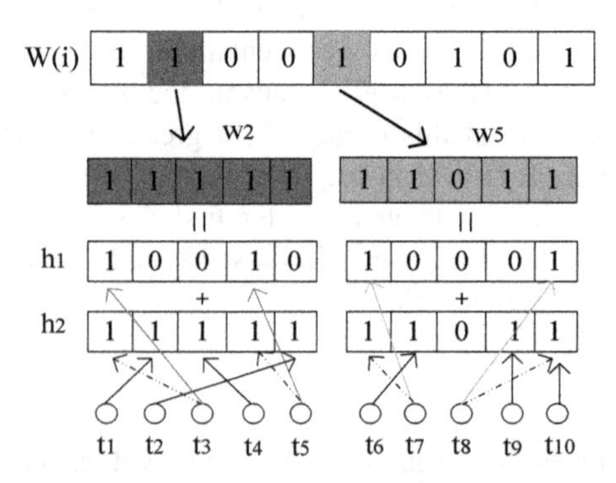

Fig. 3. Ordering phase

3.3 Collection Phase

Consider a round, the tags form w_j count the total number of "1" bits before the j-th bit in the indicating vector V_j, which is denoted as q. If the $V_j = 0$, the tag keeps silence. On the contrary, the tag calculates the corresponding slot index as $g \times q + h(id, r) \bmod g$. After selecting the slots, the tags transmit its responses to the reader in those slots. Obviously, the slots that the tags form w_i select are bounded at the range $[g \times q, g \times q + g - 1]$. At this stage, each tag needs to maintain a counter N, the initial value of the counter is set to 0, at the beginning of each slot, if the tag counter is 0, send their own information immediately, otherwise the slot does not respond. When the tag's information is successfully received, it will go into silence and will not respond to the reader commands for subsequent time slots. The tag that receives the command is calculated by the hash function with its own id segment (intercepted ID part information) and the random number r of the corresponding round, and then compare the calculated result with the corresponding bit of the vector sent by the reader to see if the tag of the corresponding bit slot is 1, if so, the position of this slot is stored in the counter N. At this time, if it is found that there is a slot that is not arranged, the counter needs to subtract the number of unscheduled slots before the corresponding bit, and waits for the counter N to become 0, then sends information to the reader. If not, the tag continues to use h_2, h_3, …, h_k calculation, until the corresponding time slot of the calculation result is the same as the hash subscript.

4 Performance Analysis

4.1 Parameter Setting and Performance Analysis

In the tag ordering stage, assuming that the number of tags is m, the number of tags to be sorted by the i-th round is m_i, the number of remaining available slots is m_i, and the number of tags which successfully assigned is m_{si}. Assuming f_i is the probability that a tag is successfully sorted in this round, in other words, f_i is the probability of a tag selects a time slot and the other tag does not select this slot. Assuming that the probability of a tag choose a slot which selected by one of the $m_i - 1$ tags in this round is $\frac{1}{m_i}$, on the contrary, the probability of they not selecting the same slot is $1 - \frac{1}{m_i}$, so we can get

$$f_i = C_{m_i}^1 \frac{1}{m_i} (1 - \frac{1}{m_i})^{m_i - 1} \tag{1}$$

For further analysis, since the mi-tag of the i-th round is tags that is not sorted successfully in the previous i − 1 round, assuming that the probability that a tag was successfully marked in the previous i − 1 round was F_{i-1}. The possibility of participating in the next round of sorting is $1 - F_{i-1}$. Then the probability that a tag of participating the i-round being successfully sorted is

$$f_i = C_{m_i}^1 \frac{1}{m_i} (1 - F_{i-1})(1 - \frac{1}{m_i}(1 - F_{i-1}))^{m_i - 1} \tag{2}$$

Since tags and time slots of the i − 1 rounds are unsuccessful sorting for the last round, through the results of the i − 1 round, we can get the probability that a tag will be successful sorting after i round

$$F_i = F_{i-1} + (1 - F_{i-1}) \times f_i = F_{i-1} + (1 - F_{i-1})^2 e^{-(1 - F_{i-1})} \tag{3}$$

When i = 1, we can get $f_1 = F_1 = e - 1$ in (1), and we can get the number of tags that can be sorted successfully by recursion. Based on the above analysis results, as shown in Fig. 4 that the curve above indicates a change in the percentage of arranged tags as the number of polling increases, the abscissa is the number of polling, and the ordinate is the percentage of the total label. The following curve is the slope of the upper curve, it can be seen that 83.9% of the tags were successfully sorted after the 6th round, and the success rate of the remaining tags was very low.

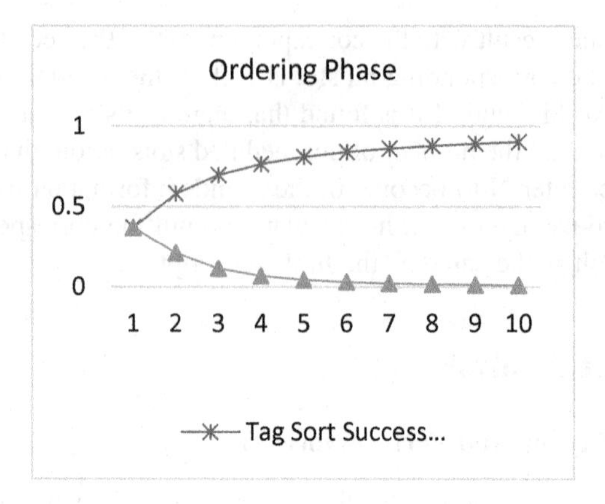

Fig. 4. The curve above indicates a change in the percentage of arranged tags as the number of polling increases, the abscissa is the number of polling, and the ordinate is the percentage of successfully ordered tags in the total tags. The following curve is the slope of the upper curve.

In the third stage, the tag receives the sort vector v generated by the second-stage and the random number r for each round, and tags check if the value calculated by the hash function is the same as the corresponding bit value of the sort vector v. The same result will be saved to the counter N, if different, continue to use the next hash function to calculate. At the same time, in order to avoid the waste of time slots, in the process of tags transmit information, make full use of each time slot for information collection. Finally, the value of the counter N = the result of the hash calculation − the number of time slots that are not sorted before the corresponding bit of the sort vector v. The counter value is reduced by one through each time slot, tags to send their own information when the value is reduced to 0 in the current time slot. In the above analysis can be drawn, after 6 rounds, the tags of successful sort is 16.1% of total tags, then the number of slots that avoid to send is $m \times (1 - F_i) = 0.16\,\text{m}$.

4.2 Expected Execution Time

To computer the expected execution time of ICCP, we need to get the time of three phases successful implementation. Since in the first stage of the tag classification, the reader and all the tags only need to use the id segment for several simple hash calculations, do not need to send other information, the execution time is very short and can be ignored. In the second stage, when the reader constructs a sorting vector for the label reply message, the total number of tags participating in the second stage is m, and the flag in each slot is y bit, y is used for the sorting phase, which is related to the number of hash function. If the tag is sorted through the six hash function, the corresponding slot is labeled by the binary representation of 110, only 3 bit. And one class tag divided into m partition. So the implementation time of the second stage is $\frac{w}{96} \times t_{id} + \frac{m}{96} \times y \times t_{id}$. In the information collection phase of the third stage, since the unused empty slot is removed, the total number of times of tag replies to the reader is $m \times F_i$. Assuming that the amount of information returned by each tag is x bit, the execution time is $\frac{x}{96} \times m \times F_i \times t_{\text{inf}}$.

Based on the above analysis, the ICCP total execution time is calculated as follows

$$T_1 = \frac{w}{96} \times t_{id} + \frac{m}{96} \times y \times t_{id} + \frac{mx}{96} \times F_i \times t_{\inf} \tag{4}$$

5 Performance Evaluation

5.1 Simulation Setting

We first set the simulation parameter, with the ID of each tags has 96 bits [9], which contains a 16-bit CRC. Any two consecutive communications, whether from the reader to the tag or from the tag to the reader, are separated by a time interval of 302 μs [10]. The reader transfer rate [11] is 26.5 kbps, so the reader transfers a tag ID number or a fragment in the bloom vector table with a time of 3927 μs, which includes a period of time [12]. Tag's transmission rate is 53 kbps, so the time of it transmits 1 bit is 18.88 μs [13]. The time used to transmit the information is the time interval plus the time of transmission of the information multiplied by the transmitted bit length. For example, when the sensor information is 8 bits long, the time to transmit information is 452 μs [14]. Because in order to reduce the collision also need to detect the channel, detection time is 321 μs [15].

At the same time we take the false positive probability 1×10^{-4}, for the same simulation we take multiple mean as the final result.

5.2 Execution Time Comparison

We assume that the recognition area of reader has a total of N tags, these tags according to the amount of different transmission information can be divided into n class, respectively b_1 bit, b_2 bit … b_c bit. And for tag number of each class, we assume partition w = xi/g, so that the w is a constant during the set of simulation. We first evaluate the performance of all protocols under different values of N, which varies from 10000 to 100000, when there is no channel error.

As shown in Fig. 5, a total of five types of information, respectively 16 bit, 24 bit, 32 bit, 40 bit, 48 bit. It can be seen that ICCP is the shortest execution time because it classifies tags according to the information quantity in each type of information collection process, there is no waste of time slots. For the MIC, in the information collection phase to ensure that the information is successfully received by the reader, the length of each slot to be greater than 48 bit. This makes the transmission of 16 bit, 24 bit, 32 bit information, resulting in a large time slot waste, with the increase in the number of tags, the implementation of the gap will be more and more, the more obvious advantages.

Figure 6 shows the transmission of information for all tags are 48 bit, compared with Fig. 5 can be seen that the minimum execution time of the information collection protocol increases as the amount of information increases, because ICCP is no need to classify the different information. Due to reduced the waste of empty time slot in the information phase, ICCP execution time is less that compared with SIC execution time, but the gap is not too big.

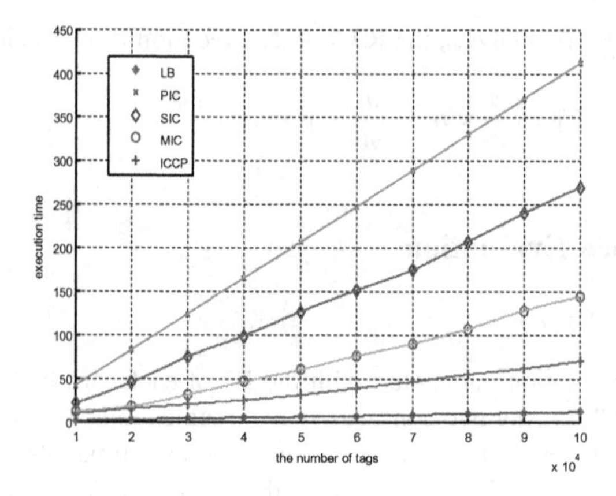

Fig. 5. Execution time of each protocol with different number of tags. And a total of five types of information, respectively 16 bit, 24 bit, 32 bit, 40 bit, 48 bit.

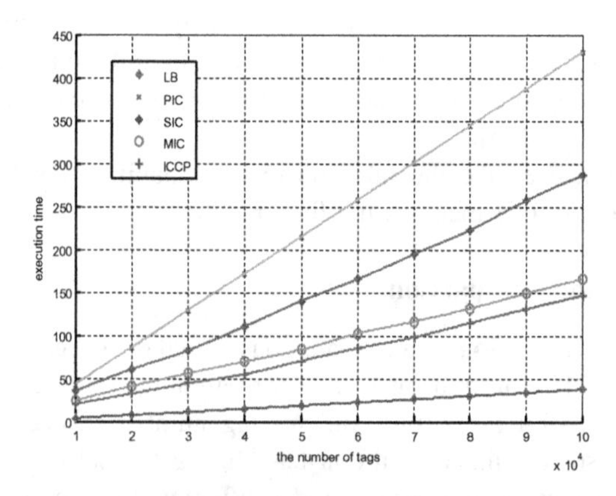

Fig. 6. Execution time of each protocol with different number of tags, and all of information is 48 bit.

The channel error rate p is defined as the percentage of slots that is corrupted. That is, in our simulations, each slot has a probability of p to be corrupted. Suppose the sensor information of every class is 16 bit, 24 bit, and 32 bit long, respectively. Figures 7 and 8 present the execution time comparison when channel error rate is 1% and 10%. With the presence of different levels of channel error, we continue to observe that ICCP performs much better than MIC, which in turn performs better than SIC, which is better than PIC. For example, there are 50000 tags, when the channel error is 1%, the execution time of protocol PIC is 232.1 s, which is about three times of the SIC and about five times of the MIC, however, the execution time of protocol ICCP only is about half of the protocol MIC. When the channel error is 10%, protocol PIC increase the execution time to 242.8 s, and our best protocol ICCP is 41.5 s, which is one third of MIC.

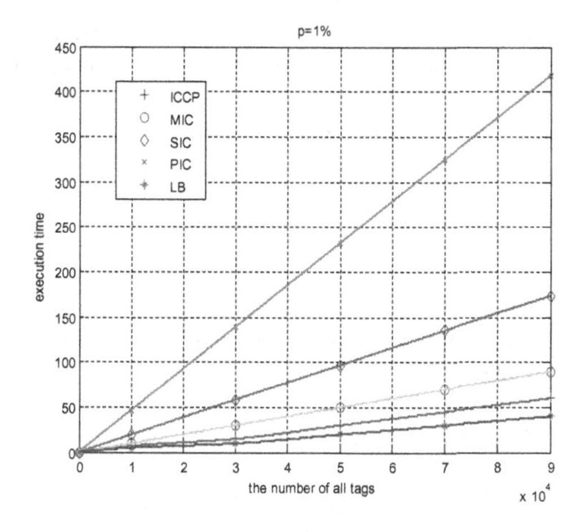

Fig. 7. Execution time comparison (in seconds) when the channel error is 1%

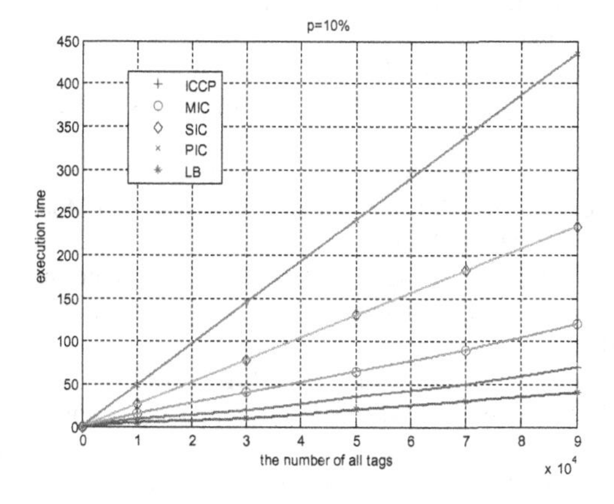

Fig. 8. Execution time comparison (in seconds) when the channel error is 10%

6 Conclusion

This paper studies the problem of how to collect multiple types information from the tags in a large-scale RFID system. Different from the previous information collection protocol, we consider that different tags need to send the amount of information is different, and propose an information classification collection protocol. The simulation results further show that the ICCP protocol is better than the existing protocol and is closer to the minimum lower limit of the information collection time.

Acknowledgments. The General Object of National Natural Science Foundation under Grants (61572346): The Key Technology to Precisely Identify Massive Tags RFID System With Less Delay; The General Object of National Natural Science Foundation (61772358): Research on the key technology of BDS precision positioning in complex landform; International Cooperation

Project of Shanxi Province under Grants (No. 201603D421012): Research on the key technology of GNSS area strengthen information extraction based on crowd sensing; The General Object of National Natural Science Foundation under Grants (61572347): Resource Optimization in Large-scale Mobile Crowdsensing: Theory and Technology.

References

1. Liu, J., Chen, S., Xiao, B., Wang, Y., Chen, L.: Category information collection in RFID systems. In: 2017 IEEE 37th International Conference on Distributed Computing Systems (ICDCS), pp. 2220–2225. IEEE (2017)
2. Xie, X., Liu, X., Li, K., Xiao, B., Qi, H.: Minimal perfect hashing-based information collection protocol for RFID systems. IEEE Trans. Mob. Comput. **16**(10), 2792–2805 (2017)
3. Li, T., Chen, S., Ling, Y.: Identifying the missing tags in a large RFID system. In: Proceedings of the Eleventh ACM International Symposium on Mobile Ad Hoc Networking and Computing, pp. 1–10. ACM (2010)
4. Bu, K., Liu, J., Xiao, B., Liu, X., Zhang, S.: Intactness verification in anonymous RFID systems. In: 2014 20th IEEE International Conference on Parallel and Distributed Systems (ICPADS), pp. 134–141. IEEE (2014)
5. Chen, S., Zhang, M., Xiao, B.: Efficient information collection protocols for sensor-augmented RFID networks. In: 2011 Proceedings IEEE INFOCOM, pp. 3101–3109. IEEE (2011)
6. Qiao, Y., Chen, S., Li, T., Chen, S.: Energy-efficient polling protocols in RFID systems. In: Proceedings of the Twelfth ACM International Symposium on Mobile Ad Hoc Networking and Computing, p. 25. ACM (2011)
7. Yue, H., Zhang, C., Pan, M., Fang, Y., Chen, S.: A time-efficient information collection protocol for large-scale RFID systems. In: 2012 Proceedings IEEE INFOCOM, pp. 2158–2166. IEEE (2012)
8. Philips Semiconductors: I-CODE smart label RFID tags (2004)
9. Tang, T., Du, G.-H.: A thin folded dipole uhf RFID tag antenna with shorting pins for metallic objects. KSII Trans. Internet Inf. Syst. **6**(9), 2253–2265 (2012)
10. Brady, M.J., Duan, D.-W., Kodukula, V.S.R.: Radio frequency identification system, 8 August 2000. US Patent 6,100,804
11. Vyas, S., Chinmay, V., Thakare, B.: State of the art literature survey 2015 on RFID. Int. J. Comput. Appl. **131**(8), 11–14 (2015)
12. Atzori, L., Iera, A., Morabito, G.: The internet of things: a survey. Comput. Netw. **54**(15), 2787–2805 (2010)
13. Priyanka, D.D., Jayaprabha, T., Florance, D.D., Jayanthi, A., Ajitha, E.: A survey on applications of RFID technology. Indian J. Sci. Technol. **9**(2) (2016)
14. Mohammed, F.H., Esmail, R.: Survey on IoT services: classifications and applications. Int. J. Sci. Res. **4**, 2124–2127 (2015)
15. Liukkonen, M.: RFID technology in manufacturing and supply chain. Int. J. Comput. Integr. Manufact. **28**(8), 861–880 (2015)

An Efficient Energy-Aware Probabilistic Routing Approach for Mobile Opportunistic Networks

Ruonan Zhao[1], Lichen Zhang[1], Xiaoming Wang[1(✉)], Chunyu Ai[2], Fei Hao[1], and Yaguang Lin[1]

[1] School of Computer Science, Shaanxi Normal University, Xi'an 710119, China
wangxm@snnu.edu.cn

[2] Division of Mathematics and Computer Science, University of South Carolina Upstate, Spartanburg 29303, USA

Abstract. Routing is a concerning and challenging research hotspot in Mobile Opportunistic Networks (MONs) due to nodes' mobility, connection intermittency, limited nodes' energy and the dynamic changing quality of the wireless channel. However, only one or several of the above factors are considered in most current routing approaches. In this paper, we propose an efficient energy-aware probabilistic routing approach for MONs. Firstly, we explore and exploit the regularity of nodes' mobility and the encounter probability among nodes to decide the time when to forward messages to other nodes. Secondly, by controlling the energy fairness among nodes, we try to prolong the network lifetime. Thirdly, by fully taking the dynamic changing quality of the wireless channel into consideration, we effectively reduce the retransmission number of messages. Additionally, we adopt a forwarding authority transfer policy for each message, which can effectively control the number of replicas for each message. Simulation results show that the proposed approach outperforms the existing routing algorithms in terms of the delivery ratio and the overhead ratio.

Keywords: Mobile Opportunistic Networks · Routing
Nodes' mobility · Encounter probability · Energy-aware

1 Introduction

Nowadays, with the development of wireless communication technology and the widespread popularity of mobile intelligent devices such as smart phones and smart watches, Mobile Opportunistic Networks (MONs) have been rapidly popularized and applied [4,12,14]. In MONs, a large number of stationary or mobile nodes (i.e., mobile intelligent devices carried by people) can conduct Device-to-Device (D2D) message transmission in a self-organizing and multi-hop manner via Bluetooth or WiFi without the support of communication infrastructures

S. Chellappan et al. (Eds.): WASA 2018, LNCS 10874, pp. 671–682, 2018.
https://doi.org/10.1007/978-3-319-94268-1_55

[1,5,6]. Consequently, the current pressure of mobile big data faced by the communication infrastructures can be effectively alleviated [2,3,7]. However, due to nodes' mobility and the lack of a complete communication path between the source and the destination, the traditional TCP/IP protocols are no longer available in MONs. Thus, routing is a concerning research hotspot in MONs.

Nevertheless, there are many challenges to implement an efficient message transmission approach for MONs [9]. Firstly, the energy of nodes (i.e., the battery power of devices) is often finite. The contact detection, forwarding and receiving messages are bound to consume some energy, respectively [10]. Therefore, energy would be a very precious resource when nodes' energy cannot be supplied in a timely manner. Although some routing algorithms, e.g. [11,14], have taken the energy factor into consideration, the energy fairness among nodes is ignored. Consequently, some nodes which have higher social status will frequently exchange messages with other nodes. Although the probability for some messages being successfully delivered would increase, these nodes will become unavailable soon due to the high energy consumption. The routing performance of the whole network will be declined as well. Thus, it is particularly important to take the energy fairness among nodes into consideration.

Secondly, in MONs, when a node transmits a message in a wireless broadcast manner, it may often face with the dynamic changing quality of the wireless channel, due to nodes' mobility and the connection between any two encountered nodes is time-varying and uncertain [13]. Thus, the predetermined relay nodes for the next hop may not receive the message, and then the message needs to be retransmitted again, which may increase the energy consumption and shorten the lifetime of the whole network. The message transmission delay is also extended.

Thirdly, it is difficult to decide the time when to forward messages due to the high mobility of nodes. A user is likely to encounter other different users at different times and may be isolated at some time, which may further increase the difficulty for designing an efficient routing algorithm.

Fortunately, people usually move in an MON according to their interests instead of moving randomly, and people who have the same interest usually have higher probability to meat each other. Therefore, we can utilize these information to improve the routing performance. Motivated by the aforementioned discussion, in this paper, we propose an efficient energy-aware probabilistic routing approach for MONs, called EEAP approach. The main contributions are summarized as follows:

- We explore and exploit the regularity of nodes' mobility and the encounter probability among nodes to decide the time when to forward messages to other nodes. Only when a node's neighbors satisfy the message forwarding conditions, the node could broadcast the corresponding message, otherwise, the node would wait for a better forwarding opportunity.
- By taking the effective utilization of nodes' energy and the energy fairness among nodes into consideration, a longer network lifetime can be achieved. Moreover, the proposed EEAP approach can effectively reduce the retransmission number of messages through considering the wireless channel quality.

- In order to improve the message delivery probability and meanwhile effectively control the number of message replicas, we adopt a forwarding authority transfer policy for each message, which also can reduce the overhead ratio. Simulation results show that the proposed approach outperforms the existing routing algorithms.

2 System Model

We consider the following MON environment in this paper. There are a large number of mobile nodes existing in the network, which is denoted by V. Each mobile node v_i ($v_i \in V$) could encounter some other nodes, when it moves in the network. Moreover, any two encountered nodes in a pair have a link, on which the weight value is to represent the encounter closeness relationship between them. We use W to represent the link weight set, and each weight value $w_{i,j}$ ($w_{i,j} \in W$) indicates the contact frequency between nodes v_i and v_j.

We assume that the inter-encounter time between any two nodes v_i and v_j in a pair follows an exponential distribution with the parameter $\lambda_{i,j}$ (e.g., $w_{i,j} = \lambda_{i,j}$). Thus, the inter-encounter time between nodes v_i and v_j is $1/\lambda_{i,j}$. Note that this assumption has been validated by realistic dataset analysis, and widely adopted in many existing researches, e.g., [8,14].

Moreover, each node v_i has an initial energy E, which is non-renewable. When the energy of node v_i is exhausted, it would become invalid and then move out from the network. Each valid node can generate messages. Let us consider that $m.sou$, $m.dst$, $m.ttl$ and $m.data$ and P_m separately denote the source, destination, initial time-to-live (TTL), data to be delivered and the required delivery probability of a message m, $\forall m \in M$. The TTL value of a message is usually considered to be a large, but finite value, which automatically decreases with time. Furthermore, a message will be invalid and removed from nodes' buffer until its TTL value decreases to 0.

Obviously, the energy consumption in MONs mainly includes three aspects: scanning channel for contact detection and communication for transmitting and receiving messages respectively. The energy consumed in communication has a lot to do with the message size and the network bandwidth. More specifically, the energy that a node v_i whose current energy is E_{cur} will consume within an observed time interval itv can be described with the following three aspects.

One is the *scanning energy consumption* for detecting the contact opportunities with other nodes, and consumes e_{sca} energy at each time. The scanning period is T, which is the time interval between any two adjacent scanning for any node. Thus, the scanning energy consumption E_{sca} can be expressed as:

$$E_{sca} = e_{sca} \times \frac{itv}{T} \tag{1}$$

Another is the *transmitting energy consumption* for forwarding messages, which consumes e_{tsm} energy per second. Let S_{tsm} denote the total size of messages that node v_i transmits during the time interval itv, thus the transmitting energy consumption E_{tsm} can be expressed as:

$$E_{tsm} = e_{tsm} \times \frac{S_{tsm}}{B}, \tag{2}$$

where B is the channel bandwidth.

The other is the *receiving energy consumption* for receiving messages, which consumes e_{rcv} energy per second. Let S_{rcv} denote the total size of messages that node v_i receives during the time interval itv, thus the receiving energy consumption E_{rcv} can be expressed as:

$$E_{rcv} = e_{rcv} \times \frac{S_{rcv}}{B}, \tag{3}$$

where B is the channel bandwidth.

Consequently, the current residual energy E_{cur} of a node v_i can be updated by the following equation after the time interval itv.

$$E_{cur} = E_{cur} - (E_{sca} + E_{tsm} + E_{rcv}) \tag{4}$$

$$= E_{cur} - (e_{sca} \times \frac{itv}{T} + e_{tsm} \times \frac{S_{tsm}}{B} + e_{rcv} \times \frac{S_{rcv}}{B}), \tag{5}$$

where the right side E_{cur} of the Eq. (5) represents the old residual energy value of node v_i before the observed time interval itv at this time, and the left side E_{cur} of the Eq. (5) denotes the new updated residual energy value of node v_i after the observed time interval itv at this time.

3 The EEAP Routing Approach

In this section, based on the presented system model, we mainly introduce the basic idea of the proposed EEAP approach. This approach consists of two phases: the initialization phase and the routing phase. The initialization phase aims to collect some useful information, and then the routing phase makes message forwarding decisions based on the collected information. The detailed description of the two phases is elaborated as follows.

3.1 The Initialization of EEAP

In an MON, time is divided into infinitely repeated time cycles with the same length l_{tc}, and each time cycle is partitioned into several discrete time slots with equal size l_{ts}, as shown in Fig. 1. For example, a week can be considered as the time cycle with the same length 7 days ($l_{tc} = 7\ days$), and a day can be regarded as the time slot whose size equals 24 h ($l_{ts} = 24$ h). Here, we assume that the encounter behavior between any two nodes is similar within a fixed time slot of each time cycle. For example, during a semester, the courses that a student should attend on every Wednesday are the same according to the schedule. Thus, he would meet same teachers and classmates on every Wednesday. Therefore, we can accumulate some useful encounter information for forwarding messages.

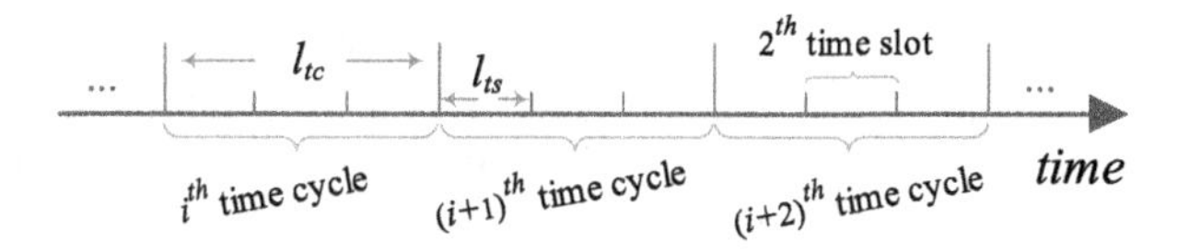

Fig. 1. An example about the time cycle and time slot

In order to simplify the expression, we use D to represent the destination $m.dst$ of message m. There are three stages in the initialization phase, which can be listed as follows:

(1) Each node v_i ($v_i \in V$) records the encounter frequency $\lambda_{i,D}$ between itself and destination D. (2) Each node calculates the probability of the number that it meets other nodes in each time slot respectively. We use $P_i(k,n)$ to express the probability that node v_i simultaneously encounters n nodes within the current k-th time slot, in which $P_i(k,n) \in [0, 1.0]$, both k and n are finite positive integers, and we set that a time cycle totally has K time slots, $1 \leq k \leq K$. (3) Each node calculates the average encounter frequency that all the other nodes it meets encounter the destination D within each time slot. Here, we use $\lambda_i(k)$ to represent the calculated average encounter frequency that all the other nodes which v_i meets encounter the destination D within the k-th time slot, that is,

$$\lambda_i(k) = \left[\frac{1}{|Ne(i,k)|} \times \sum_{v_j \in Ne(i,k)} \lambda_{j,D} \right], \tag{6}$$

where $Ne(i,k)$ represents a set of nodes that v_i meets within the k-th time slot, v_j is a node belonging to $Ne(i,k)$, $|Ne(i,k)|$ denotes the number of nodes in $Ne(i,k)$ and $\lambda_{j,D}$ is the encounter frequency between v_j and destination D.

3.2 The Implementation of EEAP

After the completion of the initialization phase, some useful information has been accumulated. This subsection will introduce the detailed implementation of the EEAP approach based on the accumulated information.

3.2.1 Determining Whether to Broadcast Message m at Present

This subsection mainly investigates determining whether to broadcast message m at present, and the main process is listed in Algorithm 1. Detailedly, in algorithm 1, lines 2–3 and lines 4–5 separately show what should node v_i do, if it encounters the destination of message m at present, or it not encounters the destination of message m at present and meanwhile the required delivery probability $P_m \leq 0$. Lines 6–19 decide whether to broadcast message m at present, if v_i not encounters the destination of message m at this moment and the required delivery probability $P_m > 0$. The detailed decision-making process is shown as follows.

When a node v_i carrying a message m encounters several other nodes, if the destination of message m not exists in the current encountered nodes, node v_i should determine whether to forward the message to the current encountered nodes at this time or not. Firstly, node v_i calculates the discrete time slot k_t that it is currently in by the following formula, that is,

$$k_t = \lfloor (t \mod l_{tc})/l_{ts} \rfloor + 1, \tag{7}$$

where t is the current time, l_{tc} is the length of a time cycle and l_{ts} is the length of a discrete time slot.

Next, node v_i determines whether the message m could be broadcast by the following two inequalities Eqs. (8) and (9). If one of the two inequalities holds, the message m would be broadcast by v_i at this time. Otherwise, v_i would give up this chance and still carry m to wait for the next broadcast chance.

$$1 - \prod_{v_j \in N_i(t)} (1 - p_j(R_{TTL})) \geq \delta, \tag{8}$$

$$1 - \prod_{v_j \in N_i(t)} (1 - p_j(R_{TTL})) \geq \max_{n=1,\ldots,N} \left\{ [1 - (1 - \hat{p}_i(k_t))^n] \cdot P_i(k_t, n) \right\}, \tag{9}$$

where δ denotes the required delivery threshold for forwarding message m at this time and $\delta \in [0,\ 1]$, $N_i(t)$ is the set of nodes that node v_i currently encounters, t is the current time, $p_j(R_{TTL})$ represents the probability that v_j would meet the destination within the remaining TTL of message m, $p_j(R_{TTL}) \in [0,\ 1]$, $P_i(k_t, n)$ denotes the probability that node v_i simultaneously encounters n nodes within the current k_t time slot, N is the maximum number of nodes that v_i can simultaneously encounter within a time cycle and N is a finite positive integer as well, $\hat{p}_i(k_t)$ represents the average encounter probability of all the other nodes that node v_i encounters within the k_t time slot meeting the destination D. As the aforementioned assumption, the inter-encounter time between any node v_i and its destination D follows an exponential distribution with the parameter $\lambda_{i,D}$. The probability that node v_j has a contact with the destination D at time t follows the exponential distribution, i.e., $\lambda_{j,D} \cdot e^{-\lambda_{j,D} \cdot t}$. Thus, the contact probability $p_j(R_{TTL})$ between node v_j and the destination D within the remaining TTL of message m can be calculated as follows:

$$p_j(R_{TTL}) = \int_0^{R_{TTL}} \lambda_{j,D} \cdot e^{-\lambda_{j,D} \cdot t} dt, \tag{10}$$

$$= 1 - e^{-\lambda_{j,D} \cdot R_{TTL}}, \tag{11}$$

where $\lambda_{j,D}$ is the encounter frequency that node v_j meets the destination and R_{TTL} is the residual TTL of message m ($m \in M$) which can be calculated by the Eq. (12).

$$R_{TTL} = m.ttl - (t - m.created), \tag{12}$$

where $m.ttl$ is the initial TTL value of message m, $m.created$ denotes the time instant when m was created and t is the current time.

Algorithm 1. Determining whether to broadcast message m at present

Input: node v_i; message m stored in the node v_i; the destination D of message m; current time t; the length of a time cycle l_{tc}; the length of a discrete time slot l_{ts}; the required delivery threshold value δ; node v_i's current neighbor node set $N_i(t)$; the current required delivery probability P_m for message m.

Begin:

1: **for all** (message m stored in the node v_i) **do**
2:   **if** ($D \in N_i(t)$)) **then**
3:     node v_i forwards message m to the destination D and finishes the transmission;
4:   **else if** (($P_m \leq 0$) & ($D \notin N_i(t)$))) **then**
5:     node v_i still carries message m and waits for the destination D;
6:   **else if** (($P_m > 0$) & ($D \notin N_i(t)$))) **then**
7:     node v_i determines whether to broadcast message m at this time by the following steps;
8:     node v_i calculates the time slot k_t that it is currently in by the Eq. (7) and the residual TTL value R_{TTL} of message m by the Eq. (12) respectively;
9:     node v_i calculates the Eq. (8);
10:     **if** (Eq. (8) holds) **then**
11:       node v_i decides to broadcast message m at present;
12:     **else**
13:       node v_i calculates the Eq. (9);
14:       **if** (Eq. (9) holds) **then**
15:         node v_i decides to broadcast message m at present;
16:       **end if**
17:     **else**
18:       node v_i decides not to broadcast message m at this time;
19:     **end if**
20:   **end if**
21: **end for**

Similarly, based on the recorded information about $\lambda_i(k)$, v_i can calculate the average encounter probability that all the other nodes it meets encounter the destination D within each time slot. We use $\hat{p}_i(k_t)$ to represent the average encounter probability of all the other nodes that node v_i encounters within the k_t time slot meeting the destination D. Thus, the average encounter probability $\hat{p}_i(k_t)$ can be calculated by

$$\hat{p}_i(k_t) = 1 - e^{-\lambda_i(k_t) \cdot R_{TTL}}, \tag{13}$$

where R_{TTL} is the remaining TTL value of message m.

The left side of Eq. (8) represents the delivery probability of message m with the assistance of v_i's current neighbor nodes. Thus, the establishment of Eq. (8) indicates that the current required broadcast threshold can be satisfied by the assistance of v_i's current neighbor nodes. Therefore, v_i can broadcast message m to its neighbor nodes.

The meaning of the Eq. (9)'s left side is similar with the Eq. (8)'s left side. The right side of Eq. (9) stands for the expected maximum delivery probability with the assistance of all neighbor nodes that node v_i simultaneously encounters within the current k_t time slot can achieve. Thus, the establishment of Eq. (9) represents that the delivery probability that message m can achieve with the assistance of v_i's current neighbor nodes is better than the later expected maximum delivery probability within the k_t time slot. Consequently, v_i does not need to wait for better neighbor nodes, and meanwhile can broadcast message m at this time. In this way, we can avoid a worse situation in the future, which means the delivery ability of the future encountered neighbor nodes is worse than the current encountered neighbor nodes.

Additionally, if the node v_i carrying message m not encounters the destination currently and the current required delivery probability $P_m \leq 0$, node v_i still carries message m until it meets the destination of message m.

3.2.2 Deciding Which Node Can Continuously Forward Message m to Other Nodes

If v_i decides to forward message m in a broadcast way, it often faces with the changing quality of wireless channel, and thus the predetermined forwarding nodes for the next hop will probably not successfully receive message m. Therefore, v_i should find out which neighbor nodes have successfully received message m after m being broadcast.

After v_i broadcasts message m to its current neighbors. If one of v_i's neighbor node v_j has received message m, it will send a confirming message to v_i. This confirming message includes v_j's ID number, its residual energy and message m's ID number. Later, node v_i calculates and updates the new required delivery probability P_m of message m by the following formula, that is,

$$P_m = P_m - \left\{1 - \prod_{v_j \in S_i} (1 - p_j(R_{TTL}))\right\}, \tag{14}$$

where the right side P_m of Eq. (14) represents the original P_m value of message m, the left side P_m of Eq. (14) denotes the new updated value, S_i is the set of v_i's neighbor nodes which have successfully received m, $S_i \subseteq N_i(t)$, $N_i(t)$ is the set of v_i's current neighbor nodes, and $p_j(R_{TTL})$ is the probability that node v_j would meet the destination within the remaining TTL of message m.

If the updated $P_m \leq 0$, v_i would always carry message m until it meets m's destination, and update P_m with the new calculated value. Because $P_m \leq 0$ indicates that v_i can satisfy the required delivery probability of message m with the assistance of v_i's current neighbor nodes. Thus, m does not need to be forwarded any more, all the nodes carrying the message m only need to delivery m to its destination, instead of forwarding it to other nodes for help. In this way, the quantity that m is forwarded can be reduced under the premise of meeting the requirements of m's delivery probability, and thus the energy consumption can be decreased and the network lifetime can be prolonged as well.

Otherwise, if the updated $P_m > 0$, it indicates that the required delivery probability of message m still cannot be guaranteed after this broadcast, and node v_i should select a new appropriate node from itself and all its current neighbors which have successfully received message m to continuously forward message m at a right time. Correspondingly, node v_i finds out a node which has the highest residual energy from the node set S_i and itself. If the node v_i is the selected node which has the highest residual energy, it will still carry m with the new updated delivery probability P_m. Otherwise, if a neighbor node v_j has the highest residual energy, v_i would send an update message including the updated P_m value to the selected neighbor node v_j. When the neighbor node v_j receives the update message sent by v_i, it will update the required delivery probability P_m of the message m stored in it with the received new P_m value. Meanwhile, v_i sets the P_m value of message m stored in it to 0, because the unsatisfied delivery probability would continuously be done by the selected neighbor node v_j. In this way, the node with the highest residual energy is selected as the forwarding node. Consequently, the residual energy of nodes can be balanced.

Additionally, if a neighbor node v_k ($v_k \in S_i$) not receives an update message, it would set the P_m value of message m stored in it to 0, and always carry message m until it meets the destination of message m. This operation ensures that each message can only be forwarded by a specified node, which can effectively reduce the replicas for each message and further decrease the network overhead ratio under the condition of ensuring the message delivery probability.

4 Performance Evaluation

In this section, we evaluate the performance of the proposed approach on ONE network simulator compared with three other routing algorithms. The simulation time is 60000 s, message TTL is 20000 s, message size $\in$ [200, 300] KB, transmitting speed is 2 MBps, buffer size is 100 MB, initial energy is 3000 J, scanning energy consumption at a time is 0.05 J, transmitting energy consumption is 2 J/s and receiving energy consumption is 3 J/s.

4.1 The Impact of Nodes' Initial Energy on Routing Performance

From Figs. 2(a) and 4(a), we can clearly see that the delivery ratio of the spray-and-wait algorithm and our approach is ascending first and then smooth with the increase of nodes' initial energy value. Nevertheless, the delivery ratio of the epidemic approach and the prophet algorithm is ascending with the increase of nodes' initial energy value. The main reasons for these phenomena are that a large number of nodes will exhaust their energy quickly because of forwarding and receiving messages when the initial energy is very small. Thus, many nodes become useless and the messages carried by them cannot be forwarded any more. Consequently, the delivery ratio of each routing algorithm is relatively low. With the increase of nodes' initial energy, the lifetime of each node becomes longer, and more messages can be forwarded. Thus, the delivery ratio of each routing algorithm increases at different extend.

Moreover, the spray-and-wait algorithm and the proposed approach can effectively control the number of replicas for each message, so the energy consumption

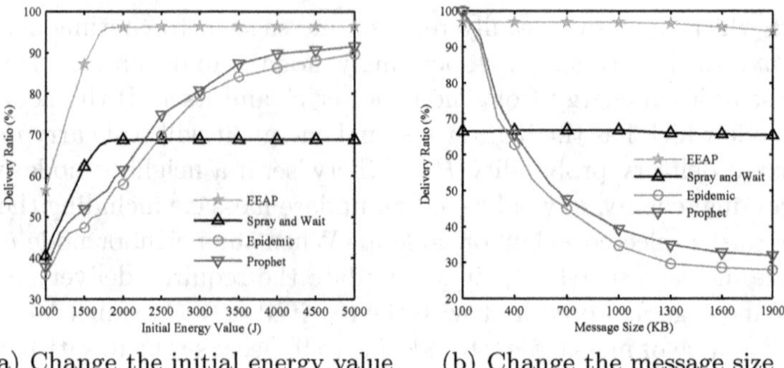

(a) Change the initial energy value (b) Change the message size

Fig. 2. The result of the delivery ratio when $|V| = 400$

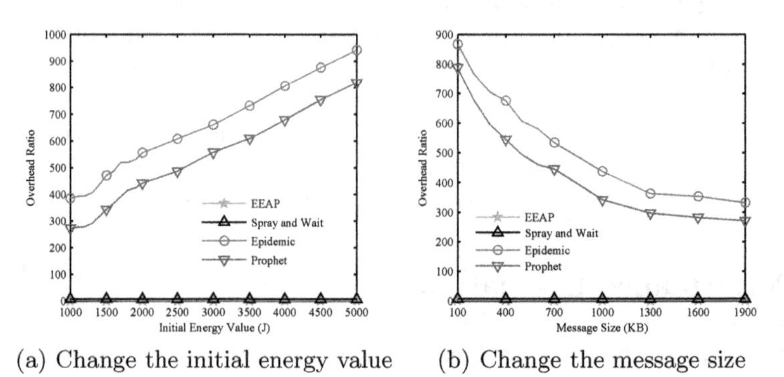

(a) Change the initial energy value (b) Change the message size

Fig. 3. The result of the overhead ratio when $|V| = 400$

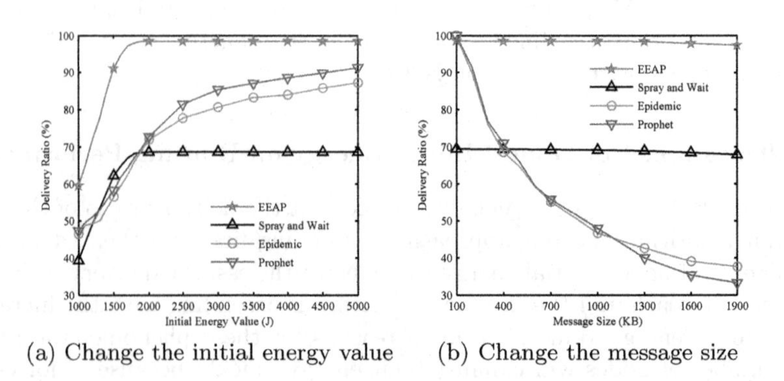

(a) Change the initial energy value (b) Change the message size

Fig. 4. The result of the delivery ratio when $|V| = 800$

of each node is less. Therefore, when the initial energy of nodes increases to a satisfactory level which can meet the requirement for forwarding and receiving messages in the network, the initial energy of nodes is no longer the most important factor affecting the delivery ratio of the two approaches. As a result, the delivery ratio of the two approaches shows a smooth trend in the later stage.

From Fig. 3(a), we can find that the spray-and-wait approach and the proposed approach have a lower overhead ratio, while the epidemic approach and the prophet approach both have a higher overhead ratio with an ascending trend. The main reasons are that the spray-and-wait approach and the proposed approach can effectively control the replicas of each message by setting the upper bound number of a message's replicas and the conditions for message broadcasting respectively. As for the epidemic approach, nodes exchange their messages without any limitation, and there are many unnecessary replicas existing in the network. Consequently, the successfully delivered messages are increased more slowly than the relayed messages with the increase of nodes' initial energy.

4.2 The Impact of Message Size on Routing Performance

From Figs. 2(b) and 4(b), we can clearly see that the delivery ratio of each routing algorithm decreases with the increase of message size. There are three main reasons for this phenomenon. (1) The energy consumption for forwarding or receiving a message would be raised with the increase of message size. Thus, the energy of nodes consumes faster and the network lifetime becomes shorter. Consequently, the number of being successfully delivered messages decreases, and the delivery ratio descends as well. (2) The number of messages that a node can store decreases with the increase of message size. (3) The time for transmitting a message will be longer with the increase of message size. Two encountered nodes are more likely to leave the communication range of each other when exchanging a message, which may interrupt the message transmission. Consequently, the number of the messages failed to be successfully transmitted increases. Thus, the delivery ratio of each routing algorithm decreases.

From Fig. 3(b), we can see that the overhead ratio of each routing algorithm is descending with the increase of message size. The main reason is that the probability for successfully forwarding a message decreases with the increase of message size. Thus, both the number of being successfully relayed messages and the number of being successfully delivered messages decrease. Consequently, the overhead ratio of each routing algorithm decreases.

5 Conclusion

In this paper, we propose an efficient energy-aware probabilistic routing approach for MONs. Each node first accumulates some useful contact information in initialization phase by exploring the regularity of nodes' mobility and the encounter relationship with other nodes. Afterwards, based on the collected information, the routing process is presented, which includes determining whether to broadcast a message m at present and deciding which node can continue to forward message m to other relay nodes two phases. Simulation has been done, and the results show that the proposed approach outperforms the previous routing algorithms.

Acknowledgment. This work is partly supported by the National Natural Science Foundation of China (No. 61702317), the Natural Science Basis Research Plan in Shaanxi Province of China (Nos. 2017JM6060, 2017JM6103), and the Fundamental Research Funds for the Central Universities of China (Nos. GK201801004, GK201603115, GK201703059 and 2018CSLZ008).

References

1. Chen, Q., Gao, H., Cheng, S., Fang, X., Cai, Z., Li, J.: Centralized and distributed delay-bounded scheduling algorithms for multicast in duty-cycled wireless sensor networks. IEEE/ACM Trans. Netw. **25**(6), 3573–3586 (2017)
2. Cheng, S., Cai, Z., Li, J., Fang, X.: Drawing dominant dataset from big sensory data in wireless sensor networks. In: Proceedings of INFOCOM 2015, Hong Kong, China, pp. 531–539, 26 April–01 May 2015
3. Cheng, S., Cai, Z., Li, J., Gao, H.: Extracting kernel dataset from big sensory data in wireless sensor networks. IEEE Trans. Knowl. Data Eng. **29**(4), 813–827 (2017)
4. Han, M., Yan, M., Cai, Z., Li, Y.: An exploration of broader influence maximization in timeliness networks with opportunistic selection. J. Netw. Comput. Appl. **63**, 39–49 (2016)
5. He, Z., Cai, Z., Cheng, S., Wang, X.: Approximate aggregation for tracking quantiles and range countings in wireless sensor networks. Theor. Comput. Sci. **607**(3), 381–390 (2015)
6. Li, J., Cheng, S., Cai, Z., Yu, J., Wang, C., Li, Y.: Approximate holistic aggregation in wireless sensor networks. ACM Trans. Sens. Netw. **13**(2), 11.1–11.24 (2017)
7. Lin, Y., Wang, X., Hao, F., Wang, L., Zhang, L., Zhao, R.: An on-demand coverage based self-deployment algorithm for big data perception in mobile sensing networks. Future Gener. Comput. Syst. **82**, 220–234 (2018)
8. Pu, L., Chen, X., Xu, J.: Crowd foraging: a QoS-oriented self-organized mobile crowdsourcing framework over opportunistic networks. IEEE J. Sel. Areas Commun. **35**(4), 848–862 (2017)
9. Saha, B.K., Misra, S., Pal, S.: SeeR: simulated annealing-based routing in opportunistic mobile networks. IEEE Trans. Mob. Comput. **16**(10), 2876–2888 (2017)
10. Wang, E., Yang, Y., Wu, J.: Energy efficient beaconing control strategy based on time-continuous markov model in DTNs. IEEE Trans. Veh. Technol. **66**(8), 7411–7421 (2017)
11. Zhang, F., Wang, X., Li, P., Zhang, L.: Energy-aware congestion control scheme in opportunistic networks. IEEJ Trans. Electr. Electron. Eng. **12**, 412–419 (2017)
12. Zhang, L., Cai, Z., Lu, J., Wang, X.: Mobility-aware routing in delay tolerant networks. Pers. Ubiquit. Comput. **19**(7), 1111–1123 (2015)
13. Zhang, L., Wang, X., Lu, J., Ren, M., Duan, Z., Cai, Z.: A novel contact prediction-based routing scheme for DTNs. Trans. Emerg. Telecommun. Technol. **28**(1), e2889.1–e2889.12 (2017)
14. Zhao, R., Wang, X., Zhang, L., Lin, Y.: A social-aware probabilistic routing approach for mobile opportunistic social networks. Trans. Emerg. Telecommun. Technol. **28**(12), e3230.1–e3230.19 (2017)

On Association Study of Scalp EEG Data Channels Under Different Circumstances

Jingyi Zheng[1], Mingli Liang[1], Arne Ekstrom[1], Linqiang Ge[2], Wei Yu[3], and Fushing Hsieh[1](✉)

[1] University of California, Davis, Davis, USA
{jgzheng,lmliang,adekstrom,fhsieh}@ucdavis.edu
[2] Georgia Southwestern State University, Americus, USA
linqiang.ge@gsw.edu
[3] Towson University, Towson, USA
wyu@towson.edu

Abstract. Electroencephalography (EEG) is an electrophysiological monitoring method to record electrical activity of the brain using different electrodes, which are considered as the EEG channels that are placed on scalp. In this paper, we propose an effective information processing approach to explore the association among EEG channels under different circumstances. Particularly, we design four different experimental scenarios and record the EEG signals under motions of eye-opening and body-movement. With sequences of data collected in time order, we first compute the mutual conditional entropy to measure the association between two electrodes. Using the hierarchical clustering tree and data mechanics algorithm, we could effectively identify the association between particular EEG channels under certain motion scenarios. We also implement the weighted random forest to further classify the classes (experimental scenarios) of the EEG time series. Our evaluation results show that we could successfully classify the particular motions with given EEG data series.

Keywords: EEG channels · Information processing
Mutual conditional entropy · Weighted random forest
Algorithm design

1 Introduction

Electroencephalography (EEG) is an electrophysiological measurements recording electrical activity of the brain from electrodes on the scalp. The EEG data typically interprets the impulses among nearby neurons [14,17]. In other words, the neural oscillations (also known as brain wave patterns) can be observed in EEG signals in the frequency domain. EEG signal normally includes several subbands such as delta (0.5–3 Hz), theta (4–7 Hz), alpha (8–13 Hz), beta (14–26 Hz), and gamma (>30 Hz) [13]. There are many existing research works speculating that the electroencephalographic activity or other electrophysiological measures

S. Chellappan et al. (Eds.): WASA 2018, LNCS 10874, pp. 683–695, 2018.
https://doi.org/10.1007/978-3-319-94268-1_56

of brain activities might be considered as a new non-muscular channel for sending messages [4,8,9,12,15].

One of the major challenges in EEG analysis and interpretation is that the EEG data is normally imbalanced, non-linear and unstable [1,5,7]. It is very difficult to capture stationary EEG data since any slight mental or physical activities would apparently be reflected in collected data. Most of the traditional linear feature selection approaches that are more used for stable data set may not be applicable for EEG data. On the other hand, EEG data is complex and intricate combination of multi-frequency signals and noises. Hence, appropriately filtrating valuable data and retrieving meaningful information become essential tasks.

In this paper, we propose an effective information processing approach to explore the association among EEG channels (electrodes) under different circumstances. Biologically, it is believed that some EEG data channels would be associatively active or inactive under certain motions. Besides, Fibromyalgia syndrome (FM) delta-theta oscillations [3,10,11] are considered to be related to body movements compared to standing-still. Hence, we setup four different experimental scenarios with the participants keeping their eyes open or closed while they are either moving or standing still. Totally, we use sixty-four EEG electrode channels to record the electrical activity of the brain. Then, we have sixty-four sequences of measurements in time order.

To discover the association between those time series data, we implement combinatorial approach to discover the potential connections among all channels. Specifically, we calculate the mutual entropy between two electrodes time series. Low entropy means two electrodes are highly associated. To see which group of electrodes are associated, we implement the hierarchical clustering tree (HC-tree) and data mechanics algorithm. The block pattern of the heat map shows mutual entropy could indeed capture the associations among electrodes. Using the combinatorial approach, we could successfully find the association between different channels. To better understand the collected data, we implement the weighted random forest algorithm to learn and classify the classes (experimental scenarios) of the EEG time series. Our evaluation results show that we could successfully classify the particular motions with given EEG data series.

The remainder of the paper is organized as follows: We introduce the preliminary in Sect. 2. We present our methodologies and show results to validate our findings in Sect. 3, we conclude this paper in Sect. 4.

2 Preliminary

In this section, we first introduce our data set and illustrate the data structure. We then provide the basic ideas of mutual conditional entropy and weighted random forest that we implement in data processing.

2.1 The Data

Our data set is four-dimension scalp EEG recording collected by Spatial Cognition Lab at University of California, Davis. In the experiment, we design four scenarios, denoted as tasks, for the participants. In the first task, participants always keep their eyes open, and switch between the statuses of moving and standing still. In the second task, participants always keep their eyes closed, and switch between moving and standing still. In the third task, participants keep moving, and switch between eyes open and closed. In the fourth task, participants always stand still, and switch between eyes open and close. In each task, there are beeps to remind participants to switch the status. During the experiment, participants are wearing a helmet recording the brain EEG signals. Each task repeats several times, denoted as trials. Figure 1 shows the experiment setting and our color code, which represents each task in the later classification results. Recall that EEG data is a combination of multi-frequency signals and noises. We dissociate the EEG signals in different frequencies and separately analyze the EEG time series in frequency domain.

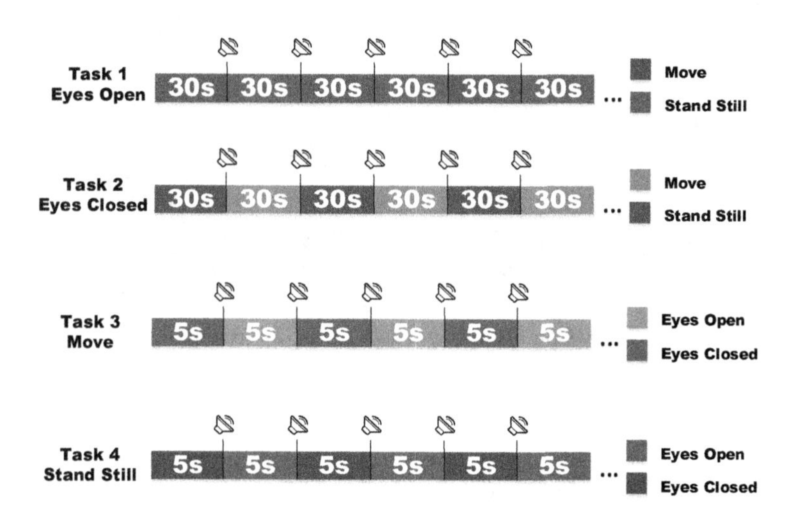

Fig. 1. EEG task setting (Color figure online)

Table 1 lists the overall data sets. Totally, we have 8 data sets with four dimensions. For example, the data set for task 1 (eyes open and move) is [64, 20, 13000, 8], where:

- 64 represents the 64 electrodes placed on the scalp.
- 20 is the frequency. As neurologists are particularly interested in delta-theta oscillations, our data contains 20 frequencies from 2 Hz to 30 Hz as in Table 2.
- 13000 is the recording time length. The first two tasks are 13000, and the other two are 1999.

Table 1. Data dimension

	Task setting	Number of trials	Time length	Data dimension
Task 1	Eyes open, move	8	13,000	[64, 20, 13000, 8]
Task 1	Eyes open, stand	8	13,000	[64, 20, 13000, 8]
Task 2	Eyes close, move	8	13,000	[64, 20, 13000, 8]
Task 2	Eyes close, stand	4	13,000	[64, 20, 13000, 4]
Task 3	Move, eyes open	24	1,999	[64, 20, 1999, 24]
Task 3	Move, eyes close	24	1,999	[64, 20, 1999, 24]
Task 4	Stand, eyes open	12	1,999	[64, 20, 1999, 12]
Task 4	Stand, eyes close	12	1,999	[64, 20, 1999, 12]

Table 2. Frequency (Hz)

1	2	3	4	5	6	7	8	9	10
2.0000	2.3064	2.6597	3.0671	3.5369	4.0788	4.7036	5.4241	6.2550	7.2132
11	12	13	14	15	16	17	18	19	20
8.3181	9.5923	11.0618	12.7563	14.7104	16.9638	19.5624	22.5591	26.0149	30.0000

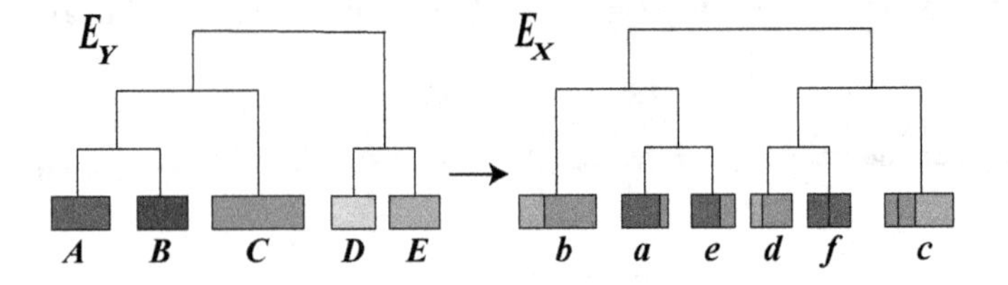

Fig. 2. Conditional entropy (Color figure online)

- 8 is the number of trials for each task. Note that EEG data itself is imbalanced and the neurologists in Spatial Cognition Lab already sift out the usable data. Hence, the number of trials for each task is different.

2.2 Combinatorial Approach for Synchrony

Consider two channels of EEG recording time series, say E_X and E_Y. We use hierarchical clustering algorithm and build a hierarchical clustering (HC) tree. As shown in Fig. 2, HC tree is cut at a certain height, then E_X and E_Y are partitioned into 5 and 6 clusters separately. We use alphabet $\{a, b, \ldots, f\}$ to denote 6 clusters for E_X and alphabet $\{A, B, \ldots, E\}$ to denote 5 clusters for E_Y.

Conditional Entropy $H[X|Y]$. Given two events X and Y taking values x_i and y_i respectively, then conditional entropy $H(X|Y)$ is calculated as:

$$H(X|Y) = -\sum_{i,j} p(x_i, y_j) log \frac{p(x_i, y_j)}{p(y_j)},$$

where $p(x_i, y_j)$ is the joint probability.

In our case, (E_X, E_Y) is bivariate categorical variable based on HC tree cut. Then, the conditional entropy $H[E_Y|E_X = a]$ is the Shannon entropy as illustrated via the color-coding in Fig. 2. Consequently, the expected value of conditional entropy of E_Y given E_X is calculated as:

$$H[E_Y|E_X] = w_a H[E_Y|E_X = a] + w_b H[E_Y|E_X = b] + ... + w_f H[E_Y|E_X = f].$$

where w_k with $k \in \{a, b, \dots, f\}$ is the proportion of E_X in cluster k. $H(E_Y|E_X)$ represents the amount of randomness in E_Y given E_X, and is a directed association measure between E_X and E_Y.

Mutual Conditional Entropy $\tilde{I}[Y \Leftrightarrow X]$. Given two events (X, Y), mutual conditional entropy is calculated as the following:

$$\tilde{I}[Y \Leftrightarrow X] = \frac{H[Y|X]}{H[Y]} + \frac{H[X|Y]}{H[X]},$$

where $H[Y]$ and $H[X]$ are Shannon entropies. $\tilde{I}[Y \Leftrightarrow X]$ measures the amount of uncertainty that cannot be mutually explained by X and Y. In another word, $\tilde{I}[Y \Leftrightarrow X]$ evaluates the association between X and Y.

In our case, $\tilde{I}[E_Y \Leftrightarrow E_X]$ represents the association between E_X and E_Y. Low value in mutual entropy means two series are highly associated. To see this, we consider the extreme case. When $\tilde{I}[E_Y \Leftrightarrow E_X]$ is 0, then $H(E_X|E_Y)$ and $H(E_Y|E_X)$ are both 0 since entropy is non-negative. $H(E_X|E_Y) = 0$ means that there is no randomness in E_X given E_Y, and same for $H(E_Y|E_X)$.

Mutual entropy measures the association between two events, either linear or non-linear, while correlation can only capture linear relationship between two events under normality. Thus, in this paper, we use mutual entropy to evaluate the association among EEG channels.

2.3 Weighted Random Forest

To better understand the collected data, we implement the weighted random forest algorithm to learn and classify the EEG time series. Random forest (RF) is an ensemble classifier that learns from tree bagging with each tree grown using randomly selected subset of features. In RF, a divide-and-conquer approach is used to improve performance, so that the classification (or regression) result is made by aggregating the results of all decision trees. More importantly, random forests is capable of dealing with large number of features, even more than observations. In this paper, the classification problem we have is to classify 100 trials using 2016 features.

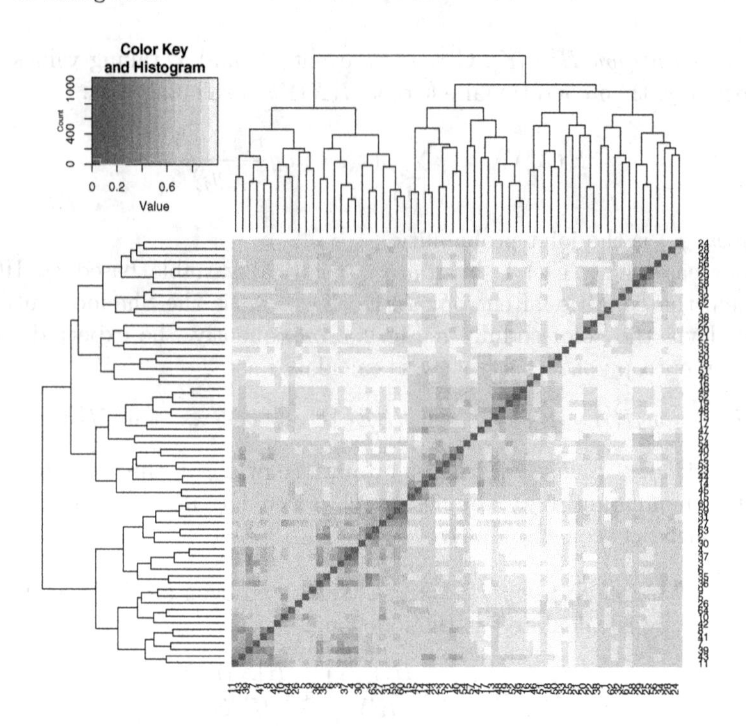

Fig. 3. Entropy matrix for task 1 stand trial 1 (Color figure online)

Our data is imbalance, like other classifier, RF would tend to focus more on the majority class causing low accuracy for minority class. To solve this problem, we implement the weighed random forest [2,16,18] to train the classifier. Specifically, we assign a weight to each class while the minority class is given a larger weight. The class weights are then incorporated into the RF algorithm. For each decision tree, Gini index is weighted using class weight. The final classification results are determined by aggregating the weighted results from each individual tree, where we consider weighting class 'votes' from each tree such that better performing trees are weighted more heavily.

3 Our Approach and Results

In this section, we first measure the associations among 64 electrode channels using mutual conditional entropy. Then, we develop a method to locate the most active channel groups and find the difference in channel activity patterns under different tasks. Based on the channel activity pattern difference, we implement weighted random forest to classify four tasks.

3.1 Electrodes Activity Pattern

Recall that brain waves can be observed from EEG recordings in the frequency domain, and the EEG is a combination of multi-frequency signals. We first dis-

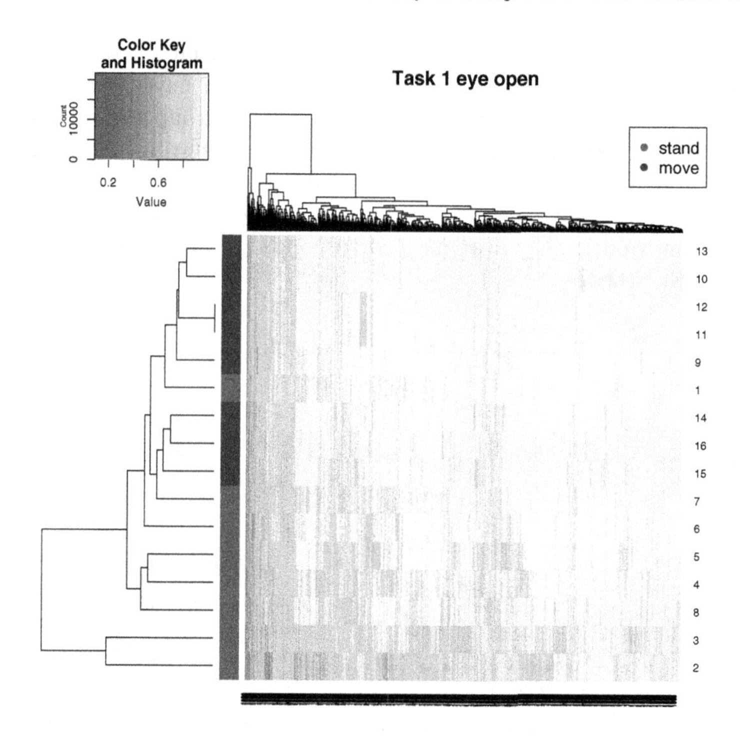

Fig. 4. Entropy matrix for task 1 with HC tree

sociate EEG data series into multiple data flows based on the frequency. Then, we analyze each data flow separately. To introduce the whole process, we use the EEG data at 20 Hz in task 1 to illustrate our method.

The data set for task 1 has 16 trials, and each trial contains 64 time series with length of 13,000. To measure the association between two electrodes, we compute the mutual conditional entropy between two electrodes time series. For each trial, we transform the original $64 * 13000$ data into a $64 * 64$ entropy matrix, which represents the association between two electrode channels. Figure 3 shows the heat map of entropy matrix with the color key indicating entropy value. Low entropy means that two electrodes are highly associated. To see which group of electrodes are associated, we arrange rows and columns based on hierarchical clustering tree (HC-tree). The block pattern of the heat map shows that there exists group of electrodes, which are associated with each other.

In order to discover the general electrode channel pattern for task 1, we need to compute and compare 16 mutual entropy matrices and find the common block pattern in the heat map. To make the comparison among 16 matrices easier, we vectorize each entropy matrix and stack them into one larger matrix. Since mutual entropy matrix is symmetric, we take the upper triangular part of matrix and vectorize it. Then, for each trial, the $64 * 64$ mutual entropy matrix is transformed into a vector with length of 2016. After that, we stack the 16 entropy vectors together, and get a $16 * 2016$ entropy matrix with each row representing one trial and elements being the mutual entropy between two electrodes. If there is a group of electrode pairs which have low entropy value among all trials, it would be the common electrodes group that are associated in task 1. To achieve that, we apply hierarchical clustering algorithm on the entropy matrix for task 1. Figure 4 shows the entropy matrix with hierarchical clustering tree (HC tree) superimposed on row and column axis. Notice that the block with low value in the heat map corresponds to a group of electrodes that are highly associated.

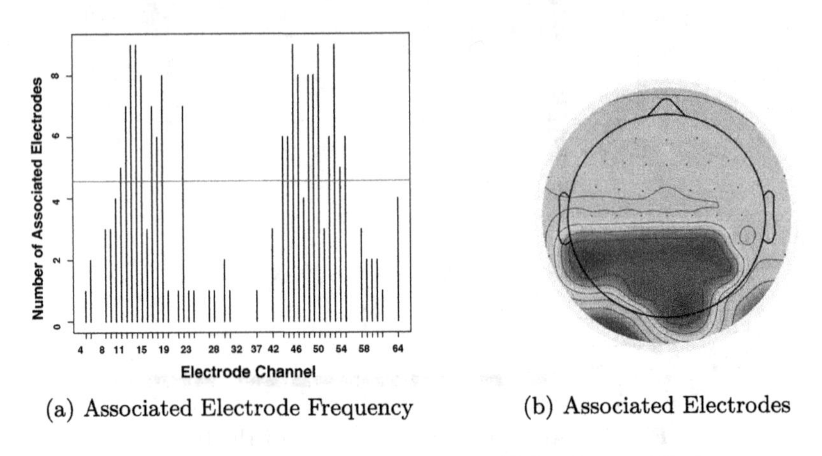

(a) Associated Electrode Frequency (b) Associated Electrodes

Fig. 5. Task 1 associated electrode (Color figure online)

To better show the block pattern, we implement data mechanics algorithm [6], which belongs to unsupervised learning strategy designed to bring out multi-scale block patterns of the target matrix. The algorithm aims to decreases the total variation of matrix by permuting row and columns, and uses iterative procedure to drive the variation further down until the block patterns converge. The number of iterations is usually between 2 to 5 depending on the size of target matrix. The detailed procedure of Data Mechanics algorithm is shown as the following:

Data Mechanics Algorithm for a $n * m$ matrix:

DM1: Denote $d_R^{(0)}$ as the Euclidean distance among m-dim row vectors. Build a hierarchical clustering tree $\mathcal{T}_R^{(0)}$ based on the row distance matrix.

DM2: Denote $d_C^{(0)}$ as the Euclidean distance among n-dim column vectors, $d_C^{(0)}$ is the initial column distance. Then, we add layers based on the row tree clusters. Choose $\mathcal{L}_R$ levels from row tree $\mathcal{T}_R^{(0)}$. Each tree level $i (i = 1, 2, \ldots, \mathcal{L}_R)$ corresponds to $G(i)$ clusters and each element in the matrix corresponds to one cluster $g(g = 1, 2, \ldots, G(i))$ at tree level i. The updated column distance $d_C^{(1)}(c_i, c_j)$ between column i and column j is defined as following:

$$d_C^{(1)}(c_i, c_j) = d_C^{(0)}(c_i, c_j) + \sum_{i=1}^{\mathcal{L}_R} \sum_{g=1}^{G(i)} (s(c_i|g, i) - s(c_j|g, i))^2$$

where $s(c_i|g, i)$ is the sum of components in column i that are in cluster $g(i)$ at tree level i.

DM3: Based on the updated column distance $d_C^{(1)}$, construct a distance matrix among all column vectors and build a HC tree $\mathcal{T}_C^{(1)}$ on column axis.

Dm4: Update row distance $d_R^{(0)}$ the same way as in [DM2] using new column tree $\mathcal{T}_C^{(1)}$. Based on the updated row distance $d_R^{(1)}$, construct the updated row tree $\mathcal{T}_R^{(1)}$.

DM5: Repeat step [DM2] to [DM4] until $(\mathcal{T}_R^{(k)}, \mathcal{T}_C^{(k)})$ pair sequence converges.

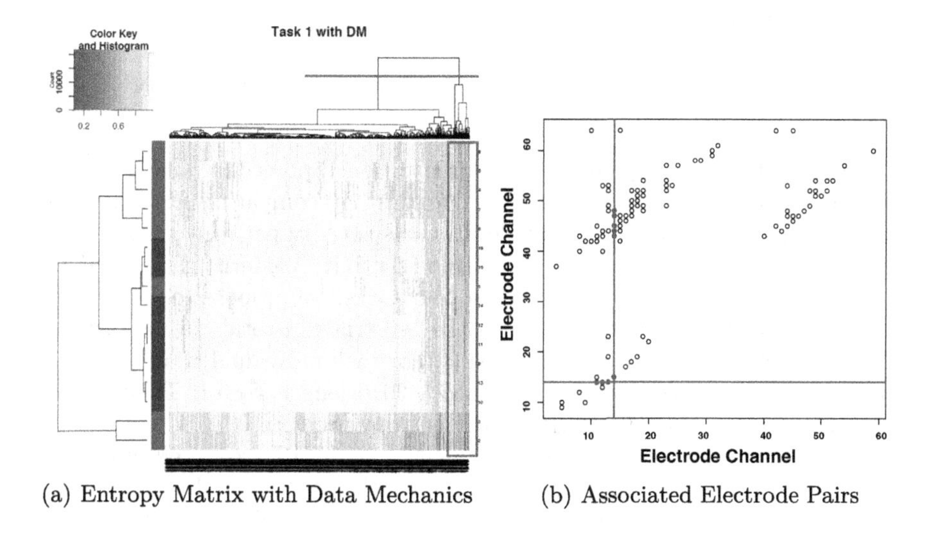

(a) Entropy Matrix with Data Mechanics (b) Associated Electrode Pairs

Fig. 6. Task 1 (Color figure online)

Figure 6 panel (a) shows the heat map of mutual entropy matrix of task 1 with updated row and column trees from data mechanics. As we can see from the figure, the block pattern in the heat map is more distinct using data mechanics. The block circled in red contains relatively low entropy value, meaning that the corresponding electrodes pairs in the column tree are highly associated across all 16 trials in task 1.

To identify the common associated electrodes group, we cut the column tree from data mechanics, and select the branch with low entropy value as in the red block, and then find out the corresponding electrodes pairs in the branch. Figure 6 panel (b) shows the highly associated electrode channel pairs. As we can see from the figure, some channels are associated to more than one channel, for example, Channel 14, which is marked in red, is associated with another 9 channels. In Fig. 5 panel (a), we summarize the number of associated electrodes for each single electrode. The mean value of number of associated electrodes is 4.45. Hence, we consider the electrodes with more than 4 associated channels as the highly active channels. To provide the intuitionistic expression, for channels which are associated to more than 4 channels, we mark them in red on electrode location map in Fig. 5 panel (b). These channels are actually part of occipital.

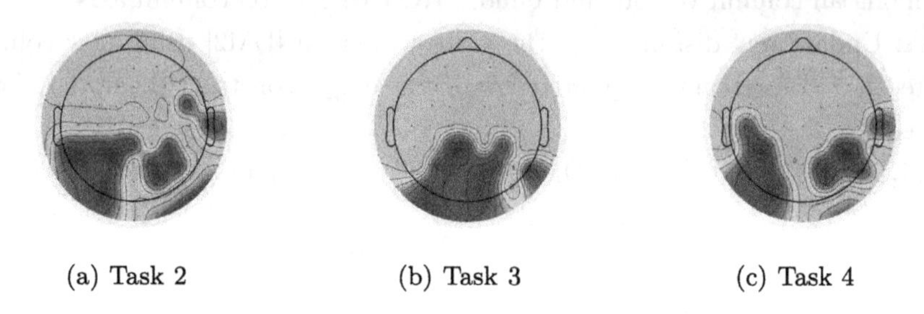

(a) Task 2 (b) Task 3 (c) Task 4

Fig. 7. Associated electrodes channels for task 2, task 3, task 4

For each task, we implement the aforementioned procedure and provide the location map of associated electrodes channels. By comparing Fig. 5 panel (b) and Fig. 7, we find that the electrode channels activity patterns are indeed different among four tasks. Since the channel activity patterns discovered from mutual entropy matrix are different for four tasks, we propose to classify trials based on their mutual entropy. There are 100 trials in total, 16 for task 1, 12 for task 2, 48 for task 3, and 24 for task 4. For each individual trial, the $64 * 64$ mutual entropy matrix is vectorized into a 2016 length vector. Then we stack 100 vectors and get a $100 * 2016$ overall entropy matrix.

Figure 8 shows the overall entropy matrix with HC tree superimposed on row and columns. The color bar indicates which task the trial belongs to. We select 4 Hz as a representative of low frequency and 30 Hz as a representative of high frequency. For low frequency like 4 Hz in Fig. 8 panel (a), the overall mutual entropy matrix shows a clear pattern difference between the first two tasks and the last two tasks. Nonetheless, the difference between task 3 and task 4 is hard to tell since most of channels are associated based on mutual entropy and the channel activity patterns look similar. Same for task 1 and task 2. For high frequency like 30 Hz, it is easier to see the difference in channel activity patterns for four tasks based on mutual entropy. As shown in panel (b), four tasks are well separated using hierarchical clustering algorithm, with some error in the bottom. To build a better classifier, especially for low frequency, we apply weighted random forest in next section.

3.2 Weighted Random Forest

The overall entropy matrix for 100 trials is used to differentiate four tasks, and the 2016 electrodes mutual entropy are used as features in the classification

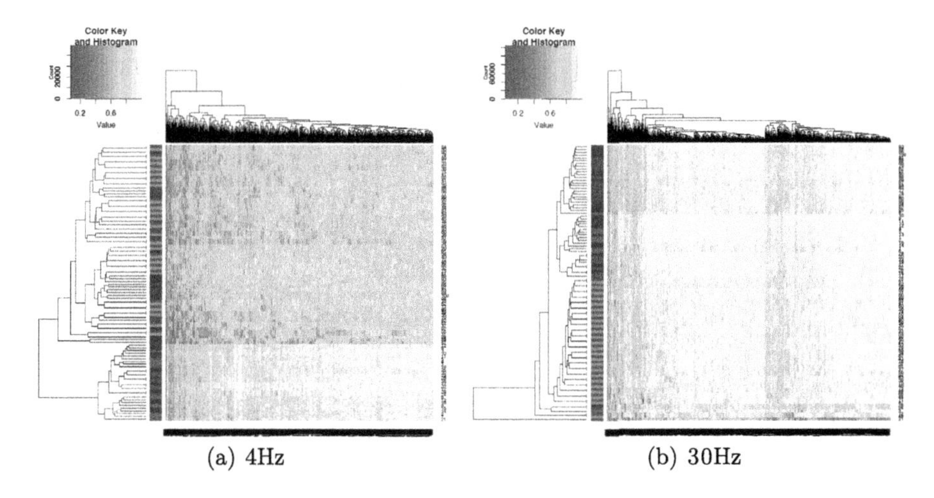

(a) 4Hz (b) 30Hz

Fig. 8. Mutual entropy matrix for 100 trials (Color figure online)

model. Hence, we have a high-dimensional spaces with limited data occurrences. A good method to solve this problem is to randomly select a subset of features and combine with ensemble learning method. Thus, random forest is our top choice. Since our data is imbalanced and random forest is sensitive to class imbalance, we implement weighted random forest instead to achieve better classification accuracy.

We first split the data, randomly choosing 80% data from each task as training set and the rest being test set. Note that we choose the 80% training ratio to demonstrate our classification process. To optimize the model parameters, we use grid search to train the weighted random forest that fit the training set as well as possible, then test the model performance on test set. To evaluate the performance, we use classification accuracy, which is the probability of correctly classifying the task that the trial belongs to and is the ratio of the number of trials correctly classified vs. the total number of trials. We also conduct the same analysis for all frequencies to compare the performances. Figure 9 shows the accuracy performances for all frequencies. As we can see from the figure, the curve appears as a growing trend when the frequency is higher. It is because that in low frequency, more electrodes are highly associated and difficult to distinguish them. Nonetheless, we still can achieve the accuracy as 85% to classify the trials in low frequencies.

4 Final Remarks

In this paper, we propose a new approach to process EEG data and discover the electrode channels activity patterns. In our approach, mutual conditional entropy is used to measure the association between two EEG time series. By vectorizing entropy matrix and stacking, $3D$ data for a certain frequency is transformed

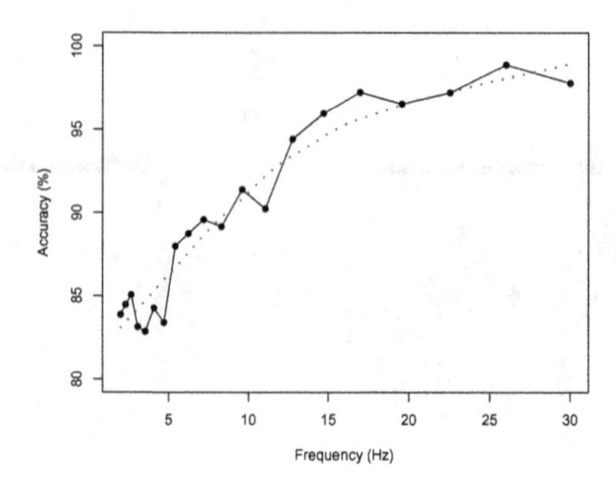

Fig. 9. Classification accuracy

into a 100 ∗ 2016 matrix. The association among 64 channels is further explored using data mechanics. The difference in channel association patterns enable us to train a weighted random forest classifier that gives 98.88% accuracy.

Our ongoing work is building the network among 64 electrode channels. In addition, the data that we analyze in this paper is pre-processed using independent component analysis (ICA) by Spatial Cognition Lab. In consideration of the complexity and originality of EEG data, the next phase of our work is to develop new methodology to effectively and efficiently extract and analyze the raw EEG data.

References

1. Acharya, U.R., Molinari, F., Sree, S.V., Chattopadhyay, S., Ng, K.H., Suri, J.S.: Automated diagnosis of epileptic EEG using entropies. Biomed. Signal Process. Control **7**(4), 401–408 (2012)
2. Booth, A., Gerding, E.H., McGroarty, F.: Performance-weighted ensembles of random forests for predicting price impact. Quant. Financ. **15**(11), 1823–1835 (2015)
3. Choi, S., Yu, E., Hwang, E., Llinás, R.R.: Pathophysiological implication of CaV3.1 T-type Ca2+ channels in trigeminal neuropathic pain. Proc. Nat. Acad. Sci. U.S.A. **113**(8), 2270–2275 (2016)
4. Delorme, A., Makeig, S.: EEGLAB: an open source toolbox for analysis of single-trial EEG dynamics including independent component analysis. J. Neurosci. Methods **134**(1), 9–21 (2004)
5. Esmaeili, V., Assareh, A., Shamsollahi, M.B., Moradi, M.H., Arefian, N.M.: Estimating the depth of anesthesia using fuzzy soft computation applied to EEG features. Intell. Data Anal. **12**(4), 393–407 (2008)
6. Fushing, H., Chen, C.: Data mechanics and coupling geometry on binary bipartite networks. PLoS ONE **9**(8), 1–11 (2014)
7. Gajic, D., Djurovic, Z., Gennaro, S.D., Gustafsson, F.: Classification of EEG signals for detection of epileptic seizures based on wavelets and statistical pattern recognition. Biomed. Eng.: Appl. Basis Commun. **26**(2), 1450021 (2014)

8. Guay, S., Beaumont, L.D., Drisdelle, B.L., Lina, J.M., Jolicoeur, P.: Electrophysiological impact of multiple concussions in asymptomatic athletes: a re-analysis based on alpha activity during a visual-spatial attention task. Neuropsychologia **108**, 42–49 (2018)
9. Holroyd, C.B., Coles, M.G.H.: The neural basis of human error processing: reinforcement learning, dopamine, and the error-related negativity. Psychol. Rev. **109**(4), 679–709 (2002)
10. Klimesch, W.: EEG alpha and theta oscillations reflect cognitive and memory performance: a review and analysis. Brain Res. Rev. **29**, 169–195 (1999)
11. Pandey, A.K., Kamarajan, C., Manz, N., Chorlian, D.B., Stimus, A., Porjesz, B.: Delta, theta, and alpha event-related oscillations in alcoholics during Go/NoGo task: neurocognitive deficits in execution, inhibition, and attention processing. Prog. Neuropsychopharmacol. Biol. Psychiatry **65**, 158–171 (2016)
12. Pfurtscheller, G., da Silva, F.L.: Event-related EEG/MEG synchronization and desynchronization: basic principles. Clin. Neurophysiol. **110**(11), 1842–1857 (1999)
13. Sanei, S., Chambers, J.: EEG Signal Processing, p. 1. Wiley, Hoboken (2007)
14. Niedermeyer, E., da Silva, F.H.L.: Electroencephalography: Basic Principles, Clinical Applications, and Related Fields. Lippincott Williams & Wilkins, Philadelphia (2005)
15. Vecchio, F., Di Iorio, R., Miraglia, F., Granata, G., Romanello, R., Bramanti, P., Rossini, P.M.: Transcranial direct current stimulation generates a transient increase of small-world in brain connectivity: an EEG graph theoretical analysis. Exp. Brain Res. **236**, 1117–1127 (2018)
16. Winham, S.J., Freimuth, R.R., Biernacka, J.M.: A weighted random forests approach to improve predictive performance. Stat. Anal. Data Min. **6**(6), 496–505 (2013)
17. Wolpaw, J.R., Birbaumer, N., McFarland, D.J., Pfurtscheller, G., Vaughan, T.M.: Brain-computer interfaces for communication and control. Clin. Neurophysiol. **113**(6), 767–791 (2002)
18. Xu, R.: Improvements to random forest methodology (2013)

A Privacy-Preserving Networked Hospitality Service with the Bitcoin Blockchain

Hengyu Zhou[1,2], Yukun Niu[1], Jianqing Liu[3], Chi Zhang[1(✉)], Lingbo Wei[1,2], and Yuguang Fang[3]

[1] School of Information Science and Technology, University of Science and Technology of China, Hefei 230027, People's Republic of China
chizhang@ustc.edu.cn
[2] State Key Laboratory of Information Security, Institute of Information Engineering, Chinese Academy of Sciences, Beijing 100093, People's Republic of China
[3] Department of Electrical and Computer Engineering, University of Florida, Gainesville, FL 32611, USA

Abstract. In recent years, we have witnessed a rise in the popularity of networked hospitality services (NHSs), an online marketplace for short-term peer-to-peer accommodations. Such systems, however, raise significant privacy concerns, because service providers such as Airbnb and 9flats can easily collect the precise and personal information of millions of participating hosts and guests through their centralized online platforms. In this paper, we propose PrivateNH, a privacy-enhancing and practical solution that offers anonymity and accountability for NHS users without relying on any trusted third party. PrivateNH leverages the recent progress of Bitcoin techniques such as Colored Coins and CoinShuffle to generate and maintain anonymous credentials for NHS participants. The credential holders (NHS hosts or guests) can then lease or rent short-term lodging and interact with the service provider in an anonymous and accountable manner. An anonymous and secure reputation system is also introduced to establish the trust between unfamiliar hosts and guests in a peer-to-peer fashion. The proposed scheme is compatible with the current Bitcoin blockchain system, and its effectiveness and feasibility in NHS scenario are also demonstrated by security analysis and performance evaluation.

Keywords: Networked hospitality services · Bitcoin blockchain
Anonymity and accountability

1 Introduction

Over the last few years, the popularity of networked hospitality services (NHSs), such as Airbnb and 9flats, has significantly increased, serving millions of users

S. Chellappan et al. (Eds.): WASA 2018, LNCS 10874, pp. 696–708, 2018.
https://doi.org/10.1007/978-3-319-94268-1_57

in hundreds of cities [1]. These services provide an efficient online marketplace where users can register themselves as hosts (to lease short-term lodging) and/or guests (to rent lodging); the service provider (SP) matches guest requests with available accommodations. In general, NHS can provide more diversified and personalized choices in accommodations at lower costs or with lower transactional overhead, and shows great advantages over traditional hotel industry. Moreover, the accountability provided by NHSs (e.g., identity verification mechanism and reputation system adopted by Airbnb) is a key feature that contributes to NHS's widespread acceptance, as it makes hosts and guests feel safer.

Despite the popularity, NHSs come with significant privacy concerns. To offer such services, SPs in NHSs collect the details of each lodging, together with real identities of the host and the guest. Note that other forms of accommodations, such as traditional hotels, also leak private information. However, with the help of centralized online platforms, data collection in NHSs is more efficient, aggressive and large-scale [2]. As a result, the SP, or any entity with access to this data, can infer privacy-sensitive information about hosts or guests, such as where they live, work, and socialize.

In this paper, we analyze the privacy threats in the current form of NHSs and propose PrivateNH, a practical solution that enhances privacy for the guests w.r.t. the SP and privacy for the hosts w.r.t. malicious outsiders, while preserving the convenience and functionality offered by the current system. PrivateNH relies on the recent progress of Bitcoin techniques such as Colored Coins [7] and CoinShuffle [8] and well-known cryptographic primitives like blind signatures [5] and private information retrieval [6]. We utilize the unmodified Bitcoin blockchain as the powerful platform to create and manage anonymous credentials for NHS participants without relying on any trusted third party. The credential holders (NHS hosts or guests) can then lease or rent short-term lodging and interact with the SP in an anonymous and accountable manner. An anonymous and secure reputation system is also introduced to establish the trust between unfamiliar hosts and guests in a peer-to-peer fashion.

In summary, our main contributions are:

- We present the first general privacy analysis of NHSs. By analyzing currently deployed NHSs, we formalize the security and privacy objectives of the next-generation NHSs.
- We propose PrivateNH, a practical system that offers enhanced privacy for hosts and guests, without affecting the convenience of these services. To facilitate adaption, PrivateNH relies exclusively on the unmodified Bitcoin blockchain system and some well-established cryptographic primitives.
- We analyze and evaluate PrivateNH, showing its effectiveness and feasibility in practical NHS scenario.

2 Preliminaries

2.1 Background on the Bitcoin Blockchain

Bitcoin is a peer-to-peer digital cash system that allows miners to mint coins called bitcoins and exchange them without authorized parties. Bitcoin uses a novel permissionless consensus protocol known as proof-of-work [4] to make all nodes agree on a log of transactions and to prevent attacks such as double-spending. This log is the Bitcoin blockchain and is managed by all nodes in the network [4,9].

The Bitcoin blockchain is an append-only public ledger which tracks all transactions in the system. A special set of participants, called miners, runs the proof-of-work protocol to extend the blockchain by appending newly generated block to the existing blockchain. A block consists of a block header and a set of transactions. The block header in a block contains a hash pointer to the previous block. The transactions in a block are hashed in a Merkle tree [4,9], and the tree's root hash is stored in the block header. Bitcoin Simplified Payment Verification (SPV) [4] is a method for verifying if particular transactions are included in a block without downloading the entire block. This method is used by some lightweight Bitcoin clients. The *blockchain* mentioned in this paper all refers to the Bitcoin blockchain.

A transaction consists of inputs and outputs. An output contains two fields: a value field which indicates the number of transferred bitcoins and a locking script that specifies what conditions must be fulfilled for those number of bitcoins to be further spent. An input contains two fields: an outpoint that references the previous output and an unlocking script to spend the bitcoins locked in the previous output. All unspent transaction outputs are called UTXO. For a valid transaction, the sum of the spent values in inputs should be greater than or equal to the sum of the values in outputs. The difference between these two sums is the mining fee for miners. The mining fee is optional, and a transaction creator can specify the amount of fee on their will.

Bitcoin allows embedding data in transactions through a particular kind of transaction output called OP_RETURN. One can specify up to 83 bytes of arbitrary data in an OP_RETURN output [3]. PrivateNH uses it to store application-specific data in the blockchain.

Due to the inherently public nature of the blockchain, users' privacy is severely restricted to linkable anonymity. Various mixing protocols have been introduced to mitigate this drawback. Ruffing et al. [8] proposed *CoinShuffle*, a fully decentralized Bitcoin mixing protocol that allows users to utilize Bitcoin in a truly anonymous manner. PrivateNH builds a credential-mixing method based on this protocol.

2.2 Cryptographic Primitives

A blind signature as introduced by Chaum [5] is a form of digital signature in which the signature requester blinds their message before sending it to the signer.

The blinded signature can, in turn, be "unblinded", to obtain a valid signature for the original message. PrivateNH uses the key property, which a signer who is asked to verify the signature of an unblinded message cannot relate this message back to the blinded version they signed.

A private information retrieval (PIR) as introduced by Benny Chor et al. [6] is a protocol that allows a user to retrieve an item from a server in possession of a database without revealing which item is retrieved. PrivateNH takes advantage of the PIR protocol to make the SP cannot link the reputation to a user when they retrieve their reputation.

3 Models, Assumptions, and Design Goals

3.1 System Model

An NHS includes three parties: hosts, guests, and the SP. The SP handles incoming querying requests from guests and matches guests with available accommodations based primarily on their locations and dates. The SP also offers essential functionalities such as accountability and reputation ratings.

3.2 Adversarial Assumptions

We assume the SP is honest-but-curious that strives to protect its business and maximize its interests. It has incentives to mining sensitive information about its guests, to either improve its quality of service or to monetize harvested data. We assume hosts are honest and will provide accurate accommodation information.

We assume hosts will not collude with the SP after the guest checks in. But some hosts may want to infer guests' identities during the online booking process.

We assume most guests want to protect their privacy. But some guests may collude with the SP to infer other guests' identities. We also assume that guests are rational, and do not misbehave if the cost of misbehaviors is significant.

We assume that the network and upper-layer protocols do not leak users identifiable information to the SP. In practice, users can use anonymous network systems (e.g., Tor) to conceal their IP addresses. We also assume that users can generate secure asymmetric key pairs and maintain the confidentiality of their secret keys.

We assume that cryptographic building blocks used in the underlying blockchain, the blind signature scheme, and the private information retrieval protocol are secure. We also assume the Bitcoin network is secure and robust.

The outsiders are active adversaries who try to collect hosts' and guests' private information and infer their identities.

3.3 Design Goals

This section describes the design goals of PrivateNH. That is, if PrivateNH satisfies these goals, it is robust against the adversarial assumptions described in Sect. 3.2.

(1) *Authentication:* Hosts and guests should be mutually authenticated to prevent criminals from participating in online booking. Together with the reputation mechanism, both the host and the guest can ensure that they are authenticated and trustful.
(2) *Guest anonymity:* The SP cannot infer guests' identities. During the online booking process, the host cannot infer their guest's identity. Moreover, Even the SP and some guests collude, they still cannot learn any knowledge about the identity of a particular guest.
(3) *Guest unlinkability:* The SP cannot tell whether two accommodations were booked by the same guest. This means the unlinkability of guests have to be preserved throughout the operations provided by the SP, including bookings and reputation ratings.
(4) *Accountability:* The SP can blacklist misbehaving users (e.g., a guest who damages a host's house). Blacklisted users are no longer able to join future online bookings.
(5) *Anonymous reputation:* It is computationally difficult for hosts and guests to misbehave during reputation ratings. It is computationally difficult for hosts and guests to show a tampered reputation without been discovered. It is computationally difficult for the SP to know whether two ratings are reviewed by the host and the guest in the same booking.
(6) *Efficiency:* All the above properties consume low communication and storage overhead.

4 PrivateNH Overview

As shown in Fig. 1, there are three types of participants in the system: the SP, hosts, and guests. We use *users* to indicate both hosts and guests in this paper.

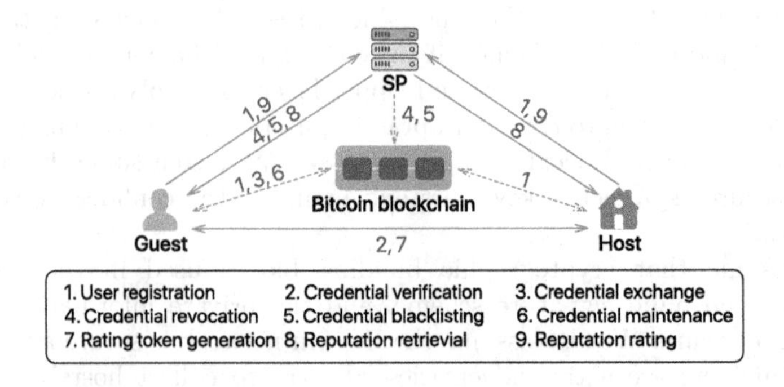

Fig. 1. System overview

A user registers themselves in the system and owns an anonymous credential which can be used to prove the validity of their identity. A credential is the hash of an ECDSA public key which can be deemed as a Bitcoin address. It can be verified using the signature signed by the corresponding secret key.

Hosts and guests can verify each other's credential without the SP. Due to the publicity of the blockchain, a host can verify whether a guest owns a valid credential by checking the blockchain, and vice versa for the guest.

Guests can exchange their credentials with the others using CoinShuffle protocol. Since CoinShuffle is a peer-to-peer protocol, there is no need for the SP to involve.

A guest can fetch their reputation from the SP using the PIR protocol and shows it to the host before the booking. Also, they can generate reputation token using the blind signature scheme to give a review for the host after the booking. These two operations are vice versa for the host.

5 PrivateNH Design and Implementation

5.1 User Registration

An individual can become a user and holds a credential by creating a registration transaction. As soon as the transaction is on the blockchain, the newly created credential becomes valid and everyone can verify it. To generate this transaction,

- First, an individual U sends a signing request $Sign(Addr_U)$ to the SP. In this step, the real identity of U is sending along with the request to the SP to resist the abuse of registrations;
- Next, after making sure the identity is valid and not been registered before, the SP signs $Addr_U$ and sends the result $Sig_{sk_{SP}}\{Addr_U\}$ back to U;
- After verifying the correctness of the signature, U generates a valid transaction and embeds the signature in it;
- Last, U broadcasts the transaction and waits for the Bitcoin network accept it.

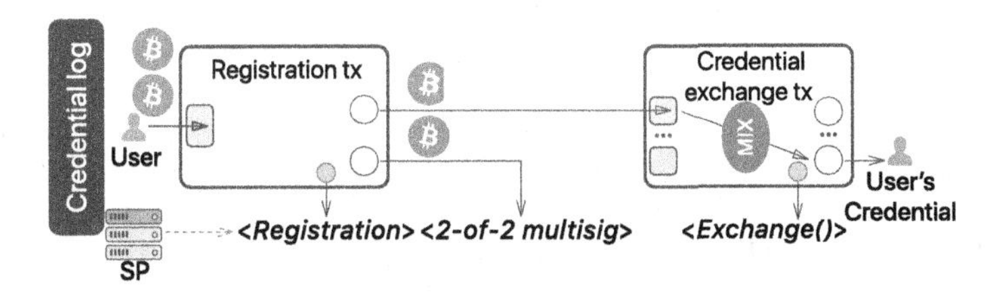

Fig. 2. Registration and credential log

The registration transaction contains three outputs. As shown in Fig. 2, they are represented in circles. We define outputs in order: the first is a genesis credential, the second is a deposit output, and the last is a registration output.

The genesis credential can be later transformed into a new and anonymous credential through *credential exchange protocol*. We discuss this in detail in Sect. 5.2.

The deposit output is a 2-of-2 multi-signature output where two signatures are required to unlock bitcoins in this output. One signature is signed by the SP, and the other is signed by the creator of the registration transaction. During the registration phase, a user transfers a fixed number of bitcoins to the output. This increases the cost for a user to misbehave. We discuss this in more detail in Sect. 5.4.

The registration output is an *OP_RETURN* output which contains a valid signature from the SP. The signature proves the user is authenticated.

5.2 Credential Exchange

Guests follow *credential exchange protocol* to get new credentials and invalidate their old ones. Our protocol works as follows. First, all participants need to verify each other. We discuss this in detail in Sect. 5.3. Then, each member provides one input address and one output address. The input is equivalent to their old credential, and the output will be their new credential. Next, all members use the decentralized mixing protocol *CoinShuffle* to construct a new transaction which randomized the mapping pattern between inputs and outputs provided by all members. Finally, all members broadcast the new transaction to the Bitcoin network.

When the transaction is added to the blockchain, every participant's new credential becomes valid, and the original one turns invalid automatically. Thanks to the mixing technique, nobody can tell which one is the corresponding old credential with a given new credential.

The values of all outputs must conform to a uniform distribution to make all outputs identical to observers. Suppose m guests engage in the protocol. The input value of the guest G_i is v_{in_i}, and the output value of G_i is v_{out_i}. The mining fee is $f(>=0)$. Then we have the following equations.

$$v_{out_1} = v_{out_2} = \ldots = v_{out_m} = k, \tag{1}$$

$$k = \frac{\sum_{i=1}^{m} v_{in_i} - f}{m}. \tag{2}$$

5.3 Credential Verification

Whether a guest requests an accommodation booking or participates in a credential exchange transaction, they are required to prove the validity of their credential. First, they use the private key to create a signature to show they own the credential. Then, they construct a proof to demonstrate the credential they hold is valid.

We introduce *verification path* to construct such a proof. As noted in Sect. 2, an input is always linked to one specific output of a preceding transaction. Through this, transactions become linked. By defining a mapping between the inputs and the outputs in the same transaction, a path is formed among the

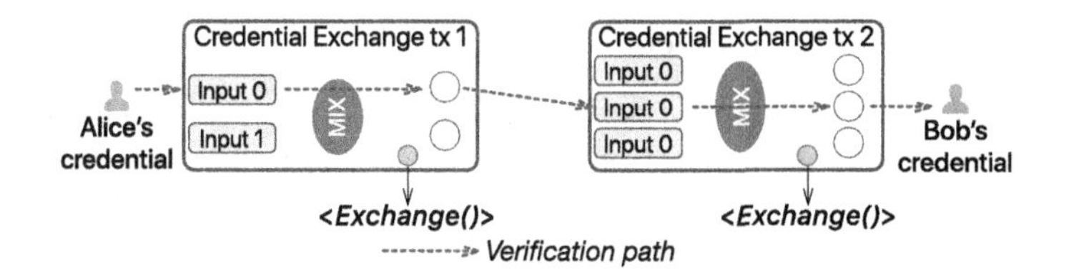

Fig. 3. Credential verification

transaction graph. In our scheme, we define the mapping with respect to the ordering of inputs and outputs. To clarify, $input_i$ is mapped to the $output_i$ within the same transaction. We specify the resulting path as the *verification path.* As depicted in Fig. 3, Bob's credential is validated using a path containing two credential exchange transactions. And the initial credential may be held by another guest, e.g., Alice.

Given a credential, a verifier can check whether there is a verification path within the blockchain by verifying:

- The output containing the credential is a UTXO;
- The credential is not on the blacklist (we discuss this in Sect. 5.4);
- There is a verification path between the credential and the checkpoint.

5.4 Credential Revocation

We introduce *credential revocation* to invalidate credentials. As shown in Fig. 4, the SP creates a blacklist transaction to invalidate credentials by embedding these credentials into the transaction's *OP_RETURN* output. A credential is revoked for two reasons: (1) Active leaving: A user leaves the system (e.g. delete the account), and (2) Passive leaving: The SP punishes a misbehaving user (e.g. blacklist the user).

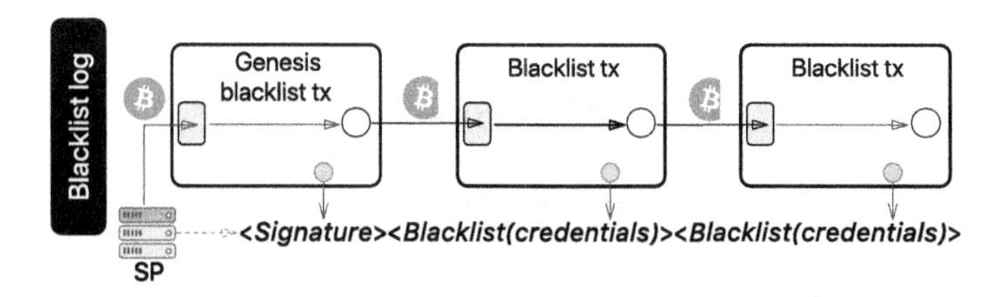

Fig. 4. Blacklist log

If a user leaves the system, they can reclaim the deposit that is locked in the registration transaction. Since the registration transaction is related to the user's identity, the SP can infer who is requesting the refunds. Thus for a passive leaving user (a.k.a. a misbehaving user), they can choose to either reclaim their deposit while revealing their identity to the SP or lost their deposit.

5.5 Credential Maintenance

If the value of a user's credential becomes too small for the user to involve in the next credential exchange transaction, the user can: (1) request the SP to revoke the credential and generate a new one, or (2) charge the credential.

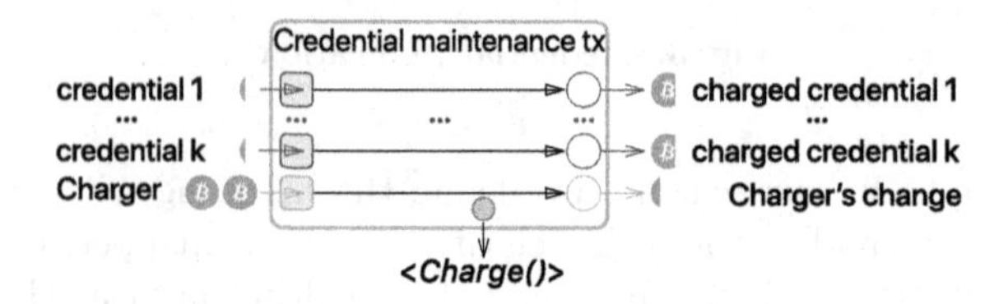

Fig. 5. Credential maintenance transaction

A user creates a credential maintenance transaction to charge their credential. As shown in Fig. 5, some users may create one credential maintenance transaction together to share the expense of the transaction fee. We define the same mapping pattern as in Sect. 5.3 in compliance with the verification path rule. The charger can be the user or a charge service provider.

5.6 Reputation Rating

We show the online booking workflow of PrivateNH in Fig. 6. We detail the steps which are the main concerns of the reputation rating.

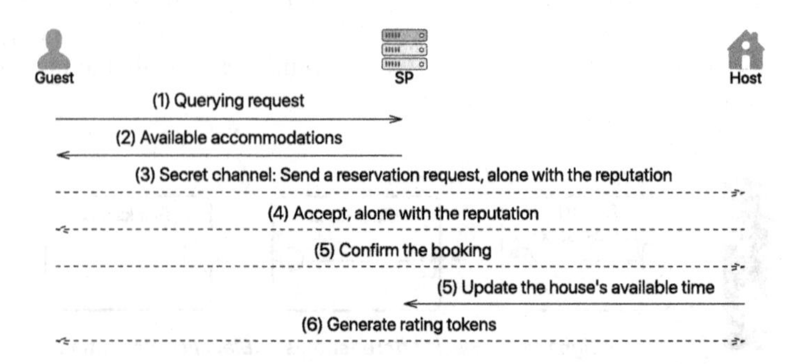

Fig. 6. PrivateNH's online booking workflow

In step 3 and 4, the guest and the host retrieves reputation from the SP using the PIR protocol without revealing which reputation they retrieve. The SP returns the reputation along with a signature $Sig_{SP}\{reputation\}$ which can be used to prove the completeness and correctness of the reputation. The SP can also return the signature with a timestamp, e.g., $Sig_{SP}\{reputation, timestamp\}$, to bring freshness to the reputation.

In step 6, the host H and the guest G use the blind signature scheme to generate rating tokens. The operations are as follows: (1) H and G generate a random value r_h and r_g respectively; (2) H sends the hash $c_h = hash(r_h)$ to G, and G sends the hash $c_g = hash(r_g)$ to H; (3) H requests a blind signature from the SP: $BlindSig_{SP}\{cert_g, sig_h\{c_g\}\}$, and G requests a blind signature from the SP: $BlindSig_{SP}\{cert_h, sig_g\{c_h\}\}$; (4) Both of them unblinds the signature to reveal the rating tokens: $RT_g = Sig_{SP}\{cert_g, sig_g\{c_h\}\}$ and $RT_h = Sig_{SP}\{cert_h, sig_h\{c_g\}\}$; and (5) G sends RT_g to H so that H can review G after the checking out, and vice versa.

The guest sends to the SP the rating token RT_h, the random value r_g used to generate the hash c_g, and a review of the host. When the SP receives this token, it checks the correctness of the signature, the correctness of the certificate $cert_h$, whether c_g has not been used before and whether c_g is the hash of r_g. And vice versa for the host. Note that, to avoid time-correlation attacks by the SP, the rating should not take place right after the checking out. This can be implemented by imposing some random delay before the rating.

6 Analysis and Evaluation

6.1 Authentication, Anonymity, Unlinkability, and Accountability

Authentication. During the authentication process, both the host and the guest have credentials, which can be validated by each other as described in Sect. 5.3. Specifically, the prover shows their credential has a verification path which originates from the credential in a registration transaction. The verification path cannot be tampered and is publicly verifiable. Since the underlying blockchain is a global append-only ledger and is assumed secure and robust, the verifier can believe the authenticity of the prover.

Anonymity. In this scheme, anonymity is enabled through credential exchange protocol. We suppose a genesis credential has a backward anonymity of 1. It is usually increased after every exchange since the backward anonymity set after an exchange is the union of all anonymity sets of the participating credentials. Given a credential exchange transaction with the set of participating credentials G and the anonymity sets A_g for all $g \in G$, the anonymity increases Δa_g for a $g \in G$ can be written as:

$$\Delta a_g = |\bigcup_{p \in G} A_p \backslash A_g|. \tag{3}$$

We have: (1) a user change their credential more frequently comes to a bigger anonymity set for themselves, and (2) more user participant in one exchange brings larger anonymity set.

Unlinkability. PrivateNH uses credential exchange protocol to make credentials unlinkable. Credential exchange protocol takes advantage of CoinShuffle. We define the *unlinkability probability* that quantifies the probability where the credential exchange protocol has at least two honest participants. It means that honest guests can get an unlinkable credential in such a probability after participating in credential exchange protocols. We assume that colluded guests are randomly selected to participate in the protocol. Let N denotes the number of all guests, m denotes the number of colluded guests, K_p denotes colluded guest rate. Then, $K_p = \frac{m}{N}$. Let n denotes the number of guests in every credential exchange protocol, r denotes the credential exchange rounds for a credential, P_r denotes the unlinkability probability after r exchanges. Then we have:

$$P_r = 1 - \frac{C_m^{n-1} C_{N-m}^1}{C_N^n} \times (\frac{C_m^{n-1} C_1^1}{C_{N-1}^{n-1} C_1^1})^{r-1}. \quad (4)$$

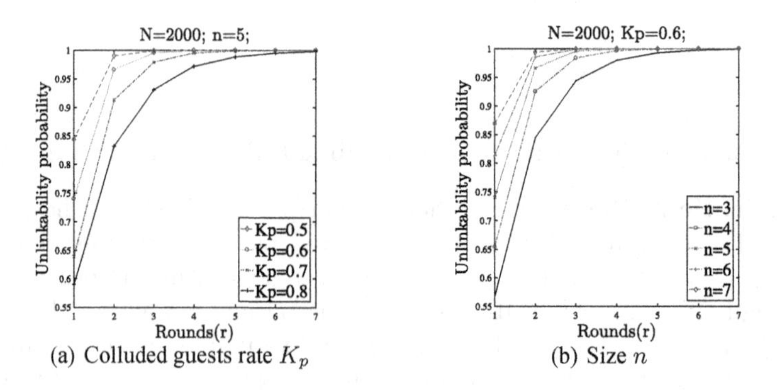

(a) Colluded guests rate K_p (b) Size n

Fig. 7. Parameter impact on the unlinkability probability

Figure 7a shows honest guests get high unlinkability through several credential exchanges even though most guests collude. Figure 7b shows the more guests in every round, the higher unlinkability probability for honest guests.

PrivateNH uses the blind signature schemes to achieve unlinkability for reputation ratings. In a reputation rating, the host and the guest *do not* provide their identifying information to the SP during the reputation rating. That is, the IP addresses and real credentials of the hosts and guests are invisible to the SP. This, together with the fact that there is no identity information included in the token, guarantees unlinkability between the host and the guest.

Accountability. PrivateNH achieves accountability without any trusted third party. First, the SP can revoke misbehaving credentials while it cannot link the credentials to users' identities. Second, users will refuse blacklisted credential holders participating in the credential exchange protocol since their new credentials may be the blacklisted ones after the protocol. Finally, the deposit raises the expense of misbehaviors. When users withdraw their deposits, their identities are disclosed and misbehaving credential holders will be identified.

6.2 Performance Evaluation

Communication Overhead. A user needs to download some system transactions to verify the others' credentials. Given n users, all users change their credentials after every booking (for example every one week), the maximum length of validation path is l, and users need to charge credentials after l changes. For the worst case, a user needs to download $n * (registration + deposit + l * maintenance)$ data in bytes. The size of different types of transactions is given in Table 1. If the system has 10000 users, and a credential exchanges 20 times before its charging. The maximum communication overhead for a user in 20 weeks is: $10000 * (351 + 333 + 224 * 20) = 51.64\,Mb = 6.455\,MB$. This is only 369 $bytes$ on average per day. And if there are too many users in the system, they do not need to download all the system created transactions. They can either download the transactions on demand or use a trust-but-verifiable server who only pushes data they need.

Table 1. Size evaluation of different system defined transactions

	Registration	Deposit	Credential exchange	Blacklist update	Credential maintenance
Size (bytes)	351	333	224	255	224

Storage Overhead. A user needs to store some public and private key pairs. Our system uses ECDSA with the secp256k1 curve. A pair of public and private key has $65+32 = 97\,bytes$. A user stores 2 pairs: one for withdrawing the deposit; another is for proving they own their latest credential. Also, a user stores all users' latest credentials, which is $(n-1) * output = (n-1) * 56\,bytes$. Given n = 10000, the storage overhead in total is $2 * 97 + 9999 * 56 = 0.56\,Mb = 0.07\,MB$. Like the discussion before, it is not necessary for a user to store all credentials of the others in the system. A user can store the other users' credentials on demand.

7 Conclusion

In this paper, we analyzed the privacy threats in the current form of NHSs. We also proposed PrivateNH, a practical solution that enhances privacy for the guests w.r.t. the SP and privacy for the hosts w.r.t. malicious outsiders, while preserving the convenience and functionality offered by the current system. The proposed PrivateNH is compatible with the current Bitcoin blockchain system, and its effectiveness and feasibility in NHS scenario are also demonstrated by the security analysis and performance evaluation.

Acknowledgments. This work was supported by the National Key Research and Development Program of China under grant 2017YFB0802202 and by the Natural Science Foundation of China (NSFC) under grant 61702474. The work of Y. Fang was partially supported by US National Science Foundation under grant IIS-1722791.

References

1. Airbnb about us. https://www.airbnb.com/about/about-us/. Accessed Mar 2018
2. Airbnb engineering & data science. https://medium.com/airbnb-engineering/tagged/data-science/. Accessed Mar 2018
3. Bartoletti, M., Pompianu, L.: An analysis of bitcoin OP_RETURN metadata. In: Brenner, M., et al. (eds.) FC 2017. LNCS, vol. 10323, pp. 218–230. Springer, Cham (2017). https://doi.org/10.1007/978-3-319-70278-0_14
4. Bonneau, J., Miller, A., Clark, J., Narayanan, A., Kroll, J.A., Felten, E.W.: SoK: research perspectives and challenges for bitcoin and cryptocurrencies. In: 2015 IEEE Symposium on Security and Privacy, pp. 104–121 (2015)
5. Chaum, D.: Blind signatures for untraceable payments. In: Chaum, D., Rivest, R.L., Sherman, A.T. (eds.) Advances in Cryptology, pp. 199–203. Springer, Boston (1983). https://doi.org/10.1007/978-1-4757-0602-4_18
6. Chor, B., Goldreich, O., Kushilevitz, E., Sudan, M.: Private information retrieval. In: Proceedings of IEEE 36th Annual Foundations of Computer Science, pp. 41–50 (1995)
7. Rosenfeld, M.: Overview of colored coins (2012). https://bitcoil.co.il/BitcoinX.pdf/. Accessed Mar 2018
8. Ruffing, T., Moreno-Sanchez, P., Kate, A.: CoinShuffle: practical decentralized coin mixing for bitcoin. In: Kutyłowski, M., Vaidya, J. (eds.) ESORICS 2014. LNCS, vol. 8713, pp. 345–364. Springer, Cham (2014). https://doi.org/10.1007/978-3-319-11212-1_20
9. Tschorsch, F., Scheuermann, B.: Bitcoin and beyond: a technical survey on decentralized digital currencies. IEEE Commun. Surv. Tutor. **18**(3), 2084–2123 (2016)

Trust-Distrust-Aware Point-of-Interest Recommendation in Location-Based Social Network

Jinghua Zhu(✉), Qian Ming, and Yong Liu(✉)

School of Computer Science and Technology,
Heilongjiang University, Harbin 150080, Heilongjiang Province, China
{zhujinghua,liuyong001}@hlju.edu.cn

Abstract. Point-of-Interest (POI) recommendation is an important personalized service in location-based social network (LBSN) which has wide applications. Traditional Collaborative Filtering methods suffer from cold-start and data sparsity problem. They also ignore connections among users and lose the opportunity to provide more accurate and personalized recommendations. In this paper, we propose a hybrid approach which incorporates user preference, geographic influence and social trust into POI recommendation system. In contrast to other trust-aware recommendation works, our approach exploits distrust links and investigates their propagation effects. We use a modified normalized Jaccard coefficient to measure the trust and distrust score. Several series of experiments are conducted and the results show that our approach perform better than the traditional Collaborative Filtering in terms of accuracy and user satisfaction.

Keywords: LBSN · POI recommendation · Collaborative filtering
Trust relationship

1 Introduction

With the rapid development of wireless networks, Web 2.0 technology and mobile smart devices, a number of location based social networking services, such as Gowalla, Foursquare and Facebook Places have become popular in recent ten years. Until June 2016, Foursquare has collected more than 8 billion check-ins and more than 65 million place shapes mapping businesses around the world; over 55 million people in the world use the service from Foursquare each month [1]. Social networks that include geographic information into shared contents are called Location Based Social Networks (LBSN). LBSN collects users' check-in information including locations and tips, and also allow users to make friends or share their information. Figure 1 demonstrates a typical LBSN, exhibiting the interactions between users and Point-of-Interest (POI), and the interactions (trust and distrust relation) among users. There are three kinds of relationships in LBSN: social relations among users, check-in behaviors between users and

S. Chellappan et al. (Eds.): WASA 2018, LNCS 10874, pp. 709–719, 2018.
https://doi.org/10.1007/978-3-319-94268-1_58

POIs, geographical links between POIs. The directed edges from users to POIs indicate the check-in behavior of users, the directed edges among users indicate their trust and distrust relationships. From Fig. 1, we can observe the following information: u_1 has visited POI l_1, l_2 and l_3; POI l_3 has been visited by u_1,u_3 and u_4; u_1 trust in u_3, u_3 trust in u_4, u_1 distrust in u_2 and u_4 distrust in u_2.

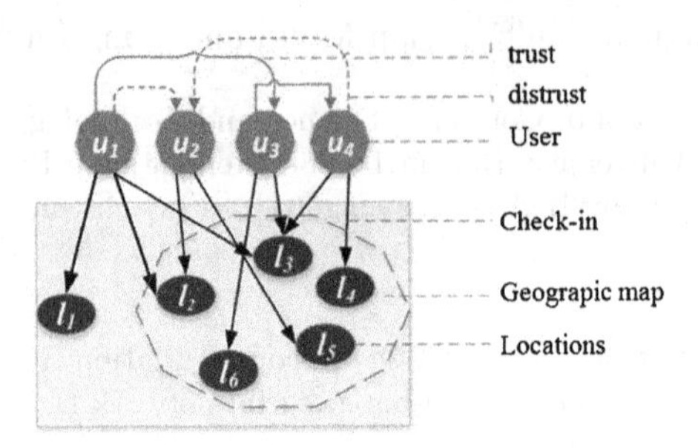

Fig. 1. Users trust-distrust relations and check-in activities in a LBSN

There are a lot of research problems in LBSN such as influence propagation [2–5], privacy protection [6,7] and POI recommendation [8–10]. In order to improve user experience in LBSN, POI recommendation is developed to suggest new locations to users by mining users' check-in records and social relations. POI recommendation is one of the most important services in LBSN since it can help users to find new potential interesting places. POI recommendation not only provides users with location-based service, but also brings benefits to advertisement agencies by publishing advertisements to the potential consumers. For example, users of Foursquare can visit the cinemas they are interested in, at the same time, the merchants are able to make the users to easily find them through POI recommendation. Because of the convenience to users and business benefits to merchants, facilitating POI recommendations in LBSN becomes a promising and interesting research problem.

POI recommendation in LBSN is a branch of recommendation system, thus we can make use of traditional recommendation methods, e.g., Collaborative Filtering (CF) technology by treating POIs as the "items" in many CF-based recommendation systems [8]. The intuitive idea is that users' preferences can be derived from other users who have similar check-in histories. Thus, user-based or item-based CF methods may be applicable to POI recommendations in LBSN. However, traditional CF-based recommendation methods suffer from several problems. First, the sparseness of the user check-in matrix seriously affects the recommendation accuracy. Second, traditional CF-based methods ignore the social relations among users. Third, traditional CF-based recommendation systems are easily under attacks from malicious recommendation. Hence, in order to provide more accurate and personalized recommendations to users, researchers start to study the social-aware and trust-aware recommendation techniques. Social friends tend to share common interests and thus can be used in POI recommendations. Several trust-aware methods have been proposed to address the data sparsity and recommendation accuracy problems. Most of the trust-aware recommendation methods only take consider of positive trust relations among users and ignore the important distrust relations among users. In contrast to trust relations that has been used in many studies, there are very few works that explore the explicit use of distrust relations in recommendation. In addition, due to the geographical characteristic of LBSN, the geographical influence between users and POIs are as important as the social influence amongst users, which may play a positive role for supporting POI recommendations in LBSN.

In this paper, aiming at providing more accurate and personalized recommendations, we propose a hybrid POI recommendation method that combines user preference, social trust-distrust and geographical influence for making better collaborative recommendations. We study the propagation characteristics of trust and distrust, give the calculation method of trust and distrust. And then we combine user similarity, geographical influence and trust relationship to build a hybrid POI recommendation framework. The experimental results show that the accuracy of the recommendation result that fusing trust-distrust relationship is much higher than that of the traditional user-similarity method, it not only can improve the credibility, but also can effectively resist the malicious recommendation.

2 Related Works

In this section, we review the POI recommendation techniques in LBSN from the following three aspects according to different influential factors: geographical influence, social influence and context-based recommendation. Because the check-in behavior depends on locations' geographical characteristics, some researchers explore geographical influence to improve POI recommendation. Ye et al. [11] employ a power law distribution model to capture the geographical influence and combine the geographical influence with collaborative filtering techniques for POI recommendation.

Based on the idea that friends often share common interests than non-friends, thus social influence can be used to improve POI recommendation. Wang et al. [12] propose a trust-based probabilistic recommendation model for social networks. They consider the attributes of products to determine the similarity among users. Forsati et al. [13] propose a matrix factorization based model for recommendation in social rating networks that incorporates both trust and distrust relationships to improve the quality of recommendations. Lee and Ma [14] exploit distrust information and investigate their propagation effects. They combine the k-nearest neighbors and the matrix factorization methods to maximize the advantages of both rating and trust information.

Compared with the check-in activity, the context information usually provide explicit preference to improve POI recommendation. Gao et al. [15] study the content information w.r.t POI properties, user interests, and sentiment indications in POI recommendation system.

3 Preliminary

3.1 Recommendation Based on User Similarity

Based on the observations that similar users usually have similar POI preferences, collaborative filtering techniques deduce the potential POI preferences of target users according to similar users' check-in behaviors. Let U and L denote user set and POI set in LBSN, $c_{i,j}$ is the check-in behavior of user $u_i \in U$ at POI $l_j \in L$. $c_{i,j} = 1$ represents u_i has checked at l_j and $c_{i,j} = 0$ means that there is no record of u_i at l_j. The check-in history of similar users can be used to discover users implicit preference of POI, which can be represented as probability to predict how likely the user would check-in at an unvisited POI. The check-in probability of u_i to l_j ($\widehat{c}_{i,j}$) can be computes in the following formula.

$$\widehat{c}_{i,j} = \frac{\sum_{u_k} w_{i,k} \cdot c_{k,j}}{\sum_{u_k} w_{i,k}} \tag{1}$$

where $w_{i,k}$ is the similarity weight between users u_i and u_k.

We use cosine similarity to calculate the weight between u_i and u_k denoted as $w_{i,k}$, as shown in the following formula.

$$w_{i,k} = \frac{\sum_{l_j \in L} c_{i,j} \cdot c_{k,j}}{\sqrt{\sum_{l_j \in L} c_{i,j}^2} \cdot \sqrt{\sum_{l_j \in L} c_{k,j}^2}} \tag{2}$$

3.2 Recommendation Based on Geographical Influence

Tobler first law of geography pointed out: "Everything is related to everything else, but near things are more related than distant things". It implies that people would like to visit nearby places close to their residential and work places. It also implies that people have high probability to visit the nearby locations of the locations they are interested in, even if those locations are far away from them.

In this paper, we explore collaborative recommendation method based on the naive Bayesian method to realize POI recommendation in LBSN. For a given user u_i and his visited POI set L_i, the likelihood of u_i check-in at all location in L_i is determined by the pair-wise distances of POI in L_i as follows.

$$Pr[L_i] = \prod_{l_m, l_n \in L_i \wedge m \neq n} Pr[d(l_m, l_n)] \tag{3}$$

where $d(l_m, l_n)$ denotes the distance between l_m and l_n, and $Pr[d(l_m, l_n)] = a \times d(l_m, l_n)^b$ follows the power law distribution model. We assume that the distances of POI pairs are independent. For a given candidate POI l_j, user u_i, and his visited POI set L_i, the likelihood for u_i to check in l_j as follows.

$$Pr[l_j | L_i] = \prod_{l_y \in L_i} Pr[d(l_j, l_y)] \tag{4}$$

4 Recommendation Based on Social Trust

Social trust network can be expressed as a directed graph $G(V, E)$, As shown in Fig. 2, each node in the network represents a user, and each edge represents the trust relationship between users.

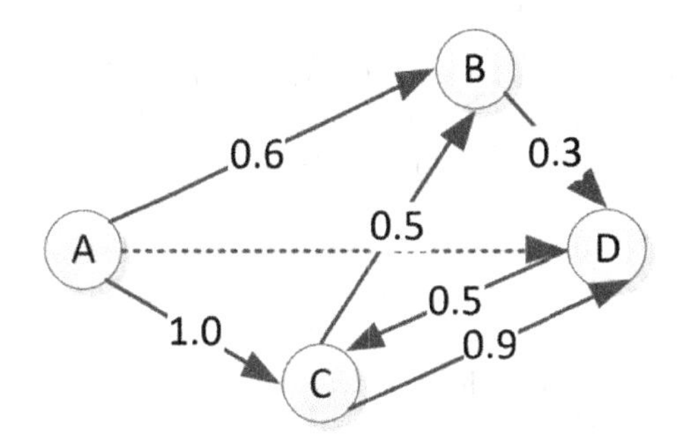

Fig. 2. User trust-distrust relation and check-in graph

Trust can be divided into two types, one is direct trust (solid line in Fig. 2, direct trust among users); the other is indirect trust (dotted line in Fig. 2, obtained through the propagation of direct trust). As shown in Fig. 2, user A has direct trust relationship with user B and user C respectively, and through certain propagation rules, we can calculate the indirect trust between user A and user D.

4.1 Local Trust Propagation

According to the scope of other users involved in the evaluation of trust relationship, trust can be divided into local trust and global trust. Local trust can reflect more personalized characteristics of trust. The calculation of local trust should be done between any two users, whereas global trust only need to be calculated once per user, so the calculation of local trust is more complex. From the aspect of system security, although both trust may be maliciously attacked, due to that the propagation of local trust is initiated by users themselves, the local trust attack decreases a lot.

Local trust has the property of transitivity, it is propagated from user to user within the same social network. Through trust propagation, we can find more implicit relationships among users to enrich information and can thus make more precise evaluations of users. In this paper, we include directed links into the equation of Jaccard coefficient to calculate the local trust and its propagation effect. Here, the trust value of a user u to the other user k without a direct link between them is described as:

$$w_{u,v} = max w_{u,k} \times \frac{|O(u) \cap O(v)|}{|O(u)|}, k \in O(u) \tag{5}$$

In the above equation, $O(u)$ is the set of direct neighbors of user u, that is those who user u trusts, and $O(v)$ is the set of neighbors with direct links connected to user v (those who trust user v). The function max provides the largest trust weight of all neighbors u trusts.

4.2 Local Distrust Propagation

In recent years, the research about trust in social network mainly focuses on trust and its propagation, there is relatively little research on the issue of distrust, especially lack the research that use distrust for recommendation. There are two main reasons: (1) Distrust information is not easy to get; (2) There is no consensus on how to deal with propagation issues of distrust and how to apply distrust to recommendation. However, the actual application system such as Epinions, Ebay and Taobao operation experiences show that the trust model that integrats distrust is an effective way to deal with user's dishonesty and malicious attacks.

The second type of local trust considered in our approach is the distrust relationship between users. Measuring distrust propagation involves the evaluation of both trust and distrust links. The complexity of measuring the distrust propagation dramatically grows along with the increase of the propagation path length, while the influence of trust decreases. Several relevant studies [9] show that length of 2 or 3 is suitable for aggregating propagation effect. Therefore, in our approach we choose to calculate the distrust propagation up to a maximum of two levels.

Based on the above analysis, to conduct the distrust propagation from user A to user C, we first select the qualified paths and then adopt the Jaccard coefficient to calculate the distrust value of each link as below:

$$\begin{aligned} W_{A,C} = & \min_{x \in O_{out}(A)} W_{A,x} \times \frac{|O_{out}(A^-) \cap O_{in}(x^+)|}{|O_{out}(A^-)|} \\ & - \max_{x \in O_{out}(A)} W_{A,x} \times \frac{|O_{out}(A^+) \cap O_{in}(x^+)|}{|O_{out}(A^+)|} \end{aligned} \tag{6}$$

In the above equation, the first term represents the distrust aggregation over all paths of the first type (a distrust link followed by a trust link); and the second term, trust over all paths of the second type (a trust link followed by a trust link). $O_{out}(A), O_{out}(A^+), O_{out}(A^-)$ denote the sets of direct neighbors, trust neighbors, and distrust neighbors of user A, respectively, and $O_{in}(x^+)$ is the set of users who trust user x. The function min in the first term provides the minimal value on the links of user A to his distrusted neighbors, and the function max in the second term provides the maximal value meaning to the highest trust relationship. Thus, only a negative $W_{A,C}$ enables a distrust propagation from A to C. The distrust propagation is performed accordingly.

In summary, under the influence of trust friends, the probability of user u_i check-in at l_j is calculated as follows:

$$\widehat{c}_{i,j} = \frac{\sum_{u_k \in F_i} W_{i,k} \cdot c_{k,j}}{\sum_{u_k \in F_i} W_{i,k}} \tag{7}$$

In the above equation, the set F_i is user i trust friends, $W_{i,k}$ is the trust value of u_i to u_k, $c_{k,j}$ is the check-in record of user k at the POI j.

5 Hybrid Collaborative Recommendation System

5.1 Hybrid Framework

As mentioned above, each of the factors such as user similarity, geographic influence, and trust relationship can be used for point-of-interest recommendation, so we can implement three recommendation system based on different factors, and then linearly integrated three recommended results to get the final sequence.

Let $S_{i,j}$ denote the check-in probability score of user u_i at POI l_j. Let $S^s_{i,j}$, $S^g_{i,j}$ and $S^t_{i,j}$ denote the check-in probability scores of user u_i at POI l_j, corresponding to recommender systems based on user similarity, geographical influence and trust, respectively. We have $S_{i,j}$ as follows.

$$S_{i,j} = (1 - \alpha - \beta) S^s_{i,j} + \alpha S^g_{i,j} + \beta S^t_{i,j} \tag{8}$$

where the two weighting parameters α and β ($0 \leq \alpha + \beta \leq 1$) denote the relative importance of geographical influence and trust relationship comparing to user similarity. Here $\alpha = 1$ states that $S_{i,j}$ depends completely on the prediction based on geographical influence; $\beta = 1$ states that $S_{i,j}$ depends completely on the predication based on trust relationship.

5.2 Check-in Probability Estimation

According to the above fusion framework, in order to estimate the check-in probability $S_{i,j}$, we need to predict the check-in probability of $S^s_{i,j}$, $S^g_{i,j}$ and $S^t_{i,j}$ corresponding to user similarity, geographical influence and trust. According to Eqs. (1), (4) and (7), we get the check-in probability $p^s_{i,j}$, $p^g_{i,j}$ and $p^t_{i,j}$ for a user u_i to visit POI l_j.

After we get the check-in probability estimation, we obtain the corresponding scores as follows.

$$S^s_{i,j} = \frac{p^s_{i,j}}{Z^s_i}, where\, Z^s_i = max_{l_j \in L - L_i} \{p^s_{i,j}\} \tag{9}$$

$$S^g_{i,j} = \frac{p^g_{i,j}}{Z^g_i}, where\, Z^g_i = max_{l_j \in L - L_i} \{p^g_{i,j}\} \tag{10}$$

$$S^t_{i,j} = \frac{p^t_{i,j}}{Z^t_i}, where\, Z^t_i = max_{l_j \in L - L_i} \{p^t_{i,j}\} \tag{11}$$

6 Performance Evaluation

6.1 Dataset Description

We conduct experiments on two real LBSN Foursquare and Gowalla. The dataset contains user information, check-in records, friendships and POI information. Foursquare contains 9800 users, 4626 POI and 45711 check-in records. Gowalla contains 3112 users, 3298 POI and 27149 check-in records. We randomly select 80% data as the training set, the remaining 20% used to test.

6.2 Performance Metrics

To evaluate the prediction accuracy, we are interested in finding out how many POI previously marked off in the preprocessing step recovered in the returned POI. More specifically, we examine two metrics: (1) the ratio of recovered POI to the recommended POI, and (2) the ratio of recovered POI to the set of POI deleted in preprocessing. The former is *precision@N* while the latter is *recall@N*, we test the performance when $N = 5,\ 10,\ 20$ with 5 as the default value.

6.3 Performance Comparison

Three factors, namely trust relationship (T), user similarity (S) and geographical influence (G), are incorporated in our unified collaborative recommendation system, denoted as TSG in our experiments. We compare the performance of TSG with the following three algorithms.

(1) User-based CF (U): This is traditional collaborative filtering recommendation algorithm based on user similarity [8].
(2) Geographic Distance (GD): This algorithm examines the importance of geographic factor [16]. The closer the POI is, the more probably it will be recommended to the user.
(3) User Similarity and Social trust (USG): This algorithm combines the traditional collaborative filtering technology with user similarity and social trust [14].

Next, we compare the effectiveness of the above algorithms. Figure 3 shows the performance of all approaches in terms of their best performance. The experiments used both Foursquare and Gowalla. The precision and recall are plotted in Fig. 3(a), (b), (c) and (d), respectively.

As can be seen from Fig. 3, the hybrid recommendation algorithm can get a better performance than the other two algorithms. And as the N increase, the accuracy of all algorithms will decrease. However, the TSG algorithm in this paper always shows the best recommend performance. This also verifies that trust-distrust relationship can resist malicious recommendation, and the accuracy and reliability of the recommendation are also improved.

Figure 4 compares two hybrid recommendation algorithms to verify the role of distrust relationship in recommendation result. TSG represents the recommendation algorithm only with trust relationship, TSG-all considers both trust

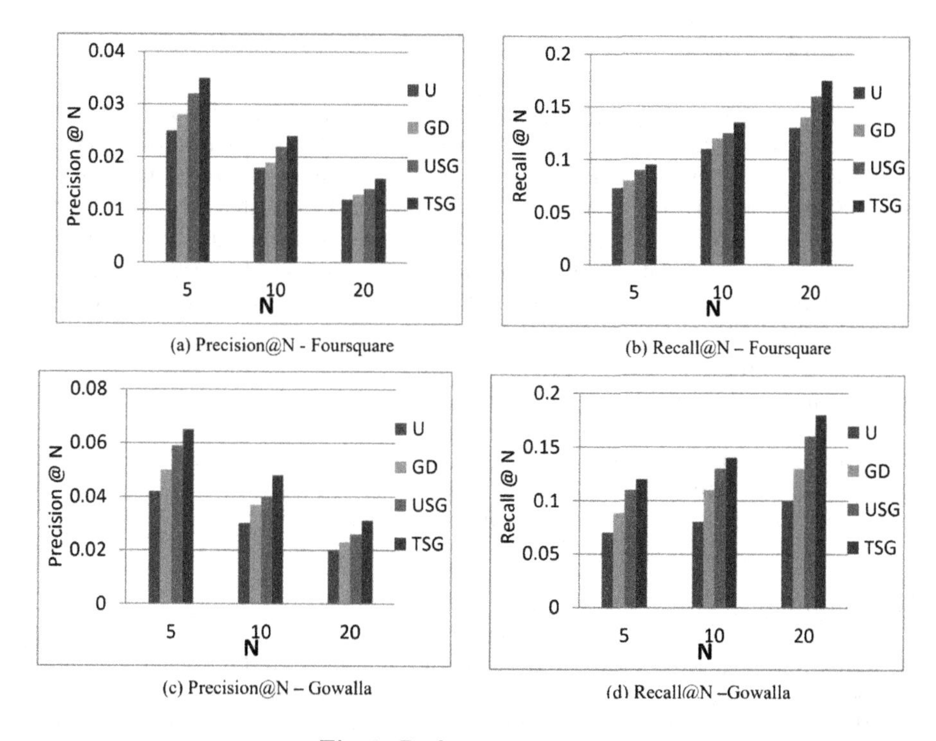

(a) Precision@N - Foursquare

(b) Recall@N – Foursquare

(c) Precision@N – Gowalla

(d) Recall@N –Gowalla

Fig. 3. Performance compare

and distrust relationship. It can be seen from Fig. 4 that the accuracy and recall of TSG-all are better than TSG. It is proved that the distrust relationship can improve the recommendation quality. To research and utilize the distrust relationship, we can discover more potential user information to improve recommendation performance.

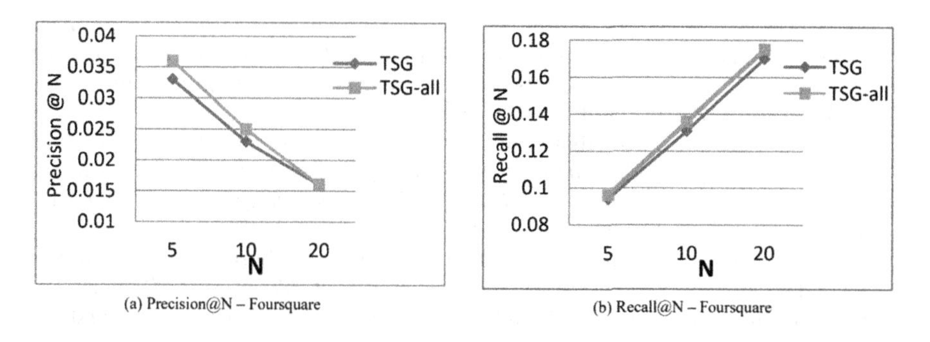

(a) Precision@N – Foursquare

(b) Recall@N – Foursquare

Fig. 4. The distrust influence on recommendation

7 Conclusion

In this paper, we focus on the problem of POI recommendation in location-based social network. We find that social influence among users can be refined into trust relationship and can be used to improve the performance of recommendation. We give the computation method of local trust and local distrust. Finally, we integrate user preference, social trust-distrust and geographical influence into a hybrid POI recommendation framework to make better collaborative recommendation. The experiments results on real location-based social network show that compared with the traditional user-based collaborative recommendation methods, our trust-distrust aware hybrid recommendation method can make more accurate and satisfying recommendation.

Acknowledgment. This work was supported in part by the National Science Foundation of China (61100048), the Natural Science Foundation of Heilongjiang Province (F2016034), the Education Department of Heilongjiang Province (12531498).

References

1. Zhao, S., King, I., Lyu, M.R.: A survey of point-of-interest recommendation in location-based social networks. Springer, New York (2016)
2. Zhu, J., Liu, Y., Yin, X., et al.: A new structure hole-based algorithm for influence maximization in large online social networks. IEEE Access, 1 (2017)
3. He, Z., Cai, Z., Wang, X.: Modeling propagation dynamics and developing optimized countermeasures for rumor spreading in online social networks. In: Proceedings of International Conference on Distributed Computing Systems, pp. 205–214. IEEE (2015)
4. Han, M., Li, Y., Li, J., et al.: Maximizing influence in sensed heterogenous social network with privacy preservation. Int. J. Sens. Netw. **1**(1), 1 (2017)
5. Li, J., Cai, Z., Yan, M., Li, Y.: Using crowdsourced data in location-based social networks to explore influence maximization. In: Proceedings of the 35th Annual IEEE International Conference on Computer Communications (INFOCOM 2016) (2016)
6. Li, J., Cai, Z., Wang, J., et al.: Truthful incentive mechanisms for geographical position conflicting mobile crowdsensing systems. IEEE Trans. Comput. Soc. Syst.
7. Han, M., Li, L., Xie, Y., et al.: Cognitive approach for location privacy protection. IEEE Access **PP**(99), 1 (2018)
8. Zhang, D., Xu, C.: A collaborative filtering recommendation system by unifying user similarity and item similarity. In: Wang, L., Jiang, J., Lu, J., Hong, L., Liu, B. (eds.) WAIM 2011. LNCS, vol. 7142, pp. 175–184. Springer, Heidelberg (2012). https://doi.org/10.1007/978-3-642-28635-3_17
9. Guo, G., Zhang, J., Thalmann, D.: Merging trust in collaborative filtering to alleviate data sparsity and cold start. Knowl. Based Syst. **57**(2), 57–68 (2014)
10. Guo, G.: Integrating trust and similarity to ameliorate the data sparsity and cold start for recommender systems. In: ACM Conference on Recommender Systems, pp. 451–454 (2013)

11. Ye, M., Yin, P., Lee, W.-C., Lee, D.-L.: Exploiting geographical influence for collaborative point-of-interest recommendation. In: Proceedings of the 34th International ACM SIGIR Conference on Research and Development in Information Retrieval, pp. 325–334 (2011)
12. Wang, Y., Yin, G., Cai, Z., et al.: A trust-based probabilistic recommendation model for social networks. J. Netw. Comput. Appl. **55**, 59–67 (2015)
13. Forsati, R., Mahdavi, M., Shamsfard, M., et al.: Matrix factorization with explicit trust and distrust side information for improved social recommendation. ACM Trans. Inf. Syst. **32**(4), 17 (2014)
14. Lee, W.P., Ma, C.Y.: Enhancing collaborative recommendation performance by combining user preference and trust-distrust propagation in social networks. Knowl. Based Syst. **106**, 125–134 (2016)
15. Gao, H., Tang, J., Hu, X., Liu, H.: Content-aware point of interest recommendation on location-based social networks. In: Proceedings of the Twenty-Ninth AAAI Conference on Artificial Intelligence, pp. 1721–1727. AAAI Press (2015)
16. Lian, D., Zhao, C., Xie, X., Sun, G., Chen, E., Rui, Y.: GeoMF: joint geographical modeling and matrix factorization for point-of-interest recommendation. In: ACM SIGKDD International Conference on Knowledge Discovery and Data Mining, pp. 831–840 (2014)

Retrieving the Relative Kernel Dataset from Big Sensory Data for Continuous Query

Tongxin Zhu[1], Jinbao Wang[2](✉), Siyao Cheng[1], Yingshu Li[3], and Jianzhong Li[1]

[1] School of Computer Science and Tech, Harbin Institute of Technology, Harbin, China
zhutongxinhit@126.com, {csy,lijzh}@hit.edu.cn

[2] The Academy of Fundamental and Interdisciplinary Sciences, Harbin Institute of Technology, Harbin, China
wangjinbao@hit.edu.cn

[3] Department of Computer Science, Georgia State University, Atlanta, GA, USA
yili@gsu.edu

Abstract. With the rapid development of Wireless Sensor Networks (WSNs), the amount of sensory data manifests an explosive growth. Currently, the sensory data generated by some WSNs is more than terabytes or petabytes, which has already exceeded the computation and transmission abilities of a WSN. Fortunately, the volume of valuable data for a given query is usually small. For a given query Q, the dataset which is highly related to it is called the relative kernel dataset $\mathcal{K}^Q$ of Q. In this paper, we study the problem of retrieving relative kernel dataset from big sensory data for continuous queries. The theoretical analysis and simulation results show that our proposed algorithms have high performance in term of accuracy and resource consumption.

Keywords: Wireless Sensor Networks · Big sensory data
Relative kernel dataset

1 Introduction

The Wireless Sensor Networks (WSNs) provide an efficient way to observe the complicated physical world. Benefiting from the wireless telecommunications, embedded systems and sensing techniques, the WSNs have rapidly developed and are widely used. According to the annual report of Gartner, 20.8 billion connected things will be in use worldwide in 2020 [1]. Besides, it is estimated that more than 250 things will connect each second by 2020, and more than 50 billion things will be connected to the Internet by 2020 [2]. Such large scale of WSNs causes the explosive growth of sensory data.

S. Chellappan et al. (Eds.): WASA 2018, LNCS 10874, pp. 720–732, 2018.
https://doi.org/10.1007/978-3-319-94268-1_59

As an important instance of Big Data, Big Sensory Data also has four V's characters. (1) **Volume**. The amount of sensory data in some WSNs has already exceeded terabyte or petabyte [1,2], so that the volume of Big Sensory Data is extremely huge. (2) **Variety**. As the applications of WSNs are becoming more and more complex, different types of sensors are applied in one WSN, so that the big sensory dataset contains a huge variety of sensory data. (3) **Velocity**. In consequence of the high sampling frequency of the large amount of sensors in a WSN, the generating velocity of big sensory data is quite fast. (4) **Value**. The existence of noises and redundancies in big sensory data diminishes its quality. Although the total value brought by the big sensory data is high, the value-scale ratio is quite low.

The above properties bring many challenges for data processing in WSNs, and make the existing sensory data acquisition, routing [3–6], data collection [7–9] and data computation [10,11] techniques no longer applicable. Taking the *Volume* property as an example, the current amount of data has already exceeded the transmission and computation ability of WSNs. Therefore, a series of new in-network processing algorithms with much lighter transmission and computation overloads should be considered.

Fortunately, the volume of valuable data for a given query Q is usually small although the volume of all sensory data in the WSN is quite huge. As the function of a WSN becomes complex, a WSN containing with n different types of sensors can support a variety of queries. Therefore, a given query Q is usually highly related to the sensory data generated by only k types of sensors, where $k < n$. The sensory data generated by that k types of sensors is denoted as the relative kernel dataset $\mathcal{K}^Q$ of Q.

Most WSNs support continuous queries since the function of the WSNs is to monitor the physical world in real time. Therefore, continuous queries are frequent in a WSN. The energy consumption of the WSNs can be reduced a lot if the energy consumption for processing the continuous queries is reduced. Processing continuous queries with the relative kernel datasets are energy efficient in both communication and computation in consequence of their quite smaller amount of sensory data compared with the raw big sensory data.

Although there exists some data reduction algorithms for reducing the amount of sensory data, they are not suitable for processing a given continuous query energy efficiently. Firstly, the simplest data reduction methods are based on sampling [12,13]. However, the sampling methods are only suitable for simple statistic queries. For other queries, the valuable data of the given query may not be sampled due to the limitation of sampling frequency. On the other hand, the sampled data may not be necessary for the given query. Secondly, the compressed sensing technique is another classical method for in-network data reduction [14,15]. However, the compressed sensing technique only considers the correlations between sensory data while ignoring the correlation between the sensory data and the given query. Then, the data reduction for a given query is not enough. Thirdly, to the best of our knowledge, the work proposed by Cheng et al. is the only one that discussed the big sensory data processing problem up to now [16,17]. However, the dominant dataset defined in [16] is completely irrelevant to the given query. That is, all queries in the WSN share one dominant dataset, which is not energy efficient enough for a given query.

Due to the above reasons, we investigate the relative kernel dataset retrieving problem for continuous queries in this paper. Two algorithms, the Relative Kernel Dataset Retrieving (RKDR) Algorithm and the Piecewise Linear Fitting Based Relative Kernel Dataset Retrieving (PLF-RKDR) Algorithm, are proposed to solve this problem. The RKDR algorithm retrieves the relative kernel dataset for a given query by the linear correlation analysis. Then a method for answering the query by the relative kernel dataset is given. The PLF-RKDR algorithm improves the accuracy of the retrieved relative kernel dataset for a given query by applying the piecewise linear fitting to retrieve the relative kernel dataset for a given query. Besides, we also provide a method to answer the query approximately by the retrieved relative kernel dataset. The major contributions of this paper are as follows.

1. The formal definition of the relative kernel dataset for a given query is firstly proposed, considering the redundancies between different types of sensory data and the correlations between sensory data and the given query.
2. Two approximate algorithms, the RKDR algorithm and the PLF-RKDR algorithm, are proposed to retrieve the relative kernel dataset for a given query.
3. Extensive simulations on both simulation datasets and real datasets are carried out to verify the accuracies of the algorithms.

The rest of the paper is organized as follows. Section 2 provides the problem definition. Section 3 proposes the relative kernel dataset retrieving algorithm for a given query Q and provides a method to estimate the result of Q by the sensory data in that relative kernel dataset. Section 4 analyzes the performance of the proposed algorithms. Section 5 shows the simulation results. Section 6 discusses the related work. Finally, Sect. 7 concludes the whole paper.

2 Problem Definition

2.1 The Wireless Sensor Networks Model

A wireless sensor work has n categories of sensors, which generates n types of sensory data from the monitored environment. These n types of sensory data are described as n attributes of the monitored environment, denoted as $x_1, x_2, \cdots, x_n$. The attribute set is denoted as $A = \{x_1, x_2, \cdots, x_n\}$. For each given query Q from users, the WSN returns a query result y_Q according to the sensory data. y_Q is described as the target value of query Q. Apparently, the target value y_Q is correlated with partial or all n attributes. Taking the atmospheric environmental monitoring WSN as an example, there are a variety of sensors collecting data of sulfur dioxide concentration, nitrogen dioxide concentration, *etc*, which are attributes of this WSN. When the user's query is the air quality, the WSN returns an air pollution index. In this application, the query Q is the air quality and the target value y_Q is the returned air pollution index.

2.2 The Correlation Model

A training set is applied to explore the correlations between n attributes and the given query Q. A training set contains m training examples $\{t_1, t_2, \cdots, t_m\}$. A training example $t_j (1 \leq j \leq m)$ is presented by the values of n attributes and the corresponding target value of query Q, i.e. $\{x_{1j}, \cdots, x_{nj}, y_{Qj}\}$.

Linear correlation coefficient is a common metric for the correlation analysis, especially in WSNs. In most applications of WSNs, the sensory data can reflect the statement of the monitored physical world intuitively. That is, the simple linear correlation is usually adequate to reflect the relationship between the sensory data and the query of a WSN. Therefore, we apply linear correlation as the correlation metric in this paper. For each attribute $x_i (1 \leq i \leq n)$, the linear correlation R_i^Q between x_i and y_Q is calculated by the following formula,

$$R_i^Q = \frac{\sum_{j=1}^{m}(x_{ij} - \overline{x_i})(y_{Qj} - \overline{y_Q})}{\sqrt{\sum_{j=1}^{m}(x_{ij} - \overline{x_i})^2}\sqrt{\sum_{j=1}^{m}(y_{Qj} - \overline{y_Q})^2}} \tag{1}$$

where m is size of the training set, and x_{ij} denotes the value of attribute x_i in the jth training example t_j, y_j denotes the target value in t_j. Besides, $\overline{x_i}$ is the average value of attribute x_i in the training set, and $\overline{y}$ is the average value of target values in the training set. R_i^Q has a value between 1 and -1, where 1 is total positive linear correlation, 0 is no linear correlation, and -1 is total negative linear correlation. That is, the greater absolute value of R_i^Q presents the stronger linear correlation between attribute x_i and query Q.

2.3 Problem Statement

In this paper, we aimed at retrieving the relative kernel dataset $\mathcal{K}^Q$ for a given query Q of a WSN. The relative kernel dataset $\mathcal{K}^Q$ is a subset of attribute set A. The definition of the relative kernel dataset $\mathcal{K}^Q$ is described as follows.

Definition 1 *(β-compatible). For any $0 < \beta < 1$, two attributes x_i and x_j is β-compatible if $|r_{ij}| \leq \beta$, where r_{ij} is the linear correlation coefficient between x_i and x_j defined as the following equation.*

$$r_{ij} = \frac{\sum_{k=1}^{m}(x_{ik} - \overline{x_i})(x_{jk} - \overline{x_j})}{\sqrt{\sum_{k=1}^{m}(x_{ik} - \overline{x_i})^2}\sqrt{\sum_{k=1}^{m}(x_{jk} - \overline{x_j})^2}} \tag{2}$$

Definition 2 *(β-compatible set). For any $0 < \beta < 1$, attribute set $S \subseteq A$ is a β-compatible set if any two attributes in S are β-compatible with each other.*

Definition 3 *(weight of set). For any attribute set $S \subseteq A$, the weight of S, denoted as $w(S)$, is defined as $w(S) = \sum_{x_i \in S} |R_i^Q|$, where R_i^Q is the correlation coefficient between attribute x_i and query Q.*

Definition 4 *((k,β)-Relative Kernel Dataset $\mathcal{K}^Q$). Given the attribute set A, query Q, the compatible parameter β, and the size of the relative kernel dataset k. The (k,β)-Relative Kernel Dataset of query Q is a subset of A, denoted as $\mathcal{K}^Q$, satisfying the following three conditions.*

(1) $\mathcal{K}^Q$ is a β-compatible subset of attribute set A,

(2) the size of $\mathcal{K}^Q$ is k, and

(3) $w(\mathcal{K}^Q) > w(S)$ for any set attribute set S satisfying the conditions (1) (2).

The sensory data in (k,β)-Relative Kernel Dataset $\mathcal{K}^Q$ is transmitted and computed each time query Q is issued by users. The smaller k indicates that less sensory data is transmitted and computed in a WSN, which saves more energy. However, the approximate result of query Q is estimated by the sensory data in $\mathcal{K}^Q$. Therefore, the smaller k will make the approximate answer less accurate.

Problem Statement. We formulate the **R**elative **K**ernel **D**ataset **R**etrieving (RKDR) problem in a WSN, as follows.

Input:
1. A query Q;
2. A training set with m training examples;
3. The required size of relative kernel dataset k;
4. The compatible parameter β.

Output:

1. The (k,β)-Relative Kernel Dataset $\mathcal{K}^Q$ of query Q;

3 Algorithms for the RKDR Problem

In this section, two algorithms, the Relative Kernel Dataset Retrieving (RKDR) Algorithm and the Piecewise Linear Fitting Based Relative Kernel Dataset Retrieving (PLF-RKDR) Algorithm, are proposed to retrieve the relative kernel dataset $\mathcal{K}^Q$ for a given query Q. We also provide two approximate methods to estimate the result of query Q by the sensory data in $\mathcal{K}^Q$.

Each attribute x_i of a WSN can be regarded as a vertex i with weight R_i^Q in a weighted graph $G(V,E)$. If attributes x_i and x_j are not β-compatible, there is an edge $(i,j) \in E$. Then, the RKDR problem can be reduced to the Weighted Maximum Independent Set (WMIS) problem. However, the WMIS problem is NP-hard. Therefore, we design heuristic algorithms for the RKDR problem.

3.1 Relative Kernel Dataset Retrieving Algorithm

The RKDR Algorithm retrieves (k,β)-Relative Kernel Dataset for a given query Q based on the linear correlation analysis. It contains the following two steps.

Step 1. Calculate the candidate relative kernel dataset $\mathcal{X}$.

At first, the candidate relative kernel dataset $\mathcal{X}$ includes all n attributes.

Step 1.1. Calculate the linear correlation coefficient R_i^Q between each attribute x_i in $\mathcal{X}$ and the target value y_Q of query Q by Eq. (1).

Step 1.2. Calculate the linear correlation coefficient r_{ij} between any two attributes x_i and x_j in $\mathcal{X}$ by Eq. (2). If x_i and x_j is not β-compatible (i.e. $|r_{ij}| > \beta$), indicating that attributes x_i and x_j are redundant in $\mathcal{X}$. If $|R_i^Q| < |R_j^Q|$, remove x_i from $\mathcal{X}$. Otherwise, remove x_j from $\mathcal{X}$. This step reduces the redundancy of the relative kernel dataset.

Step 2. Retrieve (k, β)-Relative Kernel Dataset $\mathcal{K}^Q$ from $\mathcal{X}$.

Sort the absolute values of the linear correlation coefficients of all attributes in $\mathcal{X}$, i.e. $\{|R_i^Q||x_i \in \mathcal{X}\}$, in descending order. The top-k coefficients are $\{|R_{a_1}^Q|, \cdots, |R_{a_k}^Q|\}$ and the corresponding k attributes are $\{x_{a_1}, \cdots, x_{a_k}\}$. Therefore, $\mathcal{K}^Q = \{x_{a_1}, \cdots, x_{a_k}\}$.

After the above two steps, the (k, β)-Relative Kernel Dataset for query Q is obtained. The detailed algorithm is shown in Algorithm 1.

Then once the query Q is issued by users again, we provide an approximate method to estimate the result of Q by the sensory data in $\mathcal{K}^Q$. For each attribute x_{a_i} in $\mathcal{K}^Q$, the linear function f_i between x_{a_i} and y_Q can be calculated by least-squares method. Then, each linear function f_i is assigned a weight according to $R_{a_i}^Q$. Therefore, $\mathcal{F}$ is an approximate function based on each linear function f_i and its weight to approximate the target value y_Q of query Q, shown as follows.

$$\mathcal{F} = \sum_{i=1}^{k} \left(\frac{|R_{a_i}^Q|}{\sum_{j=1}^{k} |R_{a_j}^Q|}\right) f_i \tag{3}$$

Algorithm 1. Relative Kernel Dataset Retrieving Algorithm

Input: query Q, a training set $\{t_1, \cdots, t_m\}$; the compatible parameter β; the required size k
Output: (k, β)-Relative Kernel Dataset $\mathcal{K}^Q$
1 $\mathcal{X} = \{x_1, x_2, \cdots, x_n\}$;
2 **for** *each arttribute x_i in $\mathcal{X}$* **do**
3 $\quad R_i^Q = \frac{\sum_{j=1}^{m}(x_{ij} - \overline{x_i})(y_j - \overline{y})}{\sqrt{\sum_{j=1}^{m}(x_{ij} - \overline{x_i})^2}\sqrt{\sum_{j=1}^{m}(y_j - \overline{y})^2}}$;
4 **for** *each pair of attributes x_i and x_j in $\mathcal{X}$* **do**
5 $\quad$ **if** $|r_{ij}| > \beta$ **then**
6 $\quad\quad$ If $|R_i^Q| < |R_j^Q|$, remove x_i from $\mathcal{X}$; Otherwise, remove x_j from $\mathcal{X}$;
7 Sort $\{|R_i^Q||x_i \in \mathcal{X}\}$ in descending order, and the top-k of them are $\{|R_{a_1}^Q|, \cdots, |R_{a_k}^Q|\}$;
8 $\mathcal{K}^Q = \{x_{a_1}, \cdots, x_{a_k}\}$;
9 Return $\mathcal{K}^Q$.

3.2 Piecewise Linear Fitting Based Relative Kernel Dataset Retrieving Algorithm

In the RKDR algorithm, only linear correlation is considered between each attribute and the target value. When the correlations between x_i and y_Q are exponential, quadric, logarithmic, or *etc*, the obtained relative kernel dataset

$\mathcal{K}^Q$ may incur non-negligible error for estimating the result of query Q by $\mathcal{K}^Q$. Therefore, we improve the RKDR algorithm in this section. Instead of linear correlation coefficient, the PLF-RKDR algorithm applies the piecewise linear fitting method to estimate the correlation coefficients between attributes and the target value of Q. The PLF-RKDR algorithm has the following two steps.

Step 1. Calculate the candidate relative kernel dataset $\mathcal{X}$.

At first, the candidate relative kernel dataset $\mathcal{X}$ includes all n attributes.

Step 1.1. Calculate the correlation coefficient between each attribute $x_i \in \mathcal{X}$ and the target value y_Q by the piecewise linear fitting.

We applied the method in [18] to recursively retrieve the optimal segment points in piecewise linear fitting. For attribute x_i, the training examples are sorted by the increasing order of x_i, denoted as $(x_{i1}, y_{Q1}), \cdots, (x_{im}, y_{Qm})$. $F_i^{(s,e)}(x)$ is the linear fitting function of $(x_{is}, y_{Qs}), (x_{i(s+1)}, y_{Q(s+1)}), \cdots,$ (x_{ie}, y_{Qe}) obtained by the least-squares method. The error sum of squares of linear fitting function $F_i^{(s,e)}(x)$ is $E(s,e)$. The optimal segment point (x_{ij}, y_{Qj}) satisfies that $j = \arg\min_{s<k<e}(E(s,k) + E(k,e))$. This process is conducted recursively until $E(s,e) - (E(s,j) + E(j,e)) \leq \gamma$. The detailed algorithm is shown in Algorithm 3, which returns the set of segment points SP_i. The size of SP_i is denoted as p_i.

According to the segment points in SP_i, training examples are divided into $p_i - 1$ subsets. Each subset is denoted as $S_j = \{(x_{ij^s}, y_{Qj^s}), \cdots, (x_{ij^e}, y_{Qj^e})\}$, where $1 \leq j < p_i$. The linear correlation coefficient R_{ij} for S_j and the linear function f_{ij} for S_j is calculated by the least-squares method. Therefore, the correlation coefficient R_i^Q of x_i is presented as the weighted average correlation coefficient of all $R_{ij}(1 \leq j < p_i)$. The piecewise linear function f_i is as follows.

$$f_i = \begin{cases} f_{i1}, & x_{i1^s} < x_i < x_{i1^e} \\ \cdots, & \cdots \\ f_{ip_i}, & x_{ip_i{}^s} < x_i < x_{ip_i{}^e} \end{cases} \tag{4}$$

Step 1.2. This step is the same as the Step 1.2 of the RKDR algorithm.

Step 2. Retrieve the relative kernel dataset $\mathcal{K}^Q$ from $\mathcal{X}$.

This step is the same as the Step 2 of the RKDR algorithm in Sect. 3.1.

Therefore, (k, β)-Relative Kernel Dataset $\mathcal{K}^Q$ for query Q is obtained. The detailed PLF-KDDR Algorithm is shown in Algorithm 2.

Then once the query Q is issued by users again, we provide an approximate method to estimate the result of Q by the sensory data in $\mathcal{K}^Q$. For each attribute x_{a_i} in $\mathcal{K}^Q$, the corresponding piecewise linear function f_i is assigned a weight according to $R_{a_i}^Q$. $\mathcal{F}$ is a new piecewise linear function combining all piecewise linear function $f_i(1 \leq i \leq k)$ with weight $R_{a_i}^Q$. Therefore, the approximate result of query Q is obtained by put sensory data in $\mathcal{K}^Q$ into the function $\mathcal{F}$.

4 The Performance Analysis

For RKDR Algorithm, the computation complexity of computing linear correlation coefficients is $O(mn)$ for n attributes and m training examples. Besides, the

Algorithm 2. PLF-Relative Kernel Dataset Retrieving Algorithm

Input: query Q, training set $\{t_1, \cdots, t_m\}$; compatible parameter β; required size k; threshold γ
Output: (k, β)-Relative Kernel Dataset $\mathcal{K}^Q$
1 $\mathcal{X} = \{x_1, x_2, \cdots, x_n\}$;
2 **for** *each arttribute x_i in $\mathcal{X}$* **do**
3 Sort training examples by increasing order of x_i, i.e. $(x_{i1}, y_{Q1}), \cdots, (x_{im}, y_{Qm})$;
4 $SP_i = \{1, m\} \cup$ **LS-PLF**$(1, m, \infty, \gamma)$; $p_i = |SP_i|$;
5 $(x_{i1}, y_{Q1}), \cdots, (x_{im}, y_{Qm})$ are divided into $p_i - 1$ subsets $S_1, \cdots, S_{p_i-1}$;
6 **for** *each subset $S_j (1 \leq j < p_i)$* **do**
7 $R_{ij} = \frac{\sum_{k=j^s}^{j^e}(x_{ik}-\overline{x_{ij}})(y_{Qk}-\overline{y_{ij}})}{\sqrt{\sum_{k=j^s}^{j^e}(x_{ik}-\overline{x_{ij}})^2}\sqrt{\sum_{k=j^s}^{j^e}(y_{Qk}-\overline{y_{ij}})^2}}$;
8 Calculate piecewise linear function f_{ij} by the least-squares method;
9 $R_i^Q = \sum_{j=1}^{p_i-1} R_{ij} \times \frac{j^e - j^s}{x_{im} - x_{i1}}$;
10 **for** *each pair of attributes x_i and x_j in $\mathcal{X}$* **do**
11 **if** $|r_{ij}| > \beta$ **then**
12 If $|R_i^Q| < |R_j^Q|$, remove x_i from $\mathcal{X}$; Otherwise, remove x_j from $\mathcal{X}$;
13 Sort $\{|R_i^Q| | x_i \in \mathcal{X}\}$ in descending order, and the top-k of them are $\{|R_{a_1}^Q|, \cdots, |R_{a_k}^Q|\}$;
14 Return $\mathcal{K}^Q = \{x_{a_1}, \cdots, x_{a_k}\}$.

Algorithm 3. Least-Squares Based Piecewise Linear Fitting (LS-PLF)

Input: Two endpoints s and e, $E(s, e)$, the threshold γ
Output: The set of segment points SP
1 $E_{min} = \infty$, $J_{min} = 0$;
2 **if** $e - s > 1$ **then**
3 $j = \arg\min_{s<k<e}(E(s, k) + E(k, e))$;
4 **if** $E(s, e) - (E(s, j) + E(j, e)) > \gamma$ **then**
5 $SP = SP \cup \{j\}$;
6 **LS-PLF**$(s, j, E(s, j), \gamma)$;
7 **LS-PLF**$(j, e, E(j, e), \gamma)$;
8 Return SP.

computation complexity of calculating candidate relative kernel dataset in Step 1 is $O(mn^2)$, since there are $O(n^2)$ pairs of attributes. In Step 2, the computation complexity of sorting correlation coefficients is $O(n \log n)$. In conclusion, the computation complexity for RKDR algorithm is $O(mn^2)$.

For PLF-RKDR Algorithm, the worst computation complexity of calculating optimal segment points for each attribute is $O(m^3)$. Then the computation complexity of Step 1.1 is $O(nm^3)$. For Step 1.2, the computation complexity of calculating candidate relative kernel dataset is $O(mn^2)$. In Step 2, the computation complexity of sorting correlation coefficients is $O(n \log n)$. Therefore, the computation complexity for PLF-RKDR Algorithm is $O(nm^3)$.

5 Simulation Results

This section evaluates the performance of our proposed RKDR algorithm and PLF-RKDR algorithm by extensive simulations. Simulations on both simulation dataset and real dataset are carried out.

For simulation dataset, we generate two functions among $n = 15$ attributes and the target value, denoted as $y_Q = f_1(x_1, \cdots, x_{15})$ and $y_Q = f_2(x_1, \cdots, x_{15})$.

Each training example contains fifteen randomly generated values of attributes $x_1, \cdots, x_{15}$ and a target value generated by function f_1 or f_2.

The real dataset is collected from a mobile device when a person held it making 7 types of motions, which are numbered as motion 1 to 7. For motion $i(1 \leq i \leq 7)$, y_Q is set as 1 when it happens, otherwise, it is set as 0. The attributes are the X-, Y- and Z-axis of a three-axis accelerometer and a three-axis gyroscope collected by the mobile device. Each training example contains 6 values of attributes and a target value indicating whether motion i happens.

In the following simulations, the relative error er is applied to evaluate the performance of our algorithms. In fact, $er_1 = |\frac{y'_Q - y_Q}{y_Q}|$ is the absolute error, where y'_Q is the target value estimated by sensory data in our (k, β)-Relative Kernel Dataset $\mathcal{K}^Q$ and y_Q is the true target value of query Q. However, y_Q is unknown or inaccessible in real physical world. Therefore, y_Q is estimated by $\widehat{y_Q}$, which is the target value estimated by sensory data of all n available attributes. That is, the relative error er is defined as $er = |\frac{y'_Q - \widehat{y_Q}}{\widehat{y_Q}}|$.

Firstly, comparison experiments are carried out to compare the performances of our RKDR Algorithm and PLF-RKDR Algorithm. Secondly, only PLF-RKDR algorithm is evaluated in both simulation dataset and real dataset.

5.1 The Comparison Experiments of RKDR Algorithm and PLF-RKDR Algorithm

A group of comparison experiments are carried out to compare RKDR Algorithm and PLF-RKDR Algorithm on simulation dataset. The simulation dataset is generated by two functions f_1 and f_2, each with $m = 3000$ training examples. Each training example contains $n = 15$ attributes and a target value of the given query. We compare the relative error er of our RKDR Algorithm and PLF-RKDR Algorithm with k increases, where k is the size of the (k, β)-Relative Kernel Dataset. The simulation results are presented in Fig. 1. Each data point presented in Fig. 1 is the average of simulation results on 500 times of query.

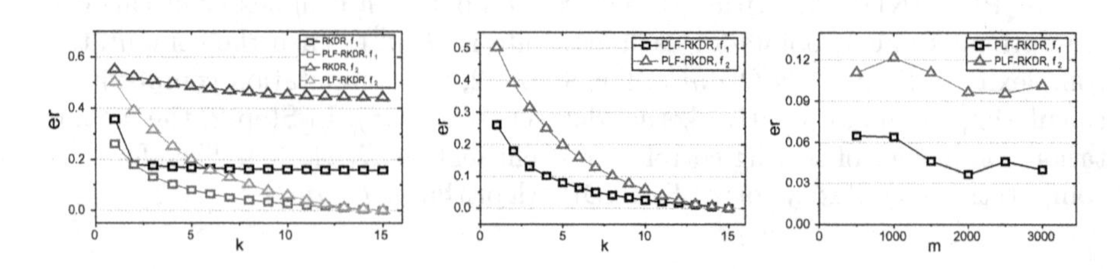

Fig. 1. RKDR vs PLF-RKDR

Fig. 2. The impact of k

Fig. 3. The impact of m

Figure 1 shows that the relative error of PLF-RKDR Algorithm is much smaller than that of RKDR Algorithm no matter how much k is. Particularly, the relative error of RKDR Algorithm is almost twice as much as that of PLF-RKDR

Algorithm when $k > 5$. Therefore, the performance of PLF-RKDR Algorithm is better than RKDR Algorithm. Besides, the relative error of both algorithms on simulation dataset of f_1 is much less than that of f_2, indicating that linear correlation coefficient may not be suitable for function f_2.

5.2 The Performance of PLF-RKDR Algorithm on Simulation Dataset

The first group of simulations investigate the impact of k, the size of t(k, β)-Relative Kernel Dataset, on the performance of PLF-RKDR Algorithm on simulation dataset. We evaluate the relative error er with the increase of k. The simulation results are presented in Fig. 2. Each data point presented in Fig. 2 is the average of simulation results on 500 times of query.

Figure 2 presents that the relative error er of PLF-RKDR Algorithm on both simulation datasets of function f_1 and f_2 decreases with k increases. Besides, the performances on simulation dataset with function f_1 and function f_2 are different with the same k since different query has different Relative Kernel Dataset. That explains why we retrieve relative kernel dataset for a given continuous query. Furthermore, Fig. 2 also presents that, the relative error reduces to 0.05 when k is only half of n, this error is pretty small in practice. That is, the network can save a half of energy when sacrifices only a few accuracy.

The second group of simulations evaluate the impact of m, the size of training examples, on the performance of PLF-RKDR Algorithm on simulation dataset. Different scales of training set is applied, i.e. m is set as 500, 1000, 1500, 2000, 2500 and 3000. The size of (k, β)-Relative Kernel Dataset is set to $k = 8$.

Figure 3 presents the performance of our PLF-RKDR algorithm on the impact of m. Each data point presented in Fig. 3 is the average of simulation results on 500 times of query. Figure 3 presents that the relative error er decreases slowly with m increases. It is worth noting that even m is relatively small, the relative error is still under 0.15 when k is no less than a half of n.

5.3 The Performance of PLF-RKDR Algorithm on Real Dataset

As motioned above, the real dataset contains 6 attributes of X-, Y- and Z-axis of a three axis accelerometer and a three axis gyroscope. The target value of motion $i (1 \leq i \leq 7)$ is set as 1 when motion i happens, otherwise, it is set as 0. If the estimated target value is closer to 1, we judge that the motion happens.

The first group of simulations show the motion judgement precisions for each motion affected by k. The simulation results of $m = 4796$ training examples are presented in Fig. 4(a). Each data point presented in Fig. 4(a) is the average of the simulation results on 1511 times of query. It shows that the precision for each motion increases with the increase of k. As the figure shows, 80% of the motions are correctly judged by our PLF-RKDR Algorithm.

The second group of simulations study the motion judgement precisions under different sizes of training examples, which are shown in Fig. 4(b). The size of (k, β)-Relative Rernel Dataset is set as $k = 4$. Each data point presented in Fig. 4(b) is the average of simulation results on 585 times of query. Figure 4(b) presents that more than 80% of the motions are correctly judged. It also shows that there was no significant relationship between the precision and m.

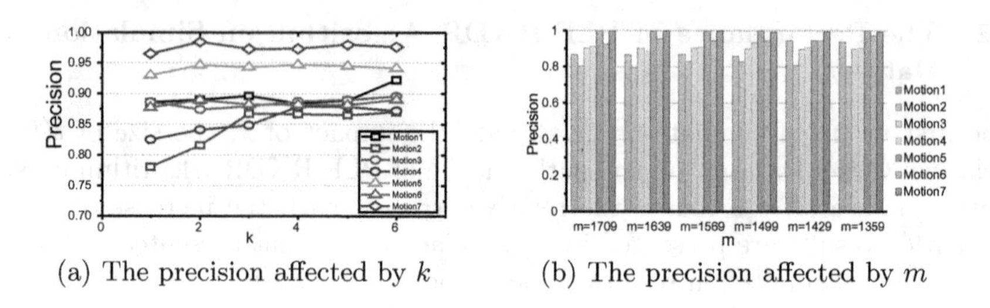

(a) The precision affected by k (b) The precision affected by m

Fig. 4. The Performance of PLF-RKDR algorithm in real dataset

6 Related Works

Most of the related works are about principal components analysis in WSN, which reduces data transmission, and then reduces the energy consumption. The work in [19] combines the compressive sensing and principal component analysis to efficiently recover the whole dataset through a small subset of data. Another work in [20] designs two distributed consensus-based algorithms to process principal component analysis in WSN. The authors in [21] propose algorithm to aggregate data through principal component analysis, which is a powerful technique for dimensionality reduction. However, these works only focus on retrieving principal components for the network and the given query has not been taken into account yet. Therefore, these methods are not suitable to retrieve relative kernel dataset from big sensory data for a given query.

The authors provide the centralized and distributed algorithms to retrieve dominant dataset from big sensory data under the condition that the information loss rate required by users is guaranteed in [16,17]. However, the dominant dataset is a general one for all queries instead of a specific one for a given query. This work cannot retrieve relative kernel dataset for a given query.

7 Conclusion

This paper studies retrieving relative kernel dataset for continuous queries from big sensory data in WSNs. The RKDR Algorithm and PLF-RKDR Algorithm are proposed to retrieve (k, β)-Relative Kernel Dataset $\mathcal{K}^Q$ for a given query Q. Then we provide methods to estimate the approximate result of query Q by sensory data in $\mathcal{K}^Q$. Extensive simulations are conducted, which demonstrate that (k, β)-Relative Kernel Dataset $\mathcal{K}^Q$ retrieved by our algorithm can estimate the result of query Q with high accuracy.

Acknowledgment. This work is partly supported by the National Natural Science Foundation of China under Grant NO. 61632010, 61502116, U1509216, 61370217, the National Science Foundation (NSF) under grant NO.1741277.

References

1. Says, G.: 6.4 billion connected things will be in use in 2016, up 30 percent from 2015. Gartner Inc. (2015)
2. Tillman, K.: How many internet connections are in the world? Right. now. CISCO (2013). http://blogs.cisco.com/news/cisco-connections-counter
3. Yu, J., Qi, Y., Wang, G., Gu, X.: A cluster-based routing protocol for wireless sensor networks with nonuniform node distribution. AEUE Int. J. Electron. Commun. **66**(1), 54–61 (2012)
4. Yu, J., Wang, N., Wang, G.: Constructing minimum extended weakly-connected dominating sets for clustering in ad hoc networks. J. Parallel Distrib. Comput. **72**, 35–47 (2012)
5. Yu, J., Ning, X., Sun, Y., Wang, S., Wang, Y.: Constructing a self-stabilizing CDS with bounded diameter in wireless networks under SINR. In: 2017 Proceedings of IEEE INFOCOM, pp. 1–9 (2017)
6. Shi, T., Cheng, S., Cai, Z., Li, J.: Adaptive connected dominating set discovering algorithm in energy-harvest sensor networks. In: 2016 Proceedings of IEEE INFOCOM, pp. 1–9 (2016)
7. Li, J., Cheng, S., Cai, Z., Yu, J., Wang, C., Li, Y.: Approximate holistic aggregation in wireless sensor networks. TOSN **13**(2), 11:1–11:24 (2017)
8. He, Z., Cai, Z., Cheng, S., Wang, X.: Approximate aggregation for tracking quantiles and range countings in wireless sensor networks. Theor. Comput. Sci. **607**, 381–390 (2015)
9. Chen, Q., Gao, H., Cai, Z., Cheng, L., Li, J.: Energy-collision aware data aggregation scheduling for energy harvesting sensor networks. In: 2018 Proceedings of IEEE INFOCOM (2018)
10. Cheng, S., Cai, Z., Li, J.: Curve query processing in wireless sensor networks. IEEE Trans. Veh. Technol. **64**(11), 5198–5209 (2015)
11. Zheng, X., Cai, Z., Li, J., Gao, H.: A study on application-aware scheduling in wireless networks. IEEE Trans. Mob. Comput. **16**(7), 1787–1801 (2017)
12. Cheng, S., Li, J., Ren, Q., Yu, L.: Bernoulli sampling based (ε, δ)-approximate aggregation in large-scale sensor networks. In: Proceedings of the 29th Conference on Information Communications, pp. 1181–1189. IEEE Press (2010)
13. Huang, Z., Wang, L., Yi, K., Liu, Y.: Sampling based algorithms for quantile computation in sensor networks. In: Proceedings of the 2011 ACM SIGMOD International Conference on Management of data, pp. 745–756. ACM (2011)
14. Zheng, H., Xiao, S., Wang, X., Tian, X., Guizani, M.: Capacity and delay analysis for data gathering with compressive sensing in wireless sensor networks. IEEE Trans. Wirel. Commun. **12**(2), 917–927 (2013)
15. Wang, H., Zhu, Y., Zhang, Q.: Compressive sensing based monitoring with vehicular networks. In: 2013 Proceedings of IEEE INFOCOM, pp. 2823–2831. IEEE (2013)
16. Cheng, S., Cai, Z., Li, J., Fang, X.: Drawing dominant dataset from big sensory data in wireless sensor networks. In: 2015 Proceedings of IEEE INFOCOM, pp. 531–539. IEEE (2015)

17. Cheng, S., Cai, Z., Li, J., Gao, H.: Extracting kernel dataset from big sensory data in wireless sensor networks. IEEE Trans. Knowl. Data Eng. **29**(4), 813–827 (2017)
18. Zong-tian, T.L.L.: Least-squares method piecewise linear fitting. Comput. Sci., S1 (2012)
19. Masiero, R., Quer, G., Munaretto, D., Rossi, M., Widmer, J., Zorzi, M.: Data acquisition through joint compressive sensing and principal component analysis. In: IEEE GLOBECOM 2009, pp. 1–6. IEEE (2009)
20. Macua, S.V., Belanovic, P., Zazo, S.: Consensus-based distributed principal component analysis in wireless sensor networks. In: 2010 IEEE Eleventh International Workshop on SPAWC, pp. 1–5. IEEE (2010)
21. Rooshenas, A., Rabiee, H.R., Movaghar, A., Naderi, M.Y.: Reducing the data transmission in wireless sensor networks using the principal component analysis. In: 2010 Sixth International Conference on ISSNIP, pp. 133–138. IEEE (2010)

KrackCover: A Wireless Security Framework for Covering KRACK Attacks

Tommy Chin[1] and Kaiqi Xiong[2](✉)

[1] Department of Computing Security, Rochester Institute of Technology, Rochester, USA
tommy.chin@ieee.org

[2] Florida Center for Cybersecurity, University of South Florida, Tampa, USA
xiongk@usf.edu

Abstract. This paper proposes KrackCover, a wireless security framework that assists in the detection of a KRACK attack with suggestions to enhance the privacy of end-users. Within KrackCover, Wi-Fi end users get an alert with suggestions when a KRACK attack is launched. Furthermore, we conduct real-world experiments by collecting data in open Wi-Fi environments including coffee shops and public libraries. Moreover, our experimental data analysis shows the applicability and effectiveness of KrackCover in the wild.

Keywords: KRACK vulnerability · Key Reinstallation Attack Internet of Things (IoT) · Wireless security · Privacy

1 Introduction

In this paper, we propose a solution called *KrackCover*. It is an adoptable framework that provides end users an alert or indication that their wireless communication has suspicion of being susceptibility to privacy evasive attackers using the Key Reinstallation Attack (KRACK) [1] or that the presence of a threat is known. KrackCover consists of a wireless sensor that evaluates 802.11 messages explicitly for a potential KRACK attack followed by network manipulation to inform the end-user of a potential hazard to their privacy. Furthermore, we conduct real-world experiments by using KrackCover. By considering a variety of scenarios, we collect data typical open Wi-Fi environments including coffee shops and public libraries at different days in a week and times on a day. Based on our experimental data analysis, we show the applicability and effectiveness of KrackCover.

We summarize the contributions of this paper as follows: (1) We investigate KRACK and propose KrackCover, a wireless security framework that help end users detect a KRACK attack with suggestions to enhance the privacy of end-users. Such suggestions should make practical impacts on wireless communities and they should be useful to wireless vendors. (2) We implement our proposed

S. Chellappan et al. (Eds.): WASA 2018, LNCS 10874, pp. 733–739, 2018.
https://doi.org/10.1007/978-3-319-94268-1_60

framework - KrackCover in real-world environments. Our approach in KrackCover consisted of two virtual machines, a mobile phone, and multiple wireless antennas for analysis where we launched the attack in the wild from one system while another captured all 802.11 wireless messages within the given area. (3) Based on our real-world experiments in open Wi-Fi environments including library and Starsbuck on different days in a week and at different times in a day, we demonstrate the applicability and effectiveness of KrackCover as a viable solution to conquer KRACK attacks.

2 Related Work

Many studies have been devoted to wireless security including Wi-Fi security [2–4]. While the risks of insecure Wi-Fi are well known, there are only a few studies found in the current literature that provides architectural solutions without having to modify the behavior of users [5]. Many of these solutions focus on user identification systems, such as measuring user signal sequences to determine if the users should be labeled as "legitimate" or "malicious" [6]. Other methods rely on using an authentication-string-based key agreement protocol [7] or other forms of identification.

A recent breaking news was the discovery of KRACK attacks [1,8,9]. KRACK attacks have been a very serious threat to the wireless communities and gives attackers to view all network traffic for a particular network system easily. Many vendors have tried to fix this issue by providing patching, but these fixes present major concerns discussed before.

3 Design

An attacker leveraging the KRACK vulnerability ultimately satisfies the following goals: (1) the ability to read and interpret network communication between a victim and a Wireless Access Point (WAP) follow by (2) obtaining leveragable data which allows the attacker to inflict further damage to both the victim and the targeted network. Mitigating the vulnerability by itself cannot be satisfied through networking alone as the end-user would need to install a software patch on their device. However, the KrackCover framework implements sensory systems that evaluate 802.11 messages such that an end-user would receive an alert or prompt for the detection of the attack.

Network security systems such as an IDS often evaluate network traffic after the WAP receives data from a client. The location in which the device detects a network threat is usually in an in-line (physical cable) configuration. The administrator designates the position of the IDS in a network including the type of traffic to inspect, however, the KRACK vulnerability applies in a wireless configuration which an in-line IDS would not be able to identify. Lastly, if a threat were to occur, the end-user would not be in a position to react to the threat in a manner to defend themselves and that any risk or compromise that may arise from an attack would ultimately be successful.

The KrackCover framework consists of a two-component approach. In the first segment, sensory systems in a wireless network are used to evaluate 802.11 (layer 2 of the OSI model) messages with actions that mimic an IDS. The second, if a threat were detected by KrackCover, the end-user would receive a warning message for the detection of the attack by temporarily redirecting their connection to an intermediate location such as a splash portal and eliminates the need for end-users to install additional software on their computer systems to receive alerting. Figure 1 depicts a high-level view of KrackCover operation and deployment strategy where integers in the figure stand for the following meaning. *Numeric 1* represents normal traffic, *Numeric 2* represents an attacker launching a KRACK attack, *Numeric 3* gives a detection alert, and *Numeric 4* indicates a temporary traffic redirection to a splash portal for the end-user. Moreover, the "Control" portion represents a network which has the ability to redirect end-user traffic in a seamless manner such that clients on the wireless network would not necessary need to negotiate a new network connection.

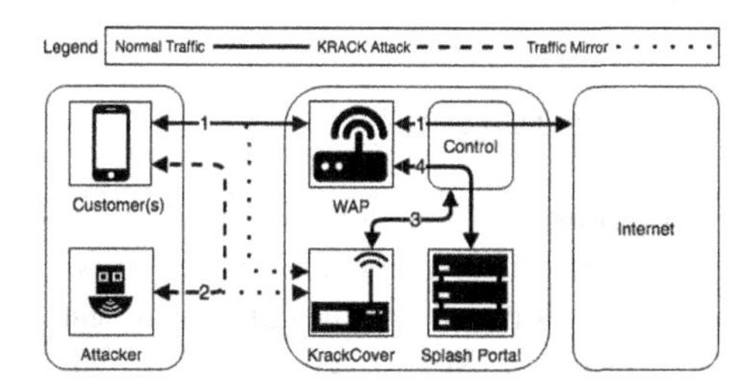

Fig. 1. A high-level overview of KrackCover.

The adaptability and manipulation of the network in the "Control" method vary between an organization and the infrastructure which KrackCover resides. In a pure physical barebone system environment, a web proxy configuration would easily suffice the goal of the redirection, but the capabilities would not encompass non-web traffic as a web-proxy would only filter web-based traffic. Moreover, if an end-user utilizes a full tunnel VPN service, the exposure of an attacker viewing plaintext traffic from a successful KRACK attack would have a minimal effect to the end-user as the attacker would only be able to see local-LAN based traffic such as ARP, discovery protocols, and other broadcast/multicast transmission. IoT devices, however, have minimal protection to a KRACK attack as much of their programming is static and uncontrolled by a user unless a vendor releases a firmware patch or update.

Evaluating wireless communication in an enclosed environment, such as a Faraday cage does not effectively reflect variable noises and other impacts in a real-world environment. Testbed solutions are widely abundant for Wi-Fi communication but lack the scalability and unaccountable impacts from a real-world configuration. Figure 2 depicts our implementation of KrackCover for our experimental evaluation which we further discuss in Sect. 4.

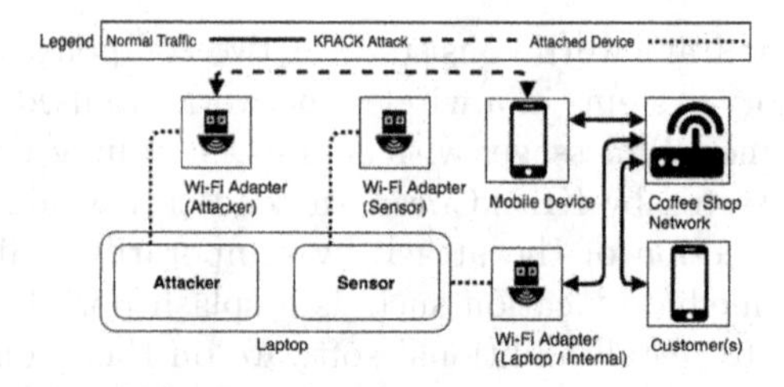

Fig. 2. A topology diagram representing the method for our experimental evaluation.

Setting up a real-world testing situation presents one significant challenge, the ability to rapidly setup, deploy, evaluate, and execute experiments without disrupting both customers, the local establishment, and nearby Wi-Fi networks. Overall, we decided to utilize the portability of a laptop, a couple of wireless antennas, a mobile phone, virtualization solutions, and situate our experimental evaluation at a local coffee shop.

4 Experimental Evaluation

Using a public space provides the realism that a wireless sensor would observe such as unassociated devices, uncertain network transmission, and environmental changes which would be difficult to emulate on a testbed. We did not ask customers at the coffee shop or post signage indicating our analysis for the concern that our evaluation would ward participants from being in the range of our data collector. One constraint in our analysis was the fact that the coffee shop we selected had a fixed operation time (open/close), which by itself created a challenge in determining the best time to conduct experiments. Fortunately, crowdsourced information from social media and geolocational data offers an approximate time in which a business has the most customers as shown in Fig. 3.

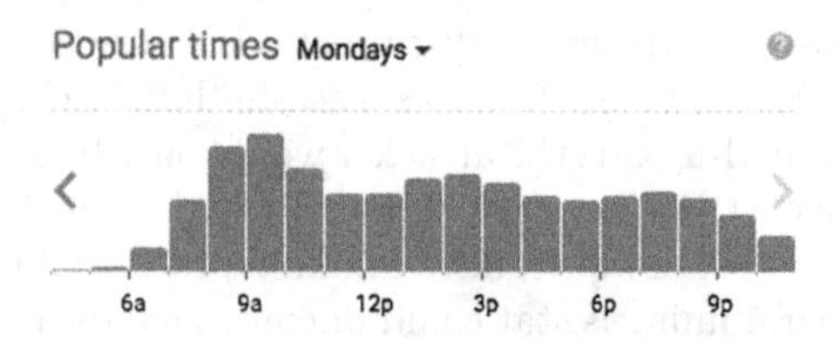

Fig. 3. Crowdsourced data offers optimum time for real-world analysis indicating where a business has the most customers. The selected service for our analysis was Google Maps.

Selecting the best time at the coffee shop for analysis showed that for a given Monday, 8 AM to 11 AM local time offered the most customers. However, it does not necessarily mean that it would correlate to the most active wireless users.

To address this concern, we, therefore, conducted and collected our experimental analysis over multiple periods between three and six-hour intervals, varying between morning, afternoon, and evenings.

Calculating the number of active devices in a given area presents a challenge without soliciting customers at the coffee shop and that a user could possess more than one wireless system. Additionally, the coffee shop by itself may have wireless devices such as Point-Of-Sale (POS) systems, Heating Ventilation Air Conditioning (HVAC) sensors, or IoT devices, which would also attribute to the number of devices in our analysis. The solution to identifying the number of devices in an area for our experimental analysis relies on a real-time full packet capture service that would collect all 802.11 orientated traffic in an area for the entire duration of our evaluation. Specifically, the communication of a Wi-Fi device utilizes a message configuration of 802.11 whether a beacon, probe, acknowledgment, etc., traffic would be visible within the range of our sensor.

4.1 Hardware Configuration

Constructing our testing environment involved many constraints, with the first being that the coffee shop has business hours and that when the store closes, the team would need to leave. We, therefore, utilized a laptop with 16 GB of memory, and a quad-core processor. The laptop hosted two virtual machines, one an attack machine running Ubuntu with 1 GB of memory, and a single core processor, and the other machine a victim machine running Debian, with two core processor and 4 GB of memory. Hardware-wise, we utilize three wireless antennas, an Alfa AWUS036NHA High Gain, a TP-Link WN722n v1, and the integrated wireless adapter on the laptop for both the attacker, KrackCover, and a victim machine, respectively. Lastly, we utilized a Samsung Galaxy S5 to host a Wi-Fi network with WPA2 encryption for our testing environment as it can rapidly deploy a Wi-Fi network with cellular tethering. Figure 4 depicts the hardware described.

Fig. 4. Experimental setup and hardware configuration for KrackCover

4.2 Results

Evaluating the performance and effectiveness of KrackCover involved real-time network analysis and full packet capture. Table 1 describes time measurements observed in our experimental analysis where each segment represents over 300 successful KRACK attacks.

Table 1. Time measurements in milliseconds on KrackCover detection where (A) represents the total execution time of a successful attack, (B) the initial time between executing the attack to the first KRACK-based packet being sent, (C) total time to transmit the necessary KRACK attack packets, (D) is the detection time between the transmission of the first packet and when the sensor records the value, and (E) the time between the first and last packet in the KRACK attack detection.

	Mean	Median	Mode	Min	Max
A	34.0690	27.1440	10.3810	3.6810	131.0800
B	196.4937	162.9040	158.7610	4.5300	10848.6720
C	199.9331	164.7290	186.2380	2.9280	10844.1860
D	230.5627	187.8250	28.4370	28.4370	10877.5670
E	2.3313	1.4260	0.0550	0.0390	19.8430

Specifically, our results demonstrated that a successful KRACK attack required an average of 34.0690 s while the fastest detected was 3.6810 s. We believe the time variations of the KRACK attack had influences by either environmental factors or software interpreting the packets. The virtual machine running as our client did not show signs of performance loss on CPU or memory and that the majority of the time, it was at idle. Detecting the KRACK attack averaged at 230.5627 ms where we capture the initial time the attack node transmitted a packet belonging to the KRACK attack. Propagation delay in the transmission was uncontrollable due to environmental constraints which would express potential reason for higher detection time.

5 Conclusions and Future Work

In this research, we have proposed a wireless security framework called KrackCover that help end users to detect a KRACK attack where an alert is given and suggestions are provided to enhance the privacy of end-users. That is, KrackCover issues an alert with suggestions to Wi-Fi end users when a KRACK attack is launched. We have further conducted the real-world experimental evaluation of KrackCover and collected experimental data in a variety of open Wi-Fi environments such as coffee shops and public libraries on different days in a week and different times on a day. Based on our experimental data analysis, we have sufficiently shows that KrackCover is applicable and effective to detect KRACK attacks in the wild. In the further work, we will investigate the scalability of KrackCover and conduct real-world experiments to evaluate its scalability.

References

1. Vanhoef, M., Piessens, F.: Key reinstallation attacks: forcing nonce reuse in WPA2. In: Proceedings of the ACM SIGSAC Conference on Computer and Communications Security, pp. 1313–1328. ACM, New York (2017)
2. Zou, Y., Zhu, J., Wang, X., Hanzo, L.: A survey on wireless security: technical challenges, recent advances, and future trends. Proc. IEEE **104**(9), 1727–1765 (2016)
3. Lashkari, A.H., Danesh, M.M.S., Samadi, B.: A survey on wireless security protocols (WEP, WPA and WPA2/802.11i). In: Proceedings of the 2nd IEEE International Conference on Computer Science and Information Technology. IEEE (2009)
4. Chin, T., Xiong, K.: MPBSD: a moving target defense approach for base station security in wireless sensor networks. In: Proceedings of the IEEE International Conference on Wireless Algorithms, Systems, and Applications. IEEE (2016)
5. Pisa, C., Caponi, A., Dargahi, T., Bianchi, G., Blefari-Melazzi, N.: WI-FAB: attribute-based WLAN access control, without pre-shared keys and backend infrastructures, pp. 31–36 (2016)
6. Cheng, L., Wang, J.: How can I guard my AP?: non-intrusive user identification for mobile devices using WiFi signals. In: Proceedings of the 17th ACM International Symposium on Mobile Ad Hoc Networking and Computing, pp. 91–100. ACM (2016)
7. Shen, W., Yin, B., Cao, X., Cai, L.X., Cheng, Y.: Secure device-to-device communications over WiFi direct. IEEE Netw. Mag. **30**(5), 4–9 (2016)
8. Gallagher, S.: How the KRACK attack destroys nearly all Wi-Fi security. In: Arts Technica (2017)
9. TP-Link: WPA2 security (KRACKs) vulnerability statement (2018). https://www.tp-link.com/us/faq-1970.html. Accessed 18 Feb 2018

Node Deployment of High Altitude Platform Based Ocean Monitoring Sensor Networks

Jianli Duan[1,2], Yuxiang Liu[1], Bin Lin[1(✉)], Yuan Jiang[3], and Fen Hou[4]

[1] Dalian Maritime University, Dalian 116026, China
duanjianli@qut.edu.cn, lyx285edu@163.com, binlin@dlmu.edu.cn
[2] Qingdao University of Technology, Qingdao 266033, China
[3] University of Chinese Academy of Science, Beijing 100049, China
[4] University of Macau, Macau 999078, China

Abstract. Territorial ocean safety and ocean development make it necessary to establish a large-scale, long-term, and low-energy integrated ocean monitoring sensor networks (OMSNs). The High Attitude Platform based OMSN (HAP-OMAN) has provided a promising solution for ocean monitoring. In this paper, we study the node deployment problem in a HAP-OMSN architecture and formulate the problem as a multi-objective linear programming (MOLP) to maximize network lifetime, minimize energy consumption and ensure network's connectivity, reliability and coverage of target points. Finally, we solve the MOLP by Gurobi, which is a powerful linear programming solver. Numerical analysis through case studies is conducted to verify the feasibility and scalability of the proposed optimization framework in different scales of network scenarios.

Keywords: High attitude platform · Wireless Sensor Network
Gurobi

1 Introduction

With the rapid development of ocean resources industries, people pay more and more attention to the protection of the marine environment and maritime security [1–3], and the demands for ocean monitoring network (OMN) increase rapidly. There are two traditional monitoring networks: shore-based network for offshore monitoring, and satellite-based open-ocean monitoring network. Due to the small coverage, the shore-based networks can only cover the limited offshore area, and the sea area beyond the reach of the shore-based networks can only be monitored through satellite-based open-ocean monitoring networks [4, 5]. However, the satellite is far away from the earth, thus the OMN nodes with limited battery capacity must consume a lot of energy to communication with the satellite and will reduce the network lifetime greatly. In addition, the satellite communication has low bandwidth,long transmission delay and high costs. Accordingly, an ocean monitoring network with larger coverage, better energy-efficiency, wider bandwidth and lower cost is urgently needed. The high attitude platform-based ocean monitoring sensor network (HAP-OMSN) is a promising

S. Chellappan et al. (Eds.): WASA 2018, LNCS 10874, pp. 740–747, 2018.
https://doi.org/10.1007/978-3-319-94268-1_61

solution. At the altitude of 17–22 km above the ground, HAP system is composed of airships which carry wireless broadband communication equipment. HAP combines the advantages of ground stations and those of satellites: large coverage, low latency, low power consumption and low cost [6]. Underwater Wireless Sensor Networks (UWSNs) are deployed underwater which can be used in different applications such as ocean ecosystem environmental monitoring, especially pollution monitoring. The underwater sensor nodes are interconnected to sinks by wireless acoustic links [7]. The HAP-OMSN can reduce the cost, improve the coverage of the network and increase data transmission speed compared with satellite-based OMN.

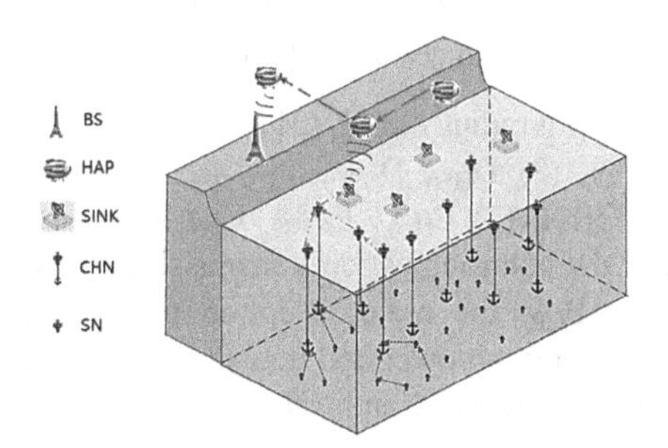

Fig. 1. HAP-OMSN architecture

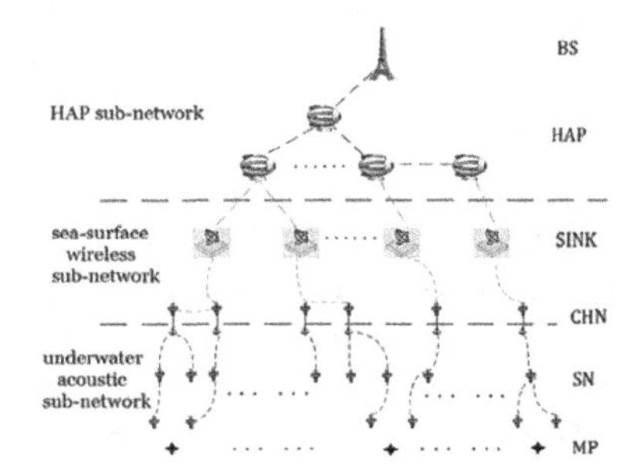

Fig. 2. The multi-tiered architecture of HAP-OMSN

In this paper, we present a merged HAP-OMSN architecture composed of high altitude wireless optical networks, surface radio networks and underwater acoustic networks for ocean monitoring as shown in Fig. 1. We investigate the multi-types node deployment (MTND) problems to maximize network lifetime and minimize energy consumption. We adopt a dual-link strategy to ensure the reliability of the network and use data fusion to remove the resulting redundant data. In addition, we formulate the MTND problem as a multi-objective linear programming (MOLP), and then solve it by Gurobi, which is a powerful integer linear programming solvers.

2 Network Model and Problem Formulation

2.1 Network Model

As shown in Fig. 2, the HAP-OMSN architecture contains three parts: the underwater acoustic subnetwork, the sea-surface wireless subnetwork and the HAP subnetwork. From the bottom to the top, the HAP-OMSN is composed of Monitoring Points (MP), Sensor Nodes (SN), Cluster-Head Nodes (CHN), SINKs, HAPs and base station (BS) on land. SN and CHN belong to the underwater acoustic subnetwork. SN can sense the required data. SINK which belongs to the sea-surface wireless subnetwork is in charge of collecting local information. HAP can complete the last step of the information transmission. MP should be within the monitoring region of two or more SNs.

HAP-OMSN can be denoted as the directed graph $\vec{G} = (\Omega, \vec{E})$. Ω represents the set of points and $\vec{E}$ represents the set of directed edges. Ω is divided into five parts and marked as Ω_{BS}, Ω_{HAP}, Ω_{SINK}, Ω_{CHN} and Ω_{SN}. i.e. $\Omega = \Omega_{BS} \cup \Omega_{HAP} \cup \Omega_{SINK} \cup \Omega_{CHN} \cup \Omega_{SN}$.

2.2 Problem Statement

The inputs, constant parameters, variables and constraints of HAP-OMSN node deployment problem are as follows.

Inputs: (1) The location of the BS, HAPs, MPs and the candidate points (CPs) for the SINKs, CHNs and SNs; (2) g_t: the amount of data produced by MPs per day.

Constant Parameters: (1) R_S: the effective sensing radius of SN; (2) D_{WH}, D_{SC}, D_{CW}: the receiving radius of HAP, SINK and CHN; (3) D_{HH}, D_{SS}, D_{CC}: the maximum communication distance of HAP, SINK and CHN; (4) E_{HAP}, E_{CHN}, E_{SN}: the initial energy of HAP, CHN and SN; (5) P_{HTR}, P_{HR}: the communication transmission power and radio reception power of HAP; (6) P_{CT}, P_{CR}: the radio reception power and transmission power of CHN; (7) P_{ST}, P_{SR}: the reception power and transmission power of SN; (8) K_1, K_2, $\alpha(f)$, β: the sending parameters of CHN and SN.

Variables: (1) b_m- if a SN is placed at CP_m, b_m is 1, otherwise bm is 0; (2) e_{ij}- if there is a communication link from $node_i$ to $node_j$, e_{ij} is 1, otherwise e_{ij} is 0; (3) l_{ij}- if there is a flow from $node_i$ to $node_j$, l_{ij} is 1, otherwise l_{ij} is 0; (4) w_{ij}- if $node_j$ can perceive the data at MP_i, w_{ij} is 1, otherwise w_{ij} is 0; (5) c_m- if a CHN is placed at CP_m, c_m is 1, otherwise c_m is 0; (6) a_m is defined: if a SINK is placed at CP_m, a_m is set to 1, otherwise a_m is 0; (7) u_{ij}^k- if the data at MP_i is received by $node_j$ via a link with $node_k$ as the SN, u_{ij}^k is 1, otherwise u_{ij}^k is 0; (8) u_{ij}- if the data at MP_i can be received from $node_j$, u_{ij} is 1, otherwise u_{ij} is 0; (9) f_{ij} is the amount of data transferred from $node_i$ to $node_j$.

Constraints: (1) Coverage: each MP should be covered by the SN; (2) Connection: HAP-OMSN should transmit the perceived data to BS; (3) Topology: HAP-OMSN should keep the tree structure. Loop is not allowed; (4) Delay: The maximum number of multi-hops from MP to BS is upper bounded.

2.3 Problem Formulation

Minimize Multi-objectives: I. F; II. $\sum_{i \in HAP} E_i F_i + \sum_{j \in \Omega_{CHN}} E_j F_j + \sum_{k \in \Omega_{SN}} E_k F_k$.

The objective function I represents that the primary goal of optimization is to minimize F, which is defined as the reciprocal of network lifetime. The objective function II is to minimize the energy consumption of the SNs, CHNs and HAPs per day, and E_i is the energy capacity of $node_i$.

Subject to

$$\sum\nolimits_{j\in\Omega_{SN}} w_{ij} \geq COVER_{MIN}\ \forall i \in \Omega_{MP} \tag{1}$$

$$e_{ij} \leq b_j,\ w_{kj} \leq b_j\ \forall i, j \in \Omega_{SN}\ \forall k \in \Omega_{MP} i \neq j \tag{2}$$

$$\sum\nolimits_{j\in\Omega_{SN}\cup\Omega_{CHN}} e_{mj} = b_m, \sum\nolimits_{i\in\Omega_{SN}} e_{in} + \sum\nolimits_{k\in\Omega_{TP}} w_{kn} \geq b_n,\ \forall m, n \in \Omega_{SN} i \neq n,\ m \neq j \tag{3}$$

$$l_{ij} \leq b_j\ \forall i, j \in \Omega_{SN} \quad i \neq j | \forall i \in \Omega_{SN}, \forall j \in \Omega_{SN} \quad \cup_{\grave{U}CHN} i \neq j \tag{4}$$

$$\sum\nolimits_{j\in\Omega_{SINK}} l_{ij} = b_i,\ \forall i \in \Omega_{SN} \tag{5}$$

$$e_{ij} \leq c_j, \sum\nolimits_{i\in\Omega_{SN}\cup\Omega_{CHN}} e_{ij} \geq c_j\ \forall i \in \Omega_{SN} \cup \Omega_{CHN},\ j \in \Omega_{CHN} i \neq j \tag{6}$$

$$\sum\nolimits_{i\in\Omega_{CHN}\cup\Omega_{SINK}} e_{ij} = c_i\ \forall i \in \Omega_{CHN}\ i \neq j \tag{7}$$

$$l_{ij} \leq c_j\ \forall i \in \Omega_{SN} \cup \Omega_{CHN},\ j \in \Omega_{CHN}\ | \forall i \in \Omega_{CHN},\ j \in \Omega_{CHN}\ i \neq j \tag{8}$$

$$\sum\nolimits_{j\in\Omega_{SINK}} l_{ij} = c_i,\ \forall i \in \Omega_{CHN} \tag{9}$$

The constraint (1) stipulates that each data source must be covered by at least $COVER_{MIN}$ SNs to ensure the reliability of the perceived data. The constraints (2)–(3) ensure that if the SN at CP exists, it should cover one DS at least or there is at least one input communication connection from other SNs. The constraints (4)–(5) constrain the relationship between the SNs and the input/output streams. The constraints (6)–(7) specify the relationship between the CHNs and the input/output paths. The constraints (8)–(9) formulate the relationship between the CHNs and the input/output flows.

$$\sum\nolimits_{i\in\Omega_{HAP}} e_{ji} = a_j,\ e_{ij} \leq a_j \quad \forall i_j \in \Omega_{CHN}, j \in \Omega_{SINK} \tag{10}$$

$$\sum\nolimits_{i\in\Omega_{CHN}} e_{ij} \geq a_j \quad \forall j \in \Omega_{SINK}\ i \neq j \tag{11}$$

$$l_{ij} \leq a_j\ \forall i \in \Omega_{SN} \cup \Omega_{CHN},\ \forall j \in \Omega_{SINK} \tag{12}$$

$$\sum\nolimits_{i\in\{\Omega_{SINK}\,\Omega_{HAP}\}} e_{ij} \geq 1 \quad \forall j \in \{\Omega_{HAP}|\Omega_{SINK}\} \tag{13}$$

$$\sum\nolimits_{j\in\Omega_{HAP}\cup\Omega_{SINK}} e_{ij} = 1,\ \forall i \in \Omega_{HAP}\ i \neq j \tag{14}$$

$$e_{ij} + e_{ji} \leq 1 \qquad i, j \in \{\Omega_{SN}|\Omega_{CHN}|\Omega_{HAP}\}\ i \neq j \tag{15}$$

$$e_{ij} d_{ij} \leq D_{\{HH|SS|CC\}} \quad \forall i, j \in \{\Omega_{SN}|\Omega_{CHN}|\Omega_{HAP}\}\ i \neq j \tag{16}$$

$$e_{ij}d_{ij} \leq D_{WH}\ \forall i \in \Omega_{SINK},\ j \in \Omega_{HAP} \tag{17}$$

$$e_{ij}d_{ij} \leq D_{CW}\ \forall i \in \Omega_{CHN},\ j \in \Omega_{SINK}\ i \neq j \tag{18}$$

$$e_{ij}d_{ij} \leq D_{SC},\ w_{ki}d_{ki} \leq R_S\ \forall k \in \Omega_{MP},\ \forall i \in \Omega_{SN},\ j \in \Omega_{CHN}\ i \in j \tag{19}$$

The constraints (10)–(12) constrain the relationship between SINKs and the input-output links. The constraints (13)–(14) stipulate the connection of HAPs. The constraints (15)–(19) define the distance between two nodes and the relationship of the communication links.

$$l_{ji} + l_{ij} \leq 1 \qquad \forall i, j \in \{\Omega_{SN}|\Omega_{CHN}\} \tag{20}$$

$$l_{ij} \geq e_{ij} \qquad \forall i \in \Omega_{SN},\ j \in \Omega_{CHN} \cup \Omega_{SN}|\forall i \in \Omega_{CHN},\ j \in \Omega_{SINK} \cup \Omega_{CHN} \tag{21}$$

$$l_{ij} + l_{jk} - 1 \leq l_{ik}\ \forall i, j \in \Omega_{SN}|\Omega_{CHN},\ k \in \Omega_{CHN} \cup \Omega_{SN}|\Omega_{SINK} \cup \Omega_{CHN} i \neq j \neq k \tag{22}$$

$$l_{ij} + l_{jk} - 1 \leq l_{ik}\ \forall i \in \Omega_{SN},\ j \in \Omega_{CHN},\ k \in \Omega_{CHN} \cup \Omega_{SINK}\ j \neq k \tag{23}$$

$$e_{ik} + e_{jk} + l_{ti} + l_{tj} \leq 3\ \forall i, j, t \in \Omega_{SN},\ \forall k \in \Omega_{SN} \cup \Omega_{CHN}\ t \neq i \neq j \neq k \tag{24}$$

$$e_{ik} + e_{jk} + l_{ij} + l_{ji} \leq 2\ \forall i, j \in \Omega_{SN},\ \forall k \in \Omega_{SN} \cup \Omega_{CHN} \quad i \neq j,\ i \neq k,\ j \neq k \tag{25}$$

$$e_{ik} + e_{jk} + l_{ti} + l_{tj} \leq 3\ \forall j, t \in \Omega_{SN},\ \forall i, k \in \Omega_{CHN} t \neq i \neq j \neq k \tag{26}$$

$$e_{ik} + e_{jk} + l_{ij} \leq 2 \quad \forall j \in \Omega_{SN},\ \forall i, k \in \Omega_{CHN} i \neq j,\ i \neq k,\ j \neq k \tag{27}$$

$$e_{ik} + e_{jk} + l_{ti} + l_{tj} \leq 3\ \forall i, j \in \Omega_{CHN} \forall k \in \Omega_{SINK} \cup \Omega_{CHN},\ t \in \Omega_{SN} \cup \Omega_{CHN},\ t \neq i \neq j \neq k \tag{28}$$

$$e_{ik} + e_{jk} + l_{ij} + l_{ji} \leq 2 \quad \forall i, j \in \Omega_{CHN},\ \forall k \in \Omega_{SINK} \cup \Omega_{CHN}\ i \neq j,\ i \neq k,\ j \neq k \tag{29}$$

$$w_{ik} \geq u_{ij}^k,\ l_{kj} \geq u_{ij}^k,\ w_{ik} + l_{kj} - 1 \leq u_{ij}^k \forall i \in \Omega_{MP},\ \forall j \in \Omega_{CHN} \cup \Omega_{SN},\ \forall k \in \Omega_{SN} \tag{30}$$

$$\sum\nolimits_{k \in \Omega_{SN}} u_{mn}^k \geq u_{mn},\ \sum\nolimits_{k \in \Omega_{SN}} u_{ij}^k + w_{ij} \geq u_{ij} \forall i,\ m \in \Omega_{MP},\ \forall j \in \Omega_{SN},\ \forall n \in \Omega_{CHN} \tag{31}$$

$$u_{ij}^k \leq u_{ij},\ w_{ik} \leq u_{ik} \quad \forall i \in \Omega_{MP},\ \forall j \in \Omega_{CHN} \cup \Omega_{SN},\ \forall k \in \Omega_{SN} \tag{32}$$

The constraints (20)–(29) constrain all flows of the entire path through communication links among nodes. The constraints (30)–(32) establish the relationship between DSs and CHNs, and it ensures that the CHN can complete the data fusion successfully.

$$\sum\nolimits_{i \in \Omega_{SN}} f_{ij} + \sum\nolimits_{k \in \Omega_{MP}} g_k w_{kj} = \sum\nolimits_{l \in \Omega_{SN} \cup \Omega_{CHN}} f_{jl}\ \forall j \in \Omega_{SN} i \neq j,\ j \neq l \tag{33}$$

$$\sum\nolimits_{k \in \Omega_{MP}} g_t u_{ij} = \sum\nolimits_{k \in \Omega_{CHN} \cup \Omega_{SINK}} f_{ij}\ \forall j \in \Omega_{CHN} j \neq k \tag{34}$$

$$\sum\nolimits_{k \in \Omega_{CHN}} f_{ij} = \sum\nolimits_{k \in \Omega_{HAP}} f_{jk} \qquad \forall j \in \Omega_{SINK} \tag{35}$$

$$f_{ij} \leq 100000 e_{ij},\ f_{ij} \geq e_{ij}\ \forall i \in \Omega_{SN},\ \forall j \in \Omega_{SN} \cup \Omega_{CHN} | \forall i \in \Omega_{CHN},\ \forall j \in \Omega_{CHN} \cup \Omega_{SINK} | \forall i \in \Omega_{SINK},\ \forall j \in \Omega_{HAP} | \forall i \in \Omega_{HAP},\ \forall j \in \Omega_{HAP} \cup \Omega_{BS} \tag{36}$$

$$P_{HTR}\left(\sum\nolimits_{i \in \Omega_{hap}} e_{ij} + \sum\nolimits_{k \in \Omega_{BS} \cup \Omega_{hap}} e_{jk}\right) + \sum\nolimits_{l \in \Omega_{SINK}} P_{HR} f_{lj} \leq (E_{HAP} - E_d d_{js}) F_i\ \forall j \in \Omega_{HAP},\ \forall s \in \Omega_{BS}\ i \neq j \neq k \tag{37}$$

$$\sum\nolimits_{i \in \Omega_{SN}} f_{ij} P_{SR} d_{ij}^{k_2} 10^{\frac{\alpha(f)}{10} d_{ij}} + \sum\nolimits_{l \in \Omega_{CHN}} P_{CR} f_{lj} + \sum\nolimits_{k \in \Omega_{SINK} \cup \Omega_{CHN}} (P_{CT} + \beta d_{ij}^{k_1}) f_{lj} \leq E_{CHN} F_i\ \forall j \in \Omega_{CHN}\ j \neq k \neq l \tag{38}$$

$$\sum\nolimits_{i \in \Omega_{SN}} f_{ij} P_{SR} d_{ij}^{k_2} 10^{\frac{\alpha(f)}{10} d_{ij}} + \sum\nolimits_{k \in \Omega_{SN} \cup \Omega_{CHN}} f_{ij} P_{ST} d_{jk}^{k_2} 10^{\frac{\alpha(f)}{10} d_{jk}} \leq E_{SN} F_i \forall j \in \Omega_{SN},\ i \neq j \neq k \tag{39}$$

The constraints (33)–(35) specify a balance between the input data and output data. The constraint (36) stipulates data transfer between two nodes by data connection, and also shows if the communication connection is built from node i to node j, data transmission between them should exist. The constraints (37)–(39) stipulate Fi of each node by the input and output power of the SNs, CHNs and HAPs.

3 Numerical Analysis

To test the performance of our HAP-OMSN formulation in terms of validation, feasibility and scalability, a series of simulations with different scales are carried out. Table 1 lists some representative simulations.

Firstly, we take scenario 0 as an example to validate the formulation where the global optimal solutions can be obtained by exhaustive search. As shown in Fig. 3, all constraint conditions are met in the deployment results, for instance, MP0 is covered by SN4 and SN5, which means the distance constraints are met. Meanwhile, the network lifetime and energy consumption are the same as those obtained by the exhaustive search. Thus our MTND formulation is validated.

Based on the formulation validation, we expand the scale of the scenes to verify the solvability and scalability of MTND. The results in Table 1 show that the MTND model can solve the deployment problem in middle and large-scale network scenarios, and the optimization time of Gurobi increases with the growth of the network scale.

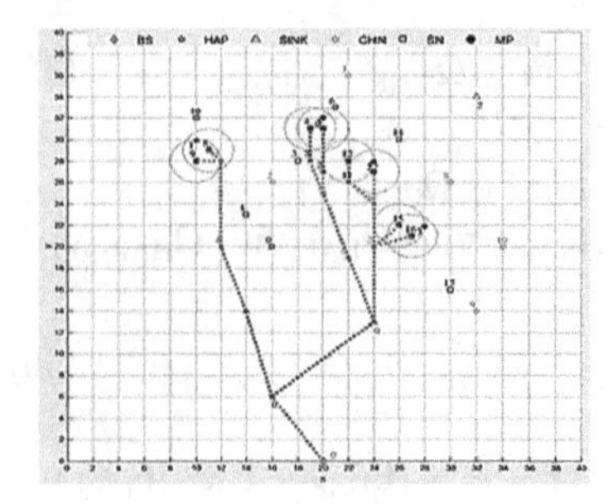

Fig. 3. the layouts of Scenario 0

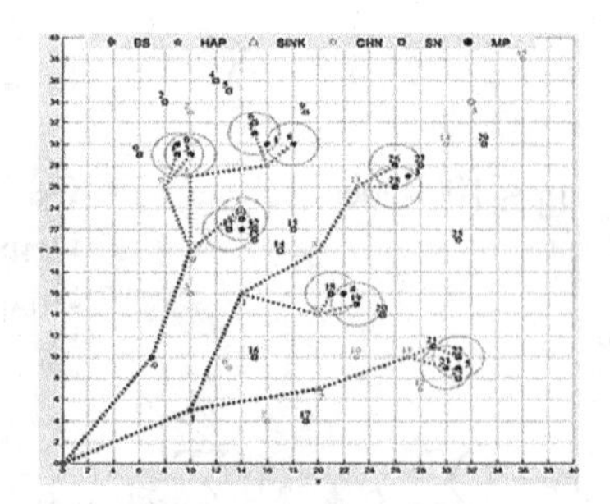

Fig. 4. the layout results of Scenario 1

Take Scenario 1 as an example, Fig. 4 shows the simulation result of Scenario 1, and we can see that the MNTD is solvable and scalable when we increase the size of the sensor network. Therefore, the results have demonstrated that the proposed optimization framework is feasible, solvable and scalable for the deployment of HAP-OMSN networks in practice.

Table 1. The results of four scenarios

Scenario	Node ID			Run time (S)
	CPs for SINKs, CHNs, SNs	BSs, HAPs, MPs	Selected SINKs, CHNs, SNs	
0	2, 11, 18	1, 1, 4	2, 6, 11	0.55
1	4, 16, 30	1, 2, 6	3, 8, 12	5.89
2	6, 24, 52	1, 3, 10	4, 13, 30	61.17
3	8, 36, 79	1, 3, 16	6, 19, 44	1226.9
4	9, 38, 88	1, 3, 18	5, 19, 53	202436.7

4 Conclusion

In this paper, we present the HAP-OMSN architecture to meet the increasing demands for maritime monitoring network firstly. We formulate the MTND problem as an MOLP to maximize lifetime, minimize energy consumption and ensure network's connectivity, reliability and coverage of target points. A serious of case studies are conducted to validate the MTND formulation, and the simulation results have demonstrated the solvability and scalability. With the proposed optimization framework, a guideline can be provided for the deployment of OMSN in practice for future ocean monitoring.

Acknowledgement. This study is sponsored by National Science Foundation of China (NSFC) No. 61771086 and Fundamental Research Funds for Central Universities under grant No. 3132016318.

References

1. Cheng, S., Cai, Z., Li, J., et al.: Drawing dominant dataset from big sensory data in wireless sensor networks. In: Computer Communications, pp. 531–539 (2015)
2. Cheng, S., Cai, Z., Li, J.: Curve query processing in wireless sensor networks. IEEE Trans. Veh. Technol. **64**(11), 5198–5209 (2015)
3. Cheng, S., Cai, Z., Li, J., et al.: Extracting kernel dataset from big sensory data in wireless sensor networks. IEEE Trans. Knowl. Data Eng. **29**(4), 813–827 (2017)
4. Duan, J., Zhao, T., Lin, B.: Optimal topology design of high altitude platform based maritime broadband communication networks. In: Gao, X., Du, H., Han, M. (eds.) COCOA 2017. LNCS, vol. 10627, pp. 462–470. Springer, Cham (2017). https://doi.org/10.1007/978-3-319-71150-8_38
5. Zhao, T.: Deployment and optimization of maritime broadband communication system based on high altitude platform. Dalian Maritime University, Dalian (2017)
6. Grace, D., Mohorĉiĉ, M.: Broadband communications via high altitude platforms. IEEE Commun. Surv. Tutor. **7**(1), 2–31 (2010)
7. Ayaz, B., Allen, A., Wiercigroch, M.: Improving routing performance of underwater wireless sensor networks. In: OCEANS 2017, Aberdeen, pp. 1–9 (2017)

LAMP: Lightweight and Accurate Malicious Access Points Localization via Channel Phase Information

Liangyi Gong[1], Chundong Wang[1], Likun Zhu[1], Jian Zhang[1], Wu Yang[2], Yiyang Zhao[3], and Chaocan Xiang[4(✉)]

[1] Tianjin Key Laboratory of Intelligence Computing and Novel Software Technology, Ministry of Education, Tianjin University of Technology, Tianjin 300384, China
gongliangyi@gmail.com

[2] Information Security Research Center, Harbin Engineering University, Harbin, China

[3] AI Laboratory, Simple Edu Co., Beijing, China

[4] The Department of Computer Science, Army Logistics University of PLA, Beijing, China
xiang.chaocan@gmail.com

Abstract. Malicious access points (APs) have emerged as a serious security problem. In this paper, we propose LAMP, a lightweight and accurate malicious access points localization via channel phase information. We utilize a commercial wireless card for malicious APs localization with little human effort. The localization framework consists of three components: feature extraction, direction determination and localization. We first extract the available channel phase information by a linear transformation and purify eigenvalues by denoising. Then, the direction of an AP can be accurately detected by our developed self-adaptive MUSIC algorithm. Finally, the localization of the AP is realized by a triangulation algorithm with only one device placed at multiple locations. Experimental results show the median detection error of angle is about 5° and median localization error is about 35 cm.

Keywords: Malicious access point · Channel phase information
MUSIC algorithm · Triangulation

1 Introduction

With the rapid development of mobile computing, WiFi plays an increasingly important role in our daily lives. Some criminal offenders utilize the laptop or the embedded board to construct a malicious AP, which brings great threats for personal privacy and fund security. However, active work has been done to identify a malicious AP but less work is done to locate the malicious AP's position after detecting its presence.

In our work, we take the basic view point on how to locate a malicious AP's position. The state-of-the-art malicious access point localization approaches can

S. Chellappan et al. (Eds.): WASA 2018, LNCS 10874, pp. 748–753, 2018.
https://doi.org/10.1007/978-3-319-94268-1_62

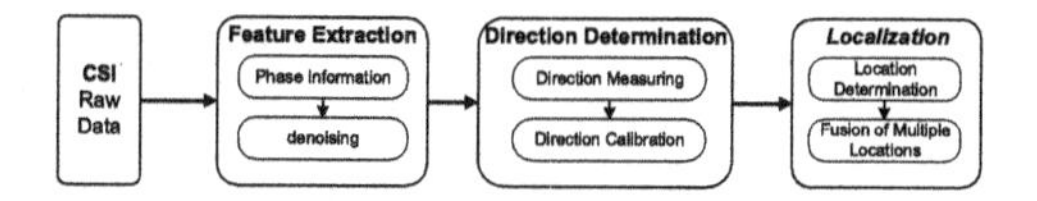

Fig. 1. System framework

be classified into two categories. The first category of the work leverages professional hardware to locate APs [6,7]. However, the work involves higher hardware cost and is neither scalable nor portable. The other category of the work utilizes human effort for direction determination. One commonly used method is to hold a wireless sniffer and walk along the direction with decreasing signal power to reach the AP [1]. One work [5] builds a signal contour map with multiple wireless sniffers in an area of interest to locate the malicious AP. However, the RSSI collected from the sniffer is seriously affected by indoor multipath effect [8]. Besides, they involve either heavy labor or high hardware cost.

In this paper, we design LAMP, a **L**ightweight and **A**ccurate **M**alicious access **P**oints localization via channel phase information. Our main idea is to leverage the beneficial phase information to determinate the direction of malicious APs and then to locate their locations. The LAMP contains three blocks to enable the malicious APs localization, as illustrated in Fig. 1. First, we obtain the transformed channel phase information and utilize the filter and sanitization to purify it. Second, the direction determination block utilizes the self-adaptive MUSIC algorithm to estimate the angle of the malicious AP. Third, we tackle this malicious AP localization based on triangulation with only one device. Extensive experiments show that the average detection error of angle is about 5° and the median error of localization is about 35 cm.

We summarize the main contribution of our works as following:

- We put forward a lightweight and accurate malicious APs localization via channel phase information, called as LAMP. It can achieve high accurate the malicious APs localization by only one ubiquitous wireless device. Compared with previous works, the LAMP relies on minimal infrastructural cost and human effort.
- We develop a novel self-adaptive MUSIC algorithm which can estimate the arrival angle more accurately. Moreover, based on the triangulation principle, we realize the malicious APs localization with only one wireless device by constructing numerous virtual devices.

2 System Design

2.1 Feature Extraction

Channel Phase Information. The measured phase $\tilde{\phi}$ of CFR for the kth subcarrier can be expressed as: $\tilde{\phi_k} = \phi_k + 2\pi\frac{k}{N}\tau_\epsilon + \lambda + n$, where ϕ_k is real phase of kth subcarrier, N is the hits of FFT demodulator, τ_ϵ is the clock synchronization error that is proportional to the clock offset, λ is unknown constant phase error,

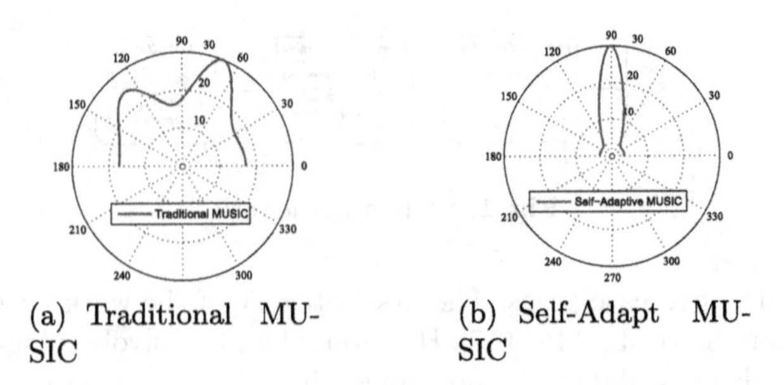

(a) Traditional MUSIC (b) Self-Adapt MUSIC

Fig. 2. Direction measuring results of different algorithms

n is some measurement noise. Because it is impossible to accurately measure and correct synchronization error of a transmitter and a receiver, the raw phase information obtained from the WiFi NICs behaves extremely random over the all feasible field.

To obtain real phases of CFR, we employ a linear transformation [3] on the raw phase to mitigate the random error (τ_ϵ, λ). The final processed phase can be written as: $\hat{\phi_{k_i}} = \phi_{k_i} - \frac{k_i}{k_i - k_1}(\phi_{k_n} - \phi_{k_1}) - \frac{1}{m}\sum_{j=1}^{m} \phi_{k_j}$.

Denoising. Though we have obtained the stable channel phase information, the phase information is still affected by environmental noises and hardware noises. The abnormal data can seriously affect the accuracy of direction determination. In the paper, we employ Savitzky-Golay filter [2] to filter the phase differences.

2.2 Direction Determination of Malicious APs

Direction Measuring: As shown in Fig. 2(a), a client in a line of sight (LOS) angle of 90° is located in 69°. The reason may be that the radio frequency (RF) oscillator brings phase offset in CSI, so we propose a self-adapt MUSIC algorithm based on traditional MUSIC algorithm.

If the true phases when the signal reaches our receiver with three antennas are respectively ϕ_0, ϕ_1, ϕ_2, the observed phases processed by the oscillators are actually measured as $\phi_0 + \psi_0, \phi_1 + \psi_1, \phi_2 + \psi_2$, where $\psi 0, \psi 1, \psi 2$ are the phase offset errors generated by the oscillators. In the self-adapt MUSIC algorithm, we are concerned with the phase offset errors of receiving signals between antennas, that is $\phi_1 - \phi_0, \phi_2 - \phi_1$. Let $\delta_0 = \psi_0 - \psi_1$, $\delta_1 = \psi_1 - \psi_2$. We assume that the phase offset errors between the antennas are $\langle \delta_0, \delta_1 \rangle$. If we can obtain the values of δ_0 and δ_1, we can gain real phase offset errors among antennas from the observations.

However, $\langle \delta_0, \delta_1 \rangle$ is a hidden random variable, so we can not directly get the size of these two variables and only select all combinations of $\langle \delta_0, \delta_1 \rangle$ to find the best combination of the reception effect. We estimate the best deviation for each packet and then calculates the offset of all packets. We will generate two phase offset values $\langle \delta_0, \delta_1 \rangle$ in the multiple clustering algorithm and select the most frequent phase offset combination as our final estimation.

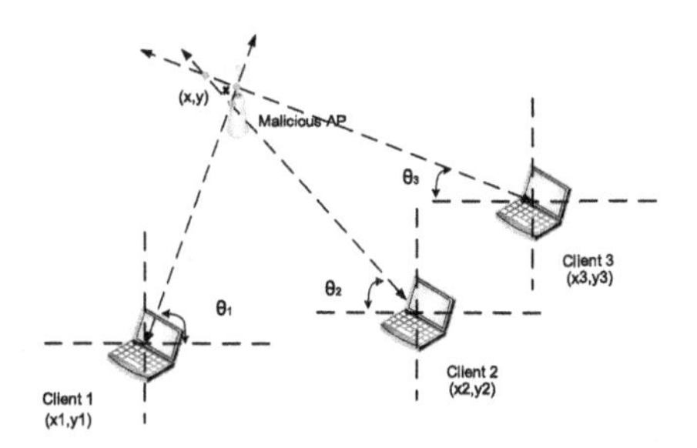

Fig. 3. Localization of malicious AP through geometric relationship.

Direction Calibration: In the self-adapt MUSIC algorithm, we group three antennas into two groups {<antenna1, antenna2>, <antenna2, antenna3>}. In each group, we calculate the AoA based on only main eigenvalue of feature vector E. At last, we obtain the accurate LOS angle $\hat{x} = \frac{1}{2}(\dot{x}_1 + \dot{x}_2)$, where $\dot{x}_i$ is the output detected angle of each group. Testing the same situation as the traditional algorithm, that is, the 90° AP on the LOS, our self-adaptive algorithm can accurately detect the LOS angle as 89°, as shown in Fig. 2(b).

2.3 Localization of Malicious APs

Location Determination: A simple localization system employs at least two receivers to locate the locations of the AP, as shown in Fig. 3. The localization needs two inputs: the physical locations where the sensing device is placed, (x_i, y_i) for i^{th} position, and the angle θ_i derived at i^{th} position towards the malicious AP. For i^{th} position, a straight line: $y = a_i x + b_i$ can be uniquely determined by the two inputs, where $a_i = -tan(\theta_i)$ and $b_i = y_i + x_i tan(\theta_i)$.

If we employ two receivers, two straight lines can intersect at one point. The intersection point $(\hat{x}, \hat{y})$ is the position of the malicious AP. Then we can achieve the coordinates of estimated location of the malicious AP: $\hat{x} = -\frac{b_1 - b_2}{a_1 - a_2}$, $\hat{y} = \frac{a_1 b_2 - a_2 b_1}{a_1 - a_2}$, where $a_1 = -tan(\theta_1)$ and $b_1 = y_1 + x_1 tan(\theta_1)$, $a_2 = -tan(\theta_2)$ and $b_2 = y_2 + x_2 tan(\theta_2)$. However, it is high-cost to utilize two clients. It is expected to realize the LAMP only relying on one device. So, we put forward a scheme: we derive the direction of the AP at multiple locations instead of multiple clients.

Fusion of Multiple Locations: In order to reduce the influence of direction determination errors on the localization accuracy, it is better to repeat direction determination at more than two locations. So, we can obtain multiple intersections $(\hat{x_i}, \hat{y_i})$, $i = 1, 2, \ldots, n$, where n is the total number of intersections. Then we can derive the location of the malicious AP by calculating the centroid of these intersections: $(x, y) = (\frac{1}{n}\sum_{i=1}^{n} \hat{x_i}, \frac{1}{n}\sum_{i=1}^{n} \hat{y_i})$. The geometric relationship based our method is shown in Fig. 3. The client is placed at three different locations to perform direction determination and locate the malicious AP.

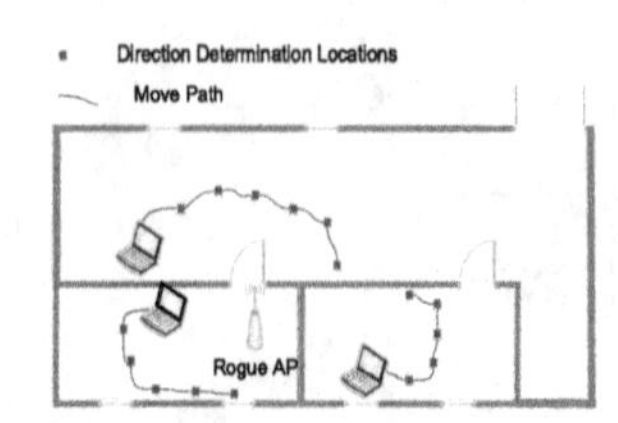

Fig. 4. Experimental testbed.

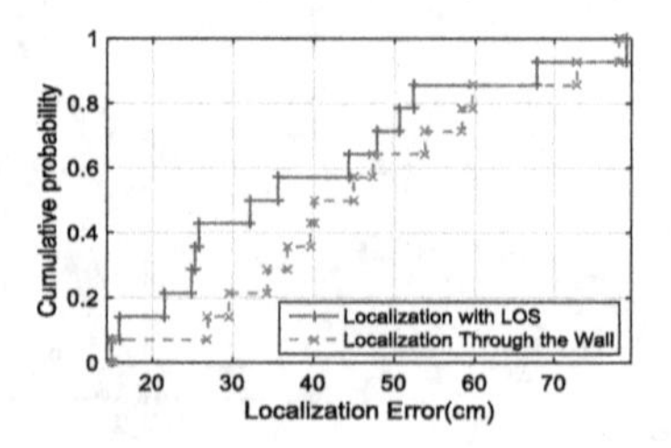

Fig. 5. CDF of localization error.

3 Experimental Evaluation

3.1 Experimental Setting

As shown in Fig. 4, our experimental environment is a small floor with two office rooms. The experimental facilities are two mini PCs equipped with Intel 5300 WiFi NICs and 3 antennas. We installed CSI tools on the mini PCs [4]. We set one as a rogue AP and put it in the 6.35 m × 8.5 m office room. In order to eliminate the interference of surrounding signals, we set the wireless card at the monitor mode and 5.32 GHz band. As shown in Fig. 4, we put the receiver at multiple places and calculate the angle of AP.

3.2 Performance Evaluation

(1) Localization Performance: We present the localization estimation results in the Fig. 5 with two types of experiments. The localization error is related to the direction determination accuracy. From the figure, we can observe that the median localization error with LOS path is about 35 cm, and the median localization error through the wall is about 45 cm. Because the number of multipath signals increases after the WiFi signals go through the wall, the error of direction determination increases affected by indoor multiple effect.

(2) Direction Determination: In the Fig. 6, we present the performance of direction determination with LOS path in the office room. From the Fig. 6(a), it can be observed that the performance of direction determination based on self-adapt MUSIC in different degrees is better than that based on original MUSIC algorithm. The detected angles of our algorithm in degrees with average error 6.3° are closer to the real values than the results of MUSIC algorithm. As shown in the Fig. 6(b), the median error of direction determination based on the self-adapt MUSIC algorithm is about 5°, but the median error of direction determination based on original MUSIC algorithm is about 11°.

We evaluate the time window with different numbers of packages on the performance of direction determination. The result is shown in Fig. 6(c) and proves that the error in degrees decreases with the increasing of number of packets. The 80% of the APs detection errors with 300 packets don't exceed 10°.

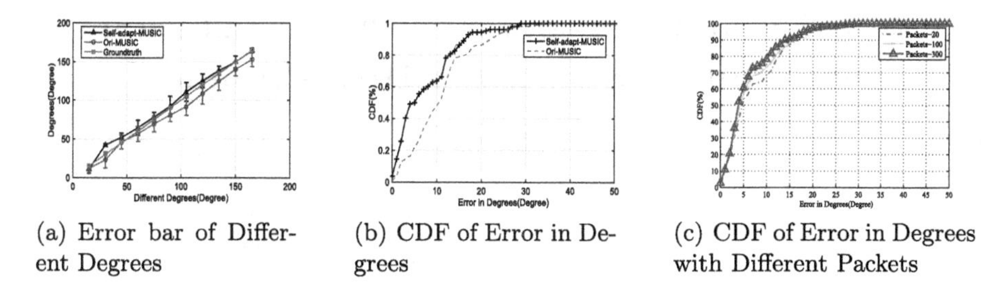

(a) Error bar of Different Degrees

(b) CDF of Error in Degrees

(c) CDF of Error in Degrees with Different Packets

Fig. 6. Direction determination performance

4 Conclusion

In this paper, we propose LAMP, a lightweight and accurate malicious access points localization system. We utilize purified channel phase information for accurate direction determination of malicious APs with little human effort. The location of malicious AP can be located with only one device. Experimental results prove that our system can achieve high performance of localization.

Acknowledgments. Our work is supported by NSF China (Grants No. 61502520, 61472098, 61672038 and 61572366), National Key R&D Plan (2016YFB0800805), Chongqing Research Program of Basic Research and Frontier Technology: No. CSTC2016JCYJA0053.

References

1. Adelstein, F., Alla, P., Joyce, R., Iii, G.G.R.: Physically locating wireless intruders. In: Proceedings of the International Conference on Information Technology: Coding and Computing, ITCC, vol. 1, pp. 482–489 (2004)
2. Fearn, T.: Savitzky-Golay filters, pp. 14–15 (2000)
3. Qian, K., Wu, C., Yang, Z., Liu, Y., Zhou, Z.: PADS: passive detection of moving targets with dynamic speed using PHY layer information. In: IEEE International Conference on Parallel and Distributed Systems, pp. 1–8 (2015)
4. Qian, K., Wu, C., Zhou, Z., Zheng, Y., Yang, Z., Liu, Y.: Inferring motion direction using commodity Wi-Fi for interactive exergames. In: CHI Conference, pp. 1961–1972 (2017)
5. Schweitzer, D., Brown, W., Boleng, J.: Using visualization to locate rogue access points. Consortium for Computing Sciences in Colleges (2007)
6. Shah, S.F.A., Srirangarajan, S., Tewfik, A.H.: Implementation of a directional beacon-based position location algorithm in a signal processing framework. IEEE Trans. Wirel. Commun. **9**(3), 1044–1053 (2010)
7. Subramanian, A.P., Deshpande, P., Jie, G., Das, S.R.: Drive-by localization of roadside WiFi networks. In: The Conference on Computer Communications, INFOCOM 2008, pp. 718–725. IEEE (2008)
8. Yang, Z., Zhou, Z., Liu, Y.: From RSSI to CSI: indoor localization via channel response. ACM Comput. Surv. **46**(2), 1–32 (2013)

NASR: NonAuditory Speech Recognition with Motion Sensors in Head-Mounted Displays

Jiaxi Gu[1(✉)], Kele Shen[2], Jiliang Wang[2], and Zhiwen Yu[1]

[1] School of Computer Science and Engineering, Northwestern Polytechnical University, Xi'an 710072, People's Republic of China
gujiaxi@mail.nwpu.edu.cn
[2] School of Software, Tsinghua University, Beijing 100084, People's Republic of China

Abstract. With the growing popularity of Virtual Reality (VR), people spend more and more time wearing Head-Mounted Display (HMD) for an immersive experience. HMD is physically attached on wearer's head so that head motion can be tracked. We find it can also detect subtle movement of facial muscles which is strongly related to speech according to the mechanism of phonation. Inspired by this observation, we propose NonAuditory Speech Recognition (NASR). It uses motion sensor for recognizing spoken words. Different from most prior work of speech recognition using microphone to capture audio signal for analysis, NASR is resistant to acoustic noise of surroundings because of its nonauditory mechanism. Without using microphone, it consumes less power and requires no special permissions in most operating systems. Besides, NASR can be seamlessly integrated into existing speech recognition systems. Through extensive experiments, NASR can get up to 90.97% precision with 82.98% recall rate for speech recognition.

Keywords: Head-Mounted Display · Motion sensor
Speech recognition · Machine learning

1 Introduction

With the growing popularity of Virtual Reality (VR), more and more people put their heads in various Head-Mounted Display (HMD) for an immersive experience, especially for watching 360° videos and playing VR games. An HMD is attached on one's head to track his/her head movement to provide dynamic visual content or interaction. We find that the subtle head movement caused by phonation can also be recorded and then used for speech recognition. Inspired by this observation, we propose NonAuditory Speech Recognition (NASR).

In comparison with traditional audio-based methods, NASR is low-cost and loose-constraint. It is based on the anatomic structure motion caused by phonation. When one speaks, his/her facial muscles moves to shape the sound and air

S. Chellappan et al. (Eds.): WASA 2018, LNCS 10874, pp. 754–759, 2018.
https://doi.org/10.1007/978-3-319-94268-1_63

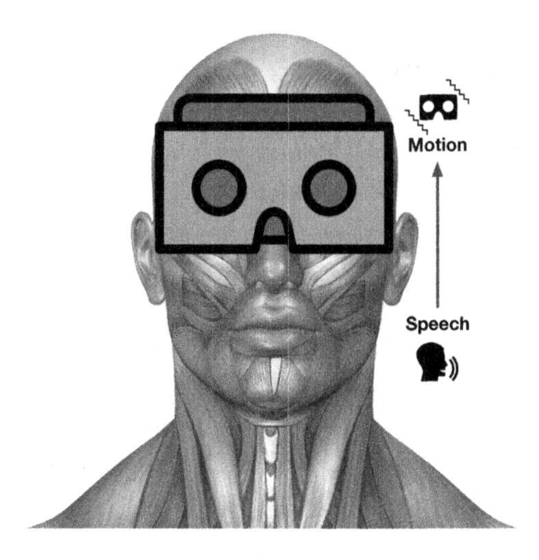

Fig. 1. Human speech is produced with the movement of relevant facial muscles. Through the motion sensors in HMD, speech can be recognized.

stream into recognizable speech [1]. As Fig. 1 shows, the movement of facial muscles can be captured by the sensors of HMD. Different from microphones, motion sensors are always available without consuming much power. Besides, NASR is inherently resistant to acoustic noise of surroundings. It can be seamlessly integrated into existing speech recognition models. To the best of our knowledge, NASR is the first work to merely use motion signal to do speech recognition.

It is however nontrivial to implement NASR because of the following challenges. Firstly, as speech is produced mainly by the muscles near lips, the motion of muscles near HMD is relatively weak. Secondly, different persons have more or less diverse accents. It makes the motion patterns of facial muscles hard to recognize. Last but not least, the speech tone and speed make a difference even for the same individual. Therefore, multiple variables have to be considered to make NASR robust and accurate for speech recognition.

In summary, we have the following contributions in this paper:

1. We propose a creative speech recognition method for HMD by merely using motion sensors without touching any audio signal.
2. We present a multidimensional motion data structure and extract relevant features for machine learning.
3. We implement NASR in the real world using an HMD equipped with a smart phone and evaluate it with extensive experiments.

2 Methodology

The application scenario of NASR is speech recognition with motion sensors in HMDs. The source data is motion signal recorded during the speaking of an

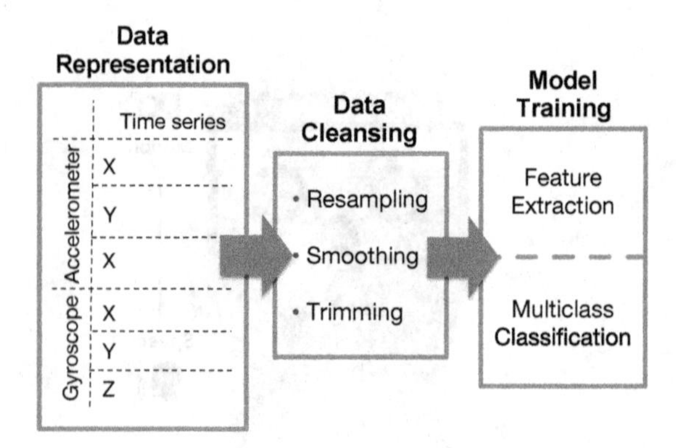

Fig. 2. The whole process of NASR is divided into three main phases.

HMD wearer. The objective is to recognize the speech content. For this objective, we use a machine learning method to build a classifier. By extracting the relevant features from the motion signal, we can build a classifier for recognize the incoming utterance. As Fig. 2 shows, the whole process of NASR can be divided into three main phases: *data representation*, *data cleansing* and *model training*.

2.1 Data Representation

The accelerometer and gyroscope are two typical motion sensors in consumer HMDs. Either sensor has 3 axes so we have total 6 time series. We denote it as:

$$\mathbf{S} = \begin{array}{c} \\ acce_x \\ acce_y \\ acce_z \\ gyro_x \\ gyro_y \\ gyro_z \end{array} \begin{array}{c} \begin{array}{cccccc} 1 & 2 & \dots & t & \dots & n \end{array} \\ \begin{pmatrix} S_{1,1} & S_{1,2} & \dots & S_{1,t} & \dots & S_{1,n} \\ S_{2,1} & S_{2,2} & \dots & S_{2,t} & \dots & S_{2,n} \\ S_{3,1} & S_{3,2} & \dots & S_{3,t} & \dots & S_{3,n} \\ S_{4,1} & S_{4,2} & \dots & S_{4,t} & \dots & S_{4,n} \\ S_{5,1} & S_{5,2} & \dots & S_{5,t} & \dots & S_{5,n} \\ S_{6,1} & S_{6,2} & \dots & S_{6,t} & \dots & S_{6,n} \end{pmatrix} \end{array} \in \mathbb{R}^{6\times n}$$

The rows of $\mathbf{S}$ represent different axes of motion sensors. The columns of $\mathbf{S}$ represent samples of sensor readings on the timeline. The n is determined by the sampling rate and duration. For simplicity, we limit a separate word utterance into a fixed time span, i.e., 2 s. Thus, the sampling rate is a critical factor. As different types of motion sensors or sensors made by different manufacturers may have different sampling rate, we use a linear interpolation to do resampling to make all the six sensor data have the same number of samples in the fixed sampling time. As a result, it is reasonable to represent the motion pattern with a matrix $\mathbf{S}$.

2.2 Data Cleansing

First, for making the motion signal uniform and reasonable, we resample the data from axes of different sensors in a specific sampling rate. Second, for removing the interference from physiological activities such as heartbeat and respiration, we use a low-pass filter to remove the high-frequency component for smoothing. Last but not least, the time span of an utterance is uncertain. To generate valid training data for our model, we need to trim each sample to contain exactly one spoken word. Ideally, the time span of a sample needs to start and stop at the edges of the spoken word. We use a threshold of motion amplitude and remove the period of time in which any motion is barely detected.

2.3 Model Training

Feature engineering on time series is a classic but still challenging process. Possible relevant features include basic features of the time series such as the mean value, the standard deviation, the number of peaks or more complex features such as the time reversal symmetry statistics. After feature extraction, we use a subset of samples as a training set. Since the target words are predefined, it is a typical multiclass classification. There are various classification algorithms for this purpose such as Decision Tree and Random Forest. We will evaluate the performance of several different classification algorithms.

3 Implementation and Evaluation

The HMD kit of our experiments consists of two components including a smart phone and a plastic VR headset. The smart phone we use is Samsung Galaxy S7 running Android 8.0 and the VR headset is a plastic version of Google Cardboard which costs about $10. We develop a VR application using Google VR SDK to get motion sensor data in real time. For model training, we generate speech samples from five adults including three males and two females. Our predefined word set contains three common English words: "private", "secret" and "password". For each word, we collect 280 samples in total from different people. We also collect unknown-word samples while the wearer is irrationally speaking for 4 min. Therefore, we have 4 labels in total.

For feature extraction, we use a development tool called TSFRESH which stands for *Time Series Feature extraction based on scalable hypothesis tests*. It can automatically extract features from time series. In our experiments, 500 relevant features are used. Basically, we adopt Random Forest for multiclass classification and other popular machine learning methods are evaluated too.

In the preliminary experiments, we use a constrained version of samples. Each sample lasts exact 2 s and contains at most one spoken word. We use 50% of samples as training instances and the other 50% as testing instances. We start with the subsets containing samples generated by every single person. The model training and testing are both based on the subsets. The recognition accuracy of

Table 1. The performance of NASR on different words.

Word	Private	Secret	Password	Unknown-word
Precision	81.25%	96.43%	96.55%	94.83%
Recall	83.87%	71.05%	84.85%	96.49%
F1-score	82.54%	81.82%	90.32%	95.65%

Table 2. The performance of NASR on different persons.

Person	#1	#2	#3	#4	#5
Precision	80.00%	80.00%	88.24%	100%	96.30%
Recall	100%	87.50%	83.33%	88.24%	92.86%
F1-score	88.89%	83.58%	85.71%	93.75%	94.55%

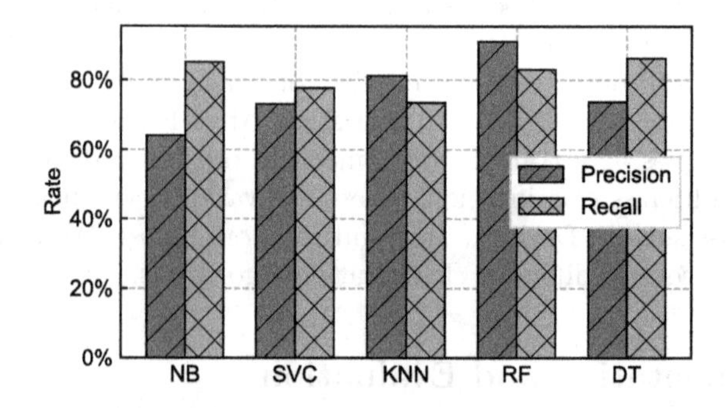

Fig. 3. The performance of NASR using different machine methods. (NB: Naive Bayes; SVC: Support Vector Classification; KNN: K-Nearest Neighbors; RF: Random Forest; DT: Decision Tree.)

different words is shown in Table 1. Then, we consider the performance of NASR between different persons. The results are shown in Table 2.

In the near-pilot experiments, we use the whole set of samples. Similarly, the 50% of samples are used for training and the rest are for testing. The results of different machine learning methods are shown in Fig. 3. The Random Forest gets the best performance and its precision is 90.97% and recall rate is 82.98%.

4 Related Work

With the development of wearable devices, motion signal can be used for various applications. Some works use accelerometers or gyroscopes on smart phones for medical purposes such as monitoring heart rate or respiration [6]. Different from using motion signal directly, Balakrishnan et al. use subtle head motion in videos to extract heart rate and beat lengths [2]. Their work points out that the Newtonian reaction to the influx of blood at each heart beat cause recognizable head motion. Hernandez et al. design a series applications for measuring physiological signs using head-mounted glasses [3], smart phone [4] and smart watch [5]. Besides, Mohamed and Youssef show that the gyroscope sensor is the most sensitive sensor for measuring the heart rate [7]. In addition to motion sensors, there are also works using wireless signal such as WiFi [8] and RFID [9] for detecting human behaviors. These works may bring more inspirations to NASR in the future.

5 Conclusion

In this paper, we propose NASR which is a nonauditory speech recognition system. Merely relying on the consequent motion during the usage of HMD for VR, the wearer's spoken words can be recognized without touching microphones. NASR shows advantages from various aspects including low battery consumption and resistance to surrounding noise. It can also help to make several potential applications such as voice command assistant like Siri and accessibility-related products. In the future work, we will test NASR on a larger scale and extend it to real-time for continuous speech recognition.

References

1. Abbs, J.H., Gracco, V.L., Blair, C.: Functional muscle partitioning during voluntary movement: facial muscle activity for speech. Exp. Neurol. **85**(3), 469–479 (1984)
2. Balakrishnan, G., Durand, F., Guttag, J.: Detecting pulse from head motions in video. In: IEEE Conference on Computer Vision and Pattern Recognition, pp. 3430–3437 (2013)
3. Hernandez, J., Li, Y., Rehg, J.M., Picard, R.W.: Bioglass: physiological parameter estimation using a head-mounted wearable device. In: International Conference on Wireless Mobile Communication and Healthcare, pp. 55–58 (2014)
4. Hernandez, J., McDuff, D.J., Picard, R.W.: Biophone: physiology monitoring from peripheral smartphone motions. In: International Conference of the IEEE Engineering in Medicine and Biology Society, pp. 7180–7183 (2015)
5. Hernandez, J., McDuff, D., Picard, R.W.: Biowatch: estimation of heart and breathing rates from wrist motions. In: Proceedings of the 9th International Conference on Pervasive Computing Technologies for Healthcare, pp. 169–176 (2015)
6. Kwon, S., Lee, J., Chung, G.S., Park, K.S.: Validation of heart rate extraction through an iPhone accelerometer. In: International Conference of the IEEE Engineering in Medicine and Biology Society, pp. 5260–5263 (2011)
7. Mohamed, R., Youssef, M.: Heartsense: ubiquitous accurate multi-modal fusion-based heart rate estimation using smartphones. Proc. ACM Interact. Mob. Wearable Ubiquit. Technol. **1**(3), 97 (2017)
8. Wu, C., Yang, Z., Zhou, Z., Liu, X., Liu, Y., Cao, J.: Non-invasive detection of moving and stationary human with WiFi. IEEE J. Sel. Areas Commun. **33**(11), 2329–2342 (2015)
9. Zhou, Z., Shangguan, L., Zheng, X., Yang, L., Liu, Y.: Design and implementation of an RFID-based customer shopping behavior mining system. IEEE/ACM Trans. Netw. **25**(4), 2405–2418 (2017)

Energy Harvesting Based Opportunistic Routing for Mobile Wireless Nanosensor Networks

Juan Xu, Jiaolong Jiang(✉), Zhiyu Wang, and Yakun Zhao

Tongji University, Shanghai 201804, China
{jxujuan,1631515,1333839,1631516}@tongji.edu.cn

Abstract. Wireless nanosensor networks (WNSN) are new type of networks that combing nanotechnology and sensor network. Aiming at the mobility problem and the energy limitation of nanonodes, a mobile WNSN opportunistic routing protocol (MWOR) based on energy harvesting is proposed. In the protocol, a piezoelectric nano energy harvesting system is introduced to break the energy bottleneck. Then the protocol establishes a new routing metric model based on the ETX measure, a node motion vector model is introduced to reflect the mobility of node, and a node survivability model is added to reflect the energy harvesting feature of the nanonodes. Simulation results show that this protocol can realize self-powered nanosensors and perpetual wireless nanosensor networks. And it has advantages in transmission success rate, energy consumption balance and effective throughput, and can be applied to mobile WNSN.

Keywords: WNSN · Opportunistic routing · Energy harvesting Mobility

1 Introduction

Wireless nanosensor networks (WNSN) are novel networks which consist of large numbers of nanosensors. These nanonodes are small in size [1], and they are expected to communicate in the Terahertz Band (0.1–10 THz). Terahertz wireless communication [2] has the characteristics of high transmission rate (Gbps–100 Gbps) and large capacity. Thus WNSNs will enable a wide range of applications in biomedical, environmental, industrial and military fields.

Routing protocol plays an important role both in WSN and WNSN. In many application scenarios, the nodes are inevitably moved by the influence of external force. In the WNSN, the movement of nodes can cause many problems. Such as, it increases the probability of packet loss, makes the terahertz channel quality changes with time and it makes the network topology changes dynamically. So this paper will study the mobile WNSN routing protocol.

The movement of the nodes will cause the dynamic change of the link, but it also increases communication opportunities with other nodes. Therefore, in

S. Chellappan et al. (Eds.): WASA 2018, LNCS 10874, pp. 760–766, 2018.
https://doi.org/10.1007/978-3-319-94268-1_64

WSN, some scholars have proposed opportunistic routing (OR) [4]. In the opportunistic routing, all the candidate nodes will cache the received data, then they find the opportunity to forward data during their moving. Normally, the data is forwarded by the node with the highest priority in the candidate set. Firstly, the routing measures based on ETX [4] are established, which can reflect the quality of the terahertz wireless channel. Besides, it introduces of mobile vector model to reflect the mobility of nodes. And the node survivability model is introduced to characterize the energy harvesting of nodes. Secondly, in order to reduce the control overhead of routing, the MWOR protocol adopts the candidate node coordination mechanism based on the competitive backoff, to avoid redundant and resource waste due to repeated forwarding of multiple candidate nodes, and establishes a reliable route in this way.

2 Design of MWOR Protocol

The main two factors that matter in opportunistic routing are, the candidate node set selection criteria, and the coordination mechanism between candidate nodes.

2.1 Route Measurement Model

Aiming at the limitation of existing routing measures, this paper proposes a opportunistic routing MWOR with diverse measures for mobile WNSN. Based on the ETX [4] measure, this protocol establishes a new routing metric model. It introduces the node motion vector model to reflect the trend of nodes' movement, and the survivability model is added to reflect the energy harvesting characteristics of the nanonodes. So the routing metric of the MWOR protocol can be expressed as:

$$M = \lambda_1(1/ETX) + \lambda_2\boldsymbol{D} + \lambda_3 RE \tag{1}$$

where M represents the measured value of a node, λ_1, λ_2 and λ_3 are the system constants.

ETX in (1) is defined as the expected transmission times the nodes successfully send a data packet to destination node, including the times of retransmissions, which reflects the quality of the wireless channel. It can be expressed as [7]:

$$ETX = \frac{1}{p_f \times p_r} \tag{2}$$

where p_f indicates the probability that the receiving end successfully receives the data packet, p_r indicates the probability that the sending end has successfully received the ACK message. The forward and reverse data transfer rates are the detecting result of a fixed-length probe packet broadcasted by the node. These two parameters can be obtained by simulating the Terahertz channel [5].

In order to reflect the moving characteristics of the nanonodes in the measure, we introduce the motion vector model $\boldsymbol{D}$ [8], which can be expressed as:

$$\boldsymbol{D} = \begin{cases} \left|\frac{d_{t_0} - d_{t_1}}{\Delta t}\right|, d_{t_0} \geq d_{t_1} \\ -\left|\frac{d_{t_0} - d_{t_1}}{\Delta t}\right|, d_{t_0} < d_{t_1} \end{cases} \tag{3}$$

where d_{t_0} and d_{t_1} respectively represent the distance between the nanonode and the control node at the time t_0 and time t_1, and Δt is the time window between the time t_0 and the time t_1. The movement vector $\boldsymbol{D}$ in Eq. (3) not only reflects the node's movement rate but also reflects the direction of movement of the node. If in the time Δt, the nanonode moves towards the control node, the value of the moving vector $\boldsymbol{D}$ is positive. If the nanonode is far from the control node, The value of the motion vector $\boldsymbol{D}$ is negative. If the moving speed of the nanonode is faster, the value of the moving vector $\boldsymbol{D}$ is larger.

As the node uses a piezoelectric energy acquisition system [6], the routing metric also need to consider the node's energy harvesting feature. The survivability model RE of the nanonode is established by combining the residual energy value of the nanonode and the rate of energy harvesting. It can be expressed as:

$$RE = \frac{r_{eh} \log \mu}{E_{max}(\mu^{\lambda} - 1)}, \lambda = \frac{E_{max} - E_{re}}{E_{max}} \tag{4}$$

where r_{eh} is the energy acquisition rate [6], E_{max} is the maximum value of the energy storage of the nanometer battery, E_{re} is the residual energy value of the node, and μ is a system constant. RE in Eq. (4) reflects the survivability of nanonodes at a certain time. In the case that all nodes have the same maximum storage capacity, the more remaining energy the node has, the faster the energy acquisition rate is, and the bigger the RE value is, which indicates that a node can work longer under current condition.

2.2 Update the Probability of Forwarding

In the MWOR protocol, node v_i uses the forwarding probability P_{v_i} as the criterion for selecting the candidate set $V_c(v_i)$. Since the time-varying characteristics of the mobile WNSN network, it is necessary to periodically update the forwarding probability in a reasonable way to reflect the state of nodes at different time.

At the initial stage of the route establishment, the forwarding probability values of all the nanonodes are set to 0. During the route establishment phase, the control node broadcasts the probe packet to all the nodes in the cluster at time interval τ. If the nanonode receives the probe packet, it's forwarding probability value increases in following way:

$$P_{v_i}(t) = P_{v_i}(t - \tau) + M_{v_i}(t) \tag{5}$$

where $P_{v_i}(t - \tau)$ and $P_{v_i}(t)$ represent the forwarding probability values of the nodes v_i at time $t - \tau$ and t, respectively, and $M_{v_i}(t)$ is the routing metric of the nanonode v_i at time t, which can be obtained from (1). As can be seen from (5), the nanonode that moves toward the control node and has stronger survivability will have a larger forwarding probability value. If the nanonode v_i has not received the probe packet after intervals $n\tau$, the forwarding probability is reduced as follows:

$$P_{v_i}(t) = \frac{P_{v_i}(t - n\tau)}{n\tau} \tag{6}$$

where $P_{v_i}(t - n\tau)$ represents the forwarding probability of the node v_i in the time interval $n\tau$. Equation (6) reflects the channel quality between the node v_i and the control node. The longer a node has not received the probe packet, the smaller forwarding probability the node has.

2.3 Coordination Mechanism Between Candidate Nodes

In order to reduce the overhead of routing, the MWOR protocol uses a coordination mechanism based competition to coordinate candidate nodes. When multiple candidate nodes cache the received data packets, they don't participate in the forwarding data packet competition immediately, but wait for a backoff time. The backoff time T_{b-o} of the candidate node v_i obeys the exponential distribution, and its probability density function can be expressed as:

$$f_{T_{b-o}}(t) = \frac{1}{\lambda(v_i)} e^{-\frac{t}{\lambda(v_i)}} \tag{7}$$

where $\lambda(v_i)$ represents the average backoff time of the nanonode v_i, which is a function related to the node forwarding probability, and can be calculated by the following formula:

$$\lambda(v_i) = K_{b-o}/P_{v_i} \tag{8}$$

where P_{v_i} represents the forwarding probability of the nanonode v_i, K_{b-o} is a system constant. As can be seen from Eq. (8), the candidate node with higher forwarding probability will have shorter backoff time, thus gets higher forwarding priority.

3 Simulation Analysis

This section we will simulate the MWOR protocol. Our simulation is based on the Nano-Sim platform proposed in [3]. We will compare MWOR with ExOR [4] protocol and flooding protocol [3]. The simulation experiment will analyze the transmission success rate, effective throughput and energy consumption of the mobile WNSN.

3.1 Simulation Results Analysis

The Success Rate of Data Packet Transmission. Figure 1 is a comparison of the transmission success rates of the three routing protocols. With the change of node density, the performance of MWOR protocol in transmission success rate is always better than the other two routes, because the measurement model in MWOR protocol takes the nodes' mobility and energy supplement characteristics into account, to a certain extent, the communication quality of the selected link is ensured.

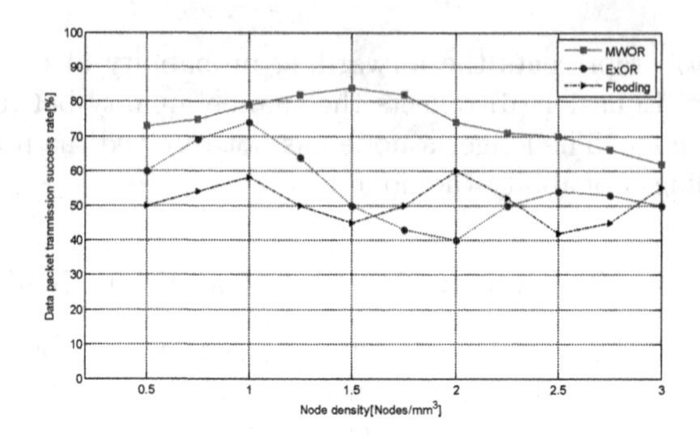

Fig. 1. Comparison of data transmission success rate.

Energy Consumption. Figure 2 compare the variance of the residual energy of the node respectively. It can be seen from Fig. 2, MWOR protocol is always better than the other two routes in the performance of the residual energy variance, which is the result of adding energy supplementation characteristics to improve the ETX measure.

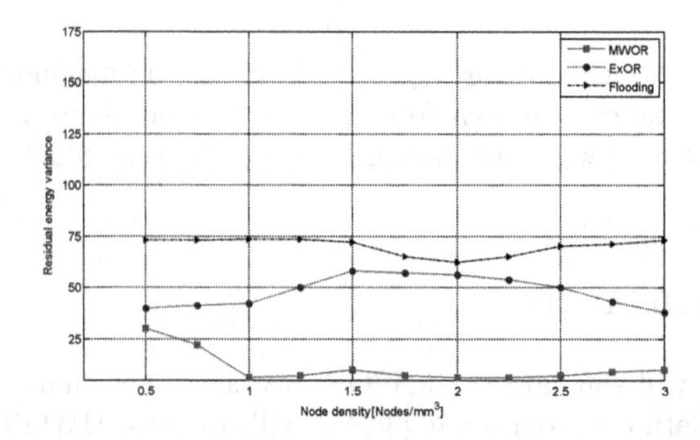

Fig. 2. Comparison of variance of residual energy.

Effective Throughput. Figure 3 shows the effective throughput of the three routes. The effective throughput is defined as the number of bits that have been successfully received per unit of time. It can be seen, when the node density increases (>1.5$Nodes/\text{mm}^3$), the MWOR protocol shows its advantages.

Although the MWOR protocol does not take into account the transmission rate, the MWOR has advantages in the transmission success rate, so gets a higher effective throughput.

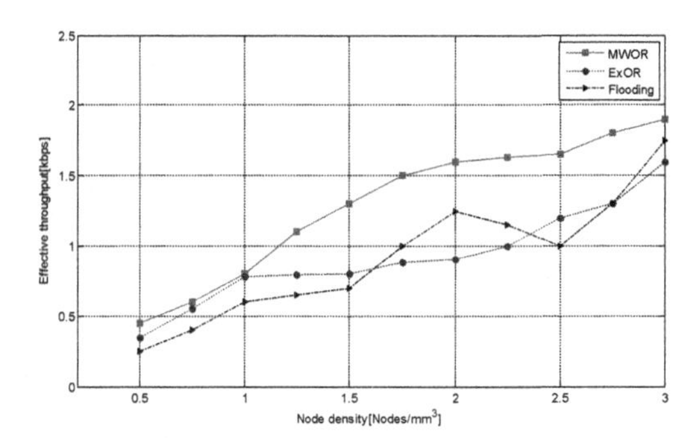

Fig. 3. Comparison of effective throughput.

4 Conclusion

In this paper, we propose a mobile WNSN opportunistic routing protocol based on energy harvesting. Based on the ETX measure, the protocol introduces the node motion vector model and the survivability model, and adopts the candidate node coordination mechanism based on the competitive backoff, so establishes the reliable route. Simulation results show that this protocol can realize self-powered nanosensors and perpetual wireless nanosensor networks. And it has advantages in transmission success rate, energy consumption balance and effective throughput, and can be applied to mobile WNSN.

References

1. Akyildiz, I.F., Brunetti, F., Blazquez, C.: Nanonetworks: a new communication paradigm. Comput. Netw. **52**(12), 2260–2279 (2008)
2. Song, H.J., Nagatsuma, T.: Present and future of Terahertz communications. IEEE Trans. Terahertz Sci. Technol. **1**(1), 256–263 (2011)
3. Piro, G., Grieco, L.A., Boggia, G.: Simulating wireless nano sensor networks in the NS-3 platform. In: IEEE International Conference on Advanced Information Networking and Applications Workshops, pp. 67–74 (2013)
4. Biswas, S., Morris, R.: ExOR: opportunistic multi-hop routing for wireless networks. In: Conference on Applications, Technologies, Architectures, and Protocols for Computer Communications, pp. 133–144 (2005)
5. Jornet, J.M., Akyildiz, I.F.: Channel modeling and capacity analysis for electromagnetic wireless nanonetworks in the Terahertz Band. IEEE Trans. Wirel. Commun. **10**(10), 3211–3221 (2011)
6. Xu, S., Hansen, B.J., Wang, Z.L.: Piezoelectric-nanowire-enabled power source for driving wireless microelectronics. Nat. Commun. **1**(7), 1–5 (2010)

7. DeCouto, D.S.J., Aguayo, D., Bicket, J.: A high-throughput path metric for multi-hop wireless routing. Wirel. Netw. **11**(4), 419–434 (2005)
8. Guangcheng, H., Xiaodong, W.: An opportunistic routing design based on RSSI in mobile sensor networks. J. Electron. **37**(3), 608–613 (2009)

Synergistic Based Social Incentive Mechanism in Mobile Crowdsensing

Can Liu[1], Feng Zeng[1(✉)], and Wenjia Li[2(✉)]

[1] School of Software, Central South University, Changsha, China
{liucan512,fengzeng}@csu.edu.cn

[2] Department of Computer Science, New York Institute of Technology, New York, USA
wli20@nyit.edu

Abstract. Most Mobile Crowdsensing (MCS) applications are large-scale and the quality of sensing for sensing tasks is interdependent. Previous incentive mechanisms have focused on quantifying participants' contribution to the quality of sensing and provide incentives directly to them, which are not applicable to the above scenario. To tackle this problem, in this article, we introduce a novel approach for MCS, called the synergistic based social incentive mechanism. The basic idea is to leverage the social ties among participants to promote cooperation. To maximize the utility of service provider, a moral hazard model is used to analyze the optimal contract between service providers and mobile users in the case of asymmetric information. Experiments show that the synergistic based social incentive mechanism can give users continuous encouragement while maximizing the utility of the principal.

Keywords: MCS · Incentive mechanism · Synergy
Social relationship

1 Introduction

Under the circumstance that smartphones integrate many sensors, the researches on MCS are becoming important and popular in recent years. Compared to traditional sensor networks, there are many advantages of data collection by MCS [1,2]. Crowdsensing with smart devices can be used for large scale sensing of the physical world at low cost by leveraging the available sensors on the devices.

When performing sensing tasks, resource consumption like power consumption and traffic consumption will reduce the enthusiasm of users to participate in the sensing tasks. Therefore, the study of incentive mechanism is of great significance in the mobile crowdsensing system. The three most dominant incentive mechanisms in MCS are the game based incentive mechanism [3], auction based incentive mechanism and contract based incentive mechanism.

S. Chellappan et al. (Eds.): WASA 2018, LNCS 10874, pp. 767–772, 2018.
https://doi.org/10.1007/978-3-319-94268-1_65

The reverse auction is the most frequently used auction method in the auction-based incentive mechanism [4–6]. Considering that users have the risk of exposing their privacy when participating in sensing tasks, [7,8] proposed a two-stage auction algorithm for privacy-preserving issues separately. The contract is a popular approach in platform-leading approaches [9–11]. Li et al. in [10] believed that the agreement on the qualities and payments in crowdsensing systems can be best modeled as a contract. Inspired by the effort-based reward from the labor market, several works [12,13] studied this problem by providing users with the amount of reward that is consistent with their performances.

The works above have provide necessary incentives for users to participate in the crowdsensing activity with independent tasks. In practice, most sensing applications are large-scale and the quality of sensing for sensing tasks are interdependent (e.g., data aggregation applications). Since the quality of sensing is interdependent among a collection of sensing tasks performed by independent participants, it is desired to stimulate cooperation among participants.

In this article, we propose a novel contract-based incentive mechanism by using the moral hazard problem from game theory, called the synergistic based social incentive mechanism. The social relationship between mobile users was applied to MCS, stimulating the cooperation between users. Then, we present detailed analysis of the conditions of cooperation and equilibrium status. Simulations are provided to demonstrate that the performance of the proposed incentive mechanism is better than traditional incentive mechanisms.

2 System Model

We consider large-scale sensing applications where the quality of sensing for sensing tasks is interdependent, and there is a synergy effects among users. Tasks in the MCS system are connected to each other due to interdependence, while participants are connected to each other by social relationships. In order to perform the task, participants need to cooperate with their social friends. It is shown in [14] that with the social relationship, participants can exert pressure on their social friends so that their behaviors will become better. We exploit this principle to promote cooperation between a participant and his/her social friends in this article. In the case of asymmetric information, we analyze the optimal contract between service providers and mobile users in order to maximize the utility of service providers.

2.1 Operation Cost

According to the commonly used method of simplifying the agent's effort cost function in the principal-agent problem, assume that the operation cost of the mobile user i is

$$C_i = \frac{1}{2}ka_i^2 + \frac{1}{2}k \sum_{j=1, j \neq i}^{n} a_{ij}^2 \tag{1}$$

The output of mobile user i is

$$x_i = a_i + \sum_{j=1, j \neq i}^{n} a_{ji} + \varepsilon_i, i = 1, 2, \ldots, n \tag{2}$$

$\varepsilon_i \sim N(0, \sigma^2)$, the sensing performance of tasks is given by Eq. 3.

$$X = \omega \sum_{i=1}^{n} x_i \tag{3}$$

2.2 Reward Package

Inspired by the effort-based reward from the labor market [15], we define the user's reward package R_i in crowdsensing as a linear combination of several rewards: (1) fixed salary, (2) performance-related reward, (3) task sharing reward, (4) spiritual motivation. The reward package of user i is written as Eq. 4.

$$R_i = R_0 + \beta x_i + \lambda X + R_p \tag{4}$$

2.3 Utility of Mobile User

All mobile users are rational and selfish. They are interested in accepting the mechanism provided by the principal only if the utility obtained under this mechanism is positive, the utility function of user i is shown in Eq. 5.

$$U_i = R_i - C_i \tag{5}$$

2.4 Utility of Principal

Assuming that the service provider is risk-neutral, the utility function of principal is expressed as Eq. 6.

$$U = E(X - \sum_{i=1}^{n} R_i) \tag{6}$$

3 Problem Formulation

With the system model, we can formulate the principal's utility maximization problem, which can be written as Eq. 7.

$$\begin{aligned} &\max_{a_i, a_{ij}, \beta, \lambda} \ U \\ &s.t (IR) U_i \geq \overline{U_i} \\ &(IC)(a_i, a_{ij}) \in \arg\max(U_i) \end{aligned} \tag{7}$$

Proposition 1. *The basic condition for achieving cooperate is $k > bn$. In this case, $a_{ij} > 0$, the mutual influence of social members is positive.*

Proof. The definition of a_{ij} shows that a_{ij} reflects the mutual relationship among users. Thus, the condition for achieving cooperation is $a_{ij} > 0$.

$$a_{ij}^* = \frac{b\beta + \lambda\omega k}{k(k - bn)} > 0 \Leftrightarrow k - bn > 0 \Leftrightarrow k > bn \tag{8}$$

Therefore, when $k > bn$, $a_{ij} > 0$, cooperation can be achieved. And there is the best strategy for users.

$$\begin{aligned} a_i^* &= \frac{(k - bn - b)\beta + \lambda\omega k}{k(k - bn)} \\ a_{ij}^* &= \frac{b\beta + \lambda\omega k}{k(k - bn)}, \forall j \neq i \end{aligned} \tag{9}$$

For the service provider, take the participation constraint into the objective function, the optimization strategy of principal is

$$\beta^* = \frac{kn^2\omega^2 (k - bn)^3 \rho\sigma^2}{-k^3 + n^2\omega \left[k + \rho(k - bn)^2\right] + \left[b(2k - bn) + (k - bn)^2 + \rho\sigma^2 k\right]}$$

$$\lambda^* = \frac{\left[b(2k - bn) + (k - bn)^2 + \rho\sigma^2 k\right]}{k + \rho\sigma^2(k - bn)^2} \left(1 - \frac{k^2\rho\sigma^2 k - bn^2}{-k^3 + n^2\omega}\right)(k - bn)\left(k + \rho k - bn^2\right)$$

4 Simulation

In the experiments, we conduct a comparison of the principal's utility among different incentive mechanisms: (1) Incentive mechanism proposed in this paper, we name it by General, (2) When mobile users's reward is only composed of material rewards, and thus we name it by Single Bonus, (3) The third one called Independent, in this case, mobile users' rewards are only related to the performance of tasks that they perform.

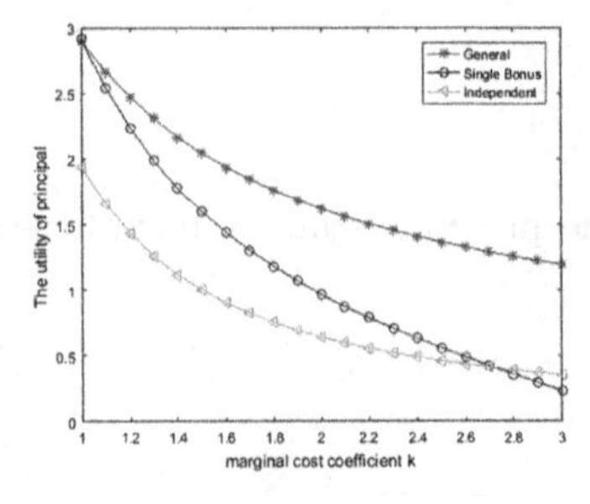

Fig. 1. The principal's utility as the marginal cost coefficient k varies

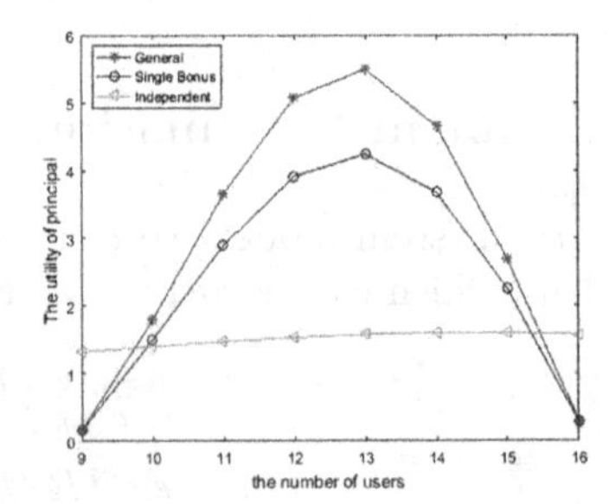

Fig. 2. The principal's utility as the number of users varie

From the the curves in Fig. 1, we see that as the cost coefficient k increases, the principal's utility is decreasing as well. In addition, we see that the principal obtains the largest utility in the General case. Followed by the Single Bonus, while the Independent gives the least utility. From the simulation results in Fig. 2, we see that as the number of users increases, the utility of the principal in the General case and the Single Bonus case increases, and then decreases after reaching the maximum value. In the Independent case, the utility of principal increases gradually with the increase of the number of users.

5 Conclusions

In this article, we have proposed a novel incentive mechanism for MCS, which leverages the power of social relationships to promote cooperation. Different from other works, we then focus on the scenario where the quality of sensing tasks are interdependent. The results show that the synergistic based social incentive mechanism is more cost-effective to promote cooperation.

References

1. He, S., Shin, D.H., Zhang, J., et al.: Near-optimal allocation algorithms for location-dependent tasks in crowdsensing. IEEE Trans. Veh. Technol. **6**(99), 1 (2017)
2. Yang, G., He, S., Shi, Z.: Leveraging crowdsourcing for efficient malicious users detection in large-scale social networks. IEEE Internet Things J. **4**(2), 330–339 (2017)
3. Nie, J., Xiong, Z., Niyato, D., et al.: A socially-aware incentive mechanism for mobile crowdsensing service market (2017)
4. Zhu, X., Yang, M., Jian, A.N., et al.: Crowdsourcing incentive method based on reverse auction model in crowd sensing. J. Comput. Appl. (2016)
5. Zhu, X., An, J., Yang, M., et al.: A fair incentive mechanism for crowdsourcing in crowd sensing. IEEE Internet Things J. **3**(6), 1364–1372 (2017)
6. Duan, Z., Li, W., Cai, Z.: Distributed auctions for task assignment and scheduling in mobile crowdsensing systems. In: IEEE International Conference on Distributed Computing Systems, pp. 635–644. IEEE (2017)
7. Wang, Y., Cai, Z., Tong, X., et al.: Truthful incentive mechanism with location privacy-preserving for mobile crowdsourcing systems. Comput. Netw. **135**, 32–43 (2018)
8. Wang, Y., Cai, Z., Yin, G., et al.: An incentive mechanism with privacy protection in mobile crowdsourcing systems. Comput. Netw. **102**, 157–171 (2016)
9. Duan, L., Kubo, T., Sugiyama, K., et al.: Motivating smartphone collaboration in data acquisition and distributed computing. IEEE Trans. Mob. Comput. **13**(10), 2320–2333 (2014)
10. Li, M., Lin, J., Yang, D., et al.: QUAC: quality-aware contract-based incentive mechanisms for crowdsensing. In: IEEE International Conference on Mobile Ad Hoc and Sensor Systems, pp. 72–80. IEEE Computer Society (2017)
11. Zhao, N., Fan, M., Tian, C., et al.: Contract-based incentive mechanism for mobile crowdsourcing networks. Algorithms **10**(3), 104 (2017)

12. Luo, T., Tan, H.P., Xia, L.: Profit-maximizing incentive for participatory sensing. In: IEEE Proceedings INFOCOM, pp. 127–135. IEEE (2014)
13. Duan, L., Kubo, T., Sugiyama, K., et al.: Incentive mechanisms for smartphone collaboration in data acquisition and distributed computing. In: Proceedings - IEEE INFOCOM, vol. 131, no. 5, pp. 1701–1709 (2012)
14. Mani, A., Rahwan, I., Pentland, A.: Inducing peer pressure to promote cooperation. Sci. Rep. **3**(3), (2013). Article number 1735
15. Bebchuk, L.A., Fried, J.M., Walker, D.I.: Managerial power and rent extraction in the design of executive compensation. Univ. Chicago Law Rev. **69**(3), 751–846 (2002)

SACP: A Signcryption-Based Authentication Scheme with Conditional Privacy Preservation for VANET

Miao Lu, Ying Wu(✉), Yuwei Xu, Yijie Yang, and Jingjing Wang

College of Computer and Control Engineering,
Nankai University, Tianjin 300350, China
{wuying,xuyw}@nankai.edu.cn

Abstract. Vehicular ad hoc network (VANET) has achieved promising performance in alleviating traffic problems. However, communications in VANET is vulnerable because of its open-access environment. To tackle this issue, we propose a conditional privacy-preserving authentication scheme (SACP) in this paper, which could protect communication security and driver privacy simultaneously. Specifically, symmetric encryption is used to encrypt messages, and the private keys and identity information of communication parties are employed to calculate the communication key. Moreover, only the recently exposed pseudonyms are maintained in SACP, which greatly alleviates the storage pressure of vehicles. Finally, a comprehensive empirical evaluation validates the superior performance of SACP.

Keywords: VANET · Communication security · Privacy
Signcryption

1 Introduction

VANET is utilized for a broad range of safety applications. However, communications in VANET are vulnerable to various attacks, such as eavesdropping, impersonation, forgery, and so forth. Authentication is feasible to verify the legality of communication parties. Yet, whereabouts of drivers would be revealed by authentication [1]. Therefore, the issue of how to verify a vehicle without revealing its privacy is crucial. However, the real identity of a malicious user should be revealed in case of any misbehavior [2].

A large number of schemes have been put forward to protect VANET. In 2007, Raya [2] proposed the first pseudonym-based scheme to protect driver privacy, which suffers large communication cost because of certificate revocation list (CRL). Instead of employing CRL, Azees [3] adopted on-board units (OBUs) to generate anonymous certificates. However, the generation time of a certificate is long and the size of a certificate is large. The identity-based signature (IBS) authentication is also used for verify pseudonyms which has smaller size than certificate. Some IBS schemes based on bilinear maps [4,5] are more time-consuming

S. Chellappan et al. (Eds.): WASA 2018, LNCS 10874, pp. 773–779, 2018.
https://doi.org/10.1007/978-3-319-94268-1_66

than those based on ECC [6,7]. Lo [6] proposed an efficient privacy-preserving authentication scheme which supports batch verification. Cui [7] incorporated ECC with hash chains to achieve authentication, but each message needs to be verified by road-side units (RSUs) before being accepted.

Besides, most of the existing schemes cannot satisfy the message confidentiality. Inspired by this, we propose the SACP in this paper. In SACP, pseudonyms are used during communications to protect the real identities of vehicles, and all of the messages are encrypted by a unique key. The main contributions of this paper are given as follows: (1) the communication key is merely accessible to communication parties, which guarantees source legality and message confidentiality simultaneously; and (2) an empirical evaluation demonstrates that SACP remains efficient after adding encryption phase.

2 Preliminaries

VANET system commonly consists of three components, which are Trust Authority (TA), RSUs and vehicles. TA is a trusted third authority at a high security level that possesses sufficient computing and storage resources. RSUs are vulnerable infrastructures fixed on the roadside and responsible for transmitting the information that is generated by TA and vehicles within the coverage areas of them. Vehicles in VANET are equipped with OBUs, which promise the communication and computation capabilities of vehicles.

Security Requirements of VANET: Due to its open-medium environment, VANET has to face various attacks. Once VANET was compromised, the benefits of users would be aggrieved, or at worse, lives of drivers may be threatened. The serious consequences compel researchers to propose more excellent schemes to protect VANET. A secure VANET should satisfy five security requirements, namely source authentication, message integrity, message confidentiality, non-repudiation and conditional anonymity.

ECC Preliminaries: In ECC, the elliptic curve equation has a variant form as $E(F_p) : y^2 = x^3 + ax + b\,(mod\,p)$ over a prime finite field F_p, where $a, b \in F_p$, $p > 3$, and $4a^3 + 27b^2 \neq 0\,(mod\,p)$. The security of ECC depends on Elliptic Curve Discrete Logarithm Problem (ECDLP) and Computational Diffie-Hellman Problem (CDHP).

3 Our Proposed Scheme

We propose SACP in this paper which contains seven stages as described below. The notations throughout this paper are listed in Table 1.

System Initialization: Let F_p be the finite field over a prime order p. G is an additive group generated by a point P on a non-singular elliptic curve $E(F_P)$ with order n; TA selects a secure one-way hash function $H()$ and a symmetric encryption algorithm $E_K()/D_K()$. The public key of TA is $PUB_{TA} = s_{TA} * P$, where $s_{TA} \in F_P$ is the secret key; TA publishes all parameters except for s_{TA}.

Table 1. Notations

Notation	Description
n, P	The order and generator of the $E(F_P)$
$H()$	A one-way hash function $H: \{0,1\}^* \rightarrow Z_p$
s_{TA}/PUB_{TA}	Private key and corresponding public key of TA
SK/PK	Private key and corresponding public key of vehicle or RSU
RID/CID	The real identity and communication identity of a vehicle or RSU
t, tt	Expiry time
c	Ciphertext
$E_K()/D_K()$	A symmetric encryption/decryption function with the secret key K
EK/DK	Communication keys
T_{cur}	Current time
θ_i	No special meaning, equal to $SK_i * PUB_{TA}$

Vehicle Registration and Pseudonym Generation: The vehicle securely submits its real identity RID_i to TA and TA is required to confirm whether the RID_i has been registered. Then, TA registers RID_i as follows:

Step 1. Choose a random number $r_i \in F_P$ and compute $R_i = r_i * P$;
Step 2. Select expiry time t_i and generate the corresponding pseudonym as

$$PID_i = RID_i \oplus H(s_{TA} * R_i \parallel t_i). \tag{1}$$

Step 3. Compute the private key SK_i and the public key PK_i for PID_i as

$$SK_i = r_i * H(PID_i \parallel R_i \parallel t_i) + s_{TA} \,(mod\, n). \tag{2}$$

$$PK_i = SK_i * P. \tag{3}$$

Step 4. Compute $\theta_i = SK_i * PUB_{TA}$;
Step 5. Send $\{PID_i, t_i, SK_i, PK_i, R_i, \theta_i\}$ to the vehicle.

RSU Registration: To register a RSU with identity RID_i, TA performs the following operations:

Step 1. Choose a random number $r_i \in F_P$ and compute $R_i = r_i * P$;
Step 2. Select expiry time t_i, then compute SK_i and PK_i as

$$SK_i = r_i * H(RID_i \| R_i \| t_i) + s_{TA} \,(mod\, n). \tag{4}$$

$$PK_i = SK_i * P. \tag{5}$$

Step 3. Compute $\theta_i = SK_i * PUB_{TA}$;
Step 4. Send $\{RID_i, t_i, SK_i, PK_i, R_i, \theta_i\}$ to the RSU.

Note that the identity information of a RSU is periodically broadcasted to facilitate the real-time connections of the newly entered vehicles with RSUs.

Message Signing: We assume that the identity of a sender and a receiver are CID_s and CID_r, respectively.

Step 1. Compute communication key EK according to its private key and the identity information of the other side as

$$EK = SK_s * H(CID_r \| R_r \| t_r) * R_r + \theta_s. \tag{6}$$

Step 2. Choose a random number $z_s \in F_P$ and compute $Z_s = z_s * p$;[1]
Step 3. Select expiry time tt and computes the signature sig_s as

$$sig_s = SK_s + z_s * H(M \| CID_s \| tt) \, (mod\, n). \tag{7}$$

Step 4. Use EK to encrypt the message M and the signature sig_s as

$$c = E_{EK}(M \| sig_s). \tag{8}$$

Step 5. Send $\{CID_s, c, tt, Z_s, R_s, t_s\}$ to the receiver.

Message Verification: After receiving $\{CID_s, c, tt, Z_s, R_s, t_s\}$, the receiver implements the following operations only if both t_s and tt are less than T_{cur}:

Step 1. Compute communication key DK and decrypt the ciphertext c with DK while it holds that $DK = EK$;

$$DK = SK_r * H(CID_s \| R_s \| t_s) * R_r + \theta_r. \tag{9}$$

$$M \| sig_s = D_{DK}(c). \tag{10}$$

Step 2. Verify sig_s as: $SK_r * sig_s * P = DK + Z_s * SK_r * H(M \| CID_s \| tt)$.

Batch Verification: Upon receiving n distinct message tuples $\{CID_i, c_i, tt_i, Z_i, R_i, t_i\}(i = 1 \dots n)$, the receiver checks the validity of tt_i and t_i. If both terms hold, the receiver computes DK_i and obtains that $h_i = H\,(M_i \| CID_i \| tt_i)\,(i = 1 \dots n)$. Finally, the receiver check whether the following equation holds:

$$(SK_r * \sum_{i=1}^{n} sig_i) * P = \sum_{i=1}^{n} DK_i + \sum_{i=1}^{n} Z_i * SK_r * h_i. \tag{11}$$

Identity Revocation: In case of a dispute, VANET should be able to reveal the real identity of the sender. Assume that the message is $\{CID_s, c, tt, Z_s, R_s, t_s\}$. Then, merely TA can compute the real identity of the sender as

$$RID_s = CID_s \oplus H(s_{TA} * R_s \parallel t_s). \tag{12}$$

[1] Step 2 could be accomplished in spare time of OBU in advance.

4 Discussion

In this section, we analyze SACP through security analysis, computation overhead and communication overhead. A comparison of SACP and several state-of-the-art schemes, i.e., CPAS [5], Lo N W [6], and SPACF [7], is also given.

Security Analysis: SACP satisfies the following security requirements:

1. Authentication and non-repudiation: according to Eq. (7), the sender computes signature with SK_s which is merely possessed by the sender, so the signature cannot be forged by any others, and this strategy contributes to mutual authentication and non-repudiation between communication parties.
2. Message integrity: the validity of a signature is verified according to Eq. (7). Once the received message M' was inconsistent with the original message M, the equation $H(M'\|CID_s\|tt) = H(M\|CID_s\|tt)$ would not hold and Eq. (7) would fail to be verified. So the equality of Eq. (7) could prove the message integrity.
3. Message confidentiality: message is encrypted by a symmetric encryption, and the key can be calculated by communication parties. In addition, ECDH, a key agreement protocol, is employed to exchange the communication key. In this manner, only communication parties can obtain the communication key.
4. Conditional privacy-preserving: we preserve the privacy of vehicles by distributing pseudonyms to them. In case of a dispute, TA can obtain the real identity of the malicious vehicle according to Eq. (12).

Computation Overhead Analysis: We adopt the same computation evaluation for VANET as in [7]. The bilinear paring and the ECC are on the security level of 80 bits. Let T_p be the execution time of a pairing operation, T_{pm} be execution time of a scale multiplication, T_{pa} be the execution time of a point addition operation and T_h be the execution time of a one-way hash function. T_p is 4.245 ms, T_{pm} is 0.442 ms, T_{pa} is 0.0018 ms and T_h is 0.0001 ms. Table 2 presents the comparisons of four schemes.

Table 2. Computation overhead of four schemes

Scheme	Sign one message	Verify one message	Batch verification of n messages
CPAS [5]	$2T_{pm} + T_h$	$3T_p + 2T_{pm} + 2T_h + T_{pa}$	$3T_p + (n+1)T_{pm} + 2nT_h + 3nT_{pa}$
Lo [6]	T_h	$3T_{pm} + 2T_{pa} + 2T_h$	$(n+2)T_{pm} + 2nT_{pa} + 2nT_h$
SPACF [7]	$2T_{pm} + 2T_h$	$2T_{pm} + T_{pa} + T_h$	$(n+2)T_{pm} + nT_{pa} + nT_h$
SACP	$T_{pm} + T_{pa} + T_h$	$3T_{pm} + 2T_{pa} + 2T_h$	$(2n+1)T_{pm} + 2nT_{pa} + 2nT_h$

Figure 1 illustrates that the total computation costs of the compared schemes are 14.4031 ms, 1.3299 ms, and 1.7701 ms, respectively. Our scheme costs 1.7737 ms to generate the communication key, which is 0.4438 ms slower than [6]. Figure 2 demonstrates the variations of the batch verification that is used for verifying n messages at the same time. In fact, there is a shortcoming in batch verification, that is, a bad message would cause the failure of entire verification, and thus, the message number should be limited. Besides, we observe that SPACF is not efficient, which may attribute to the strategy that it employs RSU to verify all of the messages transmitted within its coverage area.

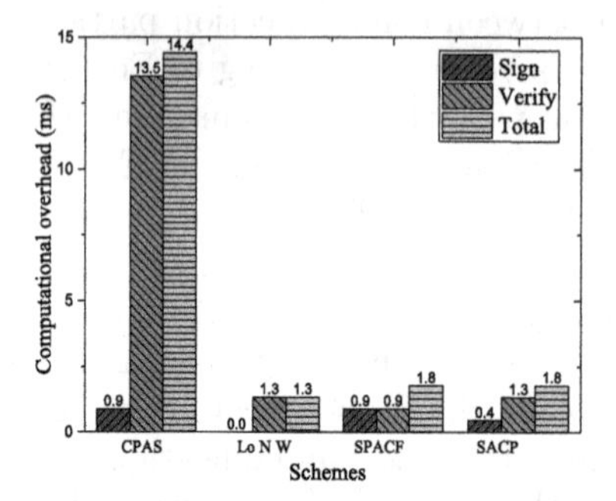

Fig. 1. Comparison of computational overhead

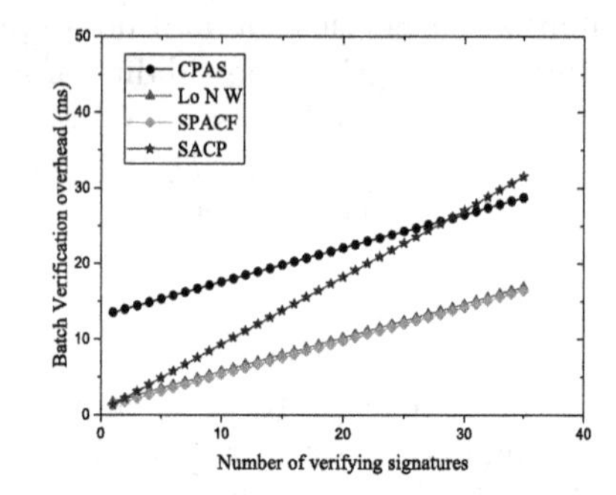

Fig. 2. Overhead variation of batch verification

Communication Overhead Analysis: As stated in [7], the sizes of the points in bilinear pairing and ECC are 128B and 40B, respectively. The size of a timestamp is set to 4B and the size of output of $H()$ is set to 20B [7]. The comparisons in terms of communication cost are illustrated in Table 3.

Table 3. Communication overhead of four schemes

Scheme	CPAS [5]	Lo N W [6]	SPACF [7]	SACP
Size	540B	168B	84B	128B

As showed in Table 3, the communication overhead of SACP is 128B, which is acceptable in VANET.

5 Conclusion

This paper proposes a novel conditional privacy-preserving authentication scheme SACP, which is designed specific to VANET. The security analysis demonstrates that SACP satisfies security requirements of VANET. Moreover, we employ the identities of communication parties to obtain the communication key, and this operation is also a component of the message signature, which promises our scheme to be effective and efficient. Extensive evaluations validate the superior performance of our scheme under different measurements.

Acknowledgments. This work was supported by the National Natural Science Foundation of China (No. 61702288) and the Natural Science Foundation of Tianjin in China (No. 16JCQNJC00700).

References

1. Qu, F., Wu, Z., Wang, F.Y., et al.: A security and privacy review of VANETs. IEEE Trans. Intell. Transp. Syst. **16**(6), 2985–2996 (2015)
2. Raya, M., Hubaux, J.P.: Securing vehicular ad hoc networks. In: International Conference on Pervasive Computing and Applications, pp. 424–429 (2007)
3. Azees, M., Vijayakumar, P., Deboarh, L.J.: EAAP: efficient anonymous authentication with conditional privacy-preserving scheme for vehicular ad hoc networks. IEEE Trans. Intell. Transp. Syst. **18**(9), 2467–2476 (2017)
4. Tzeng, S.F., Horng, S.J., Li, T., et al.: Enhancing security and privacy for identity-based batch verification scheme in VANETs. IEEE Trans. Veh. Technol. **66**(4), 3235–3248 (2017)
5. Shim, K.A.: $\mathcal{CPAS}$: an efficient conditional privacy-preserving authentication scheme for vehicular sensor networks. IEEE Trans. Veh. Technol. **61**(4), 1874–1883 (2012)
6. Lo, N.W., Tsai, J.L.: An efficient conditional privacy-preserving authentication scheme for vehicular sensor networks without pairings. IEEE Trans. Intell. Transp. Syst. **17**(5), 1319–1328 (2016)
7. Cui, J., Zhang, J., Zhong, H., Xu, Y.: SPACF: a secure privacy-preserving authentication scheme for VANET with Cuckoo filter. IEEE Trans. Veh. Technol. **66**(11), 10283–10295 (2017)

DFTinker: Detecting and Fixing Double-Fetch Bugs in an Automated Way

Yingqi Luo(✉), Pengfei Wang, Xu Zhou, and Kai Lu

National University of Defense Technology, Changsha, Hunan, People's Republic of China
nudtlyq@163.com

Abstract. The double-fetch bug is a situation where the operating system kernel fetches the supposedly same data twice from the user space, whereas the data is unexpectedly changed by the user thread. It could cause fatal errors such as kernel crashes, information leakage, and privilege escalation. Previous research focuses on the detection of double-fetch bugs, however, the fix of such bugs still relies on manual efforts, which is inefficient. This paper proposes a comprehensive approach to automatically detect and fix double-fetch bugs. It uses a static pattern-matching method to detect double-fetch bugs and automatically fix them with the support of the transactional memory (Intel TSX). A prototype tool named DFTinker is implemented and evaluated with prevalent kernels. Compared with prior works, it can automatically detect and fix double-fetch bugs at the same time and owns a high code coverage and accuracy.

1 Introduction

The wide use of multi-core hardware is making concurrent programs increasingly pervasive, especially in operating systems, network systems, and even IoT devices. However, the reliability of such system is severely threatened by the notorious concurrency bugs [5,12]. Among all the concurrency bugs, the double-fetch bug is one of the most special and significant types.

Previous research focuses on the detection of double-fetch bugs. Dynamic approaches [3,10] detect double-fetch bugs by tracing memory accesses. However, such approaches are limited by the path coverage. They cannot be applied to code that needs corresponding hardware to be executed, so device drivers cannot be analyzed without access to the device or a simulation of it. Static approaches detect double-fetch bugs based on the identification of transfer functions [9, 11], however, the accuracy and efficiency of such approaches are undesirable as they lack runtime information and still rely on manual efforts to confirm the bug. In addition, none of the previous works provides a practical solution on automatically fixing double-fetch bugs except some prevention suggestions. Thus, the fix of double-fetch bugs still relies on manually locating and rewriting the source code, and an automatic solution is in urgent need.

This paper proposes a comprehensive approach to automatically detect and fix double-fetch bugs. In the first phase, a static pattern-matching method based

S. Chellappan et al. (Eds.): WASA 2018, LNCS 10874, pp. 780–785, 2018.
https://doi.org/10.1007/978-3-319-94268-1_67

on the Coccinelle engine is used to identify double-fetch bugs. In the second phase, the identified bug is automatically fixed based on the support of the Intel Transactional Synchronization Extension (TSX). In summary, the main contribution of this paper is as follows:

- This paper proposes a comprehensive approach to automatically detect and fix double-fetch bugs at one time. The approach can cover all architectures in one detect execution, need no manual involvement, and achieve a more accurate result than previous research.
- A prototype tool named DFTinker is implemented. We have made it publicly available, hoping it can be useful for future study.
- DFTinker is evaluated with prevalent real kernels. Results show that it is effective and efficient in automatically detecting and fixing double-fetch bugs.

2 Background

In modern operating systems, the kernel space is always separated from the user space for safety [8]. Kernel code run in the kernel space and get data from users if needed, it will use specific functions, termed *transfer functions*. In Linux kernel, there are four typical transfer functions, `get_user()`, `put_user()`, `copy_from_user()`, `copy_to_user()`. All their effects are fetching data or transferring data between the kernel space and the user space. Malicious changes between two fetches many cause fatal errors in kernels, termed double-fetch bugs.

Coccinelle [7] engine is a program matching and transformation engine. It uses language SmPL (Semantic Patch Language) as rules to perform matching and transformations in C code. Coccinelle was initially targeted towards performing collateral evolutions in Linux, and it is widely used for finding and fixing bugs in system code now. One of the advantages of Coccinelle engine is path-sensitive, it is specially optimized for traversing paths.

Traditionally, transactional memory [1,2] is used to simplify concurrent programming. It allows executing load and store instructions in an atomic way. Transactional memory systems provide high-level instructions to developers so as to avoid low-level coding, and this achieves a better access model to shared memory in concurrent programming. Hardware transactional memory achieves transactions by processors, caches, and bus protocol. It provides opportunities to implement dynamic schedules according to specific CPU instructions. We choose Intel TSX to ensure data consistency.

3 Design

3.1 Detection of Double-Fetch Bugs

As Sect. 2 states, double-fetch bugs and transfer functions are closely related, and each fetch indicates an invocation of a transfer function. However, since there are many complex situations in the kernel code, such as pointer change and aliasing, double-fetch bug detection needs further and thorough analysis.

Wang *et al.* and Xu *et al.* all focus on four transfer functions to detect double-fetch bugs, i.e., `get_user()`, `__get_user()`, `copy_from_user()`, and `__copy_from_user()`, the functionality of which is transferring data from user space to kernel space.

Table 1. Expanded transfer functions.

No.	Name	Type	Parameter
1	unsafe_get_user	macro	dst, src, err
2	__copy_in_user	macro	des, src, len
3	__copy_user	function	dst, src, len
4	__copy_user_zeroing	function	dst, src, len
...			

However, these rules are not strong enough to cover all double-fetch bugs. We improve the rules as follows.

Add More Transfer Functions. Wang *et al.* used only four transfer functions in his experiment, `get_user()`, `__get_user()`, `copy_from_user()`, and `__copy_from_user()`. However, there are also many other functions containing transfer functions, and their targets are transferring data from user space to kernel space as well, such as `memdup_user()` mentioned above. Table 1 shows 4 of 15 functions (and macros) we include to detect double-fetch bugs.

Fix Incomplete Rules. Rules which Wang *et al.* proposed are theoretically correct. However, in the implementation phase, they were achieved incompletely, which leaded false negatives. Many cases are missed because of careless implementation. We fixed these rules and reduced the false negative rate.

Remove More Non-double-Fetch Bugs. Wang *et al.* used his pattern rules find 90 candidates files in total, it's still a little heavy for technicians to check manually. To lower false positive rate, more situations are added to remove those non-double-fetch bugs:

1. **The procedure returns after the first fetch.** The first situation is when the first fetch is in a `IF` statement. This case will be matched with prior rules apparently. However, there is a `RETURN` statement after the first fetch, that means, the second fetch will never be executed if the first fetch is executed. Thus, this is not a double-fetch bug actually. There are many cases like this in the kernel code.
2. **Two fetches are in the different branches.** Another situation is when the two fetches are in the different branches, just like a `SWITCH` statement. These conditions can never be satisfied at the same time, so this situation is also a non-double-fetch bug situation.

3.2 Automated Fixing with Intel TSX

Previous research only proposed suggestions on preventing double-fetch bugs [9, 11], such as don't copy the header twice. However, we need a practical solution to automatically fix the bug.

We implement fixing function with Intel's Restricted Transactional Memory (RTM) software interface. RTM defines three new instructions: `XBEGIN`, `XEND`, and `XABORT`. Programmers can use `XBEGIN` and `XEND` to specify the begin and end of a hardware transaction and use `XABORT` to explicitly abort a transaction.

As Coccinelle engine is accurate in locating lines of double-fetch bugs in the code, it is easy to fix `LOCK()` and `UNLOCK()` operations to the code. According to the feature of transaction memory, all operations between `LOCK()` and `UNLOCK()` will be executed in a transaction, execution results will be committed if there are no conflicts. In other words, if there is a malicious user changes the data in the user space after the first fetch, all operations in the transaction will be aborted and rerun from `LOCK()`, which guarantees the consistency of the data.

4 Implementation

DFTinker consists of three parts, a detector, a patcher and a supervisor. The detector is used for detecting double-fetch bugs. It is implemented as SmPL files and the Coccinelle engine. The Coccinelle engine will use these SmPL files as pattern rules to find double-fetch bugs and filter out non-double-fetch bugs cases. The patcher is used for fixing the double-fetch bugs, it is made up of header files and SmPL files. Header files are used for providing prevention interfaces and SmPL files are used for providing rules for fixing. The last part is the supervisor, which consists of Linux shell scripts. The supervisor is used for supervising the Coccinelle engine and sorting out the results of the experiments, so as to leave the process fully automated.

5 Evaluation

5.1 Detection of Double-Fetch Bugs

The experiments are conducted on a Linux laptop running Ubuntu 16.04 x64, with one Intel i7-7700HQ 2.6 GHz processor, 8 GB of memory, 250 GB SSD. We use Linux 4.14.10, OpenBSD 6.2, FreeBSD 11.1, Android 7.0.0 (kernel version 3.18), and Darwin 10.13.3, which were the relatively newer version when the experiments were conducted.

DFTinker is applied to the five prevalent open source kernels. The statistical result[1] is shown in Table 2. We find 24 double-fetch bugs in the Linux kernel, 12 bugs in the FreeBSD kernel, 41 bugs in the Android kernel, 4 bugs in the Darwin kernel. Note that DFTinker can identify all known double-fetch bugs in the Linux kernel including Wang *et al.*'s work and Xu *et al.*'s work, however, our approach is more concise in contrast, which proves DFTinker's efficiency.

[1] Due to the space limitation of the page, the full detailed results of the double-fetch bugs are available at https://github.com/luoyyqq.

Table 2. Statistical results of detection of double-fetch bugs.

Kernel	Version	Files	Size checking	Type selection	Validity checking	Reacquisition	Total bugs
Linux	4.14.10	45614	13	5	4	2	24
FreeBSD	11.1	38811	7	2	2	1	12
OpenBSD	6.2	29704	0	0	0	0	0
Android	7.0.0 (3.18)	30479	14	7	5	15	41
Darwin	10.13.3	49105	3	1	0	0	4

5.2 Automated Fixing with Intel TSX

In our experiments, we fix target functions using DFTinker and modify the data between two fetches using a user thread. The results show that the data fetched at the second time is same as the first time, which proves that DFTinker is effective in protecting double-fetch bugs.

6 Discussion

Jurczyk and Coldwind [3,4] used a dynamic approach in their Bochspwn project to study double-fetch bugs in Windows. By tracing memory accesses, they successfully found double-fetch bugs in the Windows kernel. However, such dynamic approaches can not test code under strict conditions. Our static approach has a better code coverage and can detect double-fetch bugs in the drivers, where the dynamic approaches are incapable of.

Schwarz *et al.* [6] proposed a method using cache-attack and kernel-fuzzing techniques to detect, exploit, and eliminate double-fetch bugs in Linux syscalls. However, their approach is limited to Linux syscalls, whereas large numbers of the double-fetch bugs occur in non-syscall functions, such as functions in drivers, are missed. Thus, their approach suffers from a low code coverage, whereas our approach is free from that. As for DropIt they implement, our approach can fix double-fetch bugs automatically instead of manual fixing.

Xu *et al.* [11] proposed a formal definition of double-fetch bugs and used a static analysis based on LLVM IR and symbolic execution to detect such bugs. However, their definition takes all the potential situations into consideration, which are not currently buggy but only have the potential to turn into bugs when the code is updated. Besides, their approach needs to compile the source code to LLVM IR and specify the target architecture. Thus, it detects only one architecture at one time, leading to the miss of bugs such as CVE-2016-6130. Our approach has a better code coverage, which can analyze the source code of all the architecture at one time.

Although DFTinker achieves a decent performance in detecting and fixing double-fetch bugs, it relies on the availability of the source code, which is suitable for in-house testing. We will take the binary situations into consideration in the future work.

7 Conclusion

This paper proposes an approach to automatically detect and fix double-fetch bugs. We implement a prototype named DFTinker and evaluate it with real kernels. Experiments show that DFTinker is effective and efficient in automatically detecting and fixing double-fetch bugs. DFTinker detected 81 cases in prevalent kernels and succeed in defending malicious data tampering.

Acknowledgements. This work is partially supported by The National Key Research and Development Program of China (2016YFB0200401), by program for New Century Excellent Talents in University, by National Science Foundation (NSF) China 61402492, 61402486, 61379146, 61472437, by the laboratory pre-research fund (9140C810106150C81001).

References

1. Hammond, L., Wong, V., Chen, M., Carlstrom, B.D.: Transactional memory coherence and consistency. In: Proceedings of the International Symposium on Computer Architecture, pp. 102–113 (2004)
2. Herlihy, M., Moss, J.E.B.: Transactional memory: architectural support for lock-free data structures, pp. 289–300 (1993)
3. Jurczyk, M., Coldwind, G.: Bochspwn: identifying 0-days via system-wide memory access pattern analysis. https://media.blackhat.com/us-13/us-13-Jurczyk-Bochspwn-Identifying-0-days.pdf
4. Jurczyk, M., Coldwind, G.: Identifying and exploiting windows kernel race conditions via memory access patterns. Technical report, Google Research (2013). http://research.google.com/pubs/archive/42189.pdf
5. Ma, X., Wang, Y., Qiu, Q., Sun, W., Pei, X.: Scalable and elastic event matching for attribute-based publish/subscribe systems. Future Gener. Comput. Syst. **36**(7), 102–119 (2014)
6. Schwarz, M., Gruss, D., Lipp, M., Maurice, C., Schuster, T., Fogh, A., Mangard, S.: Automated detection, exploitation, and elimination of double-fetch bugs using modern CPU features (2017)
7. Stuart, H.: Hunting bugs with Coccinelle. Masters thesis (2008)
8. Swift, M.M., Bershad, B.N., Levy, H.M.: Improving the reliability of commodity operating systems. ACM Trans. Comput. Syst. **23**(1), 77–110 (2005)
9. Wang, P., Krinke, J., Lu, K., Li, G., Dodier-Lazaro, S.: How double-fetch situations turn into double-fetch vulnerabilities: a study of double fetches in the Linux kernel. In: Usenix Security Symposium (2017)
10. Wilhelm, F.: Tracing privileged memory accesses to discover software vulnerabilities. Master's thesis, Karlsruher Institut für Technologie (2015)
11. Xu, M., Qian, C., Lu, K., Michael, B., Taesoo, K.: Precise and scalable detection of double-fetch bugs in OS kernels. http://www-users.cs.umn.edu/~kjlu/papers/deadline.pdf
12. Yu, B., Yang, L., Wang, Y., Zhang, B., Cao, Y., Ma, L., Luo, X.: Rule-based security capabilities matching for web services. Wirel. Pers. Commun. **73**(4), 1349–1367 (2013)

A Quality-Validation Task Assignment Mechanism in Mobile Crowdsensing Systems

Xingyou Xia, Lin Xue, Jie Li, and Ruiyun Yu(✉)

Northeastern University, Shenyang, China
xiaxy_neu@163.com, yury@mail.neu.edu.cn

Abstract. Participant selection or task allocation is a key issue in Mobile Crowdsensing (MCS) systems. Previous participant assignment approaches mainly focus on selecting a proper subset of users for MCS tasks, but however how to ensure that users devote effort on their tasks is a challenging problem that arises in these approaches. This paper studies the task quality control issue, and proposes a quality-validation task assignment mechanism (QTAM) in MCS systems. We theoretically model this mechanism as a Stackelberg Game, in which the users' instinct of maximizing their payoff and the validation workforce limitation are taken into account. An efficient approximation algorithm is designed to find a Strong Stackelberg Equilibrium (SSE) in the QTAM game. Extensive simulations demonstrate the efficiency and effectiveness of QTAM, which shows that QTAM can prevent untrustworthy behaviors and achieve optimal quality validation for sensing tasks.

Keywords: Mobile Crowdsensing · Quality control · Task assignment

1 Introduction

Mobile Crowdsensing [1] (MCS) is a recent sensing technology which is based on the power of users-companioned devices, uses' regular social mobility and "crowdsourcing wisdom". For MCS, tasks can be completed very quickly by accessing a large and relatively cheap workforce, but the results are not reliable. Participants are not always willing to devote effort on their tasks, and maybe provide low quality or even faked data to the system. Thus, the lack of quality control mechanisms will lead to information credibility problems in source validity and data accuracy.

To address this issue, we propose a quality-validation task assignment mechanism (QTAM) in location-based MCS systems. Different from the incentive [2, 3] or reputation [4] mechanisms, our goal is to present a task assignment approach to detect cheating users using participators for quality validation, then a punishment method is designed to alleviate the users' cheating or selfish behaviors. Some other multi-worker based approaches [5, 6] hold the idea of allocating multiple workers to a single sensing

---

This work is supported by the National Natural Science Foundation of China (61672148, 61502092), the Ministry of Education - China Mobile Research Fund (MCM20160201), and the Hundred, Thousand and Ten Thousand Talents Project of Liaoning Province (201514).

S. Chellappan et al. (Eds.): WASA 2018, LNCS 10874, pp. 786–792, 2018.
https://doi.org/10.1007/978-3-319-94268-1_68

task so as to provide the required sensing quality, but it will generate additional costs and reduce the platform's benefit to recruit extra participators.

2 System Model and Problem Formulation

We define the sensing task performers who submit invalid task results as cheating users (Cheater). Meanwhile, sensing quality measurement and assurance is called a quality-validation task in this paper. The quality-validation task performers are called checking users (Checker), and the checkers should be chosen from the reputable sensing task performers. Generally, checkers are honest because a quality-validation task is easier to be completed than a common task. Furthermore, an invalid checking result will be complained by the task performer, and this will influences the reputation of the checker (the related work can be referred in the paper [7]).

The procedure of QTAM is illustrated in Fig. 1. The checker C receives a quality validation task. P_1 exchanges sensing validation information with checker C through Device-to-Device (D2D) connection, and the information includes location check-in records, samples of sensing data and so on. The checker C gives a report on the status of P_1's task in exchange of monetary rewards. If P_1's results are validated as correct ones, the platform will grant task rewards P_1, otherwise, P_1 will be punished.

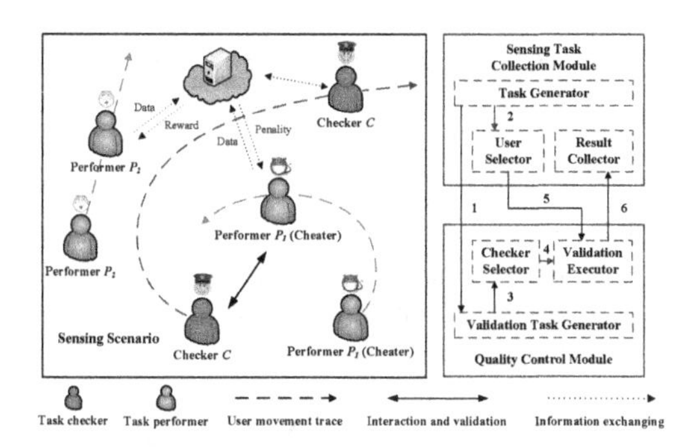

Fig. 1. Illustration of task validation procedure

The primary question is how to efficiently allocate limited checkers to protect against cheating behaviors, because users with high reputation are scarce, and recruiting extra checkers will increase cost for the platform. The main idea of QTAM is that randomly deploying checker workforce instead of allocating one checker for each task. The platform must commit to an allocating strategy for checkers before the cheaters choose their own strategies, which are known as Stackelberg games [8].

In our game, the platform is the *leader* and the task performers are *followers*. Each *player* has a set of possible pure strategies. Let $\mathcal{T} = \{T_1, \ldots, T_m\}$ be a set of tasks to be performed in a round, and a task performer chooses one task for cheating as its pure strategy, denoted as $s_P \in S_P$. The platform has a set of k ($k < m$) checkers $\mathcal{C} = \{C_1, \ldots, C_k\}$ available, and each pure strategy of the platform is an allocation schedule of checkers $\mathcal{C}$ over tasks $\mathcal{T}$, denoted as $s_C \in S_C$. A mixed strategy allows a player to play a probability distribution over pure strategies, denoted $p_P \in P_P$ and $p_C \in P_C$. Payoffs for each player are defined over all possible joint pure strategy outcomes: $S_C \times S_P \rightarrow$ *Payoff* for the platform and similarly for the task performers.

Now the quality validation problem is changed to find a SSE in the QTAM game, and the pair of strategies $\langle p_C, p_P \rangle$ is the final output of the quality-validation task assignment mechanism.

3 Algorithm Design

QTAM involves two steps: (1) the MILP program [9] is exploited to achieve an optimal *coverage* strategy for the checkers; (2) a greedy task assignment method converts MILP outputs into near-optimal quality-validation task assignment strategies.

Coverage Strategy Representation: Given a checker's allocation strategy set P_C, we can summarize it to a *coverage vector* $\mathcal{Q} = \langle \theta_i \rangle, i = 1, \ldots, m$, where θ_i represents the probability of covering T_i. A cheating vector $\Re = \langle \gamma_i \rangle$ gives the probability of cheating tasks of a cheater.

Optimal Coverage Strategy Calculation: For a strategy profile $\langle \mathcal{Q}, \Re \rangle$, the expected utilities for the platform and cheater are given by:

$$U_C^{ij} = \gamma_j(\theta_i U_C^i(z_i = 1) + (1 - \theta_i)U_C^i(z_i = 0))|y_i = 1 \quad (1)$$

$$U_P^{ij} = \gamma_j(\theta_i U_P^i(z_i = 1) + (1 - \theta_i)U_P^i(z_i = 0))|y_i = 1 \quad (2)$$

which means the checker covers task T_i and the cheater cheats task T_j. z_i is the indicator of validation on T_i: $z_i = 1$ if C_i is on T_i; $z_i = 0$ otherwise.

We first fix a coverage strategy $\mathcal{Q}$ of checkers. The cheater's optimal response to $\mathcal{Q}$ can be found using the following linear program:

$$\Re_{\mathcal{Q}}^{\max} = \arg\max_{\Re} \sum_{T_j \in T} U_P^{ij} \,|\mathcal{Q} \;\; s.t. \;\; \sum\nolimits_{T_j \in T} \gamma_j = 1 \,,\; \gamma_j \in \{0, 1\} \quad (3)$$

The objective function (3) maximizes the cheater's expected utility given $\mathcal{Q}$, while the constraints force the cheating vector to assign a single task with probability 1.

The platform seeks an allocation solution that maximizes its own utility, given that the cheater uses an optimal response $\Re_{\mathcal{Q}}^{\max}$. Therefore the platform solves the following problem:

$$\mathcal{Q}_{\Re}^{\max} = arg \max_{\mathcal{Q}} \sum_{T_i \in T} U_C^{ij} \,|\Re = \Re_{\mathcal{Q}}^{\max} \;\; s.t. \;\; \sum\nolimits_{T_i \in T} \theta_i \le k, \;\; \theta_i \in [0, 1] \quad (4)$$

The objective function (4) maximizes the platform's utility with cheater's best response $\Re_{\mathcal{Q}}^{\max}$. The conditions constraints the *coverage vector* to probabilities in the range [0, 1], and restricts the coverage by the number of available checkers k.

For simplicity, after including the characterization of $\Re_{Q}^{\max}$ through linear programming optimality conditions, the optimal coverage strategy calculation becomes:

$$\begin{aligned} \max_{Q} & \sum_{T_i \in \mathcal{T}} U_C^{ij} \\ s.t. \quad & \sum_{T_i \in \mathcal{T}} \theta_i \leq k \\ & \theta_i \in [0,\ 1],\ \ \forall T_i \in \mathcal{T} \\ & \sum_{T_j \in \mathcal{T}} \gamma_j = 1 \\ & \gamma_j \in \{0,\ 1\},\ \ \forall T_j \in \mathcal{T} \\ & 0 \leq (d - \sum_{T_i \in \mathcal{T}} U_P^{ij}) \leq (1 - \gamma_j)M \end{aligned} \tag{5}$$

In function (5), the first and second constraints enforce a feasible mixed allocation strategy for checkers, and the third and fourth constraints enforce a feasible pure strategy for the cheater. The fifth condition is used to characterize the optimal cheater's response.

Quality-Validation Task Assignment: The MILP outputs should be converted into a corresponding mixed strategy for checkers using the function $\Phi : Q \rightarrow p_C$.

Algorithm 1 The Greedy Task Assignment Algorithm

Input: the *coverage vector* $Q = \langle \theta_i \rangle, i = 1, \ldots, m$
Output: a mixed strategy p_C^{out}
$p_C^{out} \leftarrow \varnothing$
while $Q \neq O_{1 \times n}$ **do**
$\mathcal{T}^{tmp} \leftarrow \mathcal{T}$ sorted by θ_i in descending order
$s_C \leftarrow$ select first-k tasks from $\mathcal{T}^{tmp}$
$\theta^{\min} \leftarrow \min \theta_i \quad s.t.\ T_i \in s_C$
add $(s_C, \theta^{\min})$ to p_C^{out}
$Q^{tmp} = \langle \theta_i^{tmp} \rangle, \theta_i^{tmp} = \begin{cases} 0 & \text{if } T_i \notin s_C \\ \theta^{\min} & \text{if } T_i \in s_C \end{cases}$
$Q = Q - Q^{tmp}$
end while

Here, we use a greedy task assignment algorithm, as shown in Algorithm 1, to provide the function for this mapping.

4 Experimental Analyses

Baseline: We compare QTAM against the greedy based task assignment algorithm (GA) and the quality-aware online task assignment algorithm (QAOTA), which recruits two users to perform the same task in order to guarantee the quality.

QTAM must consider three fundamental elements: the allocation efficiency η (η = the number of checked cheaters/the number of cheaters), the checkers' coverage rate $\tau(\tau = k/m)$, and the task performers' cheating ratio $\delta(\delta = l/m)$.

Figure 2 shows the impact of η against the cheating behavior. When the average of η is more than 0.55, QTAM gains a better performance than GA and QAOTA. We study the average allocation efficiency η against the different cheating strategy, and the average η of QTAM can reach to 73.47%.

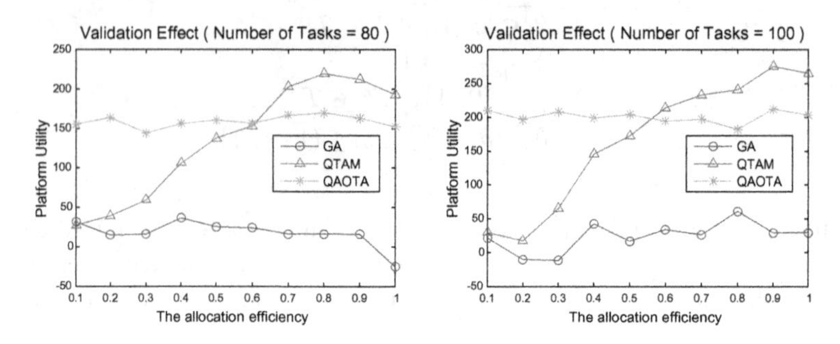

Fig. 2. The impact of the allocation efficiency η

We then perform simulations to show the impact of τ on QTAM, GA, and QAOTA. We show the results in Fig. 3. As we can see, QTAM outperforms the other two algorithms when $\tau \geq 0.3$, and this is because GA is lack of considering the tasks' quality assurance, while QAOTA recruits too many users and generates extra cost.

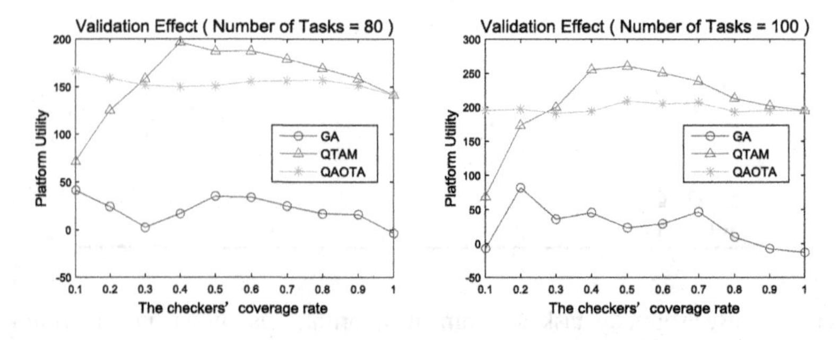

Fig. 3. The impact of the checkers' coverage rate τ

Finally, we show the impact of δ on QTAM, GA, and QAOTA in Fig. 4. It is clear that the total utility of the platform decreases with the increase of δ. In Fig. 6, GA meets a more significant impact on the factor of δ. For QTAM, it shows a relatively moderate downward trend due to the limit number of checkers, while QTAM gains the best performance when $\delta \leq 0.5$. The effect of δ on QAOTA is not very obvious.

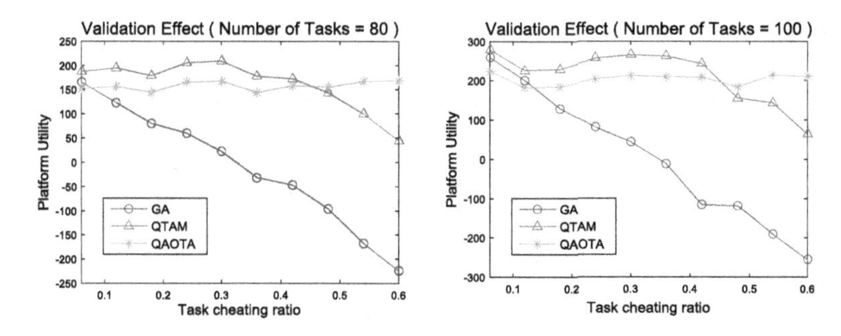

Fig. 4. The impact of the cheating ratio δ

To evaluate the effectiveness of the QTAM, we compare it against the normal Stackelberg game solution (NSGS), and the results are plotted in Figs. 5 and 6.

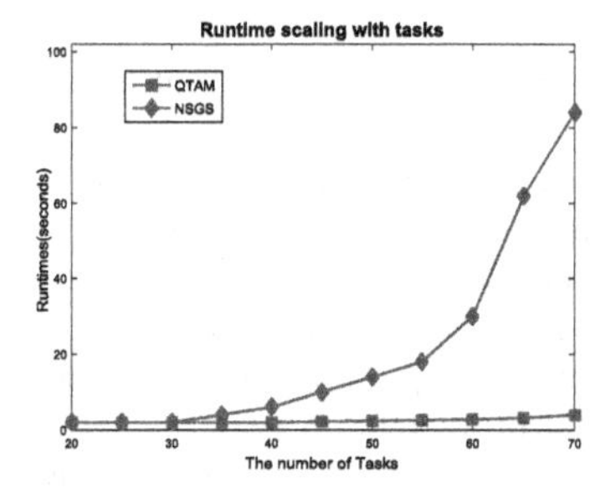

Fig. 5. The runtime comparison

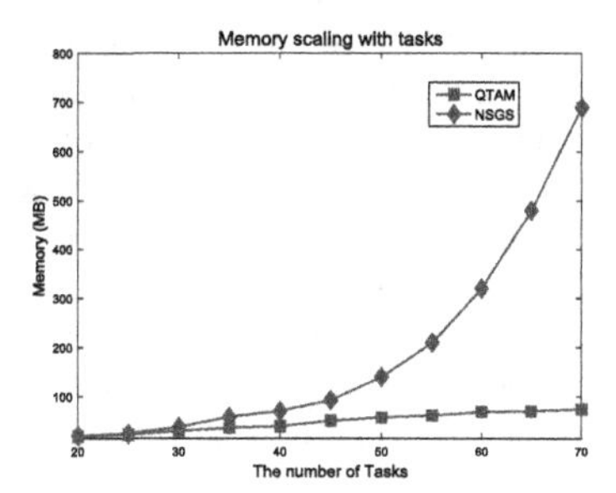

Fig. 6. The memory comparison

In this paper, QTAM introduces the *coverage vector* for checkers allocation in order to reduce requirements on both space and time for the larger case. The differences between both NSGS and QTAM are statistically significant for the larger situation, which imply the effectiveness of QTAM.

5 Conclusion

The mainstream approach to guarantee task quality in Mobile Crowd Sensing (MCS) systems is to recruit enough users to perform a task in order to reduce the impact of individual uses' cheating behaviors. In this paper, we have proposed a new quality-validation task assignment mechanism (QTAM) that enables the platform to achieve a good task completion quality with fewer users, and the basic idea of QTAM is employing participators to detect cheating users, and then using punishment mechanisms to forbid the users' cheating or selfish behaviors. Hence, the QTAM is designed based on a game-theoretic framework accounting for the fact that task cheaters can observe the platform's task validation strategy and change their targets dynamically. A checkers' allocation algorithm based on Strong Stackelberg Equilibrium (SSE) is proposed to output near-optimal validation task assignment strategies. Extensive simulation results are provided to demonstrate the efficiency of QTAM.

References

1. Guo, B., Wang, Z., Yu, Z., et al.: Mobile crowd sensing and computing: the review of an emerging human-powered sensing paradigm. ACM Comput. Surv. (CSUR) **48**(1), 7 (2015)
2. Zheng, Z., Wu, F., Gao, X., et al.: A budget feasible incentive mechanism for weighted coverage maximization in mobile crowdsensing. IEEE Trans. Mob. Comput. **16**(9), 2392–2407 (2017)
3. Peng, D., Wu, F., Chen, G.: Pay as how well you do: a quality based incentive mechanism for crowdsensing. In: Proceedings of the 16th ACM International Symposium on Mobile Ad Hoc Networking and Computing, pp. 177–186. ACM (2015)
4. Huang, K.L., Kanhere, S.S., Hu, W.: Are you contributing trustworthy data?: The case for a reputation system in participatory sensing. In: Proceedings of the 13th ACM International Conference on Modeling, Analysis, and Simulation of Wireless and Mobile Systems, pp. 14–22. ACM (2010)
5. Kang, Y., Miao, X., Liu, K., et al.: Quality-aware online task assignment in mobile crowdsourcing. In: 2015 IEEE 12th International Conference on Mobile Ad Hoc and Sensor Systems, MASS, pp. 127–135. IEEE (2015)
6. Pouryazdan, M., Kantarci, B., Soyata, T., et al.: Anchor-assisted and vote-based trustworthiness assurance in smart city crowdsensing. IEEE Access **4**, 529–541 (2016)
7. Zhang, X., Xue, G., Yu, R., et al.: Keep your promise: mechanism design against free-riding and false-reporting in crowdsourcing. IEEE Internet Things J. **2**(6), 562–572 (2015)
8. Kiekintveld, C., Jain, M., Tsai, J., et al.: Computing optimal randomized resource allocations for massive security games. In: Proceedings of The 8th International Conference on Autonomous Agents and Multiagent Systems-Volume 1, pp. 689–696. International Foundation for Autonomous Agents and Multiagent Systems (2009)
9. Jain, V., Grossmann, I.E.: Algorithms for hybrid MILP/CP models for a class of optimization problems. INFORMS J. Comput. **13**(4), 258–276 (2001)

A Novel Capacity-Aware SIC-Based Protocol for Wireless Networks

Fangxin Xu, Qinglin Zhao(✉), Shumin Yao, Hong Liang, and Guangcheng Li

Faculty of Information Technology, Macau University of Science and Technology, Avenida Wai Long, Taipa, Macao
fzxy002763@gmail.com, zqlict@hotmail.com, roller44@163.com, coolboom@126.com, guangcheng.li@hotmail.com
http://www.must.edu.mo/fi

Abstract. Successive interference cancellation (SIC) enables a receiver to receive multiple frames concurrently, thereby improving the physical-layer (PHY) capacity significantly. However, a larger PHY capacity does not necessarily imply a larger MAC throughput. To solve this problem, we design a novel capacity-aware SIC-based MAC protocol. In the contention process of our MAC protocol, we adopt a new user-selection method to enable a maximum PHY capacity, while in the transmission process, we adopt a novel in-frame rate adaption mechanism to make full use of the PHY capacity. As a result, our protocol can achieve far higher throughput than conventional 802.11 networks. Extensive simulations verify the efficiency of our protocol.

Keywords: User-selection · Rate adaption · Capacity-aware
Successive interference cancellation

1 Introduction

In 5G networks, the bandwidth requirement of each node is up to 1 Gbps. To satisfy the high bandwidth requirement, successive interference cancellation (SIC), is introduced in the draft of 5G [1]. SIC allows a receiver to receive multiple data frames concurrently, thereby achieving a high physical-layer (PHY) capacity. Figure 1 shows the basic idea of SIC. Assume that node 1 transmits Data1 to one receiver using the transmission power S_1, while node 2 transmits Data2 to the same receiver using S_2. With SIC, the receiver may first decode Data1 from the superposed signal, and then decode Data2 from the remaining signal (i.e., the signal after we cancel Data1's signal from the superposed signal).

However, a high PHY capacity does not necessarily imply a high MAC throughput. In the example of Fig. 1, assume that the power S_1 and S_2 will lead to a maximum PHY capacity. According to the Shannon capacity theorem, to decode Data1 and Data2 successfully, we require that Data1 should be transmitted at rate $R_1 \leq C_1 = B \times \log_2(1 + \frac{S_1}{S_2+N})$, and Data should be transmitted at rate $R_2 \leq C_2 = B \times \log_2(1 + \frac{S_2}{N})$, where B is the channel bandwidth and N

S. Chellappan et al. (Eds.): WASA 2018, LNCS 10874, pp. 793–798, 2018.
https://doi.org/10.1007/978-3-319-94268-1_69

represents the noise power. Let L_1 and L_2 denote the length of Data1 and Data2, respectively. If $L_1/R_1 \gg L_2/R_2$, it implies that after node 2 finishes its transmission, only node 1 continues its transmission even if the channel can support simultaneous transmission of more nodes, thereby causing a waste of channel recourse. Therefore, to maximize the MAC throughput, we need to design an appropriate MAC-layer protocol that well employs the SIC gain.

In this paper, to achieve and make full use of the maximum PHY capacity, we propose a novel capacity-aware SIC-based MAC protocol. Our MAC protocol mainly includes two processes: the contention and data transmission process. In the contention process, we adopt a new user-selection method that selects appropriate nodes and make them transmit in parallel so as to maximize the PHY capacity. In the transmission process, we adopt a novel in-frame rate adaption mechanism. In this mechanism, when the nodes with highest transmission rate finishes transmitting data frames, the receiver will inform nodes with lower rate to adopt a higher rate to transmit the remaining part of the transmitting data. In this way, the MAC-layer protocol can take advantage of the SIC gain, while maximizing the MAC throughput.

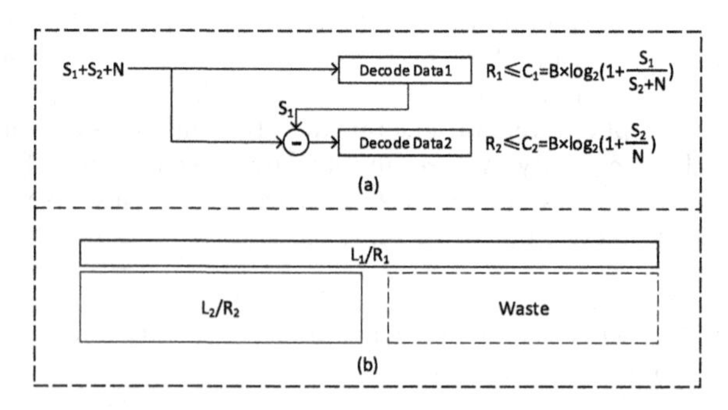

Fig. 1. (a) An example of SIC-based decoding ($R_2 \geq R_1$), (b) Transmission time of two data frames for SIC-based MAC

There are some related works [2–8] on SIC-based MAC design. The difference between us to them is that our design adopts a user-selection method and a rate adaption mechanism at the same time that makes full use of the maximum PHY capacity, thereby improving the MAC-layer throughput efficiency.

2 Capability-Aware User Selection

We consider an infrastructure-based wireless LAN (WLAN). It consists of one AP and n nodes, who are uniformly distributed around the AP. Below, we focus on uplink transmission and specify our PHY/MAC design.

2.1 PHY Design

We assume that each node and the AP are equipped with two antennas: one for data transmission/reception (Tx/Rx), the other for correlatable symbol sequences (CSS) Tx/Rx.

Like 802.11ec [6], we adopt Gold codes to generate CSSs for contention and rate adaption. Each node is assigned with one unique CSS (let CSS_A_ID donate the identity CSS of Node A) for contention, which is used as the identity of the node. Each node is also assigned with m CSSs for rate adaption, where each CSS represents one data rate (let CSS_A_R denote one of node A's m CSSs representing its data rates). For the sake of security, each node only knows its own CSSs and the AP knows all CSSs.

The AP has n correlators and adopts the successive interference cancellation (SIC) and the self-cancellation technologies. It can decode multiple simultaneously arrived data frames using SIC, and can simultaneously transmit CSS and receive data using the self-cancellation technology, and can simultaneously identify different nodes via correlation when these nodes send their identity CSSs in contention. Each node has m correlators. When the AP sends one CSS representing the data rate, the node detects the CSS via correlation and then the CSS maps to a data rate.

2.2 MAC Design

Data Frame Format. Aiming at making full use of the SIC capacity, our MAC design adopts a novel in-frame rate adaption mechanism, that is, during a frame transmission, we may adopt one transmission rate for the first part of the frame (Data-Part1) and later adopts a new transmission rate for the second part of the frame (Data-Part2). To this end, as shown in Fig. 2, we design a new data frame format consisting of 6 fields: CSS_ID is used for frame and frequency synchronizations; SIG1 and SIG2 specifying the transmission rate of the corresponding data parts; EOF is used to identify the end of a frame.

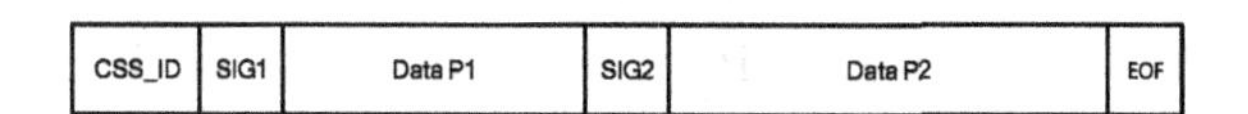

Fig. 2. Data frame format

MAC Protocol. The MAC-layer transmission is executed in a time division multiple access (TDMA) manner. Each round of transmission can be categorized into three processes: Contention processes, Transmission process and ACK process. We only consider the scenario that there are two nodes transmitting at the same time and each node only transmits one data frame in each round. We also only consider the scenario that each data frame has identical length. Below, with the example shown in Fig. 3, we specify each process.

Contention Process. At the beginning of this process, nodes which have buffered data transmit their own identity CSSs to the AP concurrently. Through correlation, the AP detects theses identity CSSs (i.e., detect these nodes) and obtain the Received Signal Strength (RSS) of each identity CSS (i.e., the correlation

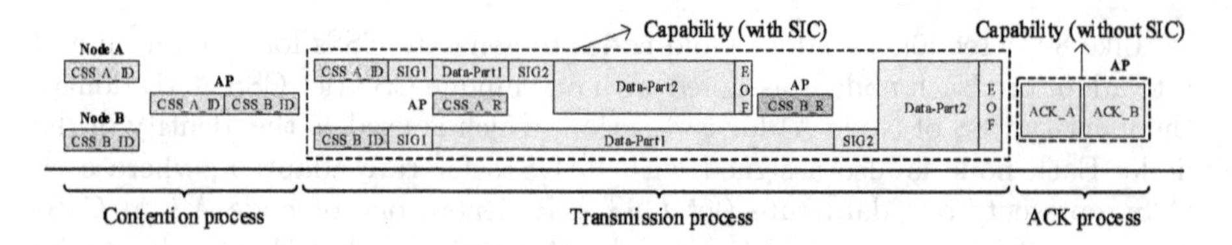

Fig. 3. An overview of MAC protocol

result). Then, the AP selects the top two nodes in the RSS ranking (correspondingly these two nodes have the highest signal-to-noise ratio therefore can make full of PHY capacity) and allows them to transmit in this round. To inform these selected nodes and their transmission orders, the AP broadcasts these selected identity CSSs sequentially, where the sequence in the broadcast represents the transmission order of nodes. These nodes detect their identity CSSs and know about transmission sequences via demodulation (instead of correlation) to ensure reliable transmissions. As shown in Fig. 3, Node A and Node B transmit their own identity CSSs (i.e., CSS_A_ID and CSS_B_ID) simultaneously. The AP detects these identity CSSs via correlation and broadcast them.

Remark: To balence the fairness and the throughput gain of SIC, we may give lower transmission probability to those nodes with higher RSS and give higher transmission probability to those nodes with lower RSS.

Transmission Process. With the help of Fig. 3, we now explain the transmission process. In this example, we assume that only nodes A and B contend for the channel. In the transmission process, the AP indicates that nodes A and B transmit their data frames at the lowest rate sequentially. Then in the transmission process, after calculating the remaining PHY capacity (which is equal to the difference between the maximum PHY bandwidth and the already used bandwidth), the AP sends CSS_A_R to ask Node A to increase its transmission rate. After that, Node A sends SIG2 with the previous rate and then sends DATA-Part2 and EOF with the data rate (that CSS_A_R maps). After Node A finishes its transmission, AP sends CSS_B_R to ask Node B to increase its transmission rate. After sending SIG2 with the previous rate, Node B sends Data-Part2 and EOF fields of its data frame with the data rate (that CSS_B_R maps).

ACK Process. When AP successfully receives the data frames, it will send ACK one by one. Like 802.11, when an ACK timeout occurs, the node will start retransmission.

3 Performance Evaluation

In this section, we evaluate the MAC-layer throughput of our design via extensive simulations. The simulator is written using Matlab. The default parameter settings are shown in Table 1.

Table 1. Parameter settings

m	10	DIFS	50 µs
R_d	11 Mbps	SIFS	10 µs
S_1	20 mW	SIG1	4 bit
S_2	20 mW	SIG2	4 bit
N	1 mW	EOF	16 bit
k_d	64	MAC Header	28 Byte
N_{us}	2	ACK	14 Byte
Bandwidth	20 MHz	DATA	1500 Byte

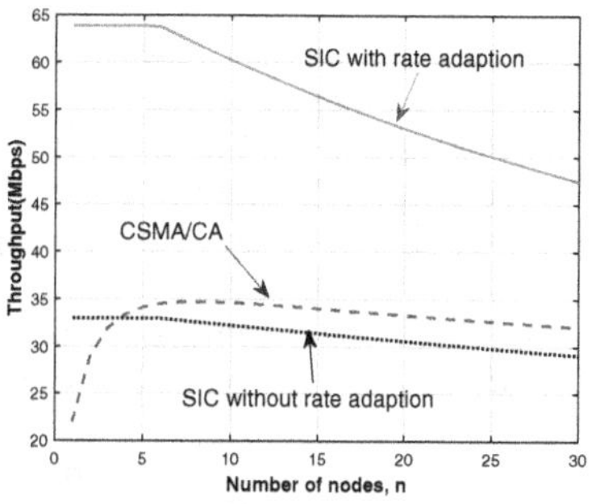

(a) Throughput vs. the number of nodes

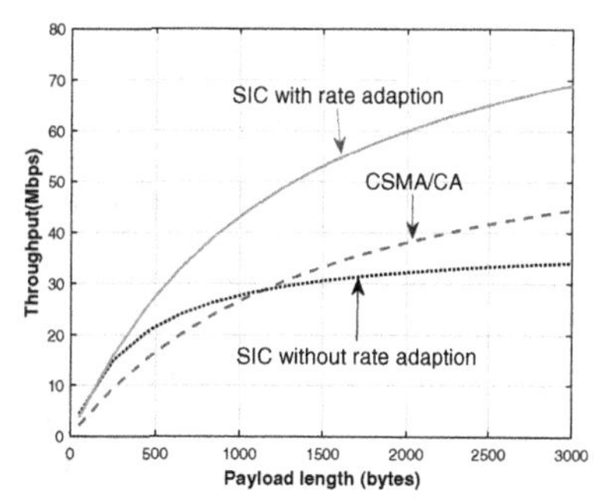

(b) Throughput vs. payload length.

Fig. 4. Simulation results

Figure 4(a) shows the comparison of the throughput of CSMA/CA, SIC without rate adaption and SIC with rate adaption. The X axis indicates the number of nodes n and the Y axis indicates throughput of the network. As shown in the figure, the throughput of SIC without rate adaption is lower then CSMA/CA when $n > 4$ since nodes in SIC without rate adaption transmit their data in the lowest data rate and therefore the channel utilization is very low. And the throughput of SIC with rate adaption is much higher than CSMA/CA overall. The main reason throughput of SIC both with and without rate adaption are decrease only when $n > 6$ is that when $n \leq 6$, the CSS length is short that results in negligible overhead.

Figure 4(b) compares the throughput of CSMA/CA, SIC without rate adaption and SIC with rate adaption as the payload varies from 50 bytes to 2000 bytes, where the number of nodes, n, is set to 20. From this figure, we can see that the throughput of all protocols increases as the length of payload increases. This is because the MAC throughput efficiency increase as the length of payload increase. Also, when the payload length is larger than a certain value, the throughput of SIC without rate adaption is lower than that of CSMA/CA, as explained in Fig. 4(a).

4 Conclusion

In this paper, we propose a novel capacity-aware SIC-based protocol. Our design can select appropriate nodes that maximizes the PHY capacity and at the same time adopt an in-frame rate adaption mechanism that makes full use of the maximum PHY capacity. This study will help better design the SIC-based protocol.

Acknowledgments. This work is supported by the Macao FDCT-MOST grant 001/2015/AMJ, and Macao FDCT grants 056/2017/A2 and 005/2016/A1.

References

1. Saito, Y., Kishiyama, Y., Benjebbour, A., Nakamura, T., Li, A., Higuchi, K.: Non-orthogonal multiple access (NOMA) for cellular future radio access. In: 2013 IEEE 77th Vehicular Technology Conference (VTC Spring), pp. 1–5. IEEE (2013)
2. Sen, S., Santhapuri, N., Choudhury, R.R., Nelakuditi, S.: Successive interference cancellation: carving out MAC layer opportunities. IEEE Trans. Mob. Comput. **12**(2), 346–357 (2013)
3. Gudipati, A., Pereira, S., Katti, S.: AutoMAC: rateless wireless concurrent medium access. In: Proceedings of the 18th Annual International Conference on Mobile Computing and Networking, pp. 5–16. ACM (2012)
4. Uddin, M.F., Mahmud, M.S.: Carrier sensing based medium access control protocol for WLANs exploiting successive interference cancellation. IEEE Trans. Wirel. Commun. **16**, 4120–4135 (2017)
5. Sen, S., Choudhury, R.R., Nelakuditi, S.: CSMA/CN: carrier sense multiple access with collision notification. IEEE/ACM Trans. Netw. (ToN) **20**(2), 544–556 (2012)
6. Magistretti, E., Gurewitz, O., Knightly, E.W.: 802.11ec: collision avoidance without control messages. IEEE/ACM Trans. Netw. **22**(6), 1845–1858 (2014)
7. Choi, W., Lim, H., Sabharwal, A.: Power-controlled medium access control protocol for full-duplex WiFi networks. IEEE Trans. Wirel. Commun. **14**(7), 3601–3613 (2015)
8. Qu, Q., Li, B., Yang, M., Yan, Z., Zuo, X., Guan, Q.: FuPlex: a full duplex MAC for the next generation WLAN. In: 2015 11th International Conference on Heterogeneous Networking for Quality, Reliability, Security and Robustness (QSHINE), pp. 239–245. IEEE (2015)

A Supervised Learning Approach to Link Prediction in Dynamic Networks

Shuai Xu, Kai Han(✉), and Naiting Xu

School of Computer Science and Technology/Suzhou Institute for Advanced Study, University of Science and Technology of China, Hefei, People's Republic of China
{sa615527,sa615237}@mail.ustc.edu.cn, hankai@ustc.edu.cn

Abstract. Link prediction, as one of fundamental problems in social network, has aroused the vast majority of research on it. However, most of existing methods have focused on the static networks, although there exist some machine learning methods for the dynamic networks, they regard either link structures or node attributes captured from a single snapshot of the network as the features, thus cannot achieve high accuracy. In this paper, following the supervised learning framework, we innovatively propose a new approach to this problem in dynamic networks. In particular, our features are captured from the variation of the structural properties and a lot of important metrics considering the long-term graph evolution of network, instead of a single snapshot. For each feature, we use an optimization algorithm to calculate the corresponding weight of each classifier, and then can determine whether there is a connection between a pair of nodes. In addition, we execute our method on two real-world dynamic networks, which indicate that our method works well and significantly outperforms the prior methods.

Keywords: Dynamic network · Link prediction
Supervised learning · Social network analysis

1 Introduction

With the rapid development of the social network, many relevant problems have raised along with it, *link prediction* is a typical one among them. In particular, the topology of network is always changing, new nodes and edges are added, meanwhile, old nodes may be deleted. Due to this highly dynamic nature, link prediction on the online social network becomes such an extremely challenging and urgent research issue.

In this field, many research works have concentrated on proposing novel and efficient methods for link prediction. Until now, most of the existing methods no matter based on the unsupervised learning or the supervised learning only focus on the static networks. In other words, they barely take account of the long-term graph evolution of the networks, which cannot achieve good performance when applied in the real dynamic networks. In view of this situation, we

S. Chellappan et al. (Eds.): WASA 2018, LNCS 10874, pp. 799–805, 2018.
https://doi.org/10.1007/978-3-319-94268-1_70

creatively put forward some new techniques to construct an approach to the link prediction problem in dynamic networks. The main contributions of this work are summarized as below:

- We innovatively propose an efficient approach with high accuracy to solve link prediction problem for online dynamic social network.
- To improve the prediction accuracy, we propose a different feature selection which fully takes account of the long-term graph evolution of social networks.
- We execute our method on two real-world dynamic social networks and the related experimental evaluations indicate that our method works surprisingly well and significantly outperforms the prior methods.

2 Related Work and Problem Definition

2.1 Related Work

Motivated by the similarity-based methods, Liben-Nowell and Kleinberg [7] originally proposed a link prediction method based on dynamic topology of networks. Later, several approaches to link prediction have been subsequently proposed either regarding the keyword distance as the property between a pair of nodes [4], or based on the user interests [3], or combining both profile similarity and social network topology [2]. In particular, in the work of Zhou et al. [12], experimental evaluations show that the link prediction algorithm based common neighbors achieves the best performance among others.

Based on the idea of Al Hasan et al. [5] that treating the link prediction problem as a binary classification problem, Da Silva Soare and Prudncio [10] proposed a link prediction algorithm which is confirmed to achieve the better performance than the traditional similar methods. Subsequently, Richard et al. [9] proposed a different method by evaluating the evolution process of dynamic network, which shows the better performance when compared with the traditional heuristics approaches. Besides, the approaches of [11] resolve the link prediction problem in dynamic/static networks, whereas they predict the future relationship by only using the link structure information but ignoring the node attributes.

2.2 Problem Definition

Given an evolving social network $G^t = (\boldsymbol{V}, E^t, P^t)$ at time t, then a sequence of graph snapshots $\{G^t\}_{t=1,\dots,T} = \{G^1, G^2, \dots, G^T\}$ can be produced at following time stamps. The link prediction aims to predict the most likely link state for a future time $t'(t' > T)$, where $\boldsymbol{V}$ remain the same across all time steps but E^t and P^t change for each time. For each pair of nodes (v, w), we compute their probability $P^{t+1}(v, w)$ of establishing link and predict their link state $L^{t+1}(v, w)$, where $P^{t+1}(v, w)$ is a element of a $n \times n$ matrix P^{t+1}, and $L^{t+1}(v, w)$ denotes the label of establishing link between v and w at time $t + 1$. In detail, if $e(v, w) \in E^{t+1}$, then $L^{t+1}(v, w) = 1$, otherwise $L^{t+1}(v, w) = 0$. Finally the network snapshot G^{t+1} can be obtained in the next time step $t = t + 1$.

3 Method Formulation

3.1 Feature Selection

In this paper, in order to improve the prediction accuracy, we propose a different feature selection which fully takes account of the long-term graph evolution of social networks. Moreover, these features are captured from the variation of the structural properties and a lot of important metrics including the change degree of common neighbors, the time-varied weight and the intimacy between common neighbors, instead of a single snapshot. In detail, these structure properties used are: Preferential Attachment (*PA*) [8], Common Neighbor (*CN*) [8], Adamic Adar (*AA*) [1] and Jaccards Coefficient (*JC*) [10], where $PA(v,w) = |N(v)| \times |N(w)|$, $CN(v,w) = |N(v) \cap N(w)|$, $AA(v,w) = \sum_{u \in N(v) \cap N(w)} \frac{1}{\log(|N(u)|)}$ and $JC(v,w) = |\frac{N(v) \cap N(w)}{N(v) \cup N(w)}|$. In addition, three important metrics used are: Time Weight (*TW*), Change Degree of CN (*CD*) and Closeness Between CNs (*CB*), where $TW(t) = \exp(-\psi(T-t))$, $CD_t(cn) = \frac{t}{\sum_{i=2}^{t} ed_{i-1,i}}$ and $CB_t(v,w) = \ln(|ACN_t|)$.

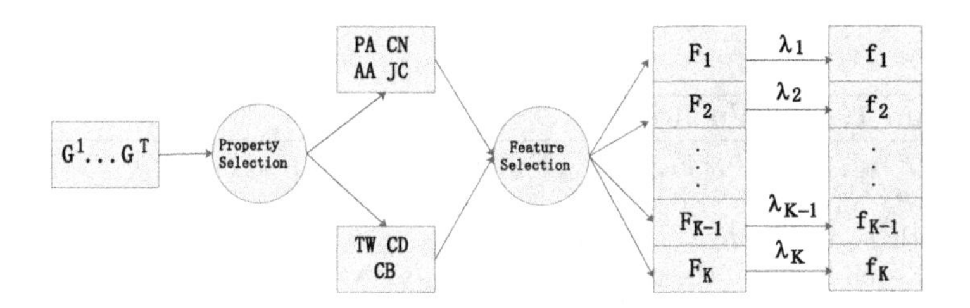

Fig. 1. The model of *SLM-Vp*.

3.2 Model of *SLM-Vp*

To address this problem, we innovatively propose a new approach *SLM-Vp*, which is a supervised learning method based on flat SVM classification methods, and these features are captured both from the variation of the structural properties and a lot of important metrics considering the long-term graph evolution of network, instead of a single snapshot. In detail, the model of *SLM-Vp* is shown in Fig. 1, where $(\lambda_1, \lambda_2, \ldots, \lambda_k)$ is a weight vector of classifiers $(f_1, f_2, \ldots, f_k)$.

Firstly, for each $G^t \in \{G^t\}_{t=1,\ldots,T}$, we can represent G^t as a $n \times n$ adjacency matrix M^t, where M^t is symmetric and each element $M^t(v,w)$ equals $L^t(v,w)$ or $P^t(v,w)$. Moreover, for an arbitrary pair of node v and w, a $K \times T$ matrix $M^t(v,w)$ can be calculated, where each element equals $F_k^t(v,w)$. Then we can calculate the change values of all its features, thus a $K \times (T-1)$ matrix $VM^{T-1}(v,w)$ can be calculated, where each element equals $\triangle F_k^t(v,w)$. Secondly, for each structure feature F_k, we can train a classifier by using a set of data composed of instances annotated with structural features and the label

$L(v, w)$ of establishing link between v and w. We denote this set of instances as $VM_k^{T-1}(v, w)$ and select both the negative and positive training instances by using a random selection algorithm. Finally, we employ an optimization algorithm to calculate the weight of each classifier. After the above three steps, we can eventually obtained a probability of establishing a link between node v and w. In this paper, we use seven classifiers for seven features respectively, which are captured from the variation of four structural properties and three important metrics.

3.3 Implementation of *SLM-Vp*

The pseudo code of *SLM-Vp* algorithm is described as in Algorithm 1 below, where TS_k^T is a training set which is generated for each F_k, and TS_k^T can be calculated in Eq. (1). Then we train a data instance TS_k^T to learn a classifier f_k^T. Note $\overrightarrow{\lambda}$ is a vector of weights and can be represented as $\overrightarrow{\lambda} = (\lambda_1, \lambda_2, \ldots, \lambda_k)$, where λ_k is the weight of classifier f_k^T. We can give an *example*[1] to illustrate.

$$TS_k^T = \{(\{\triangle F_k(v, w)\}_1^T, L_T(v, w))\}. \tag{1}$$

Algorithm 1. SLM-Vp algorithm

Input: $\{G^t\}_{t=1,\ldots,T}, P^T, \{F_1, F_2, \ldots, F_K\}, (v, w)$;
Output: $P^{T+1}(v, w)$;
1: **for** each F_k in $\{F_1, F_2, \ldots, F_K\}$ **do**
2:     **for** each (v, w) in the train set **do**
3:         Calculate $F_k^1(v, w), F_k^2(v, w), \ldots, F_k^T(v, w)$;
4:         Get $\{\triangle F_k^t(v, w)\}_1^{T-1}$
5:     **end for**
6:     Train a classifier f_k^T by using a new training set TS_k^T;
7: **end for**
8: Optimize $\min \sum(\sum \lambda_k \cdot f_k^T(v, w) - L_{T+1}(v, w))^2$ and get weight $\overrightarrow{\lambda}$;
9: **return** $P^{T+1}(v, w) = \Sigma \lambda_k \cdot f_k^{T+1}(v, w)$;

4 Experiment and Analysis

In this section, we conduct extensive experiments to validate our proposed methods on two specific co-authorship networks, i.e., *hep-th*, *hep-lat* from two sections of *Arxiv* (www.arxiv.org) which are also used by lots of prior works (e.g. [7,11]). The *hep-th*[2] network is formed by authors in theoretical high energy physics area, and the *hep-lat*[3] is composed by authors in high energy physics-lattice area.

[1] https://github.com/ustcxs/wasa/blob/master/SLM-Vp.pdf.
[2] http://arxiv.org/archive/hep-th.
[3] http://arxiv.org/archive/hep-lat.

For convenience of experiment setup, we extracted data sets from year 1995 to 2015 for *hep-th* and *hep-lat*. In detail, *hep-th* has 19258 authors and 132568 collaborations, *hep-lat* has 5136 authors and 72354 collaborations. We select training instances and testing instances by running random algorithm with about 1:1 ratio. In detail, *hep-th* has 8236 training instances and 7998 testing instances, *hep-lat* has 2125 training instances and 2204 testing instances.

For verifying our proposed method *SLM-Vp*, we make a comparison among four baseline algorithms: *RA* [12], *Tw-CN* [6], *Static-Lp* [5], *EA-Lp* [9].

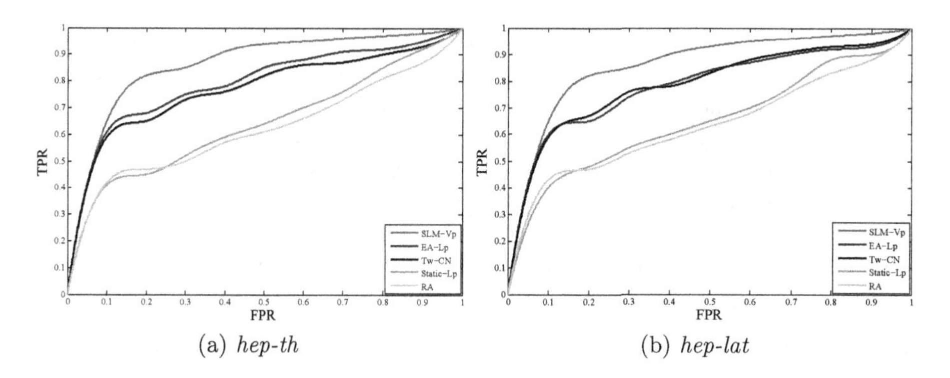

(a) *hep-th* (b) *hep-lat*

Fig. 2. The ROC curves of all the methods in two networks

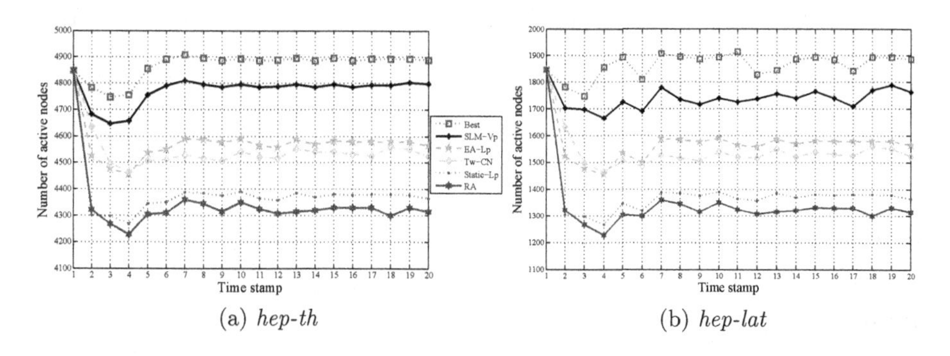

(a) *hep-th* (b) *hep-lat*

Fig. 3. The maximization influence of all methods in two networks

We test all the methods in two real-life social networks, and all the settings are decided by experimentation. As shown in Fig. 2, we demonstrate the *ROC* curves of all the methods in two networks. Table 1 lists the *AUC* values of all the methods in two networks. In Fig. 3, we predict the link state of all nodes in graph by using above five methods, then we obtain five predicted graph snapshots and select top-k most influential node sets where $k = 50$. Finally, we calculate the

Table 1. The AUC values of all the methods in two networks

Network	RA	Tw-CN	Static-Lp	EA-Lp	SLM-Vp
hep-th	0.6523	0.7885	0.6768	0.7968	**0.8585**
hep-lat	0.6456	0.7778	0.6623	0.7883	**0.8503**

influence diffusion of these node sets in a real graph snapshot. Note that the Best method shown in Fig. 3 selects top-k most influential node sets and calculates the influence diffusion in a real graph snapshot without the prediction.

The experimental results show that our *SLM-Vp* algorithms can achieve better performance than other four methods. Moreover, the *ROC* curves and the maximization influence indicate that *SLM-Vp* can achieve better performance than other existing methods in the following graph snapshots. These results show that both supervised learning and unsupervised learning can be employed to link prediction problem. However, the link prediction methods under supervised learning usually achieve better performance. Thus supervised machine leaning is a key research direction for link prediction problem in dynamic networks.

5 Conclusion

In this paper, we propose a supervised learning method called *SLM-Vp*, which is based on flat SVM classification methods to deal with link prediction problem in dynamic networks. In particular, we produce a distinctive feature selection way which fully considers the long-term graph evolution of network and achieves high accuracy. Moreover, experimental evaluations on two social networks show that our proposed method result in better performance compared with other four methods. We believe that our results are meaningful contribute to the research.

Acknowledgment. This work is partially supported by National Natural Science Foundation of China (NSFC) under Grant No. 61472460, No. 61772491, Natural Science Foundation of Jiangsu Province under Grant No. BK20161256. Kai Han is the corresponding author.

References

1. Adamic, L.A., Adar, E.: Friends and neighbors on the web. Soc. Netw. **25**(3), 211–230 (2003)
2. Akcora, C.G., Carminati, B., Ferrari, E.: User similarities on social networks. Soc. Netw. Anal. Min. **3**(3), 475–495 (2013)
3. Anderson, A., Huttenlocher, D.P., Kleinberg, J.M., Leskovec, J.: Effects of user similarity in social media. In: Proceedings of the Fifth International Conference on Web Search and Web Data Mining, pp. 703–712 (2012)
4. Bhattacharyya, P., Garg, A., Wu, S.F.: Analysis of user keyword similarity in online social networks. Soc. Netw. Anal. Min. **1**(3), 143–158 (2011)

5. Al Hasan, M., Chaoji, V., Salem, S., Zaki, M.: Link prediction using supervised learning. In: Proceedings of SDM 06 Workshop on Link Analysis, Counterterrorism and Security (2006)
6. Huang, S., Tang, Y., Tang, F., Li, J.: Link prediction based on time-varied weight in co-authorship network. In: Proceedings of the 18th International Conference on Computer Supported Cooperative Work in Design, pp. 706–709 (2014)
7. Liben-Nowell, D., Kleinberg, J.M.: The link-prediction problem for social networks. JASIST **58**(7), 1019–1031 (2007)
8. Newman, M.: Clustering and preferential attachment in growing networks. Phys. Rev. E - Stat. Nonlinear Soft Matter Phys. **64**(2), 251021–251024 (2001)
9. Richard, E., Gaïffas, S., Vayatis, N.: Link prediction in graphs with autoregressive features. J. Mach. Learn. Res. **15**(1), 565–593 (2014)
10. Da Silva Soares, P.R., Prudncio, R.B.C.: Time series based link prediction. In: The 2012 International Joint Conference on Neural Networks, pp. 1–7 (2012)
11. Wang, D., Cui, P., Zhu, W.: Structural deep network embedding. In: Proceedings of the 22nd ACM SIGKDD International Conference on Knowledge Discovery and Data Mining, San Francisco, CA, USA, 13–17 August 2016, pp. 1225–1234 (2016)
12. Zhou, T., Lü, L., Zhang, Y.C.: Predicting missing links via local information. Eur. Phys. J. B **71**(4), 623–630 (2009)

Mining Mobile Users' Interests Through Cellular Network Browsing Profiles

Fan Yan[1], Yunpeng Ding[1], and Wenzhong Li[1,2](✉)

[1] State Key Laboratory for Novel Software Technology, Nanjing University, Nanjing, China
{151220139,dingyunpeng}@smail.nju.edu.cn, lwz@nju.edu.cn
[2] Collaborative Innovation Center of Novel Software Technology and Industrialization, Nanjing, China

Abstract. Mining mobile users' interest is very important for numerous of commercial applications such as product recommendation, personalized advertisement, precision marketing, etc. In this paper, we proposed a novel clustering approach for semantic mining from cellular network browsing profiles based on the topic model. We treat each URL as a word and the user's browsing history as a document, and adopt the Latent Dirichlet Allocation (LDA) model to represent the web browsing interest of mobile users. We further used K-means to cluster the users into several groups according to their topic similarities, and apply a feature ranking approach to explain the sematic meaning of the clustering results. The performance of the proposed approach is verified on a dataset from a telecom operator, which explains users' interests well in the clusters.

Keywords: LDA · Clustering · User persona · Interest mining

1 Introduction

With the rapid development of mobile network, people are getting used to do more daily works and entertainments on their smartphones. Thus a growing number of analyses in the recent years have sought to explore mobile users related data to analyze their behavior and preference.

App using trace is one of the most popular source materials for researchers. Zhao et al. [1] collected one month of app usage from 106,762 Android users. They found that users are heterogeneous in smartphone usage, and proposed a 2-step clustering method to cluster users into 382 distinct types. Xu et al. [2] got app usage at a national level and analyzed how, when and where they are used. Some interesting patterns of using preference were revealed in this paper.

Comparing with app using trace, users' browsing profiles made up with URLs provide more semantic for researchers to mine, since URLs show what users really get in detail. Traditional method to mine user interests from surfing profiles is

S. Chellappan et al. (Eds.): WASA 2018, LNCS 10874, pp. 806–812, 2018.
https://doi.org/10.1007/978-3-319-94268-1_71

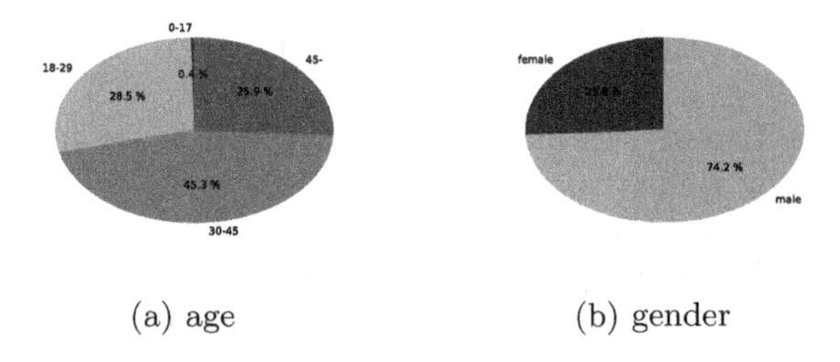

(a) age (b) gender

Fig. 1. Age and gender distribution of users.

keywords mining. Researchers try to find specific keywords in URLs to classify a record to a certain category. In that way, the keywords vocabulary should be large enough to deal with the complexity of reality, and building such vocabulary is labor-intensive and time-consuming.

To overcome those shortcomings, Giri et al. [3] proposed the unsupervised topic model (LDA) to mine the interests of the cellular users based upon their browsing profiles. However, they did not dig more from the results of LDA nor mining from time information of clicks. Extracting click stream is also an useful method to solve this problem, such as Zhang et al. [4] and Wang et al. [5].

In our research, we cooperated with a telecommunication company to get anonymous surfing profiles. To mining abundant semantic meanings, we proposed a novel clustering approach for semantic mining based on the topic model, which is quite useful to discover the hidden semantic structures in a text body. Specifically, we treat each URL as a word and each user's browsing history as a document, and adopt a famous topic model, Latent Dirichlet Allocation (LDA), to represent the web browsing interest of mobile users. To better understand the behaviors of similar users, we used K-means to cluster the users into several groups. Then we apply a feature ranking approach to explain the sematic meaning of the clustering results.

2 Dataset Overview

The dataset used in this paper is from a telecom operator of China, which contains 945 users' browsing profiles from March 15th to March 21st in 2016, together with some basic information, such as their genders and ages. The users are randomly draw from the Jiangsu province. As shown in Fig. 1, users' age varied from 10 to 79, and most of them are between 18 and 45. The gender ratio is not balanced in our dataset as the male users account for 3 fourths. Take a unique URL request as a record, the whole dataset contains about 3148110 records, about 3310 records per user in average. The data is totally anonymous with the virtual number to identify users. Those records contain 12792 different URLs. Except for URL and identifier, session start time, session end time, user agent, target IP and network traffic-related information are also included.

3 LDA Model and Implementation

LDA is short for Latent Dirichlet Allocation, which is an unsupervised topic model presented by Blei et al. in 2003 [7]. It assumes each document in the corpus as a mixture of topics and find out the possibility of the documents to be in each topic. In our case, LDA regards each users' browsing profile as a document. The model will find the hidden topic of each URL automatically.

We reviewed the work of Giri et al. [3], and used the same LDA tool–Mr. LDA [6]. It's an open-source package for scalable, multilingual topic modeling using variational inference in MapReduce.

LDA treats a document as a bag of words. Thus the order of words will not be taken into analyzing process. To capture time information, we propose to divide the users' profiles into four intervals according to its sending time. We divide the whole day into four stages: 0:00 to 5:59 is early morning, 6:00 to 11:59 is morning, 12:00 to 17:59 is afternoon, and 18:00 to 23:59 is evening. We assume that users' interests varied in different periods of time.

Stop Words and Informed Prior. Stop words are those URLs which are nearly visited by everyone. Those URLs make topics less distinctive as they may appears in every topic. We sorted all the URLs by frequency and then picked the top 100 URLs as stop URLs out. Ignoring the stop words in corpus indeed helps to get a better result, but we have to admit that we sacrificed some semantic.

After removing the stop words, how to determine the semantic meaning of each topic is still challenging. Mr. LDA provides an extension block named Informed Prior, which helps to force some similar words to be grouped into the same topic. For example, if we want URLs related to Reading/News to be grouped into the same topic, we collect some famous websites in this area and picked their URLs for topic one, like [hupu.com, sina.com, ifeng.com, baidunews.com, toutiao.com, ucweb.com ...]. When a URL β belongs to the list, we set a higher initialization value as its informed prior to this topic. In that case, we abandon the default symmetric prior provided by the model and get a better initialization.

Results of LDA. Now we have divided users' browsing profiles into four parts according to time, cleaned them up by removing stop words, and added enough common URLs for each topic. The only parameter we need to assign is the number of topics, which we chose to set as 7 after comparing the likelihood with some other parameters. From topmost URLs of each topic, we can find the semantic meanings and tick labels for users. Here is a table shows these top URLs with the score representing the probability of locating in this topic (Table 1).

As each user's profile is divided into 4 parts on different periods, and each part has the probability of locating in different topics, we finally got four 7-dimension vectors for each user. We combined them as the user feature vector. We also normalized the vector to make it easier for clustering.

Table 1. The top URLs in different topics.

Topic	Top URLs in the topic
Education/working	Imail.qq (6.92), kmail.com (6.91), yuansouti (6.90), dsp-impr2.youdao (6.91), log.yex.youdao (6.90)
Ecommecial	Talaris2-lbseleme (7.47), m.360buy (6.95), wp.360buyimg (6.93), wq.jd (6.92), pic.alipay.objects (6.92)
Entertainment	Video.qpic (6.96), vhot.hdnion.videocdn.qq (6.95), cm.passport.iqiyi (6.93), p1.music126 (6.92), ccstream.qqmusic.qq (6.91)
Game	Report.game.center (7.27), api.coolmart.net (7.11), games.qq (6.93), game11h5 (6.93), gift.gamecenter.qq (6.93)
Reading/news	Zzm.sohu (7.09), p2p.statp (7.06), d.ifengimg (7.03), stadig.ifeng (6.99), vs3.wxctu3.ucweb (6.98)
Social communication	Cn.battleofballs (7.00), ossweb-img.qq (6.91), oth.strmdt.qq (6.91), api.place.weibo (6.91), cgi.connect.qq (6.91)
Junk/ad	Router.g0.push.leancloud (6.92), hispace.clthicloud (6.91), api-webrp-analytics.cloudtoast (6.91), icsnssdk (6.91), imagebox.xiaomi (6.91)

4 User Clustering

The LDA model analyzes the original user data and provides us with normalized user vectors. To depict the similarities and personalities of users better, we choose to use K-means, a famous clustering method which based on measuring distances between samples, to cluster users into different clusters.

We combined several metrics to measure the performance of different value of K. The following equation is used to calculate the score of a certain K's clustering result.

$$Score = SE + DI + \frac{UserNum - MaxClusterNum}{UserNum} \tag{1}$$

In the equation, SE means Shannon Entropy and DI means Dunn index. We take the number of users in each cluster and divide it by the total number of users as the possibility of each cluster to calculate Shannon Entropy. Dunn index is a typical internal index used to evaluate clustering performance. Notice that the last item of the equation is not a typical index, we use the total number of users minus the number of users in the biggest cluster to penalize results which most users are grouped into few clusters.

According to our experiments, 29 was chosen as the best K for clustering. Figure 2(a) and (b) show the distribution of users in clusters. Centroids of each cluster are the intuitive materials to describe clusters' features. We visualize 29 centroids in the heat map to see how we distinguish one cluster to another. The

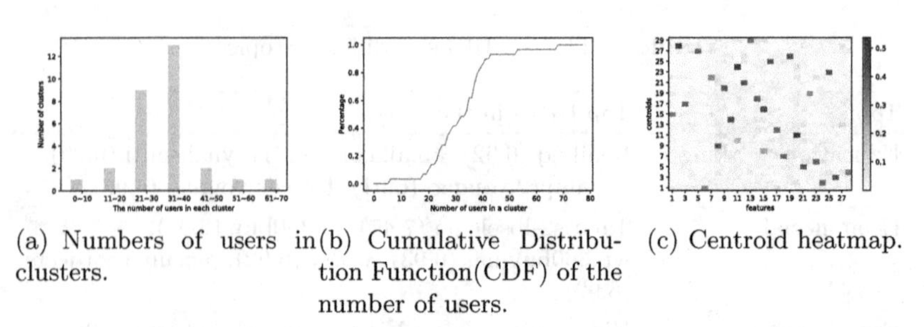

(a) Numbers of users in clusters. (b) Cumulative Distribution Function(CDF) of the number of users. (c) Centroid heatmap.

Fig. 2. The distribution of users in clusters. (Color figure online)

result is showed in Fig. 2(c). We can easily tell the difference by the highest dimension of each centroid.

5 Analysis of Clustering Results

In this part, we will explain our approach to select the most notable features and take the biggest cluster as an example to show how we give a cluster a label as personas. The following features we defined will help us to analyze the result.

Salient Features. As shown in the heat map Fig. 2(c), each centroid contains one or two dimension shown as deep blue, which means the value is significantly larger than other dimensions. Thus we chose those features with highest values as salient features.

Idiosyncratic Features. We got inspiration from the work of Zhao et al. [1], especially the part used to distinguish clusters. They call it idiosyncratic features.

$$r_{i,j} = \frac{|c_{i,j} - u_j|}{\max_{j \in J}(|c_{i,j} - u_j|)} \quad (2)$$

In this equation, $c_{i,j}$ is the value of jth dimension of ith cluster's centroid, u_j means j^{th} dimension of the "average" user. The largest $r_{i,j}$ means j^{th} feature (or dimension of the vector), is the main contributor for the cluster to be distinctive.

Example of Cluster Results: The 12th cluster is the largest cluster of the result, which contains 68 users. By visualizing the age and gender distribution of this cluster (Fig. 3(c)), we found the gender distribution is similar to the whole dataset. While the ratio of users aged from 18 to 29 in this cluster is significantly larger than average. That is to say, users are younger than average age in this cluster. The top 4 salient features for this cluster is Social/h18–24, Ad/junk/h12–18, Entertainment/h6–12 and E-commerce/h12–18. Only the first feature, Social/h18–24, is significantly larger than the average value. And top 4 Idiosyncratic Features are Social/h18–24, education/working/h12–18, Ad/junk/h12–18 and Game/h6–12.

The salient features only depict they love using social applications as other dimensions are similar to the average user. According to the idiosyncratic features, we can see these users are not fond of games and education/working applications since they are the main features distinguish them from other clusters, and values of these features are low. We can stick the label "social lovers" to this cluster.

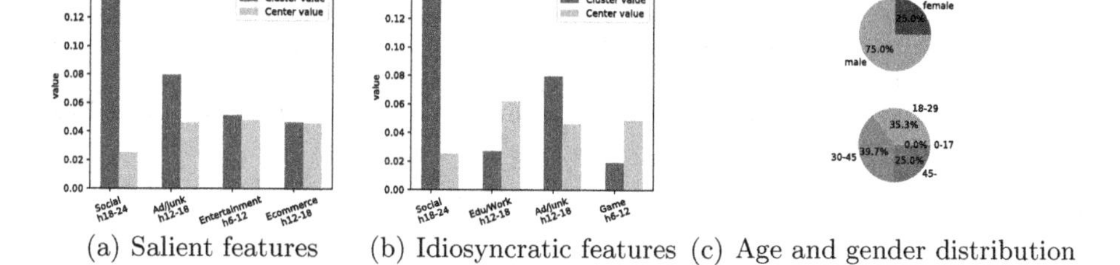

(a) Salient features (b) Idiosyncratic features (c) Age and gender distribution

Fig. 3. Analysis of the 12th cluster (68 users).

6 Conclusion

In this paper, we proposed a method to mine users' interests from cellular network browsing profiles. We used the topic model (LDA) to extract the interest and find the semantic meaning hidden behind the URLs. Using LDA, the interest of each user can be represented by a vector, which is used for clustering similar users using the K-means algorithm. The methods we provide in this paper is useful for telecommunication companies to understand their users and build personas, which is helpful for product recommendation and personalized advertisement.

Acknowledgements. This work was partially supported by the National Key R&D Program of China (Grant No. 2017YFB1001801), the National Natural Science Foundation of China (Grant Nos. 61672278, 61373128, 61321491), the science and technology project from State Grid Corporation of China (Contract No. SGSNXT00YJJS1800031), the Collaborative Innovation Center of Novel Software Technology and Industrialization, and the Sino-German Institutes of Social Computing.

References

1. Zhao, S., Ramos, J., Tao, J., Jiang, Z., Li, S., Wu, Z., Pan, G., Dey, A.K.: Discovering different kinds of smartphone users through their application usage behaviors. In: Proceedings of UbiComp 2016, pp. 498–509. ACM (2016)
2. Xu, Q., Erman, J., Gerber, A., Mao, Z., Pang, J., Venkataraman, S.: Identifying diverse usage behaviors of smartphone apps. In: Proceedings of IMC 2011, pp. 329–344. ACM (2011)

3. Giri, R., Choi, H., Hoo, K.S., Rao, B.D.: User behavior modeling in a cellular network using latent Dirichlet allocation. In: Corchado, E., Lozano, J.A., Quintián, H., Yin, H. (eds.) IDEAL 2014. LNCS, vol. 8669, pp. 36–44. Springer, Cham (2014). https://doi.org/10.1007/978-3-319-10840-7_5
4. Zhang, X., Brown, H.F., Shankar, A.: Data-driven personas: constructing archetypal users with clickstreams and user telemetry. In: Proceedings of CHI 2016, pp. 5350–5359. ACM (2016)
5. Wang, G., Zhang, X., Tang, S., Zheng, H., Zhao, B.Y.: Unsupervised clickstream clustering for user behavior analysis. In: Proceedings of CHI 2016, pp. 225–236. ACM (2016)
6. Zhai, K., Boyd-Graber, J., Asadi, N., Alkhouja, M.L.: Mr. LDA: a flexible large scale topic modeling package using variational inference in mapreduce. In: Proceedings of WWW 2012, pp. 879–888. ACM (2012)
7. Blei, D.M., Ng, A.Y., Jordan, M.I.: Latent Dirichlet allocation. J. Mach. Learn. Res. **3**(Jan), 993–1022 (2003)

A Gradient-Boosting-Regression Based Physical Health Evaluation Model for Running Monitoring by Using a Wearable Smartband System

Lan Yang, Junqi Guo(✉), Yazhu Dai, Di Lu, and Rongfang Bie

College of Information Science and Technology, Beijing Normal University, Beijing, China
guojunqi@bnu.edu.cn

Abstract. Due to insufficiency in data collection and analysis by using the existing wearable devices in physical fitness tests for adolescents, this paper presents a machine learning based physical health evaluation model for running activity monitoring, in which a gradient boosting regression algorithm is employed to process physiological data collected from a set of smartbands developed by ourselves. First, we collect two kinds of dynamic data including heart rate and acceleration when students wear our smartbands in a normal running test. Next, several key features closely related to the physical health status are extracted from the dynamic data. A gradient boosting regression (GBR) algorithm is then utilized to train a physical health evaluation model and calculate out a comprehensive score representing physical health status of each student for reference. Experiment results show that not only does the proposed model with GBR achieve higher evaluation accuracy than the one with another typical algorithm—support vector regression (SVR), but it also provides a promising solution for future physical health evaluation by using a machine-learning-model based intelligent computing instead of traditional empirical-model based manual calculation.

Keywords: Physical health evaluation · Wearable smartband
Feature extraction · Gradient boosting regression · Support vector regression

1 Introduction

Physical fitness is an important component of human health, and its level in adolescents is closely associated with health status in adulthood. Through testing some items of physical fitness and comprehensively assessing their results, we can provide beneficial help for adolescents and track their body changes for a long time [1]. Wenhua, Zhang and others used the health monitoring data obtained from traditional physical fitness tests to analyze the physical health status of adolescents [2, 3]. However, traditional physical fitness tests lack of the ability to collect and analyze data in real time and the work of evaluating adolescents' physical fitness is complicated and inefficient, which can't reflect the real-time physical condition of young students. As for the evaluating

S. Chellappan et al. (Eds.): WASA 2018, LNCS 10874, pp. 813–819, 2018.
https://doi.org/10.1007/978-3-319-94268-1_72

the physical fitness, some researchers made a comprehensive evaluation of the students' physical health status by using fuzzy mathematics and entropy method in recent years [4, 5]. Gradient boosting algorithm has achieved excellent results in data mining competitions such as Kaggle and KDD Cup, and has also been widely used in the Internet industry. For instance, Facebook predicted the online click-through rate using gradient boosting algorithm [6].

In order to improve the performance of the traditional physical fitness tests, this paper presents a physical health evaluation model based on GBR algorithm for running monitoring by using a self-developed wearable smartband system. First of all, we collect two real-time dynamic data including heart rate and acceleration. Next, three features closely related to physical health evaluation are extracted from the dynamic data, namely resting heart rate, heart rate recovery and running time. Finally, we use the GBR algorithm to calculate out the students' physical fitness score.

2 Methodology

2.1 A Wearable Smartband System

The self-developed smartband system includes data acquisition, data transmission and data analysis as shown in Fig. 1.

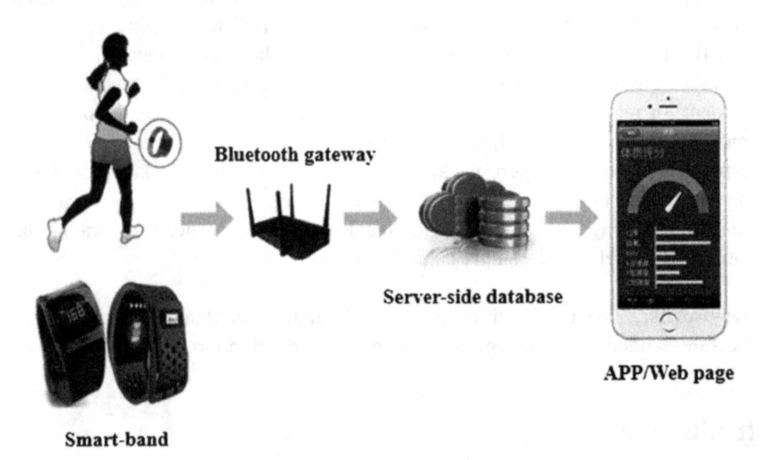

Fig. 1. A wearable smartband system

2.2 Physical Health Evaluation Model

The goal of proposed physical health evaluation model is to convert the dynamic data collected from the smartbands into the physical fitness score. The GBR based physical health evaluation model was trained by three features extracted from dynamic data, and the average scores of two items of traditional physical fitness tests (vital capacity and 50 m dash) as the data labels (Fig. 2).

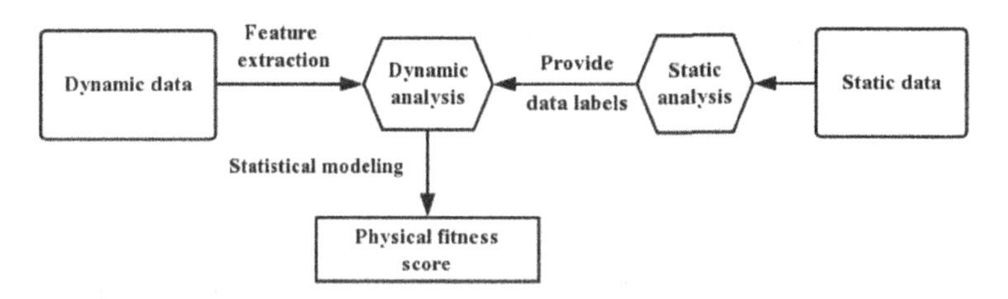

Fig. 2. Physical health evaluation model

Experimental Paradigm

Running is a very simple and popular sport and there are many features that reflect physical fitness status during running activity, therefore we design a running experiment to collect the dynamic data. When wearing the smartbands, firstly subjects do some warm-up exercises for three minutes. Then boys are requested to run 1000 m dash and girls run 800 m dash. After running, relaxing at least two minutes before handing the smartbands (Fig. 3).

Feature Extraction

The two kinds of dynamic data including heart rate and acceleration are still the original physiological data, which features are not obvious, so we extract several key features which reflect the physical health status significantly [7, 8]. Heart rate (HR) during exercise directly reflect the exercise intensity, it shows three different states from the beginning to the end of the exercise [7]. Figure 4 shows the three states of HR when running, where the green region represents the heart rate value and the red curve represents the distance variation in the running experiment. Based on the three states of HR, we extract three key features include resting heart rate, running time and heart rate recovery and assign the average heart rate value before running to resting heart rate as using acceleration to get running time, and the slope of the HR during 1 min rest after running is taken as the heart rate recovery.

Fig. 3. Running experiment

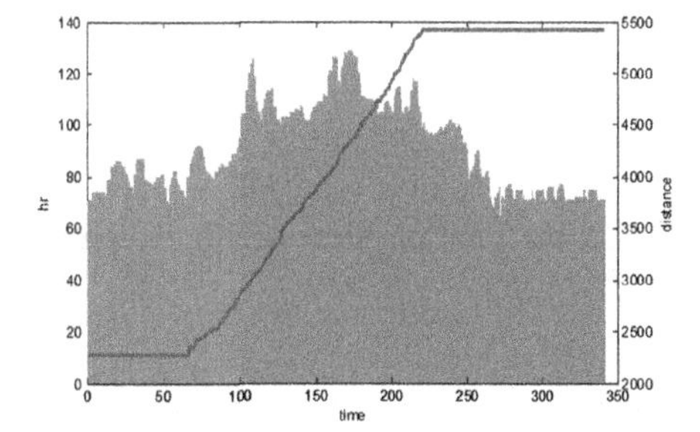

Fig. 4. The heart rate curve while running (Color figure online)

Gradient Boosting Regression

Gradient boosting is a regression technique [9], it aims at finding an additive model that minimizes the loss function. The best way to minimize the loss function is to make the loss function decrease in the direction of its gradient, the loss function is defined by the Eq. (1).

$$L(y, f(x)) = \sum_{i=1}^{N} (y_i - \tilde{y}_i)^2 \tag{1}$$

where y is the average score of the 50 m dash and vital capacity measured by traditional physical fitness tests, $\tilde{y}$ is score predicted by the proposed model.

The proposed physical health evaluation model with GBR is designed as follows:

1. Initialize the physical health evaluation model:

$$F_0(x) = \text{argmin}_{\rho} \sum_{i=1}^{N} L(y_i, \rho) \tag{2}$$

2. For m = 1, 2,... M, do the following:
   a. Calculate the negative gradient

$$g_i = -\left[\frac{\partial L(y_i, F(x_i))}{\partial F(x_i)}\right]_{F(x)=F_{m-1}(x)} \quad i = 1, 2, \ldots, N \tag{3}$$

   b. Using scores of the 3 main features x_i and the negative gradient g_i to fit a regression tree $h(x_i, a_m)$ to get a_m, which are the parameters of regression tree.
   c. Use the negative gradient of the loss function evaluated at the current evaluation model F_{m-1}, the new step length of the model ρ_m will be obtained.

$$\rho_m = \text{argmin}_{\rho} \sum_{i=1}^{N} L(y_i, F_{m-1}(x_i) + \rho h(x_i, a_m)) \tag{4}$$

   d. Update the physical health evaluation model:

$$F_m(x) = F_{m-1}(x) + \rho_m h(x, a_m) \tag{5}$$

3. Get the final physical health evaluation model:

$$F(x) = F_M(x) = \sum_{m=1}^{M} \rho_m h(x, a_m) \tag{6}$$

To avoid over-fitting and at the same time provide the best physical health evaluation accuracy, we use search grid method to choose the optimal combination of hyper-parameters in the proposed model. It is concluded that proposed model will produce best results for boosting stages set to 130, learning rate set to 0.05, maximum depth of tree set to 3 and the minimum number of samples required to be at a leaf node set to 3.

3 Experiments

Our experimental data collected in Tongzhou NO. 6 middle school of Beijing. There are 502 sets of dynamic data in the data set and we divided it into a training set and a test set randomly with a ratio 7:3. The training set is used to establish the GBR based physical health evaluation model, as well as the SVR [10] based model for comparison.

We choose the three metrics, Mean Absolute Error (MAE), Root Mean Squared Error (RMSE), R-squared (R^2), to evaluate the accuracy of the model. MAE and RMSE both indicate the error between the predicted value and the true value, the R^2 value corresponds to the students' health evaluation variance explained by the model. The R^2 value ranges between 0 and 1, the value is close to 1, indicating that the model is more effective and can explain more the output variability. MAE, RMSE and R^2 are defined as Eq. (7), the meaning of y and $\tilde{y}$ as explained above.

$$MAE = \frac{1}{n}\sum_{i=1}^{n} |\tilde{y}_i - y_i|, RMSE = \sqrt{\frac{1}{n}\sum_{i=1}^{n} (\tilde{y}_i - y_i)^2}, R^2 = 1 - \frac{\sum_{i=1}^{n} (y_i - \tilde{y}_i)^2}{\sum_{i=1}^{n} (y_i - \bar{y})^2} \quad (7)$$

Table 1 shows the MAE, RMSE and R^2 obtained by two methods:

Table 1. Three metrics results of two methods.

Metric	MAE	RMAE	R^2
GBR	2.6409	3.2081	0.8553
SVR	6.4054	8.1052	0.0765

GBR has shown to be around 60.42% more accurate as compared to SVR for physical health evaluation and can explain more the output variability in Table 1.

The difference between the results evaluated by two models and the real values on the test set is reflected in Fig. 5. The X axis is student number, and the Y axis is the corresponding students' physical fitness score. It is obvious that the score evaluated by GBR based model was close to the real value.

Figure 6 is the residual graph of the two methods. The data should be distributed at the centerline as closely as possible, indicating that the prediction values are close to the real values. The horizontal band width near the centerline of the GBR is narrower than that of SVR and the exception data of the GBR is less than that of the SVR, which proves that the proposed model with GBR achieves higher evaluation accuracy than the one with SVR.

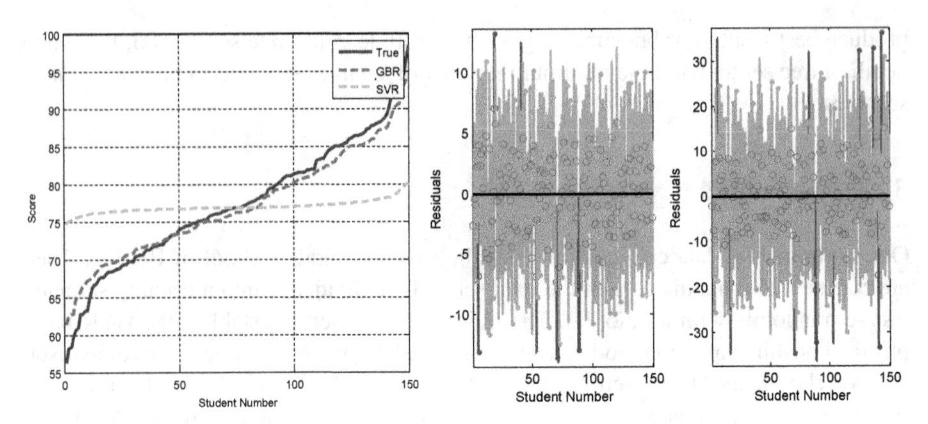

Fig. 5. Score comparison of evaluation methods

Fig. 6. Residual comparison of evaluation methods (Left: GBR, Right: SVR)

4 Conclusions

In this paper, we have proposed a physical health evaluation model based on the GBR algorithm. By using this model, the original data collected from self-developed smartbands can be converted into physical fitness score. This paper describes each part of the model, including data acquisition, feature extraction and the GBR based model training and testing. Finally, experiment results prove that the proposed model with GBR achieves higher evaluation accuracy than the one with SVR. Moreover, our work also provides an efficient solution with superior performance for future physical health evaluation, in which intelligent computing with a machine-learning model replaces traditional empirical formula and manual calculation.

Acknowledgments. This research is sponsored by National Natural Science Foundation of China (No.61401029) and Beijing Advanced Innovation Center for Future Education (BJAICFE2016IR-004).

References

1. Zhen-Wang, B.I.: Physical fitness test of adolescents in Europe and USA. Foreign Med. Sci. (2005)
2. Wen-hua, Z.H.U.: Analysis of physical health condition of primary and middle school students in Taiyuan City from 1985 to 2014. Chin J Sch. Doct. **31**(10), 727–731 (2017)
3. Zhang, Y., He, L.: Dynamic analysis on physique and health condition of Chinese adolescents based on four national fitness monitoring data from 2000 to 2014. China Youth Study **6**, 4–12 (2016)
4. He, D.: Research and development of public sports information processing system. Chongqing University (2004)

5. Yu, W.: Research on comprehensive evaluation of the test data of "national student physical health standard" in Hubei Province based on entropy method. Huazhong University of Science and Technology (2016)
6. He, X., Pan, J., Jin, O., et al.: Practical lessons from predicting clicks on ads at Facebook. In: Proceedings of the Eighth International Workshop on Data Mining for Online Advertising, pp. 1–9. ACM (2014)
7. Buchheit, M.: Monitoring training status with HR measures: do all roads lead to Rome? Front. physiol. **5**, 73 (2014)
8. Bosquet, L., Gamelin, F.X., Berthoin, S.: Reliability of postexercise heart rate recovery. Int. J. Sports Med. **29**(03), 238–243 (2008)
9. Friedman, J.H.: Greedy function approximation: a gradient boosting machine. Ann. stat. 1189–1232 (2001)
10. Ajmera, S., Singh, A.K., Chauhan, V.: An approach towards medium term forecasting based on support vector regression. In: Power India International Conference, pp. 1–6. IEEE (2016)

The Research of Spam Web Page Detection Method Based on Web Page Differentiation and Concrete Cluster Centers

Mei Yu[1,2,3,4], Jie Zhang[2,3,4], Jianrong Wang[1,2,3,4], Jie Gao[1,3,4], Tianyi Xu[1,3,4], and Ruiguo Yu[1,2,3,4](✉)

[1] School of Computer Science and Technology, Tianjin University, Tianjin, China
{yumei,wjr,gaojie,tianyi.xu,rgyu}@tju.edu.cn
[2] Tianjin International Engineering Institute, Tianjin University, Tianjin, China
tjuzhangj@tju.edu.cn
[3] Tianjin Key Laboratory of Advanced Networking (TANK Lab), Tianjin, China
[4] Tianjin Key Laboratory of Cognitive Computing and Application, Tianjin, China

Abstract. To improve the PageRank algorithm's disadvantage of assigning link weights evenly and ignoring the authority of web page, we propose an improved PageRank algorithm based on web page differentiation (DPR) which evaluate pages authority according it's links' numbers and assign corresponding weights according to its authoritativeness when assigning PR values. To improve the cluster's stability and accuracy of the K-Means algorithm, we combine DPR with K-Means, design a differentiation page-based K-Means (DPK-Means) algorithm. This algorithm will sort the pages according to the PR value obtained by the DPR algorithm and then concrete cluster centers according to the current sorting result. Experiments show that in spam detection, the DPR is superior to PageRank in terms of pages numbers, recall rate, accuracy, and F-Measure value and DPK-Means has better performance than the K-Means.

Keywords: Web page differentiation · Concrete cluster center
Spam web page detection · PageRank algorithm · K-Means algorithm

1 Introduction

Nowadays, search engines have become important for people. Search engine [1] optimization makes pages get better page rank [2]. The report stated that 88% of users only browse the 1/3 pages of the return and many pages increase ranks by cheating. Web spam makes lots of spam pages [3]. Giving the spam pages good positions will bring harm to users, society and the country. So, it is important to improve the search engine's page sorting results and effectively detect spam pages. Many researchers contribute their work in this. Maehara [4] proposed to use the graph structure to calculate personalized PR values. Xie [5] proposed an algorithm that assigns different

S. Chellappan et al. (Eds.): WASA 2018, LNCS 10874, pp. 820–826, 2018.
https://doi.org/10.1007/978-3-319-94268-1_73

values based on different edges. Lofgren [6] proposed a two-way analysis method and realized the analysis of pages based on the pages' features.

In this paper, we propose the DPR algorithm which measure the authoritative value of a page according to the proportion of webpage links' number and assign the corresponding PR value according to its authoritative value. We propose a Differentiated Page-based K-Means [7] (DPK-Means) algorithm based on differentiated pages. Experiment shows that DPK-Means has more accurate clustering effect in the spam web pages detection. And it will play a more important role in the classified search field.

2 DPR Algorithm

We propose the DPR algorithm to improve the PageRank algorithm's disadvantage of assigning link weights evenly and ignoring the authority of web page. And we explain the basic idea of DPR algorithm with the following four-node network (see Fig. 1).

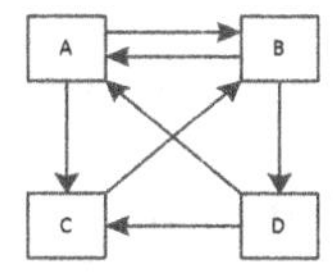

Fig. 1. Basic idea of DPR algorithm

In DPR algorithm, a hyperlink to a web page counts as a vote of support. The more votes a web page has, the more authoritative it is and the weights is larger when calculating PR values. As in Fig. 1, page B points to page A and page D. But page B shouldn't assign A and D the same PR value because the authoritative values of A and D are not the same. Based on this idea, we propose the DPR algorithm. The DPR has two steps optimization over the PageRank, referred to as DPR-A and DPR-B.

2.1 DPR-A Algorithm

In DPR-A, we use the number of incoming links of a web page to measure its authority according to the thought that the more page is recommended, the more important it is and in web graph, the times of recommended is equal to node's incoming links. In Fig. 1, the number of incoming links of A is 2, and D is 1, so A gets the $2/(1+2)$ authoritative value of B, while D gets $1/(1+2)$. The equation is shown in (1).

$$PR(p) = \frac{1-d}{N} + d\sum\nolimits_{q \in B(p)} \frac{Inlinks(p)}{\sum\nolimits_{x \in A(q)} Inlinks(x)} PR(q) \tag{1}$$

In the (1), N is the sum number of the pages, d is the damping factor, $PR(p)$ is the PR value of web page p, $B(p)$ is the set of web pages that reference p, $A(q)$ is the set of outgoing links of page q, and $Inlinks(p)$ is the number of times that p is pointed to

$\sum_{x \in A(q)} Inlinks(x)$ counts the pages that page q links out at first, and then adds all the incoming links of each page together. $Inlinks(p)/\sum_{x \in A(q)} Inlinks(x)$ is the proportion of incoming links' number of node p in the number of incoming links of the node pointed by p, and p gets the PR value from q according to this proportion.

Now, we can get the calculation procedure of DPR-A. First, by the link structure in the network, we can obtain the adjacency matrix A of the network node, shown in (2).

$$A_{ij} = \begin{cases} 1, & \text{connection exist from node j to node i} \\ 0, & \text{others} \end{cases} \tag{2}$$

Next, the transition matrix T is calculated by the adjacency matrix A, shown in (3).

$$T_{ij} = \begin{cases} In_i / \sum_{k=1}^{m} In_k, & if \quad A_{ij} = 1 \\ 0, & if \quad A_{ij} = 0 \end{cases} \tag{3}$$

The description of the algorithm for calculating T by DPR-A is shown in Table 1.

Table 1. DPR-A algorithm calculate transition matrix T description

DPR-A algorithm calculate transition matrix *T* description

Input: Adjacency matrix *A*
Output: Transition matrix *T*
Begin:
$Out_{degree} = sum(A)$; // out_{degree} save each node out degree add each row
$N = length(A)$; // *N* is the length of matrix A
$For\ j = 1\ to\ N\ do$
    $EE(j,1) = j$;
    $EE(j,2) = out_{degree(j)}$;
$G = D * A'$; // *D* is a sparse matrix with element $out_{degree(i)}$ in the i_{th} row andi_{th} column;
$Out_{degree1} = sum(G)$;
$For\ j = 1\ to\ N\ do$
    $EE1(j,1) = j$;
    $EE1(j,2) = 1/out_{degree1(j)}$;
$T = (D1 * G')'$; //$D1$ is a sparse matrix with element$\frac{1}{out_{degree1(i)}}$ in the i_{th} row and i_{th} column;
Return *T*

Finally, an initial PR value $1/N$ is assigned to each node. N is the total number of web pages. Then we can obtain the DPR-A value by using T and the iteration times M.

2.2 DPR-B Algorithm

In DPR-B, we use the number of links of a web page to measure its authority because it's can more comprehensively assess the importance of pages. In Fig. 1, A's number of links are 4, which includes 2 incoming and 2 outgoing links, and B's are 3, including 1 incoming and 2 outgoing links. According to the DPR-B, A obtains $1/(4+3)$ authoritative value of B, and D obtains $3/(4+3)$. The equation is shown in (4).

$$PR(p) = \frac{1-d}{N} + d\sum\nolimits_{q \in B(p)} \frac{Links(p)}{\sum\nolimits_{x \in A(q)} Links(x)} PR(q) \tag{4}$$

The (4) is similar to (1), we only change the *Inlinks* in (1) to *Links* in (4), which represents the link numbers rather than the incoming links. And the calculation steps of DPR-B are as same as DPR-A except for the transition matrix which we need to change the $In_i / \sum_{k=1}^{m} In_k$ in (3) to $Links_i / \sum_{k=1}^{m} Links_k$. The method to calculate the transition matrix by DPR-B is similar to Table 2, we just need to change the Out_{degree} to the *Degree* which is the sum of the out degree and the in degree.

Table 2. DPK-Means algorithm description

DPK-Means **algorithm description**
Input: Transfer matrix T ,Iteration number M, Initial matrix d, The number of clusters k, The dataset D containing n objects
Output: The k clusters with the smallest squared errors
Begin:
$T' = change(T)$;// The improved transfer matrix T' is calculated from the transfer matrix T
$Rank = PageRank(d, T', M)$; // According to T', M and d to sort the pages
$max = Max(Rank(:,1))$; // Select the maximum value of DPR as the center of non-span page
$min = Min(Rank(:,1))$; // Select the minimum value of DPR as the center of spam page
$center1 = find(D(:,1) == max)$; // In the data set, the DPR maximum page position
$center2 = find(D(:,1) == min)$; // In the data set, the DPR minimum page position
$result = K\text{-}Means(center1, center2, D, k)$; // Clustering the dataset
Return $result$

3 DPK-Means Algorithm

The main ideas of the DPK-Means are as follows. Firstly, using DPR to rank the pages, the higher the DPR value, the higher the authority of the page, and it's probability belongs to the non-spam page increases. Then, the pages with the highest and lowest DPR value are identified as the two centers of K-Means clustering. Finally, using the distance between dataset objects as the measurement standard and the error criterion as the identification of the end of algorithm to cluster pages. Show in Table 2.

The DPR algorithm makes the page differentiation more obvious, and the sorting results become more accurate. This helps the K-Means algorithm to determine the initial clustering center concretely. So, the DPK-means algorithm can not only accurately detect spam web pages, but also optimize the clustering effect of web pages.

4 Experiments

The data set used for the experiment is the WEBSPAM-UK2007 collection. The base data is a set of pages in 114,529 hosts. We use the number of spam pages detected, recall rate, precision, and F-Measure values as evaluation criteria.

4.1 Experiment Result and Comparison

Web Page Differentiation Analysis Between PageRank and DPR Algorithm. We did experiments to verify whether the DPR can increase web page differentiation. The initial PR value is 1/114529 and iterative numbers are 50. The results of top five pages, their PageRank's and DPR's PR values are show in Table 3. The last ten pages are the same in here, so we don't show in this paper.

Table 3. The top five pages of PageRank and DPR algorithm

PageRank	PR page number	PR value	DPR-A page number	DPR-A value	DPR-B page number	DPR-B value
1	81635	0.0051	83000	0.0242	83000	0.0251
2	41000	0.0034	87407	0.0195	87407	0.0226
3	5170	0.0028	41000	0.0180	41000	0.0138
4	19505	0.0026	5170	0.0106	81635	0.0105
5	58632	0.0025	19505	0.0104	5170	0.0105

The Table 3 shows that PageRank and DPR get different ranking. And DPR's PR value is higher than PageRank's. This indicates that DPR has increased the important web pages' authoritative value and makes the web page differentiation more obvious. This verifies that DPR can better express different attributes and facilitate clustering.

Spam Web Page Detected Analysis Between PageRank and DPR Algorithms. When we set different thresholds, the number of spam pages, recall rate, precision rate and F value are shown in Fig. 2.

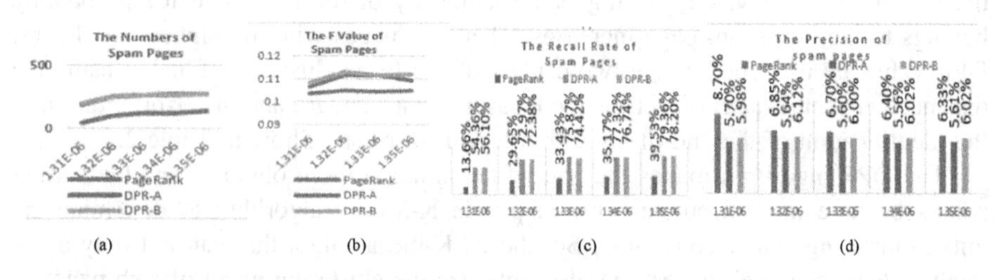

Fig. 2. (a) represents numbers of spam pages, (b) is F value, (c) is recall rate, (d) is precision.

In Fig. 2, we can see in (a) that spam pages' numbers detected by DPR are more than PageRank. The recall rate in (c) is similar to (a). Though the precision of DPR is lower than the PageRank, the F value of DPR-B, which is the harmonic mean of recall rate and precision in (b) is higher than PageRank. To sum up, DPR is better.

Cluster Analysis of DPK-Means Algorithm. We analyze the clustering effects of DPK-Means, PK-Means and K-Means algorithm when extracting different features of pages. The page feature vectors' numbers are 15, 35, 55, 75, 95, 115, 135, 139. The results of 135 are show in Fig. 3 as representative. The red squares represent non-spam webpage clusters, and the green circles represent spam webpage clusters.

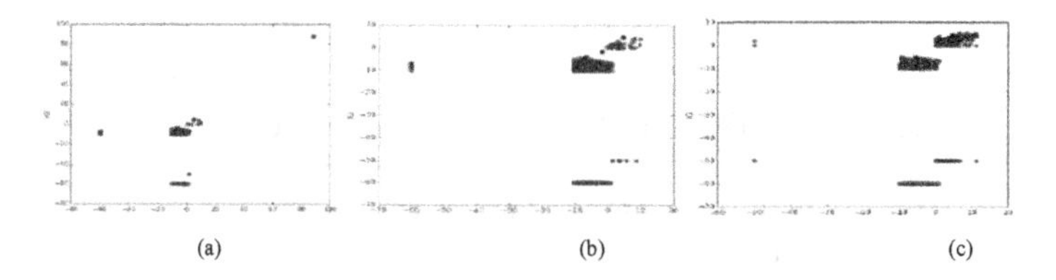

Fig. 3. The results of clustering by using different algorithm. (a) is the result of K-Means algorithm, (b) is the results of PK-Means, (c) is the results of DPK-Means. (Color figure online)

From Fig. 3, it can be seen that the K-Means detects the smallest number of spam web pages. The PK-Means works well for web page clustering, but the cluster center is at the edge of the cluster. The DPK-Means can distinguish the two types of web pages, and the number of spam pages detected is the largest. It has the best clustering effect.

Experimental Analysis of DPK-Means Algorithm Detecting Spam Web Pages. We compare the spam pages numbers detected by three algorithms: K-means, PK-Means and DPK-Means. The results are shown in Table 4.

Table 4. The number of spam pages detected in different number of feature vectors

	15	35	55	75	95	115	135	139
K-Means	212	4	4	5	5	12	7	1
PK-Means	538	482	260	542	307	28	45	45
DPK-Means	881	775	771	780	775	721	695	686

From Table 4, it can be seen that the K-Means detects the least spam pages, followed by the PK-Means. When feature vectors' number is greater than 95, the spam pages' number detected by the PK-Means is less. In summary, the DPK-Means algorithm is significantly better than the K-Means algorithm and the PK-Means algorithm.

5 Conclusion

In this paper, we propose DPR algorithm which makes the web page differentiation more obvious by assign different authorities according the links of web pages. And then we propose DPK-Means algorithm based on differentiated web pages. Experiment shows that DPR algorithm has get page differentiation more obvious in PR values and help the DPK-Means algorithm make a more accurate clustering effect in the spam web pages detection. And it will play a more important role in the field of related classified search and greatly improve work efficiency and classification accuracy.

References

1. Luh, C.-J., Yang, S.-A., Huang, T.-L.D.: Estimating Google's search engine ranking function from a search engine optimization perspective. Online Inf. Rev. **40**(2), 239–255 (2016)
2. Patel, K.M., Chauhan, P.: Analysis of spam link detection algorithm based on hyperlinks. Int. J. Data Warehous. Min. **4**(1), 67–72 (2014)
3. Castillo, C., Donato, D., Gionis, A., et al.: Know your neighbors: web spam detection using the web topology. In: International ACM SIGIR Conference on Research and Development in Information Retrieval 2007, pp. 423–430. ACM, Netherlands (2007)
4. Maehara, T., Akiba, T., Iwata, Y., et al.: Computing personalized PageRank quickly by exploiting graph structures. In: Proceedings of the VLDB Endowmen 2014, Hangzhou, pp. 1023–1034 (2014)
5. Xie, W., Bindel, D., Demers, A., et al.: Edge-weighted personalized PageRank: breaking a decade-old performance barrier. In: ACM SIGKDD International Conference on Knowledge Discovery and Data Mining 2015, pp. 1325–1334. ACM, Sydney (2015)
6. Lofgren, P., Banerjee, S., Goel, A.: Personalized PageRank estimation and search: a bidirectional approach. In: Proceedings of the Ninth ACM International Conference on Web Search and Data Mining 2016, pp. 163–172. ACM, San Francisco (2016)
7. Suci, A.M.Y.A., Sitanggang, I.S.: Web-based application for outliers detection on hotspot data using K-Means algorithm and shiny framework. In: IOP Conference Series: Earth and Environmental Science, vol. 31, no. 1 (2016)

Network Control for Large-Scale Container Clusters

Weiqi Zhang(✉), Baosheng Wang, Wenping Deng, and Hao Zeng

College of Computer, National University of Defense Technology,
Changsha Hunan, China
zhangweiqi12@nudt.edu.cn

Abstract. The recent rise of container systems like Docker has created a lot of excitement in data center. Its ability to package, transfer and run application code across many different environments enables new levels of fluidity in how we manage applications. However, container's easy-to-manage and second-boot features increase the degree of network dispersion and management difficulties, which causes the networking and security issues in container network. Aiming at the lack of control in container network, this paper designs a network control architecture for large-scale container clusters to solve the key issue of large-scale container clusters deployment in the network adapter and isolation control. Specifically, we design two different container network models and a policy-based security isolation by using VLAN partition and iptables. The experimental results show that our network control architecture could achieve rapid VLAN division and accurate isolation of node-to-node communication.

Keywords: Data center · Container network control
Network model · Security isolation

1 Introduction

Data center can build virtual pools of resources for cloud computing through virtualization. Cloud computing can realize resource deployment automatically, expansion dynamically, distribution on-demand and other functions through the data center virtual resource management technique to meet the needs of users on resources. Virtualization is one of the key technologies which is essential to building a cloud infrastructure. The initial virtualization of resources in the data center is mainly implemented by virtual machines. However, the current virtual resource management by virtual machine still has many problems, such as low utilization rate of cloud computing resources, incompatibility of applications and platforms, limitation of application operating environment, and a decrease of control ability of maintenance personnel, etc. [1,2].

With the development and popularization of containerization technologies such as Docker [3–5], container-based resource virtualization is gradually replacing virtual machines, and more and more containers are used in data center

S. Chellappan et al. (Eds.): WASA 2018, LNCS 10874, pp. 827–833, 2018.
https://doi.org/10.1007/978-3-319-94268-1_74

to implement large-scale cluster deployment [6]. The characteristics of container such as lightweight management and second-level start-up can simplify the management and solve the problem of virtual resource management currently facing in the data center. It has become an inevitable trend to use container instead of virtual machine for large-scale cluster deployment in the data center [7–9]. However, with the gradual maturation of container technologies, the problems of large volume and complex deployment are gradually exposed [10].

Container network model is the primary question in network control. Since the development of container technology, the network has been lagging behind and there is no standard interface at the network level for a long time. Although there come up with two network standard, CNM (Container Network Model) and CNI (Container Network Interface) [11,12], it is also difficult to determine an appropriate network access model [13,14]. As a result, a large number of start-up companies and open source organizations have developed a wide variety of network implementations, resulting in many complex solutions which makes it difficult to adapt container network models and manage the network [15].

Security is a guarantee of normal operation of the container network. The container depends on the operating system which can achieve basic security and isolation in Linux system. However, the safety and isolation control still can not meet the actual needs. A complete set of security isolation methods is needed in the container network urgently in order to control the container network and ensure basic network security. The realization of the network control for large-scale container clusters become a serious problem in the transformation of container-based data center [16,17].

In this paper, we analyze three aspects of the network model, network selection, and security isolation, respectively, to realize network model intelligent adaptation and network security isolation control. The main contributions of this paper are:

(1) Building a network control architecture in a large-scale container cluster environment and solving the key problems of network selection and security isolation in container network deployment.
(2) Designing container network model and achieving the node's flexible migration in overlay network model based on VxLAN [18,19] and macvlan-based [20] network model.
(3) Using VLAN partitioning and iptables management technology to realizes the container networks isolation and control.

The remainder of this paper is organized as follows. Section 2 provides the overall network control architecture. Sections 3 and 4 introduce the network model and security isolation respectively. Section 5 conducts the experimental testing and the result analysis. The conclusion and future work are presented in Sect. 6.

2 Container Network Control Architecture

Based on the implementation of network deployment, the centralized control module is added for achieving the overall controllable network. Throughout the architecture, the entire process of network deployment is implemented through the server-side based on the network model selection. The architecture is shown in Fig. 1.

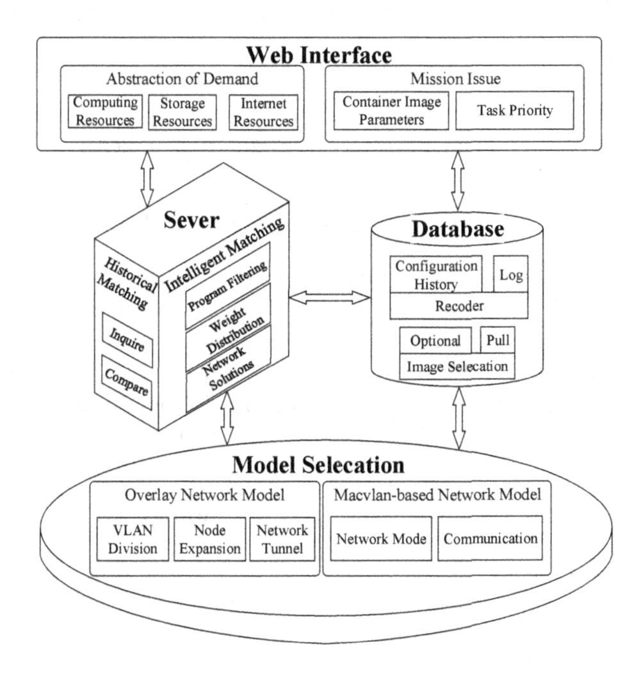

Fig. 1. Large-scale container cluster network control architecture

3 Network Model Intelligent Adaptation Technology

The design of container network control architecture gives full play to the container's deployment and removal characteristics with rapid implementation. According to different networking requirements for network security and network communication efficiency, overlay network model based on VxLAN and macvlan-based container network model are respectively designed.

In the overlay network model based on VxLAN, we design network isolation and node join/delete strategies according to the characteristics of the network structure. When the packages are transformed in, the VTEP will finish the identification, forwarding, encapsulation/decapsulation and other operations for the data frame to achieve the directional transmission and isolation of network information.

In the macvlan-based container network model, we design a container network model which can improve the network transmission efficiency and ensure the network communication quality. Different from other container network models, macvlan does not need to create bridges for network communications. In this way, it can guarantee the performance of the container network and improve the network adaptability.

4 Policy-Based Container Network Isolation Control Technology

In this part, we use VLAN isolation technology on OVS bridges and iptables to access container services which are based on different policies.

4.1 Access Control Mechanisms Based on Iptables

In large scale container cluster network, the main way to isolate tiers logically for containers of different functions, applications, and users is to configure access control. We use iptables to realize the main access control in the current Linux environment. Net filter and iptable is the main part of iptables which form the Linux platform packet filter firewall. Iptables can also be used to complete packet filtering, packet redirection and network address translation (NAT).

4.2 VLAN-Based Network Isolation Technology

The access control method based on VLAN isolation is widely used in today's virtual network. VLAN is a logical segment of network nodes on a two-tier switch port, which is not limited by the physical location. The group of VLAN is flexible. Using VLAN technology in the platform can increase network connectivity flexibility for container clusters. A virtual network environment is formed by combing different locations, networks, and users. In this way, a single broadcast domain will be composed and the network isolation will be increased. VLAN can ensure isolation between containers, which improves the network utilization and enhances the security and confidentiality of the container network.

5 Experiments and Results Analysis

We analysis the correctness and feasibility of the scheme design in the experiment. And we also verify the network isolation technology in the container network.

5.1 Container Networking Efficiency Verification

Container networking experiment is mainly for verifying the effectiveness of container networking. Our experiment is based on Huawei FusionServer RH2288 server. Because the busybox image integrates hundreds of commonly used commands and tools in Linux with less storage overhead, we use docker.io/busybox:latest as the basic mirror in the network performance testing. In the test, 200, 400, 600, 800 and 1000 containers were created respectively in the macvlan-based network model to reflect the time of networking in different container scale. Scatter plots are drawn with the container size as the main reference, as shown in Fig. 2(a) and (b).

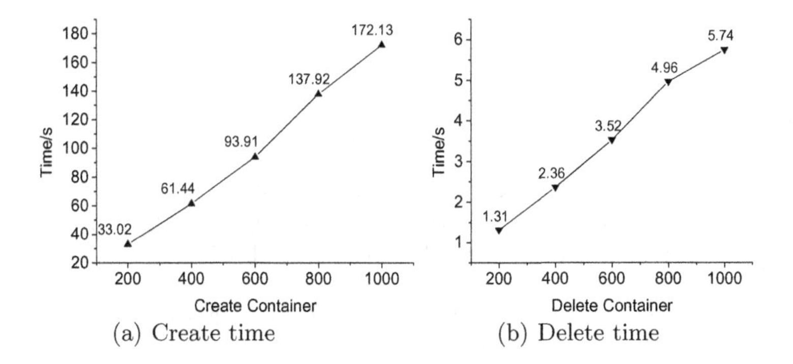

(a) Create time (b) Delete time

Fig. 2. Different sizes of container networking efficiency

The results show that when the host limit is not exceeded, the time consumption of start/delete is positively related to the number of containers. The number of containers created does not affect the subsequent container's startup and deletion. In the current environment, the average container creation time is less than 200 ms.

5.2 Container Network Isolation Test

In this experiment, we use OVS bridge as the main means of communication to test the performance of network isolation policy, based on the VxLAN network tunneling protocol. In the global network rules, we can restrict communication between different networks by iptables filter and nat table.

Several basic strategies are tested and displayed in the form of network topology. We can achieve the control of the global network by defining the filter table with minimal impact on the network. The network isolation scheme designed in this paper can achieve the basic isolation control of the network with minimal impact on network communication. However, due to the higher requirement of configuration strategy, the reasonable configuration of network rules requires a certain mastery of network communication rules.

6 Conclusion

In this paper, the development of container network technology in the large-scale cluster environment is investigated, and the important role of container network control technology in the current large-scale cluster is analyzed. According to the comparison of container network and traditional network, the container network control should focus on the breakthrough of flexible networking and security isolation. We design the security isolation module to realize the security control of the container network by dividing VLANs and adding filtering strategies. This paper plans to gradually complete the construction of container network control architecture in the future. We are going to develop a dedicated control platform to provide guidance for application developers, system administrators, and researchers to better deploy and control container networks on large-scale container clusters.

References

1. Dua, R., Raja, A.R., Kakadia, D.: Virtualization vs containerization to support PaaS. In: IEEE International Conference on Cloud Engineering, pp. 610–614 (2014)
2. Pahl, C.: Containerization and the PaaS cloud. IEEE Cloud Comput. **2**(3), 24–31 (2015)
3. Merkel, D.: Docker: lightweight Linux containers for consistent development and deployment. Linux J. **2014**(239), 2 (2014)
4. Fink, J.: Docker: a software as a service, operating system-level virtualization framework. Code4Lib J. **25**, 29 (2014)
5. Anderson, C.: Docker [software engineering]. IEEE Softw. **32**(3), 102–c3 (2015)
6. Seo, K.T., Hwang, H.S., Moon, I.Y., Kwon, O.Y., Kim, B.J.: Performance comparison analysis of Linux container and virtual machine for building cloud. Network. Commun. **66**, 105–111 (2014)
7. Xavier, M.G., Neves, M.V., Rossi, F.D., Ferreto, T.C., Lange, T., De Rose, C.A.F.: Performance evaluation of container-based virtualization for high performance computing environments, pp. 233–240 (2017)
8. Bhimani, J., Yang, J., Yang, Z., Mi, N., Xu, Q., Awasthi, M., Pandurangan, R., Balakrishnan, V.: Understanding performance of I/O intensive containerized applications for NVMe SSDs. In: Performance Computing and Communications Conference, pp. 1–8 (2017)
9. Boettiger, C.: An introduction to docker for reproducible research. ACM SIGOPS Oper. Syst. Rev. **49**(1), 71–79 (2015)
10. Felter, W., Ferreira, A., Rajamony, R., Rubio, J.: An updated performance comparison of virtual machines and Linux containers. In: 2015 IEEE International Symposium on Performance Analysis of Systems and Software (ISPASS), pp. 171–172. IEEE (2015)
11. de Bruijn, N.: eBPF based networking (2017)
12. DockerInfo: Container network spotlight: CNM and CNI. http://www.dockerinfo.net/3772.html
13. Kratzke, N.: About microservices, containers and their underestimated impact on network performance. arXiv preprint arXiv:1710.04049 (2017)

14. Bernstein, D.: Containers and cloud: from LXC to docker to kubernetes. IEEE Cloud Comput. **1**(3), 81–84 (2014)
15. Morabito, R.: A performance evaluation of container technologies on internet of things devices. In: 2016 IEEE Conference on Computer Communications Workshops (INFOCOM WKSHPS), pp. 999–1000. IEEE (2016)
16. Bui, T.: Analysis of docker security. arXiv preprint arXiv:1501.02967 (2015)
17. Dusia, A., Yang, Y., Taufer, M.: Network quality of service in docker containers. In: 2015 IEEE International Conference on Cluster Computing (CLUSTER), pp. 527–528. IEEE (2015)
18. Mahalingam, M., Dutt, D., Duda, K., Agarwal, P., Kreeger, L., Sridhar, T., Bursell, M., Wright, C.: Virtual extensible local area network (VXLAN): a framework for overlaying virtualized layer 2 networks over layer 3 networks. Technical report (2014)
19. Zhang, H.: System and method for VXLAN inter-domain communications. US Patent 9,036,639, 19 May 2015
20. Cizixs: Linux network virtualization: macvlan. http://cizixs.com/2017/02/14/network-virtualization-macvlan

Joint Optimization of Flow Entry Aggregation and Routing Selection in Software Defined Wireless Access Networks

Zhipeng Zhao[1], Bin Wu[1(✉)], Jie Xiao[2], and Zhenyu Hu[3]

[1] School of Computer Science and Technology, Tianjin University, Tianjin, China
{zzpeng, binw}@tju.edu.cn

[2] Department of Electrical and Electronic Engineering, The University of Hong Kong, Hong Kong, China

[3] Henan Electric Power Survey & Design Institute, Henan, China

Abstract. Software Defined Wireless Access Networks (SD-WAN) produce large amounts of traffic, which has caused limited flow table space in the TCAM. To solve the issue, this paper enables routing to overlap as many as possible with a minimum transmission delay, and thus traffic can be allocated to common paths to achieve flow entry aggregation. However, overlapping routing and traffic delay are two conflicting factors. An Integer Linear Program (ILP) is first proposed to minimize the total cost by tradeoff of the two factors. Through numerical experiments, this paper shows the effectiveness of the proposed method.

Keywords: ILP · SD-WAN · Flow entry aggregation
TCAM (ternary content addressable memory)

1 Introduction

Wireless Access Network (WAN) can efficiently expand the coverage and capacity of networks by easily accessing access nodes [1]. According to a Cisco report, the proportion of worldwide wireless access users increased 120% in 2016 and had reached 96.5% by 2017, so large numbers of nodes and large amount of traffic degrade the network performance [2]. Software Defined Network (SDN) is a new type of network innovation architecture, which was first proposed by the Clean Slate research group in Stanford University [3]. It separates control plane from data plane by Openflow technology, thus realizing flexible control of network traffic and providing a good platform for the innovation of WAN [4]. In SD-WAN, control layer is mainly responsible by the SDN controller for scheduling resource in data plane and maintaining state information of the network. Data layer is made of Openflow switches which are responsible for data processing, data forwarding and status collection based on flow tables [5].

To achieve high-performance data forwarding and processing, current commercial switch is usually implemented by ternary content addressable memory (TCAM), which

S. Chellappan et al. (Eds.): WASA 2018, LNCS 10874, pp. 834–839, 2018.
https://doi.org/10.1007/978-3-319-94268-1_75

can parallelly match all rules formed by aggregating routes through a certain method [6, 7]. TCAM is very expensive due to the characteristics of 'high cost', 'high power consumption', and 'large silicon space occupation' [8, 9]. In addition, the fine-grained definition of SDN traffic increases the number of rules. Therefore, current Openflow switches face a severe problem of insufficient flow table space, which restricts the application of SD-WAN [10–12].

Existing methods to solve the above problem can be classified into three categories: (1) Considering the differences in the occurrence time, duration time and traffic of different hosts, and offloading the processing rules that are not currently used to reduce size. Typical works include: TFO [2], Cacheflow [3], AHTM [4] and TimeoutX [5]. (2) Analyzing redundant rules in the flow table and aggregating some of the rules without changing the semantics of the original rules. Typical works include ORTC [6], SMALTA [7], and bitweaving [8]. (3) Using decentralized routing among different switches to ensure that the flow table is not overflowed. Typical works include: OBS [9], Palette [10] and vCRIB [11]. The first method involves offload overhead. The second method separates routing from flow aggregation without achieving a joint design, and thus leads to a poor performance. The third method can minimize average traffic delay and achieve load balance through decentralized routing, but cannot easily generate overlapping routes.

In this paper, we jointly consider flow entry aggregation and routing selection to reduce the number of rules in flow table and average traffic delay. An Integer Linear Program (ILP) is formulated to find the optimal solution by leveraging the two factors.

The rest of this paper is organized as follows. System model is proposed in Sect. 2. ILP is formulated in Sect. 3 to achieve optimal joint design. Performance of the proposed algorithm is shown in Sect. 4. Finally, we conclude this paper in Sect. 5.

2 System Model

In this paper, we consider a SD-WAN system model, which consists of a SDN controller and a number of Openflow switches. Assuming that one wireless access user with a given amount of traffic is required to be transmitted from one source to its destination switch. In this case, no direct wireless transmission paths between the source switch and the destination switch exist, so the optimal routing algorithm should be designed. Moreover, under the assumption that the flow table spaces of all switches are the same, the routing algorithm should be jointly designed with flow entry aggregation to further reduce the system cost. We denote s_i as the ith relay switch, $1 \leq i \leq n$, where n is the number of relay switches. For convenience, s_1 and s_n are denoted as the source switch and the destination switch. We introduce binary connection variables to characterize the connection status between SDN switches in this paper. c_{ij} is denoted as the connection variable between s_i and s_j. We set $c_{ij} = 1$ representing that s_i is the adjacent node of s_j, and set $c_{ij} = 0$, when $0 \leq i, j \leq n + 1, i \neq j$. In this paper, we assume that the network topology of the SDN is given, so c_{ij} is a given constant. Figure 1 shows the system model in this paper.

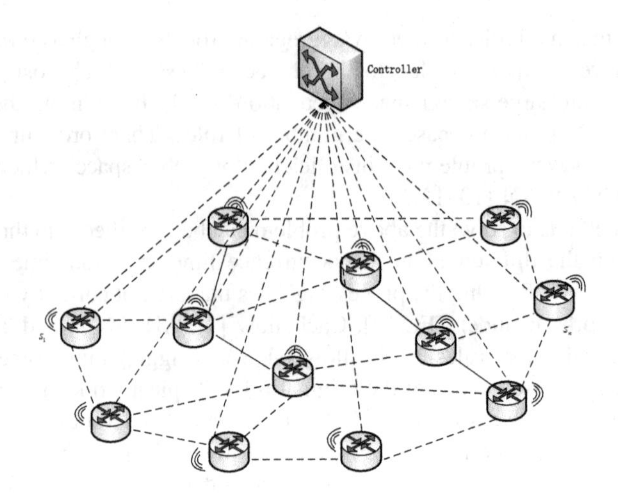

Fig. 1. A SD-WAN system model.

3 ILP Formulation

The number of rules in the flow tables and the traffic delay lead to the total cost at a SD-WAN network. To reduce the number of rules in the flow table, it is inevitable to increase the traffic delay. Therefore, we set a traffic delay cost scaling factor λ which represents the total delay at a SD-WAN network. In our design, the number of rules and the transmission delay can be properly controlled by the variable λ.

3.1 Notation List

Input:

N: The set of all nodes in a SD-WAN network *G* (*N*, *L*).
L: The set of all bidirectional edges in a SD-WAN network *G* (*N*, *L*).
d_{ij}: Wireless traffic delay between nodes *i* and *j* in the SD-WAN network. In this paper, it is regarded as the distance between nodes *i* and *j*, and $d_{ij} = d_{ji}$.
q_{sd}: The requested traffic from switch *s* to switch *d*.
λ: The cost scaling factor representing the transmission delay in SD-WAN network.
δ: The cost of each entry in the flow table.

Variables:

X_{ij}^{sd}: Binary variables. It is defined by $\{i,j,s,d \in N|i<j,s<d\}$. If some traffic start from the source switch *s* to the destination switch *d* and cross the wireless transmission path from the node *i* to the node *j*, it takes 1. Otherwise, it takes 0.
Y_{ij}^{s}: Binary variables. It is defined by $\{i,j,s,d \in N|i<j,s<d\}$. If some traffic start form the source switch *s* and cross the wireless transmission path from the node *i* to the node *j*, it takes 1. Otherwise, it takes 0.

3.2 ILP Formulation

$$\min\{\lambda\sum_{s\in n}\sum_{d\in n}\sum_{i\in n}\sum_{j\in n}X_{ij}^{sd}d_{ij}q_{sd}+\delta\sum_{i\in n}\sum_{j\in n}\sum_{s\in n}Y_{ij}^{s}\} \tag{1}$$

Subject to

$$\sum_{j\in n}X_{ij}^{sd}-\sum_{j\in n}X_{ij}^{sd}=1,s,d\in N,i=s,i\neq j \tag{2}$$

$$\sum_{j\in n}X_{ij}^{sd}-\sum_{j\in n}X_{ij}^{sd}=-1,s,d\in N,j=d,i\neq j \tag{3}$$

$$\sum_{j\in n}X_{ij}^{sd}-\sum_{j\in n}X_{ij}^{sd}=0,i,s,d\in N,i\neq s\neq d \tag{4}$$

$$\sum_{s\in n}\sum_{d\in n}\sum_{i\in n}\sum_{j\in n}X_{ij}^{sd}-\sum_{i\in n}\sum_{j\in n}\sum_{s\in n}Y_{ij}^{s}\leq 0 \tag{5}$$

Objective (1) minimizes the total cost. The first term is the cost of traffic delay, and the delay of all traffic is computed by using λ. The second term represents the cost of the flow table. In this paper, we assume these routes forwarded from the same output port with the same source node can be aggregated as a rule. Therefore, we define the sum number of entries after aggregating as the flow table cost, which is computed by using δ. Constraints (2)–(4) formulate only one wireless multi-hop transmission path between any two nodes in the network. Constraint (5) formulates the relationship between two variables.

4 Performance Evaluation

Experiments are carried out based on the topology with 11 nodes and 26 edges as shown in Fig. 2. We define the transmission delay between an arbitrary pair of nodes in Table 1. Our simulations are run in MATLAB 7.11 in an 8 GHz computer with 8 GB memory. In order to efficiently verify the performance of the proposed algorithm, we obtain the average result from randomly generated 300000 data samples which are denoted by the traffic matrix and compare the rules variation and the total cost, with the increase of the total number of flows.

In Fig. 3, we compare the total number of rules of Min Hop and ILP with the increase of the total number of flows. When the number of flow is less than 5, Min Hop algorithm is slightly better than STR, and once the number of flow surpasses 5, the average increment percentages of aggregation performance of the ILP is 40% better than Min Hop. When the number of flows is 15, the number of rules of the ILP does not change, but the number of flows of the Min Hop algorithm need to reach 50, implying that our proposed ILP is more suitable for the high performance network containing large traffic. Figure 4 shows the total cost with the increase of the total number of flows

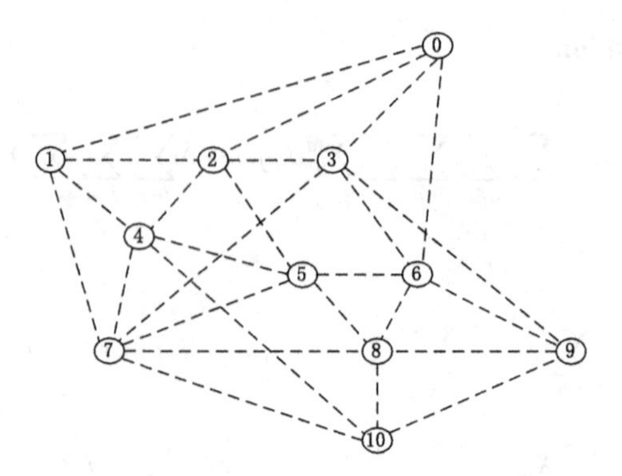

Fig. 2. A network topology G (11, 26).

Table 1. Transmission delay.

Traffic	Delay	Traffic	Delay	Traffic	Delay	Traffic	Delay
(0, 1)	131	(2, 3)	660	(4, 7)	330	(7, 8)	600
(0, 2)	760	(2, 4)	210	(4, 10)	730	(7, 10)	720
(0, 3)	390	(2, 5)	390	(5, 6)	730	(8, 9	730
(0, 6)	740	(3, 6)	340	(5, 7)	400	(8, 10)	320
(1, 2)	550	(3, 7)	109	(5, 8)	350	(9, 10)	720
(1, 4)	390	(3, 9)	660	(6, 8)	565		
(1, 7)	450	(4, 5)	220	(6, 9)	320		

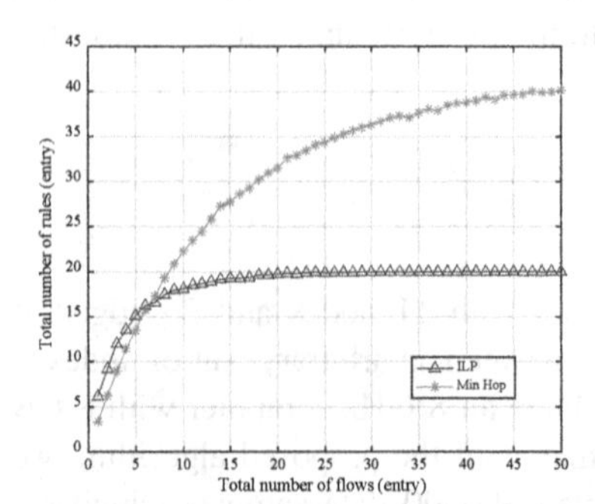

Fig. 3. The total number of rules vs the total number of flows.

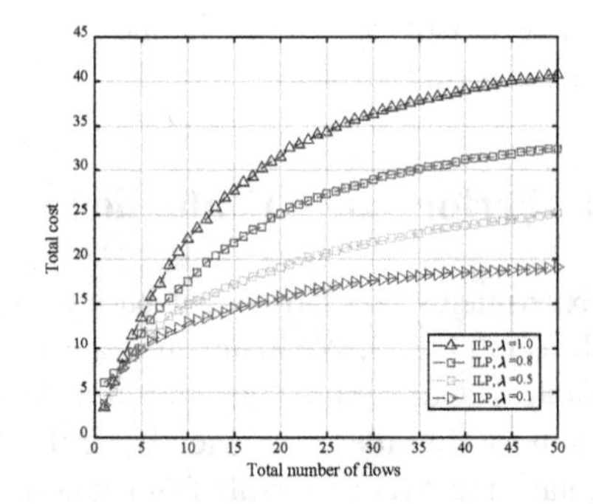

Fig. 4. The total cost vs the total number of flows under different λ.

character with different values of parameter λ and the total cost increases with the increase of parameter λ, indicating the parameter λ can adjust the total cost by the tradeoff of the delay cost and hardware cost.

5 Conclusion

We addressed the flow entry shortage issue in SD-WAN. By jointly considering flow entry aggregation and routing selection, we formulated an ILP to minimize the total cost with a proper set of parameters. Through numerical experiments, we showed that our proposed method can reduce the number of flow entries while reducing traffic delay.

Acknowledgment. This research is supported by the National Key R&D Program (Grant No. 2016YFB0201403), the Natural Science Fund of China (NSFC project No. 61701054). It is also supported by Tianjin Key Laboratory of Advanced Networking (TANK), School of Computer Science and Technology, Tianjin University, Tianjin, P. R. China.

References

1. Abujoda, A., Dietrich, D., Papadimitriou, P., et al.: Software-defined wireless access networks for internet access sharing. Comput. Netw. **93**(P2), 359–372 (2015)
2. Sarrar, N., Uhlig, S., Feldmann, A., et al.: Leveraging Zipf's law for traffic offloading. SIGCOMM Comput. Commun. Rev. **42**(1), 16–22 (2012)
3. Katta, N., Alipourfard, O., Rexford, J., et al.: Infinite cache flow in software-defined networks. In: 3rd ACM SIGCOMM Workshop on Hot Topics in Software Defined Networking (HotSDN), Chicago, Illinois, USA, pp. 175–180 (2014)
4. Zhang, L., Lin, R., Xu, S., et al.: AHTM: achieving efficient flow table utilization in software defined networks. In: IEEE Global Communications Conference (GLOBECOM), Austin, Texas, USA, pp. 1897–1902 (2014)
5. Zhang, L., Wang, S., Xu, S., et al.: Timeoutx: an adaptive flow table management method in software defined networks. In: IEEE Global Communications Conference (GLOBECOM), San Diego, CA, USA, pp. 1–6 (2015)
6. Draves, R., King, C., Venkatachary, S., et al.: Constructing optimal IP routing tables. In: IEEE INFOCOM, New York, USA, vol. 1, pp. 88–97 (1999)
7. Uzmi, Z.A., Nebel, M., Tariq, A., et al.: Smalta: practical and near-optimal FIB aggregation. In: 7th International Conference on Emerging Networking EXperiments and Technologies (CoNEXT), Tokyo, Japan, pp. 29:1–29:12 (2011)
8. Meiners, C., Liu, A., Torng, E.: Bit weaving: a non-prefix approach to compressing packet classifiers in TCAMs. IEEE/ACM Trans. Netw. **20**(2), 488–500 (2012)
9. Kang, N., Liu, Z., Rexford, J., et al.: Optimizing the "one big switch" abstraction in software-defined networks. In: 9th International Conference on Emerging Networking EXperiments and Technologies (CoNEXT), Santa Barbara, California, USA, pp. 13–24 (2013)
10. Kanizo, Y., Hay, D., Keslassy, I.: Palette: distributing tables in software-defined networks. In: IEEE INFOCOM, Turin, Italy, pp. 545–549 (2013)
11. Moshref, M., Yu, M., Sharma, A., et al.: Scalable rule management for datacenters. In: 10th USENIX Symposium on Networked Systems Design and Implementation (NSDI), Lombard, IL, USA, pp. 157–170 (2013)
12. Sheu, J.P., Chen, Y.C.: A scalable and bandwidth-efficient multicast algorithm based on segment routing in software-defined networking. In: IEEE International Conference on Communications, pp. 1–6. IEEE (2017)

Loc-Knock: Remotely Locating Survivors in Mine Disasters Using Acoustic Signals

Jiaxi Zhou[1(✉)], Zhonghu Xu[2], Shuo Zhang[2], and Jing Xu[2]

[1] The 38th Research Institute of China Electronics Technology Group Corporation, Hefei, China
18019961203@189.cn

[2] School of Computer Science, University of Science and Technology of China, Hefei, China
{xzhh,zshuo,jxu125}@mail.ustc.edu.cn

Abstract. In mine disasters, the location information of trapped miners is extremely important in rescue. In this paper, we propose a novel localization approach called Loc-Knock that remotely locates miners through acoustic signals, i.e., ***knocking sounds***. By applying an implicit synchronization method to the acoustic signals of indeterminate phases, frequencies and speeds, Loc-Knock can collaboratively reconstruct remote signals from unsynchronized sensors and locate the miners through a synchronization-free TDoA (Time Difference of Arrival) based localization method. To evaluate our approach, we conduct extensive experimental study in an operating coal mine. The results in the given 1 km field-test tunnel proved that Loc-Knock can precisely locate the miner over the tunnel.

Keywords: Post-disaster remote localization · Mine disasters · TDoA

1 Introduction

As known, mine disasters, e.g., gas explosion, landslide, mine flooding, etc., occur from time to time all over the world and took away a lot of lives of miners in past decades. In mine disasters, a key information in rescue is *the locations of miners*, since practically we must know the exact locations of the miners before providing the life supply or making rescue effort. Great efforts have been made in recent decade to address the mine localization issues. Most state-of-the-art solutions mainly rely on wired or wireless based technologies to provide underground localization and environment monitoring. Unfortunately, *in underground disasters during which these networking and sensing approaches should have been most needed, these approaches lose their power*. They either cease to function due to device breakdown, power cutoff, etc., or fail to send miners' locations to the ground due to the consequent harsh conditions, e.g., cable breaks, collapsed tunnel blocks the signal transmission, etc. In this paper we propose a novel approach that leverages acoustic signals (e.g., knocking sounds generated by the miners) for remote localization of miners.

S. Chellappan et al. (Eds.): WASA 2018, LNCS 10874, pp. 840–845, 2018.
https://doi.org/10.1007/978-3-319-94268-1_76

2 An Overview of System Design

The goal of Loc-Knock is to determine the locations of miners in mine disasters. Its acoustic design and the rail (conveyor) based signal propagation way allow Loc-Knock be able to overcome the difficulties traditional approaches meet. In Loc-Knock, we use TDoA (time difference of arrival) based localization approach. As shown in Fig. 1, M represents the location of the miner, $S = \{S_0, S_1, \ldots, S_{K_1-1}\}$ and $S' = \{S_0^{'}, S_1^{'}, \ldots, S_{K_2-1}^{'}\}$ represents the randomly deployed sensors on both sides of M. R_1, $R_1^{'}$ and R_2 are the locations where the reference signals are generated. Ideally, after S and S' receive the knocking signal generated at M, the location of the miner can be calculated based on the time difference of signal arrivals and signal propagation speed v.

Fig. 1. An example illustration of Loc-Knock

Note that measuring the time difference of signal arrivals requires that S and S' be tightly synchronized, while the sensors are actually unsynchronized. To solve this problem, we use a reference signal generated at the location R_2, as shown in Fig. 1. This reference signal could provide implicitly synchronization by comparing the time differences of the signal arrivals (More details are illustrated in Sect. 5.

The positioning accuracy of Loc-Knock highly depends on the accuracy of the measured signal arrival time. Obviously, the higher the SNR of a signal, the better the accuracy of the signal arrival detection. However, after a long-distance propagation, the received signal at each sensor has been overwhelmed by background noise and is too weak to be recognized. To tackle this issue, we first use a reference signal generated at R_1 (R_1') to implicitly synchronize the sensors of S (S'), based on which we further propose a collaborative beamforming approach to reconstruct the remote signal.

Furthermore, the signal arrival time could possibly not be accurately detected after beamforming due to the overwhelming noise. Besides, the overwhelmed parts are different among sensors since the background noise at different sensors (locations) is always different. To tackle this issue, we apply frequency domain analysis for featured signal selection and propose a differential based signal arrival detection method that detects the time point when the signal reaches a maximum power change rate.

Besides, the signal propagation speed v also plays an important role in the localization. v is usually unknown and varies among different environments due to the variances of chemical composition of rail tracks or surrounding environmental parameters. To deal with this uncertainty, we again apply the reference signal generated at R_1 (R_1'). By measuring the signal arrival time over air and rail (conveyor) at S, we can calculate the signal propagation speed v onsite.

3 Collaborative Reconstruction of Indeterminate Signals

In order to collaboratively reconstruct the remote signals, we randomly deploy the sensors along the rail (conveyor) to form a linear sensor array, and apply beamforming-based technology to collaboratively reconstruct the signal with unsynchronized sensors.

Suppose a randomly spaced linear array consists of K unsynchronized sensors, labeled as $S_i, i = 0, 1, 2, ..., K-1$. The inter-sensor spacing between sensor S_{i-1} and sensor S_i is denoted by d_i. If a wave impinges upon the linear array at an angle θ, and by setting the phase of the signal at the origin arbitrarily to zero, the wavefront arrives at sensor S_i sooner than at sensor S_{i-1} by $\Delta\omega_i$, since the differential distance along the two paths is $d_i \sin\theta$ [1]. $\Delta\omega_i$ is defined as $\Delta\omega_i = 2\pi f \frac{d_i \sin\theta}{v}$ where f, v is the frequency and the transmission speed of the signal, respectively. Let each sensor weighted with a complex weight W_i. Adding all the outputs of the sensors together gives the array factor F:

$$F(\theta) = W_0 + W_1 e^{j\omega_1} + \cdots + W_{K-1} e^{j\sum_{i=1}^{K-1} \Delta\omega_i} = W_0 + \sum_{i=1}^{K-1} W_i e^{j\sum_{k=1}^{i} 2\pi f \frac{d_i \sin\theta}{v}}$$

Let $W_i = A_i e^{ji\alpha_i}$, where A_i denotes the signal amplitude of sensor S_i, and α_i denotes the phase shift between sensors S_{i-1} and S_i. Therefore, $F = \sum_{i=0}^{K-1} A_i e^{ji(2\pi f \frac{d_i}{v} + \alpha)}$. By setting $\alpha = -2\pi f \frac{d_i}{v}$ we have the maximum output of array factor F.

4 Differential Based Signal Recognition

Before the knocking signal arrives, the waveshape of the signal represents background noise. When the knocking signal hits at time t_i, the waveshape starts to change and correspondingly the distribution of the observe signal begins to change. However, in underground environments the background noise overwhelms a piece of the signal arrival part, and the overwhelmed parts are different among sensors since the background noise at different sensors (locations) is always different. To solve this problem, we propose a differential based approach. The basic idea is given as follows: when a hammer knocking on the rail, the kinetic momentum of the hammer reaches a maximum change rate at a specific time t', which corresponds to the time point that the signal reaches a maximum power change rate. The time difference between t' and t is a fixed value to the same signal. Therefore, we detect the time t' when the signal has the maximum power change rate. Since $t' - t$ is a fixed value, it would not incur any additional error when we calculate time differences of signal arrivals.

5 Synchronization-Free TDoA Based Localization

In Loc-Knock, we use TDoA based localization method to locate the miner. Let M represents the location of the miner (the source of the knocking signal),

S and S' represent the locations of two sensing arrays distant from M. Note that in TDoA based approach the time clock of S and S' should be tightly synchronized. However, during sensor deployment it is likely that S and S' have different clocks and time drifting behaviors. To tackle this issue, we apply another reference signal R_2 for implicit synchronization of S and S'. Suppose the location of S is the origin of the one-dimensional coordinate. Let Δt_1 denote the time difference of arrival of the reference signal R_2 at S and S', and Δt_2 denote the time difference of the knocking signal arrival at S and S', i.e., $\Delta t_1 = \frac{R_2}{v} - \frac{L-R_2}{v}$ and $\Delta t_2 = \frac{M}{v} - \frac{L-M}{v}$ where L is the distance between S and S'.

Let $\Delta t = \Delta t_2 - \Delta t_1$. We get $M = R_2 + \frac{1}{2}(\Delta t_2 - \Delta t_1)v = R_2 + \frac{1}{2}\Delta tv$ where M is the computed location of the miner who generates the acoustic signal.

6 Experimental Study

6.1 Experiment Settings

The experiment is conducted in an operating coal mine. The field-test environments approved for us to use is a 24-h operating tunnel, which is about 1000 m long and 300 m deep underground. Figure 2 shows a snapshot of the experiment scenarios. We deploy 17 sensors on the rail and conveyor in the tunnel, with 16 sensors at one end of the tunnel and 1 sensor at the other end. The distance between the two set of sensors is set to 850 m, 750 m, 650 m, respectively. The distance between the miner and the 16-sensors set is set to 800 m, 700 m, 600 m, respectively. The distance between the miner and the single sensor is fixed to 50 m. The sensitivity of the sensors we used in the experiment is 4 mV/Pa and its frequency response spread from 20 Hz to 2.5 KHz. The sampling rate of each sensor is set to 128 K sps. The purpose we use oversampling is to reduce noise.

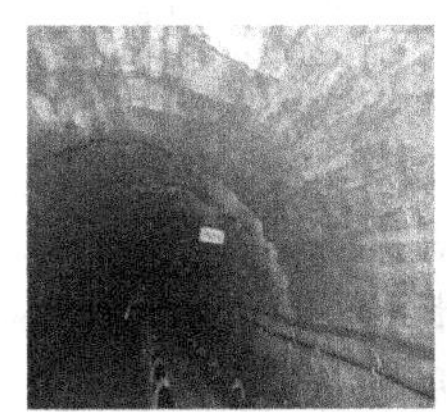

(a) At 950m of the tunnel

(b) Deployed sensors

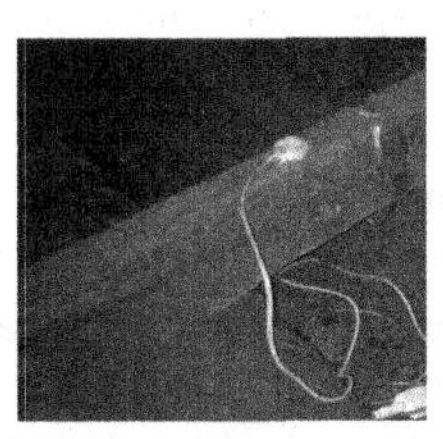

(c) The close-up of a sensor and the rail

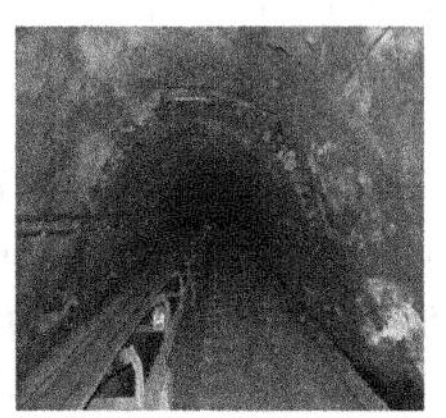

(d) Rail and belt conveyor

Fig. 2. Snapshot of the field-test environment

6.2 The Reconstructed Signals

Figure 3(a), (d) and (g) show the frequency distribution of the original signals that are recorded around the miner in the 600 m, 700 m and 800 m experiments, respectively. The knocking signal is overwhelmed by noise in the low frequency domain, while has obvious high energy response around 1000 Hz.

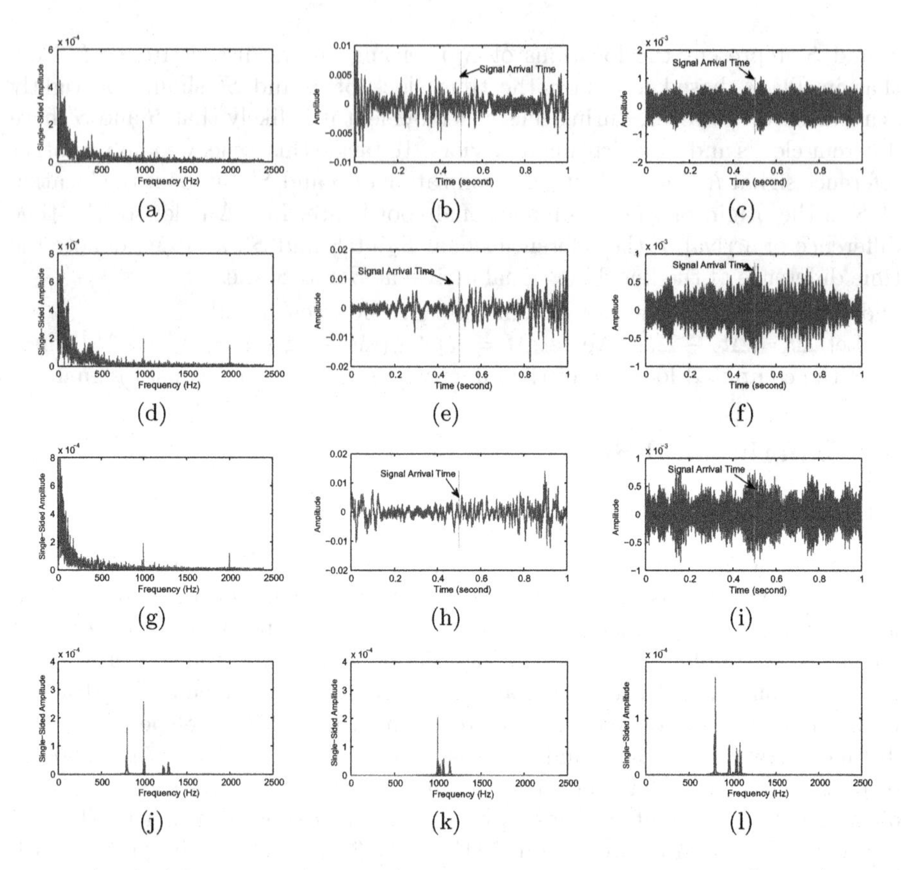

Fig. 3. (a, d, g) Spectrum chart of signal recorded around the miner; (b, e, h) signal received at a sensor of the 16-sensors set; (c, f, i) signal after beamforming at the 16-sensors set; (j, k, l) spectrum chart of signal after beamforming and filtering

Figure 3(b), (e) and (h) show the original signals received at a sensor of the 16-sensors set in the 600 m, 700 m and 800 m experiments, respectively. We can see that the knocking signal is completely overwhelmed in the background noise and cannot be recognized at all after a long-distance propagation.

Figure 3(c), (f) and (i) show the reconstructed signals after collaborative beamforming at the 16-sensors set in the 600 m, 700 m and 800 m experiments, respectively. According to the figures, it is hard to tell whether the reconstructed signals in the 700 m and 800 m experiments are enhanced or note. While the signal arrival point of the reconstructed signal in the 600 m experiments seems still be overwhelmed by noise. This is because there is a subterranean stream that generating strong babbling noise beside the deployed 16-sensors between the 600 m and 700 m experiments. At this stage, the beamforming results are not straightforward.

Figure 3(j), (k) and (l) are the spectrum charts of the signals after the reconstructed signals are processed by the band-pass filter and frequency selection in the 600 m, 700 m and 800 m experiments, respectively. Multiple frequencies (e.g., frequencies around 1000 Hz) with a maximal energy level of the signals stands out in the figures and constitute the signal for signal arrival detection. At this stage, Loc-Knock usually reproduce the remote knocking signal with the SNR as high as 20 db in the 600 m, 700 m and 800 m experiments, respectively. According to existing theoretical and experimental results [2,3], this suggests that Loc-Knock could work at a longer distance with a smaller SNR, e.g., 2 km at 12 db, 4 km at 6 db, which could be sufficiently long to cover the range of most mine disasters.

Figure 4(a), (b) and (c) show the cumulative distribution of localization error in the 600 m, 700 m and 800 m experiments, respectively. In 80% measurements, the estimated location is within 20 m from the miner in the 600 m and 700 m experiments, and in 90% measurements the estimated location is within 10 m from the miner in the 800 m experiments.

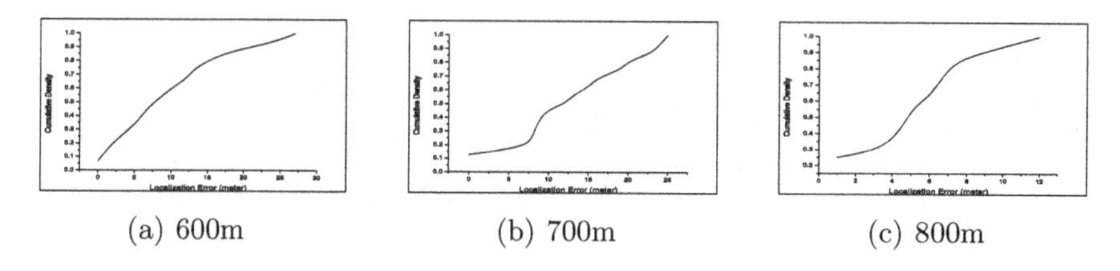

(a) 600m (b) 700m (c) 800m

Fig. 4. Cumulative distribution of localization error at 600 m, 700 m and 800 m

7 Conclusion

In this paper, we develop a new localization approach called Loc-Knock, which leverages acoustic signals for remote localization of miners. By collaboratively reconstructing the remote acoustic signals, Loc-Knock can well extract the knocking signal and accurately detect the signal arrival through a differential based approach, based on which Loc-Knock further provide a synchronization free TDoA based localization approach. According to the extensive experimental study in an operating coal mine, it is shown that in the given 1 km field-test tunnel Loc-Knock can remotely and precisely locate the miners over the tunnel.

References

1. Litva, J., Lo, T.K.: Digital Beamforming in Wireless Communications, 1st edn. Artech House, Inc., Norwood (1996)
2. Maev, R.G.: Acoustic Microscopy: Fundamentals and Applications. Wiley-VCH, Weinheim (2008)
3. Attenuation of Sound, Kaye and Laby Online. National Physical Laboratory, UK

A Recognition Approach for Groups with Interactions

Weiping Zhu(✉), Jiaojiao Chen, Lin Xu, and Yan Gu

School of Computer Science, Wuhan University, Wuhan, People's Republic of China
{wpzhu,jiaojiaochen,yangu}@whu.edu.cn,
cathyxl@outlook.com

Abstract. People often participate in activities in groups, such as buying goods in a shopping mall or walking around in a park. Interactive groups refer to the groups whose members have interactions such as shaking hands, embracing, which are not uncommon occurrences in our daily life. Existing group recognition approaches are based on the similarity of the individuals' locations or signal features. The interactions among people are probably regarded as dissimilar and affect the recognition accuracy. Moreover, when not all group members perform the interactions, group recognition is even more difficult to achieve. In this paper, we propose an approach called Interactive Group Recognizing (IGR) for recognizing groups with interactions among their members. The actions of individuals are inferred based on the sensing data, and the disparity between two individuals is computed using the sliding window technique. After that, groups are recognized using a majority-voting based method. Experimental results show that compared with the existing approach, the average group recognition accuracy of IGR is improved by 6.9%.

Keywords: Group recognition · Interactions · IGR

1 Introduction

Group recognition refers to the identification of a group of people such as classmates, colleagues, and family members who conduct activities together. It is important for evacuation management, crowd control, socially aware recommendations and so on. During such activities, it is quite common that the group members perform interactions such as waving hands, shaking hands and embracing. We call groups with such type of interactions *interactive groups*.

Existing group recognizing approaches are based on the similarity of sensing data [7,8] or locations [1,5,9] of individuals in a group, the interactions among people are probably regarded as dissimilar, and hence affect the group recognition accuracy. For example, when Alice and Blair come across their friends, Clark and Danny, 20 m away, they wave hands and then walk toward each other; during this process, their locations are different and their signal features (based on their walking and wave directions) are in fact opposite. Moreover,

S. Chellappan et al. (Eds.): WASA 2018, LNCS 10874, pp. 846–852, 2018.
https://doi.org/10.1007/978-3-319-94268-1_77

not all group members perform the same interactions simultaneously. For example, Alice shakes hands with Clark, and at the same time Blair embraces with Danny. Such inconsistent actions make the recognition of groups even more difficult. Therefore, there is a significant demand to improve existing approaches to recognize such interactive groups.

In this paper, we propose an algorithm called Interactive Group Recogning (IGR) to recognize interactive groups. We first collect the sensing data of persons, and then infer their interactive actions. We then split the sequence of actions of each person into several non-overlapped windows. A disparity matrix is computed based on their action sequences in each window, and the group affiliation is then determined using a threshold based approach [4] or a joint-density based clustering approach [5]. The final group affiliation is obtained through the use of majority voting to handle the inconsistent interactions of group members. We invited eleven persons to conduct real tests to evaluate the effectiveness of the IGR algorithm. Compared with a state-of-the-art approach, DBAD [4], the average group recognition accuracy of IGR is improved by 6.9%.

2 System Model and the Proposed Solution

We assume that there are several groups of people present in a large spatial space. These groups conduct their activities concurrently, and may affect each other. During an activity, some group members perform common actions such as walking, running, sitting down, standing up, or remaining stationary, as well as interactions such as waving hands, shaking hands, embracing, holding hands, or hooking arms. The acceleration data of each person are collected using mobile devices carried by the person. All data are transmitted to a central server for processing. All the people in a group are within the scope of the communication. Our purpose is to recognize all groups with maximum accuracy based on the collected acceleration data.

We propose an algorithm called Interactive Group Recognizing (IGR) to recognize interactive groups. The algorithm includes following four phases.

Interaction Recognition: First, the sensing data are segmented in sliding windows. Because most of the interactions last no more than 2 s, we set the size of the sliding window to 2 s, and the overlapped time to 1 s [10], to guarantee that a complete interaction is included in a window. Next, features are extracted from the processed data, and machine learning approaches can be used to recognize the actions. In this study, we use the features in the time and frequency domains [3]. The time domain features include the mean value, standard deviation, maximum and minimum value. The frequency domain features include the mean value, standard deviation, skewness, and kurtosis [2]. Various types of approaches including support vector machine, random forests, and k-nearest neighbor can be used to deduce the actions based on the features of the sensing data. Finally, we obtain a sequence of actions for each person.

Disparity Matrix Generation: We further split the action sequences of each person into several non-overlapping windows (called action sequence windows), and generate a disparity matrix M^t for the tth windows of all persons. The window size represents the number of actions of a person included in a window, and M^t_{ij} denotes the disparity between the action sequences of persons i and j ($i, j \in N$) at window t. The disparity matrix reflects the similarity of actions performed by two persons. In the specific implementation, the edit distance [6] is used to compute the action disparity between two persons. We smooth the tth disparity matrix by applying a low-pass filter $\overline{M^t}_{ij} = (1/b) \times \sum_{\tau=0}^{b-1} M^{t-\tau}_{ij}$ where b is the length of the low-pass filter, and denotes the number of consecutive windows used in the filter [4,7].

Group Affiliation: We deduce the group affiliation matrix V^t based on the disparity matrix $\overline{M^t}$ and threshold ϕ. If $\overline{M^t}_{ij}$ is equal to or less than ϕ, V^t_{ij} is set to *1*, denoting that persons i and j are in the same group at window t, whereas if $\overline{M^t}_{ij}$ is greater than ϕ, V^t_{ij} is set to 0, denoting that persons i and j are not in a group at window t. The optimal value of ϕ dependents on the specific activities conducted by the persons [4].

Majority-Voting Based Group Recognition: After the group affiliation matrix of each window is acquired, the final group result can be computed. Existing studies [5], finalize the results of group affiliations when the groups remain for a sufficient number of successive time windows. These methods perform well when the members of a group stay relatively close to each other or perform similar actions simultaneously. However, this is not always true for interactive groups. We propose a majority-voting based method to solve this problem. Suppose that the number of total windows is m, and g_{ij} denotes the number of windows in which persons i and j are classified into the same group. If $g_{ij} \geq \alpha \times m$, then we consider i and j to be in the same group and set the final group affiliation matrix $v_{ij} = 1$; otherwise, $v_{ij} = 0$. In addition, α is a constant having different values in different environments. According to our experiment results, it should be set to around 0.7. The final group affiliation can be obtained based on the final group affiliation matrix.

3 Experiment Evaluation

We invited 11 volunteers to conduct experiments in a laboratory hall of 15 m × 10 m in size. Each person had a mobile phone in hand but their orientations had no constraints. We developed a data acquisition application and installed it on all of the devices. The volunteers performed four experiments. In experiment 1, the individuals form two groups, including 5 and 6 persons, respectively. In experiment 2 and 4, the individuals form three groups, each including 4 persons. In experiment 3, the individuals form four groups, including 2, 3, 2 and 4 persons, respectively. The persons performed the activities following a combination of three stages: meeting, merging, and separating. In the meeting stage, all members of the group walk to a gathering place. When walking to the gathering place, their walking speeds and directions have no constraints and hence are probably different. The group members perform different interactions when walking. In the

Table 1. Group affiliation accuracy of DBAD, cross-correlation based approach (corr.) and IGR (100%)

Method	Method	Exp. 1	Exp. 2	Exp. 3	Exp. 4	Average
Threshold	DBAD	0.611	0.815	0.817	0.77	0.753
	Corr.	0.593	0.725	0.825	0.732	0.718
	IGR	0.741	0.84	0.888	0.848	0.829
Joint-density clustering	DBAD	0.574	0.653	0.666	0.509	0.600
	Corr.	0.504	0.388	0.361	0.36	0.403
	IGR	0.689	0.744	0.846	0.801	0.77

Table 2. Group recognition accuracy of different methods (100%)

Method	Exp. 1	Exp. 2	Exp. 3	Exp. 4	Average
DBAD	0.916	0.96	0.933	0.916	0.931
Corr.	0.583	0.75	0.822	0.777	0.733
IGR	1.0	1.0	1.0	1.0	1.0

merging stage, the members are merged into a group, and then move together. While moving, the members of the group interact with each other freely. In the separating stage, the group is split into several subgroups, and the members in different subgroups perform different interactions. And there are no restrictions to the sequence and number of group activities during an experiment.

We compared the accuracy of the group affiliation of IGR with those of cross-correlation based approach [8] and DBAD [4]. Because the collected data are processed in windows, we first compared their performance at the window level and then compared the final results.

With regard to the comparison at the window level, IGR, DBAD and cross-correlation were applied to obtain a disparity matrix of each window. Next, either a threshold-based method or joint-density based clustering can be used to obtain the group affiliation. For the threshold-based method, the value of threshold was iterated to achieve the maximal accuracy. The average accuracy were averaged over all windows and all experiments for comparison. The results are shown in Table 1. It can be seen that the group affiliation accuracy of IGR are higher than those of DBAD and the cross-correlation based approach for the four experiments regardless of whether a threshold-based approach or joint-density based clustering approach were applied. When using a threshold-based method, the average accuracy of IGR is improved by 7.6% and 11.05% compared with DBAD and cross-correlation based approach. When using joint-density based clustering, the average accuracy of IGR is improved by 16.95% and 36.68% compared with DBAD and cross-correlation based approach. The performance threshold based processing is better than joint-density based clustering, and is therefore adopted in IGR. We then use majority voting to handle the difference in

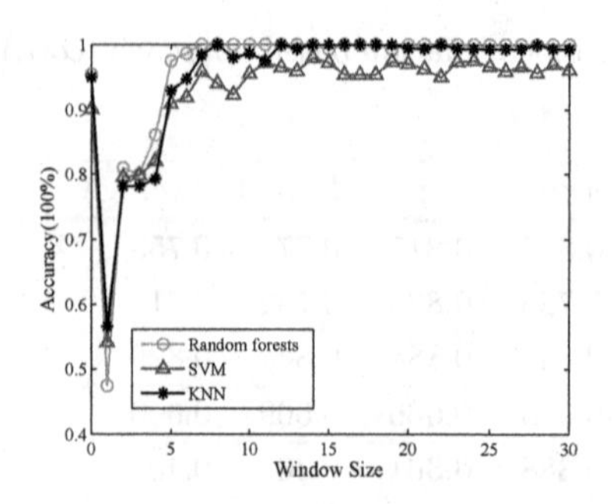

Fig. 1. Group recognition accuracy versus the window size of action sequence

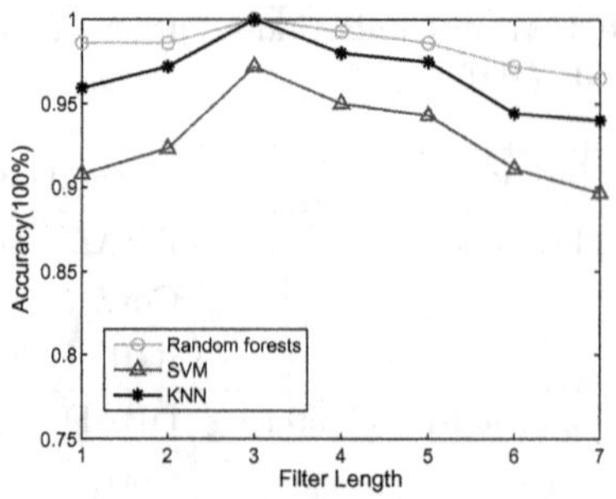

Fig. 2. Group recognition accuracy versus the low-pass filter length

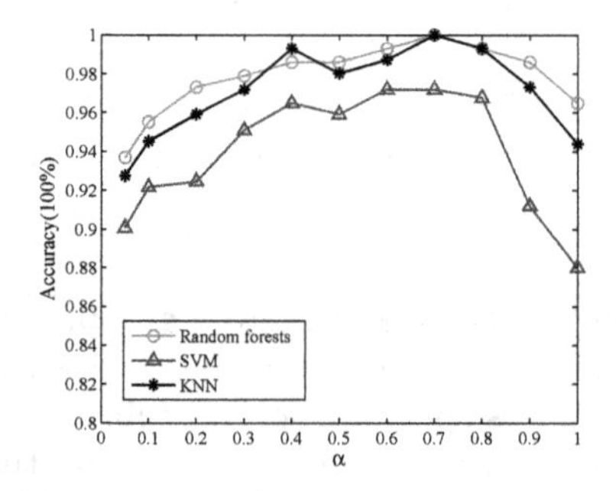

Fig. 3. Group recognition accuracy versus α used in majority voting.

the window results, and obtain the final group affiliation. The group recognition accuracy of the three approaches are shown in Table 2. It can be seen that the average accuracy of IGR for the four experiments is improved by 6.9% compared with DBAD, and 26.7% compared with cross-correlation based approach.

We further examined how the parameters of IGR affect its performance. We first set the low-pass filter length to 3 and α to 0.7, and change the window size from 0 to 30. The results are shown in Fig. 1. According to the figure, when the window size is less than 5, the group recognition accuracy is relatively low. This is due to the inconsistency of the interactions in a group. When the window size increases to more than 10, the group recognition accuracy is close to 1. This occurs because the increased time window can include more interactions, and hence the inconsistencies are more likely to be eliminated. We then set the window size to 14 and α to 0.7, and change the filter length from 1 to 7, the results of which are shown in Fig. 2. According to the figure, all three approaches achieve their maximum group recognition accuracy when the length of the filter is 3. When the length of the filter is less than or more than 3, the accuracy decreases because the disparity matrixes are smoothed by the low-pass filter;

therefore, a smaller filter length cannot eliminate the noise in the collected data, and a larger one probably causes the loss of useful information in the current window. Next, we set the window size to 14 and the filter length to 3, and change α from 0.05 to 1, the results of which are shown in Fig. 3. According to the figure, all three approaches achieve their maximum group recognition accuracy when α is 0.7. When α is small, it is likely that several groups are recognized as a single group, and thus the accuracy is reduced. When α is large, the members in a group are likely separated into different groups due to the inconsistency of the interactions in a group, and thus the accuracy is affected.

4 Conclusion

This paper investigated the recognition of groups with interactions. To achieve such recognition, an algorithm called Interactive Group Recognizing (IGR) is proposed, through which we collect the sensing data from individuals and deduce their interactions. The disparity between two individuals is obtained by calculating the difference of their interactions. We further proposed a majority-voting based group recognition method to handle the problem in which not all group members perform the interactions simultaneously. Compared with the group recognition results from DBAD and cross-correlation based approach, IGR improves the average group recognition accuracy by 6.9% and 26.7%, respectively.

Acknowledgment. This research is supported in part by National Natural Science Foundation of China No. 61502351, Luojia Young Scholar Funds of Wuhan University No. 1503/600400001, and Chutian Scholars Program of Hubei, China.

References

1. Anagnostopoulos, C., Hadjiefthymiades, S., Kolomvatsos, K.: Time-optimized user grouping in location based services. Comput. Netw. **81**, 220–244 (2015)
2. Chen, Y., Zhao, Z., Wang, S., Chen, Z.: Extreme learning machine-based device displacement free activity recognition model. Soft. Comput. **16**(9), 1617–1625 (2012)
3. Figo, D., Diniz, P.C., Ferreira, D.R., Cardoso, J.M.P.: Preprocessing techniques for context recognition from accelerometer data. Pers. Ubiquit. Comput. **14**(7), 645–662 (2010)
4. Gordon, D., Wirz, M., Roggen, D., Beigl, M.: Group affiliation detection using model divergence for wearable devices. In: Proceedings of ACM International Symposium on Wearable Computers, pp. 19–26 (2014)
5. Kjærgaard, M.B., Wirz, M., Roggen, D., Tröster, G.: Mobile sensing of pedestrian flocks in indoor environments using WiFi signals. In: Proceedings of IEEE International Conference on Pervasive Computing and Communications, pp. 95–102 (2012)
6. Ristad, E.S., Yianilos, P.N.: Learning string-edit distance. IEEE Trans. Pattern Anal. Mach. Intell. **20**(5), 522–532 (1998)

7. Roggen, D., Wirz, M., Tröster, G., Helbing, D.: Recognition of crowd behavior from mobile sensors with pattern analysis and graph clustering methods. Netw. Heterogen. Media **6**(3), 521–544 (2011)
8. Wirz, M., Roggen, D., Troster, G.: Decentralized detection of group formations from wearable acceleration sensors. In: Proceedings of International Conference on Computational Science and Engineering, vol. 4, pp. 952–959 (2009)
9. Yu, N., Han, Q.: Grace: recognition of proximity-based intentional groups using collaborative mobile devices. In: Proceedings of IEEE International Conference on Mobile Ad Hoc and Sensor Systems, pp. 10–18 (2014)
10. Zhao, Z., Chen, Z., Chen, Y., Wang, S., Wang, H.: A class incremental extreme learning machine for activity recognition. Cogn. Comput. **6**(3), 423–431 (2014)

Security Enhancement of Over-the-Air Update for Connected Vehicles

Akshay Chawan[1], Weiqing Sun[1(✉)], Ahmad Javaid[1], and Umesh Gurav[2]

[1] College of Engineering, University of Toledo, Toledo, OH 43606, USA
Weiqing.Sun@utoledo.edu
[2] Tech Mahindra Americas Inc., Schaumburg, IL 60173, USA

Abstract. Similar to wireless software updates for the smartphones, over the air (OTA) update is used to update the software and firmware for various electronic control units (ECUs) in the connected vehicles. It is an efficient and convenient approach to update the software in the car and it will save the customers the visiting time to repair small bugs in the software. However, OTA updates will open a new attack vector for the hackers. They can possibly exploit the OTA channel to steal OEM firmware, to reprogram ECUs and even control the vehicle remotely. In this paper, we perform a comprehensive security analysis for the current OTA mechanism to understand its associated threats. We also propose an approach to secure the original OTA software update method by incorporating biometric iris scan and cryptographic checksum before updating the software and firmware over the air. These security enhancements can help to mitigate the threats by preventing unwanted and potentially malicious software updates through the OTA channel.

1 Introduction

It is the century of exciting technology with different types of smart devices. Traditional accustomed electronic devices are now equipped with an extraordinary level of intelligence. Mobile phones, as a representative daily used electronic device, have become "smarter" in a way that they can perform more functions than just making calls. And most recently normal cars are becoming smart cars, which are capable of more than just being used as a mode of transportation. All these electronic devices bring great benefits to an individual with increasing levels of comfort, convenience, and efficiency. And modern automobiles are transforming into "smartphones-on-wheels" [1] which uninterruptedly generate, process, exchange and store large amounts of data.

The vehicles are evolving into a computerized system as they are going through a rapid evolution by replacing most of the mechanical structures with electrical structures. In addition, automobile systems can connect to the external networks such as the Internet by using their wireless interfaces, and can enhance the consumer experience by empowering new features and services [2]. However, with the connection of the car to the external networks, it also makes the connected cars susceptible to the hackers who can attack the car by seeking and manipulating weaknesses in its computer systems or networks.

S. Chellappan et al. (Eds.): WASA 2018, LNCS 10874, pp. 853–864, 2018.
https://doi.org/10.1007/978-3-319-94268-1_78

As an important security measure, it is crucial to update automobile computerized systems regularly as the loss of its integrity and confidentiality can have an adverse effect to the users. The new trends for car automation and security, deemed more viable as recognized globally in this modern era, have replaced the traditional ways such as using various locking and anti-theft systems. Moreover, the research community has developed efficient software algorithms to enable deployment of general biometric based platforms. To remain vibrant in this technology reformation, the development of different car security applications has generated a substantial amount of excitement among researchers. It results in the shifting of car safety mechanisms from a traditional way of securing one's car to advanced technologies because of the immediate benefits of accessibility, mobility, feasibility, spontaneity, and real-time communication.

In this paper, we make use of biometric iris authentication technology on OTA to ensure that only legitimate users can install the available updates from OEM (original equipment manufacturer) servers to the respective ECUs and sensors. We also use the cryptographic checksum verification to ensure the integrity of the software updates before they can be installed in the car.

2 Related Work

2.1 Threats to Connected Vehicles

Presently, encryption is uncommon, and, if accessible, regularly utilizes comparative keys over a progression of vehicles and ECUs. The traditional information transmissions are not encoded or validated and authentication is used while reinventing ECUs. McCoy et al. [3] demonstrated that ECUs can be reconstructed by obtaining premodified security keys from the automobile tuning group, and outlined the security issues in contemporary networked vehicles.

The greater part of these attacks has been performed with direct associations with the vehicle, yet in [4] attacks by means of external interfaces like integral telematics unit have also been reported and the same was demonstrated by Checkoway. As of late, as per an analysis of automotive networks and control units by Miller and Valasek in 2013 [5], various types of vehicles have been evaluated and two vehicles have been attacked widely. Miller and Valasek made use of either vehicle's OBD port or the vehicle's networks to execute these attacks.

In the attacks demonstrated by Paverd et al. [6], they performed the attack through a cellular connection and have broadened the purpose of the attack. Hoppe in his research [7] discussed the attacks particularly related to the Controller Area Network (CAN), and their countermeasures. In addition, Othmane in his case study [8] did a contextual analysis exploring the probability of attacks on vehicles in light of expert knowledge. Garcia et al. performed a case of a Rainbow table assault to break the 96-bit Megamos Crypto calculation utilized by numerous vehicle producers. Building the 1.5 Terabyte rainbow table took short of one week, however the comprehensive pursuit just took seconds [9]. As it illustrates that security instruments can be compromised and the vehicle can be stolen without much of a stretch, this can have a potential ramifications for vehicle proprietors. Zhang et al. from Cisco Systems demonstrated that with the use of embedded web programs, on-board diagnostic ports, removable ports and media players, numerous potential instruments can be tainted by malware. He also described the critical danger postured by malware to the vehicle's control framework and the potential ways by which vehicle's security keys can be compromised by the attackers to update the ECU programs [10].

2.2 Authentication Frameworks Used in Automobiles

The TESLA (Timed Efficient Stream Loss-Tolerant Authentication) framework has been designed for low-execution communication frameworks [6]. In TESLA, the sender delivers another symmetric key and calculates the MACs for at least one message. In the wake of accepting the messages, the sender communicates the key on the bus, allowing each recipient to confirm the sender of the earlier message. As the keys are sent alongside the information of the following message, the corresponding overhead of TESLA is insignificant. In any case, the inevitable time delay amongst receiving and confirming a message constrains TESLA's real-time configurations. Furthermore, TESLA just provisions fractional recipient validation in communication frameworks without sender identifications, and does not convey stream approval or encryption.

For the vehicle domain, focus has been given on Lightweight Broadcast Authentication Protocol for CAN in the research work known as LiBra-CAN developed by Groza et al. [11]. It validates senders at the receipting ECUs by means of Mixed Message Authentication Codes (M-MACs) in which keys are administered to constellations of ECUs. LiBra-CAN does not fret about key trades and needs pre-shared keys.

The lightweight authentication mechanism CANAuth has been proposed by Singelee et al. [12]. The keys are allocated for message clusters and it permits broadcast authentication. Before the authentication can be achieved for message clusters, it calls for pre-shared keys similar to LiBra-CAN and CANAuth.

VeCure is another authentication protocol developed by Wang and Sawhney [13] in 2014. Centered on trust, the ECUs are split into various classes and then keys are allocated to these classes. It suggests programming these keys at the initial setup of the vehicle and is dependent on pre-programmed keys.

Most of the systems explained in the above research are not satisfactory or operational, as these strategies rely on the basic trust of pre-programmed keys. Due to the long lifetime of vehicles, keys must be updated occasionally in order to dodge different attacks. In addition, the safe generation and programming of keys isn't a minor issue. Thus, the above approaches may not be reasonable for real-world applications.

3 Proposed OTA Software Update

We will describe the proposed overall secure OTA software update process in this section. In addition, we will discuss the two key enhancements: checksum comparison and iris based authentication.

3.1 Process of Secure OTA Software Update

Figure 1 shows the flow of the proposed software update process by downloading and installing the software update from the remote OEM server to the corresponding ECU in the vehicle. Initially, we will create the checksum from the software package and this checksum will be encrypted using OEM's private key. Then the software package along with the encrypted checksum and the OEM's digital signature will be sent to the OEM remote server. The telematics unit on the car receives the update files from OTA servers through the radio links and then transfers them to Central Gateway. In the normal situation, the central gateway directly transfers the updates to the body control module (BCM) and then to the respective ECUs. The central gateway is connected to the firmware control unit that informs the central gateway when a specific update should be sent to the ECU in case of updating multiple ECUs simultaneously. A database is connected to the central gateway as well as to the iris scan authentication unit. It stores the firmware/software updates securely until they are required to be sent

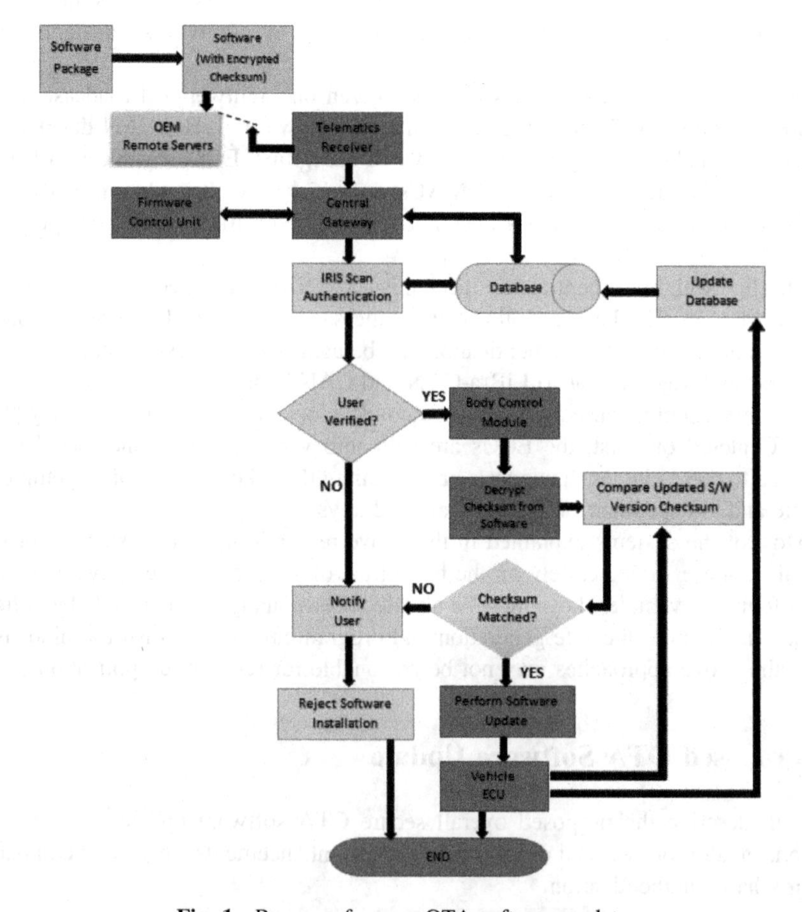

Fig. 1. Process of secure OTA software update

to their corresponding ECUs. Before downloading the software update files, the central gateway also verifies the version of the software by comparing it with the information stored in the database.

In our proposed method, it is required to go through an additional authentication and verification module before the updates actually arrive at the ECU from central gateway. This authentication module will store car owner's iris scan and sits between central gateway and body control module. The owner will have the capability to select which updates to install and at what time slots to install them to the ECU. In this case, even if the attacker hacks the central gateway, he/she will not be able to take control of or manipulate the ECU and the sensors connected to them through OTA updates.

If the central gateway receives an update and the owner of the car is not available, the updates will be stored in the database. These stored software/firmware update files would be secured using encryption and authentication protection to avoid any modification. The firmware control unit will hold a table of each ECU containing the information such as the serial number and version of the firmware update received and already installed in the vehicle. Once the owner starts the car it will prompt him/her about the update and with the iris scan camera fitted on the steering wheel he/she can authenticate and give his/her consent, before the BCM receives the updates.

Before the BCM sends the software update files to ECU for installation, it will verify whether the software package is legitimate by checking the attached digital signature. Once the legitimacy of the software update file is confirmed, its checksum value will be decrypted using OEM's public key stored in the BCM. After that, the vehicle ECU will compare the checksum value of the updated version with the version of the software that was previously installed in the car. If the checksum values are not equal, the user will be notified an error condition and then the installation will be terminated. Otherwise, the software update process continues and if the newly triggered software update installs successfully, the ECU approves the updated software version.

After the above steps, the ECU will set the newly installed software update as active, and the formerly active storage memory back to the inactive status. It also conveys the same information to the database to populate it with the information of the latest version of the software. After this, the whole software update procedure ends.

3.2 Checksum Comparison Implementation

Automotive service tool is a programming tool that provides abilities to analyze, repair, debug or monitor a system or product. The type of service tool varies with different manufacturers. A few examples of automotive service tools are CANOE and CanAnalyzer. Service tool is connected to ECU with the help of two buses namely CAN High bus and CAN Low bus.

As shown in Fig. 2, every module continuously sends heart beat messages to other modules in order to check whether any other module is live at the given time. Each node of the module has its own unique source address. With the use of the source addresses, the module can send out the heart beat messages globally to other modules.

At the communication initiation phase, the service tool requests the module to send its ID. After receiving the module ID, the service tool compares the received ID to see if it matches with the ID present in the service tool database.

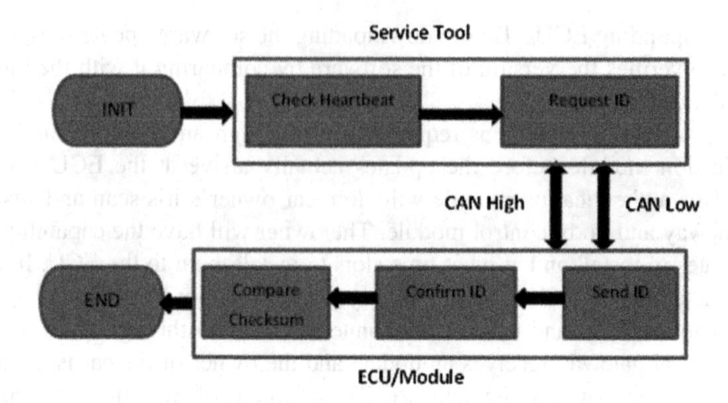

Fig. 2. Checksum comparison mechanism

Now as the communication is established, the service tool requests the specific module to enter the programming mode. The module processes the request and sends the acknowledgement to enter the programming mode.

Once the service tool receives a positive acknowledgement signal from the module, it requests the controller to enter the programming mode. This is the time when the checksum of the available update or firmware is calculated. The module needs to calculate the checksum of every program to be installed. This calculated checksum is then compared with the OEM's checksum value which is stored in the CAN database.

If both the checksums are the same, the flashing of the program updates will begin until the successful installation of the available update or firmware. On the other hand, if the third party is trying to install a program into the module with its checksum not matching with OEM's checksum, it will inform the module to abort the installation process and throw an error and/or a warning message.

3.3 Iris Based Authentication Implementation

In our proposed approach, the security of the OTA software update process will depend on the iris authentication of the car owner. As soon as a person enters the vehicle, the CCD digital camera scanning the iris will get activated immediately and then will use both visible and near-infrared light to take a clear, high-contrast picture of that person's iris.

As shown in Fig. 3, either the person looks into an iris scanner, or the camera focuses automatically, a mirror or audible feedback from the system are also used to determine that person's position accurately. An individual's eye should be approximately 4–12 in. away from the camera to obtain the clear image from the iris, the camera then clicks a picture of the subject, and the computing system locates the center of the pupil, edge of the pupil, edge of the iris, eyelids, and eyelashes [15]. A computing system then removes the eyelids and eyelashes from the image and analyzes only the patterns of iris. The core iris pattern will then be encoded compared with the already scanned iris data stored in the database [16] that is connected to the central

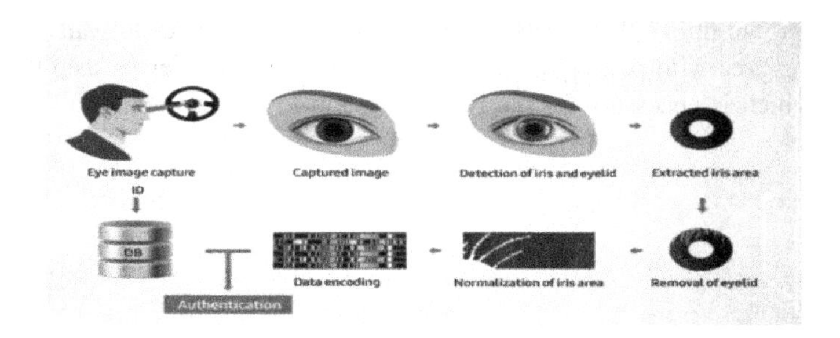

Fig. 3. Procedure to detect and authenticate iris [14]

gateway and iris authentication unit. If the code matches with the iris of the person sitting in the driver's seat with any of the code from the database, it will activate the required system and then proceed to software updates installation over the air.

4 Simulation and Evaluation

For the evaluation, we make use of CANOE software to simulate our proposed OTA security enhancement method. In particular, we created nodes using the protocol CAN-J1939 and used the CAPL scripting language.

4.1 Condition 1: OTA Upgrade Mode

We have taken Ignition key, Vehicle speed and Park brake as input signals and we get the corresponding Iris approval as the output signal. Ignition key and Park brake are digital inputs to the BCM. The value of these signals is transmitted periodically on the CAN bus by the BCM. Generally, these signals are broadcasted every 100 ms on the CAN network. In our setup, these signals are simulated using BCM node simulation.

Scenario. Assume the car is in a parking area with parking brakes ON or is stationary without the parking brakes ON. It shows that irrespective of the status of the parking brakes, the iris module turns ON and will start authenticating the user, provided the Ignition key is ON and Vehicle speed is zero.

The condition represented in Fig. 4 is a manual iris approval system. This condition will be specifically used during the OTA software update process. In this state, the driver can activate the iris system manually when the ignition is ON, irrespective of the parking brake status (set or not set). As shown in Fig. 4, the iris system is set to OTA mode, hence at time point of around 118th second after the ignition is turned ON, the iris system is activated irrespective of the park break is set at 124th second and at 138th second. At 145th second, the iris system matches the iris code from the database with the iris of the driver and sends a positive acknowledgement to OTA system to upgrade the required software. This will ensure that the software can be upgraded only with the car owner's consent. The car owner will be the only one to decide which software to upgrade at what time. This prevents the car from being attacked by an external source and making changes to the car firmware, which can be fatal to the driver.

The driver can utilize this condition even when he/she decides to wait in the car in some parking area with parking break set. This will help the user to keep the ignition turned on which is necessary for the HVAC system to be activated.

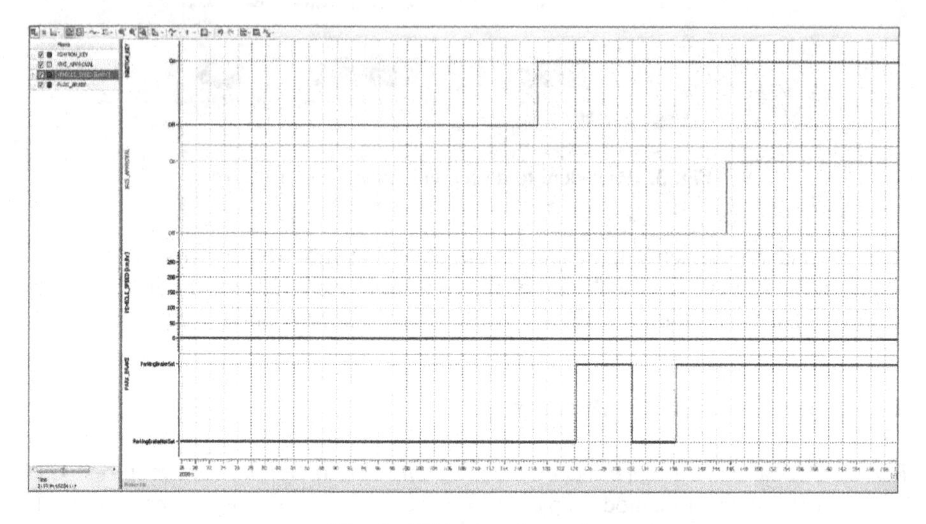

Fig. 4. OTA upgrade mode

4.2 Condition 2: Program Installation Using Checksum Comparison

The graphs in Figs. 5, 6 and 7 represent the different states of the software update process. It determines whether the update will be installed successfully.

- HVAC_UPDATE signal is a checksum value of the received software update.
- HVAC_CHKSM signal is a checksum value of the software that is already install in the HVAC ECU.
- Iris_Approval signal is a signal received from the iris module.
- Ignition signal is a signal received from BCM.

With the above inputs, we get a corresponding output Program_OK signal which can be in one of the four states (Active, Stop, Inactive and Not available). The update can be installed successfully only after satisfying the following two conditions:

1. It receives a positive Iris Approval.
2. The checksum of the available update or firmware is the same as that of the respective module checksum already stored in the non-volatile memory of the CAN stack.

We will demonstrate the process by using an example of updating the HVAC module in the following three different conditions.

4.3 Condition 3A: Idle Mode

Scenario. When an unauthorized user turns ON ignition, even though the iris module is ON, the Iris approval will fail, as it will not match with iris data from the database. This will ensure that an unknown person cannot install any update in the vehicle.

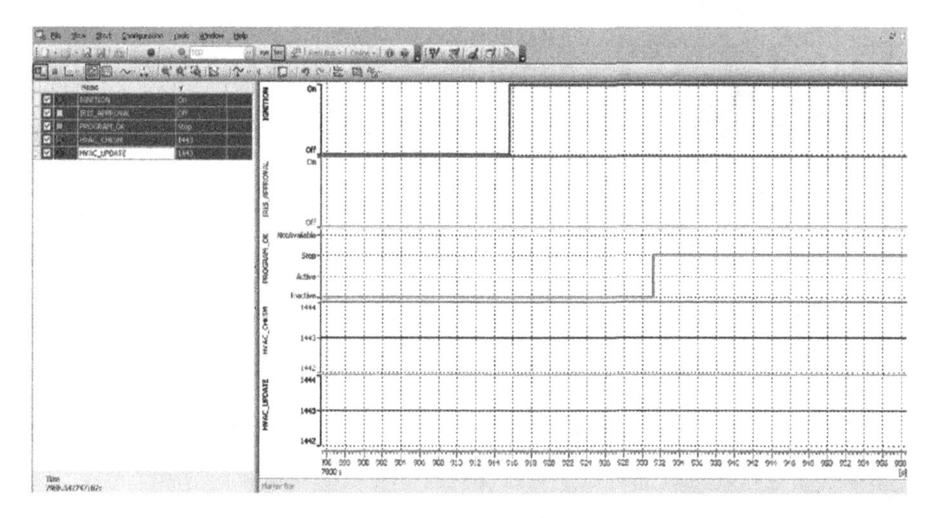

Fig. 5. Idle mode

As shown in Fig. 5, the ignition key is set ON at the time of around 7916th second. The checksum of the available HVAC update is determined to be 1443, and it is the same as the one stored in the database and represented by HVAC_CHKSM. In this scenario, the Iris approval bit is off through the experiment as the driver is not in front of the iris scanner or an unauthorized individual's iris is scanned. The graph shows that the download process has not started yet and hence the Program_OK bit is in inactive status. However, as the ignition is set ON, the system tries to download the update automatically after a few seconds. In this condition, even if the firmware tries to download the update it will not let the system install the update and the Program_OK bit will change into the Stop status as it happens at 7931st second as shown in Fig. 5. This indicates that the firmware installation is unsuccessful irrespective of having the same checksum value.

This will prevent installation of unauthorized applications or firmware to your vehicle by an unauthorized user.

4.4 Condition 3B: Iris Enabled Stop to Active Mode

Scenario. The authorized driver turns ON the ignition key, the car is stationary and connected to the Internet. This activates the iris module and then the iris of the driver will be scanned and then compared with the iris data stored in the database. As the scanned iris is from an authorized person, it moves on to checksum comparison module. In addition, the received update is genuine and from the trusted OEM, the checksum will match with the checksum of the software already installed in the car. Hence, the update will be installed successfully.

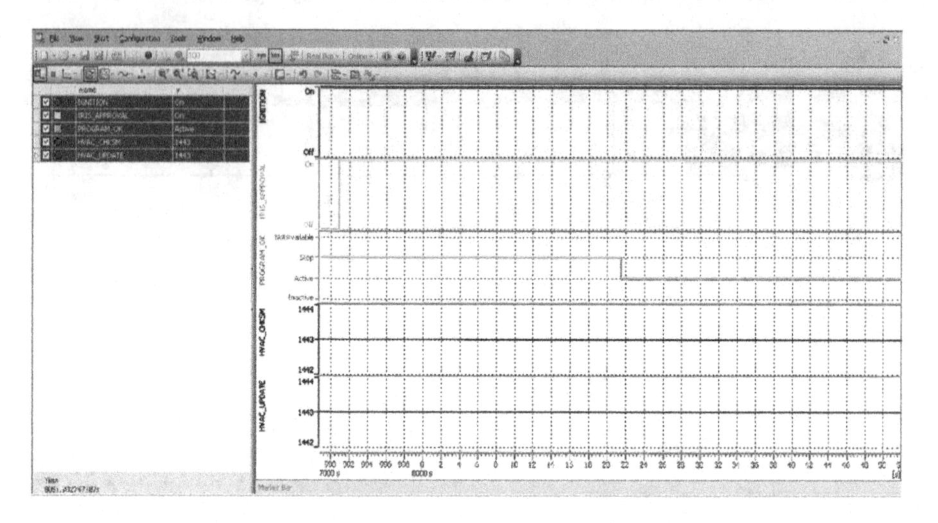

Fig. 6. Iris enabled stop to active mode

The graph in Fig. 6 shows that the ignition key is set ON throughout the experiment. As the ignition is set ON, the update is downloaded and ready to be installed but it is waiting for the Iris approval, hence the Program_OK bit is still in the Stop status. In this scenario, the Iris approval bit is set to ON status at the 7991st second as the scanned iris matches with the iris data originally stored in the database. Now the system will compare the checksum of the available update with the respective HVAC checksum stored in the system. As the checksum of the HVAC_update available is the same as the HVAC_CHKSM that is sent by the ECU i.e. 1443, the Program_OK bit transforms from Stop status to Active status at the 8022nd second. This helps to complete the installation of the required update without any error.

Assumption: We have assumed both HVAC_update and HVAC_CHKSM value as '1443' for experimental purpose.

4.5 Condition 3C: Iris Enabled Inactive to Active Mode

Scenario. The driver is already driving the car and then receives a message of available software update. The software will not be downloaded at this moment, as the car is moving. As the driver stops the car and connects to the Internet, the software update will be downloaded. As the authorized driver is driving and the update available is genuine, both the conditions are satisfied and hence the update is successfully installed in the car.

The ignition bit is set ON at 8154th second as shown in Fig. 7. In this graph, it shows that the Program_OK bit is in Inactive status at the initial stage. This means the system has not downloaded the update yet but it is available for downloading. At around 8186th second, the Iris approval bit is set high. At this time, the system starts downloading the update and compares the checksum of the update file with the respective module (for example, the HVAC module) checksum already stored in the system. Once it authorizes that both the checksums are the same, the Program_OK bit will transform directly to the Active state and will enable the system to install the update successfully.

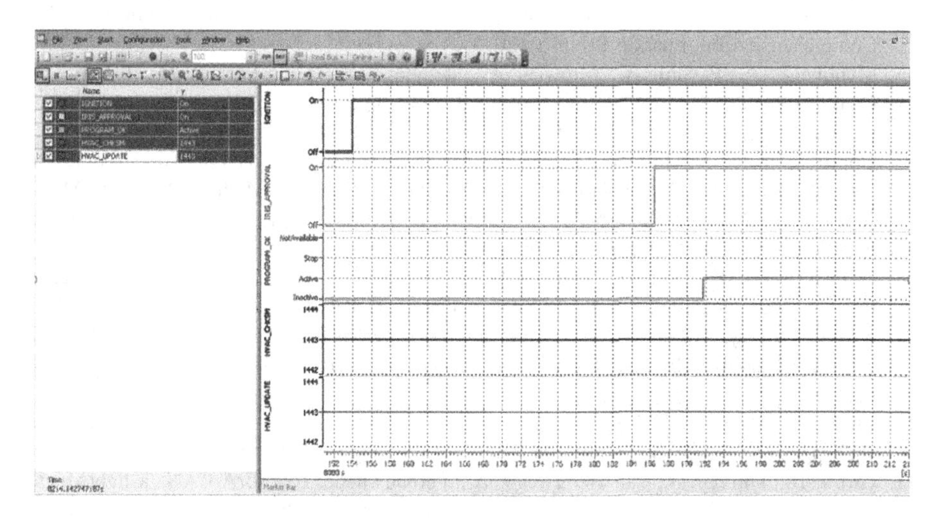

Fig. 7. Iris enabled inactive to active mode

5 Conclusion

In this paper, we analyze the cyber threats against OTA on connected vehicles. This suggests that we should stress over them and think about their potential impact. Having the capacity to compromise a car's ECU is, however, only half the story. The rest of the worry is what an attacker is able to do with those competencies. The method of securely installing software updates through OTA, proposed in this paper, incorporates a 2-step security verification process including iris recognition and checksum comparison, which will help to mitigate the potentially threats associated with OTA software updates. For the future work, we plan to implement and evaluate the proposed approach on real vehicles to gain more insights on its efficiency and usability.

References

1. van Roermund, T.: Secure connected cars for a smarter world, NXP Semiconductors (2015)
2. Birnie, A., van Roermund, T.: A multi-layer vehicle security framework, NXP Semiconductors (2016)
3. Checkoway, S., McCoy, D., Kantor, B., Anderson, D., Shacham, H., Savage, S., Koscher, K., Czeskis, A., Roesner, F., Kohno, T.: Comprehensive experimental analyses of automotive attack surfaces. In: USENIX Security, San Francisco (2011)
4. Koscher, K., Czeskis, A., Roesner, F., Patel, S., Kohno, T., Checkoway, S., McCoy, D., Kantor, B., Anderson, D., Shacham, H., Savage, S.: Experimental security analysis of a modern automobile. In: Proceedings of Symposium on Security and Privacy (SP) (2010)
5. Miller, C., Valasek, C.: Adventures in automotive networks and control units. In: IOActive Comprehensive Information Security (2013)
6. Mundhenk, P., Paverd, A., Mrowca, A., Steinhorst, S., Lukasiewycz, M., Fahmy, S., Chakraborty, S.: Security in automotive networks: lightweight authentication and authorization. ACM Trans. Des. Autom. Electron. Syst. (TODAES) **22**, 25:1–25:27 (2017)
7. Hoppe, T., Kiltz, S., Dittmann, J.: Security threats to automotive CAN networks – practical examples and selected short-term countermeasures. In: Harrison, Michael D., Sujan, M.-A. (eds.) SAFECOMP 2008. LNCS, vol. 5219, pp. 235–248. Springer, Heidelberg (2008). https://doi.org/10.1007/978-3-540-87698-4_21
8. Othmane, L.B., Fernando, R., Ranchal, R., Bhargava, B., Bodden, E.: Likelihood of threats to connected vehicles. Int. J. Next-Gener. Comput. (IJNGC) **5**, 290–303 (2014)
9. Verdult, R., Ege, B., Garcia, F.D.: Dismantling megamos crypto: wirelessly lockpicking a vehicle immobilizer. In: 22nd USENIX Security Symposium, Washington, DC, pp. 703–718 (2013)
10. Zhang, T., Antunes, H., Aggarwal, S.: Defending connected vehicles against malware: challenges and a solution framework. IEEE Internet Things J. **1**, 10–21 (2014)
11. Groza, B., Murvay, S., van Herrewege, A., Verbauwhede, I.: LiBrA-CAN: a lightweight broadcast authentication protocol for controller area networks. In: Pieprzyk, J., Sadeghi, A.-R., Manulis, M. (eds.) CANS 2012. LNCS, vol. 7712, pp. 185–200. Springer, Heidelberg (2012). https://doi.org/10.1007/978-3-642-35404-5_15
12. Van Herrewege, A., Singelee, D., Verbauwhede, I.: CANAuth-a simple, backward compatible broadcast authentication protocol for CAN bus. In: Proceedings of the 2011 ECRYPT Workshop on Lightweight Cryptography (2011)
13. Wang, Q., Sawhney, S.: VeCure: a practical security framework to protect the CAN bus of vehicles. In: International Conference on the Internet of Things, Cambridge (2014)
14. Samsung news. http://samsungnews.net/samsung-giai-thich-cach-hoat-dong-cua-may-quet-mong-mat-tren-galaxy-note-7/
15. Wilson, T.V.: How biometrics work. http://science.howstuffworks.com/biometrics4.htm
16. Determan, G.: Security alarm notification using iris detection systems. Patent No. US2006 0072793 (2006)

Resilient SDN-Based Communication in Vehicular Network

Kamran Naseem Kalokhe[1], Younghee Park[1(✉)], and Sang-Yoon Chang[2]

[1] Computer Engineering Department, San Jose State University, San Jose, USA
younghee.park@sjsu.edu
[2] Computer Science Department, University of Colorado Colorado Springs, Colorado Springs, USA

Abstract. Vehicular ad-hoc network (VANET) is the key component of intelligent transportation system (ITS) for various services like road safety and traffic efficiency. However, current VANET architectures provide less flexibility and scalability for vehicle-to-vehicle communication because of static underlying network infrastructure. Based on the software defined networking (SDN), open programmable networks through logically centralized control, this paper proposes a VANET monitoring system to provide a resilient and efficient routing for better services for communication between vehicles. It also provides the secure channels between the SDN controller and the network devices for reliable communication. Experimental results show that our proposed system selects the best path according less network latency in VANET.

Keywords: VANET · Reactive routing · Software defined networking

1 Introduction

Vehicular Ad hoc Network (VANET) has become the main infrastructure to make cars connected in each other over different communication technologies, such as Wi-Fi, LTE, ZigBee, Wi-Max, GSM and UMTS [2, 18, 19]. A full-connected vehicle network improves vehicle road safety, enhanced traffic and travel efficiency. This connection includes Vehicle to Vehicle (V2V) and Vehicle to Infrastructure (V2I) in VANET. V2V communication enables services ranging from accident avoidance and resource sharing in traffic management. In V2I communication is established between two vehicles via Road Side Unit (RSU).

Even though VANET provides many benefits for drivers and passengers, it is still confronting a great deal of difficulties, for example, inefficient resource utilization, absence of QoS, security and protection issues and uneven traffic distribution [1]. Moreover, the current VANET architectures lack of flexibility and scalability in order to support dynamic mobility in vehicles. To resolve these underlying issues, researchers are focusing on incorporating SDN into VANET [2, 3]. SDN provides programmability to the network by extracting the control plane from the data plane. SDN have been successfully implemented to efficiently control the datacenter traffic.

S. Chellappan et al. (Eds.): WASA 2018, LNCS 10874, pp. 865–873, 2018.
https://doi.org/10.1007/978-3-319-94268-1_79

SDN has advanced in the wired network domain, especially for datacenters. SDN involvement is also progressive in the wireless and ad-hoc domain [5].

This paper proposes a resilient and efficient routing system to react to the real-time network status in VANET by using the concept of SDN. The system consists of two main components: network monitoring, and path selection. The network monitoring keeps collecting network data related to switches and road side units (RSUs) in VANET by using SNMP (simple network management protocol). The path selection provides the best path from source to destination by selecting the best set of RSUs through a modified shortest path algorithm. Since traffic in the VANET is passed through the first available path, which leads to the underutilized network resources, high latency and increase drop rate in the network. By using our proposed system, we can achieve low network latency while our proposed system keeps governing the entire VANET based on the SDN controller.

Our contributions in this paper is followed. First, we present a traffic engineering technique for VANET by integrate SDN into the VANET. Second, we implement a new path selection system by keep monitoring network status in VANET based on SNMP. Lastly, we evaluate our proposed system in Mininet Wi-Fi [18] which has capability to extend virtualized Wi-Fi stations and access points.

The remainder of the paper is structured as follows. Section 2 discusses previous works. In Sect. 3, we present our proposed architecture. Section 4 shows the performance evaluation, and finally, Sect. 5 concludes this paper.

2 Related Work

Vehicular ad-hoc network (VANET) aims to support vehicles-to-vehicles (V2V) communication with road side units (RSUs) [2, 18]. Generally, as VANET efficiently manages vehicles and road traffic, it is used for traffic management and safety services for vehicles. These services have reduced the number of car accidents and saved life of people traveling in vehicles by controlling traffic and by providing useful information for drivers and passengers. SDN-based traffic engineering can improves quality of services for end users to maximize the performance of VANET.

Software Defined Networks (SDN) gives us high flexibility and programmability to control routing paths according to the current network status [15–17]. The controller has an intelligence to control the entire network. The routing module in the controller is used to push flow rules in the network devices and receive network statistical information from switches or routers for tracking network conditions. In this paper, OpenFlow is integrated into a wireless VANET environment. VANET utilizes SDN to improve usage of channels and to optimize wireless resources while reducing interference in data transmission in VANET [2, 3, 20, 21]. Multiple controllers established a hierarchical controlling mechanism to control VANET as a master-slave model [20]. Ramon practiced a SDN model into VANET to manage data communication among vehicles and RSUs while the controller communicated with OpenFlow switches [21].

3 The Proposed Method

3.1 System Architecture

Figure 1 shows our SDN-based VANET infrastructure including vehicles, RSU (Roadside Units), and an SDN controller. The SDN controller continuously monitors network dynamics and infrastructure to determine the best path for vehicle-to-vehicle communication. The controller installs flow rules into switches according to a selected routing path for the end-to-end communication. The RSU is equipped with programmability in the data plane by enabling the software of Open vSwitch in order to dynamically handle flow rules in the data plane. The communication among vehicles (V2V) and V2I (Vehicle to Infrastructure) in VANET is performed by Openflow-enabled RSU while the controller communicates with RSU to control the traffic.

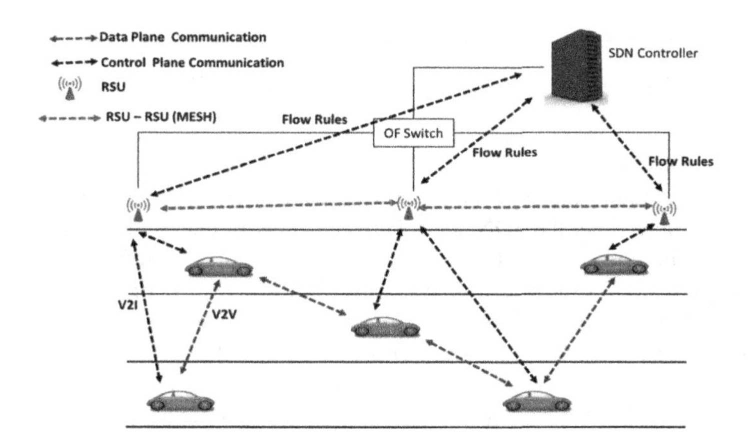

Fig. 1. A software-defined networking based VANET

The SDN controller consists of two main subsystems: a network monitoring system and a routing path selection system. The network monitoring system in the controller continuously monitors traffic to collect statistical information in VANET and to track physical device status. The path selection system chooses the best routing path between two end vehicles by using the input from the network monitoring system. Compared to a wired network, VANET has limitations to monitor individual vehicles due to high mobility and inaccessibility. Thus, the controller keeps monitoring VANET by focusing on RSUs, and reroute traffic based on the monitoring results of RSUs (Fig. 2).

The networking monitoring system collects information about VANET by monitoring RSUs and vehicles using SNMP (Simple Network Management Protocol). SNMP aims to gather information about legacy devices like RSUs and vehicles while the SDN controller also collects data in OpenFlow switches. The SDN controller keeps polling data including traffic statistics, interface information, CPU utilization and memory utilization in OpenFlow switches. The SDN controller using SNMP also collects information about interface status, interface bandwidth, CPU and memory

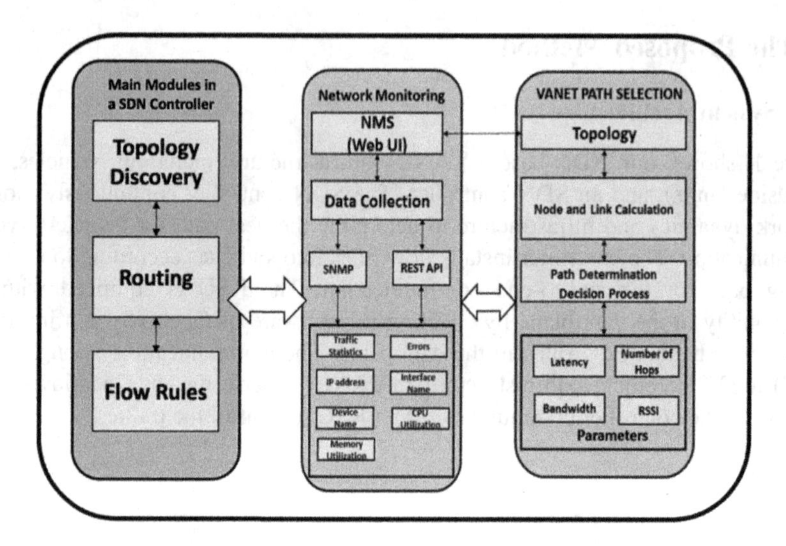

Fig. 2. Our proposed system architecture in an SDN controller

utilization, packet drops in RSUs. About the vehicles, the controller identifies end-to-end communication between RSUs rather than vehicles because of no direct accessible vehicles from the SDN controller. The network monitoring system identifies failures by using errors or CPU, and memory utilization. The traffic statistical information is used for a path selection system to find the best routing path of RSUs to support an efficient vehicle-to-vehicle communication.

The path selection system in the controller decides on the best routing path of RSUs based on the information collected by the network monitoring system. The path selection system calculates the path cost (C_{ij}^{Pr}) for all the possible paths ($P_r = \{P_1, P_2, ..., P_r\}$) from source RSU($N_i$) to destination RSU($N_j$) and then selects the lowest path cost for routing. All the paths are determined by the depth first search algorithm from source N_i to destination N_j. The topology of RSUs is represented as a graph G. Note that G_i is the set of RSUs (M is the total number of RSUs). As like a mesh network, the graph G is a complete connected graph from node N_i to node N_j. The path selection system computes the path cost of all the possible paths between N_{ij} according to Eq. 1. From all the path costs for each path P_r, the path selection system selects the best path for routing traffic between two vehicles.

The path cost (C_{ij}^{Pr}) is determined by latency, bandwidth, the number of hops, and a signal strength (RSSI) between an RSU and a vehicle for a path. Here is the path cost (C_{ij}^{Pr}) from source RSU (called a node) to destination RSU for a particular path P_r. Note that L_{ij}^{Pr} is latency between a node i and a node *j for each path*, P_r. The latency is computed by RTT (round trip time) between two nodes. H_{ij}^{Pr} is the total number of hops for all the possible paths (P_r) between the two nodes N_{ij}. $B = \sum_{k=1}^{n} (b_k)$ ($k = \{1, ..., n\}$) is the average bandwidth of the total bandwidth after summing each b_k. Note that b_k is the link bandwidth between a node N_i and its neighboring node (not a destination node). The RSSI (Received Signal Strength) is a value to indicate the signal quality between an RSU and a vehicle. The closer between an RSU and a vehicle is, the higher value RSSI is. Thus, if a vehicle is close to an RSU, RSSI has a better signal strength. The $RSSI_{final}$ value is compute by rounding off the absolute value of $RSSI$ divided by 10 ($RSSI_{final} = round\ (abs\ (\frac{RSSI}{10}))$ to normalize the final RSSI value. Since the range of RSSI is from −26 to −62, we need the final value which lies in the first quadrant and is near to the origin.

From Eq. 1, the path selection system chooses the lowest path cost value for a routing path between two RSUs to support end-to-end vehicle communication with small latency. Note that C_{ij}^{Pr} is affected by normalized values (α, β, γ) depending on importance. For example, if, γ is a very small value, the path cost is impacted by only latency, bandwidth, and the number of hops.

$$C_{ij}^{Pr} = \alpha(L_{ij}^{Pr} * B) + \beta(H_{ij}^{Pr}) + \gamma(RSSI_{final}) \quad (i.e.\, \alpha + \beta + \gamma \leq 1) \qquad (1)$$

4 Evaluation

4.1 Implementation

We used Mininet Wi-Fi [18] to set up VANET for our evaluation. We implemented our proposed system using Floodlight version 1.2 [19]. We also implemented a server application to collect data using SNMP. All the collected data are saved in MySQL database in the controller. The path selection system is implemented by a Python program with REST APIs to extract real-time traffic information. We created a VANET topology using five RSUs and four vehicles with random speed in the Mininet Wi-Fi. Figure 3 shows the topology of the five RSUs with a number. A connection between two RSUs is indicated by the solid line from p_1 to p_{10}. The final routing path to support V2V communication will be determined by the path selection system depending on the real-time network status of RSUs. The four vehicles keep moving with a random speed in Mininet Wi-Fi through other scripts by managing the four vehicles as a JSON format.

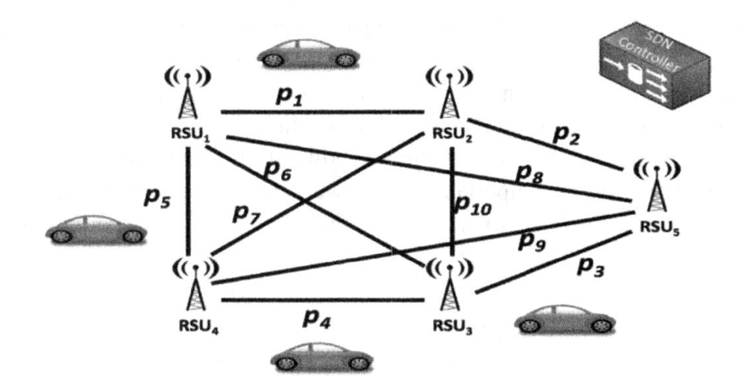

Fig. 3. Our experiment topology in Mininet Wi-Fi.

4.2 Experimental Results

We evaluate our design system in Mininet Wi-Fi with a given topology in Fig. 3. RSU_1 is a source and RSU_5 is a destination in Fig. 3. Between the two nodes (N_{15}, $i = 1$ and $j = 5$), there are ten difference paths for routing P_r: $P_1 = \{p_8\}$, $P_2 = \{p_1, p_2\}$, $P_3 = \{p_5, p_9\}$, $P_4 = \{p_6, p_3\}$, $P_5 = \{p_5, p_4, p_3\}$, $P_6 = \{p_5, p_7, p_2\}$, $P_7 = \{p_1, p_{10}, p_3\}$, $P_8 = \{p_6, p_{10}, p_2\}$, $P_9 = \{p_5, p_4, p_{10}, p_2\}$, and $P_{10} = \{p_5, p_4, p_{10}, p_7, p_9\}$. Each path P_r ($r = 10$) is computed by Eq. 1 in Sect. 3. We first test each parameter that affects to compute a path cost. Figure 4 shows the number of hops between the two nodes(N_{15}), RSU_1 and RSU_5. P_1 has only one hop for N_{15}, but P_{10} has five hops for N_{15}. Figure 4 also shows a signal quality value (RSSI) between an RSU and a vehicle depending on a distance. If the distance between an RSU and a vehicle is greater, the RSSI value decreases.

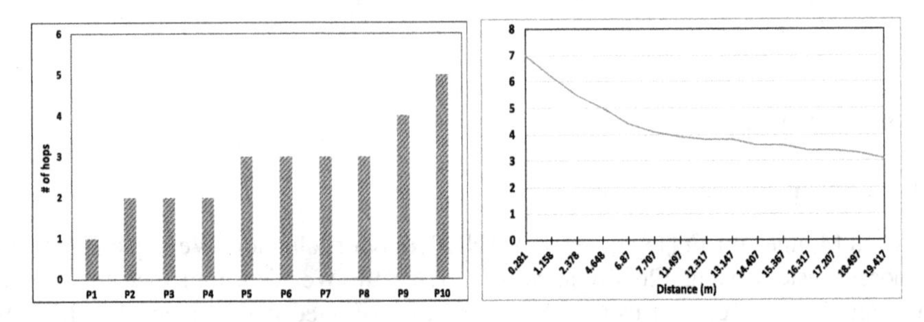

Fig. 4. The number of hops between two nodes (RSU_1 and RSU_5) and signal quality ($RSSI_{final}$) according to a different distance between an RSU and a vehicle.

Figure 5 shows the latency of each link and the average bandwidth of each link in the topology of Fig. 3. Generally, the latency increases as the number of hops increases. This result can be also affected by available bandwidth.

Figure 5 shows the latency of each link and the average bandwidth of each path in the topology of Fig. 3. Generally, the latency increases as the number of hops increases. This result can be also affected by available bandwidth. Figure 6 presents the result of the total path cost for each path P_r. In this experiment, we didn't consider the RSSI value by $\gamma = 0$. We test our path selection system for two difference cases. The first case (case 1) has $\alpha = 0.5$ and $\beta = 0.5$ and the second case (case 2) is set to $\alpha = 0.8$ and $\beta = 0.2$. As shown in Fig. 6, for both cases P_1 is selected as the best path between the two nodes RSU_1 and RSU_5. P_{10} has the highest path cost for both cases since it has the highest latency and the low bandwidth. We can adjust these parameters (α, β, γ) depending on the importance when we determine the best routing path. We evaluated each factor in Eq. 1 to analyze the path cost variation according to different settings.

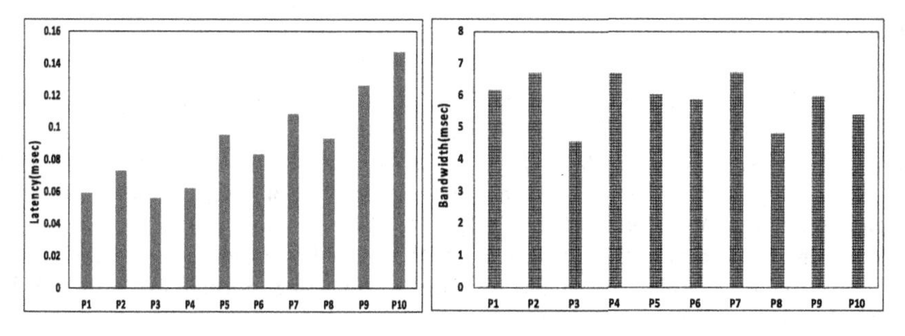

Fig. 5. The latency and the average bandwidth for each path P_r (r = 1, 2, 3, …, 10)

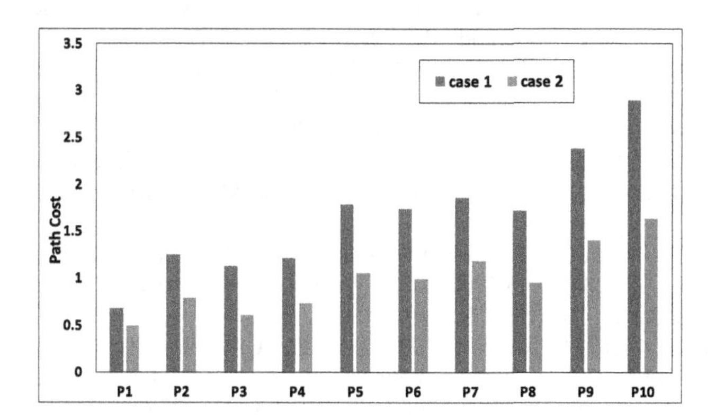

Fig. 6. The path cost for each path P_r (r = 1, 2, 3, …, 10)

5 Conclusion

Vehicular ad-hoc network (VANET) has gained a lot of interests for an intelligent transportation system for high mobility. VANET is autonomously created over a wireless network for data transfer among vehicles. However, the unstable wireless ad-hoc networks make VANET less flexible and efficient regarding high mobility and availability. This paper develops an SDN-based routing algorithm to provide the best path for reliable and efficient end-to-end vehicle communication through RSU (Roadside Units). The routing path is determined by the main features of VANET, such as latency, bandwidth, the number of hops, and RSSI (Received Signal Strength). The experimental results showed that SDN can control VANET and improve the utilization of VANET through reactive routing based on RSU.

Acknowledgement. This work is supported by NSF CNS #1637371.

References

1. Ali, G.N., Chong, P.H.J., Samantha, S.K., Chan, E.: Efficient data dissemination in cooperative multi-RSU vehicular ad hoc networks. J. Syst. Softw. **117**, 508–527 (2016)
2. Ku, I., You, L., Gerla, M., Ongaro, F., Gomes, R.L., Cerqueira, E.: Towards software-defined VANET: architecture and services. In: 13th Annual Mediterranean Ad Hoc Networking Workshop (MED-HOCNET) (2014)
3. Zongjian, H., Jiannong, C., Xuefeng, L.: SDVN: enabling rapid network innovation for heterogeneous vehicular communication. IEEE Netw. Mag. Spec. Issue Softw. Defined Wirel. Netw. (2015)
4. Duan, P., Peng, C., Zhu, Q., Shi, J., Cai, H.: Design and analysis of software defined vehicular cyber physical systems. In: 2014 20th IEEE International Conference on Parallel and Distributed Systems (ICPADS), December 2014
5. Abolhasan, M., Lipman, J., Ni, W., Hagelstein, B.: Software-defined wireless networking: centralized, distributed, or hybrid? IEEE Netw. **29**(4), 32–38 (2015)
6. Truong, N.B., Lee, G.M., Ghamri-Doudane, Y.: Software defined networking-based vehicular adhoc network with fog computing. In: 2015 IFIP/IEEE International Symposium on Integrated Network Management (IM), May 2015
7. Liyanage, M., Ylianttila, M., Gurtov, A.: Securing the control channel of software-defined mobile networks. In: World of Wireless, Mobile and Multimedia Networks (WoWMoM) (2014)
8. Anderson, J., Martin, J.: Towards a system for controlling client-server traffic in virtual worlds using SDN. In: 2013 12th Annual Workshop on Network and Systems Support for Games (NetGames), Denver, CO, USA (2013)
9. de Oliveira, R.L.S., Schweitzer, C.M., Shinoda, A.A., Prete, L.R.: Using mininet for emulation and prototyping software-defined networks. In: 2014 IEEE Colombian Conference on Communications and Computing (COLCOM), Bogota (2014)
10. ONF: Open networking foundation (2014). https://www.opennetworking.org/
11. Mininet: An Instant Virtual Network on your Laptop (or other PC) - Mininet. http://mininet.org/
12. Raghavan, B., Casado, M., Koponen, T., Ratnasamy, S., Ghodsi, A., Shenker, S.: Software-defined internet architecture: decoupling architecture from infrastructure. In: Proceedings of the 11th ACM Workshop on Hot Topics in Networks, Ser. HotNets-XI. ACM, New York (2012)
13. Mckeown, N.: How SDN will Shape Networking, October 2011. http://www.youtube.com/watch?v=c9-K5OqYgA
14. Kim, H., Feamster, N.: Improving network management with software defined networking. IEEE Commun. Mag. **51**(2), 114–119 (2013)
15. McKeown, N., Anderson, T., Balakrishnan, H., Parulkar, G., Peterson, L., Rexford, J., Shenker, S., Turner, J.: OpenFlow: enabling innovation in campus networks. SIGCOMM Comput. Commun. Rev. **38**(2), 69–74 (2008)
16. Haleplidis, E., Denazis, S., Pentikousis, K., Denazis, S., Salim, J.H., Meyer, D., Koufopavlou, O.: SDN layers and architecture terminology. Internet Draft, Internet Engineering Task Force, September 2014. http://www.ietf.org/id/draft-irtfsdnrg-layer-terminology-02.txt
17. Big Switch Networks: Project Floodlight (2013). http://www.projectfloodlight.org/
18. Mininet-WiFi: SDN emulator supports WiFi networks. http://mininet.org
19. Floodlight Project. http://www.projectfloodlight.org/floodlight

20. Erel, M., Teoman, E., Ozçevik, Y., Seçinti, G., Canberk, B.: Scalability analysis and flow admission control in mininet-based SDN environment. In: IEEE Conference on Network Function Virtualization and Software Defined Networks, pp. 18–19 (2015)
21. Fontes, R.D.R., Campolo, C., Rothenberg, C.E., Molinaro, A.: From theory to experimental evaluation: resource management in software-defined vehicular networks. IEEE Access **5**, 3069–3076 (2017)

Authentication Protocol Using Error Correcting Codes and Cyclic Redundancy Check

C. Pavan Kumar[1] and R. Selvakumar[2(✉)]

[1] School of Computer Science and Engineering (SCOPE), VIT University, Vellore 632014, Tamil Nadu, India
pavankumarc@ieee.org

[2] School of Advanced Science (SAS), VIT University, Vellore 632014, Tamil Nadu, India
rselvakumar@vit.ac.in

Abstract. Authenticating devices in communication system is an important and challenging task. With many diverse devices getting connected to communicate, establishing authentication of such devices among themselves (or with a central server) is essential to overcome possible attacks in the communication channel and by adversaries. In this paper, an authentication protocol is proposed based on linear error correcting codes, pseudo random numbers and cyclic redundancy check function. General protocol is provided in this paper and can be used for any specific linear error correcting codes defined over finite field. The proposed protocol is resistant against replay attack, man-in-the-middle and impersonation kind of attacks. One of the advantages of the proposed protocol is that it can be incorporated within the framework of any communication system that uses linear error correction system to achieve reliability or can be implemented independently to achieve security in terms of authentication.

Keywords: Authentication protocol · Error correcting codes
Cyclic redundancy check

1 Introduction

With diverse devices getting connected to internet and communicate with each other (or with central server), it is necessary to have authentication protocols that ensure that the messages received from senders are trustworthy and genuine, and are not altered from intruders of network or jammers or attackers present in the communication channel [1,2]. In an environment that is prone to attacks, it is very important to establish authenticity of devices in communication system. Shannon's influential paper on Mathematical Theory of Communication has immensely influenced research in many directions [3]. The problems of achieving

S. Chellappan et al. (Eds.): WASA 2018, LNCS 10874, pp. 874–882, 2018.
https://doi.org/10.1007/978-3-319-94268-1_80

reliability and security in digital communication paradigm are addressed separately in literature. To achieve reliability error correcting codes are used rather than redundantly transmitting the same packets until an acknowledgment message is received from receiver. Additional parity bits are added to the message to be transmitted so that errors if any will be corrected by employing suitable decoding techniques. Adding such parity bits are little computationally expensive but provides better throughput over transmitting redundant packets repeatedly which incurs additional overhead. Encoding E can be given as an injective map as follows [4]:

$$E : \Sigma^k \rightarrow \Sigma^n \tag{1}$$

where, message m of length k is encoded into a message of length n over an alphabet Σ. The additional bits r added to message of length k are called parity bits or redundancy bits and $n = k + r$. Similarly, decoding D can be given as a function as follows [4]:

$$D : \Sigma^{n+\eta} \rightarrow \Sigma^k \tag{2}$$

where, η is the noise added by the communication channel.

To achieve security, efficient cryptographic techniques are used which are broadly classified into public key cryptosystem and private key cryptosystem. In public key cryptosystem, sender will possess a public key using which he/she encrypts the message and transmits. User upon receiving the encrypted message uses his/her private key to decrypt the message. In private key cryptosystem, both sender and receiver will agree upon a mutual set of keys that act as key to encrypt and decrypt. Sender using his/her private key, encrypts the message and transmits whereas receiver will use his/her private key to decrypt the message.

Encryption Enc can be given as a mapping from actual message defined over an alphabet to another message defined over different alphabet as follows:

$$Enc : \Sigma^n \rightarrow \sigma^n \tag{3}$$

where, Σ and σ are the alphabets over which plain text and cipher text are defined respectively. Similarly, decryption Dec is given as a map as $Dec : \sigma^n \rightarrow \Sigma^n$.

In this paper, we propose a multi purpose light weight authentication protocol based on coding theory, i.e., by exploiting the error correction capability of the code. Maurya et al. [5] have used such coding theory based authentication protocol to authenticate Radio-Frequency Identification (RFID) tags with the RFID readers with the help of a trusted server. In our proposed protocol the necessity of such server assistance is removed. This makes the proposed protocol suitable for use in diverse communication scenarios such as Device to Device (D2D), Machine-to-Machine (M2M), Vehicle-to-Vehicle (V2V), Cognitive Radio Networks (CRNs) and Cyber Physical Systems (CPS) that uses error correction codes for achieving reliability. In these communication setup, i.e., Device to Device (D2D), Machine-to-Machine (M2M), Vehicle-to-Vehicle (V2V), both sender and receiver will have almost same capabilities (can be termed broadly as Ubiquitous computing [6]). The protocol is constructed using the techniques of coding theory especially linear error correcting codes that are defined over Field $\mathbb{F}_q^n$, pseudo random numbers and Cyclic Redundancy Check functions. General protocol is given in this paper and it can be used with any linear error correcting codes defined over $\mathbb{F}_q^n$. Here, the emphasis is on achieving authentication in communication systems exploiting error correction capability of the code employed to achieve reliability. Proposed method is also useful in achieving security in terms of authentication in communication systems that has only physical layer in the communication stack and not the upper layers [1,7] (in scenarios such as using TV spectrum as opportunistic cognitive radio [8,9]).

Present paper is organized as follows: In Sect. 2 related works and attack models are discussed. Section 3 discusses the proposed authentication protocol. Analysis of the proposed protocol is made in Sect. 4. Section 5 deals with conclusions and future work.

2 Related Works and Attack Models

2.1 Related Works

To overcome these challenges coding theory and cryptography techniques can be efficiently combined. The idea of employing error correcting codes for authenticating messages were introduced by Gilbert et al. [10]. The mathematical formulation of such schemes and a survey on construction of unconditionally secure authentication schemes from error correcting codes was given by Simmons [11,12]. There after many authentication schemes were proposed in the literature [13,14] to achieve security. Kacewicz [15] has analyzed few error correcting codes that are suitable to achieve reliability as well as security in the context of wireless communication systems.

Tsimbalo et al. [16–18] have exploited the Cyclic Redundancy Check (CRC) function bits of Bluetooth Low Energy (BLE) and IEEE 802.15.4 standards meant for resource constrained IoT devices intended to detect errors (not for correcting, as error correction or high level encoding involves processing overhead for the resource constrained devices), and proposed Forward Error Correction (FEC) over CRC which was purposefully disabled in those two protocols for saving energy at sender side. Tsimbalo et al. [16] proposed forward error correcting codes over such CRC redundant bits and checked the performance of such codes in communication paradigms where receiver can process received codewords from constrained devices and employ decoding to correct errors instead of discarding received packets and requesting sender to resend discarded packets.

Ez-zazi et al. [19] have proposed coding based reliable communication scheme for constrained IoT devices. The scheme uses Low Density Parity Check codes and CRC to achieve reliability. Sender will encode data sensed using LDPC codes and CRC of encoded data is computed, such encoding and CRC computation is termed as joint FEC/CRC by authors. Such encoded data is transmitted to the next hop or base station, upon receiving the joint FEC/CRC encoded package, first CRC will be checked to find if there are any errors in the received packets, if not it will be transmitted further. If any errors are detected by CRC, then the received packet will be decoded using belief propagation technique of decoding LDPC codes and errors will be corrected. If such packet's error correction is performed at intermediate hops, then the message will be further encoded using LDPC scheme and further CRC is computed before transmitting it further.

Alabady et al. [20] have proposed novel Low Complexity Parity Check (LCPC) codes for the resource constrained IoT devices. The proposed scheme encodes the data using LCPC codes at sender and uses three stage decoding algorithm to correct up to two bit errors if any and decode the message. This scheme corrects only one bit and two bit errors and beyond that discards the received packet and request the sender to retransmit the discarded packet.

2.2 Attack Models

The common attacks both cryptographic and coding theory methods used to achieve reliability and security subjected to are - eavesdropping, Denial-of-Service attacks, intrusion attacks, man-in-the-middle attacks, replay kind of attacks. By combining or selectively using techniques from coding theory as well as cryptography, attacks such as intruder, Man-in-the-Middle, replay and Denial-of-Service can be effectively mitigated.

Intrusion Attack. In intrusion attack, the adversaries will join the network or the group of existing communication entities posing as legitimate user and then use the network resources or information similar to legitimate users or sometimes even dominating legitimate users. Mitigating or overcoming or detecting such intrusion is necessary to make network resources available only for legitimate users.

Man-in-the-Middle Attack. Communication between sender and receiver is through a channel that is subjected to noise and attacks. Additive White Gaussian Noise (AWGN) is considered in the channel that changes bits and the channel is assumed to be insecure. Any intruder posing as legitimate user can send messages to receiver and thwart communication. Identifying the source of received message at receiver is an important task to mitigate such attack.

Replay Attack. The communication between sender and receiver can be copied (copying session keys) and used at later instances. Such attacks should be avoided in real time communication systems so that entities will not function adversely.

Denial-of-Service Attack. Denial-of-Service attack will make legitimate users deprive of services. One instance of it is to use powerful signal and entirely change the routing of packets in the network to other unintended nodes than to legitimate receiver present in the network. Also, making the receiver deprive of legitimate data over network is an instance of Denial-of-Service attack.

3 Authentication Protocol

The proposed protocol works in two phases, namely, initialization phase and execution phase. Initialization phase involves setting up of the environment necessary for execution of the protocol. Notations used in the protocol are given in Table 1.

Table 1. Notations

Notation	Description
K	Shared secret key
$\mathbb{C}$	Linear error correcting code over $\mathbb{F}_q^n$
c_i	Codeword $c_i \in \mathbb{C}$
CRC	Cyclic redundancy check sum
R_s	Random number generated at sender
R_r	Random number generated at receiver
$\|\|$	Concatenation operation
$\oplus$	XOR operation

3.1 Assumptions

The following assumptions are made in defining the protocol.

1. $[n, k, d]_q$ error correcting code is assumed, where n is the length of the codeword, k is the information to be encoded, d is the minimum Hamming distance and q is the alphabet over which code is defined [21]. The error correcting code $\mathbb{C}$ is assumed to have 2^k distinct codewords.
2. It is assumed that the random number generated at the sender R_s is such that $wt(Rs) \leq t$, where t is the error correction capacity of the code. It is to ensure that the errors that occur in this can only be corrected (Hamming bound).
3. The length of pseudo random numbers R_s and R_r are assumed to be n, which is same as the length of the codeword $c \in \mathbb{C}$.

3.2 Initialization Phase

In the initialization phase, both sender and receiver will compute functions that are necessary for the working of the protocol. Both sender and receiver will share a secret key K, also called as shared secret key. Both sender and receiver will have respective pseudo random number generators that can generate pseudo random numbers indicated by R_s and R_r respectively. But the pseudo random generator at the sender can generate random number which satisfies the condition that $wt(R_s) \leq t$. Both sender and receiver will have their respective Cyclic Redundancy Check (CRC) functions that computes the CRC sums for the inputs provided. The encoder function of sender will encode message m of length k into codeword c of length n.

3.3 Execution Phase

Both sender and receiver will compute S_s and S_r respectively by XOR-ing random number generated with the shared secret key K. $S_s = R_s \oplus K$ and $S_r = R_r \oplus K$. Both sender and receiver will exchange S_s and S_r. At receiver, random number of sender R_s will be computed by XOR-ing the received S_s values with the shared secret key K, i.e., $R_s = S_s \oplus K$. Similarly, sender will compute $R_r = S_r \oplus K$ to get R_r. Thus, both sender and receiver will get R_r and R_s respectively without being explicitly sharing it.

Sender will concatenate the codeword c_i to be transmitted with the random number generated by both sender and receiver and compute CRC sum of the concatenated message, i.e., $CRC(c_i||R_s||R_r)$. Also, the sender will compute $c_i \oplus R_s$. Sender will transmit both $CRC(c_i||R_s||R_r)$ and $c_i \oplus R_s$ to the receiver. Let $E_1 = CRC(c_i||R_s||R_r)$ and $E_2 = c_i \oplus R_s$. E_2 will be similar to codeword.

Receiver will receive messages E_1 and E_2 from the sender. It will decode the received message E_2 employing Maximum Likelihood decoding as $R_s \leq t$ and produce codeword c_i'. Further, the receiver will compute E_3 such that $E_3 = CRC(c_i'||R_s||R_r)$. If E_1 computed at sender is equivalent to E_3 computed at receiver, i.e., $E_1 = E_3$ then the communication is authentic and it is sent from genuine sender, if not, i.e., $E_1 \neq E_3$ then message is being received from other sources than the intended sender. E_1 can be shared with the receiver as private key similar to that of Private Key Cryptosystem. Overview of the proposed Authentication protocol is given in Fig. 1.

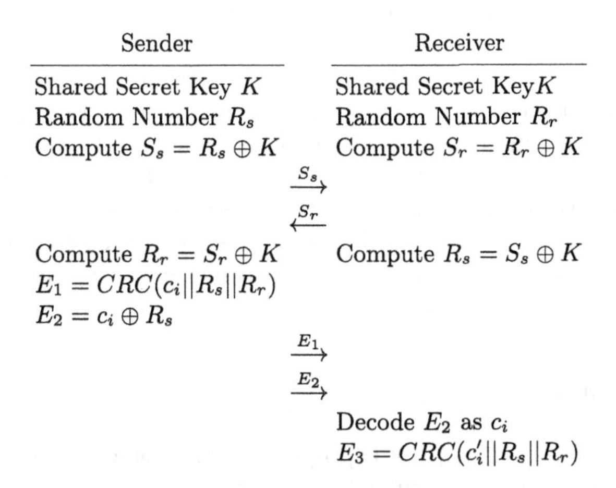

Fig. 1. Proposed authentication protocol

4 Analysis of Proposed Authentication Protocol

4.1 Replay Attack

Adversary or intruder of the channel tries to send the same session details in a future instance to communicate with the receiver. But in the proposed protocol even if shared secret key K is known to the intruder, it is difficult to guess the random numbers R_r and R_s generated at receiver and sender respectively, as they will be generated for each session. CRC computed at both sender as well as at receiver further makes it difficult even if intruder knows either R_s or R_r as they will be freshly generated.

4.2 Impersonation Attack

If intruder stores E_2 and tries to use it in future instance proposed protocol will reject it as values of R_r and R_s will be generated for sessions or as and when required. Further, CRC will be computed at both sender and receiver. That makes it difficult for intruders to impersonate the sessions. Even though E_2 look like codeword it will be XOR-ed with R_s thus impersonating codeword is also difficult.

4.3 Man-in-the-Middle Attack

Intruder in the channel can alter the message E_2 in the channel and transmit. But due to the nature of generating S_s, S_r, E_2 and E_1 it will be easy to detect unauthenticated transmissions received and reject them.

4.4 Computation Cost

Three $\oplus -$ operations are performed at sender to compute S_s, R_r and E_1. Similarly, two $\oplus -$ operations are performed at receiver to compute S_r and R_s respectively. One time CRC operation is performed at sender as well as receiver. Both sender and receiver will perform two $||$ operation. If computation time to perform $\oplus$ operation is indicated by $T_{\oplus}$, CRC operation by T_{CRC} and $||$ operation by $T_{||}$, then computation time taken at the sender to implement the proposed protocol is $3T_{\oplus} + 2T_{||} + 1T_{CRC}$. This computation is in addition to the computation time required at the sender to compute codeword $c \in \mathbb{C}$. Similarly at the receiver it takes a total of $2T_{\oplus} + 2T_{||} + 1T_{CRC}$ to implement the proposed protocol in addition to the cost involved at receiver to decode the received codeword c' using Maximum Likelihood Decoder.

5 Conclusion

A simple authentication protocol is proposed in this paper based on linear error correcting codes, pseudo random number generators and CRC function. The proposed protocol provides resistance against replay attack, impersonation attack and man-in-the-middle kind of attacks. The protocol can also be employed in any communication setups that uses linear error correcting codes (to achieve reliability) as discussed in paper to achieve security in terms of authentication. Further, it is interesting to incorporate this protocol in real time systems and analyze its performance.

References

1. Harrison, W.K., Almeida, J., Bloch, M.R., McLaughlin, S.W., Barros, J.: Coding for secrecy: an overview of error-control coding techniques for physical-layer security. IEEE Signal Process. Mag. **30**(5), 41–50 (2013)
2. Mukherjee, A., Fakoorian, S.A.A., Huang, J., Swindlehurst, A.L.: Principles of physical layer security in multiuser wireless networks: a survey. IEEE Commun. Surv. Tutor. **16**(3), 1550–1573 (2014)
3. Shannon, C.E.: A mathematical theory of communication. Bell Syst. Tech. J. **27**(3), 379–423 (1948)
4. Sudan, M.: Coding theory: tutorial & survey. In: Proceedings of the 42nd IEEE Symposium on Foundations of Computer Science, pp. 36–53. IEEE (2001)
5. Maurya, P.K., Pal, J., Bagchi, S.: A coding theory based ultralightweight RFID authentication protocol with CRC. Wirel. Pers. Commun. **97**(1), 967–976 (2017)
6. Friedewald, M., Raabe, O.: Ubiquitous computing: an overview of technology impacts. Telemat. Inform. **28**(2), 55–65 (2011)
7. Liu, Y., Chen, H.H., Wang, L.: Physical layer security for next generation wireless networks: theories, technologies, and challenges. IEEE Commun. Surv. Tutor. **19**(1), 347–376 (2017)
8. Nekovee, M.: Cognitive radio access to TV white spaces: spectrum opportunities, commercial applications and remaining technology challenges. In: 2010 IEEE Symposium on New Frontiers in Dynamic Spectrum, pp. 1–10. IEEE (2010)
9. Rempe, D., Snyder, M., Pracht, A., Schwarz, A., Nguyen, T., Vostrez, M., Zhao, Z., Vuran, M.C.: A cognitive radio TV prototype for effective TV spectrum sharing. In: 2017 IEEE International Symposium on Dynamic Spectrum Access Networks, DySPAN, pp. 1–2. IEEE (2017)
10. Gilbert, E.N., MacWilliams, F.J., Sloane, N.J.: Codes which detect deception. Bell Labs Tech. J. **53**(3), 405–424 (1974)
11. Simmons, G.J.: Authentication theory/coding theory. In: Blakley, G.R., Chaum, D. (eds.) CRYPTO 1984. LNCS, vol. 196, pp. 411–431. Springer, Heidelberg (1985). https://doi.org/10.1007/3-540-39568-7_32
12. Simmons, G.J.: A survey of information authentication. Proc. IEEE **76**(5), 603–620 (1988)
13. Moulin, P., Koetter, R.: Data-hiding codes. Proc. IEEE **93**(12), 2083–2126 (2005)
14. Schillewaert, J., Thas, K.: Construction and comparison of authentication codes. SIAM J. Discret. Math. **28**(1), 474–489 (2014)
15. Kacewicz, A.: Coding Theory for Security and Reliability in Wireless Networks. Cornell University, Ithaca (2010)
16. Tsimbalo, E., Fafoutis, X., Piechocki, R.J.: CRC error correction in IoT applications. IEEE Trans. Ind. Inf. **13**(1), 361–369 (2017)
17. Tsimbalo, E., Fafoutis, X., Piechocki, R.: Fix it, don't bin it!-CRC error correction in Bluetooth low energy. In: 2015 IEEE 2nd World Forum on Internet of Things, WF-IoT, pp. 286–290. IEEE (2015)

18. Tsimbalo, E., Fafoutis, X., Piechocki, R.J.: CRC error correction for energy-constrained transmission. In: 2015 IEEE 26th Annual International Symposium on Personal, Indoor, and Mobile Radio Communications, PIMRC, pp. 430–434. IEEE (2015)
19. Ez-zazi, I., Arioua, M., El Oualkadi, A., El Assari, Y.: Joint FEC/CRC coding scheme for energy constrained IoT devices. In: Proceedings of the International Conference on Future Networks and Distributed Systems, p. 18. ACM (2017)
20. Alabady, S.A., Salleh, M.F.M., Al-Turjman, F.: LCPC error correction code for IoT applications. Sustain. Cities Soc. (2018). https://doi.org/10.1016/j.scs.2018.01.036
21. Moon, T.K.: Error Correction Coding: Mathematical Methods and Algorithms. Wiley, Hoboken (2005)

Understanding Data Breach: A Visualization Aspect

Liyuan Liu, Meng Han(✉), Yan Wang, and Yiyun Zhou

Kennesaw State University, Kennesaw, GA, USA
{lliyuan,ywang63,yzhou20}@students.kennesaw.edu, menghan@kennesaw.edu

Abstract. Data breaches happen daily, in too many places, result in data loss including personal, health, financial information that are crucial, sensitive, and private. The cost of the data breach is not only considered as potentially damaging to the monetary penalty, but also regarding other more severe problems such as consumer confidence, social trust and personal safety. In this paper, we brush up the world's most significant data breaches in which amount of lost records are more than 30,000 records from 2004 to 2017. From many different aspects, the data visualization technique is used to demonstrate the fact and the hidden information of the data breach phenomenon. We also employ a case study which includes the income data of the residents in the United States and the public transportation data in New York to point out the potential risk of the data breach. Based on the case study, we once again exhibit the real dangers of data breach are out of hands. Based on the analysis and visualization, we find some interesting insights which seldom researchers focus on before and it is apparently the real dangers of data breach are beyond the common imagination.

Keywords: Data breach · Data visualization · Danger of data breach

1 Introduction

Data breaches may happen every day, in numerous places at once to keep count. In the United States, the number of data breaches is 8,741 and the number of records exposed is 1,069,914,088 from January 2005 to March 2018 based on Identity Theft Resource Center (ITRC) report [1]. Recently, Facebook is facing the heat after Cambridge Analytica (a British consulting company) and it was accused of harvesting data of up to 50 million Facebook users without permission [2]. This data breach hurts facebook's stock price precariously. Since the New York Times first published a story over the weekend detailing this breach, Facebook's worth as a company has fallen by $61 billion to $476.4 billion [3]. Investors are gradually decreasing their trust and confidence in Facebook and social media companies.

Actually, since 2005, data breaches have more than tripled as advances in technology and they have made the collection and sharing of information easier and more efficient. According to the Identity Theft Resource Center's report [1],

S. Chellappan et al. (Eds.): WASA 2018, LNCS 10874, pp. 883–892, 2018.
https://doi.org/10.1007/978-3-319-94268-1_81

the number of breaches is 1,500 in 2017, which has increased by 37% compared to 2016. The biggest data breach record in 2017 happened in Equifax. As one of the major credit reporting agencies, Equifax revealed that cybercriminals had penetrated their network. The breach bared the data of 143 million Americans' privacy information [4].

With the volume of data growing exponentially year after year in the world nowadays, cybercriminals have received a more significant opportunity to expose vast volumes of data in a single breach. Based on the 12th annual Cost of Data Breach Study sponsored by IBM in 2017, the global average cost of a data breach is $3.62 million [5]. In fact, the cost of the data breach is not only considered as potentially damaging to the monetary penalty, but also regarding other more severe problems such as consumer confidence, social trust and personal safety. Therefore, data breach is a hot and valuable topic in the research area. Baker *et al.* published the Data Breach Investigations Report in 2011 and analyzed data breach such as threat agencies, threat action and other aspects [6]. Sen and Borle used an empirical approach to estimate the Contextual Risk of Data Breach. They employed the opportunity theory of crime, the institutional anomie theory, and institutional theory to recognize factors that could increase or decrease the contextual risk of data breach [7,8]. Quick *et al.* collected the world's biggest data breaches that lost more than 30,000 records as a public dataset [9]. Later, Tayan used this dataset to summarize the worldwide impact of security of breaches on various it-based industries in his research [9–11]. With the increasing of the severity in the cost of data breach, some researchers are paying more attention to the dangers of data breach. Delgado *et al.* described the dangers of data breaches that the companies are facing [12]. Gatzlaff and McCullough found that data breaches could not only lose the personal information of customers or employees but also may affect the shareholder wealth of the company [13]. There are some researchers employed the case study in their data breach related researches. Acquisto *et al.* employed an event study that is beyond the data breach and found that there exists a negative and statistically significant effect of data breaches on a company's market value on the announcement day for the breach [14].

However, data breach is still happening every day. The data breach related datasets are hard to collect and most of them are too messy to analysis. Surprisingly, even though more and more researchers focus on data breach, most of them ignore to dig in why to originated data breach and what's the characteristics of each data breach record. Even though there are many data breaches incident reports published [4,15,16], few researchers focus on analysis and conclude the other unusual pattern of the data breach from the organization, data breach types *et al.* aspects. Besides only making conclusions on the data breach incidents, we find other interesting findings that few researchers focused before. In addition, many researchers are neglecting the dangers of data breach. In fact, the dangers of data breach are not only cost the monetary problems that most research focus on, but they will also cause more severe issues such as social trust and personal safety that beyond our imagination. Consequently, our contributions are concluded as follows:

- We are the first to re-organize and analyze the details of the biggest data breach records in the world from 2004 to 2017 that use visualization method to our best knowledge.
- Based on the data visualization, we find some novel and interesting insights. We point out some commonalities and differences of the data breach records to draw on the experience of other researchers further study.
- We employ a case study, using two independent datasets [17,18]. After the phases of data preprocessing and implementing deep learning neural network algorithm, we are shocked by the fact that we can locate people's address, track their moving activity route and predict their income. More surprisingly, we can use two independent datasets by matching them together to find some critical and private information. This kind of potential data breach dangers can appear everywhere. Behind the data breach, there are more dangers cost beyond our vision.

It is worth to mention that we are not using the leaky data. Instead, we only collect and analyze two datasets that seem individualistic and normal. It is apparent that if it is a real data breach, it will cause more severe problems.

2 Related Work

There has been abundant literature related to data breach. Baker *et al.* published the Data Breach Investigations Report in 2011. In their research, they concluded 2010 Threat Event Overview and analyzed the threat agents, threat actions and other aspects [6]. Sen and Borle used an empirical approach to estimate the Contextual Risk of Data Breach. They employed the opportunity theory of crime, the institutional anomie theory, and institutional theory to recognize factors that could increase or decrease the contextual risk of data breach [7,19,20]. Harris *et al.* published California data breach report in 2016. In the report, they present their findings on the nature of the breaches that are occurring and what can be learned from them about threats and vulnerabilities. They also make some reasonable recommendations aiming at reducing the risk of data breaches and decreasing the harms that are resulted from data breach [21]. Quick *et al.* collected the world's biggest data breaches that losses greater than 30,000 records as a public dataset and there are several researchers used this public dataset in their research [9]. Tayan used this dataset to summarize the worldwide impact of security of security breaches on various it-based industries in his research [10]. Cai and Zhang *et al.* did a lot of work such as developing a novel privacy preserving approach for smartphones [22–26].

With the data breach growing exponentially year after year in the world, there are more and more researchers paying attention to the cost of data breach. Ponemon Institute and IBM released the annual research report to demonstrate the average monetary cost of a data breach [5]. Delgado and Beranek described the dangers of data breaches in apartment companies facing [12]. They indicated that apartment companies face challenges in keeping personal information safe, and the information of apartment companies owned are very private including subject's name, address, social security number and driver's license number, as well as additional information found in leases, financial records, insurance records and other documentation [12]. Gatzlaff and McCullough found the effect of data breaches in not only losing the personal information of customers or employees but also affecting shareholder wealth of the company [13]. They employed a cross-sectional analysis of the cumulative abnormal returns and found that firm size and subsidiary status mitigate have negative effect of a data breach on the firm's stock price. There are some researchers employed the case study in their data breach related researches to find the hidden data breach cost [27]. Acquisto *et al.* employed an event study for the findings beyond the data breach [14]. Ayyagari and Tyks applied a case study in University to prove losing students personally identifiable information can cause a significant financial burden on the university as it was not prepared to handle an information security disaster [28,29].

3 Data Source and Data Visualization

In this paper, after data cleaning and data preprocessing, there are three datasets can be used:

- World's Biggest Data Breaches [9]: Including the data breach record that losses greater than 30,000 records from 2004 to 2017 all over the world.
- US Household Income Statistics [17]: Including Household and geographic statistics data and geographic location data. In this dataset, there are 8 variables and 32,526 observations.
- New York City Transport Statistics [18]: This dataset is about the public transportation stream service data in New York City. In this dataset, there are 9 variables and 984,914 observations.

Based on the data visualization, we find more profound insights of data breach records from many different aspects. Firstly, as Fig. 1 shows, web, government, financial are the top three types of organizations that lost the most records.

Secondly, there are 5 data breach methods including lost/stolen device or media, hacked, accidentally published, inside job, and poor security. Figure 3 shows the top organization types that lost the most records in each data leak method. For the method of lost/stolen device or media, government, healthcare and financial are the top three types. Web, financial, and retail are the top three types through hacked. Web and government mostly have data breach by accidentally published. inside job: Financial and web are the top two types through inside job. And web mostly lose information by poor security. In 2017, the number of records stolen is soaring than 2016. The number of records lost of government and web is increasing dramatically in recent years. Government's data breach is more specific and serious than others. The two primary data leak methods of government data leak are accidentally published and stolen device. Thirdly, as Fig. 2 displays, Yahoo is the company that has the most substantial numbers of data breach. Figure 2 displays the total numbers of lost records,

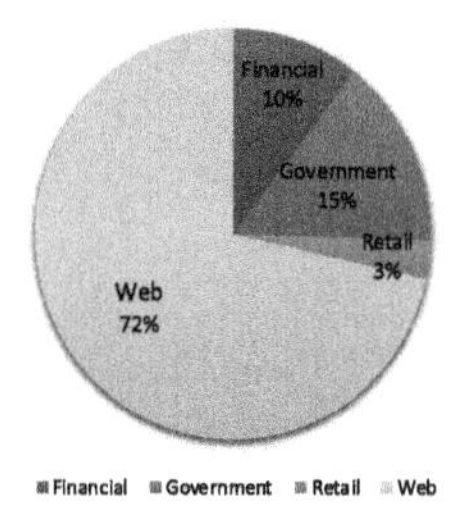

Fig. 1. The number of data lost by organization

Entity	Organization	Method Of Leak	Records Lost
Yahoo	web	hacked	1,532,000,000
River City Media	web	accidentally published	1,370,000,000
Aadhaar	government	accidentally published	1,000,000,000
Spambot	web	poor security	711,000,000
Friend Finder Network	web	hacked	412,000,000
Court Ventures	financial	inside job	200,000,000
Deep Root Analytics	web	poor security	198,000,000
MySpace	web	hacked	164,000,000
Massive American busine..	financial	hacked	160,000,000
Ebay	web	hacked	145,000,000

Fig. 2. The biggest data breach by entity

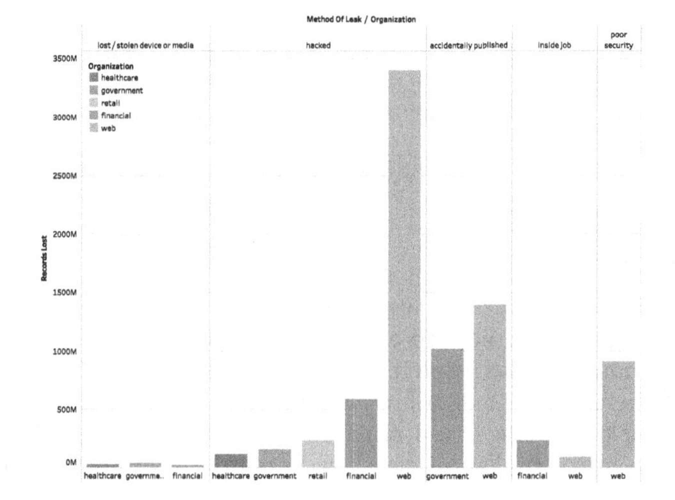

Fig. 3. The number of data lost by leak method and organization

method of leak and organization type of biggest data breaches by entity from 2004 to 2017. The primary method of these data breaches is hacked. Although accident published is not the frequent method of the data breach, the number of lost records is enormous in each record.

Figure 4 performs the numbers of lost records in each organization type by year. Since 2016, government and web organizations' data breaches are increasing quickly. The data breach of Healthcare and retail are decreasing in recent years. The most severe two methods of data leak are hacked and accidentally published. With respect to the trend, the data leak results from accidentally published and poor security are increasing rapidly in recent years. Figure 5 shows that legal, financial, government, military and healthcare are the top types of the organizations that own the most sensitive data. The data breach problem of financial companies and government are more severe than others since number of the data they lost are more than 10 times than healthcare companies and military.

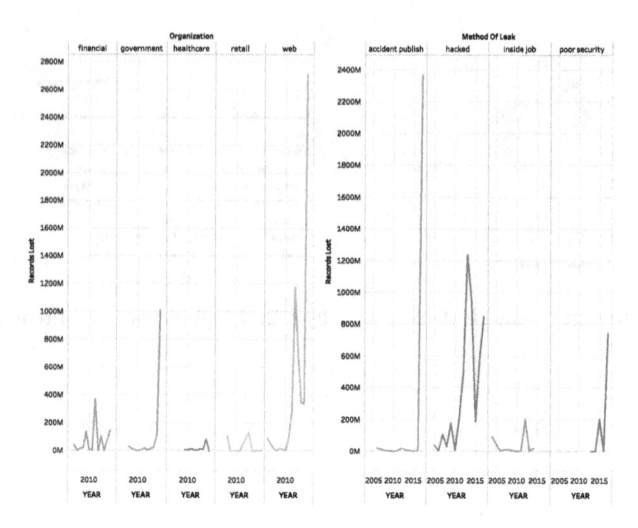

Fig. 4. The number of data lost by organization and method of leak in time series

4 Case Study

Now is the big data era. Big data gives unprecedented chances for business, innovation social engineering and other areas. However, it also brings new challenges for in-house counsel when it comes to security, privacy and what crosses people's life behavior. The data breach gives criminals weapons through which they can steal the messages and bring even more severe scams. Therefore, we employ a case study to demonstrate how harmful a data breach could be. We use two independent datasets. One is the U.S residents income census data and the other is the New York City public transportation data [17,18,30].

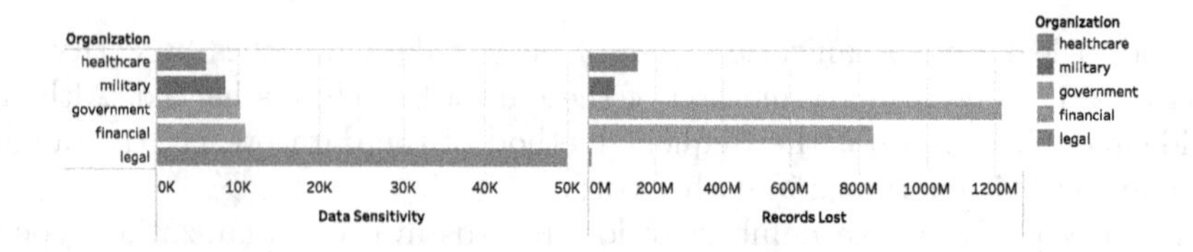

Fig. 5. Lost data sensitivity and number of lost records by organization

Apparently, these two datasets are individualistic at first glance. With this case study, we get the conclusion that the hidden dangers of data breaches are beyond our imagination.

Firstly, we find the hidden correlation between these two datasets. Based on the longitude and latitude information, we locate each residents' address from U.S income dataset and further select New York city residents. Figure 6 shows an example of the address location results in Manhattan area based on our data.

In Fig. 6, red dots represent people's location and blue dots represent the public transportations' station.

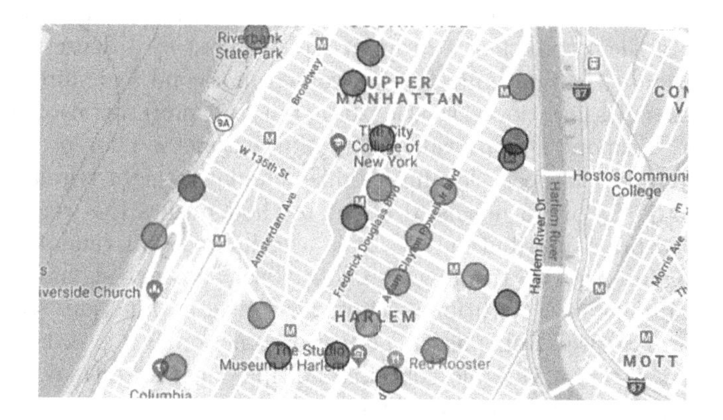

Fig. 6. Location and public transportation station in Manhattan area (Color figure online)

Secondly, we use the hidden relationship between the two datasets to track people's possible moving activity route and then find the possible transportation lines people potential to take. Based on the transportation lines information, it is easy to collect the arrival time of each station. Besides, we also know the income and other private information of those people. Figure 7 displays an example of our findings while combining two independent datasets.

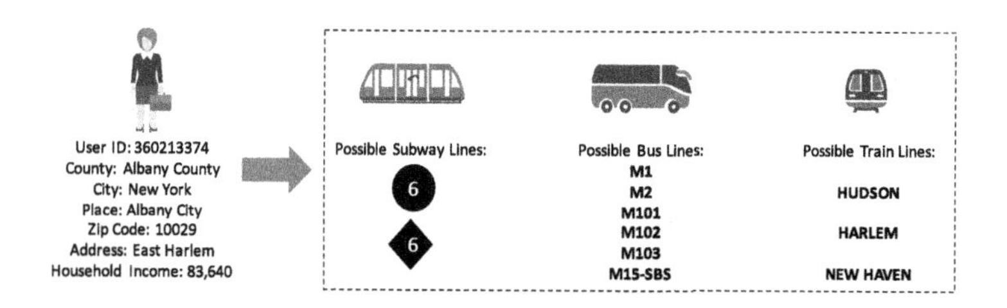

Fig. 7. Example of moving track

Thirdly, with the boosting growth of advanced technology such as artificial intelligence and deep learning, using developed models to predict unknown information is widespread. One of the principal advantages of Deep learning is that it can extrapolate new features from a limited set of features contained in a training set. In this case study, we implement deep neural network to predict residents' income to see how dangerous if criminals apply advanced technology to the leaky data. The definition and architecture of our deep neural network model are demonstrated as follows: At first, we discretize our target variable to three categories based on the distributions. 80% observations was randomly chosen as training data and the rest of 20% was used as testing data. We use a deep feed-forward neural network to classify the data with four layers. The first two layers include 128 node layers while the third layer includes 64 node layers. Both of them consist of a dropout and followed by a dense layer. A linear combination activation unit (ReLU) worked in second and third layers and multiple classes classification activation function (softmax) was used in the prediction layer. Dropout is a prevalent and efficient method for preventing the over-fitting problem in deep learning. Unlike standard weight regularizers such as based on the L1 or L2 norms, dropout method pushes the weights toward some expected prior distribution [31]. It is a technique that randomly selected neurons are ignored during training. ReLU has been shown to significantly speed up network training over traditional sigmoidal activation functions such as tanh [32]. Therefore, we used ReLU as the activation function in the second and third layers. Since the target variable contains multiple classes, softmax is a proper activation function to classify the numerous classes variables. Let the network layers as $L = \{1, 2, 3, 4\}$, $L = 1$ is the input layer and $L = 4$ is the output layer. $Y^{(L)}$ denotes the values of the output layer, $Y^{(L-1)}$ denotes the incoming values into the layer, $W^{(L)}$ is the weights of the layer that linearly transforms n input values into m output values, b^L denotes the bias, and $F^{(L)}$ denotes the activation vector function. For each layer when $L = 2, 3, 4$, the equation can be defined by:

$$D^{(L)} = Y^{(L-1)} * R^{(L)} \tag{1}$$

$$O^{(L)} = W^{(L)} D^{(L)} + b^{(L)} \tag{2}$$

$$Y^{(L)} = F(O^{(L)}) \tag{3}$$

When $L = \{2, 3\}$, the ReLU activation function is:

$$F(O_i{}^{(L)}) = (Y_1{}^{(L)}, \ldots, Y_i{}^{(L)}, \ldots, Y_m{}^{(L)}) \tag{4}$$

When L = 4, the softmax activation function is defined as:

$$Y^* = \frac{e^{O_j}}{\sum_{K=1}^{3} e^{O_3}}, j = 1, 2, 3 \tag{5}$$

It denotes the output of the model. The sum of the cross-entropy is used to evaluate the loss for each batch sample. Training using backpropagation and Adam optimizer is applied because they are significantly faster than the standard stochastic gradient descent [33]. With this method, we can estimate people's income even though we don't have the exact income information in the future.

This case study is a warning light that the hidden dangers of data breaches should not be underestimated. In the real world, leaky data are more private and sensitive than the data we used in this case.

5 Conclusion

In this paper, we re-organize and investigate the details of the biggest data breach records in the world from 2004 to 2017 from the visualization perspective which few researchers did previously. Based on the visualization analysis, we find some interesting insights that were ignored by most of the previous researchers. From the analysis of the case study, we once again exhibit the real dangers of data breach are out of hands. In this case study, we are not using the real leaky data. Instead, we only analyze two independent and public datasets. In fact, our findings can be referred by other researchers in the future studies. Data breach is a world-wide and noteworthy problem since its cost is not only considered as potentially damaging to the monetary penalty, but also regarding other more severe problems such as consumer confidence, social trust and personal safety. It deserves more attention from researchers and data holder organizations.

Due to the serious potential damage from the data breaches, there are a lot of efforts required in the future research. We plan to focus on applying deep learning algorithms to data breaches detection, since the deep learning method will be able to not only show stealthy data breach but also advance the detection accuracy as well as achieve timely protection.

References

1. Interstate Technology & Regulatory Council: ITRC sponsors and supporters (2018). https://www.idtheftcenter.org/data-breaches.html
2. Cadwalladr, C., Graham-Harrison, E.: Revealed: 50 million Facebook profiles harvested for Cambridge Analytica in major data breach (2018)
3. Shen, L.: Here's who is winning after Mark Zuckerberg's response to the Facebook data breach (2018). https://www.yahoo.com/news/winning-mark-zuckerberg-response-facebook-180549803.html
4. Armerding, T.: The 17 biggest data breaches of the 21st century (2018). https://www.csoonline.com/article/2130877/data-breach/the-biggest-data-breaches-of-the-21st-century.html
5. Ponemon Institute: 2017 Ponemon Cost of Data Breach Study (2018)
6. Baker, W., et al.: 2011 data breach investigations report. Verizon RISK Team, pp. 1–72 (2011)
7. Sen, R., Borle, S.: Estimating the contextual risk of data breach: an empirical approach. J. Manage. Inf. Syst. **32**(2), 314–341 (2015)
8. Li, W., et al.: Secure multi-unit sealed first-price auction mechanisms. Secur. Commun. Netw. **9**(16), 3833–3843 (2016). SI on cyber security, crime, and forensics of wireless networks and applications
9. Quick, M., et al.: World's biggest data breaches. Informationisbeautiful **16** (2017). http://www.informationisbeautiful.net/visualizations/worlds-biggest-data-breaches-hacks/
10. Tayan, O.: Concepts and tools for protecting sensitive data in the it industry: a review of trends, challenges and mechanisms for data-protection. Int. J. Adv. Comput. Sci. Appl. **8**(2) (2017)
11. Han, M., et al.: Privacy reserved influence maximization in GPS-enabled cyber-physicaland online social networks. In: 2016 IEEE International Conferences on Social Computing and Networking (SocialCom), pp. 284–292. IEEE (2016)

12. Delgado, J.M., Beranek, A.J.: The dangers of data breach. Multi Hous. News **47**(4), 42 (2012)
13. Gatzlaff, K.M., McCullough, K.A.: The effect of data breaches on shareholder wealth. Risk Manage. Insur. Rev. **13**(1), 61–83 (2010)
14. Acquisti, A., Friedman, A., Telang, R.: Is there a cost to privacy breaches? An event study. In: ICIS 2006 Proceedings, p. 94 (2006)
15. Edwards, B., Hofmeyr, S., Forrest, S.: Hype and heavy tails: a closer look at data breaches. J. Cybersecur. **2**(1), 3–14 (2016)
16. Hu, C., et al.: Privacy-preserving combinatorial auction without an auctioneer. EURASIP J. Wirel. Commun. Netw. **38**, 1–8 (2018)
17. Golden Oak Research Group Ltd.: U.S. Income Database Kaggle (2017). https://www.kaggle.com/goldenoakresearch/us-household-income-stats-geo-locations
18. MichaelStone: New York City Transport Statistics (2017)
19. Han, M., et al.: Cognitive approach for location privacy protection. IEEE Access **6**, 13466–13477 (2018)
20. Sooksatra, K., et al.: Solving data trading dilemma with asymmetric incomplete information using zero-determinant strategy. In: International Conference on Wireless Algorithms, Systems and Applications (2018)
21. Harris, K.D., General, A.: California data breach report, 7 Aug 2016
22. Zhang, L., Cai, Z., Wang, X.: FakeMask: a novel privacy preserving approach for smartphones. IEEE Trans. Netw. Serv. Manage. **13**(2), 335–348 (2016)
23. Cai, Z., et al.: Collective data-sanitization for preventing sensitive information inference attacks in social networks. IEEE Trans. Dependable Secure Comput., 1 (2016). https://doi.org/10.1109/TDSC.2016.2613521
24. Cai, Z., Zheng, X.: A private and efficient mechanism for data uploading in smart cyber-physical systems. IEEE Trans. Netw. Sci. Eng., 1 (2018). https://doi.org/10.1109/TNSE.2018.2830307
25. Liang, Y., Cai, Z., Yu, J., Han, Q., Li, Y.: Deep learning based inference of private information using embedded sensors in smart devices. IEEE Netw. Mag. (2018)
26. Zheng, X., Cai, Z., Li, Y.: Data linkage in smart IoT systems: a consideration from privacy perspective. IEEE Commun. Mag. (2018)
27. Han, M., et al.: Mining public business knowledge: a case study in SEC's EDGAR. In: 2016 IEEE International Conferences on Social Computing and Networking (SocialCom), pp. 393–400. IEEE (2016)
28. Ayyagari, R., Tyks, J.: Disaster at a university: a case study in information security. J. Inf. Technol. Educ. **11**, 85–96 (2012)
29. Li, J., Cai, Z., Wang, J., Han, M., Li, Y.: Truthful incentive mechanisms for geographical position conflicting mobile crowdsensing systems. IEEE Trans. Comput. Soc. Syst. **5**, 324–334 (2018). https://doi.org/10.1109/TCSS.2018.2797225
30. Han, M., Duan, Z., Li, Y.: Privacy issues for transportation cyber physical systems. In: Sun, Y., Song, H. (eds.) Secure and Trustworthy Transportation Cyber-Physical Systems. SCS, pp. 67–86. Springer, Singapore (2017). https://doi.org/10.1007/978-981-10-3892-1_4
31. Franklin, J.: The elements of statistical learning: data mining, inference and prediction. Math. Intell. **27**(2), 83–85 (2005)
32. Krizhevsky, A., Sutskever, I., Hinton, G.E.: ImageNet classification with deep convolutional neural networks. In: Advances in neural information processing systems, pp. 1097–1105 (2012)
33. Kinga, D., Adam, J.B.: A method for stochastic optimization. In: International Conference on Learning Representations (ICLR) (2015)

A Novel Recommendation-Based Trust Inference Model for MANETs

Hui Xia(✉), Benxia Li, Sanshun Zhang, Shiwen Wang, and Xiangguo Cheng

College of Computer Science and Technology, Qingdao University, Qingdao 266071, People's Republic of China
xiahui@qdu.edu.cn

Abstract. Over the last few years, trust, security, and privacy in mobile ad hoc networks have received increasing attention. The proposed trust-based countermeasures are considered to be promising approaches, which play an important role for reliable data transmission, qualified services with context-awareness, and information security. The foundation of these countermeasures is trust computation. In order to address this issue, we first study trust properties, and subsequently abstract a novel recommendation-based trust inference model. Two trust attributes called the subjective trust and the recommendation trust, are selected to quantify the trust level of a specific entity. Recommendations provide an effective way to build trust relationship, by making use of the information from others rather than exclusively relying on one's own direct observation. To compute the subjective trust and the recommendation trust precisely, some comprehensive factors are introduced. Furthermore, the concept of belief factor is proposed to integrate these two trust attributes. The aim of this trust model is, the network can itself detect, prevent and exclude the misbehaving entities, and obtain strong resistibility to malicious attacks as well. The effectiveness and resistibility of the model are analyzed theoretically and evaluated experimentally. The experimental results show that this new mechanism outperforms existing mechanisms.

Keywords: Mobile ad hoc networks · Trust-based countermeasure
Trust computation · Trust inference model · Recommendation trust
Belief factor

1 Introduction

A mobile ad hoc network can be seen as an implementation of the perception layer, which is at the most front-end of information collection and plays a fundamental role in the internet of things (IoT). This type of network can offer a high quality of service, and a high level of flexibility for data perception, collection, transmission, process, analysis and utilization. The recent proliferation of mobile entities (e.g. mobile phones) has given rise to numerous applications. Various data collected by underlying mobile entities can be further processed, mined and analyzed for the sake of multifarious promising services with intelligence. As we become increasingly reliant on intelligent and interconnected devices in every aspect of our lives, critical security issues involved

S. Chellappan et al. (Eds.): WASA 2018, LNCS 10874, pp. 893–906, 2018.
https://doi.org/10.1007/978-3-319-94268-1_82

in this type of wireless network are raised as well, and remain serious impediments to widespread adoption. In recent years, trust, security and privacy in mobile ad hoc networks have received increasing attention [1]. The existence of malicious entities can make a seriously damage on the availability and correctness of network services. In order to address the abovementioned problem, several security extensions and detection systems have been proposed in the literature to counter various types of malicious attacks [2, 3].

Cryptography-based solutions (PKI) is based on a centralized server [3], while have clear deficiencies: (*a*) these solutions cannot recognize the malicious entities since they all have been authorized identities, thus are vulnerable to suffer from DoS/DDoS attacks, internal attacks, privacy breach and impersonation attacks; (*b*) most cryptographic operations consume a significant amount of bandwidth, and cause a high network overhead to be incurred, and are therefore not suitable for resource-constrained networks. In contrast, trust-based countermeasures are considered to be more acceptable as promising approaches [4, 5], which play an important role for reliable data transmission and fusion, qualified services with context-awareness and information security. For instance, these countermeasures have been widely used to neutralize packet dropping attack, exclude misbehaving entities from the network and guarantee security interactions between entities.

The trust computation is the foundation of those countermeasures, which plays an important role to initialize a trusted network system (e.g. the network can itself detect, prevent and exclude the misbehaving nodes [6, 7], and obtain strong resistibility to attacks as well [8, 9]). And the feedback of trust information can be applied to traditional network services with unique security assurances at different levels [10]. The trust mechanism can help entities overcome perceptions of uncertainty and risk in consumption on network services and security applications (e.g. routing function) [11, 12]. However, the computational model is often very complicated. And the selection rules of the trust decision attributes and the calculation methods of weights have not been solved effectively. Moreover, most of these proposed models have poorly capability to resist various types of attacks. The basis of some countermeasures may be exploited to fulfill new attacks. As mentioned above, how to design an effective and efficient trust-based framework is a challenging task in such networks.

We therefore focus in this paper on the design of trust inference model. We first study trust properties, and subsequently propose a novel recommendation-based trust inference model to achieve the trust computing problem, where two trust attributes called the subjective trust and the recommendation trust, are selected to quantify the trust level of a mobile entity. The trust information can be used to classify entities as honesty or malevolence.

The remainder of this paper is organized as follows. Section 2 discusses recent works in the literature. In Sect. 3, we describe in detail a trust inference model. Section 4 presents the experimental results and analysis of the performance of this new trust model. Finally, Sect. 5 presents concluding remarks with possible extensions and directions for future research.

2 Related Works

A detailed survey of various trust computing approaches was presented in [4]. According to this survey, distributed trust computations can be classified as neighbor sensing, recommendation-based trust, and hybrid methods (based on direct experience and recommendations from other nodes). Movahedi et al. [5] presented a holistic view on various trust management frameworks geared for MANETs. Besides, they proposed taxonomy for the main identified trust-distortion attacks and they provided a holistic classification of the main evaluation metrics.

Shen and Li [6] presented a hierarchical account-aided reputation management system to effectively provide cooperation incentives. A hierarchical locality-aware distributed hash table infrastructure is employed to globally collect all node reputation information in the system, which is used to calculate more accurate reputations and to detect abnormal reputation information. To complement the insufficiency of identity authentications, Chen et al. [7] presented a novel trust management scheme based on the information from behavior feedback. The successors generate verified feedback packets for each positive behavior and, consequently, the 'behavior-trust' relationship is formed for slow-moving nodes. Shabut et al. [8] proposed a recommendation-based trust model with a defense scheme, which utilized the clustering technique to dynamically filter out attacks related to dishonest recommendations between certain times based on the number of interactions, the compatibility of information and the distance between nodes.

Due to the characteristics of a group communication system (e.g. the existence of selfish nodes and high survival time requirements), Cho and Chen [9] summarised a detailed analysis of trust management for this system. Barnwal and Ghosh [10] proposed an efficient trust estimation scheme to detect the errant/malicious nodes that disseminate incorrect kinematics information to vehicular cyber-physical systems. An attack-resistant trust management scheme was proposed for VANETs in [11], which was not only able to detect and cope with malicious attacks, but it also evaluated the trustworthiness of both data and mobile nodes. Specially, node trust is assessed in two dimensions, i.e. functional trust and recommendation trust.

To secure the data plane of ad-hoc networks, Tan et al. [12] proposed a novel trust management system. In this system, fuzzy logic was employed to formulate imprecise empirical knowledge, which was used to evaluate the path trust value. Together with fuzzy logic, graph theory was adopted to build a novel trust model for calculating the node trust value. Chen and Wang [13] demonstrated a layered trust management model based on a vehicular cloud system. This model could benefit from the efficient use of physical resources (e.g. computing, storage and communication costs) and the exploration of its deployment in a vehicular social network scenario based on a three-layer cloud computing architecture.

Yao et al. [14] proposed the concept of incorporating social trust in the routing decision process and the design of a trust routing protocol based on the social similarity (TRSS) scheme. TRSS is based on the observation that nodes move around and contact each other according to their common interests or social similarities. Based on direct and recommended trust, those untrustworthy nodes will be detected and purged from the trusted list. Because only trusted node packets will be forwarded, the selfish nodes have incentives to behave well again.

However, there are some problems with the abovementioned trust mechanisms in mobile ad hoc environments, such as the selection of trust attributes, the calculation of their weights and lack of direct interaction experience. Besides, the quality of recommended information is not guaranteed. Moreover, they are weak resistibility to various types of malicious attacks.

3 Trust Inference Model

The concept of trust has appeared in many academic works and can be used to achieve certain missions and system goals. Yan and Wang [15] utilized two dimensions of trust levels to control data access to pervasive social networks in a heterogeneous manner using attribute-based encryption. The trust levels can be evaluated either by a trusted server or by individual entities or by using them both.

3.1 Concept of Trust

In accordance with prior related studies, 'trust' helps humans overcome perceptions of uncertainty and risk when engaging in social activities and sharing social resources. But what is 'trust'? How should one describe and quantify it? Trust is a very complicated concept that relates to the confidence, belief, reliability, honesty, integrity, security, dependability, ability and other characters. According to the method for establishing trust relationships in a human society, building trust relationships in mobile ad hoc networks has a similar process. Therefore, it is a challenging issue to model, manage and maintain trust. For a specific network environment, a high level of trust for a service provider denotes that this provider does not only follow the willingness of a requesting entity, but it also effectively provides a mutually agreed upon service (e.g. to transmit information efficiently).

The concept of trust involves two kinds of participants (i.e. trustor that giving the evaluation and trustee that being evaluated), and the policies are designed by the trustor in the decision-making process.

3.2 Overall Design of Trust Model

In the variety of trust management trust models, the recommendation-based method accounts for a large percentage, which is a common evaluating criterion based on the recommended experiences from reliable recommenders. In order to gain the trust of a specific entity, the two most important steps are: the selection and the synthesis of trust attributes.

We abstract a novel trust inference model to mobile ad hoc networks, where two attributes, called subjective trust (ST) and recommendation trust (RT), are selected to quantify the trust level for a specific entity. The subjective trust of the trustee relative to the trustor is calculated based on the evaluation of historical interactions between them, whereas the recommendation trust of the trustee is computed based on all referenced reliable evaluations from neighbors in numerous interactions. Recommendations provide an effective way to build trust relationship, by making use of the information from others rather than exclusively relying on one's ow direct observation. The aim of this trust model is that the network can detect, prevent and exclude the misbehaving nodes and it can establish a trusted network system.

3.3 Synthesis of Trust

In the trust system of human society, after each interaction, both participants will make an evaluation for this interaction to the other. A lower interaction evaluation (*IE*) makes decrement in the trust level and vice versa. Two basic factors, called the interaction period and the interaction amount, are introduced to estimate the quality of an evaluation for a specific interaction.

Interaction Period (*IP*): The attenuation of trust is well known. The researches based on economic theory show that, when calculating the rating of an object, the evaluations of historical interactions should be properly attenuated. Using a similar way used by human societies, the evaluations for the recent interactions are more credible, and should be given greater weights on the synthesis of multiple evaluations. In order to achieve this purpose, we will introduce a proper time attenuation function.

Interaction Amount (*IA*): Similar to the vector *IP*, the interaction amount is another noteworthy vector. A larger scale interaction makes a bigger impact on the trust evaluation and reflects an entity's performance more exactly, which should also be given a greater weight for the synthesis of multiple evaluations. A trustor will pay more attention to a larger scale interaction in the future.

In a mobile ad hoc network, the metric TV_{ij} represents the trust value of a specific node j from the perspective of an evaluating node i. This metric can be calculated via synthesizing the abovementioned trust attributes using the following equation:

$$TV_{ij} = \alpha ST_{ij} + \beta RT_{ij} \tag{1}$$

where ST_{ij} and RT_{ij} represent the subjective trust and the recommendation trust levels for node j as derived from node i. The weights α and β (α, $\beta \geq 0$, $\alpha > \beta$ and $\alpha +$ $= 1$) are called confidence factors. In general, subjective trust should be given a greater weight unless there are rarely interactions between two sides. The detailed calculation process will be shown in the following Subsect. 3.3.3.

3.3.1 Calculation of Subjective Trust

You can find that this trust attribute appears in all trust models. There is variety of evaluation methods (e.g. a statistical approach) to characterize this attribute from different aspects. However, the most central part is still the subjective measure of trustee's historical behaviors using mathematical methods.

To simplify the discussion and implementation, we establish a simple inference method. Assume that there are n-th number of interactions between couple of nodes, i.e. node i and node j. And IE_k, IP_k and IA_k, $(\forall k(1 \leq k \leq n))$ represent the interaction evaluation, the interaction period and the interaction amount of the k-th interaction, respectively. Then we put forward a quality model to synthesize the abovementioned interaction factors.

$\forall k(1 \leq k \leq n)$, ST_{ij}^{k} can be calculated by using (2):

$$\begin{cases} ST_{ij}^{0} = Threshold, & if\ k = 0 \\ ST_{ij}^{k} = \sum_{m=1}^{k} W_m IE_m, & if\ 1 \leq k \leq n \end{cases} \tag{2}$$

where ST_{ij}^{k} represents the subjective trust for node j, which can be derived from node i after the k-th interaction and W_m represents the weight of IE_m. As shown in Eq. (2), the final subjective trust is a weighted average value of all the interaction evaluations that occurred during different interaction periods. We set a value interval for the variables, IE_m and ST_{ij}^{k} (i.e. $0 \leq IE_k, ST_{ij}^{k} \leq 1$). At the beginning of an experimental simulation, we initially set the subjective trust value to 0.5 (i.e. the threshold value).

The two factors (i.e. *IP* and *IA*) are involved into calculating the weight W_m. A rational solution is carried out using the following equation:

$$W_m = \frac{\rho_{m,k} IA_m}{\sum_{m=1}^{k} \rho_{m,k} IA_m} \tag{3}$$

where $\rho_{m,k}$ represents the time attenuation function. An effective attenuation approach could be used to accelerate the convergence rate of a computing process and to guarantee that this process reaches a stable state. To effectively calculate the subjective trust, the recent interaction should be given a bigger weight. In other words, the value of the attenuation function increases as the interaction period becomes closer to the recent time, i.e. $\sum_{1 \leq m \leq k, m=1}^{k} \rho_{m,k} = 1$ and $\rho_{1,k} < \rho_{2,k} < \ldots < \rho_{k,k}$.

To address the above issues, we introduce a simple model that considers the interaction period through the following equation:

$$\rho_{m,k} = \frac{T_m}{\sum_{m=1}^{k} T_m} \tag{4}$$

where T_m denotes the beginning time of the m-th interaction period.

Then, Eq. (3) can be transformed via Eq. (4). We can obtain two conclusions: (*a*) if the interaction amount for a specific interaction period is small, then it is difficult to obtain a high weight is obtained for this interaction evaluation in the calculation of subjective trust. The benefit of this new approach is that this design could prevent a dishonest node from cheating in a future large interaction if it has obtained a high trust level based on a number of small, honest interactions; (*b*) compared to the historical interactions, the evaluation of the latest interaction is more important and credible. In other words, the calculation of subjective trust is more likely to reflect the recent behaviors of a specific node.

3.3.2 Calculation of Recommendation Trust

If a node wants to interact with another node in an unfamiliar environment, it should first estimate a trust level for this other node. The calculation of trust should rely on the experience-based recommendations from all the referenced and reliable third parties. Even there are direct interactions between a pair of nodes; the trust estimation should also consider the reliable recommendations to overcome the subjectivity. In other words, an effective trust mechanism should use trusted neighbors to scout the behaviors of a specific node. Due to a large number of false recommendations and dishonest recommendations derived from variety of malicious attacks, like ballot stuffing, bad mouthing, conflict and collusion, how to establish a recommendation-based trust model is a challenging issue in such a network.

Each physical neighbor of a specific node can obtain a subjective trust of this node, and the credibility of recommended information derived from such neighbors will directly affect the accuracy of the composition of the recommendation trust. The computational method is based on the following intuitive ideas:

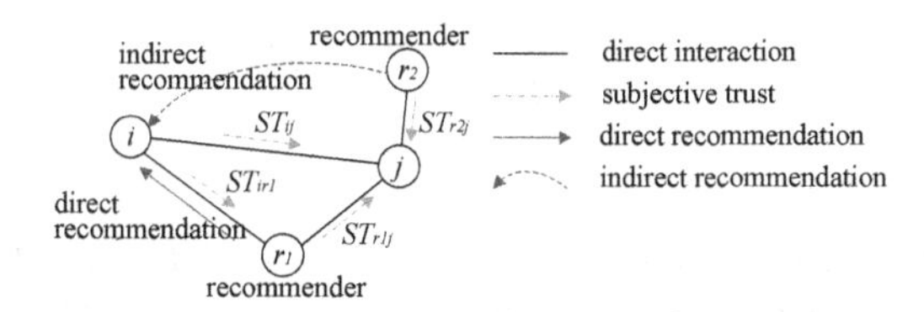

Fig. 1. Calculation of recommendation trust

As shown in Fig. 1, it is assumed that, after *k*-th interaction between node *i* and node *j*, node *i* can will calculate a value of subjective trust ST_{ij}^{k} for node *j* based on the *ST* model. Nodes r_1 and r_2 are two types of neighboring nodes that have previous interactions with node *j* and their historical evaluations of subjective trust for node *j* are $ST_{r_1 j}^{n}$ and $ST_{r_2 j}^{m}$ respectively.

Based on the conditions of the recommender(s) (i.e. whether there are previous interactions with the monitored node), the recommended experience can be divided into directly recommended experience d_RE_{ij} and indirectly recommended experience in_RE_{ij}.

This concept of credibility factor reflects the credibility of the recommended experience given by a recommender. Two computational approaches are presented to estimate the credibility factor of these two different types of recommended experiences, respectively. Moreover, we set a recommended threshold value *d*, which makes this recommendation mechanism more reasonable.

Directly Recommended Experience

If the recommender has historical interactions with the monitoring node, according to the transferable property of trust, this monitor can use recommended experience directly.

The value of the directly recommended experience for node j is an average of recommended experiences provided by multiple recommenders (e.g. node r_1 as shown in Fig. 1.). This operation is carried out using the following equation:

$$d\_RE_{ij} = \sum_{l=1}^{n} \left(\frac{ST_{ir_l}}{\sum_{l=1}^{n} ST_{ir_l}} \times ST_{r_l j} \right) \tag{5}$$

Indirectly Recommended Experience
If the recommender has no historical interactions with the monitor, while they have a common set of evaluated nodes, then this monitor can still effectively use recommended experience.

It is assumed that a monitor node i and a recommender node r have a consistent view if they have a higher similarity rating on a public node set (denoted by $Set(i, r)$). The deviation of their trust evaluation about a public node set is defined using the following equation:

$$Diff_{ir} = \frac{\sum_{k \in Set(i,r)} |ST_{ik} - ST_{rk}|}{|Set(i, r)|} \tag{6}$$

where $|Set(i, r)|$ represents the number of public node sets. The credibility of this type of recommended experience can be obtained as $CR_{ir} = 1 - Diff_{ir}$. The value of this variable also needs to satisfy the recommended threshold for the synthesizing operation. Then, we can obtain the following equation:

$$in\_RE_{ij} = \sum_{s=1}^{m} \left(\frac{CR_{ir_s}}{\sum_{s=1}^{n} CR_{ir_s}} \times ST_{r_s j} \right) \tag{7}$$

The concept of the recommended threshold can be used to counter slander attacks from malicious nodes. Furthermore, the interaction amount (*IA*) between a recommender and a specific node makes a great impact on the estimation of recommendation trust. A larger interaction amount results in a more credible recommendation. The *IA* can be considered as another weighting factor and is used to optimize the calculation of recommendation trust. Equations (5) and (7) can therefore be derived into the following equations:

$$d\_RE_{ij} = \sum_{l=1}^{n} \left(\frac{ST_{ir_l} \times IA_{r_l j}}{\sum_{l=1}^{n} ST_{ir_l} \times IA_{r_l j}} \times ST_{r_l j} \right) \tag{8}$$

$$in\_RE_{ij} = \sum_{s=1}^{m} \left(\frac{CR_{ir_s} \times IA_{r_s j}}{\sum_{s=1}^{m} CR_{ir_s} \times IA_{r_s j}} \times ST_{r_s j} \right) \tag{9}$$

Such assignments and the abovementioned recommended credibility mechanism can effectively prevent malicious nodes from providing false recommended experience, which is used to raise or decrease the trust level for a specific node.

Finally, we can calculate the recommendation trust for a specific node from the monitor's point of view using the following equation:

$$RT_{ij} = \omega_1 d\_RE_{ij} + \omega_2 in\_RE_{ij} (\omega_1 + \omega_2 = 1) \tag{10}$$

where ω_l and ω_2 are the weighting factors for the two types of recommended experiences.

$$\omega_1 = \frac{\sum_{l=1}^{n} IA_{r_l j}}{\sum_{l=1}^{n} IA_{r_l j} + \sum_{s=1}^{m} IA_{r_s j}} \quad \omega_2 = \frac{\sum_{s=1}^{m} IA_{r_s j}}{\sum_{l=1}^{n} IA_{r_l j} + \sum_{s=1}^{m} IA_{r_s j}} \tag{11}$$

Besides, we find a phenomenon that, the number of recommenders is also an important factor for calculating the recommended trust on the basis of a variety of researches. A larger number of recommenders yielded a more accurate recommended experience. We can design a function $\phi(x)$ that considers the number of recommenders as follows:

$$\phi(n+m) = e^{-1/(n+m)}, \lim_{n \to \infty} \phi(n+m) = 1 \tag{12}$$

where $(n + m)$ represents the number of recommenders. Without considering the abovementioned factor, multiple malicious nodes can easily raise each other's trust through complicity. Therefore, this mechanism can be used to counter collision attacks from malicious nodes.

Then, Eq. (10) should be optimized and derived as follows:

$$RT_{ij} = \phi(n+m) \times (\omega_1 d\_RE_{ij} + \omega_2 in\_RE_{ij}) \, (\omega_1 + \omega_2 = 1) \tag{13}$$

It hints that if a node wants to raise its recommendation trust, it must deal with a large number of honest nodes.

3.3.3 Calculation of Confidence Factors α and β

The primary remaining problem in calculating the trust is how to legitimately calculate the confidence factors α and β. The process of calculating trust must consider the confidence level of the subjective trust and the recommendation trust.

In general, if a monitor is familiar with the behavioral performance of a specific node, then it is supposed to be more convinced with the calculation of subjective trust. On the contrary, if there are rarely interactions between them, then the synthesis of trust is supposed to mainly rely on the recommended experience.

Based on the abovementioned analysis and taking the influence of the interaction amount into consideration, a simplified approach based on game theory is proposed to determine the values of confidence factors α and β using the following equation:

$$\alpha = \frac{IA_{ij}}{IA_{ij} + \sum\limits_{p \in SetC} IA_{ip}}, \beta = \frac{\sum\limits_{p \in SetC} IA_{ip}}{IA_{ij} + \sum\limits_{p \in SetC} IA_{ip}} \tag{14}$$

SetC includes two types of nodes: (1) direct recommenders. In other words, for a specific interaction period, these nodes not only directly interact with a monitor i, but they also interact with a specific node j. (2) Set(i, r), which was discussed in Subsect. 3.3.2, for node r represents an indirect recommender.

3.4 Trust Table

Each node can calculate the trust values of its neighbours and establish a trust table to record and update this information as shown in Table 1.

Table 1. The trust table for a special node i

Neighbour ID(i)	TV	Property flag
k	0.92	0
m	0.73	0
j	0.23	1
...	...	...

The vector *TV* denotes a neighbor's trust value on the node i's point of view, and *Property Flag* indicates whether this neighbor is a malicious node or not. With the help of the trust model, if a particular node is detected as a malicious node by all its neighbors, then this node will be logically excluded from the local network and prevented from engaging in any activity with the local network (e.g. data forwarding).

4 Trust Model Performance

To verify the validity and accuracy of our trust mechanism, compared with the BFT [7], the RT [8], the Tan [12] and the TRSS [14], we have conducted a comprehensive test using NetLogo [16]. NetLogo is an agent-based programming language and integrated modeling environment, which can simulate the large-scale dynamic environment and interaction process between agents.

In this simulator, 200 agents were arranged, and a simulation time of 3000 steps was used for simulation scenario. Two types of behaviors were launched by agents (i.e. trustworthy behaviors and malicious behaviors). In trustworthy behaviors, a specific agent provides reliable service in an interaction. While in malicious behaviors, a specific agent provides untrustworthy service (i.e. launching various types of attacks). The dynamic environment contained 20% malicious agents and each malicious agent provided bad services (e.g. launched malicious behaviors) in the entire simulation time over a rate of 80% in the following experiments. Figure 2, 3, 4 and 5 show the simulation results during the simulation.

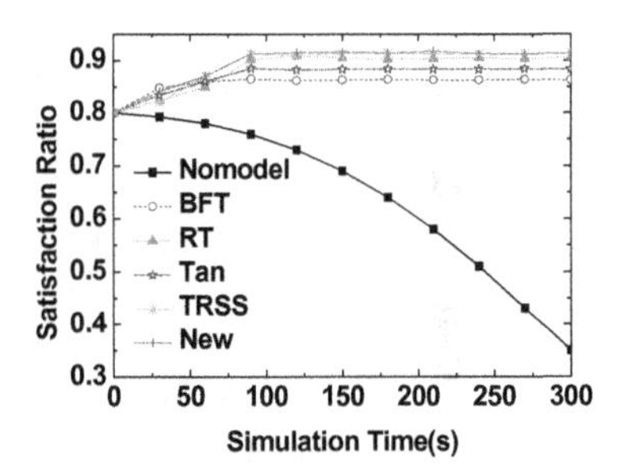

Fig. 2. Satisfaction ratios of network interaction

As shown in Fig. 2, the satisfaction ratios of network interaction perform as a function of the simulation time. The x-axis of this figure stands for the number of simulation steps. The satisfaction ratios for all trust models rise in the initial stage of simulation, after that, remain stable. While this ratio of no-model decreases gently in the entire simulation time. Trust models can take advantage of the trust concept, gradually detect malicious agents in the simulation process, and guarantee the interactions occurring only between the trustworthy agents. A longer the existing time of malicious node will lead a greater damage on the simulation environment. Our model has a better performance.

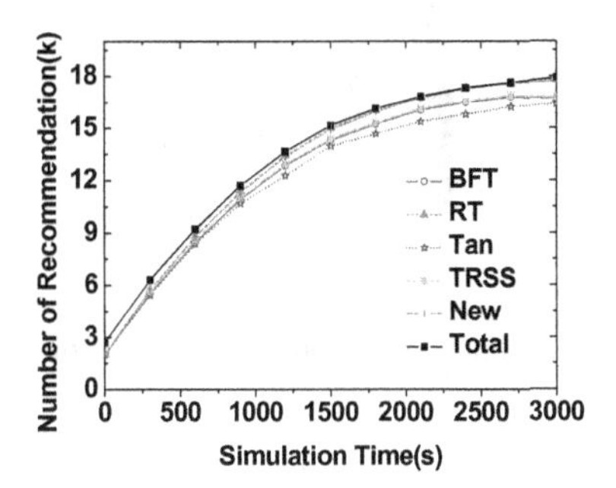

Fig. 3. Number of good recommended experience

Figure 3 illustrates the correction rate of the recommended experience, where y-axis is the number of the recommended experience. We can see that, with the increase of total number of recommended experience in network, the number of bad recommended experience is very small, and decreases as time passes. With the help of trust mechanisms, the network can identify its inherent malicious nodes. When calculating a specific node's trust level, the proportion of good recommended experience provided by malicious nodes will be neglected, and the malicious nodes will be slowly removed from the set of recommending nodes. Almost all the recommended experience is good recommended experience, so the plot representing the good recommended experience is almost coincident with the one representing the total recommended experience. Due to the effective and consummate recommendation trust evaluation rules in our new model, the performance outperforms existing mechanisms.

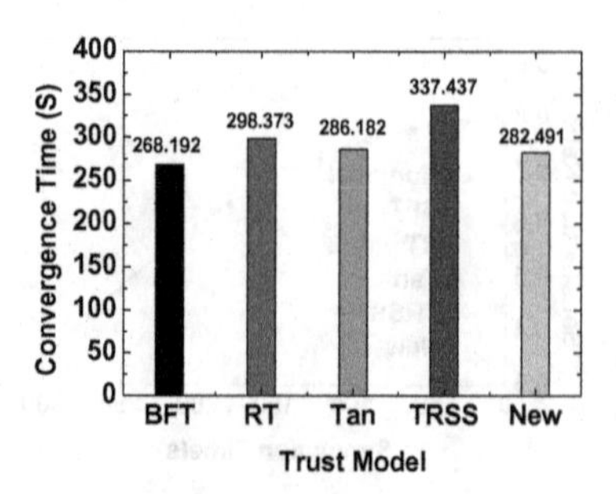

Fig. 4. Convergence time

Figure 4 shows the comparison of the convergence time of different trust models. Our new trust mechanism has a faster convergence time. The reason is that the new proposed mechanism adopts an iterative calculation method, and the update process of trust level is only related to the current trust information.

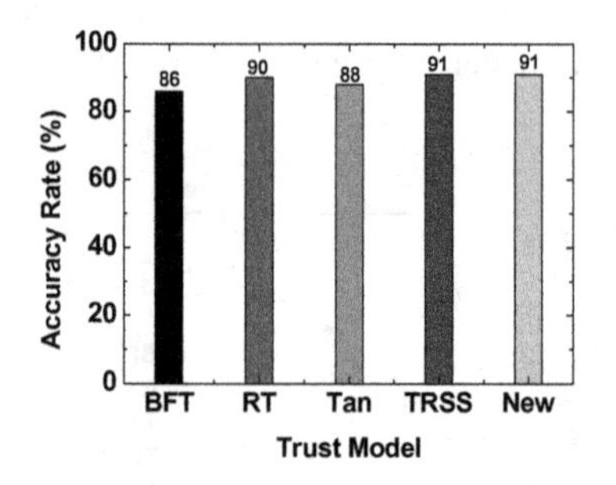

Fig. 5. Accuracy detection ratio of malicious agents

Figure 5 shows the comparison of the accuracy detection rate of malicious agents. The simulation environment can gradually identify its inherent malicious agents benefit from the trust mechanism. In the process of trust calculation, the proportion of recommendation experience provided by malicious agents will be neglected. These malicious agents will be slowly removed from the local network area. Due to the effective and consummate trust evaluation rules in our new trust model, its performance is better than the others.

5 Conclusions and Future Work

Unlike wired networks, which have a higher level of security for gateways and routers, mobile ad hoc networks are vulnerable to various attacks due to their inherent features. It is relatively easy for multiple malicious entities to bring down the whole network in several network services. Trust-based countermeasure is considered to be more acceptable as a promising approach. Therefore, we carry out a detailed study of the various trust-based countermeasures and abstract a novel recommendation-based trust inference model which is used to establish a trusted network. Two trust attributes called the subjective trust and the recommendation trust, are selected to quantify the trust level of a specific entity. The subjective trust of a specific entity is calculated based on its historical behaviors, whereas the recommendation trust is computed based on all the reliable recommendations. The convincing experimental results show that our new model performs better than the other trust-based countermeasures.

In future work, we plan to conduct an in-depth study of trust-based strategies, taking into account the requirements for deployment area issues, network applications and security levels. Moreover, trust computations and management can be an attractive target for attackers, since major decisions can be taken based on these trust computations. A malicious node may behave well towards one group of nodes and badly towards another group; this is known as a conflicting behaviour attack. Hence, defence mechanisms are required that can ensure trusted information, confidentiality and integrity in order to enable and support the most secure routing decisions.

Acknowledgment. This work is sponsored by the Natural Science Foundation of China (NSFC) under Grant No. 61402245, the Project funded by China Postdoctoral Science Foundation under Grand No. 2015T80696 and 2014M551870, the Shandong Provincial Natural Science Foundation No. ZR2014FQ010, the Project of Shandong Province Higher Educational Science and Technology Program No. J16LN06, the Qingdao Postdoctoral Application Research Funded Project.

References

1. Sicari, S., Rizzardi, A., Grieco, L.A., et al.: Security, privacy and trust in Internet of Things: the road ahead. Comput. Netw. **76**, 146–164 (2015)
2. Liu, Y.X., Dong, M.X., Ota, K.: ActiveTrust: secure and trustable routing in wireless sensor networks. IEEE Trans. Inf. Forensics Secur. **11**(9), 2013–2027 (2016)
3. Yao, J.P., Feng, S.L., Zhou, X.Y.: Secure routing in multihop wireless ad-hoc networks with decode-and-forward relaying. IEEE Trans. Commun. **64**(2), 753–764 (2016)
4. Govindan, K., Mohapatra, P.: Trust computations and trust dynamics in mobile adhoc networks: a survey. IEEE Commun. Surv. Tutor. **14**(2), 279–298 (2012)
5. Movahedi, Z., Hosseini, Z., Bayan, F., Pujolle, G.: Trust-distortion resistant trust management frameworks on mobile ad hoc networks: a survey. IEEE Commun. Surv. Tutor. **18**(2), 1287–1309 (2016)
6. Shen, H.Y., Li, Z.: A hierarchical account-aided reputation management system for MANETs. IEEE-ACM Trans. Netw. **23**(1), 70–84 (2015)
7. Chen, X., Sun, L., Ma, J.F., Ma, Z.: A trust management scheme based on behavior feedback for opportunistic networks. China Commun. **12**(4), 117–129 (2015)
8. Shabut, A.M., Dahal, K.P., Bista, S.K., Awan, I.U.: Recommendation based trust model with an effective defence scheme for MANETs. IEEE Trans. Mob. Comput. **14**(10), 2101–2115 (2015)
9. Cho, J., Chen, I.: On the tradeoff between altruism and selfishness in MANET trust management. Ad Hoc Netw. **11**(8), 2217–2234 (2013)

10. Barnwal, R.P., Ghosh, S.K.: KITE: an efficient scheme for trust estimation and detection of errant nodes in vehicular cyber-physical systems. Secur. Commun. Netw. **9**(16), 3271–3281 (2016)
11. Li, W.J., Song, H.B.: ART: an attack-resistant trust management scheme for securing vehicular ad hoc networks. IEEE Trans. Intell. Transp. Syst. **17**(4), 960–969 (2016)
12. Tan, S.S., Li, X.P., Dong, Q.K.: A trust management system for securing data plane of ad-hoc networks. IEEE Trans. Veh. Technol. **65**(9), 7579–7592 (2016)
13. Chen, X., Wang, L.M.: A cloud-based trust management framework for vehicular social networks. IEEE Access **5**, 2967–2980 (2017)
14. Yao, L., Man, Y.M., Huang, Z., Deng, J., Wang, X.: Secure routing based on social similarity in opportunistic networks. IEEE Trans. Wirel. Commun. **15**(1), 594–605 (2016)
15. Yan, Z., Wang, M.J.: Protect pervasive social networking based on two-dimensional trust levels. IEEE Syst. J. **11**(1), 207–218 (2017)
16. Wilensky, U.: NetLogo (1999). http://ccl.northwestern.edu/netlogo/

Author Index

Printed in the United States
By Bookmasters